volume 3

reviews in NUMBER THEORY

as printed in
MATHEMATICAL REVIEWS
1940 through 1972
volumes 1—44 inclusive

edited by

WILLIAM J. LeVEQUE

AMERICAN MATHEMATICAL SOCIETY · PROVIDENCE · RHODE ISLAND · 1974

Reviews reprinted from
MATHEMATICAL REVIEWS
volumes 1—44 published during 1940—1972

AMS (MOS) classification numbers (1970).
Primary 10 – XX, 12 – XX; Secondary 00 – XX, 01 – XX.

Library of Congress Cataloging in Publication Data **CIP**

LeVeque, William Judson, comp.
Reviews in number theory.

1. Numbers, Theory of—Abstracts. I. Mathematical reviews. II. Title.
QA241.L577 512′.7′08 74-11335
ISBN 0-8218-0205-4 (v. 3)

Printed in the United States of America

SERIES CONTENTS

CONTENTS

H. GEOMETRY OF NUMBERS

For the number of points with integral coordinates in special regions (frequently, but not necessarily, large regions), see **P20** ***ff.***

See also Sections E32, J04, J08, J12, J16, J20, J24, J28, J32, J36, J40, J44, J48, J52.

H02 BOOKS AND SURVEYS

See also reviews H05-52, H05-89, H15-10, H25-40, H30-27, H30-68, H30-69, H30-91.

H02-1 (1, 67c)

Hancock, Harris. Development of the Minkowski Geometry of Numbers. The Macmillan Company, New York, 1939. xxiv+839 pp. $12.00.

In this book the author gives a faithful account of all the work of Minkowski on the geometry of numbers [Gesammelte Abhandlungen, 1911; Geometrie der Zahlen, 1910; Diophantische Approximationen, 1907] in the form of a translation and commentary. He does not connect Minkowski's results and methods with newer researches on the subject; and many simplifications and extensions of the theory, obtained during the last 25 years, are not mentioned. The book will be useful to those who are unable to read German or find Minkowski's own writings too difficult. It is well printed and contains a large number of excellent figures. Contents: (1) Preliminary notions (including Minkowski's inequality $M^3J\leqq 8$ for 3 dimensions, and some applications). (2) Surfaces that are nowhere concave. (3) The volume of bodies. (4) Bodies which with respect to their volumes have more than one point with integral coordinates. (5) Applications. (6) Algebraic numbers. (7) Arithmetical theory of a pair of lines, etc. (The chapters 2–6 correspond to Kapitel 1–4 in Geometrie der Zahlen.) (8) Shorter papers of Minkowski (historical remarks on quadratic forms, discriminant theorem, etc.). (9) A criterion for algebraic numbers. (10) The theory of continued fractions (containing among other things Blichfeldt's proof of the Tchebycheff-Minkowski theorem, and recent work of Dr. Pepper on an algorithm of Minkowski). (11) Periodic approximation of algebraic numbers. (12) On the approximation of a real quantity through rational numbers. (13) A further analytic-arithmetic inequality. (14) The arithmetic of the ellipsoid. (15) Computation of a volume through successive integrations. (16) Proof of the new analytic-arithmetic inequality. (17) The extreme standard bodies. (The chapters 13–17 correspond to Kapitel 5 of the Geometrie der Zahlen.) (18) Densest placement of congruent homologous bodies. (19) Miscellany. (20) New theory of quadratic forms, etc. (Minkowski's reduction theory; his asymptotic formula for the mean value of the class number of quadratic forms in n variables). *K. Mahler.*

Referred to in H02-2.

H02-2 (30# 64)

Hancock, Harris

Development of the Minkowski geometry of numbers. Vols. One, Two.

Dover Publications, Inc., New York, 1964. *Vol. One:* xix+*pp.* 1–452. $2.50; *Vol. Two:* ix+*pp.* 453–839. $2.50.

This is an unabridged and unaltered republication, in two volumes, of the work first published in 1939 by Macmillan, New York [MR **1**, 67]. *E. S. Barnes* (Adelaide)

Citations: MR 1, 67c = H02-1.

H02-3 (8, 502e)

Mordell, L. J. Geometry of numbers. Proc. First Canadian Math. Congress, Montreal, 1945, pp. 265–284. University of Toronto Press, Toronto, 1946. $3.25.

Report on recent work in the geometry of numbers, especially on various results obtained by Davenport, Mahler, Mordell and Ollerenshaw. Most of the results and methods reported concern domains which are not convex.

V. Jarník (Prague).

H02-4 (9, 271a)

Davenport, H. The geometry of numbers. Math. Gaz. **31**, 206–210 (1947).

This is the report of an address, the purpose of which was to introduce the theory of the geometry of numbers to non-specialists. Minkowski's proof of the fundamental lattice point theorem for convex bodies is given. This is followed by applications among which is the Hermite proof of the Lagrange result that every integer can be represented as the sum of at most four integral squares. *D. Derry.*

H02-5 (13, 919f)

Davenport, H. Recent progress in the geometry of numbers. Proceedings of the International Congress of Mathematicians, Cambridge, Mass., 1950, vol. 1, pp. 166–174. Amer. Math. Soc., Providence, R. I., 1952.

This address gives an account of a selection of some of the developments in the geometry of numbers since 1936. The results mentioned are explained clearly and appropriate references are given. A few rather deep problems are mentioned. *C. A. Rogers* (London).

Referred to in H02-11.

H02-6 (14, 454b)

Chabauty, Claude. Sur des problèmes de géométrie des nombres. Algèbre et Théorie des Nombres. Colloques Internationaux du Centre National de la Recherche Scientifique, no. 24, pp. 27–28. Centre National de la Recherche Scientifique, Paris, 1950.

H02-7 (16, 117c)

Hlawka, Edmund. Grundbegriffe der Geometrie der Zahlen. Jber. Deutsch. Math. Verein. **57**, Abt. 1, 37–55 (1954).

This is an excellent exposition of the concepts, methods, and results of the modern geometry of numbers. A rather complete list of references is given, and some of the open questions in the subject are noted. *W. J. LeVeque.*

Referred to in H02-11.

H02-8 **(16, 451d)**

Keller, Ott-Heinrich. Geometrie der Zahlen. Enzyklopädie der mathematischen Wissenschaften mit Einschluss ihrer Anwendungen. I 2, 27. Band I. Algebra und Zahlentheorie. 2. Teil. C. Reine Zahlentheorie. Heft 11, Teil III. B. G. Teubner Verlagsgesellschaft, Leipzig, 1954. 84 pp. DM 8.80.

This is an account of results in the theory of numbers obtained by geometric methods up to 1951. Among the chapter headings are convex bodies in lattices, star bodies, linear forms, minima of homogeneous forms, inhomogeneous forms, definite quadratic forms, continued fractions and algebraic numbers. Some proofs of fundamental results are sketched. The results on star bodies which have been obtained in the last two decades are listed with considerable completeness. Included among these is a paragraph of conjectures made by Mahler. The chapter on definite quadratic forms contains a clear account of the work done on the reduction problem. *D. Derry.*

Referred to in H02-11.

H02-9 **(19, 57h)**

Люстерник, Л. А. [Lyusternik, L. A.] Выпуклые фигуры и многогранники. [Convex figures and polyhedra.] Gosudarstv. Izdat. Tehn.-Teor. Lit., Moscow, 1956. 212 pp. 2.95 rubles.

The six chapter-headings are: convex figures and bodies and their support lines and planes; centrally symmetric convex figures; convex polyhedra; linear systems of convex bodies; theorems of Minkowski and Alexandrov (this chapter was written by Alexandrov); supplementary material. Visualizability is emphasized throughout. The entire book can be read with very little mathematical background, the first three chapters with almost none at all.

H02-10 **(19, 124e)**

Hlawka, Edmund. Das inhomogene Problem in der Geometrie der Zahlen. Proceedings of the International Congress of Mathematicians, 1954, Amsterdam, vol. III, pp. 20–27. Erven P. Noordhoff N.V., Groningen; North-Holland Publishing Co., Amsterdam, 1956. $7.00.

Let T be a set of points in n dimensional space and let $\mathscr{G}$ be a lattice in the space. If for every point $\boldsymbol{p}$ of space there is a lattice-vector $\boldsymbol{g}$ such that $\boldsymbol{p}+\boldsymbol{g}$ lies in T, we say that $\mathscr{G}$ is a covering lattice for T. Many problems, both of a general and a special nature, can be related to this concept, and the author's extensive survey includes much work that was previously scattered and unconnected. His collection of references will be valuable to all workers in the field. A few more recent references are: B. J. Birch, Proc. Cambridge Philos. Soc. **53** (1957), 269–272; J. W. S. Cassels, Introduction to Diophantine approximation, Cambridge, 1957; C. A. Rogers, Mathematika **4** (1957), 1–6; G. L. Watson, Rend. Circ. Mat. Palermo (2) **5** (1956), 93–100 [MR **17**, 1235]. *H. Davenport* (London).

Citations: MR 17, 1235b = H30-35.

H02-11 **(19, 124f)**

Minkowski, Hermann. Diophantische Approximationen. Eine Einführung in die Zahlentheorie. Chelsea Publishing Co., New York, 1957. viii+235 pp. $4.50.

This is a reprint of Minkowski's classical book [Teubner, Leipzig, 1907] with some minor corrections. The book was written to give an elementary exposition of his work on the Geometry of Numbers and of its applications to the theories of Diophantine Approximation and of Algebraic Numbers. The work is now mainly interesting for historical reasons and for some of its details. It seems irritating nowadays to find general n-dimensional theorems proved first in 2dimensions and then again in later chapters in 3dimensions and in 4dimensions. Readers interested in finding the way the subject has developed are referred to the books: J. F. Koksma, Diophantische Approximationen [Springer, Berlin, 1936]; O.-H. Keller, Geometrie der Zahlen, Enzykl. Math. Wiss. I 2, 27 [Teubner, Leipzig, 1954; MR **16**, 451]; J. W. S. Cassels, Introduction to Diophantine approximation, [Cambridge, 1957]; and to the review articles: H. Davenport, Proc. Internat. Congress Math., Cambridge, Mass., 1950, v. 1, pp. 166–174 [Amer. Math. Soc., Providence, R. I., 1952; MR **13**, 919]; E. Hlawka, Jber. Deutsch. Math. Verein. **57** (1954), Abt. 1, 37–55 [MR **16**, 117]; and the paper reviewed above. *C. A. Rogers* (Birmingham).

Citations: MR 13, 919f = H02-5; MR 16, 117c = H02-7; MR 16, 451d = H02-8.

Referred to in J02-19, R54-41.

H02-12 **(40# 2515)**

Minkowski, Hermann

Geometrie der Zahlen.

Bibliotheca Mathematica Teubneriana, Band 40.

Johnson Reprint Corp., New York-London, 1968. vii+256 *pp.* $15.00.

This work appeared originally in complete form in 1910 [Teubner, Leipzig; reprint, Chelsea, New York, 1953]. The first 240 pages appeared as the first section in 1896.

H02-13 **(23# A1609)**

Bambah, R. P.

Some problems in the geometry of numbers.

J. Indian Math. Soc. (*N.S.*) **24** (1960), 157–172 (1961).

This is a useful summary of solved and unsolved problems in the geometry of numbers. Altogether 20 problems are mentioned; 13 of these are classified as homogeneous, the remainder being inhomogeneous. Since the paper was written, several improvements have been made in certain of the problems mentioned, notably in the problem of packings and coverings of spheres. *R. A. Rankin* (Glasgow)

H02-14 **(25# 3909)**

Chabauty, Claude

Introduction à la géométrie des nombres.

Enseignement Math. (2) **8** (1962), 41–53.

Expository paper. *J. W. S. Cassels* (Cambridge, England)

Referred to in Z10-17.

H02-15 **(28# 1175)**

Cassels, J. W. S.

An introduction to the geometry of numbers.

Die Grundlehren der mathematischen Wissenschaften in Einzeldarstellungen mit besonderer Berücksichtigung der Anwendungsgebiete, Bd. 99.

Springer-Verlag, Berlin-Göttingen-Heidelberg, 1959. viii+344 *pp.* *DM* 69.00.

Seit den klassischen Werken von Minkowski ist das hier vorliegende Werk das erste Lehrbuch der Geometrie der Zahlen. Verfasser hat selbst Bedeutendes zu diesem Gebiet beigetragen. Nun eine kurze Inhaltsübersicht: Ch. I—Lattices—enthält die grundlegenden Sätze über Gitter. Ch. II—Reduction—enthält die Reduktionstheorie der positiv definiten Formen, angewendet auf die Theorie der quadratischen Formen. Es wird bei den binären und ternären quadratischen Formen, sowohl definit wie indefinit, das erste Minimum bestimmt (im binären Fall werden auch einseitige Probleme behandelt), ebenso bei den binären kubischen Formen mit positiver Diskriminante nach Davenport und Chalk. Für negative Diskriminante wird nur eine Beweisskizze gegeben, die dann im 3. Kapitel ergänzt wird. Ch. III—Theorems of Blichfeldt and Minkowski. Es werden hier auch die Verallgemeinerungen von Siegel und R. Rado behandelt, die Gitterkonstante einer Menge (in der Literatur auch Determinante einer Menge genannt) eingeführt und nach einer Mitteilung von Mahler die Gitterkonstante Δ eines Simplex bestimmt. Dann wird die Methode von Mordell behandelt, um Δ in gewissen Fällen zu bestimmen (das ist wohl die beste Darstellung dieser Methode), zunächst an zwei einfachen Fällen dargestellt und dann auf die binären kubischen Formen mit negativer Diskriminante angewendet. Am Schluß dieses Kapitels wird gezeigt, wie die Geometrie der Zahlen zum Beweis klassischer Sätze der Zahlentheorie verwendet werden kann. Ch. IV—Distance functions. Hier wird die Theorie der Sternkörper entwickelt, konvexe Körper und ihre Polarkörper werden eingeführt und die zugehörigen Sätze von Mahler entwickelt. Ch. V—Mahler's compactions theorem. Der Beweis dieses Satzes wird nach Chabauty geführt, kritische Gitter werden eingeführt und insbesondere beschränkte Sternkörper in Betracht gezogen und gezeigt, daß die Bestimmung des zugehörigen Δ auf eine endliche Menge von gewöhnlichen Minimalproblemen zurückgeführt werden kann. Die Theorie der reduziblen Sternkörper wird am Kreis erläutert. Dann wird für konvexe Körper der Satz von Swinnerton-Dyer gebracht und die Kugel für $n=3$ behandelt. Den Abschluß bildet in Anwendung von diophantischen Approximationen der Satz von Davenport, welcher einen Satz von Furtwängler vertieft. Ch. VI—The theorem of Minkowski-Hlawka. Hier wird unter anderem die Verschärfung des Satzes von W. Schmidt und der Beweis einer Vermutung von Rogers, welcher vom Verfasser herrührt, gebracht. Im Fall der Ebene wird ein allgemeines Kriterium dafür gebracht, daß ein Sternbereich von endlichem Typus ist. Ch. VII—The quotient space. Hier wird der Summensatz von Macbeath, der ein Analogon des Satzes von H. B. Mann darstellt, bewiesen. Ch. VIII—Successive minimas—bringt die Sätze von Rogers und Chabauty, von Minkowski und den Übertragungssatz von Mahler bei polaren Körpern. Ch. IX—Packings—enthält die Theorie der Wabenzellen und im Fall der Ebene die Sätze von Fejes Tóth und Rogers und die Bestimmung der Gitterkonstanten für konvexe Zylinder. Den Abschluß bildet die Abschätzung der Gitterkonstanten für Kugeln und Produkt von Linearformen nach Blichfeldt. Ch. X—Automorphs—bespricht die Isolierungsmethode. Sätze von Mordell, welche die Bestimmung der Gitterkonstanten von n-dimensionalem Bereiche auf $n-1$-dimensionale Probleme zurückführen, zerlegbare ternäre kubische Formen und die Sätze von Davenport und Rogers über die Existenz von unendlich vielen Gitterpunkten in Sternkörpern. Ch. XI—Inhomogeneous problems—enthält die Übertragungssätze, Produkt von Linearformen und die Sätze von Minkowski und Remak. Dabei wird auf den Algorithmus der geteilten Zelle von Delaunay eingegangen. Die Darstellung ist sehr klar und durchsichtig. Hervorzuheben sind die zahlreichen Literaturangaben und die Fülle von originellen Beweisführungen. *E. Hlawka* (Zbl **86**, 262)

Referred to in H02-16, H02-20, H30-86.

H02-16 (31# 5841)

Kassels, Dž. V. S. [Кассельс, Дж. В. С.]
[Cassels, J. W. S.]

An introduction to the geometry of numbers [Введение в геометрию чисел].

Translated from the English by A. N. Andrianov and I. V. Bogačenko. Edited by A. V. Malyšev.

Izdat. "Mir", Moscow, 1965. 421 *pp.* 1.68 *r.*

This is a translation of the book published by Springer, Berlin, 1959 [MR **28** #1175].

Citations: MR 28# 1175 = H02-15.

H02-17 (30# 2405)

Rogers, C. A.

Packing and covering.

Cambridge Tracts in Mathematics and Mathematical Physics, No. 54.

Cambridge University Press, New York, 1964. viii + 111 *pp.* $5.50.

This monograph, based on the work of the author and his collaborators, gives a very elegant and readable account of many important results in the theory of packings and coverings, lattice as well as non-lattice. The author has managed in most cases to include proofs of best known results without affecting the readability of this book, which is entirely self-contained. Although the main text deals with estimates from above and below for the densities of best lattice and general packings and coverings for a set K, where K is an n-dimensional convex body, sphere or simplex and n is large, the introduction contains a lucid and complete survey of the present knowledge of many other problems also. Because of the many unsolved problems in the field, this introduction is very valuable. The introduction is supplemented by an exhaustive bibliography. It may be worth remarking that the bibliography contains references to more than 125 contributions between the years 1944 and 1964 by more than forty mathematicians. A chapter by chapter description follows.

In Chapter 1, the author gives alternative definitions (and proves their equivalence) for densities of a set, a lattice packing, a periodic packing, a general packing, a lattice covering, a periodic covering and a general covering, and proves also the obvious upper bound for packing densities and the lower bound for covering ones. He then proves the invariance of the best packing and covering densities under affine transformations, as well as the equality of the best general and best periodic densities. In the rest of the review $\delta_L(S)$, $\delta(S)$, $\vartheta_L(S)$ and $\vartheta(S)$ will denote the densities, respectively, of the best lattice packing, the best general packing, the best lattice covering and the best general covering for S. In Chapter 2, using a modification of a method of his note on coverings and packing [J. London Math. Soc. **25** (1950), 327–331; MR **13**, 323], the author obtains lower bounds for $\delta(K)$ when K is an open n-dimensional set, a convex body with center at 0 or a general convex body. If $\mu(S)$ denotes the volume of S and DS its "difference" set, then $\delta(K)\geqq 2\mu(K)/\mu(DK)$ if K is an open set and if K and DK have volumes. For symmetrical convex K in n dimensions, the estimates are $\delta(K)\geqq 1/2^{n-1}$ and $\delta(K)\geqq\vartheta(K)/2^n$. He then proves his result with G. C. Shephard [Arch. Math. **8** (1957), 220–233; MR **19**, 1073] on an upper bound for $\mu(DK)/\mu(K)$ when K is convex, and combining this with the earlier estimate, obtains the result $\delta(K)\geqq 2(n!)^2/(2n!)$ for all n-dimensional convex K.

Chapter 3 is based on the author's paper [Mathematika **4** (1957), 1–6; MR **19**, 877], and the principal result is $\vartheta(K) \leqq n \log n + n \log \log n + 5n$ for all n-dimensional convex bodies K. Before the appearance of this paper all estimates were of the type $\vartheta(K) \leqq c^n$, c a constant greater than 1. An interesting part of the proof is that "if we take a large n-dimensional cube C and throw at random a fixed number of translates of K in C, the translates cover a fairly good proportion of C". In Chapter 4, denoting by $\Lambda(\alpha_1, \cdots, \alpha_{n-1}, \eta)$ a lattice of determinant 1, depending on parameters $\alpha_1, \cdots, \alpha_{n-1}, \eta$ in a special way, the author proves that under some simple conditions

$$\lim_{\eta \to +0} \frac{1}{\eta} \int_0^\eta d\eta \int_0^1 \cdots \int_0^1 \sum_{\substack{x \in \Lambda(\alpha_1, \cdots, \alpha_{n-1}, \eta) \\ x \neq 0}} f(x)\, d\alpha_1 \cdots d\alpha_{n-1} = \int_{-\infty}^{\infty} \cdots \int_{-\infty}^{\infty} f(x)\, dx_1 \cdots dx_n.$$

As a consequence one gets the Minkowski-Hlawka theorem for sets with volumes and the lower bound $\delta_L(K) \geqq 2\mu(K)/\mu(DK)$ if K and DK have volumes. The author remarks that it has become "clear" to him that every lattice of determinant 1 can be "approximated arbitrarily closely" by a suitable $\Lambda(\alpha_1, \cdots, \alpha_{n-1}, \eta)$. Using methods of Chapters 3 and 4, a result of A. M. Macbeath on the approximation of convex bodies by cylinders and a result of his own on the overlapping of translates of a convex body [J. London Math. Soc. **33** (1958), 208–212; MR **20** #7247], all proved here, the author obtains the estimates

$$\vartheta_L(K) \leqq n^{\log_2 n + \log \log n}, \qquad \vartheta_L(s) \leqq c_1(n \log_2 n)^{\log_2 \sqrt{(2\pi e)}},$$

where $n \geqq 3$, K is an n-dimensional convex body and S is an n-dimensional sphere. These results are great improvements over earlier estimates of the type c^n, but are weaker than the sharper results of the author [Mathematika **6** (1959), 33–39; MR **23** #A2130], whose proofs are too difficult to include in the book. Chapter 6, based on results of Minkowski, the author and G. C. Shephard [loc. cit.], gives an upper bound for $\delta(s)$ when s is a simplex. In Chapter 7, he uses a dissection of space with the help of Voronoi polyhedra to prove that for n-dimensional spheres K, $\delta(K) \leqq \sigma_n$, where σ_n is the ratio of the volume of a certain part of a regular simplex to that of the whole simplex. An asymptotic formula of H. E. Daniels shows that $\delta(K) \leqq \sigma_n \sim (n/e)(1/\sqrt{2})^n$. This estimate is better than the Blichfeldt result, $\delta(K) < \frac{1}{2}(n+2)(1/\sqrt{2})^n$, and its earlier improvements. This chapter is based on the author's paper in Proc. London Math. Soc. (3) **8** (1958), 609–620 [MR **21** #847]. Chapter 8, based on a paper by H. S. M. Coxeter, L. Few and the author [Mathematika **6** (1959), 147–157; MR **23** #A2131], uses a dissection of space due to Delaunay, which is in some sense dual to the Voronoi dissection, to prove

$$\vartheta_L(K) \geqq \vartheta(K) \geqq \tau_n = \left(\frac{2n}{n+1}\right)^{n/2} \sigma_n \sim \frac{n}{e\sqrt{e}}$$

for n-dimensional spheres K. This is a great improvement over earlier results $\vartheta_L(K) \geqq c_n \sim 4/3$ (the reviewer and H. Davenport) and $\vartheta(K) \geqq c_n{}^* \sim 16/15$ (P. Erdős and the author). *R. P. Bambah* (Chandigarh)

Citations: MR 13, 323b = H30-15; MR 19, 1073f = H05-57; MR 20# 7247 = H30-41; MR 21# 847 = H30-42; MR 23# A2130 = H30-49.

Referred to in H02-18, H02-20, H05-81, H15-78, H30-68, H30-80, H30-81.

H02-18 **(41# 4390)**

Роджерс, К. [**Rodžers, K.**] [**Rogers, C. A.**]

Укладки и покрытия. (Russian) [**Packings and coverings**]

Translated from the English by B. Z. Moroz and O. M. Fomenko. Edited by A. V. Malyšev.

Izdat. "Mir", Moscow, 1968. 134 *pp.* 0.41 *r.*

The English original has been reviewed [Cambridge Univ. Press, London, 1964; MR **30** #2405].

Citations: MR 30# 2405 = H02-17.

Referred to in H02-20.

H02-19 **(42# 202)**

Grasselli, Jože

Fundamental concepts of the geometry of numbers. (Slovenian)

Obzornik Mat. Fiz. **15** (1968), 1–9.

Introductory expository article.

H02-20 **(42# 5915)**

Lekkerkerker, C. G.

Geometry of numbers.

Bibliotheca Mathematica, Vol. VIII.

Wolters-Noordhoff Publishing, Groningen; North-Holland Publishing Co., Amsterdam-London, 1969. ix+510 *pp.* *Hfl.* 90.00; 210*s.*; $25.20.

This book is the product of a great deal of labour. The author describes almost all work available up to 1964 in the classical geometry of numbers over the n-dimensional Euclidean space. All basic results are proved. Others are described lucidly and references are given to the sources where complete proofs are available. The book deals with the homogeneous problem, in its various formulations, arithmetical and geometrical in terms of critical determinants as well as packings, and with both types of non-homogeneous problems, those corresponding to non-homogeneous determinants and those corresponding to coverings.

In the preface to his book *An introduction to the geometry of numbers* [Springer, Berlin, 1959; MR **28** #1175], J. W. S. Cassels remarked: "When I first took an interest in the geometry of numbers, I was struck by the absence of any book which gave the essential skeleton of the subject as it was known to the experienced workers in the subject." Cassels' own book, C. A. Rogers' book *Packing and covering* [Cambridge Univ. Press, Cambridge, 1964; MR **30** #2405; Russian translation, Izdat. "Mir", Moscow, 1968; MR **41** #4390] and the present book admirably fulfil the need pointed out by Cassels. These three books are in many ways complementary to each other. The bibliography at the end of the book under review is another good feature.

The book contains seven chapters. Chapter 1 gives preliminaries on convex bodies, star sets, star bodies, lattices and algebraic number fields. Chapter 2, besides giving an account of the classical work of Minkowski on existence of lattice points in symmetrical convex bodies (Minkowski's fundamental theorem, his second theorem) and its generalizations by Blichfeldt, Mordell and others, has a good account of polar reciprocal convex bodies and Mahler's compound convex bodies. There is also a discussion of extremal bodies and some discussion of the non-homogeneous problem. Chapter 3 starts with Mahler's compactness theorem and existence of critical lattices. The Minkowski-Hlawka theorem is proved and its improvements by Rogers and Schmidt discussed. Lattice packings and coverings in R_n and R_2 are also studied. Chapter 4 deals with the critical determinants of star bodies, automorphic star bodies, reducible star bodies, etc. on the lines of the general theory developed by K. Mahler and extended by Davenport and C. A. Rogers. Chapter 5 deals with methods of Mahler and Mordell for critical determinants of two dimensional star domains, Minkowski's method of critical determinants of three-dimensional convex bodies, critical determinants of cylinders, and other special bodies, Blichfeldt's method of estimating densities of sphere packings, Macbeath's theorem on non-homogeneous minima and Mordell's method of connecting the critical determinant of an n-dimensional body with an $(n-1)$-dimensional one. Chapter 6 deals with Voronoï's theory of extreme quadratic forms and discusses the later results of Coxeter and others. This chapter also contains results on minima of various forms, including products of n homogeneous linear forms, and binary cubic forms. There is also a discussion of isolated minima and asymmetric inequalities. The chapter ends with application to Diophantine approximations. The last chapter deals with non-homogeneous results, including products of non-homogeneous linear forms, covering densities for spheres, and indefinite quadratic forms. The chapter ends with a discussion of the infinity of solutions for non-homogeneous inequalities. *R. P. Bambah* (Chandigarh).

Citations: MR 28# 1175 = H02-15; MR 30# 2405 = H02-17; MR 41# 4390 = H02-18.

H05 LATTICES AND CONVEX BODIES: GENERAL THEORY

See also reviews H15-52, H30-29, H30-57, J20-39, J56-6, P28-9, P28-11, Q25-1.

H05-1 (1, 202b)

Mahler, Kurt. Ein Übertragungsprinzip für lineare Ungleichungen. Časopis Pěst. Mat. Fys. **68**, 85–92 (1939).

The author proves the following generalization of Khintchine's "Übertragungssatz" [Rend. Circ. Mat. Palermo **50**, 170–195 (1926)]. Let $f_1, \cdots, f_n$ be linear forms in $x_1, \cdots, x_n$ with real coefficients and determinant $d\neq 0$, and let $g_1, \cdots, g_n$ be the contragredient linear forms in $y_1, \cdots, y_n$, so that $\sum f_i g_i = \sum x_i y_i$. Suppose there exist integral $x_1, \cdots, x_n$, not all zero, for which

$$|f_1|=t_1, \quad |f_2|\leqq t_2, \cdots, \quad |f_n|\leqq t_n.$$

Then there exist integral $y_1, \cdots, y_n$, not all zero, for which

$$|g_1|\leqq (n-1)\lambda/t_1, \quad |g_2|\leqq \lambda/t_2, \cdots, \quad |g_n|\leqq \lambda/t_n,$$

where $\lambda^{n-1}=t_1t_2\cdots t_n/|d|$. The proof is very simple, and uses only Minkowski's classical theorem on linear forms. It follows as a corollary that if $\Pi|f_i|$ can be made arbitrarily small, so also can $\Pi|g_i|$. The author also proves an analogous "Übertragungssatz" for linear forms with p-adic coefficients. *H. Davenport* (Manchester).

Referred to in J16-6.

H05-2 (1, 202c)

Mahler, Kurt. Ein Übertragungsprinzip für konvexe Körper. Časopis Pěst. Mat. Fys. **68**, 93–102 (1939).

The author generalizes the "Übertragungsprinzip" for linear forms [see preceding review] to relations between an arbitrary convex body K with center at the origin and the polar reciprocal body K'. Let $F(\mathbf{x})$ ($\mathbf{x}$ any point in n-dimensional space) be the distance-function corresponding to K, so that K is defined by $F(\mathbf{x})\leqq 1$. The distance-function $G(\mathbf{x})$ for K' is given by

$$G(\mathbf{x})=\max_{\mathbf{y}\neq 0}\frac{|\mathbf{x}\cdot\mathbf{y}|}{F(\mathbf{y})}, \quad \mathbf{x}\cdot\mathbf{y} \text{ the scalar product.}$$

Let $\sigma_1=F(\mathbf{x}_1), \cdots, \sigma_n=F(\mathbf{x}_n)$ be the successive lattice-point minima associated with K [see Minkowski, Geometrie der Zahlen, 1910, Kap. 5], and $\tau_1=G(\mathbf{y}_1), \cdots, \tau_n=G(\mathbf{y}_n)$ be those associated with K'. The author proves that

$$(1) \qquad \sigma_h\tau_{n-h+1}\geqq 1, \qquad h=1, 2, \cdots, n.$$

The proof depends on the existence of $i\leqq h, j\leqq n-h+1$ with $|\mathbf{x}_i\mathbf{y}_j|\geqq 1$. In the proof of this, at the bottom of p. 99, the phrase following "also auch" should read "jeder Punkt der durch 0, $x^{(1)}, \cdots, x^{(h)}$ gelegten linearen h-dimensionalen Mannigfaltigkeit $L^{(h)}$ auf jedem Punkt der durch. . . ."

The author also proves by elementary geometrical considerations that the volumes J, J' of K, K' satisfy

$$(2) \qquad 4^n/(n!)^2\leqq JJ'\leqq 4^n,$$

and conjectures the more precise inequality

$$4^n/n!\leqq JJ'\leqq \pi^n/\Gamma(n/2+1)^2,$$

the two sides corresponding to the parallelepiped and to the ellipsoid. From (1), (2) and Minkowski's inequality

$$2^n/n!J\leqq \sigma_1\cdots\sigma_n\leqq 2^n/J,$$

it follows that

$$(1') \qquad 1\leqq \sigma_h\tau_{n-h+1}\leqq (n!)^2.$$

H. Davenport (Manchester).

Referred to in H05-46, H05-53, H15-60, J20-1.

H05-3 (2, 35h)

Weyl, Hermann. Theory of reduction for arithmetical equivalence. Trans. Amer. Math. Soc. **48**, 126–164 (1940).

Minkowski's investigations in the geometry of numbers started from the problem of reduction for positive quadratic forms of n variables, but he found later some considerable difficulties in the application of his geometrical method to this problem and finally solved it directly by arithmetical ideas. The author sets forth a more geometrical theory of reduction which depends upon Minkowski's fundamental inequality (1) $S_1\cdots S_nV\leqq 2^n$ concerning a symmetric convex body in relationship to a lattice. In this inequality appear n vectors $v_1, \cdots, v_n$ of the lattice which generally do not constitute a basis. The relationship (1) is replaced by another inequality in which the vectors $v_1, \cdots, v_n$ form a basis of the lattice. It is remarked that this result had already been obtained independently by K. Mahler [Quart. J. Math. **9**, 259–262 (1938)]. The new inequality gives rather strong estimates for certain quantities in the theory of reduction. In the second chapter the convex body is specialized to the case of an ellipsoid. The results of the first chapter lead to a finite number of conditions which define a reduced positive quadratic form. This gives Minkowski's theorem that the domain of reduced positive quadratic forms is a convex polyhedron Z with a finite number of faces. By any unimodular substitution S of the variables, Z is carried into an equivalent cell Z_S, and these equivalent cells cover without gaps and overlappings the domain G of all positive quadratic forms. Minkowski had proved that Z borders on not more than a finite number of equivalent cells Z_S. The author obtains the sharper result that into any compact subset of G penetrate only a finite number of cells Z_S.

The whole theory is extended to lattices and forms in which complex numbers or quaternions take the place of real numbers, under the essential conditions that every ideal in the corresponding ring of integers is a principal ideal. *C. L. Siegel* (Princeton, N. J.).

Referred to in H05-5.

H05-4 (2, 350c)

Mahler, Kurt. An analogue to Minkowski's geometry of numbers in a field of series. Ann. of Math. (2) **42**, 488–522 (1941).

Notation: $\mathfrak{P}$ an arbitrary field, z an indeterminate, $\mathfrak{T}$ the ring of all polynomials in z with coefficients in $\mathfrak{P}$, $\mathfrak{K}$ the field of all formal Laurent series $x=\alpha_f z^f+\alpha_{f-1}z^{f-1}+\cdots$ with coefficients $\alpha_f, \alpha_{f-1}, \cdots$ in $\mathfrak{P}$, P_n the n-dimensional space of all points $X=(x_1, \cdots, x_n)$ with coordinates $x_1, \cdots, x_n$ in $\mathfrak{K}$, Λ_n the set of all lattice-points in P_n (that is, that of all points with coordinates in $\mathfrak{T}$). The non-Archimedean valuation $|x|$ is defined as $|0|=0$ and $|x|=e^f$ if $\alpha_f\neq 0$. A distance function is any function satisfying $F(0)=0$, $F(X)>0$ if $X\neq 0$, $F(tX)=|t|F(X)$ for all t in $\mathfrak{K}$, $F(X-Y)\leqq\max(F(X), F(Y))$; it is further assumed that $F(X)$ is an integral power of e, for all $X\neq 0$ in P_n. A convex body $C(\tau)$ is defined by the inequality $F(X)\leqq\tau$, for any $\tau>0$. Similar to the arithmetical definition of the volume of a body in the ordinary real space, a certain positive constant V may be defined as the volume of $C(1)$.

The following theorem is proved: There are n independent points $X^{(1)}, \cdots, X^{(n)}$ in Λ_n with the following properties: (1) $F(X^{(1)})$ is the minimum of $F(X)$ in all lattice points $X\neq 0$, and, for $k\geqq 2$, $F(X^{(k)})$ is the minimum of $F(X)$ in all lattice points X which are independent of $X^{(1)}, \cdots, X^{(k-1)}$. (2) The determinant of the points $X^{(1)}, \cdots, X^{(n)}$ is 1. (3) The numbers $F(X^{(k)})=\sigma^{(k)}$ satisfy the formulae $0<\sigma^{(1)}\leqq\sigma^{(2)}\leqq\cdots\leqq\sigma^{(n)}$, $\sigma^{(1)}\sigma^{(2)}\cdots\sigma^{(n)}=V^{-1}$. The result is in analogy to a well-known theorem of Minkowski [Geometrie der Zahlen, Leipzig, 1910, §§ 50–53], and the proofs in the paper are also based on the methods of Minkowski. Some special Diophantine problems in P_n are considered as examples. *C. L. Siegel* (Princeton, N. J.).

Referred to in H05-75, J64-17, K40-14, R58-26, R58-43, R58-44.

H05-5 (3, 272d)

Weyl, Hermann. Theory of reduction for arithmetical equivalence. II. Trans. Amer. Math. Soc. **51**, 203–231 (1942).

[The first part appeared in the same Trans. **48**, 126–164 (1940); cf. these Rev. **2**, 35.] Let $\mathfrak{F}$ be a field of finite degree f over the field of rational numbers, with the basis $\sigma_1, \cdots, \sigma_f$, and $[\mathfrak{F}]$ an order in $\mathfrak{F}$. Any n-uple $(\xi_1, \cdots, \xi_n)$ of numbers ξ_k $(k=1, \cdots, n)$ in $\mathfrak{F}$ is a vector. A set of vectors constitutes a lattice L belonging to the order $[\mathfrak{F}]$ if $\mathfrak{x}+\mathfrak{y}$ and $\lambda\mathfrak{x}$ are contained in L for all $\mathfrak{x}, \mathfrak{y}$ in L and all λ in $[\mathfrak{F}]$. We assume that L is discrete and contains n linearly independent vectors. The unit lattice $I=I^{(n)}$ consists of all vectors $(\xi_1, \cdots, \xi_n)$ with ξ_k $(k=1, \cdots, n)$ in $[\mathfrak{F}]$. Two lattices L and Λ are called equivalent if one is carried into the other by a nonsingular linear transformation $\xi_k=\sum_{l=1}^n\delta_{kl}\eta_l$ $(k=1, \cdots, n)$; all linear transformations with $L=\Lambda$ form the modular group. If Λ contains I, then $j=[\Lambda:I]$ denotes the number of vectors in Λ which are incongruent modulo I; more generally, $j_k=[\Lambda^{(k)}:I^{(k)}]$, where $\Lambda^{(k)}$ consists of all vectors $(\xi_1, \cdots, \xi_k, 0, \cdots, 0)$ in Λ.

For arbitrary real variables $t_1, \cdots, t_f$, let $\tau^*=t_1\sigma_1+\cdots+t_f\sigma_f$, $\tau^{(l)}=t_1\sigma_1^{(l)}+\cdots+t_f\sigma_f^{(l)}$, where $\sigma_k^{(l)}$ denotes the lth conjugate of σ_k, and $|\tau^*|=\max|\tau^{(l)}|$. The f numbers $\tau^{(l)}$ (for the r real conjugates, $l=1, \cdots, r$) and $\frac{1}{2}(\tau^{(l)}+\tau^{(l+s)})$, $(1/2i)(\tau^{(l)}-\tau^{(l+s)})$ (for the s pairs of conjugate complex conjugates, $l=r+1, \cdots, r+s$; $r+2s=f$) are called the splitting coordinates of τ^*. Let $\xi_k^*=x_{k1}\sigma_1+\cdots+x_{kf}\sigma_f$, $\mathfrak{x}^*=(\xi_1^*, \cdots, \xi_n^*)$. A continuous function $F(\mathfrak{x}^*)=F(\xi_1^*, \cdots, \xi_n^*)$ of the nf real variables x_{kl} $(k=1, \cdots, n; l=1, \cdots, f)$ is called a gauge function if (1) $F(\mathfrak{x}^*)>0$ except for $\mathfrak{x}^*=0$, (2) $F(\tau^*\mathfrak{x}^*)\leqq|\tau^*|F(\mathfrak{x}^*)$, (3) $F(\mathfrak{x}_1^*+\mathfrak{x}_2^*)\leqq F(\mathfrak{x}_1^*)+F(\mathfrak{x}_2^*)$. There exist n linearly independent vectors $\theta_1, \cdots, \theta_n$ in L such that $F(\mathfrak{x})\geqq F(\theta_k)=M_k$ $(k=1, \cdots, n)$ for every vector $\mathfrak{x}$ in L outside $[\theta_1, \cdots, \theta_{k-1}]$. The mapping $\mathfrak{x}^*=\eta_1^*\theta_1+\cdots+\eta_n^*\theta_n\to(\eta_1^*, \cdots, \eta_n^*)$ carries $F(\mathfrak{x}^*)$ into a function $g(\eta_1^*, \cdots, \eta_n^*)$ and L into an equivalent lattice Λ which contains the unit lattice I for $[\mathfrak{F}]$. Let V_k be the volume defined by $g(\mathfrak{x}^*)<1$, $\xi_{k+1}^*=\cdots=\xi_n^*=0$, computed in terms of the splitting coordinates of $\xi_1^*, \cdots, \xi_k^*$, and let Δ denote the absolute value of the determinant of the $\sigma_k^{(l)}$ $(k, l=1, \cdots, f)$; then $(M_1\cdots M_k)^f v_k j_k\leqq(2^f\Delta)^k$ and $j_k\leqq(kf)!(4/\pi)^{ks}(\Delta/f!)^k$ $(k=1, \cdots, n)$. The main theorem of Minkowski's "Geometry of Numbers" is contained in these two inequalities, for the special case $f=1$.

The Hermitian forms $\gamma^*(\mathfrak{x}^*)=\sum_{k,l=1}^n\xi_k^*\gamma_{kl}^*\bar{\xi}_l^*$ constitute a linear space of $\frac{1}{2}n(n+1)f-ns$ dimensions and the positive ones a convex cone G in that space. The two inequalities are applied to the gauge function F introduced by $F^2=\mathrm{tr}(\gamma^*(\mathfrak{x}^*))$; they lead to the construction of a fundamental domain in G for the modular group of an arbitrary lattice L belonging to an arbitrary order $[\mathfrak{F}]$. For $f=1$, this contains the result of Minkowski [J. Reine Angew. Math. **129**, 220–274 (1905)], and for $L=I$, $[\mathfrak{F}]$=principal order, the result of P. Humbert [Comment Math. Helv. **12**, 263–306 (1940); cf. these Rev. **2**, 148].

The whole theory goes through also in the case of a quaternion algebra $\mathfrak{F}$ over a totally real field if the norm is totally positive; the results are analogous to those in the case of a field $\mathfrak{F}$. *C. L. Siegel* (Princeton, N. J.).

Citations: MR 2, 35h = H05-3; MR 2, 148a = E12-3.

H05-6 (3, 273a)

Weyl, Hermann. On geometry of numbers. Proc. London Math. Soc. (2) **47**, 268–289 (1942).

Let $f(\mathfrak{x})$ be any continuous function in the n-dimensional space of vectors $\mathfrak{x}=(x_1, \cdots, x_n)$ enjoying the following properties: $f(\mathfrak{x})>0$ except for the origin $\mathfrak{x}=\mathfrak{o}=(0, \cdots, 0)$; $f(t\mathfrak{x})=|t|f(\mathfrak{x})$ for any real factor t; $f(\mathfrak{x}+\mathfrak{x}')\leqq f(\mathfrak{x})+f(\mathfrak{x}')$. Let $2M_1$ be the minimum of $f(\mathfrak{a})$ for all lattice points $\mathfrak{a}\neq\mathfrak{o}$ and $\mathfrak{d}_1$ the corresponding value of $\mathfrak{a}$; let $2M_k$ $(k=2, \cdots, n)$ be the minimum of $f(\mathfrak{a})$ for all lattice points outside the $(k-1)$-dimensional linear manifold spanned by $\mathfrak{d}_1, \cdots, \mathfrak{d}_{k-1}$ and $\mathfrak{d}_k$ the corresponding value of $\mathfrak{a}$. Minkowski [Geometrie der Zahlen, Teubner, Leipzig, 1910, pp. 211–218] proved the inequality $M_1\cdots M_nV\leqq 1$, where V denotes the volume of the convex solid $\mathfrak{K}$ defined by $f(\mathfrak{x})<1$. In the first part of his paper, the author derives a more general theorem concerning functions $\phi(\mathfrak{x})$ which satisfy the following 3 conditions: $0\leqq\phi(\mathfrak{x})\leqq 1$; $\phi(\mathfrak{x})=0$ outside a finite region; the locus of the points $\mathfrak{x}$ where $\phi(\mathfrak{x})\geqq r$ is a convex set for any $r>0$. Let

$$\phi^{(k)}(\mathfrak{x})=\phi(\mathfrak{x})\int_0^1\prod_{\mathfrak{a}}\{1-t\phi(\mathfrak{x}-\mathfrak{a})\}dt,$$

the product extending to all vectors $\mathfrak{a}\neq\mathfrak{o}$ of a k-dimensional discrete lattice $(1\leqq k\leqq n)$, and $\phi_q(\mathfrak{x})=\phi(q^{-1}\mathfrak{x})$ $(q>0)$,

$$J^{(k)}[\phi]=\int\phi^{(k)}(\mathfrak{x})dx_1\cdots dx_n,$$

where the integral extends over the whole $\mathfrak{x}$-space; then $q^{k-n}J^{(k)}[\phi_q]$ is a monotonically increasing function of q. This theorem leads to Minkowski's inequality if ϕ is chosen as the characteristic function of the convex solid $\mathfrak{K}$, that is, $\phi=1$ inside $\mathfrak{K}$ and $\phi=0$ outside.

It seems that Minkowski did not notice a simple idea connecting his inequality with the problem of reduction of the lattice in terms of the given convex solid $\mathfrak{K}$. This idea was found by K. Mahler [Quart. J. Math., Oxford Ser. **9**, 259–262 (1938)]. In the second part, the author obtains independently again Mahler's result and applies it in particular to the reduction of quadratic forms. *C. L. Siegel.*

Referred to in H05-78.

H05-7 (7, 244i)

Pipping, Nils. Zur Geometrie der Zahlen. Acta Acad. Aboensis **14**, no. 13, 8 pp. (1944).

If S_1, S_2, S_3 are the successive minima associated with a convex body of volume J in 3-dimensional space, a simple proof is given that $S_1S_2S_3J\leqq 3!2^3$, an inequality that is weaker than Minkowski's [see Geometrie der Zahlen, Teubner, Berlin, 1910, chapter 5]. *H. Davenport* (London).

H05-8 **(8, 317i)**

van der Corput, J. G., and Davenport, H. On Minkowski's fundamental theorem in the geometry of numbers. Nederl. Akad. Wetensch., Proc. **49**, 701–707 = Indagationes Math. **8**, 409–415 (1946).

Let L be an n-dimensional lattice of determinant 1, the origin O being a lattice-point; K denotes a closed convex body, symmetrical about O and containing no lattice-point other than O in its interior; $V(M)$ denotes the volume of a set M. (I) There exists a polyhedron K' with the following properties: K' is convex, symmetrical about O, K' contains K and has not more than $2(2^n-1)$ faces, K' contains no lattice-point other than O in its interior, but each face of K' contains at least one lattice-point in its interior. Clearly $V(K')\leqq 2^n$ and so $V(K)\leqq 2^n-(V(K')-V(K))$, which may lead to improvements of the Minkowski inequality $V(K)\leqq 2^n$. (II) Let $n=2$ and let us suppose that the boundary of K consists of a curve with a continuous radius of curvature ρ; let $\rho\geqq\rho_0$, where ρ_0 is a positive number. Then $V(K)\leqq 4-(2\sqrt{3}-\pi)\rho_0^2$; here $2\sqrt{3}-\pi$ is the best possible constant. (III) The n-dimensional case. Let R be any $(n-1)$-dimensional plane having at least one point in common with K; we denote by $d(R)$ the distance of R from the nearest parallel tangent plane to K and by $U(R)$ the $(n-1)$-dimensional volume of the common part of R and K. If there is a number $\rho_1\geqq 1$ such that $U(R)\leqq(\rho_1 d(R))^{(n-1)/2}$ for every R, then $V(K)<2^n-c_1\rho_1^{-1}$; if there is a number $\rho_2(0<\rho_2\leqq 1)$ such that $U(R)\geqq(\rho_2 d(R))^{(n-1)/2}$ for every R, then $V(K)<2^n-c_2\rho_2^n$. The numbers $c_1>0$, $c_2>0$ depend only on n.

V. Jarník (Prague).

Referred to in H05-12, H30-12, H99-3.

H05-9 **(8, 565f)**

Estermann, T. Note on a theorem of Minkowski. J. London Math. Soc. **21**, 179–182 (1946).

The author gives a new simple proof of the following theorem. Let $0<\lambda_1\leqq\cdots\leqq\lambda_n$ and let S be a convex set of points in n-dimensional space with the following property: for any integer k from 1 to n and any two points $(x_1,\cdots,x_n)$ and $(y_1,\cdots,y_n)$ of $\lambda_k S$, where $x_1-y_1,\cdots,x_n-y_n$ are integers, we have $x_k=y_k$, $x_{k+1}=y_{k+1},\cdots,x_n=y_n$. Then $\lambda_1\lambda_2\cdots\lambda_n V(S)\leqq 1$. Here $V(S)$ is the n-dimensional volume of S and λS denotes the set of the points $(\lambda x_1,\cdots,\lambda x_n)$ corresponding to points $(x_1,\cdots,x_n)$ of S. *V. Knichal.*

H05-10 **(9, 10h)**

Mahler, K. On the minimum determinant and the circumscribed hexagons of a convex domain. Nederl. Akad. Wetensch., Proc. **50**, 692–703 = Indagationes Math. **9**, 326–337 (1947).

Let K denote a plane convex domain, symmetric with respect to the origin, $V(K)$ its area and $\Delta(K)$ the lower bound of determinants of lattices whose only point interior to K is the origin. Such a lattice with determinant equal to $\Delta(K)$ is known as a critical lattice. The author shows easily that a hexagon H, symmetric with respect to the origin, exists, enclosed by three pairs of parallel tangents to K, so that $\Delta(H)=\Delta(K)$. As H has a single critical lattice, namely the lattice containing the midpoints of the sides of H, a determination of a circumscribed hexagon H, with $\Delta(H)$ a minimum, provides a method of determining $\Delta(K)$ as well as the lattice which defines it. By use of these hexagons a tangent is shown to exist at every point of K which is a point of a lattice critical to K. The method is applied mainly to $2n$-sided polygons Π_n symmetric with respect to the origin. Let Q be the lower bound of $V(K)/\Delta(K)$, where K runs through all plane convex domains, and Q_n the lower bound of the same quantities, where K runs through all polygons Π_n. The existence of a polygon Π_n with $V(\Pi_n)/\Delta(\Pi_n)=Q_n$ is proved and if $n=4$ such a polygon is shown to be a regular octagon. Within a regular octagon Π_4 an irreducible convex domain K is found with $\Delta(\Pi_4)=\Delta(K)$. The quantity $V(K)/\Delta(K)$ is then computed for this domain, thus giving an upper bound for the constant Q. *D. Derry* (Vancouver, B. C.).

Referred to in H05-17, H05-90, H10-15, H15-84.

H05-11 **(9, 501b)**

Mahler, K. On lattice points in polar reciprocal convex domains. Nederl. Akad. Wetensch., Proc. **51**, 482–485 = Indagationes Math. **10**, 176–179 (1948).

Let k be a plane convex domain, symmetric with respect to the origin, and K the polar reciprocal of k with respect to the circle $x^2+y^2=1$. Let $\Delta(k)$ and $\Delta(K)$ denote the lower bounds of the determinants of the k-admissible and K-admissible lattices, respectively. The author shows by elementary methods that $\frac{1}{2}\leqq\Delta(k)\Delta(K)\leqq\frac{3}{4}$. By computation of the constants for special domains the inequalities are shown to be the best possible. *D. Derry.*

H05-12 **(10, 102b)**

Jarník, Vojtěch. On the main theorem of the Minkowski geometry of numbers. Časopis Pěst. Mat. Fys. **73**, 1–8 (1948). (English. Czech summary)

Let K be a convex area in the plane, symmetrical about the origin, bounded by a curve with a continuous radius of curvature which is never less than ρ_1. Let Λ be a plane lattice of determinant 1, and let τ_i, for $i=1$ or 2, be the least number such that K, when magnified by a factor τ_i, contains at least i independent points of Λ. Then van der Corput and Davenport [Nederl. Akad. Wetensch., Proc. **49**, 701–707 = Indagationes Math. **8**, 409–415 (1946); these Rev. **8**, 317] proved a result equivalent to $\tau_1^2\leqq(1+\delta\rho_1^2)^{-1}$, where $\delta=\frac{1}{2}3^{\frac{1}{2}}-\frac{1}{4}\pi$. The author now proves that

$$\tau_1\tau_2(1+\delta\rho_1^2)(1+\epsilon\rho_1^2)\leqq 1+(1-\tau_1/\tau_2)\delta\rho_1^2+(\tau_1/\tau_2)\epsilon\rho_1^2,$$

where $\epsilon=1-\frac{1}{4}\pi$; this is more precise if $\tau_1<\tau_2$.

H. Davenport (London).

Citations: MR 8, 317i = H05-8.

H05-13 **(10, 284g)**

Chalk, J. H. H., and Rogers, C. A. The critical determinant of a convex cylinder. J. London Math. Soc. **23**, 178–187 (1948).

Let D be a plane convex domain, symmetric with respect to the origin O, and K the convex cylinder of all points (x,y,z) for which $|z|\leqq 1$ and (x,y) is a point of D. Let $\Delta(D)$ and $\Delta(K)$ denote the lower bounds of the determinants of all the D-admissible lattices and K-admissible lattices, respectively. The object of this paper is to show $\Delta(D)=\Delta(K)$. The authors prove that a critical lattice of K may always be deformed continuously into a critical lattice with three linearly independent boundary points for which $|z|=1$. These points are shown to generate the lattice and a study of their projections on the plane $z=0$ leads to the result. This result is then used to show that the density of the closest lattice packing of 3-space by bodies congruent to K is the same as the corresponding density of the closest lattice packing of the plane by domains congruent to D. The same problem is the subject of the paper reviewed below.

D. Derry (Vancouver, B. C.).

Referred to in H05-15, H15-38, H15-56, H15-78.

H05-14 **(10, 285a)**

Yeh, Yenchien. Lattice points in a cylinder over a convex domain. J. London Math. Soc. **23**, 188–195 (1948).

The author discusses the same problem as that of the paper reviewed above. By use of a result of Minkowski the author shows it is sufficient to consider those K-admissible lattices generated by the three vectors OP_1, OP_2, OP_3 with P_1, P_2, P_3 on K and for which either the three z-coordinates are all 1 or all nonnegative and less than 1. The projections of these three vectors on the plane $z=0$ are shown to define certain D-admissible lattices by means of which the lower bound for the volume of the fundamental parallelepiped generated by P_1, P_2, P_3 is obtained. *D. Derry.*

Referred to in H15-56.

H05-15 (11, 83e)

Chalk, J. H. H., and Rogers, C. A. Corrigendum: The critical determinant of a convex cylinder. J. London Math. Soc. **24**, 240 (1949).

The paper appeared in the same J. **23**, 178–187 (1948); these Rev. **10**, 284.

Citations: MR 10, 284g = H05-13.

H05-16 (10, 511e)

Jarník, Vojtěch. On Estermann's proof of a theorem of Minkowski. Časopis Pěst. Mat. Fys. **73**, 131–140 (1949). (English. Czech summary)

Let M be a convex closed bounded n-dimensional set of points, symmetrical with respect to the origin, and having interior points. Let V denote the volume of M, and τ_i the least positive number τ for which the set τM contains at least i linearly independent lattice points (i.e., points with integer coordinates). A theorem of Minkowski states that (1) $\tau_1\tau_2\cdots\tau_n V\leqq 2^n$. Another theorem of Minkowski [Geometrie der Zahlen, part 1, Teubner, Leipzig, 1896, pp. 235–236] contains necessary and sufficient conditions for the equality sign to hold in (1). The author gives a new proof of the latter theorem. *T. Estermann* (London).

H05-17 (10, 593d)

Mahler, Kurt. Sui determinanti minimi delle sezioni di un corpo convesso. Atti Accad. Naz. Lincei. Rend. Cl. Sci. Fis. Mat. Nat. (8) **5**, 251–252 (1948).

Let K be a convex body in the 3-dimensional space, symmetric about the x_3-axis. Let π_z be the plane $x_3=z$ and let K_z be the intersection of K and π_z. Let $\Delta(K_z)$ be the lower bound of the determinants of all plane lattices situated in π_z and containing the point $O_z=[0, 0, z]$ which are K_z-admissible (i.e., which have no point in the interior of K_z except O_z). Theorem: suppose $K(z_1)\neq 0$, $K_{z_2}\neq 0$, $0\leqq t\leqq 1$, $z=(1-t)z_1+tz_2$; then $\Delta^{\frac{1}{2}}(K_z)\geqq(1-t)\Delta^{\frac{1}{2}}(K_{z_1})+t\Delta^{\frac{1}{2}}(K_{z_2})$. This result follows from an analogous inequality of Brun (where $\Delta(K_z)$ is replaced by the area of K_z) and from a theorem of Reinhardt [Abh. Math. Sem. Hamburg. Univ. **10**, 216–230 (1934)] and Mahler [Nederl. Akad. Wetensch., Proc. **50**, 692–703 = Indagationes Math. **9**, 326–337 (1947); these Rev. **9**, 10] according to which $4\Delta(K_z)$ is the minimum of the areas of all hexagons situated in π_z, containing K_z and symmetric about O_z. *V. Jarník* (Prague).

Citations: MR 9, 10h = H05-10.

H05-18 (11, 160e)

Laub, Josef. Über Punktgitter. Veröffentlichungen Math. Inst. Tech. Hochschule Braunschweig **1946**, no. 1, i+51 pp. (1946).

Let $E_1, \cdots, E_n$ be n independent points in R_n, and let Λ be the lattice generated by them. If $A_1, \cdots, A_n$ are n independent points of Λ, then every point Y of Λ may be written as $Y=M^{-1}(y_1A_1+\cdots+y_nA_n)$, where $y_1, \cdots, y_n$ are integers, and M is a positive integer depending only on $A_1, \cdots, A_n$. Denote by C the 2^n-cell consisting of all points $X=\lambda_1A_1+\cdots+\lambda_nA_n$, where $|\lambda_1|+\cdots+|\lambda_n|\leqq 1$, and by π the parallelepiped consisting of all points $X=\lambda_1A_1+\cdots+\lambda_nA_n$, where $0\leqq\lambda_1\leqq 1, \cdots, 0\leqq\lambda_n\leqq 1$. Assume that no point of Λ different from $0, \mp A_1, \mp A_2, \cdots, \mp A_n$ belongs to C. Then π contains no points of Λ different from its vertices if, and only if, $M=1$. For the lowest dimensions, the author shows in a very simple way that only the following values are possible: $n=2$, $M=1$; $n=3$, $M=1$ or 2; $n=4$, $M=1$ or 2 or $\cdots$ or 5; $n=5$, $M=1, 2, \cdots, 26, 29$, but not $27, 28, 30, 31, \cdots, 37$, and there may be further possible values. For all these cases, the configuration of the lattice points in π is determined. It is also proved that for $n\geqq 5$ all cases $M=1, 2, \cdots, n+2$ are possible, and further results are given for odd n. [See the following papers: Ph. Furtwängler, Math. Ann. **99**, 71–83 (1928); N. Hofreiter, Monatsh. Math. Phys. **40**, 181–192 (1933); E. Brunngraber, Über Punktgitter, Dissertation, Wien, 1944.] *K. Mahler.*

H05-19 (11, 331d)

Chalk, J. H. H., and Rogers, C. A. The successive minima of a convex cylinder. J. London Math. Soc. **24**, 284–291 (1949).

Davenport [Nederl. Akad. Wetensch., Proc. **49**, 822–828 = Indagationes Math. **8**, 525–531 (1946); these Rev. **8**, 565] has conjectured that if $\lambda_1, \cdots, \lambda_n$ are the successive minima for a lattice Λ of a convex region S in n-dimensional space and $\Delta(S)$ is the critical determinant of S, then $\lambda_1\cdots\lambda_n\Delta(S)\leqq d(\Lambda)$, the determinant of Λ. This has been proved true when $n=2$, when S is a generalized ellipsoid, or when the volume of S is $2^n\Delta(S)$. Here it is proved when $n=3$ and S is a convex symmetrical cylinder. *L. Tornheim* (Ann Arbor, Mich.).

Citations: MR 8, 565d = J36-9.

H05-20 (12, 45j)

Schüler, Hans. Vereinfachter Beweis eines Minkowskischen Satzes über konvexe Körper mit Mittelpunkt. Arch. Math. **2**, 202–204 (1950).

The author makes an error in attempting to give a short proof of Minkowski's theorem on the successive minima of a bounded convex body symmetric in the origin. The difficulty centers around the following erroneous statement on page 204: "$\mathfrak{B}^{(j)}$ ist . . . ganz enthalten in dem Körper $\mathfrak{B}^{(j)*}$." *P. T. Bateman* (Urbana, Ill.).

H05-21 (12, 46a)

Schneider, Theodor. Verallgemeinerung einer Minkowskischen Ungleichung über konvexe Körper mit Mittelpunkt. Math. Ann. **122**, 35–36 (1950).

This paper contains the same fallacy as the paper reviewed above. *P. T. Bateman* (Urbana, Ill.).

H05-22 (12, 320c)

Rédei, L. Endlich-projektivgeometrisches Analogon des Minkowskischen Fundamentalsatzes. Acta Math. **84**, 155–158 (1950).

The author points out that Minkowski's fundamental theorem has the following corollary. Suppose that $m_1, \cdots, m_k$ are positive integers and that $\mathfrak{K}$ is a convex body in n-dimensional Euclidean space symmetrical in the origin and having volume not less than $2^n m_1\cdots m_k$. Let $L_i(x_1, \cdots, x_n)$, $i=1, \cdots, k$, be homogeneous linear forms with integral coefficients (or, more generally, functions on ordered n-tuples of integers taking integral values and having the property that $L_i(x_1, \cdots, x_n)\equiv L_i(y_1, \cdots, y_n) \pmod{m_i}$ implies $L_i(x_1-y_1, \cdots, x_n-y_n)\equiv 0 \pmod{m_i}$). Then there exists a nonzero lattice point $(x_1, \cdots, x_n)$ in $\mathfrak{K}$ such that $L_i(x_1, \cdots, x_n)\equiv 0 \pmod{m_i}$, $i=1, \cdots, k$. The title of the paper refers to the special case of the preceding result in which $0\leqq k\leqq n-1$, $m_1, \cdots, m_k$ are all equal to the same prime number p, and $\mathfrak{K}$ contains no nonzero lattice point all coordinates of which are divisible by p. The further specialization in which $k=1$ and $n=2$ is a generalization of a theorem of Thue [Christiania Vid. Selsk. Forh. **1902**, no. 7]. In another paper [Nieuw Arch. Wiskunde (2) **23**, 150–162 (1950); these Rev. **11**, 417] the author has given some applications of this case to power residues mod p.

P. T. Bateman (Urbana, Ill.).

Citations: MR 11, 417i = N72-2.
Referred to in A10-6.

H05-23 (13, 115b)

Störmer, Horand, und Walter, Gerhard. Verschärfung eines Satzes von Mahler über konvexe Körper in inhomogener Lage. Arch. Math. **2**, 346–348 (1950).

Let $0<t<1$, $n\geqq 2$. Given a distance function $F(\mathfrak{x})$ in n-space. Suppose $F(\mathfrak{x})\leqq 1$ defines a centrally symmetric convex body of volume J, and suppose $\mathfrak{x}=\mathfrak{o}$ is the only integral vector satisfying $F(\mathfrak{x})\leqq 2t\cdot J^{-1/n}$. Then to every vector $\mathfrak{a}$ there is an integral vector $\mathfrak{g}$ such that

$$(1)\qquad F(\mathfrak{g}+\mathfrak{a})<\frac{(n-1)t^{n/2}+1}{t^{n-1}J^{1/n}}.$$

The proof given by the authors is very simple. By Minkowski's theorem there are n linearly independent integral vectors $\mathfrak{y}_1, \cdots, \mathfrak{y}_n$ such that $F(\mathfrak{y}_1)\leqq F(\mathfrak{y}_2)\leqq\cdots\leqq F(\mathfrak{y}_n)$ and $\prod_1^n F(\mathfrak{y}_\nu)\leqq 2^nJ^{-1}$. Thus $F(\mathfrak{y}_k)^{n-k+1}\prod_1^{k-1}F(\mathfrak{y}_\nu)\leqq 2^nJ^{-1}$. Since $F(\mathfrak{y}_\nu)>2tJ^{-1/n}$, this implies

$$(2)\qquad F(\mathfrak{y}_k)<2t^{-(k-1)/(n-k+1)}J^{-1/n}\quad (k=1, 2, \cdots, n).$$

Let $\mathfrak{a}=\sum a_k\mathfrak{y}_k$. Choose $\mathfrak{g}=\sum g_k\mathfrak{y}_k$ such that the g_k's are integers satisfying $|g_k+a_k|\leqq\frac{1}{2}$. Then

$$F(\mathfrak{g}+\mathfrak{a})=F(\sum(g_k+a_k)\mathfrak{y}_k)\leqq\sum|g_k+a_k|F(\mathfrak{y}_k)\leqq\tfrac{1}{2}\textstyle\sum_1^n F(\mathfrak{y}_k),$$

and (2) yields (1). *P. Scherk* (Saskatoon, Sask.).

Referred to in H05-28.

H05-24 (13, 212e)

Cohn, Harvey. On finiteness conditions for a convex body. Proc. Amer. Math. Soc. **2**, 544–546 (1951).

The following theorem is proved. Let K be a convex body in d-dimensional Euclidean space containing as its interior the hypersphere $\Sigma_d(r)$ of radius r and centre at the origin. Then if K fails to contain in its interior those lattice points (other than the origin) in a second sphere $\Sigma_d(C_dr^{1-d})$ it will contain in its interior no lattice points at all except the origin. In fact K will then lie entirely within the second sphere. This is proved by applying Minkowski's theorem to the body formed by joining the two points at the ends of a diameter of $\Sigma_d(\rho)$ to a circle of radius r centred at the origin and lying in a plane perpendicular to the diameter. This method gives $C_d=2^{d-1}d\pi^{-\frac{1}{2}(d-1)}\Gamma\{\frac{1}{2}(d+1)\}$. It is shown how this value of C_d may be improved slightly, and it is proved by means of Perron's transferal principle [J. F. Koksma, Diophantische Approximationen, Springer, Berlin, 1936, p. 67] that numbers C_d for which the theorem is true possess a positive lower bound C_d^* which is such that, as $d\to\infty$, $\log C_d^*/(d\log d)$ lies between $\frac{1}{2}$ and -1. *R. A. Rankin.*

H05-25 (13, 212f)

Cohn, Harvey. On the finite determination of critical lattices. Proc. Amer. Math. Soc. **2**, 547–549 (1951).

Let K be a finite convex body symmetric about the origin. The critical lattices for K are determined by minimising a determinant subject to an infinity of inequalities. Minkowski's finite basis theorem is used, in conjunction with the theorem proved in the paper reviewed above, to show that this infinite set of inequalities can be reduced to a finite set whose number depends only on the dimension of the space.

R. A. Rankin (Birmingham).

H05-26 (14, 540f)

Swinnerton-Dyer, H. P. F. Extremal lattices of convex bodies. Proc. Cambridge Philos. Soc. **49**, 161–162 (1953).

A neat proof that a critical (or, more generally, extremal) lattice of an n-dimensional convex body has at least $\frac{1}{2}n(n+1)$ points on the boundary. For $n=2$ or 3 this was proved by Minkowski [Ges. Abh., Bd. II, Teubner, Leipzig-Berlin, 1911, pp. 3–42] and for spheres and all n by Korkine and Zolotareff [Math. Ann. **11**, 242–292 (1877)]; but the general result is apparently novel.

J. W. S. Cassels (Cambridge, England).

H05-27 (14, 540g)

Hlawka, Edmund. Zur Theorie des Figurengitters. Math. Ann. **125**, 183–207 (1952).

In this memoir many of the known results of the geometry of numbers, including theorems of Minkowski, Blichfeldt, Siegel, Khintchine, and Koksma, are extended to more general situations. The most general situation contemplated is that in which the n-dimensional Euclidean space is replaced by a locally compact space R having a measure theory and satisfying certain other conditions, and a lattice is replaced by a discontinuous group Γ of measure-preserving automorphisms of R, with a bounded and measurable fundamental region. By a "Figurengitter" the author means the system of sets γA obtained from a given measurable set A in R by the application of all automorphisms γ of Γ. The various extensions of the classical theorems require the imposition of various additional restrictions on R and Γ and on the functions introduced. Some new results for the classical case are also given. One is the following [(28) of §4]: let $f(\mathbf{x})\leqq 1$ define a convex body of volume V in n-dimensional space, symmetrical about 0, and let $M_1, \cdots, M_n$ denote the successive Minkowski minima of $f(\mathbf{x})$ for a lattice of determinant Δ. Then for every point $\mathbf{x}_0$ of space there is a lattice point $\mathbf{x}$ such that

$$f(\mathbf{x}-\mathbf{x}_0)\leqq\tfrac{1}{2}M_n\left\{\frac{2^n\Delta}{VM_1\cdots M_n}\right\},$$

where $\{t\}$ denotes the least integer $\geqq t$. *H. Davenport.*

H05-28 (14, 541a)

Scherk, Peter. Convex bodies off center. Arch. Math. **3**, 303 (1952).

Let $F(\mathfrak{x})$ be a distance function in n-space such that $F(\mathfrak{x})\leqq 1$ defines a symmetric convex body of volume J. Suppose that $\mathfrak{x}=\mathfrak{o}$ is the only integral vector satisfying $F(\mathfrak{x})\leqq 2t\,J^{-1/n}$. Then it is shown that to every vector $\mathfrak{a}$ there is an integral vector $\mathfrak{g}$ such that

$$F(\mathfrak{a}-\mathfrak{g})<\frac{1+(n-1)t^n}{t^{n-1}J^{1/n}}.$$

This improves further a result due to Mahler recently improved by H. Störmer and G. Walter [Arch. Math. **2**, 346–348 (1949); these Rev. **13**, 115]. A slightly stronger form of the author's result is contained in the paper of E. Hlawka reviewed above. *C. A. Rogers* (London).

Citations: MR 13, 115b = H05-23.

H05-29 (14, 624a)

Macbeath, A. M. A theorem on non-homogeneous lattices. Ann. of Math. (2) **56**, 269–293 (1952).

Let R be a closed convex set in n-dimensional space. Suppose R has an interior and that it has n tac-planes having one and only one point in common. For a lattice Λ, let $d(\Lambda)$ denote its determinant. The critical determinant $\delta(K)$ of a set K symmetrical about a point u is inf $d(\Lambda)$ for all Λ such that $\Lambda\cap K=u$. The main result is that if $\Lambda\cap R\neq 0$, there is a point x in $\Lambda\cap R$ such that $\delta(R\cap R')<d(\Lambda)$, where $R'=2x-R$ is the set of all points $2x-r$ with r in R. Further, since the volume $f(x)$ of $R\cap R'$ is at most $2^n\delta(R\cap R')$, $f(x)<2^nD(\Lambda)$. The result and its proof are a considerable generalization of some work of Čebotarev and of Chalk [Quart. J. Math., Oxford Ser. **18**, 215–227 (1947); these Rev. **9**, 413]. The proof is long, requiring most of the paper, and is mostly geometric, using results from the theory of convex sets. It involves obtaining a sequence of points of Λ for which $\delta(R\cap R')$ decreases ultimately to less than $d(\Lambda)$. Particular applications of the theorem give results of Ollerenshaw [J. London Math. Soc. **20**, 22–26 (1945); these Rev. **7**, 417] and Blaney [ibid. **23**, 153–160 (1948); these Rev. **10**, 511]. Another consequence is that every lattice Λ has a point x such that

$$0\leqq x_1-x_2^2-\cdots-x_n^2\leqq\{2^{n-2}(n+1)d(\Lambda)/A\}^{2/(n+1)},$$

where A is the volume of the unit sphere in $n-1$ dimensions.
L. Tornheim (Ann Arbor, Mich.).

Citations: MR 7, 417d = H20-8; MR 9, 413b = J40-6; MR 10, 511c = J48-4.
Referred to in H05-31.

H05-30 (14, 624b)

Macbeath, A. M. Non-homogeneous lattices in the plane. Quart. J. Math., Oxford Ser. (2) **3**, 268–281 (1952).

A much simpler proof of the theorem of the paper reviewed above is given for the case of the plane. The theorem is then used to prove that if $k>2$ and $d(\Lambda)=a(k-2)(k^2-4)^{\frac12}$, then there is a point of Λ in the set $a\leqq xy<(k-1)^2a$, $x\geqq 0$, $y\geqq 0$. That this result is best possible for integral k is proved after first showing that the critical determinant of the set $x^2-\lambda y^2+(\lambda+1)|y|\leqq 1$, $|y|\leqq 1$ is $\frac12(2-\lambda)^{\frac12}$ when $0\leqq\lambda<1$. *L. Tornheim* (Ann Arbor, Mich.).

H05-31 (15, 941c)

Rogers, C. A. A note on the theorem of Macbeath. J. London Math. Soc. **29**, 133–143 (1954).

A point x of a set G in n-dimensional space is called exterior if there is a half space H such that $H\cap G=x$. Let Γ be a grid of determinant $d(\Gamma)$ and R be an open convex set not an infinite cylinder. If $\Gamma\cap R$ is not empty it has an exterior point and every exterior point x satisfies (1) $\Delta(S(R,x))\leqq d(\Gamma)$, where $(R,x)=\{R-x\}\cap\{-R+x\}$ and $\Delta(S)$ is the critical determinant of S. This generalizes a result of Macbeath [Ann. of Math. (2) **56**, 269–293 (1952); these Rev. **14**, 624]. The proof is simple. Conditions when an infinitude of points of $\Gamma\cap R$ satisfy (1) are given. Let $V(R,x)$ be the volume of (R,x) and $V(R,\Gamma)$ the g.l.b. of $V(R,x)$ for $x\in\Gamma\cap R$. There exists a constant $\lambda_n<1$ and dependent only on the dimension n such that if $x\in\Gamma\cap R$ and $V(R,x)<V(R,\Gamma)/\lambda_n$, then x is an exterior point of $\Gamma\cap R$. *L. Tornheim* (Ann Arbor, Mich.).

Citations: MR 14, 624a = H05-29.
Referred to in J32-55.

H05-32 (14, 624f)

Rankin, R. A. The anomaly of convex bodies. Proc. Cambridge Philos. Soc. **49**, 54–58 (1953).

Let K be a convex symmetrical body in 3-dimensional space with critical determinant $\Delta(K)$. For any lattice Λ let $\mu_r=\mu_r(K,\Lambda)$ be the lower bound of the positive numbers μ such that the expanded body μK contains r linearly independent points of Λ (for $r=1,2,3$). It has been conjectured that $\mu_1\mu_2\mu_3\Delta(K)\leqq d(\Lambda)$. The author proves that $\mu_1\mu_2\mu_3\Delta(K)\leqq\beta d(\Lambda)$ where $\beta=1.226\cdots$ satisfies

$$\beta(\beta+6)^2=64.$$

The proof depends on a geometrical result of the author [Proc. Cambridge Philos. Soc. **49**, 44–53 (1953); these Rev. **14**, 678]. *C. A. Rogers* (London).

H05-33 (15, 780e)

Sawyer, D. B. The lattice determinants of asymmetrical convex regions. J. London Math. Soc. **29**, 251–254 (1954).

Let K be an open convex region in Euclidean n-space which contains the origin O. If $P'OP$ be a straight line segment joining the boundary points P', P of K, the coefficient of asymmetry λ of K is defined to be the bound PO/OP'. The author shows that a region N exists in K, symmetric with respect to the origin, for which

$$V(N)\geqq\left(\frac{2\lambda}{\lambda+1}\right)^n\{\lambda^n-(\lambda-1)^n\}^{-1}V(K),$$

where $V(K)$, $V(N)$ represent the n-dimensional volumes of K and N respectively. This leads to an upper bound of bound $V(K)/\Delta(K)$ for a given λ, $\Delta(K)$ being the critical determinant of the admissible lattices of K. This relationship reduces to the Minkowski inequality $V(K)/\Delta(K)\leqq 2^n$ if $\lambda=1$ and K thus becomes symmetric. A lower bound of bound $V(K)/\Delta(K)$, for a given λ, $\lambda\geqq 2$, is obtained by construction of a special asymmetric region K. *D. Derry.*

Referred to in H05-41.

H05-34 (16, 341d)

Wolff, Karl H. Über kritische Gitter im vierdimensionalen Raum (R_4). Monatsh. Math. **58**, 38–56 (1954).

In her thesis [Wien, 1944] E. Brunngraber has shown that, if Λ is a critical lattice of a convex symmetrical body K in 4-dimensional space, then it is possible to choose a coordinate system, so that the points $e_1=(1,0,0,0)$, $e_2=(0,1,0,0)$, $e_3=(0,0,1,0)$, $e_4=(0,0,0,1)$ form a basis of Λ while the first three of these points, together with at least one of the points $p_1=(0,0,0,1)$, $p_2=(0,1,1,2)$, $p_3=(1,1,1,2)$, $p_4=(1,1,1,3)$, $p_5=(1,1,2,4)$, $p_6=(1,2,2,5)$, lie on the boundary of K. The author gives another proof of this result. He is then able to list, in each of the cases $k=1,2,\cdots,6$, a finite number of points with integral coordinates, having the property, that, if none of these points lie in a convex symmetrical body with e_1, e_2, e_3, p_k on its boundary, then no point with integral coordinates, other than the origin, lies in the body. This extends to 4-dimensions classical results of Minkowski [Nachr. Ges. Wiss. Göttingen. Math.-Phys. Kl. **1904**, 311–355] proved in explicit form only in 2 and 3 dimensions. The method is applied, in a refined form to the example when K is a sphere; and the critical lattices are completely determined in this case. *C. A. Rogers* (Birmingham).

Referred to in H05-65, H05-77.

H05-35 (16, 574b)

Ehrhart, Eugène. Une généralisation du théorème de Minkowski. C. R. Acad. Sci. Paris **240**, 483–485 (1955).

A short proof of the following analogue of Minkowski's fundamental theorem in 2 dimensions is given: Let K be a plane closed convex body with area $\geqq 4.5$ and let Λ denote any lattice of determinant 1 having a point at the centroid G of K. Then there exist at least two points of Λ, other than G, in K.

The result is best possible in two ways. (i) The constant

4.5 is exact, and is attained when K is a triangle. (ii) When K is a triangle, there is a lattice Λ of determinant 1 having only two points, other than G, in K. In a footnote, the author states a somewhat similar result for convex solids of revolution in 3 dimensions. *J. H. H. Chalk* (London).

Referred to in H05-37, H05-40, H05-44, H05-45, H05-80, P28-21.

H05-36 (16, 680b)

Hlawka, Edmund. Zur Theorie der Überdeckung durch konvexe Körper. Monatsh. Math. **58**, 287–291 (1954).

Four estimates are obtained for the inhomogeneous minimum M_I of a convex body K in n-dimensional space relative to a lattice G. If K is given by $f(\mathbf{x}) \leqq 1$, where f is homogeneous of degree 1, then M_I is defined by

$$M_I = \max_{\mathbf{p}} \min_{\mathbf{g}} f(\mathbf{p}+\mathbf{g}),$$

where $\mathbf{p}$ is any point and $\mathbf{g}$ any point of G. One of the estimates is expressed in terms of the number B of pairs of points of G, other than O, in the interior of K; it is

$$2M_I \leqq w^{1/n}\{w\}^{1-1/n}, \quad w = 2^n(B+1)\Delta V^{-1},$$

where $\{w\}$ denotes the least integer $\geqq w$ and $\Delta = \det G$, V = volume of K. Another estimate, which follows from this, is

$$2M_I \leqq v^{1/n}\{v\}^{1-1/n}\mu_1, \quad v = 2^n \Delta V^{-1}\mu_1^{-n},$$

where $\mu_1 = \min f(\mathbf{g})$ over $\mathbf{g} \neq O$. All the estimates are proved by applying Minkowski's fundamental theorem, or a similar theorem, to a suitable cylindrical body in $n+1$ dimensions, constructed in terms of K and an arbitrary point $\mathbf{p}$. *H. Davenport* (London).

Referred to in H05-59.

H05-37 (16, 740b)

Ehrhart, Eugène. Sur les ovales et les ovoïdes. C. R. Acad. Sci. Paris **240**, 583–585 (1955).

This note is mainly concerned with non-central convex bodies. Methods similar to those developed by the author in two previous articles [Rev. Math. Spéc. **64**, 61–62, 85–87 (1953); C. R. Acad. Sci. Paris **240**, 483–485 (1955); MR **16**, 574] here lead to results both geometrical and arithmetical in character. Among them is the following analogue of Minkowski's fundamental theorem on central convex bodies. Theorem. Let K denote a closed convex solid of revolution in 3-dimensional space. Let Λ denote any lattice of determinant 1 having a point at the centroid G of K. Then, if $V(K) \geqq 4^4/3^3$, K contains at least one point of Λ, other than G. *J. H. H. Chalk* (London).

Citations: MR 16, 574b = H05-35.

Referred to in H05-40, H05-44.

H05-38 (16, 802a)

Mahler, Kurt. On a problem in the geometry of numbers. Rend. Mat. e Appl. (5) **14**, 38–41 (1954).

Let K be a bounded convex body, symmetric with respect to the origin, of points $(x_1, x_2, \cdots, x_n)$ in Euclidean n-space. For $s>0$, let K_s denote the convex body of all the points of K for which $|x_n| \leqq s$ and $\Delta(s)$ the critical determinant of K_s. The author conjectures that $\Delta(s)/s$ is a decreasing function of s and proves this for $n=2$. *D. Derry.*

H05-39 (16, 898c)

Kneser, Martin. Ein Satz über abelsche Gruppen mit Anwendungen auf die Geometrie der Zahlen. Math. Z. **61**, 429–434 (1955).

Let A and B be two non-null finite subsets of an abelian group G with $n(A)$, $n(B)$ elements respectively. The author proves that $A+B=G$ if $n(A)+n(B)>n(G)$, while $n(A+B) \geqq n(A)+n(B)-n(H)$ if $n(A)+n(B) \leqq n(G)$. Here H is a proper subgroup of G. This is a weak form of a stronger result stated by the author which can be proved by methods similar to those he has used in an earlier paper [Math. Z. **58**, 459–484 (1953); MR **15**, 104], but is all the author needs to prove certain results concerning the n-dimensional torus group T. Thus, if A and B are two non-null subsets of T, then $A+B=T$ when $i(A)+i(B)>i(T)$, while $i(A+B) \geqq i(A)+i(B)$ when $i(A)+i(B) \leqq i(T)$. Here the function i denotes Jordan content. As remarked by the author, this result is included in a paper of A. M. Macbeath [Proc. Cambridge Philos. Soc. **49**, 40–43 (1953); MR **15**, 110] where it is proved similarly.

Applications to the geometry of numbers are given and results such as the following, which includes as particular cases theorems of Hlawka [Math. Z. **49**, 285–312 (1943); MR **5**, 201] and Schneider [Arch. Math. **2**, 81–86 (1950); MR **11**, 331] are obtained. Let $0<k \leqq k+h \leqq l$; suppose that the set A of Euclidean n-dimensional space contains not more than k, and the set B not more than l inner points which are congruent modulo I, where I is an n-dimensional lattice of determinant $d(I)$. If

$$(1-hl^{-1})k^{-1}i(A)+l^{-1}i(B)>d(I),$$

then for every vector x there exist at least $k+h$ lattice vectors $g \in I$ such that $x-g$ lies in $A+B$. *R. A. Rankin* (Glasgow).

Citations: MR 5, 201c = H25-1; MR 11, 331e = H25-9; MR 15, 104c = B08-42; MR 15, 110a = B08-44.

Referred to in H05-51.

H05-40 (16, 908a)

Ehrhart, Eugène. Sur les ovales en géométrie des nombres. C. R. Acad. Sci. Paris **240**, 935–938 (1955).

This article is a continuation of previous notes [same C. R. **240**, 483–485, 583–585 (1955); MR **16**, 574, 740] on properties of non-central convex bodies K in the geometry of numbers. As before, lattices are assumed to have a fixed point at the centroid of K. *J. H. H. Chalk* (London).

Citations: MR 16, 574b = H05-35; MR 16, 740b = H05-37.

Referred to in H05-44.

H05-41 (16, 1090b)

Sawyer, D. B. The lattice determinants of asymmetrical convex regions. II. Proc. London Math. Soc. (3) **5**, 197–218 (1955).

[For part I see J. London Math. Soc. **29**, 251–254 (1954); MR **15**, 780.] Let K be a plane convex region of which the origin O is an interior point. Let $\Delta(K)$, $A(K)$ denote the lattice determinant and area of K respectively. The coefficient of symmetry λ of K is defined to be the upper bound of PO/OP' where P, P' are the boundary points of K on a chord POP'. Where $\lambda_0 = (8+\sqrt{15})/7$,

$$\begin{aligned}
\gamma(\lambda) &= 3\lambda^2-2\lambda+3-2(\lambda-1)(2\lambda^2+2)^{1/2} && \text{for } 1 \leqq \lambda \leqq \lambda_0,\\
\gamma(\lambda) &= \tfrac{9}{2}-\tfrac{3}{2}(\lambda-2)^2 && \text{for } \lambda_0 \leqq \lambda \leqq 2,\\
\gamma(\lambda) &= \tfrac{1}{2}(\lambda+1)^2/\lambda-1 && \text{for } 2 \leqq \lambda \leqq 3,\\
\gamma(\lambda) &= 2\lambda-(2\lambda^2-4\lambda-2)^{1/2} && \text{for } 3 \leqq \lambda.
\end{aligned}$$

The purpose of the paper is to prove $A(K) \leqq \gamma(\lambda)\Delta(K)$. This is a generalization of the Minkowski result for symmetric regions for which $\lambda=1$. The problem is divided into a considerable number of subcases. The work is elementary but too detailed to be described in a review. The regions K for which $A(K)=\gamma(\lambda)\Delta(K)$ are shown to be convex polygons and are completely determined. *D. Derry.*

Citations: MR 15, 780e = H05-33.

H05-42 (16, 1090c)

Sawyer, D. B. The lattice determinants of asymmetrical convex regions. III. Quart. J. Math., Oxford Ser. (2) **6**, 27–33 (1955).

Let K be a convex body in Euclidean n-space symmetric with respect to one of its points, and of which the origin O is an interior point. The coefficient of symmetry λ of K is defined to be the upper bound of PO/OP', where P, P' are the boundary points of K on a chord POP'. Where λ_0 is defined as a root of a certain equation $q(\lambda)=2\lambda(\lambda+1)^{-1}$ for $1\leqq\lambda_0\leqq\lambda_0$, and $q(\lambda)=2^n\{1+n(\lambda-1)\}(\lambda+1)^{-n}$ for $\lambda_0\leqq\lambda$. The principal result of the above paper is that a convex body L exists within K, symmetric with respect to the origin, for which $V(L)\geqq q(\lambda)V(K)$, where $V(L)$, $V(K)$ denote the volumes of L and K respectively. This is a best possible result, as for each λ regions K and L are shown to exist for which $V(L)=q(\lambda)V(K)$. *D. Derry* (Vancouver, B. C.).

H05-43 (16, 1145f)

Bambah, R. P. Polar reciprocal convex bodies. Proc. Cambridge Philos. Soc. **51**, 377–378 (1955).

Let K and K' be two convex bodies in the n-dimensional Euclidean space which are polar reciprocal to each other with respect to the unit sphere J_n centered at the origin 0. Let $V(K)$, $V(K')$ and J_n be the volumes of K, K' and J_n respectively. The author proves the inequalities: a) $V(K)V(K')\geqq n^{-n/2}J_n^2$ for K having 0 as centre of symmetry; b) $V(K)V(K')\geqq 4^n/(n!)^2$ for K having 0 as an inner point. *L. A. Santaló* (Buenos Aires).

H05-44 (17, 350e)

Ehrhart, Eugène. Propriétés arithmo-géométriques des ovales. C. R. Acad. Sci. Paris **241** (1955), 274–276.

This is a continuation of three previous notes [same C. R. **240** (1955), 483–485, 583–585, 935–938; MR **16**, 574, 740, 908]. Let K denote a plane non-central convex body (i.e. "oval") with area >6 and let Λ denote any lattice of determinant 1 having a point at the centroid G of K. Then, under certain restrictions which are probably spurious, it is shown that K contains a point of Λ, other than G, together with its image in G. An example in which K is a triangle suffices to show that the constant 6 is best possible. *J. H. H. Chalk* (London).

Citations: MR 16, 574b = H05-35; MR 16, 740b = H05-37; MR 16, 908a = H05-40.

H05-45 (17, 350f)

Ehrhart, Eugène. Propriétés arithmogéométriques des polygones. C. R. Acad. Sci. Paris **241** (1955), 686–689.

The critical cases in the author's extension of Minkowski's fundamental theorem [same C. R. **240** (1955), 483–485; MR **16**, 574] are characterized. *J. H. H. Chalk* (London).

Citations: MR 16, 574b = H05-35.
Referred to in P28-20, P28-21.

H05-46 (17, 589a)

Mahler, Kurt. On compound convex bodies. I. Proc. London Math. Soc. (3) **5** (1955), 358–379.

Let $1\leqq p<n$ and let $X^{(\pi)}=(x_{\pi 1},\cdots,x_{\pi n})$ $[\pi=1,2,\cdots,p]$ denote p points in euclidean n-space R_n. They determine the p-vector (1) $\Xi=[X^{(1)},\cdots,X^{(p)}]$. The $N=\binom{n}{p}$ components $\xi_1,\cdots,\xi_N$ of Ξ are the minors of order p of the matrix $(x_{\pi\nu})$ arranged in some definite order. Ξ can be interpreted as a point on an algebraic manifold $\Omega(n,p)$ in R_N. Let $K^{(1)},\cdots,K^{(p)}$ denote convex bodies in R_n symmetric with respect to its origin. If each $X^{(\pi)}$ ranges over $K^{(\pi)}$, Ξ will range over a closed bounded set on $\Omega(n,p)$ whose convex hull $\mathrm{K}=[K^{(1)},\cdots,K^{(\pi)}]$ is a convex body in R_N symmetric with respect to its origin. If each $K^{(\pi)}$ has the distance function $F^{(\pi)}(X)$, that of K will be $\Phi(\mathrm{H})=\inf\sum_\varrho\prod_\pi F^{(\pi)}(X_\varrho^{(\pi)})$, where the lower bound is extended over all finite decompositions $\mathrm{H}=\sum_\varrho[X_\varrho^{(1)},\cdots,X_\varrho^{(p)}]$. Here H may be any point in R_N.

Put $P=\binom{n-1}{p-1}$ and let $c_1,\cdots$ denote positive constants which depend only on n and p. Let $V(K),\cdots$ denote the volume of $K,\cdots$. In this paragraph let $K^{(1)}=\cdots=K^{(p)}=K$. Theorem 1: There are c_1, c_2 such that $c_1\leqq V(\mathrm{K})V(K)^{-P}\leqq c_2$ for every K. Let $\Delta(K),\cdots$ denote the lattice determinant of $K,\cdots$. Theorem 1 combined with Minkowski's and Hlawka's inequalities $2\zeta(n)\Delta(K)\leqq V(K)\leqq 2^n\Delta(K)$ yields Theorem 2: There are c_4, c_5 such that $c_4\leqq\Delta(\mathrm{K})\Delta(K)^{-P}\leqq c_5$ for every K. — If $X^{(1)},\cdots,X^{(p)}$ range in dependently over the points of a lattice L in R_n, the points (1) form a set $\Pi\subset\Omega(n,p)$. The finite sums $\sum_\varrho\Xi_\varrho$ with $\Xi_\varrho\subset\Pi$ form a lattice $\Lambda=[L]^{(p)}$ in R_N. Let $0<m_1\leqq m_2\leqq\cdots\leqq m_n$ $[0<\mu_1\leqq\mu_2\leqq\cdots\leqq\mu_N]$ denote the successive minima of K in L [of K in Λ]. Number the N products $M_\varkappa=m_{\nu_1}m_{\nu_2}\cdots m_{\nu_p}$ $[1\leqq\nu_1<\nu_2<\cdots<\nu_p\leqq n]$ in the order of increasing size. Theorem 3: There is a c_7 such that $c_7M_\varkappa\leqq\mu_\varkappa\leqq M_\varkappa$ $[\varkappa=1,\cdots,N]$ for every K and L. — The convex body K^{-1} consists of all those $Y=(y_1,\cdots,y_n)$ for which $|XY|=|\sum_1^n x_\nu y_\nu|\leqq 1$ for all $X=(x_1,\cdots,x_n)\subset K$. Theorem 4: Let $p=n-1$ [thus $\mathrm{K}\subset R_n$]. Then there are c_9, c_{10} such that $c_9V(K)K^{-1}\subset\mathrm{K}\subset c_{10}V(K)K^{-1}$ for every K. The lattice L^{-1} consists of those Y for which XY is an integer for all $X\subset L$. It is similar to $[L]^{(n-1)}$. Let $0\leqq m_1'\leqq\cdots\leqq m_n'$ denote the successive minima of K^{-1} in L^{-1}. Theorems 3 and 4 imply Theorem 5: There are c_{11}, c_{12} such that $c_{11}\leqq m_k m_{n-k+1}'\leqq c_{12}$ $[k=1,\cdots,n]$. A direct proof of this theorem with $c_{11}=1$, $c_{12}=(n!)^2$ was given by Mahler [Časopis Pěst. Mat. Fys. **68** (1939), 93–102; MR **1**, 202].

The lattice L_0 $[\Lambda_0]$ consists of all points in R_n $[R_N]$ with integral coordinates. Let (a_{hk}) be an nth-order square matrix. The elements of the square matrix $(\alpha_{\eta\varkappa}^{(p)})$ of order N are the N^2 minors of order p of (a_{hk}), the ordering of its rows and columns being the same as that of the $\xi_\varkappa$. Let

$$m=\min_{x\subset L_0}\max_{h=1,\cdots,n}\left|\sum_{k=1}^n a_{hk}x_k\right|$$

and

$$\mu=\min_{\Xi\subset\Lambda_0}\max_{\eta=1,\cdots,N}\left|\sum_{\varkappa=1}^N\alpha_{\eta\varkappa}^{(p)}\xi_\varkappa\right|.$$

Theorem 6: There are c_{15}, c_{16} such that $\mu\leqq c_{15}m^{(n-p)/(n-1)}$ and $m\leqq c_{16}\mu^{1/p}$ for all unimodular (a_{hk}). *P. Scherk*.

Citations: MR 1, 202c = H05-2.
Referred to in H05-50, H05-56, J68-65.

H05-47 (17, 589b)

Mahler, Kurt. On compound convex bodies. II. Proc. London Math. Soc. (3) **5** (1955), 380–384.

[Cf. the preceding review]. Let $p=r+s$; $r>0$, $s>0$. Let $K^{(1)}=K^{(2)}=\cdots=K^{(r)}=K_1$; $K^{(r+1)}=\cdots=K^{(p)}=K_2$; $\mathrm{K}=[K^{(1)},\cdots,K^{(p)}]$. Put $s=V(\mathrm{K})\{V(K_1)^rV(K_2)^s\}^{-P/p}$. Let $n\geqq 3$; $2\leqq p\leqq n-1$. Then S has no upper bound which depends only on n and p. There is a $c>0$ such that $S\geqq c$ for all K_1, K_2. *P. Scherk* (Saskatoon Sask.).

Referred to in H05-56.

H05-48 (17, 589c)

Mahler, Kurt. On the minima of compound quadratic forms. Czechoslovak Math. J. **5(80)** (1955), 180–193. (Russian summary)

[The second last review is quoted as I]. Let (a_{hk}) be symmetric and let $F(X)=\sum_{h,k=1}^n a_{hk}x_hx_k$ be positive definite. Then $(\alpha_{\eta\varkappa}^{(p)})$ also is symmetric and $\Phi^{(p)}(\Xi)=\sum_{\eta,\varkappa=1}^n\alpha_{\eta\varkappa}^{(p)}\xi_\eta\xi_\varkappa$ is also positive definite. Let $0<m_1\leqq m_2\leqq\cdots\leqq m_n$ $[0<\mu_1^{(p)}\leqq\cdots\leqq\mu_N^{(p)}]$ denote the successive minima of $F(X)$ in L_0 [of $\Phi^{(p)}(\Xi)$ in Λ_0]. Define $M_\varkappa$ as in I.

Let G_n denote the unit sphere in R_n. Theorem 1: $\Delta(G_n)^{2P}M_k \leqq \mu_k^{(p)} \leqq M_k$ $[k=1, 2, \cdots, N]$. It implies Theorem 2: $\Delta(G_n)^{2P}m_1^p \leqq \mu_1^{(p)} \leqq \Delta(G_n)^{-2}m_1^{(n-p)/(n-1)}$ if $|a_{hk}|=1$. Theorem 3: $\Delta(G_n)^{2N} \leqq \mu_\eta^{(p)}\mu_{N-\eta+1}^{(n-p)} \leqq \Delta(G_n)^{-2}$ $[\eta=1, 2, \cdots, N]$. The final Theorem 4 shows that Theorem 3 of I holds with $c_7=n^{-p/2}\Delta(G_n)^P$ if $L=L_0$, $\Lambda=\Lambda_0$.

P. Scherk (Saskatoon, Sask.).

Referred to in H05-50.

H05-49 (17, 402c)

Mahler, Kurt. The p-th compound of a sphere. Proc. London Math. Soc. (3) **5** (1955), 385–391.

Let Y be a vector in R_N, $N=\binom{n}{p}$, the components of which are the determinants of the $p\times p$ matrices in the matrix $[X_1, \cdots, X_p]$, where $X_1, \cdots, X_p$ are vectors from R_n. Γ_n^p denotes the pth compound of the unit n-sphere G_n, i.e., the convex hull of all points Y for which $X_1, \cdots, X_p \in G_n$. The author shows $\Gamma_n^p = G_N \cap \Omega(n, p)$, where G_N is the unit sphere in R_N and $\Omega(n, p)$ is the Grassmann manifold in R_N. It follows from this that Γ_n^p is the convex hull of the intersection of the surface of G_N with $\Omega(n, p)$. The second of these results is used to construct the support and distance functions of Γ_4^2. From the latter it follows that Γ_4^2 consists of all points $Y=(\xi_1, \cdots, \xi_6)$ for which $\sum_{H=1}^{H=3}(\xi_H+\xi_{H+3})^2 \leqq 1$, $\sum_{H=1}^{H=3}(\xi_H-\xi_{H+3})^2 \leqq 1$. *D. Derry.*

H05-50 (18, 196c)

Mahler, K. Invariant matrices and the geometry of numbers. Proc. Roy. Soc. Edinburgh. Sect. A. **64** (1956), 223-238.

This is a generalization of earlier results [Proc. London Math. Soc. (3) **5** (1955), 358–379; MR **17**, 589].

Let $R_n=\{X\}$ denote affine n-space with the unit vectors $U_1=(1, 0, \cdots, 0)$, $U_n=(0, \cdots, 0, 1)$. Given a p-linear mapping

$$M: X^{(1)}, \cdots, X^{(p)} \to [X^{(1)}, \cdots, X^{(p)}]$$

of the p-tuples of R_n into affine N-space R_N with the following properties: (i) There are N sets of indices $\nu_{\eta 1}, \cdots, \nu_{\eta p}$ such that the points $Y_\eta=[U_{\nu_{\eta 1}}, \cdots, U_{\nu_{\eta p}}]$ span R_N $(\eta=1, 2, \cdots, N)$. (ii) To every non-singular affine transformation T of R_n there is an affine transformation T^* of R_N such that $[TX^{(1)}, \cdots, TX^{(p)}]=T^*[X^{(1)}, \cdots, X^{(p)}]$. (iii) There is a constant P such that $|\det T^*|=|\det T|^P$ for all T. (iv) M maps the p-tuples of rational points onto rational points. — If $T=T(t_1, \cdots, t_n)$ is the transformation $TU_\lambda=t_\lambda U_\lambda$, then

$$T^*Y_\eta=t_1^{\alpha_{\eta 1}} \cdots t_n^{\alpha_{\eta n}}\cdot Y_\eta,$$

where the $\alpha_{\eta\nu}$ are non-negative integers with $\alpha_{\eta 1}+\cdots+\alpha_{\eta n}=p$. Put $q=q(M)=\max_{\eta,\nu}\alpha_{\eta\nu}$. — The positive constants $c_1, c_2, \cdots$ depend only on M.

A "convex body" is a closed bounded convex set with interior points which is symmetric with respect to the origin. Theorem 1: If $X^{(1)}, \cdots, X^{(p)}$ range through the convex body K in R_n, the convex hull of the points $[X^{(1)}, \cdots, X^{(p)}]$ is the associated convex body K. Let $V(K), \cdots$ denote the volume of $K\cdots$. Theorem 3: There are c_1, c_2 such that $c_1V(K)^p \leqq V(\mathrm{K}) \leqq c_2V(K)^p$. — Let $F(X)$ and $\Phi(\Xi)$ denote the distance functions of K and K respectively. Theorem 4:

$$\Phi([X^{(1)}, \cdots, X^{(p)}]) \leqq F(X^{(1)}) \cdots F(X^{(p)})$$

for all $X^{(\pi)} \subset R_n$. — Let $L_0(\Lambda_0)$ be the lattice of the integral points of R_n (of R_N). Let $0<m_1 \leqq m_2 \cdots \leqq m_n$ $(0<\mu_1 \leqq \cdots \leqq \mu_N)$ be the successive minima of K in L_0 (of K in Λ_0). Let $M_1, \cdots, M_N$ denote the N products $m_{\nu_{\eta 1}} \cdots m_{\nu_{\eta p}}$ arranged in order of increasing size. Theorem 5: $c_3M_\eta \leqq \mu_\eta \leqq c_4M_\eta$ for some c_3, c_4 $(\eta=1, \cdots, N)$. Theorem 6: $m_1^p \leqq c_5\mu_1$, $\mu_1^{n-1} \leqq c_6V(K)^{q-p}m_1^{nq-p}$ for some c_5, c_6. — Let $T=(a_{hk})$ be a unimodular affine transformation; $T^*=(\alpha_{\eta\kappa})$. They define distance functions

$$F(X)=\max_{h=1,\cdots,n} |\sum_{k=1}^{n} a_{hk}x_k|, \quad \Phi(\Xi)=\max_{\eta=1,\cdots,N} |\sum_{\kappa=1}^{N} a_{\eta\kappa}\xi_\kappa|$$

of the points $X=(x_1, \cdots, x_n) \subset R_n$ and $\Xi=(\xi_1, \cdots, \xi_N) \subset R_N$. Put $m_1=\min_{X\subset L_0} F(X)$, $\mu_1=\min_{\Xi\subset\Lambda_0}\Phi(\Xi)$. Theorem 6 leads to Theorem 7: $m_1^p \leqq c_1\mu_1$; $\mu_1^{n-1} \leqq c_8m_1^{nq-p}$ [cf. Mahler, Czechoslovak Math. J. **5(80)** (1955), 180–193; MR **17**, 589]. — The paper is concluded "with some short remarks on the connections to representation theory."

P. Scherk (Philadelphia, Pa.).

Citations: MR 17, 589a = H05-46; MR 17, 589c = H05-48.

H05-51 (18, 114e)

Birch, B. J. A transference theorem of the geometry of numbers. J. London Math. Soc. **31** (1956), 248–251.

Let A be a convex body in n-space with volume V and distance function $F(x)$. Where I is the lattice of all points with integer coordinates, let Λ be the lower bound of the numbers t for which the bodies tA with centers on the points of I cover the whole space. If $\lambda_1, \cdots, \lambda_n$ be the successive minima of $F(x)$, defined by Minkowski, let $Q_k=2^n(\lambda_1\cdots\lambda_{k-1}\cdot\lambda_k^{n-k-1}V)^{-1}$. The principal result of the paper is that $\Lambda \leqq \frac{1}{2}(Q_k\lambda_k)$, $1\leqq k\leqq n$, if $Q_k \geqq n!$ Examples are given to show that both the above estimate as well as the estimate for Λ given by Kneser [Math. Z. **61** (1955); 429–434; MR **16**, 898] valid for $n=2$, $Q_2<2$, cannot be improved. Thus the best possible estimate for Λ is known for all 2-dimensional convex regions. *D. Derry.*

Citations: MR 16, 898c = H05-39.

Referred to in H05-59.

H05-52 (18, 875g)

Müller, Claus. Die Grundprobleme der Geometrie der Zahlen. Math.-Phys. Semesterber. **5** (1956), 63–70.

This article is mainly expository. It contains two proofs of Minkowski's Fundamental Theorem, one of which appears to be a new variant of C. L. Siegel's analytical proof [Acta Math. **65** (1935), 307–323], and a discussion of the number of lattice points in a large circle.

C. A. Rogers (Birmingham).

H05-53 (19, 125e)

Birch, B. J. Another transference theorem of the geometry of numbers. Proc. Cambridge Philos. Soc. **53** (1957), 269–272.

Let λ and Λ be respectively the homogeneous and the inhomogeneous minimum of a convex body of volume V with respect to a lattice of unit determinant. The author gives a short and elegant proof that $\Lambda^{n-1}\lambda V \geqq 2/n$ and a rather longer proof of a slightly stronger result. He shows by an example that these results cannot be much improved. The proofs are entirely from first principles. Similar but weaker results have been given by Mahler using "successive minima" [Časopis Pěst. Mat. Fys. **68** (1939), 93–102; MR **1**, 202]. *J. W. S. Cassels.*

Citations: MR 1, 202c = H05-2.

H05-54 (19, 164b)

Woods, A. C. The anomaly of convex bodies. Proc. Cambridge Philos. Soc. **52** (1956), 406–423.

Let K be a closed convex set in E^n containing the origin as interior point, and Λ a lattice with determinant $d(\Lambda)$. For each i $(i=1, \cdots, n)$ we define $\mu_i(\Lambda)$ as the greatest lower bound of those positive μ for which μK contains i linearly independent lattice points, put

$$\Delta(K)=\inf_\Lambda \mu_1^{-n}(\Lambda)d(\Lambda)$$

and define the anomaly of K as

$$M(K)=\sup_{\Lambda} \mu_1(\Lambda)\cdots\mu_n(\Lambda)\Delta(K)/d(\Lambda).$$

It is known that for K with the origin as center $1\leqq M(K)\leqq 2^{(n-1)/2-1/n}$, but no K with $M(K)>1$ is known (star shaped K with $M(K)>1$ are known). It is proved here that $M(K)=1$ for $n=3$ if K has the origin as center, and $M(K)=1$ for $n=2$ and arbitrary K. *H. Busemann.*

H05-55 (19, 877d)

Rogers, C. A. The compound of convex bodies. J. London Math. Soc. **32** (1957), 311–318.

Let $1\leqq p\leqq n-1$ and let $X^{(\pi)}=(x_{\pi 1},\cdots,x_{\pi n})$ $(\pi=1,2,\cdots,p)$ denote p points in euclidean R_n. They determine the p-vector $\Xi=[X^{(1)},\cdots,X^{(p)}]$. Its $N=\binom{n}{p}$ components are the minors of order p of the matrix $(x_{\pi\nu})$. If they are arranged in some definite order, Ξ can be interpreted as a point in R_N. Let $K^{(1)},\cdots,K^{(p)}$ be symmetric convex bodies in R_n. If each $X^{(\pi)}$ ranges over $K^{(\pi)}$, Ξ will range over a point set in R_N whose convex hull $\mathrm{K}=[K^{(1)},\cdots,K^{(p)}]$ is a symmetric convex body. Put $P=\binom{n-1}{p-1}$ and let c_1, c_2 denote positive constants which depend only on n and p. Let $V(K)$, $\cdots$ denote the volume of K, $\cdots$ and put $S=V(\mathrm{K})\{\prod_{\pi=1}^{p} V(K^{(\pi)})\}^{-P/p}$. Mahler showed that S has no upper bound if $p>1$ and that (*) $S>c$, if there are only two distinct convex bodies among the $K^{(\pi)}$'s. The author proves (*) in the general case by symmetrizing each $K^{(\pi)}$ with respect to all the coordinate $(n-1)$-spaces. His proof uses the following lemma: Let $H^{(\pi)}$ be the convex body obtained from $K^{(\pi)}$ by Steiner symmetrization with respect to a given $(n-1)$-space through the origin; $\pi=1,2,\cdots,p$. Then $V([H^{(1)},\cdots,H^{(p)}])\leqq c_2V(\mathrm{K})$. — In the last section upper bounds for the successive minima of K and for their product are given without proofs. [Cf. Mahler, Proc. London Math. Soc. (3) **5** (1955), 358–379, 380–384; MR **17**, 589.] *P. Scherk.*

H05-56 (19, 877e)

Lekkerkerker, C. G. On the volume of compound convex bodies. Nederl. Akad. Wetensch. Proc. Ser. A. **60**= Indag. Math. **19** (1957), 284–289.

Let $K^{(1)},\cdots,K^{(p)}$ be p symmetric convex bodies in R_n where $1\leqq p\leqq n-1$; let $N=\binom{n}{p}$ and $P=\binom{n-1}{p-1}$; and let $\mathrm{K}=[K^{(1)},\cdots,K^{(p)}]$ be the compound convex body of $K^{(1)},\cdots,K^{(p)}$ in R_N [K. Mahler, Proc. London Math. Soc. (3) **5** (1955), 358–379, 380–384; MR **17**, 589]. Put $Q=V(\mathrm{K})\prod_{\pi=1}^{p} V(K^{(\pi)})^{-P/p}$, where V denotes the volume. In his paper, Mahler stated the conjecture that $Q\geqq c$ where $c>0$ depends only on n and p. This conjecture was proved by C. A. Rogers [see paper reviewed above] by means of Steiner symmetrisation. The author, without knowledge of Rogers' work, gives a second and completely different proof which is based on Minkowski's theorem on the successive minima of a convex body in the lattice of points with integral coordinates. Let $m_1^{(\pi)},\cdots,m_n^{(\pi)}$, for $\pi=1,\cdots,p$, be the successive minima of $K^{(\pi)}$, and let $A_1^{(\pi)},\cdots,A_n^{(\pi)}$ be n independent lattice points at which these minima are attained. Then N systems of p suffixes

$$(\mu_{1i},\cdots,\mu_{pi})\quad(i=1,2,\cdots,N)$$

can be constructed such that the compound points

$$\Xi^{(i)}=[A_{\mu_{1i}}{}^{(1)},\cdots,A_{\mu_{pi}}{}^{(p)}]$$

in R_N are independent and that further

$$\prod_{i=1}^{N}(m_{\mu_{1i}}{}^{(1)}\cdots m_{\mu_{pi}}{}^{(p)})\leqq\left(\prod_{\pi=1}^{p}\prod_{\nu=1}^{n} m_{\nu}{}^{(\pi)}\right)^{P/p}.$$

The 2^N points

$$\mp(m\mu_{1i}{}^{(1)}\cdots m\mu_{pi}{}^{(p)})^{-1}\Xi^{(i)}\quad(i=1,2,\cdots,N)$$

form now the vertices of a generalised "octahedron" contained in K, and the lower bound for Q is a consequence of Minkowski's theorem. *K. Mahler* (Manchester).

Citations: MR 17, 589a = H05-46; MR 17, 589b = H05-47.

H05-57 (19, 1073f)

Rogers, C. A.; and Shephard, G. C. The difference body of a convex body. Arch. Math. **8** (1957), 220–233.

Let K be a convex body in Euclidean n-space, and denote by DK its difference (or vector) body $K-K$. The volumes of these bodies satisfy an inequality of the form $V(DK)\leqq c_nV(K)$, where c_n depends only on n. For $n=2$ and 3 the smallest possible c_n have been known for a long time. For $n>3$ the known value of c_n turned out to be far from the best one. Using a lemma due to Th. Bang, the authors prove in an elegant manner that the inequality holds with $c_n=\binom{2n}{2}$, and that this is the best possible value. It is further shown that equality holds only for simplexes. Finally, from the result, a lower estimate, considerably better than those previously known, for the density of packings by the translates of a non-central convex body is deduced. {Reviewer's remark: C. Godbersen, Dissertation, Göttingen, 1938, should be added to the list of references. In this paper, the above results have been conjectured, and proofs are given for particular classes of bodies.} *W. Fenchel* (Copenhagen).

Referred to in H02-17.

H05-58 (20# 39)

Woods, A. C. On a theorem of Minkowski's. Proc. Amer. Math. Soc. **9** (1958), 354–355.

Let K be a closed strictly convex body in R_n, symmetric about the origin O and let Λ be an arbitrary lattice in R_n. The successive minima $\mu_1(\Lambda)$, $\mu_2(\Lambda)$, $\cdots$, $\mu_n(\Lambda)$ of K are defined as the least upper bounds of numbers $c_1,c_2,\cdots,c_n$, respectively, which have the property that c_iK contains at most $i-1$ linearly independent points of Λ within its interior $(i=1,2,\cdots,n)$. Let $X_1,X_2,\cdots,X_n$ denote n linearly independent points of Λ such that $\mu_i(\Lambda)K$ contains $X_1,X_2,\cdots,X_i$ for $i=1,2,\cdots,n$, and let $Z_1,Z_2,\cdots,Z_n$ denote a basis for Λ satisfying

$$X_i=\sum_{j=1}^{i} g_{ij}Z_j\quad(i=1,2,\cdots,n),$$

where the coefficients g_{ij} are integers with $g_{ii}>0$.

Denote by $X_{i1},X_{i2},\cdots,X_{ir_i}$ all the points of $\Lambda\cap\mu_i(\Lambda)K$ of the form

$$X_{is}=\sum_{j=1}^{i} g_{ij}{}^{(s)}Z_j,$$

where the coefficients $g_{ij}{}^{(s)}$ are integers with $g_{ii}{}^{(s)}>0$. Then the main result of the note is the inequality

$$\sum_{i=1}^{n} r_i\leqq 2^n-1.$$

This clearly implies that if Λ is admissible for K, then there are at most 2^n-1 pairs of points $\pm X$ of Λ lying on the boundary of K, which is a familiar theorem of Minkowski. *J. H. H. Chalk* (Hamilton, Ont.)

H05-59 (20# 3120)

Bambah, R. P. Some transference theorems in the geometry of numbers. Monatsh. Math. **62** (1958), 243–249.

The usual set of constants must be introduced: $f(X)=$ the gauge function for a convex body K in n-space of volume $V(K)$; Λ=given lattice of determinant $d(\Lambda)$; $\lambda_1/2$=sup of all t for which Λ is a packing lattice of tK; μ=inf of all t for which Λ is a covering lattice of tK; $\delta(\Lambda, \frac{1}{2}\lambda K)$=density of packing of $\frac{1}{2}\lambda K$ into Λ, or $(\lambda/2)^n V(K)/d(\Lambda)$; $\delta_1=\delta(\Lambda, \frac{1}{2}\lambda_1 K)$; $Q_1=1/\delta_1$; δ=density of best lattice packing of K; $Q=1/\delta$.

The author shows that if Λ is a packing lattice of $\frac{1}{2}\lambda K$, then $\mu<\frac{1}{2}\lambda\cdot 3^{m+1}$, where $3^m \leqq Q\delta < 3^{m+1}$. This result is an improvement of the result of Birch [J. London Math. Soc. **31** (1956), 248–251; MR **18**, 114], and it overlaps the result of Hlawka for non-lattice packings [Monatsh. Math. **58** (1954), 287–291; MR **16**, 680]. The author obtains similar results with inequalities

$$\mu<\tfrac{1}{2}\lambda 3^{m+1}(Q\delta/3^{m+1})^{1/n},\ \mu<\tfrac{1}{2}\lambda 2^{M+2},$$
$$\mu<\tfrac{1}{2}\lambda 2^{M+2}(Q\delta/2^{M+1})^{1/n},$$

where $2^M \leqq Q\delta < 2^{M+1}$. The method is due to Rogers [J. London Math. Soc. **25** (1950), 327–331; MR **13**, 323], but the proofs here are self-contained.

H. Cohn (Tucson, Ariz.)

Citations: MR 13, 323b = H30-15; MR 16, 680b = H05-36; MR 18, 114e = H05-51.

H05-60 (20# 5768)

Woods, A. C. On two-dimensional convex bodies. Pacific J. Math. **8** (1958), 635–640.

Consider a lattice L of points in E^2 with coordinates x^1, x^2. For a given closed convex set K with interior points let $\mu_i(K)$ $(i=1, 2)$ be the supremum of the numbers c_i such that c_iK contains at most $i-1$ independent points of L. Put $\Delta(K)=\inf d(L')$, where $d(L')$ is the determinant of L', and L' traverses all lattices such that K contains at most one point of L'. It is proved here that $\mu_1(L)\mu_2(L) \times\Delta(K)\leqq d(L)$, for any K and L. If $K(t)$ denotes the subset $|x^2|\leqq t$ of K then $\Delta[K(t)]/t$ decreases for $t>0$. These results were proved previously for K with center, respectively, by C. Chabauty [Ann. Sci. Ecole Norm. Sup. (3) **66** (1949), 367–394; MR **11**, 418] and by Mahler [Rend. Mat. e Appl. (5) **13** (1954), 38–41].

H. Busemann (Cambridge, Mass.)

Citations: MR 11, 418e = H15-34.

H05-61 (21# 41)

Groemer, Helmut. Eine Bemerkung über Gitterpunkte in ebenen konvexen Bereichen. Arch. Math. **10** (1959), 62–63.

Theorem: Let S be a closed 2-dimensional convex set symmetric about the origin. Suppose that the boundary of S is a curve with continuously varying curvature and affine-length λ. Then S contains a point with integral co-ordinates other than the origin provided that $\lambda^3\geqq 288$. The constant 288 is the best possible. The proof depends on the lower bound $\lambda^3/72$ for the area of any circumscribed hexagon [Fejes Tóth, *Lagerungen in der Ebene, auf der Kugel und im Raum*, Springer, Berlin–Göttingen–Heidelberg, 1953; MR **15**, 248] and the characterization of the lattice-constant of S in terms of circumscribed hexagons [Reinhardt, Abh. Math. Sem. Hansische Univ. **10** (1934), 216–230].

J. W. S. Cassels (Cambridge, England)

Citations: MR 15, 248b = H30-27.

H05-62 (21# 318)

Rogers, C. A.; and Shephard, G. C. Convex bodies associated with a given convex body. J. London Math. Soc. **33** (1958), 270–281.

Let K be an n-dimensional convex body with volume $V(K)$. Then symmetrical convex bodies associated with K are defined as follows. The "difference body" $DK=K-K$. The "reflexion body" R_aK for each $a\in K$ is the minimal convex body containing K which is centrally symmetric in a; and if a is any point such that R_aK has minimum volume, then the body R_aK is designated RK. The associated $(n+1)$-dimensional body CK is the convex hull of the two sets determined by $(X_1, X_2, \cdots, X_n, 0)=(X, 0)$ and $(-X, 1)$ for $X\in K$. This paper establishes inequalities among the volumes $V(K)$ and the associated convex bodies of K. Theorems 2 and 3 are stated as examples. Theorem 2: $V(K)\leqq V(CK)\leqq 2^nV(K)/(n+1)$, with equality on the left if and only if K is centrally symmetric, and with equality on the right if and only if K is an n-dimensional simplex. Theorem 3: If $a\in K$, then $V(K)\leqq V(R_aK)\leqq 2^nV(K)$, with equality on the left if and only if K is centrally symmetric with a as centre, and with equality on the right if and only if K is a simplex with a as a vertex.

P. C. Hammer (Madison, Wis.)

H05-63 (21# 3403)

Woods, A. C. A counter-example in the geometry of numbers. Quart. J. Math. Oxford Ser. (2) **10** (1959), 46–47.

The example concerns the critical lattices of a plane convex body K and those of a body K^* obtained by rotating K.

C. G. Lekkerkerker (Amsterdam)

H05-64 (21# 6568)

Melzak, Z. A. Minkowski's theorem with curvature limitations. I. Canad. Math. Bull. **2** (1959), 151–158.

The object of the research begun in this paper is to investigate how far the constant 4 can be improved in Minkowski's fundamental theorem on lattice points in a plane convex body K, if a curvature limitation is imposed on K. Only the integer lattice is considered. The convex bodies considered are the "r-regions", with the property that through each boundary point there passes a circle of radius r which contains K completely.

The paper, stated to be preliminary to further work, consists of qualitative results about maximal r-regions. A generalization to n dimensions is considered, and it is shown that a maximal r-region in n-space is the intersection of at most $f(n)$ solid spheres, where $f(n)$ is an even integer and $f(n)<2^n n(n-2)!+n-2/(n-1)$.

A. M. Macbeath (Dundee)

H05-65 (22# 3730)

Mordell, L. J. Lattice octahedra. Canad. J. Math. **12** (1960), 297–302.

Let A_i $(i=1, 2, \cdots, n)$ be a basis of the real vector space R^n, and K the convex hull of $\{\pm A_i\}$. The determination of all the lattices Λ of R^n (i.e., the discrete subgroups of R^n) such that $K\cap\Lambda=\{0, \pm A_i\}$ was known for $n\leqq 4$ [E. Bruungraber, Dissertation, Wien, 1944; K. H. Wolff, Monatsh. Math. **58** (1954), 38–56; MR **16**, 341]. The author gives a simpler proof for the result.

I. G. Amemiya (Tokyo)

Citations: MR 16, 341d = H05-34.

Referred to in H05-69.

H05-66 (22# 4015)

Woods, A. C. On the irreducibility of convex bodies. Canad. J. Math. **11** (1959), 256–261.

The first result is similar to results of Mahler [Nederl. Akad. Wetensch. Proc. **49** (1946), 331–343; MR **8**, 12] and Rogers [ibid. **50** (1947), 868–872; MR **9**, 228] for irreducible star bodies. An example shows that a convex body may be irreducible among the convex bodies, but not among the star bodies. The terminology "the lattice Λ is free at the point X" is not elegant, as in the present paper it is a qualification of X, but not of Λ.

C. G. Lekkerkerker (Amsterdam)

Citations: MR 8, 12b = H15-8; MR 9, 228a = H15-19.

H05-67 (22# 5626)

Groemer, Helmut. Über lineare homogene diophantische Approximationen. Arch. Math. **11** (1960), 188–191.

Let $F(P)$ denote a distance-function defined for all points $P=(x_1, x_2, \cdots, x_n)$ of the Euclidean space R_n, i.e., $F(P)>0$ for $P\neq(0, 0, \cdots, 0)$, $F(tP)=|t|F(P)$ for any real number t, $F(P+Q)\leq F(P)+F(Q)$. Let $G(P)$ denote a distance-function in R_m. Furthermore let $L_i(x_j)=\sum_{j=1}^m a_{ij}x_j$ $(i=1, 2, \cdots, n)$ denote n linear forms with real coefficients. Then the author proves the following theorem.

If F, G and L_i $(i=1, 2, \cdots, n)$ are given, there exists an infinity of systems of integers x_j $(j=1, 2, \cdots, m)$, and for every system $\{x_j\}$ integers y_i $(i=1, 2, \cdots, n)$, such that

$$F^n(L_i(x_j)-y_i)G^m(x_j) < \left(\frac{n}{n+m}\right)^n\left(\frac{m}{n+m}\right)^m\binom{n+m}{m}\frac{2^{n+m}}{V(F)V(G)},$$

where $V(F)$ and $V(G)$ denote the volumes of the convex bodies given by $F\leq 1$ and $G\leq 1$.

The author gives a short direct proof using Minkowski's lattice theorem. He observes that this problem is connected with a more general theory developed by Mullender [Nederl. Akad. Wetensch. Proc. **51** (1948), 874–884; MR **10**, 285]. In fact it is possible to give also a simple proof using Mullender's ideas.

Finally, the author gives some applications.

Corrections: The right-hand member of the inequality in Satz 3 (p. 191) is erroneous: The factors

$$\left(\frac{n}{n+m}\right)^n\left(\frac{m}{n+m}\right)^m\binom{n+m}{m}\left(\frac{2}{\pi}\right)^{n+m}$$

should be replaced by

$$\left(\frac{n}{n+m}\right)^{2n}\left(\frac{m}{n+m}\right)^{2m}\binom{2n+2m}{2m}\left(\frac{4}{\pi}\right)^{n+m}.$$

Moreover, in the next inequality, due to Minkowski, the right-hand member should read

$$\left(\frac{n}{n+1}\right)^{2n}\frac{2n+1}{n+1}\left(\frac{4}{\pi}\right)^{n+1}.$$

J. Popken (Amsterdam)

Citations: MR 10, 285c = H15-25.

H05-68 (24# A1257)

Lekkerkerker, C. G.

A theorem on the distribution of lattices.

Nederl. Akad. Wetensch. Proc. Ser. A **64** = *Indag. Math.* **23** (1961), 197–210.

Arising from the simultaneous approximation to zero of r real linear forms in s integral variables, the following theorem provides relevant information about bases of a lattice. Roughly, it tells us that a lattice Λ in n-space has a basis $\{\mathbf{b}_1, \cdots, \mathbf{b}_n\}$ such that, in the first r coordinates, the vectors $\mathbf{b}_1, \cdots, \mathbf{b}_r$ are approximately fixed large multiples of $\mathbf{e}_1, \cdots, \mathbf{e}_r$ (where $\mathbf{e}_i$ denotes the ith unit vector) while the remaining vectors $\mathbf{b}_{r+1}, \cdots, \mathbf{b}_n$, if taken modulo the subspace spanned by $\mathbf{b}_1, \cdots, \mathbf{b}_r$, are approximately fixed small multiples of $\mathbf{e}_{r+1}, \cdots, \mathbf{e}_n$. Theorem: Let Λ be a lattice in n-space and let r, s be positive integers with $r+s=n$. Let $\varepsilon>0$ and let N, M be sufficiently large positive numbers with $N^rM^{-s}=d(\Lambda)$. Then Λ has a basis $\{\mathbf{b}_1, \cdots, \mathbf{b}_n\}$ such that $\mathbf{b}_i=N(\mathbf{e}_i+\boldsymbol{\delta}_i)$, for $i=1, \cdots, r$, and $\mathbf{b}_i=\theta_{1i}\mathbf{b}_1+\cdots+\theta_{ri}\mathbf{b}_r+M^{-1}(\mathbf{e}_i+\boldsymbol{\delta}_i)$, for $i=r+1, \cdots, n$, where θ_{hi} $(h=1, \cdots, r;\ i=r+1, \cdots, n)$ are real numbers, and where the coordinates d_{hi} of the vectors $\boldsymbol{\delta}_i$ satisfy $d_{hi}=O\{\max(N^{-1+\varepsilon}, M^{-1+\varepsilon})\}$, for $h, i=1, \cdots, r$ and $h, i=r+1, \cdots, n$, and $d_{hi}=0$, for $h=1, \cdots, r;\ i=r+1, \cdots, n$.

A similar approximation theorem, for the special cases $(r, s)=(n-1, 1)$ and $(1, n-1)$, is due to H. Davenport [J. London Math. Soc. **30** (1955), 186–195; MR **16**, 803], who also gives information about the last coordinates of $\mathbf{b}_1, \cdots, \mathbf{b}_r$. *J. H. H. Chalk* (Toronto)

Citations: MR 16, 803a = J12-28.

Referred to in H15-76.

H05-69 (25# 3908)

Bantegnie, R.

À propos d'un problème de Mordell sur les octaèdres latticiels.

J. London Math. Soc. **37** (1962), 321–328.

Let M denote a lattice in n-space with basis A_i $(i=1, 2, \cdots, n)$. If Λ is any lattice containing $\pm A_i$ $(i=1, 2, \cdots, n)$ but having no other point $\neq 0$ in common with the convex cover Ω of $\pm A_i$ $(i=1, 2, \cdots, n)$, a general theorem of Minkowski asserts that the index of Λ/M is bounded by $(n!)$.

An associated problem is the determination of the points of Λ modulo M, particularly, the maximum order of an element of Λ/M. This question has two variants according as Ω is taken to be open or closed, so let the maximum order in each case be denoted by p_n, p_n', respectively. Recently, Mordell [Canad. J. Math. **12** (1960), 297–302; MR **22** #3730] considered the second problem and found $p_4'=5$. The new result is $p_4=16$, and the proof incorporates $p_2=2$, $p_3=4$ and $p_4'=5$. *J. H. H. Chalk* (Toronto)

Citations: MR 22# 3730 = H05-65.

Referred to in H05-79, H05-85.

H05-70 (26# 670)

Andrews, George E.

A lower bound for the volume of strictly convex bodies with many boundary lattice points.

Trans. Amer. Math. Soc. **106** (1963), 270–279.

In a previous article of the author [same Trans. **99** (1961), 272–277; MR **22** #10979], it was shown that if a strictly convex body C in n-space contains N non-coplanar lattice points (i.e., points with integral coordinates) on its boundary, then

$$S(C) > k(n)N^{(n+1)/n},$$

where $S(C)$ denotes the surface area of C and $k(n)>0$ is a constant depending only on n. If $V(C)$ denotes the volume C, a corresponding result is shown to hold: If C has N non-coplanar lattice points on its boundary, then $V(C)>K(n)N^{(n+1)/(n-1)}$, where $K(n)>0$ depends only on n.

J. H. H. Chalk (Toronto)

Citations: MR 22# 10979 = P28-44.

H05-71 (26# 674)

Sawyer, D. B.

Lattice points in rotated convex sets.

Quart. J. Math. Oxford Ser. (2) **13** (1962), 221–228.

Let K denote a convex bounded plane region symmetric with respect to the origin. The author determines a constant α with the property that if the area of K is less than α, then K can be rotated about the origin to a position in the plane which contains no point with integer coordinates other than the origin. An example is constructed to show that α is the best possible such constant.

D. Derry (Vancouver, B.C.)

H05-72 (27# 4138)

White, G. K.

A refinement of van der Corput's theorem on convex bodies.

Amer. J. Math. **85** (1963), 320–326.

The following theorem is proved: Let K be an open convex body in n-space, centrally symmetric about the origin 0, and with volume $V(K)$. Let Λ be any lattice with det $\Lambda = \Delta \neq 0$. Then if there are at most $k = 1, 2, \cdots$, distinct points $\{0 = P_0, P_1, \cdots, P_{k-1}\}$ of Λ such that $P_i \in K$ $(i = 0, 1, \cdots, k-1)$ and $P_i - P_j \in K$ $(i, j = 0, 1, \cdots, k-1)$, then $V(K) \leqq k2^n\Delta$. Further $V(K) = k2^n\Delta$ if and only if every grid of the form $\Gamma = X + \Lambda$ (where X is an arbitrary point), which has no point on the boundary of K, has exactly $k2^n$ points in K.

This result extends a theorem of van der Corput [Acta Arith. **1** (1935), 62–66; ibid. **2** (1936), 145–146] and gives the additional information on when the inequality for the volume becomes an equality.

G. C. Shephard (Birmingham)

H05-73 (28# 1176)

Bergmann, G.

Theorie der Netze.

Math. Ann. **149** (1963), 361–418.

Minkowski's classical 3-dimensional parallelepiped algorithm (for finding units in a cubic field) is extended in this paper to higher dimensions. The task is carried out with great thoroughness and generality; the principal generalization (apart from arbitrariness of dimension) concerns the notion of extremal figures.

A "vacancy" (Vakanz) is defined as a property V of finite sets of lattice points which is fulfilled by at least one such set but not by the set consisting of the origin $\mathfrak{o}$ alone. A V-set is a finite set of lattice points with property V. In the ordinary Minkowski case V is the property of containing at least one lattice point $\neq \mathfrak{o}$, but other vacancies play an equally important role, for example, the property V_B of containing a lattice basis.

The cylinder $Z: |x_k| < \zeta_k\ (>0)$, $k = 1, \cdots, s$, where x_k is real for $1 \leqq k \leqq r_1$, complex for $r_1 < k \leqq r_1 + r_2 = s$, $r_1 + 2r_2 = n$, is called extremal with respect to a given lattice G and vacancy V if (1) each face of Z contains at least one lattice point (not necessarily in its interior), (2) there is a V-set in Z, (3) there is no V-set in the interior of Z. If only conditions (1) and (2) are fulfilled, the cylinder is called normed.

Minkowski's notion of a neighbour (Nachbar) is replaced by that of "companion" (Begleiter). A companion of a given extremal (or normed) Z is an extremal (normed) cylinder obtained from Z through the operations of compression or expansion; these are straightforward generalizations of Minkowski's neighbour-producing operations. The author proves Minkowski's completeness theorem for arbitrary vacancies; from a given extremal cylinder Z it is possible to reach every extremal cylinder by a chain of successive companions.

The notion of companion is extended to couples $\{A, \mathfrak{A}\}$ where A is a normed cylinder (essentially with respect to V_B) and $\mathfrak{A}$ is a matrix corresponding to an ordered lattice basis contained in A and chosen according to a specified selection rule. If $\{B, \mathfrak{B}\}$ is a companion of $\{A, \mathfrak{A}\}$, then the transition $\{A, \mathfrak{A}\} \to \{B, \mathfrak{B}\}$ is characterised by (i) an index $+p$ or $-p$, indicating whether the transition is due to expansion or compression in the x_p-direction, and (ii) a unimodular integral transformation U from the basis $\mathfrak{A}$ to the basis $\mathfrak{B}$. Starting from a given couple $\{A, \mathfrak{A}\}$ one thus obtains the "net expansion" of $\{A, \mathfrak{A}\}$, consisting essentially of a directed graph whose edges are labelled by an index $\pm p$ and a transformation U. By the completeness theorem, every extremal Z will appear among the vertices $\{Z, \mathfrak{Z}\}$ of the net expansion of a given $\{A, \mathfrak{A}\}$.

If G is derived in the usual manner from the conjugate bases of an ideal of a number field K of degree n (a K-matrix for brevity) and θ is the diagonal transformation $\vartheta_1, \cdots, \vartheta_n$, where ϑ_i are conjugates of a unit of K, then the net expansion of $\{Z, \mathfrak{Z}\}$ with extremal Z will contain the vertex $\{Z\theta, \mathfrak{Z}\theta\}$. Thus the net expansion will eventually lead to a set of basic units of the field.

A net is called periodic if among the net expansions of its vertices there are only a finite number of distinct ones. The net of a couple $\{A, \mathfrak{A}\}$ with real $\mathfrak{A}$ is periodic if and only if $\mathfrak{A}$ is a totally real K-matrix, or one which can be derived from such a matrix by a trivial diagonal transformation. This is a far-reaching generalization of the theorem of periodic continued fractions.

Another important result concerns the class number of K. Provided that K is not totally complex, the number of essentially distinct nets which can be derived from K-matrices is equal to the class number of K. This again is a generalization a well-known result in continued fractions.

The actual determination of a net expansion seems to be a major undertaking. As an illustration the author describes a periodic net expansion in the field of $x^3 - 3x + 1$. Its complete specification requires 82 indexed transformations and 23 diagonal unit transformations. {At least one misleading error was noted in the labelling of the net in Figure 2. There are two $+2$ edges going out from the vertex $\mathfrak{o}$, an obvious impossibility.}

G. Szekeres (Kensington)

Referred to in R22-29, R22-32.

H05-74 (28# 1177)

Ennola, V.

On the lattice constant of a symmetric convex domain.

J. London Math. Soc. **36** (1961), 135–138.

Let C be an irreducible convex body in the euclidean plane, symmetric in the origin. The author shows that the octagon with vertices $(\pm 1, 0)$, $(\pm\frac{1}{2}, \pm\frac{1}{2}\sqrt{3})$, $(0, \pm 1)$ may be inscribed in some linear transform K of C in such a manner that the points $(\pm 1, 0)$, $(\pm\frac{1}{2}, \pm\frac{1}{2}\sqrt{3})$ belong to one critical lattice of K. With special choice of a boundary point R of K relative to such an octagon the author concludes that the supremum over all C of the ratio of the volume of C to its critical determinant exceeds $3.5252\cdots$, thus improving the known bound $3.4641\cdots$ due to Mahler [Duke Math. J. **13** (1946), 611–621; MR **8**, 444]. In a footnote the author remarks that a refinement of his method yields the bound 3.57.

A. C. Woods (Columbus, Ohio)

Citations: MR 8, 444e = H25-5.

Referred to in H05-90.

H05-75 (30# 3092)

Aggarwal, Satish Kumar; Kaur, Gurnam; Manocha, J. N.

On Mahler's convex sets.

Res. Bull. Panjab Univ. (*N.S.*) **14** (1963), 87–91.

In n-dimensional affine space over a field k complete with respect to a nontrivial non-archimedean valuation, it is shown that Mahler's definitions of a convex set and of a convex body in terms of gauge functions [Mahler, Ann. of Math. (2) **42** (1941), 488–522; MR **2**, 350] are equivalent to geometric definitions similar to the conventional geometric definitions in real affine space.

J. W. S. Cassels (Cambridge, England)

Citations: MR 2, 350c = H05-4.

H05-76 (32# 1173)

Sudan, G.; Bucur, C.

Remarques sur deux théorèmes arithmétiques.

Bull. Math. Soc. Sci. Math. Phys. R. P. Roumaine (*N.S.*) **6** (**54**) (1962), 235–238 (1964).

The authors show that Minkowski's theorem on the existence of an integral point in a symmetrical convex plane region may be strengthened for parallelograms of a special type, and deduce a well-known result on rational approximation to an irrational number.

{The article contains several confusing misprints.}

E. S. Barnes (Adelaide)

H05-77 **(32# 2380)**

Cleaver, F. L.

On a theorem of Voronoï.

Trans. Amer. Math. Soc. **120** (1965), 390–400.

Let L denote a lattice in E_n and H a hypersphere with the points O (origin), $X_1, X_2, \cdots, X_m$ of L on its boundary and no points of L in its interior. In case $n=3$, assume that the set S of points $X_1, \cdots, X_m$ contains 3 linearly independent points. It is proved that if there exists a choice of 3 linearly independent points in S that is not a basis for L, then L is a rectangular lattice, H has 8 points of L on its boundary, and the center of H is at the center of one of the cells of L. In the case $n=4$, assume that the set S contains 4 linearly independent points. If $m=4$, then the set S is a basis of L. If $m>4$, there exist parallel hyperplanes K and K' such that all the points of $L \cap H$ lie in these two hyperplanes. If $m \geqq 9$, then there exists a hyperplane with 8 points of $L \cap H$, and consequently a 3-dimensional rectangular sublattice of L. We omit the lengthier details of the conclusions in case $m=8$, $n=4$. The results in E_4 are based on a result of E. Brunngraber; cf. K. H. Wolff [Monatsh. Math. **58** (1954), 38–56; MR **16**, 341].

I. Niven (Eugene, Ore.)

Citations: MR 16, 341d = H05-34.

H05-78 **(32# 5602)**

Bambah, R. P.; Woods, Alan; Zassenhaus, Hans

Three proofs of Minkowski's second inequality in the geometry of numbers.

J. Austral. Math. Soc. **5** (1965), 453–462.

Although the proof of Minkowski's first theorem is intuitive, there is no similarly intuitive proof of the second theorem (about the product of the successive minima of a lattice with respect to a lattice). In the present paper the three authors each give a proof: one a variant of that of Weyl [Proc. London Math. Soc. (2) **47** (1942), 268–289; MR **3**, 273], and two of that of Davenport [Quart. J. Math. Oxford **10** (1939), 119–121].

J. W. S. Cassels (Cambridge, England)

Citations: MR 3, 273a = H05-6.

Referred to in H05-82.

H05-79 **(32# 5603)**

Bantegnie, Robert

Sur l'indice de certains réseaux de R^4 permis pour un octaèdre.

Canad. J. Math. **17** (1965), 725–730.

Let G be any lattice in R^n containing Z^n (the lattice of points in R^n with integer coordinates) and having the property: $G \cap \Omega_n = (O)$, where Ω_n is the "octahedron" defined by $\sum_{i=1}^n x_i < 1$. In a previous article [J. London Math. Soc. **37** (1962), 321–328; MR **25** #3908], the author found the maximum value (in fact, 16) for the index of Z^n in G, in the special case when $n=4$ and G/Z^4 is cyclic. Here the investigation for $n=4$ is completed by showing that the corresponding bound is 18 when G/Z^4 is not cyclic.

J. H. H. Chalk (Toronto, Ont.)

Citations: MR 25# 3908 = H05-69.

Referred to in H05-85.

H05-80 **(33# 6506)**

Hammer, Joseph

Some relatives of Minkowski's theorem for two-dimensional lattices.

Amer. Math. Monthly **73** (1966), 744–746.

Let K be a convex domain in two-dimensional Euclidean space, $A(K)$ its area, and $P(K)$ its perimeter. The following results are proved. (I) If $A(K) \geqq P(K)$ and K is symmetric about a lattice point, then K contains at least four more lattice points. (II) If $A(K) \geqq 2^r P(K)$, where r is a positive integer, then K contains at least $2^{r+2}-1$ lattice points.

These results generalise the classical theorem of Minkowski and also results of E. Ehrhart [C. R. Acad. Sci. Paris **240** (1955), 483–485; MR **16**, 574] and of E. A. Bender [Amer. Math. Monthly **69** (1962), 742–744].

G. C. Shephard (Birmingham)

Citations: MR 16, 574b = H05-35.

Referred to in H99-9.

H05-81 **(34# 147)**

Woods, A. C.

A note on dense subsets of lattices.

J. London Math. Soc. **41** (1966), 742–744.

The author gives a proof of the fact, stated by C. A. Rogers [*Packing and covering*, Cambridge Univ. Press, New York, 1964; MR **30** #2405], that every lattice of determinant 1 can be approximated arbitrarily closely by lattices with a basis of the form $(x, 0, 0, \cdots, 0)$, $(0, x, 0, \cdots, 0)$, $(0, 0, x, \cdots, 0)$, $\cdots$, $(0, \cdots, 0, x, 0)$, $(a_1, a_2, \cdots, a_{n-1}, x^{1-n})$, where $a_1, a_2, \cdots, a_{n-1}, x$ are real numbers with $x \geqq 0$.

I. G. Amemiya (Sapporo)

Citations: MR 30# 2405 = H02-17.

H05-82 **(34# 1276)**

Woods, A. C.

A generalisation of Minkowski's second inequality in the geometry of numbers.

J. Austral. Math. Soc. **6** (1966), 148–152.

Let L be a discrete point set in the n-dimensional Euclidean space R_n. Suppose $0 \in L$ and L contains n linearly independent points. For each parallelepiped P and real number $k>0$, denote by $L(kP)$ the number of points of L in kP and by $V(kP)$ the volume of kP. Define $D(L, P) = \liminf V(kP)/L(kP)$ as $k \to \infty$, and $d(L) = \inf d(L, P)$ over all non-degenerate P. ($1/d(L)$ is an "upper density" of L.)

Let K be a convex body in R_n with center 0 and volume $V(K)$. Let L be a set for which $d(L)$ is defined. Suppose L has the property that if $X, Y \in L$, then $X-Y$ or $Y-X \in L$. For each $i=1, 2, \cdots, n$, define $m_i = \inf u_i$ such that $u_i \geqq 0$ is real and $u_i K$ contains i linearly independent points of L. The author proves that $m_1 m_2 \cdots m_n V(K) \leqq 2^n d(L)$. When L is an n-dimensional lattice, this result coincides with the famous second theorem of Minkowski.

The proof is an adaptation of Davenport's proof of Minkowski's theorem as amended by the reviewer, the author and H. Zassenhaus [same J. **5** (1965), 453–462; MR **32** #5602]. The author remarks that Minkowski's own proof or its simplifications by Weyl, Cassels, or the reviewer, the author and Zassenhaus do not seem to apply without further restrictions on L.

R. P. Bambah (Columbus, Ohio)

Citations: MR 32# 5602 = H05-78.

H05-83 **(37# 160)**

Bambah, R. P.; Woods, A. C.

Convex bodies with a covering property.

J. London Math. Soc. **43** (1968), 53–56.

A subset S of euclidean n-space R_n is said to have the covering property if every set congruent to S contains at least one point with integer coordinates. Let K be the collection of closed convex sets in R_n. A set $C \in K$ with the covering property is said to be irreducible if no proper subset $C' \in K$ has the covering property. It is proved that any member of K with the covering property contains an irreducible member of K with the covering property. Say that a point P on the boundary of a set $C \in K$ with the covering property is irreducible if no subset C' of C such that $C' \in K$ and $P \notin C'$ has the covering property. Then a set C is irreducible with the covering property if and only if every extreme point on the boundary of C is irreducible. A construction is given of an irreducible member of K with the covering property contained in the rectangle with sides of length 1 and $\sqrt{2}$. *I. Niven* (Eugene, Ore.)

H05-84 (37# 2693)

Hammer, Joseph

On some analogies to a theorem of Blichfeldt in the geometry of numbers.

Amer. Math. Monthly **75** (1968), 157–160.

For the unit square lattice in Euclidean 2-space the author shows the following theorems. (1) Any central-symmetric (closed bounded) convex region of area greater than π, which is symmetric about a lattice point, can be brought into such a position by rotating it around the centre that it contains, besides the centre, at least two other lattice points. (2) Given any convex region of area greater than $(9/8)\pi$ whose centre of gravity is a lattice point, by rotating it around this point, the region can be brought into such a position that it covers, besides the centre, two further lattice points. (3) A convex region whose area is greater than $\pi/2$, by rotation around an arbitrary inner point, can be brought into such a position that it covers at least one lattice point. (4) In any convex region whose area is greater than $\pi/18$, a point can be found inside the region so that by rotating it around that point, the region can be brought into a position that makes it cover at least one lattice point. An application on the approximation of a given irrational by rationals is discussed. The n-dimensional versions of Theorems 1, 3 and 4 are indicated.

F. Supnick (New York)

H05-85 (37# 5154)

Bantegnie, R.

Le "problème des octaèdres" en dimension 5.

Acta Arith. **14** (1967/68), 185–202.

Let $B=\{e_i\}$ be a basis for a lattice M in n-space, and Ω the open convex cover of $B'=\{\pm e_i\}$. Denote by $\overline{\mathfrak{M}}$, $\mathfrak{M}$ the families of sublattices $\Lambda \supset$M such that, respectively, (i) $\Lambda \cap \bar{\Omega}=\{O, B'\}$, (ii) $\Lambda \cap \Omega=\{O\}$. Then Ω is an (n-dimensional) octahedron, and the "octahedron problem" concerns the lattices in (i) $\overline{\mathfrak{M}}$, (ii) $\mathfrak{M}$, and, in particular, $\bar{m}_n$, m_n, the maximum indices Λ:M for $\Lambda \in \overline{\mathfrak{M}}, \mathfrak{M}$. Let $\bar{\mathfrak{L}}$, $\mathfrak{L}$ be those subsets of $\overline{\mathfrak{M}}$, $\mathfrak{M}$ for which Λ/M is cyclic, and $\bar{p}_n$, p_n the corresponding maximum indices. Then $\bar{m}_n$, m_n, $\bar{p}_n$, p_n are known for $n \leqq 4$ (see table, p. 186), p_4 and m_4 having been previously found by the author [J. London Math. Soc. **37** (1962), 321–328; MR **25** #3908; Canad. J. Math. **17** (1965), 725–730; MR **32** #5603]. He now goes on to find $\bar{p}_5=41$, $p_5=48$, examples being Λ generated by $e_1e_2e_3e_4$ and (i) $x=(1/41)(18e_1+16e_2+10e_3+4e_4+e_5)$, (ii) $x=(1/48)(23e_1+13e_2+7e_3+4e_4+e_5)$. In addition he supplies lists which essentially give $\overline{\mathfrak{M}}$ for $n=5$, and $\mathfrak{M}$ (and thence, easily, $\overline{\mathfrak{M}}$) for $n=4, 3$; a partial list only is given for $\mathfrak{M}$ for $n=5$.

To obtain his results the author needed a large electronic computer plus considerable analysis to keep computer time within reason; according to him larger $n=6, 7, \cdots$, are at present beyond the capabilities of even today's giant computers. *G. K. White* (Vancouver, B.C.)

Citations: MR 25# 3908 = H05-69; MR 32# 5603 = H05-79.

H05-86 (38# 2094)

Chalk, J. H. H.; Rogers, C. A.

A study of convex cones from the point of view of the geometry of numbers.

J. Reine Angew. Math. **230** (1968), 139–166.

Let K be an n-dimensional closed convex cone which does not contain any complete straight line and has $\mathbf{o}$ as its vertex. Let $D=K \cup (-K)$. For $\mathbf{k} \in K$ define the Macbeath region $M(\mathbf{k})=K \cap (2\mathbf{k}-K)$; and the regions $S(\mathbf{k})$, $C(\mathbf{k})$ by $S(\mathbf{k})=(K-2\mathbf{k}) \cap (2\mathbf{k}-K)$, $C(\mathbf{k})=\mathrm{Con}\{M(\mathbf{k}) \cup (-M(\mathbf{k}))\}$. (Con denotes the convex hull.) Let H be a half space containing $\mathbf{o}$ which meets K in a bounded set. Define $A(H)=\mathrm{Con}\{(K \cap H) \cup (-(K \cap H))\}$. The authors say a point $\mathbf{x}$ is "near" the boundary of D if it lies in a set $M(\mathbf{k})$, $C(\mathbf{k})$, $S(\mathbf{k})$ or $A(H)$ of small volume. For $\mathbf{x} \in K$ all these definitions are equivalent. For $\mathbf{x} \notin K$, the definition in terms of $M(\mathbf{k})$ is meaningless while the other three are equivalent. The authors' "first contribution (Theorem 5) to the geometry of numbers is to show that there will be points of any given lattice outside D reasonably close to its boundary. Further, if there should be an especially 'large' subset of K that is devoid of all lattice points, then there will be a lattice point outside D and extra close to its boundary (Theorem 6). On the other hand, if there is an especially 'large' convex region of a suitable type that contains no lattice points outside D, then there is necessarily a point of the lattice other than $\mathbf{o}$ in K and very 'close' to its boundary (Theorem 7). ...If an especially large region of K is devoid of lattice points close to the boundary of K, then there will be an associated point of the lattice outside D and extra close to its boundary."

The exact statement of Theorem 5 is as follows: If Λ is a lattice with determinant $d(\Lambda)$, there is a half-space H^*, containing $\mathbf{o}$, with $H^* \cap K$ bounded and $V(A(H^*))=2^n d(\Lambda)$, such that the successive minima $\lambda_1(0), \cdots, \lambda_{n-1}(0)$ of Λ for the star-set $0=(\mathbf{o}) \cup \{A(H^*)\backslash D\}$ satisfy $\lambda_1(0)\lambda_2(0)\cdots\lambda_{n-1}(0) \leqq 2^{n-1/n}$. The statements of other theorems are similar.

The authors also discuss the existence points of Λ "near" the boundary of K which lie near the tac-planes of K in the Euclidean sense. Further, let K^* be the cone reciprocal to K and Λ^* the lattice polar to Λ. The authors relate the existence of points of Λ near the boundary of $D=K \cup (-K)$ to the existence of sets of independent points of Λ^* outside $D^*=K^* \cup (-K)^*$ but "near" its boundary.

"These results, taken together, give very detailed information concerning the distribution of lattice points near the boundary of K." All quotations are from the authors' well-written introduction.

R. P. Bambah (Chandigarh)

H05-87 (40# 90)

Danicic, I.

An elementary proof of Minkowski's second inequality.

J. Austral. Math. Soc. **10** (1969), 177–181.

Soit K un ouvert convexe de l'espace euclidien à n dimensions de volume V. Si K admet l'origine comme centre de symétrie et si $\lambda_1, \cdots, \lambda_n$ sont les minima successifs de K par rapport au réseau Λ des points à coordonnées entières, on a (1) $\lambda_1\lambda_2\cdots\lambda_n V \leqq 2^n$. C'est la seconde inégalité de Minkowski. L'auteur en donne une preuve brève et élémentaire. Pour cela il remarque que si $N(l)$ est le nombre des points $(u_1, \cdots, u_n)$ de Λ pour lesquels

$$X = (2u_1/l, (\lambda_1/\lambda_2)2u_2/l, \cdots, (\lambda_1/\lambda_n)2u_n/l)$$

est dans $\lambda_1 K$ on a $\lim_{l\to\infty} N(l)/l^n = 2^{-n}\lambda_1 \cdots \lambda_n V$ et que, par suite, (1) est prouvé par (2) $N(l) \leqq l^n(1+o(1))$ pour $l \to \infty$. Si $P_1, \cdots, P_n$ sont des points linéairement indépendants de Λ tels que pour chaque entier j avec $1 \leqq j \leqq n$ les points de $\Lambda \cap \lambda_j K$ dépendent linéairement de $P_1, \cdots, P_{j-1}$ et que pour $\lambda > \lambda_j$, $P_1, \cdots, P_j$ sont dans λK, il commence par se ramener au cas où $(u_1', \cdots, u_n') \in \Lambda \cap \lambda_r K$ entraîne $u_r' = u_{r+1}' = \cdots = u_n' = 0$. Il distingue alors les bons X et les mauvais X ; il montre que les seconds contribuent dans (2) au plus pour $o(l^n)$ et les premiers pour l^n (il dit qu'un point est mauvais si l'on peut trouver un entier r, $1 \leqq r \leqq n-1$, tel que $c_r = (\lambda_{r+1}/\lambda_r - 1)(\lambda_r/\lambda_1)$ soit différent de 0 et des entiers $v_1, \cdots, v_n$ tels que l'hypercube $2v_i/c_r l \leqq x_i \leqq 2(v_i+1)/c_r l$ $(i = 1, \cdots, n)$ contienne ce point et ait des points à la fois dans $\lambda_1 K$ et au dehors; un point est bon s'il n'est pas mauvais). *R. Bantegnie* (Besançon)

H05-88 (42# 5914)

Jones, A. J.
Cyclic overlattices. I.
Acta Arith. **17** (1970), 303–314.

Let Λ be a k-dimensional lattice and M a sublattice of Λ. Suppose Λ is "cyclic over M", i.e., the quotient graph Λ/M is cyclic. An element a of Λ is said to be a generator of Λ over M if the coset $a+M$ generates Λ/M. Let F be the distance function of a bounded convex body with center at 0. Let Λ^*, M^*, F^* be the polar reciprocals of Λ, M, F. Suppose M^* is admissible for $F^*(x) < 1$. Let $D > 1$. Let $S_D = \{x: F^*(x) < D, x \in \Lambda^*\}$. Let W_D^* be the space generated by S_D and Λ_D^*, M_D^* be the parts of Λ^*, M^* in W_D^*. If $\Lambda_D^* \neq M_D^*$, Theorem 1, due to Cassels, states that $F(a) > 1/D$ for all generating points a of Λ over M. On the other hand, if $\Lambda_D^* = M_D^*$, the author proves the existence of generating points a of Λ over M such that $F(a)$ is bounded by a suitable function of D. Another theorem is also proved. The results are motivated by an application to appear later. The author uses the words convex body to mean a convex body symmetrical about 0.
R. P. Bambah (Chandigarh)

H05-89 (43# 1926)

Mahler, Kurt
A lecture on the geometry of numbers of convex bodies.
Bull. Amer. Math. Soc. **77** (1971), 319–325.

A discussion of some general results in the theory of successive minima of convex bodies.
C. G. Lekkerkerker (Amsterdam)

H05-90 (44# 2707)

Tammela, P.
An estimate of the critical determinant of a two-dimensional convex symmetric domain. (Russian)
Izv. Vysš. Učebn. Zaved. Matematika **1970**, *no.* 12 (103), 103–107.

Let $Q(K) = V(K)/\Delta(K)$, where V and Δ are the volume and critical determinant and K is a plane convex body, symmetric with respect to the origin. K. Reinhardt [Abh. Math. Sem. Univ. Hamburg **10** (1934), 216–230] and K. Mahler [Nederl. Akad. Wetensch. Proc. Ser. A **50** (1947), 692–703; MR **9**, 10] found a smoothed octagon K_0 for which $Q(K_0) = 3.6096\ldots$, and they conjectured that K_0 minimized Q. V. Ennola [J. London Math. Soc. **36** (1961), 135–138; MR **28** #1177] showed that $Q(K) \geqq 3.5252\ldots$, and remarked that elaborations of his method would give a better estimate. The author accomplishes that task by showing that $Q(K) \geqq 3.570624\ldots$.
W. J. Firey (Corvallis, Ore.)

Citations: MR 9, 10h = H05-10; MR 28# 1177 = H05-74.

H10 LATTICES AND CONVEX BODIES: SPECIAL PROBLEMS

See also reviews H20-3, H20-20, J68-56, J68-61, P28-20, P28-22, P28-26, P28-38, P28-42, P28-61, P28-73, P28-82, P28-88, R28-1.

H10-1 (4, 36b)

Heinhold, Josef. Zur Geometrie der Zahlen. Math. Z. **47**, 199–214 (1941).

Let $f(x, y)$ be the distance function of a convex domain and consider the lower bound $k = k(f)$ of all positive constants c, such that for arbitrary real numbers ξ, η the inequality $f(x+\xi, y+\eta) \leqq c$ has an integral solution x, y. It is proved that there exists a unimodular transformation $x = \alpha x' + \beta y'$, $y = \gamma x' + \delta y'$ and two real numbers u, v satisfying $0 \leqq u \leqq v \leqq \frac{1}{4}$ and $f(x_r, y_r) = k$ $(r = 1, 2, 3)$, where $x_1' = \frac{1}{4} + u$, $y_1' = \frac{1}{4} + v$, $x_2' = \frac{3}{4} - u$, $y_2' = -\frac{1}{4} - v$, $x_3' = \frac{1}{4} + u$, $y_3' = -\frac{3}{4} + v$. The result is applied in the special case $f^p = |Ax+By|^p + |Cx+Dy|^p$ with real A, B, C, D, $AD - BC \neq 0$, $p \geqq 1$.
C. L. Siegel (Princeton, N. J.).

H10-2 (7, 51b)

Hua, Loo-Keng. A remark on a result due to Blichfeldt. Bull. Amer. Math. Soc. **51**, 537–539 (1945).

Let $\xi_1, \cdots, \xi_n$ be n linear forms in $x_1, \cdots, x_n$ of unit determinant, of which $n-2s$ are real and $2s$ are complex conjugate in pairs $(n \geqq 3, 2s \leqq n)$. The convex body

$$F(x_1, \cdots, x_n) \equiv |\xi_1|^\sigma + \cdots + |\xi_n|^\sigma \leqq 1, \qquad \sigma \geqq 1,$$

is of volume

$$V(\sigma) = 2^{n-(1+2/\sigma)s}\pi^s\Gamma^{n+2s}(1+1/\sigma)\Gamma^s(1+2/\sigma)\Gamma^{-1}(1+n/\sigma),$$

and so, by Minkowski's theorem [Geometrie der Zahlen, Leipzig, 1910, p. 76], there are integers $(x_1, \cdots, x_n) \neq (0, \cdots, 0)$ such that $F(x_1, \cdots, x_n) \leqq r^\sigma$ if $r^n \geqq 2^n V(\sigma)^{-1}$. Van der Corput and Schaake [Acta Arith. **2**, 152–160 (1936)] improved this, by means of Blichfeldt's method [Trans. Amer. Math. Soc. **15**, 227–235 (1914)], to

$$r^n \geqq \epsilon(\sigma)^{n/\sigma}(n+\sigma)\sigma^{-1}V(\sigma)^{-1},$$

where $\epsilon(\sigma)$ is the smallest constant such that, for all k and all complex $z_1, \cdots, z_n$,

$$\sum_{p,q=1}^{k} |z_p - z_q|^\sigma \leqq \epsilon(\sigma)k\sum_{p=1}^{k} |z_p|^\sigma.$$

They also proved that $\epsilon(\sigma) = 2^{\sigma-1}$ if $\sigma \geqq 2$. The author deduces this result in a few lines from M. Riesz's convexity theorem [Hardy, Littlewood and Pólya, Inequalities, Cambridge University Press, 1934, p. 219, theorem 296]. He also shows that $\epsilon(\sigma) = 2$ if $1 \leqq \sigma \leqq 2$, and gives another slight improvement if $1 \leqq \sigma \leqq 1.865 \cdots$. *K. Mahler.*

Referred to in H10-12, H10-13.

H10-3 (7, 368b)

Mahler, K. On lattice points in a cylinder. Quart. J. Math., Oxford Ser. **17**, 16–18 (1946).

If K denotes a convex body in 3-space symmetric with respect to the origin, a lattice Λ whose only point interior to K is the origin is known as a K-admissible lattice. Let $d(\Lambda)$ be the determinant of such a lattice. Let $\Delta(K)$ be the lower bound of $d(\Lambda)$ for all K-admissible lattices Λ. This paper shows that $\Delta(K)=\frac{1}{2}\sqrt{3}$ when K is the cylinder $x_1^2+x_2^2\leqq 1$, $|x_3|\leqq 1$. The proof depends on a bound for the number of nonoverlapping circular cylinders of the same dimensions which can be placed in a cube with axes perpendicular to a face of the cube. This is a consequence of an analogous result for a plane from an earlier paper of Segre and Mahler [Amer. Math. Monthly **51**, 261–270 (1944); these Rev. **6**, 16]. *D. Derry* (Vancouver, B. C.).

Referred to in H10-9, H10-19, H10-30.

H10-4 (8, 196b)

Rédei, L. Über Gitterparallelogramme. Mat. Fiz. Lapok **49**, 73–75 (1942). (Hungarian. German summary)

In der mit einem rechtwinkligen Koordinatensystem versehenen Ebene heissen die Punkte mit ganz rationalen Koordinaten Gitterpunkte, die Parallelogramme mit solchen Ecken Gitterparallelogramme. Zählt man zu diesen zwei benachbarte Seiten nicht mit, so gilt bekanntlich, dass die Anzahl der enthaltenen Gitterpunkte dem Inhalt gleich ist. Neben dem üblichen Beweis mit Grenzübergang und zwei neuerdings veröffentlichte elementare Beweisen werden zwei weitere elementare Beweise hingestellt.

From the author's summary.

H10-5 (8, 196c)

Hajós, G. Über Gitterparallelogrammen. Mat. Fiz. Lapok **48**, 398–400 (1941). (Hungarian. German summary)

The author gives a simple geometrical proof of the well-known theorem that a triangle whose vertices have integer coordinates and which has no lattice point in its interior or boundary has area $\frac{1}{2}$. *P. Erdös* (Syracuse, N. Y.).

H10-6 (8, 196d)

Makai, E. Über Gitterdreiecke und Gitterparallelogramme. Mat. Fiz. Lapok **50**, 47–50 (1943). (Hungarian. German summary)

Folgender bekannte Satz wird elementargeometrisch bewiesen: der Inhalt eines jeden elementaren Gitterdreiecks ist $\frac{1}{2}$. Der Beweis beruht auf folgenden beiden Tatsachen: (1) ein jedes elementares Gitterdreieck ist entweder kongruent mit dem Dreieck (0, 0), (1, 0), (1, 1), oder es hat einen stumpfen Winkel und (2) zu einem jeden stumpfwinkligen elementaren Gitterdreieck $A_1A_2A_3$ kann man ein anderes elementares Gitterdreieck $A_1A_2A_4$ mit demselben Inhalt, aber von kleinerem Umfang konstruieren.

From the author's summary.

H10-7 (9, 500h)

Hlawka, Edmund. Über Potenzsummen von Linearformen. Akad. Wiss. Wien, S.-B. IIa. **154**, 50–58 (1945).

Let $L_1, \cdots, L_n$ be linear forms in $x_1, \cdots, x_n$ of determinant $D\neq 0$, consisting of r forms with real coefficients and s pairs of forms with conjugate complex coefficients ($r+2s=n$). Let $q\geqq 2$ be real. Minkowski, in his Geometrie der Zahlen, gave an inequality of the form $|L_1|^q+\cdots+|L_n|^q\leqq A|D|^{q/n}$, which always has a solution in integers $x_1, \cdots, x_n$, not all zero. Blichfeldt [Math. Ann. **101**, 605–608 (1929)] gave two improvements on this when $q=2$. The first of these was extended to $q>2$ by van der Corput and Schaake [Acta Arith. **2**, 152–160 (1936)]. The author now extends the second and more precise result of Blichfeldt. His conclusion is that Minkowski's A can be replaced by $\frac{1}{2}(1+n/q)^{q/n}A/(1+r)^{q/n}$, where

$$r=\frac{q}{q+2n}2^{-n/q}-K^{n+\frac{1}{2}q}\left\{\frac{n}{q}K^{\frac{1}{2}q}+\frac{n+q}{q}K^{-\frac{1}{2}q}-\frac{4n(n+q)}{q(q+2n)}\sqrt{2}\right\},$$

and $K=2^{1-1/q}-1$. *H. Davenport* (London).

Referred to in H10-8, H10-11.

H10-8 (10, 236b)

Hlawka, Edmund. Über Potenzsummen von Linearformen. II. Österreich. Akad. Wiss. Math.-Nat. Kl. S.-B. IIa. **156**, 247–254 (1948).

The author considers here the minimum of $|L_1|^p+\cdots+|L_n|^p$ for $1\leqq p\leqq 2$ [see a previous paper on the case $p\geqq 2$, same S.-B. **154**, 50–58 (1945); these Rev. **9**, 500]. By an appropriate modification of Blichfeldt's method, he obtains an improvement on Minkowski's estimate for the minimum.

H. Davenport (London).

Citations: MR 9, 500h = H10-7.

Referred to in H10-13, H30-24.

H10-9 (10, 236c)

Hlawka, Edmund. Über Gitterpunkte in Zylindern. Österreich. Akad. Wiss. Math.-Nat. Kl. S.-B. IIa. **156**, 203–217 (1948).

Consider a cylinder in n-dimensional space, whose base is an ellipsoid in m-dimensional space ($2\leqq m<n$). The author proves that, if the volume exceeds $(m+2)2^{n-1-\frac{1}{2}m}$, the cylinder contains a point with integral coordinates other than the origin. The proof is by the author's form of Blichfeldt's method [Math. Z. **49**, 285–312 (1943); these Rev. **5**, 201], and a knowledge of this is assumed. The theorem relates, more generally, to a body defined by $g(x_1, \cdots, x_m)\leqq 1$, $h(x_{m+1}, \cdots, x_n)\leqq 1$, where h is a convex function and g has a specified measure of convexity, in a sense defined by the author. Later sections of the paper are concerned with extensions of the author's "Alternativsatz." [Note by the reviewer: for the cylinder with $m=2$, see K. Mahler, Quart. J. Math., Oxford Ser. **17**, 16–18 (1946); these Rev. **7**, 368.]

H. Davenport (London).

Citations: MR 5, 201c = H25-1; MR 7, 368b = H10-3.

H10-10 (10, 433d)

Ollerenshaw, Kathleen. The critical lattices of a sphere. J. London Math. Soc. **23**, 297–299 (1948).

This paper gives an elementary proof that the determinants of admissible lattices of the unit sphere are not less than $1/\sqrt{2}$ and determines all the admissible lattices with determinant $1/\sqrt{2}$ (i.e., the critical lattices).

D. Derry (Vancouver, B. C.).

Referred to in H10-14.

H10-11 (11, 160b)

Rankin, R. A. On sums of powers of linear forms. I. Ann. of Math. (2) **50**, 691–698 (1949).

Let $\beta\geqq 2$, $\alpha\beta=1$; let

$$L_j(P)=L_j(x_1, \cdots, x_k)=\sum_{k=1}^{n} a_{jk}x_k \quad (1\leqq j\leqq n)$$

be n linear forms with determinant $D\neq 0$; we suppose that $L_1, \cdots, L_r$ are real and that L_{r+v} is the complex conjugate of L_{r+s+v} ($r+2s=n$). Put

$$g(P)=\left\{\sum_{j=1}^{n}|L_j|^\beta\right\}^{1/\beta}.$$

Let $M(g)$ be the minimum of $g(P)$ for integral x_k not all zero, $M_\beta = \sup M(g)$ for all systems of forms L_j with fixed r, s, D; let T_β be the volume of the body $g(P) \leqq 1$. Minkowski's theorem on convex bodies gives the inequality $M_\beta \leqq 2T_\beta^{-1/n} = \mu_1$. This result has been improved for certain values of β by van der Corput and Schaake [Acta Arith. **2**, 152–160 (1936)] and more recently by Hlawka [Akad. Wiss. Wien, S.-B. IIa. **154**, 50–58 (1945); these Rev. **9**, 500]. The author gives sharper results for $2 < \beta \leqq 2n$, viz.,

$$M_\beta \leqq \mu_1 2^{-\alpha}(1+\alpha n)^{1/n}(1+R_\beta)^{-1/n},$$

$$R_\beta = \frac{2^{-\alpha n}}{1+2\alpha n} - (\sqrt{2}-1)^{1+2\alpha n}\left(1+\frac{\sqrt{2}}{1+2\alpha n}\right).$$

The method is similar to that of the author's paper [same Ann. (2) **48**, 1062–1081 (1947); these Rev. **9**, 226]. Further improvements are possible; the author gives an example for $\beta=4$, $n=3$ and for $\beta=4$, n large. *V. Jarník* (Prague).

Citations: MR 9, 226d = H30-8; MR 9, 500h = H10-7.

H10-12 (11, 160c)

Rankin, R. A. On sums of powers of linear forms. II. Ann. of Math. (2) **50**, 699–704 (1949).

Same notations as in part I [see the preceding review], but we suppose now that $1 \leqq \beta \leqq 2$. The author proves the inequality (1) $M_\beta \leqq 2^\alpha(1+(1-\alpha)n)^{1/n} T_\beta^{-1/n} = \nu_\beta$ and, more precisely, the inequality (2) $M_\beta \leqq (1+R)^{-1/n}\nu_\beta$, where $R=(1-\alpha)n(\sqrt{2}-1)^{2n(1-\alpha)+2}$. Inequality (1) is an improvement of a result of Hua [Bull. Amer. Math. Soc. **51**, 537–539 (1945); these Rev. **7**, 51]. A further improvement of (2) is given for $1 < \beta \leqq \frac{4}{3}$. The proofs, similar to those of part I, use a convexity theorem of M. Riesz [see, e.g., Hardy, Littlewood, and Pólya, Inequalities, Cambridge University Press, 1934, theorem 296]. Another improvement can be obtained, in some cases, from the obvious inequality $M_\beta \leqq n^{\alpha-\alpha_0} M_{\beta_0}$ $(0<\beta\leqq\beta_0,\ \alpha_0\beta_0=1)$ if we use any upper bound of $n^{\alpha-\alpha_0}M_{\beta_0}$ and then choose $\beta_0 \geqq \beta$ so as to make this upper bound a minimum. *V. Jarník* (Prague).

Citations: MR 7, 51b = H10-2.

H10-13 (10, 284e)

Rankin, R. A. On sums of powers of linear forms. III. Nederl. Akad. Wetensch., Proc. **51**, 846–853 = Indagationes Math. **10**, 274–281 (1948).

Let $1 \leqq \beta \leqq 2$, $n \geqq 2$; let $L_j = \sum_{k=1}^n a_{jk}x_k$ $(j=1, \cdots, n)$ be n real linear forms with determinant $D \neq 0$. Let $g_\beta(x_1, \cdots, x_n) = (\sum_{j=1}^n |L_j|^\beta)^{1/\beta}$ and let $M(g_\beta)$ be the lower bound of g_β for all systems of integers $x_1, \cdots, x_n$ other than $0, \cdots, 0$; then (1) $M(g_\beta) \leqq n^\alpha |D|^{1/n}/A_\delta'(\alpha, \delta)$, where

$$A_\delta'(\alpha,\delta) = 2^{-\alpha n^\delta}\left(\frac{2-\delta}{1-\delta}\right)^{\alpha-\delta}\left(\frac{(1+n\delta)(\alpha+\gamma-2)}{(1-\delta)I_\gamma|D|}\right)^{-1/n};$$

here $\alpha=\beta^{-1}$, $\gamma=\delta^{-1}$ and δ is any number such that $\frac{1}{2} \leqq \delta \leqq \alpha \leqq 1$, $\delta \leqq \frac{1}{3}(1+\alpha)$; further $I_\gamma = 2^n\Gamma^n(1+\delta)|D|^{-1}\Gamma^{-1}(1+n\delta)$. For large values of n and for $\delta=\frac{1}{2}$, (1) has asymptotically the same degree of precision as an analogous formula given by E. Hlawka [Österreich. Akad. Wiss. Math.-Nat. Kl. S.-B. IIa. **156**, 247–254 (1948); these Rev. **10**, 236]. But if $\alpha > 0.77673 \cdots$, there are values of δ which are more advantageous than $\delta=\frac{1}{2}$. The improvement is about 0.7% for $\alpha=0.9$ and about 2% for $\alpha=1$. See also Loo-Keng Hua [Bull. Amer. Math. Soc. **51**, 537–539 (1945); these Rev. **7**, 51], P. Mullender [Amsterdam thesis, 1945; these Rev. **9**, 335] and paper II of the present series [above].
[In the last formula on p. 853 replace $1/n$ by $-1/n$.]
V. Jarník (Prague).

Citations: MR 7, 51b = H10-2; MR 9, 335b = J56-6; MR 10, 236b = H10-8.
Referred to in H30-24.

H10-14 (11, 160d)

Ollerenshaw, Kathleen. The critical lattices of a four-dimensional hypersphere. J. London Math. Soc. **24**, 190–200 (1949).

Let $\Delta(H)$ denote the lower bound of the determinants of the lattices whose only point interior to the hypersphere $x_1^2+x_2^2+x_3^2+x_4^2 \leqq 1$ is the origin. The paper gives a geometrical proof of the fact that $\Delta(H)=\frac{1}{2}$ and constructs those lattices with determinant $\frac{1}{2}$. The proof follows along the lines of the author's treatment of the 3-dimensional case [same J. **23**, 297–299 (1948); these Rev. **10**, 433] and starts from the fact that lattices of minimum determinant have 4 linearly independent lattice points on the boundary of H. The results are obtained by a study of certain tetrahedra in the 3-dimensional lattice defined by 3 of these lattice points with the use of the fact that the radii of their circumspheres cannot be less than that of the sphere obtained by cutting H by the hyperplane through the fourth lattice point and parallel to the hyperplane of the above 3-dimensional lattice. *D. Derry* (Vancouver, B. C.).

Citations: MR 10, 433d = H10-10.
Referred to in H10-21.

H10-15 (11, 233c)

Ledermann, Walter, and Mahler, Kurt. On lattice points in a convex decagon. Acta Math. **81**, 319–351 (1 plate) (1949).

Among all symmetric convex decagons those are extreme for which the densest lattice packing covers the least area, i.e., for which V/Δ is least, V being the area of the decagon and Δ the determinant of a critical lattice. The extreme decagons are determined and are affinely equivalent but not equivalent to the regular decagon although the regular octagon is extreme among symmetric octagons. The method depends on the fact for convex bodies that 4Δ is the lower bound of the areas of the circumscribing hexagons; cf. K. Reinhardt [Abh. Math. Sem. Hamburg. Univ. **10**, 216–230 (1934)] and K. Mahler [Nederl. Akad. Wetensch., Proc. **50**, 692–703 = Indagationes Math. **9**, 326–337 (1947); these Rev. **9**, 10]. The proof is mostly analytic rather than geometric. The extreme decagon can be "smoothed" giving a smaller value for V/Δ, but not as small as that for the smoothed regular octagon [cf. op. cit.]; this supports the conjecture that the latter is extreme among all 2-dimensional convex bodies. *L. Tornheim* (Ann Arbor, Mich.).

Citations: MR 9, 10h = H05-10.

H10-16 (11, 582e)

Skolem, Th. Proof of a theorem on 3-lattices. Norske Vid. Selsk. Forh., Trondheim **21**, no. 44, 197–200 (1949).

Another proof is given of the theorem that in a 3-dimensional lattice there is always a fundamental parallelepiped having the product of its sides not exceeding $\sqrt{2}$ times its volume. Let $\overline{OX}$ be the length of the vector OX. The proof uses only the inequalities that if $\overline{OA}$, $\overline{OB}$, $\overline{OC}$ generate a parallelepiped for which $\overline{OA}\cdot\overline{OB}\cdot\overline{OC}$ is minimal then $\overline{OA \pm OB} \geqq \overline{OB}$ while $\overline{OA \pm OC}$, $\overline{OB \pm OC}$, and $\overline{OA \pm OB \pm OC}$ all are not less than $\overline{OC}$. *L. Tornheim.*

H10-17 (11, 716e)

Cohn, Harvey. Minkowski's conjectures on critical lattices in the metric $(|\xi|^p+|\eta|^p)^{1/p}$. Ann. of Math. (2) **51**, 734–738 (1950).

Continuing work started by C. S. Davis [J. London Math. Soc. **23**, 172–175 (1948); these Rev. **10**, 512, 856], the author shows that Minkowski's conjecture about the critical lattices of the convex region $|\xi|^p+|\eta|^p \leqq 1$ where $p \geqq 1$ [Minkowski,

Diophantische Approximationen, 1st ed., Teubner, Leipzig, 1907, pp. 51–58] is true for sufficiently large positive p, but false in certain small finite intervals for p. *K. Mahler.*

Citations: MR 10, 512f = H20-14.

Referred to in H10-25, H10-34, H10-35.

H10-18 (12, 161d)

Hlawka, Edmund. Über Gitterpunkte in Parallelepipeden. J. Reine Angew. Math. **187**, 246–252 (1950).

The author's main result is as follows. Suppose we are given n independent directions in n-dimensional space, and a positive number V. Then there exists a parallelepiped P, with centre O and volume V, and with its faces normal to the given directions, such that the number of pairs of opposite points with integral coordinates lying in P (not counting O) is less than $A_n V$. Here A_n is the constant $(1/n)(n!)^2 2^{\frac{1}{2}n(n-1)}$. The proof is by an extension of the method by which Siegel [see Davenport, Acta Arith. **2**, 262–265 (1937)] proved that such a parallelepiped exists, with volume depending only on n, which contains no point with integral coordinates except O. A new feature is the definition (§ 3) of successive minima depending on a number l. The author also proves a theorem relating to any convex body with centre O. This could in fact be deduced from the earlier result, and with a better constant, by circumscribing a parallelepiped to the convex body [see Dvoretzky and Rogers, Proc. Nat. Acad. Sci. U. S. A. **36**, 192–197 (1950), theorem 5A; these Rev. **11**, 525]. *H. Davenport.*

Referred to in H15-52.

H10-19 (12, 161f)

Chalk, J. H. H. On the frustrum of a sphere. Ann. of Math. (2) **52**, 199–216 (1950).

Let K_λ be the truncated sphere

$$\max\left((x^2+y^2+z^2)^{\frac{1}{2}},\ |z|/\lambda\right)\leqq 1,$$

where $0<\lambda\leqq 1$. Then the critical determinant of K_λ is (1) $\Delta(K_\lambda)=\frac{1}{2}\lambda(3-\lambda^2)^{\frac{1}{2}}$. The only critical lattices (i.e., lattices with determinants equal to $\Delta(K_\lambda)$ having no point other than (0, 0, 0) in the interior of K_λ) are the lattices which may be obtained, by means of rotations about the z-axis and reflexions in the coordinate planes, from the lattice

$$x=v-\tfrac{1}{2}w,\quad y=\tfrac{1}{2}w(3-\lambda^2)^{\frac{1}{2}},\quad z=\lambda u+\tfrac{1}{2}\lambda w$$

($u, v, w=0, \pm1, \pm2, \pm3, \cdots$). As a simple corollary we get the following result of K. Mahler [Quart. J. Math., Oxford Ser. (1) **17**, 16–18 (1946); these Rev. **7**, 368]: The critical determinant of the cylinder $\max\left((x^2+y^2)^{\frac{1}{2}}, |z|\right)\leqq 1$ is equal to $3^{\frac{1}{2}}/2$. Another corollary of (1): Let $F(x, y, z)$ be a positive definite quadratic form of determinant D. Let α, β, γ be real numbers satisfying $\bar{F}(\alpha, \beta, \gamma)>D$, where $\bar{F}$ is the adjoint form; put $\bar{F}(\alpha, \beta, \gamma)=\lambda^{-2}D$, $\lambda>0$. Then there are integers x, y, z, not all zero, such that

$$\max\left(F(x, y, z),\ |\alpha x+\beta y+\gamma z|^2\right)\leqq 2^{\frac{2}{3}}D^{\frac{1}{3}}\lambda^{-\frac{2}{3}}(3-\lambda^2)^{-\frac{1}{3}}.$$

V. Jarník (Prague).

Citations: MR 7, 368b = H10-3.

H10-20 (13, 114b)

Whitworth, J. V. The critical lattices of the double cone. Proc. London Math. Soc. (2) **53**, 422–443 (1951).

Let K be the convex body obtained by rotating a square of side $\sqrt{2}$ about one of its diagonals. The object of this paper is to determine the critical lattices of K, i.e. the lattices of minimum determinant whose only point interior to K is (0, 0, 0) which is at the centre of K. These lattices, of determinant $\frac{1}{8}\sqrt{6}$, are shown to have seven points on each of the two component cones of K. Furthermore, any two of the lattices may be rotated into each other. The author starts from the three possible types of critical lattices for convex bodies in 3-space determined by Minkowski [Gesammelte Abhandlungen, vol. II, Teubner, Leipzig and Berlin, 1911, pp. 1–42]. The points of a critical lattice on one of the component cones of K are shown to belong to one of five different configurations. Four of these configurations are shown to lead either to incompatible relations or to lattices whose determinant exceeds $\frac{1}{8}\sqrt{6}$. The fifth configuration leads to the critical lattice which is completely determined, except for a rotation, by the Minkowski relations when the convex body is specialized to K. *D. Derry.*

H10-21 (13, 444b)

Ollerenshaw, Kathleen. Addendum: On the critical lattices of a sphere and four-dimensional hypersphere. J. London Math. Soc. **26**, 316–318 (1951).

In this note it is proved that if a lattice is admissible for a sphere and contains three linearly independent lattice points on the boundary of the sphere then these lattice points generate the lattice. This fact is used to meet an objection to the author's determination of the critical lattices of a 4-dimensional hypersphere [same J. **24**, 190–200 (1949); these Rev. **11**, 160]. The analogous result for 4-space is shown not to be true but only when the lattice involved is critical in which case it can be generated by four suitably chosen lattice points on the surface of the hypersphere.

D. Derry (Vancouver, B. C.).

Citations: MR 11, 160d = H10-14.

H10-22 (14, 851a)

Eichler, Martin. Note zur Theorie der Kristallgitter. Math. Ann. **125**, 51–55 (1952).

Let R_0 be an n-dimensional vector space over the rational field k_0, with a positive definite metric (scalar product) with the properties (1) $\xi\eta=\eta\xi \in k_0$, (2) $(x\xi)\eta=x(\xi\eta)$, (3) $(\xi+\eta)\zeta=\xi\zeta+\eta\zeta$, (4) $\xi^2>0$ for $\xi\neq0$, where $\xi, \eta \in R$, $x \in k$. A crystal-lattice (lattice) in R_0 is a $\mathfrak{o}_0$-module $\mathfrak{J}$ of rank n (additive abelian group) with respect to the order (integral set) $\mathfrak{o}_0$ of all rational integers. If $\mathfrak{J}$ has mutually orthogonal sublattices $\mathfrak{J}_1, \mathfrak{J}_2, \cdots$, of ranks $r_1, r_2, \cdots, r_1+r_2+\cdots=n$, such that every $\iota \in \mathfrak{J}$ is uniquely expressible in the form $\iota=\iota_1+\iota_2+\cdots$, $\iota_\nu \in \mathfrak{J}_\nu$, then $\mathfrak{J}$ is said to be the direct sum $\mathfrak{J}=\mathfrak{J}_1+\mathfrak{J}_2+\cdots$, of $\mathfrak{J}_1, \mathfrak{J}_2, \cdots$. The author proves (Theorem 1) that if $\mathfrak{J}=\mathfrak{J}_1'+\mathfrak{J}_2'=\mathfrak{J}_1+\mathfrak{J}_2$, both sums direct, then $\mathfrak{J}_1=\mathfrak{J}_1\cap\mathfrak{J}_1'+\mathfrak{J}_1\cap\mathfrak{J}_2'$, and similarly for $\mathfrak{J}_2, \mathfrak{J}_1', \mathfrak{J}_2'$. From this follows (Theorem 2) that if

$$\mathfrak{J}=\mathfrak{J}_1+\mathfrak{J}_2+\cdots=\mathfrak{J}_1'+\mathfrak{J}_2'+\cdots,$$

both sums direct, with directly indecomposable summands, then the $\mathfrak{J}_\nu$ and $\mathfrak{J}_\nu'$ are identical apart from order. Bases of $\mathfrak{J}$, etc., relative to $\mathfrak{o}_0$ are employed in the proof of Theorem 1, and hence the proof does not go in general for an algebraic number field k. Also, an example shows that the definiteness (property 4) of the metric is essential. However, the author obtains Theorems 1 and 2 for a field k of degree h over k_0, and an arbitrary order $\mathfrak{o}$ in k (i.e., not necessarily the set of all integers of k), provided k has an involutory automorphism $x\to\bar{x}$, such that $x+\bar{x}$ and $x\bar{x}$ are totally real for all x in k. Then $\xi\to\bar{\xi}$, with $\overline{x\xi}=\bar{x}\bar{\xi}$, is an extension of the automorphism to an n-dimensional vector space R over k. The scalar product is required to satisfy (1) $\xi\bar{\eta}=\overline{\bar{\xi}\eta} \in k$, (2) $(x\xi)\bar{\eta}=x(\xi\bar{\eta})$, (3) $(\xi+\eta)\bar{\zeta}=\xi\bar{\zeta}+\eta\bar{\zeta}$, (4) $\xi\bar{\xi}$ totally positive for $\xi\neq0$. The author gets around the basis difficulty by extending R over k to a space of dimension hn over k_0, with scalar product $\xi\circ\eta=\mathrm{trace}_{k/k_0}(\xi\bar{\eta})=\mathrm{trace}_{k/k_0}(\eta\bar{\xi})$, which has properties (1), $\cdots$, (4) above. *R. Hull.*

Referred to in H10-23.

H10-23 **(15, 780b)**

Kneser, Martin. Zur Theorie der Kristallgitter. Math. Ann. **127**, 105–106 (1954).

The author gives a shorter and more constructive proof of Theorems 1 and 2 of a paper by Eichler [Math. Ann. **125**, 51–55 (1952); these Rev. **14**, 851]. *R. Hull.*

Citations: MR 14, 851a = H10-22.

H10-24 **(14, 1065f)**

Watson, G. L. Minkowski's conjectures on the critical lattices of the region $|x|^p+|y|^p\leqq 1$. I. J. London Math. Soc. **28**, 305–309 (1953).

Minkowski's problem to find the critical lattices for the regions $|x|^p+|y|^p\leqq 1$ $(p>1)$ is unsolved as yet. However, any critical lattice has six points on the boundary and no points inside (apart from the origin). Special lattices of this type are (i) the one which contains the point $x=1$, $y=0$, and (ii) the one containing $x=y=2^{-p}$. Their determinants are denoted by Δ_0, Δ_1, respectively. Probably the critical lattice is either of these, so it is desirable to know which of Δ_0, Δ_1 is the smaller. Minkowski conjectured that sgn $(\Delta_0-\Delta_1)=$sgn $(p-2)$ $(p>1)$; this was disproved by C. S. Davis [same J. **23**, 172–175 (1948); these Rev. **10**, 512, 856]. Davis also conjectured that there is a number p_0 (about 2.5725) such that sgn $(\Delta_0-\Delta_1)=$sgn $(p-2)(p_0-p)$ $(p>1)$. The present author proves this, with $2.57<p_0<2.58$. The discussion depends on the sign of

$$f(p)=2^{-p}(H+1)^p+2^{-p}(H-1)^p-2,$$

where $H^p=4(1-2^{-p})$. *N. G. de Bruijn* (Amsterdam).

Citations: MR 10, 512f = H20-14.

Referred to in H10-25, H10-34, H10-35.

H10-25 **(15, 203d)**

Watson, G. L. Minkowski's conjectures on the critical lattices of the region $|x|^p+|y|^p\leq 1$. II. J. London Math. Soc. **28**, 402–410 (1953).

In part I of this paper [ibid. **28**, 305–309 (1953); these Rev. **14**, 1065] the author considered two special symmetrical lattices, and determined which of their determinants is the smaller. The critical determinant is $\min_P \Delta(P)$, where P runs through the lattices that have six points on the boundary and none inside (apart from the origin). In the present part the author shows that min (Δ_0, Δ_1) gives a local minimum of $\Delta(P)$ for all p. In a small region around $p_0=2.57\cdots$ both give local minima, so that there is an unsymmetrical maximum. This disproves a conjecture of Minkowski stating that the only local extremals are Δ_0 and Δ_1. A similar result was given by H. Cohn [Ann. of Math. (2) **51**, 734–738 (1950); these Rev. **11**, 716], but the present author indicates an essential error in Cohn's paper.

N. G. de Bruijn (Amsterdam).

Citations: MR 11, 716e = H10-17; MR 14, 1065f = H10-24.

Referred to in H10-34, H10-35.

H10-26 **(15, 203e)**

Ap Simon, H. G. The critical lattices of the off-centre hypercube. Quart. J. Math., Oxford Ser. (2) **4**, 204–209 (1953).

There is exactly one critical lattice of the hypercube A: $|y_r-1|\leqq c_r$, where $c_r>1$ $(r=1,\cdots,n)$ and it is given by $y_r=(c_r-1)\xi_r+2\sum\xi_i$, the ξ_i being integral. It is determined as the unique critical lattice of a convex hyperbody D which lies in A and the symmetric image of A with respect to the origin. The uniqueness is proved by first showing that the lattice provides a space-filling lattice-packing of $\frac{1}{2}D$.

L. Tornheim (Ann Arbor, Mich.).

H10-27 **(15, 607e)**

Sawyer, D. B. On the covering of lattice points by convex regions. Quart. J. Math., Oxford Ser. (2) **4**, 284–292 (1953).

Let K be a closed central convex region in the euclidean plane, such that, however it is displaced in the plane, at least one lattice point is covered. Here displacements include both translations and rotations. The author proves that the area of K is at least 4/3, and that this minimum value is attained only if K is congruent to the region K^*, defined by $|y|\leqq\frac{3}{4}-x^2$, $|x|\leqq\frac{1}{2}$. The special position of K^* is connected to the fact that to every point A on one of the vertical sides there can be found points B and C on the curved sides such that AB and AC are perpendicular and both of length 1. *N. G. de Bruijn* (Amsterdam).

Referred to in H10-28.

H10-28 **(17, 351e)**

Sawyer, D. B. On the covering of lattice points by convex regions. II. Quart. J. Math. Oxford Ser. (2) **6** (1955), 207–212.

The author shows that the theorem of the first part of the paper [same J. (2) **4** (1953), 284–292; MR **15**, 607] remains true if the condition that the set K is closed and central is omitted entirely. This result was conjectured by J. L. Massera and J. J. Schäffer [Fac. Ingen. Montevideo. Publ. Inst. Mat. Estadíst. **2** (1951), 55–74; MR **13**, 768]. It is pointed out in a note that the same result has been obtained independently by Shäffer, whose work is in course of publication. *N. G. de Bruijn.*

Citations: MR 15, 607e = H10-27.

H10-29 **(18, 22e)**

Szüsz, P. Beweis eines zahlengeometrischen Satzes von G. Szekeres. Acta Math. Acad. Sci. Hungar. **7** (1956), 75–79. (Russian summary)

A simple proof based on continued fractions is given of the following theorem of G. Szekeres [J. London Math. Soc. **12** (1937), 88–93]: For any two real numbers α, β $(\alpha\neq\beta)$ there exists a central-symmetric parallelogram of area $>K=2(1+5^{-\frac{1}{2}})$, having sides with slopes α, β, and containing no lattice point besides the origin in its interior. *L. Tornheim* (Berkeley, Calif.).

H10-30 **(21# 1299)**

Woods, A. C. The critical determinant of a spherical cylinder. J. London Math. Soc. **33** (1958), 357–368.

Let $F(x_1, x_2, \cdots, x_n)$ be a distance function in R_n. Then

$$\underline{F}(x_1, x_2, \cdots, x_n, x_{n+1}) = \text{Max}\,(F(x_1, x_2, \cdots, x_n), |x_{n+1}|)$$

defines a distance function in R_{n+1}. Assume the starbody K: $F\leqq 1$ is of finite type and has critical determinant $\Delta(K)$. Then the corresponding starbody $\underline{K}$: $\underline{F}\leqq 1$ in R_{n+1} (the cylinder over the base K) is again of finite type and $\Delta(\underline{K})\leqq\Delta(K)$. Mahler [Quart. J. Math. Oxford Ser. **17** (1946), 16–18; MR **7**, 368] proved that if K is the 2-dimensional sphere $x_1^2+x_2^2\leqq 1$, then $\Delta(\underline{K})=\Delta(K)$; and Chalk, Rogers and Yeh extended that result to all convex bodies K in R_2. Varnavides showed that $\Delta(\underline{K})=\Delta(K)$ also remains true for the star domain $|x_1x_2|\leqq 1$. Rogers and Davenport, however, constructed starbodies K in R_2 for which $\Delta(\underline{K})<\Delta(K)$. Now the authors prove that, if K: $x_1^2+x_2^2+x_3^2\leqq 1$ is the unit sphere in R_3, then $\Delta(\underline{K})=\Delta(K)$. Corollary: Call a system of non-overlapping spheres a regular packing if the centres form a lattice, and call it a semiregular packing if the centres form the union of a lattice and a translation of the lattice, which does not form a new lattice. It follows from the theorem that there is no semiregular packing of spheres with a density closer than that of the closest regular packing. *J. F. Koksma* (Amsterdam)

Citations: MR 7, 368b = H10-3.

H10-31 (28# 5040)

Ohnari, Setsuo

On the lattice constant of a regular n-gon ($n \equiv 0 \bmod 6$). (Japanese)

Sûgaku **14** (1962/63), 236–238.

Let φ be a point set on a plane containing the origin 0. A lattice Λ is said to be φ-admissible if $\Lambda \cap \varphi = \{0\}$. The lattice constant $\Delta(\varphi)$ is defined by the infimum of the determinants $d(\Lambda)$ for all φ-admissible lattices Λ. The purpose of the present paper is the computation of the lattice constant of a regular n-gon when $n \equiv 0 \pmod 6$. Denote by φ_n the interior of a regular n-gon whose vertices are ω^ν, $0 \leqq \nu < n$, where $\omega = \exp(2\pi i \cdot n^{-1})$. Then the result is

$$\Delta(\varphi_n) = \frac{\sqrt{3}}{2}\cos^2\frac{\pi}{n}.$$

The proof is due to several results on the geometry of numbers, and the author explains why the assumption $6|n$ is necessary. *S. Hitotumatu* (Tokyo)

H10-32 (28# 5041)

White, G. K.

Lattice tetrahedra.

Canad. J. Math. **16** (1964), 389–396.

Let $\{x\}$ denote the fractional part of x. It is proved that if r_i ($i = 1, 2, 3$) are 3 rational numbers with $r_i + r_j$ not an integer for $i \neq j$, then there exists an integer u for which $0 < \sum_{i=1}^3 \{r_i u\} \leqq 1$. From this, the following theorem is deduced. If T is a closed tetrahedron and Λ is a lattice which contains the vertices of T, then the two following conditions are equivalent: (1) The only points of Λ in T other than the vertices lie on a pair of opposite edges of T; (2) There is a pair of parallel lattice planes of Λ through a pair of opposite edges of T such that no points of Λ lie between these planes.

References to related problems are given. {The letters u and w are missing in equation (18) and on p. 395, line 3, respectively.} *I. Danicic* (London)

H10-33 (29# 522)

Ehrhart, Eugène

Une généralisation probable du théorème fondamental de Minkowski.

C. R. Acad. Sci. Paris **258** (1964), 4885–4887.

The following conjecture is proposed by the author, and established for $n = 2$: If the center of gravity G of an n-dimensional convex body K is a point of the n-dimensional lattice of points with integral coordinates, and if the n-content of K is greater than $(n+1)^n/n!$, then K contains in its interior a lattice point different from G. It is shown by examples that, for all n, the conjecture is not valid for bodies with n-content equal to $(n+1)^n/n!$. *B. Grünbaum* (Jerusalem)

H10-34 (34# 146)

Kuharev, V. G.

The critical determinant of the region $|x|^p + |y|^p \leqq 1$. (Russian)

Dokl. Akad. Nauk SSSR **169** (1966), 1273–1275.

The author describes a method by which the truth or falsity of Minkowski's conjecture [*Diophantische Approximationen*, p. 47, Teubner, Leipzig, 1907] about the value of the critical determinant in question can be decided at least for most p; he has verified the truth of the conjecture for $p = 1.3$, 1.4, 1.5, 1.6, 1.7, 2.2, 2.3, 3, 4, and 5. For earlier related work, see L. J. Mordell [J. London Math. Soc. **16** (1941), 152–156; MR **3**, 167], H. Cohn [Ann. of Math. (2) **51** (1950), 734–738; MR **11**, 716], C. S. Davis [J. London Math. Soc. **23** (1948), 172–175; MR **10**, 512] and G. L. Watson [ibid. **28** (1953), 305–309; MR **14**, 1065; ibid. **28** (1953), 402–410; MR **15**, 203].

{This article has appeared in English translation [Soviet Math. Dokl. **7** (1966), 1090–1093].} *J. W. S. Cassels* (Cambridge, England)

Citations: MR 3, 167c = H20-3; MR 10, 512f = H20-14; MR 11, 716e = H10-17; MR 14, 1065f = H10-24; MR 15, 203d = H10-25.

H10-35 (38# 4416)

Kuharev, V. G.

Investigations on Minkowski's conjecture concerning the critical determinant of the region $|x|^p + |y|^p \leqq 1$. (Russian. English summary)

Vestnik Leningrad. Univ. **23** (1968), *no.* 13, 34–50.

Minkowski stated a conjecture on the critical determinant and the critical lattices of the region $|x|^p + |y|^p \leqq 1$, $p \geqq 1$. In its original version the conjecture was not correct [C. S. Davis, J. London Math. Soc. **23** (1948), 172–175; MR **10**, 512], but there can be given a more refined and analytical variant; furthermore, Minkowski's conjecture was shown to be correct for sufficiently large p [cf. H. Cohn, Ann. of Math. (2) **51** (1950), 734–738; MR **11**, 716; G. L. Watson, J. London Math. Soc. **28** (1953), 305–309; MR **14**, 1065; ibid. **28** (1953), 402–410; MR **15**, 203]. In this paper the author proves a new equivalent version of the refined conjecture, adapted to the use of electronic computers. An algorithm is described and by means of this algorithm the conjecture is proved for $p = 1.3$; 1.4; 1.5; 1.6; 1.7; 2.2; 2.3; 3; 4; 5. *F. Schweiger* (Vienna)

Citations: MR 10, 512f = H20-14; MR 11, 716e = H10-17; MR 14, 1065f = H10-24; MR 15, 203d = H10-25.

H10-36 (39# 1408)

Schmidt, Wolfgang M.

A problem of Schinzel on lattice points.

Acta Arith. **15** (1968/69), 199–203.

Let E^n denote Euclidean n-space, E^+ its non-negative orthant, Λ a point lattice in E^n, $\Lambda^+ = \Lambda \cap E^+$. It is proved that, if Λ is a sublattice of the integer lattice, there exists a finite set $S \subset \Lambda^+$ such that every $\mathbf{g} \in \Lambda^+$ is expressible as a non-negative integral linear combination of n points of S. The case $n = 2$ of this result was proved incidentally by A. Schinzel [same Acta **11** (1965), 1–34; MR **31** #4783] by means of continued fractions. By an easy compactness argument, the author deduces the complete result from the following more general proposition: Every $\mathbf{x} \neq \mathbf{0}$ in E^+ is contained in a neighbourhood $N(\mathbf{x})$ in E^+ which is open with respect to E^+ and which is contained in a finite union of cones $C(\mathbf{B})$. Here $\mathbf{B} = (\mathbf{u}_1, \cdots, \mathbf{u}_n)$ denotes generally a basis of Λ for which all $\mathbf{u}_i \in \Lambda^+$; and $C(\mathbf{B})$ is then the cone of points $\sum_1^n \lambda_i \mathbf{u}_i$ with all $\lambda_i \geqq 0$. *E. S. Barnes* (Adelaide)

Citations: MR 31# 4783 = C10-34.

H10-37 (44# 156)

Kuharev, V. G.

Critical determinant of the region $|x|^p + |y|^p \leqq 1$. (Russian)

Izv. Vysš. Učebn. Zaved. Matematika **1971**, *no.* 2 (105), 62–70.

Minkowski enunciated a conjecture for the critical determinant of the region in the title. It is a special case of one of Minkowski's general theorems that the conjecture is equivalent to the statement that the determinants of a set of admissible lattices (parametrized by a continuous variable, say τ) are all not less than the conjectural value. The author describes some rather elaborate inequalities that, with the aid of the "Promin'-M" computer of the Togliatti Polytechnic Institute, have enabled him to verify that this is so for a certain subset of the required values of τ and for $1.3 \leqq p \leqq 1.301$.

J. W. S. Cassels (Cambridge, England)

H15 NONCONVEX BODIES: GENERAL THEORY

See also review J12-28.

H15-1 (1, 265d)

Santaló, L. A. Integral geometry of unbounded figures. Publ. Inst. Mat. Univ. Nac. Litoral 1, no. 2, 58 pp. (1939). (Spanish)

Given a group of translations in the Euclidean plane with a two dimensional fundamental domain. A "uniform net" (u.n.) is a set of congruent figures in the plane which is transformed into itself by each translation of the group. The "area" and "length" of such a u.n. are the area and length of that part of it which lies in a fundamental domain. The total curvature c of the u.n. is defined in the following way: The part of the u.n. which lies in the domain formed by $\mu \cdot \nu$ fundamental domains, μ in one direction and ν in the other, may have the total curvature $c(\mu, \nu)$. Then

$$c = \lim_{\substack{\mu\to\infty \\ \nu\to\infty}} \frac{c(\mu, \nu)}{\mu \cdot \nu},$$

when it exists.

With these definitions, the author determines mean values and geometrical probabilities connected with the conception of kinematic measure [cf. Blaschke, Vorlesungen über Integralgeometrie I, Hamburg Math. Einzelschr. **1936**], considering positions of figures which can be transformed into each other by the translations of the group as identical; the discussed configurations consist of two or three figures, one or two of which are u.n., fixed or movable. Most of the many results, including two translations of Blaschke's "Fundamental Formula," are the same as in the ordinary integral geometry. [This seems not surprising, for the integral geometry of u.n. is essentially integral geometry in a finite Euclidean 2-space.] An interesting new result is the determination of the mean value of the sum of curvatures in the points of intersection of a movable curve with a fixed one which may be the element of a u.n. The means of translating ordinary integral formulas to those on u.n. is a theorem similar to the mentioned extension of the notion of total curvature.

The second part of this paper treats analogous problems with analogous methods and results in the Euclidean 3-space. An appendix gives an elegant simple proof of the following theorem of Blichfeldt [A new principle in the geometry of numbers with some applications, Trans. Amer. Math. Soc. **15**, 227–235 (1914)]: Given a u.n. in the Euclidean m-space consisting of k isolated points in each of the fundamental domains. Then any closed body of volume V can be placed by a translation in such a way that it has more than $kV\alpha^{-1}$ points in common with the u.n. (α is the volume of a fundamental domain). The proof follows almost at once from the fact that the mean value of the number of points of intersection of the u.n. with the movable body equals $kV\alpha^{-1}$ (no rotations). The appendix also contains some related results. *P. Scherk* (New York, N. Y.).

H15-2 (4, 212d)

Mahler, K. Note on lattice points in star domains. J. London Math. Soc. **17**, 130–133 (1942).

This note is a summary of results of which proofs will appear in the Proceedings of the London Mathematical Society. The author, generalizing Minkowski's original methods, obtains theorems on lattice points in nonconvex domains. He treats the general star region K, a closed bounded point set satisfying (a) K contains the origin O of the coordinate system (x, y) in its interior; (b) the boundary L of K is a Jordan curve consisting of a finite number of analytic arcs; (c) every radius vector from O intersects L in one and only one point. A lattice Λ of points P

$$(x, y) = (\alpha h + \beta k, \gamma h + \delta k), \quad h, k = 0, \pm 1, \pm 2, \cdots,$$

is said to be K-admissible if the origin O is the only point of Λ interior to K. Let $d(\Lambda) = |\alpha\delta - \beta\gamma|$, and let $\Delta(K)$ be the lower limit of $d(\Lambda)$ for all K-admissible lattices. Then $\Delta(K) > 0$, and the author states that there exists at least one K-admissible lattice Λ such that $d(\Lambda) = \Delta(k)$ (a critical lattice in Mordell's terminology). Furthermore, he has found an algorithm by means of which all critical lattices of K can be determined in a finite number of steps, and he lists a few special cases in which the algorithm has been applied in detail. *D. C. Spencer.*

Referred to in H20-4, H20-7.

H15-3 (6, 257d)

Mordell, L. J. On the geometry of numbers in some non-convex regions. Proc. London Math. Soc. (2) **48**, 339–390 (1945).

Let $f(x, y)$ be defined for $x \geqq 0$, $y \geqq 0$ and have the following properties: (1) $f(x, y)$ is continuous and nonnegative; (2) $f(x, y) = f(y, x)$ and $f(x, y) = xf(1, y/x)$; (3) $f(1, y)$ is steadily increasing for all y, and $f(1+y, 1-y)$ is steadily decreasing for $0 \leqq y \leqq 1$ and is not a constant; (4) the region $x \geqq 0$, $y \geqq 0$, $f(x, y) \geqq f(1, 1)$ is convex. Then unique positive numbers a, b, c exist such that $f(a+b, a-b) = f(a, b) = cf(1, 1)$, $a^2 + b^2 = 2$, $a > b$; moreover $a < 2c < 2$. From the properties of $f(x, y)$, the region $\mathfrak{R} : f(|x|, |y|) \leqq cf(1, 1)$ is symmetrical in the origin and in the four lines $x=0$, $y=0$, $x+y=0$, $x-y=0$; it contains the origin as an inner point, and it is not convex. The author proves the following general theorem, which is the first of its kind.

Let P be any point on the arc C of the first quadrant boundary of $\mathfrak{R}$ between the two points (b, a) and (a, b); there then exist two unique points Q and R on the second quadrant boundary of $\mathfrak{R}$ such that the four points $OPQR$ form the vertices of a parallelogram, $\Pi(P)$ say. If for all P on C this parallelogram is at least of area 2, then every lattice

$$x = \alpha\xi + \beta\eta, \quad y = \gamma\xi + \delta\eta \qquad (\xi, \eta = 0, \pm 1, \pm 2, \cdots)$$

of determinant $|\alpha\delta - \beta\gamma| = 2$ contains at least one point of $\mathfrak{R}$ different from 0, i.e. a point $(x, y) \neq (0, 0)$ such that $f(|x|, |y|) \leqq cf(1, 1)$. This result is further the best possible one (i.e. the constant $cf(1, 1)$ cannot be replaced by any smaller number) if there exists at least one position of P on C for which the lattice generated by the vertices of $\Pi(P)$ contains no inner points of $\mathfrak{R}$ except O. To prove this theorem, point sets are constructed abutting on $\mathfrak{R}$ and containing at least one point of every lattice of determinant 2 without inner points of $\mathfrak{R}$; the assertion then follows by combining these points.

When applying this theorem, the discussion of the area of $\Pi(P)$ presents difficulties. The author has carried out the calculations in a number of cases of particular interest, and he finds that each of the following regions contains a point different from O of every lattice of determinant 1: (1) the region

$$|x|+|y|+\{(\lambda^2-1)(x^2+y^2)+2|xy|\}^{\frac{1}{2}} \leqq \{\lambda^2+2+(-\lambda^4+12\lambda^2-4)^{\frac{1}{2}}\}^{\frac{1}{2}},$$

where $1\leqq\lambda<\sqrt{2}$; (2) the region

$$\min(|x|+m|y|, |y|+m|x|)\leqq(1+2m-m^2)(5-4m+m^2)^{-\frac{1}{2}},$$

where $1/\sqrt{3}\leqq m\leqq 1$; (3) the region $|x|^p+|y|^p\leqq 2^{1-\frac{1}{2}p}c^p$, where $0.329\cdots\leqq p<1$, and where c has the same meaning as before; (4) the region $|xy|(x^2+y^2)\leqq 2/\sqrt{17}$, or the equivalent region $|x^4-y^4|\leqq 4/\sqrt{17}$; (5) the region $|xy|(|x|+|y|)\leqq 30/(13)^{\frac{1}{2}}$. The results (1)–(4) are the best possible ones, but (5) is not, in the opinion of the reviewer.

Both the methods and the results of this paper go back to an earlier paper [Proc. London Math. Soc. (2) **48**, 198–228 (1943); these Rev. **5**, 172] of the author, where he studied the two regions $|x^3+x^2y-2xy^2-y^3|\leqq 1$ and $|x^3-xy^2-y^3|\leqq 1$. The paper shows that nonconvex lattice point problems form a very promising subject of research in the geometry of numbers. *K. Mahler* (Manchester).

Citations: MR 5, 172d = J52-3.

Referred to in H20-7, H20-9, H20-22, J52-16.

H15-4 (7, 417e)

Jarník, Vojtěch. Zwei Bemerkungen zur Geometrie der Zahlen. Věstník Královské České Společnosti Nauk. Třída Matemat.-Přírodověd. **1941**, no. 24, 12 pp. (1942) (Czech. German summary)

Proofs of the following theorems. (1) Let N be a bounded closed set in n-dimensional Euclidean space with at least one inner point, $V(N)$ the set of all points $X-Y$, where X, Y lie in N, and $J(N)$ the inner Jordan measure of N. Denote by τ_1 the smallest positive number such that $\tau_1 V(N)$ contains a lattice point $X_1\neq 0$ and, for $k=2, 3, \cdots, n$, by τ_k the smallest positive number such that $\tau_k V(N)$ contains a lattice point X_k independent of $X_1, \cdots, X_{k-1}$. Then $\tau_1\tau_2\cdots\tau_n J(N)\leqq 2^{n-1}$. The author states that V. Knichal has proved that the constant 2^{n-1} cannot, in general, be replaced by 1; for convex bodies N, this is, however, allowed by Minkowski's classical theorem [Geometrie der Zahlen, Leipzig, 1896–1910, § 53; see also H. Davenport, Quart. J. Math., Oxford Ser. **10**, 119–121 (1939)]. (2) Let now M be a convex body and $N=\frac{1}{2}M$, so that $V(N)=M$. Denote by σ the smallest positive number such that the point set $X+\sigma M$ contains for every point X at least one lattice point. Then $\frac{1}{2}\tau_n\leqq\sigma\leqq(n/2)\tau_n$, and this is the best possible result. (3) Let $L_1(X), \cdots, L_n(X)$ be n real linear forms in $x_1, \cdots, x_n$ of determinant 1, and let $\epsilon>0$. Then there exists a lattice point $X\neq 0$ such that $|L_1(X)\cdots L_n(X)|\leqq 2^{-(n-1)/2}+\epsilon$. [This is not the best possible result; in an as yet unpublished paper, H. Davenport has shown that approximately 12^{-n} may be taken as the right-hand side.] *K. Mahler.*

Referred to in H15-15, H15-23.

H15-5 (7, 506g)

Mordell, L. J. Further contribution to the geometry of numbers for non-convex regions. Trans. Amer. Math. Soc. **59**, 189–215 (1946).

Minkowski's theory on lattice points in convex regions led him to the following problem. If $f(x, y)\leqq 1$ denotes a convex, symmetrical domain with its center at the origin, to find a parallelogram of minimum area, one of whose vertices is at the origin and whose other three vertices are on the boundary $f(x, y)=1$. The author extends this problem to a quite general class of nonconvex infinite domains $f(x, y)\leqq 1$, very similar to the special region R: $|x^n+y^n|\leqq 1$ (more precisely written: $|\operatorname{sgn} x|x|^n+\operatorname{sgn} y|y|^n|\leqq 1$, where n denotes a real positive number), develops a method for solving the problem and discusses the application to the special region R in full detail for $n\geqq 3$. *J. F. Koksma.*

H15-6 (8, 12d)

Mahler, Kurt. Lattice points in two-dimensional star domains. I. Proc. London Math. Soc. (2) **49**, 128–157 (1946).

A star domain K is defined to be a closed bounded set, containing the origin O as an inner point, which is symmetric with respect to O and such that its boundary is cut in exactly two points by every straight line through O. A lattice is defined to be K-admissible if its only point interior to K is O. Let $\Delta(K)$ be the lower bound of the determinants of all K-admissible lattices. If the determinant of a lattice is $\Delta(K)$, the lattice is said to be critical. The boundary of K is assumed to consist of a finite number of analytic arcs.

The purpose of the paper is to determine a process whereby $\Delta(K)$ may be found for a given star domain K in a finite number of steps. The author shows easily that a critical lattice contains at least two independent boundary points P_1, P_2 of K and then that there is only a finite number of K-admissible lattices containing both P_1 and P_2. Critical lattices are defined to be singular or regular according as they contain exactly four or more than four boundary points of K. An inner tac line is a straight line through a boundary point P of K such that all its points within a neighborhood of P lie in K. If a lattice containing P_1, P_2 is singular then the line through P_2 parallel to OP_1 is shown to be an inner tac line. This leads to two equations connecting P_1 and P_2 which, in general, determine a finite number of possible positions for the two points because of the analytic character of the boundary. This in turn leads to the determination of every possible K-admissible lattice with these two points and so to all the singular lattices. A somewhat similar argument leads to the determination of all regular lattices. The relations used in this case are the minimal condition for the determinant of the lattices and the equations derived from the boundedness of the number of lattices with three or more given boundary points. Several special cases arise because of lattice points located at the end points of the component arcs of the boundary and because of the coincidence of certain analytic functions in the solution of the equation systems.

An example of a domain with a singular lattice is given but actual applications of the method are left for part II. The author points out that the lattice problem for nonsymmetric domains and for infinite star domains K containing a bounded star domain K' with a critical K-admissible lattice can easily be reduced to the problem for bounded star domains. *D. Derry* (Vancouver, B. C.).

Referred to in H15-10, H15-42, H15-59, J28-19, J36-10.

H15-7 (8, 195f)

Mahler, K. On lattice points in n-dimensional star bodies. I. Existence theorems. Proc. Roy. Soc. London. Ser. A. **187**, 151–187 (1946).

Let $F(x_1, \cdots, x_n)=F(X)$ be a continuous nonnegative function over real Euclidean n-space which does not vanish everywhere and with the property that $F(tX)=|t|F(X)$ for all real t. The set of points X for which $F(X)\leqq 1$ is a star body K. A lattice Λ whose only point interior to K is the origin is defined to be K-admissible; $\Delta(K)$ denotes the lower bound of the determinants of all K-admissible lattices. A lattice with determinant $\Delta(K)$ is defined to be critical. The purpose of this paper is to study critical lattices of a star body K and especially the points of such lattices which are on or near the boundary of K. A number of fundamental results are first derived. The author introduces a definition for a bounded set of lattices and a definition for the convergence of a lattice sequence which enable him to establish a lattice point adaptation of the Bolzano-Weierstrass theorem. Much use of this is made throughout the paper. Examples of star bodies are given with finite and with infinite $\Delta(K)$. Every critical lattice of a bounded star body K is shown to contain n linearly independent points on the boundary of K. Examples are given (1) of a bounded K which has exactly one critical lattice whose only points on the boundary of K are n linearly independent points and their reflections with respect to the origin and (2) of an unbounded star body K with a critical lattice without points on the boundary of K. For every positive ϵ a critical lattice is shown to contain a point P with $1 \leqq F(P) < 1+\epsilon$. These results lead to the problem: does every critical lattice of a star body K, for all $\epsilon > 0$, contain n linearly independent points P for all of which $1 \leqq F(P) < 1+\epsilon$? Most of the results of the paper contribute to partial solutions of this problem.

Let Γ be a group of linear transformations of Euclidean n-space which map the star body K into itself, such that for a given positive c and a given point X of K an element Ω of Γ always exists with $|\Omega X| < c$. If K admits such a group it is defined to be automorphic. Such star bodies are shown to possess critical lattices which contain at least one boundary point of K. If a body K has a group Γ so that for each positive c and each element $X \neq 0$ of K an element Ω of Γ exists with $|\Omega X| > c$ then infinitely many points P exist in every critical lattice so that $1 \leqq F(P) < 1+\epsilon$ for each positive ϵ. A set of conditions is given which if imposed on Γ lead to results regarding the minimum number of linearly independent points P of a critical lattice with $1 \leqq F(P) < 1+\epsilon$ for each positive ϵ. From this follows the affirmative answer for the above problem if K is defined by $F(x_1, \cdots, x_n) = (x_1 \cdots x_n)^{1/n}$. *D. Derry* (Vancouver, B. C.).

Referred to in H15-10, H15-18, H15-42, H15-43, H15-59, H15-62, H15-64, H15-70, H20-19, J28-19, J48-7.

H15-8 (8, 12b)

Mahler, K. Lattice points in n-dimensional star bodies. II. Reducibility theorems. I, II. Nederl. Akad. Wetensch., Proc. **49**, 331–343, 444–454 = Indagationes Math. **8**, 200–212, 299–309 (1946).

[Part I is in the course of publication in Proc. Roy. Soc. London. Ser. A.]

Let $F(x_1, \cdots, x_n) = F(X)$ be a continuous function over Euclidean n-space which assumes positive values except at the origin, where it vanishes, and has the property that $F(tX) = |t| F(X)$ for all real t. Let H be the star body of all points X for which $F(X) \leqq 1$. Let $\Delta(H)$ denote the lower bound of the determinants of all H-admissible lattices (that is, lattices whose only point interior to H is the origin). An H-admissible lattice with determinant $\Delta(H)$ is said to be critical. A star body H which contains a star body K as a proper subset such that $\Delta(H) = \Delta(K)$ is said to be reducible, otherwise irreducible.

If a star body is irreducible the author shows that a critical lattice goes through every boundary point. Sufficient conditions are given under which a bounded star body is irreducible. This second set of conditions is used to show that some simple convex bodies in 2 and 3-space are irreducible, among them the square, sphere and cube, and also to develop a construction whereby additional convex irreducible bodies in 2-space may be constructed. Some non-convex irreducible star bodies are listed without proofs of irreducibility. The constant ω_H is defined to be the maximum of $F(X_1+X_2)$, where X_1 and X_2 are points in n-space for which $F(X_1)+F(X_2)=1$. Some inequalities involving the critical lattices of a star body and ω_H are developed. The author derives some results on a class of unbounded star bodies and uses these to give a simple construction for a star body which has critical lattices without points on its boundary. Unsolved problems are proposed on every topic discussed, among them the following. Does every bounded star body H contain an irreducible star body K with $\Delta(H) = \Delta(K)$? *D. Derry* (Vancouver, B. C.).

Referred to in H05-66, H15-10, H15-13, H15-19, H15-28, H15-42, H15-50, H15-59, H15-82, H20-12, H30-79, J28-19, J36-10.

H15-9 (8, 12a)

Mahler, K. Lattice points in n-dimensional star bodies. II. Reducibility theorems. III, IV. Nederl. Akad. Wetensch., Proc. **49**, 524–532, 622–631 = Indagationes Math. **8**, 343–351, 381–390 (1946).

This is the concluding part of the paper reviewed above. The author defines an infinite star body K to be boundedly reducible if it contains a bounded star body H with $\Delta(H) = \Delta(K)$. If no such body H exists, K is said to be boundedly irreducible. An example of a boundedly irreducible body of dimension n is given. A critical lattice Λ of K is defined to be strongly critical if a bounded star body K^* exists within K such that, for every K^*-admissible lattice Λ^* sufficiently close to Λ, $d(\Lambda^*) \geqq d(\Lambda)$. The author shows that, if every critical lattice of K is strongly critical, K is boundedly reducible. This criterion is applied to show that the domains $|x_1x_2x_3| \leqq 1$, $(x_1^2+x_2^2)|x_3| \leqq 1$ are each boundedly reducible. The proofs use the determination of the critical lattices of these domains given by H. Davenport [Proc. London Math. Soc. (2) **44**, 412–431 (1938); **45**, 98–125 (1939)]. As a consequence of the second of these results the author derives the following approximation theorem. If α_1, α_2 are irrational numbers, arbitrarily large integers u_1, u_2, u_3 exist so that

$$(u_1/u_3-\alpha_1)^2+(u_2/u_3-\alpha_2)^2 \leqq (2/\sqrt{23})|u_3|^{-3}.$$

The final result of the paper is that, if $F_r(X) \rightarrow F(X)$ as $r \rightarrow \infty$ uniformly for X on the unit hypersphere $|X|=1$, then $\liminf \Delta(K_r) \geqq \Delta(K)$, where K_r is the star body of all points X for which $F_r(X) \leqq 1$. An example is given for which $\lim \Delta(K_r)$ exists and exceeds $\Delta(K)$. *D. Derry.*

Referred to in H15-10, H15-59, H20-10, J36-10.

H15-10 (8, 566a)

Mahler, K. Lattice points in n-dimensional star bodies. Univ. Nac. Tucumán. Revista A. **5**, 113–124 (1946).

This is a report of a lecture delivered in May, 1945. It contains an informal account of work done by the author recently in n-dimensional lattice point theory. The results described have, in the meantime, appeared in two papers [Proc. Roy. Soc. London. Ser. A. **187**, 151–187 (1946); Nederl. Akad. Wetensch., Proc. **49**, 331–343, 444–454, 524–532, 622–631 (1946); these Rev. **8**, 195, 12]. The descriptions are illustrated by special examples. No complete proofs are given. *D. Derry* (Vancouver, B. C.).

Citations: MR 8, 12a = H15-9; MR 8, 12b = H15-8; MR 8, 12d = H15-6; MR 8, 195f = H15-7.

H15-11 (8, 369h)

Mordell, L. J. Lattice points in some n-dimensional non-convex regions. I, II. Nederl. Akad. Wetensch., Proc. **49**, 773–781, 782–792 = Indagationes Math. **8**, 476–484, 485–495 (1946).

(A) Let R be the n-dimensional region defined by $|x_n| \leqq f(|x_1|, \cdots, |x_{n-1}|)$, where (for $x_1 > 0, \cdots, x_{n-1} > 0$) $f \geqq 0$ is steadily decreasing and differentiable and $\partial f/\partial x_i$ are steadily increasing functions of each variable separately. Let $x_1/p_1+x_2/p_2+\cdots+x_n/p_n=1$ be the tangent plane to $x_n = f(x_1, \cdots, x_{n-1})$ at a point $(\xi_1, \cdots, \xi_n)$, $\xi_i > 0$; we suppose that $p_n \leqq 2\xi_n$ and that $f(x_1, \cdots, x_{n-1})/p_n \geqq 1 - x_1/p_1 - \cdots - x_{n-1}/p_{n-1}$ for all $x_i > 0$ [this condition is not stated in the paper, but it is necessary in order to ensure the correctness of the conclusion on p. 486 = 783, line 3 from above; in the case $n=2$, it is a consequence of the other conditions]. Denote by V_n the volume of the region $x_1/p_1+\cdots+x_n/p_n\geqq\frac{1}{2}$, $x_n+\frac{1}{2}p_n\leqq f(x_1,\cdots,x_{n-1})$, $x_1\geqq\xi_1,\cdots,x_{n-1}\geqq\xi_{n-1}$, $x_n\geqq 0$. Let Λ be an n-dimensional lattice of determinant Δ, where $0<\Delta\leqq(p_1\cdots p_n)/n!+2^nV_n$. Then R contains a lattice point other than the origin. (B) Results analogous to (A); however, the x_i's are no longer coordinates, but sums of coordinates $t_1, t_2, \cdots$: $x_1=t_1+\cdots+t_\alpha$, $x_2=t_{\alpha+1}+\cdots+t_{\alpha+\beta}$, $\cdots$. [The denominator on p. 488 = 785, line 9 from below, and p. 489 = 786, line 6 from above, should be $n!$ instead of $\alpha!\beta!\cdots$.]

Applications. (C) Simultaneous approximation. Proof of a result of Blichfeldt [Trans. Amer. Math. Soc. **15**, 227–235 (1914)] in the form of an integral and a better evaluation, due to Davenport, of this integral. (D) Proof of a theorem of Koksma and Meulenbeld [Nederl. Akad. Wetensch., Proc. **45**, 256–262, 354–359, 471–478, 578–584 (1942); these Rev. **5**, 256]. (E) Evaluation of $|x_1|^\lambda+\cdots+|x_n|^\lambda$ $(0<\lambda<1)$ and of the symmetrical function $\sum|x_1x_2\cdots x_r|$. [Concerning pp. 491–492 = 788–789: in order to ensure the correctness of these considerations, we must suppose that the least value of $f(x_1,\cdots,x_{n-1})$, the value of the sum $x_1+\cdots+x_{n-1}$ being given, is taken at the point $x_1=x_2=\cdots=x_{n-1}$. This condition is satisfied in the applications given in the paper. In the formula for V_2 on p. 495 = 792 the factor $(n-1)^{r-1}$ is omitted in the first term of the curly bracket.]

V. Jarník (Prague).

Citations: MR 5, 256e = J24-6.

Referred to in H15-25, H15-26, H15-35.

H15-12 (8, 444d)

Rado, R. A theorem on the geometry of numbers. J. London Math. Soc. **21**, 34–47 (1946).

Der Verfasser gibt eine Verallgemeinerung des Minkowskischen Satzes, der auch die Verallgemeinerungen von Mordell [Compositio Math. **1**, 248–253 (1934)] und van der Corput [Acta Arith. **1**, 62–66 (1935)] noch als Spezialfälle umfasst. Es seien $x, y, \cdots$ Punkte des n-dimensionalen Euklidischen Raumes; ξ durchlaufe ein Gitter mit der Determinante D; es enthalte den Nullpunkt. Es sei λ eine (n, n)-reihige Matrix. Eine Punktmenge S heisse eine λ-Menge, wenn mit x und y jedesmal auch $\lambda(x-y)$ der Menge angehört. Eine Funktion $f(x)$ heisse eine λ-Funktion, wenn für jede Zahl a die Menge derjenigen Punkte, für die $f(x)\geqq a$, eine λ-Menge bildet. Ist insbesondere $\lambda=\frac{1}{2}E$ gleich der halben Einheitsmatrix, so ist jede λ-Menge konvex und symmetrisch in bezug auf den Nullpunkt, und umgekehrt. Der Verfasser beweist den folgenden Satz. Ist $f(x)$ eine λ-Funktion, S eine λ-Menge und bedeutet $\sum_S'$ die Summierung über alle der Menge S angehörigen Gitterpunkte mit Ausnahme des Nullpunktes, so ist

$$f(0)+\tfrac{1}{2}\textstyle\sum_S' f(\xi)\geqq|D|^{-1}\|\lambda\|\displaystyle\int_S f(x)dx.$$

Setzt man $f(x)\equiv 1$ und $\lambda=\frac{1}{2}E$, so erhält man den Minkowskischen Satz. *O.-H. Keller* (Dresden).

Referred to in H15-21, H15-44.

H15-13 (8, 445a)

Mahler, K. On irreducible convex domains. Nederl. Akad. Wetensch., Proc. **50**, 98–107 = Indagationes Math. **9**, 73–82 (1947).

Let K be a plane convex domain which is symmetric with respect to the origin. A lattice whose only point interior to K is the origin is known as K-admissible. Let $\Delta(K)$ denote the lower bound of the determinants of all the K-admissible lattices and let H be a star domain, symmetric with respect to the origin, with $H<K$. If, for every such H, $\Delta(H)<\Delta(K)$, K is defined to be irreducible. The purpose of this paper is to show that an irreducible convex domain H exists within every domain K with $\Delta(H)=\Delta(K)$. Using general results he established in an earlier paper [same Proc. **49**, 331–343 = Indagationes Math. **8**, 200–212 (1946); these Rev. **8**, 12], the author shows that every reducible convex domain K contains a smaller convex domain H with $\Delta(H)=\Delta(K)$ and then expresses the required irreducible convex domain by means of a convergent sequence of the smaller domains H. *D. Derry.*

Citations: MR 8, 12b = H15-8.

H15-14 (8, 445b)

Mahler, K. On the area and the densest packing of convex domains. Nederl. Akad. Wetensch., Proc. **50**, 108–118 = Indagationes Math. **9**, 83–93 (1947).

If K and $\Delta(K)$ are defined as in the paper reviewed above, let $V(K)$ denote the area of K. The author studies the lower bound Q of $V(K)/\Delta(K)$ over all possible K. A domain K with $V(K)/\Delta(K)=Q$ is defined to be an extreme domain. The existence of extreme domains is established and they are proved to be irreducible. The determination of these domains is reduced to a problem in the calculus of variations. However, the integral of the resulting Euler equations, an ellipse, is shown not to be an extreme domain. An alternative formulation of the problem of the determination of the constant Q is given in terms of sets of nonoverlapping domains congruent to K whose centers form a lattice. *D. Derry* (Vancouver, B. C.).

H15-15 (8, 565g)

Jarník, Vojtěch, and Knichal, Vladimír. On the main theorem of the geometry of numbers. Rozpravy II. Třídy České Akad. **53**, no. 43, 15 pp. (1943). (Czech)

The authors consider the extension of Minkowski's theorem to an arbitrary set A in r-dimensional space. Let $J(A)$ be the inner Lebesgue measure of A and let $\mathfrak{B}(A)$ be the set of all points $P-Q$ for $P\varepsilon A$, $Q\varepsilon A$, the subtraction being defined in an obvious way. Further, let $\tau_j''(N)$ denote the lower bound of all numbers $\alpha>0$ for which the set $\sum_{0<\beta\leqq\alpha}\beta N$ contains at least j independent lattice points. Here βN is the set obtained from an arbitrary set N by expansion about the origin in the ratio β. The most important part of the authors' first theorem is that, if $0<J(A)<\infty$ and $\tau_1''(\frac{1}{2}\mathfrak{B}(A))\neq 0$, then $\mu_1\mu_2\cdots\mu_rJ(A)\leqq 2^r$, where $\mu_1,\mu_2,\cdots,\mu_r$ are any numbers such that $0<\mu\leqq\tau_j''(\frac{1}{2}\mathfrak{B}(A))$ $(1\leqq j\leqq r)$ and $\mu_2/\mu_1,\cdots,\mu_r/\mu_{r-1}$ are integers. From this it follows that

$$T(A)=\tau_1''(\tfrac{1}{2}\mathfrak{B}(A))\tau_2''(\tfrac{1}{2}\mathfrak{B}(A))\cdots\tau_r''(\tfrac{1}{2}\mathfrak{B}(A))\leqq 2^{2r-1},$$

and it is shown that the upper bound 2^{2r-1} can be improved slightly. For a bounded convex set A, $T(A)\leqq 2^r$ by Minkowski's theorem. However, 2^r is not the correct upper bound for an arbitrary set A as is shown by the authors' second theorem where an example is given of a set A for which $T(A)>2^r$. The main theorem is an extension of a result due to Jarník [Věstník Královské České Společnosti Nauk. Třida Matemat.-Přírodověd., **1941**; these Rev. **7**, 417] who proved a less general result for a bounded closed set A. Use is made in the proof of a generalisation of a theorem of Blichfeldt due to Jessen [cf. Fenchel, Acta Arith. **2**, 230–241 (1937)]. [Results of a similar but less general type have been given independently by Rogers in the paper reviewed below.] *R. A. Rankin.*

Citations: MR 7, 417e = H15-4.

Referred to in H15-17, H15-23, H15-30.

H15-16 (8, 565h)

Rogers, C. A. A note on a theorem of Blichfeldt. Nederl. Akad. Wetensch., Proc. **49**, 930–935 = Indagationes Math. **8**, 589–594 (1946).

Let L be an n-dimensional lattice with determinant $\Delta>0$. Let S be a closed bounded set with inner volume $V(S)>0$. If $\lambda>0$, we denote by λS the set of all points λx, where $x\varepsilon S$.

We define $2n$ numbers $0<\lambda_1\leqq\cdots\leqq\lambda_n$, $0<\mu_1\leqq\cdots\leqq\mu_n$ $(\lambda_k\leqq\mu_k)$ in the following way: λ_1 is the smallest number for which there exists a nonvanishing lattice-vector A_1 which is contained in $\lambda_1 S$ (i.e., $A_1=\overline{xy}$, where $x\varepsilon\lambda_1 S$, $y\varepsilon\lambda_1 S$); λ_2 is the smallest number for which $\lambda_2 S$ contains a lattice-vector A_2, linearly independent of A_1, etc. (e.g., $\lambda_3 S$ contains a lattice-vector A_3, linearly independent of A_1, A_2); μ_k is the smallest number such that the set μS contains, for every $\mu\geqq\mu_k$, at least k linearly independent lattice-vectors. If S is convex and symmetrical about the origin, we have $\lambda_k=\mu_k$, $\lambda_1\lambda_2\cdots\lambda_n V(S)\leqq\Delta$ [Minkowski]. Results. (I) We have $(\lambda_2{}^n/k)V(S)\leqq\Delta$, $(\mu_2{}^n/k')V(S)\leqq\Delta$, where k, k' are integers, $\lambda_2/\lambda_1\leqq k<\lambda_2/\lambda_1+1$, $\mu_2/\lambda_1\leqq k'<\mu_2/\lambda_1+1$. (II) If $n=2$ then to every $\epsilon>0$ there corresponds an S such that $\lambda_1\lambda_2 V(S)>(\frac{7}{6}-\epsilon)\Delta$. [For more general results (except that concerning μ_2) see the paper of Jarník and Knichal reviewed above.] *V. Jarník* (Prague).

H15-17 (12, 677e)

Jarník, Vojtěch, et Knichal, Vladimír. Sur le théorème de Minkowski dans la géométrie des nombres. Acad. Tchèque Sci. Bull. Int. Cl. Sci. Math. Nat. **47** (1946), 171–185 (1950).

This is a French version of a paper which appeared earlier in Czech [Rospravy II. Třídy České Akad. **53**, no. **43** (1943); these Rev. **8**, 565]. For later literature on these questions, see C. A. Rogers [Proc. London Math. Soc. (2) **51**, 440–449 (1949); these Rev. **10**, 683]. *H. Davenport.*

Citations: MR 8, 565g = H15-15; MR 10, 683b = H15-36.

H15-18 (9, 227c)

Rogers, C. A. A note on a problem of Mahler. Proc. Roy. Soc. London. Ser. A. **191**, 503–517 (1947).

This paper answers the following problem proposed by Mahler [same Proc. Ser. A. **187**, 151–187 (1946), p. 167; these Rev. **8**, 195]: does every critical lattice of a star body $F(x)\leqq 1$, for given $\epsilon>0$, contain n linearly independent lattice points P for all of which $1\leqq F(P)<1+\epsilon$? The author shows that the star body $\min\{|x^2-xy-y^2|,\ |y^2-xy-x^2|\}\leqq 1$ possesses the single critical lattice of all points with integral coordinates and that its only admissible lattices Λ with $|\Delta(\Lambda)|<1.6$ are obtained by magnifying the critical lattice. The proofs depend on some elementary results from the theory of minima of indefinite binary quadratic forms. Star bodies are then constructed by adding certain narrow rectangular strips, centered on the x-axis, to the above star body. The single critical lattice of each of these bodies is shown to be a magnification of the lattice of points of integral coordinates and has the property that its only points in the star body are the origin and the two points on the x-axis nearest to the origin. Thus Mahler's problem is answered in the negative for the 2-dimensional case. An induction proof of some length (seven pages) is given showing that star bodies with similar properties exist for all dimensions. Examples are given illustrating the lack of continuity of $\Delta(K)$ for certain variable star bodies K. *D. Derry* (Vancouver, B. C.).

Citations: MR 8, 195f = H15-7.

H15-19 (9, 228a)

Rogers, C. A. A note on irreducible star bodies. Nederl. Akad. Wetensch., Proc. **50**, 868–872 = Indagationes Math. **9**, 379–383 (1947).

The principal results of a paper of Mahler [same Proc. **49**, 331–343 = Indagationes Math. **8**, 200–212 (1946); these Rev. **8**, 12] are obtained by a similar but simpler treatment. This simplification is effected by introducing the notion of an irreducible point, a point P on the boundary of a star body H with the property that, for every star body K not containing P with $K<H$, $\Delta(K)<\Delta(H)$. The results are easy consequences of the following theorem and its converse. For every irreducible point P a lattice Λ can always be constructed which is H-admissible except for lattice points in given neighborhoods of P and $-P$ and with $d(\Lambda)<\Delta(H)$. *D. Derry* (Vancouver, B. C.).

Citations: MR 8, 12b = H15-8.

Referred to in H05-66, H20-20.

H15-20 (9, 334f)

Cassels, J. W. S. A theorem on star domains. Quart. J. Math., Oxford Ser. **18**, 236–243 (1947).

Let K be a plane star body with a set of continuously changing K-admissible lattices Λ_t, $0\leqq t\leqq 1$, of constant positive determinant δ, which satisfies the following conditions. Each lattice Λ_t contains a parallelogram $OP_tQ_tR_t$, where O is the origin and P_t, Q_t, R_t boundary points of K such that OP_t, OQ_t move monotonically in the same direction while $R_t=Q_t-P_t$ and the triangles OQ_tR_t, $O(-R_t)P_t$ are within K. For star bodies with such lattice systems, the author shows that $\Delta(K)=\delta$ and that each critical lattice contains at least a point P_t or a point Q_t. The proof is obtained by showing that the critical lattices of K are the critical lattices of a parallelogram the midpoints of whose sides are $\pm P_t$, $\pm Q_t$. The result is applied to determine the critical lattices of some special bounded star bodies. *D. Derry.*

Referred to in H15-59.

H15-21 (10, 19d)

Cassels, J. W. S. On a theorem of Rado in the geometry of numbers. J. London Math. Soc. **22** (1947), 196–200 (1948).

Sei $f(x)$ eine beschränkte nicht-negative Vektorfunktion der n reellen Veränderlichen $x=[x_1,\cdots,x_n]$ im R_n und es verschwinde $f(x)$ ausserhalb eines beschränkten Bereiches des R_n. Es sei λ eine nicht-singuläre Matrix; ξ durchlaufe alle Gitterpunkte eines gegebenen Gitters der Determinante $D>0$. Ist nun $f(\lambda x-\lambda y)\geqq\min\{f(x), f(y)\}$, so hat R. Rado [J. London Math. Soc. **21**, 34–47 (1946); diese Rev. **8**, 444] gezeigt, dass die Ungleichung

$$f(0)+\tfrac{1}{2}\sum_{\xi\neq 0}f(\xi)\geqq(\|\lambda\|/D)\int_{R_n}f(x)dx=(\|\lambda\|/D)v$$

besteht. Hier wird bewiesen: Gelten für $f(x)$ die Bedingungen $f(\lambda x-\lambda y)\geqq k\{f(x)+f(y)\}$ für positives $k\leqq\frac{1}{2}$ und alle Vektoren x und y im R_n und ferner $f(0)=\max f(x)$, so ist

$$f(0)+\frac{1}{2k}\sum_{\xi\neq 0}f(2\xi)\geqq\frac{\|\lambda\|v}{D}+\frac{1}{\|\lambda\|Dv}\sum_{\xi\neq 0}\left|\int f(\lambda^{-1}x)e^{2\pi i(x\xi)}dx\right|^2$$

erfüllt. Die Beweise benutzen eine Methode von C. L. Siegel [Acta Math. **65**, 307–323 (1935)], die auf der Besselschen Ungleichung bei n-fachen Fourier-Reihen beruht. *T. Schneider* (Göttingen).

Citations: MR 8, 444d = H15-12.

Referred to in H15-44.

H15-22 (10, 102a)

Mahler, K. On the admissible lattices of automorphic star bodies. Acad. Sinica Science Record **2**, 146–148 (1948).

A star body K of points X for which $F(X)\leqq 1$ is defined to be automorphic if a group Γ of linear substitutions exists, each of which maps K into itself, together with a hypersphere Σ, $|X|\leqq c$, such that each point of K may be mapped into a point of Σ by an appropriate member of Γ. Let d_F denote the set of K-admissible lattices Λ each of which has the property: lower bound of $F(P)=1$, P in Λ, $P\neq 0$. The author proves that the set of determinants of lattices of d_F is closed if K is automorphic. *D. Derry.*

H15-23 (10, 102c)

Jarník, Vojtěch. On the successive minima of arbitrary sets. Časopis Pěst. Mat. Fys. **73**, 9–15 (1948). (English. Czech summary)

Let M be any set of points in n-dimensional space, and M_1 the set of all points $\frac{1}{2}(x-y)$, where x and y are points of M. The author studies various definitions of successive minima, possible generalizations of Minkowski's definition [Geometrie der Zahlen, Leipzig, 1896–1910, chap. 5] for a convex set, symmetrical about the origin. He defines λ_i, μ_i, ν_i, π_i for $i=1, 2, \cdots, n$ as the lower bounds of all positive numbers α such that (i) the union of all sets βM_1 with $\beta \leqq \alpha$, (ii) the set αM_1, (iii) every set βM_1 with $\beta > \alpha$, (iv) the common part of all sets βM_1 with $\beta \geqq \alpha$, respectively, contains at least i linearly independent points with integral coordinates. Here αM_1 denotes the set of all points αx, where x is in M_1. It follows from these definitions that each of the four sets of numbers is monotonic increasing (in the wide sense), and that $\lambda_i \leqq \mu_i \leqq \nu_i \leqq \pi_i$. Various inequalities are known which generalize partially Minkowski's inequality for the product of the successive minima [Jarník, Věstník Královské České Společnosti Nauk. Třida Matemat.-Přírodověd., **1941**; Jarník and Knichal, Rozpravy II. Třídy České Akad. **53**, no. 43 (1943); these Rev. **7**, 417; **8**, 565; C. A. Rogers, forthcoming paper]. The aim of the present paper is to show that some of these cannot be essentially improved. For example, it is known that

$$\mu_1 \cdots \mu_{i-1}\nu_i\mu_{i+1} \cdots \mu_n L(M) \leqq 2^{2n-1},$$

where $L(M)$ denotes the inner Lebesgue measure of M; and the author constructs, for any T, i, j a set M for which

$$\lambda_1 \cdots \lambda_{i-1}\nu_i\lambda_{i+1} \cdots \lambda_{j-1}\nu_j\lambda_{j+1} \cdots \lambda_n L(M) > T.$$

Other results relate to the set M_p, obtained by "differencing" p times. *H. Davenport* (London).

Citations: MR 7, 417e = H15-4; MR 8, 565g = H15-15.

Referred to in H15-36.

H15-24 (10, 236a)

Cassels, J. W. S. On two problems of Mahler. Nederl. Akad. Wetensch., Proc. **51**, 854–857 = Indagationes Math. **10**, 282–285 (1948).

This paper answers two questions proposed by Mahler by constructing a plane star domain K which has a critical lattice with no points on the boundary of K and which possesses a group Γ of affinities which map K into itself. The group Γ has the properties that an appropriate substitution will map a point of the plane other than O into a point of the plane beyond an arbitrary distance from the origin while another appropriate substitution will map a point of K into a point within a certain fixed circle $|X| < C$. *D. Derry* (Vancouver, B. C.).

H15-25 (10, 285c)

Mullender, P. Lattice points in non-convex regions. I. Nederl. Akad. Wetensch., Proc. **51**, 874–884 = Indagationes Math. **10**, 302–312 (1948).

The author gives [as an application of a well-known theorem of Blichfeldt] the following generalization of a theorem of L. J. Mordell [same Proc. **49**, 773–781, 782–792 = Indagationes Math. **8**, 476–484, 485–495 (1946); these Rev. **8**, 369] concerning nonconvex regions. Let S_1 and S_2 denote the p- and q-dimensional spaces of points $x=(x_1, \cdots, x_p)$ and $y=(y_1, \cdots, y_q)$, respectively, and let S denote the $n=(p+q)$-dimensional space of points $(x, y)=(x_1, \cdots, x_p, y_1, \cdots, y_q)$. Let M and N be convex regions which are closed, bounded and symmetric about the origin O in the spaces S_1 and S_2, respectively. Let P and Q denote the p- and q-dimensional volumes of the regions M and N, respectively. Let $f(\mu)$ be a positive, steadily decreasing function of μ, defined for $0<\mu<a$, with steadily increasing derivative. Put $f(0)=\lim_{\mu\to 0+} f(\mu)$ and let $f(a)=\lim_{\mu\to a-} f(\mu)=0$, where $f(0)$ and a may be infinite. Let ξ be any number for which $0<\xi<a$, $f(\xi)+\xi f'(\xi)>0$. Let $g(t)$ be a differentiable function of t, defined for $0\leqq t\leqq \frac{1}{2}\xi$, with $g(\frac{1}{2}\xi)=\frac{1}{2}\xi$, $g'(t)<0$. Let $\beta=g(0)<a-\frac{1}{2}\xi$. Let $t=h(\bar{t})$ be the inverse function of $\bar{t}=g(t)$. Put $\varphi(\mu)=\frac{1}{2}f(\xi)-\int_\mu^{\frac{1}{2}\xi} f'(t+g(t))dt$ for $0\leqq\mu\leqq\frac{1}{2}\xi$; $\varphi(\mu)=\frac{1}{2}f(\xi)+\int_{\frac{1}{2}\xi}^{\mu} f'(h(t)+t)dt$ for $\frac{1}{2}\xi\leqq\mu\leqq\beta$; $\varphi(\mu)=f(\mu)-\varphi(0)$ for $\beta\leqq\mu\leqq a$. Let R be the set of the points $(x, y)\varepsilon S$ for which there exist two numbers μ, ν such that $0\leqq\mu\leqq a$, $0\leqq\nu\leqq f(\mu)$, $x\varepsilon\mu M$, $y\varepsilon\nu N$. Let $0\neq|\Delta|\leqq pPQ\int_0^\alpha(\varphi(t))^q t^{p-1}dt$, where α is defined by $\varphi(\alpha)=0$. Then R contains a point other than O of any lattice with determinant Δ. In the second part of this paper the author gives some applications of this theorem. *V. Knichal* (Prague).

Citations: MR 8, 369h = H15-11.

Referred to in H05-67, H15-26, J12-18.

H15-26 (10, 593c)

Mullender, P. Lattice points in non-convex regions. II. Nederl. Akad. Wetensch., Proc. **51**, 1251–1261 = Indagationes Math. **10**, 395–405 (1948).

[For part I cf. same vol., 874–884 = Indagationes Math. **10**, 302–312 (1948); these Rev. **10**, 285.] The author considers the following problem. Given a closed region R in n-dimensional Euclidean space E, construct a region $K\subset E$, with the (largest possible) volume V_K, such that all points P_1-P_2, with P_1 and P_2 in K, are points of R. The author calls such a region K suitable. By a theorem of H. F. Blichfeldt [Trans. Amer. Math. Soc. **15**, 227–235 (1914)] any lattice Λ, with $0<|\Delta|\leqq V_K$ (Δ denotes the determinant of Λ), then has a point other than the origin contained in R. In part A of this paper $R\subset E$ is the set of the points $(x_1, \cdots, x_n)$, with $|x_1 \cdots x_n|\leqq 1$. Using results of J. F. Koksma and B. Meulenbeld [same Proc. **45**, 256–262, 354–359, 471–478, 578–584 (1942); these Rev. **5**, 256] and L. J. Mordell [same Proc. **49**, 773–781, 782–792 = Indagationes Math. **8**, 476–484, 485–495 (1946); these Rev. **8**, 369] the author constructs a suitable region K (which is larger than an almost trivial one defined by $|x_1|+\cdots+|x_n|\leqq n/2$) and gives its volume. In part B the author constructs a suitable region K for the regions R defined as follows: let $F(x)\equiv F(x_1, \cdots, x_n)$ be a single-valued, twice differentiable function of $x_1, \cdots, x_n$ defined for all nonnegative values of the variables; let $F(x)>0$, $\partial F(x)/\partial x_j>0$ ($j=1, \cdots, n$), $F(x)+F(y)\leqq F(x+y)$, $F(tx)=tF(x)$ for $x_1, \cdots, x_n>0$; $y_1, \cdots, y_n>0$; $t\geqq 0$; then R is defined as the set of points (x) satisfying the inequality $F(|x_1|, \cdots, |x_n|)\leqq F(1, \cdots, 1)$. *V. Knichal* (Prague).

Citations: MR 5, 256f = J16-2; MR 5, 256g = J16-2; MR 5, 256h = J16-2; MR 5, 256i = J16-2; MR 8, 369h = H15-11; MR 10, 285c = H15-25.

Referred to in H15-27.

H15-27 (11, 418d)

Mullender, P. Lattice points in non-convex regions. III. Nederl. Akad. Wetensch., Proc. **52**, 18–28 = Indagationes Math. **11**, 50–60 (1949).

Continuing two previous notes the author gives the proof of the theorem contained in part II B [same Proc. **51**, 1251–1261 = Indagationes Math. **10**, 395–405 (1948); these Rev. **10**, 593] and applies them to the following cases [see the notation in the cited review]:

$$(1)\qquad F(x_1, \cdots, x_n)=(X_1^{p_1} \cdots X_r^{p_r})^{1/(p_1+\cdots+p_r)},$$

where $X_i=p_i^{-1}\sum_{k=1}^{p_i}|x_{p_1+\cdots+p_{i-1}+k}|$, with $p_1+\cdots+p_r=n$;

$$(2)\qquad F(x_1, \cdots, x_n)=(|x_1|^\sigma+\cdots+|x_n|^\sigma)^{1/\sigma},$$

where $0<\sigma<1$. *V. Knichal* (Prague).

Citations: MR 10, 593c = H15-26.

H15-28 **(10, 355c)**

Mullineux, N. On two problems of K. Mahler on irreducible star domains. Ann. Mat. Pura Appl. (4) **26,** 375–382 (1947).

To answer two problems proposed by Mahler [Nederl. Akad. Wetensch., Proc. **49**, 444–454 = Indagationes Math. **8**, 299–309 (1946), problems 3 and 4, pp. 449–450; these Rev. **8**, 12] the author constructs plane bounded star bodies K_d, depending on a real parameter d (the boundary curves are either hyperbolas or straight lines), which are irreducible, i.e., $\Delta(K)<\Delta(K_d)$ if $K<K_d$, and which have the following two properties. (1) Boundary points X, X_1 and X_2 of K_d exist so that the origin, X and X_1+X_2 are collinear and $|X_1+X_2|/|X|$ becomes infinite with d; (2) $V(K_d)/\Delta(K_d)$ becomes infinite with d, where $V(K_d)$ is the area of K_d.

D. Derry (Vancouver, B. C.).

Citations: MR 8, 12b = H15-8.

H15-29 **(10, 355d)**

Mahler, K. On the critical lattices of arbitrary point sets. Canadian J. Math. **1**, 78–87 (1949).

Let S be any point set within Euclidean n-space. A lattice is defined to be S-admissible if its only point interior to S is the origin. Let $\Delta(S)$ be the lower bound of the determinants of all the S-admissible lattices. If $0<\Delta(S)<\infty$, a lattice with determinant $\Delta(S)$ is defined to be critical. The author shows easily that S has a critical lattice if a sphere with center at the origin exists together with an infinite sequence of S-admissible lattices whose determinants are bounded and none of which have points other than the origin in the sphere. The main part of the paper deals with the construction of a set S which has admissible but no critical lattices. This set is the part of Euclidean n-space whose coordinates exceed 1 after a certain infinite set of open parallelepipeds has been removed. All the S-admissible lattices are shown to be generated by centers of the parallelepipeds not in S. Their determinants are shown to be greater than 1 while $\Delta(S)=1$. Hence no critical lattices can exist. This system of S-admissible lattices is enumerable. Another example of a space S is given whose only admissible lattices are the lattice of all points with integral coordinates and its sublattices. This space consists of all points $(x_1, \cdots, x_n)$ for which $\max[|x_1-u_1|, \cdots, |x_n-u_n|]\geqq\frac{1}{6}$, where $(u_1, \cdots, u_n)$ runs through all vectors with integral components.

D. Derry (Vancouver, B. C.).

H15-30 **(10, 511f)**

Chabauty, Claude. Géométrie des nombres d'ensembles non convexes. C. R. Acad. Sci. Paris **227**, 747–749 (1948).

Let S be a set of points in n-dimensional space, which is a star set relative to the origin, and let Δ denote the critical determinant of S. Let G be a lattice of determinant D. The author considers the successive minima $\mu_1, \cdots, \mu_n$ defined as follows: μ_h is the upper bound of all numbers $\mu\geqq 0$ for which the points of G in the set μS have a linear dimension less than h. His main theorem is that $\Delta\mu_1\cdots\mu_h\leqq D\psi(n)$. Here $\psi(n)$ is a function of n only, defined as follows. For any real $x_1, \cdots, x_n$ with $0<x_1\leqq x_2\leqq\cdots\leqq x_n$, take numbers $y_1, \cdots, y_n$ with $0<y_h<x_h$ for which y_{h+1}/y_h is an integer, such that $y_1\cdots y_n$ is as large as possible. Then $\psi(n)$ is the upper bound of $(x_1\cdots x_n)/(y_1\cdots y_n)$ for all sets x. The author notes that $\psi(2)=\sqrt{2}$, and that $\psi(n)<2^{n-C}/n^{\log 2}$, where C is Euler's constant. The method of proof of the main result is essentially that of Jarník and Knichal [Rozpravy II. Třídy České Akad. **53**, no. 43 (1943); these Rev. **8**, 565]. Errata: in the definition of $\gamma(S)$, "Sup" should stand first on the right; in the definition of $\alpha(S)$, the symbol A should be replaced by S, and $\alpha(S)$ should strictly be $\alpha(S, G)$; on page 749, $O(n)$ should be $o(1)$. [See the three following reviews.]

H. Davenport (London).

Citations: MR 8, 565g = H15-15.

Referred to in H15-36.

H15-31 **(10, 511g)**

Rogers, C. A. The product of the minima and the determinant of a set. Nederl. Akad. Wetensch., Proc. **52**, 256–263 = Indagationes Math. **11**, 71–78 (1949).

Let S be any set of points in n-dimensional space, and let $\Delta(S)$ be its critical determinant, i.e., the lower bound of the determinants of all lattices with no point different from O in S. For any lattice Λ, of determinant $d(\Lambda)$, define the successive minima $\mu_1, \cdots, \mu_n$ as follows: μ_k is the lower bound of the positive numbers μ such that the set μS contains at least k linearly independent points of Λ. [If S is a star set, this definition agrees with that of Chabauty in the preceding review.] The author's main theorem is that

$$(*)\qquad \mu_1\cdots\mu_n\Delta(S)\leqq 2^{\frac{1}{2}(n-1)}d(\Lambda).$$

The author states that this refinement of a result of Jarník is implicit in an earlier paper [submitted to Proc. London Math. Soc. in 1946 but not yet published]. The inequality is strict when S is a bounded star body. [See the two following reviews.]

H. Davenport (London).

Referred to in H15-36, H15-52.

H15-32 **(10, 511h)**

Chabauty, Claude. Géométrie des nombres d'ensembles non convexes. C. R. Acad. Sci. Paris **228**, 796–797 (1949).

The author proves that the function $\psi(n)$, defined in his previous paper [see the second preceding review], has the value $2^{\frac{1}{2}(n-1)}$, thus obtaining the inequality (*) above. The only difference between this result and that of Rogers is that here the set is supposed to be a star set. The author also gives a construction for a star set S in n-dimensional space for which the inequality becomes an equality. Let C_k $(k=1, \cdots, n)$ be the set of all points $(tg_1, \cdots, tg_k, 0, \cdots, 0)$ for which $g_1, \cdots, g_k$ are integers with H.C.F. unity, $g_k\neq 0$, and $t\geqq 2^{(1-k)/n}$. The set S consists of all points which are not in any of $C_1, \cdots, C_n$, and the lattice is that of all points with integral coordinates. By modifying S, a bounded star body is constructed for which the ratio of the two sides of (*) is arbitrarily near to 1. The proofs are naturally somewhat condensed. [See also the following review.]

H. Davenport (London).

Referred to in H15-34, H15-52.

H15-33 **(10, 512a)**

Mahler, K. On the minimum determinant of a special point set. Nederl. Akad. Wetensch., Proc. **52**, 633–642 (1949).

The author constructs a set S, as described in the preceding review, and so proves that the inequality (*) is best possible. There is also a modification of the construction, so as to obtain a bounded star body for which the ratio of the two sides of (*) is arbitrarily near to 1. The proof is detailed, and is accompanied by an investigation of the critical lattices of the set S.

H. Davenport (London).

Referred to in H15-34, H15-52.

H15-34 **(11, 418e)**

Chabauty, Claude. Sur les minima arithmétiques des formes. Ann. Sci. École Norm. Sup. (3) **66**, 367–394 (1949).

Let $f(x)$ be a function, defined at all points $x=(x_1, \cdots, x_n)$ of the Euclidean space R^n, and satisfying $f(tx)=|tf(x)|$ for all real t and for all points x. Let G be a lattice in R^n with determinant $m(G)$. For $h=1, \cdots, n$, the hth arithmetical minimum $\mu_h=\mu_h(f, G)$ of f for G is defined to be the lower bound of the numbers μ such that there are at least h linearly independent points x of G with $f(x)\leqq\mu$. Write $\gamma(f)=\sup \mu_1(f, G)\{m(G)\}^{-1/n}$, the upper bound being taken over all lattices G. [Then in Mahler's notation

$$\Delta(S)=\{\gamma(f)\}^{-n},$$

where S is the set defined by $f(x)\leqq 1$.] The author gives detailed proofs of his main results that (a) for every such function f and every lattice G,

$$(*) \qquad \mu_1 \cdots \mu_n \leqq 2^{\frac{1}{2}(n-1)}m(G)\{\gamma(f)\}^n,$$

and (b) that this inequality is satisfied with equality in certain special cases. He had previously given concise proofs of these results [C. R. Acad. Sci. Paris **228**, 796–797 (1949); these Rev. **10**, 511]. See also papers by the reviewer [Nederl. Akad. Wetensch., Proc. **52**, 256–263 = Indagationes Math. **11**, 71–78 (1949); these Rev. **10**, 511] and by K. Mahler [ibid., 633–642 = Indagationes Math. **11**, 195–204 (1949); these Rev. **10**, 512], giving independent proofs of (a) and of (b), respectively.

The convexity coefficient $\omega(f)$ of the function f is defined to be the upper bound of $f(x+y)/\{f(x)+f(y)\}$ taken over all pairs of points, not both coinciding with the origin. Let $V(f)$ be the interior Lebesgue measure of the set S given by $f(x)\leqq 1$. Let φ be a positive definite homogeneous polynomial in $x_1, \cdots, x_n$ of degree d with integral coefficients. By application of (*) to the function $f=|\varphi|^{1/d}$ it is shown that the number of arithmetically inequivalent polynomials φ of this form, having $n+d+\omega(f)+\{V(f)\}^{-1}$ less than any given constant, is finite.

The result (*) is used to generalize results of Khintchine [Math. Ann. **113**, 398–415 (1936)] and Mahler [Nederl. Akad. Wetensch., Proc. **41**, 634–637 (1938)] by establishing a connection between the homogeneous and nonhomogeneous problems for bounded star bodies. Among the other topics discussed are results corresponding to (*) for star bodies and for arbitrary sets. *C. A. Rogers.*

Citations: MR 10, 511h = H15-32; MR 10, 512a = H15-33.

Referred to in H05-60.

H15-35 (10, 513a)

Macbeath, A. M. Non-convex regions in three and more dimensions. Proc. Cambridge Philos. Soc. **45**, 161–166 (1949).

Let R be the region of points $(x_1, \cdots, x_n)$ in Euclidean n-space for which $f(|x_1|, \cdots, |x_n|)\leqq k$, where $f(|x_1|, \cdots, |x_n|)$ is a function which is strictly increasing in each of the variables and has the property that $f(px+qy)\geqq k$, $p>0$, $q>0$, $p+q=1$, $f(x)=f(y)=k$. If $\lambda_1x_1+\cdots+\lambda_nx_n=1$ is a tac-hyperplane for a boundary point of R with positive coordinates, let $u=\varphi(c)$ be the nonnegative function defined for positive c by the relations $f(z_1, \cdots, z_{n-1}, c)\geqq k$, $u=\min(\lambda_1z_1+\lambda_2z_2+\cdots+\lambda_{n-1}z_{n-1})$. The author constructs a nonconvex subset R_2 of R whose volume V may be expressed as an integral involving the function $\varphi(c)$ and which has the property that, if a and b are vectors representing the points of R_2, the point defined by $a-b$ is in R. This follows readily from a variation of a theorem of Mordell [Nederl. Akad. Wetensch., Proc. **49**, 773–781, 782–792 = Indagationes Math. **8**, 476–484, 485–495 (1946); these Rev. **8**, 369] for plane domains. The principal result of the paper is to show that a lattice with determinant not exceeding V has a point, other than the origin, in R. This follows from the above results by applying a theorem of Blichfeldt [Trans. Amer. Math. Soc. **15**, 227–235 (1914)]. The result is applied to the region of points $(x_1, \cdots, x_n)$ for which $(|x_1|+c)(|x_2|+c) \cdots (|x_n|+c)\leqq(1+c)^n$, c nonnegative. *D. Derry* (Vancouver, B. C.).

Citations: MR 8, 369h = H15-11.

H15-36 (10, 683b)

Rogers, C. A. The successive minima of measurable sets. Proc. London Math. Soc. (2) **51**, 440–449 (1949).

Let S be any set of points in n-dimensional space, of positive Lebesgue measure $V(S)$. Let $\mathbf{D}S$ denote the difference set, consisting of all points $(x_1-y_1, \cdots, x_n-y_n)$, where $(x_1, \cdots, x_n)$ and $(y_1, \cdots, y_n)$ are points of S. With $\mathbf{D}S$, the author associates successive minima $\nu_1, \cdots, \nu_n$, in the sense of Minkowski, defining ν_k to be the lower bound of all positive numbers ν for which the set $\mathbf{D}\nu S$ contains k linearly independent points with integral coordinates. One result which he proves is that $\nu_1 \cdots \nu_n V(S)\leqq 2^{\frac{1}{2}(n-1)}$. Some of the results relate to alternative definitions of the successive minima. The paper [submitted in 1946] has now been largely superseded by later work of Rogers [Nederl. Akad. Wetensch., Proc. **52**, 256–263 = Indagationes Math. **11**, 71–78 (1949); these Rev. **10**, 511] and of Chabauty [C. R. Acad. Sci. Paris **227**, 747–749 (1948); **228**, 796–797 (1949); these Rev. **10**, 511]. See also V. Jarník [Časopis Pěst. Mat. Fys. **73**, 9–15 (1948); these Rev. **10**, 102]. *H. Davenport* (London).

Citations: MR 10, 102c = H15-23; MR 10, 511f = H15-30; MR 10, 511g = H15-31.

Referred to in H15-17.

H15-37 (11, 13a)

Mahler, Kurt. On the successive minima of a bounded star domain. Ann. Mat. Pura Appl. (4) **27**, 153–163 (1948).

Let $F(X)$ be a continuous distance function, defined for the points of a plane, which defines a bounded star domain K of points X for which $F(X)\leqq 1$. Let P, Q run through all point pairs of the lattice Λ which generate a lattice and for which $F(P)\leqq F(Q)$. Let $\mu_1(\Lambda)$, $\mu_2(\Lambda)$ be defined to be the minima of $F(P)$ and $F(Q)$, respectively, while $\mu(\Lambda)$ is defined to be the product $\mu_1(\Lambda)\mu_2(\Lambda)$. Let $M(K)$ be the upper bound of $\mu(\Lambda)$, where Λ runs through all lattices whose determinants are equal to $\Delta(K)$. A lattice Λ for which $\mu(\Lambda)=M(K)$ is defined to be extreme. The principal result of the paper is that an extreme lattice exists for every plane bounded symmetric star body K. The author also discusses some of the relationships between extreme and critical lattices. For the latter it follows easily that $\mu(\Lambda)=1$ and so if $M(K)=1$ every critical lattice is extreme. This is always the case if K is convex. A star body K is constructed for which $M(K)>1$ which means that, in this case, the critical lattices are not extreme. *D. Derry.*

H15-38 (11, 83d)

Rogers, C. A. On the critical determinant of a certain nonconvex cylinder. Quart. J. Math., Oxford Ser. **20**, 45–47 (1949).

If C is a plane star body of points (x, y), let K be the 3-dimensional cylindrical star body of all points (x, y, z), $|z|\leqq 1$. Let $\Delta(S)$ denote the lower bound of all determinants of lattices with no interior points of the star body S other than the origin. The author constructs a nonconvex star body C, bounded by straight lines, with $\Delta(C)<\Delta(K)$. This is in contrast to the earlier result that $\Delta(C)=\Delta(K)$ if C is convex [J. H. H. Chalk and C. A. Rogers, J. London Math. Soc. **23**, 178–187 (1948); these Rev. **10**, 284]. *D. Derry* (Vancouver, B. C.).

Citations: MR 10, 284g = H05-13.

Referred to in H15-39.

H15-39 **(12, 161e)**

Davenport, H., and Rogers, C. A. On the critical determinants of cylinders. Quart. J. Math., Oxford Ser. (2) **1**, 215–218 (1950).

Let K be a plane star body, symmetric with respect to the origin, and C the cylindrical star body of points (x, y, z), where $|z| \leqq 1$ and (x, y) is a point of K. Let $\Delta(K)$, $\Delta(C)$ denote the critical determinants of K and C respectively. It is clear that $\Delta(C) \leqq \Delta(K)$. Recently Rogers [same J., Oxford Ser. (1) **20**, 45–47 (1949); these Rev. **11**, 83] constructed a star body K with $\Delta(C) < \Delta(K)$. In this note another star body K is constructed so that $\Delta(C)/\Delta(K)$ is arbitrarily small. *D. Derry* (Vancouver, B. C.).

Citations: MR 11, 83d = H15-38.

Referred to in H15-56, H15-78.

H15-40 **(11, 233a)**

Davenport, H., and Rogers, C. A. A note on the geometry of numbers. J. London Math. Soc. **24**, 271–280 (1949).

Let $G(x_1, \cdots, x_r)$, $E(x_1, \cdots, x_r)$ and $H(x_{r+1}, \cdots, x_n)$ be distance functions (not necessarily convex), where $1 \leqq r \leqq n-1$. It is proved that the star bodies in n dimensions defined by $G^{r/n}H^{(n-r)/n} \leqq 1$ and $G^{r/n}(E+H)^{(n-r)/n} \leqq 1$ have the same critical determinant, possibly infinite. Seven direct applications are given; e.g., the critical determinant of the star body $|x|(x^2+|yz|) \leqq 1$ is 7 and of $|x|(x^2+y^2+z^2) \leqq 1$ is $\frac{1}{2}(23)^{\frac{1}{2}}$. These functions are closely related to $|x(x^2+yz)|$ and $|x(-x^2+y^2+z^2)|$. It is shown that the critical determinant of the former is at least 9.1 and that of the latter is finite but exceeds $\frac{1}{2}(23)^{\frac{1}{2}}$. *L. Tornheim.*

H15-41 **(11, 241a)**

Tsuji, Masatsugu. On Blichfeld's theorem in the geometry of numbers. Jap. J. Math. **19**, 427–431 (1948).

Let $\mu(e)$ be a nonnegative additive set function defined on Borel sets e in n-dimensional Euclidean space. Suppose that the mean value

$$\lim_{T\to\infty} T^{-n} \int_a^{a+T} \cdots \int_a^{a+T} d\mu(x_1, \cdots, x_n)$$

exists for all $a \geqq 0$ and has a value κ independent of a. If E is a bounded Borel set of positive measure $m(E)$, and if $\epsilon > 0$ is given, then by translating E into suitable positions one can obtain sets E' and E'', satisfying

$$\kappa m(E) - \epsilon \leqq \mu(E') \leqq \mu(E'') \leqq \kappa m(E) + \epsilon.$$

When a lattice with density κ is given, and $\mu(e)$ is taken to be the number of lattice points in e, the theorem reduces to a result of Blichfeldt [Trans. Amer. Math. Soc. **15**, 227–235 (1914)]. *C. A. Rogers* (Princeton, N. J.).

No. 4 of vol. 19 of the Japanese Journal of Mathematics was erroneously paged. Nos. 1–3 are paged consecutively from 1–516; no. 4 is paged 233–547. This paper is in no. 4.

H15-42 **(12, 394b)**

Davenport, H., and Rogers, C. A. Diophantine inequalities with an infinity of solutions. Philos. Trans. Roy. Soc. London. Ser. A. **242**, 311–344 (1950).

Let $F(x_1, \cdots, x_n)$ be a continuous real function of n real variables satisfying $F(tx_1, \cdots, tx_n) = t^h F(x_1, \cdots, x_n)$ for all real t, where h is a positive integer. Let $x_1, \cdots, x_n$ denote any real forms in $u_1, \cdots, u_n$ of determinant 1. Now if the inequality $|F(x_1, \cdots, x_n)| \leqq \lambda$ has at least one solution in integers $(u_1, \cdots, u_n) \neq (0, \cdots, 0)$ one may ask whether for each $\lambda' > \lambda$ the inequality (1) $|F(x_1, \cdots, x_n)| \leqq \lambda'$ has infinitely many such solutions. Carrying further Mahler's work on star bodies [Proc. Roy. Soc. London. Ser. A. **187**, 151–187 (1946); Nederl. Akad. Wetensch., Proc. **49**, 331–343, 444–454, 524–532, 622–631 (1946); these Rev. **8**, 195, 12], which enables one to prove under certain general conditions that the answer to the above question is affirmative, the authors prove a number of theorems concerning inequalities of the type (1) having an infinity of integer solutions.

The inequality (2) $|F(x_1, \cdots, x_n)| \leqq 1$ defines a starbody K in R_n. This body is called automorphic, if F possesses automorphisms, i.e., linear transformations with real coefficients which leave $|F|$ invariant, such that every point $(x_1, \cdots, x_n)$ of the body K can be transformed by an appropriate automorphism into a point of a bounded region. The body is called fully automorphic if moreover every point except the origin can be transformed by appropriate automorphisms into points arbitrarily far from the origin.

Now consider a lattice defined by n linear forms $x_1, \cdots, x_n$ of the integer variables $u_1, \cdots, u_n$ with determinant $\Delta > 0$. This lattice is said to be admissible for K if none of its points except the origin is strictly inside K. The lower bound of the determinants Δ of all lattices which are admissible for K is denoted by $\Delta(K)$. Further, the definitions of critical lattices for K, reducible starbodies, and boundedly reducible starbodies are used [the reader can find these in the reviews of the cited papers by Mahler]. Moreover, they give a new definition: K is said to be fully reducible if there exists a bounded starbody H contained in K such that $\Delta(H) = \Delta(K)$, whereas H has exactly the same critical lattices as K. If K is fully reducible, it is certainly boundedly reducible. The authors then prove theorem 1. Suppose K is fully automorphic and suppose Λ is a lattice with determinant $\Delta > 0$. Then (a) if $\Delta < \Delta(K)$, there are an infinity of solutions of (2) with strict inequality; (b) if K is boundedly reducible and $\Delta \leqq \Delta(K)$, there are an infinity of solutions of (2); (c) if K is fully reducible and $\Delta \leqq \Delta(K)$, there are an infinity of solutions of (2) with strict inequality, unless Λ is a critical lattice of K, in which case there are an infinity of solutions of (2) with equality.

The authors prove the fully-reducibility for the following special starbodies; calculating the corresponding values of $\Delta(K)$ they apply theorem 1 to them: $|x_1x_2| \leqq 1$ $(\Delta(K) = \sqrt{5})$; $-1 \leqq x_1x_2 \leqq k$ $(\Delta(K) = (k^2+4k)^{\frac{1}{2}}$ for any positive integer $k)$; $|x_1x_2x_3| \leqq 1$ $(\Delta(K) = 7)$; $|(x_1^2+x_2^2)x_3| \leqq 1$ $(\Delta(K) = \frac{1}{2}\sqrt{23})$; $|x_1^2+x_2^2-x_3^2| \leqq 1$ $(\Delta(K) = \sqrt{\frac{2}{3}})$; $|x_1^2+x_2^2+x_3^2-x_4^2| \leqq 1$ $(\Delta(K) = \sqrt{\frac{7}{4}})$; $|x_1^2+x_2^2-x_3^2-x_4^2| \leqq 1$ $(\Delta(K) = \frac{3}{2})$.

Further, the authors investigate the possible distributions of the lattice points within given bodies. A starbody K_0 contained in K is said to generate K if for every bounded part K' of K there is an automorphism Ω of K, such that ΩK_0 contains K'. If K_0 generates K, then obviously ΩK_0 also generates K for every automorphism Ω of K. Now the authors prove: theorem 4. Suppose K_0 generates K, K being a fully automorphic starbody. Suppose Λ is a lattice with determinant Δ. Then (a) if $\Delta < \Delta(K)$ there are an infinity of points of Λ strictly inside K_0; (b) if K is boundedly reducible, and $\Delta \leqq \Delta(K)$, there are an infinity of points of Λ in K_0; (c) if K is fully reducible and $\Delta \leqq \Delta(K)$, then there are an infinity of points of Λ strictly inside of K_0, unless Λ is a critical lattice of K, in which case there are an infinity of points of Λ on the boundary of K_0. Theorem 8. Suppose K_0 generates K, K not necessarily being fully automorphic. Assume that Θ is an automorphism of K which reduces K_0. Suppose that Λ is a lattice with determinant Δ which has no point other than the origin in $C(K_0, \Theta)$. Then the conclusions (a), (b), (c) of theorem 4 are valid. Remark. The set $C(K_0, \Theta)$ is defined to consist of all points which are common to all starbodies $\Theta^m K_0$ $(m = 1, 2, \cdots)$. The authors give various applications of the above theorems which cannot be explicitly quoted here, and moreover deduce "isolation theorems": If for a class of functions F the value λ_0 is the greatest lower bound of all λ' for which (1) has infinitely many solutions for all functions F of the class, one may ask which "critical" functions F may be excluded in order to improve the inequality (1) for the remaining functions of the class [cf. Markoff's results for binary quadratic forms, Math. Ann. **15**, 381–407 (1879)]. Thus, e.g., new results are deduced for the case $F(x, y, z) = (x^2+y^2)z$, where x, y, z denote linear forms in u, v, w, of determinant 1. The paper closes with an interesting conjecture concerning the simultaneous approximation of two real numbers θ_1, θ_2 by rationals p_1/q, p_2/q. *J. F. Koksma* (Amsterdam).

Citations: MR 8, 12b = H15-8; MR 8, 12d = H15-6; MR 8, 195f = H15-7.

Referred to in H15-58, H15-67, J36-22.

H15-43 (12, 479f)

Chabauty, Claude. Limite d'ensembles et géométrie des nombres. Bull. Soc. Math. France **78**, 143–151 (1950).

The author extends a theorem of Mahler [Proc. Roy. Soc. London. Ser. A. **187**, 151–187 (1946); these Rev. **8**, 195] concerning lattices in Euclidean space to lattices in a locally compact topological group, by a direct group-theoretical method which does not depend on inequalities of Minkowski and Hermite, as does that of Mahler. Let G be a σ-compact locally compact group; a discrete subgroup H is called a "lattice" if $m(G/H)$ is finite, where m is the unique G-invariant regular measure on the left coset space G/H, normalized in the usual way (relative to the obvious measure on H). A sequence $\{F_n\}$ of lattices converges to a lattice F if for every compact set K and neighborhood N of the identity e in G, and for sufficiently large n, there exists for each $x\varepsilon F_n\cap K$ an element y in F such that $y^{-1}x\varepsilon N$, and for each $y\varepsilon F\cap K$ an $x\varepsilon F_n$ such that $y^{-1}x\varepsilon N$. Then if (1) G has a basis for the open sets consisting of sets whose boundaries are of Haar measure zero, (2) $\{H_n\}$ is a sequence of lattices in G such that (a) there exists a neighborhood N of the identity e in G such that $H_n\cap N = e$ $(n=0, 1, \cdots)$, and (b) $m(G/H_n)$ is bounded as a function of n $(n=0, 1, \cdots)$, the following holds: (3) there exists a subsequence of the H_n which converges to a lattice H, and $G\cap H=e$, $m(G/H)\leqq \liminf_{n\to\infty} m(G/H_n)$. If the hypothesis is strengthened by requiring also (c) there exists an open set S in G of finite measure such that $G=S\cdot H_n$ $(n=0, 1, \cdots)$, then (3) can be strengthened to assert that $m(G/H)=\lim_{n\to\infty} m(G/H_n)$. In the course of the proof a number of lemmas concerning the quotient measure m in G/H, where H is a lattice, are obtained.

I. E. Segal (Chicago, Ill.).

Citations: MR 8, 195f = H15-7.

Referred to in H15-64, H15-65.

H15-44 (12, 678f)

Hlawka, Edmund. Bemerkungen zu einem Satz von R. Rado. Anz. Öster. Akad. Wiss. Math.-Nat. Kl. **1950**, 219–226 (1950).

Let $f(x)$ be a bounded nonnegative function vanishing outside a bounded set and integrable in the Riemann sense over n-dimensional space R_n. Suppose that there is a nonsingular matrix A so that $f(A(x-y))\geqq \min\{f(x), f(y)\}$ for all x, y. Let Γ be any lattice with determinant 1. A theorem of R. Rado [J. London Math. Soc. **21**, 34–47 (1946); these Rev. **8**, 444] asserts that

$$(*)\qquad f(o)+\tfrac{1}{2}\textstyle\sum' f(g)\geqq |A|\,V, \quad \text{where} \quad V=\displaystyle\int f(x)dx,$$

the sum being taken over all the points g of Γ other than the origin o and the integral being over R_n. The author discusses this theorem and its improvement by J. W. S. Cassels [J. London Math. Soc. **22**, 196–200 (1948); these Rev. **10**, 19]. He shows that if equality holds in (*) then every point of R_n can be represented in the form $x+A^{-1}g$, where g is in Γ and x is in the closure of the set of points y with $f(y)>0$. The author remarks that by use of one of his earlier theorems [Math. Z. **49**, 285–312 (1943); these Rev. **5**, 201] it follows that if f, A, and $\epsilon>0$ are given it is possible to find a lattice Γ with determinant 1 so that

$$f(o)+\tfrac{1}{2}\textstyle\sum' f(g)\leqq \tfrac{1}{2}V+f(o)+\epsilon.$$

Let f_1 and f_2 be two functions satisfying the same conditions as f (with the same matrix A) and let Γ be a lattice with determinant 1. Then, if $|A|(V_1+V_2)^2>V_1\sum_1+V_2\sum_2$ where $\sum_i=f_i(o)+\frac{1}{2}\sum' f_i(g)$ and $V_i=\int f_i(x)dx$, every point of R_n is of the form $\pm(y_1-y_2)+A^{-1}g$, where g is in Γ and $f_i(A^{-1}y_i)>0$, $i=1, 2$.

A p-adic form of Rado's theorem is given.

C. A. Rogers (London).

Citations: MR 5, 201c = H25-1; MR 8, 444d = H15-12; MR 10, 19d = H15-21.

H15-45 (12, 806a)

Schneider, Theodor. Über einen Blichfeldtschen Satz aus der Geometrie der Zahlen. Arch. Math. **2**, 349–353 (1950).

Let $\mu=\mu(E)$ be a nonnegative countably additive set function defined for Borel sets E in n-dimensional space R_n. Let $\rho=\liminf_{w\to\infty} w^{-n}\mu(W)$, $\rho'=\limsup_{w\to\infty} w^{-n}\mu(W)$, where W is an arbitrary n-dimensional cube with side w; and suppose that $0<\rho\leqq\rho'<+\infty$. Let $f=f(z)$ be a bounded nonnegative function vanishing outside a bounded set in R_n, the integral $\int f dz$ over R_n being positive. It is shown that, provided certain integrability conditions are satisfied, it is possible, for each $\epsilon>0$, to find points ζ and ζ' such that

$$\int f(z-\zeta)d\mu>\rho\int f dz-\epsilon,$$

$$\int f(z-\zeta')d\mu<\rho'\int f dz+\epsilon.$$

If μ is periodic in each of the coordinates, then $\rho=\rho'$, and the points ζ, ζ' may be chosen to satisfy

$$\int f(z-\zeta)d\mu\geqq\rho\int f dz\geqq\int f(z-\zeta')d\mu.$$

C. A. Rogers (London).

H15-46 (13, 114c)

Bambah, R. P. On the geometry of numbers of non-convex star-regions with hexagonal symmetry. Philos. Trans. Roy. Soc. London. Ser. A. **243**, 431–462 (1951).

Let R be a plane star region, symmetric with respect to each of the three diagonals l_1Ol_4, l_2Ol_5, l_3Ol_6 of a regular hexagon and also to each of the bisectors of the angles l_1Ol_2, l_2Ol_3, l_3Ol_4, for which the boundary points are finite except possibly those on the hexagon diagonals, and such that each of the six sets defined as the points within an angular region l_1Ol_2, l_2Ol_3, etc., complementary to R is convex. Let T_1, T_2 be the boundary points of R for which OT_1, OT_2 bisect the angles l_2Ol_1, l_1Ol_6 respectively. Then if the points of R within the angular region T_1OT_2 are included within the parallelogram with sides OT_1, OT_2 the author shows easily that the single critical lattice of R is generated by this parallelogram. Such regions R are said to be of type I. For regions R other than of type I a unique parallelogram $OA_1A_2(A_1+A_2)$ is shown to exist for which A_1, A_2, A_1+A_2 are boundary points of R such that $\angle A_1OA_2=60°$ and A_1 is within the angular region bounded by Ol_1 and the bisector of the angle l_2Ol_1. Let $\mathfrak{L}_1$ be the lattice generated by OA_1 and OA_2, $\mathfrak{L}_2$ the reflection of $\mathfrak{L}_1$ about the bisector of the angle l_2Ol_1 and Δ the determinant of these lattices. These star regions are now subdivided into types II, III, and IV according to the manner in which the boundary of R cuts certain lines which join points of $\mathfrak{L}_1$ and $\mathfrak{L}_2$. The principal result of the paper shows that, for regions R of types II and III, $\Delta(R) \geqq \Delta$ with equality if and only if $\mathfrak{L}_1$ and $\mathfrak{L}_2$ are admissible, in which case these are the only critical lattices, and that for regions of type IV a somewhat similar result holds but which is conditioned by the fact that at least one of an infinite set of parallelograms be not less than Δ. The proofs of these results, while of elementary character, involve considerable geometric detail. As an easy consequence of his results for type II regions, the author obtains the critical lattices for the region $|f(x, y)| \leqq 1$, where $f(x, y)$ is a binary cubic form with positive discriminant, which were first determined by Mordell [Proc. London Math. Soc. (2) **48**, 198–228 (1943); these Rev. **5**, 172]. In addition, the results for regions of types I, II and III are applied to regions R in which the boundary arcs are (1) circular, (2) parabolic and (3) hyperbolic. The results for type IV regions are applied to star shaped dodecagons. *D. Derry.*

Citations: MR 5, 172d = J52-3.

Referred to in H20-26.

H15-47 (13, 444a)

Macbeath, A. M. The finite-volume theorem for non-homogeneous lattices. Proc. Cambridge Philos. Soc. **47**, 627–628 (1951).

If the volume of a point set K in Euclidean n-space is finite then there exists a non-homogeneous lattice, of arbitrarily small determinant, having no point in common with K. More generally, if the volume of that part of K not lying between some pair of hyperplanes is finite, the same result holds. *L. Tornheim* (Ann Arbor, Mich.).

H15-48 (14, 253a)

Sawyer, D. B. The number of non-homogeneous lattice points in n-dimensional point sets. Proc. Cambridge Philos. Soc. **48**, 735–736 (1952).

Let R be a set in Euclidean n-space and let $\Delta'(R)$ be the g.l.b. of determinants of non-homogeneous lattices with no point in R. Suppose $m \geqq 2$ and integral. If Λ is a non-homogeneous lattice of non-zero determinant $d(\Lambda)$ such that $\Delta'(R) > md(\Lambda)$, then R contains at least $m+n-1$ points of Λ. An example is presented to show that for $m=2$ this result is best possible. The proof, of a geometrical nature, is simple. *L. Tornheim* (Ann Arbor, Mich.).

H15-49 (14, 253b)

Sawyer, D. B. The number of non-homogeneous lattice points in n-dimensional point sets. Proc. Cambridge Philos. Soc. **49**, 156–157 (1953).

Same as the paper reviewed above.

H15-50 (14, 253c)

Rogers, C. A. The reduction of star sets. Philos. Trans. Roy. Soc. London. Ser. A. **245**, 59–93 (1952).

A set S of points in Euclidean n-space is defined to be a star set if $X \in S$ implies $\lambda X \in S$, $-1 \leqq \lambda \leqq 1$. If, in addition to this, the points λX, $-1 < \lambda < 1$, are interior points of S, then S is defined to be a star body. $\Delta_M(S)$ is defined to be the lower bound of the determinants of the lattices whose only point interior to a star body S is the origin. A star body S is defined to be reducible in the sense of Mahler if it contains a star body T as a proper subset with $\Delta_M(S) = \Delta_M(T)$. The present paper starts from the following problem of Mahler [Nederl. Akad. Wetensch., Proc. **49**, 444–454 (1946); these Rev. **8**, 12]. Does every bounded star body S contain an irreducible star body T with $\Delta_M(S) = \Delta_M(T)$? In answer to this problem a plane, bounded, reducible star body S is constructed which contains no irreducible star body T with $\Delta_M(S) = \Delta_M(T)$.

The author introduces the following modifications of the basic concepts. A lattice is defined to be admissible for a star set S if it contains no points of S except the origin. $\Delta(S)$ is the lower bound of the determinants of admissible lattices. A critical lattice is a limit lattice of admissible lattices with determinant $\Delta(S)$; $\Delta_M(S) \leqq \Delta(S)$. Examples of star sets, one bounded by straight lines, are given for which $\Delta_M(S) < \Delta(S)$. The principal result of the paper is that every bounded star set S, with $\Delta(S) > 0$, contains an irreducible star set T with $\Delta(T) = \Delta(S)$. In other words the analogue of Mahler's problem with the modified concepts has an affirmative answer.

In the course of the proof a point X of S is defined to be primitively irreducible if X is a point of a lattice Λ with $d(\Lambda) < \Delta(S)$ and such that the only lattice points of S are of the form mX, $m = 0, \pm 1, \pm 2, \cdots$. A point X of S is defined to be an outer boundary point of S if λX is not in S for $\lambda > 1$. The author shows first that a star set is irreducible if every outer boundary point is primitively irreducible or the limit point of such points. The final irreducible star subset T of S is obtained by progressively eliminating, in a finite number of steps, those outer boundary points of S which are not irreducible. Results analogous to those of Mahler are obtained. Among these is that every outer boundary point of an irreducible star set S with $\Delta(S) < \infty$ is a point of a critical lattice of S.

In a second section of the paper the author considers bounded star sets S whose boundary points $(x_1, x_2, \cdots, x_n)$ satisfy an equation

$$G(x_1, x_2, \cdots, x_n) = 0$$

where $G(x_1, x_2, \cdots, x_n)$ is a polynomial. By methods analogous to those of the previous section he shows that S contains an irreducible star set T with $\Delta(T) = \Delta(S)$ where the boundary points of T also satisfy a polynomial equation. *D. Derry* (Vancouver, B. C.).

Citations: MR 8, 12b = H15-8.

Referred to in H30-79.

H15-51 (14, 253d)

Ap Simon, H. A method of finding the critical lattices of spheres containing the origin. Quart. J. Math., Oxford Ser. (2) **3**, 91–93 (1952).

Let K be a closed three-dimensional star body which is the union of a strictly convex body K_1 containing the origin as an inner point and its reflection K_2 in the origin. A grid G is the set of 27 points $aA+bB+cC$, where A, B, C are independent points and each of a, b, c takes the values $-1, 0, 1$. The determinant of G is the determinant of the lattice generated by A, B, C. A grid G is called admissible for K if A, B, C lie on K and no grid point except 0 is an inner point of K. In the set of admissible grids a grid is critical if its determinant is minimal. The problem of determining critical lattices is transformed to that of finding critical grids by a result that the equivalent lattice of a critical grid of K is critical if it is admissible. The author outlines how this result can be applied to the case of K_1 a sphere to find the critical lattices of K and refers to his thesis for details. *L. Tornheim* (Ann Arbor, Mich.).

H15-52 (14, 1066c)

Prachar, Karl. Über höhere zahlengeometrische Minima. Arch. Math. **4**, 39–42 (1953).

Let S be an n-dimensional set with the origin $\mathbf{o}$ as centre. For any positive integer l let $\Delta_l(S)$ be the lower bound of the determinants $d(\Lambda)$ of the lattices Λ having less than l point pairs $\pm\mathbf{x}$ in S. The rth successive minimum $\sigma_{rl} = \sigma_{rl}(S, \Lambda)$ of order l of S for a lattice Λ is defined to be the lower bound of the positive numbers σ such that the set σS contains, not only at least r linearly independent points of Λ, but also at least l point pairs of Λ. If S is a convex body and has volume V, it was proved by Minkowski that

$$\sigma_{11}\sigma_{21}\cdots\sigma_{n1}V \leqq 2^n d(\Lambda)$$

and by Hlawka [J. Reine Angew. Math. **187**, 246–252 (1950); these Rev. **12**, 161] that $\sigma_{1l}\sigma_{2l}\cdots\sigma_{nl}V \leqq l2^n d(\Lambda)$. Chabauty [C. R. Acad. Sci. Paris **228**, 796–797 (1949); these Rev. **10**, 511] and Rogers [Nederl. Akad. Wetensch., Proc. **52**, 256–263 (1949); these Rev. **10**, 511; see also Mahler, ibid. **52**, 633–642 (1949); these Rev. **10**, 512] prove that in the general case $\sigma_{11}\sigma_{21}\cdots\sigma_{n1}\Delta_1(S) \leqq 2^{(n-1)/2}d(\Lambda)$. The author in this paper proves the result

$$\sigma_{1l}\sigma_{2l}\cdots\sigma_{nl}\Delta_l(S) \leqq 2^{(n-1)/2}d(\Lambda),$$

bearing the same relationship to Hlawka's result as that of Chabauty's and Rogers' bears to that of Minkowski. He further shows that $\sigma_{1l}\sigma_{2l}\cdots\sigma_{nl}\Delta_l(S) \leqq d(\Lambda)$ in the cases when (a) S is convex and two-dimensional, or (b) S is an n-dimensional sphere, or (c) S is an n-dimensional cube.

C. A. Rogers (London).

Citations: MR 10, 511g = H15-31; MR 10, 511h = H15-32; MR 10, 512a = H15-33; MR 12, 161d = H10-18.

H15-53 (14, 1066d)

Cohn, Harvey. Stable lattices. Canadian J. Math. **5**, 261–270 (1953).

Let $\varphi(\mathbf{x})$ be a real continuously differentiable function, homogeneous of positive degree h, defined on the vectors of n-space. Let $\mathfrak{L}$ be a lattice with determinant Δ. Write $M(\mathfrak{L}) = \inf_{\mathbf{x}\in\mathfrak{L},\, \mathbf{x}\neq(0)} |\varphi(\mathbf{x})|\Delta^{-h/n}$. Further, $M_0 = \sup_{\mathfrak{L}} M(\mathfrak{L})$ ($M_0^{-n/h}$ is what is usually called the determinant of the region defined by $\varphi(\mathbf{x}) \leqq 1$). A lattice $\mathfrak{L}_0$ is called critical if $M(\mathfrak{L}_0) = M_0$; in that case $\mathfrak{L}_0$ gives the absolute maximum of $M(\mathfrak{L})$. Relative maxima can also occur. The author defines a special type of relative maxima, which are furnished by what he calls stable lattices. The definition is somewhat complicated: Write $F = \varphi(\mathbf{x})\Delta^{-h/n}$; if the lattice point $\mathbf{x}^{(k)}$ of $\mathfrak{L}$ varies continuously with $\mathfrak{L}$, the variations of $\mathfrak{L}$ give rise to a differential $d|F^{(k)}|$. Let Q be the minimum number of lattice points $\mathbf{x}^{(1)}, \cdots, \mathbf{x}^{(Q)}$ with the property that for every lattice point $\mathbf{x}^{(k)}$ the differential $d|F^{(k)}|$ is a linear combination of $d|F^{(1)}|, \cdots, d|F^{(Q)}|$. ($Q$ is called the free dimension, and is shown to be independent of $\mathfrak{L}$.) Now $\mathfrak{L}$ is called stable if $\mathbf{x}^{(1)}, \cdots, \mathbf{x}^{(Q+1)}$ can be chosen in such a way that (i) $|F^{(i)}| = M(\mathfrak{L}) > 0$ $(i = 1, \cdots, Q+1)$, and (ii) in every non-trivial relation $A_1 d|F^{(1)}| + \cdots + A_{Q+1} d|F^{(Q+1)}| = 0$ all A_i are positive.

The author investigates, with $\varphi = x_1\cdots x_n$, the stability of the lattices furnished by the numbers of a module in a totally real algebraic number field of degree n, relating the vector of its n conjugates to any number of the field (hence φ is the norm). Complex fields are also considered.

N. G. de Bruijn (Amsterdam).

Referred to in H15-54.

H15-54 (15, 687c)

Cohn, Harvey. Stable lattices. II. Canadian J. Math. **6**, 265–273 (1954).

In part I of this paper [same J. **5**, 261–270 (1953); these Rev. **14**, 1066] the author defined the notion of stability of a lattice with respect to a norm function, and gave criteria for stability of the norm in algebraic number fields. In the present part further results are obtained: (i) The integer-module of R ($\cos 2\pi/N$) has stable norm if $N \neq 1, 2, 3, 4, 6, 12$; (ii) a necessary and sufficient condition for stability of the norm in complex modules; (iii) the norm in the integer-module of the cyclotomic field $R(\exp 2\pi i/N)$ is stable if and only if N is square-free.

N. G. de Bruijn.

Citations: MR 14, 1066d = H15-53.

H15-55 (15, 106f)

Rogers, K. The minima of some inhomogeneous functions of two variables. J. London Math. Soc. **28**, 394–402 (1953).

Let l, m be two lines through the origin dividing the plane into four quadrants. Let K be a closed point set, symmetric in 0, containing $(\frac{1}{2}, 0)$, $(0, \frac{1}{2})$, $(\frac{1}{2}, \frac{1}{2})$, and such that the part of each quadrant not in K is convex. Then for every (x_0, y_0) there is in K a $(x, y) \equiv (x_0, y_0) \pmod 1$. As a consequence, for a certain class of functions $f(x, y)$ described by geometric properties and for any (x_0, y_0) there exists $(x, y) \equiv (x_0, y_0)$ (mod 1) such that

$$|f(x, y)| \leqq \max [f(0, \tfrac{1}{2}), f(\tfrac{1}{2}, 0), \min \{f(\tfrac{1}{2}, \tfrac{1}{2}), f(\tfrac{1}{2}, -\tfrac{1}{2})\}].$$

An example of such an f is $|\xi\eta|$ where $\xi = ax + by$, $\eta = cx + dy$ $(ad - bc \neq 0)$ and for it the result had been proved by Barnes [Quart. J. Math., Oxford Ser. (2) **1**, 199–210 (1950); these Rev. **13**, 16]. Other examples given are $|\xi\eta|(|\xi| + |\eta|)$ and $|\xi^n\eta^n|$.

L. Tornheim (Ann Arbor, Mich.).

Citations: MR 13, 16a = R12-23.

Referred to in H15-61.

H15-56 (15, 292c)

Few, L. The critical determinant of a displaced convex cylinder. J. London Math. Soc. **29**, 26–30 (1954).

Let K be a star-region in the (x, y)-plane about O, and C the three-dimensional region $(x, y) \in K$, $|z| \leqq 1$. Then $\Delta(C) \leqq \Delta(K)$ and $\Delta(K)/\Delta(C)$ can be arbitrarily large, where $\Delta(C)$, $\Delta(K)$ denote the respective lattice constants for lattices with a point at O [Davenport and Rogers, Quart. J. Math., Oxford Ser. (2) **1**, 215–218 (1950); these Rev. **12**, 161]. If K is convex with centre O then $\Delta(C) = \Delta(K)$ [Yeh, J. London Math. Soc. **23**, 188–195 (1948); these Rev. **10**, 285; Chalk and Rogers, ibid. **23**, 178–187 (1948); these Rev. **10**, 284]. In this paper the author generalizes the proof of Chalk and Rogers to show that $\Delta(C) = \Delta(K)$ if K is any centrally symmetric convex region and O is any inner point of K.

W. S. Cassels (Cambridge, England).

Citations: MR 10, 284g = H05-13; MR 10, 285a = H05-14; MR 12, 161e = H15-39.

H15-57 (15, 406d)

Swinnerton-Dyer, H. P. F. Inhomogeneous lattices. Proc. Cambridge Philos. Soc. **50**, 20–25 (1954).

Let K be an open non-empty set of points in Euclidean n-space and Λ an inhomogeneous lattice $x_i = \sum_{j=1}^{j=n} a_{ij}u_j + x_i^0$, $1 \leqq i \leqq n$, and $u_1, u_2, \cdots, u_n$ integral. Λ is defined to be K-admissible if it does not contain any point of K. A boundary point P of K is defined to be exterior if it is the limit of points of K on the segment OP. Let R_0 denote the closure of the set of these points and where $\epsilon > 0$ let R_ϵ denote the set of points $(1+\lambda)P$, $P \in R_0$, $0 \leqq \lambda \leqq \epsilon$. The author shows, where $\Delta(K)$ is the lower bound of the determinants of the K-admissible lattices, that $\Delta(K)$ is the lower bound of those K-admissible lattices which have a point in R_ϵ where ϵ is an arbitrary positive number. K is defined to be automorphic if a number c exists so that, for X within K, a homogeneous linear unimodular substitution Ω exists with $\Omega K = K$ and $|\Omega X| < c$. By use of his first result the author shows that if K is automorphic and has the property that every line through a point of R_0 contains points of K then if $\Delta(K)$ is finite a K-admissible lattice exists with determinant $\Delta(K)$ which contains a point of R_0. It is stated that examples exist which show that the result cannot be improved.

D. Derry (Paris).

H15-58 (15, 406e)

Rogers, C. A. Almost periodic critical lattices. Arch. Math. **4**, 267–274 (1953).

The lattice Λ is said to be periodic for a group G of linear transformations if there exists a compact set H contained in G with the property that, if Ω is any linear transformation of G, there is a Θ in H such that $\Theta\Omega\Lambda=\Lambda$. The lattice Λ is said to be almost periodic for a group G of linear transformations if, for every neighborhood of Λ in the space of lattices, there exists a compact set H contained in G, with the property that, if Ω is any linear transformation of G, there is a Θ in H such that $\Theta\Omega\Lambda$ lies in the given neighborhood of Λ. The author shows by a direct application of results of Gottschalk [Ann. of Math. (2) **47**, 762–766 (1946); these Rev. **8**, 159] that every automorphic star body S with $\Delta(S)<\infty$ possesses a critical lattice which is almost periodic for its group of automorphisms. Further results in case S or G satisfy certain conditions. [For the above terminology cf. H. Davenport and C. A. Rogers, Philos. Trans. Roy. Soc. London. Ser. A. **242**, 311–344 (1950); these Rev. **12**, 394.] *J. F. Koksma* (Amsterdam).

Citations: MR 12, 394b = H15-42.

H15-59 (16, 117d)

Pötzl, Hans. Über Sternkörper im dreidimensionalen Raum. Monatsh. Math. **58**, 91–102 (1954).

The author generalizes work of Mahler on two-dimensional regions [Proc. London Math. Soc. (2) **49**, 128–157 (1946); these Rev. **8**, 12] to show that lattice constants of bounded star bodies in three-dimensional space whose surface is composed of a finite number of analytic pieces may be found in a finite number of steps. He is apparently unaware of Mahler's later work [Proc. Roy. Soc. London. Ser. A. **187**, 151–187 (1946); Nederl. Akad. Wetensch., Proc. **49**, 331–343, 444–454, 524–532, 622–631 (1946); these Rev. **8**, 195, 12] in n-dimensions. The author also analyses a method of the reviewer [Quart. J. Math., Oxford Ser. **18**, 236–243 (1947); these Rev. **9**, 334] for finding the lattice constants of certain two-dimensional regions and shows why it is unreasonable to expect it to generalize to three dimensions. *J. W. S. Cassels* (Cambridge, England).

Citations: MR 8, 12a = H15-9; MR 8, 12b = H15-8; MR 8, 12d = H15-6; MR 8, 195f = H15-7; MR 9, 334f = H15-20.

H15-60 (16, 802f)

Hlawka, Edmund. Inhomogene Minima von Sternkörpern. Monatsh. Math. **58**, 292–305 (1954).

This paper generalizes a group of theorems found by the reviewer and afterwards proved more simply by Cassels [Proc. Cambridge Philos. Soc. **48**, 72–86, 519–520 (1952); MR **13**, 919]. Let S be a star body in n-dimensional space, symmetrical about the origin, with distance-function $f(\mathbf{x})$. The inhomogeneous minimum of S relative to a lattice Λ is defined by

$$E(S,\Lambda)=\sup_{\mathbf{x}}\inf_{\mathbf{l}} f(\mathbf{l}+\mathbf{x}),$$

where $\mathbf{l}$ is any point of Λ and $\mathbf{x}$ any point of space. The object of the paper is to obtain a lower bound for $E(S,\Lambda)$ when S is the direct product of a bounded star body S_1 in n_1 dimensions and a bounded star body S_2 in n_2 dimensions $(n_1+n_2=n)$, so that $f=(f_1^{n_1}f_2^{n_2})^{1/n}$. Let λ_1, λ_2 be positive parameters satisfying $\lambda_1^{n_1}\lambda_2^{n_2}=1$, and let S_{λ_1,λ_2} denote the (bounded) n-dimensional star body defined by

$$\max(\lambda_1 f_1, \lambda_2 f_2)\leqq 1.$$

The first general result given is that if S_1 and S_2 are convex, then

$$(*)\qquad E(S,\Lambda)\geqq c_n^{-1}\inf_{\lambda_1,\lambda_2}\mu_n(S_{\lambda_1,\lambda_2},\Lambda),$$

where μ_n denotes the nth successive minimum of the star body S_{λ_1,λ_2} relative to Λ, and $c_n=128n^{7/2}(n!)^5$. The earlier results of the reviewer arise as the special cases when S_1 and S_2 are either intervals in one dimension or circles in two dimensions, and n is accordingly 2 or 3 or 4. The author does not prove (*) explicitly, but proves a theorem which is still more general, in that it is postulated only that one of S_1, S_2 is convex. He introduces the polar bodies $H_1\leqq 1$ and $H_2\leqq 1$ of S_1 and S_2 in their respective spaces, in the sense of Mahler [Časopis Pěst Mat. Fys. **68**, 93–102 (1939); MR **1**, 202], and denotes by U_{λ_1,λ_2} the (bounded) star body defined by $\max(\lambda_1 H_1, \lambda_2 H_2)\leqq 1$. Then the more general theorem asserts an inequality similar to (*) but with $\mu_n(S_{\lambda_1,\lambda_2},\Lambda)$ replaced by $1/\mu_1(U_{\lambda_1,\lambda_2},\Lambda')$, where Λ' is the n-dimensional lattice reciprocal to Λ. The proof is stated to be based on the methods of Cassels. *H. Davenport.*

Citations: MR 1, 202c = H05-2; MR 13, 919a = J32-31.

H15-61 (16, 1089j)

Bambah, R. P., and Rogers, K. An inhomogeneous minimum for nonconvex star-regions with hexagonal symmetry. Canad. J. Math. **7**, 337–346 (1955).

For certain functions $f(x, y)$ it is well known that there exist real numbers x, y taking assigned residues mod 1 and satisfying an inequality of the type

$$|f(x,y)|\leqq\max\{|f(\tfrac12,0)|, |f(0,\tfrac12)|, |f(\tfrac12,\tfrac12)|, |f(\tfrac12,-\tfrac12)|\}.$$

L. J. Mordell [Duke Math. J. **19**, 519–527 (1952); MR **14**, 540] investigated the conditions on such a function when the region $|f(x,y)|\leqq 1$ has one asymptote, while K. Rogers [J. London Math. Soc. **28**, 394–402 (1953); MR **15**, 106] considered the case of two asymptotes. Here the authors generalize the work of R. P. Bambah [Proc. Cambridge Philos. Soc. **47**, 457–460 (1951); MR **13**, 114] on the special region with three asymptotes which arose in connection with a problem on the inhomogeneous minimum of a binary cubic form with the three real linear factors.

J. H. H. Chalk (London).

Citations: MR 13, 114d = J52-19; MR 14, 540d = J52-22; MR 15, 106f = H15-55.

Referred to in H15-81.

H15-62 (17, 465d)

Rogers, K. On the generators of an ideal, with an application to the geometry of numbers in unitary space U_2. Amer. J. Math. **77** (1955), 621–627.

The author defines an n-dimensional lattice in the space of n complex co-ordinates, the integers of some complex quadratic field $\mathfrak{K}$ taking on the rôle of the rational integers in the ordinary case. For $n=2$ he shows that the analogue of Mahler's compactness theorem [Proc. Roy. Soc. London. Ser. A. **187** (1946), 151–187; MR **8**, 195] holds, and so the analogue of Mahler's existence theorem for critical lattices may be deduced. It is stated that proofs for general n will be published jointly with Swinnerton-Dyer. An application is given in the paper reviewed below. The proofs depend on two results valid for any algebraic number field $\mathfrak{K}$. (I) If p, q, r, s are integers (of $\mathfrak{K}$) and the greatest common ideal divisors of p, q and r, s are equal, then to every integer u of $\mathfrak{K}$ there is a 2×2 integer matrix A^* such that $(p,q)A=(r,s)$. The case $u=1$ is Satz 35 of Hilbert's Zahlbericht. (II) For any rational integer $e>0$ there is a finite set of integral 2×2 matrices U_m such that every integral 2×2 matrix A with $|\text{norm}(\det A)|\leqq e$ is of the type $A=U_mT$, where T and T^{-1} are both integral. *J. W. S. Cassels.*

*of determinant u

Citations: MR 8, 195f = H15-7.

H15-63 (17, 465e)

Rogers, K. Complex homogeneous linear forms. Proc. Cambridge Philos. Soc. **52** (1956), 35–38.

Let $\omega=\frac12(-1+i\sqrt7)$; then 1 and ω form a basis for the ring $Z(\omega)$ of integers in the quadratic field $Q(i\sqrt7)$. Using results of the author and H. P. F. Swinnerton-

Dyer (not yet published), the following theorem is proved: Let $\xi=ax+by$, $\eta=cx+dy$, where a, b, c, d are complex numbers and $ad-bc=\Delta\neq 0$. Then there exist $x, y\in Z(\omega)$, not both zero, such that

$$\max(|\xi|, |\omega|)\leqq \left(\frac{2|\Delta|}{21^{\frac{1}{2}}-3}\right)=\mu.$$

The sign of equality is necessary only for forms which, apart form unimodular factors, can be reduced by transformations with coefficients in $Z(\omega)$ and determinant ± 1 to $\mu\xi_0$, $\mu\eta_0$, where

$$\xi_0=x-\frac{1+i\sqrt{3}}{2}y,$$

$$\eta_0=\frac{3-\sqrt{21}+i(\sqrt{7}-\sqrt{3})}{4}x+\frac{\sqrt{3}-\sqrt{7}}{2}iy.$$

W. J. LeVeque (Ann Arbor, Mich.).

H15-64 (20# 1666)

Rogers, K.; and Swinnerton-Dyer, H. P. F. The geometry of numbers over algebraic number fields. Trans. Amer. Math. Soc. **88** (1958), 227–242.

The authors extend Mahler's compactness theorem on ordinary lattices [Proc. Royal Soc. London, Ser. A **187** (1946), 151–187; C. Chabauty, Bull. Soc. Math. France **78** (1950), 143–151; MR **8**, 195; **12**, 479] to lattices over any finite algebraic number field and deduce applications similar to his. They also find a lower bound for the number of lattice points of their type on the boundary of a convex body, and they sketch the proof of an analogue to Minkowski's linear form theorem for forms in two variables in the quadratic field generated by $\sqrt{-5}$.

K. Mahler (Manchester)

Citations: MR 8, 195f = H15-7; MR 12, 479f = H15-43.

H15-65 (17, 1059f)

Santaló, L. A. On geometry of numbers. J. Math. Soc. Japan **7** (1955), 208–213.

The author essentially proves the following lemma. Let m be a bi-invariant countably additive measure defined on a group G_1. Let F be a countable group of one-one transformations of G_1 onto itself leaving the measure invariant. Suppose that the unit x_0 of F is the identity transformation and let $x_1, x_2, \cdots$ be the other elements of F. Suppose that D_0 is a fundamental domain for G_1 with respect to F, which is measurable and which has a finite positive measure. Let $P_1, \cdots, P_N$ be N fixed elements in D_0. Define the element P_{hN+i} to be x_hP_i for $i=1, \cdots, N$ and $h=1, 2, \cdots$. Let f be a function defined so that $f(P_{j+N})=f(P_j)$, $j=1, 2, \cdots$. Then if H is a measurable set of G_1 the mean value over all x in D_0 of the sum $\sum f(P_i)$, taken over all P_i in xH is

$$\frac{m(H)}{m(D_0)}\sum_{i=1}^{N} f(P_i).$$

This lemma is used to prove two rather complicated theorems which reduce in one special case to results of Blichfeldt and Minkowski [see Blichfeldt, Trans. Amer. Math. Soc. **15** (1914), 227–235] and in another special case to results of M. Tsuji [J. Math. Soc. Japan **4** (1952), 189–193; MR **14**, 623]. Other comparable generalizations of the theorems of Blichfeldt and Minkowski have been given by A. M. Macbeath [Thesis, Princeton, 1950] and by C. Chabauty [Bull. Soc. Math. France **78** (1950), 143–151, footnote 6; MR **12**, 479]. *C. A. Rogers* (Birmingham).

Citations: MR 12, 479f = H15-43; MR 14, 623f = H99-2.

H15-66 (17, 1060a)

Armitage, J. V. On a method of Mordell in the geometry of numbers. Mathematika **2** (1955), 132–140.

Let K be an n-dimensional star body with distance function $F(x_1, x_2, \cdots, x_n)$ and r an integer $1\leqq r\leqq n$ for which (1) the adjoint of an automorph (a linear transformation $x_1\to x_1', \cdots, x_n\to x_n'$ for which $F(x_1, x_2, \cdots, x_n)=F(x_1', x_2', \cdots, x_n')$ is an automorph. (2) Each point $(x_1, x_2, \cdots, x_n)$ can be mapped, by a suitably chosen automorph, into a point $(\alpha, \alpha, \cdots, \alpha, 0, 0, \cdots, 0)$, where r is the number of non-zero coordinates. K_{n-1} is the intersection of K by a certain hyperplane.

A lower bound is determined for the critical determinant $\Delta(K)$ in terms of $\Delta(K_{n-1})$ and a constant dependent on K. This result is applied to obtain two results due to Mordell, one involving the region $|x_1\cdot x_2\cdot\cdots\cdot x_n|\leqq 1$ [Mat. Sb. N.S. **12(54)** (1943), 273–276; MR **5**, 201] and the second involving the n-dimensional sphere [J. London Math. Soc. **19** (1944), 6–12; MR **6**, 57, 334].

The methods are used to obtain a result concerning more specialized star bodies. By means of this result an upper bound is found for the constant C_3 defined as follows. If C is a constant such that for any three real numbers $\theta_1, \theta_2, \theta_3$ not all rational, infinitely many rational approximations p_1/q, p_2/q, p_3/q exist with

$$\left(\theta_1-\frac{p_1}{q}\right)^2+\left(\theta_2-\frac{p_2}{q}\right)^2+\left(\theta_3-\frac{p_3}{q}\right)^2<\left(\frac{C}{q}\right)^{2/3},$$

then C_3 is the greatest lower bound of all the constants C.

D. Derry (Vancouver, B.C.).

Citations: MR 5, 201b = J36-8; MR 6, 57d = J52-9.

Referred to in H15-73, H15-77.

H15-67 (18, 21c)

Hejtmanek, Johann. Über eine Klasseneinteilung der Sternkörper. Monatsh. Math. **60** (1956), 11–20.

Let k be a given non-negative integer. The author constructs a star body in Euclidean n-space, symmetric with respect to the origin, which admits a critical lattice with exactly k point pairs on the boundary of the star body, provided (1) $n=2$, $k\geqq 2$, in which case the constructed star body is bounded, or (2) a star body K_n exists which is both completely reducible and fully automorphic. Star bodies K_n, $2\leqq n\leqq 4$, which satisfy the necessary conditions have been given by Davenport and Rogers [Philos. Trans. Roy. Soc. London. Ser. A. **242** (1950), 311–344; MR **12**, 394]. Thus the problem is solved for these values of n.

It is shown that the Minkowski-Hlawka theorem implies that a star body of finite volume must be of the finite type, i.e. it must admit at least one lattice. *D. Derry*.

Citations: MR 12, 394b = H15-42.

H15-68 (19, 127a)

Rogers, C. A. A single integral inequality. J. London Math. Soc. **32** (1957), 102–108.

When ρ is a bounded non-negative Lebesgue-integrable function on Euclidean n-space, the spherical symmetrization ρ^* of ρ is the function defined for each point X in n-space (except $X=0$, at which point ρ^* is defined to be $\sup\{\rho(X):X \text{ in } n\text{-space}\}$) as the supremum of numbers t such that the measure of the set of points Y with $\rho(Y)>t$ is not less than the measure of the set of points Y with $|Y|<|X|$. Generalizing results of an earlier paper of his [same J. **31** (1956), 235–38; MR **18**, 757], the author proves the following theorem: if $\rho_i, \cdots, \rho_k$ are bounded non-negative Lebesgue-integrable functions and $\rho_1^*, \cdots,$

ρ_k^* are their respective spherical symmetrizations, then for any constants c_{ij} ($i=1, \cdots, k$; $\lambda=1, \cdots, m$),

$$\int\cdots\int \prod_{i=1}^{k} \rho_i\Big(\sum_{j=1}^{m} c_{ij}X_j\Big)dX_1\cdots dX_m \leqq \int\cdots\int \prod_{i=1}^{k} \rho_i^*\Big(\sum_{j=1}^{m} c_{ij}X_j\Big)dX_1\cdots dX_m.$$

An application of interest in the geometry of numbers is noted. *T. A. Botts* (Charlottesville, Va.).

H15-69 (22# 8004)

Schmidt, Wolfgang. A metrical theorem in geometry of numbers. Trans. Amer. Math. Soc. **95** (1960), 516–529.

Let S be a Borel set in the n-dimensional space R_n not containing the origin. Let $V=V(S)$ be its volume, $L(S)$ the number of lattice points g (i.e., points g with integral coordinates) in S, and $P(S)$ the number of primitive lattice points in S. The author considers the discrepancies $D(S)=|L(S)V(S)^{-1}-1|$, $E(S)=|P(S)\zeta(n)V(S)^{-1}-1|$. His first theorem runs as follows.

Let $\psi(s)$ ($s\geqq 0$) be a positive, nondecreasing function with $\int_0^\infty \psi(s)^{-1}\,ds<\infty$. Suppose $n\geqq 3$. Then, for almost all $n\times n$-matrices A,

$$D(AS) = O(V^{-1/2}\log V\psi^{1/2}(\log V)),$$
$$E(AS) = O(V^{-1/2}\log V\psi^{1/2}(\log V)),$$

if S runs through any family Φ which is totally ordered (by inclusion), such that all $V(S)$ are finite and $V(S)\to\infty$ if S runs through Φ.

The proof uses integration of expressions like

$$\Big(\sum_g \rho(Ag)-V(S)\Big)^2$$

over certain fundamental regions of matrices, ρ denoting the characteristic function of S. Results of the above type are not true for $n=1$. A slightly weaker result is proved in the case $n=2$, by the same method. But here the details are essentially more complicated.

C. G. Lekkerkerker (Amsterdam)

Referred to in H25-39.

H15-70 (22# 12100)

Groemer, Helmut. The number of lattice points on the boundary of a star body. Proc. Amer. Math. Soc. **11** (1960), 757–762.

Suppose that the set S of all points X in Euclidean n-space E_n, defined by $f(X)\leqq 1$, is a bounded star body [cf. K. Mahler, Proc. Roy. Soc. London Ser. A **87** (1946), 151–187; MR **8**, 195]. Then there exists a least number k such that, for all X, Y in E_n,

$$f(X+Y) \leqq k\{f(X)+f(Y)\},$$

and this k is known as the concavity coefficient of S. By the properties of the distance function $f(X)$ it follows that $k\geqq 1$, and moreover, that $k=1$ if and only if S is convex. Define $A_n(k)$ as the greatest number with the property that there exists an n-dimensional star body S with concavity coefficient k and a lattice having $A_n(k)$ points on the boundary of S but none, other than the origin 0, in the interior of S. A classical result of Minkowski [*Geometrie der Zahlen*, Teubner, Leipzig, 1910; p. 79] is $A_n(1)=3^n-1$. His method has been adapted by the author to give the following estimates for $A_n(k)$. Theorem: Let $m_i=[2k]+i$ ($i=1, 2, \cdots, n$) and put

$$P_n(k) = \inf m_i^n \sum_{d|m_i} \mu(d)d^{-n},$$

$$Q_n(k) = \sum_{i=1}^{\infty} \mu(i)\{(2[k/i]+1)^n-1\}.$$

Then $Q_n(k)\leqq A_n(k)\leqq P_n(k)$. In particular, it may be remarked that, since $\sum\mu(d)d^{-n}<1$, a star body S (of concavity coefficient k) having more than $m_1^n-1=[2k+1]^n-1$ lattice points on its boundary has at least one lattice point not 0 in its interior. Thus Minkowski's result 3^n-1 is valid not only for $k=1$, but for all $k<\frac{3}{2}$. For large k, the following estimates for $P_n(k)$ and $Q_n(k)$ are given:

$$P_n(k) = 2^n\zeta^{-1}(n)k^n+o(k^n),$$
$$Q_n(k) = 2^n\zeta^{-1}(n)k^n+o(k^n),$$

as $k\to\infty$. Thus $A_n(k)=2^n\zeta^{-1}(n)k^n+o(k^n)$ as $n\to\infty$.

J. J. H. Chalk (Toronto)

Citations: MR 8, 195f = H15-7.

H15-71 (24# A1256a; 24# a1256b)

Lekkerkerker, C. G.

Lattice points in unbounded point sets. I. The one-dimensional case.

Nederl. Akad. Wetensch. Proc. Ser. A **61** = *Indag. Math.* **20** (1958), 197–205.

Lekkerkerker, C. G.

Lattice points in unbounded point sets. II. The n-dimensional case.

Nederl. Akad. Wetensch. Proc. Ser. A **61** = *Indag. Math.* **20** (1958), 206–216.

Im ersten Teil seiner Arbeit gibt Verfasser kurze Beweise von vier Sätzen, die bereits von J. F. Koksma, de Bruijn und H. Kersten erhalten wurden. Diese Sätze beschränken sich auf den eindimensionalen Fall. Satz 1: Ist $f(x)$ eine nichtnegative meßbare Funktion und $\int_0^\infty f(x)dx<\infty$, so gilt für fast alle x $F(x)=\sum_{n=1}^\infty f(nx)<\infty$. Satz 2: Ist $f(x)$ Riemann-integrierbar in jedem endlichen Intervall und $\int_0^\infty f(x)dx=\infty$, dann existiert ein $x>0$ mit $F(x)=\infty$. Zu jeder Lebesgue-meßbaren Menge V aus $[0,\infty)$ lassen sich folgende Mengen bilden: W_1 aus allen $x>0$, so daß $kx\in V$ für unendlich viele positive ganze Zahlen k gilt, W_2 mit $kx\in V$ für höchsten endlich viele k und W_3 für kein k. Satz 3: Es existiert eine Jordan-meßbare Menge V mit $\mu(V)=\infty$, so daß $\mu(W_1)=0$ und $\mu(W_3)=\infty$. Satz 4: ρ und d sind positive Zahlen mit $\rho<1$. Wenn V Dichte $\geqq\rho$ in jedem Intervall einer Folge von punktfremden Intervallen der Länge d hat, so ist $\mu(W_2)=0$. Im zweiten Teil werden diese vier Sätze mit Hilfe der Theorie des Siegelschen Maßes in Gitterräumen auf den n-dimensionalen Fall übertragen und können folgendermaßen formuliert werden: Satz 1: Ist $f(x)$ eine nichtnegative Riemann-integrierbare Funktion in R_n und $\int_{R_n} f(x)dx<\infty$, dann gilt für fast alle Gitter Λ $\sum_{x\in\Lambda} f(x)<\infty$. Satz 2: Ist dagegen $\int_{R_n} f(x)dx=\infty$, dann ist die Menge der Gitter Λ mit $\sum_{x\in\Lambda} f(x)=\infty$ überall dicht. Satz 3: Es existiert eine Jordan-meßbare Menge V in R_n mit $\mu(V)=\infty$, sodaß fast alle Gitter nur endlich viele Punkte in V haben. Satz 4: ρ und d sind positive Zahlen mit $\rho<1$. Wenn V Dichte $\geqq\rho$ in jedem Würfel einer Folge von punktfremden Würfeln $g^{(k)}+dE$, $g^{(k)}\in R_n$, E: $0\leqq x_i\leqq 1$, $i=1, \cdots, n$ hat, dann haben fast alle Gitter unendlich viele Punkte in V.

H. Hejtmanek (Zbl **86**, 38)

Referred to in K99-2.

H15-72 (26# 672)

Hejtmanek, Johann

Ausgezeichnete Gitter eines Sternkörpers.

Monatsh. Math. **64** (1960), 22–34.

Sei ein Sternkörper mit $\Delta(S)<\infty$, $f(x)$ seine Distanzfunktion, Λ ein Gitter, $f(\Lambda)=\inf f(X)$, $X\in\Lambda$, $X\neq 0$. Gibt es zu einem Wert Δ ein Gitter Λ mit $d(\Lambda)=\Delta$, $f(\Lambda)=1$, so heisst Δ ein M-Wert (Markoffscher-Wert). Der Verfasser beweist unter anderen folgende Sätze: Ist S unbeschränkt, so ist die Menge der zulässigen Gitter abgeschlossen und nirgends dicht im Raum aller Gitter; besitzt S ein endliches Inhalt $V(S)$, so gibt es zu fast allem Gitter Λ mit $d(\Lambda)=1$ ein $\lambda>0$ mit $f(\lambda\Lambda)=1$; ist $V(S)=\infty$, so ist die Menge der Gitter mit $d(\Lambda)=\Delta(S)$, $0<f(\Lambda)\leqq 1$ vom Mass Null. Weiterhin wird der Sternkörper K untersucht, der als Vereinigungsmenge von $|x_1x_2|\leqq 1$, $|x_1+x_2|\leqq\sqrt{5}$ einerseits (in der Arbeit steht fehlerhaft $|x_1+x_2|\leqq 1$) und von $\max(|x_1{}^\alpha x_2|, |x_1x_2{}^\alpha|)\leqq 1$ $(\alpha\geqq 1)$ andererseits entsteht. Die M-Werte von K bilden für genügend grosses α das Intervall $[\sqrt{5}, \infty)$. Mit denselben Methoden wird auch ein neuer Beweis für Davisschen Satz gegeben [Quart. J. Math. Oxford Ser. (2) **1** (1950), 241–242; MR **12**, 393] dass das offene Intervall $(\sqrt{12}, \sqrt{13})$ kein M-Wert von $|x_1x_2|\leqq 1$ enthält. {Im Beweis, Seite 28, sind g_1, g_2 durch $x_2=x_1-\sqrt{12}$ bzw. $x_2=x_1-\sqrt{13}$, die Punkte $P\pm 2Q_1$ durch $2P\pm Q_1$, $P\pm 2Q_2$ durch $2P-Q_2$, $P+Q_2$, G_1 durch $2P-Q_1$, G_2 durch $2P-Q_2$ zu ersetzen. Die Arbeit ist nicht leicht zu lesen, da sie sehr knapp und auch nicht immer genau gefasst ist. Der Beweis des Satzes 2, 2 — mit Δ ist auch Δ/n (n natürliche Zahl) kein M-Wert — ist falsch, wie es auch Verfasser bestetigte; folgt aber leicht aus Satz 2, 3 — mit Δ ist auch $n\Delta$ ein M-Wert. Wäre nähmlich Δ/n ein M-Wert, sollte das auch für $n(\Delta/n)=\Delta$ gelten.} *J. Surányi* (Budapest)

Citations: MR 12, 393b = J28-13.

H15-73 (26# 2401)

Lekkerkerker, C. G.

Eine Mordell'sche Methode in der Geometrie der Zahlen.

J. Reine Angew. Math. **206** (1961), 20–25.

If Y is the lattice of points $y=(y_1, \cdots, y_n)$ in Euclidean n-space with integral coordinates and A is an arbitrary non-singular linear transformation of the space, let $\Lambda=AY$ denote the general lattice. The polar lattice Λ^* corresponding to Λ is defined by $\Lambda^*=A^*Y$, where A^* is the transpose of the inverse of A. Let S be a star body in the space with distance function $F(x)$. Then the minimum $\lambda(S, \Lambda)$ of S with respect to Λ is $\lambda(S,\Lambda)=\inf_{0\neq x\in\Lambda}F(x)$ and the absolute minimum $\lambda(S)$ of S is

$$\lambda(S) = \sup_{\Lambda}\{\lambda(S, \Lambda)\cdot d(\Lambda)^{-1/n}\},$$

where $d(\Lambda)$ is the determinant of Λ. S is said to be of finite type or of infinite type, according as $\lambda(S)>0$ or $\lambda(S)=0$, respectively. A linear transformation Ω of space with the property $\Omega S=S$ is said to be an automorphism of S.

The object of this paper is to present the ideas of Mordell and others, which in special instances have led to an inequality between $\lambda(S)$ and $\lambda(\Sigma)$, where Σ is some $(n-1)$-dimensional plane section of S, in an appropriately general setting. The first attempt at this was made by J. V. Armitage [Mathematika **2** (1955), 132–140; MR **17**, 1060], who incidentally neglected to mention an obvious restriction on the sections Σ of S (see author's remarks on p. 20). We quote Satz 1 as being typical of the main results. Satz 1: Let S be a star body of the finite type admitting a group Γ of automorphisms Ω of S with the following properties: (1) the normals through 0 to the planes H with $\lambda(S\cap H)=0$ are not everywhere dense in space; (2) to each H with $\lambda(S\cap H)>0$, there is an $\Omega_H\in\Gamma$ such that $\Omega_H H=H_0$, where H_0 is some fixed plane; (3) $\Omega\in\Gamma\Rightarrow\Omega^*=A^*\Omega\in\Gamma$. Then there is a point b^0 on the boundary of S, which is normal to H_0, for which

$$[\lambda(S)]^{n-2} \leqq [\lambda(S\cap H_0)]^{n-1}|b^0|.$$

This result is capable of certain minor extensions, and some 8 special cases are considered. *J. H. H. Chalk* (Toronto)

Citations: MR 17, 1060a = H15-66.

Referred to in H15-77.

H15-74 (26# 4972)

Bombieri, Enrico

Sulla dimostrazione di C. L. Siegel del teorema fondamentale di Minkowski nella geometria dei numeri. (English summary)

Boll. Un. Mat. Ital. (3) **17** (1962), 283–288.

Let Λ be a lattice and S a closed bounded domain in n-dimensional space. For $\mathbf{v}\in R^n$, put

$$V_S(\mathbf{v}) = \sum_{\mathbf{u}\in\Lambda} V((S+\mathbf{v})\cap(S+\mathbf{u})).$$

Here, $V(T)$ denotes the volume of T. Applying an old method used by Siegel in his proof of Minkowski's fundamental theorem [Acta Math. **65** (1935), 307–323], the author deduces the following result. If $b_0, b_1, \cdots, b_N$ are non-negative real numbers such that $\sum_{h=0}^N b_h\cos(hx)\geqq 0$ for each real x, then $d(\Lambda)\sum b_hV_S(h\mathbf{v})\geqq\sum b_hV^2(S)$.

A first application is the following refinement of Blichfeldt's theorem: If S does not contain two points $\mathbf{x}$, $\mathbf{x}'$ with $\mathbf{x}\neq\mathbf{x}'$, $\mathbf{x}-\mathbf{x}'\in\Lambda$, then $d(\Lambda)\geqq 3V^2(S)\cdot(V(S)+2V_S(\tfrac{1}{3}\mathbf{u}))^{-1}$ for each $\mathbf{u}\in\Lambda$. A second application, which concerns the product of n inhomogeneous linear forms is announced. *C. G. Lekkerkerker* (Amsterdam)

Referred to in H15-75, J40-34.

H15-75 (26# 3646)

Bombieri, Enrico

Un principio generale della Geometria dei Numeri e sue applicazioni ai problemi non omogenei.

Atti Accad. Naz. Lincei Rend. Cl. Sci. Fis. Mat. Nat. (8) **33** (1962), 45–48.

Exposition of some consequences of a theorem proved earlier [Boll. Un. Mat. Ital. (3) **17** (1962), 283–288; MR **26** #4972]. These consequences concern, in particular, the minimum of the product of n inhomogeneous linear forms. *C. G. Lekkerkerker* (Amsterdam)

Citations: MR 26# 4972 = H15-74.

Referred to in J40-37.

H15-76 (28# 2087)

Lekkerkerker, C. G.

Homogeneous simultaneous approximations.

Nederl. Akad. Wetensch. Proc. Ser. A **66** = *Indag. Math.* **25** (1963), 578–586.

An approximation property of matrices, developed in an earlier article [same Proc. **64** (1961), 197–210; MR **24** #A1257], is applied to a general problem of simultaneous Diophantine approximations. If $\phi(\mathbf{x})$, $\psi(\mathbf{y})$ are distance functions of bounded symmetric star bodies in, respectively, r- and s-dimensional spaces, let $K_{\phi,\psi}$ denote the (unbounded) star body in n-dimensional space $(n=r+s)$ defined by $\phi(\mathbf{x})^r\psi(\mathbf{y})^s\leqq 1$, and let $\Delta(K_{\phi,\psi})$ denote its critical (lattice) constant. It is easily shown that $\Delta(K_{\phi,\psi})$ is positive and finite. Now, let Θ be an $r\times s$ matrix of real numbers θ_{ij} and let $\mathbf{u}$, $\mathbf{v}$ be integral vectors in r- and s-dimensional spaces, respectively. Then we define

$$C_{\phi,\psi}(\Theta) = \liminf_{\psi(\mathbf{v})\to\infty}\phi(\Theta\mathbf{v}-\mathbf{u})^r\psi(\mathbf{v})^s,$$

$$C_{\phi,\psi} = \sup_{\Theta} C_{\phi,\psi}(\Theta).$$

The main theorem (which is a generalization of earlier

work of Furtwängler and Davenport) is

$$C_{\phi,\psi} = \{\Delta(K_{\phi,\psi})\}^{-1}$$

which has, as a corollary, $C_{\phi,\psi} = C_{\psi,\phi}$.

J. H. H. Chalk (Toronto, Ont.)

Citations: MR 24# A1257 = H05-68.

H15-77 **(29# 5792)**

Mullender, P.

Some remarks on a method of Mordell in the Geometry of Numbers.

Acta Arith. **9** (1964), 301–304.

The "method" of the title leads to an inequality connecting the critical determinant of an n-dimensional star body with that of a related $(n-1)$-dimensional star body. Several authors [see, e.g., C. G. Lekkerkerker, J. Reine Angew. Math. **206** (1961), 20–25; MR **26** #2401] have attempted to analyse the underlying principles of the method; the present paper seeks to elucidate it further.

Let $F(x)$, $G(x)$ be the distance functions of two star bodies K_F, K_G of finite type in n-dimensional Euclidean space, X. Suppose that K_F possesses a group Ω of automorphs such that the adjoint of an automorph of K_F is an automorph of K_G; that is, there is a group of non-singular matrices, A, such that $F(Ax) = F(x)$, $G(\tilde{A}x) = G(x)$ for all $x \in X$, where $\tilde{A}$ is the transposed inverse of A. The main theorem reads as follows. Suppose that there are in X a k-dimensional linear sub-space R and an $(n-k)$-dimensional linear sub-space S, perpendicular to R, such that (a) the k-dimensional and $(n-k)$-dimensional star bodies $R \cap K_F$ and $S \cap K_G$ are of finite type in R, S, respectively; (b) to any k-dimensional linear sub-space R' of X such that the k-dimensional star body $R' \cap K_F$ is of finite type in R' there corresponds an automorph of K_F in Ω transforming R' into R. Then, for $k = n-1$,

$$(\Delta_{K_F})^k \geqq \left(\frac{\Delta_{R\cap K_F}}{\Delta_{S\cap K_G}}\right)^n \cdot (\Delta_{K_G})^{n-k},$$

where Δ_{K_F} denotes the critical determinant of K_F, etc.

The integer k is introduced in order to indicate possible generalizations of the theorem. However, it is not easy to see what the right generalization should be, and it may be a difficult problem to solve, if indeed there is a solution at all. For an example of a particular case in which the method can be carried further, see a paper by the reviewer [Mathematika **2** (1955), 132–140; MR **17**, 1060].

J. V. Armitage (Durham)

Citations: MR 17, 1060a = H15-66; MR 26# 2401 = H15-73.

H15-78 **(30# 3067)**

Groemer, Helmut

Über gewisse dichteste Anordnungen von Punkten in der Ebene.

Arch. Math. **15** (1964), 385–387.

Let S be a plane bounded symmetrical star domain with distance function $f(x)$. A set Σ in the plane is said to be S-admissible if $f(P-Q) \geqq 1$ for all P, Q in Σ. The density $d(\Sigma)$ of Σ is defined as $\liminf_{\lambda\to\infty} \lambda^{-2}N(\Sigma, \lambda)$, where $N(\Sigma, \lambda)$ is the number of points of Σ in the square of side λ and centre 0. Let $\delta(S) = \sup d(\Sigma)$ over all S-admissible sets Σ, and $\delta_L(S) = \sup d(\Sigma)$ over all S-admissible lattices Σ. If S is convex, it is known (C. A. Rogers and L. Fejes Tóth) that $\delta(S) = \delta_L(S)$ [see C. A. Rogers, *Packing and covering*, p. 12, Cambridge Univ. Press, New York, 1964; MR **30** #2405; see also J. H. Chalk and C. A. Rogers, J. London Math. Soc. **23** (1948), 178–187; MR **10**, 284; C. A. Rogers, Acta Math. **104** (1960), 305–306; MR **23** #A2132]. M. R. Von Wolff [Acta Math. **108** (1962), 53–60; MR **26** #2400] gave an example of a star S with explicitly determined $\delta_L(S)$ and $\delta(S) > \delta_L(S)$. The author shows how one can deduce from a result of H. Davenport and C. A. Rogers [Quart. J. Math. Oxford Ser. (2) **1** (1950), 215–218; MR **12**, 161] that given any $\rho > 0$, there exists a star S with $\delta(S) > \rho\delta_L(s)$.

R. P. Bambah (Columbus, Ohio)

Citations: MR 10, 284g = H05-13; MR 12, 161e = H15-39; MR 23# A2132 = H30-18; MR 26# 2400 = H25-35; MR 30# 2405 = H02-17.

H15-79 **(31# 2213)**

White, George K.

Grid bases in n-space.

Quart. J. Math. Oxford Ser. (2) **16** (1965), 126–134.

If S is a set of points in n-space and $D(S)$ denotes the greatest lower bound of $d(\Gamma)$, taken over all grids Γ without points in S, it has been conjectured that, if Γ is any grid with $d(\Gamma) < \frac{1}{2}D(S)$, then S contains a basis of Γ. In this paper the author proves a result of this kind about the more extensive family of partial bases. A set of $n+1$ points of Γ is called a partial basis if the simplex spanned by them contains no other points of Γ. Let R denote an open non-empty convex subset of n-space which contains no line, let $V(X)$ be the volume of the set $R \cap (2X - R)$, and, for each real positive α let $T(\alpha)$ be the subset of R consisting of points X such that $V(X) \leqq \alpha$. The main result is that every n-dimensional grid Γ in n-space with determinant $d(\Gamma) \leqq 2^{-n}\lambda$ has a partial basis contained in $T(2\lambda)$.

A. M. Macbeath (Birmingham)

H15-80 **(31# 3934)**

Oler, Norman

Zassenhaus' lemma on sectorial norm-distances.

Illinois J. Math. **9** (1965), 338–342.

Let the plane E_2 be divided into a finite number $2r \geqq 4$ of sectors S_i by lines through the origin. Let F be a real-valued function on E_2 such that $F(X) < 1$ is a symmetrical star body whose complement in each sector is convex. A set E is said to be F-admissible if, for all A, B in E with $A \neq B$, $F(A-B) \geqq 1$. An F-admissible triangle ABC is of type I if, with a suitable choice of signs, $\pm(A-B)$, $\pm(B-C)$, $\pm(C-A)$ lie in the same sector. All other F-admissible triangles are of type II. Let $\Delta = 2 \inf A(T)$, where $A(T)$ is the area of T and T runs over all triangles of type II. If Π is a Jordan polygon with vertices $X_1, \cdots, X_k$ and $E = \{X_1, \cdots, X_k, Y_1, \cdots, Y_j\}$ is an F-admissible set with $Y_1, \cdots, Y_j$ inside Π, H. Zassenhaus [*Statistical geometry of numbers*, Prentice Hall (in press)] has proved the remarkable theorem: $A(\Pi)/\Delta + F(\Pi)/2 + N - 1 \geqq 0$, where $F(\Pi) = \sum_{i=1}^k F(X_{i+1} - X_i)$, $X_{k+1} = X_1$, $A(\Pi)$ is the area of Π and N is the number of points in E. The proof depends on the following lemma: "If no side of Π is the largest side of a type I triangle, then Π can be triangulated into triangles of type II with vertices at points of E." The author gives a new proof of this lemma in a wider context, in that he makes his triangles independent of F. {There seems to be a small misprint in conditions 4(b), 4(c), page 340, where $\overrightarrow{BD}$ (in 4(b)) and $\overrightarrow{AD}$ (in 4(c)) should be replaced by $\overrightarrow{DB}$ and $\overrightarrow{DA}$, respectively.}

R. P. Bambah (Columbus, Ohio)

H15-81 **(31# 4767)**

Chander, Vishwa; Hans, Rajinder Jeet

An inhomogeneous minimum of a class of functions.

Monatsh. Math. **69** (1965), 193–207.

L. J. Mordell [Duke Math. J. **19** (1952), 519–527; MR **14**, 540], E. S. Barnes [Quart. J. Math. Oxford Ser. (2) **1**

(1950), 199–210; MR **13**, 16], J. H. H. Chalk [Proc. Cambridge Philos. Soc. **48** (1952), 392–401; MR **13**, 919], K. Rogers [J. London Math. Soc. **28** (1953), 394–402; MR **15**, 106], and the reviewer [Proc. Cambridge Philos. Soc. **47** (1951), 457–460; MR **13**, 114] have proved theorems of the following type: "Let $f(x, y)$ be a function of a given form. Then for given real x_0, y_0, there exist integers x, y such that $|f(x+x_0, y+y_0)| \leq \max[|f(\frac{1}{2}, 0)|, |f(0, \frac{1}{2})|, |f(\frac{1}{2}, \pm\frac{1}{2})|]$." In particular, the reviewer and K. Rogers [Canad. J. Math. **7** (1955), 337–346; MR **16**, 1089] proved the theorem for functions corresponding to "regions with hexagonal symmetry" and remarked that the analogous result is not true for functions corresponding to regions with four asymptotes. The authors prove that the theorem becomes true for functions corresponding to "regions with four asymptotes and octogonal symmetry" if the right-hand side is replaced by the maximum of $|f|$ over six points, which can be determined as soon as f is given. Like earlier results, their theorem is best possible in the sense that the equality sign is needed for certain special forms. *R. P. Bambah* (Columbus, Ohio)

Citations: MR 13, 16a = R12-23; MR 13, 114d = J52-19; MR 13, 919c = J52-20; MR 14, 540d = J52-22; MR 15, 106e = D76-5; MR 16, 1089j = H15-61.

H15-82 (37# 1319)

Mullender, P.

On lattice coverings and inextensible star bodies.

J. London Math. Soc. **43** (1968), 99–104.

K. Mahler [Nederl. Akad. Wetensch. Proc. **49** (1946), 331–343; MR **8**, 12] introduced the notion of reducibility of star bodies and proved that if a is a point on the boundary of the irreducible star body K, then there exists at least one critical lattice of K containing a. In the present paper a similar result is proved for the inhomogeneous case; if the star body K_f and the star body K_g containing K_f have the same critical covering determinant, then K_f is said to be extensible. For example, a regular hexagon and its circumscribing circle have the same critical covering determinant and so the hexagon is extensible. If K is bounded and not extensible and a is a point on the boundary of K, then it is proved that there is at least one critical covering grid (inhomogeneous lattice) of K containing a and not containing any point of K. The author also remarks on some differences between the homogeneous and inhomogeneous cases. *H. J. Godwin* (Swansea)

Citations: MR 8, 12b = H15-8.

H15-83 (39# 128)

Bantegnie, Robert

Réseaux multipermis et multicritiques.

Math. Z. **108** (1969), 173–190.

Let S be an open star body in the space R^n, and let Λ be an arbitrary lattice. For $m=1, 2, \cdots$, put $\Phi_m=\Phi_m(S)=\{\Delta : \text{card}(\Lambda \cap S) \leq m\}$, $\Delta_m(S)=\inf\{d(\Lambda) : \Lambda \in \Phi_m\}$. If $\Lambda \in \Phi_m$ and $d(\Lambda)$ is an absolute [extreme] minimum on the set Φ_m, then call Λ an m-S-critical [m-S-extreme] lattice.

The author extends various results known for $m=1$. Among other things, he finds: (i) If Φ_m is not empty, then there exist m-S-critical lattices. (ii) If S is bounded, then each m-S-extreme lattice has n independent points on the boundary $\bar{S}$; if S is bounded, convex and symmetric about the origin, then such a lattice has at least $\frac{1}{2}n(n+1)$ pairs of points $\pm x$ on the boundary. (iii) If K is the cube $|x_i|<1$ $(i=1, \cdots, n)$, then $\Delta_m(K)=m^{-1}$; the m-K-critical lattices can be determined explicitly. (iv) Inequalities for the quotient $\Delta_m(S)/\Delta(S)$ in case S is a centrally symmetric bounded convex body in the plane; for $m=3, 5, 7$, the possible configurations, with respect to S, of the points of an m-S-extreme lattice are determined. In the proofs extensive use is made of a result of J. W. S. Cassels [J. London Math. Soc. **33** (1958), 281–284; MR **20** #5803] on sublattices not containing a given set of points of a given lattice. *C. G. Lekkerkerker* (Amsterdam)

Citations: MR 20# 5803 = H25-31.

H15-84 (39# 130)

Hans, Rajinder Jeet

On the Reinhardt-Mahler theorem.

Proc. Amer. Math. Soc. **20** (1969), 391–396.

Soit dans l'espace euclidien à n dimensions R_n un ensemble donné S de volume $V(S)$ et de constante critique $\Delta(S)$. De façon indépendante, K. Reinhardt [Abh. Math. Sem. Hamburg Univ. **10** (1934), 216–230] et K. Mahler [Nederl. Akad. Wetensch. Proc. **50** (1947), 692–703; MR **9**, 10] ont prouvé que, pour $n=2$, si K est un domaine convexe symétrique, alors $\Delta(K)=H(K)/4$, où $H(K)$ est l'aire du plus petit "hexagone" symétrique contenant K. Une généralisation peut être: Si K est un corps convexe symétrique de R_n, K est contenu dans un corps symétrique et pavant $\mathscr{P}$ et l'on a $\Delta(K)=\Delta(\mathscr{P})=V(\mathscr{P})/2^n$. L'auteur montre que cette généralisation n'est pas valable pour $n \geqq 3$. Si $\mathscr{K}$ désigne la classe de tous les corps ouverts symétriques de R_n, on dit que $K \in \mathscr{K}$ est $\mathscr{K}$-maximal si $H \in \mathscr{K}$, $H \not\supseteq K$ implique $\Delta(H)>\Delta(K)$. L'octaèdre $|x|+|y|+|z|>1$ est $\mathscr{K}$-maximal sans être partie pavante. Pour $n>2$, il existe dans R_n des polyèdres qui sont $\mathscr{K}$-maximaux sans être pavants: l'impossibilité de la généralisation citée en résulte. En effet, l'auteur montre que si $K \in \mathscr{K}$ est une partie pavante polyédrale, alors toute face à $n-1$ dimensions de K est une union finie de corps convexes symétriques disjoints; or on peut trouver un polyèdre $\mathscr{K}$-maximal ayant pour face à $n-1$ dimensions un simplexe, et un simplexe n'est pas union finie de corps convexes symétriques disjoints. {Remarque du rapporteur: Il résulte des travaux de H. Groemer [Math. Z. **79** (1962), 364–375; MR **25** #5454] que si $K \in \mathscr{K}$ pave R_n les faces à $n-1$ dimensions de K sont centrées; il suffirait donc de montrer qu'on peut trouver dans $\mathscr{K}$ un polyèdre K $\mathscr{K}$-maximal et dont les faces ne sont pas centrées.}

{Les références à Ollernshaw doivent être lues Ollerenshaw.} *R. Bantegnie* (Besançon)

Citations: MR 9, 10h = H05-10.

H20 NONCONVEX BODIES: SPECIAL PROBLEMS

See also reviews H15-3, H15-5, H15-11, H15-20, H15-27, H15-40, H15-42, H15-51, H15-66, H15-72, J08-1, J08-5, J08-14, J12-14, J12-27, J28-2, J28-7, J28-14, J28-22, J52-16.

H20-1 (3, 167a)

Watson, G. N. Proof of a conjecture stated by Mordell. J. London Math. Soc. **16**, 157–166 (1941).

Mordell remarked that a proof of the following conjecture is important in the discussion of lattice points connected with $f=x^p+y^p$: When $0<p<1$, the curve

$$\{(x+1)^p-1\}\{(y+1)^p-1\}=(1-x^p)(1-y^p)$$

possesses the property that any line parallel to $x+y=0$ cannot meet it at more than two points lying in the positive quadrant. The author demonstrates that the coefficients a_n $(n>0)$ of the power series

$$\{(1+\lambda+\xi)^p-1\}\{(1+\lambda-\xi)^p-1\} - \{1-(\lambda+\xi)^p\}\{1-(\lambda-\xi)^p\}=\sum_{n=0}^{\infty} a_n\xi^{2n}$$

are positive, when $\frac{1}{2} \leqq \lambda \leqq \frac{1}{2}(-1+\sqrt{5})$. The proof of the conjecture follows then quite simply. *C. L. Siegel.*

H20-2 (3, 167b)

Mordell, L. J. Some results in the geometry of numbers for non-convex regions. J. London Math. Soc. 16, 149–151 (1941).

Let L be a lattice $x=\alpha\xi+\beta\eta$, $y=\gamma\xi+\delta\eta$, where $\alpha, \beta, \gamma, \delta$ are real numbers, $\alpha\delta-\beta\gamma=1$, and ξ, η run through all integer values. Let $f=f(x, y)$ be a homogeneous function of x, y, of positive dimension. The results are concerned with the best possible value of κ, independent of $\alpha, \beta, \gamma, \delta$, such that a point of L, other than (0, 0), satisfies the inequality $|f(x, y)| \leqq \kappa$. Five applications are given: $f=x+y+\{\mu(x^2+y^2)+2xy\}^{\frac{1}{2}}$ $(0 \leqq \mu<1)$; $f=\min(x+my, y+mx)$ $(3^{-\frac{1}{2}} \leqq m<1)$; $f=x^p+y^p$ $(p=\frac{1}{2}$ or $\frac{1}{3})$; $f=x^4-y^4$; $f=xy(x+y)$. The proofs are to be published in the Proc. London Math. Soc. *C. L. Siegel* (Princeton, N. J.).

Referred to in H20-3.

H20-3 (3, 167c)

Mordell, L. J. Lattice points in the region $|Ax^4+By^4| \leqq 1$. J. London Math. Soc. 16, 152–156 (1941).

In the case $f=x^4+y^4$, the best possible value of κ is $(6+4\sqrt{6})/15$ [cf. the preceding review]. By a geometric idea of Minkowski [Diophantische Approximationen, Teubner, Leipzig, 1907, pp. 47–58], the proof is reduced to the determination of the minimum value of $\alpha\delta-\beta\gamma$ under the conditions $\alpha^4+\gamma^4=1$, $\beta^4+\delta^4=1$, $(\alpha-\beta)^4+(\gamma-\delta)^4=1$, and the solution of this problem is then simplified by the theory of invariants of the binary quartic. *C. L. Siegel.*

Citations: MR 3, 167b = H20-2.

Referred to in H10-34, H20-7, H20-14.

H20-4 (6, 119b; 6, 119c; 6, 119d)

Mahler, K. On lattice points in the domain $|xy| \leqq 1$, $|x+y| \leqq \sqrt{5}$ and applications to asymptotic formulae in lattice point theory. I. Proc. Cambridge Philos. Soc. 40, 107–116 (1944).

Mahler, K. On lattice points in the domain $|xy| \leqq 1$, $|x+y| \leqq \sqrt{5}$ and applications to asymptotic formulae in lattice point theory. II. Proc. Cambridge Philos. Soc. 40, 116–120 (1944).

Mahler, K. On lattice points in an infinite star domain. J. London Math. Soc. 18, 233–238 (1943).

The author defined [J. London Math. Soc. 17, 130–133 (1942); these Rev. 4, 212] a finite star domain K in the (x, y)-plane as a closed set of points symmetric in $O=(0, 0)$, containing the origin O in its interior and bounded by a Jordan curve which does not intersect more than once any radius vector from O. A domain K is said to be an infinite star domain if its points in every circle of radius r about O form a finite star domain K_r. The lattice Λ of points $(x, y)=(\alpha h+\beta k, \gamma h+\delta k)$, where $h, k=0, \pm 1, \cdots$, of determinant $d(\Lambda)=|\alpha\delta-\beta\gamma|$ is called K-admissible if the origin O is the only point of Λ interior to K. The lower bound of $d(\Lambda)$ for all K-admissible Λ is denoted by $\Delta(K)$. A K-admissible Λ for which $d(\Lambda)=\Delta(\Lambda)$ is called a critical lattice.

In the first paper of the above series, the author proves (an extension of a result of Hurwitz) that $\Delta(K_1)=\sqrt{5}$ if K_1 is the finite star domain $|xy| \leqq 1$, $|x+y| \leqq \sqrt{5}$, determines all the critical lattices for this domain, and shows that, if a star domain H is a proper part of K_1, then $\Delta(H)<\sqrt{5}=\Delta(K)$. In the second paper he considers Mordell's domain G: $|x|^\alpha+|y|^\alpha \leqq 1$, and, employing the above results, derives the simple asymptotic formula $\Delta(G) \sim 2^{-2/\alpha}\sqrt{5}$ for small α. As another application of these results, he obtains an asymptotic expression for the upper bound of the minima of positive quartic forms f of discriminant unity, when the absolute invariant J tends to infinity.

In the third paper the author considers infinite plane star domains K and shows that such domains possess admissible lattices if and only if $\lim_{r\to\infty}\Delta(K_r)$ is finite, in which case there exists a critical lattice Λ and $d(\Lambda)=\Delta(K)=\lim_{r\to\infty}\Delta(K_r)$. Employing a deep result of Markoff on the minima of indefinite binary quadratic forms, he shows that, unlike the case of finite star domains, there exist infinite star domains K^* whose critical lattices have no points on the boundary of K^*. A note at the end of this paper indicates that the author has extended his results to spaces of more than two dimensions. *A. E. Ross* (St. Louis, Mo.).

Citations: MR 4, 212d = H15-2.

Referred to in J56-9.

H20-5 (6, 257c)

Mordell, L. J. Lattice points in the region $|x^3+y^3| \leqslant 1$. J. London Math. Soc. 19, 92–99 (1944).

The author gives a simple proof that every lattice Λ of determinant $(\frac{243}{7})^{1/6}$ contains a point $\neq (0, 0)$ of the set $\mathfrak{R}$: $|x^3+y^3| \leqq 1$. It contains only boundary points of this set, if it is either of basis (a, b), (c, d), or of basis (b, a), (d, c), where a, b, c, d are the real numbers satisfying $b<0$,

$$a^3+b^3=c^3+d^3=(a+c)^3+(b+d)^3=(c-a)^3+(d-b)^3=1.$$

If Λ is different from these two "critical lattices," it has (1) a point P in the region $\mathfrak{R}_1$ external to $\mathfrak{R}$ and bounded by the line from (c, d) to $(c-a, d-b)$; (2) a point P_1 in the region $\overline{\mathfrak{R}}_1$ symmetrical to $\mathfrak{R}_1$ in $x+y=0$; (3) a point P_2 in the region $\mathfrak{R}_2$ external to $\mathfrak{R}$ and bounded by the line through (a, b), $(a+c, b+d)$, and the tangent to $x^3+y^3=1$ at (c, d); (4) a point P_3 in the region $\overline{\mathfrak{R}}_2$ symmetrical to $\mathfrak{R}$ in $x+y=0$. The midpoint of one of the line segments joining two of the points (0, 0), P, P_1, P_2, P_3 must then be a point of Λ; a simple discussion shows that then at least one point of Λ different from (0, 0) is an inner point of $\mathfrak{R}$. The result is equivalent to one of the author's for the region $|x^3-xy^2-y^3| \leqq 1$, which he applied in the theory of cubic binary forms [cf. the same J. 17, 107–115 (1942); 18, 201–210, 210–217, 218–221 (1943); Proc. London Math. Soc. (2) 48, 198–228 (1943); these Rev. 4, 131; 6, 37, 38; 5, 172; and, for related work of Davenport, Proc. Cambridge Philos. Soc. 39, 1–21 (1943); same J. 18, 168–176 (1943); these Rev. 4, 212; 5, 254]. *K. Mahler* (Manchester).

Citations: MR 4, 131d = J36-5; MR 4, 212a = J36-7; MR 5, 172d = J52-3; MR 5, 254f = J52-4; MR 6, 37h = D32-15; MR 6, 37i = J52-5; MR 6, 38a = J52-7.

Referred to in H20-7.

H20-6 (7, 146b)

Ollerenshaw, Kathleen. The critical lattices of a square frame. J. London Math. Soc. 19, 178–184 (1944).

Let F_μ be the region between the squares $(\pm 1, \pm 1)$, $(\pm\mu, \pm\mu)$, where $0<\mu<1$. A lattice is defined to be F_μ-admissible if none of its points is in the interior of F_μ. If $\Delta(F_\mu)$ is the lower bound of the determinants of F_μ-admissible lattices, a lattice whose determinant is actually equal to $\Delta(F_\mu)$ is known as a critical lattice. The author proves that $\Delta(F_\mu)=[1/(1-\mu)]^{-2}$ and constructs all critical lattices. The result is applied to derive a lattice-point theorem for an arbitrary lattice with determinant Δ by magnifying F_μ so that its magnification has a critical lattice with determinant Δ. *D. Derry* (Saskatoon, Sask.).

H20-7 (7, 368a)

Delaunay, B. Local method in the geometry of numbers. Bull. Acad. Sci. URSS. Sér. Math. [Izvestia Akad. Nauk SSSR] 9, 241–256 (1945). (Russian. English summary)

In recent years, the problem of the lattices of smallest determinant satisfying certain conditions (for example, to contain no inner points of a given region except the origin) has been studied particularly by H. Davenport, L. J. Mordell, K. Mahler and B. Segre [see Davenport, J. London Math. Soc. **16**, 98–101 (1941); **18**, 168–176 (1943); Mordell, J. London Math. Soc. **16**, 152–156 (1941); **17**, 107–115 (1942); **18**, 201–210, 210–217 (1943); **19**, 92–99 (1944); Proc. London Math. Soc. (2) **48**, 198–228 (1943); 339–390 (1945); Mahler, J. London Math. Soc. **17**, 130–133 (1942); **18**, 233–238 (1943); Proc. Cambridge Philos. Soc. **40**, 107–116, 116–120 (1944); Duke Math. J. **12**, 367–371 (1945); Segre, Duke Math. J. **12**, 337–365 (1945); these Rev. **3**, 70; **5**, 254; **3**, 167; **4**, 131; **6**, 37, 257; **5**, 172; **6**, 257; **4**, 212; **6**, 119, 258]. The author tries to solve problems of this kind by what he calls the "local method." He replaces the variable lattice by the configuration of only a finite number of points in the lattice and satisfies the extremal conditions by varying the configuration; if the lattice determined by the configuration is admissible, then the problem is solved. This is carried out in detail for the region $|xy|\leqq 1$ connected with Hurwitz's theorem on $|\alpha-p/q|<1/(q^2\sqrt{5})$ and for the region $|xy(x+y)|\leqq 1$ considered by Mordell. [The local method is closely connected with the method of Mordell. Moreover, there are simple regions like $x^2(x^2+y^2)\leqq 1$ for which this method cannot lead to a solution since the critical lattices have no points on the boundary of the region.]

K. Mahler (Manchester).

Citations: MR 3, 70f = J36-2; MR 3, 167c = H20-3; MR 4, 131d = J36-5; MR 4, 212d = H15-2; MR 5, 172d = J52-3; MR 5, 254f = J52-4; MR 6, 37h = D32-15; MR 6, 37i = J52-5; MR 6, 119a = N12-1; MR 6, 257c = H20-5; MR 6, 257d = H15-3; MR 6, 258a = J08-1.

H20-8 (7, 417d)

Ollerenshaw, Kathleen. Lattice points in a hollow n-dimensional hypercube. J. London Math. Soc. **20**, 22–26 (1945).

For positive μ let H_μ be the hypercube $|x_r|\leqq\mu$, $r=1, \cdots, n$, in Euclidean n-space. For $\mu<1$ let F_μ be the set of points in H_1 which are not interior points of H_μ. Let $m=[1/(1-\mu)]$. The author shows that a point of every lattice with determinant not greater than $(1/m)^n$, other than the origin, exists in F_μ. If $\mu<\frac{1}{2}$ and H_μ contains a lattice point other than the origin then F_μ is shown to contain an interior lattice point and so the result is equivalent to Minkowski's linear form theorem for H_1. For $\mu\geqq\frac{1}{2}$ the lattice point in F_μ is interior to it except when the lattice assumes the special form $x_r=\xi_r/m$, $r=1, \cdots, n$, while $\xi_1, \cdots, \xi_n$ run through all integers. The result is a consequence of the recently proved Minkowski conjecture on the boundary case of the linear form theorem [G. Hajós, Math. Z. **47**, 427–467 (1941); these Rev. **3**, 302]. A lattice point theorem for an arbitrary lattice is obtained by a magnification of the n-space. *D. Derry* (Vancouver, B. C.).

Citations: MR 3, 302b = H30-4.
Referred to in H05-29.

H20-9 (7, 506h)

Ollerenshaw, Kathleen. Lattice points in a circular quadrilateral bounded by the arcs of four circles. Quart. J. Math., Oxford Ser. **17**, 93–98 (1946).

Those arcs of the four circles $x^2+y^2=\pm\lambda x$ and $x^2+y^2=\pm\mu y$ which do not pass through the origin form the boundary of a nonconvex star domain K. The author determines $\Delta(K)$ and the critical lattices of K by means of Mordell's method [Proc. London Math. Soc. (2) **48**, 339–390 (1945); these Rev. **6**, 257]. *K. Mahler* (Manchester).

Citations: MR 6, 257d = H15-3.
Referred to in H30-94.

H20-10 (8, 195d)

Mahler, Kurt. Lattice points in two-dimensional star domains. II. Proc. London Math. Soc. (2) **49**, 158–167 (1946).

This paper applies methods developed in the first paper of the series [same vol., 128–157 (1946); these Rev. **8**, 12] to determine the constant $\Delta(H)$ and to find all the critical lattices for three simple, convex, nonsymmetric domains H which contain the origin O. The three domains considered, the ellipse, the parallelogram, and the triangle, are transformed into a circle, a square and an isosceles right-angled triangle, respectively, by suitable affinities. Each domain H is then replaced by a symmetric domain K consisting of all the points of H and of all the points of the reflection of H with respect to O. As the critical lattices of H and K are the same, $\Delta(H)=\Delta(K)$. The nonexistence of singular lattices is established easily and the results for the first two domains are then obtained by finding the parallelograms of minimum area with a vertex at O and three vertices simultaneously on the boundary of the domain H and the domain K. The third result is a consequence of Minkowski's linear form theorem. *D. Derry* (Vancouver, B. C.).

Citations: MR 8, 12a = H15-9.

H20-11 (8, 195e)

Mahler, Kurt. Lattice points in two-dimensional star domains. III. Proc. London Math. Soc. (2) **49**, 168–183 (1946).

This paper, like the one reviewed above, applies the methods the author has developed in the first paper of this series to find the critical lattices and the constant $\Delta(K)$ for a special star domain K. Here K is defined to be the set of points which are within at least one of the ellipses $a_1x^2+2b_1xy+c_1y^2=1$, $a_2x^2+2b_2xy+c_2y^2=1$, $a_1c_1-b_1^2=1$, $a_2c_2-b_2^2=1$. By means of an affinity which maps one of the ellipses into a circle it is shown that $\Delta(K)$ is completely determined by the simultaneous invariant $J=a_1c_2-2b_1b_2+c_1a_2$ and so may be regarded as a function $D(J)$ of J. A simple proof is given that every critical lattice Λ has determinant $d(\Lambda)\geqq\sqrt{\frac{3}{4}}$ and contains a pair of lattice points on each ellipse, both of which pairs form a basis of Λ. The author shows that for each J there are only a finite number of lattices satisfying the two conditions above. He constructs all such lattices for $2\leqq J\leqq 25$ and tabulates them. By use of the table the actual value of $D(J)$ is found within the interval considered. This part of the paper involves considerable detail. Upper and lower bounds for $D(J)$ for all J are stated without proof. *D. Derry* (Vancouver, B. C.).

H20-12 (8, 317j)

Ollerenshaw, Kathleen. The critical lattices of a circular quadrilateral formed by arcs of three circles. Quart. J. Math., Oxford Ser. **17**, 223–239 (1946).

Let K_r denote the star domain of all points (x, y) which satisfy at least one of the inequalities $x^2+y^2\leqq r^2$ $(0<r<1)$, $x^2+y^2-|x|\leqq 0$. A lattice whose only point interior to K_r is the origin is defined to be admissible. The lower bound of the determinants of all admissible lattices is designated by $\Delta(K_r)$, while an admissible lattice with determinant $\Delta(K_r)$ is known as a critical lattice. The purpose of the paper is to find $\Delta(K_r)$ and to construct all critical lattices. The author shows that admissible lattices of minimum determinant, containing a point of intersection of the circle $x^2+y^2=r^2$ and one of the circles $x^2+y^2\pm x=0$, are critical lattices and that these are the only critical lattices except when $\frac{1}{2}3^{\frac{1}{2}}<r<1$, in which case admissible lattices with three points on the circle $x^2+y^2=r^2$ are also critical. The methods used are elementary. The author uses a theorem of Mahler [Nederl. Akad. Wetensch., Proc. **49**, 331–343 (1946); these

Rev. **8**, 12] to construct star domains R within K_r with $\Delta(R)=\Delta(K_r)$ but for which $\Delta(R')<\Delta(R)$ for every star domain R' within R but different from R. *D. Derry.*

Citations: MR 8, 12b = H15-8.

H20-13 (10, 285b)

Varnavides, P. On lattice points in a hyperbolic cylinder. J. London Math. Soc. **23**, 195–199 (1948).

Let K be the cylinder of all points (x, y, z) for which $|xy|\leqq 1$, $|z|\leqq 1$. It is shown that the lower bound $\Delta(K)$ of the determinants of admissible lattices of K is $\sqrt{5}$, which is the same as the corresponding number for the plane domain $|xy|\leqq 1$. *D. Derry* (Vancouver, B. C.).

H20-14 (10, 512f)

Davis, C. S. Note on a conjecture by Minkowski. J. London Math. Soc. **23**, 172–175 (1948).

Minkowski [Minkowski, Diophantische Approximationen, 2d ed., Teubner, Leipzig, 1927, pp. 51–58] was led to the problem of finding the minimum possible area of a parallelogram with one vertex in the origin and the three remaining vertices on the curve $|\xi|^p+|\eta|^p=1$ $(p\geqq 1)$. This problem is easy for $p=1$, $p=2$ and $p=\infty$. For $1<p<2$ Minkowski conjectured that the solution is given by a parallelogram with one vertex on a coordinate axis; its area is $\Delta_0=(1-2^{-p})^{1/p}$. For $p>2$ he conjectured that the minimum is given by a parallelogram with one vertex on one of the lines $\xi=\pm\eta$, this area being Δ_1 given by the equation $(2^{2/p}\Delta_1+1)^p+(2^{2/p}\Delta_1-1)^p=2^{p+1}$. Mordell proved the truth of this conjecture for $p=4$ [J. London Math. Soc. **16**, 152–156 (1941); these Rev. **3**, 167], and now the author proves that the conjecture is false in general. Minkowski remarked that Δ_0 and Δ_1 are clearly extrema as is seen from the symmetry of the curve; he assumed that there are no further extrema, and finally he made a statement which is equivalent to: $\Delta_0<\Delta_1$ for $1<p<2$; $\Delta_1<\Delta_0$ for $p>2$. The author proves that these inequalities are false and he conjectures: $\Delta_1<\Delta_0$ for $1<p<2$ and for $p>p_0$; $\Delta_0<\Delta_1$ for $2<p<p_0$, where p_0 is a certain value, about 2.5725. *J. F. Koksma* (Amsterdam).

Citations: MR 3, 167c = H20-3.

Referred to in H10-17, H10-24, H10-34, H10-35.

H20-15 (12, 678e)

ApSimon, H. On the critical lattices of the 'quadrifoil.' Quart. J. Math., Oxford Ser. (2) **2**, 17–25 (1951).

Let R_a be the set of all points (x, y) of the plane interior to at least one of the circles $(x\pm a)^2+y^2=1$, $x^2+(y\pm a)^2=1$, $0<a\leqq 1$. The above paper determines the critical lattices of the regions R_a. For $1/\sqrt{5}<a\leqq 1$ there is only one such lattice and this is generated by a boundary point of R_a which is on the line $x+y=0$ and the reflection of this point about the x-axis. For $0<a<1/\sqrt{5}$ each critical lattice contains one of the above points P and two boundary points Q, T of R_a with $T=Q+P$, while for $a=1/\sqrt{5}$ lattices of both types exist. The proofs make use of the fact that critical lattices of subregions of R_a which are admissible for R_a are likewise critical for R_a. *D. Derry* (Vancouver, B. C.).

H20-16 (13, 322d)

Mullineux, N. Lattice points in the star body K:

$$|x_1^2+x_2^2-x_3^2|\leqq 1,\ |x_3|\leqq\sqrt{2}.$$

Proc. London Math. Soc. (2) **54**, 1–41 (1951).

Let K be the 3-dimensional star body of points (x_1, x_2, x_3) for which $|x_1^2+x_2^2-x_3^2|\leqq 1$, $|x_3|\leqq\sqrt{2}$. The purpose of this paper is to determine the critical lattices of K. A construction for these lattices is given. Moreover they are shown to be either the critical lattices of the infinite star body $|x_1^2+x_2^2-x_3^2|\leqq 1$ or lattices which can be transformed into certain of these lattices by an affine transformation: $x_1=x_1'+\tau_1x_3'$, $x_2=x_2'+\tau_2x_3'$, $x_3=x_3'$. Their determinant is $\frac{1}{2}\sqrt{6}$. By use of the group of affine transformations which leave the form $x_1^2+x_2^2-x_3^2$ invariant it is shown that critical lattices pass through every boundary point of K.

To prove his results the author considers a certain plane through the origin which contains a two-dimensional sublattice of a given critical lattice of K and which cuts the boundary of K in points on the hyperboloid $x_1^2+x_2^2-x_3^2=1$. Now K is subjected to an affine deformation so that the above plane cuts the deformed K in a circle. Any plane parallel to this plane which cuts the deformation of K will do so in a region bounded by line segments and circular arcs. About half the paper deals with plane lattices without points interior to such regions. These results depend on a study of the critical lattices of symmetric plane regions which consist of the set-theoretic join of the points of a circle and the points interior to two intersecting circles of equal radius.

Results analogous to the main result are stated for the star body of points (x_1, x_2, x_3) for which $-2\leqq x_1^2+x_2^2-x_3^2\leqq 1$, $|x_3|\leqq\sqrt{6}$. *D. Derry* (Vancouver, B. C.).

H20-17 (13, 443d)

Macbeath, A. M. A new sequence of minima in the geometry of numbers. Proc. Cambridge Philos. Soc. **47**, 266–273 (1951).

Consider only polynomials $f(x, y)$ of the form

$$k+px+qy-(rx+sy)^2,$$

where $|ps-qr|=1$. For every $\epsilon>0$ there is a finite set of polynomials $f_1, \cdots, f_N$ such that if f is not equivalent to a polynomial of the form f_i+c $(i=1, \cdots, N)$ then integers x, y exist such that $0\leqq f(x, y)\leqq\epsilon$. In particular, if $\epsilon=(13/8)^{1/3}$, then $N=3$ and f_1, f_2, f_3 with their corresponding "minima" are given. This problem is related to that of finding admissible non-homogeneous lattices for $a\leqq y-x^2\leqq b$. *L. Tornheim* (Ann Arbor, Mich.).

H20-18 (14, 850d)

Ollerenshaw, Kathleen. An irreducible non-convex region. Proc. Cambridge Philos. Soc. **49**, 194–200 (1953).

Let K be the region of points (x, y) for which $|xy|\leqslant 1$, $|x^2-y^2|\leqq 2$, and let $\Delta(K)$ denote the lower bound of the determinants of the admissible lattices of K. A star domain H is said to be irreducible if, for every star domain H' which is a proper subset of H, $\Delta(H')<\Delta(H)$. The author determines the irreducible star domain H within K for which $\Delta(H)=\Delta(K)$ and proves it to be unique. This is the first known example of a reducible star domain Q which contains a unique irreducible star domain P for which $\Delta(P)=\Delta(Q)$. Where $A(H)$ is the area of H, it is pointed out that $A(H)/\Delta(H)$ is less than the corresponding ratio for any other known star domain. A domain G is constructed which is not a star domain for which $A(G)/\Delta(G)$ takes the least known value of this quotient. *D. Derry.*

H20-19 (14, 850f)

Mahler, K. On the lattice determinants of two particular point sets. J. London Math. Soc. **28**, 229–232 (1953).

If K is a domain in the plane, then its determinant $\Delta(K)$ is defined as the inf of the determinants of all lattices which do not contain inner points of K, apart from the origin. Consider the following regions: $K(|xy|\leqq 1)$; $R_t(|xy|\leqq 1, x\geqq t)$; $K_t(|xy|\leqq 1, |y|\leqq t)$; S_t is the union of R_t and the strip $0\leqq x\leqq t$ $(t>0)$. From the facts that $\Delta(K)=\sqrt{5}$ (Hurwitz) and $\Delta(K_t)=\sqrt{5}$ [K. Mahler, Proc. Roy. Soc. London. Ser.

A. **187**, 151–187 (1946); these Rev. **8**, 195], the author derives that $\Delta(R_t)=\sqrt{5}$. Furthermore, he shows that $\Delta(S_t)=\frac{1}{2}(3+\sqrt{5})$, which is deduced from a theorem of A. V. Prasad [J. London Math. Soc. **23**, 169–171 (1948); these Rev. **10**, 513], which asserts that to every real number ξ at least one fraction u/v can be found such that

$$|\xi-u/v|\leqq 2v^{-2}(3+\sqrt{5})^{-1}.$$

It is trivial that $\Delta(R_t)$ and $\Delta(S_t)$ do not depend on t (if $t>0$), as all regions R_t are equivalent under linear transformations of determinant 1, and the same holds for S_t. Nevertheless, it is surprising to notice the fact that under the regions with determinant 1 some are "arbitrary small", and some are "arbitrary large". *N. G. de Bruijn.*

Citations: MR 8, 195f = H15-7; MR 10, 513d = J04-6.

H20-20 (15, 780a)

Ollerenshaw, Kathleen. Irreducible convex bodies. Quart. J. Math., Oxford Ser. (2) 4, 293–302 (1953).

A star body S is defined to be reducible if a star body T exists which is properly contained within S and for which the critical determinants $\Delta(S)$ and $\Delta(T)$ are the same. Let $f(x_1, x_2)\leqq 1$, define an irreducible convex plane domain. The author shows that both the n-dimensional sphere, $n\leqq 5$, and the generalized cylinder $f(x_1, x_2)\leqq 1$, $|x_3|\leqq 1, \cdots, |x_n|\leqq 1$ are irreducible star bodies. Rogers [Nederl. Akad. Wetensch., Proc. 50, 868–872 (1947); these Rev. 9, 228] has proved that a star body S is irreducible if each of its boundary points P satisfies the following condition. For every ϵ, $\epsilon>0$, a lattice Λ exists with $\Delta(\Lambda)=\Delta(S)$ whose only points interior to S are 0, $\pm P^*$ where $|P-P^*|<\epsilon$. The results in the paper reviewed are obtained by constructing lattices Λ for boundary points of the given star bodies. *D. Derry.*

Citations: MR 9, 228a = H15-19.

H20-21 (18, 21d)

Mahler, K. A property of the star domain $|xy|\leq 1$. Mathematika 3 (1956), 80.

The author shows that each star polygon with finitely many sides which is contained in the star domain K: $|xy|\leq 1$ is of smaller lattice determinant than K.

C. G. Lekkerkerker (Amsterdam).

H20-22 (20# 3123)

Clarke, L. E. The critical lattices of a star-shaped octagon. Acta Math. **99** (1958), 1–32.

In a classical paper, L. J. Mordell [Proc. London Math. Soc. (2) **48** (1945), 339–390; MR **6**, 257] introduced a method for determining the critical lattices and determinants of a general class of non-convex star domains in the plane. One of his examples was the star-shaped octagon of vertices

$$(1,0),\ (l/(l-1), l/(l-1)),\ (0,1),\ -(l/(l-1), l/(l-1)),$$
$$(-1,0), -(l/(l-1), -l/(l-1)), (0,-1), (l/(l-1), -l/(l-1)),$$

where $l=\tan(45°+\theta)>1$, for which he solved the problem when $30°\leqq\theta\leqq 45°$. The author now extends Mordell's method and deals with the octagon when $15°\leqq\theta\leqq 30°$, the result being that

$$\Delta(K)=\begin{cases} 1+\dfrac{1}{2l} & \text{if } 22\tfrac{1}{2}°\leqq\theta\leqq 30°,\\[2mm] \dfrac{2l^2(l+1)(3l+1)}{(3l^2-1)^2} & \text{if } \theta_0\leqq\theta\leqq 22\tfrac{1}{2}°,\\[2mm] \dfrac{l^2(l^2+4l+5)}{(l^2+2l-1)^2} & \text{if } 15°\leqq\theta\leqq\theta_0.\end{cases}$$

Here θ_0, roughly $18\frac{1}{2}°$, is defined by the equation

$$3l^6+4l^5-7l^4-24l^3-7l^2+4l+3=0.$$

The author also shows that, if $n=2, 3, 4, \cdots$, then $\frac{1}{2}(1+\sqrt{(n^2-2n+2)})\leqq\Delta(K)\leqq\frac{1}{2}(n+1)$ if $[(n+1)/(n-1)]^{\frac{1}{2}}\leqq l\leqq(n-1)^{-1}[1+\sqrt{(n^2-2n+2)}]$.

K. Mahler (Manchester)

Citations: MR 6, 257d = H15-3.

H20-23 (25# 3422)

Cohen, A. M.

Numerical determination of lattice constants.

J. London Math. Soc. **37** (1962), 185–188.

It had been conjectured that the star body

$$S: |xyz| \leqq 1,\ (x-y)^2+(y-z)^2+(z-x)^2 \leqq 14,$$

has the lattice constant $\Delta(S)=7$. Using the Mercury computer at Manchester, the author disproves this conjecture and shows that (i) $\Delta(S)\leqq 6.260638$, and (ii) $\Delta(S)<7$ even if the constant 14 is replaced by 15.15625. *K. Mahler* (Canberra)

H20-24 (29# 2228)

Foster, D. M. E.

On generating points of a lattice in the region

$$|x_1^2+x_2^2-x_3^2| \leqq 1.$$

Proc. Glasgow Math. Assoc. **6**, 141–155 (1964).

If Λ is a lattice of determinant Δ in 3-space, it is of interest to know for what values of k there is a basis for Λ contained in the set

$$|x_1^2+x_2^2-x_3^2| \leqq k.$$

In this paper the value $k=\Delta^{2/3}$ is found for the case when Λ has points other than 0 on the cone

$$x_1^2+x_2^2-x_3^2 = 0$$

and the value $k=(27\Delta^2/25)^{1/3}$ for the case when Λ has no points on the cone other than 0. The results are best possible in the sense that in each case a lattice can be found which does not satisfy the conclusion of the theorem if k is replaced by any smaller value k'. The proof makes use of continued fraction theory, classical theory of quadratic forms and geometric methods. Subsidiary results about bases for two-dimensional lattices are obtained. *A. M. Macbeath* (Birmingham)

H20-25 (38# 6469)

Dauenhauer, Mary Hughes

The densest irregular packing of the Mordell cubic norm-distance.

Illinois J. Math. **12** (1968), 660–681.

Let $\mathscr{S}$ be the star-domain $f(x, y)=|y(3x^2-y^2)|<1$. A set $\mathscr{P}\subset R_2$ is said to be $\mathscr{S}$-admissible if (x_1, y_1), $(x_2, y_2)\in\mathscr{P}$ imply $f(x_1-x_2, y_1-y_2)\geqq 1$. When $\mathscr{P}$ is a lattice the definition coincides with the standard definition of admissible lattices. The author proves that no $\mathscr{S}$-admissible set $\mathscr{P}$ can be denser than the critical lattices of $\mathscr{S}$. She also proves that the "critical" $\mathscr{S}$-admissible sets $\mathscr{P}$ are in some sense like the critical lattices of $\mathscr{S}$. The proof is based on the Zassenhaus method of domains of action and the convex body used for the domains of action is the regular hexagon inscribed in $\mathscr{S}$. Although the proof is elementary, it needs a lot of careful work.

R. P. Bambah (Chandigarh)

H20-26 (39# 5478)

Rehman, M.

On the critical determinant of an unbounded star domain of hexagonal symmetry.

Panjab Univ. J. Math. (*Lahore*) **1** (1967), 79–85.

Let R be the region in the euclidean plane determined by the inequalities, either $2|y|((\sqrt{3})|x|-|y|)\leqq 1$ and $|y|\leqq(\sqrt{3})|x|$, or $y^2-3x^2\leqq 1$ and $|y|\geqq(\sqrt{3})|x|$. R is an unbounded star domain whose boundary consists of the hyperbola $y^2-3x^2=1$ together with the two rotations of this that yield hexagonal symmetry. The author applies R. P. Bambah's method [Philos. Trans. Roy. Soc. London Ser. A **243** (1951), 431–462; MR **13**, 114] to show that the critical determinant of R is given by $(2/\sqrt{3})(a^2+ab+b^2)$, where $4ab=1$ and $b^2=2a^2$.

A. C. Woods (Columbus, Ohio)

Citations: MR 13, 114c = H15-46.

H25 THE MINKOWSKI-HLAWKA THEOREM; SIEGEL'S MEAN VALUE THEOREM

See also reviews H02-17, H05-39, H15-67.

H25-1 (5, 201c)

Hlawka, Edmund. Zur Geometrie der Zahlen. Math. Z. **49**, 285–312 (1943).

We use small gothic letters for points or vectors in n-dimensional Euclidean space R_n ($n\geqq 2$) and gothic capitals for square matrices of order n. The letter $\mathfrak{g}$ denotes a lattice point. All point sets are assumed closed, bounded and Jordan-measurable. If the set S consists of the points $\{\mathfrak{x}\}$, then $S+\mathfrak{a}$ and $\mathfrak{A}S$ denote the sets of points $\{\mathfrak{x}+\mathfrak{a}\}$ and $\{\mathfrak{A}\mathfrak{x}\}$, respectively. The letters k_0 and r are integers satisfying $k_0\geqq 1$, $0\leqq r\leqq k_0-[k_0/2]-1$.

(A) Let K be a set of volume V such that (a) the origin $\mathfrak{o}$ is an inner point of K and (b) every radius vector from $\mathfrak{o}$ meets the boundary of K in just one point. A theorem of Minkowski [Gesammelte Abhandlungen, vol. 1, Teubner, Leipzig, 1911, pp. 265, 269–270, 276] states that, if $V<\zeta(N)$ (or $V<2\zeta(n)$ if K is symmetrical in $\mathfrak{o}$), then there exists a matrix $\mathfrak{A}$ of determinant 1 such that $\mathfrak{A}K$ contains no lattice point $\mathfrak{g}\neq\mathfrak{o}$. This theorem has an important application to the minimum of definite quadratic forms in n variables [ibid., p. 270] but no proof seems to have been published. The author gives a proof based on the following theorem which is of interest in itself. "Let $\varphi(\mathfrak{x})$ be real, bounded and Riemann-integrable, and let $\varphi(\mathfrak{x})=0$ if $\mathfrak{x}$ lies outside a finite cube. Then to every $\epsilon>0$ there is a unimodular $\mathfrak{A}$ such that

$$\sum_{\mathfrak{g}\neq\mathfrak{o}}\varphi(\mathfrak{A}\mathfrak{g})\leqq\int_{-\infty}^{+\infty}\cdots\int_{-\infty}^{+\infty}\varphi(\mathfrak{x})dx_1\cdots dx_n+\epsilon."$$

This inequality implies not only Minkowski's theorem, but also the following one. "If S is of volume less than k_0, then there exists a matrix $\mathfrak{A}$ of determinant 1 such that $\mathfrak{A}S$ contains at most k_0-1 lattice points $\mathfrak{g}\neq\mathfrak{o}$."

(B) Let $\mathfrak{K}$ be a diagonal matrix with the positive numbers $k_1,\cdots,k_n$ in the diagonal. For every set M, denote by $M(\mathfrak{K})$ the set of all points $\mathfrak{K}^{-1}(\mathfrak{x}-\mathfrak{y})$, where $\mathfrak{x}$ and $\mathfrak{y}$ run over M. The author uses methods of C. Siegel [Acta Math. **65**, 307–323 (1935)] and J. G. van der Corput [Acta Arith. **1**, 62–66 (1935)] to prove the following existence and "alternative" theorems. "If M is of volume $V\geqq k_0k_1\cdots k_n$, then $M(\mathfrak{K})$ contains at least k_0 pairs of lattice points $\mp\mathfrak{g}\neq\mathfrak{o}$." "If M is of volume $V=k_0k_1\cdots k_n$ and if $M(\mathfrak{K})$ contains at most k_0-1 pairs of inner lattice points $\mp\mathfrak{g}\neq\mathfrak{o}$, then every point of R_n lies in at least one and in at most k_0 of the sets $M+\mathfrak{K}\mathfrak{g}$, where $\mathfrak{g}$ runs over all lattice points." "If M is of volume $V\geqq\frac{1}{2}k_0k_1\cdots k_n$ and if at most k_0-r-1 pairs of lattice points $\mp\mathfrak{g}\neq\mathfrak{o}$ are inner points of $M(\mathfrak{K})$, then every point in R_n belongs to at least $r+1$ sets $M(\mathfrak{K})+\mathfrak{g}$, where $\mathfrak{g}$ runs over all lattice points." The last theorem can be further generalized.

(C) Let c and σ be positive numbers, and let $f(\mathfrak{x})$ be a function in R_n with the following properties. (a) $f(\mathfrak{x})>0$ for $\mathfrak{x}\neq\mathfrak{o}$, but $f(\mathfrak{o})=0$. (b) $f(t\mathfrak{x})=|t|f(\mathfrak{x})$ for all real t. (c) If $\mathfrak{x}_1,\cdots,\mathfrak{x}_k$ are any number of points in R_n, and if $1\leqq k_0<k$, then there are k_0+1 of these points, say $\bar{\mathfrak{x}}_0,\bar{\mathfrak{x}}_1,\cdots,\bar{\mathfrak{x}}_{k_0}$, such that

$$(k-k_0)f(\bar{\mathfrak{x}}_i-\bar{\mathfrak{x}}_0)^\sigma\leqq c\sum_{j=1}^{k}f(\mathfrak{x}_j)^\sigma,\quad i=1,2,\cdots,k_0.$$

[This condition may be replaced by

$$\sum_{i=1}^{k}\sum_{j=1}^{k}f(\mathfrak{x}_i-\mathfrak{x}_j)^\sigma\leqq ck\sum_{j=1}^{k}f(\mathfrak{x}_j)^\sigma$$

for all k]. (d) $f(\mathfrak{x})$ has a positive lower bound on $\mathfrak{x}^2=1$. The author proves the following existence and "alternative" theorems [for the existence theorem, see J. G. van der Corput and G. Schaake, Acta Arith. **2**, 152–160 (1936)]. "If $\mathfrak{J}=\int\cdots\int dx_1\cdots dx_n$, the integral extending over $f(\mathfrak{x})\leqq 1$, and $t>((k_0/\sigma\mathfrak{J})(n+\sigma))^{1/n}$, then at least k_0 pairs of lattice points $\mp\mathfrak{g}\neq\mathfrak{o}$ lie in the set $f(\mathfrak{x})\leqq c^{1/\sigma}t$." "If at most k_0-r-1 pairs of lattice points $\mp\mathfrak{g}\neq\mathfrak{o}$ lie in the set

$$f(\mathfrak{x})\leqq c^{1/\sigma}((k_0/2\sigma\mathfrak{J})(n+\sigma))^{1/n},$$

then to every point $\mathfrak{x}_0$ in R_n there are at least $r+1$ lattice points $\mathfrak{g}_i$ such that

$$f(\mathfrak{x}_0-\mathfrak{g}_i)\leqq c^{1/\sigma}((k_0/2\sigma\mathfrak{J})(n+\sigma))^{1/n}."$$

K. Mahler (Manchester).

Referred to in H05-39, H10-9, H15-44, H25-2, H25-3, H25-9, H25-11, H25-14, H25-15, H25-22, H25-25, H25-28, H25-29, H30-10.

H25-2 (6, 257b)

Siegel, Carl Ludwig. A mean value theorem in geometry of numbers. Ann. of Math. (2) **46**, 340–347 (1945).

Let Ω_1 be the set of n-rowed square matrices with real elements and determinant $+1$. The author starts from Minkowski's fundamental region K of reduced positive quadratic forms. Using K he defines a fundamental region F for Ω_1 with respect to the proper unimodular group T_1 of all matrices of Ω_1 with integral elements. The Ω_1 space is endowed with a measure ω, so that $\int_F d\omega=1$. Let R be a bounded set in the real Euclidean n-space, measurable in the Jordan sense, and let $f(x)$ be a bounded function integrable in R and zero elsewhere. The main purpose of the paper is to prove the theorem

$$\int_F\sum_{g\neq 0}f(Ag)d\omega=\int_R f(x)\{dx\},$$

where g is an integral vector in Euclidean n-space and $A\epsilon\Omega_1$.

The proof consists of an analysis of the integrals

$$\psi(\lambda)=\int_F\lambda^n\sum_{g\neq 0}f(\lambda Ag)d\omega_1,\qquad 0<\lambda\leqq 1,$$

where the integration is performed with respect to a specially defined measure in Ω_1. The author shows that $\psi(\lambda)$ is independent of λ and derives the result from $\psi(1)=\lim_{\lambda\to 0}\psi(\lambda)$.

An immediate deduction from this result is the following theorem of Hlawka [Math. Z. **49**, 285–312 (1943); these Rev. **5**, 201]. A matrix A exists in Ω_1 so that

$$\sum_{g\neq 0}f(Ag)\leqq\int_R f(x)\{dx\}+\epsilon.$$

Two simple consequences of the analysis are given. One is a proof of Minkowski's assertion that, if B is an n-dimensional star domain of volume less than $\zeta(n)$, then there exists a lattice of determinant 1 such that B does not contain any lattice point other than 0. The other is an evaluation of the volume of that part of Minkowski's fundamental region K of reduced positive quadratic forms for which the determinant is less than or equal to 1. The paper closes by suggesting generalizations of the main result either by imposing fewer restrictions on $f(x)$ or by using other spaces and groups in place of Ω_1 and T_1, respectively.

D. Derry (Saskatoon, Sask.).

Citations: MR 5, 201c = H25-1.

Referred to in E12-54, F99-3, H25-4, H25-5, H25-7, H25-12, H25-17, H25-18, H25-19, H25-23, H25-26, H25-32, H25-33.

H25-3 (7, 244h)

Mahler, K. On a theorem of Minkowski on lattice points in non-convex point sets. J. London Math. Soc. **19**, 201–205 (1944).

A star body K in n-dimensional Euclidean space is a closed bounded point set which contains the origin O as an inner point and whose frontier is met by every radius vector in just one point. Denote by V the volume of K and put $E=2$ if K is symmetrical with respect to O, $E=1$ otherwise. The following theorem was stated by H. Minkowski [Gesammelte Abhandlungen, vol. I, Teubner, Leipzig-Berlin, 1911, pp. 265, 270, 277] and proved for the first time by E. Hlawka [Math. Z. **49**, 285–312 (1943); these Rev. **5**, 201]. Let $n>1$, $V<E\zeta(n)$; then there exists a lattice of determinant 1 which contains no point of K other than O. The author presents a simple geometrical proof of the less precise statement where $\zeta(n)$ is replaced by n^{-1}.

C. L. Siegel (Princeton, N. J.).

Citations: MR 5, 201c = H25-1.

Referred to in H25-5.

H25-4 (7, 411g)

Weil, André. Sur quelques résultats de Siegel. Summa Brasil. Math. **1**, 21–39 (1946).

This paper exploits homogeneous spaces to simplify some results of Siegel. The first (and longer) part of the paper proves a formula of Siegel [Ann. of Math. (2) **46**, 340–347 (1945); these Rev. **6**, 257] concerning the group G of $n\times n$ matrices (real elements) of determinant 1, the subgroup Γ of those with integer elements, and the set of lattices in Euclidean n-space E^n. A lattice R in E^n is any transform by an element of G of the set of points in E^n all of whose coordinates are integers. Siegel's formula asserts, for $f(x)$ (Riemann integrable) defined on E^n, that $\int f(x)dx=\int\sum f(x)dR$, where the summation is over all $x\neq 0$ in R, so that this sum is a function of R.

The author first proves a general integration formula involving groups $G\supseteq g\supseteq\gamma$, where g,γ are closed in the locally compact group G and all three are unimodular (left and right Haar measures coincide). It asserts that the integral over G/γ equals the iterated integral over g/γ and G/g, the integrals involved being those defined on a general homogeneous space by Weil [L'intégration dans les groupes topologiques et ses applications, Actual. Sci. Ind., no. 869, Hermann, Paris, 1940; these Rev. **3**, 198]. Then he considers still another subgroup Γ (closed and unimodular) with $G\supseteq\Gamma\supseteq\gamma$ and as a special case of the above obtains (in case g/γ has finite measure) for $f(x)$ defined on G/g

$$\int_{G/g} f(x)dx=c\int_{G/\Gamma}d\bar{x}\int_{\Gamma/\gamma}f(x\xi)d\xi,$$

where c is a constant. This is the general formula which yields Siegel's, by taking G, Γ as in the first paragraph above, choosing g properly so that $G/g=E^n$, and defining $\gamma=\Gamma\cap g$. To obtain Siegel's formula it remains to (1) show that g/γ has finite measure (easy), (2) turn the integral over Γ/γ into $\sum f(x)$ (done following Siegel) and (3) eliminate the constant in Weil's formula (difficult). This last includes proving that G/Γ has finite measure.

The second part of the paper improves other theorems of Siegel [Ann. of Math. (2) **44**, 674–689 (1943); these Rev. **5**, 228]. A continuous mapping is called proper if the inverse image of a compact set is compact. For a locally compact group G with closed subgroups Γ and g the natural representation of Γ on G/g is called proper if each transformation of the representation is proper. It is proved that, for G, Γ unimodular, and G/Γ of finite measure, Γ is proper on G/g if and only if g is compact. *W. Ambrose.*

Citations: MR 6, 257b = H25-2.

Referred to in H25-32.

H25-5 (8, 444e)

Mahler, Kurt. The theorem of Minkowski-Hlawka. Duke Math. J. **13**, 611–621 (1946).

Il résulte d'un théorème de géométrie intégrale, dû à C. L. Siegel [Ann. of Math. (2) **46**, 340–347 (1945); ces Rev. **6**, 257] que, si K est un corps de volume non supérieur à 1 dans R^n, il existe un réseau unitaire (i.e., transformé, par une substitution unimodulaire réelle, du réseau des points à coordonnées entières) dont aucun point, sauf peut-être O, ne soit intérieur à K; si K est étoilé symétrique de centre O, il résulte du même théorème qu'il suffit de supposer K de volume non supérieur à $2\zeta(n)$ (c'est le théorème "de Minkowski-Hlawka"). L'auteur montre que, si K est convexe symétrique de centre O, la constante $2\zeta(n)$ n'est pas la meilleure possible. Sa démonstration (indépendante de celles de Siegel et de Hlawka) procède par récurrence sur n, et repose (a) sur une démonstration élémentaire directe du résultat pour $n=2$; (b) sur un lemme applicable à tout corps K symétrique de centre O [sur ce lemme, cf. K. Mahler, J. London Math. Soc. **19**, 201–205 (1944); ces Rev. **7**, 244]; (c) sur le théorème de Brunn-Minkowski.

A. Weil (São Paulo).

Citations: MR 6, 257b = H25-2; MR 7, 244h = H25-3.

Referred to in H05-74, H25-6.

H25-6 (9, 11a)

Davenport, H., and Rogers, C. A. Hlawka's theorem in the geometry of numbers. Duke Math. J. **14**, 367–375 (1947).

The authors give some improvements of the results of K. Mahler [same J. **13**, 611–621 (1946); these Rev. **8**, 444] and C. A. Rogers [to appear in Ann. of Math.]. (I) Let c_n denote the upper bound of V/Δ for all convex bodies K in n dimensions, symmetrical about the origin O, where Δ is the determinant of any lattice with no point (except O) in K and V the volume of K. Then, for $n\geqq 3$, $c_n\geqq 2n^{-1}(c_{n-1}^{n/(n-1)}-1)/(c_{n-1}^{1/(n-1)}-1)$ and $\lim\inf_{n\to\infty}c_n\geqq c$, where $\log c=2(1-1/c)$, $c>1$ (the numerical value of c is $4.921\cdots$). (II) Let Q be any positive definite quadratic form in n variables of determinant 1 and let $M(Q)$ denote the minimum value of Q for integral values of the variables, not all zero. Let γ_n be the least number such that $M(Q)\leqq\gamma_n$ for all such Q and let $\Theta=0.596\cdots$ denote the minimum, for $0<\eta<\pi$, of $2(\log\pi/\eta)^{-1}\sum_{t=1}^{\infty}e^{-\eta t^2}$. Then $\lim\inf_{n\to\infty}n^{-1}T(n)\gamma_n^{n/2}\geqq 1/\Theta$, where $T(n)=\pi^{n/2}/\Gamma(1+n/2)$. *V. Knichal* (Prague).

Citations: MR 8, 444e = H25-5.

Referred to in H25-7, H25-13, H25-24.

H25-7 (9, 270h)

Rogers, C. A. Existence theorems in the geometry of numbers. Ann. of Math. (2) **48**, 994–1002 (1947).

Let S be a closed bounded n-dimensional star body ($n\geqq 2$), symmetrical about the origin O, with a volume $V(S)$. The author gives a simple proof of a theorem of C. L. Siegel (except for an ϵ term) [same Ann. (2) **46**, 340–347 (1945); these Rev. **6**, 257]. This theorem is used (1) to deduce a Minkowski's assertion: if $V(S)<2\zeta(n)$ (Riemann zeta function), it is possible to find a lattice $\mathfrak{G}$ of unit determinant such that O is the only point of $\mathfrak{G}$ in S; (2) to prove a new theorem: if $V(S)<2n\zeta(n)l^{-1}(1-l^{-n})^{-1}=\sigma(n)$, it is possible to find a lattice $\mathfrak{G}$ of unit determinant such that $\lambda_1\cdots\lambda_n>1$, where λ_i is the least value of λ such that the body λS contains at least i linearly independent points of $\mathfrak{G}$. This theorem is used to establish the following result: there exists a positive definite quadratic form $Q(u)$ of unit determinant in n variables u_i such that $Q(u)\geqq\pi^{-1}\{\sigma(n)\Gamma(1+n/2)\}^{2/n}$ for all the lattice points $u\neq O$. See also Davenport and Rogers [Duke Math. J. **14**, 367–375 (1947); these Rev. **9**, 11].

V. Knichal (Prague).

Citations: MR 6, 257b = H25-2; MR 9, 11a = H25-6.
Referred to in H25-11, H25-12, H25-22.

H25-8 (10, 683a)

Hlawka, Edmund. Über eine Verallgemeinerung des Satzes von Mordell. Ann. of Math. (2) **50**, 314–317 (1949).

The following result was proved by a method of Siegel [Davenport, Acta Arith. **2**, 262–265 (1937)]. For any lattice in n-dimensional space, of determinant 1, there exist positive numbers $\lambda_1, \cdots, \lambda_n$, with

$$\lambda_1\cdots\lambda_n\geqq(1\cdot 2^2\cdot 3^3\cdot\cdots\cdot(n-1)^{n-1}n!)^{-1}$$

such that the "box" defined by $|x_1|<\lambda_1, \cdots, |x_n|<\lambda_n$ contains no lattice point except the origin O. The author gives another result of a similar kind, which he enunciates as follows. Let $f(x)$ be the distance-function of a convex body, and A a matrix of determinant 1. Then there exists an integral unimodular matrix U, independent of A, and a diagonal matrix D, such that the body $f(UDU^{-1}Ax)\leqq t$ contains no point with integral coordinates [except O] if its volume is less than $2^n(n!)^{-4n^2-n-1}$. The proof is complicated, and not easy to follow.

H. Davenport (London).

H25-9 (11, 331e)

Schneider, Theodor. Über einen Hlawkaschen Satz aus der Geometrie der Zahlen. Arch. Math. **2**, 81–86 (1950).

Let $k_1, \cdots, k_n$ be positive numbers, M a bounded, closed, Jordan measurable set, and $M(K)$ the set of points $((x_1-y_1)/k_1, \cdots, (x_n-y_n)/k_n)$, where $(x_1, \cdots, x_n)$ and $(y_1, \cdots, y_n)$ are in M. If M has a volume not less than $\frac{1}{2}k_0k_1\cdots k_n$, k_0 being a positive integer, and if $M(K)$ has at most k_0-1 pairs of symmetric lattice points distinct from the origin in its interior, then for every point $r=(r_1, \cdots, r_n)$ there exist at least k_0 lattice points g such that $r+g$ is in $M(K)$. This improves a theorem of Hlawka [Math. Z. **49**, 285–312 (1943); these Rev. **5**, 201]. The proof is essentially arithmetical. A refinement of a theorem of van der Corput [Acta Arith. **2**, 145–146 (1936)] on the number of lattice points in $M(K)$ is made and then applied to the set M' of points $(x_1', \cdots, x_{n+1}')$ where $x_i'=x_i+(r_ik_i/k_{n+1})x_{n+1}$ $(i=1, \cdots, n)$ and $x_{n+1}'=x_{n+1}$ with $(x_1, \cdots, x_n)$ in M and $|x_{n+1}|\leqq k_{n+1}$.

L. Tornheim (Ann Arbor, Mich.).

Citations: MR 5, 201c = H25-1.
Referred to in H05-39.

H25-10 (12, 806b)

Rogers, C. A. On theorems of Siegel and Hlawka. Ann. of Math. (2) **53**, 531–540 (1951).

The main theorem proved is the following. Let K be any convex body in n-dimensional space, symmetrical about the origin O, whose volume does not exceed a certain number depending only on n; let Λ be any lattice of determinant 1. Then there exists a transformation of the special form $x_i=\omega_ix_i'$, $i=1, 2, \cdots, n$, with $\omega_1\cdots\omega_n=1$, such that the transformed lattice Λ' has no point other than O in K. This theorem is a powerful generalization of a result of Siegel [see Davenport, Acta Arith. **2**, 262–265 (1937)], Siegel's result being the case when K is a sphere. The whole difficulty of the proof lies in extending Siegel's result from the case of a sphere to that of an arbitrary ellipsoid. The essential lemma (lemma 2) is one which in effect permits the introduction of a single cross-product term into the inequality defining the sphere; once this has been achieved, the further extension to an arbitrary ellipsoid is not difficult.

H. Davenport (London).

H25-11 (13, 919d)

Sanov, I. N. A new proof of Minkowski's theorem. Izvestiya Akad. Nauk SSSR. Ser. Mat. **16**, 101–112 (1952). (Russian)

The theorem in question is that generally known as the Minkowski-Hlawka theorem; it was stated without proof by Minkowski and proved by Hlawka in a paper which appeared in 1943 [Math. Z. **49**, 285–312 (1943); these Rev. **5**, 201]. The author gives a proof of this theorem, based on another paper of his [Leningrad. Gos. Univ. Učenye Zapiski **111**, 32–46 (1949)], which the reviewer has not been able to consult. The author states that this earlier paper was presented to a seminar at Leningrad in 1941, and that Lemma 1 in it represents an "insignificantly weaker" form of the theorem. There is no reference to the other proofs of the Minkowski-Hlawka theorem which have since been given [see, e.g., Rogers, Ann. of Math. **48**, 994–1002 (1947); these Rev. **9**, 270]. In view of these later proofs, the paper of Sanov is probably now only of historical interest.

H. Davenport (London).

Citations: MR 5, 201c = H25-1; MR 9, 270h = H25-7.

H25-12 (14, 143c)

Malyšev, A. V. On the Minkowski-Hlawka theorem concerning a star body. Uspehi Matem. Nauk (N.S.) **7**, no. 2(48), 168–171 (1952). (Russian)

The author extends the Minkowski-Hlawka theorem on bounded star bodies to unbounded star bodies. I.e., he proves that if the volume of an n-dimensional symmetric star body, centred at the origin, is less than $2\zeta(n)$, then it is possible to find a unimodular transformation transforming the body into one containing no lattice point other than the origin. This is proved by expanding the body about one axis Ox_1 and contracting it about the others so that the sum of the areas of sections at distance unity apart and perpendicular to Ox_1 is approximately equal to the volume. A line not in the plane $x_1=0$ is then chosen as the new x_1 axis in such a way that the new body contains no lattice point other than those in the plane $x_1=0$. This process is carried out for each of the n axes, the body being finally transformed into one containing no lattice-point other than the origin. Similar extensions of the Minkowski-Hlawka theorem valid under different or additional conditions have been mentioned by Siegel [Ann. of Math. **46**, 340–347 (1945), p. 346; these Rev. **6**, 257] and Rogers [ibid. **48**, 994–1002 (1947), p. 1000; these Rev. **9**, 270].

R. A. Rankin (Birmingham).

Citations: MR 6, 257b = H25-2; MR 9, 270h = H25-7.

H25-13 (14, 624c)

Cassels, J. W. S. A short proof of the Minkowski-Hlawka theorem. Proc. Cambridge Philos. Soc. **49**, 165–166 (1953).

The theorem in question is that if a bounded star body K in n-dimensional space, symmetrical about the origin 0, has volume $<2\sum_{m=1}^{\infty}m^{-n}$, then there exists a lattice of determinant 1 having no point except 0 in K. The method of proof is essentially that devised by Rogers and used by Davenport and Rogers [Duke Math. J. **14**, 367–375 (1947); these Rev. **9**, 11] to prove a more precise result when K is convex.

H. Davenport (London).

Citations: MR 9, 11a = H25-6.

H25-14 (14, 624e)

Rogers, C. A. The number of lattice points in a star body. J. London Math. Soc. **26**, 307–310 (1951).

For abbreviation, by "star" we denote a set in n-dimensional space which is star-shaped with respect to the origin 0 and also symmetric with respect to 0. A "lattice" means a lattice of which 0 is a lattice point. If S is a star, then $\Delta(S)$ denotes the lower bound of the determinants $d(\Lambda)$ of the lattices Λ with no points other than 0 in S. $V(S)$ denotes the Jordan measure of S. If S is convex, then a well-known theorem of Minkowski asserts that $2^n\Delta(S)\geqq V(S)$, whereas van der Corput [Acta Arith. **2**, 145–146 (1936)] proved more generally that $V(S)>m2^n d(\Lambda)$ (S convex) implies that S contains at least m distinct pairs $\pm A_1, \cdots, \pm A_m$ of points of Λ.

The author states the following conjecture (which, if true, shows among other things that van der Corput's result follows from Minkowski's). Let m be a positive integer, S a star, Λ a lattice. Then, if $\Delta(S)>md(\Lambda)$, S contains m distinct pairs $\pm A_1, \cdots, \pm A_m$ of points of Λ. He states that he is able to prove this when m is of certain special forms, and gives the proof for the case that m is a prime. In that case he even shows that either there is a point $A_1 \varepsilon \Lambda$ ($A_1\neq 0$) such that $pA_1 \varepsilon S$, or there are $p+1$ distinct pairs of primitive points of Λ in S.

As his conjecture is proved for a set of numbers $m_1, m_2, \cdots$ for which $m_{n+1}/m_n\to 1$, the author is able to give a simple proof for the inequality $2\zeta(n)\Delta(S)\leqq V(S)$, which was stated by Minkowski and proved by Hlawka [Math. Z. **49**, 285–312 (1943); these Rev. **5**, 201].

N. G. de Bruijn (Amsterdam).

Citations: MR 5, 201c = H25-1.

Referred to in H25-21, H25-31.

H25-15 (15, 607c)

Churchhouse, R. F. An extension of the Minkowski-Hlawka theorem. Proc. Cambridge Philos. Soc. **50**, 220–224 (1954).

Let R be an n-dimensional region, symmetric about the origin, of content $V(R)$ and critical determinant $\Delta(R)$. Put $V(R)/\Delta(R)=Q(R)$. If R is convex, the Minkowski-Hlawka theorem [Hlawka, Math. Z. **49**, 285–312 (1943); these Rev. **5**, 201] asserts that $Q(R)\geqq 2\zeta(n)$. The non-convex two-dimensional case is considered in the present paper. Let g be any real-valued function such that $g(0)=1$, $g(1)=0$, g is non-increasing and concave upward for $0<x<1$, and $g(x)=g^{-1}(x)$ for $0<x<1$. Let α be the unique solution in $0<x<1$ of the equation $x=g(x)$. Then if R is the region containing the origin and bounded by the arcs $|y|=g(|x|)$, and $V(R)\geqq 2\alpha$, then $Q(R)\geqq 4$. The constant 4 is best possible, since equality holds when R is the square $|x|+|y|\leqq 1$.

W. J. LeVeque (Ann Arbor, Mich.).

Citations: MR 5, 201c = H25-1.

H25-16 (16, 680a)

Rogers, C. A. The Minkowski-Hlawka theorem. Mathematika **1**, 111–124 (1954).

Let Λ be a discrete set of points in real Euclidean n-space which does not contain the origin. Where $\delta(\Sigma(r))$ denotes the number of points of Λ within a sphere Σ of radius r divided by the volume of Σ, let $\delta(\Lambda)$ be

$$\limsup_{r\to\infty}\sup_{\Sigma(r)}\delta(\Sigma(r)).$$

The purpose of this paper is to prove the following result. If S is a set with outer Jordan content $V(S)$ for which $\delta(\Sigma)V(S)<1$. Then a linear transformation α with unit determinant exists so that $\alpha\Lambda$ has no point in common with S. After specializing Λ to be a lattice, the Minkowski-Hlawka theorem follows simply from this result, thus showing that this theorem depends largely on the density properties of the lattice.

Let u be a unit vector and $C(u,\theta)$ be the cone with vertex at the origin whose generators make an angle θ with u. If

$$\delta(C(u,\theta))=\limsup_{r\to\infty}\sup_{\Sigma(r)\subset C(u,\theta)}\delta(\Sigma(r)),$$

then $\delta(u)$ is defined as $\lim_{\theta\to 0}\delta(C(u,\theta))$. Most of the paper is concerned with the following theorem which is stronger than the author's main result. If u is a unit vector for which $\delta(u)$, $\delta(-u)$ are finite and $\rho(x)$ is a function, integrable in the Riemann sense over the whole space, then, for $\epsilon>0$, a linear transformation α exists with unit determinant for which

$$\sum_{x\varepsilon\Lambda}\rho(\alpha x)<\delta(u)\int_P\rho(x)\,dx+\delta(-u)\int_N\rho(x)\,dx+\epsilon,$$

where P and N are the half-spaces $x_n=0$, $x<0$.

The proof makes use of transformations

$$\alpha x=(\omega x_1+\omega\Phi_1 x_n,\ \cdots,\ \omega x_{n-1}+\omega\Phi_{n-1}x_n,\ \omega^{-n+1}x_n).$$

$I(x)$ is defined as

$$\int_{-\Phi}^{+\Phi}\cdots\int_{\Phi}^{+\Phi}\rho(\alpha x)\,d\Phi_1\cdots d\Phi_{n-1}.$$

By a series of elementary estimates which exploit the density properties of Λ, it is shown that

$$\sum_{x\varepsilon\Lambda}I(x)\leqq(2\Phi)^{n-1}\left[\delta(u)\int_P\rho(x)\,dx+\delta(-u)\int_N\rho(x)\,dx+\epsilon\right].$$

As

$$\sum_{x\varepsilon\Lambda}I(x)=\int_{-\Phi}^{+\Phi}\cdots\int_{-\Phi}^{+\Phi}\sum_{x\varepsilon\Lambda}\rho(\alpha x)\,d\Phi_1\cdots d\Phi_{n-1}$$

it follows that at least one set of values $\Phi_1,\cdots,\Phi_{n-1}$ exist which define an appropriate α. Proofs are indicated of generalizations of this result in which $\rho(x)$ is assumed to be integrable in the Lebesgue sense. A generalization of the Minkowski-Hlawka theorem is given in which the hypothesis involves the $\delta(u)$ concept.

D. Derry.

H25-17 (17, 241d)

Macbeath, A. M., and Rogers, C. A. A modified form of Siegel's mean-value theorem. Proc. Cambridge Philos. Soc. **51** (1955), 565–576.

If Λ is a lattice in Euclidean n-space with determinant 1, and $\varrho(x)$ a real function defined for the points of Λ, let $\varrho(\Lambda)=\sum_{x\in\Lambda,x\neq 0}\varrho(x)$. In his proof of the Minkowski-Hlawka theorem, Siegel [Siegel, Ann. of Math. (2) **46** (1945), 340–347; MR **6**, 257] defines a measure $\mu(\gamma)$ for the space of all linear transformations γ of determinant 1 and considers

$$\int_F\varrho(\gamma\Lambda_0)d\mu(\gamma)\Big/\int_F d\mu(\gamma),$$

where Λ_0 is the lattice of all points with integer coordinates and F a fundamental region defined by the use of Minkowski's theory of reduced quadratic forms. The region F is introduced because the complete space of all γ does not have finite measure. In the present paper Siegel's region F is replaced by a region $\|\gamma\|\leqq K$ where $\|\gamma\|=\sup_{|x|=1}|\gamma x|$. This region is a compact subset within the region of all γ and has a finite measure. Λ_0 is replaced by a more general discrete point set for which $\lim N(r)/V(r)=d$, where $N(r)$ denotes the number of points of Λ_0 in the sphere $|x|\leqq r$ and $V(r)$ the volume of the sphere. The author's principal result is that

$$\lim_{K\to\infty}\int_{\|\gamma\|\leqq K}\varrho(\gamma\Lambda_0)d\mu(\gamma)\Big/\int_{\|\gamma\|\leqq K}d\mu(\gamma)=\int\varrho(x)dx,$$

where $\varrho(x)$ is assumed to be integrable in the Riemann sense and vanishes outside a bounded region. This result can be used in place of Siegel's analogous result in the proof of the Minkowski-Hlawka theorem. *D. Derry.*

Citations: MR 6, 257b = H25-2.

Referred to in H25-30.

H25-18 (17, 242a)

Rogers, C. A. The moments of the number of points of a lattice in a bounded set. Philos. Trans. Roy. Soc. London. Ser. A. **248** (1955), 225–251.

The author improves the Minkowski-Hlawka Theorem ($V_0=2\zeta(n)$) for $n\geqq 6$ to $V_0=2+2/(3[1+633\times 2^{-n}])$: If a bounded symmetric set S, not containing the origin, has Jordan content $V(S)<V_0$, there will exist an n-dimensional unimodular lattice with no point in S. The proof involves defining the lattices Λ generated by $A_t=\omega(\delta_{1t},\cdots,\delta_{(n-1)t},\alpha_t\omega^{-n})$ $(1\leqq t\leqq n;\ \alpha_n=1)$, and then, with ϱ equal to the number of lattice points of Λ in S, defining the moment $\mu_k=\lim_{\omega\to 0}\int_0^1\cdots\int_0^1\varrho^k d\alpha_1\cdots d\alpha_{n-1}$, as suggested by the invariant measure of Siegel [Ann. of Math. (2) **46** (1945), 340–347; MR **6**, 257]. For if S always has two lattice points of Λ_1, the discriminant inequality may be applied to the form $\int_0^1\cdots\int_0^1(\varrho-2)(\xi\varrho+\eta)^2 d\alpha_1\cdots d\alpha_{n-1}$. The proof further involves explicit formulas for μ_k, in which the "principal" term is roughly $V(S)^k$ and the "error" terms are estimated by decomposing Λ into generating sub-lattices. *H. Cohn* (Detroit, Mich.).

Citations: MR 6, 257b = H25-2.

Referred to in H25-25.

H25-19 (17, 715d)

Rogers, C. A. Mean values over the space of lattices. Acta Math. **94** (1955), 249–287.

Let F denote the fundamental region for real $n\times n$ matrices of determinant 1 defined by Minkowski in his theory of the reduction of quadratic forms. Siegel [Ann. of Math. (2) **46** (1945), 340–347; MR **6**, 257] by use of the Lebesgue measure of n^2-space defined a measure $\mu(\Omega)$ for the matrices Ω of F which is normalized so that $\int_F d\mu(\Omega)=1$. For a lattice Λ of determinant 1, let Ω be the uniquely determined matrix of F for which $\Lambda=\Omega\Lambda_0$, Λ_0 being the lattice of all points with integral coordinates. If $\varrho(\Lambda)=\varrho(\Omega\Lambda_0)$ be a function defined for all lattices Λ of determinant 1 which is Borel measurable over F

$$\int_F\varrho(\Omega\Lambda_0)d\mu(\Omega)$$

is defined to be the Siegel mean of $\varrho(\Lambda)$. If $\Lambda(\theta_1,\cdots,\theta_{n-1})$ be the lattice generated by the points $(\omega,\cdots,\theta_1\omega^{-n+1})$, $\cdots$, $(0,\cdots,\omega,\theta_{n-1}\omega^{-n+1})$, $(0,\cdots,0,\omega^{-n+1})$, then

$$\lim_{\omega\to 0}\int_0^1\cdots\int_0^1\varrho[\Lambda(\theta_1,\cdots,\theta_{n-1})]\,d\theta_1\cdots d\theta_{n-1}$$

is designated by $M_\Lambda[\varrho(\Lambda)]$.

The paper consists of a number of results dealing with $M_\Lambda[\varrho(\Lambda)]$. Conditions are given so that $M_\Lambda[\varrho(\Lambda)]=\int_F\varrho(\Omega\Lambda_0)d\mu(\Omega)$. Among the other results are the following. If Y is a point of the mn-space defined by the m n-vectors $X_1,\cdots,X_m$, $0<m\leqq n-1$, and $\varrho(Y)=\varrho(X_1,\cdots,X_m)$ is a function for which $\int\varrho(Y)dY$ exists, let

$$\varrho(\Lambda)=\sum\varrho(X_1,\cdots,X_m),$$

where $X_1,\cdots,X_m$ run through all lattice points of Λ subject to certain auxiliary conditions. It is shown that $M_\Lambda[\varrho(\Lambda)]=K\int\varrho(Y)dY$, where K is determined in terms of the auxiliary conditions.

A combination of the above two results leads to a proof that

$$\int_F\sum\varrho(X_1,X_2)d\mu(\Omega)=\frac{1}{\zeta(n)^2}\int\int\varrho(X_1,X_2)dX_1dX_2$$
$$+\frac{1}{\zeta(n)}\int\varrho(X_1,X_1)dX_1+\frac{1}{\zeta(n)}\int\varrho(X_1,-X_1)dX_1,$$

where the n-vectors X_1, X_2 of the sum run through all the primitive points of the lattice $\Omega\Lambda_0$. The proofs consisting mostly of elementary manipulations of the integrals are too detailed to be described here.

D. Derry (Vancouver, B.C.).

Citations: MR 6, 257b = H25-2.

Referred to in H25-20, H25-26.

H25-20 (19, 1041a)

Schmidt, Wolfgang. On the convergence of mean values over lattices. Canad. J. Math. **10** (1958), 103–110.

The reviewer [Acta Math. **94** (1955), 249–287; MR **17**, 715] recently obtained a very complicated (but nevertheless useful) formula for certain mean values taken over the space of all lattices of determinant 1, but he had to adopt the convention that both sides might take the value $+\infty$. The author now shows that under appropriate conditions, both sides are finite; he also shows how the infinite sums involved can be rearranged in a more convenient form.

C. A. Rogers (Birmingham).

Citations: MR 17, 715d = H25-19.

H25-21 (17, 1188f)

Schmidt, Wolfgang. Über höhere kritische Determinanten von Sternkörpern. Monatsh. Math. **59** (1955), 274–304.

Let S be a star body which is symmetrical in the origin O. If k is a positive integer, let $\Delta^k(S)$ denote the lower bound of the determinants of the lattices Λ that have less than k pairs $\pm X$ of lattice points in S (we assume the existence of one such lattice). The reviewer conjectured [J. London Math. Soc. **26** (1951), 307–310; MR **14**, 624]

$$k\Delta^k(S)\geqq\Delta^1(S),$$

but only proved the conjecture when k is a prime. The author gives a series of theorems, showing that the conjecture holds when k is of various forms, culminating in a theorem (Satz 5) showing that the conjecture holds for all sufficiently large integers k. He also discusses the more general inequality

$$k\Delta^{kl}(S)\geqq\Delta^l(S),$$

showing that this holds when $l=2$ and k is prime, but that it fails for suitable $S=S_{k,l}$ when $l\geqq 3$, $k>1$. However he shows that, if S is bounded, then

$$l\Delta^l(S)\leqq\lim_{k\to\infty}k\Delta^k(S),$$

for each positive integer l, the limit on the right having a finite value. This limit is determined in the case when S is a convex plane domain. The inequality

$$k\Delta^k(S)\leqq\frac{V(S)}{2\zeta(n)}\sum_{\nu=1}^{k}\frac{1}{\nu^n},$$

generalizing the Minkowski-Hlawka theorem, is established.

The reviewer remarks that, using the results of A. E. Western's [ibid. **9** (1934), 276–278] examination of J. W. L. Glaisher's factor tables and D. N. Lehmer's list of prime numbers, it is easy to verify that the author's first four theorems prove the conjecture for all $k\leqq 10^7$.

C. A. Rogers (Birmingham).

Citations: MR 14, 624e = H25-14.

Referred to in H25-31.

H25-22 (18, 21a)

Schmidt, W. Eine neue Abschätzung der kritischen Determinante von Sternkörpern. Monatsh. Math. **60** (1956), 1–10.

Let S be a finite star body in n-dimensional space which is symmetric with respect to the origin. Let $\Delta(S)$ denote the critical determinant of S and $V(S)$ its volume. In the above paper a system of constants $A(n, K)$ is constructed dependent on the first K prime numbers for which $V(S)/\Delta(S) \geq 2\zeta(n)A(n, K)$. For $n>2$ $A(K, n)$ exists which is greater than 1. This shows that the bound $2\zeta(n)$ given by the Minkowski-Hlawka theorem [Hlawka, Math. Z. **49** (1943), 285–312; MR **5**, 201] is not the best possible. An essential link in the derivation of the result is a limit obtained by Rogers by analysis [Ann. of Math. (2) **48** (1947), 994–1002; MR **9**, 270]. Apart from this the proofs depend only on elementary congruence computations. *D. Derry* (Vancouver, B.C.).

Citations: MR 5, 201c = H25-1; MR 9, 270h = H25-7.

Referred to in H25-25.

H25-23 (18, 21b)

Rogers, C. A. The number of lattice points in a set. Proc. London Math. Soc. (3) **6** (1956), 305–320.

Let $\varrho(x_1, x_2, \cdots, x_n)=\varrho(x)$ be a non-negative measurable function and Λ a lattice defined in n-space. A function $\varrho^*(x)$, called the spherical symmetrization of $\varrho(x)$, is defined as the greatest lower bound of the numbers ϱ with the property that the measure of the set of points x' with $\varrho(x')>\varrho$ does not exceed the measure of the set of points y with $|y|\leq|x|$. Let $\mu(\Lambda)$ be the measure defined for the space of lattices with determinant by Siegel [Ann. of Math. (2) **46** (1945), 340–347; MR **6**, 257] and $\varrho(\Lambda)=\sum_{0\neq x\in\Lambda}\varrho(x)$. The author shows that

$$\int\{\varrho(\Lambda)\}^k d\mu(\Lambda)\leq\int\{\varrho^*(\Lambda)\}^k d\mu(\Lambda), \text{ for } k=1, 2, 3,$$

and provided that the set of points x with $\varrho(x)>c$ is convex for each constant c, for $k=1, 2, 3, \cdots$.

If $\varrho(x)$ is specialized to be the characteristic function of a Borel set with measure V, symmetric with respect to the origin, then it is shown that for $n\geq n(k)$,

$$2^k e^{-\frac{1}{2}V}\sum_{r=0}^{\infty}\frac{r^k}{r!}(\tfrac{1}{2}V)^r\leq\int\{\varrho(\Lambda)\}^k d\mu(\Lambda)\leq$$
$$2^k e^{-\frac{1}{2}V}\sum_{r=0}^{\infty}\frac{r^k}{r!}(\tfrac{1}{2}V)^r+f(n, k, V),$$

where $\lim_{n\to\infty} f(n, k, V)=0$. This result is applied to obtain the improvement of the Minkowski-Hlawka theorem that

$$\frac{V(S)}{\Delta(S)}\geq\frac{\sqrt{n}}{3},$$

provided n is sufficiently large. This result, for large n, is stronger than another improvement of the same theorem recently given in the paper reviewed above. *D. Derry* (Vancouver, B.C.).

Citations: MR 6, 257b = H25-2.

Referred to in H25-24.

H25-24 (18, 287a)

Lekkerkerker, C. G. On the Minkowski-Hlawka theorem. Nederl. Akad. Wetensch. Proc. Ser. A. **59**=Indag. Math. **18** (1956), 426–434.

The author shows that if K is a symmetric convex body of volume V in n dimensions then for $n\geq5$ there exists a K-admissible lattice of determinant not exceeding V/c, where $c=4.921\cdots$ is the solution of $c\log c=2(c-1)$. Further, for $n\geq6$ the basis may be chosen to lie in the cube $|x_i|\leq2.13(V/\kappa_n)^{1/n}$, where κ_n is the volume of the unit sphere, after a suitable rotation of the coordinate system. The second result remains valid for $n=2, 3, 4, 5$ provided that 3, 3.82, 4.41, 4.80 respectively are substituted for c. The first result improves one of Davenport and Rogers [Duke Math. J. **14** (1947), 367–375; MR **9**, 11]: the proofs here are an elaboration of theirs and involve some detailed computation. [For results for general starbodies which are sometimes stronger see Rogers, Proc. London Math. Soc. (3) **6** (1956), 305–320; MR **18**, 21.] *J. W. S. Cassels* (Cambridge, Mass.).

Citations: MR 9, 11a = H25-6; MR 18, 21b = H25-23.

H25-25 (18, 382h)

Schmidt, Wolfgang. Eine Verschärfung des Satzes von Minkowski-Hlawka. Monatsh. Math. **60** (1956), 110–113.

Let S be a bounded Jordan-measurable set not containing the origin. When he gave a proof of a similar statement of Minkowski, E. Hlawka [Math. Z. **49** (1943), 285–312; MR **5**, 201] proved that there will be a lattice with determinant 1, having no point in S, provided the volume of S satisfies $V(S)<1$. The condition $V(S)<1$ has been relaxed by the author [Monatsh. Math. **60** (1956), 1–10; MR **18**, 21] and by the reviewer [Philos. Trans. Roy. Soc. London. Ser. A. **248** (1955), 225–251; MR **17**, 242]. Here the author proves the statement under the condition $V(S)<2(1+2^{1-n})^{-1}(1+3^{1-n})^{-1}$. For small values of n this is the best result obtained so far. The proof is simple, but is concise and is best read in conjunction with the paper by the author referred to above. *C. A. Rogers* (Birmingham).

Citations: MR 5, 201c = H25-1; MR 17, 242a = H25-18; MR 18, 21a = H25-22.

H25-26 (20# 1672)

Schmidt, Wolfgang. Mittelwerte über Gitter. Monatsh. Math. **61** (1957), 269–276.

Let $f(X_1, X_2, \cdots, X_k)$ be a function of the k points $X_1, X_2, \cdots, X_k$ in n-dimensional Euclidean space. Let F be a fundamental domain for the space $\mathscr{A}$ of $n\times n$ matrices A of determinant 1 relative to the integral unimodular matrices. C. L. Siegel [Ann. of Math. **46** (1945), 340–347; MR **6**, 257] introduced an invariant volume element $d\omega$ in $\mathscr{A}$ with $\int_F d\omega=1$ and proved the case $k=1$ of the formula $(1\leq k\leq n-1)$

$$\int_F\sum f(Ag_1, Ag_2, \cdots, Ag_k)\,d\omega=$$
$$\int\int\cdots\int f(X_1, X_2, \cdots, X_k)\,dX_1dX_2\cdots dX_k,$$

the sum being taken over all sets of k linearly independent vectors $g_1, g_2, \cdots, g_k$ with integral elements. He also stated the general result without proof. The author now gives a proof of Siegel's assertion and uses it to give a much simpler and more direct proof of an apparently more general result proved recently by the reviewer [Acta Math. **94** (1955), 249–287; MR **17**, 715]. *C. A. Rogers* (Birmingham)

Citations: MR 6, 257b = H25-2; MR 17, 715d = H25-19.

Referred to in H25-27, H25-33.

H25-27 (20# 5769)

Schmidt, Wolfgang. Mittelwerte über Gitter. II. Monatsh. Math. **62** (1958), 250–258.

In the first section the author corrects the proof of theorem 2 of part I [Monatsh. Math. **61** (1957, 269–276; MR **20**#1672]. In the second part he evaluates the integral

$$\int\sum f(Ag_1, \cdots, Ag_n)d\mu(A),$$

where μ is Siegel's measure in the group L of all linear transformations of determinant 1, where the sum is taken over all sets $g_1, \cdots, g_n$ of vectors which form a basis for the lattice of points with integral coordinates, and where the integral is taken over a fundamental region in the group L relative to the sub-group of all integral unimodular matrices. *C. A. Rogers* (Birmingham)

Citations: MR 20# 1672 = H25-26.

Referred to in H25-33.

H25-28 **(20# 3121)**

Schmidt, Wolfgang. The measure of the set of admissible lattices. Proc. Amer. Math. Soc. **9** (1958), 390–403.

Let S be a Borel set in n-dimensional space with the property that there is no point X in S for which $-X$ is also in S. The author's main result shows that, if the volume V of S satisfies $V \leqq \frac{1}{8}n \log(4/3) - \frac{1}{2}\log 3$, and $n \geqq 13$, then the proportion (measured by use of Siegel's invariant measure) of the lattices of determinant 1 which have no point in S lies strictly between 0 and $2e^{-V}$. In particular, this gives an improvement of the Minkowski-Hlawka theorem [E. Hlawka, Math. Z. **49** (1943), 285–312; MR **5**, 201]. The main result is also used to show that, under certain conditions, it is possible to choose a lattice of determinant 1 having no lattice point in such a set S, but having lattice points in other sets $S_1, S_2, \cdots, S_m$. The author's method has since been developed and refined by the reviewer [#3122 below] and the author [not yet published]. *C. A. Rogers* (Birmingham)

Citations: MR 5, 201c = H25-1.

H25-29 **(20# 3122)**

Rogers, C. A. Lattice covering of space: The Minkowski-Hlawka theorem. Proc. London Math. Soc. (3) **8** (1958), 447–465.

The two main results are as follows. (I) Let K be an n-dimensional convex body, where n is sufficiently large. There exists a lattice Λ such that the union of the sets $K+g$ with $g \in \Lambda$ covers the whole of space, the density $\vartheta(K, \Lambda)$ of the covering satisfying

$$\vartheta(K, \Lambda) \leqq \text{const. } n^3\left(1+\sqrt[4]{\frac{16}{27}}\right)^n < (1.8774)^n.$$

The previous best result, due to the author, was the upper bound 2^n. (II) Let S be an n-dimensional Borel set of finite positive measure V such that for no point x do both x and $-x$ belong to S. If n is sufficiently large and if

$$V \leqq \tfrac{1}{4}n \log \tfrac{4}{3} - \log n - \text{const.},$$

then there exists a lattice of determinant 1 containing no point of S. This is much stronger than Hlawka's original theorem [Math. Z. **49** (1943), 285–312; MR **5**, 201] and is a slight improvement on a recent result by W. Schmidt [#3121 above]. *K. Mahler* (Manchester)

Citations: MR 5, 201c = H25-1.

H25-30 **(20# 3852)**

Macbeath, A. M.; and Rogers, C. A. A modified form of Siegel's mean value theorem. II. Proc. Cambridge Philos. Soc. **54** (1958), 322–326.

Λ is defined to be a discrete point set in real Euclidean n-space for which $\lim_{r\to\infty} N(r)/r = d$, where $N(r)$ is the number of points of Λ in a sphere of radius r with centre at the origin. If γ is an $n \times n$ matrix with determinant 1, $\|\gamma\|$ is defined to be $\sup_{|x|\leq 1} |\gamma x|$, x being a point in Euclidean n-space. Where $\rho(x)$ is a function which vanishes outside a bounded region, $\rho(\gamma\Lambda)$ is defined to be $\sum \rho(x)$, the sum to be taken over all x in $\gamma\Lambda$. In part I of this paper [same Proc. **51** (1955), 565–576; MR **17**, 241] the authors defined a measure $\mu(\gamma)$ over sets of matrices γ and proved that if $\int \rho(x)dx$ exists as a Riemann integral, then

$$\int_{\|\gamma\|\leq K} \rho(\gamma\Lambda)d\mu(\gamma) \Big/ \int d\mu(\gamma) \to d\int \rho(x)dx,$$

as $K\to\infty$, where $\int \rho(x)dx$ is defined over the whole space. The present paper uses this result to establish the same limit if $\int \rho(x)dx$ is integrable in the Lebesgue sense.

D. Derry (Vancouver, B.C.)

Citations: MR 17, 241d = H25-17.

H25-31 **(20# 5803)**

Cassels, J. W. S. On the subgroups of infinite Abelian groups. J. London Math. Soc. **33** (1958), 281–284.

Let A be an additively written torsion-free abelian group; let $a_1, \cdots, a_s$ be elements none of which can be written in the form nb (n is an integer >1, $b \in A$); let $j_1, \cdots, j_s$ be positive integers and put $J = j_1 + \cdots + j_s$. Then there exists a subgroup B of A of index

(1) $$[A:B] \leqq J+1$$

which contains none of the J elements $i_\sigma a_\sigma$ ($1 \leqq i_\sigma \leqq j_\sigma$; $\sigma = 1, \cdots, s$). There can be strict inequality in (1) except when $s=1$. In the special case when A is a free abelian group of finite rank, this theorem yields a conjecture of C. A. Rogers [J. London Math. Soc. **26** (1951), 307–310; MR **14**, 624] which was partially proved earlier by Rogers and by W. Schmidt [Monatsh. Math. **59** (1955), 274–304; MR **17**, 1188]. *A. Kertész* (Debrecen)

Citations: MR 14, 624e = H25-14; MR 17, 1188f = H25-21.

Referred to in H15-83.

H25-32 **(21# 1966)**

Macbeath, A. M.; and Rogers, C. A. Siegel's mean value theorem in the geometry of numbers. Proc. Cambridge Philos. Soc. **54** (1958), 139–151.

Let T_n be the set of real $n \times n$ matrices with unit determinant, G_n the unimodular group in T_n, F_n a fundamental domain for T_n/G_n and $\bar{\mu}$ the Siegel measure in T_n. Further let LI be the set of $n \times l$ matrices with integer elements and linearly independent columns, $1 \leqq l \leqq n-1$, and μ the Lebesgue measure in X, the set of real $n \times l$ matrices. Siegel [Ann. of Math. (2) **46** (1945), 340–347; MR **6**, 257] stated without proof that for any Lebesgue integrable function defined on X,

(1) $$\int_{F_n} \sum_{q \in \text{LI}} f(tq)d\bar{\mu}(t) = \bar{\mu}(F_n) \int_X f(x)d\mu(x).$$

The authors give a proof of this result. In case $l=1$, (1) becomes the mean value theorem for lattices proved by Siegel [loc. cit.]. The authors state that their main object is a simplified version of this proof. Under the inductive assumption that $\bar{\mu}(F_m)$ is finite for $m<n$ they arrive at a mean value formula for primitive matrices. Applying elementary geometry and Lebesgue's convergence theorem to this formula they deduce the identity

$$\bar{\mu}(F_n) = \left(\frac{n-1}{n}\right) \zeta(n)\bar{\mu}(F_{n-1}).$$

The formula for primitive matrices is thus induced and the proof of (1) is completed in a simple way. With this approach and a separate proof of the existence of a fundamental domain the authors point out that Minkowski's reduction theory is unnecessary to the argument. For another approach see Weil [Summa Brasil. Math. **1** (1946), 21–39; MR **7**, 411]. *A. C. Woods* (New Orleans, La.)

Citations: MR 6, 257b = H25-2; MR 7, 411g = H25-4.

H25-33 **(22# 5627)**

Schmidt, Wolfgang. Masstheorie in der Geometrie der Zahlen. Acta Math. **102** (1959), 159–224.

Let S be a Borel set in R^n, $\Delta(S)$ its critical determinant, $V(S)$ its volume. By integrating with respect to the measure induced in the space of lattices by the Haar measure, Siegel [Ann. of Math. (2) **46** (1945), 340–347; MR **6**, 257] proved a mean value theorem about the average number of lattice points in S, from which Hlawka's theorem that $V(S)/\Delta(S) \geqq 1$ follows immediately as a corollary.

A natural method of proceeding further would seem to be to obtain estimates for the number of k-tuples of lattice points in K. For $k \leqq n-1$, methods similar to Siegel's, using integration in the same space, were used by Rogers and the author, from which considerable improvements on Hlawka's result were derived for large n.

Then [Monatsh. Math. **61** (1957), 269–276; **62** (1958), 250–258; MR **20** #1672, 5769] the author obtained a mean value formula for the n-tuples. In this case the integral involves not only the Haar measure, but also integration with respect to some extra parameters. In the present paper, this mean value theorem is generalized to k-tuples, where $k \geqq n$. With the aid of the new formula, an expression is found for the Haar measure of the set of admissible lattices. Considerable difficulties of convergence arise in the proof, since the obvious term-by-term integration leads to a divergent series (as the author actually shows for $n=2$). These convergence difficulties are overcome rather ingeniously by ordering the points in R^n. A careful study is also needed of certain sums involving the abelian factor groups modulo the sublattices generated by the k-tuples of lattice points, using the "Möbius function" of Delsarte.

In the latter part of the paper, these results are used to deduce a still further improvement of the Minkowski-Hlawka theorem: $V(S)/\Delta(S)$ is not less than $nr-2$, where $r=0.278\cdots$, for sufficiently large n. Elegant applications in 2-space include an exact value, in the form of a finite series, for the measure of the set $A(S)$ of S-admissible lattices in the case when S is an annulus. A corollary is deduced about the numbers represented by a positive binary quadratic form.

Applications are also made to the problem of lattice coverings of space, using ideas of Rogers, though the author has added a note in proof to the effect that his results in this direction have been superseded by a recent paper of Rogers himself.

Two other results worth mentioning are: (i) If S has infinite volume, then almost all lattices have infinitely many points in S; (ii) The measure of the set $A(S)$ is a continuous function of S in the topology defined by the measure of the symmetric difference of subsets of R^n.

A. M. Macbeath (Dundee)

Citations: MR 6, 257b = H25-2; MR 20# 1672 = H25-26; MR 20# 5769 = H25-27.

Referred to in H30-86.

H25-34 **(25# 3014)**

Rogers, C. A.

The chance that a point should be near the wrong lattice point.

J. London Math. Soc. **37** (1962), 161–163.

Let Λ be a lattice of determinant Δ in n-dimensional Euclidean space, and suppose that the probability that a point $\mathbf{x}=(x_1, \cdots, x_n)$ lies in the volume element $d\mathbf{x}= dx_1 \cdots dx_n$ is given by $p(\mathbf{x})d\mathbf{x}$. If $p(\Lambda)$ is the probability that $\mathbf{x}$ is nearer to some other point of Λ than the origin, an estimate is found for the mean value of $p(\Lambda)$, averaged over the space of all lattices with determinant Δ. In particular, if $p(\Lambda)$ is the characteristic function of a sphere with centre at the origin and volume V, it is shown that $p(\Lambda) \leqq V/2\Delta\zeta(n)$, $\zeta(s)$ denoting Riemann's zeta-function.

A. M. Macbeath (Birmingham)

H25-35 **(26# 2400)**

von Wolff, M. R.

A star domain with densest admissible point set not a lattice.

Acta Math. **108** (1962), 53–60.

Let $\mathfrak{S}$ be a symmetric $(-\mathfrak{S}=\mathfrak{S})$ star domain in the real affine plane. A point set $\mathfrak{P}$ is said to be "admissible for $\mathfrak{S}$" (in the sense of the Minkowski-Hlawka theory) if P_1, $P_2 \in \mathfrak{P}$ and $P_1 \neq P_2$ imply $P_2 \notin \operatorname{int}(P_1+\mathfrak{S})$. If $\mathfrak{S}$ is convex, it is known [cf. L. Fejes Tóth, Acta Sci. Math. (Szeged) **12** (1950), 62–67; MR **12**, 352; or C. A. Rogers, Acta Math. **86** (1951), 309–321; MR **13**, 768] that among the densest admissible point sets there is always at least one lattice. The author constructs a symmetric star domain (with 14 sides) containing a rectangle $\mathfrak{R}$, a lattice $\mathfrak{C}_1$, and a point L' such that $\mathfrak{P}:=\mathfrak{C}_1 \cup (L'+\mathfrak{C}_1)$ is not a lattice but admissible for $\mathfrak{S}$. Now, if $D(\mathfrak{M})$ denotes the density of a point set $\mathfrak{M}$, and if $\mathfrak{L}$ is a densest lattice admissible for $\mathfrak{R}$, it is easily checked that $D(\mathfrak{L})=D(\mathfrak{P})=2$ and that no such $\mathfrak{L}$ is admissible for $\mathfrak{S}$. So the assertion of the title is proven. Finally (by aid of an IBM 610 computer) the critical lattice $\mathfrak{L}_1$ for $\mathfrak{S}$ is determined; it is $D(\mathfrak{L}_1)=3024/1583$.

L. W. Danzer (Göttingen)

Citations: MR 13, 768i = H30-17.

Referred to in H15-78.

H25-36 **(26# 3675)**

Schmidt, Wolfgang M.

On the Minkowski-Hlawka theorem.

Illinois J. Math. **7** (1963), 18–23.

If $V(S)$ is the volume and $\Delta(S)$ the critical determinant of a bounded Borel set in R_n, the author makes a further contribution to the problem of determining the greatest lower bound of the quotient $Q(S)=V(S)/\Delta(S)$, known to exceed unity by the Minkowski-Hlawka Theorem. Improved estimates have been found for $n \geqq 3$ by Rogers and the author.

It is shown here that for $n=2$, $Q(S) \geqq 16/15$. This is a considerable improvement on the known result $Q(S)>1$. The proof makes elegant use of the sublattices of index 2 and 3.

A similar use of sublattices of index 2 is combined with the known methods in a study of dimension $n \geqq 3$. Unfortunately, the detailed result is not established rigorously, since, on page 20, line 15, the statement is made: "There will be a minimal dependent set of at least three lattice points, since every minimal dependent set mod 2 of two points consists of two identical points mod 2." There would seem to be no objection to a minimal dependent set mod 2 which contains only one lattice point in 2Λ. The set of lattices with such a point would make a contribution $2^{-n}V$ to the integral on page 20. As this is a rather small contribution, it is probable that the final result

$$Q(S) \geqq n \log 2^{1/2} - c_1 \text{ for } n \geqq c_2$$

can still be established by a modification of the argument, but the paper as it stands is not accurate.

It should be mentioned that the calculations involved are much simpler than those in previous work on this question.

A. M. Macbeath (Birmingham)

Referred to in H25-37.

H25-37 **(27# 4772)**

Schmidt, Wolfgang M.

Correction to my paper, "On the Minkowski-Hlawka theorem".

Illinois J. Math. **7** (1963), 714.

A minor correction to formula (3) in the original paper [same J. **7** (1963), 18–23; MR **26** #3675]; the original result is still valid.

Citations: MR 26# 3675 = H25-36.

H25-38 **(26# 4973)**

Bateman, Paul T.

The Minkowski-Hlawka theorem in the geometry of numbers.

Arch. Math. **13** (1962), 357–362.

A new proof is given of Hlawka's theorem. It is based on the idea of Rogers—averaging over a certain $(n+1)$-parameter group of linear transformations (matrices which have zeros except in the diagonal and the final column). It improves on a similar proof given by Cassels, through the introduction of a smoothing process which makes the result applicable to Lebesgue measurable functions rather than merely Jordan-integrable.

A. M. Macbeath (Birmingham)

H25-39 **(33# 1288)**

Schwald, Andreas

Über das asymptotische Verhalten der Anzahl von k-tupeln linear unabhängiger Gitterpunkte.

Österreich. Akad. Wiss. Math.-Natur. Kl. S.-B. II **173** (1965), 195–249.

Let S be a Borel measurable set in R_n $(n>1)$ with volume $V(S)<\infty$. For a fixed k, $1\leqq k<n$, let $L(S)$, $P(S)$ and $B(S)$ denote the number of k-tuples in S, of linearly independent points, of linearly independent primitive points, and points which can be completed to a basis of the fundamental lattice Λ, respectively. In order to investigate the asymptotic behaviour of these functions as $V(S)\to\infty$, the author introduces the discrepancy functions

$$D(S) = |k!L(S)V(S)^{-k}-1|,$$
$$E(S) = |k!P(S)\zeta(n)^k\cdot V(S)^{-k}-1|,$$
$$F(S) = |k!B(S)\prod_{j=0}^{k-1}\zeta(n-j)V(S)^{-k}-1|.$$

The main theorems proved are as follows. (1) Let $\psi(s)$ $(s\geqq 0)$ be a positive nondecreasing function such that $\int_0^\infty \psi^{-1}(s)\,ds$ exists. Let $n\geqq 3$, $1<k<n/2$; then, for almost all linear transformations A,

$$D(AS) = O[V^{-1/2}\log^{1/2} V\,\psi^{1/2}(\log V)],$$
$$E(AS) = O[V^{-1/2}\log^{1/2} V\,\psi^{1/2}(\log V)],$$
$$F(AS) = O[V^{-1/2}\log^{1/2} V\,\psi^{1/2}(\log V)],$$

for all S in a totally ordered (by inclusion) family Φ such that $V(S)\to\infty$ as S ranges over Φ. (2) Let $n>2$, $1\leqq k\leqq n-1$, $S\in\Phi$; then, for almost all A,

$$D(AS) = O[V^{-1/2}\log V\,\psi^{1/2}(\log V)],$$
$$E(AS) = O[V^{-1/2}\log V\,\psi^{1/2}(\log V)].$$

Theorem 2 is a generalization of Theorem 1 of Schmidt [Trans. Amer. Math. Soc. **95** (1960), 516–529; MR **22** #8004]. The proof uses integrals of the sums of the form

$$\left(\sum[g_1,\cdots,g_k\in\Lambda \text{ lin indep}]\rho(Ag_1)\cdots\rho(Ag_k)-\frac{V^k}{\|A\|^k}\right)^2$$

over suitable fundamental regions of the space of matrices. Here, $\rho(X)$ is the characteristic function of S.

$$(3)\quad \int_0^1 \nu^n\left\{\int_F \sum[g_1,\cdots,g_n\in\Lambda \text{ lin indep}]\,\rho(\nu^{1/n}Ag_1)\cdots\rho(\nu^{1/n}Ag_n)\,d\mu(A)\right\}d\nu = V^n+O(V^{n-1}\log^2 V),$$

$$\int_0^1 \nu^n\left\{\int_F \sum[g_1,\cdots,g_n\in\Lambda \text{ lin indep, primitive}]\,\rho(\nu^{1/n}Ag_1)\cdots\rho(\nu^{1/n}Ag_n)\,d\mu(A)\right\}d\nu = V^n\zeta(n)^{-n}+O(V^{n-1}\log^4 V).$$

The above is suggested as a probable replacement for the non-existing generalization of Siegel's mean value theorem for $k=n$ in order to prove Theorem 1 for $k=n/2$, Theorem 2 for $k=n$ and a possible generalization of Theorem 2 of Schmidt.

V. C. Dumir (Columbus, Ohio)

Citations: MR 22# 8004 = H15-69.

H25-40 **(34# 5763)**

Schmidt, Wolfgang M.

Masstheorie in der Geometrie der Zahlen. (French summary. With discussion)

Les Tendances Géom. en Algèbre et Théorie des Nombres, pp. 225–229. *Éditions du Centre National de la Recherche Scientifique, Paris,* 1966.

In this expository article the author reports results obtained in the geometry of numbers by himself, Rogers and others.

H. J. Godwin (Swansea)

H25-41 **(37# 1320)**

Sawyer, D. B.

Lattice points in rotated star sets.

J. London Math. Soc. **43** (1968), 131–142.

The Minkowski-Hlawka theorem can be stated in the following form. Let U be a bounded symmetric star body in E^n with volume $V(U)<2\zeta(n)$, then there exists an element g of the group of linear transformations of E^n of determinant 1, such that Ug contains no point of the integer lattice other than O. In the paper under review U is a star set, measurable but not necessarily bounded or symmetric, g is restricted to the group of proper rotations, and $2\zeta(n)$ is replaced by J_n/c_n, where J_n is $\pi^{n/2}/\Gamma(1+\frac{1}{2}n)$ (the volume of the unit ball in E^n): the conclusion of the theorem is then the same. If $C_n=\inf c_n$ then $C_n\leqq 2n$ $(n=2,\ n\geqq 4)$ and $C_3=9/\sqrt{2}=6.36\cdots$; lower bounds are given for the C_n but $C_2=4$ is the only result known to be best-possible. A corollary on successive minima is obtained, and also some sharpening of the results for the cases when U is a cylinder, or U is convex and symmetrical in O.

H. J. Godwin (Swansea)

H30 LATTICE PACKING AND COVERING

See also reviews E12-140, H02-10, H02-17, H05-13, H05-57, H10-26, H10-30, H25-29, H25-33, H99-3, J44-9.

H30-1 **(2, 153e)**

Perron, Oskar. Über lückenlose Ausfüllung des n-dimensionalen Raumes durch kongruente Würfel. Math. Z. 46, 1–26 (1940).

Bei seiner Untersuchung der Minkowskischen Vermutung über den Grenzfall des Minkowskischen Satzes über homogene Linearformen, hat O. H. Keller diese Vermutung verallgemeinert zu einer Aussage über das Auftreten gewisser Würfelpaare bei lückenloser Ausfüllung des R_n durch kongruente Würfel [J. Reine Angew. Math. **163**, 231–248 (1930)]. Die Kellersche Vermutung wird vom Verfasser eingehend untersucht. Gezeigt wird, dass für jedes feste n die Entscheidung über Richtigkeit oder Falschheit der Vermutung durch endlich viele Versuche erzwungen werden kann. *J. F. Koksma* (Amsterdam).

Referred to in H30-4.

H30-2 (2, 153f)

Perron, Oskar. Modulartige lückenlose Ausfüllung des R_n mit kongruenten Würfeln. I. Math. Ann. **117**, 415–447 (1940).

A cube with center $(a^1, \cdots, a^n)$ in Euclidean n-space R_n is the set of points $(x^1, \cdots, x^n)$ with $-\frac{1}{2} \leqq x^\nu - a^\nu < \frac{1}{2}$ $(\nu = 1, \cdots, n)$. An M-covering of R_n is a set of cubes such that each point of R_n lies in one and only one of them, and such that the vectors from the origin to the centers of the cubes form a modul. Minkowski conjectured that every M-covering contains n cubes with the respective centers $(a_i^1, \cdots, a_i^n)$ such that for conveniently numbered coordinates $a_i^i = 1$ and $a_i^k = 0$ if $i > k$ $(i, k = 1, 2, \cdots, n)$. According to a theorem of the author every M-covering contains n cubes with the respective centers $(\alpha_i^1, \cdots, \alpha_i^n)$ such that $\alpha_i^i = 1$, $0 \leqq \alpha_i^k < 1$ for $i \neq k$ $(i, k = 1, \cdots, n)$ [Math. Z. **46**, 1–26 (1940), theorem 5; cf. the preceding review]. In this paper he shows that Minkowski's hypothesis is equivalent to the conjecture that all the products

$$\alpha_{\lambda_1}^{\lambda_2} \alpha_{\lambda_2}^{\lambda_3} \cdots \alpha_{\lambda_{r-1}}^{\lambda_r} \alpha_{\lambda_r}^{\lambda_1}$$

vanish $(r = 2, \cdots, n;\ \lambda_i \neq \lambda_k$ for $i \neq k;\ i, k = 1, \cdots, r)$. He proves that conjecture for $n \leqq 8$. His main tool is his theorem that, if the center of a cube of an M-covering is not the origin, at least one of its coordinates is an integer not equal to 0. *P. Scherk* (New Haven, Conn.).

Referred to in H30-3, H30-5.

H30-3 (3, 253a)

Perron, Oskar. Modulartige lückenlose Ausfüllung des R_n mit kongruenten Würfeln. II. Math. Ann. **117**, 609–658 (1941).

In part I [Math. Ann. **117**, 415–447 (1940); these Rev. **2**, 153] the author established Minkowski's conjecture for Euclidean space of $n \leqq 8$ dimensions. The present part consists of the extension to nine dimensions. No fresh principles are involved; but the multitude of cases and sub-cases renders the work formidable indeed, and seems to indicate that the extension to still higher values of n would not be humanly feasible by this method. *H. S. M. Coxeter.*

Citations: MR 2, 153f = H30-2.

Referred to in H30-5.

H30-4 (3, 302b)

Hajós, Georg. Über einfache und mehrfache Bedeckung des n-dimensionalen Raumes mit einem Würfelgitter. Math. Z. **47**, 427–467 (1941).

In this paper Minkowski's conjecture on the boundary case of his theorem on homogeneous linear forms is proved. We define a cube with center $(a_1, \cdots, a_n)$ in Euclidean n-space R_n as the set of points $(x_1, \cdots, x_n)$ with $-\frac{1}{2} \leqq x_\nu - a_\nu < \frac{1}{2}$ $(\nu = 1, \cdots, n)$, and we define a k-covering of R_n as a set of cubes such that each point of R_n belongs to precisely k of them and such that their centers form a lattice L. Then Minkowski's conjecture can be formulated geometrically as the special case $k = 1$ of the following hypothesis: The boundaries of two suitable cubes of a k-covering of R_n have an $(n-1)$-dimensional cube in common.

The author first proves by a method similar to that of Perron's proof of a theorem of Keller [Math. Z. **46**, 1–26 (1940), in particular, pp. 6–12; these Rev. **2**, 153] that we can assume the coordinates of all the lattice points to be rational $(k \geqq 1)$. Then L can be considered as a sublattice of another one L' which is generated by certain points $(1/m_1, 0, \cdots, 0), (0, 1/m_2, 0, \cdots, 0), \cdots, (0, 0, \cdots, 0, 1/m_n)$, where $m_1, m_2, \cdots, m_n$ are rational integers [of course, L' is not uniquely determined]. The group of L' (mod L) is a finite Abelian group G. If L belongs to a 1-covering, there are n elements $A_1, \cdots, A_n$ of G and n positive integers $a_1, \cdots, a_n$ such that each element A of G can be written in one and only one way in the form $A = \alpha_1 A_1 + \cdots + \alpha_n A_n$ $(0 \leqq \alpha_\nu < a_\nu;\ \nu = 1, \cdots, n)$. Minkowski's conjecture is equivalent to the (purely group theoretical) contention that $a_\kappa A_\kappa = 1$ for a suitable κ. Replacing n by a larger number, this theorem can readily be reduced to the case that all the a's are prime numbers. The remainder of its proof is complicated and (19 pages) long. The more general conjecture is proved to be true for $n \leqq 3$ and all k, but false for $n \geqq 4$ and suitable $k > 1$. *P. Scherk.*

Citations: MR 2, 153e = H30-1.

Referred to in H20-8, H30-22, H30-44, J16-8, J16-9.

H30-5 (6, 16d)

Hofreiter, N. Gitterförmige lückenlose Ausfüllung des R_n mit kongruenten Würfeln. Monatsh. Math. Phys. **50**, 48–64 (1941).

This is an attempt to simplify the work of Perron on Minkowski's conjecture [Math. Ann. **117**, 415–447, 609–658 (1940–1941); these Rev. **2**, 153; **3**, 253]. The conjecture is again established for Euclidean space of not more than nine dimensions. Some of Perron's lemmas are used, along with the following. Given $2n$ numbers $\alpha_1, \cdots, \alpha_n; \beta_1, \cdots, \beta_n$ $(0 \leqq \alpha_i < 1;\ 0 \leqq \beta_i < 1)$, we can choose $\epsilon_i = \pm 1$ $(i = 1, \cdots, n)$ so that $|\sum \epsilon_i \alpha_i| < 1$, $|\sum \epsilon_i \beta_i| < 1$. *H. S. M. Coxeter.*

Citations: MR 2, 153f = H30-2; MR 3, 253a = H30-3.

H30-6 (6, 259c)

Misra, Rama Dhar. On lattice sums for closest packing crystals. Proc. Benares Math. Soc. (N.S.) **4**, 109–112 (1943).

Let ν_q denote the number of ways of expressing an even number q as the sum of three squares (or as the sum of squares of four numbers whose sum is zero); for example, $\nu_4 = 6$ from the permutations of $(\pm 2)^2 + 0^2 + 0^2$ or of $1^2 + 1^2 + (-1)^2 + (-1)^2$, but $\nu_{28} = 0$. The author tabulates ν_q for $q \leqq 50$ in order to investigate the sums

$$S_n = 2^{-\frac{1}{2}n}\nu_2 + 4^{-\frac{1}{2}n}\nu_4 + 6^{-\frac{1}{2}n}\nu_6 + \cdots,$$

which he tabulates for $n = 4, 5, \cdots, 15$. He remarks that, for large values of n, the first term alone (namely, $12/2^{\frac{1}{2}n}$) provides a good approximation to S_n.

Geometrically, ν_q is the number of spheres distant $(2q)^{\frac{1}{2}}$ radii from a given sphere (center to center) in any closest packing. [In the first paragraph on page 110, the descriptions of the cubic and hexagonal packings have unhappily been interchanged [cf. W. W. R. Ball, Mathematical Recreations and Essays, 11th ed., Macmillan, New York, 1939, p. 149].] *H. S. M. Coxeter* (Toronto, Ont.).

H30-7 (8, 169g)

Fejes, László. Eine Bemerkung über die Bedeckung der Ebene durch Eibereiche mit Mittelpunkt. Acta Univ. Szeged. Sect. Sci. Math. **11**, 93–95 (1946).

The author proves the following theorem. Let B be a convex region with a center, which is not an ellipse. Then there exists a covering of the plane by regions congruent to B whose centers form a lattice and whose density is less

than $2\sqrt{3}\pi/9$. In case B is an ellipse the density is at least $2\sqrt{3}\pi/9$. Several similar problems are investigated.
P. Erdös (Syracuse, N. Y.).

H30-8 **(9, 226d)**

Rankin, R. A. On the closest packing of spheres in n dimensions. Ann. of Math. (2) **48**, 1062–1081 (1947).

The purpose of this paper is to extend a result of Blichfeldt's [Math. Ann. **101**, 605–608 (1929)]. Let C_n be an n-dimensional hypercube of edge L in n-dimensional Euclidean space; let $N(L)$ be the maximum number of hyperspheres of unit radius and content K_n which can be placed in C_n without overlapping each other or the sides of C_n. Then the packing constant ρ_n is defined to be $\rho_n = \lim_{L\to\infty} K_n N(L)/L^n$. The packing of the hyperspheres in C_n need not be regular, i.e., their centers need not form a lattice. The regular packing constant $\rho_n' = \lim_{L\to\infty} K_n N'(L)/L^n$, where $N'(L)$ is the maximum number of unit hyperspheres which can be placed in C_n without overlapping in any regular packing.

Proofs are given that both ρ_n and ρ_n' exist as unique limits; and that ρ_n' is related to the number γ_n which occurs in the theory of quadratic forms. A series of lemmas is developed and used to find expressions leading to the evaluation of bounds to the packing constant. The results are: $\rho_2 \leqq 0.92998 \cdots$, $\rho_3 \leqq 0.82711 \cdots$ (which is lower than Blichfeldt's two estimates) and $\rho_4 \leqq 0.71197 \cdots$. For $n=2$ the closest packing is known to be regular hexagonal, so that $\rho_2 = \rho_2' = \pi/\sqrt{12} = 0.90689968 \cdots$. When $n>2$ it is not known whether the closest packing is regular, but it is shewn that $\rho_3 \geqq \rho_3' = \pi/\sqrt{18} = 0.74048 \cdots$, and $\rho_4 \geqq \rho_4' = \pi^2/16 = 0.61685 \cdots$.
S. Melmore (York).

Referred to in H10-11, H30-19, H30-27, H30-59, N52-10.

H30-9 **(9, 226e)**

Durrieu, Mauricio. A geometrical demonstration of the amount of empty space in a collection of spheres of equal diameter, arranged regularly, tangent in rows and layers in quincunxes. Ciencia y Tecnica **109**, 351–355 (1947). (Spanish. French summary)

H30-10 **(9, 501a)**

Hlawka, Edmund. Über einen Satz aus der Geometrie der Zahlen. Akad. Wiss. Wien, S.-B. IIa. **155**, 75–82 (1947).

The theorem in question is of an "alternative" type, and generalizes results of Mordell and of the author [Math. Z. **49**, 285–312 (1942); these Rev. **5**, 201]. Let M_1, M_2 be bounded, closed sets of points in n-dimensional space with Jordan contents V_1, V_2. Let $k_0, k_1, \cdots, k_n$ be positive integers and suppose that $V_1+V_2>k_0k_1\cdots k_n$. Let M_1' be the set of points $((x_1-y_1)/k_1, \cdots, (x_n-y_n)/k_n)$, where $x=(x_1, \cdots, x_n)$ and $y=(y_1, \cdots, y_n)$ are points of M_1; and similarly for M_2'. Let M_3 be the set of all points given by the same formula where x now belongs to M_1 and y to M_2 or vice versa. Suppose there are at most k_0-r_i-1 points, other than 0, with integral coordinates in M_i', where $i=1$ or 2, and where $r_i \geqq 0$ and $k_0-r_i>V_i/k_1\cdots k_n$. Then the theorem asserts that the set M_3, after all integral translations, provides a covering of space in which every point is covered at least $r+1$ times, where $r=[\frac{1}{2}(r_1V_1+r_2V_2)(V_1V_2)^{-\frac{1}{2}}]$. The proof is elementary and arithmetical. *H. Davenport.*

Citations: MR 5, 201c = H25-1.

H30-11 **(10, 60a)**

Fejes Tóth, László. On the densest packing of convex domains. Nederl. Akad. Wetensch., Proc. **51**, 544–547 = Indagationes Math. **10**, 188–192 (1948).

Let C be a plane convex domain symmetric with respect to the origin. The plane is covered by nonoverlapping domains each of which is a displacement of C whose centers form a lattice. Let d denote the ratio of the area so covered to the total area of the plane. The paper gives an elementary proof that for a given C the above lattice can always be chosen so that $d>\frac{1}{2}\sqrt{3}$. *D. Derry.*

H30-12 **(11, 12b)**

Hlawka, Edmund. Ausfüllung und Überdeckung konvexer Körper durch konvexe Körper. Monatsh. Math. **53**, 81–131 (1949).

Let $K=K(r, f)$ be a convex body in n-dimensional space, with the origin 0 as centre, defined by an inequality $f(x)=f(x_1, \cdots, x_n)\leqq r$. Let

$$M=\sup_{\Gamma}\inf_{0\neq x\in\Gamma} f(x), \quad \epsilon=\inf_{\Gamma}\sup_{z}\inf_{x\in\Gamma} f(x+z),$$

where Γ is an arbitrary lattice with determinant 1 and x and z are arbitrary points. The author proves that

$$1\leqq[1-\tfrac{1}{2}(1-\tfrac{1}{2}J^{1/n}M)^n]^{-1}\leqq\epsilon^nJ\leqq(\tfrac{1}{2}n)^nM^nJ\leqq n^n,$$

where J is the volume of $K(1, f)$.

A set of bodies $K_i: f(x-p_i)\leqq r$, $i=1, \cdots, a$, is said to form a packing of a bodies K into a convex body B, if each K_i is contained in B and no two K_i have common inner points. A similar set K_i, $i=1, \cdots, u$, is said to form a covering of B by u bodies K if each K_i contains an inner point of B and every point of B is contained in some K_i. A packing of K into B is said to be a lattice packing, if, for some lattice Γ and some point z, (i) $p_i=g_i+z$ for some g_i of Γ, and (ii) if $p+z\epsilon\Gamma$ and the body $f(x-p)\leqq r$ is contained in B, then p is one of the p_i. [The author omits the essential condition (ii); his definitions 5, 13 and 24 need appropriate modifications, and so do the proofs of theorems 5 and 26.] A lattice covering is defined similarly. Let $a(B, K)$ and $u(B, K)$ be the upper and lower bounds of the number of bodies K in the packings and coverings of B, respectively. Let $a^*(B, K)$ and $u^*(B, K)$ be the corresponding bounds for the lattice packings and lattice coverings.

The author obtains both upper and lower bounds for both $a^*(B, K)$ and $u^*(B, K)$, which ensure that, if V is the volume of B, the limits

$$\Delta^*=\lim_{r\to 0} a^*(B, K)Jr^n/V, \quad \vartheta^*=\lim_{r\to 0} u^*(B, K)Jr^n/V$$

exist and have the values $(\frac{1}{2}M)^nJ$ and ϵ^nJ. Further it is shown that

$$\Delta=\lim_{r\to 0} a(B, K)Jr^n/V, \quad \vartheta=\lim_{r\to 0} u(B, K)Jr^n/V$$

exist and are independent of B.

It is clear that $\Delta^*\leqq\Delta\leqq 1\leqq\vartheta\leqq\vartheta^*$. Improved upper bounds for Δ and lower bounds for ϑ are found for certain classes of bodies K similar to those discussed by J. G. van der Corput and H. Davenport [Nederl. Akad. Wetensch., Proc. **49**, 701–707 = Indagationes Math. **8**, 409–415 (1946); these Rev. **8**, 317]. Other topics discussed are: upper bounds for Δ for spheres and cylinders; packings and coverings with convex bodies which do not have a centre; distributions of points $p_1, \cdots, p_s$ in B with $f(p_i-p_j)\geqq d$, if $i\neq j$; packings in which at most $k-1$ bodies may overlap; coverings in which each point of B belongs to at least k bodies; packings and coverings with bodies, which are not all the same size, or which are selected from a set of bodies, not all the same shape; selection from coverings of systems of nonoverlapping bodies. *C. A. Rogers* (Princeton, N. J.).

Citations: MR 8, 317i = H05-8.

Referred to in H30-15, H30-54, H30-59.

H30-13 **(11, 12c)**

Hlawka, Edmund. Ausfüllung und Überdeckung durch Zylinder. Anz. Öster. Akad. Wiss. Wien. Math.-Nat. Kl. **85**, 116–119 (1948).

Let $f(x_1, \cdots, x_n) = \max\{f_1(x_1, \cdots, x_m), f_2(x_{m+1}, \cdots, x_n)\}$, where f, f_1 and f_2 are the distance functions of symmetrical convex bodies K, K_1 and K_2 in n, m and $n-m$ dimensional spaces. Let Δ and ϑ, Δ_1 and ϑ_1, and Δ_2 and ϑ_2 be the densities defined, as in the preceding review, for the bodies K, K_1 and K_2. The author proves that

$$\Delta_1\Delta_2 \leqq \Delta \leqq \min(\Delta_1, \Delta_2),\ \vartheta_1\vartheta_2 \geqq \vartheta \geqq \max(\vartheta_1, \vartheta_2),$$

and discusses some simple consequences of these results.

C. A. Rogers (Princeton, N. J.).

H30-14 **(11, 12d)**

Whitworth, J. V. On the densest packing of sections of a cube. Ann. Mat. Pura Appl. (4) **27**, 29–37 (1948).

The author considers the densest lattice packing of convex bodies K of the form

$$\text{(A)}\qquad |x| \leqq 1, \quad |y| \leqq 1, \quad |z| \leqq 1, \quad |x+y+z| \leqq \tau,$$

where $0<\tau<3$, and gives, without proofs, a summary of the results which he has obtained. His method is based on Minkowski's work and, for $\tau=1$, includes Minkowski's result concerning the densest packing of octahedra [Gesammelte Abhandlungen, v. 2, pp. 1–42 = Nachr. Ges. Wiss. Göttingen **1904**, 311–355]. For such a densest packing the lattice formed by the centres of the bodies K must be K-admissible, i.e., no point of the lattice different from the origin is an inner point of (A), and must be critical, i.e., its determinant $d(\Lambda)$ equals the lower bound $\Delta(K) = \inf d(\Lambda)$ extended over all K-admissible lattices Λ. The packing constant $q(K) = \frac{1}{4}V(K)/\Delta(K)$ where $V(K)$ is the volume of K.

The investigation falls into three parts according as (a) $0<\tau\leqq\frac{1}{2}$, (b) $\frac{1}{2}\leqq\tau\leqq1$, (c) $1\leqq\tau<3$ and the critical lattices are determined in each case. The main results are as follows:

(a) $\Delta(K) = \frac{3}{4}\tau,\quad q(K) = \frac{1}{9}(9-\tau^2)$,

(b) $\Delta(K) = -\frac{1}{27}(\tau^3+3\tau^2-24\tau+1)$,
$q(K) = \frac{9}{4}\tau(9-\tau^2)/(-\tau^3-3\tau^2+24\tau-1)$,

(c) $\Delta(K) = \frac{1}{27}\tau(\tau^2-9\tau+27)$,
$q(K) = \frac{9}{8}(\tau^3-9\tau^2+27\tau-3)/\tau(\tau^2-9\tau+27)$.

In cases (b) and (c) there is only one critical lattice for each value of τ, but there are an infinity of critical lattices in case (a) for each τ. The packing constant $q(K)$ is a minimum for $\tau=1$, so that, of the bodies considered, the octahedra can be packed least densely.

R. A. Rankin.

H30-15 **(13, 323b)**

Rogers, C. A. A note on coverings and packings. J. London Math. Soc. **25**, 327–331 (1950).

Let K be a bounded symmetrical n-dimensional convex body in n-dimensional space and let δ and ϑ be the densities of the closest lattice packing of K into the whole space and of the most economical lattice covering of the whole space by K respectively. Further, δ^* and ϑ^* denote the corresponding quantities when the centres of the bodies are not restricted to form a lattice. Then $\delta\leqq\delta^*\leqq1\leqq\vartheta^*\leqq\vartheta$. The author proves that $\vartheta^*\leqq 2^n\delta^*$ and $\vartheta\leqq3^{n-1}\delta$. The second inequality is an improvement on the result $\vartheta\leqq n^n\delta$ obtained by Hlawka [Monatsh. Math. **53**, 81–131 (1949); these Rev. **11**, 12].

R. A. Rankin (Birmingham).

Citations: MR 11, 12b = H30-12.

Referred to in H02-17, H05-59, H30-20, H30-21.

H30-16 **(13, 323c)**

Davenport, Harold. Sur un système de sphères qui recouvrent l'espace à n dimensions. C. R. Acad. Sci. Paris **233**, 571–573 (1951).

If each point of n-dimensional space belongs to at least one of a system of spheres, each of the same volume V, their centres forming a lattice of determinant d, the spheres are said to form a lattice covering of space with density $\vartheta = V/d$. A lattice covering is constructed with $\vartheta < (\frac{1}{6}\pi e+\epsilon_n)^{n/2}$, where $\epsilon_n\to0$ as $n\to\infty$ (here $(\frac{1}{6}\pi e)^{\frac{1}{2}} = 1.193\cdots$). It is stated that if n is sufficiently large there is a lattice covering with $\vartheta<(1.15)^n$, but that Bambah and the author have proved that $\vartheta>\frac{4}{3}-\epsilon_n$ (where $\epsilon_n\to0$ as $n\to\infty$) for all lattice coverings.

C. A. Rogers (London).

H30-17 **(13, 768i)**

Rogers, C. A. The closest packing of convex two-dimensional domains. Acta Math. **86**, 309–321 (1951).

The main result of the present paper is the following remarkable theorem. If n similarly situated congruent convex discs of area a can be packed into a convex domain of area A, then $na/A \leqq nd/(n-1+d)$, where d denotes the density of the closest lattice packing of the discs. It follows that the density of an irregular packing of an infinite set of homothetic congruent convex discs cannot exceed the density of the closest lattice of the discs. If the discs have a center of symmetry, then by a result of the reviewer [Acta Sci. Math. Szeged **12**, Pars A, 62–67 (1950); these Rev. **12**, 352] the last proposition holds without restriction of homothetic discs.

L. Fejes Tóth (Veszprém).

Referred to in H25-35, H30-18, H30-51.

H30-18 **(23# A2132)**

Rogers, C. A.

The closest packing of convex two-dimensional domains, corrigendum.

Acta Math. **104** (1960), 305–306.

The author points out that the proof of his main theorem [Th. 1, Acta Math. **86** (1951), 309–321; MR **13**, 768] on the number of translates of a plane convex domain packed in a convex domain S is valid only for symmetrical K. He gives a revised version of the theorem for non-symmetrical K. His result that the density of the best packing of K in the plane is the same as that of the best lattice packing is unaffected.

R. P. Bambah (Chandigarh)

Citations: MR 13, 768i = H30-17.

Referred to in H15-78.

H30-19 **(13, 863e)**

Lekkerkerker, C. G. Packing of spheres. Math. Centrum Amsterdam. Rapport **ZW-1951-023**, 8 pp. (1951). (Dutch)

Let $A(w)$ be the maximum number of solid hyperspheres of unit radius and volume ω_n which can be packed into an n-dimensional hypercube of side w, and let $A^*(w)$ be the corresponding number when the centers of the spheres are required to form a lattice. The quantities

$$\rho_n = \lim_{w\to\infty} w^{-n}A(w)\omega_n \quad\text{and}\quad \rho_n^* = \lim_{w\to\infty} w^{-n}A^*(w)\omega_n$$

are called the packing constant and the regular packing constant, respectively. It is shown here that ρ_n and ρ_n^* exist for every integer $n\geqq1$, and that $\rho_n\leqq(n+2)2^{-(n+1)/2}$; this appears to be the first general non-trivial upper bound for ρ_n, although R. A. Rankin [Ann. of Math. **48**, 1062–1081 (1947); these Rev. **9**, 226] obtained bounds for ρ_3 and ρ_4, while it is known that $\rho_2=\rho_2^*=\pi/\sqrt{12}$. (Clearly $\rho_n^*\leqq\rho_n$, but only for $n=2$ is equality known to hold.) Finally, it is shown that ρ_n^* is connected with a constant occurring in the theory of positive definite n-ary quadratic forms; this was also shown by Rankin. *W. J. LeVeque.*

The published review was erroneous in attributing the results quoted to Lekkerkerker. The paper is purely expository; all the results mentioned in the review are due to R. E. Rankin [Ann. of Math. (2) **48**, 1062–1081 (1947); these Rev. **9**, 226], except for the relationship between ρ_n^* and a constant occurring in the theory of positive definite n-ary quadratic forms, which was well-known before Rankin's work.
W. J. LeVeque (Ann Arbor, Mich.).

Citations: MR 9, 226d = H30-8.

H30-20 (14, 22e)

Bambah, R. P., and Roth, K. F. A note on lattice coverings. J. Indian Math. Soc. (N.S.) **16**, 7–12 (1952).

A lattice Λ of determinant $d(\Lambda)\neq 0$ is called a covering lattice for a symmetrical convex body K in n-dimensional space if every point of the space belongs to a body of the type $K+X$ where X is a point of Λ. The density $\theta(K)$ of the most economical lattice covering of space by K is defined to be the lower bound of $V(K)/d(\Lambda)$ for all covering lattices Λ for K; here $V(K)$ is the content of K. Hlawka has shown that $\theta(K)\leqq n^n$ and this has been improved by Rogers [J. London Math. Soc. **25**, 327–331 (1950); these Rev. **13**, 323] to $\theta(K)\leqq 3^n$. The authors use the Brunn-Minkowski theorem to construct an n-dimensional space-filling body and deduce that $\theta(K)\leqq \pi n^n/(3\sqrt{3n}!)$ when K is symmetrical about its coordinate planes. This is an improvement on Rogers's inequality since the expression on the right is asymptotically equal to $e^n(\pi/54n)^{\frac{1}{2}}$ as $n\rightarrow\infty$. *R. A. Rankin.*

Citations: MR 13, 323b = H30-15.

H30-21 (14, 75h)

Davenport, H. The covering of space by spheres. Rend. Circ. Mat. Palermo (2) **1**, 92–107 (1952).

Let Λ be an n-dimensional lattice of determinant $d(\Lambda)>0$, and $V(K)$ be the content of a symmetric convex body K. Let the whole of the space be covered by bodies congruent to K centered at the points of Λ, and define $\vartheta(K)$, the density of thinnest covering, to be the lower bound of $V(K)/d(\Lambda)$ for all lattices with this property. It is proved that, for the n-dimensional sphere S_n,

$$\vartheta(S_n)<\left(\frac{11\pi e}{54\sqrt[4]{3}}+\epsilon\right)^{\frac{1}{2}n}<(1.15)^n$$

for large n, where ϵ is an arbitrary sufficiently small positive number. The method of proof is to choose a fixed integer $k>0$ and a fixed k-dimensional lattice Λ_0, to write $n=mk+h$ where $k\leqq h<2k$, and to evaluate $V(K)/d(\Lambda)$ for a particular n-dimensional lattice Λ which is defined in terms of Λ_0. Even the simple choice $k=1$ provides a good upper limit for $\{\vartheta(S_n)\}^{1/n}$, namely $(\pi e/6)^{1/2}$, but by choosing Λ_0 to correspond to the extreme form $\sum_{i=1}^k\xi_i^2+2\sum_{i<j}\xi_i\xi_j$ a better upper limit is obtained which is a minimum for $k=8$, and this gives the result stated. The paper concludes with a proof of the less sharp inequality $\vartheta(S_n)<(1.621)^n$ by a method which is a modification of one used by Rogers [J. London Math. Soc. **25**, 327–331 (1950); these Rev. **13**, 323].
R. A. Rankin (Birmingham).

Citations: MR 13, 323b = H30-15.

Referred to in H30-25, H30-35, H30-54, H30-81.

H30-22 (14, 131h)

Gericke, H. Äquivalenz des Satzes von Hajos mit einer Vermutung von Minkowski. Arch. Math. **3**, 34–37 (1952).

Earlier proofs of the equivalence of the Minkowski conjecture on linear forms (M) and Hajós's theorem on abelian groups (H) have employed the intermediate result that every simple lattice covering of n-dimensional space by hypercubes is columnated. See, for example, the paper of Hajós [Math. Z. **47**, 427–467 (1941); these Rev. **3**, 302]. The author seeks to give a direct proof, without considering lattices of hypercubes, for the case when the coefficients of the forms are rational. This he does by using the fact that the determinant of the system of defining relations of the group is equal to its order. In both parts of the proof the author appears to make the assumption that if $\mathfrak{a}_1, \mathfrak{a}_2, \cdots, \mathfrak{a}_n$ are generators of a finite abelian group having addition as its operation, then $\sum_k y_k\mathfrak{a}_k=\sum_{i,k}x_iq_{ik}\mathfrak{a}_k$ implies that $y_k=\sum_i x_iq_{ik}$. No justification for this conclusion is given. In the cases $n=2$ and 3 (H) is proved without appealing to (M), the proof in the latter case depending on the assumption mentioned above. *R. A. Rankin* (Birmingham).

In his review of this paper the reviewer stated that in two places the author appeared to make a certain assumption without justification. The reviewer has now received a letter from the author which makes it clear that the difficulty is one of distinguishing between what is being assumed and what is to be proved. The author admits that his original wording conveys the impression recorded by the reviewer and shows how, by a slight change of wording, this can be avoided in both places, thus justifying his conclusions. *R. A. Rankin.*

Citations: MR 3, 302b = H30-4.

H30-23 (14, 541b)

Chabauty, Claude. Empilement de sphères égales dans R^n et valeur asymptotique de la constante γ_n d'Hermite. C. R. Acad. Sci. Paris **235**, 529–532 (1952).

The object of this paper is to show that the Hermite constant γ_n associated with positive definite quadratic forms is asymptotically equal to $n/(2\pi e)$; this result would supersede the known inequalities

$$\frac{1}{2\pi e}\leqq\liminf n^{-1}\gamma_n\leqq\limsup n^{-1}\gamma_n\leqq\frac{1}{\pi e}.$$

Pressure of space has forced the author to condense his argument and omit certain steps in the proofs of his lemmas, and for this reason the paper is hard to read. The reviewer has been in correspondence with the author, who has informed him that it is not always possible, as stated in support of Lemma 5, to choose points A, B from a set of $n+2$ points on the unit sphere in R^n so that the segment AB meets the convex cove. of the remaining n points. Until this part of the argument can be amended, the main result of the paper must remain in doubt. A correction will be published in due course. *R. A. Rankin* (Birmingham).

H30-24 (14, 541d)

Chabauty, Claude. Nouveaux résultats de géométrie des nombres. C. R. Acad. Sci. Paris **235**, 567–569 (1952).

Utilisant les définitions et les notations d'une note antérieure [voir la note analysée ci-dessus] l'auteur démontre le théorème suivant: Si $G\subset R^n$, pour tous les entiers naturels h et k, on a

$$((n+2)^{kh}-1)D(G)\geqq\Omega_n\rho^n(h, G)((k-1)(n+1)k^{-1}n^{-1})^{n/2}$$

et en particulier

$$D^{2/n}(G)\geqq 4\pi en^{-1}\rho^2(h, G)(1-\epsilon(nh^{-1})),$$

où $\epsilon(t)$ désigne une fonction positive qui$\rightarrow 0$ avec t^{-1}.

Les applications de ce théorème au cas d'un système de formes linéaires donnent des améliorations des résultats classiques de Minkowski et des résultats de Blichfeldt [Monatsh. Math. Phys. **43**, 410–414 (1936); **48**, 531–533 (1939); ces Rev. **1**, 68], Hlawka [Österreich. Akad. Wiss.

Math.-Nat. Kl. S.-B. IIa. **156**, 247–254 (1948); ces Rev. **10**, 236], Rankin [Indagationes Math. **10**, 274–281 (1948); ces Rev. **10**, 284], et Rogers [Acta Math. **82**, 185–208 (1950); ces Rev. **11**, 501]. *J. F. Koksma.*

Citations: MR 1, 68c = R28-1; MR 10, 236b = H10-8; MR 10, 284e = H10-13; MR 11, 501e = J36-13.

H30-25 (14, 787f)

Bambah, R. P., and Davenport, H. The covering of n-dimensional space by spheres. J. London Math. Soc. **27**, 224–229 (1952).

It is proved that the density ϑ of any covering of n-dimensional space by equal spheres (whose centres form a lattice) satisfies $\vartheta > \frac{4}{3} - \epsilon_n$ where $\epsilon_n \to 0$ as $n \to \infty$. It is shown that if Π is a convex polyhedron of volume V with at most $2(2^n - 1)$ faces and is inscribed in an n-dimensional sphere of volume J_n and if the foot of the perpendicular from 0 to each face falls inside the face, then $J_n/V > 4 - \epsilon_n$. This is done by splitting Π into pyramids with apex at the centre of the sphere and considering their volumes. This result is applied to Voronoi's polyhedron Π which consists of those points which are nearer to a given point of the lattice than to any other lattice-point, and yields the inequality stated. A result in the other direction, namely that there exists a lattice-covering for which $\vartheta < (1.15)^n$ has been given by Davenport [Rend. Circ. Mat. Palermo (2) **1**, 92–107 (1952); these Rev. **14**, 75]. *R. A. Rankin* (Birmingham).

Citations: MR 14, 75h = H30-21.

Referred to in H30-26, H30-34, H30-54.

H30-26 (14, 1066b)

Erdös, Paul, and Rogers, C. A. The covering of n-dimensional space by spheres. J. London Math. Soc. **28**, 287–293 (1953).

Given a system of equal spheres which cover n-dimensional space, one can define the density of the covering as the lower limit of the density of covering of a large cube as the cube expands. Let ϑ^*, depending only on n, denote the lower bound of the densities of all such coverings. The main result of the paper is that $\vartheta^* > 16/15 - \epsilon_n$, where $\epsilon_n \to 0$ as $n \to \infty$. A similar result, with 4/3 for 16/15, has been proved by Bambah and Davenport [same J. **27**, 224–229 (1952); these Rev. **14**, 787] for coverings in which the centres of the spheres are restricted to form a lattice. The present proof is again based on the consideration of the polyhedron Π consisting of all points which are nearer to the centre of one particular sphere than to the centre of any other sphere. The first difficulty is to estimate the number of faces of Π, and the authors show by an ingenious argument that this number is at most $\vartheta^* 4^n - 1$. To complete the proof, an upper bound is needed for the proportion of the volume of an n-dimensional sphere which can be filled by a convex polyhedron with a given number of faces. A fairly simple argument on this question leads to the final result but with $(16/15)^{1/2}$ in place of 16/15. To obtain the latter, appeal is made to an inequality of Rogers concerning integrals over convex sets [same J. **28**, 293–297 (1953); these Rev. **14**, 965]. *H. Davenport* (London).

Citations: MR 14, 787f = H30-25.

H30-27 (15, 248b)

Fejes Tóth, L. Lagerungen in der Ebene, auf der Kugel und im Raum. Die Grundlehren der Mathematischen Wissenschaften in Einzeldarstellungen mit besonderer Berücksichtigung der Anwendungsgebiete, Band LXV. Springer-Verlag, Berlin-Göttingen-Heidelberg, 1953. x+197 pp. DM 24.00; bound DM 27.00.

This book is a collection of recent work done by Hadwiger, van der Waerden, the author, and many others, on problems of densest packing and thinnest covering in the plane, on the sphere, and in space, and numerous other extremal problems. The regions considered are simple and connected, in two or three dimensions.

Chapter I contains proofs of geometrical theorems which are used in the subsequent work and are also of interest in themselves. Direct and indirect proofs are given of the following theorem: Let r and R be the inradius and circumradius of a convex n-gon of area F and perimeter L. Then

$$nr^2 \tan\frac{\pi}{n} \leqq F \leqq \tfrac{1}{2}nR^2 \sin\frac{2\pi}{n}, \quad 2nr \tan\frac{\pi}{n} \leqq L \leqq 2nR \sin\frac{\pi}{n},$$

with equality in all four cases only if the polygon is regular. Thus, among all convex n-gons inscribed in (circumscribed about) a given circle, the regular n-gon has maximum (minimum) area and perimeter. Several interesting inequalities concerning triangles are proved. For example, if R_1, R_2, R_3 are the distances from an arbitrary point O to the vertices of a triangle of area Δ, then

$$R_1 + R_2 + R_3 \geqq 2 \cdot 3^{1/4} \Delta^{1/2}$$

with equality only if the triangle is equilateral and O is its centre. A short treatment of the regular and Archimedean polyhedra (which includes a proof of Euler's formula) is accompanied by clear drawings of these bodies.

Chapter II deals with the problem of approximating convex regions by inscribed and circumscribed polygons. Let T be (the area of) a convex region, and T_n the inscribed n-gon of maximum area. It is shown that

$$T_n \geqq T \frac{n}{2\pi} \sin\frac{2\pi}{n},$$

with equality only if T is an ellipse. Thus the approximation of the area of a convex region by inscribed polygons is worst when the region is an ellipse. The corresponding problem for circumscribed polygons has not been solved. Nevertheless, for large n the external curve is shown to approximate a circle.

The similar problem of approximating the boundary of a convex region by in-(circum-)scribed polygons is unsolved, but the following inequality is proved. For any convex region there are circumscribed and inscribed n-gons of perimeter L_n and l_n such that

$$\frac{L_n - l_n}{L_n} \leqq 2 \sin^2 \frac{\pi}{2n}.$$

Chapter II also contains an exposition of Blaschke's concept of the affine length of a curve.

Let a countable number of unit circles be packed (i.e., placed without overlapping) in the plane. Let $N(R)$ be the number of these circles wholly contained in a large circle of radius R and centre O. The density d of the packing is defined to be

$$d = \lim_{R\to\infty} \frac{N(R)\pi}{\pi R^2} = \lim_{R\to\infty} \frac{N(R)}{R^2}.$$

If the circles are placed so as to cover the plane (i.e., so that every point of the plane is contained inside, or on the boundary of, at least one circle), the above limit defines the density D of the covering. Chapter III contains proofs that

$$d \leqq \pi/\sqrt{12} = .9069\cdots, \quad D \geqq 2\pi/\sqrt{27} = 1.209\cdots.$$

In both cases equality is attained only when the centres of the spheres are the points of a lattice formed by equilateral triangles.

The problem of how best to cover a finite convex region is considered, and it is proved that if T is (the area of) a convex region of boundary L, then

$$\left[\frac{2}{3\sqrt{3}}T+\frac{2}{\pi\sqrt{3}}L+1\right]$$

unit circles suffice to cover T. A similar formula for the maximum number of unit circles which can be packed into a convex region is derived.

Inequalities are established for bounds of the density of a packing of circles of n different radii. Also considered is the more general problem of packing congruent convex regions into a larger convex region. The following inequality seems particularly interesting. If n congruent convex regions G are packed into a convex hexagon of area S, then $n \leqq S/s$, where s is the area of the smallest hexagon circumscribed about G.

Let P be (the area of) the fundamental parallelogram of a lattice generated by a translation group. If the regions obtained by applying the translations to a region (of area) G do not overlap, then the density of this lattice packing is G/P. If the regions arranged in this way cover the plane, the above ratio gives the density of the lattice covering. Let $d(G)$ and $D(G)$ be the densities of the densest lattice packing of G and the thinnest lattice covering by G. Chapter IV considers the problem of finding bounds for these densities. If G is convex, it is shown that

$$d(G) \geqq 2/3, \quad D(G) \leqq 3/2,$$

with equality in both cases only when G is a triangle. If G is centrally symmetrical as well as convex the situation is understandably different. In this case it is shown that $D(G) \leqq 2\pi/\sqrt{27}$, with equality only if G is an ellipse. The smallest value for $d(G)$ has not been established. For an ellipse it is $\pi/2\sqrt{3} = .9069\cdots$, but for a regular octagon it is $4(3-\sqrt{2})/7 = .90616\cdots$ and for a "smoothed" octagon (diagram on page 104) it is

$$(9-4\sqrt{2}-\log 2)/(2\sqrt{2}-1) = .9024\cdots.$$

The author agrees with Mahler's conjecture that this last density is the extreme one.

Chapter V contains numerous theorems which exhibit the extremal properties of regular and semi-regular polyhedra. For example, the following three theorems are proved. (i) If $n \geqq 3$ congruent spherical caps are packed on the surface of a sphere, and if d is the density of the packing, then

$$d \leqq \frac{n}{2}(1-\tfrac{1}{2}\operatorname{cosec} w_n), \quad \text{where} \quad w_n = \frac{n}{n-2}\frac{\pi}{6}.$$

(ii) If $n \geqq 3$ congruent spherical caps cover the surface of a sphere, and D is the density of the covering, then

$$D \geqq \frac{n}{2}(1-3^{-1/2}\cot w_n).$$

In (i) and (ii) equality occurs only when $n=3$, 4, 6, 12 and the centres of the caps are the vertices of an equilateral triangle (inscribed in a great circle), a regular tetrahedron, octahedron, or icosahedron. (iii) Let V be the volume, and e, f, and k the number of vertices, faces, and edges of a polyhedron which contains the unit sphere. Then

$$V \geqq \frac{k}{3}\sin\frac{\pi f}{k}\left(\tan^2\frac{\pi f}{2k}\tan^2\frac{\pi e}{2k}-1\right),$$

with equality only if the polyhedron is regular and circumscribed about the sphere.

The isoperimetric theorem is discussed and its solution is given in the form (due to M. Goldberg): If a convex polyhedron of n faces has volume V and surface F, then

$$\frac{F^3}{V^2} \geqq 54(n-2)\tan w_n(4\sin^2 w_n - 1).$$

How should n points be placed on the surface of a unit sphere so that the minimum distance between the $\frac{1}{2}n(n+1)$ pairs of points is maximum? Theorem (i) above solves this problem for $n=3$, 4, 6, 12. Chapter VI contains the recent work of van der Waerden, Habricht and Schütte for other values of n. For $n=5$, 7, 8, 9 the problem has been solved. The configurations are more or less irregular, and the methods used for a particular n do not necessarily help in solving the problem for other values of n. For $n=10$, 11, 13, 14, 15, 16, 32, the author gives Schütte and van der Waerden's conjectures for the best configurations.

Chapter VII contains a discussion of the problems of the densest packing of equal spheres, and the thinnest covering of space by equal spheres. If the condition is imposed that the centres of the spheres constitute a lattice, the problems have been solved: The best packing density is $\pi/\sqrt{18} = .74048\cdots$ and is attained when the centres of the spheres are the points of a face-centered cubic lattice, and the thinnest covering density is $5^{3/2}\pi/24 = 1.464\cdots$ which is attained when the centres form a body-centred cubic lattice.

Although it is widely believed that $\pi/\sqrt{18}$ is the greatest density without the lattice restriction, this has never been proved. Let d be the density of an arbitrary packing of spheres. Blichfeldt [Math. Ann. **101**, 605–608 (1929)] showed $d < .835$ and Rankin [Ann. of Math. (2) **48**, 1062–1081 (1947); these Rev. **9**, 226] improved this to $d < .828$. The author gives a proof (different from Blichfeldt's) that $d < .835$, and a non-rigorous proof that $d < .7545$, but observes that great technical difficulties must be overcome before the conjecture $d \leqq \pi/\sqrt{18}$ can be proved.

Each chapter begins with a description of the problems to be discussed and ends with a historical outline, in which the author gives the sources of the theorems and proofs. There is an excellent bibliography (which includes works as recent as 1952). The author indicates the direction of further research. The exceptionally clear treatment of the variety of problems, and the 124 beautiful diagrams, combine to make this book extremely attractive. *W. O. Moser.*

Citations: MR 9, 226d = H30-8.
Referred to in H05-61, H30-56, H30-69, H30-71.

H30-28 (15, 292d)

Bambah, R. P. On lattice coverings. Proc. Nat. Inst. Sci. India **19**, 447–459 (1953).

A homogeneous lattice Λ, of determinant $d(\Lambda) \neq 0$, in n-dimensional Euclidean space R_n is called a covering lattice for a set K if every point of R_n belongs to the set $K+\Lambda$, and the covering constant $c(K)$ is the upper bound of $d(\Lambda)$ for all covering lattices. If there exists a lattice Λ for which $d(\Lambda)=c(K)$ then Λ is called a maximal covering lattice for K. The author proves a great many results on coverings, of which the following is a selection. I. A maximal covering lattice exists when K is a closed bounded set containing O in its interior. II. When K is closed and bounded and Λ is a maximal covering lattice for K, there are n independent points on the boundary of K which are just covered. III. If K is unbounded and closed and has Lebesgue measure $V(K)$ (possibly ∞), then $c(K) \leqq V(K)$. IV. If K is unbounded, closed and contains O in its interior, and if $V(K) < \infty$, then there exists a maximal covering lattice such that the set of all points of R_n which do not lie in $K+\Lambda$ has measure zero. V. There exist star-sets K with $V(K)=\infty$ for which (i) $c(K)<\infty$, (ii) $c(K)=\infty$. VI. $\sup\{V(K)/c(K)\}=\infty$, the supremum being taken over all bounded star-sets K. VII. If $K^{(t)}$ is the set of points of an unbounded star-set K which lie within a distance t of the origin, then $\lim_{t\to\infty} c(K^{(t)}) = c(K)$. From this last result three further theorems are deduced which are analogous to certain results of Mahler and of Davenport and Rogers. *R. A. Rankin.*

H30-29 (15, 607d)

Bambah, R. P. On polar reciprocal convex domains. Proc. Nat. Inst. Sci. India **20**, 119–120 (1954).

Let K be a centrally symmetric convex body with center at the origin in Euclidean space E_n. A lattice is K-admissible if it contains no points of K except the origin. A lattice is K-covering if every point of E_n lies in a translate of K by an element of the lattice. The critical determinant $\Delta(K)$ is the minimal volume of the fundamental parallelotope for all K-admissible lattices, the covering constant $c(K)$ is the maximal volume of the fundamental parallelotope for all K-covering lattices. The author proves: If k, K are centrally symmetric convex domains in E_2 with centers at the origin, which are polar reciprocal with respect to the unit circle C with center at the origin, then $2 \leqq \Delta(K)c(k) \leqq 9/4$. These bounds are proved best possible by considering the inscribed and circumscribed square of C for the lower bound and the inscribed and circumscribed regular hexagon of C for the upper bound. Generalizations to higher dimensions are indicated. *E. G. Straus* (Los Angeles, Calif.).

Referred to in H30-30, H30-40.

H30-30 (16, 65a)

Bambah, R. P. On polar reciprocal convex domains. Addendum. Proc. Nat. Inst. Sci. India **20**, 324–325 (1954).

In this addendum to an earlier note [same Proc. **20**, 119–120 (1954); these Rev. **15**, 607] the author establishes the best possible inequalities $27/4 \leqq c(k)c(K) \leqq 9$.

E. G. Strauss (Los Angeles, Calif.).

Citations: MR 15, 607d = H30-29.
Referred to in H30-40.

H30-31 (15, 687b)

Tornheim, Leonard. Lattice packing in the plane without crossing arcs. Proc. Amer. Math. Soc. **4**, 734–740 (1953).

Let S be a set in a Euclidean plane and let $E(S)$ be the vector difference set $S-S$. Then a (trivial) necessary and sufficient condition, that a lattice Λ should be such that no two sets of the form $S+\lambda$, $S+\lambda'$ with λ, λ' in Λ should have a common point, is that no point of Λ other than the origin should lie in $E(S)$. The author investigates the problem of finding a similar condition that there should be no pair of (Jordan) arcs, one contained in $S+\lambda$ and the other in $S+\lambda'$, which cross each other in a suitably defined sense. The ways in which arcs can cross are studied in some detail. A point is called a local boundary point of $E(S)$ if the points x in its neighbourhood cannot be represented continuously in the form $x=y-z$ with y and z in S. It is shown that if the only points of Λ other than the origin belonging to $E(S)$ are local boundary points then there will be no arcs in $S+\lambda$, $S+\lambda'$ which cross each other at a single point. [The reader will be much puzzled if he fails to realize that A' should be read for A in (iii) of the definition of §1. Note also that A and B should be interchanged in the example following this definition.] *C. A. Rogers* (London).

H30-32 (15, 780c)

Bambah, R. P. On lattice coverings by spheres. Proc. Nat. Inst. Sci. India **20**, 25–52 (1954).

In this paper the author obtains, for the first time, the most economical lattice coverings of three-dimensional Euclidean space by equal spheres. Let Λ be a lattice, of determinant $d(\Lambda)$, and suppose that equal spheres K of volume $V(K)$ are centred at each point of Λ, and that each point of the space belongs to at least one such sphere. Define $\theta(K)$ to be the lower bound of $V(K)/d(\Lambda)$, taken over all Λ. Then $\theta(K)$ is the density of the most economical lattice covering by spheres K and is shown by the author to be $5^{3/2}\pi/24$. Further, a lattice Λ provides the most economical lattice covering by K if and only if, by a suitable choice of orthogonal axes, Λ is generated by the points

$$\frac{2r}{\sqrt{5}}(-1, 1, 1), \quad \frac{2r}{\sqrt{5}}(1, -1, 1), \quad \frac{2r}{\sqrt{5}}(1, 1, -1),$$

and K has the equation $x^2+y^2+z^2=r^2$. This is the body-centred cubic lattice.

This result is equivalent to the following theorem: Let $f(x, y, z) = ax^2+by^2+cz^2+2ryz+2szx+2txy$ be a positive definite quadratic form with real coefficients and determinant $D>0$. Then there exist real numbers x_0, y_0, z_0 such that for all integers we have

$$f(x+x_0, y+y_0, z+z_0) \geqq \left(\frac{125}{1024}D\right)^{\frac{1}{3}},$$

the sign of equality being necessary if and only if

$$f \sim \rho(3x^2+3y^2+3z^2-2yz-2zx-2xy), \tag{1}$$

and ρ is any positive number.

If D_f is the determinant of a positive definite form f, if $m_f(\xi, \eta, \zeta)$ is the minimum of $f(\xi-x, \eta-y, \zeta-z)$ for all integers x, y, z, and if $M_f = m_f D_f^{-1/3}$, then this theorem is equivalent to the statement $\inf M_f = m = (125/1024)^{1/3}$. To prove this f is taken in Seeber's reduced form, i.e. $0<a\leqq b\leqq c$, $2|s|\leqq a$, $2|t|\leqq a$, $2|r|\leqq b$, $a+b+2r+2s+2t\geqq 0$ and r, s, t are (i) all non-negative or (ii) all negative. Forms of type (i) are first considered and are shown to have an M_f with lower bound greater than m. This is done by considering the various possible relative orders of r, s and t. Forms of type (ii) are then considered. The associated polyhedron Π_f is the region consisting of points X for which $f(X)\leqq f(X-A)$ for each point A with integral coordinates, other than the origin. Π_f is bounded by at most seven pairs of opposite planes and M_f clearly is attained at a vertex, of which there are three types giving values f_1, f_2 and f_3 to f. By considering in detail all the possible relative orderings of f_1, f_2 and f_3, it is shown that a form f for which M_f attains its infimum must have $f_1=f_2=f_3$, and from this it is deduced that f is in fact the form (1). *R. A. Rankin* (Birmingham).

Referred to in H30-65, H30-75, H30-76, H30-81, J48-10.

H30-33 (15, 780d)

Bambah, R. P. Lattice coverings with four-dimensional spheres. Proc. Cambridge Philos. Soc. **50**, 203–208 (1954).

Let θ_n be the density of the thinnest lattice covering by spheres in n-dimensional Euclidean space R_n. The author proves that

$$1.5194 < \frac{4}{15\sqrt{3}}\pi^2 \leqq \theta_4 \leqq \frac{2}{5\sqrt{5}}\pi^2 < 1.7656.$$

This is equivalent to the following theorem which he proves: Let

$$f(x) = f(x_1, x_2, x_3, x_4) = \sum_{i,j=1}^{4} a_{ij}x_ix_j \quad (a_{ij}=a_{ji})$$

be a positive definite quadratic form with real coefficients and determinant $D=|a_{ij}|$. Then there exist real numbers u_1, u_2, u_3, u_4 such that for all integers x_1, x_2, x_3, x_4 we have

$$f(x+u) = f(x_1+u_1, x_2+u_2, x_3+u_3, x_4+u_4) \geqq \left(\frac{64}{675}D\right)^{1/4}.$$

On the other hand, there exists a positive definite quadratic form $f(x)$ such that for all real numbers u_1, u_2, u_3, u_4 we can find integers x_1, x_2, x_3, x_4 with

$$f(x+u) = f(x_1+u_1, x_2+u_2, x_3+u_3, x_4+u_4) \leqq \left(\frac{16}{125}D\right)^{1/4},$$

the equality sign in the last relation being necessary. The proof of the first part depends upon a lemma quoted from the paper reviewed above. It is shown that the last part of the theorem is satisfied by the form

$$f(x)=4\sum_{i=1}^{4}x_i^2-2\sum_{1\leq i<j\leq 4}x_ix_j,$$

and it is conjectured that, in fact, $\theta_4=2\pi^2/5\sqrt{5}$.

R. A. Rankin (Birmingham).

Referred to in H30-36, H30-75.

H30-34 **(17, 1235a)**

Davenport, H. Le recouvrement de l'espace par des sphères. Colloque sur la Théorie des Nombres, Bruxelles, 1955, pp. 139–145. Georges Thone, Liège; Masson and Cie, Paris, 1956.

Any lattice Λ in Euclidean n-space determines a partition of the whole space into congruent centrally-symmetrical polytopes whose centres are the lattice points. The interior of such a polytope Π, with centre O, consists of all points which are nearer to O than to any other lattice point. Since this "Dirichlet region" or "Voronoi polyhedron" is a fundamental region for the translation group of Λ, it has the same content $d(\Lambda)$ as the unit cell or basic parallelotope. Let s and S denote the in- and circum-spheres of Π, so that s touches the faces nearest to O while S passes through the vertices farthest from O; let $V(s)$, $V(S)$ denote their contents. Such spheres round all the lattice points form a "packing" of density $\delta(\Lambda)=V(s)/d(\Lambda)$ and a "covering" of density $\theta(\Lambda)=V(S)/d(\Lambda)$. Let δ_n denote the upper bound of $\delta(\Lambda)$ for all possible lattices, and θ_n the lower bound of $\theta(\Lambda)$. After remarking that $\delta_{n-2}<2^{-\frac{1}{2}n}n$, the author gives an illuminating sketch of the manner in which the inequality

$$\theta_n>\tfrac{4}{3}-o(1) \qquad (n\to\infty)$$

was obtained by Bambah and Davenport [J. London Math. Soc. **27** (1952), 224–229; MR **14,** 787] and

$$\theta_n<(1.107)^n$$

by G. L. Watson [see the following review].

H. S. M. Coxeter (Toronto, Ont.).

Citations: **MR 14, 787f** = **H30-25.**

H30-35 **(17, 1235b)**

Watson, G. L. The covering of space by spheres. Rend. Circ. Mat. Palermo (2) **5** (1956), 93–100.

In the notation of the preceding review, the author proves that $\theta_n<(1.107)^n$ by applying the ingenious method of Davenport [Rend. Circ. Mat. Palermo (2) **1** (1952), 92–107; MR **14,** 75] to an eight-dimensional lattice consisting of the points whose coordinates $y_1, \cdots, y_8$ are integers satisfying the congruences

$$y_i+y_{i+4}\equiv y_1+y_2+y_3+y_4 \pmod 2 \qquad (i=1, 2, 3, 4).$$

This lattice was discovered by T. Gosset [Messenger of Math. **29** (1899), 43–48, p. 48] as the final member 5_{21} of a finite family of semi-regular $(n+4)$-dimensional polytopes n_{21}, beginning with the triangular prism $(-1)_{21}$. It was rediscovered by E. Cartan [Ann. Mat. Pura Appl. (4) **4** (1927), 209–256, p. 222] in connection with the simple Lie group E_8; he took the coordinates to be nine integers, mutually congruent (mod 3), with sum zero. The same points were shown by H. F. Blichfeldt [Math. Z. **39** (1934), 1–15] to be the centres of the densest regular packing of spheres in 8-space. It also represents the integral Cayley numbers [Coxeter, Duke Math. J. **13** (1946), 561–578, p. 571; MR **8,** 370]. The author's coordinates $y_1, \cdots, y_8$ are essentially the same as those proposed by P. Du Val [Coxeter, Proc. London Math. Soc. (2) **34** (1932), 126–189, p. 185].

The essential step in the present work consists in integrating a certain function over the content of the Dirichlet region which, being the reciprocal of 4_{21} [Coxeter, Regular polytopes, Methuen, London, 1948, p. 204; MR **10,** 261], is dissected by its hyperplanes of symmetry into 192.10! congruent simplexes T_9 [ibid., p. 194]. The author further subdivides T_9 into three orthoschemes, one occupying one-seventh of its content while the other two are congruent. *H. S. M. Coxeter* (Toronto, Ont.).

Citations: **MR 8, 370d** = **R54-5; MR 14, 75h** = **H30-21.**

Referred to in H02-10, H30-75.

H30-36 **(18, 721c)**

Few, L. Covering space by spheres. Mathematika **3** (1956), 136–139.

If Λ is a lattice in three-dimensional space with the property that the spheres of radius 1 centered at the lattice points cover the whole of space, then

$$d(\Lambda)\leq\frac{32}{5\sqrt{5}}$$

as was proved by Bambah [Proc. Cambridge Philos. Soc. **50** (1954), 203–208; MR **15,** 780] and later by Barnes [Canad. J. Math. **8** (1956), 293–304; MR **17,** 1060]. The author gives a third proof which is fully elementary and which, unlike the two former proofs, makes no use of the theory of reduction of ternary quadratic forms.

J. F. Koksma (Amsterdam)

Citations: **MR 15, 780d** = **H30-33; MR 17, 1060c** = **J48-10.**

Referred to in H30-76.

H30-37 **(18, 875f)**

Ehrhart, Eugène. Sur l'empilement réticulaire d'ovales ou d'ovoïdes. C. R. Acad. Sci. Paris **244** (1957), 550–553.

The author gives simple geometrical proofs of results of Minkowski reducing the problem of finding the closest lattice packing of a given convex domain or body to the corresponding problem for its difference domain or body.

C. A. Rogers (Birmingham).

H30-38 **(19, 976g)**

Broch, E. K. Some remarks concerning the plane lattice and close-packing of equal circles. Avh. Norske. Vid. Akad. Oslo. I. **1956,** no. 3, 7 pp.

In this essay, the author describes the closest lattice packing of circles in the plane, and explains why no "twinning" of the lattice is possible. He promises to explain in a subsequent paper how simple and multiple twin-formations can occur in the closest lattice packings of spheres in three dimensional space, and how this leads to an infinite number of different close-packed structures.

C. A. Rogers (Birmingham).

H30-39 **(20# 1284)**

Bambah, R. P. Maximal covering domains. Proc. Nat. Inst. Sci. India. Part A. **23** (1957), 540–543.

Let K, K' be closed bounded symmetrical convex domains. Let $c(K)$ denote the supremum of the discriminants of those lattices Λ for which $K+\Lambda$ is the whole plane. K is called maximal if $K\subset K'\neq K$ implies $c(K)<c(K')$. For every point p on the boundary of K let $t(p)$ denote the area of the largest triangle with the vertex p and inscribed in K. If $t(p)$ is independent of p, K is maximal; in particular, parallelograms and ellipses are maximal; conversely if K is maximal and strictly convex, $t(p)$ is independent of p. The author conjectures that K then is an ellipse. *P. Scherk* (Saskatoon, Sask.)

H30-40 **(20# 3509)**

Bambah, R. P. An analogue of a problem of Mahler. Res. Bull. Panjab Univ. no. **109** (1957), 299–302.

Let K be a bounded convex region in the (x, y)-plane which is symmetric with respect to the origin. Let K_s denote the intersection of K with the strip $|y| \leqq s$ and let $T(s)$ denote the area of the largest triangle inscribed in K_s. The author proves: $T(s)/s \geqq T(t)/t$ if $0<s<t$.

The covering constant $c(K_s)$ is the supremum of the discriminants of those lattices Λ for which $K+\Lambda$ is the whole plane. Since $c(K_s)=2T(s)$, this result implies $c(K_s)/s \geqq c(K_t)/t$ [cf. Bambah, Proc. Nat. Inst. Sci. India **20** (1954), 119–120, 324–325; MR **15**, 607; **16**, 65].

P. Scherk (Saskatoon, Sask.)

Citations: MR 15, 607d = H30-29; MR 16, 65a = H30-30.

H30-41 **(20# 7247)**

Rogers, C. A. Lattice coverings of space with convex bodies. J. London Math. Soc. 33 (1958), 208–212.

It is shown that there exists a lattice covering of n-dimensional space by the translates of an arbitrary convex body with density 2^n. The main tool is an estimate of the average density of the points not covered by the lattice translates of a fixed body of volume V with $0<V\leqq 1$ where the average is extended over all lattices of determinant 1. With the help of Siegel's mean value theorem it is shown that this average is no greater than $1-V+\frac{1}{2}V^2$.

A sharpening of the density result to $(1.8774)^n$ for large n is announced for a later paper.

E. G. Straus (Los Angeles, Calif.)

Referred to in H02-17.

H30-42 **(21# 847)**

Rogers, C. A. The packing of equal spheres. Proc. London Math. Soc. (3) **8** (1958), 609–620.

In any packing of equal spheres in Euclidean n-space, we may associate with each sphere a Dirichlet region or "Voronoi polyhedron" whose interior consists of all the points that are nearer to the center of that sphere than to the center of any other sphere. Such regions, each surrounding a sphere, fit together to fill the whole space [G. L. Dirichlet, J. Reine Angew. Math. **40** (1850), 216–219]. The density of the packing may be defined as the average of the ratio of the content of a sphere to the content of the Dirichlet region that surrounds it.

The author proves that the closest possible packing of $n+1$ spheres of radius 1 is attained when their centers are at the vertices of a regular simplex of edge 2, whose content is $2^{\frac{1}{2}n}(n+1)^{\frac{1}{2}}/n!$ [cf. R. A. Rankin, Proc. Glasgow Math. Assoc. **2** (1955), 145–146; MR **17**, 523]. Part of this content is occupied by "sectors" of the $n+1$ spheres. In the notation of Schläfli [*Gesammelte Mathematische Abhandlungen*, Bd. 2, Verlag Birkhäuser, Basel, 1953; MR **14**, 833; p. 178], this part is $2^{-n}(n+1)!J_nF_n(\alpha)$, where $J_n=\pi^{\frac{1}{2}n}/\Gamma(1+\frac{1}{2}n)$ is the content of a whole sphere, $2\alpha=\sec^{-1}n$ is the dihedral angle of the regular simplex, and the "Schläfli function" F_n is defined recursively by the formulae

$$F_0(\theta) = F_1(\theta) = 1, \quad F_{n+1}(\theta) = \frac{2}{\pi}\int_\alpha^\theta F_{n-1}(\phi)d\theta,$$

$\alpha=\frac{1}{2}\sec^{-1}n, \phi=\frac{1}{2}\sec^{-1}(\sec 2\theta-2)$. Thus the local density of this packing is

$$\sigma_n = 2^{-3n/2}(n!)^2(n+1)^{\frac{1}{2}}J_nF_n(\alpha).$$

The same density could be maintained throughout the whole space if regular simplexes could be fitted together to make a regular honeycomb $\{3, 3, \cdots, 3, p\}$, where $p=\pi/\alpha$. Then a cell of the reciprocal honeycomb $\{p, 3, \cdots, 3, 3\}$ would serve as a Dirichlet region. Setting

$$g_{p,3,\cdots,3} = 2^n/F_n(\alpha), \ (-1, k) = 1-k/(n+1), \ {}_{n-1}R = 1$$

in formula 8.87 of the reviewer's *Regular polytopes* [Methuen, London, 1948, Pitman, New York, 1949; MR **10**, 261; p. 161], one verifies that J_n/σ_n is the proper expression for the content of such a cell $\{p, 3, \cdots, 3\}$ of inradius 1. When $n>2$, p is fractional, the regular honeycomb can only exist in a statistical sense, and σ_n is an unattained upper bound for the density. When $n=3$, this upper bound

$$\sigma_3 = 3\pi(2)^{\frac{1}{2}}F_3(\tfrac{1}{2}\sec^{-1}3) = (2)^{\frac{1}{2}}(3\sec^{-1}3-\pi) = 0.7797\cdots$$

is better than Rankin's $0.8271\cdots$ [Ann. of Math. (2) **48** (1947), 1062–1081; MR **9**, 226; see also the paper by Coxeter, reviewed below].

An ingenious argument, suggested by H. E. Daniels, enables the author to obtain the asymptotic formula

$$\sigma_n \sim 2^{-\frac{1}{2}n}n/e,$$

which is better than Blichfeldt's bound $2^{-\frac{1}{2}n-1}(n+2)$ [Math. Ann. **101** (1929), 605–608].

The "dual" problem, of the thinnest covering of space by equal spheres, can in fact be treated by using a polytope $\{p, 3, \cdots, 3\}$ of circumradius 1 (instead of inradius 1). Since the ratio of circumradius to inradius is $\{2n/(n+1)\}^{\frac{1}{2}}$, a lower bound for the density of a covering (attained only when $n=1$ or 2) is $\{2n/(n+1)\}^{\frac{1}{2}n}\sigma_n$. When $n=3$, this unattained lower bound $(3(3^{\frac{1}{2}})/2)\sec^{-1}3-\pi)=1.431\cdots$ is, as one would expect, slightly less than the known density $1.464\cdots$ of the thinnest lattice covering [Bambah, Proc. Nat. Inst. Sci. India **20** (1954), pp. 25–52; MR **15**, 780].

The corresponding asymptotic expression is $n/e^{3/2}$. This is better than Erdös and Rogers' 16/15 [J. London Math. Soc. **28** (1953), 287–293; MR **14**, 1066].

H. S. M. Coxeter (Toronto, Ont.)

Referred to in H02-17, H30-62, H30-63.

H30-43 **(21# 3812)**

Heppes, A. Mehrfache gitterförmige Kreislagerungen in der Ebene. Acta Math. Acad. Sci. Hungar. **10** (1959), 141–148. (Russian summary, unbound insert)

A k-fold packing of circles in the plane is a set of congruent (unit) circles in the plane such that no point lies in the interior of more than k of the circles. A lattice packing is one in which the centers of the circles form a lattice. The least upper bound of the densities of k-fold packings is denoted by d_k' and that of lattice packings by d_k. The author proves by elementary geometric arguments that $d_k=kd_1=k\pi/\sqrt{12}$ for $k=1, 2, 3, 4$ and $d_k>kd_1$ for $k\geqq 5$. Since the author had previously shown [Elem. Math. **10** (1955), 125–127; MR 17, 523] that $d_k'>kd$ for $k>1$, it follows that $d_k'>d_k$ for $k=2, 3, 4$.

E. G. Straus (Los Angeles, Calif.)

Referred to in H30-56, H30-58, H30-64.

H30-44 **(22# 2596)**

Rogers, K.; Straus, E. G. A class of geometric lattices. Bull. Amer. Math. Soc. **66** (1960), 118–123.

Hajós's theorem [Math. Z. **47** (1941), 427–467; MR **3**, 302] gives a full enumeration of the critical lattices Λ of the cube K: $|x_1|\leqq 1, |x_2|\leqq 1, \cdots, |x_n|\leqq 1$ in R^n: Let A be an $n\times n$ matrix the columns of which form a basis of Λ. Then Λ is critical if the basis can be chosen such that A has 1's in the diagonal and 0's above it. A is said to have the property P if for every vector $\mathbf{u}\neq\mathbf{0}$ with integral components at least one component of $A\mathbf{u}$ is an integer not 0. If it were true that (i) "every A with property P

has an integral row", Hajós's theorem would follow from a result by Siegel stating that every point $\neq \mathbf{0}$ of a critical lattice of K has at least one integral coordinate $\neq 0$; but (i) is false for $n \geqq 5$. The authors investigate in detail matrices A with the property P and prove in particular the following theorems. (1) Let k be an algebraic number field of class number 1; let j be the ring of integers in k; and let A be an $n \times n$ matrix with elements in k. If for every vector $\mathbf{u} \neq \mathbf{0}$ with components in j at least one component of $A\mathbf{u}$ is in j and is $\neq 0$, then det A is likewise in j and $\neq 0$. (2) If A is as in (1) and $0 < |\text{norm} (\det A)| < 1$, there exists a $\mathbf{u} \neq \mathbf{0}$ with components in j such that $A\mathbf{u}$ has no component $\neq 0$ in j. *K. Mahler* (Manchester)

Citations: MR 3, 302b = H30-4.

H30-45 (22# 9484)

Macbeath, A. M. Abstract theory of packings and coverings. I. Proc. Glasgow Math. Assoc. **4**, 92–95 (1959).

The author considers the measure space (X, S, μ) and a countable group G of permutations of X, each of which is measure-preserving. e denotes an identical permutation. A set $P \in S$ is a G-packing if $P \cap gP = \varphi$ for all $g \neq e$, $g \in G$; and a G-covering if $\bigcup\{gP: g \in G\} = X$; P is a fundamental domain if it is both a G-covering and a G-packing. The author shows that the measure of any G-covering is greater than or equal to the measure of any G-packing and uses this result to show that if X is a metric space such that the measure of a sphere of radius $r > \inf \mu(C)$ for C any G-covering then there exists $g \in G$, $g \neq e$ such that the distance between the centre of the sphere and its image under g is less than $2r$. The author also proves a somewhat similar result when X is a group and G is a sub-group acting on X by left translation. A number of applications are given which include results of Minkowski, Siegel, Blichfeldt and Santaló.

H. G. Eggleston (London)

H30-46 (22# 9485)

Świerczkowski, S. Abstract theory of packings and coverings. II. Proc. Glasgow Math. Assoc. **4**, 96–100 (1959).

The author establishes necessary and sufficient conditions for the existence of a fundamental domain, using the terminology of the previous review. The measure space (X, S, μ) has a fundamental domain if and only if the following three properties hold: (1) if $A \in S$, $g \in G$, $\mu(A) > 0$, and $g \neq e$, then there exists $B \subset A$ such that $\mu(B) > 0$ and $B \cap gB = \varphi$; (2) no $g \neq e$ has fixed points; (3) if $A \in S$ has arbitrarily small coverings then $\mu(A) = 0$.

The author also gives further results of a similar nature and an example to show that his criterion is valid only when the measure μ is σ-finite and complete.

H. G. Eggleston (London)

H30-47 (23# A120)

Macbeath, A. M.

On convex fundamental regions for a lattice.

Canad. J. Math. **13** (1961), 177–178.

F is a closed convex region of real Euclidean n-space E_n. Λ is the lattice of all points of E_n with integer coordinates. F is defined to be a fundamental region if the set of all translations $x + F$, $x \in \Lambda$, of F covers E_n and $F° \cap x + F° = \emptyset$, $x \neq 0$, where $F°$ is the set of interior points of F. It is shown, with the use of the Brunn-Minkowski theorem, that every fundamental region has a centre of symmetry. If each vertex of F is a vertex of exactly n Λ-translates of F, F is said to be primitive. The primitive fundamental regions were determined by Voronoĭ [J. Reine Angew. Math. **134** (1908), 198–287] and the above result for such fundamental regions is a consequence of this determination. The present result holds also for non-primitive regions F which are not yet characterized.

D. Derry (Vancouver, B.C.)

H30-48 (23# A862)

Few, L.

A mixed packing problem.

Mathematika **7** (1960), 56–63.

Let Λ' be a lattice in Euclidean 3-space E_3. Let P be any point in E_3 and denote by Λ the "displaced lattice" consisting of all points $X + P$, where $X \in \Lambda'$. Suppose that a sphere of radius 1 is centred at each point of Λ and a sphere of radius $R-1 (1 < R \leqq 2)$ is centred at each point of Λ'. Now if P and Λ' are such that no two spheres of the system overlap, Λ is called a "mixed packing lattice". Let Δ denote the lower bound of the determinants $d(\Lambda) = d(\Lambda')$ of all mixed packing lattices. It is easily proved that Λ is an attained bound, and any mixed packing lattice Λ, with $d(\Lambda) = \Delta$, is called a critical mixed packing lattice. The values of Δ are determined, namely:

$$\begin{aligned} \Delta &= 4\sqrt{2} && \text{if } 1 < R^2 \leqq 2, \\ &= 4(R^2-1)\sqrt{(4-R^2)} && \text{if } 2 \leqq R^2 \leqq 3, \\ &= 8\sqrt{(R^2-2)} && \text{if } 3 \leqq R^2 \leqq 4, \end{aligned}$$

and the critical mixed packing lattices identified.

J. H. H. Chalk (Toronto)

H30-49 (23# A2130)

Rogers, C. A.

Lattice coverings of space.

Mathematika **6** (1959), 33–39.

Ist $\vartheta_n(K)$ die Dichte der dünnsten gitterförmigen Überdeckung des R_n durch konvexe Körper, welche aus K durch Parallelverschiebung hervogehen, dann wird gezeigt: Für großes n ist $\log \vartheta_n \leqq (\log^2 \log n + O(1)) \log n$. Ist K die (n-dimensionale) Kugel S, so ist $\log \vartheta_n(S) \leqq \log n + \frac{1}{2} \log^2 (2\pi e) \log \log n + O(1)$. Bisher war nur bekannt, daß $\log \vartheta_n = O(n)$.

E. Hlawka (Zbl **87**, 42)

Referred to in H02-17, H30-54.

H30-50 (24# A1887)

Zassenhaus, Hans

Modern developments in the geometry of numbers.

Bull. Amer. Math. Soc. **67** (1961), 427–439.

In this report of an invited address to the American Mathematical Society, the author describes the work of his students and associates on a type of packing problem, to which he has given the name "statistical geometry of numbers". Let Γ be a plane star domain with distance function $d(x)$. Let J be a Jordan polygon with vertices $P_1, \cdots, P_n$ and let $S = \{P_1, \cdots, P_n, P_{n+1}, \cdots, P_N]$ be a finite set lying in the closed domain bounded by J. S is called a packing set for Γ if $d(P-Q) \geqq 1$ for all P, Q in S. Defining $C(J) = \sum_{i=0}^{n-1} d(P_{i+1} - P_i)$, $P_0 = P_n$, he defines the slackness $\sigma(S, J, d)$ of the packing by

$$\sigma(S, J, d) = \frac{A(J)}{\Delta} + \frac{C(J)}{2} + 1 - N,$$

where $A(J)$ is the area of J and Δ is the critical determinant of Γ.

The author conjectured in 1947 that $\sigma(S, J, d) \geqq 0$ for all symmetrical convex Γ. This result has been proved by Oler [Acta Math. **105** (1961), 19–48] *. A similar result was proved by N. E. Smith [Ph.D. Thesis, McGill Univ., 1951] for the set $|x_1 x_2| \leqq 1$.

The author also describes the so-called "domain of

action method" developed by M. Rahman [Ph.D. Thesis, McGill Univ., 1957] and used by Sr. Mary Robert von Wolff [Ph.D. Thesis, Univ. of Notre Dame, 1961] to get estimates for densities of non-lattice "packings". {The author uses the concept of a packing of Γ in a slightly different sense from the usual one of non-overlapping of translates of Γ. His Γ packings are equivalent to the non-overlapping packings of sets Σ such that Γ is the difference set of Σ.} *R. P. Bambah* (Chandigarh)

*See next review.

Referred to in H30-96.

H30-51 (24# A2900)

Oler, N.

An inequality in the geometry of numbers.

Acta Math. **105** (1961), 19–48.

Let Γ be a plane symmetrical convex domain with distance function $\mu(x)$. The pair (Π, E), where Π is a Jordan polygon bounding a closed domain Π^* and E is a finite set contained in Π^*, is called weakly admissible for Γ, if the vertices $P_1, \cdots, P_n$ of Π belong to E and if for any points P, Q, of E, such that the segment PQ lies in Π^*, the sets $\frac{1}{2}\Gamma+P$, $\frac{1}{2}\Gamma+Q$ do not overlap. Defining

$$M(\Pi) = \sum_{i=1}^{n} \mu(P_{i+1}-P_i), \quad P_{n+1} = P_1,$$

the author proves the remarkable theorem that for weakly admissible (Π, E), we have

$$\frac{A(\Pi^*)}{\Delta}+\frac{M(\Pi)}{2}+1 \geqq N,$$

where $A(\Pi^*)$ is the area of Π^*, Δ is the critical determinant of Γ, and N is the number of points in E.

The theorem confirms a conjecture of Zassenhaus and generalises an inequality of C. A. Rogers [Acta Math. **86** (1951), 309–321; MR **13**, 768], which the latter used to deduce his famous theorem that the best lattice and best general packings of a symmetrical convex domain in the plane have the same density. [For a sketch of another proof, see L. Fejes Tóth, Acta Sci. Math. (Szeged) **12** (1950), 62–67; MR **12**, 352.] *R. P. Bambah* (Chandigarh)

Citations: MR 13, 768i = H30-17.

Referred to in H30-92.

H30-52 (24# A3562)

Heppes, A.

Ein Satz über gitterförmige Kugelpackungen.

Ann. Univ. Sci. Budapest. Eötvös Sect. Math. **3–4** (1960/61), 89–90.

The author proves, in a particularly simple manner, the following pleasing theorem. Given any lattice-packing of (3-dimensional Euclidean) space by equal spheres (i.e., the spheres are non-overlapping and their centers form a point-lattice), there exist three lines in linearly independent directions, each of which has no point in common with any of the spheres. *W. Moser* (Winnipeg, Man.)

H30-53 (25# 2516)

Groemer, Helmut

Über die dichteste gitterförmige Lagerung kongruenter Tetraeder.

Monatsh. Math. **66** (1962), 12–15.

The author proves that the density of the closest lattice packing of congruent tetrahedra is at least 18/49, and points out a mistake in an argument of Minkowski which was supposed to show that that density is 9/38.

T. Estermann (London)

Referred to in H30-89.

H30-54 (26# 6862)

Bleicher, M. N.

Lattice coverings of n-space by spheres.

Canad. J. Math. **14** (1962), 632–650.

In Euclidean n-space E_n, n independent vectors form a basis for a lattice Λ; $d(\Lambda)$, the absolute value of the $n \times n$ determinant whose rows are the coordinates of a basis, is called the determinant of Λ. For any Λ there is a unique minimal positive number r such that if spheres of radius r are placed with centres at all points of Λ, the entire space is covered. The density of this covering is $\theta_n(\Lambda) = J_n r^n/d(\Lambda)$, where J_n is the volume of the n-dimensional unit sphere. $\theta_n(\Lambda)$, considered as a function of Λ, has an absolute minimum θ_n, the density of the most efficient lattice covering [see Hlawka, Monatsh. Math. **53** (1949), 81–131; MR **11**, 12]. A lattice is called extreme if it yields a relative minimum of $\theta_n(\Lambda)$. In 1952 Davenport [Rend. Circ. Mat. Palermo (2) **1** (1952), 92–107; MR **14**, 75] showed that for large n, $\theta_n < (11\pi e/54\sqrt{3}+\varepsilon)^n < (1.15)^n$, and Bambah and Davenport [J. London Math. Soc. **27** (1952), 224–229; MR **14**, 787] showed $\theta_n < 4/3-\varepsilon_n$ where $\lim \varepsilon_n = 0$; in 1959 Rogers [Mathematika **6** (1959), 33–39; MR **23** #A2130] proved that $\theta_n < \beta_n = O(n(\log_e n)^{(\frac{1}{2}\log_2 2\pi e)})$ {and not, as the author states, $\theta_n = O(n(\log_e n)^{(\frac{1}{2}\log_2 2\pi e)})$}, and Coxeter, Few and Rogers [ibid. **6** (1959), 147–157; MR **23** #A2131]; (in this review, and also in MR **23** #A2130, the $\log_2 2\pi e$ of the above expression is incorrectly given as $\log^2 2\pi e$) exhibited numbers $t_n < \theta_n$ with $t_n \sim n/e\sqrt{e}$. {Bleicher's $\theta_n(\Lambda)$ is Rogers' $\theta(S_n, \Lambda)$.} In this paper the author considers the lattice Λ_n, whose matrix is

$$\begin{bmatrix} \frac{1}{\sqrt{2}} & \frac{-1}{\sqrt{6}} & \frac{-1}{\sqrt{12}} \cdots & \frac{-1}{\sqrt{k(k+1)}} & \cdots \frac{-1}{\sqrt{n(n+1)}} \\ 0 & \frac{\sqrt{2}}{\sqrt{3}} & \frac{-1}{\sqrt{12}} \cdots & \frac{-1}{\sqrt{k(k+1)}} & \cdots \frac{-1}{\sqrt{n(n+1)}} \\ \vdots & & & & \\ 0 & 0 & 0 \quad \cdots & \frac{\sqrt{k}}{\sqrt{(k+1)}} & \cdots \frac{-1}{\sqrt{n(n+1)}} \\ \vdots & & & & \\ 0 & 0 & 0 \quad \cdots & 0 & \cdots \frac{\sqrt{n}}{\sqrt{(n+1)}} \end{bmatrix} \frac{\sqrt{(n+1)2\sqrt{3}}}{\sqrt{n(n+2)}}.$$

He computes

$$\theta_n(\Lambda_n) = \frac{1}{\Gamma((n+2)/2)}\left[\frac{\pi n(n+2)}{12(n+1)^{1-1/n}}\right]^{n/2};$$

observes that $\theta_n(\Lambda_n) \leqq \theta_n$ with equality for $n=1, 2, 3$; proves that Λ_n is extreme for all n; and, observing that $\beta_n < \theta_n(\Lambda_n)$, notes that Λ_n is, for all sufficiently large n, not absolutely extreme. *W. Moser* (Winnipeg, Man.)

Citations: MR 11, 12b = H30-12; MR 14, 75h = H30-21; MR 14, 787f = H30-25; MR 23# A2130 = H30-49.

Referred to in H30-73, H30-80, J48-18.

H30-55 (26# 6863)

Erdős, P.; Rogers, C. A.

Covering space with convex bodies.

Acta Arith. **7** (1961/62), 281–285.

It is shown that for every convex body K in Euclidean n-space, E^n, with sufficiently large n, there exists a covering of E^n by translates of K whose density is less than $n \log n + n \log \log n + 4n$ and so that no point is covered more than $e(n \log n + n \log \log n + 4n)$ times. The translations considered are unions of a finite number of lattices. The error term in the estimation of $\log \mathscr{M}(\delta(F_h))$ should be $O(\log n)$ instead of $O(1)$. This does not affect the validity of the proof. *E. G. Straus* (Los Angeles, Calif.)

H30-56 **(26# 6864)**

Blundon, W. J.

Multiple packing of circles in the plane.

J. London Math. Soc. **38** (1963), 176–182.

An r-fold packing of equal circles in the plane is a set of congruent (unit) circles such that no point lies in the interior of more than r of the circles. The least upper bound of the densities of r-fold packings is denoted by $\delta^{(r)}$, and that of lattice packings (where the centres of the circles form a lattice) by $\delta_1^{(r)}$. It is well known that $\delta^{(1)}=\delta_1^{(1)}=\pi/2\sqrt{3}=0.9069\cdots$ [see L. Fejes Tóth, *Lagerungen in der Ebene, auf der Kugel und im Raum*, Ch. III, Springer, Berlin, 1953; MR **15**, 248]. A Heppes [Acta Math. Acad. Sci. Hungar. **10** (1959), 141–148; MR **21** #3812] proved that $\delta^{(r)}=r\delta^{(1)}=r\pi/2\sqrt{3}$ for $r\leqq 4$, and $\delta^{(r)}>r\delta^{(1)}$ for $r>4$. In the paper under review the author considers the cases $r=5$ and $r=6$, and proves that $\delta^{(5)}=(8/7)(\sqrt{21})\delta^{(1)}=5.237\cdots\delta$, and that the lattice generated by the points $(\frac{1}{2}\sqrt{2},\ 0)$ and $(\frac{1}{4}\sqrt{2},\ \frac{1}{4}\sqrt{14})$ provides a closest 5-fold packing; $\delta^{(6)}=(35/8)(\sqrt{2})\delta=6.187\cdots\delta$, and the lattice generated by the points $((2/35)\sqrt{105},\ 0)$ and $(0,\ (4/35)\sqrt{70})$ provides a closest 6-fold packing. *W. Moser* (Winnipeg, Man.)

Citations: MR 15, 248b = H30-27; MR 21# 3812 = H30-43.

Referred to in H30-64.

H30-57 **(27# 5731)**

Gameckiĭ, A. F.

The optimality of Voronoĭ's lattice of first type among lattices of first type of arbitrary dimension. (Russian)

Dokl. Akad. Nauk SSSR **151** (1963), 482–484.

A Voronoi basis of a lattice Γ is a set of $n+1$ vectors with zero sum and the further properties that any n form a (usual) basis while the $(n+1)!$ ordering of vectors determines the simplicial star of the origin. Such a basis is said to be of first type if the $n(n+1)/2$ scalar products of these $n+1$ vectors are equal. The author considers critical lattices for the "packing" of circumscribed (instead of inscribed) spheres. Specifically for Γ, let R_Γ be the smallest radius for which spheres of radius $\geqq R_\Gamma$ about each lattice point enclose all of E^n. Then let D_Γ denote the (Dirichlet) volume for the (fundamental parallelogram of the) lattice Γ. The critical lattice Γ_0 which maximizes $R_\Gamma{}^n/D_\Gamma$ is then shown to have a Voronoi basis of first type. Reference is made to the author's earlier work [same Dokl. **146** (1962), 991–994]. {Some further cognate references to Voronoi's work can be found in Coxeter [Canad. J. Math. **3** (1951), 391–441; MR **13**, 443], and also in Voronoi [J. Reine Angew. Math. **134** (1908), 198–287].} *H. Cohn* (Tucson, Ariz.)

Citations: MR 13, 443c = J44-11.

Referred to in H30-75, H30-80.

H30-58 **(27# 6192)**

Blundon, W. J.

Note on a paper of A. Heppes.

Acta Math. Acad. Sci. Hungar. **14** (1963), 317.

A. Heppes [same Acta **10** (1959), 141–148; MR **21** #3812] proved that if d_k is the density of the closest k-fold packing of equal circles in the plane, then $d_k\geqq kd_1$, with equality for $k<5$. Now the author observes that a trivial additional remark to Heppes' proof shows that $d_k\geqq d_1(k^2-1)/\sqrt{(k^2-4)}$ for $k\geqq 5$.

W. Moser (Winnipeg, Man.)

Citations: MR 21# 3812 = H30-43.

Referred to in H30-64.

H30-59 **(29# 525)**

Groemer, Helmut

Existenzsätze für Lagerungen im Euklidischen Raum.

Math. Z. **81** (1963), 260–278.

In recent decades there has accumulated a large body of theorems having to do with densities of packings and coverings in R^n, Euclidean n-space. The classical one is that of the densest packing and thinnest covering of R^n by congruent spheres. These have been generalized in many ways, e.g., to spheres of different radii, convex and non-convex bodies, multiple packings and coverings, and so on. Most of these results are inequalities on the magnitudes of densities, and do not prove the existence of configurations having maximum (packing) or minimum (covering) densities. Indeed, the only existence theorems of this sort known to date are due to R. A. Rankin [Ann. of Math. (2) **48** (1947), 1062–1081; MR **9**, 226] for the case of packing and coverings of R^n by congruent spheres, and E. Hlawka [Monatsh. Math. **53** (1949), 81–131; MR **11**, 12] for the more general case of packings and coverings by convex bodies. The author considers the problem in a general way and obtains such general existence theorems that this paper is an important addition to this field; indeed, it is certain to be considered a classic. A few definitions and theorems will have to suffice for a review.

A finite or infinite sequence $\Sigma=(S_i)$ of point sets S_i in R^n is a configuration provided: (a) any bounded region of R^n contains only a finite number of S_i; (b) the set of all diameters of the S_i is bounded; (c) to each S_i there corresponds a positive number γ_i. If Σ is a finite sequence, $\gamma(\Sigma)$ denotes the sum of γ_i taken over all S_i in Σ.

For any point p the ρ-section of p relative to Σ is the set of all $S_i\in\Sigma$ whose distance from p is $\leqq\rho$.

A set $\mathscr{S}=(\Sigma_\nu)$ of configurations Σ_ν is said to be proper if the following conditions (I), (II), and (III) or (III′) are satisfied. (I) If $\Sigma_\nu\in\mathscr{S}$, and Σ_ν' is obtained from Σ_ν by a translation, then $\Sigma_\nu'\in\mathscr{S}$. (II) There exists $\rho\geqq 0$ such that if Φ is a configuration whose ρ-sections coincide with the ρ-sections of some $\Sigma_\nu\in\mathscr{S}$, then $\Phi\in\mathscr{S}$. (III) If $\Sigma_\kappa\in\mathscr{S}$ and $S\in\Sigma_\kappa$, then $\Sigma_\kappa-(S)\in\mathscr{S}$. (III′) If $\Sigma_\kappa\in\mathscr{S}$ and $S\in\Sigma_\lambda\in\mathscr{S}$, then $\Sigma_\kappa\cup(S)\in\mathscr{S}$.

K will denote a bounded point set with positive volume (Jordan content) $\mu(K)$. For any positive number σ, the points σp with $p\in K$ will be denoted by σK. $\bar K$ is the closure of the convex cover of K, and $\tilde K$ the interior of K. K is starlike if for all $0\leqq\sigma<1$, $\sigma\bar K\subset K$. Σ^*K and Σ_*K denote the set of all $S\in\Sigma$ with $K\cap S\neq\varnothing$ and $S\subset K$, respectively. Let

$$\delta^*(\Sigma, K)=\frac{\gamma(\Sigma^*K)}{\mu(K)},\qquad \delta_*(\Sigma, K)=\frac{\gamma(\Sigma_*K)}{\mu(K)},$$

$$\bar\delta(\Sigma, K)=\limsup_{\sigma\to\infty}\delta^*(\Sigma,\sigma K),$$

$$\underline{\delta}(\Sigma, K)=\liminf_{\sigma\to\infty}\delta_*(\Sigma,\sigma K).$$

If $\bar\delta(\Sigma, K)=\underline{\delta}(\Sigma, K)$ for all K, Σ is said to be "gleichmässig".

Let $\mathscr{S}$ be a collection of configurations. Let $\mathscr{K}$ be the set of all K in R^n. Various constants (densities) are defined as follows.

$$\delta^0(\mathscr{S}, K)=\sup\bar\delta(\Sigma, K),\quad \delta_0(\mathscr{S}, K)=\inf\underline{\delta}(\Sigma, K)\qquad(\Sigma\in\mathscr{S}),$$

$$\delta^0(\mathscr{S})=\sup\delta^0(\mathscr{S}, K),\quad \delta_0(\mathscr{S})=\inf\delta_0(\mathscr{S}, K)\qquad(K\in\mathscr{K}).$$

Theorem: If $\mathscr{S}$ is proper, then

$$\tilde{\delta}^0(\mathscr{S}, K) = \lim_{\sigma\to\infty} \sup_{\Sigma\in\mathscr{S}} \delta^*(\Sigma, \sigma K),$$

$$\tilde{\delta}_0(\mathscr{S}, K) = \lim_{\sigma\to\infty} \inf_{\Sigma\in\mathscr{S}} \delta_*(\Sigma, \sigma K)$$

exist. Furthermore, these limits, as well as $\delta^0(\mathscr{S}, K)$ and $\delta_0(\mathscr{S}, K)$, are independent of K, and $\delta^0(\mathscr{S}) = \tilde{\delta}^0(\mathscr{S})$, $\delta_0(\mathscr{S}) = \tilde{\delta}_0(\mathscr{S})$. Furthermore, there exist "gleichmässige" configurations Σ^0, Σ_0 in $\mathscr{S}$, such that $\delta^0(\mathscr{S}) = \bar{\delta}(\Sigma^0) = \underline{\delta}(\Sigma^0)$, $\delta_0(\mathscr{S}) = \bar{\delta}(\Sigma_0) = \underline{\delta}(\Sigma_0)$. Theorem (with an error corrected): If K is arbitrary and Σ is "gleichmässig", or if K is starlike and Σ is arbitrary, then

$$\bar{\delta}(\Sigma, K) = \limsup_{\sigma\to\infty} \delta^*(\Sigma, \sigma K) = \limsup_{\sigma\to\infty} \delta_*(\Sigma, K),$$

$$\underline{\delta}(\Sigma, K) = \liminf_{\sigma\to\infty} \delta_*(\Sigma, \sigma K) = \liminf_{\sigma\to\infty} \delta^*(\Sigma, K).$$

Furthermore, if Σ' is a translation of Σ, then

$$\bar{\delta}(\Sigma', K) = \bar{\delta}(\Sigma, K), \quad \underline{\delta}(\Sigma', K) = \underline{\delta}(\Sigma, K).$$

Theorem: If $\mathscr{S}$ is an arbitrary collection of configurations, and if K is starlike, then in the case where the limits exist,

$$\tilde{\delta}^0(\mathscr{S}) = \limsup_{\sigma\to\infty}\,_{\Sigma} \delta^*(\Sigma, \sigma K) = \limsup_{\sigma\to\infty}\,_{\Sigma} \delta_*(\Sigma, \sigma K) \quad (\Sigma \in \mathscr{S}),$$

$$\tilde{\delta}_0(\mathscr{S}) = \liminf_{\sigma\to\infty}\,_{\Sigma} \delta_*(\Sigma, \sigma K) = \liminf_{\sigma\to\infty}\,_{\Sigma} \delta^*(\Sigma, \sigma K) \quad (\Sigma \in \mathscr{S}).$$

W. Moser (Montreal, Que.)

Citations: MR 9, 226d = H30-8; MR 11, 12b = H30-12.

H30-60 (29# 1581)

Coxeter, H. S. M.

An upper bound for the number of equal nonoverlapping spheres that can touch another of the same size.

Proc. Sympos. Pure Math., Vol. VII, pp. 53–71. *Amer. Math. Soc., Providence, R.I.,* 1963.

Reasons are given for believing that the number of equal nonoverlapping $(n-1)$-spheres that can touch another of the same size (in Euclidean n-space) has the upper bound $2f_{n-1}(n)/f_n(n)$, where the function f is defined recursively by

$$f_n(x) = \frac{1}{\pi}\int_{n-1}^{x} \frac{f_{n-2}(t-2)}{t(t^2-1)^{1/2}}\,dt,$$

with initial conditions $f_0(x) = f_1(x) = 1$.

The author summarizes the history of the problem (from the seventeenth century right up to the latest research) with an excellent list of references. The material is arranged so as to be useful to the specialist and interesting to the novice. Some of this information is described in the following paragraphs.

Kepler (1611, misprinted as 1911 on page 71), discovered the cubic close-packing of equal spheres in Euclidean 3-space. To the question "Can a rigid material sphere be brought into contact with 13 other such spheres of the same size?" Gregory said "Yes" and Newton said "No"; 180 years later the issue was settled in favor of Newton.

In any packing of space by equal spheres each sphere is surrounded by a Dirichlet region (or Voronoi polyhedron) consisting of all points which are as near to the center of that sphere as to the center of any other. The density of the packing is defined as the average of the ratio of the volume of a sphere to the volume of the Dirichlet region that surrounds it. Thus Kepler's cubic close-packing is seen to have density $\pi/3\sqrt{2} = 0.74048\cdots$.

Schläfli (1852) generalized spherical trigonometry to an arbitrary number of dimensions and defined the remarkable function $f_n(x)$, which enabled him to express the content of a regular simplex in spherical $(n-1)$-space in terms of its dihedral angle. He observed that, when n is odd,

$$f_n(x) = f_{n-1}(x) - \tfrac{1}{3}f_{n-3}(x) + \tfrac{2}{15}f_{n-5}(x) - \cdots,$$

where the coefficients (apart from sign) agree with

$$\tan x = x + \tfrac{1}{3}x^3 + \tfrac{2}{15}x^5 + \cdots.$$

Relations resembling this were given later by Poincaré, Sommerville, Peschl, and Guinand.

Minkowski [J. Reine Angew. Math. **129** (1905), 220–274] proved that in Euclidean n-space with $n = 3$, 4, or 5, the densest packing of equal spheres whose centres form a lattice consists of spheres of radius $1/\sqrt{2}$ whose centres have integral Cartesian coordinates with an even sum. Blichfeldt [Math. Z. **39** (1934), 1–15] obtained the corresponding results for $n = 6$, 7, and 8. The author described the centres of such spheres (of radius $3/\sqrt{2}$) as having for coordinates nine integers, mutually congruent modulo 3 and satisfying the equations

$$x_1+x_2+x_3 = x_4+x_5+x_6 = x_7+x_8+x_9 = 0 \quad (n = 6),$$

$$x_1+x_2+x_3+x_4+x_5+x_6 = x_7+x_8+x_9 = 0 \quad (n = 7),$$

$$x_1+x_2+x_3+x_4+x_5+x_6+x_7+x_8+x_9 = 0 \quad (n = 8).$$

Let $N(\phi)$ denote the maximum number of "caps" ($(n-2)$-spheres) of angular radius ϕ that can be packed on the unit $(n-1)$-sphere. Rankin [Proc. Glasgow Math. Assoc. **2** (1955), 139–144; MR **17**, 523] proved that $N(\phi) = [\sec(\pi - 2\phi)] + 1$ for $\pi - \operatorname{arcsec} n \leqq 2\phi \leqq \pi$. Davenport and Hajós [Mat. Lapok **2** (1951), 68] proved that $N(\phi)$ cannot take a value between $n+1$ and $2n$, so that $N(\phi) = n+1$ for $\frac{1}{2}\pi < 2\phi \leqq \pi - \operatorname{arcsec} n$. For smaller values of ϕ Rankin obtained inequalities for the value of $N(\phi)$. In the paper under review the author gives reasons for believing that

$$N(\phi) \leqq 2f_{n-1}(x)/f_n(x), \qquad x = \sec 2\phi + n - 2.$$

Leech computed values of $f_n(x)$, which the author uses to establish the following inequalities for $N_n = N(\phi/6)$, which is the maximum number of spheres that can touch another of the same size in Euclidean n-space: $N_1 = 2$, $N_2 = 6$, $N_3 = 12$, $24 \leqq N_4 \leqq 26$, $40 \leqq N_5 \leqq 48$, $72 \leqq N_6 \leqq 85$, $126 \leqq N_7 \leqq 146$, $240 \leqq N_8 \leqq 244$.

The author considers the problem in spherical, elliptic, and hyperbolic spaces as well, and concludes with a proof that $2^{(n-1)/2}\pi^{1/2}e^{-1}n^{3/2}$ is an asymptotic expression for the bound named in the title of the paper.

W. Moser (Montreal, Que.)

Referred to in H30-62.

H30-61 (29# 3444)

Chalk, J. H. H.

A local criterion for the covering of space by convex bodies.

Acta Arith. **9** (1964), 237–243.

Let K be a convex body in n-dimensional Euclidean space R^n, and Λ_0 the integer lattice. Let $\mu = \inf\{t : tK + \Lambda_0 = R^n\}$ be the non-homogeneous minimum, and let $\lambda_K{}^*$ be the smallest value of t such that for some $d \in R^n$ the convex body $tK + d$ contains "a K-dimensional set of $K+1$ points of Λ_0". The following inequality is proved:

$$\mu \leqq \tfrac{1}{2}(n+1)\lambda_n{}^* \qquad \text{if } n \text{ is odd},$$

$$\leqq \tfrac{1}{2}n\left(1+\frac{1}{n+1}\right)\lambda_n{}^* \quad \text{if } n \text{ is even}.$$

The value of μ is calculated for the modified octahedron.

A. M. Macbeath (Birmingham)

H30-62 **(29# 5166)**

Leech, John

Some sphere packings in higher space.

Canad. J. Math. **16** (1964), 657–682.

In the first part of this paper the densest lattice packings of equal spheres in four and eight dimensions are generalized to obtain a packing in 2^m dimensions in which each sphere is touched by $\prod_{r=1}^{m}(2^r+2)$ others.

In the second part of the paper analogues of the densest lattice packing in eight dimensions are generalised to obtain new packings in twelve and twenty-four dimensions. In the packing in twelve dimensions, each sphere is touched by 648 others, in twenty-four dimensions each sphere is touched by 98256 others. The lattices associated with these packings seem to be connected with the Steiner systems $S(5, 6, 12)$ and $S(5, 8, 24)$. Packings in spaces of lower dimension are derived by taking suitable linear sections.

The third part of the paper connects the number of spheres touching a given sphere of the packing, and the density of the packing, with the upper bounds determined by Coxeter [Proc. Sympos. Pure Math., Vol. VII, pp. 53–71, Amer. Math. Soc., Providence, R.I., 1963; MR **29** #1581] and Rogers [Proc. London Math. Soc. (3) **8** (1958), 609–620; MR **21** #847].

J. A. Todd (Cambridge, England)

Citations: MR 21# 847 = H30-42; MR 29# 1581 = H30-60.

Referred to in H30-78.

H30-63 **(29# 6395)**

Baranovskiĭ, E. P.

On packing n-dimensional Euclidean spaces by equal spheres. I. (Russian)

Izv. Vysš. Učebn. Zaved. Matematika **1964**, *no.* 2 (39), 14–24.

The author obtains an upper bound D_n for the density of packings of equal n-dimensional spheres. D_n is expressed in terms of the volume of a certain simplex and, to this extent, is equivalent to the work of C. A. Rogers [Proc. London Math. Soc. (3) **8** (1958), 609–620; MR **21** #847], although there are differences in the methods used. For $n=3$ an explicit calculation is made giving

$$D_3 = 2^{1/2}\{3\cos^{-1}(\tfrac{1}{3})-\pi\} = 0.77963\ldots,$$

in agreement with Rogers. No asymptotic evaluation for large n is made as was done by H. E. Daniels in the paper of Rogers. *R. A. Rankin* (Glasgow)

Citations: MR 21# 847 = H30-42.

H30-64 **(30# 2404)**

Blundon, W. J.

Some lower bounds for density of multiple packing.

Canad. Math. Bull. **7** (1964), 565–572.

Let d_k be the least upper bound of the densities of k-fold packings of the plane by equal circles (i.e., every point of the plane is an interior point of at most k of the circles) whose centers form a lattice. It is known that $d_1=\pi/2\sqrt{3}=.9069\ldots$, and it is easy to show that $d_k/kd_1\geqq 1$ [the author, J. London Math. Soc. **38** (1963), 176–182; MR **26** #6864], with equality only if $k<5$ [A. Heppes, Acta Math. Acad. Sci. Hungar. **10** (1959), 141–148; MR **21** #3812]. The author [ibid. **14** (1963), 317; MR **27** #6192] proved that $d_k/kd_1\geqq(k^2-1)/k(k^2-4)^{1/2}$, $k\geqq 5$. In the paper under review, the author proves the following. Theorem 1: Let $c=[k\theta]$, where $\theta=(6-\sqrt{10})/13=.21828\ldots$. Let $f(x)=(1-x^2)/(1-4x^2)^{1/2}$. Then $d_k/kd_1\geqq f(c/k)$, for $k\geqq 5$, and a reduced basis for the lattice providing this packing is given by the points $(a, 0)$ and $(0, h)$ for even k, $(a, 0)$ and $(\frac{1}{2}a, h)$ for odd k, where $a^2=12/(k^2-c^2)$ and $h^2=(k^2-4c^2)/(k^2-c^2)$. Theorem 2: For every $\varepsilon>0$, there exist arbitrarily large positive integers k such that $d_k/kd_1>f(\theta)-\varepsilon$, where

$$f(\theta) = \frac{41\sqrt{5}+20\sqrt{2}}{845}(5+16\sqrt{10})^{1/2} = 1.0585\ldots.$$

W. Moser (Montreal, Que.)

Citations: MR 21# 3812 = H30-43; MR 26# 6864 = H30-56; MR 27# 6192 = H30-58.

H30-65 **(31# 126)**

Delone, B. N.; Ryškov, S. S.

Solution of the problem on the least dense lattice covering of a 4-dimensional space by equal spheres. (Russian)

Dokl. Akad. Nauk SSSR **152** (1963), 523–524.

The problem of the least dense lattice covering by equal spheres, which is trivial in the plane, was first solved for three-dimensional Euclidean space by Bambah [Proc. Nat. Inst. Sci. India **20** (1954), 25–52; MR **15**, 780] and for the first of the three types of four-dimensional lattice by Gameckiĭ [Dokl. Akad. Nauk SSSR **146** (1962), 991–994]. Using a table of G. F. Voronoĭ [*Collected works in three volumes*, Vol. II, pp. 341–368, Izdat. Akad. Nauk Ukrain. SSR, Kiev, 1952; MR **16**, 2] on the structures of the stars and Dirichlet regions, for all three types of four-dimensional lattice, the authors solve the corresponding problem for the second and third types of four-dimensional lattice.

{This paper has appeared in English translation [Soviet Math. Dokl. **4** (1963), 1333–1334].}

R. Finkelstein (Tucson, Ariz.)

Citations: MR 15, 780c = H30-32; MR 16, 2h = Z25-11.

Referred to in H30-75, H30-80, H30-81, J48-18.

H30-66 **(31# 665)**

Ignat'ev, N. K.

On a practical method of finding dense packings of n-spheres. (Russian)

Sibirsk. Mat. Ž. **5** (1964), 815–819.

Verfasser hat in seiner früheren Arbeit [Sb. Trudov Gos. NII Min. Svjazi SSSR **8** (12) (1958), 85–145] ein Verfahren entwickelt, sehr dichte gitterförmige Kugelpackungen im R_n zu konstruieren. In diesen Gittern sind orthotope Gitter, d.h. solche mit lauter rechten Winkeln enthalten. Die Struktur hängt von der Restklasse mod 4 ab, in der die Dimensionszahl vorkommt. Für $n\leqq 10$ sind es dichteste gitterförmige Packungen. In vorliegender Arbeit illustriert er sein Verfahren für die Dimensionszahlen $n\leqq 20$. Er gibt dabei keine Beweise. Die Frage bleibt offen, ob seine Packungen maximale Dichte haben.

Auch abgesehen davon scheinen mir seine Ergebnisse wichtig. Sie haben überraschenderweise eine Anwendung in der praktischen Integration von Funktionen mehrerer Veränderlicher durch elektronische Rechenmaschinen gefunden. *O.-H. Keller* (Halle)

H30-67 **(32# 2381)**

Cleaver, F. L.

On coverings of four-space by spheres.

Trans. Amer. Math. Soc. **120** (1965), 401–416.

In E_4 a lattice L is said to be S_4-admissible if no points of L except the origin lie inside the unit hypersphere S_4 with center at the origin. Let $d(L)$ denote the determinant of L. The following results are established. If L is S_4-admissible and $|d(L)|\leqq 1$, then any hypersphere H of radius $\geqq 1$ must contain a point of L in its interior or on its boundary. If the radius of H is 1 and if H has no points of L in its interior, then $d(L)=1$ and either L is the unit cubic lattice and H has its center C at the center of one of the cells of L, or, for some choice of coordinates, L is generated by the points $(1, 0, 0, 0)$, $(\frac{1}{2}, \frac{1}{2}\sqrt{3}, 0, 0)$, $(0, 1/\sqrt{3}, \sqrt{2}/\sqrt{3}, 0)$, $(0, 0, 0, \sqrt{2})$, C is congruent to $(0, -1/\sqrt{3}, 1/\sqrt{6}, 1/\sqrt{2})$ with respect to L, and H has 12 points of L on its boundary. These results, which are shown not to extend to E_5, form a generalization of a theorem of N. Hofreiter and F. J. Dyson: cf. F. J. Dyson [Ann. of Math. (2) **49** (1948), 82–109; MR **10**, 19].

I. Niven (Eugene, Ore.)

Citations: MR 10, 19a = J40-7.

H30-68 (32# 4604)

Davenport, H.

Problems of packing and covering. (Italian summary)

Univ. e Politec. Torino Rend. Sem. Mat. **24** (1964/65), 41–48.

This is a very readable sketch of the problems that have been exhaustively treated by L. Fejes Tóth [*Regular figures*, Macmillan, New York, 1964; MR **29** #2705] and C. A. Rogers [*Packing and covering*, Cambridge Univ. Press, New York, 1964; MR **30** #2405]. The author gives a new and very simple proof that the closest packing of equal circles in the Euclidean plane is achieved by the incircles of the hexagons in the regular tessellation {6, 3}.

H. S. M. Coxeter (Toronto, Ont.)

Citations: MR 30# 2405 = H02-17.

H30-69 (32# 6322)

Lekkerkerker, C. G.

Packings and coverings of figures in the plane. (Dutch)

Math. Centrum Amsterdam Afd. Zuivere Wisk. **1964**, *ZW*-008, 7 *pp.*

In this expository lecture the author deals with the densest lattice packings and the least dense lattice coverings by convex figures in the plane, in particular, with their relations to the smallest circumscribed and largest inscribed hexagons. [See, e.g., L. Fejes Tóth, *Lagerungen in der Ebene, auf der Kugel und im Raum*, Springer, Berlin, 1953; MR **15**, 248.]

K. Mahler (Canberra)

Citations: MR 15, 248b = H30-27.

H30-70 (33# 105)

Kelly, J. B.

Polynomials and polyominoes.

Amer. Math. Monthly **73** (1966), 464–471.

If $p=(n_1, n_2, \cdots, n_k)$ is any lattice point in Euclidean k-space E_k, define $M(p)=\prod_1^k x_i^{n_i}$. For any finite set S of lattice points in E_k, define the "polynomial (in k indeterminates) of S" to be $P(S)=\sum_{p\in S} M(p)$. Conditions on the ability of the "polyomino" S to cover a specified region R in E_k (typically a rectangular parallelotope) are obtained in terms of polynomial ideal theory, and applied to settle the possibility of several specific coverings (some negatively, some affirmatively). This is an algebraic generalization of coloring arguments, just as the ordinary checkerboard coloring distinguishes the two residue classes modulo $1+i$ in the ring of Gaussian integers.

S. W. Golomb (La Canada, Calif.)

H30-71 (33# 638)

Eggleston, H. G.

A minimal density plane covering problem.

Mathematika **12** (1965), 226–234.

Let Σ be a countable set of discs C_i in the plane. Let C_i have radius r_i and $\inf r_i=r>0$. The set Σ is said to be saturated if every circle of radius r in the plane meets some C_i. Let $X(k)$ denote the circle about O of radius k, and $\mu(k)$ the proportion of $X(k)$ (in area) occupied by the discs C_i. Then $\mu(\Sigma)=\lim\inf_{k\to\infty}\mu(k)$ is said to be the lower density of Σ. L. Fejes Tóth [*Lagerungen in der Ebene, auf der Kugel und im Raum*, Springer-Verlag, Berlin, 1953; MR **15**, 248] posed the problem of finding $\inf\mu(\Sigma)$ over all saturated Σ. The author proves that over all saturated packings Σ, $\inf\mu(\Sigma)=\pi/6\sqrt{3}=\mu(\Sigma_0)$, where Σ_0 is a packing of equal circles of radius r around the points of the lattice generated by an equilateral triangle of radius $2r\sqrt{3}$.

Starting from a saturated packing Σ, he first obtains a triangulation of the plane into triangles $T(a, b, c)$ for which the vertices a, b, c are centers of discs in Σ, and the circles C_a, C_b, C_c (in the obvious notation) meet a circle of radius r. He then proves that for each $T(a, b, c)$ the proportion of T occupied by C_a, C_b, C_c is at least equal to the corresponding proportion when T is an equilateral triangle of side $2r\sqrt{3}$ and C_a, C_b, C_c all have radius r.

R. P. Bambah (Columbus, Ohio)

Citations: MR 15, 248b = H30-27.

H30-72 (33# 1289a; 33# 1289b)

Woods, A. C.

The densest double lattice packing of four-spheres.

Mathematika **12** (1965), 138–142.

Woods, A. C.

Lattice coverings of five space by spheres.

Mathematika **12** (1965), 143–150.

In these two papers, the author proves, for $n=4$ and $n=5$, respectively: If L is a lattice, in Euclidean space R_n, of determinant 1, and if there exists a sphere centred at the origin O containing n linearly independent points of L on its boundary and none (other than O) in its interior, then any sphere in R_n of radius $\frac{1}{2}\sqrt{n}$ contains a point of L. This result is well known for $n=2, 3$; the proof given here for $n=4$ is simpler than those previously given by Hofreiter [Monatsh. Math. Phys. **40** (1933), 351–392; ibid. **40** (1933), 393–406] and Dyson [Ann. of Math. (2) **49** (1948), 82–109; MR **10**, 19], and uses only the weaker hypothesis that no two points of L are within unit distance apart. The method of proof is a generalization of that used by Davenport [J. London Math. Soc. **14** (1939), 47–51] for the case $n=3$ in his proof of Remak's theorem, and rests on the Korkine-Zolotareff reduction theory of positive quadratic forms.

E. S. Barnes (Adelaide)

Citations: MR 10, 19a = J40-7.

H30-73 (33# 6509)

Baranovskiĭ, E. P.

Local density minima of a lattice covering of a four-dimensional Euclidean space by equal spheres. (Russian)

Dokl. Akad. Nauk SSSR **164** (1965), 13–15.

Let Λ be a lattice in the n-dimensional Euclidean space R_n. There exists a unique minimal number $r>0$ such that the spheres of radius r about the points of Λ cover R_n. If $d(\Lambda)$ is the determinant of Λ, the density $\theta_n(\Lambda)$ of this covering is defined to be $J_n r^n/d(\Lambda)$, where J_n is the volume of the unit sphere. It was shown by E. S. Barnes [Canad. J. Math. **8** (1956), 293–304; MR **17**, 1060] that for $n=2, 3$ there is only one locally minimum density $\theta_n(\Lambda)$. M. N. Bleicher [ibid. **14** (1962), 632–650; MR **26** #6862] and A. F. Gameckiĭ [Dokl. Akad. Nauk SSSR **146** (1962), 991–994; MR **34** #145] independently showed that the density corresponding to the lattice Γ_n (corresponding to the quadratic form $n\sum x_i^2-\sum_{i\neq j} x_ix_j$) is locally minimum for all n. The author determines all three locally minimum densities in the case $n=4$. He shows that there is one minimum in each of the three Voronoĭ classes. The proof depends on the work of Voronoĭ [*Collected works in three volumes* (Russian), Vol. II, pp. 239–368, Izdat. Akad. Nauk Ukrain. SSR, Kiev, 1952; MR **16**, 2], Delone [Uspehi Mat.

Nauk **3** (1937), 16–62; ibid. **4** (1938), 102–164], and the author and Ryškov [Proc. Second All-Union Geometr. Conf. (Kharkov, 1964) (Russian), Har'kov. Gos. Univ., Kharkov, 1964]. {The last two references are not available to the reviewer, who is not familiar with some of the terminology used in this paper.}

{This article has appeared in English translation [Soviet Math. Dokl. **6** (1965), 1131–1133].}

R. P. Bambah (Columbus, Ohio)

Citations: MR 16, 2h = Z25-11; MR 17, 1060c = J48-10; MR 26# 6862 = H30-54; MR 34# 145 = H30-75.

Referred to in H30-80, H30-82.

H30-74 (33# 7935)

Groemer, Helmut

Zusammenhängende Lagerungen konvexer Körper.

Math. Z. **94** (1966), 66–78.

Ist K ein konvexer Körper und Λ ein Gitter des R^n, so wird unter einer zum Gitter Λ gehörigen K-Lagerung $\mathfrak{L}$ die Menge aller Körper $K+u$ verstanden, wobei u alle Vektoren aus Λ durchläuft. Die Körper der Lagerung können sich dabei auch überlappen. Die Lagerung $\mathfrak{L}$ heißt zusammenhängend, wenn es zu je zwei Körpern U und V von $\mathfrak{L}$ eine U und V verbindende Kette $U=X_0, X_1, X_2, \cdots, X_k=V$ von Körpern aus $\mathfrak{L}$ gibt, bei der jedes X_{i-1} mit X_i $(i=1, 2, \cdots, k)$ Punkte gemeinsam hat. In der Note werden verschiedene Sätze über zusammenhängende K-Lagerungen bewiesen, deren Dichte $\delta(K, \Lambda)=V(K)/D(\Lambda)$ bei gegebenem K minimal ist; $V(K)$ bezeichnet dabei das Volumen von K und $D(\Lambda)$ das Volumen des Grundspats von Λ. Eine K-Lagerung mit der Dichte $\delta(K)=\inf \delta(K, \Lambda)$, wobei Λ alle Gitter mit zusammenhängenden K-Lagerungen durchläuft, heißt eine dünnste zusammenhängende K-Lagerung. K^+ bezeichnet den durch Zentralsymmetrisierung aus K hervorgehenden Körper. Es werden dann verschiedene notwendige und hinreichende Bedingungen für zu einem Gitter Λ gehörige dünnste zusammenhängende K-Lagerungen $\mathfrak{L}$ von konvexen Körpern angegeben. Z.B. die Bedingung, daß in $2K^+$ eine Basis von Λ existiert, und daß für $x_i \in 2K^+$ gilt $D(\Lambda)=\sup |\det(x_1, x_2, \cdots, x_n)|$. Je zwei Körper einer dünnsten zusammenhängenden Lagerung konvexer Körper können einander nicht überlappen; jeder Körper einer solchen Lagerung wird mindestens von 2^n und höchstens von $2(2^n-1)$ anderen Körpern der Lagerung berührt.

Für die Dichte $\delta(K)$ gibt es dabei eine nur von der Dimensionszahl n abhängige Schranke. Für deren bestmöglichen Wert gilt die folgende Ungleichung, in der S^n eine Kugel, I^n ein Simplex, H^n ein verallgemeinertes Oktaeder bedeutet: $1/n!=\delta(H^n)=\delta(I^n)\leqq\delta(K)\leqq\delta(S^n)=\pi^{n/2}/2^n\Gamma(n/2+1)$. Ist K zentralsymmetrisch, so gilt $\delta(K)=1/n!$ nur für verallgemeinerte Oktaeder.

K. Strubecker (Karlsruhe)

H30-75 (34# 145)

Gameckiĭ, A. F.

On the theory of covering Euclidean n-space by equal spheres. (Russian)

Dokl. Akad. Nauk SSSR **146** (1962), 991–994.

This paper is a contribution to the problem of finding those lattices Γ in Euclidean n-space for which the covering density, proportional to $\rho(\Gamma)=R_\Gamma{}^n/V_\Gamma$, is a minimum. Here V_Γ is the volume of the fundamental region of Γ and R_Γ is the radius for the smallest set of congruent spheres, centred over Γ, which cover the space.

Let $\Gamma_1{}^n$ denote the lattice constructed from a regular Selling frame. $\rho(\Gamma_1{}^n)$ is the absolute minimum of $\rho(\Gamma)$ for $n=2, 3$ (shown by R. Bambah [Proc. Nat. Inst. Sci. India **20** (1954), 25–52; MR **15**, 780; Proc. Cambridge Philos. Soc. **50** (1954), 203–208; MR **15**, 780]) and perhaps 4, but the estimate $\rho(\Gamma)\leqq(1.107)^n$ by G. L. Watson [Rend. Circ. Mat. Palermo (2) **5** (1956), 93–100; MR **17**, 1235] shows that $\rho(\Gamma_1{}^n)\sim(1.17)^n$ is not the absolute minimum for large n.

The main result of the paper is that $\Gamma_1{}^n$ furnishes a local minimum for $\rho(\Gamma)$. The proof is along the following lines. Let R_λ be the distance from the centre to the λth vertex of the Dirichlet (Voronoĭ) region for Γ. If the R_λ are equal for a minimizing Γ, then Γ also minimizes $\sum R_\lambda{}^2/\sqrt[n]{V_\Gamma}{}^2$. The author shows that $\Gamma_1{}^n$ does this locally (and has equal R_λ) with the aid of a representation of $\sum R_\lambda{}^2$ in terms of parameters associated with Selling frames.

{Subsequent related papers: the author [Dokl. Akad. Nauk SSSR **151** (1963), 482–484; MR **27** #5731]; B. N. Delone and S. S. Ryškov [ibid. **152** (1963), 523–524; MR **31** #126]. Bibliographical errors: reference 3 is in volume 50, not 56; reference 6 is in Mathematika **6** (1959).}

{This article has appeared in English translation [Soviet Math. Dokl. **3** (1962), 1410–1414].}

W. J. Firey (Corvallis, Ore.)

Citations: MR 15, 780c = H30-32; MR 15, 780d = H30-33; MR 17, 1235b = H30-35; MR 27# 5731 = H30-57; MR 31# 126 = H30-65.

Referred to in H30-73, H30-90.

H30-76 (34# 682)

Baranovskiĭ, E. P.

On minima of the density of a lattice covering of space by equal balls. (Russian)

Ivanov. Gos. Ped. Inst. Učen. Zap. **34** (1963), *vyp. mat.*, 71–76.

Jeder Punkt eines Punktgitters des dreidimensionalen Euklidischen Raumes E_3 sei Mittelpunkt einer Kugel vom Radius R derart, daß die Kugeln den E_3 überdecken und R unter dieser Bedingung minimal ist. Der Quotient D aus dem Volumen einer solchen Kugel und dem Volumen eines Fundamentalparallelepipeds des Gitters heißt die Dichte der Überdeckung.

R. P. Bambah [Proc. Nat. Inst. Sci. India **20** (1954), 25–52; MR **15**, 780], E. S. Barnes [Canad. J. Math. **8** (1956), 293–304; MR **17**, 1060] und L. Few [Mathematika **3** (1956), 136–139; MR **18**, 721] zeigten, daß D den kleinstmöglichen Wert $(5\sqrt{5})\pi/24$ genau im Falle des raumzentrierten Würfelgitters annimmt.

Der Autor gibt hierfür einen elementaren Beweis (Satz 2). Dabei benutzt er einen Satz über Beziehungen der lokalen Minima zweier mit den Punkgittern verbundener Funktionen (Satz 1), der in verallgemeinerter Form auch für Punktgitter des E_n gültig ist. *H. Sachs* (Ilmenau)

Citations: MR 15, 780c = H30-32; MR 17, 1060c = J48-10; MR 18, 721c = H30-36.

H30-77 (34# 3441a)

Baranovskiĭ, E. P.; Ryškov, S. S.

The second local density minimum of a lattice covering of the four-dimensional Euclidean space by equal balls. (Russian)

Sibirsk. Mat. Ž. **7** (1966), 731–739.

H30-78 (35# 878)

Leech, John

Notes on sphere packings.

Canad. J. Math. **19** (1967), 251–267.

This paper supplements an earlier one by the author [same J. **16** (1964), 657–682; MR **29** #5166]. New sphere packings in $[2^m]$ $(m\geqq 6)$, [11], [22], [23], [24], which are denser than those previously recorded, and some relations between them are discussed. *J. A. Todd* (Cambridge, England)

Citations: MR 29# 5166 = H30-62.

H30-79 **(35# 5394)**

Hans, Rajinder Jeet

Extremal packing and covering sets.

Monatsh. Math. **71** (1967), 203–213.

The author shows that any open set in euclidean n-space that is star in the origin is contained in a maximal such open star set with the same lattice packing constant. Likewise, it is shown that any bounded and closed set that is star in the origin contains a minimal such star set with the same lattice covering constant. These theorems are similar to those of K. Mahler [Nederl. Akad. Wetensch. Proc. **49** (1946), 331–343; MR **8**, 12] and C. A. Rogers [Philos. Trans. Roy. Soc. London Ser. A **245** (1952), 59–93; MR **14**, 253] for which short proofs are also included. *A. C. Woods* (Columbus, Ohio)

Citations: MR 8, 12b = H15-8; MR 14, 253c = H15-50.

H30-80 **(35# 6034)**

Barnes, E. S.; Dickson, T. J.

Extreme coverings of n-space by spheres.

J. Austral. Math. Soc. **7** (1967), 115–127.

Let J be a sphere in the n-dimensional Euclidean space R_n. If Λ is a lattice in R_n, (J, Λ) is a lattice covering by J if $R_n \subset \bigcup_{A\in\Lambda}(J+A)$. The density $\theta(J, \Lambda)$ of a lattice covering (J, Λ) is $V(J)/d(\Lambda)$, where $V(J)$ is the volume of J and $d(\Lambda)$ the determinant of Λ. The constant $\theta_n = \min \theta(J, \Lambda)$ over all lattice coverings (J, Λ) is called the density of the thinnest lattice covering by n-dimensional spheres. It is known for $n \leqq 4$. (θ_3 was determined by the reviewer, and other proofs were given by the first author and L. Few; θ_4 has been determined by B. N. Delone and S. S. Ryškov [Dokl. Akad. Nauk SSSR **152** (1963), 523–524; MR **31** #126].) Various estimates from above and below for θ_n have been obtained for other n (see, e.g., C. A. Rogers [*Packing and covering*, Cambridge Univ. Press, New York, 1964; MR **30** #2405]).

The problem of finding θ_n can be easily translated into one for quadratic forms as follows. Let $f(X) = X'AX$ be a positive definite quadratic form in n variables $X = (x_1, \cdots, x_n)$, with real coefficients and determinant $d(f) = |\det A|$. For $P \in R_n$, define $m(f; P) = \min_{X\in\Lambda_0} f(X+P)$; here Λ_0 is the fundamental lattice of all points with integral coordinates. Let $m(f) = \max_P m(f; P)$, $\mu(f) = m(f)/d^{1/n}(f)$. Then $\theta_n = \min_f J_n \mu^{n/2}(f)$, where J_n is the volume of the sphere $|X| < 1$. Thus the problem is reduced to finding $\min_f \mu(f)$. The forms where this minimum is attained are absolutely extreme and correspond to thinnest coverings. A form f is called (locally) extreme if $\mu(f) \leqq \mu(\varphi)$ for all φ near enough to f in the obvious sense. Clearly, absolutely extreme forms are (locally) extreme also. Any systematic method of finding extreme forms is of obvious interest. Extreme forms have been determined for $n \leqq 4$ ($n = 2, 3$, the first author [Canad. J. Math. **8** (1956), 293–304; MR **17**, 1060]; $n = 4$, the second author ["Lattice coverings of n-space by spheres", Ph.D. thesis, Univ. of Adelaide, Adelaide, 1966] and E. P. Baranovskiĭ [Dokl. Akad. Nauk SSSR **164** (1965), 13–15; MR **33** #6509]). Also, M. N. Bleicher [Canad. J. Math. **14** (1962), 632–650; MR **26** #6862] and A. F. Gameckiĭ [Dokl. Akad. Nauk SSSR **151** (1963), 482–484; MR **27** #5731] have shown that the form $f_0 = n \sum x_i^2 - 2 \sum x_i x_j$ is extreme for all n.

In this useful paper, the authors obtain a criterion for a form in the interior of a Voronoĭ cone to be extreme. As an application, they obtain the Bleicher-Gameckiĭ theorem. They next prove that if a Voronoĭ cone Δ has an extreme form f in its interior, then all extreme forms in the interior or the boundary of Δ consist of multiples of f. Since there is only one Voronoĭ cone in R_n for $n = 2, 3$ and the form $n \sum x_i^2 - 2 \sum x_i x_j$ lies in its interior, values of θ_2, θ_3 follow at once. From their second theorem they deduce the result that if f is an extreme form in the interior of a Voronoĭ cone Δ, then f and Δ have the same group of automorphisms. This theorem also shows that the only possible extreme forms in the principal cone are multiples of f_0.

{Reviewer's remark: In a corrigendum (presumably to be published), the authors point out that their proof of Lemma 4.2 is incomplete in that the ε there may depend on the choice of T, whereas they need one independent of T. In the corrigendum they show how this can be done.} *R. P. Bambah* (Columbus, Ohio)

Citations: MR 17, 1060c = J48-10; MR 26# 6862 = H30-54; MR 27# 5731 = H30-57; MR 30# 2405 = H02-17; MR 31# 126 = H30-65; MR 33# 6509 = H30-73.

Referred to in H30-82, H30-95.

H30-81 **(35# 7207)**

Ryškov, S. S.

Effective realization of a method of Davenport in the theory of coverings. (Russian)

Dokl. Akad. Nauk SSSR **175** (1967), 303–305.

In the study of coverings of Euclidean n-space by equal spheres, centred at the points of a lattice, the lattices Γ_1^n characterized by the metric form $(n+1)\sum_{i=1}^n x_i^2 - \sum_{i=1}^n \sum_{j=1}^n x_i x_j$ have special interest. For $n = 2, 3, 4$ these furnish the most economical coverings; see R. P. Bambah [Proc. Nat. Inst. Sci. India **20** (1954), 25–52; MR **15**, 780] and B. N. Delone and the author [Dokl. Akad. Nauk SSSR **152** (1963), 523–524; MR **31** #126]. However it is known that this is not true for sufficiently large n; see C. A. Rogers [*Packing and covering*, Cambridge Univ. Press, New York, 1964; MR **30** #2405]. By using a modification of work of H. Davenport [Rend. Circ. Mat. Palermo (2) **1** (1952), 92–107; MR **14**, 75], the author constructs lattices Γ_*^n which determine more economical coverings than those based on Γ_1^n for even $n \geqq 114$ and odd $n \geqq 201$. The metric form for Γ_*^{2n} has as its matrix $[\Gamma_*^{2n}] = [\Gamma_1^n] \otimes [\Gamma_1^2]/(n+1)$, where $\otimes$ signifies the Kronecker product. The odd-dimensional case is not explicitly described.

{This article has appeared in English translation [Soviet Math. Dokl. **8** (1967), 865–867].} *W. J. Firey* (Dunedin)

Citations: MR 14, 75h = H30-21; MR 15, 780c = H30-32; MR 30# 2405 = H02-17; MR 31# 126 = H30-65.

H30-82 **(36# 4439)**

Dickson, T. J.

The extreme coverings of 4-space by spheres.

J. Austral. Math. Soc. **7** (1967), 490–496.

The author applies results proved by E. S. Barnes and himself [same J. **7** (1967), 115–127; MR **35** #6034] to determine the three locally extreme lattice coverings of three-dimensional Euclidean space by equal spheres. The results agree with those of E. P. Baranovskiĭ [Dokl. Akad. Nauk SSSR **164** (1965), 13–15; MR **33** #6509], whose work seems to have escaped the notice of the author. The author's method is different and his proofs are clear and complete. *R. P. Bambah* (Columbus, Ohio)

Citations: MR 33# 6509 = H30-73; MR 35# 6034 = H30-80.

H30-83 **(37# 4023)**

Few, L.

Double covering with spheres.

Mathematika **14** (1967), 207–214.

Author's summary: "Let Λ be a lattice in 3-dimensional space which provides a double covering for spheres of

unit radius. By this we mean that if X is any point of space, there are at least two distinct lattice points P, Q such that $XP \leqq 1$, $XQ \leqq 1$. Let $d(\Lambda)$ be the determinant of Λ. We shall prove that $d(\Lambda) \leqq \xi$, where $12\xi^2 = 76\sqrt{6} - 159$, $\xi = 1.50447\ldots$. We shall also prove that the lattice Λ_0 generated by the points $(0, a_0, 0)$, $(2b_0, 0, 0)$, $(b_0, 0, b_0\sqrt{3})$, where $3a_0^2 = 2\sqrt{6} - 3$, $12b_0^2 = 9 - \sqrt{6}$, does provide a double covering for spheres of unit radius and has $d(\Lambda_0) = \xi$. Also, Λ_0 is the only such lattice."
J. H. H. Chalk (Toronto, Ont.)

H30-84 (37# 4714)

Chalk, J. H. H.
A note on coverings of E^n by convex sets.
Canad. Math. Bull. **10** (1967), 669–673.

Let $(x_1, x_2, \cdots, x_n)$ denote the coordinates of a point X of Euclidean n-space E^n. Let $a_0, a_1, \cdots, a_n$ be a set of $n+1$ points of E^n with the property that $b_i = a_i - a_0$ ($i = 1, 2, \cdots, n$) form a linearly independent set and define a lattice Λ of points $u_1b_1 + \cdots + u_nb_n$, u_i integral. Let K be any closed bounded centrally symmetric convex set containing the points $a_0, a_1, \cdots, a_n$. There is a constant $\lambda_n > 0$, depending only on n, such that (i) $\bigcup_{X \in \Lambda} (\lambda_n K + X) \supset E^n$. In this paper an example is provided for all odd n to verify that (i) implies $\lambda_n \geqq (n+1)/2$. *F. Supnick* (New York)

H30-85 (37# 6838)

Bambah, R. P.; Woods, A. C.
The covering constant for a cylinder.
Monatsh. Math. **72** (1968), 107–117.

Let S be a set in the n-dimensional euclidean space R_n. For $A \in R_n$ denote by $S + A$ the set of points $X + A$, $X \in S$. A lattice Λ is said to be a covering lattice for S if the sets $S + A$, $A \in \Lambda$, cover R_n completely. The covering constant $c(S)$ of S is defined by $c(S) = \sup d(\Lambda)$ where the upper bound is taken over all covering lattices Λ for S. The authors obtain a simpler proof for the covering constant for three-dimensional cylinders whose base is a symmetric convex domain. *F. Supnick* (New York)

H30-86 (38# 120)

Gruber, Peter
Zur Gitterüberdeckung des R^n durch Sternkörper.
Österreich. Akad. Wiss. Math.-Natur. Kl. S.-B. II **176** (1967), 1–7.

Let Λ be a lattice in R^n of determinant $d(\Lambda)$ and let f be the distance function of a star body in R^n. For given $\mathfrak{y} \in R^n$, put $m_k(\mathfrak{y}, f, \Lambda) = \inf\{t \mid f(\mathfrak{x} - \mathfrak{y}) \leqq t$ contains at least k lattice points$\}$, so that $m_1 \leqq m_2 \leqq m_3 \leqq \cdots \leqq m_\infty$. Let $\mu_k(f, \Lambda) = \sup\{m_k(\mathfrak{y}, f, \Lambda) \mid \mathfrak{y} \in R^n\}$. Suppose that $\mathfrak{y}_r \to \mathfrak{y}$, $\Lambda_r \to \Lambda$ and that $f_r \to f$ uniformly in $|\mathfrak{x}| \leqq 1$ for $r \to \infty$. It is proved that, for $k < \infty$, $\limsup_{r\to\infty} m_k(\mathfrak{y}_r, f_r, \Lambda_r) \leqq m_k(\mathfrak{y}, f, \Lambda)$ and that equality holds if f vanishes only at the origin. Similar results hold for $\mu_k(f, \Lambda)$. Accordingly m_k and μ_k are semi-continuous in the sense of J. W. S. Cassels [*An introduction to the geometry of numbers*, p. 305, Springer, Berlin, 1959; MR **28** #1175]. By taking $f_0(\mathfrak{x}) = |x_1x_2 \cdots x_n|^{1/n}$, it is shown that, for each $k \geqq 1$, m_k and μ_k are, however, not continuous. This uses a lower bound for $\mu_1(f_0, \Lambda)$ due to V. Ennola [Ann. Univ. Turku. Ser. AI **28** (1958); MR **20** #3825], an upper bound due to the author [Acta Arith. **13** (1967/68), 9–27; MR **36** #3729], as well as a further result of the author's, deduced from Satz 15* of W. M. Schmidt [Acta Math. **102** (1959), 159–224; MR **22** #5627], which states that, if the region $|f| \leqq 1$ has infinite volume, then for almost all linear transformations τ, $m_\infty(\mathfrak{y}, f, \tau\Lambda_0) = 0$ for almost all y, where Λ_0 is the integral lattice. *R. A. Rankin* (Glasgow)

Citations: MR 20# 3825 = J32-56; MR 22# 5627 = H25-33; MR 28# 1175 = H02-15; MR 36# 3729 = J40-37.

H30-87 (39# 4752)

Few, L.; Kanagasabapathy, P.
The double packing of spheres.
J. London Math. Soc. **44** (1969), 141–146.

Ein System von Einheitskugeln des n-dimensionalen euklidischen Raumes R_n, deren Zentren die Punkte eines n-dimensionalen Gitters Λ der Determinante d sind, heißt k-fache Packung, wenn jeder Punkt von R_n von höchstens k Kugeln überdeckt wird. Ist $\Delta_k^{(n)} = \inf d(\Lambda)$, wobei sich die Bildung der unteren Schranke über alle Gitter Λ des R_n erstreckt, die eine k-fache Packung erlauben, so ist $\Delta_2^{(n)} \leqq \frac{1}{2}\Delta_1^{(n)}$. Für $n = 2$ gilt hier Gleichheit, wie der erste Verfasser früher zeigte; es ist $\Delta_1^{(2)} = 2\sqrt{3}$, $\Delta_2^{(2)} = \sqrt{3}$. In der vorliegenden Note wird nun nachgewiesen, daß für $n > 2$ stets $\Delta_2^{(n)} < \frac{1}{2}\Delta_1^{(n)}$ ausfällt. Genauere Ergebnisse:

$$\Delta_2^{(3)} = \tfrac{3}{2}\sqrt{3} = \tfrac{1}{2}(\sqrt{27/32})\Delta_1^{(3)},$$

$$\Delta_2^{(4)} \leqq \tfrac{3}{2}\sqrt{6} = \tfrac{1}{2}(\sqrt{27/32})\Delta_1^{(4)},\ \Delta_2^{(n)} \leqq (\tfrac{3}{4})^{n/2}\Delta_1^{(n)},\ n \geqq 5.$$

Die Beweise greifen auf weiter zurückliegende Resultate zurück [erster Verfasser, dasselbe J. **28** (1953), 297–304; MR **14**, 1115], die insbesondere dazu dienen, die Hilfsaussage $\Delta_2^{(n)} \leqq (\frac{3}{4})^{(n-1)/2}\Delta_1^{(n-1)}$ sicherzustellen.
H. Hadwiger (Bern)

H30-88 (40# 1893)

Kaur, Gurnam
Extreme quadratic forms for coverings in four variables.
Proc. Nat. Inst. Sci. India Part A **32** (1966), 414–417.

Author's summary: "In this paper we prove that the polynomial $\phi_1 = \sum x_i^2 + \sum x_ix_j$, $i, j = 1, 2, \cdots, 4$, which is extreme in the packing case is not extreme in the covering case."

H30-89 (40# 3435)

Hoylman, Douglas J.
The densest lattice packing of tetrahedra.
Bull. Amer. Math. Soc. **76** (1970), 135–137.

Das Problem, im dreidimensionalen euklidischen Raum die dichteste gitterförmige Lagerung kongruenter Tetraeder zu finden, hat schon Minkowski beschäftigt. Ist J ein konvexer Körper, $V(J)$ sein Volumen, Λ ein Gitter mit $|\det \Lambda| = \Delta$ und ist $(x+J) \cap (y+J) = \varnothing$ für jedes $x \in \Lambda$, $y \in \Lambda$, dann heißt $V(J)/\Delta$ die Dichte der Lagerung. H. Groemer [Monatsh. Math. **66** (1962), 12–15; MR **25** #2516] zeigte 1962 für die dichteste Lagerung kongruenter Tetraeder, daß sie mindestens gleich 18/49 ist. Der Autor zeigt unter Anwendung zweier Lemmata von Minkowski und durch Ausrechnen von 38 Fällen, daß das Beispiel von Groemer optimal ist und die dichteste Lagerung kongruenter Tetraeder gleich 18/49 ist.
J. M. Wills (Berlin)

Citations: MR 25# 2516 = H30-53.

H30-90 (41# 1651)

Gameckiĭ, A. F.; Duhovnyĭ, S. M.
A certain extremal problem. (Russian)
Mat. Issled. **4** (1969), *vyp.* 2 (12), 131–137.

In Euclidean n-space E^n, consider a lattice Γ whose fundamental parallelepiped has volume V. Inscribe in each simplex of the partition associated with Γ a ball, and let R_Γ be the least of the radii of these balls. Finally, write

γ_n for the volume of the unit ball in E^n. The problem of the paper is to find those lattices Γ for which the associated packing density $d_n=\gamma_n(R_\Gamma)^n/V$ is maximal. The first author in an earlier paper [Dokl. Akad. Nauk SSSR **146** (1962), 991–994; MR **34** #145] considered the corresponding covering problem. By similar methods the authors prove here that the lattices $\Gamma_1{}^n$, constructed from a regular Selling frame (for which the associated metric form is $n\sum_i x_i{}^2-\sum_i\sum_j x_ix_j$), furnish a local maximum for d_n. *W. J. Firey* (Corvallis, Ore.)

Citations: MR 34# 145 = H30-75.

H30-91 (41# 6783)

Hlawka, E.

Überdeckung durch konvexe Scheiben.

S.-B. Berlin. Math. Ges. **1961–64**, 28–36.

A useful survey article describing results and open problems in lattice packings and coverings.

R. P. Bambah (Chandigarh)

H30-92 (41# 9109)

Folkman, J. H.; Graham, R. L.

A packing inequality for compact convex subsets of the plane.

Canad. Math. Bull. **12** (1969), 745–752.

Let K be a convex domain with centre 0 in the plane. Let μ be its gauge function. Let Π be a Jordan polygon and E a finite set of points consisting of the vertices of Π and other points in the interior of the polygon. Suppose the sets $\frac{1}{2}K+P$, $P\in E$, form a packing, i.e., for all P, Q in E, $\mu(P-Q)\geqq 1$. Let n be the number of points in E, $A(\Pi)$ the area of Π, $M(\Pi)$ its μ-perimeter and Δ the critical determinant of K. Then N. Oler [Acta Math. **105** (1961), 19–48; MR **24** #A2900] proved the remarkable Zassenhaus-Oler inequality: (A) $A(\Pi)/\Delta+M(\Pi)/2+1\geqq n$. Oler's proof covers thirty pages of ingenious and detailed arguments. The authors prove by simple geometrical arguments the special case of (A) when K is a circle. They deduce their result from a similar result in which Π is replaced by a simplicial complex. *R. P. Bambah* (Chandigarh)

Citations: MR 24# A2900 = H30-51.

H30-93 (42# 203)

Hans, Rajinder J.

Covering constants of some non-convex domains.

Indian J. Pure Appl. Math. **1** (1970), *no.* 2, 127–141.

Let S be a set in euclidean 2-space R^2. A lattice Λ in R^2 is said to be a covering lattice for S if every point of R^2 is in some set of the form $S+p$ with $p\in\Lambda$. The "covering constant" of S is defined by $\sup d(\Lambda)$, where $d(\Lambda)$ is the determinant of Λ and Λ varies over all covering lattices for S. In this paper the covering constants and the corresponding extremal lattices for several non-convex plane domains are found. For example, if the domain S is given by $\max\{|x|, |y|\}\leqq\frac{3}{2}$, $\min\{|x|, |y|\}\leqq 1$, the covering constant is shown to be 15/2 and the corresponding extremal lattices are of the type $T\Lambda$, where T is an automorphism of S and Λ is a lattice with basis (0, 3), (5/2, c), where $1\leqq c\leqq 2$. Other polygonal regions which are considered are of a more complicated nature. Finally, the domain which can be represented as the union of two intersecting unit circles is investigated. In this case the covering constant is a rather involved function of the distance of the two centers of the circles. The method which the author uses to obtain all these results has some similarity to Mordell's method for finding the critical determinants of plane non-convex regions. *H. Groemer* (Tucson, Ariz.)

H30-94 (42# 204)

Hans-Gill, Rajinder Jeet

Covering constant of a star domain.

J. Number Theory **2** (1970), 298–309.

Soit dans $\mathbf{R}_n$ un réseau Λ dont le déterminant a pour valeur absolue $d(\Lambda)$. Pour $S\subset\mathbf{R}_n$, on dit que Λ est pour S un réseau couvrant si tout point de $\mathbf{R}_n$ est dans le translaté de S par un point de Λ. $c(S)=\sup d(\Lambda)$, où Λ est pour S un réseau couvrant, est la constante de recouvrement de S; si on peut trouver Λ avec $d(\Lambda)=c(S)$, Λ est appelé réseau maximal couvrant. L'auteur détermine la constante de recouvrement d'un domaine étoilé plan non convexe limité par quatre cercles égaux sécants, domaine dont on connaît la constante critique [K. Ollerenshaw, Quart J. Math. Oxford Ser. **17** (1946), 93–98; MR **7**, 506]. Il s'agit de $S=\{(x, y)\colon \min((x\pm 1)^2+(y\pm 1)^2)\leqq 2\}$. Il trouve $c(S)=\max f(\theta)$ avec $\pi/4\leqq\theta\leqq\pi/2$ et $f(\theta)=8+2^{1/2}(3\sin\theta+\cos\theta)+2^{1/2}(12+2^{3/2}(3\sin\theta-\cos\theta))^{1/2}$. Il détermine aussi les réseaux maximaux couvrants.

R. Bantegnie (Besançon)

Citations: MR 7, 506h = H20-9.

H30-95 (42# 1767)

Delone, B. N.; Dolbilin, N. P.;
Ryškov, S. S.; Štogrin, M. I.

A new construction of the theory of lattice coverings of an n-dimensional space by congruent balls. (Russian)

Izv. Akad. Nauk SSSR Ser. Mat. **34** (1970), 289–298.

The lower bound r for the radius of equal balls, centred at the points of a lattice $\mathscr{E}$ in Euclidean n-space E^n and covering E^n, is the covering radius of $\mathscr{E}$. The volume of the ball of radius r, divided by the volume of the fundamental parallelepiped of $\mathscr{E}$, gives the covering density for $\mathscr{E}$. This paper outlines a method for discovering local minima for the covering density; the method uses the correspondence between $\mathscr{E}$, the associated positive definite quadratic form determined from the lattice frame, and the point of E^N, $N=n(n+1)/2$, whose components are the coefficients a_{ij}, $i\leqq j$, of the aforementioned quadratic form, in order to shift the question to one concerning properties of certain convex sets in E^N.

In more detail, the first author's method of the empty sphere [Uspehi Mat. Nauk **3** (1937), 16–62; ibid. **4** (1938), 102–164] shows the covering radius of $\mathscr{E}$ to be the radius of the circumsphere of the frame of $\mathscr{E}$; the volume of the fundamental parallelepiped is the discriminant d of the quadratic form for $\mathscr{E}$. In terms of E^N, d is called the height of the point associated with $\mathscr{E}$; the problem now consists in extremizing this height over certain sets of points in E^N which are associated with lattice frames whose circumspheres have no more than unit radius.

The positive definite quadratic forms determine a convex cone $\mathbf{K}$ in E^N which decomposes into a set $\{\Delta\}$ of convex polyhedral cones of G. F. Voronoĭ [*Collected works in three volumes* (Russian), Vol. II, pp. 239–368, Izdat. Akad. Nauk Ukrain. SSR, Kiev, 1952; MR **16**, 2]; each Δ determines a certain finite set of automorphisms of $\mathbf{K}$ and the partition $\{\Delta\}$. Let $\mathbf{V}$ be the set of points of $\mathbf{K}$ whose associated frames have a circumsphere of radius no more than one. Fundamental lemma: $\mathbf{V}$ is convex and bounded. Now form the intersection $\mathbf{W}$ of the finite set of the images of $\mathbf{V}$ under the automorphisms associated with the Voronoĭ cone Δ and, in turn, the intersection of $\mathbf{W}$ with the closure of Δ. Call this set $\mathbf{W}(\Delta)$; the fundamental lemma also applies to $\tilde{\mathbf{W}}(\Delta)$ and, using simple properties of convex sets, the authors show that, for a given Δ, the height d has a unique local extremum over $\tilde{\mathbf{W}}(\Delta)$. This is a theorem of E. S. Barnes and T. J. Dickson [J. Austral. Math. Soc. **7** (1967), 115–127; MR **35** #6034; corrigendum, ibid. **8** (1968), 638–640; MR **37** #6839].

The paper closes with a condition which is necessary and sufficient for a point of E^N to correspond to a local extremal lattice; the condition is expressed in terms of properties of certain convex sets in E^N, and its analytic form is said to be equivalent to the necessary and sufficient conditions described in the work of Barnes and Dickson cited above.

W. J. Firey (Corvallis, Ore.)

Citations: MR 16, 2h = Z25-11; MR 35# 6034 = H30-80.

H30-96 (42# 4496)

Zassenhaus, Hans

On a packing inequality for plane sectorial norm distances.

Punjab Univ. J. Math. (Lahore) **2** (1969), 1–9.

This paper has its origins in the work of the author and of his former students (see the author's survey article in Bull. Amer. Math. Soc. **67** (1961), 427–439 [MR **24** #A1887] for further details and for most references). In general terms, this work is concerned with certain types of distance function $N(P,Q)\geqq 0$, where P, Q denote arbitrary points of an n-dimensional Euclidean space E_n with origin O, which possesses the following properties: (i) $N(P,Q)=N(O,Q-P)$ for all P, Q in E_n, (ii) $N(O,\lambda P)=\lambda N(O,P)$ for all $P\in E_n$, $\lambda\geqq 0$, (iii) $N(O,-P)=N(O,P)$, together with an anti-triangle inequality of the type (iv) $N(O,P_2)\geqq N(O,P_1)+N(O,P_3)$, obtaining whenever $OP_1P_2P_3$ form a parallelogram in any one of finitely many non-overlapping closed n-dimensional cones $\{C_i\}$, $1\leqq i\leqq p$, with $\bigcup_{1\leqq i\leqq p}C_i=E_n$. The cones $\{C_i\}$ are called the sectors of the sectorial norm distance N and the "gauge-body" for N is the set of points $P\in E_n$ with $N(O,P)<1$. Thus, note that (iv) disqualifies the ordinary Euclidean distance function. The "octahedron", defined by $|x_1|+\cdots+|x_n|<1$, is the (convex) gauge-body for the N-distance function $N_1(P,Q)=\sum_{1\leqq i\leqq n}|x_i-y_i|$ and $N_2(P,Q)=\prod_{1\leqq i\leqq n}|x_i-y_i|^{1/2}$, $P=(x_1,\cdots,x_n)$, $Q=(y_1,\cdots,y_n)$, gives a (non-convex) gauge-body $|x_1\cdots x_n|<1$; both having the 2^n coordinate "octants" as sectors. A triangle with vertices at P, Q, R in E_n is said to be sectorial if the points $Q-P$, $R-Q$ are contained in one of the cones C_i. An arbitrary point-set M is said to be N-admissible if the N-distance between any two distinct points of M is at least 1.

Specializing now to the case $n=2$ (there seems to be little or no progress when $n\geqq 3$), denote by $t(N)$ the infimum of the (Euclidean) areas of the N-admissible non-sectorial triangles. For the case $N=N_2$, as defined above, it is known that $t(N_2)=\frac{1}{2}\sqrt{5}$, and this is the area of the N-admissible triangle OQR with $Q=(1,1)$, $R=(\frac{1}{2}(1+\sqrt{5}),\frac{1}{2}(1-\sqrt{5}))$. Let M denote a finite N-admissible point-set and let $\Pi=P_1\cdots P_bP_1$ denote a Jordan polygon with its vertices in M and such that all other points of M belong to the interior of Π. Then the important achievement of this paper is a proof of the following theorem. Theorem: If $A(\Pi)$ denotes the Euclidean area of Π, $|M|$ denotes the cardinality of M and $C(\Pi)=\sum_{1\leqq i\leqq b}N(P_i,P_{i+1})$, where $P_{i+1}=P_1$ denotes the N-circumference of Π, then $A(\Pi)/(2t(N))+\frac{1}{2}C(\Pi)+1\geqq |M|$, whenever $t(N)>0$.

This theorem enables one to deduce that $(2t(N))^{-1}$ is an upper bound for the irregular packing density, taken with respect to plane sectorial norm distance. In the special case $N=N_2$, for example, the bound is exact and coincides with the reciprocal of the critical determinant (originally established by N. E. Smith [see the bibliography of the article under review]).

Unfortunately, quite a few misprints are present.

J. H. H. Chalk (Toronto, Ont.)

Citations: MR 24# A1887 = H30-50.

H30-97 (42# 6725)

Leech, John; Sloane, N. J. A.

New sphere packings in more than thirty-two dimensions.

Proc. Second Chapel Hill Conf. on Combinatorial Mathematics and its Applications (Univ. North Carolina, Chapel Hill, N.C., 1970), *pp.* 345–355. *Univ. North Carolina, Chapel Hill, N.C.,* 1970.

The authors announce new sphere packings in dimensions 36, 40, 48, 60 and 2^m ($m\geqq 6$) obtained from error-correcting codes; details and proofs are to appear in a future paper (the authors, "Sphere packings and error-correcting codes", to appear). The density of their packing in E^{512} is higher by a factor of 2^{186} than that of the best previously known! *P. McMullen* (London)

H30-98 (43# 2613)

Ryškov, S. S.

The polyhedron $\mu(m)$ and certain extremal problems of the geometry of numbers. (Russian)

Dokl. Akad. Nauk SSSR **194** (1970), 514–517.

Der Autor der vorliegenden Arbeit knüpft an Untersuchungen von G. Voronoĭ an, in denen zur Behandlung von Fragen über gitterförmige Packungen n-dimensionaler Kugeln auf ein gewisses Polyeder π, welches in Verbindung zu quadratischen Formen mit n Variablen steht, zurückgegriffen wird. Ryškov betrachtet ein anderes Polyeder $\mu(m)$. Das Polyeder $\mu(m)$ ist der Durchschnitt von Halbräumen, welche quadratischen Formen von n Variablen in einem gewissen N-dimensionalen Raum E^N, $N=\frac{1}{2}n(n+1)$, zugeordnet sind. Nach einer Aufzählung der Eigenschaften von $\mu(m)$ werden Beziehungen zwischen den Polyedern $\mu(m)$ und π aufgezeigt. Es folgt unter wesentlicher Verwendung des Polyeders $\mu(m)$ die Behandlung bekannter (auf G. Voronoĭ, H. Minkowski und B. N. Delone [Uspehi Mat. Nauk **3** (1937), 16–62] zurückgehender) und zweier neuer Extremalaufgaben über gitterförmige Packungen n-dimensionaler Kugeln nach einheitlichen Gesichtspunkten. Dabei ergeben sich gewisse Vereinfachungen gegenüber der Verwendung des Polyeders π. Das Polyeder $\mu(m)$ kann als eine Brücke zwischen den Methoden Voronoĭs und den Methoden Minkowskis zur Lösung der erwähnten Extremalaufgaben angesehen werden.

{This article has appeared in English translation [Soviet Math. Dokl. **11** (1970), 1240–1244].}

O. Krötenheerdt (Halle a.d. Saale)

H30-99 (44# 6608)

Bambah, R. P.; Woods, A. C.

The thinnest double lattice covering of three-spheres.

Acta Arith. **18** (1971), 321–336.

It is known that the covering constant for lattice coverings of three-dimensional euclidean space by congruent spheres is $5\sqrt{(5)}\pi/24$. In this paper the authors generalize this result by considering sphere coverings of 3-space in which the centers of the spheres are located at the points of the union of two lattices that are related by a translation. They show that in this case $(5\sqrt{5})\pi/24$ is again the greatest lower bound of the corresponding densities.

H. Groemer (Tucson, Ariz.)

H99 NONE OF THE ABOVE, BUT IN THIS CHAPTER

See also reviews R32-5, R32-7, U10-14, Z10-16, Z10-20, Z10-30.

H99-1 (8, 502h)

Kotzig, Anton. Sur les "translations k." Časopis Pěst. Mat. Fys. **71**, 55–66 (1946). (Czech. French summary)

Let n, k be given integers with $1 \leqq k \leqq n$. A "translation k" of a point $(x_1, \cdots, x_n)$ in n-dimensional space is one which replaces it by $(x_1-a_1, \cdots, x_n-a_n)$, where $a_1, \cdots, a_n$ are nonnegative integers and the number l of a's which are different from zero satisfies $1 \leqq l \leqq k$. Let M denote the set of all points $(x_1, \cdots, x_n)$ for which $x_1, \cdots, x_n$ are nonnegative integers. The author proves that there is a unique subset A of M such that (1) any point of M can be changed into a point of A by a suitable translation k; (2) no point of A can be changed into another point of A by a translation k. The explicit construction of A is that it consists of all points $(x_1, \cdots, x_n)$ for which $b_m(x_1)+\cdots+b_m(x_n) \equiv 0 \bmod (k+1)$, where $b_m(x)=0$ or 1 denotes the coefficient of 2^m in the dyadic expansion of a nonnegative integer x.

H. Davenport (Stanford University, Calif.).

H99-2 (14, 623f)

Tsuji, Masatsugu. Theorems in the geometry of numbers for Fuchsian groups. J. Math. Soc. Japan **4**, 189–193 (1952).

The non-euclidean metric $ds=2|dz|/(1-|z|^2)$ is introduced in the unit circle $|z|<1$. By use of an analogue of a theorem of Blichfeldt [Trans. Amer. Math. Soc. **15**, 227–235 (1914)] the following analogue of Minkowski's fundamental theorem is obtained. Let G be a Fuchsian group of transformations of $|z|<1$ into itself. Let the fundamental domain D_0 of G have finite non-euclidean area $\sigma(D_0)$. Let Δ be the disc $|z| \leqq \rho$ (where $\rho<1$) with non-euclidean area $\sigma(\Delta)=4\pi\rho^2/(1-\rho^2)$. If $\sigma(\Delta) \geqq 4\sigma(D_0)+\sigma^2(D_0)/\pi$, then Δ contains a point other than $z=0$ equivalent to $z=0$ under G. Analogues of some other simple results in the geometry of numbers are given.

C. A. Rogers (London).

Referred to in H15-65.

H99-3 (24# A1258)

Groemer, Helmut

Lagerungs- und Überdeckungseigenschaften konvexer Bereiche mit gegebener Krümmung.

Math. Z. **76** (1961), 217–225.

The author considers a plane lattice of determinant 1 and a convex region K, of area $|K|$, symmetrical about one lattice point and containing no other lattice point in its interior. His chief theorems are as follows: If r is the greatest lower bound of the radius of curvature of the boundary of K, then

$$|K| \leqq 4-(2\sqrt{3}-\pi)r^2;$$

and if R is the least upper bound of the radius of curvature of the boundary of K, then

$$|K| \leqq 6R^2 \arctan(2/3R^2).$$

He points out that the former result, under the additional assumption that the curvature varies continuously, was proved another way by van der Corput and Davenport [Nederl. Akad. Wetensch. Proc. **8** (1946), 409–415; MR **8**, 317].

H. S. M. Coxeter (Toronto)

Citations: MR 8, 317i = H05-8.

H99-4 (35# 135)

Hlawka, Edmond [Hlawka, Edmund] 135

Geometrie der Zahlen und trigonometrische Interpolation bei Funktionen von mehreren Variablen. (French summary)

Les Tendances Géom. en Algèbre et Théorie des Nombres, pp. 83–86. *Éditions du Centre National de la Recherche Scientifique, Paris,* 1966.

Étant donné, dans R^s, un réseau Γ_0, N points β_i formant un groupe mod Γ_0, et une fonction f admettant le groupe de périodes Γ_0, on majore (sans démonstration), en fonction de l'oscillation de f, la différence entre f et un polynôme trigonométrique valant $f(\beta_i)$ en chaque point β_i.

M. Hervé (Paris)

H99-5 (35# 1557)

Niven, Ivan; Zuckerman, H. S.

Lattice point coverings by plane figures.

Amer. Math. Monthly **74** (1967), 353–362.

A set S in the Euclidean plane (with a given system of rectangular coordinates) is said to have the covering property if every T congruent to S contains a lattice point (point with integral coordinates). A circle has this property if and only if its radius is at least $1/\sqrt{2}$, and a rectangle if and only if it has sides a, b with $a \geqq 1$, $b \geqq \sqrt{2}$ [see the first author, *Diophantine approximations*, p. 54, Interscience, New York, 1963; MR **26** #6120]. In this paper the authors determine necessary and sufficient conditions for ellipses and parallelograms to have this property. A necessary condition is given for triangles also. A corrected proof of the sufficiency of this condition is promised.

R. P. Bambah (Columbus, Ohio)

Citations: MR 26# 6120 = J02-15.

Referred to in H99-6.

H99-6 (36# 2557)

Niven, Ivan; Zuckerman, H. S.

Correction for: "Lattice point coverings by plane figures".

Amer. Math. Monthly **74** (1967), 952.

The original article appeared in same Monthly **74** (1967), 353–362 [MR **35** #1557].

Citations: MR 35# 1557 = H99-5.

H99-7 (37# 161)

Schmidt, Wolfgang M.

Asymptotic formulae for point lattices of bounded determinant and subspaces of bounded height.

Duke Math. J. **35** (1968), 327–339.

Denote by Γ^k $(0 \leq k \leq n)$ an integral k-dimensional point-lattice in Euclidean space E^n (with a given Cartesian coordinate system); its determinant $d(\Gamma^k)$ is then to be calculated on the Euclidean space E^k which it spans. Say that Γ^k is primitive if it is not properly contained in any integral k-dimensional lattice. The author establishes, for $1 \leq k \leq n-1$, an asymptotic formula for $P(n, k, H)$, the number of primitive $\Gamma^k \subset E^n$ with $d(\Gamma^k) \leq H$, in the form $a(n, k)H^n+O(H^{n-b(n,k)})$, where $a(n, k)$ involves products and quotients of ζ- and Γ-functions and $b(n, k)=\max(1/k, 1/(n-k))$. A similar result is also obtained for $L(n, k, H)$, the number of $\Gamma^k \subset E^n$ with $d(\Gamma^k) \leq H$, for $1 \leq k \leq n$.

The author notes the interesting result: $P(n, k, H)=P(n, n-k, H)$, and relates it to the definition of the height $H(S^k)$ of a k-dimensional subspace of E^n given by him in Ann. of Math. (2) **85** (1967), 430–472 [MR **35** #4165]. When

S^k is a rational subspace (i.e., one defined by linear equations with rational coefficients), $H(S^k) = d(\Gamma^k)$, where Γ^k is the unique primitive lattice contained in S^k.

E. S. Barnes (Adelaide)

Citations: MR 35# 4165 = J99-7.

H99-8 (39# 6178a; 39# 6178b)

Maier, E. A.
On the minimal rectangular region which has the lattice point covering property.
Math. Mag. **42** (1969), 84–85.

Niven, Ivan; Zuckerman, H. S.
The lattice point covering theorem for rectangles.
Math. Mag. **42** (1969), 85–86.

Two proofs of the following theorem of I. Niven [*Diophantine approximations*, p. 54, Interscience, New York, 1963; MR **26** #6120]: A closed rectangle of sides a, b, $a \leqq b$, catches a lattice point, no matter where it is placed in the plane, if and only if $a \geqq 1$, $b \geqq \sqrt{2}$.

R. P. Bambah (Chandigarh)

Citations: MR 26# 6120 = J02-15.

H99-9 (42# 966)

Reich, Simeon
Two-dimensional lattices and convex domains.
Math. Mag. **43** (1970), 219–220.

Let A, P, d denote the area, perimeter and diameter of a plane convex domain D. The author establishes four theorems giving bounds on the minimum number of lattice points contained in D. The following is typical. Theorem: If $A \geqq 2^r(P/2+d)$, where r is any positive integer, then D contains at least $2^{r+2}-2$ lattice points. The theorems generalize earlier work by E. A. Bender [Amer. Math. Monthly **69** (1962), 742–744] and J. Hammer [ibid. **71** (1964), 534–535; MR **29** #2710; ibid. **73** (1966), 744–746; MR **33** #6506]. All the proofs are quite short.

G. T. Sallee (Davis, Calif.)

Citations: MR 33# 6506 = H05-80.

H99-10 (43# 8007)

Wills, Jörg M.
Ein Satz über konvexe Körper und Gitterpunkte.
Abh. Math. Sem. Univ. Hamburg **35** (1970), 8–13.

Let $V(K)$ and $F(K)$ denote the volume and surface area, respectively, of a convex body K in Euclidean n-space R^n. Let $s(n)$ be the supremum of $V(K)/F(K)$ as K ranges over all convex bodies in R^n that contain no lattice point. E. A. Bender [Amer. Math. Monthly **69** (1962), 742–744] showed that $s(2) = \frac{1}{2}$, and the author [Monatsh. Math. **72** (1968), 451–463; MR **39** #865] has shown that $s(3) = \frac{1}{2}$ and $s(n) \geqq \frac{1}{2}$ for $n \geqq 2$. In this paper it is proved that also $s(n) = \frac{1}{2}$. {H. Hadwiger [Math. Z. **116** (1970), 191–196; MR **42** #6721] has since proved the author's conjecture that $s(n) = \frac{1}{2}$ for all $n \geqq 2$.}

G. D. Chakerian (Davis, Calif.)

J. DIOPHANTINE APPROXIMATION

For distribution (mod 1) and for metric results on continued fractions and other algorithms, see the next chapter.

J02 BOOKS AND SURVEYS

General texts which include standard material on approximation to real numbers are to be found in **Z01** and **Z02** and are not cross-referenced here.

See also reviews H02-8, H02-10, H02-15, H02-16, H02-20, J12-33, J12-48, J20-30, J20-34, J32-56, J36-21, J40-13, J40-35, J52-15, J64-14, J68-16, J68-22, J68-29, J68-35, J68-38, J68-40, J76-5, J76-18, J76-47, J76-50, J76-52, J76-67, J84-44, J84-45, J88-35, J88-42, M50-39, Q05-43, R06-35, Z02-16, Z02-18, Z02-32, Z02-55.

J02-1 **(7, 275c)**

de Vries, G. W. A theorem of Minkowski. Mathematica, Zutphen. B. **12**, 119–125 (1944). (Dutch)

Exposition of results of Davenport [Quart. J. Math., Oxford Ser. **10**, 119–121 (1938)].

J02-2 **(10, 102e)**

Hinčin, A. Ya. Dirichlet's principle in the theory of Diophantine approximations. Uspehi Matem. Nauk (N.S.) **3**, no. 3(25), 3–28 (1948). (Russian)

An expository survey of the Dirichlet "Schubfachprinzip" and its applications in linear Diophantine analysis. In chapter 1 the principle is applied to prove various known theorems concerning solutions of sets of homogeneous linear inequalities; by using a generalised version of the standard Dirichlet principle, the treatment of limiting cases is somewhat facilitated. In chapter 2 the principle is applied to nonhomogeneous linear inequalities; by a method due originally to Mordell, theorems are proved correlating the solubility of the nonhomogeneous with that of the corresponding homogeneous inequalities. *F. J. Dyson.*

J02-3 **(11, 231c)**

Gel'fond, A. O. The approximation of algebraic numbers by algebraic numbers and the theory of transcendental numbers. Uspehi Matem. Nauk (N.S.) **4**, no. 4(32), 19–49 (1949). (Russian)

The author reports on the history and recent progress of the theory of transcendental numbers, laying particular stress on the work by Russian mathematicians beginning with Euler. After an introductory paragraph, the following subjects are discussed. [2] Liouville's theorem on the approximation of algebraic numbers, and its improvement by Thue and Siegel. [3] The author's recent further improvement of the Thue-Siegel theorem [similar to, but more general than, Dyson's work in Acta Math. **79**, 225–240 (1947); these Rev. **9**, 412; see also T. Schneider, Arch. Math. **1**, 288–295 (1949); these Rev. **10**, 592] in a paper unavailable outside the USSR, and its connection with the problem of effectiveness, i.e., of finding bounds for the solutions and not only for their number. The Thue-Siegel method does not have this property, but more recent work by N. Fel'dman and the author [also unavailable] is said to be promising in this direction. [4] Transcendency of e and π. Sketch of a proof of Lindemann's theorem without use of Hermite's explicit formulae. Siegel's work on Bessel functions. Measures of transcendency of e and π, in particular the new results by Fel'dman [see the second following review]. [5] Transcendency of α^β and $(\log \alpha)/(\log \beta)$ for algebraic α, β. The author's earlier result for e^π, and the final complete solution by him and Schneider. Sketch of his proof. Schneider's work on elliptic functions. The author's work on measures of transcendency of α^β and $(\log \alpha)/(\log \beta)$; its connection with the problem of effectiveness [C. R. (Doklady) Acad. Sci. URSS (N.S.) **7** (**1935 II**), 177–182; Bull. Acad. Sci. URSS. Sér. Math. [Izvestiya Akad. Nauk SSSR] **1939**, 509–518; Rec. Math. [Mat. Sbornik] N.S. **7**(**49**), 7–25 (1940); these Rev. **1**, 295, 292]. Sketch of a new method of the author by which he proves that if $a \neq 0, \neq 1$ is algebraic, and α is a cubic irrational, then a^α and a^{α^2} are algebraically independent over the rational field. Outlook on further applications of this method, e.g., to improve still more the measure of transcendency of $(\log \alpha)/(\log \beta)$. [6] Expressions defined by series or products, and especially the work by Morduchay-Boltovskoy. The paper ends with a bibliography. *K. Mahler* (Manchester).

Citations: MR 1, 292d = D60-1; MR 1, 295c = J80-1; MR 9, 412h = J68-5; MR 10, 592h = J68-7.

Referred to in J02-4, J02-5, J02-6, J80-11, Z10-51.

J02-4 **(13, 630a)**

Gel'fond, A. O. The approximation of algebraic numbers by algebraic numbers and the theory of transcendental numbers. Acta Math. Acad. Sci. Hungar. **1**, 229–260 (1950). (Russian)

This paper was already printed under the same title in Uspehi Matem. Nauk (N.S.) **4**, no. 4 (32), 19–49 (1949); these Rev. **11**, 231. *K. Mahler* (Manchester).

Citations: MR 11, 231c = J02-3.

J02-5 **(13, 727e)**

Gel'fond, A. O. The approximation of algebraic numbers by algebraic numbers and the theory of transcendental numbers. Amer. Math. Soc. Translation no. **65**, 45 pp. (1952).

Translated from Uspehi Matem. Nauk (N.S.) **4**, no. 4(32), 19–49 (1949); these Rev. **11**, 231.

Citations: MR 11, 231c = J02-3.

Referred to in Z10-51.

J02-6 **(15, 292e)**

Gel'fond, A. O. Transcendentnye i algebraičeskie čisla. [Transcendental and algebraic numbers.] Gosudarstv. Izdat. Tehn.-Teor. Lit., Moscow, 1952. 224 pp. 7.20 rubles.

According to the author, the aims of this book are to show the contemporary state of the theory of transcendental numbers, to exhibit the fundamental methods of this theory, to present the historical course of development of these methods, and to show the connections which exist between this theory and other problems in the theory of numbers.

The first chapter begins with an historical survey, in which results due to Liouville, Thue, Siegel, Schneider, Mahler, Morduhaĭ-Boltovskoĭ, Kuzmin, Dyson and the author are reviewed. The measure of a number ζ of an algebraic field K, of degree σ and having an integral basis $\omega_1, \cdots, \omega_\sigma$, is then defined as $\min\max(|p_1|, \cdots, |q_\sigma|)$, where the minimum is taken over all representations $\zeta=(p_1\omega_1+\cdots+p_\sigma\omega_\sigma)/(q_1\omega_1+\cdots+q_\sigma\omega_\sigma)$ of ζ. In terms of this measure, the following generalization of Dyson's theorem is proved. I. Let α and β be two arbitrary numbers in an algebraic field K_0 of degree ν. Let ζ and ζ_1 be numbers of an algebraic field K of degree s, whose measures with respect to a fixed integral basis of K are q and q_1, respectively. Let θ and θ_1 be two real numbers such that $2\leqq\theta\leqq\theta_1\leqq\nu$, $\theta\theta_1=2\nu(1+\epsilon)$, where $\epsilon>0$ is arbitrarily small but fixed. Then if the inequality $|\alpha-\zeta|<q^{-s\theta}$ has a solution ζ with measure $q>q'(K_0, K, \alpha, \beta, \epsilon, \delta)$, the inequality $|\beta-\zeta_1|<q_1^{-s\theta_1}$ cannot have solutions with measure q_1 under the condition that

$$\log q_1\geqq\left[\frac{\theta-1}{2(\sqrt{(1+\epsilon)}-1)}+\delta\right]\log q,$$

where δ is any arbitrarily small positive constant. The proof, which is given in detail and thus becomes available for the first time outside the USSR [see Gel'fond, Uspehi Matem. Nauk (N.S.) **4**, no. 4(32), 19–49 (1949); these Rev. **11**, 231], uses an improvement of the method of Thue. A p-adic analogue, and other variants, of I are proved. I is then used to prove II: Let $\alpha, \zeta_1, \cdots, \zeta_s$ be numbers of an algebraic field K_0, and suppose that no product of integral powers of $\zeta_1, \cdots, \zeta_s$ is 1. Then the inequality

$$|\alpha-\zeta_1^{x_1}\cdots\zeta_s^{x_s}|<e^{-\epsilon x},$$

where $x=\max(|x_i|)$, and the congruence $\alpha\equiv\zeta_1^{y_1}\cdots\zeta_s^{y_s} \pmod{\mathfrak{P}^m}$, where $m=[\delta y]$, $y=\max(|y_i|)$, can have, for arbitrary $\epsilon>0$, $\delta>0$, only finitely many rational integral solutions $x_1, \cdots, x_s, y_1, \cdots, y_s$. Here $\mathfrak{P}$ is a prime ideal of K_0. The chapter concludes with some variants of II, and applications of I–II, for some of which see Gel'fond, loc. cit.

The second chapter contains an exposition of Siegel's method, in the form given in the latter's monograph [Transcendental numbers, Princeton, 1949; these Rev. **11**, 330]. The first two sections of the third and last chapter are concerned with the Gel'fond-Schneider theorem, and the work of Schneider on elliptic functions, using Gel'fond's method. In the third section, this method is applied to give an effective transcendence measure for the $\mathfrak{P}$-adic logarithm, as follows. III. Let a and b be numbers of an algebraic number field K, such that $a^{n_1}\neq b^{n_2}$ for any rational integers n_1 and n_2 not both zero, and let $\mathfrak{P}$ be a prime ideal of K with $N(\mathfrak{P})=p^\sigma$, $p\equiv 0 \pmod{\mathfrak{P}^\mu}$. If $|a|_\mathfrak{P}=|b|_\mathfrak{P}=1$, then the inequalities

$$\left|\log a^\lambda-\frac{n_1}{n_2}\log b^\lambda\right|_\mathfrak{P}<\mathfrak{P}^{-m_0},\quad \lambda=p^{3\sigma\mu}-p^{(3\mu-1)\sigma},$$

$$0<|n_1|+|n_2|\leqq 2N,\quad m_0=[\log^7 N],\quad (n_1, n_2)=1,$$

cannot have solutions for $N>N_0$, where N_0 can be explicitly specified as a function of a, b, and $\mathfrak{P}$. This is followed by an outline of the proof of the author's theorem [loc. cit. pp. 32–36, and Doklady Akad. Nauk SSSR (N.S.) **64**, 277–280 (1949); these Rev. **10**, 682], that if $a\neq 0, 1$ is algebraic and α is a cubic irrationality, then a^α and a^{α^2} are algebraically independent over the rationals.

The last two sections of the third chapter include a detailed repetition of the author's work [Uspehi Matem. Nauk (N.S.) **4**, no. 5(33), 14–48 (1949); these Rev. **11**, 231; **12**, 1001] on algebraic independence of certain classes of transcendental numbers. An application to the problem of character sums is noted. *W. J. LeVeque.*

Citations: **MR** 10, 682d = J88-3; **MR** 11, 231c = J02-3; **MR** 11, 231d = J88-5; **MR** 11, 330c = J76-18.

Referred to in D60-44, D60-61, E16-84, J02-7, J02-9, J02-24, J68-59, J76-40, J80-26, J80-29, J80-38, J80-42, J88-41, Q05-63, R14-98.

J02-7 (22# 2598)

Gelfond, A. O. Transcendental and algebraic numbers. Translated from the first Russian edition by Leo F Boron. Dover Publications, Inc., New York, 1960. vii+190 pp. $1.75.

Translation of the 1952 edition [GITTL, Moscow; MR **15**, 292].

Citations: **MR** 15, 292e = J02-6.

Referred to in D60-61, E16-84, J02-24, J68-59, J76-70, J80-26, J80-29, J80-38, J80-42, J88-41, J88-45, Q05-63, Q05-89.

J02-8 (17, 17e)

Poitou, Georges. Approximations diophantiennes et groupe modulaire. Publ. Sci. Univ. Alger. Sér. A. **1** (1954), 15–21 (1955).

An exposition of results which can be obtained directly from geometric considerations relative to the modular and Picard groups. The results concern the minimum of a positive definite quadratic or Hermitian form, and the approximation of a number by fractions (rational, or from an algebraic number field having only principal ideals), the latter problem also being modified by requiring that the denominators of the fractions be not divisible by a given integer. *W. J. LeVeque.*

J02-9 (19, 252f)

Schneider, Theodor. Einführung in die transzendenten Zahlen. Springer-Verlag, Berlin-Göttingen-Heidelberg, 1957. v+150 pp. DM 21.60.

This addition to the small library of modern books on trancendental numbers [see C. L. Siegel, Transcendental numbers, Princeton, 1949; MR **11**, 330; and A. O. Gel'fond, Transcendental and algebraic numbers, Gostehizdat, Moscow, 1952; MR **15**, 292] will be welcomed by mathematicians everywhere, the more so since it contains proofs of the author's fundamental work on the transcendency of elliptic and modular functions. (His even more far-reaching results on abelian functions are also mentioned, but not proved.)

In Chapter 1, the author starts with Liouville numbers, proceeds to a slight generalisation of Roth's recent theorem on the approximation of algebraic numbers, and ends with the actual construction of special classes of transcendental numbers, both Liouville and non-Liouville. Chapter 2 on the values of periodic functions begins with a proof of the transcendency of e^α (special theorem of Lindemann) by means of interpolation series. Then a general theorem on the algebraic dependence of integral or meromorphic functions is proved and shown to imply the transcendency of e^α, of α^β, of $(\log\alpha)/(\log\beta)$, and in particular the author's results already referred to. Chapter 3 gives an account of the still unsatisfactory position in the classification of transcendental numbers and deals in particular with the methods due to K. Mahler and J. F. Koksma. In Chapter 4 measures of transcendency for numbers like e^α and α^β are proved and connected with the classification theory in Chapter 3. The final chapter deals with Siegel's method of proving the transcendency of solutions of linear differential equations. As examples, the general theorem of Lindemann on the exponential function and Siegel's theorem on the transcendency of Bessel functions are derived.

The book contains many valuable references to the literature. It will serve as an excellent introduction to different aspects of the theory. *K. Mahler.*

Citations: MR 11, 330c = J76-18; MR 15, 292e = J02-6.

Referred to in J02-10, J02-21, J68-18, J68-23, J76-47, J76-55, J76-67, J76-68, J84-20, J84-23, J84-27, Q05-61.

J02-10 (21# 5620)

Schneider, Théodor. Introduction aux nombres transcendants. Traduit de l'allemand par P. Eymard. Gauthier-Villars, Paris, 1959. viii+151 pp. Paperbound: 3500 francs; $7.30.

The original [*Einführung in die transzendenten Zahlen*, Springer, Berlin-Göttingen-Heidelberg, 1957] is reviewed in MR **19**, 252.

Citations: MR 19, 252f = J02-9.

Referred to in J02-21, J76-55, J76-68.

J02-11 (19, 396h)

Cassels, J. W. S. An introduction to Diophantine approximation. Cambridge Tracts in Mathematics and Mathematical Physics, No. 45. Cambridge University Press, New York, 1957. x+166 pp. $4.00.

Chapter headings: I. Homogeneous approximation; II. The Markoff chain; III. Inhomogeneous approximation; IV. Uniform distribution; V. Transference theorems; VI. Rational approximation to algebraic numbers; VII. Metrical theory; VIII. The Pisot-Vijayaraghavan numbers; and Appendices.

This tract is rather more than an introduction to the subject of Diophantine approximation, and is best described as a concise textbook. It contains a remarkable wealth of material, and most of the major theories of the subject are treated fairly fully. The author has taken much trouble to present the proofs in small compass, though in consequence it is not always possible to follow the general principle of a proof without mastering the details.

Chapter II contains all the main results of Markoff's theory of the minima of indefinite binary quadratic forms, proved in a mere 25 pages. Chapter III gives a full account of what is known in connection with Kronecker's theorem. Chapter IV contains the substance of Weyl's great memoir of 1916, in a modern presentation. Chapter V is based on the work of Khintchine, with recent developments by Hlawka, the author and Birch. Chapter VI contains the proof of K. F. Roth's remarkable recent theorem. Chapters VII and VIII are well-integrated accounts of work which was previously scattered in the literature.

Everyone interested in the subject has reason to be grateful to the author for this valuable work. It should be studied in conjunction with Koksma's report [Diophantische Approximationen, Springer, Berlin, 1936], which surveys a somewhat wider field, but for the most part without proofs. *H. Davenport* (London).

Referred to in J02-12, J04-42, J16-24, J20-35, J20-43, J24-42, J32-64, J68-69, R06-31, Z10-20.

J02-12 (22# 10976)

Касселс, Дж. В. С. [Cassels, J. W. S.]. Введение в теорию диофантовых приближений [An introduction to Diophantine approximation]. Translated by A. M. Polosuev; edited and supplemented by A. O. Gel'fond. Izdat. Inostr. Lit., Moscow, 1961. 213 pp. 0.84 r.

For a review of the original [Cambridge Univ. Press, New York, 1957], see MR **19**, 396.

Citations: MR 19, 396h = J02-11.

Referred to in J20-43, J24-42, J68-69.

J02-13 (24# A1249)

Inkeri, K.

On diophantine approximation. (Finnish)

Arkhimedes **1960**, *no.* 2, 8–17.

A readable, expository account is given on Diophantine approximation. Theorems of Liouville, Thue, Maillet, Dyson, Roth, Bergström, Landau, Walfisz, Ostrowski, Catala, and Siegel are surveyed, and several new proofs based on the author's work are indicated.

L. Sario (Los Angeles, Calif.)

J02-14 (26# 78)

Mahler, Kurt

Lectures on diophantine approximations. Part 1: g-adic numbers and Roth's theorem.

Prepared from the notes by R. P. Bambah of my lectures given at the University of Notre Dame in the Fall of 1957.

University of Notre Dame Press, Notre Dame, Ind., 1961. xi+188 *pp.* $2.25.

The present work constitutes an extremely valuable contribution to the literature on p-adic diophantine approximation. Up to now the only other exposition has been Lutz's *Sur les approximations diophantiennes linéaires p-adiques* [Actualités Sci. Indust., No. 1224, Hermann, Paris, 1955; MR **16**, 1003], which has little in common with the present book; Lutz's was devoted exclusively to small p-adic values assumed by linear forms, while this Part I of Mahler's lectures is concerned principally with the development of the relevant valuation and pseudo-valuation theory and its application to algebraic and transcendental numbers. A second part is promised, dealing with applications of the geometry of numbers to diophantine approximation in p-adic fields.

The author makes heavy use of his own theory of pseudo-valuations as a unifying device. Both valuations and pseudo-valuations are functions defined over some field K and satisfying the relations $w(a) \geqq 0$ (with $w(a)=0$ if and only if $a=0$) and $w(a\pm b) \leqq w(a)+w(b)$; but while a valuation is multiplicative (i.e., $w(ab)=w(a)w(b)$), a pseudo-valuation is required only to satisfy the inequality $w(ab) \leqq w(a)w(b)$. The principal reason for the importance of the notion of pseudo-valuation is that the author wants to consider several valuations simultaneously, and it is easily shown that if $w_1, \cdots, w_r$ are pseudo-valuations, then the function w_Σ defined by $w_\Sigma(a)=\max\{w_1(a), \cdots, w_r(a)\}$ is a pseudo-valuation. Four kinds of pseudo-valuations of the rational field Γ are of basic importance: the ordinary archimedean and p-adic valuations, the g-adic pseudo-valuations, and the g^*-adic pseudo-valuations. The last two are functions w_Σ as above; in the first case the component valuations are appropriate powers of p-adic valuations, for the various primes p dividing the integer $g>1$, while in the second, an archimedean valuation is combined with a g-adic pseudo-valuation. It is shown in Chapter I that every pseudo-valuation of the rational field Γ is equivalent to one of these four types, when equivalence is defined as usual. In the four cases, sequential completion of the rational field leads to the field of real numbers, a p-adic field, a g-adic ring or a g^*-adic ring, respectively.

The following important decomposition theorem gives the connection between the completions Γ_{w_Σ} and $\Gamma_{w_1}, \cdots, \Gamma_{w_r}$: There is a 1-1 correspondence $\alpha \leftrightarrow (\alpha_1, \cdots, \alpha_r)$ between the elements α of Γ_{w_Σ} and the ordered sets $(\alpha_1, \cdots, \alpha_r)$ of one element from each of $\Gamma_{w_1}, \cdots, \Gamma_{w_r}$ such that, if $\alpha \leftrightarrow (\alpha_1, \cdots, \alpha_r)$ and $\beta \leftrightarrow (\beta_1, \cdots, \beta_r)$, then $\alpha+\beta \leftrightarrow (\alpha_1+\beta_1, \cdots, \alpha_r+\beta_r)$ and $\alpha\beta \leftrightarrow (\alpha_1\beta_1, \cdots, \alpha_r\beta_r)$. This correspondence is defined by

$$\alpha = \lim_{m\to\infty} a_m(w_\Sigma) \text{ and } \alpha_h = \lim_{m\to\infty} a_m(w_h), \quad h = 1, 2, \cdots, r,$$

where $\{a_m\}$ is a sequence in Γ which is a Cauchy sequence with respect to each of $w_1, \cdots, w_r, w_\Sigma$.

Chapter II is devoted to series expansions of p-adic, g-adic and g^*-adic numbers. In Chapter III the author's necessary and sufficient condition that a real number be transcendental, which by now is classical, is extended to the present context: Let α be an element of the ring or field resulting from completion of Γ by a pseudo-valuation ω. Then α is transcendental over Γ if and only if, given any $\Lambda > 0$, there exist a positive integer m and infinitely many distinct polynomials $F(x)$ with integral coefficients, of degrees not exceeding m and of heights A tending to infinity, such that $0 < \omega\{F(\alpha)\} < A^{-\Lambda}$. In Chapter IV an analogue of the continued fraction algorithm is developed for p-adic, g-adic and g^*-adic numbers. This theory appears to be far from fully developed, and it is to be expected that further significant results will be obtained in this direction.

The above material occupies the first 72 pages. The next 90 pages are devoted to detailed proofs of four rational archimedean and non-archimedean versions of Roth's theorem, in which one is concerned with the accuracy of approximation (relative to one of the pseudo-valuations) by rational numbers, having special arithmetic structure, of an algebraic element of the corresponding ring or field. Since these theorems are rather complicated, we quote only the simplest. Let ξ be a real algebraic number and $\Xi \leftrightarrow (\xi_1, \cdots, \xi_r)$ a g-adic number, each of whose components is algebraic over Γ. Let ρ, σ, λ, μ be real constants satisfying $\rho > 0$, $\sigma > 0$, $0 \leqq \lambda \leqq 1$, $0 \leqq \mu \leqq 1$; let c_1, c_2, c_3, c_4 be positive constants; and let $g' \geqq 2$ and $g'' \geqq 2$ be fixed integers. Finally, let $\kappa_1, \kappa_2, \kappa_3, \cdots$ be an infinite sequence of rational numbers, $\kappa_k = P_k/Q_k \neq 0$, where $(P_k, Q_k) = 1$ and $H_k = \max(|P_k|, |Q_k|)$, with the following two properties.

(A_d) For all k,

$$|\kappa_k - \xi| \leqq c_1 H_k^{-\rho} \qquad \text{if } d = 1,$$

$$|\kappa_k - \Xi|_g \leqq c_2 H_k^{-\sigma} \qquad \text{if } d = 2,$$

$$|\kappa_k - \xi| \leqq c_1 H_k^{-\rho} \text{ and } |\kappa_k - \Xi|_g \leqq c_2 H_k^{-\sigma} \qquad \text{if } d = 3.$$

(B) For all k,

$$|P_k|_{g'} \leqq c_3 H_k^{\lambda-1} \text{ and } |Q_k|_{g''} \leqq c_4 H_k^{\mu-1}.$$

Then

$$\rho \leqq \lambda + \mu \qquad \text{for } d = 1,$$

$$\sigma \leqq \lambda + \mu \qquad \text{for } d = 2,$$

$$\rho + \sigma \leqq \lambda + \mu \qquad \text{for } d = 3.$$

The case $d = 1$, $\lambda = \mu = 1$ is Roth's theorem [Mathematika **2** (1955), 1–20; corrigendum, 168; MR **17**, 242]; one of the other versions contains as special cases the extensions of Roth's theorem due to Ridout [ibid. **4** (1957), 125–131; MR **20** #32; ibid. **5** (1958), 40–48; MR **20** #3851].

In Chapter 9 various applications of the above theorems are given, e.g., that if ξ is a real irrational algebraic number with continued fraction convergents p_n/q_n, then the greatest prime factor of p_n, as well as that of q_n, tends to infinity with n.

Finally, there are three appendices. The first gives an alternative proof, due to G. E. H. Reuter, of a lemma of Schneider used in the proof of Roth's theorem. The second gives a generalization of a theorem of M. Cugiani [Collectanea Mathematica, No. 169, Milano, 1958] dealing with solutions of the inequality

$$\left| \xi - \frac{P_k}{Q_k} \right| < Q_k^{-2-\varepsilon(Q_k)},$$

where $\varepsilon(Q) \to 0$ as $Q \to \infty$. In the third appendix a brief development is given of the completions of an algebraic number field K by valuations and pseudo-valuations determined by prime ideals of K. The book closes with conjectural statements of the generalizations of the four extensions of Roth's theorem mentioned above to the archimedean, $\mathfrak{p}$-adic or $\mathfrak{g}$-adic approximation of algebraic numbers by elements of a fixed set of algebraic number fields. The analogue of that one of the extensions which was quoted in full above would reduce to the archimedean theorem of the reviewer [*Topics in number theory*, Vol. 2, pp. 121–160, Addison-Wesley, Reading, Mass., 1956; MR **18**, 283] in a special case.

{The book contains a few misprints, of which the most serious is the omission of the factor

$$\prod_{j=r+1}^{r+r'} |\kappa|_{\mathfrak{p}_j}{}^{*g(\mathfrak{p}_j)}$$

in the displayed inequality of Assertion (2.I) on page 187.}

W. J. LeVeque (Ann Arbor, Mich.)

Citations: MR 16, 1003d = J64-14; MR 17, 242d = J68-14; MR 18, 283b = Z01-38; MR 20# 32 = J68-18; MR 20# 3851 = J68-20.

Referred to in K40-58.

J02-15 (26# 6120)

Niven, Ivan

Diophantine approximations.

The Ninth Annual Series of Earle Raymond Hedrick Lectures of The Mathematical Association of America. Interscience Tracts in Pure and Applied Mathematics, No. 14.

Interscience Publishers, a division of John Wiley & Sons, New York-London, 1963. viii + 68 *pp.* $5.00.

This book presents a review of the problem of approximating irrationals by rationals in both $K(1)$ and $K(i)$. The five chapters that make up the book are headed as follows: (1) The approximation of irrationals by rationals; (2) The product of linear forms; (3) The multiples of an irrational number; (4) The approximation of complex numbers; (5) The product of complex linear forms. Some of the proofs are new and many others have been revised by the author. Continued fractions are not used in this monograph, and the treatment is self-contained. Some refinements, as the author notes, are therefore necessarily omitted. At the end of each chapter is a section entitled 'Further results', giving a bibliographic account of closely related work, the sources from which proofs are drawn, and also some conjectures and their background.

G. M. Petersen (Swansea)

Referred to in H99-5, H99-8, J04-50, J20-36, J56-29.

J02-16 (30# 1981)

Turán, Pál

Diophantine approximation and applied mathematics. (Hungarian. English and Russian summaries)

Mat. Lapok **14** (1963), 264–276.

Author's summary: "The subjects touched are kinetic gas-theory, stability-theory of ordinary differential equations, retarded differential equations and approximative determination of eigenvalues of matrices."

J02-17 **(30# 4727)**

Mahler, Kurt

Transcendental numbers.

J. Austral. Math. Soc. **4** (1964), 393–396.

An expository paper.

J02-18 **(33# 1286)**

Lang, Serge

Report on diophantine approximations.

Bull. Soc. Math. France **93** (1965), 177–192.

From the author's introduction: "The theory of transcendental numbers and diophantine approximations has only few results, most of which now appear isolated. It is difficult, at the present stage of development, to extract from the literature more than what seems a random collection of statements, and this causes a vicious circle. On the one hand, technical difficulties make it difficult to enter the subject, since some definite ultimate goal seems to be lacking. On the other hand, because there are few results, there is not too much evidence to make sweeping conjectures, which would enhance the attractiveness of the subject.

"With these limitations in mind, I have nevertheless attempted to break the vicious circle by imagining what would be an optimal situation, and perhaps recklessly to give a coherent account of what the theory might turn out to be. I especially hope thereby to interest algebraic geometers in the theory."

The section headings: (1) Measure theoretic results, (2) Classical numbers, (3) The convergent case, (4) Asymptotic approximations, (5) Generalizations, and (6) Relation with transcendental numbers. *S. Ikehara* (Tokyo)

J02-19 **(34# 1266)**

Minkowski, H.

Diophantische Approximationen. Einführung in die Zahlentheorie.

Thesaurus Mathematicae, No. 2.

Physica-Verlag, Würzburg, 1961. 235 *pp.* *DM* 27.00.

This is a reprinting of the Teubner, Leipzig, 1907 edition, which was reprinted by Chelsea, New York, 1957 [MR **19**, 124].

Citations: MR 19, 124f = H02-11.

J02-20 **(35# 129)**

Lang, Serge

Introduction to diophantine approximations.

Addison-Wesley Publishing Co., Reading, Mass.-London-Don Mills, Ont., 1966. viii+83 *pp.* $6.75.

According to the author's foreword, "The quantitative aspects of the theory of diophantine approximations are, at the moment, still not very far from where Euler and Lagrange left them. Very recent work seems to have opened some fruitful lines of research . . .".

The first chapter presents the fundamental approximation properties of continued fractions: near the end of the chapter it is asserted that Perron made the first attempt to extend such results to linear forms in several variables.

In Chapter II the author defines a real irrational number α to be of type $\leqq g$, where g is a nondecreasing function $\geqq 1$, if for all sufficiently large B there are relatively prime p and q such that $|q\alpha-p|<q^{-1}$ and $B(g(B))^{-1}\leqq q<B$. It is shown that α is of bounded type if and only if it has bounded partial quotients. The principal theorem of the chapter (due to the author [Amer. J. Math. **87** (1965), 481–487; ibid. **87** (1965), 488–496; MR **31** #3383]) is that if α is of type $\leqq g$, $g(t)=O(\omega(t))$, and $\mu(t)=\omega(t)^{1/2}g(t)^{1/2}/t$ decreases, then the number of solutions of the inequalities $0<q\alpha-p<\psi(q)=\omega(q)/q$, $1\leqq q\leqq N$, is of the form $\int_1^N \psi(t)\,dt+O(\int_1^N \mu(t)\,dt)$. {Reviewer's remark: The proof by Erdős mentioned on page 27 appears to have a serious flaw; for related work see also the reviewer [J. Reine Angew. Math. **202** (1959), 215–220; MR **22** #12090].}

Chapter III is devoted to estimates of sums whose terms involve the fractional part of $n\alpha$, $n=1, 2, \cdots, N$, and quadratic exponential sums, where the hypotheses again involve a bound on the type of α. For the relation between these theorems and classical work in the area by Ostrowski, Behnke and Hecke, see the review of the author's paper in Proc. Nat. Acad. Sci. U.S.A. **55** (1966), 31–34 [MR **32** #5594].

Chapter IV contains a proof of the author's theorem that if α is a real quadratic irrationality, and if $\lambda(N)$ is the number of solutions of the inequality $0<q\alpha-p<c/q$, $1\leqq q\leqq N$, then either $\lambda(N)=O(1)$ or $\lambda(N)=c_1\log N+O(1)$ as $N\to\infty$. In Chapter V, analogous estimates are obtained when α is replaced by e.

The title of this book might lead the reader to expect somewhat broader coverage, and to feel some surprise at the names missing from the bibliography. However, it evidences the same care and scholarship as the other books of the author. *W. J. LeVeque* (Ann Arbor, Mich.)

Citations: MR 22# 12090 = J24-25; MR 31# 3383 = J24-35; MR 32# 5594 = J24-40.

Referred to in J24-49, J24-57.

J02-21 **(35# 4170)**

Lipman, Joseph

Transcendental numbers.

Queen's Papers in Pure and Applied Mathematics, No. 7.

Queen's University, Kingston, Ont., 1966. vii+83 *pp.* (*loose errata*) $2.00.

A short account of basic results on transcendental numbers. The first chapter deals with rational (and more generally with algebraic) approximations of algebraic numbers, in particular the theorems of Liouville and of Roth (the latter only quoted). These propositions are applied to the construction of transcendental numbers (Liouville numbers, $\sum_0^\infty 10^{-3^n}$). The second chapter contains a detailed proof of T. Schneider's general theorem on pairs of meromorphic functions [*Einführung in die transzendenten Zahlen*, Springer, Berlin, 1957; MR **19**, 252; French translation, Gauthier-Villars, Paris, 1959; MR **21** #5620; Math. Ann. **121** (1949), 131–140; MR **11**, 160], with applications to e^α, α^β, elliptic functions, and the modular function. In the final chapter the author proves the following weakened form of the reviewer's measure of transcendency of e: $|\sum_{h=0}^n a_h e^h|>H^{-n-(cn^3\log(n+1))/\log\log H}$, where the a's are integers not all zero, and $H=\max_h|a_h|$ [the reviewer, J. Reine Angew. Math. **166** (1931/32), 118–136; ibid. **166** (1931/32), 137–150]. The proof is different from that of the reviewer and seems unnecessarily long and complicated. {The text is disfigured by many misprints, not all listed on an enclosed sheet. For "Gelfand" read everywhere "Gelfond". The correct reference for the decimal fraction $0.f(1)f(2)f(3)\ldots$ by the reviewer is Nederl. Akad. Wetensch. Proc. **40** (1937), 421–428.}

K. Mahler (Canberra)

Citations: MR 11, 160a = J76-17; MR 19, 252f = J02-9; MR 21# 5620 = J02-10.

J02-22 **(35# 5397)**

Lang, Serge

Introduction to transcendental numbers.

Addison-Wesley Publishing Co., Reading, Mass.-London-Don Mills, Ont., 1966. vi+105 *pp.* $7.50.

Die Theorie der transzendenten Zahlen wird hier verstanden als die Aufgabe, "Transzendenz und algebraische Unabhängigkeit von Werten klassischer, geeignet normierter Funktionen nachzuweisen". Von diesem Standpunkt aus bleiben zwar manche Gebiete der Theorie ausser Betracht—so etwa die diophantischen Approximationen (diese hat der Verfasser an anderer Stelle behandelt) oder auch die metrische Theorie der transzendenten Zahlen, wo in letzter Zeit eine ganze Reihe von interessanten Ergebnissen erzielt worden sind. Jedoch wird die Darstellung dadurch in sich geschlossen und übersichtlich.

Ausgangspunkt der Untersuchung (Kapitel II) ist ein Satz, wonach bei gegebenen komplexen Zahlen β_1, β_2, z_1, z_2, z_3, wenn die β_μ und auch die z_ν jeweils über dem rationalen Zahlkörper Q linear unabhängig sind, wenigstens eine der Zahlen $e^{\beta_\mu z_\nu}$ transzendent ist. Der Beweis wird dann so verallgemeinert, dass ein Satz über die algebraische Abhängigkeit zweier meromorpher Funktionen mit gewissen Wachstumseigenschaften herauskommt. Später (Kapitel V) wird dieser Satz im Hinblick auf das (noch unerreichte) Fernziel verallgemeinert, die Transzendenz von Zahlen wie $e+\pi$ zu beweisen. Es ergibt sich ein (zuvor anscheinend noch nicht publizierter) Satz über die algebraische Abhängigkeit gegebener meromorpher Funktionen $f_1, \cdots, f_d$ mit gewissen funktionentheoretischen und arithmetischen Wachstumseigenschaften.

Im dritten Kapitel wird die vorher eingeführte Methode auf solche meromorphen Funktionen angewendet, die einer algebraischen Differentialgleichung genügen. Es wird als Verallgemeinerung von Resultaten von Gel'fond und Schneider ein Satz des Verfassers [Topology **1** (1962), 313–318; MR **28** #95] über die Anzahl der Punkte bewiesen, an denen endlich viele gegebene meromorphe Funktionen aus einem Ring mit gewissen Differenzierbarkeitseigenschaften sämtlich Werte in einem gegebenen algebraischen Körper annehmen können. Als Korollar ergibt sich der bekannte Gel'fond-Schneidersche Satz über die Transzendenz von α^β. Sodann (Kapitel IV) wird dieser Satz auf Funktionen mehrerer Variablen verallgemeinert, wie es der Verfasser in einer früheren Arbeit [ibid. **3** (1965), 183–191; MR **32** #7506] getan hat. Es folgen u.a. Anwendungen auf abelsche Mannigfaltigkeiten.

Das sechste Kapitel behandelt im wesentlichen Ergebnisse von Feldman aus der Zeit von 1951 bis 1964 über untere Schranken für Zahlen der Form $\log|P(\xi_1, \cdots, \xi_m)|$ (bei gegebenen algebraischen Zahlen ξ_ν und beliebigen Polynomen P mit ganzrationalen Koeffizienten) und verwandte Fragen.

Im letzten Kapitel wird die Siegelsche Methode für Lösungsfunktionen linearer Differentialgleichungen behandelt. Zur Illustration dieser Methode wird zunächst der Satz von Lindemann bewiesen; dann wird er unter Verwendung eines Lemmas von Šidlovskiĭ [Izv. Akad. Nauk SSSR Ser. Mat. **23** (1959), 35–66; MR **21** #1295] auf beliebige E-Funktionen verallgemeinert.

In einem Anhang wird diskutiert, inwieweit die Ergebnisse des Buches auf den p-adischen Fall übertragbar sind.

Jedes Kapitel endet mit historischen Bemerkungen, die mit heuristischen Erläuterungen der Beweisgedanken und der methodischen Schwierigkeiten wie mit Hinweisen auf offene Probleme verbunden sind.

Die Stärke des Buches liegt in dem Versuch, viele Einzelergebnisse durch Axiomatisierung der Beweise in einen einheitlichen Rahmen einzuordnen und überhaupt lehrbuchreif zu machen. Alles in allem ein gut gelungenes Werk, das bestimmt die Weiterentwicklung dieses Gebietes sehr befruchten wird. *B. Volkmann* (Stuttgart)

Citations: MR 21# 1295 = J88-15; MR 28# 95 = J76-45; MR 32# 7506 = J76-51.

Referred to in J80-23, J88-37, J88-48, J88-50.

J02-23 **(35# 5400)**

Fel'dman, N. I.; Šidlovskiĭ, A. B.

The development and present state of the theory of transcendental numbers. (Russian)

Uspehi Mat. Nauk **22** (1967), *no.* 3 (135), 3–81.

This valuable article gives an up-to-date report on the theory of transcendental numbers and discusses in detail the different methods of proofs of transcendency so far known. It thus may serve to replace the fourth chapter of J. F. Koksma's well-known book [*Diophantische Approximationen*, Ergeb. Math. Grenzgeb. (N.F.), Heft 4, Springer, Berlin, 1936]. The progress in the theory becomes evident from the list of references; more than half of the 440 titles given have appeared since 1936.

The authors stress the great contributions by Soviet mathematicians. The mathematical world will be grateful to have now a full list of Russian papers on transcendental numbers up to about 1967.

The article consists of a short historical introduction and the following six sections. (1) Rational approximations of algebraic numbers (theorems of Liouville-Thue-Siegel-Roth; the recent effective results by Baker). (2) The Hermite-Lindemann method (proofs of the transcendency of e and π and of the general theorem of Lindemann-Weierstrass; measures of irrationality and transcendency of e and π). (3) Gel'fond's methods (his paper of 1929 on e^π; the proofs of the transcendency of α^β of 1934 by Gel'fond and by Schneider; Schneider's work on elliptic, modular, and Abelian functions; measures of transcendency). (4) Siegel's method (transcendency of $J_0(\alpha)$ and more general results; Siegel's general method of 1949). (5) The extension of Siegel's method (Šidlovskiĭ's general theory of E-functions; his and his students' work on special E-functions; recent work on measures of irrationality and transcendency). (6) Conclusion (in particular, classification of transcendental numbers, and Sprindžuk's solution of Mahler's problem on S-numbers).

Koksma's report of 1936 was the source of much progress in the theory of Diophantine approximations. We may hope and expect that the present report will play a similar role in the theory of transcendental numbers.

K. Mahler (Canberra)

J02-24 **(38# 5721)**

Šmelev, A. A.

The algebraic independence of certain transcendental numbers. (Russian)

Mat. Zametki **3** (1968), 51–58.

Let α_1 and α_2 be algebraic numbers, the logarithms of which are linearly independent over the rational field, and let β be a quadratic algebraic number. Using unproved lemmas (see below), the author presents a proof that the numbers α_1^β, α_2^β, $(\ln \alpha_2)/\ln \alpha_1$ cannot be algebraically expressed in terms of one of them. The reviewer did not work through the details.

The lemmas in question appear in A. O. Gel'fond's book [*Transcendental and algebraic numbers* (Russian), GITTL, Moscow, 1952; MR **15**, 292; English translation, Dover, New York, 1960; MR **22** #2598]. The present reviewer considers Lemmas I and IV of Chapter I, § 2, and Lemmas I–III and V–VII of Chapter III, § 5.

The phrase translated as "all different from zero" should be replaced everywhere by "not all zero".

In Chapter I, the inequalities in (19) and on the following line are not strict. The right hand side of (26) should be $H_1^m/(2mn_1+1)$ with corresponding changes in the proof. In Lemma IV, f_1 and f_2 can be assumed to be polynomials of non-negative degrees with complex coefficients. Inequality (30) is not strict if f_1 and f_2 are constant.

In Chapter III, Lemmas II and II′ refer to polynomials

with complex coefficients. In equation (126), "cos" should be "sin". In equation (139) the factor h^m should occur in front of the product sign. There is similar poor notation involving product signs throughout this section. Different integration variables are needed in the display after (139).

Lemma III is very obscure and not proved. The reviewer in a forthcoming article (to appear in Trans. Amer. Math. Soc.) replaces this lemma with a simpler lemma which will have the same force.

In Lemma V, the sentence "In case $\nu=1$, the number $a=1$" is part of the definition of the quantity a. In inequalities (166) and (168) the third "α" should be "a". The words "not greater than" should be omitted, as well as the first paragraph of the proof. In Lemma VI, the inequality $H > H'(\alpha)$ should be added to the conditions (169). The second inequality of (179) is not necessarily strict.

In Lemma VII, the inequalities for $|Q_q(\alpha)|$ in (181) and (183) appear to be falsely derived, and the reviewer sees no way of repairing the proof. The proofs of Theorems I and II in § 5 depend on this lemma, as does the result of the present author. *R. Spira* (E. Lansing, Mich.)

Citations: MR 15, 292e = J02-6; MR 22# 2598 = J02-7.

J02-25 **(39# 5479)**

Popken, J.

The real numbers seen from a number-theoretic standpoint. (Dutch)

Euclides (Groningen) **41** (1965/66), 244–254.

In this expository paper, the author treats some classic theorems concerning the problem of approximating an irrational number by rationals, and deals to some extent with the notion of irrationality measure, mentioning the work of Liouville, Mahler, Thue, Roth and Jarník.

L. Kuipers (Carbondale, Ill.)

J02-26 **(40# 7207)**

Gel'fond, A. O.

On Hilbert's seventh problem. (Russian)

Hilbert's problems (Russian), pp. 121–127. *Izdat. "Nauka", Moscow,* 1969.

The author, who died recently, will forever be remembered for his great contributions to the theory of transcendental numbers. The present lecture gives an expository account of the development and the present position of this theory.

K. Mahler (Columbus, Ohio)

J02-27 **(44# 6615)**

Lang, Serge

Transcendental numbers and diophantine approximations.

Bull. Amer. Math. Soc. **77** (1971), 635–677.

The author gives an account of the theory of transcendental numbers and algebraic independence of numbers from Hermite to A. Baker and of the theory of Diophantine approximations from Dirichlet and Liouville to K. F. Roth and W. Schmidt. An extensive bibliography with 97 entries is given at the end of the paper.

In Section 1 he deals with the behavior of values of the exponential function, including the theorems of Hermite, Lindemann, Gel'fond and Schneider (Hilbert's 7th Problem), and Baker, and mentions Schanuel's conjecture that the degree of transcendentality of $Q(\alpha_1, \cdots, \alpha_n, e^{\alpha_1}, \cdots, e^{\alpha_n})$ is a least n whenever $\alpha_1, \cdots, \alpha_n$ are complex numbers that are linearly independent over Q. In Section 2 he gives a brief but very readable and lucid account of the ideas underlying the proofs of these theorems. In Section 3 he deals with the values of more general meromorphic functions at algebraic points, giving his own version on meromorphic functions satisfying algebraic differential equations with algebraic coefficients and Bombieri's generalization to meromorphic functions of several complex variables.

Section 4, entitled "General exponential functions" is concerned with work, largely initiated by the author, extending the results of Section 1 to linear, group and Abelian varieties. The limitations on the possible extensions of Schanuel's conjecture and other open problems are discussed. Finally, in Section 5 the author briefly defines Siegel's E-functions and states the Šidlovskiĭ-Siegel theorem on the algebraic independence of the values at regular algebraic points of E-functions satisfying a system of linear differential equations whose coefficients are rational functions over an algebraic number field.

In Section 6 he gives some basic metric results on Diophantine approximation including the theorems of Dirichlet, Hinčin and others on the kind of approximation possible for all, almost all or almost no real number and the expected number of times such good approximations occur. In Section 7 he gives a rather selective classification of numbers according to the types of Diophantine approximation which they admit. In Section 8 he studies Diophantine approximations of algebraic numbers, starting with Liouville and continuing to the results of Roth (including a sketch of the main ideas of the proof) and Schmidt on simultaneous approximations. Effective bounds of Baker and Stark and p-adic extensions are mentioned. In Section 9 he studies transcendence measures and Diophantine approximation results of Popken, Mahler, Gel'fond, Feldman and Baker on e, π, linear forms in logarithms of algebraic numbers, etc. In Section 10 he gives a criterion of Gel'fond for algebraicity and states that it can be used to prove algebraic independence of certain pairs of numbers without, however, explaining Gel'fond's method for doing so.

The final section is devoted to applications to Diophantine analysis. It includes Siegel's theorem on algebraically integral points on algebraically algebraic curves of genus $\geqq 1$ and its generalizations and the Baker-Coates effective bounds in the case of curves of genus 1.

E. G. Straus (Los Angeles, Calif.)

J04 HOMOGENEOUS APPROXIMATION TO ONE NUMBER

Papers in which the main emphasis is on approximative properties of convergents and semi-convergents of continued fractions, as distinguished from the approximability of real numbers, are listed in **A58**.

See also Sections J08, J24, J28.

See also reviews A02-6, A56-17, A68-6, B52-1, H05-76, H05-84, H15-42, H15-72, H20-7, J02-8, J02-15, J12-10, J12-64, J20-25, J20-41, J24-13, J24-60, J28-26, J56-11, J56-12, J56-13, J56-20, J56-22, J56-26, J56-28, J56-30, J56-31, J56-32, J64-2, J64-5, J68-39, J84-24, J99-7, K15-21.

J04-1 **(1, 203a)**

Scott, W. T. Approximation to real irrationals by certain classes of rational fractions. Bull. Amer. Math. Soc. **46**, 124–129 (1940).

Using the method developed by Humbert and Ford (depending on geometric properties of elliptic modular transformations) the author considers the inequality

$$(1)\qquad \left|\omega-\frac{p}{q}\right|<\frac{k}{q^2},$$

ω being a real irrational number and $k>0$ a constant. Each irreducible fraction p/q with (1) belongs to one of the three classes $[o/e]$, $[e/o]$ and $[o/o]$, o denoting an odd integer and e an even integer. The author shows: If $k\geqq 1$ there are infinitely many fractions p/q of each of the classes satisfying (1), regardless of the value of ω. If $k<1$ there exist irrational numbers ω, everywhere dense on the real axis, for which (1) is satisfied by only a finite number of fractions of a given one of the three classes. For the method used, cf. Kap. 3 of the reviewer's "Diophantische Approximationen" [Ergebnisse der Mathematik IV, 4, Berlin, 1936].

J. F. Koksma (Amsterdam).

Referred to in J04-2, J04-13, J04-29, J04-38.

J04-2 **(2, 149c)**

Robinson, Raphael M. The approximation of irrational numbers by fractions with odd or even terms. Duke Math. J. 7, 354–359 (1940).

The author studies the problem [A. Hurwitz, Math. Ann. **39**, 279–285 (1891)] of approximation to an irrational number ξ by rational numbers A/B subject to the condition of $|\xi-A/B|<1/\mu B^2$ for various values of μ. The results obtained extend those of Hurwitz and of W. T. Scott [Bull. Amer. Math. Soc. **46**, 124–129 (1940); these Rev. **1**, 203] and are the best of their type. The methods, unlike those of Scott [cf. L. R. Ford, Proc. Edinburgh Math. Soc. **35** (1916)], involve the traditional use of continued fractions, their convergents and their secondary convergents. The author's bibliography contains only the paper of Scott.

W. Leighton (Houston, Tex.)

Citations: MR 1, 203a = J04-1.

Referred to in J04-3, J04-13, J04-38.

J04-3 **(3, 150b)**

Oppenheim, Alexander. Rational approximations to irrationals. Bull. Amer. Math. Soc. **47**, 602–604 (1941).

The author proves that, if p and q are relatively prime, $q>0$, and if (1) $|x-p/q|<1/q^2$, then p/q is one of the three irreducible fractions p'/q', $(p'+p'')/(q'+q'')$, $(p'-p'')/(q'-q'')$, where p''/q'', p'/q' are two consecutive convergents in the regular continued fraction expansion of the irrational number x; further, one at least of the two fractions $(p'+\epsilon p'')/(q'+\epsilon q'')$, where $\epsilon=\pm 1$, satisfies (1). The editor calls attention to a paper by R. M. Robinson [Duke Math. J. 7, 354–359 (1940); cf. these Rev. **2**, 149] and to a result of P. Fatou [C. R. Acad. Sci. Paris **139**, 1019–1021 (1904)]. *W. Leighton* (Houston, Tex.).

Citations: MR 2, 149c = J04-2.

Referred to in J04-61.

J04-4 **(7, 274d)**

Huzurbazar, V. S. On a property of rational numbers. J. Univ. Bombay (N.S.) **14**, part 3, 1–3 (1945).

The property is that all rational numbers sufficiently close to a given real positive number (except perhaps for the number itself) have arbitrarily large numerators and denominators. *R. P. Boas, Jr.* (Providence, R. I.).

J04-5 **(10, 284c)**

Heilbronn, H. On the distribution of the sequence $n^2\theta$ (mod 1). Quart. J. Math., Oxford Ser. **19**, 249–256 (1948).

The general problem mentioned in the title is not treated, but rather the special question, how closely can the fractional part of $n^2\theta$ approach zero? This question was first posed by Hardy and Littlewood [Acta Math. **37**, 155–191 (1914)]; they conjectured that for all real θ it is possible to make $|n^2\theta-m|<c/N$ for every positive integer N, for suitable integers n, m with $1\leqq n\leqq N$. Here c is an absolute constant. Vinogradov [Bull. Acad. Sci. URSS [Izvestiya Akad. Nauk SSSR] (6) **21**, 567–578 (1927)] proved a general theorem, of which a special case is that this statement holds if c/N is replaced by $c(\eta)/N^{\frac{2}{3}-\eta}$, for every positive η. ($c(\eta)$ depends only on η.) The present work improves this to $c(\eta)/N^{\frac{1}{2}-\eta}$ by an adaptation of Vinogradov's method.

W. J. LeVeque (Cambridge, Mass.).

Referred to in D76-10, J04-57, J04-64, J04-68, J12-42.

J04-6 **(10, 513d)**

Prasad, A. V. Note on a theorem of Hurwitz. J. London Math. Soc. **23**, 169–171 (1948).

Generalizing Hurwitz's well-known approximation theorem, the author proves the following theorem. Let p_n/q_n denote the nth convergent to $\xi_0=\frac{1}{2}(\sqrt{5}-1)$ starting with $p_1/q_1=1$. Let $C_m=\frac{1}{2}\{\sqrt{5}+1\}+p_{2m-1}/q_{2m-1}$ $(m=1, 2, \cdots)$. Then for any irrational number ξ the inequality

$$|\xi-p/q|\leqq C_m^{-1}q^{-2}$$

has at least m solutions in relatively prime integers p, q, with $q>0$. If $\xi=\xi_0$ there are not more than m such solutions.

J. F. Koksma (Amsterdam).

Referred to in H20-19, J04-38, J56-15.

J04-7 **(11, 646f)**

Perron, Oskar. Neuer Beweis zweier klassischer Sätze über Diophantische Approximationen. Acta Sci. Math. Szeged **12**, Leopoldo Fejér et Frederico Riesz LXX annos natis dedicatus, Pars B, 125–130 (1950).

A well-known theorem of Hurwitz states that any irrational number θ has infinitely many rational approximations x/y $(y>0)$ such that $y|\theta y-x|<5^{-\frac{1}{2}}$. A closely related theorem of Korkine and Zolotareff states that a quadratic form $ax^2+bxy+cy^2$ with $b^2-4ac=1$ assumes a value numerically not exceeding $5^{-\frac{1}{2}}$. The author gives simple proofs of these theorems. The proof is on the same lines in each case, and uses the lemma that if $0<\delta<1$ and $0<\lambda<1$ then one at least of the three numbers δ, $\lambda+\delta\lambda^2$, $1-\lambda-\delta(1-\lambda)^2$ does not exceed $5^{-\frac{1}{2}}$. *H. Davenport* (London).

J04-8 **(12, 807a)**

Hartman, S. Sur une condition supplémentaire dans les approximations diophantiques. Colloquium Math. **2**, 48–51 (1949).

En utilisant la théorie des fractions continues, l'auteur montre le théorème suivant, qui donne la réponse à une question qui lui avait été posée par S. Mazur: Soit $\xi>0$ un nombre irrationnel. Quels que soient les entiers $a\geqq 0$, $b\geqq 0$ et $s>0$, il existe une infinité de couples de nombres naturels u, v, satisfaisant aux conditions

$$(1)\qquad \left|\xi-\frac{u}{v}\right|\leqq\frac{2s^2}{v^2},$$

$$(2)\qquad u\equiv a \pmod{s},\qquad v\equiv b \pmod{s}.$$

L'auteur fait remarquer que l'exposant dans le numérateur du membre droit de (1) ne peut pas être diminué; pour s'en convaincre il suffit de poser $a=b=0$ dans (2). [Le référent remarque qu'on peut diminuer un peu le coefficient 2 dans (1) en utilisant des théorèmes connus sur les approximations diophantiques linéaires non-homogènes, comme il montrera dans une note dans Simon Stevin]. Enfin l'auteur applique son résultat en montrant que

$$\liminf (\sin \alpha n)^n = \liminf (\cos \alpha n)^n = -1,$$

où α est un nombre réel incommensurable avec π.

J. F. Koksma (Amsterdam).

Referred to in J04-10, J04-11, J04-18.

J04-9 **(13, 116b)**

Kogoniya, P. G. On the set of Markov numbers. Doklady Akad. Nauk SSSR (N.S.) **78**, 637–640 (1951). (Russian)

The numbers $L(\alpha)=\liminf|q(q\alpha-p)|$, where $q>0$, p are integers, are called the Markov numbers. It is well-known that $L(\alpha)\leqq 5^{-\frac{1}{2}}$ and that there are only a denumerable set of $L(\alpha)>\frac{1}{3}$. It is shown (I) that the set of all $L(\alpha)$ has Hausdorff dimension 1 and (II) that the set of $L(\alpha)$ in $(\frac{1}{3}-\epsilon, \frac{1}{3})$ has the power of the continuum for all $\epsilon>0$. More precisely if $N\geqq 4$ and β is any number whose partial quotients do not exceed $N-2$ a simple continued fraction argument shows that there is an α whose partial quotients have upper limit N for which $L(\alpha)=(N+2\beta)^{-1}$; and analogous but more complicated results hold for $N=2, 3$. The result (I) then follows from Jarnik's estimate for the dimension of the set of β [J. F. Koksma, Diophantische Approximationen, Springer, Berlin, 1936, p. 49]. *J. W. S. Cassels.*

J04-10 **(13, 825e)**

Koksma, J. F. Sur l'approximation des nombres irrationnels sous une condition supplémentaire. Simon Stevin **28**, 199–202 (1951).

Let θ be irrational, and let $s\geqq 1$, a and b be integers. It is shown that to each $\epsilon>0$ there correspond infinitely many pairs of integers p, q with $q>0$ such that

$$|\theta-p/q|<s^2(1+\epsilon)/\sqrt{5}q^2 \text{ and } p\equiv a \pmod{s},\ q\equiv b \pmod{s}.$$

This theorem, in which the constant $\sqrt{5}$ is clearly best possible, is an improvement on the corresponding theorem of S. Hartman [Colloquium Math. **2**, 48–51 (1949); these Rev. **12**, 807] in which the above inequality is replaced by $|\theta-p/q|<2s^2/q^2$. In the case that not both of a and b are divisible by s, it is also shown that there are infinitely many pairs p, q for which $|\theta-p/q|<s^2/4|q(q-b)|$ and $p\equiv a \pmod{s}$, $q\equiv b \pmod{s}$. *W. J. LeVeque.*

Citations: MR 12, 807a = J04-8.

Referred to in J04-38.

J04-11 **(13, 825f)**

Descombes, Roger, et Poitou, Georges. Sur certains problèmes d'approximation. C. R. Acad. Sci. Paris **234**, 581–583 (1952).

Soit ξ un nombre réel positif, s un nombre naturel et soient a et b des nombres entiers. A la suite de S. Hartman [Colloquium Math. **2**, 48–51 (1949); ces Rev. **12**, 807] les auteurs à l'aide de la théorie des fractions continues, étudient la limite inférieure $K(\xi, s, a, b)$ de $|v(\xi v-u)|$ pour les fractions u/v à termes positifs congrus à a et b respectivement modulo s. Le référent, en utilisant une méthode tout à fait différente, a démontré $K(\xi, s, a, b)\leqq 5^{-1/2}s^2$ (et bien $\leqq\frac{1}{4}s^2$, si l'on admet des valeurs négatives pour u et v) [voir l'analyse ci-dessus]. Des résultats classiques de Hurwitz, il suit facilement qu'on ne peut pas améliorer ce résultat (il suffit de poser $a=b=0$ ou de prendre $s=1$). En supposant que les trois nombres a et b et s sont premiers entre eux et (si le référent a bien compris) en supposant aussi que $s\geqq 2$, les auteurs montrent que toujours $K(\xi, s, a, b)\leqq 4s^2/9$, pendant que d'autres résultats sont annoncés. Ils remarquent qu'on peut se limiter au cas $a=1$, $b=0$ quelque soit le nombre $s\geqq 2$.

J. F. Koksma (Amsterdam).

Citations: MR 12, 807a = J04-8.

Referred to in J04-12, J04-22, J20-18.

J04-12 **(13, 921c)**

Poitou, Georges, et Descombes, Roger. Sur certains problèmes d'approximation. II. C. R. Acad. Sci. Paris **234**, 1522–1524 (1952).

En utilisant la notation de la première partie [mêmes C. R. **234**, 581–583 (1952); ces Rev. **13**, 825] les auteurs en remarquant que le problème général équivaut à son cas particulier en faisant $a=1$, $b=0$, continuent à rechercher $k(\xi, s)=k(\xi, s, 1, 0)$ et $k(s)=$ la borne supérieure des nombres $k(\xi, s)$ pour tous ξ (s supposé fixe). Pour $3\leqq s\leqq 10$ les auteurs donnent une liste de $k(s)$. Ces nombres sont isolés dans l'ensemble des valeurs de $k(\xi, s)$ et ne sont atteint que pour certains nombres quadratiques (dits critiques) ξ. Quant au cas $s=2$, on a $k(2)=1$, mais le nombre $k(2)$ n'est pas isolé dans l'ensemble des $k(\xi, 2)$. Les auteurs recherchent la distribution des nombres $C(s)=s^2/k(s)$ sur l'axe réel ($s=2, 3, 4, \cdots$) et montrent $C(s)\geqq 2.5$, borne qui n'est pas la valeur la plus grande possible. Il y a une infinité de valeurs $C(s)<3$. Enfin les auteurs recherchent l'expression $\bar{K}(s)$, analogue à $K(s)$ qu'on trouve si dans la définition de $K(\xi, s)$ on admet aussi des valeurs négatives pour u et v (voir la note citée). *J. F. Koksma* (Amsterdam).

Citations: MR 13, 825f = J04-11.

Referred to in J04-22, J20-18.

J04-13 **(14, 23d)**

Kuipers, L., and Meulenbeld, B. Some properties of continued fractions. Acta Math. **87**, 1–12 (1952).

The authors consider the diophantine inequality

$$(1) \qquad |\alpha-p/q|<k/q^2,$$

where α is irrational and p/q belongs to one of the three classes, odd/odd, odd/even, even/odd, of irreducible fractions. They use properties of simple continued fractions to obtain the following results which had previously been obtained by the reviewer [Bull. Amer. Math. Soc. **46**, 124–129 (1940); these Rev. **1**, 203] using other methods: (i) For any α there exist infinitely many p/q of each of the three classes satisfying (1) when $k\geqq 1$; (ii) If $k<1$, there exist irrationals α, everywhere dense, for which (1) holds for only a finite number of p/q of a given class. The distribution of approximants of a simple continued fraction among the three classes is discussed. The authors do not cite a paper of R. M. Robinson [Duke Math. J. **7**, 354–359 (1940); these Rev. **2**, 149], in which present methods are used to obtain (i) and a strengthened form of (ii).

W. T. Scott (Evanston, Ill.).

Citations: MR 1, 203a = J04-1; MR 2, 149c = J04-2.

Referred to in A58-27.

J04-14 **(14, 359a)**

Lehner, Joseph. A diophantine property of the Fuchsian groups. Pacific J. Math. **2**, 327–333 (1952).

If α is any irrational real number and $h\leqq\frac{1}{2}\sqrt{5}$, then it is well-known that there exists an infinity of reduced fractions p_i/q_i such that

$$(1) \qquad \left|\alpha-\frac{p_i}{q_i}\right|<\frac{1}{2hq_i^2} \quad (i=1, 2, 3, \cdots).$$

The author gives a proof of this for $h\leqq\frac{1}{2}\sqrt{3}$, using the theory of the modular group, and proves the following generalisation. Let G be a Fuchsian group, with real axis as principal circle, which is generated by a finite number of substitutions. Let P, the set of parabolic points of G, be an infinite set including the point ∞. Let $\alpha \varepsilon \bar{P}-P$. Then there exists p_i/q_i $(i=1, 2, \cdots)$ for which (1) is true, where $p_i/q_i \varepsilon P$ and $h=h(G)$ is a geometrical constant depending only on the group G. The number $h(G)$ is defined during the course of the proof, which makes use of L. R. Ford's method of constructing fundamental regions by means of isometric circles. Roughly the idea of the proof is to take a line through the point α perpendicular to the real axis. This passes through an infinity of fundamental regions. These are all mapped onto the principal fundamental region R_0 so that the line is replaced by an infinity of arcs crossing R_0. It is shown that an infinity of these arcs possess points which are at a distance greater than some fixed distance h' from the real axis. These arcs correspond to transformation matrices with bottom row $(q_i, -p_i)$, say, and for these the inequality (1) is valid. In the case of the ordinary modular group $h=h'\geqq\frac{1}{2}\sqrt{3}$. *R. A. Rankin* (Birmingham).

Referred to in J04-29.

J04-15 **(14, 454e)**

Kurzweil, Jaroslav. A contribution to the metric theory of diophantine approximations. Czechoslovak Math. J. **1(76)** (1951), 149–178 (1952) = Čehoslovack. Mat. Ž. **1(76)** (1951), 173–203 (1952).

If $g(q)$ is a given positive function defined for real $q>0$, an irrational number is said to admit the approximation $g(q)$ if there exist infinitely many pairs of integers p, q $(q>0)$ such that

$$\left|x-\frac{p}{q}\right|<\frac{1}{q^2g(q)}.$$

Khintchine showed that the Lebesgue measure of the set of x in $(0, 1)$ which admit the approximation $g(q)$ is zero if $\int^\infty dx/\{xg(x)\}$ converges and unity if this integral diverges. In the case of convergence Jarník [Mat. Sbornik **36**, 371–382 (1929); Math. Z. **33**, 505–543 (1931)] investigated the set of x in $(0, 1)$ which admit the approximation $g(q)$ by means of Hausdorff measure. Here the author considers the corresponding problem of the set Q_g of x in $(0, 1)$ which do not admit the approximation $g(q)$ when the integral diverges. The functions $f_1(d)$ and $f_2(d)$ are defined by

$$f_1(d)=\exp\left\{\frac{2}{3}\int_w^{d^{-\frac{1}{2}}}\frac{dx}{xg(x)}\right\},\quad f_2(d)=\exp\left\{2\int_w^{d^{-\frac{1}{2}}}\frac{dx}{xg(x)}\right\}$$

and $g(x)$ is assumed to satisfy certain additional subsidiary conditions. The main result proved is the theorem that if $g(q)>1000$, then the Hausdorff measure of Q_g formed with respect to f_1 is zero whilst that formed with respect to f_2 is infinity. To prove this the numbers in $(0, 1)$ are expressed as continued fractions and divided into intervals each of which consists of numbers whose expansions as a continued fraction begin in a prescribed way. These intervals are used to cover sets associated with Q_g but the details of the analysis are too complicated to be described here. The theorem is used to obtain the Hausdorff dimension of Q_g for functions $g(q)$ of the form $\log^{\alpha_1} q \log_2{}^{\alpha_2} q\cdots\log_n{}^{\alpha_n} q$, where the suffix denotes a repeated logarithm and the Hausdorff measure is taken with respect to functions of the form

$$f_s(d)=\exp\left\{\log^{-s_1}\frac{1}{d}\log_2{}^{-s_2}\frac{1}{d}\cdots\log_r{}^{-s_r}\frac{1}{d}\right\}.$$

In particular, if $g(q)=\log^\alpha q$ $(0<\alpha\leqq1)$ and

$$f_s(d)=\exp(\log^{-s}1/d)\quad(-1<s\leqq0)$$

then $\dim Q_g=\alpha-1$. The case g a constant greater than 1000 is also considered and it is shown that, if $f_s(d)=d^{s-1}$ $(0<s\leqq1)$, then $1-0.99_g{}^{-1}\leqq\dim Q_g\leqq1-0.25_g{}^{-1}$.

R. A. Rankin (Birmingham).

Referred to in J16-16.

J04-16 **(14, 851c)**

Descombes, Roger. Sur un problème d'approximation non homogène. C. R. Acad. Sci. Paris **236**, 1401–1403 (1953).

Let s be a positive integer and for irrational ξ and integers a, b with $(a, b, s)=1$ put

$$(*)\qquad H(\xi, a, b)=\liminf|v(v\xi-u)|$$

taken over integers v, u with $u\equiv a$, $v\equiv b$ (s). The upper bound of $H(\xi, a, b)$ over ξ is independent of a, b and is denoted by $H(s)$. If s is even, then $H(2)=1$, $H(s)=\frac{1}{4}s(s-2)$ $(s>2)$ and it is non-isolated. If s is odd, then $H(s)$ is isolated; the values for $3\leqq s\leqq11$ are given and

$$H(s)=\tfrac{1}{4}(s^2-2s-1)(1-2/s)^{-1/2}$$

for $s\geqq13$. The proofs depend on showing that for (*) one need consider only (u, v) related to the successive convergents in a certain way. *J. W. S. Cassels.*

Referred to in J04-22.

J04-17 **(14, 851d)**

Descombes, Roger. Sur un théorème classique d'Hurwitz. C. R. Acad. Sci. Paris **236**, 1460–1462 (1953).

For irrational ξ and integer s define

$$c_s(\xi)=\limsup|q(q\xi-p)|^{-1}$$

taken over integers p, q with $s\nmid q$. Let c_s, γ_s be the upper bound of $c_s(\xi)$ and its greatest point of accumulation, respectively. Then $c_2=\gamma_2=2$, but $c_s=5^{1/2}$ for $s>3$. For $s>2$ the value of γ_s depends on the index j of the first term in the Fibonacci sequence $0, 1, 1, 2, \cdots$ divisible by s. If j is even, then $\gamma_s=1+3.5^{-1/2}$ and $c_s(\xi)$ takes no value between c_s and γ_s. If j is odd, then γ_s is given by a complicated expression depending on j and there are infinitely many ξ with $c_s(\xi)>\gamma_s$. Proofs depend on considering best "s-approximations" p_n/q_n with $s\nmid q_n$ and such that $|q\xi-p|\geqq|q_n\xi-p_n|$ if $|q|\leqq|q_n|$ and $s\nmid q$. *J. W. S. Cassels.*

Referred to in J04-22.

J04-18 **(15, 688a)**

Černý, Karel. Remark on Diophantine approximation. Časopis Pěst. Mat. **77**, 241–242 (1952). (Czech)

The author proves that if ξ is any irrational number and a, b, s, are arbitrary integers $(s>0)$, then there exist infinitely many pairs of integers u, v such that $u\equiv a$ (mod s), $v\equiv b$ (mod s), and

$$\left|\xi-\frac{u}{v}\right|<\frac{(1+\epsilon)s^2}{5^{\frac{1}{2}}v^2}.$$

Here ϵ is an arbitrary positive number and the result is best possible in the sense that $5^{\frac{1}{2}}$ cannot be replaced by a larger number.

This is an improvement on a result of S. Hartman [Colloquium Math. **2**, 48–51 (1949); these Rev. **12**, 807], where a similar result with the right-hand number replaced by $2s^2/v^2$ is proved (see also remark by the reviewer). The proof is extremely simple, following immediately from Khintchine's result that if ξ is irrational, α real and $\epsilon>0$, then there exist infinitely many integers p, $q>0$ such that

$$|q\xi-p-\alpha|<\frac{1+\epsilon}{5^{\frac{1}{2}}q}.$$

R. A. Rankin (Birmingham).

Citations: MR 12, 807a = J04-8.

J04-19 **(15, 781e)**

Kuipers, L., and Meulenbeld, B. On a certain classification of the convergents of a continued fraction. II. Nieuw Arch. Wiskunde (3) **2**, 32–39 (1954).

The authors continue the investigation begun in part I

[same Arch. (3) **1**, 199–211 (1953); these Rev. **15**, 510], in which irreducible fractions P/Q are separated into 8 classifications determined by the ordered pair $(r(P), r(Q))$ of remainders mod 3 of P, Q. It is shown that for any irrational α the inequality $|\alpha - P/Q| < k/Q^2$ is satisfied by infinitely many P/Q of at least four classes if $k \geqq 1$, and of all eight classes if $k > 3/\sqrt{5}$. There exist irrationals α for which the inequality is satisfied by infinitely many fractions of only four classes when $k=1$; moreover, there exist irrationals α for which the inequality does not hold for infinitely many fractions of all of the eight classes when $k < 3/\sqrt{5}$.
W. T. Scott (Evanston, Ill.).

Citations: MR 15, 510f = A54-18.

J04-20 (16, 224g)

Meulenbeld, B. On the approximating decimal fractions of decimals. Simon Stevin **30**, 65–78 (1954).

Let α be a positive irrational number less than one, and let the successive "decimal" approximations (base g) in reduced form be P_m/Q_m. These approximating decimal fractions (adf) are classified by the parities of P_m and Q_m, so that there are the classes (e/o) = even/odd, (o/o), and (o/e). Whether there are adf of a given class depends in part on g; for example, there are no (o/o) or (e/o) if g is a power of 2. The author proves various assertions about the existence of irrational numbers for which there are one, several successive, or infinitely many adf of a given class, or pair of classes.
W. J. LeVeque (Ann Arbor, Mich.).

J04-21 (16, 451e)

Kogoniya, P. G. On the structure of the set of Markov numbers. Akad. Nauk Gruzin. SSR. Trudy Tbiliss. Mat. Inst. Razmadze **19**, 121–133 (1953). (Russian. Georgian summary)

Let α be any irrational number satisfying $0<\alpha<1$ and with partial quotients $a_1, a_2, \cdots$, so that α has the continued fraction representation $[0, a_1, a_2, \cdots]$. Let $L(\alpha)$ be inf c subject to the condition that $|\alpha - p/q| < c/q^2$ has infinitely many solutions for integers p, q with $q>0$. The set of all positive $L(\alpha)$ is denoted by M_L. In the range $L(\alpha) > \frac{1}{3}$, M_L contains only a countable infinity of values which form a decreasing sequence, decreasing from $1/\sqrt{5}$ to the limit $\frac{1}{3}$. The author is chiefly concerned with that part of M_L in the range $0 < L(\alpha) < \frac{1}{3}$. If $M(N)$ denotes the set of all α for which lim sup $a_k = N$ and $M_L(N)$ is the set of all $L(\alpha)$ corresponding to α in $M(N)$, then M_L is the union of all sets $M_L(N)$ for $N=1, 2, 3, \cdots$. It is proved that, for $N \geqq 2$, $M_L(N)$, and consequently M_L, has the power of the continuum. This is done by constructing a set of numbers $L(\alpha)$ of this power belonging to $M_L(N)$, the construction being easier for $N \geqq 4$ than for $N=2$ and 3.

It is also shown that $\frac{1}{3}$ is the maximum point of condensation of the set M_L; in fact, the set of all points of $M_L(2)$ lying in the interval $1/(3+\epsilon) < L(\alpha) < 1/3$ has the power of the continuum for every $\epsilon > 0$. Finally it is shown that the set M_L has Hausdorff dimension 1. The proof uses Jarnik's inequalities for the Hausdorff dimension of the set of α for which $a_k \leqq N$ for all k.
R. A. Rankin (Glasgow).

Referred to in J04-32.

J04-22 (16, 803d)

Descombes, R. Etude diophantienne de certaines formes linéaires non homogènes. Bull. Soc. Math. France **82**, 197–299 (1954).

The author presents amplifications and detailed proofs of results announced earlier [Descombes and Poitou, C. R. Acad. Sci. Paris **234**, 581–583, 1522–1524 (1952); MR **13**, 825, 921; Descombes, ibid. **236**, 1401–1403, 1460–1462 (1953); MR **14**, 851]. The proofs use ingenious modifications of the continued fraction process but require much detailed numerical argumentation.
J. W. S. Cassels.

Citations: MR 13, 825f = J04-11; MR 13, 921c = J04-12; MR 14, 851c = J04-16; MR 14, 851d = J04-17.

Referred to in Z10-3.

J04-23 (16, 908f)

Tornheim, Leonard. Approximation to irrationals by classes of rational numbers. Proc. Amer. Math. Soc. 6, 260–264 (1955).

(I) Let r, s, m be integers without common factor and $m \geqq 1$. For any real ξ there are infinitely many pairs of integers a, b $(b>0)$ such that $sa \equiv rb \pmod{m}$ and (*) $b|b\xi - a| < 5^{-1/2}m$. (II) If in addition a, b are constrained to be coprime, then (I) remains true provided that m is of the form p^e $(p \neq 2)$, 2^e, $p^e q^f$ for primes p, q and integers $e>0$, $f>0$ and that $5^{-1/2}$ in (*) is replaced by $5^{-1/2}$, $\frac{1}{2}$, 1 respectively. (III) For any fixed r, s, m the constants in (I), (II) are the best possible. The proofs are elementary, simple and straightforward.
J. W. S. Cassels.

J04-24 (17, 590e)

Djerasimović, B. Beitrag zur Untersuchung der Perron'schen Modularfunktion $M(\gamma)$ einer Irrationalzahl. Bull. Soc. Math. Phys. Serbie **6** (1954), 86–92. (Serbo-Croatian. German summary)

Perron's function $M(\gamma)$ is defined as follows. Let γ be any irrational and let A_n/B_n be the nth convergent to the regular continued fraction of γ. The numbers λ_n defined by

$$\left|\gamma - \frac{A_n}{B_n}\right| = \frac{1}{\lambda_n B_n^2} \quad (n=0, 1, 2, \cdots)$$

have a limit superior as $n \to \infty$, which is denoted by $M(\gamma)$. Any other irrational γ' equivalent to γ in the sense that $\gamma' = (a\gamma+b)/(c\gamma+d)$ $(ad-bc=1)$ is such that $M(\gamma') = M(\gamma)$. The author is concerned with the limit points of $M(\gamma)$. He shows first that every integer greater than 2 is such a limit point, as are also infinitely many rational numbers. At such points ϱ the numbers γ for which $M(\gamma) = \varrho$ have the power of the continuum. The same is true of certain cases in which ϱ is a quadratic surd. Every integer greater than 3 is a limit point of such points ϱ.
D. H. Lehmer (Berkeley, Calif.).

J04-25 (17, 829g)

Cugiani, Marco. Sopra una questione di approssimazione diofantea non lineare. Boll. Un. Mat. Ital. (3) **10** (1955), 489–497.

The author considers the set $I(\alpha)$ of points $p^2 - q^2\alpha$, where α is a positive real number and p, q are integers by considering the continued fraction development of $\alpha^{\frac{1}{2}}$. He proves the following results. (I) For almost all α the point 0 is a point of accumulation of $I(\alpha)$ both on the left and right. Then $I(\alpha)$ is dense on the real axis. (II) There exist α for which 0 is an isolated point of $I(\alpha)$ and also α for which 0 is a point of accumulation only on left (or right). (III) $I(\alpha)$ has points of accumulation γ, δ in $0 < \gamma \leqq 2\alpha^{\frac{1}{2}} + 1$ and $0 > \delta > -2\alpha^{\frac{1}{2}} - 1$. The proofs are quite straightforward on noting that if $\beta \in I(\alpha)$ then $n^2\beta \in I(\alpha)$ for all integers n.
J. W. S. Cassels.

Referred to in J04-27, J68-17.

J04-26 (19, 121d)

Hartman, S.; und Knapowski, S. Bemerkungen über die Bruchteile von $p\alpha$. Ann. Polon. Math. 3 (1957), 285–287.

The authors discuss the solubility of

(*) $$|\alpha-r/q|<B/q^b$$

in (positive) integers r, q, of which one or both may be required to be prime. Here α, B, b are given positive numbers (α irrational). In Satz 1 it is proved that, if c is a (Linnik) constant such that every arithmetical progression $l+jq$ ($j=1, 2, \cdots$; $0<l<q$; $(l, q)=1$) contains a prime $p<q^c$, and if α is such that (*) has an infinity of integer solutions with $(r, q)=1$ when $b=c+1$, $B=1$, then (*) has an infinity of solutions with q prime when $b=1+c^{-1}$, $B=2$. In Satz 2 the known fact that there exists a prime between x and $x+x^{5/8+\varepsilon}$ (ε an arbitrarily small positive number; $x>x_0(\varepsilon)$) is made to yield the consequence that (*) has an infinity of solutions with r, q both prime when $b=3/8-\varepsilon$, $B=1$. These theorems are elementary deductions from the basic facts about primes. The authors also quote an observation of Jarník that it follows from a theorem of Vinogradov on the distribution of the fractional parts of $p\alpha$ [Trudy Mat. Inst. Steklov. **23** (1947), p. 177; MR **10**, 599; **15**, 941] that (*) has an infinity of solutions with q prime when $b=6/5-\varepsilon$, $B=1$.

A. E. Ingham (Cambridge, England).

Citations: MR 10, 599a = L02-2; MR 15, 941b = L02-3.

J04-27 (19, 124c)

Lekkerkerker, Cornelis Gerrit. Una questione di approssimazione diofantea e una proprietá caratteristica dei numeri quadratici. I. Atti Accad. Naz. Lincei. Rend. Cl. Sci. Fis. Mat. Nat. (8) **21** (1956), 179–185.

Following Cugiani [Boll. Un. Mat. Ital. (3) **10** (1955), 489–497; MR **17**, 829] the author considers the set $I(\alpha)$ of values of $r^2-s^2\alpha$, where $\alpha>0$ is a fixed irrational and r, s run through all integers. He puts $\alpha=\theta^2$ and considers also the much more natural set $J(\theta)$ of values of $s(s-r\theta)$. In an obvious sense it follows at once that $I'(\alpha)=2\theta J'(\theta)$, where $I'(\alpha)$, $J'(\theta)$ are the derived sets. The author now shows easily that $J'(\theta)$ is discrete if and only if θ is a quadratic irrationality. Using a result of Marshall Hall on the expression of a number as the sum of two numbers with bounded partial quotients [Ann. of Math. (2) **48** (1947), 966–993; MR **9**, 226], he constructs a number θ whose partial quotients are at most $k-1$, where k is any integer greater than 6, and such that $J(\theta)$ is dense in the interval $(-1/k, 1/k)$.

J. W. S. Cassels.

Citations: MR 9, 226b = A54-7; MR 17, 829g = J04-25.
Referred to in J04-28.

J04-28 (19, 533f)

Lekkerkerker, Cornelis Gerrit. Una questione di approssimazione diofantea e una proprietà caratteristica dei numeri quadratici. II. Atti Accad. Naz. Lincei. Rend. Cl. Sci. Fis. Mat. Nat. (8) **21** (1956), 257–262.

Completes the proofs of results already enunciated in the first part [same Rend. (8) **21** (1956), 179–185; MR **19**, 124].

J. W. S. Cassels (Cambridge, England).

Citations: MR 19, 124c = J04-27.

J04-29 (19, 533g)

Rankin, R. A. Diophantine approximation and horocyclic groups. Canad. J. Math. **9** (1957), 277–290.

Let Γ be a real zonal horocyclic group of the first kind (i.e., a Fuchsian group of the first kind with a finite number of generators and real axis as principal circle, and with ∞ as a parabolic fixed point), and let $Tz=(az+b)/(cz+d)$ $(ad-bc=1)$, a, b, c, d real. The principal problem considered is the diophantine approximation problem of determining h so that the inequality (*) $|\omega-a/c|=|\omega-T\infty|<1/hc^2$ will have infinitely many solutions $a/c=T\infty$, where $T\in\Gamma$ and ω is not a parabolic fixed point for Γ. If T denotes the coefficient matrix of Tz, if U^λ denotes the matrix for $U^\lambda z=z+\lambda$, where λ is the cusp width of Γ at ∞, and if Γ_U denotes the cyclic subgroup of Γ generated by $U^\lambda z$, then an equivalent statement of the problem is the determination of h so that there will be infinitely many left cosets of Γ_U in Γ whose members T satisfy (*). A lower bound, h_Γ, and two upper bounds, h_Γ' and h_Γ'', are determined for $h(E, \Gamma)=\sup h$ in the set of all h for which (*) holds for transformations T belonging to infinitely many left cosets of Γ_U in Γ, and it is shown that $h=h_\Gamma$ is a solution of the problem. A lower bound for $h(E, \Gamma)$, and also a solution of the problem, had previously been obtained by Lehner [Pacific J. Math. **2** (1952), 327–333; MR **14**, 359]. For the corresponding problem of approximating ω by $ST\infty$, where S is an arbitrary but fixed transformation having real axis as fixed circle, it is shown that $h(S, \Gamma)=h(E, \Gamma)$. For the modular group $\Gamma(1)$ and for ω irrational the inequality $\sqrt{3}\leqq h(E, \Gamma(1))\leqq\sqrt{5}$ is obtained. Ford [Proc. Edinburgh Math. Soc. **35** (1917), 59–65], in a detailed study that was restricted to $\Gamma(1)$, showed that $h=\sqrt{5}$ is the maximal solution of the corresponding problem. It is remarkable that, for the principal congruence group $\Gamma(2)$ of level 2, the present general treatment yields $h=h(S, \Gamma(2))=1$, $S\in\Gamma(1)$, which had been shown by the reviewer [Bull. Amer. Math. Soc. **46** (1940), 124–129; MR **1**, 203] to be the maximal solution of the corresponding diophantine approximation problem. The proofs depend on Hermite's geometric formulation of the problem and Ford's determination of fundamental regions by the method of isometric circles.

W. T. Scott (Evanston, Ill.).

Citations: MR 1, 203a = J04-1; MR 14, 359a = J04-14.

J04-30 (20# 2324)

Obrechkoff, Nikola. Sur l'approximation diophantienne des nombres réels. C. R. Acad. Sci. Paris **246** (1958), 31–32.

L'auteur démontre le théorème suivant: Soit a un nombre natural et soit n un nombre naturel plus grand que a. Alors pour chaque nombre réel ω, satisfaisant à la condition $0<\omega\leqq a$, il existe aux moins deux nombres entiers, x, y, non négatifs et tels qu'on ait

(1) $$|\omega x-y|\leqq\left\{\left[\frac{n-a}{a+1}\right]+2\right\}^{-1}\quad(0<x+y\leqq n).$$

Le signe d'égalité dans (1) est atteint.

La démonstration du théorème utilise les propriétés de la suite de Farey.

J. Popken (Amsterdam)

J04-31 (20# 4544)

Negoescu, N. L'ordination de quelques fractions continues, doubles. An. Şti. Univ. "Al. I. Cuza" Iaşi. Sect. I (N.S.) **3** (1957), 11–17. (Romanian. Russian and French summaries)

It is known that every irrational number θ can be developed uniquely in a simple infinite continued fraction of the form $\theta=[a_0, a_1, a_2, \cdots]$, where a_0 is an integer and the $a_i\geqq1$, $i=1, 2, \cdots$, are also integers. One designates by $[a_1, \cdots, a_r, [a_{r+1}], a_{r+2}, \cdots]$ the sum $[a_{r+1}, a_{r+2}, \cdots]+[0, a_r, a_{r-1}, \cdots, a_1]$. This expression is named, after R. Robinson [Bull. Amer. Math. Soc. **53** (1947), 351–361; **54** (1948), 693–705; MR **8**, 566; **10**, 235], a double continued fraction. Using a generalization of the method of Robinson, the author proves, in addition to a number of relations concerning these continued fractions, the following theorem. Let $a_0, a_1, \cdots, a_{r-1}, s$ be integers $\geqq n$; $\alpha, \beta, \gamma, \delta$ real numbers $>n$; B_0 the block $a_{r-1}, \cdots, a_1, [a_0], a_1, \cdots, a_{r-1}$; and r an even number. Then $[\gamma, n, s, B_0, s-1, n, \delta]<[\alpha, n, s, B_0, s, n, \beta]$, $[\gamma, n, s, B_0, s+1, n, \delta]>[\alpha, n, s, B_0, s, n, \beta]$, where $\alpha, \beta<(n^2+1)^2s+n(n^2+n+1)$.

E. Frank (Chicago, Ill.)

Citations: MR 8, 566b = J08-4; MR 10, 235c = J08-8.
Referred to in J04-33.

J04-32 **(20# 6410)**

Kogoniya, P. G. Condensation points of the set of Markov numbers. Dokl. Akad. Nauk SSSR (N.S.) **118** (1958), 632–635. (Russian)

Let $M(N)$ be the set of all irrational numbers α $(0<\alpha<1)$ whose continued fraction partial quotients a_k satisfy $\lim\sup_{k\to\infty} a_k=N$, where $N=1, 2, 3, \cdots$. Let $L(\alpha)$ be the lower bound of the set of all positive numbers c for which

$$\left|\alpha-\frac{p}{q}\right|<\frac{c}{q^2}$$

has infinitely many solutions in integers p, $q>0$, and let $M_L(N)$ be the set of all values $L(\alpha)$ for $\alpha\in M(N)$. It is shown that $(N^2+4N)^{-\frac{1}{2}}$, which is the smallest number in $M_L(N)$, is a point of condensation of $M_L(N)$ $(N\geqq 2)$ and that $22/(65+9\sqrt{3})$ is the greatest point of condensation of $M_L(3)$. [For earlier work of the author on related problems see Akad. Nauk Gruzin. SSR. Trudy Tbiliss. Mat. Inst. Razmadze **19** (1953), 121–133; MR **16**, 451.]

R. A. Rankin (Glasgow)

Citations: MR 16, 451e = J04-21.

J04-33 **(21# 3399)**

Negoescu, N. Inégalités pour certaines fractions continues multiples. An. Şti. Univ. "Al. I. Cuza" Iaşi. Secţ. I. (N.S.) **4** (1958), 1–9. (Romanian. Russian and French summaries)

Let $[a_0, a_1, \cdots, a_v, \cdots]$ represent the continued fraction expansion of an irrational number θ, and $[a_0, a_1, \cdots, a_v]$ be the approximant p_v/q_v of order v. Let

$$[a_{v+1}, a_{v+2}, \cdots]+[0, a_v, a_{v-1}, \cdots, a_1]$$

be the double continued fraction

$$[a_1, \cdots, a_v, [a_{v+1}], a_{v+2}, \cdots],$$

introduced by R. Robinson [Bull. Amer. Math. Soc. **53** (1947), 351–361; **54** (1948), 693–705; MR **8**, 566; **10**, 235]. This concept is generalized to that of a multiple continued fraction. Let

$$[A_0; a_{v,1}; a_{v,2}; \cdots; a_{v,m}]$$

be the m-tuple continued fraction which designates the sum

$$\sum_{i=1}^{m} [a_0, a_1, a_2, \cdots, a_{v-1}, a_{v,i}]$$

of m regular continued fractions. A number of inequalities are proved here on multiple continued fractions, based on the author's previous work [An. Şti. Univ. "Al. I. Cuza" Iaşi. Secţ. I (N.S.) **3** (1957), 11–17; Gaz. Mat. Fiz. Ser. A (N.S.) **10** (**63**) (1958), 482–491; MR **20** #4544, #5988]. The principal theorem is as follows: Let a_0, $a_1, \cdots, a_{v-1}$ be integers $\geqq n$, m and s integers $\geqq 2$, γ_i and δ_i real numbers $>n$. Let A_0 be the block $a_0, a_1, \cdots, a_{r-1}$, and v an even number. Then

(a) $[A_0; s-1, n, \gamma_1; \cdots; s-1, n, \gamma_{m-1}; s, n, \gamma_m] < [A_0; s, n, \delta_1; \cdots; s, n, \delta_m]$,

(b) $[A_0; s+1, n, \gamma_1; \cdots; s+1, n, \gamma_{m-1}; s, n, \gamma_m] > [A_0; s, n, \delta_1; \cdots; s, n, \delta_m]$,

where

$$\delta_i < (m-1)(n^2+1)^2 s+(3-m)n(n^2+n+1) \quad (i = 1, 2, \cdots, m).$$

E. Frank (Chicago, Ill.)

Citations: MR 8, 566b = J08-4; MR 10, 235c = J08-8; MR 20# 4544 = J04-31; MR 20# 5988 = A54-26.

J04-34 **(22# 36)**

Sudan, Gabriel. Sur un théorème de Obrechkoff. C. R. Acad. Sci. Paris **249** (1959), 2700–2701.

Obrechkoff used Farey series in order to prove the following theorem: Given $\omega>0$ and two integers a and n such that $0<\omega\leqq a<n$, there exists at least one pair of non-negative integers x, y, satisfying simultaneously $|\omega x-y|\leqq\left(\left[\frac{n-a}{a+1}\right]+2\right)^{-1}$ and $0<x+y\leqq n$. The present paper proves the statement using elementary geometric considerations.

E. Grosswald (Princeton, N.J.)

J04-35 **(22# 706)**

Obreškov, N. Sur l'approximation diophantienne des formes linéaires. Bŭlgar. Akad. Nauk Izv. Mat. Inst. **3**, no. 2, 3–18 (1959). (Bulgarian. Russian and French summaries)

Résumé de l'auteur: "Nous démontrons les théorèmes suivants. (1) Soit a un nombre positif et entier et n un nombre entier plus grand que a. Alors pour chaque nombre réel ω, $0<\omega\leqq a$, ils existent au moins deux nombres entiers non négatifs x et y tels que l'on a

$$(1) \qquad |\omega x-y| \leqq \left(\left[\frac{n-a}{a+1}\right]+2\right)^{-1}, \quad 0 < x+y\leqq n.$$

Le signe d'égalité dans (1) est atteint. (2) Soit p et d deux nombres entiers et positifs, $d\leqq p$, et soit $n>0$ un nombre entier arbitraire mais multiple de $d/(p, d)$. Soit $\omega_1, \cdots, \omega_p$ p nombres réels arbitraires. Alors ils existent p nombres entiers $x_1, \cdots, x_p$, non négatifs, desquels d au moins sont différents de 0 et tels que l'on a (y entier),

$$(2) \qquad |\omega_1x_1+\cdots+\omega_px_p-y| \leqq \frac{1}{(np/d)+1},$$
$$x_s \leqq n \quad (s = 1, \cdots, p).$$

Le signe d'égalité dans (2) est atteint."

J04-36 **(22# 3724)**

Hartman, S. A feature of Dirichlet's approximation theorem. Acta Arith. **5**, 261–263 (1959).

Let p be a prime. The author considers the inequalities (1) $0<x\leqq ct$, $|\alpha x-y|<t^{-1}$, where α is irrational and x, y are integers with $p\nmid x$. He constructs an α such that for any $c>0$ and some $t=t(c)>1$ there is no solution of (1). Using an ergodic theorem for continued fractions proved by Ryll-Nardzewski [Studia Math. **12** (1951), 74–79; MR **13**, 757] he shows that, in the case $p=2$, this is true for almost all α.

C. G. Lekkerkerker (Amsterdam)

Citations: MR 13, 757b = K50-5.

J04-37 **(22# 3727)**

Flor, Peter. Inequalities among some real modular functions. Duke Math. J. **26** (1959), 679–682.

Let $\alpha=(a_0, a_1, a_2, \cdots)$ be the regular continued fraction of an irrational α and put $\alpha_n=(a_n, a_{n+1}, \cdots)$, $\beta_n=(a_{n-1}, \cdots, a_1)$, $M(\alpha)=\lim\sup(\alpha_n+\beta_n{}^{-1})$, $T(\alpha)=\lim\sup\alpha_n\beta_n$, $k(\alpha)=\lim\sup a_n$. The author derives and discusses some simple inequalities relating the functions $M(\alpha)$, $T(\alpha)$, $k(\alpha)$, such as $T(\alpha)^{1/2}+T(\alpha)^{-1/2}\leqq M(\alpha)\leqq T(\alpha)-T(\alpha)^{-1}$.

C. G. Lekkerkerker (Amsterdam)

J04-38 **(22# 12096)**

Eggan, L. C. On Diophantine approximations. Trans. Amer. Math. Soc. **99** (1961), 102–117.

In the sequel θ denotes an irrational number. A classical result of Hurwitz states that there are infinitely many pairs of integers p, q with $q^2|\theta-p/q|<5^{-1/2}$. A. V. Prasad [J. London Math. Soc. **23** (1948), 169–171; MR **10**, 513], among others, has shown that if only one solution (p, q)

is demanded, the appropriate constant is less than $5^{-1/2}$ and is, in fact, $\frac{1}{2}(3-\sqrt{5})$.

In the first part of this paper the author generalizes Prasad's result as follows: "Let n be a positive integer and let $\theta_n=((n^2+4)^{1/2}-n)/2$. For any positive integer m, let $c_m=\theta_n+n+P_{2m-1}/Q_{2m-1}$, where P_j/Q_j is the jth convergent to the continued fraction expansion $[0, n, n, \cdots]$ of θ_n. Then if $\theta=[a_0, a_1, a_2, \cdots]$ and if $a_j\geqq n$ for infinitely many values of j, there are at least m solutions in relatively prime integers p, q $(q>0)$ to the inequality $q^2|\theta-p/q|\leqq 1/c_m$. Moreover, the constant c_m cannot be improved, since for $\theta=\theta_n$ there are exactly m solutions and equality is attained." For $n=1$ this result is due to Prasad. For $n=2$ it provides the second link in the Markoff chain.

In the second part of this paper the author applies his method to other results in the theory of diophantine approximations. W. T. Scott [Bull. Amer. Math. Soc. **46** (1940), 124–129; MR **1**, 203] has shown that the inequality $q^2|\theta-p/q|<1$ has an infinity of solutions (p, q) if we restrict the fractions p/q to those chosen from any one of the following three classes: (a) p, q both odd; (b) p odd and q even; (c) p even and q odd. The author shows that the constant 1 on the right is the best possible even if only one solution is demanded instead of infinitely many; in a similar way he treats a related result due to R. M. Robinson [Duke Math. J. **7** (1940), 345–359; MR **2**, 149]. A. Khintchine [Math. Ann. **111** (1935), 631–637] proved that for any real number α and for $\sigma>0$ the inequality $v|v\theta-u-\alpha|<5^{-1/2}+\sigma$ has an infinity of solutions in integers (u, v) with $v>0$. In the paper reviewed here it is shown that the constant $5^{-1/2}$ cannot be replaced by any smaller number, even if only one solution is demanded.

Moreover, the author also attacks with his method the problem of asymmetric approximations, introduced by B. Segre [Duke Math. J. **12** (1945), 337–365; MR **6**, 258] and treated later by a number of other authors.

In the last part of this paper the author extends the following theorem of J. F. Koksma [Simon Stevin **28** (1951), 199–202; MR **13**, 825]. Let $\sigma>0$ and let $s\geqq 1$, a, b denote arbitrary integers. Then there exists an infinity of solutions of the inequality

$$q^2|\theta-p/q| < 5^{-1/2}(1+\sigma)s^2$$

such that $p\equiv a \pmod{s}$, $q\equiv b \pmod{s}$.

For the numbers θ equivalent to $\frac{1}{2}(5^{1/2}-1)$ the constant $5^{-1/2}$ on the right is the best possible. The author sharpens this result by disregarding the numbers θ equivalent to $\frac{1}{2}(5^{1/2}-1)$. *J. Popken* (Amsterdam)

Citations: MR 1, 203a = J04-1; MR 2, 149c = J04-2; MR 6, 258a = J08-1; MR 10, 513d = J04-6; MR 13, 825e = J04-10.

Referred to in J04-39, J04-59, J08-21, J08-22.

J04-39 (32# 4088)

Sinha, T. N.

On Diophantine approximation.

J. Indian Math. Soc. (*N.S.*) **28** (1964), 139–144 (1965).

The theorem proved in this paper is the case $n=2$ of Theorem 2.3 of the reviewer's paper [Trans. Amer. Math. Soc. **99** (1961), 102–117; MR **22** #12096].

L. C. Eggan (Tacoma, Wash.)

Citations: MR 22# 12096 = J04-38.

J04-40 (23# A855)

Schmidt, Asmus L.

Simple proofs of an approximation theorem of Hurwitz. (Danish. English summary)

Nordisk Mat. Tidskr. **7** (1959), 157–162, 190.

The theorem of Hurwitz is that if ξ is irrational, then there are infinitely many rational numbers p/q such that $|\xi-p/q|<(5^{1/2}q^2)^{-1}$. The author gives a new and simple proof, and sketches a proof by the reviewer [*Topics in number theory, Vol. I*, Addison-Wesley, Reading, Mass., 1956, p. 154; MR **18**, 283].

W. J. LeVeque (Ann Arbor, Mich.)

Citations: MR 18, 283b = Z01-38.

J04-41 (23# A1598)

Marcus, Solomon

Les approximations diophantiennes et la catégorie de Baire.

Math. Z. **76** (1961), 42–45.

A theorem of M. K. Fort [Amer. Math. Monthly **58** (1951), 408–410] reads: Let the real function f be defined on an arbitrary real interval. Let A be the set of points at which f is discontinuous, B the set at which f is differentiable. If A is everywhere dense, then B is a set of the first category in the sense of Baire. The author applies this theorem to the theory of Diophantine approximations. In this way he finds: Let $f(x)$ be a real function positive for every $x>0$. Then the real numbers ξ, such that the inequality $|\xi-p/q|<f(q)$ has only a finite number of solutions in integers p, q $(q>0)$, form a set of the first category. From this result it follows easily: Let $M(x)$ denote the "modular function of Perron", i.e., the least upper bound of positive numbers c such that $|x-p/q|<c^{-1}q^{-2}$ has an infinity of solutions in integers p, q $(q>0)$. Then the set E of real numbers x such that $M(x)$ is finite is of the first category. This last result is an analogue of the well-known theorem that E has measure zero. The author shows that the two theorems stated above can also be derived by other methods. *J. Popken* (Amsterdam)

J04-42 (23# A1602)

Delaunay, B. N.; Vinogradov, A. M.

Über den Zusammenhang zwischen den Lagrangeschen Klassen der Irrationalitäten mit begrenzten Teilnennern und den Markoffschen Klassen der extremen Formen. (Russian summary)

Sammelband zu Ehren des 250. *Geburtstages Leonhard Eulers, pp.* 101–108. *Akademie-Verlag, Berlin,* 1959.

For irrational θ put

$$\lambda_\theta = \limsup_{|q|\to\infty} |q(q\theta-p)|^{-1} \quad (p, q \text{ integers}).$$

Further, if $f=f(x, y)=ax^2+2bxy+cy^2$ is any indefinite binary quadratic form for which one has $\inf|f(p, q)|=1$ (p, q integers not both zero), then put $\lambda_f=2(b^2-ac)^{1/2}$. As is well known, the values of λ_θ and λ_f which are $\leqq 3$ were determined by Markov [see, e.g., J. W. S. Cassels, *An introduction to Diophantine approximation*, Cambridge Univ. Press, New York, 1957, Ch. II; MR **19**, 396]. The present authors prove inter alia that the set of all possible values λ_θ coincides with the set of values λ_f; to this end they use the so-called polygon of Klein connected with the continued fraction expansion of θ or with a similar algorithm for f due to F. Klein. In the summary it is stated that, as a consequence of a result of M. Hall [Ann. of Math. (2) **48** (1947), 966–993; MR **9**, 226], the quantity λ_f takes each value >6.1. *C. G. Lekkerkerker* (Amsterdam)

Citations: MR 9, 226b = A54-7; MR 19, 396h = J02-11.

J04-43 (23# A2358)

Sudan, Gabriel

Démonstrations géometriques de quelques théorèmes sur les fractions continues. (Russian, English, and German summaries)

Bul. Inst. Politehn. București **22** (1960), *no.* 2, 73–83.

Using Klein's representation of continued fractions, the author proves the following three theorems, due originally to Seidel, Legendre and Perron, respectively: (I) A necessary and sufficient condition that the continued fraction $\{a_0; a_1, a_2, \cdots\}$, with arbitrary positive denominators $a_1, a_2, \cdots$, be convergent is that $\sum a_n$ be divergent. (II) A necessary and sufficient condition that the irreducible fraction p/q be a convergent to the real number α is that $|\alpha - p/q| \leqq q^{-1}(q+q_1)^{-1}$, where p_1/q_1 is the penultimate convergent to p/q. (III) The Markov constant $M(\alpha)$ is finite if and only if the irrational number α has bounded denominators in its regular continued-fraction expansion. *W. J. LeVeque* (Ann Arbor, Mich.)

J04-44 **(24# A3132)**

Kogonija, P. G.

The set of condensation points of the set of Markov numbers. (Russian)

Akad. Nauk Gruzin. SSR Trudy Tbiliss. Mat. Inst. Razmadze **26** (1959), 3–16.

If α is an irrational number, its Markov number $L(\alpha)$ is the infimum of all numbers $c>0$ such that the inequality $|\alpha-p/q|<cq^{-2}$ has infinitely many integral solutions $q>0$, p. Let $M(N)$ be the set of irrational α for which the continued fraction expansions $[0; a_1, a_2, \cdots]$ have the property that $\limsup_{k\to\infty} a_k = N$, and let $M_L(N)$ be the set of Markov numbers of all elements of $M(N)$. Theorem 1: The minimal point of condensation of $M_L(N)$, for $N \geqq 2$, is equal to the minimal element $(N^2+4N)^{-1/2}$ of the set. Theorem 2: The maximal element $(N^2+4)^{-1/2}$ of $M_L(N)$ is an isolated point. *W. J. LeVeque* (Ann Arbor, Mich.)

Referred to in N44-24.

J04-45 **(24# A3133)**

Kogonija, P. G.

The maximal condensation point of a subset of the set of Markov numbers. (Russian)

Akad. Nauk Gruzin. SSR Trudy Tbiliss. Mat. Inst. Razmadze **26** (1959), 17–22.

Preserve the notation of the preceding review [#A3132]. Theorem 1: The maximal point of condensation of $M_L(3)$ is $22(65+9\sqrt{3})^{-1}$. Theorem 2: The minimal point $(N^2+4N)^{-1/2}$ of $M_L(N)$ belongs to $M_L(N+1)$ for $N \geqq 3$. *W. J. LeVeque* (Ann Arbor, Mich.)

J04-46 **(27# 1416)**

Kogonija, P. G.

On the set of "generalized" Markov numbers. (Russian. Georgian summary)

Tbiliss. Gos. Univ. Trudy Ser. Meh.-Mat. Nauk **84** (1962), 143–149.

Let p_k/q_k be the sequence of convergents to the real number α. Consider only those α for which

$$\limsup \frac{\log q_{k+1}}{\log q_k} = J-1,$$

where $J>2$ is fixed. Put $L_J(\alpha) = \liminf q_k^J|\alpha - p_k/q_k|$. It is shown that $L_J(\alpha)$ can take infinitely many values and that 0 and ∞ are limiting points of the set of values taken. *J. W. S. Cassels* (Cambridge, England)

J04-47 **(29# 5788)**

Davenport, H.

A remark on continued fractions.

Michigan Math. J. **11** (1964), 343–344.

It is shown that, for any irrational number θ and for any prime P, at least one of the numbers

$$P^2\theta,\ \theta,\ \theta+1/P,\ \cdots,\ \theta+(P-1)/P$$

has a simple continued fraction whose partial denominators a_n satisfy $a_n > P-2$ for infinitely many indices n. *W. T. Scott* (Tempe, Ariz.)

J04-48 **(31# 4766a; 31# 4766b)**

Kogonija, P. G.

On the connection between the spectra of Lagrange and Markov. II, III. (Russian. Georgian summary)

Tbiliss. Gos. Univ. Trudy Ser. Meh.-Mat. Nauk **102** (1964), 95–104; *ibid.* **102** (1964), 105–113.

Kogonija, P. G.

On the connection between the spectra of Lagrange and Markov. IV. (Russian. Georgian summary)

Akad. Nauk Gruzin. SSR Trudy Tbiliss. Mat. Inst. Razmadze **29** (1963), 15–35 (1964).

Part I appeared in Tbiliss. Gos. Univ. Trudy **76** (1959), 161–171. Readers acquainted only with Georgian should note that the summaries in that language are lamentably perfunctory and give only an inadequate idea of the wealth of theorems and details in the Russian text.

Let $\theta=[0; a_1, a_2, \cdots]$ (continued fraction), where the a_j are nonnegative integers, and put

$$\lambda(\theta) = \limsup_k\{[0; a_{k-1}, a_{k-2}, \cdots, a_1]+a_k+[0; a_{k+1}, a_{k+2}, \cdots]\}$$

(which determines the closeness of rational approximations to θ by the theory of continued fractions). The Lagrange spectrum is the set of values that can be taken by $\lambda(\theta)$ as θ varies. Similarly, let $M=[a_j]_{-\infty<j<\infty}$ be a doubly infinite sequence of positive integers and put

$$\mu(M) = \sup_k\{[0; a_{k-1}, \cdots]+a_k+[0; a_{k+1}, \cdots]\}$$

(which is suggested by the classical theory of binary quadratic forms). The set of values taken by $\mu(M)$ as M varies is called the Markov spectrum. It is well known, and the author gives another proof, that the Lagrange spectrum is a subset of the Markov spectrum. The main purpose of this series of papers is to establish conditions under which an element of the Markov spectrum is in the Lagrange spectrum. The theorems are too numerous to quote in detail, a typical one being that if M has max $a_j=2$ and the sequence 121 of digits occurs only finitely often, then there is a θ with $\lambda(\theta)=\mu(M)$. The fundamental idea of the proofs is to find a θ whose continued fraction mimics the critical portion of M arbitrarily closely arbitrarily often, without containing subsequences which would lead to a higher value of θ. The author also determines all the θ [M] with $\limsup a_j=N$ [$\sup a_j=N$] for which $\lambda(\theta)$ [$\mu(M)$] takes the extreme values $(N^2+4)^{1/2}$ and $(N^2+4N)^{1/2}$. She also gives a proof that the Markov spectrum is a closed point set. *J. W. S. Cassels* (Cambridge, England)

Referred to in J04-55, J04-60.

J04-49 **(32# 2376)**

Güting, Rainer

Über den Zusammenhang zwischen rationalen Approximationen und Kettenbruchentwicklungen.

Math. Z. **90** (1965), 382–387.

If γ is a real number in $(0, 1)$, put

$$w_1(h,\gamma) = \min_{b_0,b_1}|b_0\gamma+b_1|,$$

$$\vartheta_1(\gamma) = \limsup_{h\to\infty}\left(-\frac{\log w_1(h,\gamma)}{\log h}\right),$$

where b_0 and b_1 run over the integers satisfying $|b_0|\leqq h$, $|b_1|\leqq h$, $b_0\gamma+b_1\neq 0$. Let $\gamma=[a_0; a_1, a_2, \cdots]$ be the continued fraction for γ, and let p_n/q_n be its convergents. The author proves that

$$\vartheta_1(\gamma) = 1+\limsup_{n\to\infty} \frac{\log a_n}{\log q_{n-1}} = \limsup_{n\to\infty} \frac{\log q_n}{\log q_{n-1}},$$

and deduces that the set of all γ satisfying

$$\limsup_{n\to\infty} \frac{\log a_n}{\log q_{n-1}} = d \qquad (0 \leqq d \leqq \infty)$$

has the Hausdorff dimension $2/(2+d)$. If further, $g_n = \sqrt[n]{(a_1 a_2 \cdots a_n)}$, $c = \limsup_{n\to\infty} \log g_n$, $d = \liminf_{n\to\infty} \log g_n$, then

$$1 \leqq \vartheta_1(\gamma) \leqq 1+(c-d)\min\left(\frac{1}{d}, \frac{1}{\log \frac{1}{2}(1+\sqrt{5})}\right),$$

and hence $\vartheta_1(\gamma)=1$ if $c=d$. *K. Mahler* (Canberra)

Referred to in J84-43.

J04-50 (32# 2377)

Negoescu, N.

Remarks on diophantine approximations. (Romanian and Russian summaries)

An. Şti. Univ. "Al. I. Cuza" Iaşi Secţ. I a Mat. (N.S.) **10** (1964), 257–263.

Since 1940, numerous results have appeared which state that irrational numbers have good rational approximations which also satisfy other conditions. The author obtains many of these as special cases of his results. A typical theorem is as follows: If $r_1 < m_1$, $r_2 < m_2$ are non-negative integers, r_1, r_2 not both zero, then any irrational number θ has infinitely many rational approximations x/y such that $|\theta - x/y| < (1+\varepsilon)m_1 m_2/4y^2$, with $x \equiv r_1 \pmod{m_1}$ and $y \equiv r_2 \pmod{m_2}$. Corresponding results for complex approximations are also obtained. The proofs involve rather straightforward applications of the theorems of Minkowski (real case) and Hlawka (complex case) on the product of two linear forms [see, for example, Ivan Niven, *Diophantine approximations*, Interscience, New York, 1963; MR **26** #6120]. *L. C. Eggan* (Tacoma, Wash.)

Citations: MR 26# 6120 = J02-15.

J04-51 (32# 5596)

Ćetković, Simon

Approximation of transcendental numbers by an arbitrary everywhere dense set of real numbers. (Serbo-Croatian. French summary)

Bull. Soc. Math. Phys. Serbie **11** (1959), 81–87.

The following theorem is proved. Let $\{a_n\}$ be an arbitrary sequence of real numbers, everywhere dense in a fixed interval (d_0, e_0), let g and h be functions such that for $n = 1, 2, \cdots, 0 \leqq g(n) < h(n) \leqq n^{-\alpha}$, for some $\alpha > 1$, and let B be the set of real numbers c for which the inequality $g(n) \leqq |c - a_n| \leqq h(n)$ has infinitely many solutions n. Then B is of measure 0, and $B \cap (d, e)$ has the power of the continuum, for each $(d, e) \leqq (d_0, e_0)$.

J04-52 (33# 4011)

Samušenok, I. N.

An analytic relation between arithmetic characteristics of irrational numbers. (Russian)

Volž. Mat. Sb. Vyp. 1 (1963), 164–168.

Let $\lambda(\theta)$ be the g.l.b. of the set of values $0 < c < 1$ for which, for a given irrational number θ and an infinite set of values t, there exist pairs of integers p and $q > 0$ such that the inequalities $|q\theta - p| < 1/t$, $q < ct$ hold. Let $\Lambda(\theta)$ be the g.l.b. of values $c < 1$ for which the above inequalities are solvable for a sufficiently large t. Theorem: If $\theta = \frac{1}{2}(\sqrt{5}-1)$ and numbers equivalent to θ are excluded from the set of all irrational numbers, then $\Lambda \geqq \frac{1}{2}+\lambda$, $\Lambda \geqq 1-\lambda$, $\Lambda \geqq (\sqrt{5}+1)/(2\sqrt{5}) + (\sqrt{15}-1)(1/\sqrt{5}-\lambda)/7$. *G. Biriuk* (Ann Arbor, Mich.)

Referred to in J04-56.

J04-53 (33# 5563)

Kinney, John R.; Pitcher, Tom S.

The Hausdorff-Besicovich dimension of the level sets of Perron's modular function.

Trans. Amer. Math. Soc. **124** (1966), 122–130.

For a real number ξ denote, as usual, $\limsup q\|q\xi\|$ by $\{M(\xi)\}^{-1}$, where q tends to infinity through integral values and $\|q\xi\|$ denotes the distance between $q\xi$ and the nearest integer. The authors use the results of their recent paper [#6670 below] to estimate the dimension of the set $L(\gamma)$ of ξ for which $M(\xi) = \gamma$, both from above and below. In particular, the dimension is positive for all $\gamma \geqq (7/2) + (7/4)2^{1/2}$. The proofs depend heavily on the theory of continued fractions and, in particular, on Marshall Hall's theorem about representation as the sum of continued fractions with partial quotients at most 4 [Ann. of Math. (2) **48** (1947), 966–993; MR **9**, 226]. *J. W. S. Cassels* (Cambridge, England)

Citations: MR 9, 226b = A54-7; MR 33# 6670 = K55-29.

Referred to in K55-29.

J04-54 (34# 5757)

Freĭman, G. A.; Judin, A. A.

On the Markov spectrum. (Russian. Lithuanian and English summaries)

Litovsk. Mat. Sb. **6** (1966), 443–447.

Für jede zweiseitig unendliche Folge $M = \{a_n\}$ $(n = 0, \pm 1, \pm 2, \cdots)$ von natürlichen Zahlen sei $\mu_k(M) = [0; a_{k-1}, a_{k-2}, \cdots] + a_k + [0; a_{k+1}, a_{k+2}, \cdots]$ $(k = 1, 2, \cdots)$ und $\mu(M) = \sup_{k=1}^{\infty} \mu_k(M)$. ([ ] bedeutet Kettenbruch.) Die Menge aller solchen Zahlen $\mu(M)$ wird als Markov-Spektrum S bezeichnet. Sei μ_0 das Supremum der Komplementärmenge $\bar{S}$. Dann wird durch Verfeinerung einer Methode von M. Hall, Jr. [Ann. of Math. (2) **48** (1947), 966–993; MR **9**, 226] gezeigt, dass $\mu_0 < 5.118$ ist. Die beste bekannte Schranke dieser Art lag bisher bei 6.1. *B. Volkmann* (Stuttgart)

Citations: MR 9, 226b = A54-7.

J04-55 (34# 7456)

Kogonija, P. G.

Certain questions of rational approximation. (Russian. Georgian summary)

Thbilis. Saḫelmc. Univ. Šrom. Mekh.-Math. Mecn. Ser. **117** (1966), 45–62.

The author continues her investigations [see, e.g., same Šrom. **102** (1964), 95–104; ibid. **102** (1964), 105–113; MR **31** #4766a] into the structure of the Lagrange and Markov spectra. (Let $[0; a_1, a_2, \cdots]$ be the continued fraction expansion of an irrational number $\theta \in (0, 1)$, and put $\lambda(\theta) =$

$$\limsup_k\{[0; a_{k-1}, a_{k-2}, \cdots, a_1] + a_k + [0; a_{k+1}, a_{k+2}, \cdots]\};$$

the Lagrange spectrum is defined to be the set of values taken by $\lambda(\theta)$ as θ varies. The Markov spectrum is the set of values taken by $\mu(M) = \sup_k\{[0; a_{k-1}, a_{k-2}, \cdots] + a_k + [0; a_{k+1}, a_{k+2}, \cdots]\}$ as M varies over all doubly infinite sequences of positive integers $[a_j]_{-\infty<j<\infty}$.) A finer partitioning of these spectra into subsets yields the following results: there are no points of the Markov spectrum, and hence of the Lagrange spectrum, in the intervals $(4(\sqrt{30})/7, \sqrt{10})$ and $(\sqrt{10}, (7\sqrt{10}+4\sqrt{30}+2)/14)$; if $M_1' = \{\cdots, 1, 2, 1, 1, 2, 1, \cdots\}$, then $\mu(M_1') = \sqrt{10}$, and

M_1' is the only doubly infinite sequence with this property. The author also establishes a new sufficient condition that an element of the Markov spectrum be contained in the Lagrange spectrum. *S. E. Schuur* (E. Lansing, Mich.)

Citations: MR 31# 4766a = J04-48.

Referred to in J04-62.

J04-56 (35# 5396)

Samušenok, I. N.

On the question of estimation of the lower bound for an arithmetic characteristic of irrational numbers. (Russian)

Volž. Mat. Sb. Vyp. 4 (1966), 160–162.

Let $\lambda(\theta)$ be the g.l.b. of the set of values $0<c<1$ for which, for a given irrational number θ and an infinite set of values t, there exist pairs of integers p and $q>0$ such that $|q\theta-p|<t^{-1}$ and $q<ct$. Let $\Lambda(\theta)$ be the g.l.b. of values $0<c<1$ for which the above inequalities are satisfied for all sufficiently large t. Theorem: For the set of all irrational numbers, $\Lambda\geqq\frac{1}{2}(1-2\lambda+\sqrt{(1+4\lambda^2)})$. If $\theta=\frac{1}{2}(\sqrt{5}-1)$ and numbers equivalent to it are deleted from the set of all irrational numbers, then $\Lambda\geqq\frac{1}{2}(1-2\lambda+\sqrt{(1+4\lambda^2)})$, $\Lambda\geqq\frac{1}{2}+\lambda$.

This improves an earlier result of the author [same Sb. Vyp. 1 (1963), 164–168; MR **33** #4011].

S. E. Schuur (E. Lansing, Mich.)

Citations: MR 33# 4011 = J04-52.

J04-57 (36# 6355)

Davenport, H.

On a theorem of Heilbronn.

Quart. J. Math. Oxford Ser. (2) **18** (1967), 339–344.

H. Heilbronn [same J. **19** (1948), 249–256; MR **10**, 284] proved that if θ is real, and $\|x\|$ denotes the distance of x from the nearest integer, then given $N>0$, $\varepsilon>0$, there exists an integer n such that $1\leqq n\leqq N$, $\|n^2\theta\|<c(\varepsilon)N^{-1/2+\varepsilon}$, where $c(\varepsilon)$ is a constant depending only on ε. The author generalizes this result as follows. Let $f(n)$ be a polynomial with real coefficients, degree k and $f(0)=0$. Let $R=2^k-1$. Then given $\varepsilon>0$, $N>0$, there exists an integer n such that $1\leqq n\leqq N$, $\|f(n)\|<c(\varepsilon,k)N^{-1/R+\varepsilon}$, where $c(\varepsilon,k)$ depends only on ε and k. Taking $f(n)=n^2\theta$, one gets the Heilbronn result with the weaker exponent $-\frac{1}{3}+\varepsilon$ instead of $-\frac{1}{2}+\varepsilon$.

R. P. Bambah (Columbus, Ohio)

Citations: MR 10, 284c = J04-5.

Referred to in J04-64.

J04-58 (37# 2695)

Freĭman, G. A.

Non-coincidence of the spectra of Markov and of Lagrange. (Russian)

Mat. Zametki **3** (1968), 195–200.

It is known that the Lagrange spectrum (related to the minima of binary indefinite quadratic forms) is a subset of the Markov spectrum (related to the approximation of irrational by rationals). In this note the author constructs explicitly an isolated element of the Markov spectrum which does not belong to the Lagrange spectrum and shows the existence of a countable infinity of elements of this kind. *J. W. S. Cassels* (Cambridge, England)

Referred to in J04-63.

J04-59 (39# 2707)

Gyapjas, F.

On diophantine approximation.

Ann. Univ. Sci. Budapest. Eötvös Sect. Math. **10** (1967), 143–154.

Let m be a positive integer or ∞. The author examines closely when for a constant $C_m>0$ and for a quadratic irrational δ the inequality $q\|q\delta\|\leqq C_m^{-1}$ has at least m solutions in integers $q>0$ ($\|x\|$ denotes the distance from x to the closest integer). He uses his results to deduce a result of L. C. Eggan [Trans. Amer. Math. Soc. **99** (1961), 102–117; MR **22** #12096], and a result on the attainment of the minimal value of a binary quadratic form with integer coefficients. {The article is so full of misprints and unclear and inaccurate statements that it is difficult to understand.} *W. W. Adams* (College Park, Md.)

Citations: MR 22# 12096 = J04-38.

J04-60 (40# 1343)

Kinney, J. R.; Pitcher, T. S.

On the lower range of Perron's modular function.

Canad. J. Math. **21** (1969), 808–816.

Let ξ be an irrational number. O. Perron [S.-B. Heidelberger Akad. Wiss. Natur. Kl. **12** (1921), 3–17] defined the modular function $M(\xi)$: the inequality $|\xi-p/q|<(1+d)/(M(\xi)q^2)$ is satisfied by an infinity of relatively prime pairs (p,q) for positive d, but by at most a finite number of such pairs for negative d. Using continued fractions, one writes

$$\xi=[x_1,x_2,\cdots]=\frac{1}{x_1+}\frac{1}{x_2+}\cdots\quad(\xi\in(0,1)),$$

$$[y_1,\cdots,y_k]=\frac{1}{y_1+}\frac{1}{y_2+}\cdots\frac{1}{+y_k},$$

where $y_1,\cdots,y_k$ are positive integers. Now Perron proved $M(\xi)=\limsup_k M_k(\xi)$, where

$$M_k(\xi)=x_k+[x_{k-1},x_{k-2},\cdots,x_1]+[x_{k+1},x_{k+2},\cdots].$$

The range of M is known as the Lagrange spectrum. If ξ^* is a doubly infinite sequence of positive integers, $\xi^*=(\cdots,x_1,x_0,x_1,\cdots)$, we set $M^*(\xi^*)=\sup_k M_k^*(\xi^*)$, where $M_k^*(\xi^*)=x_k+[x_{k-1},x_{k-2},\cdots]+[x_{k+1},x_{k+2},\cdots]$. The range of M^* is known as the Markov spectrum. P. Kogonija proved that the Markov spectrum is closed and contains the Lagrange spectrum [see Tbiliss. Gos. Univ. Trudy Ser. Meh.-Mat. Nauk **102** (1964), 95–104; ibid. **102** (1964), 105–113; MR **31** #4766a; Akad. Nauk Gruzin. SSR Trudy Tbiliss. Mat. Inst. Razmadze **29** (1963), 15–35 (1964); MR **31** #4766b]. The authors investigate the range of $M(\xi)$ for $\xi\in\{\xi\|x_i\leqq 2\}$. They show that the values of $M(\xi)$ in the neighbourhood of 3 form a nowhere dense set, in fact, a set of Hausdorff dimension less than 1, and also that there exist intervals $[a_i,b_i]$, $b_i\to 2\sqrt{3}$ in which no values of $M(\xi)$ occur. *L. Kuipers* (Carbondale, Ill.)

Citations: MR 31# 4766a = J04-48; MR 31# 4766b = J04-48.

J04-61 (40# 5548)

Negoescu, N.

Approximation to certain classes of real irrational numbers by rational fractions x/y with $(x,y)=1$ and $(x,m)=1$ or $(y,m)=1$. (Romanian summary)

An. Şti. Univ. "Al. I. Cuza" Iaşi Secţ. I a Mat. (N.S.) **14** (1968), 277–286.

Let Θ be real irrational, $[a_0,\cdots,a_n,\cdots]$ its (regular) continued fraction expansion and p_n/q_n its nth convergent. Arithmetically characterized integer solutions x,y $(y\geqq 1)$ of (1): $|\Theta-x/y|\leqq 1/\xi y^2$ are investigated for certain classes of Θ's. It is proved: (i) If $a_{n-1}\geqq a_n$, p a prime and k a natural number, then (at least) one of the p_{n-2+j}/q_{n-2+j} $(j=0,1)$ satisfies (1) with $\xi=a_n$ and $(x,p^k)=1$. (ii) If $a_{n+i}\geqq a_n$ $(i=1,2)$, $p\geqq a_n+1$, then one of the p_{n-1+j}/q_{n-1+j} $(j=0,1,2,3)$ satisfies (1) with $\xi=a_n+1/a_n$ and $(x,p^k)=1$. (iii) If $a_{n+i}\geqq a_n$ $(i=1,2,3,4)$, $p>a_n^2+1$, then one of the p_{n-1+j}/q_{n-1+j} $(j=0,\cdots,5)$ satisfies (1) with $\xi=(a_n^2+4)^{1/2}$ and $(x,p^k)=1$. Symmetric theorems for y are stated and some simple corollaries are given; furthermore, a theorem on solutions of (1) for which $(x, p^k q^l) = 1$, q a prime $\neq p$. It is interesting to note: (iv) One of the two interconvergents $(p_n + \varepsilon p_{n-1})/(q_n + \varepsilon q_{n-1})$, $\varepsilon = \pm 1$, satisfies (1) with $\xi = ((a_n+1)^2+1)/((a_n+1)^2-1)$. From this theorem a sharpening of a result of A. Oppenheim [Bull. Amer. Math. Soc. **47** (1941), 602–604; MR **3**, 150] is obtained. All proofs of the paper are elementary, only the simplest properties of continued fractions are used. *P. Bundschuh* (Freiburg)

Citations: MR 3, 150b = J04-3.

J04-62 (40# 7205)

Kogonija, P. G.

Certain questions of rational approximation. II. (Russian. Georgian summary)

Thbilis. Sahelmç. Univ. Šrom. Mekh.-Math. Mecn. Ser. **129** (1968), 267–274.

Let $M = \{\cdots, a_{-2}, a_{-1}, a_0, a_1, a_2, \cdots\}$ be a doubly infinite sequence of positive integers, and define $\mu(M) = \sup_{-\infty<k<\infty}\{[0; a_{k-1}, a_{k-2}, \cdots] + a_k + [0; a_{k+1}, a_{k+2}, \cdots]\}$ (square brackets denote continued fractions). The author proves that if $M_{m,n} = \{\cdots, 1_m, 2, 2, 1_m, 2, 2, 1_n, 2, 2, 1_n, \cdots\} = \{(1_m, 2, 2)_{-\infty}, (1_n, 2, 2)_\infty\}$, where 1_n stands for a sequence of n consecutive 1's, then $\mu(M_{m,n}) < 3$ if $m = n =$ even number; otherwise $\mu(M_{m,n}) > 3$. {Cf. Part I of this paper in same Šrom. **117** (1966), 45–62 [MR **34** #7456].}

S. E. Schuur (E. Lansing, Mich.)

Citations: MR 34# 7456 = J04-55.

J04-63 (41# 3403)

Berštein, A. A.

Necessary and sufficient conditions for the occurrence of points of the Markov spectrum in the Lagrange spectrum. (Russian)

Dokl. Akad. Nauk SSSR **191** (1970), 971–973.

The author gives a rather elaborate set of conditions which are together necessary and sufficient to ensure that a number λ of the Markov spectrum in the interval $4(30)^{1/2}7^{-1} < \lambda < (689)^{1/2}2^{-3}$ should also belong to the Lagrange spectrum. The argument is said to be a development of ideas of G. A. Freĭman [Mat. Zametki **3** (1968), 195–200; MR **37** #2695].

{This article has appeared in English translation [Soviet Math. Dokl. **11** (1970), 463–466].}

J. W. S. Cassels (Cambridge, England)

Citations: MR 37# 2695 = J04-58.

J04-64 (42# 206)

Liu, Ming-chit

On a theorem of Heilbronn concerning the fractional part of θn^2.

Canad. J. Math. **22** (1970), 784–788.

Als Verschärfung eines Satzes von H. Heilbronn [Quart. J. Math. Oxford Ser. **19** (1948), 249–256; MR **10**, 284] beweist der Verfasser: Zu jedem reellen θ und jeder natürlichen Zahl N gibt es eine natürliche Zahl n mit $1 \leqq n \leqq N$ und $\|\theta n^2\| < AN^{-1/2+\varepsilon(N)}$. Dabei ist $\|x\|$ der Abstand von x von der nächsten ganzen Zahl; A ist eine Konstante; $\varepsilon(N) = 1/(\log\log N)$. Für alle $N \geqq N_1$ kann $A = 1$ gewählt werden. Der Beweis stützt sich auf vier Lemmas; er beruht auf einer Verschärfung einer Methode von H. Davenport [dieselben J. (2) **18** (1967), 339–344; MR **36** #6355]. *H. J. Kanold* (Braunschweig)

Citations: MR 10, 284c = J04-5; MR 36# 6355 = J04-57.

J04-65 (42# 4497)

Cusick, T. W.

Measures of diophantine approximation. (Italian summary)

Boll. Un. Mat. Ital. (4) **3** (1970), 761–767.

Let $\sigma(\theta)$ be defined for irrational θ by $\sigma(\theta) = \sum_{j=1}^{\infty} \|q_j\theta\|$, where q_j denotes the denominator of the jth convergent in the continued fraction expansion of θ. The author studies the function $\lambda(\theta) = \liminf_{n\to\infty} q_n(\sigma(\theta) - \sum_{j=1}^{n}\|q_j\theta\|)$. It is shown that $\lambda(\theta)$ behaves similarly to the function $M(\theta) = \liminf_{n\to\infty} q_n\|q_n\theta\|$, namely: If θ is equivalent to θ', then $\lambda(\theta) = \lambda(\theta')$ and $\lambda(\theta) > 0$ if and only if the partial quotients of θ are bounded. *F. Schweiger* (Salzburg)

J04-66 (42# 5916)

Baker, A.; Schmidt, Wolfgang M.

Diophantine approximation and Hausdorff dimension.

Proc. London Math. Soc. (3) **21** (1970), 1–11.

For $\lambda > 0$ real and $n \geqq 1$ an integer, set $K_n(\lambda)$ equal to the set of all real numbers ξ such that the inequality $|\xi - \alpha| < H(\alpha)^{-(n+1)\lambda'}$ has an infinite number of solutions in algebraic numbers of degree $\leqq n$ for all $\lambda' < \lambda$ ($H(\alpha)$ denotes the height of α). Denote by $K_n'(\lambda)$ the set of all $\xi \in K_n(\lambda)$ which are not in $K_n(\lambda')$ for all $\lambda' > \lambda$. By "dim" we mean the Hausdorff dimension. The authors prove that $\dim K_n(\lambda) = \dim K_n'(\lambda) = 1/\lambda$. In an analogous problem, let $M_n(\lambda)$ equal the set of real numbers ξ such that the inequality $|P(\xi)| < H(P)^{-(n+1)\lambda'+1}$ has an infinite number of solutions in polynomials P, with integer coefficients, of degree $\leqq n$, for all $\lambda' < \lambda$ ($H(P)$ denotes the maximum of the absolute values of the coefficients of P). Then it is shown that $1/\lambda \leqq \dim M_n(\lambda) \leqq 2/\lambda$. Similar results are shown concerning the Koksma (respectively, Mahler) classification of transcendental numbers. The case $n = 1$ of the above results is an old theorem of V. Jarník [Mat. Sb. **36** (1929), 371–382] and A. S. Besicovitch [J. London Math. Soc. **9** (1934), 126–131]. The second result follows from the first by using theorems concerning the relations between the two problems which were established by E. Wirsing [J. Reine Angew. Math. **206** (1960), 67–77; MR **26** #79]. The first theorem uses, as an essential tool, Baker's refinement of Sprindzuk's theorem settling Mahler's conjecture [A. Baker, Proc. Roy. Soc. London Ser. A **292** (1966), 92–104].

W. W. Adams (College Park, Md.)

Citations: MR 26# 79 = J84-20.

J04-67 (42# 7603)

Davenport, H.; Schmidt, Wolfgang M.

Dirichlet's theorem on diophantine approximation.

Symposia Mathematica, Vol. IV (INDAM, Rome, 1968/69), pp. 113–132. Academic Press, London, 1970.

The paper contains the proofs of three theorems. The first refines Dirichlet's theorem in the one-dimensional case; it is shown that if

$$\alpha = a_0 + \frac{1}{a_1+}\frac{1}{a_2+}\cdots$$

and

$$\gamma(\alpha) = \liminf\left(\frac{1}{a_{n+1}+}\frac{1}{a_{n+2}+}\cdots\right)\left(\frac{1}{a_n+}\frac{1}{a_{n-1}+}\cdots+\frac{1}{a_1}\right),$$

then the inequalities $|\alpha x - y| < cN^{-1}$, $1 \leqq x \leqq N$, are solvable for all sufficiently large N if $c > (1+\gamma(\alpha))^{-1}$, but not if $c < (1+\gamma(\alpha))^{-1}$. (For all irrational α, $0 \leqq \gamma(\alpha) \leqq \frac{1}{2}(3-\sqrt{5}) = 0.382\ldots$.)

The two-dimensional case is considered in the other theorems. A pair of numbers α, β such that $1, \alpha, \beta$ are linearly independent over the rationals is said to be badly approximable if for some $C>0$, $|\alpha x+\beta y+z|>C\{\max(|x|, |y|)\}^{-2}$ for all integers x, y, z with $x, y\neq 0, 0$. Theorem 2 gives an analogue of Theorem 1: If (α, β) is a badly approximable pair, then there is a $c<1$ such that for all large N the inequalities $|\alpha x+\beta y+z|<cN^{-2}$, $1\leqq\max(|x|, |y|)\leqq N$ are solvable, and the same is true of the inequalities $|\alpha z-x|<cN^{-1/2}$, $|\beta z-y|<cN^{-1/2}$, $1\leqq z\leqq N$. Theorem 3 concerns the pairs (α, α^2): If $0<\kappa<\frac{1}{4}$, then for almost all α there are infinitely many N for which the inequalities $|\alpha^2x+\alpha y+z|<\kappa N^{-2}$, $1\leqq\max(|x|, |y|)\leqq N$ are unsolvable in integers x, y, z.

W. J. LeVeque (Claremont, Calif.)

Referred to in J12-64.

J04-68 (43# 4772)

Liu, Ming-Chit

On the fractional parts of θn^k and ϕn^k.

Quart. J. Math. Oxford Ser. (2) **21** (1970), 481–486.

The author proves the theorem: For an integer $k\geqq 2$ let $K=2^{k-1}$. Let δ be any arbitrary positive number; for any $\varepsilon>0$ there exist positive constants $C(k, \varepsilon)$, $C(\delta, \varepsilon)$ such that, for any real numbers θ and ϕ and integers $N\geqq 1$, there exists an integer m satisfying $1\leqq m\leqq N$, and $\|\theta m^2\|<C(\delta, \varepsilon)N^{(-1/(7+\delta))+\varepsilon}$, $\|\phi m^2\|<C(\delta, \varepsilon)N^{(-1/(7+\delta))+\varepsilon}$, $\|\theta m^k\|<C(k, \varepsilon)N^{(-1/(3K+1))+\varepsilon}$, $\|\phi m^k\|<C(k, \varepsilon)N^{(-1/(3K+1))+\varepsilon}$, where $k=3, 4, \cdots$. This result improves several extensions of a result of H. Heilbronn [same Quart. (2) **19** (1948), 249–256; MR **10**, 284]: For each $\varepsilon>0$ there exists $C(\varepsilon)>0$ such that for any real number θ and any $N\geqq 1$ there is an integer m satisfying $1\leqq m\leqq N$, $\|\theta m^2\|<C(\varepsilon)N^{-1/2+\varepsilon}$. Above $\|t\|$ means the distance from t to the nearest integer.

A. R. Freedman (Burnaby, B.C.)

Citations: MR 10, 284c = J04-5.

J04-69 (43# 6162)

Wassef, Pierre

Une application de la théorie ergodique à la théorie métrique des fractions continues.

Séminaire Delange-Pisot-Poitou: 1969/70, *Théorie des Nombres, Fasc.* 1, *Exp.* 8, 10 *pp. Secrétariat mathématique, Paris,* 1970.

The author presents a proof of the following theorem on diophantine approximation: Let $f(q)$ be positive for $q\geqq 1$; if $qf(q)$ is not increasing and the series $\sum f(q)$ diverges, then the inequality $|qx-p|<f(q)$ has infinitely many integer solutions $q>0$, p for almost all x; if $\sum f(q)$ converges, the inequality has infinitely many integer solutions for almost no x.

The proof follows the pattern of the known ergodic method. *F. Schweiger* (Salzburg)

J04-70 (44# 2709)

Diviš, Bohuslav; Novák, Břetislav

A remark on the theory of Diophantine approximations.

Comment. Math. Univ. Carolinae **12** (1971), 127–141.

Let β be an irrational number and $(b_0; b_1, b_2, \cdots)$ its simple continued fraction expansion. For $t\geqq 1$ put $\psi_\beta(t)=\min\{|q\beta-p|: p, q \text{ integral}, 0<q\leqq t\}$, $\mu_\beta=\lim\sup t\psi_\beta(t)$, $R_\beta=\lim\sup(b_k; b_{k-1}, \cdots, b_1)\cdot(b_{k+1}, b_{k+2}, \cdots)$. Then $\mu_\beta=(1+R_\beta^{-1})^{-1}$. The authors investigate the set $\mathfrak{M}(N)$ of values of R_β, if β runs through the irrationals with $\lim\sup b_k=N$ $(N=1, 2, \cdots)$. By direct computation, they show that $\mathfrak{M}(N)$ has an isolated minimum and a maximum which is a condensation point. Using a result of M. Hall on the product of continued fractions [Ann. of Math. (2) **48** (1947), 966–993; MR **9**, 226] they further prove that $\bigcup_1^\infty\mathfrak{M}(N)$ contains an interval $[R^*, \infty)$, where $R^*\leqq 12+8\sqrt{2}$. *C. G. Lekkerkerker* (Amsterdam)

Citations: MR 9, 226b = A54-7.

J08 ASYMMETRIC HOMOGENEOUS APPROXIMATION TO ONE NUMBER

See also reviews A58-12, H15-42, J04-38, J04-39, J04-61, J32-23, Z10-4.

J08-1 (6, 258a)

Segre, B. Lattice points in infinite domains and asymmetric Diophantine approximations. Duke Math. J. **12**, 337–365 (1945).

Let $K_{a,b}$ be the set of points (x, y) for which $-a<xy<b$, $a>0$, $b\geqq 0$. A lattice Λ is defined to be $K_{a,b}$-admissible if it has no point within $K_{a,b}$ except possibly the origin. The area of a fundamental parallelogram of Λ is called the determinant of Λ and is designated by $d(\Lambda)$. Let $\Delta(K_{a,b})$ be the lower bound of $d(\Lambda)$ for all $K_{a,b}$-admissible lattices. The author deduces, by simple geometric methods, that $\Delta(K_{a,b})\geqq(a^2+4ab)^{\frac{1}{2}}$ and that the equality sign occurs if and only if a/b is integral or $b=0$. When the equality sign holds, he constructs all lattices Λ for which $d(\Lambda)=\Delta(K_{a,b})$. The results are applied to derive inequalities for $\Delta(K)$ when K belongs to some general classes of domains for which $K_{a,b}\subseteq K$. A second application is the following approximation theorem. Let τ be any nonnegative number. Then, for any irrational number ξ, infinitely many rationals m/n exist for which

$$-\frac{1}{n^2(1+4\tau)^{\frac{1}{2}}}<\frac{m}{n}-\xi<\frac{\tau}{n^2(1+4\tau)^{\frac{1}{2}}}.$$

For $\tau=1$ this reduces to a classical result of Hurwitz [Math. Ann. **39**, 279–284 (1891)]. When τ is integral the author determines, by continued fraction methods similar to those of Hurwitz, the class of numbers ξ for which this is the best possible approximation. *D. Derry* (Saskatoon, Sask.).

Referred to in H20-7, J04-38, J08-3, J08-4, J08-5, J08-7, J08-9, J08-10, J08-11, J08-12, J08-13, J08-14, J08-19, J08-20, J08-21, J08-22, J08-23, J08-24, J40-6.

J08-2 (6, 258b)

Mahler, K. A theorem of B. Segre. Duke Math. J. **12**, 367–371 (1945).

The author uses methods developed by Mordell [cf. J. London Math. Soc. **18**, 201–210 (1943); these Rev. **6**, 37] to prove the result of the paper reviewed above that the set of points (x, y) for which $-1\leqq xy<0$ contains a point of the lattice Λ if $d(\Lambda)\leqq 1$. He discusses the connection of this result with continued fractions used in the theory of binary quadratic forms. *D. Derry* (Saskatoon, Sask.).

Citations: MR 6, 37h = D32-15.

Referred to in J40-6.

J08-3 (8, 196e)

Olds, C. D. Note on an asymmetric Diophantine approximation. Bull. Amer. Math. Soc. **52**, 261–263 (1946).

By means of geometric methods B. Segre [Duke Math. J. **12**, 337–365 (1945); these Rev. **6**, 258] deduced the following theorem on "asymmetric" Diophantine approximation. Let τ be any nonnegative number. Then, for any irrational number θ, infinitely many rationals x/y exist, such that

$$\frac{-1}{y^2(1+4\tau)^{\frac{1}{2}}}<\frac{x}{y}-\theta<\frac{\tau}{y^2(1+4\tau)^{\frac{1}{2}}}, \qquad y>0.$$

The author gives a very short and purely arithmetical proof for the special case $\tau \geqq 1$. *J. Popken* (Groningen).

Citations: MR 6, 258a = J08-1.

Referred to in J08-4, J08-9, J08-10, J08-20.

J08-4 **(8, 566b)**

Robinson, Raphael M. Unsymmetrical approximation of irrational numbers. Bull. Amer. Math. Soc. **53**, 351–361 (1947).

By means of geometrical methods B. Segre [Duke Math. J. **12**, 337–365 (1945); these Rev. **6**, 258] deduced a theorem on asymmetric approximation and C. D. Olds [same Bull. **52**, 261–263 (1946); these Rev. **8**, 196] gave a proof making use of Farey series. The author makes use of continued fractions and proves, among other results, that for any ξ and any $\epsilon > 0$ the inequality

$$-\frac{1}{(5^{\frac{1}{2}}-\epsilon)B^2} < \frac{A}{B} - \xi < \frac{1}{(5^{\frac{1}{2}}+1)B^2}$$

has infinitely many integer solutions. *J. F. Koksma.*

Citations: MR 6, 258a = J08-1; MR 8, 196e = J08-3.

Referred to in A58-12, J04-31, J04-33, J08-7, J08-8, J08-10, J08-20.

J08-5 **(9, 335a)**

Cassels, J. W. S. The lattice properties of asymmetric hyperbolic regions. I. On a theorem of Khintchine. Proc. Cambridge Philos. Soc. **44**, 1–7 (1948).

Let $k \geqq 0$, $l \geqq 0$; let $K > 0$, $L > 0$ be such that

$$K+L \geqq \max(k, l); \quad K^2+2(L+l)K+2(L-l) \geqq 1;$$
$$L^2+2(K+k)L+2(K-k) \geqq 1.$$

(I) If $\theta > 0$, α are real numbers, θ irrational, $\alpha \neq m - n\theta$ for every pair of integers m, n and if there is an infinity of pairs of positive integers p, q such that (1) $-k < q(p-\theta q) < l$, then there is also an infinity of pairs of positive integers p, q such that (2) $-K < q(p-q\theta-\alpha) < L$. An easy corollary of (I) is (II): Let Λ be a unimodular lattice in the plane, having a point at the origin but having no other point on the y-axis. Suppose that, for each $u > 0$, there is a point of Λ in the region $-k < xy < l$, $y > 0$, $|x| < u$. Then, to each $v > 0$, and to every point V of the plane, there is a point W in the region $-K < xy < L$, $y > 0$, $|x| < v$ such that $W - V$ is a point of Λ. Many results about admissible values of k, l in (1) are known; e.g., for every θ we may take $k = l = 5^{-\frac{1}{2}}$ and so $K = L = 5^{-\frac{1}{4}}$ [Khintchine, Math. Ann. **111**, 631–637 (1935); Jogin, Uchenye Zapiski Moskov. Gos. Univ. Matematika **73**, 37–40 (1944); these Rev. **7**, 273]. Asymmetric results (i.e., with $k \neq l$) concerning (1) are due to B. Segre [Duke Math. J. **12**, 337–365 (1945); these Rev. **6**, 258]. *V. Jarník* (Prague).

Citations: MR 6, 258a = J08-1; MR 7, 273j = J20-1.

Referred to in J32-16.

J08-6 **(9, 569c)**

Negoescu, Nicolae. Sur des approximations asymétriques. C. R. Acad. Sci. Paris **226**, 1495–1497 (1948).

Le problème de la meilleure approximation asymétrique d'un nombre irrationnel θ par des rationnels consiste de trouver la valeur maximum de ξ telle qu'il existe une infinité de fractions p/q qui satisfassent à

(1) $$-1/(\xi q^2) < p/q - \theta < \tau/(\xi q^2), \qquad \tau > 0.$$

La borne supérieure de l'ensemble (ξ) des nombres ξ pour lesquels (1) soit satisfaite par une infinité de fractions rationnelles p/q soit designée par $M(\theta, \tau)$. L'auteur donne quelques considérations sur ce nombre $M(\theta, \tau)$ à l'aide de la théorie des fractions continues et en donne une répresentation géométrique. *J. F. Koksma* (Amsterdam).

Referred to in J08-7, J08-15.

J08-7 **(10, 102d)**

Negoescu, Nicolae. Théorèmes sur des approximations asymétriques. C. R. Acad. Sci. Paris **226**, 1664–1666 (1948).

Continuation d'une note antérieure [mêmes C. R. **226**, 1495–1497 (1948); ces Rev. **9**, 569]. L'auteur énonce quelques théorèmes qui sont des généralisations des théorèmes de Perron et d'autres [voir Koksma, Diophantische Approximationen, Ergebnisse der Math., v. 4, no. **4**, Springer, Berlin, 1936, chap. III, § 2] et de B. Segre [Duke Math. J. **12**, 337–365 (1945); ces Rev. **6**, 258] et qui sont liés à quelques théorèmes de R. M. Robinson [Bull. Amer. Math. Soc. **53**, 351–361 (1947); ces Rev. **8**, 566]. *J. F. Koksma* (Amsterdam).

Citations: MR 6, 258a = J08-1; MR 8, 566b = J08-4; MR 9, 569c = J08-6.

J08-8 **(10, 235c)**

Robinson, Raphael M. The critical numbers for unsymmetrical approximation. Bull. Amer. Math. Soc. **54**, 693–705 (1948).

If ξ is an irrational number, let $M^+(\xi)$ denote the least upper bound of the values of μ for which infinitely many rational numbers A/B satisfy the inequality $0 < A/B - \xi < 1/(\mu B^2)$. Similarly $M^-(\xi)$ is defined measuring the approximability from the left. The number ξ is called critical if there is no other irrational number ξ' for which simultaneously $M^+(\xi') < M^+(\xi)$, $M^-(\xi') < M^-(\xi)$. The author gives a necessary and sufficient condition that ξ be critical. This condition requires that the expansion of ξ as a continued fraction has the form

$$\xi = [q_0, q_1, \cdots, q_{n-1}, 1, r_1, 1, r_2, 1, r_3, \cdots],$$

where the sequence $r_1, r_2, \cdots$ is of a certain type. [Cf. same Bull. **53**, 351–361 (1947); these Rev. **8**, 566.] *J. F. Koksma* (Amsterdam).

Citations: MR 8, 566b = J08-4.

Referred to in A54-26, J04-31, J04-33.

J08-9 **(10, 235d)**

Negoescu, Nicolae. Quelques précisions concernant le théorème de M. B. Segre sur des approximations asymétriques des nombres irrationnels par les rationnels. Bull. École Polytech. Jassy [Bul. Politehn. Gh. Asachi. Iaşi] **3**, 3–16 (1948).

B. Segre [Duke Math. J. **12**, 337–365 (1945); ces Rev. **6**, 258] a démontré géométriquement le théorème suivant: Tout nombre irrationel θ a une infinité d'approximations rationnelles p/q telles que l'on ait

(1) $$-1/(q^2\xi) < p/q - \theta < \tau/(q^2\xi)$$

où $\tau \geqslant 0$ est réel et $\xi = (1+4\tau)^{\frac{1}{2}}$. C. D. Olds en donna une démonstration arithmétique pour le cas $\tau \geqq 1$ [Bull. Amer. Math. Soc. **52**, 261–263 (1946); ces Rev. **8**, 196]. L'auteur en donne une troisième démonstration, généralisant la méthode de Borel qui fait usage de la théorie des fractions continues et qui permet l'auteur (a) de remplacer la condition $\xi = (1+4\tau)^{\frac{1}{2}}$ par $\xi = \max((1+4\tau)^{\frac{1}{2}}, (\tau^2+4\tau)^{\frac{1}{2}})$, (b) de la remplacer par une autre condition encore meilleure, si l'on exclut les nombres θ équivalent à $(1+\sqrt{5})/2$ et (c) de démontrer des précisions sur les réduites p_n/q_n qui réalisent l'approximation (1). *J. F. Koksma* (Amsterdam).

Citations: MR 6, 258a = J08-1; MR 8, 196e = J08-3.

Referred to in J08-10, J08-20.

J08-10 **(13, 630e)**

Negoescu, Nicolae. Note on a theorem of unsymmetric approximation. Acad. Repub. Pop. Române. Bul. Şti. A. **1**, 115–117 (1949). (Romanian)

B. Segre [Duke Math. J. **12**, 337–365 (1945); these Rev. **6**, 258] proved by a geometrical method that if θ is a given irrational real number there are infinitely many rational fractions p/q for which

$$(*)\qquad -\frac{1}{q^2\xi}<\frac{p}{q}-\theta<\frac{\tau}{q^2\xi},$$

where τ is any real non-negative number and $\xi=\sqrt{(1+4\tau)}$. Proofs of this result have also been given by others including the author. Thus Olds [Bull. Amer. Math. Soc. **52**, 261–263 (1946); these Rev. **8**, 196] has given a simple arithmetical proof valid for $\tau\geqq 1$. The author shows how the argument of Olds can be modified so as to apply for all $\tau>0$. He uses continued fractions in place of the Farey series used by Olds. The author also states that by using the method of Humbert [J. Math. Pures Appl. (7) **2**, 155–167 (1916)] it can be shown that one of every three consecutive convergents to θ satisfies (*). This the reviewer has only been able to check for $\tau\geqq 1$, since otherwise it does not seem to be necessarily true that $\mu_{2n+2}\geqq(1+\xi)/2\tau$. The theorem is also proved by a method similar to that used by Fujiwara [Jap. J. Math. **1**, 15–16 (1924)]. It is stated that it is possible to replace ξ by $\max(\sqrt{(1+4\tau)},\sqrt{(\tau^2+4\tau)})$ as proved by Robinson [Bull. Amer. Math. Soc. **53**, 351–361 (1947); these Rev. **8**, 566] and the author [Bull. École Polytech. Jassy [Bul. Politehn. Gh. Asachi. Iaşi] **3**, 3–16 (1948); these Rev. **10**, 235]. *R. A. Rankin.*

Citations: MR 6, 258a = J08-1; MR 8, 196e = J08-3; MR 8, 566b = J08-4; MR 10, 235d = J08-9.

Referred to in J08-12.

J08-11 (15, 106b)

Sawyer, D. B. The minima of indefinite binary quadratic forms. J. London Math. Soc. **28**, 387–394 (1953).

A geometric proof is given of a result of Segre [Duke Math. J. **12**, 337–365 (1945); these Rev. **6**, 258]. For the indefinite form $a\xi^2+b\xi\eta+c\eta^2$ of discriminant D, integers ξ, η not both zero exist such that $-\lambda\phi_0(\lambda)\leqq f/D^{1/2}\leqq\lambda^{-1}\phi_0(\lambda)$, where $\lambda>0$ and $\phi_0(\lambda)=\min[(4+\lambda^2)^{-1/2},(4+\lambda^{-2})^{-1/2}]$. Strict inequalities hold unless λ^2 is integral and f is equivalent to a multiple of $f_0=\xi^2+\lambda^2\xi\eta-\lambda^2\eta^2$. The author also proves that if f is not equivalent to kf_0, then ϕ_0 may be replaced by $(4+(\lambda+\lambda^{-3})^2)^{-1/2}$ and strict inequality holds unless $\lambda=1$ and f is equivalent to a multiple of $\xi^2+2\xi\eta-\eta^2$.

L. Tornheim (Ann Arbor, Mich.).

Citations: MR 6, 258a = J08-1.

J08-12 (16, 18b)

LeVeque, W. J. On asymmetric approximations. Michigan Math. J. **2**, 1–6 (1954).

It is known that every irrational ξ has infinitely many approximations u/v such that

$$(*)\qquad -\frac{\tau}{\alpha\xi^2}<\xi-\frac{u}{v}<\frac{1}{\alpha v^2},$$

where u, v are rational integers and

$$\alpha=\max\{(1+4\tau)^{1/2},(\tau^2+4\tau)^{1/2}\}$$

[B. Segre, Duke Math. J. **12**, 337–365 (1945); these Rev. **6**, 258]. When $\tau=1$ a classical theorem states that indeed one of three consecutive convergents p_n/q_n to ξ satisfies (*) and N. Negoescu alleged that this contines to hold in the general case [Acad. Repub. Pop. Române. Bul. Şti. A. **1**, 115–117 (1949); these Rev. **13**, 630]. The author disproves this with an example but shows that one of p_{n-1}/q_{n-1}, p_n/q_n, p_{n+1}/q_{n+1} satisfies (*) with $(\tau^2+4\tau)^{1/2}$ if n is odd but with $(1+4\tau)^{1/2}$ if n is even. *J. W. S. Cassels* (Cambridge, England).

Citations: MR 6, 258a = J08-1; MR 13, 630e = J08-10.

Referred to in J08-20.

J08-13 (16, 1003a)

Tornheim, Leonard. Asymmetric minima of quadratic forms and asymmetric Diophantine approximation. Duke Math. J. **22**, 287–294 (1955).

Let $f(x,y)=ax^2+bxy+cy^2$ be an indefinite quadratic form not representing 0 and normalized to $b^2-4ac=1$. Let p^{-1}, n^{-1} be respectively the infima of the positive values of $f(x,y)$, $-f(x,y)$ for integral x, y and put $B=B(f)=\max(p^2,k^2n^2)$, where $k\geqq 1$ is fixed. Segre [same J. **12**, 337–365 (1945); MR **6**, 258] has shown that $B\geqq B_1=k^2+4k$ with equality only when k is an integer and f is equivalent to $x^2-kxy-ky^2$. The author shows that for $k\geqq 2$, k an integer, the next highest value is

$$B_2=\left[\frac{k^2+k+(3k+1)(k^2+4k)^{1/2}}{2(2k-1)}\right]^2,$$

taken only for forms equivalent to

$$x^2-(k-1)xy-\tfrac{1}{2}[3k-(k^2+4k)^{1/2}]y^2.$$

But B_2 is a limit point on the right of $B(f)$ values, so there is no B_3, in striking contrast to the well-known Markoff chain for $k=1$. The author investigates in more detail the $B(f)$ values in a neighbourhood on the right of B_2. He also proves the analogous results for diophantine approximation. The proof shows first that if $B(f)$ is fairly small the continued fraction associated with f has the shape $(\cdots,1,X_1,1,X_2,1,X_3,\cdots)$ where $X_j=k$ or $k-1$ and then considers the structure of the X_j in detail. [For non-integral k see Barnes and Swinnerton-Dyer, Acta Math. **92**, 199–234 (1954); MR **16**, 802.] *J. W. S. Cassels.*

Citations: MR 6, 258a = J08-1; MR 16, 802c = J32-48.

Referred to in J44-42, J44-43.

J08-14 (17, 133a)

Lekkerkerker, C. G. On the determinant of an asymmetric hyperbolic region. Ann. Mat. Pura Appl. (4) **38** (1955), 253–266.

Let $K_{a,b}$ denote the plane domain defined by $-a\leqq xy\leqq b$, $a>0$, $b>0$; and let $\Delta(a,b)$ denote its critical determinant with respect to lattices having a fixed point at the origin. Observing that there is no loss of generality in taking $a=1$, $b\geqq 1$, the author proves that

$$\Delta(1,b)\geqq\min\{\sqrt{(\beta^2+4b)},\ \sigma\sqrt{(b^2+4b\sigma)}\},$$

where $\beta=-[-b]$, $\sigma=b/[b]$; the equality sign holding if and only if, either $\sigma\sqrt{(b^2+4b\sigma)}\leqq\sqrt{(\beta^2+4b)}$, or $(\beta+2)/(\beta+1-b)$ is an integer. This result is a sharper form of a theorem of B. Segre [Duke Math. J. **12** (1945), 337–365; MR **6**, 258]. Defining b_0 by the relations $b_0^4=4(1+b_0)/5$, $b_0>1$, it is shown that the (fully automorphic) star body $K_{1,b}$ is boundedly irreducible for $1<b<b_0$. *J. H. H. Chalk* (London).

Citations: MR 6, 258a = J08-1.

J08-15 (18, 468d)

Negoescu, Nicolae. Approximation asymétrique des nombres irrationnels par des rationnels. An. Şti. Univ. "Al. I. Cuza" Iaşi. Secţ. I. (N.S.) **1** (1955), 21–30. (Romanian. Russian and French summaries)

Let θ be an irrational number having the regular continued fraction expansion $[a_0,a_1,\cdots,a_n,\cdots]$ and, for any integer r set

$$\lambda_r=[a_{r+1},a_{r+2},\cdots]+[0,a_r,a_{r-1},\cdots,a_1].$$

Given $\tau>0$, let μ be such that the inequality

$$-1/\mu q^2<p/q-\theta<\tau/\mu q^2$$

is satisfied by infinitely many convergents p/q of θ. If $M(\theta,\tau)=\limsup\mu$, then the author has shown [C. R.

Acad. Sc. Paris **226** (1948), 1495–1497; MR **9**, 569] that $M(\theta, \tau) = \max[\limsup \lambda_{2\nu}, \tau \limsup \lambda_{2\nu+1}]$. For every given irrational θ, $M = M(\theta)$ represents a broken line with vertex P at $\tau = \limsup \lambda_{2\nu}/\limsup \lambda_{2\nu+1}$, $M = \limsup \lambda_{2\nu}$ and the set $\mathfrak{M}$ of all such points P has the power of the continuum. Properties of $\mathfrak{M}$ and of some of its subsets, corresponding to quadratic irrationalities are studied.

E. Grosswald (Philadelphia, Pa.).

Citations: MR 9, 569c = J08-6.

J08-16 (18, 468e)

Negoescu, Nicolae. Une méthode arithmétique pour le problème des approximations asymétriques. An. Ști. Univ. "Al. I. Cuza" Iași. Secț. I. (N.S.) **1** (1955), 31–38. (Romanian. Russian and French summaries)

Let θ be an irrational number and denote by $[a_0, a_1, \cdots, a_n, \cdots]$ its expansion in a regular continued fraction. Denote the denominator of the convergent $[a_{n+1}, a_{n+2}, \cdots, a_m]$ by $r_{n,m}$ and set $\varepsilon_n = \frac{1}{2}(1-(-1)^n)$. Then, for integers $n<m<l$, $m-n \equiv l-m \equiv 1 \pmod 2$ and arbitrary $\tau > 0$, at least one of the three convergents p_n/q_n, p_m/q_m, p_l/q_l of θ satisfies $-1/\xi q^2 < p/q - \theta < \tau/\xi q^2$ with

$$\xi = \left[\left(\frac{\tau^{\varepsilon_l} r_{n,m}^2 + \tau^{\varepsilon_n} r_{m,l}^2 + \tau^{\varepsilon_m} r_{n,l}^2}{r_{n,m} r_{m,l} r_{n,l}}\right)^2 - 4\frac{\tau^{\varepsilon_n+\varepsilon_l}}{r_{n,l}^2}\right]^{1/2}.$$

This theorem generalizes a result by Fujiwara [Proc. Imp. Acad. Tokyo **2**, (1926), 1–3] and contains as particular cases theorems of Vahlen, Hurwitz, Borel, Khintchine, Humbert and Segre on the (symmetric and asymmetric) approximations of irrational numbers by rationals. The method used is an extension of that of Fujiwara.

E. Grosswald (Philadelphia, Pa.).

J08-17 (26# 1283)

Negoescu, Nicolae

Méthode de Borel pour le problème des approximations asymétriques. (Romanian. Russian and French summaries)

Acad. R. P. Romîne Fil. Iași Stud. Cerc. Ști. Mat. **12** (1961), 195–204.

Author's summary: "Dans cette note l'auteur démontre le théorème suivant: Si $n < m < l$, $m-n$ et $l-m$ sont des nombres impairs ≥ 1, l'une au moins des réduites p_n/q_n, p_m/q_m, p_l/q_l du nombre irrationel θ satisfait aux inégalités:

$$-\frac{1}{\xi q^2} < \frac{p}{q} - \theta < \frac{\tau}{\xi q^2}, \quad \tau > 0,$$

où

$$\xi = \left[\left(\frac{\tau^{\varepsilon_l} r_{n,m}^2 + \tau^{\varepsilon_n} r_{m,l}^2 + \tau^{\varepsilon_m} r_{n,l}^2}{r_{n,m} r_{m,l} r_{n,l}}\right)^2 - 4\frac{\tau^{\varepsilon_n+\varepsilon_l}}{r_{n,l}^2}\right]^{1/2},$$

$r_{n,m}$ est le dénominateur de la réduite $[a_{n+1}, a_{n+2}, \cdots, a_m]$ du nombre θ_n et $\varepsilon_n = (1-(-1)^n)/2$. Ce théorème constitue une généralisation d'un théorème de Fujiwara [voir, par exemple, Fujiwara, Proc. Imp. Acad. Japan **2** (1926), 97–99]. La démonstration est donnée par une extension de la méthode arithmétique utilisée par Borel."

J08-18 (22# 5622)

Negoescu, N. Nombres et points critiques pour les ensembles N_{n+} et M_{n+} dans le problème des approximations asymétriques. I. Acad. R. P. Romîne. Fil. Iași. Stud. Cerc. Ști. Mat. **10** (1959), 1–12. (Romanian. Russian and French summaries)

Here is introduced the notion of a critical number for the ensemble of numbers N_{n+} that have a development in a simple infinite continued fraction of the form $\theta = [a_0, a_1, a_2, \cdots]$ with $a_i \geq n$ $(i \geq i_0)$, where θ is an irrational number. The following theorem is proved: A necessary condition in order that a number $\theta \in N_{n+}$ be a critical number is that the development of θ in a continued fraction be of the form $[a_0, a_1, a_2, \cdots, a_r, n, b_1, n, b_2, n, b_3, \cdots]$, where the sequence $\{b_i\}$ satisfies one of the three following conditions: (i) $b_i \to \lambda \geq n$; (ii) $b_i \to +\infty$; (iii) $\liminf b_i = \lambda$ and $\limsup b_i = \lambda + 1$, with λ an integer $\geq n$.

E. Frank (Chicago, Ill.)

J08-19 (24# A1250)

Eggan, L. C.; Niven, Ivan

A remark on one-sided approximation.

Proc. Amer. Math. Soc. **12** (1961), 538–540.

The best possible values of c, such that for every irrational ξ, the unequalities

$$0 < \frac{a}{b} - \xi < \frac{1}{cb^2}$$

have (i) infinitely many rational solutions a/b, (ii) at least one rational solution a/b, are determined. The answer is $c=1$ in both cases, the result for (i) being due, essentially, to B. Segre [Duke Math. J. **12** (1945), 337–365; MR **6**, 258]. The results, attributed to Robinson in the course of the proofs, appear to be due to J. H. Grace [Proc. London Math. Soc. (2) **17** (1918), 247–258].

J. H. H. Chalk (Toronto)

Citations: MR 6, 258a = J08-1.

J08-20 (25# 2032)

Niven, Ivan

On asymmetric Diophantine approximations.

Michigan Math. J. **9** (1962), 121–123.

The reviewer [Duke Math. J. **12** (1945), 337–365; MR **6**, 258] has proved geometrically the following theorem. If τ denotes any non-negative real number, every irrational number θ has infinitely many rational approximations h/k satisfying

$$-[(1+4\tau)^{1/2}k^2]^{-1} < \theta - h/k < \tau[(1+4\tau)^{1/2}k^2]^{-1}.$$

Subsequently, a proof under the restriction $\tau > 1$ has been given by C. D. Olds [Bull. Amer. Math. Soc. **52** (1946), 261–263; MR **8**, 196] using Farey sequences, and various proofs by continued fractions were given by N. Negoescu [Bull. École Polytech. Jassy **3** (1948), 3–16; MR **10**, 235], R. M. Robinson [Bull. Amer. Math. Soc. **53** (1947), 351–361; MR **8**, 566] and W. J. LeVeque [Michigan Math. J. **2** (1954), 1–6; MR **16**, 18]. Here a nice short proof of the theorem is obtained using Farey sequences.

B. Segre (Rome)

Citations: MR 6, 258a = J08-1; MR 8, 196e = J08-3; MR 8, 566b = J08-4; MR 10, 235d = J08-9; MR 16, 18b = J08-12.

J08-21 (31# 122)

Eggan, L. C.

On asymmetric Diophantine approximations.

Atti Accad. Naz. Lincei Rend. Cl. Sci. Fis. Mat. Natur. (8) **36** (1964), 814–815.

The author announces the following theorem. "Let m and n be positive integers and let $\theta_n = [0, \overline{m, mn}]$ and $\varphi_n = [0, \overline{mn, m}]$ have convergents P_j/Q_j and H_j/K_j, respectively. If $\theta = [a_0, a_1, a_2, \cdots]$ is any irrational number such that $a_{2j+1} \geq m$ for infinitely many j, then for any positive integer k there exist at least k rational numbers p/q with $q > 0$ satisfying

$$\frac{-1}{(c(m,n,k)q^2)} \leq \frac{p}{q} - \theta \leq \frac{1}{(nc(m,n,k)q^2)},$$

where $c(m, n, k) = \varphi_n + m + H_{2k-1}/K_{2k-1}$. Moreover, the constant $c(m, n, k)$ cannot be improved since equality is attained on the right hand side for $\theta=\theta_n$. If $a_{2j}\geqq m$ for infinitely many j, the statement holds with $p/q-\theta$ replaced by $\theta-p/q$, and the left hand side is best possible for $\theta=\varphi_n$."

A proof will be given elsewhere, but the author asserts that this proof is similar to the one given by him for the special case $n=1$ [Trans. Amer. Math. Soc. **99** (1961), 102–117; MR **22** #12096]. The theorem generalizes results of B. Segre [Duke Math. J. **12** (1945), 337–365; MR **6**, 258] and of M. Müller [Arch. Math. **6** (1955), 253–258; MR **16**, 1090]. *J. Popken* (Amstelveen)

Citations: MR 6, 258a = J08-1; MR 16, 1090f = A58-7; MR 22# 12096 = J04-38.

J08-22 (31# 2205)

Eggan, L. C.

Finitely many asymmetric Diophantine approximations.
J. London Math. Soc. **40** (1965), 509–517.

If θ is an irrational positive number, denote by $\theta=[a_0, a_1, a_2, \cdots]$ its continued fraction expansion, where the a's are integers and $a_0\geqq 0$, $a_j\geqq 1$ for $j\geqq 1$. It is immediately seen that, for any positive integers x and y, the periodic continued fraction $[0, \overline{x, y}]$ equals $(\sqrt{(x^2y^2+4xy)}-xy)/(2x)$. In particular, if m and n are positive integers, consider the continued fractions $\theta_n=[0, \overline{m, mn}]$, $\varphi_n=[0, \overline{mn, m}]$, and denote by H_j/K_j the convergents of φ_n.

The main result of the present paper is that, if θ is such that $a_{2j+1}\geqq m$ for infinitely many values of j, then for any positive integer k there exist at least k rational numbers p/q (with $q>0$) satisfying $-1/(cq^2)\leqq p/q-\theta\leqq 1/(cnq^2)$, where $c=c(m, n, k)=\varphi_n+m+H_{2k-1}/K_{2k-1}$; here the constant c cannot be improved, and equality is attained on the right-hand side for $\theta=\theta_n$. Likewise, if $a_{2j}\geqq m$ for infinitely many j, the first part of the statement holds with $p/q-\theta$ replaced by $\theta-p/q$, and the left-hand side is best possible for $\theta=\varphi_n$.

Some interesting corollaries are drawn in the case when $k=1$. Moreover, the case when $n=1$ coincides with Theorem 2.3 of the author [Trans. Amer. Math. Soc. **99** (1961), 102–117; MR **22** #12096]. Finally, by taking the limit as $k\to\infty$, Theorems 11 and 12 of the reviewer [Duke Math. J. **12** (1945), 337–365; MR **6**, 258], as well as a generalization of them given by E. A. Maier [Atti Accad. Naz. Lincei Rend. Cl. Sci. Fis. Mat. Natur. (8) **37** (1964), 237–241], are immediately obtained.
B. Segre (Rome)

Citations: MR 6, 258a = J08-1; MR 22# 12096 = J04-38.

J08-23 (32# 5595)

Maier, Eugene Alfred

On asymmetric diophantine approximation.
Atti Accad. Naz. Lincei Rend. Cl. Sci. Fis. Mat. Natur. (8) **37** (1964), 237–241 (1965).

Theorem: Let m be a positive integer and let θ be an irrational number with continued fraction expansion $\langle a_0, a_1, a_2, \cdots\rangle$ such that $a_{2j+1}\geqq m$ for infinitely many j. Then, if $t\geqq 0$, there exist infinitely many fractions h/k with $k>0$ such that

$$-(m^2+4t)^{-1/2} < k^2\left(\frac{h}{k}-\theta\right) < t(m^2+4t)^{-1/2}.$$

Setting $m=1$ one gets a known theorem of Segre [Duke Math. J. **12** (1945), 337–365; MR **6**, 258].
C. G. Lekkerkerker (Amsterdam)

Citations: MR 6, 258a = J08-1.

J08-24 (32# 1167)

Negoescu, N.

On asymmetric Diophantine approximations. (**Romanian and Russian summaries**)
An. Şti. Univ. "Al. I. Cuza" Iaşi Secţ. I a Mat. (N.S.) **10** (1964), 1–20.

This paper continues the study of asymmetric Diophantine approximations as initiated by B. Segre [Duke Math. J. **12** (1945), 337–365; MR **6**, 258] who proved the basic theorem that to every irrational θ and every real $\tau\geqq 0$ there correspond infinitely many rationals p/q such that (*) $-(q^2\xi)^{-1}<p/q-\theta<\tau(q^2\xi)^{-1}$, where $\xi=(1+4\tau)^{1/2}$, and a similar result with $p/q-\theta$ replaced by $\theta-p/q$. In addition to a new proof of this theorem, the author establishes several new results of which the following can serve as an example. Let a/b and c/d be adjacent fractions in the Farey sequence of any fixed order such that $a/b<(a+c)/(b+d)<\theta<c/d$. Let h and m be the positive integers defined by $(a+hc)/(b+hd)<\theta<(a+hc+c)/(b+hd+d)$ and

$$\frac{(a+hc)(m+1)+c}{(b+hd)(m+1)+d} < \theta < \frac{(a+hc)m+c}{(b+hd)m+d}.$$

Then (*) holds with $\xi=(m^2+4\tau)^{1/2}$ and p/q replaced by at least one of the three fractions

$$\frac{(a+hc)}{(b+hd)},\quad \frac{(a+hc)m+c}{(b+hd)m+d},\quad \frac{c}{d}.$$

A fairly extensive bibliography is given, especially of papers treating asymmetric approximations by use of Farey sequences, as contrasted with the continued fractions approach. *I. Niven* (Eugene, Ore.)

Citations: MR 6, 258a = J08-1.

Referred to in J08-25.

J08-25 (33# 4010)

Negoescu, N.; Olaru, E.

On some general theorems of diophantine approximation. (**Romanian. Russian and French summaries**)
An. Şti. Univ. "Al. I. Cuza" Iaşi Secţ. I a Mat. (N.S.) **11** (1965), 13–20.

By specialising results in a previous paper of the first author [same An. (N.S.) **10** (1964), 1–20; MR **32** #1167], eight theorems on the rational solutions of the unsymmetric inequality (*) $-1/(\xi q^2)<p/q-\theta<\tau/(\xi q^2)$, where $\tau>0$, are established. By way of example, let $\theta=[a_0, a_1, a_2, \cdots]$ be an irrational number with the convergents p_n/q_n; let k, m, and n be positive integers such that $m=n+2k-1$, and let

$$\frac{p}{q} = [0, a_{n+1}, a_{n+2}, \cdots, a_{m-1}, a_m'],$$

where $0<a_m'\leqq a_m$. Then, according as n is even or odd, one of the three fractions p_{n-1}/q_{n-1}, p_n/q_n, and $[a_0, a_1, \cdots, a_{m-1}, a_m']$ satisfies (*) with

$$\xi = \left[\left(\frac{q}{p}+\tau\frac{p}{q}+\frac{\tau}{pq}\right)^2-\frac{4\tau^2}{q^2}\right]^{1/2}$$

or $\xi=[(\tau q/p+p/q+1/pq)^2-4/q^2]^{1/2}$, respectively. {Cf. S. Fukasawa [Japan. J. Math. **2** (1925), 101–114].}
K. Mahler (Canberra)

Citations: MR 32# 1167 = J08-24.

J12 SIMULTANEOUS HOMOGENEOUS APPROXIMATION; ONE LINEAR FORM

See also Section A60.

See also reviews A04-4, A58-3, A60-15, A60-17, H15-42, H15-66, H15-76, J04-35, J04-67, J16-2, J16-18, J16-27, J24-13, J24-36, J36-25, J56-2, J56-3, J56-4, J64-1, J64-6, J64-8, J64-15, J64-18, J64-28, J68-30, J68-39, J68-44, J68-46, J68-52, J68-56, J68-62, J84-41, J99-7, M50-7, R04-4, Z10-4, Z10-20.

J12-1 **(2, 253a)**

Koksma, J. F. und Meulenbeld, B. Ueber die Approximation einer homogenen Linearform an die Null. Nederl. Akad. Wetensch., Proc. **44**, 62–74 (1941).

Let S_n be any system of n real numbers $\alpha_1, \alpha_2, \cdots, \alpha_n$ ($n \geqq 1$, fixed); and let

$$L_n = \alpha_1 x_1 + \alpha_2 x_2 + \cdots + \alpha_n x_n - y,$$

where $(x_1, x_2, \cdots, x_n, y)$ is a lattice-point, and

$$X = \max(|x_1|, |x_2|, \cdots, |x_n|) \geqq 1.$$

L_n is said to admit the approximation $\phi(X)$ if there are infinitely many systems $(x_1, x_2, \cdots, x_n, y)$ for which

$$|\alpha_1 x_1 + \alpha_2 x_2 + \cdots + \alpha_n x_n - y| < \phi(X).$$

It is well known that L_n admits the approximation $1/X^n$ ($n \geqq 1$). The authors show that L_n admits the approximation $1/c_n X^n$ if

$$c_n \leqq (1+1/n)^n \{1+((n-1)/(n+1))^{n+3}\}.$$

The method is geometrical. By means of a lemma the authors are able to translate to linear forms L_n the volume calculations used by H. F. Blichfeldt [Trans. Amer. Math. Soc. **15**, 227–235 (1914)] to prove the analogous result concerning the simultaneous rational approximation of S_n.

D. C. Spencer (Cambridge, Mass.).

Referred to in J12-18, J56-3, J56-6.

J12-2 **(2, 350b)**

Robinson, Raphael M. On the simultaneous approximation of two real numbers. Bull. Amer. Math. Soc. **47**, 512–513 (1941).

Using Dirichlet's "pigeonhole principle" the author proves that for every pair of real numbers ξ_1 and ξ_2 and for every positive integer s there exist three integers a_1, a_2 and b such that $0 < b \leqq s$ and

$$|b\xi_k - a_k| \leqq \max\left(\frac{[s^{\frac{1}{2}}]}{s+1}, \frac{1}{[s^{\frac{1}{2}}]+1}\right), \qquad k=1, 2.$$

This pair of simultaneous approximations is the best possible in the sense that for every s there exist values of ξ_1 and ξ_2 such that one of these inequalities would not be satisfied if the sign of equality were removed. The corresponding problem for the simultaneous approximation of more than two real numbers "appears more difficult."

D. H. Lehmer (Berkeley, Calif.).

J12-3 **(7, 245b)**

Linés Escardó, Enrique. Aplicaciones de la Teoria de Redes Regulares al Estudio de las Funciones Cuasiperiodicas. [Applications of the Theory of Regular Nets to the Study of Quasiperiodic Functions]. Consejo Superior de Investigaciones Cientificas, Madrid, 1943. 79 pp. (Spanish)

Let $f(x_0, \cdots, x_n)$ be a real continuous function of period 1 in each of the variables $x_0, \cdots, x_n$; let the real numbers $\alpha_0, \cdots, \alpha_n$, 1 be linearly independent over the rational field; and let $k_0, \cdots, k_n$ be arbitrary real numbers. The author shows by means of Kronecker's theorem (for which a proof is given) that $\varphi(t) = f(\alpha_0 t + k_0, \cdots, \alpha_n t + k_n)$ is an almost periodic function of t, and that its Fourier series is

$$\varphi(t) \sim \sum_{\mu_0, \cdots, \mu_n} A_{\mu_0, \cdots, \mu_n} \exp\{2\pi i(\mu_0 k_0 + \cdots + \mu_n k_n)\} \times \exp\{2\pi i(\mu_0\alpha_0 + \cdots + \mu_n\alpha_n)t\},$$

where $A_{\mu_0, \cdots, \mu_n}$ are the Fourier coefficients of $f(x_0, \cdots, x_n)$. Hence the characteristic exponents of $\varphi(t)$ form a module of finite base. The author deduces properties of $\varphi(t)$ from those of $f(x_0, \cdots, x_n)$, and also shows how generalized almost periodic functions can be obtained when $f(x_0, \cdots, x_n)$ is subjected to less restrictive conditions. The monograph concludes with applications to special ergodic problems and with a constructive method, by means of continued fractions, for solving inequalities

$$|\xi_1 - a\eta| < \epsilon_1, \quad |\xi_2 - b\eta| < \epsilon_2, \cdots, |\xi_n - l\eta| < \epsilon_n$$

in integers $\xi_1, \cdots, \xi_n, \eta$. *K. Mahler* (Manchester).

J12-4 **(7, 274b)**

Pipping, Nils. Approximation mehrerer reellen Zahlen durch rationale Zahlen mit gemeinsamem Nenner. Acta Acad. Aboensis **13**, no. 9, 12 pp. (1942).

Dirichlet proved that, if $\xi_1, \cdots, \xi_n$ denote $n \geqq 1$ arbitrary real numbers and m denotes any positive integer, n fractions $p_1/q, \cdots, p_n/q$ ($q \geqq 1$) exist, such that $|\xi_\nu - p_\nu/q| \leqq q^{-m}$ ($\nu = 1, \cdots, n$), $q \leqq m^n$ [cf. the reviewer's Diophantische Approximationen, Ergebnisse der Math. **4**, no. 4, Springer, Berlin, 1936, chapters 1, 4]. By a refinement of Dirichlet's argument ("Schubfachprinzip") the author proves that, if $m \geqq 2$, the inequality $q \leqq m^n$ can be replaced by $q \leqq m^n - 2^n + 1$. By a geometrical interpretation of the method he proves a further improvement in the case $n=2$, $m=3$, namely $q \leqq 5$. Finally, he considers the approximation $\sum_{\nu=1}^{n} |q\xi_\nu - p_\nu| \leqq 1$. From the above theorem it follows that this approximation always can be realized with integers $p_1, \cdots, p_n, q$ such that $1 \leqq q \leqq n^n - 2^n + 1$. Improvements are given in special cases.

J. F. Koksma (Amsterdam).

J12-5 **(7, 365a)**

van der Pol, Balth. Music and elementary theory of numbers. Music Review **7**, 1–25 (1946).

J12-6 **(7, 506f)**

Davenport, H., and Mahler, K. Simultaneous Diophantine approximation. Duke Math. J. **13**, 105–111 (1946).

Theorem 1(a). If $c > 2/\sqrt{23}$ and α, β are any two irrational numbers, then there exist an infinity of fractions p/r, q/r for which $(\alpha - p/r)^2 + (\beta - q/r)^2 < c/r^3$. (b) This is false if $c < 2/\sqrt{23}$. Theorem 2(a). If $c > 2/\sqrt{23}$ and α, β are any two real numbers for which $1, \alpha, \beta$ are linearly independent, then there exist an infinity of sets of integers p, q, r, satisfying $|\alpha p + \beta q + r| < c/(p^2+q^2)$, $p^2+q^2 > 0$. (b) This is false if $c < 2/\sqrt{23}$.

This result is interesting because the problem of finding the best possible value of $c > 0$ for which the simultaneous inequalities

$$|\alpha - p/r| < c/r^{\frac{3}{2}}, \quad |\beta - q/r| < c/r^{\frac{3}{2}}$$

have an infinity of rational solutions p/r, q/r ($r > 0$) is still unsolved. The proof of (a) in both cases is based on investigations on star bodies by Mahler [in course of publication] and on a theorem of Davenport on three dimensional lattices [Proc. London Math. Soc. (2) **45**, 98–125 (1939)]. The proof of (b) in each case is based on the properties of the cubic field $k(\phi)$, where ϕ is the real root of $t^3-t-1=0$.

J. F. Koksma (Amsterdam).

Referred to in J12-18, J12-24, J12-27.

J12-7 (9, 135c)

Stoll, A. Das Proportionalwahl-Problem als diophantische Näherungsaufgabe. Vierteljschr. Naturforsch. Ges. Zürich **92**, 204–212 (1947).

With elections under a many party system one faces the problem of a fair apportionment of seats to the various parties [or states in the U. S. House of Representatives]. Let p_k be the fraction of votes cast for the kth party [population in the kth state]. The problem consists in finding integers n_k such that $\sum n_k=n$ and that the system $\{n_k\}$ comes as near as possible to the system $\{np_k\}$. The deviation of n_k from np_k is measured by a convex function $f_k(n_k)$ and the total deviation of $\{n_k\}$ from $\{np_k\}$ is defined as the sum of the individual deviations. Typical examples are $f_k(n_k)=|n_k-np_k|$, $(n_k-np_k)^2$ and $(n_k-np_k)^2/np_k$. The optimal solution is found for these and other cases.

W. Feller (Ithaca, N. Y.).

J12-8 (10, 284b)

Barbour, J. M. Music and ternary continued fractions. Amer. Math. Monthly **55**, 545–555 (1948).

This paper discusses the problem of the division of the octave into equally tempered intervals by means of continued fractions. There is an interesting historical account of the early and modern attempts to achieve a perfect tuning system. Objections to the 12 tone equally tempered system in common use have been raised by many writers. The chief trouble with the system is the sharpness of the major third. In fact with equal temperament the ratio $(\log \frac{5}{4})/\log 2=.3219281$ is to be compared with $\frac{4}{12}=.333\cdots$. The latter is too high by more than 3 percent. For the fifth the ratio $(\log \frac{3}{2})/\log 2=.5849625$ is to be compared with $\frac{7}{12}=.58333\cdots$. A division of the octave into more than 12 intervals is necessary to improve these approximations. Previous attempts have employed ordinary continued fractions rational approximations to $(\log \frac{3}{2})/\log 2$, obtained as usual from convergents and semi-convergents. The approximate third is then made to fit as well as possible.

The author submits that this unsymmetrical procedure is opposed to the fact that today's music is based on the major triad and holds that the irrational ratios (1) $\log \frac{5}{4}:\log \frac{3}{2}:\log 2$ should be approximated simultaneously without favoring one ratio over another. This calls for the use of ternary continued fractions. However, the ordinary Jacobi algorithm converges too rapidly. This means that one very soon reaches approximations involving the division of the octave into more than 100 parts. The author gives two modifications of the Jacobi algorithm which reduce its rate of convergence and thus obtains as many as 16 best approximations $A:B:C$ to (1) with $C<100$. The tenth of these is the standard 4:7:12. Others are 6:11:19, 7:13:22, 10:18:31. It is suggested that these modifications of Jacobi's algorithm have other applications. *D. H. Lehmer.*

Referred to in J12-19, J12-20, J12-37.

J12-9 (10, 512e)

Hinčin, A. Ya. On the fractional parts of a linear form. Izvestiya Akad. Nauk SSSR. Ser. Mat. **13**, 3–8 (1949). (Russian)

Let $\{z\}$ denote the distance of a real number z from the nearest integer. Let $S(x)=\theta_1x_1+\cdots+\theta_nx_n$ be a linear form with real coefficients in n integer variables x_i, components of an n-dimensional vector x. Let $N(x)=\max(|x_1|,\cdots,|x_n|)$. Then a classical theorem of Dirichlet [see J. F. Koksma, Diophantische Approximationen, Springer, Berlin, 1936, p. 5] states that for every $t\geqq 1$ there exists an integral vector x such that $\{S(x)\}<t^{-1}$, $0<N(x)\leqq t^{1/n}$. The author proves the following extension of Dirichlet's theorem. For every $t\geqq 1$ there exist n linearly independent integral vectors $x^1,\cdots,x^n$ such that

$$\{S(x^i)\}<t^{-1},\qquad i=1,\cdots,n,$$

$$0<\prod_{i=1}^{n}N(x^i)\leqq c_nt,$$

where c_n is a constant depending only on n. The proof is elementary, making repeated use of the Dirichlet "Schubfachprinzip." *F. J. Dyson* (London).

J12-10 (10, 513c)

Obrechkoff, Nikola. Sur l'approximation des nombres irrationnels. C. R. Acad. Sci. Paris **228**, 352–353 (1949).

Si $0<\omega<1$ et si n est un nombre entier positif, il existe deux nombres entiers x et y tels que $|\omega x-y|<1/(n+1)$, $1\leqq x\leqq n$. Cette amélioration du théorème de Dirichlet se trouve déjà chez Minkowski [Koksma, Diophantische Approximationen, Springer, Berlin, 1936, pp. 5–6]. Si ω se trouve en dehors de quelques intervalles indiquées par l'auteur (dépendant de n) on peut remplacer dans l'inégalité mentionnée $1/(n+1)$ par $1/(n+2)$ (démontré à l'aide d'une suite spéciale de Farey). L'auteur donne un énoncé semblable pour l'inégalité

$$|\omega_1x_1+\cdots+\omega_mx_m-z|\leqq(n+1)^{-m},\qquad |x_i|\leqq n.$$

J. F. Koksma (Amsterdam).

J12-11 (10, 514a)

Todd, H. On Diophantine approximation to certain exponential and Bessel functions. Proc. London Math. Soc. (2) **50**, 550–559 (1949).

It is shown that solutions of a certain type of linear difference equation of order $s+1$ provide simultaneous rational approximations to s irrationals, which are determined by the coefficients of the difference equation and the initial values chosen for the formation of the solutions. Applications are given for the case in which the irrationals thus defined are combinations of exponential or of Bessel functions. Theorems and examples are too complicated to be quoted here in detail. *J. F. Koksma* (Amsterdam).

J12-12 (11, 82i)

Obrechkoff, Nikola. Sur l'approximation diophantique linéaire. Atti Accad. Naz. Lincei. Rend. Cl. Sci. Fis. Mat. Nat. (8) **6**, 283–285 (1949).

By a modification of the Schubfachprinzip the following sharp form of Dirichlet's theorem is proved. Let $\omega_1,\cdots,\omega_m$ be arbitrary real numbers and let n be a positive integer. Then there exist integers $x_1,\cdots,x_m$ not all zero and an integer y such that $|x_i|\leqq n$ for $i=1,\cdots,n$ and

$$\left|\sum_{i=1}^{n}\omega_ix_i-y\right|\leqq(n+1)^{-m};$$

the equality sign holds if and only if the numbers ω_i are all rational and have, in some order, the denominators $n+1,(n+1)^2,\cdots,(n+1)^m$. The author also shows that, if $a_1,\cdots,a_n$ are integers, a necessary and sufficient condition that the form $\sum_1^m a_ix_i$ represent every integer (mod n^m) when the x_i vary independently over a complete residue system (mod n) is that the a's, in some order, are numbers b_in^i, $i=0,\cdots,m-1$, where the numbers b_i are all prime to n.

W. J. LeVeque (Ann Arbor, Mich.).

J12-13 (11, 583e)

Jarník, Vojtěch. Une remarque sur les approximations diophantiennes linéaires. Acta Sci. Math. Szeged **12**, Leopoldo Fejér et Frederico Riesz LXX annos natis dedicatus, Pars B, 82–86 (1950).

Let $S=\{\theta_1, \theta_2, \cdots, \theta_r\}$ be a system of r real numbers, and let $t\geqq 1$. Put $\psi_s(t)=\min |a_1\theta_1+\cdots+a_r\theta_r+a_0|$ where the minimum extends over all integers a_i for which

$$0<\max(|a_1|, \cdots, |a_r|)\leqq t.$$

Assume $\psi_s(t)>0$ for all $t\geqq 1$, so that $1, \theta_1, \theta_2, \cdots, \theta_r$ are linearly independent over the rational field. The author proves the following results. (1) If $A=\liminf_{t\to\infty} t^r\psi_s(t)=A>0$, $B=\limsup_{t\to\infty} t^r\psi_s(t)$, then $(B/A)^{2^r}\geqq 2$. (2) If $r=2$, then $A\leqq 36B^3$, which is better than (1) when B is small. The inequality (2) is an immediate consequence of the following result. (3) Let $r=2$. Denote by $\varphi(t)$ a continuous steadily decreasing function of $t\geqq 1$ such that $t\varphi(t)\to 0$ as $t\to\infty$. If $\psi_s(t)<\varphi(t)$ for all sufficiently large t, then there exist arbitrarily large t such that $\psi_s(t)<\varphi(1/6t\varphi(t))$.

K. Mahler (Manchester).

J12-14 (12, 82e)

Prasad, A. V. Simultaneous Diophantine approximation. Proc. Indian Acad. Sci., Sect. A. **31**, 1–15 (1950).

Let $K_n^{(X)}$ be the region in $(n+1)$-dimensional space defined by $n^{-1}\sum_{r=1}^{n} x_r^2|x_0|^{2/n}\leqq 1$, $n^{-1}\sum_{r=1}^{n} x_r^2\leqq X^2$. It is shown for certain numbers C_n'' with $C_n''\sim n^{\frac{1}{2}}/(\pi e)^{\frac{1}{2}n}\ 3.763\ldots$ that, provided X is sufficiently large, every lattice with determinant less than $(C_n'')^{-1}$ has a point other than the origin in $K_n^{(X)}$. In this result the constant $3.763\ldots$ is larger than the constant $2e^{\frac{1}{2}}=3.297\ldots$ which can be obtained by applying a result of Blichfeldt [Trans. Amer. Math. Soc. **15**, 227–235 (1914), theorem II] to an $(n+1)$-dimensional sphere inscribed in $K_n^{(X)}$. By simple applications of his result the author obtains a theorem on the simultaneous Diophantine approximations to n irrational numbers and a theorem on the approximate solution in integers of a homogeneous linear equation in $n+1$ variables.

C. A. Rogers (London).

J12-15 (12, 82f)

Raisbeck, Gordon. Simultaneous Diophantine approximation. Canadian J. Math. **2**, 283–288 (1950).

Given any set of real numbers $z_1, z_2, \cdots, z_n$ and an integer t we can find an integer q, $1\leqq q\leqq t^{n-1}(t-1)-2^{n-1}+1$, and a set of integers $p_1, p_2, \cdots, p_n$ such that $|qz_j-p_j|\leqq 1/t$, $j=1, 2, \cdots, n$. This result is sharper than the classical results of Dirichlet-Minkowski and than a theorem of Pipping [Acta Acad. Aboensis **13**, no. 9 (1942); these Rev. **7**, 274]. For $n=2$ the author proves that his result is the best possible.

J. F. Koksma (Amsterdam).

J12-16 (12, 163b)

Černý, Karel. Contribution à la théorie des approximations diophantiques simultanées. Acta Fac. Nat. Univ. Carol., Prague no. **188**, 27 pp. (1948). (French. Czech summary)

The author's main result is as follows. Let s be a positive integer greater than 1. Let $\omega(x)$ be positive and continuous and $\omega^s(x)x^{s+1}$ monotonic for $x\geqq 1$. Suppose that $\int_1^\infty \omega^s(x)x^s dx$ converges and $\int_1^\infty \omega^{s-1}(x)x^{s-1}dx$ diverges. Let $\tau(x)\geqq x$ be defined for $x\geqq 1$, with $\tau(x)/x\to\infty$ as $x\to\infty$. Then there exists a real number Θ_s such that, for almost all real numbers $\Theta_1, \cdots, \Theta_{s-1}$, the system $(\Theta_1, \cdots, \Theta_s)$ admits simultaneous Diophantine approximation with the function $\omega(x)$, but does not admit it with the function $\omega(\tau(x))$. This result is an extension of one due to Jarník [cf. Koksma, Diophantische Approximationen, Springer, Berlin, 1936, chapter 5, theorem 13], which asserted the existence of a system $(\Theta_1, \cdots, \Theta_s)$ with these properties. The construction of Θ_s is effected by a continued fraction, and the set-theoretic part of the proof is less complicated than was the case with Jarník's original proof.

H. Davenport (London).

Referred to in J16-28.

J12-17 (12, 163d)

Obreškov, N. On Diophantine approximations of linear forms for positive values of the variables. Doklady Akad. Nauk SSSR (N.S.) **73**, 21–24 (1950). (Russian)

The author proves the following simple but elegant variation on a well known result on Diophantine approximation. Let $\omega_1, \cdots, \omega_k$ be real numbers, and n a positive integer. Then there exist integers $x_1, \cdots, x_k$ (not all zero) and y, such that $0\leqq x_i\leqq n$ and (*) $|\omega_1x_1+\cdots+\omega_kx_k+y|\leqq N^{-1}$, where $N=kn+1$. The proof is by Dirichlet's principle. Give $x_1, \cdots, x_k$ all sets of values satisfying $0\leqq x_i\leqq n$, subject to the condition that if $x_i>0$ then $x_1=\cdots=x_{i-1}=n$. There are N such sets of values, and if they are arranged in lexicographical order, then $x_i'-x_i\geqq 0$ for $i=1, \cdots, k$ if the set x_i precedes the set x_i'. Considering the N values of $\omega_1x_1+\cdots+\omega_kx_k$ (mod 1), it is plain that there must be two sets x_i and x_i' for which the corresponding values of the linear form differ (mod 1) by at most $1/N$. This gives the result. The author further proves that the sign of equality in (*) is necessary if and only if (**) $\omega_1\equiv\cdots\equiv\omega_k\equiv\lambda/N$ (mod 1), for some integer λ relatively prime to N. He gives an extension to m linear forms $\varphi_i=a_{i1}x_1+\cdots+a_{ik}x_k$, $i=1, 2, \cdots, m$, the result being that there exist integers $x_1, \cdots, x_k$ (not all zero) and $y_1, \cdots, y_m$ such that $0\leqq x_i\leqq n$ and $|\varphi_i+y_i|\leqq N^{-1/m}$ for $i=1, \cdots, m$. The previous argument extends at once to this case, on considering points $(\varphi_1, \cdots, \varphi_m)$ whose coordinates are treated mod 1. Another extension of the first result is given to the case when the inequalities $0\leqq x_i\leqq n$ are replaced by $0\leqq x_i\leqq n_i$, and N is taken to be $n_1+\cdots+n_i+1$. The author states that the cases of equality are again those given by (**). This does not seem obvious to the reviewer, as the proof given in the previous case does not now apply.

H. Davenport (London).

Referred to in J12-22.

J12-18 (12, 245c)

Mullender, P. Simultaneous approximation. Ann. of Math. (2) **52**, 417–426 (1950).

The author deals with the following problems: (1) to give the lower bound of the numbers $c>0$ for which the simultaneous inequalities $|x|\geqq 1$, $|\alpha-y/x|<c^{\frac{1}{2}}/|x|^{\frac{3}{2}}$, $|\beta-z/x|<c^{\frac{1}{2}}/|x|^{\frac{3}{2}}$ for any real α and β have an infinity of integral solutions x, y, z; and (2) to give the lower bound of the numbers $c>0$ for which the simultaneous inequalities $X=\max(|y|, |z|)\geqq 1$, $|\alpha y+\beta z-x|<c/X^2$ for any real α and β have an infinity of integral solutions x, y, z. These problems still seem far from a solution, but several authors have given estimates for admissible values of c [Minkowski, Geometrie der Zahlen, Teubner, Leipzig-Berlin, 1910; Blichfeldt, Trans. Amer. Math. Soc. **15**, 227–235 (1914); Koksma and Meulenbeld, Nederl. Akad. Wetensch., Proc. **44**, 62–74 (1941); Mullender, ibid. **50**, 173–185 (1947); **51**, 874–884 (1948) = Indagationes Math. **9**, 136–148 (1947); **10**, 302–312 (1948); Davenport and Mahler, Duke Math. J. **13**, 105–111 (1946); these Rev. **2**, 253; **9**, 335; **10**, 285; **7**, 506]. By an ingenious method the author now improves all known estimates, proving that in both problems, each value of $c>2^7/3^4\cdot 3^{\frac{1}{2}}\cdot 23^{\frac{1}{4}}=1/2.400$ is admissible. This result is remarkable in view of a related problem for which Davenport and Mahler [loc. cit.] proved the value $2/23^{\frac{1}{2}}=1/2.398$ to be the exact lower bound of c. By using results of Furtwängler [Math. Ann. **96**, 169–175 (1926); **99**, 71–83 (1928)] and the author [first reference cited] it is shown that the exact lower bound of c in any case is not less than $23^{-\frac{1}{2}}$.

J. F. Koksma (Amsterdam).

Citations: MR 2, 253a = J12-1; MR 7, 506f = J12-6; MR 9, 335c = J56-7; MR 10, 285c = H15-25.

J12-19 (12, 675e)

Rosser, J. B. Generalized ternary continued fractions. Amer. Math. Monthly **57**, 528–535 (1950).

This paper illustrates the application of the author's algorithm [same Monthly **48**, 662–666 (1941); these Rev. **3**, 161] for solving the linear Diophantine equation to the problem of simultaneous approximation of irrationals involved in the tempering of musical scales treated previously by Barbour [ibid. **55**, 545–555 (1948); these Rev. **10**, 284]. The method consists in taking simple linear combinations of previously obtained approximations in such a way as to minimize the absolute values of the coefficients involved. As a result, the author obtains 9 possible methods of tempering a scale with an "octave" of n notes with $12<n\leqq 118$.

D. H. Lehmer (Berkeley, Calif.).

Citations: MR 3, 161b = D04-3; MR 10, 284b = J12-8.

J12-20 (12, 675f)

Brun, Viggo. Music and ternary continued fractions. Norske Vid. Selsk. Forh., Trondheim **23**, 38–40 (1950).

This paper is concerned with a problem discussed by J. M. Barbour [Amer. Math. Monthly **55**, 545–555 (1948); these Rev. **10**, 284] of the division of the octave into equally tempered intervals. The problem is that of simultaneously approximating the irrational ratios $\log \frac{5}{4} : \log \frac{3}{2} : \log 2$ by integers $x:y:z$. The author points out that an algorithm given by him in 1919 gives a simple solution of the problem and gives the results of Barbour (and a little more) for this particular problem. The algorithm is as follows: Given a set of three real numbers a, b, c, each with an associated triplet of integers

$$\begin{array}{ll} a & (\alpha_1, \alpha_2, \alpha_3) \\ b & (\beta_1, \beta_2, \beta_3) \\ c & (\gamma_1, \gamma_2, \gamma_3) \end{array}$$

we replace the row corresponding to the greatest of a, b, c by a row whose number is the difference between the greatest and the midmost of a, b, c and whose triplet is formed by adding the corresponding elements of the triplets belonging to the greatest and midmost of a, b, c. The process is now repeated. Initially, the numbers a, b, c, are given and have triplets $(1, 0, 0)$, $(0, 1, 0)$, $(0, 0, 1)$. Quotations are given from correspondence between Stieltjes and Hermite pointing out defects in the Jacobi ternary continued fraction method of solving the problem. *D. H. Lehmer.*

Citations: MR 10, 284b = J12-8.
Referred to in A60-20.

J12-21 (13, 727g)

Obrechkoff, N. Sur l'approximation diophantique linéaire. C. R. Acad. Bulgare Sci. **3**, no. 2–3 (1950), 1–4 (1951). (French. Russian summary)

Let $\omega_1, \cdots, \omega_m$ be real, and let n be a positive integer. It is well known that there exist integers $x_1, \cdots, x_m$ (not all 0) and y such that $|x_i|\leqq n$ and

$$|\omega_1 x_1+\cdots+\omega_m x_m-y|\leqq(n+1)^{-n}.$$

The author proves that the sign of equality is necessary in the last inequality if and only if $\omega_1, \cdots, \omega_m$ are a permutation of $\lambda_1/(n+1), \lambda_2/(n+1)^2, \cdots, \lambda_m/(n+1)^m$, where $\lambda_1, \cdots, \lambda_m$ are integers relatively prime to $n+1$.

H. Davenport (London).

J12-22 (13, 921a)

Obrechkoff, Nikola. Sur l'approximation diophantique des formes linéaires pour des valeurs positives des variables. Annuaire [Godišnik] Univ. Sofia. Fac. Sci. Livre 1. **46**, 343–356 (1950). (Bulgarian. French summary)

The present paper contains the detailed proofs of the results announced, and proved in outline, in an earlier note [Doklady Akad. Nauk SSSR (N.S.) **73**, 21–24 (1950); these Rev. **12**, 163], together with a further slight generalization. The doubt concerning one particular point, expressed in the review of the earlier note, is removed by the detailed proof now given. *H. Davenport* (London).

Citations: MR 12, 163d = J12-17.
Referred to in J12-23.

J12-23 (14, 1067a)

Obrechkoff, N. Sur l'approximation diophantique linéaire pour des valeurs positives des variables. C. R. Acad. Bulgare Sci. **4** (1951), no. 1, 1–4 (1953). (Russian summary)

This is the same in substance as Annuaire [Godišnik] Univ. Sofia. Fac. Sci. Livre 1. **46**, 343–356 (1950); these Rev. **13**, 921. *H. Davenport* (London).

Citations: MR 13, 921a = J12-22.

J12-24 (14, 956d)

Davenport, H. Simultaneous Diophantine approximation. Proc. London Math. Soc. (3) **2**, 406–416 (1952).

Let α, β be given real numbers. What is the greatest lower bound c_0 of all constants $c>0$ for which the simultaneous inequalities $r(p-\alpha r)^2<c$, $r(q-\beta r)^2<c$ have infinitely many integral solutions p, q, r $(r>0)$? It was shown by Furtwängler that $c_0\geqq 1/23^{1/2}$ and by Davenport and Mahler that $c_0\leqq 2/23^{1/2}$ [Duke Math. J. **13**, 105–111 (1946); these Rev. **7**, 506], whereas Mullender proved

$$c_0\leqq 2^7/3^{9/2}23^{1/2}=1/2.4003\cdots$$

by means of the geometry of numbers. Now using in principle Mullender's method the author proves

$$c_0\leqq 1/46^{1/4}=1/2.6043\cdots.$$

At the end of the paper the author makes some remarks concerning the possibilities of improving his estimate.

J. F. Koksma (Amsterdam).

Citations: MR 7, 506f = J12-6.

J12-25 (15, 857e)

Černy, Karel. Sur les approximations diophantiennes. Čehoslovack. Mat. Ž. **2**(77), 191–220 (1952). (Russian. French summary)

The author proves, with the aid of the theory of Lebesgue measure, a result of Jarník based on difficult considerations involving Hausdorff measure [Math. Z. **33**, 505–543 (1931)]. Suppose that $s>1$, $\omega(x)$ is a positive continuous function for $x\geqq 1$, and $\omega^s(x)x^{s+1}$ is monotonic for $x\geqq 1$. Suppose also that $\int_1^\infty \omega^s(x)x^s dx$ and $\int_1^\infty \omega^{s-1}(x)x^{s-1}dx$ are respectively convergent and divergent. Finally, let $\tau(x)$ be a function defined for $x\geqq 1$ such that $\tau(x)/x\geqq 1$, and $\tau(x)/x\to\infty$ as $x\to\infty$. Then there exists a number Θ_s such that, for almost all points $(\Theta_1, \Theta_2, \cdots, \Theta_{s-1})$ of Euclidean $(s-1)$-dimensional space, $(\Theta_1, \Theta_2, \cdots, \Theta_s)$ is a proper system and admits the approximation $\omega(x)$ but not $\omega(\tau(x))$. Here $(\Theta_1, \Theta_2, \cdots, \Theta_s)$ is said to be proper if $k_0+k_1\Theta_1+k_2\Theta_2+\cdots+k_s\Theta_s=0$, for integral k_i, implies $k_0=k_1=\cdots=k_s=0$. In the proof Θ_s is constructed as a continued fraction whose convergents satisfy certain elaborate conditions.

From this result the following is deduced: Let $s\geqq 1$, and let $\omega(x)$, $\lambda(x)$ be two functions which are positive and continuous for $x\geqq 1$. Let the functions $\omega(x)x^2, \omega^2(x)x^3, \cdots, \omega^s(x)x^{s+1}, \omega(x)x^{2+\epsilon}$ (for a certain $\epsilon>0$), $\lambda(x)$ be monotonic and let $\lambda(x)\to 0$ for $x\to\infty$. Also suppose that $\int_1^\infty \omega^s(x)x^s dx$ is convergent. Then there exists a proper system

$$(\Theta_1, \Theta_2, \cdots, \Theta_s)$$

which admits the approximation $\omega(x)$, but not $\omega(x)\lambda(x)$.

R. A. Rankin (Birmingham).

Referred to in J16-28.

J12-26 (16, 223a)

Davenport, H. Simultaneous Diophantine approximation. Mathematika **1**, 51–72 (1954).

If θ and ϕ are irrationals, then there are infinitely many pairs of rational numbers p/r, q/r, such that

$$|\theta-p/r|<r^{-3/2}, \quad |\phi-q/r|<r^{-3/2} \quad (r>0),$$

as may be proved by Dirichlet's "Schubfachprinzip" [for references cf. Koksma, Diophantische Approximationen, Springer, Berlin, 1936]. Perron proved that for a suitable constant $c>0$ there exist pairs θ, ϕ, such that for all fractions p/r, q/r the inequalities

$$(1) \qquad |\theta-p/r|>cr^{-3/2}, \quad |\phi-q/r|>cr^{-3/2} \quad (r>0)$$

hold. (By a theorem of Khintchine it is a priori clear that the set of all such pairs (θ, ϕ) in the cartesian plane has the Lebesque measure 0.) The author now proves that for a suitable constant $c>0$ there exist continuum-many distinct pairs θ, ϕ of the last named kind. In his proof, which uses sequences of unimodular substitutions, he first considers the inequality

$$(2) \qquad (p^2+q^2)|\theta p+\phi q+r|>c',$$

constructing pairs θ, ϕ for which (2) holds for all integers p, q, r with $p^2+q^2>0$, even for $c'=1/200$. Then by a well known "Übertragungsprinzip" it can easily be proved that for each such couple also (1) holds (with suitable c).

J. F. Koksma (Amsterdam).

Referred to in J12-31, J16-29, J84-41.

J12-27 (16, 574d)

Cassels, J. W. S. Simultaneous Diophantine approximation. J. London Math. Soc. **30**, 119–121 (1955).

Let C denote the upper bound of those positive constants c for which the inequalities

$$\left|\alpha-\frac{p}{r}\right|\leqq c^{-1/2}r^{-3/2}; \quad \left|\beta-\frac{q}{r}\right|\leqq c^{-1/2}r^{-3/2}$$

for all couples of real numbers α, β have an infinity of integer solutions $r>0$, p, q. It is shown that $C\leqq 7/2$. First it is remarked that the critical determinant of the domain $|X|\max(Y^2, Z^2)\leqq 1$ is C and then that for any three linear forms L_i $(i=1, 2, 3)$ giving a critical lattice for $|XYZ|\leqq 1$, so that they have determinant 7, the related forms

$$M_1=L_1, \quad M_2=\tfrac{1}{2}(L_2+L_3), \quad M_3=\tfrac{1}{2}(L_2-L_3)$$

have determinant 7/2 and give an admissible lattice for the above mentioned domain. The proof of the first remark depends on a modification of an argument due to Davenport and Mahler [Duke Math. J. **13**, 105–111 (1946); MR **7**, 506].

J. F. Koksma (Amsterdam).

Citations: MR 7, 506f = J12-6.

Referred to in J12-31.

J12-28 (16, 803a)

Davenport, H. On a theorem of Furtwängler. J. London Math. Soc. **30**, 186–195 (1955).

Let $F(x_1, \cdots, x_{n-1})$ be the distance function of a bounded star body in $n-1$ dimensions, so that F is continuous, homogeneous of degree 1, and positive except at $x_1=\cdots=x_{n-1}=0$. Let $C(F)$ be the inf of the constants C such that for any real $\theta_1, \cdots, \theta_{n-1}$ there are infinitely many approximations $q_1/q, \cdots, q_{n-1}/q$ satisfying

$$F(q\theta_1-q_1, \cdots, q\theta_{n-1}-q_{n-1})<(C/q)^{1/(n-1)}.$$

Finally, let K be the n-dimensional region defined by

$$\{F(x_1, \cdots, x_{n-1})\}^{n-1}|x_n|\leqq 1.$$

The main theorem is that $C(F)=1/\Delta(K)$, where $\Delta(K)$ is the critical determinant of K. The method is similar to, but simpler than, that devised by P. Furtwängler [Math. Ann. **96**, 169–175 (1926); **99**, 71–83 (1928)], who proved a consequence of the above theorem, namely that if

$$F(x_1, \cdots, x_{n-1})=\max(|x_1|, \cdots, |x_{n-1}|),$$

then $C(F)\geqq |d|^{-1/2}$, where d is the numerically smallest discriminant of any real algebraic number field of degree n. It is also shown that the inf of the numbers C' such that the inequality

$$|\theta_1 p_1+\cdots+\theta_{n-1}p_{n-1}+p|<C'\{F(p_1, \cdots, p_{n-1})\}^{-n+1}$$

has infinitely many solutions, is again $1/\Delta(K)$.

W. J. LeVeque (Ann Arbor, Mich.).

Referred to in H05-68.

J12-29 (16, 1003b)

Mahler, Kurt. On a problem in Diophantine approximations. Arch. Math. **6**, 208–214 (1955).

Let $L(x)$ be a linear form in n variables $x_1, \cdots, x_n$; assume that $n\geqq 2$, and that the coefficients of this form are real numbers of absolute value not greater than $a\geqq 1$. Denote by v_{n-1} the volume of the $(n-1)$-dimensional polyhedron defined by the inequalities

$$\max(|x_i|)\leqq 1, \quad |x_1+\cdots+x_n|\leqq 1.$$

Then for every real number $N\geqq 1$, there are integers $x_1, \cdots, x_n$ not all zero such that $|L(x)|\leqq N^{-1}$ and

$$\max(|x_i|)\leqq 2(aN/v_{n-1})^{1/(n-1)}.$$

This is an improvement of the usual Dirichlet theorem, in which the right-hand side of the last inequality is $2(anN)^{1/(n-1)}$, since $v_{n-1}\geqq 2$ and $v_{n-1}\sim(3/2\pi n)^{1/2}2^n$ as $n\to\infty$. It is obtained by applying Minkowski's theorem on convex bodies, rather than the Schubfachprinzip.

W. J. LeVeque (Ann Arbor, Mich.).

J12-30 (16, 1090e)

Obrechkoff, N. Sur l'approximation des formes linéaires. Bŭlgar. Akad. Nauk. Izv. Mat. Inst. **1**, no. 2, 35–46 (1954). (Bulgarian. Russian and French summaries)

Let $\omega_1, \cdots, \omega_k$ be any k real numbers $(k\geqq 2)$, and let n be a positive integer. The author proves that there exist integers $x_1, \cdots, x_k$, not all zero, satisfying

$$|x_i|\leqq n \quad (i=1, \cdots, k), \quad |\omega_1 x_1+\cdots+\omega_k x_k|\leqq\frac{n\Omega}{(n+1)^k-1},$$

where $\Omega=|\omega_1|+\cdots+|\omega_k|$. He also shows that the second inequality can be strict unless $\omega_1, \cdots, \omega_k$ have values proportional to $\pm 1, \pm(n+1), \cdots, \pm(n+1)^{k-1}$ in some order. Both results are easily proved. *H. Davenport.*

J12-31 (17, 715a)

Cassels, J. W. S. Simultaneous diophantine approximation. II. Proc. London Math. Soc. (3) **5** (1955), 435–448.

[For part I see J. London Math. Soc. **30** (1955), 119–121; MR **16**, 574.] It is proved that for every integer $N\geqq 1$ there are continuum-many N-tuples of real numbers $\alpha_1, \cdots, \alpha_N$ such that

$$r^{1/N}\max|r\alpha_n-p_n|\geqq\gamma_N \quad (1\leqq n\leqq N)$$

for all integers $r>0$, $p_1, \cdots, p_N$, where $\gamma_N>0$ depends only on N. The case $N=2$ was proved, by a different method, already by Davenport [Mathematika **1** (1954), 51–72; MR **16**, 223]. *J. F. Koksma* (Amsterdam).

Citations: MR 16, 223a = J12-26; MR 16, 574d = J12-27.

Referred to in J16-29.

J12-32 (18, 875d)

Cassels, J. W. S. On a result of Marshall Hall. Mathematika 3 (1956), 109–110.

It has been shown by the reviewer [Ann. of Math. (2) **48** (1947), 966–993; MR **9**, 226] that any real number is the sum (or difference) of two continued fractions with partial quotients at most 4. Given any two real numbers β_1 and β_2, let γ_1 and γ_2 be continued fractions with partial quotients at most 4 such that $\gamma_1-\gamma_2=\beta_1-\beta_2$. Then with $\alpha=\gamma_1-\beta_1=\gamma_2-\beta_2$ we will have $x|(\alpha+\beta_1)x-y|>C$ and $x|(\alpha+\beta_2)x-y|>C$ for all integers $x>0$ and y, where we may take $C=1/5$. This result is generalized here to any number of real numbers β_i. It is shown that if $\beta_1, \cdots, \beta_r$ are any real numbers, then there exists a real number α such that $x|(\alpha+\beta_j)x-y|>C_r$ $(j=1, \cdots, r)$ for all integers $x>0$ and y, where we can take $C_r=\frac{1}{8}(r+1)^2$.

Marshall Hall, Jr. (Columbus, Ohio).

Citations: MR 9, 226b = A54-7.

Referred to in J12-40, J12-41, J16-29.

J12-33 (19, 18g)

Davenport, H. Simultaneous Diophantine approximation. Proceedings of the International Congress of Mathematicians, 1954, Amsterdam, vol. III, pp. 9–12. Erven P. Noordhoff N.V., Groningen; North-Holland Publishing Co., Amsterdam, 1956. $7.00.

A brief survey of the little that is known about the approximation to a set of irrationals $\theta_1, \cdots, \theta_n$ by rational fractions $p_1/q, \cdots, p_n/q$ with a common denominator, other than the so-called "metrical theory".

J. W. S. Cassels (Cambridge, Mass.).

J12-34 (19, 124b)

Barbour, J. M. A geometrical approximation to the roots of numbers. Amer. Math. Monthly **64** (1957), 1–9.

The author considers various methods of approximation to the numbers $2^{r/12}(0<r<12)$ which have been proposed since 1581 [V. Galilei, Dialogo della musica antica e moderna, Florence, 1581], this being of interest as a practical method of achieving equal temperament of the musical scale. An ingenious geometric method put forward by Strähle [Kongl. Swenska Wetenskaps Acad. Handl. **4** (1743), 281–286] has not received its due credit since, when its exact implications were worked out by Faggot [ibid. **4** (1743), 286–291], he made a serious numerical error which implied that the maximum error of the approximation was about 1.7% instead of the correct amount of 0.15%. The author corrects this error and generalizes Strähle's method to give approximation to fractional powers N^m of any positive number N. The method consists essentially of replacing N^m by a bilinear transformation in m with coefficients adjusted to give the correct values at $m=0$, $\frac{1}{2}$, and 1, namely

$$N^m=\frac{Nm+N^{\frac{1}{2}}(1-m)}{m+N^{\frac{1}{2}}(1-m)}.$$

R. A. Rankin (Glasgow).

J12-35 (20# 849)

Obrechkoff, N. Sur une question de l'approximation diophantique des formes linéaires. C. R. Acad. Bulgare Sci. **9** (1956), no. 4, 1–4. (Russian summary)

A proof of the following theorem: "Let α, β, and $r\geqq 1$ be arbitrary real numbers. There exist integers x, y, z not all zero such that

$$|\alpha x+\beta y+z|<\frac{\theta(1+\alpha^2+\beta^2)^{\frac{1}{2}}}{r^2},\ \max(|x|, |y|, |z|)\leqq r$$

if $\theta\geqq 1$; but this is not always true if $\theta<1$." This improves a much weaker result by E. Borel [J. Math. Pures Appl. (5) **9** (1903), 329–375]. *K. Mahler* (Manchester)

J12-36 (20# 2325)

Obrechkoff, Nikola. Sur l'approximation diophantienne des formes linéaires. C. R. Acad. Sci. Paris **246** (1958), 204–205.

An inequality is obtained concerning the approximation of a homogeneous linear form in variables x_μ, divided into p sets, with the condition that in each set the x_μ are of equal sign. The result is best possible.

C. G. Lekkerkerker (Amsterdam)

J12-37 (24# A705)

Brun, Viggo

Music and Euclidean algorithms. (Norwegian. English summary)

Nordisk Mat. Tidskr. **9** (1961), 29–36, 95.

In four problems from music theory, the first and last of which are

$$\frac{\log 2}{x}\approx\frac{\log \frac{3}{2}}{y}$$

and

$$\frac{\log 2}{x}\approx\frac{\log \frac{14}{8}}{y}\approx\frac{\log \frac{12}{8}}{x}\approx\frac{\log \frac{11}{8}}{u}\approx\frac{\log \frac{10}{8}}{v},$$

the simultaneous approximation of certain real numbers by rational numbers with the same denominator is wanted. The author has earlier introduced and studied generalizations of Euclid's algorithm for more than two numbers, in the form of alternating subtractions, whereby the largest number at each step is replaced by the difference between the largest and the second largest number. Application to the problems stated yields sets of increasing pairs, triples, quadruples, and quintuples of integers, respectively, for $x, y, \cdots$ [cf. #A706 below].

The number pairs for the first problem begin (2, 1), (3, 2), (5, 3), (7, 4), (12, 7), $\cdots$; they have already been derived by Euler [*Opera omnia, Series tertia, Vol. I*, Teubner, Leipzig, 1926; p. 197]. The last pair shown is the basis of our musical scale in which twelve fifths equal seven octaves. The second problem, the author points out, has been studied by I. M. Barbour [Amer. Math. Monthly **55** (1948), 545–555; MR **10**, 284] by means of Jacobi's generalization of continued fractions. For the third problem (the second relation given above without the term $\frac{\log \frac{11}{8}}{u}$), A. D. Fokker, in 1947, obtained the quadruple (31, 25, 18, 10) and remarked: "Nous retrouvons le tempérament de Huygens avec trente-et-un cinquièmes de tons dans l'octave" [cf. Huygens, *Œuvres complètes, Vol. XX*, 1940, pp. 139–173]. Fokker, who communicated the last problem to the author, is also experimenting with a special organ based on this scale (cf. the article by M. Vogel in Euler, *Opera omnia, Series tertia, Vol. XI*, Teubner, Leipzig, 1960; pp. xliv–lx [MR **23** #A1492], as well as reference to the *Opera omnia* above). *C. J. Scriba* (Toronto)

Citations: MR 10, 284b = J12-8.

Referred to in A60-20.

J12-38 **(24# A706)**

Selmer, Ernst S.

Continued fractions in several dimensions. (Norwegian. English summary)

Nordisk Nat. Tidskr. **9** (1961), 37–43, 95.

From the summary: "In connection with the preceding article by V. Brun [cf. #A705 above], another subtraction algorithm is introduced, whereby the largest number at each step is replaced by the difference between the largest and the smallest number. This means slower convergence and consequently more convergents, as illustrated for Brun's problem.

"In three dimensions, the two methods are compared with respect to the problem of periodicity of the expansion for $\sqrt[3]{D^2}:\sqrt[3]{D}:1$. Such a periodicity is obtained for $D=2$ with Brun's method, for $D=2$, 3, 4 and 5 with the author's. The cases $D=3$ (Brun) and $D=6$, 7 (Selmer) do at least not show any periodicity within the first 100 steps."

C. J. Scriba (Toronto)

J12-39 **(26# 3670)**

Karimov, B.

Remarks on the Dirichlet principle in the theory of linear Diophantine approximations. (Russian. Uzbek summary)

Izv. Akad. Nauk UzSSR Ser. Fiz.-Mat. Nauk **1962**, *no.* 5, 20–24.

The author uses the Dirichlet "Schubfachprinzip" to prove the existence of solutions of some elementary Diophantine inequalities.

S. Knapowski (New Orleans, La.)

J12-40 **(26# 3671)**

Davenport, H.

A note on Diophantine approximation.

Studies in mathematical analysis and related topics, pp. 77–81. *Stanford Univ. Press, Stanford, Calif.*, 1962.

Let $\|x\|$ denote the distance from the real number x to the nearest integer. Cassels [Mathematika **3** (1956), 109–110; MR **18**, 875] proved: Let $\lambda_1, \cdots, \lambda_r$ be given real numbers, then there exists a real number α such that all the numbers $\alpha+\lambda_1, \cdots, \alpha+\lambda_r$ are "badly approximable"; more precisely, they satisfy $\|(\alpha+\lambda_q)u\|>C/u$ $(q=1, \cdots, r)$ for all positive integers u, where $C^{-1}=8(r+1)^2$. In the paper reviewed here this result is generalized to simultaneous Diophantine approximation to k-tuples of real numbers. In the simplest case (simultaneous approximation to pairs of real numbers) the author obtains: "For any real numbers $\lambda_1, \cdots, \lambda_r$; $\mu_1, \cdots, \mu_r$ there exist real numbers α and β such that

$$\max(\|(\alpha+\lambda_q)u\|^2, \|(\beta+\mu_q)u\|^2) > C/u \qquad (q = 1, \cdots, r)$$

for all positive integers u, where $C=2^{-4r-7}$. Moreover, the set α, β which have this property when C is replaced by $C'=2^{-4r-11}$ has the cardinal of the continuum." The author gives a short and very elegant geometrical proof for these assertions. Finally, he gives simple geometrical interpretations of his results.

J. Popken (Berkeley, Calif.)

Citations: MR 18, 875d = J12-32.

Referred to in J12-41, J12-46, J16-30.

J12-41 **(29# 3432)**

Davenport, H.

A note on Diophantine approximation. II.

Mathematika **11** (1964), 50–58.

Let $\|\alpha\|$ denote the distance from the real number α to the nearest integer. Cassels [Mathematika **3** (1956), 109–110; MR **18**, 875] proved the following. Let $\lambda_1, \cdots, \lambda_r$ be given real numbers, then there exists a real number α such that all numbers $\alpha+\lambda_1, \cdots, \alpha+\lambda_r$ are "badly approximable"; more precisely, they satisfy $\|(\alpha+\lambda_q)u\|>C/u$ $(q=1, \cdots, r)$ for all positive integers u, where $C=C(r)>0$, i.e., the partial quotients of their continued fractions are bounded.

In the paper reviewed here this result is generalized in different directions. One of the simplest is the following: "Suppose $f_1(\alpha), \cdots, f_r(\alpha)$ are functions of a real variable α such that their first derivatives are continuous in some interval containing α_0, and suppose that none of these derivatives is zero for $\alpha=\alpha_0$. Then there exists α such that $\|f_q(\alpha)u\|>C/u$ $(q=1, \cdots, r)$ for all positive integers u, where $C=C(f_1, \cdots, f_r)>0$. The set of these α has the cardinal of the continuum".

This theorem can be extended to simultaneous Diophantine approximation of k-tuples of real numbers. In the two-dimensional case this result reads: "Let $f_1(\alpha, \beta), \cdots, f_r(\alpha, \beta)$ and $g_1(\alpha, \beta), \cdots, g_r(\alpha, \beta)$ be real functions with continuous partial derivatives of the first order in a neighbourhood of some point (α_0, β_0). Suppose that the Jacobian of each pair (f_q, g_q) does not vanish for (α_0, β_0) $(q=1, \cdots, r)$. Then there exists a set of pairs (α, β) with the cardinal of the continuum, for which

$$\max(\|f_q(\alpha, \beta)u\|^2, \|g_q(\alpha, \beta)u\|^2) > C/u$$

for all integers $u>0$ and for $q=1, \cdots, r$, where

$$C = C(f_1, \cdots, f_r, g_1, \cdots, g_r) > 0".$$

Finally, the author shows that one can go even further and find values of functions which possess properties of being badly approximable of different dimensions at the same time. The exact wording of these results is omitted in this review.

The author uses a geometrical method of proof already used by himself in a special case in Part I [*Studies in mathematical analysis and related topics*, pp. 77–81, Stanford Univ. Press, Stanford, Calif., 1962; MR **26** #3671].

J. Popken (Amstelveen)

Citations: MR 18, 875d = J12-32; MR 26# 3671 = J12-40.

Referred to in J12-46.

J12-42 **(29# 3438)**

Danicic, I.

On the fractional parts of θx^2 and ϕx^2.

J. London Math. Soc. **34** (1959), 353–357.

Nach Heilbronn [Quart. J. Math. Oxford Ser. **19** (1948), 249–256; MR **10**, 284] gibt es zu jedem $\varepsilon>0$ ein $C(\varepsilon)$, so daß für jedes reelle θ und jedes $N\geqq 1$ eine ganze Zahl x existiert, für die $1\leqq x\leqq N$, $\|\theta x^2\|<CN^{-1/2+\varepsilon}$ gilt. Für die simultane Approximation zweier beliebiger reeller Zahlen θ, ϕ gilt: Zu jedem $\varepsilon>0$ gibt es ein $C(\varepsilon)$ und ein ganzes x, so daß $1\leqq x\leqq N$, $\|\theta x^2\|<CN^{-1/8+\varepsilon}$, $\|\phi x^2\|<CN^{-1/8+\varepsilon}$ gilt, wobei wieder $N\geqq 1$ gegeben ist. Der Beweis erfolgt mit Hilfe von trigonometrischen Summen. Man vergleiche die neue Arbeit des Verfassers [Mathematika **5** (1958), 30–37; MR **20** #3103], in der eine Verallgemeinerung auf beliebige quadratische Formen durchgeführt wurde.

N. Hofreiter (Zbl **88**, 257)

Citations: MR 10, 284c = J04-5; MR 20# 3103 = D76-10.

J12-43 **(29# 4718)**

Brun, Viggo

Euclidean algorithms and musical theory.

Enseignement Math. (2) **10** (1964), 125–137.

J12-44 **(30# 3065)**

Kaširskiĭ, Ju. V.
Non-linear transference theorems. (Russian)
Uspehi Mat. Nauk **19** (1964), *no.* 6 (120), 175–181.

Classical transference theorems show how information about some problem for a given set of linear forms gives information about the corresponding problem for the transposed set. The present author deals with transference theorems for a system of quadratic forms. One of his theorems is as follows. "If, for a $\sigma>0$, arbitrary $b_1, b_2, \cdots, b_n$, arbitrary $T>0$, and arbitrary $\tau_1, \tau_2, \cdots, \tau_n$ with $\tau_1\tau_2\cdots\tau_n=cT^{1-\sigma}$ and $\tau_i\leqq\frac{1}{2}$, there exists a solution of $\|\alpha_i t^2-b_i\|\leqq\tau_i$, $|t|\leqq T$ $(i=1, 2, \cdots, n)$, then for arbitrary $S>0$ and b there exists a solution of $\|\alpha_1 t_1{}^2+\cdots+\alpha_n t_n{}^2-b\| < c_1 S^{-n\sigma/(2n+8)+\varepsilon}$, $|t_i|\leqq S$ $(i=1, 2, \cdots, n)$." The method of trigonometric sums is used throughout.

S. Knapowski (Gainesville, Fla.)

J12-45 **(31# 4763)**

Sudan, G.; Bucur, C. D.
On the approximation of linear ternary forms. (Romanian. Russian, English, French and German summaries)
Bul. Inst. Politehn. București **25** (1963), *no.* 4, 13–17.

Authors' summary: "On montre comment le théorème de Borel-Obreškov [N. Obreškov, C. R. Acad. Bulgare Sci. **9** (1956), no. 3, 1–3; MR **19**, 937; ibid. **9** (1956), no. 4, 1–4; MR **19**, 397], concernant l'approximation d'une forme linéaire ternaire, peut être deduite très simplement dans le cas où N est un nombre naturel, du théorème de Minkowski [*Diophantische Approximationen*, p. 60, Teubner, Leipzig, 1907]."

J12-46 **(31# 5838)**

Schmidt, Wolfgang M.
On badly approximable numbers.
Mathematika **12** (1965), 10–20.

An n-tuple $(\beta_1, \cdots, \beta_n)$ of real numbers is said to be badly approximable if there is a number $C>0$, depending on the n-tuple, such that $\max(\|\beta_1 q\|, \cdots, \|\beta_n q\|)>Cq^{-1/n}$ for all positive integers q. (Here $\|\theta\|$ denotes the difference between θ and the nearest integer.) The author proves the following remarkable theorem: There are continuum many sets of n-tuples $(\alpha_1, \cdots, \alpha_n)$ of real numbers such that any set $(\beta_1, \cdots, \beta_n)$ of algebraically independent real numbers from the algebraic closure of $\mathbf{Q}(\alpha_1, \cdots, \alpha_n)$ is badly approximable. (Here $\mathbf{Q}$ denotes the rationals.) This is a special case of a theorem about the bad approximability of a sequence of vector functions of $(\alpha_1, \cdots, \alpha_n)$ which is too elaborate to quote here. The $(\alpha_1, \cdots, \alpha_n)$ are obtained as the intersection of an ingenious sequence of inductively constructed nested intervals, which is said to have been suggested by work of Davenport, who proved a weaker result [*Studies in mathematical analysis and related topics*, pp. 77–81, Stanford Univ. Press, Stanford, Calif, 1962; MR **26** #3671; Mathematika **11** (1964), 50–58; MR **29** #3432]. *J. W. S. Cassels* (Cambridge, England)

Citations: MR 26# 3671 = J12-40; MR 29# 3432 = J12-41.

J12-47 **(33# 1287)**

Osgood, Charles F.
Some theorems on diophantine approximation.
Trans. Amer. Math. Soc. **123** (1966), 64–87.

Let Q be the rational field, n a positive integer, and K the quadratic field $Q(\sqrt{-n})$. Denote by $\theta(z)$ and $\phi(z)$ monic polynomials of positive degrees k and l in $K[z]$ and $Q[z]$, respectively, where $\theta(z)$ has no positive integral zero. Put $f(z)=\sum_{d=0}^{\infty} z^d \prod_{e=1}^{d} \theta(e)^{-1}$, so that $(\theta(z\, d/dz)-z)f(z)=\theta(0)$. Assume that $\phi(z)=z^\delta\psi(z)^k$, where $\psi(z)$ is monic of degree t, thus $l=kt+\delta$, where, further, $\psi(z)$ has only simple zeros $\omega_0, \cdots, \omega_{t-1}$, none a positive integer, and where δ is 0 or 1, with the second case excluded if at least one ω_j is irrational. Denote by h and m integers satisfying $0\leqq h\leqq l-1$, $m\geqq 1$, and put $N=lm-k$ and

$$P_N(\delta) = \frac{1}{2\pi i}\int_C f(z)z^h\phi(z)^{-m}\,dz,$$

where the contour C contains the zeros of $\phi(z)$ in its interior. Next, let S be the set of all values $f^{(r)}(\omega_j)$, where $0\leqq r\leqq k-1$, $0\leqq j\leqq t-1$; $c_1, c_2, c_3, \cdots$ positive constants independent of N (i.e., of m); and let $K^*=K(\omega_0, \cdots, \omega_{t-1})$. From the integral, and from the differential equation for $f(z)$,

$$P_N(\delta) = \varepsilon+\sum_{r=0}^{k-1}\sum_{j=0}^{t-1}\varepsilon_{rj}f^{(r)}(\omega_j),$$

where the ε_{rj} and ε are in K^*, have absolute values $<c_1{}^m$ and a least common denominator $v(m)<m^{km}c_2{}^m$; further, $|P_{N+\alpha}(\delta)|<(c_3 m^{-k})^{lm}$ for $0\leqq\alpha\leqq(k+2)l$. Put $g_N(\delta)=v(m)P_N(\delta)$, so that $g_N(\delta)$ is a linear polynomial in the elements of S with integral coefficients in K^*. Again, for $0\leqq\alpha\leqq(k+2)l$, the height of $g_{N+\alpha}(\delta)$ is $<(c_4 m)^{km}$, and for every $\eta>0$ and all sufficiently large N, $|g_{N+\alpha}(\delta)|<[(c_4 m)^{km}]^{-(l-1)+\eta}$. Further, there are representations $f^{(r)}(\omega_j)=\sum_{\alpha=0}^{(k+2)l}\gamma_{rj\alpha}g_{N+\alpha}(\delta)$ for $f^{(r)}(\omega_j)\in S$, and if $\delta=1$, or $\theta(0)\neq 0$, or both, $1=\sum_{\alpha=0}^{(k+2)l}\gamma_\alpha g_{N+\alpha}(\varepsilon)$, with the coefficients $\gamma_{rj\alpha}$ and γ_α in K^*. In the remaining case $\delta=\theta(0)=0$ the constant terms in all $g_{N+\alpha}(\delta)$, where $0\leqq\alpha\leqq(k+2)l$, are zero.

From these properties, the author deduces the following result. Let $\omega_0, \cdots, \omega_{t-1}$ be t elements of K that are $\neq 0$ and non-conjugate, and of which the last t_2, but not the first t_1, are in Q; here $t=t_1+t_2$. Put $t_3=2t_1+t_2$, and denote by J the ring of algebraic integers in K. If $\theta(z)$ has only rational zeros, then there exists, for every $\eta>0$, a $c(\eta)>0$ such that

$$\max_{\substack{0\leqq r\leqq k-1\\ 0\leqq j\leqq t-1}}\left|f^{(r)}(\omega_j)-\frac{p_{rj}}{q}\right| > c(\eta)q^{-(1+[kt_3]^{-1}+\eta)}$$

for all positive integers q and all p_{rj} in J. An analogous result holds for the approximations to zero of linear polynomials in the values $f^{(r)}(\omega_j)$. By generalising the method, the author proves several further theorems which are more complicated.

This very interesting paper is unfortunately written in a difficult style, and the notation is not very well chosen.

K. Mahler (Canberra)

J12-48 **(33# 4009)**

Davenport, Harold
Quelques problèmes d'approximation diophantienne. (Italian and English summaries)
Rend. Sem. Mat. Fis. Milano **35** (1965), 125–130.

Author's summary: "An account is given of recent results which establish the existence of irrational numbers, of various types, which do not admit good rational approximations. Extensions to problems of simultaneous approximation are briefly indicated." *C. Karanikolov* (Sofia)

J12-49 **(35# 1551)**

Osgood, Charles F.
A method in diophantine approximation.
Acta Arith. **12** (1966/67), 111–129.

The problem treated in this paper is a generalization of the following situation: Let $y(x)$ be a real-valued function satisfying a functional equation of the form (1) $y=\sum_{i=1}^{l} g_i T^i y$, in which T denotes a linear functional operator of a very special class (say a differential operator of the form $\psi_1(x)\, d/dx+\psi_2(x)$) and in which the g_i are real-valued functions of x. The g_i also belong to a particular class and satisfy moreover for a given real value x_1 certain conditions of rationality. Then one wants to study the simultaneous approximation of the set of real numbers $\{T^i y(x_1)\}_{i=0}^{l-1}$ by a set of rationals $\{p_i/q\}_{i=0}^{l-1}$.

In earlier work of a similar type, information about the asymptotic behavior of $T^n y(x_1)$ as $n\to\infty$ has been used. Here, however, the author considers $T^{-n}y(x_1)$ instead.

Although the main ideas in this paper are simple and clear, the paper is difficult to read, mainly because the author aims at great generality. For instance, the paper starts with a master theorem containing 13 conditions, too long to reproduce here. Therefore we give here only the simplest of the two corollaries following the theorem (all functions considered have real values): Let ψ_1, ψ_2 be functions on $[0, a]$ such that $1/\psi_1$ and ψ_2 belong to C^{l-1} ($l\geqq 2$). Set $T=\psi_1\, d/dx+\psi_2$. Furthermore, let the functions g_i belong to C^i on $[0, a)$ and let them be such that $(\psi_1\, d/dx)^i g_i\equiv 0$ ($i=1, 2, \cdots, l$). Now $\deg g_i$ is defined as the smallest integer j such that $(\psi_1\, d/dx)^{j+1} g_i\equiv 0$. Let x_1 in $(0, a)$ be such that $g_i(x_1)\neq 0$ and $(\psi_1\, d/dx)^j g_i(x_1)$ is rational for $i=1, 2, \cdots, l$, $j\geqq 0$. Further, let $y=y(x)$ satisfy (1) and, moreover, let (2) $(\psi_1\, d/dx)^j g_i\cdot T^{i-j-1-k}y\to 0$ as $x\to 0$, for each set of integers i, j, k with $0\leqq j<i$, $k\geqq 0$, $i-j-1-k\geqq 0$. Then either $T^i(x_1)=0$ for all $i=0, 1, \cdots, l-1$, or for every $\varepsilon>0$ there exists $C(\varepsilon)$ such that

$$\max_{0\leqq i\leqq l-1} |T^i y(x_1)-p_i/q|\geqq C(\varepsilon)q^{-(1+d+\varepsilon)},$$

where $d=\max_{1\leqq i\leqq l}\{\deg g_i/(i-\deg g_i)\}$. (The condition (2) is needed to ensure an inequality of the type $|T^{-n}y(x_1)|\leqq (C(x_1)/n)^n$.)

In the second corollary, an extension of the first corollary to matrix-valued functions is given. Several applications to solutions of differential equations are included.

J. Popken (Amstelveen)

Referred to in J12-50, J12-56.

J12-50 (36# 6356)

Osgood, Charles F.

A method in diophantine approximation. II.

Acta Arith. **13** (1967/68), 383–393.

By a slight alteration of the method used in the previous part of this paper [same Acta **12** (1966/67), 111–129; MR **35** #1551] the author obtains the principal result of the present part (Theorem III, p. 391). This result, however, is quite abstractly stated and is too extensive to quote here. Therefore we give only a special case.

"Let each $g_j(z)$ ($1\leqq j\leqq l$; $l>1$) be a polynomial of degree less than j with coefficients in the Gaussian field. Suppose that X is a simply connected region of the complex plane where $a(z)$ is analytic, and that $y_1, \cdots, y_n$ are $n\geqq 1$ solutions of the differential equation (1) $y=\sum_{j=1}^{l} g_j(z)D^j y+a(z)$, which are analytic in some open disk $N\subseteq X$ with center z_0. Suppose that the Gaussian rational $z_1\in N$ and that $0\notin g_l(N)$. Let $\mathscr{C}$ be a differentiable path in X with endpoints at z_0 which does not pass through any of the zeros of $g_l(z)$. Suppose that $\tilde{y}_1, \cdots, \tilde{y}_n$ are the function elements obtained by extending $y_1, \cdots, y_n$, respectively, about $\mathscr{C}$ and that the $y_j-\tilde{y}_j$ are linearly independent. Let $d=\max_{1\leqq i\leqq l}\{\deg g_i/(i-\deg g_i)\}$. Let $\|\alpha\|$, for any complex α, denote the distance from α to the nearest Gaussian integer.

"Then for each $\varepsilon>0$ there exists a $c(\varepsilon)>0$ such that $\max_{0\leqq i<l}\|\sum_{j=1}^{n} A_j D^i(y_j-\tilde{y}_j)(z_1)\|\geqq c(\varepsilon)\min_j|A_j|^{-(d+\varepsilon)}$ for all Gaussian integers $A_1, \cdots, A_n$, not all zero."

The author gives also two elegant applications of this theorem.

There are a few misprints; e.g., the symbol $\mathscr{C}$ in the lemma on p. 389 should be replaced by C.

J. Popken (Amstelveen)

Citations: MR 35# 1551 = J12-49.

Referred to in J12-56, J12-63.

J12-51 (37# 2696)

Sudan, G.; Bucur, C. D.

Sur un problème de Barbour et Brun.

Bull. Math. Soc. Sci. Math. R. S. Roumanie **9** (**57**) (1965), 227–233 (1966).

The problem referred to in the title (arising from musical theory) is that of finding integers x, y, z such that (*) $x^{-1}\log 2\simeq y^{-1}\log 3/2\simeq z^{-1}\log 4/3$. Using a method of F. Lettenmeyer for the simultaneous approximation to two (or more) real numbers [Deutsche Math. **3** (1938), 89–108], the author computes several acceptable solutions of (*) and of two other relations of a similar type.

C. G. Lekkerkerker (Amsterdam)

J12-52 (38# 122)

Spohn, W. G.

Blichfeldt's theorem and simultaneous diophantine approximation.

Amer. J. Math. **90** (1968), 885–894.

Sei R die Menge der reellen Zahlen und $\alpha=(\alpha_1, \cdots, \alpha_n)\in R^n$. Zu einem $x\in R$ sei $\|x\|$ der Abstand zur nächsten ganzen Zahl. Das bekannteste Problem der simultanen homogenen diophantischen Approximation ist die Frage nach $s_n=(C_n)^{-1/n}=\sup_{\alpha\in k^n}\liminf_{q\to\infty} q^{1/n}\max_{1\leqq k\leqq n}\|q\alpha_k\|$. Hurwitz zeigte 1891: $C_1=5^{1/2}$; alle weiteren C_n sind bis heute unbekannt. Eine umfangreiche Literatur befaßt sich mit oberen und unteren Schranken für die C_n. Verfasser zeigt $C_n\geqq c_n=n2^{n+1}\int_0^1 (1+u)^{-n}(1+u^n)^{-1}u^{n-1}\,du$ und $c_n\sim\pi$. Für $n>2$ ist das die bisher beste Abschätzung der C_n nach unten. Verfasser geht aus von Blichfeldts Satz der Minkowskischen Geometrie der Zahlen. Wesentlicher Punkt der interessanten Arbeit ist die Verschärfung eines Satzes von Mullender mittels Variationsrechnung.

J. M. Wills (Berlin)

J12-53 (38# 2095)

Wills, Jörg M.

Zur simultanen homogenen diophantischen Approximation. I.

Monatsh. Math. **72** (1968), 254–263.

Let R be the set of reals, Z the set of integers, Q the set of rationals, I the set of irrationals, F the set of non-integer reals. For $\alpha\in R$, let $\|\alpha\|$ denote the distance of α from the nearest integer. For $\alpha=(\alpha_1, \cdots, \alpha_n)\in R^n$, $q\in Z$, define $\mu(q, \alpha)=\min_{1\leqq i\leqq n}\|q\alpha_i\|$, $\mu'(q, \alpha)=\max_{1\leqq i\leqq n}\|q\alpha_i\|$. Let $\lambda_1(\alpha)=\sup_{q\in Z}\mu(q, \alpha)$, $\lambda_2(\alpha)=\limsup_{|q|\to\infty}\mu(q, \alpha)$, $\lambda_3(\alpha)=\liminf_{|q|\to\infty}\mu(q, \alpha)$, $\lambda_4(\alpha)=\inf_{q\in Z}\mu(q, \alpha)$. Let $\lambda_i'(\alpha)$, $i=1, 2, 3, 4$, denote the analogous concepts for $\mu'(q, \alpha)$. For $\alpha\in R^n$ it is shown that $\lambda_1(\alpha)=\lambda_2(\alpha)$; $\lambda_3(\alpha)=\lambda_3'(\alpha)=\lambda_4(\alpha)=\lambda_4'(\alpha)=0$. For $\alpha\in R^n-Q^n$, $\lambda_1'(\alpha)=\frac{1}{2}$. For $\alpha\in Q^n-Z^n$, $\frac{1}{3}\leqq\lambda_1'(\alpha)\leqq\frac{1}{2}$ and both bounds are attained. Further, if $\omega(n)=\inf_{\alpha\in F^n}\lambda_1(\alpha)$ then $\omega(1)=\frac{1}{3}$. Let $x(n)=\inf_{\alpha\in I^n}\lambda_1(\alpha)$. Then it is proved that $\chi(1)=\frac{1}{2}$, $\chi(2)=\frac{1}{3}$, and $1/(2n)\leqq\chi(n)\leqq 1/(n+1)$. The proofs use Kronecker's approximation theorem. The author conjectures that $\chi(n)=1/(n+1)$.

V. C. Dumir (Chandigarh)

J12-54 **(38# 2096)**

Wills, Jörg M.

Zur simultanen homogenen diophantischen Approximation. II.

Monatsh. Math. **72** (1968), 368–381.

For various definitions and notations see Part I [#2095 above]. Let $F = R - Z$ and define

$$\omega(n) = \inf_{\alpha \in F^n} \sup_{q \in Z} \min_{1 \leqq i \leqq n} \|q\alpha_i\|.$$

For $m \geqq 2$ define $h(m)$ to be 0 if m is a prime and h otherwise, where h is the number of distinct prime factors of m. Let $w(n) = \max\{m | \frac{1}{2}\varphi(m) + h(m) \leqq n\}$. It is shown that (i) $w(1) = 3$, $w(2) = 5$, $w(3) = 8$ and $6(n-2) \leqq w(n) \leqq n^2 - 4$ for $n \geqq 4$, (ii) $1/(2n^2) \leqq \omega(n) \leqq 1/w(n)$, for $n \geqq 2$. $\omega(1) = \frac{1}{3}$ was proved in Part I. *V. C. Dumir* (Chandigarh)

Referred to in J12-65.

J12-55 **(39# 6828)**

Adams, William W.

Simultaneous diophantine approximations and cubic irrationals.

Pacific J. Math. **30** (1969), 1–14.

Sei c_0 das Infimum aller c mit der Eigenschaft, daß es für jedes Paar reeller β_1, β_2 unendlich viele $q > 0$ ganz gibt mit $(+)$ $q^{1/2}\|q\beta_i\| < c^{1/2}$, $i = 1, 2$ ($\|x\| = \min_{g \text{ ganz}} |x - g|$). Nach J. W. S. Cassels und H. Davenport ist $2/7 \leqq c_0 \leqq 46^{-1/4}$. Der Autor führt für reelle kubische Zahlkörper mit der Basis $1, \beta_1, \beta_2$ eine gewisse Konstante $C_0 > 0$ ein und zeigt: Theorem 1: Sei $c > C_0$. Dann hat $(+)$ für alle β_1, β_2 mit $1, \beta_1, \beta_2$ Basis eines reellen kubischen Zahlkörpers unendlich viele Lösungen. Ist $c < C_0$, dann gibt es ein Paar β_1, β_2 so, daß $(+)$ nur endlich viele Lösungen hat. Theorem 2: $2/7 \leqq C_0 \leqq 46^{-1/4}$.

Aus dem 2. Teil von Theorem 1 folgt $C_0 \leqq c_0$. Der Autor vermutet $C_0 = 2/7$. *J. M. Wills* (Berlin)

J12-56 **(40# 1367)**

Osgood, Charles F.

A method in diophantine approximation. III, IV.

Acta Arith. **16** (1969/70), 5–22; *ibid.* **16** (1969/70), 23–40.

The author continues his studies on linear differential equations which satisfy certain conditions of a mixed arithmetical and function-theoretical character [same Acta **12** (1966/67), 111–129; MR **35** #1551; ibid. **13** (1967/68), 383–393; MR **36** #6356]. This time, however, he is mainly interested in the algebraic structure of their solution-spaces.

Let X be a bounded region in the complex plane, which is star-shaped about zero. Let Y denote the Riemann surface over X belonging to $\log z$. Further, a function is said to have property A if it is analytic on X and to have property B if it is analytic on Y and bounded on every finite angular sector $\alpha < \arg z < \beta$ in Y. If f_1 has property A and f_2 has property A [B], then the convolution $f_1 * f_2(z) = \int_0^z f_1(z-t) f_2(t)\,dt$ has property A [B]. Here the path of integration is the ray from 0 to z. The author then introduces the set R_y of all functions $r(z)$ with property A which satisfy an equation of the form

$$(Dz)^\alpha r(z) = \sum_{j=0}^{\alpha-1} h_j(D)(Dz)^j r(z),$$

where $\alpha > 0$, D means differentiation and where the $h_j(D)$ are polynomials in D with Gaussian rationals as coefficients. Finally, R_y' denotes the set of all functions with property A which satisfy a linear homogeneous differential equation that has a regular singular point at infinity (at worst) and with coefficients which are polynomials in z with Gaussian rationals as their coefficients. Then the author proves (a) R_y is a ring under $*$ and $+$, (b) R_y' is a subring of R_y, (c) R_y' is also a ring under multiplication of functional values and $+$, (d) if we replace the field of Gaussian rationals by any finite extension of this field in the definitions of R_y and R_y', then we obtain the same sets.

Now let M_y be the set of all functions with property B which satisfy an equation $m(z) = \sum_{j=1}^{l} g_j(z) D^j m(z) + a(z)$, where $a(z)$ has property A and where each $g_j(z)$ is a polynomial of degree $< j$ with Gaussian rationals as coefficients. It is shown that M_y is an R_y module under $*$.

A few examples are given as applications of these general theorems combined with some results of Part II. Moreover, in Proposition II (p. 6) a remarkable approximation theorem of Part II is generalized.

Part IV contains generalizations of some of the results discussed above to the case of partial differential equations. *J. Popken* (Amstelveen)

Citations: MR 35# 1551 = J12-49; MR 36# 6356 = J12-50.

J12-57 **(40# 2613)**

Žogina, L. I.

The approximation of cubic irrationalities by rational fractions. (Russian)

Sverdlovsk. Gos. Ped. Inst. Učen. Zap. **54** (1967), 83–87.

The author shows that for A an integer (not a perfect cube), any positive triple satisfying $x^3 + Ay^3 + A^2z^3 - 3Axyz = 1$ satisfies $|A^{1/3} - y/z| < A^{-1/2}z^{-3/2}$, $|A^{2/3} - x/z| < z^{-3/2}$. He considers recurrence formula based on the fundamental solution (i.e., the ring unit). The author cites a reference to R. P. McKeon and H. H. Goldstine, Trudy Tbiliss. Mat. Inst. **8** (1940), 165–172 [MR **3**, 65], which ignored work of G. B. Mathews [Proc. London Math. Soc. **24** (1893), 319–327]. *Harvey Cohn* (Tucson, Ariz.)

Citations: MR 3, 65e = D32-4.

J12-58 **(40# 4209)**

Pjartli, A. S.

Diophantine approximations of submanifolds of a Euclidean space. (Russian)

Funkcional. Anal. i Priložen. **3** (1969), *no.* 4, 59–62.

The point $(\omega_1, \cdots, \omega_n)$ in R^n is called ν-normal if there is a constant C such that

$$|k_0 + k_1\omega_1 + \cdots + k_n\omega_n| > C(\sum_{i=1}^n |k_i|)^{-\nu}$$

for all integers k_i for which the right hand sum is not 0. The following theorem is proved: almost all points of a non-degenerate submanifold of R^n are ν-normal for $\nu > n^2 + n - 1$. The submanifolds, (I^m, γ, R^n), under consideration are so defined that the conditions guarantee the existence of non-flat curves of suitable dimension through each point of the surface $\omega = \gamma(x)$. This reduces the theorem in question to the corresponding theorem for non-flat curves, which, in turn, is proved by making use of properties of functions whose derivatives are both bounded and bounded away from zero.

J. B. Roberts (Halifax, N.S.)

J12-59 **(41# 5325)**

Libkind, L. M.

A problem in the theory of arrangement. (Russian)

Uspehi Mat. Nauk **24** (1969), *no.* 4 (148), 203–204.

Verfasser berichtet ohne Beweise über eine Reihe von Ergebnissen über diophantische Approximationen von folgendem Typus: Es sei L ein Gitter des n-Raumes mit dem Minimalabstand 2. Die Matrix von L hänge noch von einigen Parametern ab. Diese sollen so bestimmt werden, daß die Dichte $\delta_n(L)$ möglichst groß wird, daß sich also alle n-tupel reeller Zahlen möglichst gut durch L approximieren lassen.

Z.B. sei

$$L = \begin{pmatrix} 1 & & & & \\ & 1 & & & \\ & & \cdot & & \\ & & & \cdot & \\ & & & & \cdot \\ \xi_1 & \xi_2 & \cdots & \xi_{n-1} & \Delta \end{pmatrix}.$$

Δ sei gegeben und $\Delta \leqq \pi^{n/2}/(2^n\Gamma(n/2+1))$. Dann kann man die ξ_i so bestimmen, daß $\delta_n(L) \geqq \xi(n)/2^n$ wird. Ein weiterer Satz gibt die Möglichkeit, mit Hilfe eines n-Gitters, das in der Umgebung des Nullpunktes besonders dünn ist, ein $(n+1)$-Gitter mit großer Dichte aufzubauen.

O.-H. Keller (Halle a.d. Saale)

J12-60 **(41# 8371)**

Slesoraĭtene, R. [Sliesoraitienė, R.]

An analogue of the Mahler-Sprindžuk theorem for polynomials of the second degree in two variables. (Russian. Lithuanian and English summaries)

Litovsk. Mat. Sb. **9** (1969), 627–634.

Let k, m, n be positive integers such that $n \leqq \binom{k+m}{k}$, and let $(i_{1l}, i_{2l}, \cdots, i_{kl})$, where $l = 1, 2, \cdots, n$, be n distinct systems of k non-negative integers of sum $\leqq m$. Further, let $x_1, x_2, \cdots, x_k$ be k independent real variables. On generalising his solution of Mahler's problem [V. G. Sprindžuk, *Mahler's problem in metric number theory* (Russian), Izdat. "Nauka i Tehnika", Minsk, 1967; MR **39** #6832; English translation, Amer. Math. Soc., Providence, R.I., 1969; MR **39** #6833], Sprindžuk raised the problem whether, for every $\varepsilon > 0$, the inequality $|\sum_{l=1}^{n} a_l x_1^{i_{1l}} x_2^{i_{2l}} \cdots x_k^{i_{kl}}| < \{\max(|a_1|, |a_2|, \cdots, |a_n|)\}^{-(n-1)-\varepsilon}$ has for almost all $(x_1, x_2, \cdots, x_k)$ in the sense of Lebesgue only finitely many solutions in integers $(a_1, a_2, \cdots, a_n)$. By means of trigonometric sums he gave a positive answer for $m = 2$ and arbitrary k [Litovsk. Mat. Sb. **2** (1962), no. 1, 147–152; MR **28** #2088]. The author gives a proof independent of trigonometric sums in the special case when $m = 2$, $k = 1$. This proof depends on estimates for integrals.

K. Mahler (Columbus, Ohio)

Citations: MR 28# 2088 = J24-32; MR 39# 6832 = J84-44; MR 39# 6833 = J84-45.

Referred to in J84-49, J84-50.

J12-61 **(42# 1769)**

Mack, J. M.

A note on simultaneous approximation.

Bull. Austral. Math. Soc. **3** (1970), 81–83.

Let a_{ij} (not all zero), b_i, p_j and q be integers and $\theta_1, \cdots, \theta_n$ real numbers which satisfy $\sum_{j=1}^{n} a_{ij}\theta_j = b_i$ $(i = 1, 2, \cdots, r)$. The author notes that there exists a constant $c > 0$, depending on the a_{ij}, such that if $q > 0$, $p_1, \cdots, p_n$ satisfy $\max_{1 \leqq j \leqq n} |q\theta_j - p_j| < c$, then $\sum_{j=1}^{n} a_{ij}(p_j/q) = b_i$ $(i = 1, \cdots, r)$. Thus, good simultaneous rational approximations to a finite set of rationally dependent real numbers must satisfy the same rational dependence relations. The result relates to the problem of constructing satisfactory higher dimensional generalizations of the continued fraction algorithm (cf. O. Perron "Grundlagen für eine Theorie des Jacobischen Kettenbruchalgorithmus" [Math. Ann. **64** (1907), 1–76]).

L. C. Eggan (Normal, Ill.)

J12-62 **(42# 7619)**

Kaufman, R. [Kaufman, Robert P.]

Probability, Hausdorff dimension, and fractional distribution.

Mathematika **17** (1970), 63–67.

The following theorem is proved. Theorem: Let $\psi(q)$ be a function defined for natural q's satisfying the following conditions: $q^m\psi^\beta(q)$ decreases monotonically to zero and $\sum_{q=1}^{\infty} q^m\psi^\beta(q) = \infty$ with some $0 < \beta < m$; here m is a natural number. Further let $X_i(q)$ $(i = 1, 2, \cdots, m)$ be m sequences of independent random variables uniformly distributed in $(0, 1)$. If F is any compact set in the m-dimensional space with positive measure, E' the set of points for which the inequality $\max_i \|(q + X_i(q))x_i\| \leqq q\psi(q)$ has infinitely many solutions, then $E' \cap F$ has almost surely infinite Hausdorff β-measure.

This theorem is related to a theorem of V. Jarník [Math. Z. **33** (1931) 505–543] who proved that the set of points for which the inequality $\max_i \|qx_i\| \leqq q\psi(q)$ has an infinity of solutions, has infinite Hausdorff β-measure. The proof is carried out in a geometric way, computing length of intervals and intersections of intervals; it is detailed only for $m = 1$.

Because of the random variables $X_i(q)$ and the phrase "almost surely", the theorem of the author does not contain Jarník's theorem. It would contain it if the statement of the theorem held also for any given sequence of bounded m-dimensional vectors $X(q) = (X_1(q), \cdots, X_m(q))$ instead of the random variables. The reviewer thinks that even this stronger statement is true.

P. Szüsz (Stony Brook, N.Y.)

J12-63 **(43# 3214)**

Osgood, Charles F.

On the diophantine approximation of values of functions satisfying certain linear q-difference equations.

J. Number Theory **3** (1971), 159–177.

Der Verfasser setzt seine Untersuchungen über die simultane Approximation spezieller Werte von Lösungen gewisser Differentialgleichungen [Acta Arith. **13** (1967/68), 383–393; MR **36** #6356] fort. Hier wird die Approximierbarkeit von Funktionswerten von Lösungen bestimmter q-Differenzengleichungen untersucht. Als interessante Anwendung ergibt sich eine Aussage über die diophantische Approximation der Kettenbrüche

$$\eta(z) = 1 + \frac{zq^{-1}}{1} + \frac{zq^{-2}}{1} + \cdots + \frac{zq^{-n}}{1} + \cdots$$

Dabei ist q eine ganze Gaußsche Zahl mit $|q| > 1$, und z eine von Null verschiedene Gaußsche Zahl. Zu jedem $\varepsilon > 0$ gibt es dann ein $N(\varepsilon)$ derart, daß für alle ganzen Gaußschen Zahlen p und r mit $r \neq 0$, $|r| > N(\varepsilon)$, $|\eta^{(z)} - pr^{-1}| \geqq |r|^{-(2+\varepsilon)}$ gilt.

R. Wallisser (Freiburg)

Citations: MR 36# 6356 = J12-50.

J12-64 **(43# 4766)**

Davenport, H.;

Schmidt, W. M. [Schmidt, Wolfgang M.]

Dirichlet's theorem on diophantine approximation. II.

Acta Arith. **16** (1969/70), 413–424.

By Dirichlet's theorem is meant either of the following two statements, for $\mu = 1$: (a) for each positive integer N there exist integers $x_1, \cdots, x_n, y$ not all zero, satisfying $|\alpha_1 x_1 + \cdots + \alpha_n x_n + y| < \mu N^{-n}$, $\max(|x_1|, \cdots, |x_n|) \leqq \mu N$; (b) for each positive integer N there exist integers $x_1, \cdots, x_n, y$, not all zero, with $\max(|\alpha_1 y - x_1|, \cdots, |\alpha_n y - x_n|) < \mu N^{-1}$, $|y| \leqq \mu N^n$. $(\alpha_1, \cdots, \alpha_n$ are real.) The authors' principal theorem is that for $\mu < 1$, each of these statements is false for almost all n-tuples $\alpha_1, \cdots, \alpha_n$, even if $N \geqq 1$ is replaced by $N \geqq N_0(\mu)$. This is proved for (a) directly, and (b) is deduced by a transfer theorem.

The proof depends on the following result, of some interest in itself. Suppose $1 \leqq m \leqq l$ and write points of lm-dimensional space as $\mathbf{X} = (\mathbf{x}_1, \cdots, \mathbf{x}_m)$, where $\mathbf{x}_1, \cdots, \mathbf{x}_m$ are in l-dimensional space. Let S be a bounded Jordan measurable set in lm-dimensional space, of volume $V(S)$. Then as $t\to\infty$, the number of integer points $\mathbf{X}$ in tS such that $\mathbf{x}_1, \cdots, \mathbf{x}_m$ are part of a basis of the integer lattice of l-dimensional space is asymptotically equal to

$$t^{ml}V(S)\{\zeta(l)\zeta(l-1)\cdots\zeta(l-m+1)\}^{-1}.$$

{Part I appeared in Symposia Mathematica, Vol. IV (INDAM, Rome, 1968/69), pp. 113–132; Academic Press, London, 1970 [MR **42** #7603].}

W. J. LeVeque (Claremont, Calif.)

Citations: MR 42# 7603 = J04-67.

Referred to in J12-66.

J12-65 **(44# 161)**

Wills, Jörg M.

Zur simultanen homogenen diophantischen Approximation. III.

Monatsh. Math. **74** (1970), 166–171.

{Part II appeared in same Monatsh. **72** (1968), 368–381 [MR **38** #2096].} R is the set of real numbers, Z the set of integers, $\|x\|$ is the distance from $x\in R$ to the nearest integer. We denote by F the set $R-Z$. Let

$$\omega(n) = \inf_{\alpha\in F^n} \sup_{q\in Z} \min_{1\leq i\leq n}\|\alpha_{iq}\|.$$

The author proves that $\lim\sup_{n\to\infty}\omega(n)n\log\log n\leq\frac{1}{2}e^{-C}$ and $\lim\inf_{n\to\infty}\omega(n)n\log\log n\geq\frac{1}{2}e^{-C}$, where C is Euler's constant. In the proof he uses the asymptotic behavior of the prime numbers. *S. Kotov* (RŽMat **1970** #10 A112)

Citations: MR 38# 2096 = J12-54.

J12-66 **(44# 3960)**

Schmidt, Wolfgang M.

Diophantine approximation and certain sequences of lattices.

Acta Arith. **18** (1971), 195–178. (*errata insert*)

Let $\alpha_1, \cdots, \alpha_n$ be real numbers. For positive integers N let $\Lambda(\alpha_1, \cdots, \alpha_n; N)$ be the lattice in $n+1$ space $\mathbf{R}^{n+1}$, with basis $(N^{-1}, 0, \cdots, 0, \alpha_1 N)$, $(0, N^{-1}, 0, \cdots, 0, \alpha_2 N), \cdots$, $(0, \cdots, 0, N^{-1}, \alpha_n N), (0, \cdots, 0, N^n)$. So Dirichlet's theorem for one linear form asserts that there is a non-zero lattice point of $\Lambda(\alpha_1, \cdots, \alpha_n, N)$ in the region $|x_1|\leq 1, \cdots$, $|x_{n+1}|\leq 1$, for all N. The author studies this sequence of lattices in the space X of all lattices of determinant 1. The space X is a topological space if two lattices are defined to be close when one lattice can be transformed into the other by a linear transformation that is close to the identity. The author shows first that the sequence $\Lambda(\alpha; N)$ $(N=1, 2, \cdots)$ is everywhere dense in X if and only if every block of positive integers occurs infinitely often in the sequence of partial quotients of the expansion of α as a simple continued fraction. Thus $\Lambda(\alpha, N)$ is dense for almost all α. In the general case the author shows that the lattices $\Lambda(\alpha_1, \cdots, \alpha_n; N)$ $(N=1, 2, \cdots)$ are everywhere dense for almost all n-tuples $(\alpha_1, \cdots, \alpha_n)$ in $\mathbf{R}^n$. This work generalizes earlier work of H. Davenport and the author [same Acta **16** (1969/70), 413–424; MR **43** #4766].

W. W. Adams (College Park, Md.)

Citations: MR 43# 4766 = J12-64.

J16 SYSTEMS OF HOMOGENEOUS LINEAR FORMS

See also reviews A04-4, H05-67, H05-68, H15-63, J12-17, J20-1, J20-6, J20-7, J24-41, J40-40, J56-5, J56-6, J56-7, J64-4, J64-7, J64-10, J64-14, J64-19, J64-27, J99-7.

J16-1 **(1, 103h)**

Derry, D. Remarks on a conjecture of Minkowski. Amer. J. Math. **62**, 61–66 (1940).

The author shows that a special case of Minkowski's hypothesis "MI" on linear forms [Mordell, C. R. Congr. Int. Math. Oslo, vol. 1, 226–238 (1936)] can be expressed in group-theoretic terms as follows. Let $\mathfrak{B}$ be an Abelian p-group of order p^{rn} with rank less than n, and let $\mathfrak{C}_1, \mathfrak{C}_2, \cdots, \mathfrak{C}_n$ be cyclic subgroups having the following property: for every subgroup $\mathfrak{D}$ whose factor group $\mathfrak{B}/\mathfrak{D}$ is cyclic, groups $\mathfrak{C}_{s_1}, \mathfrak{C}_{s_2}, \cdots, \mathfrak{C}_{s_k}$ exist such that the factors of the series

$$\mathfrak{B}=(\mathfrak{C}_{s_1}, \mathfrak{C}_{s_2}, \cdots, \mathfrak{C}_{s_k}, \mathfrak{D}), \quad \cdots, \quad (\mathfrak{C}_{s_1}, \mathfrak{D}), \quad \mathfrak{D}$$

have order not greater than p^r. Then (according to the conjecture) the groups $\mathfrak{C}_1, \mathfrak{C}_2, \cdots, \mathfrak{C}_n$, after a possible rearrangement, build a series

$$\mathfrak{B}=(\mathfrak{C}_1, \mathfrak{C}_2, \cdots, \mathfrak{C}_n), \quad \cdots, \quad (\mathfrak{C}_1, \mathfrak{C}_2), \quad \mathfrak{C}_1, \quad (E),$$

whose factors are all cyclic and of order p^r.

H. S. M. Coxeter (Toronto, Ont.).

J16-2 **(5, 256f; 5, 256g; 5, 256h; 5, 256i)**

Koksma, J. F. et Meulenbeld, B. Sur le théorème de Minkowski, concernant un système de formes linéaires réelles. I. Introduction. Applications. Nederl. Akad. Wetensch., Proc. **45**, 256–262 (1942).

Koksma, J. F. et Meulenbeld, B. Sur le théorème de Minkowski, concernant un système de formes linéaires réelles. II. Lemmes et démonstration du théorème 1. Nederl. Akad. Wetensch., Proc. **45**, 354–359 (1942).

Koksma, J. F. et Meulenbeld, B. Sur le théorème de Minkowski, concernant un système de formes linéaires réelles. III. Démonstration des lemmes 5 et 6. Nederl. Akad. Wetensch., Proc. **45**, 471–478 (1942).

Koksma, J. F. et Meulenbeld, B. Sur le théorème de Minkowski, concernant un système de formes linéaires réelles. IV. Démonstration du lemme 1 (fin). Remarque sur le théorème 1. Nederl. Akad. Wetensch., Proc. **45**, 578–584 (1942).

The authors prove the following theorem. "Let n and r be positive integers such that $1\leqq r\leqq n$. For $(n+1)/2\leqq r\leqq n$ define a number $\rho_{n,r}$ by

$$\rho_{n,r}=\frac{2^{n+1}}{(n+1)!}\left\{\frac{1}{r^r}\sum_{\mu=0}^{r}\binom{n+1}{\mu}\left(r-\frac{n+1}{2}\right)^{\mu}(n+1-r)^{r-\mu}\right.$$
$$\left.+\frac{r(n+1)}{n+1-r}\binom{n}{r}\sum_{\mu=r+1}^{\infty}\frac{1}{\mu r^{\mu}}\left(r-\frac{n+1}{2}\right)^{\mu}\right\},$$

and then put $\rho^*_{n,r}=\rho_{n,n+1-r}$ for $1\leqq r<(n+1)/2$, $\rho^*_{n,r}=\rho_{n,r}$ for $(n+1)/2\leqq r\leqq n$. Further denote by $L_1, \cdots, L_{n+1}$ a set of $n+1$ linear forms with real coefficients and determinant $\Delta\neq 0$. Then to every $t>2$ there exists a system of integers $x_1, \cdots, x_{n+1}$ not all zero for which these forms $L_\nu=L_\nu(x_1, \cdots, x_{n+1})$ satisfy the inequalities

$$\sum_{\nu=1}^{r}|L_\nu|\leqq 2\left(\frac{t^{n+1-r}|\Delta|}{\rho^*_{n,r}}\right)^{1/r}, \quad \sum_{\nu=r+1}^{n+1}|L_\nu|\leqq\frac{2}{t},$$

$$\left(\sum_{\nu=1}^{r}|L_\nu|\right)^r\left(\sum_{\nu=r+1}^{n+1}|L_\nu|\right)^{n+1-r}\leqq\frac{|\Delta|}{\rho^*_{n,r}},$$

$$|L_1\cdots L_{n+1}|\leqq\frac{|\Delta|}{\rho^*_{n,r}r^r(n+1-r)^{n+1-r}}."$$

On specializing the L's, results on the approximations of

or of

$$|a_1x_1+\cdots+a_nx_n-x_{n+1}|$$

$$\sum_{\nu=1}^{n} |a_\nu - x_\nu/x_{n+1}|$$

to zero are obtained. The proof makes use of the general method of H. F. Blichfeldt [Trans. Amer. Math. Soc. **15**, 227–235 (1914)]; it suffices to consider the case that L_ν contains only $x_1, \cdots, x_\nu$. *K. Mahler* (Manchester).

Referred to in H15-26, J56-6.

J16-3 **(5, 256j)**

Meulenbeld, B. Des approximations diophantiques d'un système de formes linéaires complexes. Nederl. Akad. Wetensch., Proc. **45**, 924–928 (1942).

The author states, without proof, results similar to those in the papers reviewed above, but now allows the forms to have complex coefficients. *K. Mahler* (Manchester).

J16-4 **(9, 271e)**

Dyson, F. J. On simultaneous Diophantine approximations. Proc. London Math. Soc. (2) **49**, 409–420 (1947).

This paper is concerned with generalizations of Khintchine's "Übertragungsprinzip" [see Koksma, Diophantische Approximationen, Ergebnisse der Math., v. 4, no. 4, Springer, Berlin, 1936, chap. 5]. Let θ_{ij} $(1\leqq i\leqq n,\ 1\leqq j\leqq m)$ be real numbers. Let

$$f_i=\theta_{i1}x_1+\cdots+\theta_{im}x_m-X_i, \qquad 1\leqq i\leqq n.$$

The matrix θ_{ij} is said to admit the (nonnegative) exponents $\alpha_1, \cdots, \alpha_n; \beta_1, \cdots, \beta_m$ if, for all $A>0$, $B>0$, the inequalities

$$|Af_i|^{\beta_i}|Bx_j|^{\alpha_i}\leqq 1, \quad 1\leqq i\leqq n,\ 1\leqq j\leqq m,$$

are simultaneously soluble in integers $x_1, \cdots, x_m, X_1, \cdots, X_n$ with not all the x's zero. Let ζ be defined by

$$\zeta+\sum_{\alpha_i>\zeta}(\alpha_i-\zeta)=\sum_{\beta_j>-\zeta}(\beta_j+\zeta).$$

Let $\rho_i=\max(0, \alpha_i-\zeta)$ for $1\leqq i\leqq n$; $\sigma_j=\max(0,\beta_j+\zeta)$ for $1\leqq j\leqq m$. The main theorem is that if the matrix θ admits the exponents (α_i, β_j) then its transpose θ' admits the exponents (σ_j, ρ_i). This includes Khintchine's original theorem and various extensions of it. The proof is based on the "Schubfachschluss." [An alternative method would be to follow Mahler's very simple proof of Khintchine's original theorem, Rec. Math. [Mat. Sbornik] N.S. **1(43)**, 961–962 (1936).] *H. Davenport* (Stanford University, Calif.).

Referred to in J16-5, J16-6, J16-10, J16-19, J20-14.

J16-5 **(9, 334d)**

Hinčin, A. Ya. A transfer theorem for singular systems of linear equations. Doklady Akad. Nauk SSSR (N.S.) **59**, 217–218 (1948). (Russian)

The author generalizes his earlier transfer theorem (Übertragungssatz) [Rend. Circ. Mat. Palermo **50**, 170–195 (1926)] to a system of several forms; his result has been anticipated by F. J. Dyson [Proc. London Math. Soc. (2) **49**, 409–420 (1947); these Rev. **9**, 271] who proves more general transfer theorems. In particular, the author's result is obtained from theorem 4 of Dyson on putting $\beta=0$. *K. Mahler* (Manchester).

Citations: MR 9, 271e = J16-4.

J16-6 **(10, 512d)**

Hinčin, A. Ya. On some applications of the method of the additional variable. Uspehi Matem. Nauk (N.S.) **3**, no. 6(28), 188–200 (1948). (Russian)

The method of the additional variable, introduced by L. J. Mordell [J. London Math. Soc. **12**, 34–36, 166–167 (1937)] is the following. Suppose that $L_1, \cdots, L_n$ are linear forms in any number of variables, and that it is required to prove that there exist integer values of the variables for which the L_i satisfy certain conditions; then one considers the set of $(n+1)$ forms $L_1-\alpha u, L_2, \cdots, L_n, u$, where u is the additional variable, and α is a suitably chosen real number. One proves in turn, (i) that the modified set of forms satisfy the required conditions for some integer u within known bounds, (ii) that the only value of u for which the conditions can in fact be satisfied is zero. Three applications of the method are made, as follows. A proof in strengthened form of a theorem of K. Mahler [Časopis Pěst. Mat. Fys. **68**, 85–92 (1939); these Rev. **1**, 202]. A simplified proof of a theorem of V. Jarník [Trav. Inst. Math. Tbilissi [Trudy Tbiliss. Mat. Inst.] **3**, 193–212 (1938)]. A greatly shortened proof, following a suggestion by Mahler, of a theorem of the reviewer [Proc. London Math. Soc. (2) **49**, 409–420 (1947); these Rev. **9**, 271]. *F. J. Dyson* (London).

Citations: MR 1, 202b = H05-1; MR 9, 271e = J16-4.
Referred to in J16-7, Z10-51.

J16-7 **(12, 12g)**

Khintchine, A. Ya. On some applications of the method of the additional variable. Amer. Math. Soc. Translation no. **18**, 14 pp. (1950).

Translated from Uspehi Matem. Nauk (N.S.) **3**, no. 6(28), 188–200 (1948); these Rev. **10**, 512.

Citations: MR 10, 512d = J16-6.
Referred to in Z10-51.

J16-8 **(10, 683c)**

Rédei, L. Vereinfachter Beweis des Satzes von Minkowski-Hajós. Acta Univ. Szeged. Sect. Sci. Math. **13**, 21–35 (1949).

The author considers the following four theorems. (I) Let R_n be the module of all real vectors $(x)=(x_1, \cdots, x_n)$ and suppose that there is a submodule U of R_n such that there is exactly one element (x) with $0\leqq x_i<1$ $(i=1, \cdots, n)$ in every residue class of R_n/U. Then U contains at least one element $(0, \cdots, 0, 1, 0, \cdots, 0)$. (Ia) Let $L_i(x)$ $(i=1, \cdots, n)$ be n real linear forms in $x_1, \cdots, x_n$ with determinant ± 1. If the system of inequalities $|L_i(x)|<1$ $(i=1, \cdots, n)$ has no integral solution other than $x_1=\cdots=x_n=0$, then at least one of the forms $L_i(x)$ has integral coefficients. (Ib) Let $\mathfrak{P}$ be a lattice and suppose that, for every given point $(x_1, \cdots, x_n)$, there is one and only one point $(\xi_1, \cdots, \xi_n)$ of $\mathfrak{P}$ such that $0\leqq x_i-\xi_i<1$ $(i=1, \cdots, n)$. Then $\mathfrak{P}$ contains a unit vector (i.e., two points whose distance apart is unity). (Ic) If the finite Abelian group G can be represented in the form $G=[\alpha_1]\cdots[\alpha_n]$, where the factors $[\alpha_i]$ are simplexes, then at least one factor is a group. By a simplex $[\alpha]$ is meant a complex of elements $1, \alpha, \cdots, \alpha^{e-1}$ of the group where e exceeds unity but does not exceed the order of the group element α.

All four theorems are equivalent and to each may be appended a corollary in which the conclusions are sharpened. Thus the corollary to (I) is that U has n basis vectors $(x_i)=(x_{i1}, \cdots, x_{in})$ $(i=1, \cdots, n)$ which may be chosen in the form $x_{ii}=1$, $x_{ij}=0$ $(j>i)$. Theorem I is new and is due to the author; (Ia) is the famous Minkowski conjecture which was recently proved by Hajós [Math. Z. **47**, 427–467 (1941); these Rev. **3**, 302] who also proved a more general theorem on k-coverings of n-dimensional space. Hajós's proof is in three parts: (i) demonstration that (Ib) need be proved for "rational" lattices only; (ii) proof of equivalence of (Ib) and (Ic); (iii) proof of (Ic). By far the hardest part is (iii), and the author states that he is publishing a simplified proof elsewhere. The purpose of his present paper is to complete the picture by giving a simplified proof of the equivalence of (Ic) with the other three theorems. The steps in the argument are as follows: (a) (Ia) and (Ib) are equivalent. (b) (I) and (Ib) are equivalent. (c) If (Ib) is true for rational lattices it is true in general. (A very simple proof of this is given.) (d) There is a correspondence between finite modules and rational lattices. (e) (Ib) and (Ic) are equivalent. *R. A. Rankin* (Cambridge, England).

Citations: MR 3, 302b = H30-4.

Referred to in J16-9, J40-13.

J16-9 (11, 318d)

Rédei, L. Kurzer Beweis des gruppentheoretischen Satzes von Hajós. Comment. Math. Helv. **23**, 272–282 (1949).

Let G be a finite Abelian group with unit element 1. A set of elements of G: $1, \alpha, \cdots, \alpha^{e-1}$, where e is greater than unity and less than the order of α, is called a simplex and is denoted by $[\alpha]_e$; if e is prime then $[\alpha]_e$ is called a prime simplex. If H and K are two complexes consisting of m and n elements of G, respectively, the product HK is defined to be the direct product of H and K when this product contains mn different elements. Hajós has proved the following theorem. (H) If G can be represented as a product

$$(1)\qquad G=[\alpha_1]_{e_1}[\alpha_2]_{e_2}\cdots[\alpha_n]_{e_n},\qquad n\geqq 1,$$

then at least one simplex $[\alpha_i]_{e_i}$ is a subgroup of G [Math. Z. **47**, 427–467 (1941); these Rev. **3**, 302; see also L. Rédei, Acta Univ. Szeged. Sect. Sci. Math. **13**, 21–35 (1949); these Rev. **10**, 683]. This theorem provides the proof of the Minkowski conjecture concerning linear forms and can be stated in several equivalent forms other than the above group theoretical one [see the above references]. The present paper contains a proof of (H) which is simpler than that given by Hajós. The main steps in the proof are as follows. (i) If (H) is true when each factor in (1) is a prime simplex it is true in general. (ii) The ("ordinary") complexes of G may be regarded as forming a subset of the group ring Γ of G over the ring of the rational integers; the members of Γ are called "general" complexes. Thus the simplex $[\alpha]_e$ corresponds to the member $\gamma=1+\alpha+\cdots+\alpha^{e-1}$ of Γ. Let $\bar{\alpha}$ denote either γ, for a prime e, or $1-\alpha$; let $\{K_1, \cdots, K_r\}$ denote the group generated by the members of the general complexes $K_1, \cdots, K_r$ and let $g\{K_1, \cdots, K_r\}$ be the number of prime factors, not necessarily different, dividing the order of this group. Then if the equation $K\bar{\alpha}_1\bar{\alpha}_2\cdots\bar{\alpha}_m=0$ holds in Γ where K is a general complex, and if no factor $\bar{\alpha}_i$ may be omitted without affecting the truth of this equation, it is shown that

$$g\{K, \alpha_1, \alpha_2, \cdots, \alpha_m\}-g\{K\}<m.$$

(iii) If the product $K[\alpha_1]_{p_1}\cdots[\alpha_r]_{p_r}$ is a group, where $p_1, \cdots, p_r$ are prime and K is an ordinary complex and not a group, then $g\{K, \alpha_1, \cdots, \alpha_r\}-g\{K\}<r$. (iv) Proof of (H) from (i), (ii) and (iii). The most difficult part of the proof is the demonstration of (ii) which is effected by induction on $g\{\alpha_1\}+g\{\alpha_2\}+\cdots+g\{\alpha_m\}$. *R. A. Rankin.*

Citations: MR 3, 302b = H30-4; MR 10, 683c = J16-8.

Referred to in J40-13.

J16-10 (12, 319c)

Maler, K. [Mahler]. On a theorem of Dyson. Mat. Sbornik N.S. **26(68)**, 457–462 (1950). (Russian)

Let m and n be positive integers with $n>m$; let x_i, y_i be two independent sets of n integer variables; let f_j be n linear forms in the x_i, and let g_i be n linear forms with integer coefficients and determinant ± 1 in the y_i. Let $\sum_{j=1}^n f_jg_j=\sum_{i,k=1}^n e_{ik}x_iy_k$, where the e_{ik} are integers. Let a_j be n positive real numbers satisfying $a_j\geqq 1$ for $j\leqq m$ and $a_j\leqq 1$ for $j>m$; let z be the unique number such that $z\prod_{j=1}^m \max(1, z^{-1}a_j)\prod_{j=m+1}^n \min(1, z^{-1}a_j)=1$. The following theorem is proved. If a set of integers x_i, not all zero, exists such that $f_j\leqq a_j^{-1}$, $j=1, \cdots, (n-1)$; $f_n=a_n^{-1}$, then a set of integers y_i, not all zero, exists such that $g_j<\max(1, z^{-1}a_j)$, $j=1, \cdots, m$; $g_j\leqq\min(1, z^{-1}a_j)$, $j=m+1, \cdots, n-1$; $g_n\leqq(n-1)\min(1, z^{-1}a_n)$. This theorem is a generalization of one proved by a more complicated method by the reviewer [Proc. London Math. Soc. (2) **49**, 409–420 (1947); these Rev. **9**, 271]. *F. J. Dyson* (Birmingham).

Citations: MR 9, 271e = J16-4.

J16-11 (12, 483d)

Chabauty, Claude, et Lutz, Élisabeth. Sur les approximations diophantiennes linéaires réelles. I. Problème homogène. C. R. Acad. Sci. Paris **231**, 887–888 (1950).

In order to generalize a theorem of Khintchine [Rend. Circ. Mat. Palermo **50**, 170–195 (1926); cf. the reviewer, Diophantische Approximationen, Springer, Berlin, 1936, p. 69, theorem 8] the authors consider a system of linear forms (1) $L_i=\sum_{j=1}^q a_{ij}x_j+u_i$ $(i=1, 2, \cdots, p;\ p+q=n)$. Putting $X=\max(|x_j|)$, they prove: (A) If $p<\frac{1}{2}n$ and if $\varphi(t)$ denotes any positive decreasing function, there exists a free system (1), such that the inequalities $|L_i|<\varphi(t)$, $i=1, 2, \cdots, p$, $1\leqq X\leqq t$, have an integer solution $x_1, \cdots, x_q, u_1, \cdots, u_p$ for each sufficiently large value of t; (B) if $\frac{1}{2}n<p<n$ and if $\psi(t)$ denotes any positive increasing function, $\psi(t)\to\infty$ for $t\to\infty$, there exists a free system (1), such that the inequalities $|L_i|<\psi(t)t^{-q(p-1)}$, $i=1, 2, \cdots, p$, $1\leqq X\leqq t$, have an integer solution $x_1, \cdots, x_q, u_1, \cdots, u_p$ for each sufficiently large value of t. In these theorems the notion "free" means that no algebraic relation (with integer coefficients which do not all vanish) holds between the pq real numbers $a_{11}, \cdots, a_{pq}$. Replacing the notion "free" by another notion, "pure", the authors state that (A) remains valid, if one replaces the condition $p<\frac{1}{2}n$ by $p\leqq\frac{2}{3}(n-1)$. The proof of (A) is a generalization of Khintchine's argument; the proof of (B) consists of an application of (A) to the "conjugate" system $L_j^*=y_j-\sum_{i=1}^p a_{ij}v_i$ and an application of Minkowski's fundamental theorem on successive dilatation of a symmetric convex body. *J. F. Koksma* (Amsterdam).

Referred to in J16-20, J20-15, J32-31, J64-11.

J16-12 (12, 595b)

Obrechkoff, N. Sur l'approximation de n formes linéaires à n inconnues. Annuaire [Godišnik] Univ. Sofia. Fac. Sci. Livre 1. **45**, 287–292 (1949). (Bulgarian. French summary)

Let $f_1, \cdots, f_m$ be real linear forms in $x_1, \cdots, x_m$, and let n be a positive integer. It follows at once from the consideration of differences (mod 1) that there exist integers $x_1, \cdots, x_m$, not all zero, with $|x_i|\leqq n$, such that $|f_i-y_i|\leqq(n+1)^{-1}$ for $i=1, \cdots, m$, where $y_1, \cdots, y_m$ are also integers. The author proves that the sign of equality is needed only if $f_i=(b_{i1}x_1+\cdots+b_{im}x_m)/(n+1)$, where the b_{ij} are integers and their determinant is relatively prime to $n+1$. *H. Davenport* (London).

J16-13 (13, 116e)

Rogers, C. A. The asymptotic directions of n linear forms in $n+1$ integral variables. Proc. London Math. Soc. (2) **52**, 161–185 (1951).

Let $x_0, x_1, \cdots, x_n$ be the $n+1$ linear forms

$$x_i=\sum_{j=0}^n a_{ij}u_j\qquad(i=0, 1, \cdots, n)$$

in the $n+1$ variables $u_0, u_1, \cdots, u_n$ with real coefficients a_{ij} and determinant $\Delta\neq 0$, such that $x_1, \cdots, x_n$ are not all zero for any integral values of $u_0, u_1, \cdots, u_n$ which are not all zero. We consider the lattice of points X with coordinates $(x_0, x_1, \cdots, x_n)$ corresponding to integral values of $u_0, u_1, \cdots, u_n$. We call $\mathbf{x}$ the vector $(x_1, \cdots, x_n)$ in n-dimensional space and $|\mathbf{x}|$ its length. A direction given by a unit vector $\mathbf{v}$ will be called an asymptotic direction for the forms $x_0, \cdots, x_n$ if there exists an infinite sequence $\{X^{(r)}\}$ of distinct lattice points $X^{(r)}$ with $x_0^{(r)}>0$, such that $\mathbf{x}^{(r)}\to 0$, $\mathbf{x}^{(r)}/|\mathbf{x}^{(r)}|\to\mathbf{v}$ as $r\to\infty$. This asymptotic direction will be said to be of order $O(f(x))$, when $f(x)=o(1)$, if for at least one such sequence $\{X^{(r)}\}$ the inequality $\mathbf{x}^{(r)}=O(f(x_0^{(r)}))$ holds. It is explained by the author that without loss of generality one always may put $x_0=u_0$. Hence one may speak of the asymptotic directions for n linear forms $x_1,\cdots,x_n$ in the $n+1$ variables $u_0, u_1,\cdots,u_n$. In this paper the author investigates to what extent such asymptotic directions of various orders are arbitrary. He proves 6 theorems. Theorem 1 gives a condition (condition (a)) such that if (a) is fulfilled by the system of linear forms, its asymptotic directions of any given order can be determined in terms of the asymptotic directions of a certain set of $n-1$ linear forms in n variables. Theorem 2 states that if (a) is not fulfilled, then every direction is an asymptotic direction of order $o(1)$. Theorem 3 states that given any direction $\mathbf{v}$ there are forms $x_1, x_2,\cdots,x_n$ such that the only asymptotic directions of order $o(1)$ are $\mathbf{v}$ and $-\mathbf{v}$. The proofs of these theorems are easy; Theorem 2 is a consequence of Kronecker's theorem, while Theorem 3 depends on a simple example. The main results are Theorems 4 and 5. Theorem 4 states that the forms $x_1,\cdots,x_n$ always have an asymptotic direction $\mathbf{v}$ of order $O(x^{-1/n})$ such that $-\mathbf{v}$ is either an asymptotic direction of order $O(x^{-1/n})$ or a limit point of asymptotic directions of order $O(x^{-1/n})$. Theorem 5 states that, given any direction $\mathbf{v}$ and any positive decreasing function $f(x)$ of order $o(1)$, then there are forms $x_1,\cdots,x_n$ which do not satisfy the condition (a), but which are such that the only asymptotic directions of order $O(f(x))$ are $\mathbf{v}$ and $-\mathbf{v}$. Theorem 6 is a further elaboration of Theorem 5. For applications see the paper reviewed below. *J. F. Koksma* (Amsterdam).

J16-14 (13, 116f)

Rogers, C. A. The signatures of the errors of simultaneous Diophantine approximations. Proc. London Math. Soc. (2) **52**, 186–190 (1951).

In this paper five theorems are proved which are applications or restatements of the general theorems of the preceding paper for the special case that the $n+1$ forms $x_0, x_1,\cdots,x_n$ mentioned there are the forms $x_0=u_0$, $x_1=u_1-\theta_1u_0,\cdots,x_n=u_n-\theta_nu_0$, where at least one of the real numbers $\theta_1,\cdots,\theta_n$ is irrational. These theorems give important information on the set of signs of the numbers $x_1, x_2,\cdots,x_n$ in connection with the closeness of the approximation of the system $\theta_1,\cdots,\theta_n$ by the system of fractions $u_1/u_0,\cdots,u_n/u_0$. *J. F. Koksma* (Amsterdam).

J16-15 (14, 359c)

Apfelbeck, Alois. A contribution to Khintchine's principle of transfer. Czechoslovak Math. J. 1(76) (1951), 119–147 (1952) = Čehoslovack. Mat. Ž. 1(76) (1951), 141–171 (1952).

The author generalises Khintchine's principle of transfer [see, for example, J. F. Koksma, Diophantische Approximationen, Springer, Berlin, 1936, Kap. V, and more recent work of Jarník, e.g., Acta Arith. **2**, 1–22 (1936)]. Let ϑ_{ij} $(1\leqq i\leqq n, 1\leqq j\leqq m)$ be given real numbers and put

$$S_i(x)=\sum_{j=1}^{m}\vartheta_{ij}x_j+x_{m+i}\quad (i=1, 2,\cdots,n),$$

$$T_j(y)=\sum_{i=1}^{n}\vartheta_{ij}y_i+y_{n+j}\quad (j=1, 2,\cdots,m),$$

where the numbers x_j, y_i are integers $(i, j=1, 2,\cdots,m+n)$. Define

$$\psi_1(t)=\min_{\substack{0<\max|x_j|\leq t\\ 1\leq j\leq m}}(\max_{1\leq i\leq n}|S_i(x)|),$$

$$\psi_2(t)=\min_{\substack{0<\max|y_i|\leq t\\ 1\leq i\leq n}}(\min_{1\leq j\leq m}|T_j(y)|),$$

$$\alpha=\sup\omega\quad\text{for which}\quad \limsup_{t\to\infty}\psi_1(t)t^{(m+\omega)/n}<+\infty,$$

$$\beta=\sup\omega'\quad\text{for which}\quad \limsup_{t\to\infty}\psi_2(t)t^{(n+\omega')/m}<+\infty.$$

The "indices" α and β indicate the degree of precision to which the systems of equations $S_i(x)=0$ and $T_j(y)=0$ can be solved in integers. Khintchine and Jarník considered the case $m=1$ (or $n=1$). The following theorem is proved:

(i) $\beta\geqq\dfrac{n\alpha}{m(m+n-1)+(m-1)\alpha}$, $\alpha\geqq\dfrac{m\beta}{n(m+n-1)+(n-1)\beta}$.

(ii) If $m\geqq 2$ and if $\alpha>2(m+n-1)(m+n-3)$, then

$$\beta\geqq\frac{n\alpha-2n(m+n-3)}{(m-1)\alpha+m-(m-2)(m+n-3)}.$$

The extent to which these results are best possible is also investigated. The method of proof is too complicated to be adequately described here. *R. A. Rankin.*

Referred to in J16-16.

J16-16 (16, 452d)

Jarník, Vojtěch. Über lineare diophantische Approximationen. Bericht über die Mathematiker-Tagung in Berlin, Januar, 1953, pp. 189–192. Deutscher Verlag der Wissenschaften, Berlin, 1953. DM 27.80.

Summary of the following two papers: Apfelbeck, Czechoslovak Math. J. 1(76), 119–148 (1952) [MR **14**, 359]; Kurzweil, ibid. 1(76), 149–178 (1952) [MR **14**, 454].

Citations: MR 14, 359c = J16-15; MR 14, 454e = J04-15.

J16-17 (17, 242e)

Jarník, Vojtěch. Contribution à la théorie des approximations diophantiennes linéaires et homogènes. Czechoslovak Math. J. **4(79)** (1954), 330–353. (Russian. French summary)

Let Θ be the $s\times r$ matrix with real elements Θ_{ji} $(j=1, 2,\cdots,s;\ i=1, 2,\cdots,r)$ and put

$$\psi(t)=\chi(\Theta, t)=\min\{\max_{1\leq j\leq s}|\Theta_{j,1}x_1+\cdots+\Theta_{j,r}x_r+x_{r+j}|\},$$

where the minimum is taken over all systems of integers $x_1, x_2,\cdots,x_{r+s}$ for which $0<\max(|x_1|,\cdots,|x_r|)\leq t$. Let $\alpha=\alpha(\Theta)$ or $\beta=\beta(\Theta)$ denote the upper bound of all real γ for which

$$\limsup_{t\to\infty}t^\gamma\psi(t)<\infty\quad\text{or}\quad\liminf_{t\to\infty}t^\gamma\psi(t)<\infty,$$

respectively. Ignoring the case when $\psi(t)=0$ for all sufficiently large t, we have $r/s\leqq\alpha\leqq\beta\leqq\infty$. The author obtains a lower bound for β as a function of α. He proves, in particular, that if $r=2$, $s\geqq 1$, $\alpha<+\infty$, then $\beta\geqq\alpha(\alpha-1)$, and that for $r>2$, $s\geqq 1$, $\beta\geqq\alpha^{r/(r-1)}-3\alpha$ if $(5r^2)^{r-1}<\alpha<+\infty$. He also proves that if $r=1$, $s\geqq 2$, $m>2$, $m^{s-1}>m^{s-2}+\sum_{k=0}^{s-2}m^k$ and

$$\alpha_0=1-\frac{1}{n}-\cdots-\frac{1}{m^{s-1}},\quad \beta_0=m\frac{m^{s-1}-m^{s-2}-\cdots-1}{m^{s-1}+m^{s-2}+\cdots+1},$$

then there exists a matrix Θ consisting of s linearly independent numbers for which $\alpha=\alpha_0$ and, if $m^s>1+2\sum_{k=1}^{s-1} m^k$, $\beta=\beta_0$. For large m, $\beta_0=\alpha_0^2/(1-\alpha_0)+O(1)$, which compares with the author's general result that when Θ contains at least two linearly independent numbers, $\beta\geqq\alpha^2/(1-\alpha)$ for $\alpha<1$ ($r=1$, $s\geqq 2$). *R. A. Rankin.*

J16-18 (21# 2639)

Obrechkoff, Nikola. Sur l'approximation diophantienne des formes linéaires. Ark. Mat. **3** (1958), 537–542.

The author uses Dirichlet's Schubfachschluss to prove the following result, and deduces some corollaries. Let $n_1, \cdots, n_p, m_1, \cdots, m_p$ be positive integers. Let a_{ij} $(1\leqq i\leqq p,\ 1\leqq j\leqq n_p)$ be real. Then there exist integers x_{ij} and y such that (1) for each i the numbers x_{ij} $(1\leqq j\leqq n_p)$ are either all non-negative or all non-positive, (2) $|x_{ij}|\leqq m_i$,

$$(3)\qquad \left|\sum_{i=1}^{p}\sum_{j=1}^{n_p} a_{ij}x_{ij}-y\right| \leqq \prod_{i=1}^{p}(n_im_i+1)^{-1}.$$

{In the array (3) on p. 538, $m_{\nu-1}$ should be $m_\nu-1$, and similarly for $m_{\nu-2}$.} *H. Davenport* (Cambridge, England)

J16-19 (22# 707)

Jarník, Vojtěch. Eine Bemerkung zum Übertragungssatz. Bŭlgar. Akad. Nauk Izv. Mat. Inst. **3**, no. 2, 169–175 (1959). (Bulgarian and Russian summaries)

Let

$$\Theta = \begin{bmatrix}\theta_{11} & \cdots & \theta_{m1}\\ \cdot & \cdots & \cdot\\ \theta_{1n} & \cdots & \theta_{mn}\end{bmatrix},\qquad H = \begin{bmatrix}\theta_{11} & \cdots & \theta_{1n}\\ \cdot & \cdots & \cdot\\ \theta_{m1} & \cdots & \theta_{mn}\end{bmatrix} = \Theta^T$$

be a real $n\times m$ matrix and its transposed; put

$$S_j(x) = \sum_{\mu=1}^{m}\theta_{\mu j}x_\mu - x_{m+j}\quad (j=1,2,\cdots,n),$$

$$T_i(x) = \sum_{\nu=1}^{n}\theta_{i\nu}x_\nu - x_{n+i}\quad (i=1,2,\cdots,m).$$

Denote by $A(\Theta)$ the least upper bound of all $a>0$ for which there are infinitely many integral solutions $x_1,\cdots,x_m$, $x_{m+1},\cdots,x_{m+n}$ of the inequalities $|S_j(x)|<X^{-a}$ $(j=1,2,\cdots,n)$, where $X=\max(|x_1|,\cdots,|x_m|)>0$, and define $A(H)$ analogously. It is known that $m/n\leqq A(\Theta)\leqq\infty$, $n/m\leqq A(H)\leqq\infty$, and a general transfer principle [J. Dyson, Proc. London Math. Soc. (2) **49** (1947), 409–420; MR **9**, 271] states that

$$A(\Theta) \geqq \frac{mA(H)+(m-1)}{(n-1)A(H)+n},$$

with a similar inequality obtained by interchanging Θ and H. The author proves that this transfer principle is best in the following sense. (1) If $n\geqq m\geqq 1$, $n/m\leqq A\leqq\infty$ or $1<n<m$, $(m-1)/(n-1)\leqq A\leqq\infty$, there exists a pair of matrices Θ and H satisfying

$$\text{(J)}\qquad A(H) = A,\quad A(\Theta) = \frac{mA+(m-1)}{(n-1)A+n}.$$

(For the excluded case $1=n<m$ see V. Jarník [Acta Arith. **2** (1936), 1–22].) This result is an immediate consequence of the following one. (2) Let $m\geqq 1$, $n>1$, $\max(n/m, (m-1)/(n-1))\leqq A\leqq\infty$ and let Θ and H be restricted to the matrices satisfying

$$A(\theta_{11}) = A,\quad \theta_{21} = \cdots = \theta_{m1} = 0.$$

Then (J) holds for almost all such matrices.

K. Mahler (Notre Dame, Ind.)

Citations: MR 9, 271e = J16-4.

J16-20 (22# 4701)

Jarník, Vojtěch. Eine Bemerkung über diophantische Approximationen. Math. Z. **72** (1959/60), 187–191.

Let $\Theta=(\theta_{ij})$ $(1\leqq i\leqq n,\ 1\leqq j\leqq m)$ be a matrix of real numbers. For $t\geqq 1$ write

$$\Psi_\Theta(t) = \min_{0<x\leqq n}\left(\max_{1\leqq i\leqq n}\left|\sum_{1\leqq j\leqq m}\theta_{ij}x_j+x_{m+i}\right|\right),$$

where $x=\max(|x_1|,\cdots,|x_m|)$ and the $x_1,\cdots,x_{m+n}$ are integers. Let $\phi(t)$ be a positive decreasing function of t, decreasing however fast, and let $\phi_1(t)$ be a positive function of t which tends to infinity, however slowly, when t tends to infinity. The author shows that there exist algebraically independent sets of θ_{ij} such that

$$(*)\qquad \begin{aligned}\Psi_\Theta(t) &= O(\phi_1(t)/t)\quad (m=1, n>1),\\ \Psi_\Theta(t) &= O(\phi(t))\quad (m>1).\end{aligned}$$

Indeed the projection on any coordinate axis of the set of such algebraically independent θ_{ij} in any given open set has the power of the continuum.

Results of this kind but with linear independence instead of algebraic independence were obtained by Hinčin [Rend. Circ. Mat. Palermo **50** (1926), 170–195]. For $n<m$ the existence of algebraically independent θ_{ij} with (*) was shown by Chabauty and Lutz [C. R. Acad. Sci. Paris **231** (1950), 887–888; MR **12**, 483] but their results were weaker for $m\leqq n$. Since the author [Trav. Inst. Math. Tbilissi **3** (1938), 193–212] has shown in the case $m=1$, $n>1$ that $\limsup t\Psi_\Theta(t)=+\infty$ when and only when at least two of the numbers $\theta_{11},\cdots,\theta_{n1}$, are linearly independent, it follows that the author's results are the best possible. *J. W. S. Cassels* (Cambridge, England)

Citations: MR 12, 483d = J16-11.
Referred to in J16-25, J16-26.

J16-21 (26# 4966)

Kaširskiĭ, Ju. V.
Transference theorems. **(Russian)**
Dokl. Akad. Nauk SSSR **149** (1963), 1019–1022.

The author is concerned with modifications of the classical transference problem: Given linear forms

$$L_j = L_j(x_1,x_2,\cdots,x_m) = \sum_{i=1}^{m}\alpha_{ij}x_i,$$

$$M_i = M_i(u_1,u_2,\cdots,u_n) = \sum_{j=1}^{n}\alpha_{ji}u_j,$$

what can be said about the non-trivial solvability of $\max\|L_j\|\leqq A$, $\max|x_i|\leqq X$, if we know that there are no non-trivial solutions of $\max\|M_i\|\leqq B$, $\max|u_j|\leqq U$? One of the author's results refers to the inequality

$$\prod_{i=1}^{n}\max(1,|x_i|)\prod_{j=1}^{m}\|L_j\| \leqq C,\quad \max|x_i|\leqq X$$

(instead of $\max\|L_j\|\leqq A$, $\max|x_i|\leqq X$). He works with trigonometric sums. *S. Knapowski* (New Orleans, La.)

J16-22 (27# 1414)

Karimov, B.
On two-dimensional Diophantine approximations. **(Russian. Uzbek summary)**
Izv. Akad. Nauk UzSSR Ser. Fiz.-Mat. Nauk **1963**, *no.* 1, 5–10.

The author generalizes a lemma of A. Khintchine [Rend. Circ. Mat. Palermo **50** (1926), 170–195] on the two-dimensional case. Let θ_1, θ_2, c denote positive numbers, p_i, q_i, r_i $(i=1,\cdots,N)$ positive integers, the pairs $(q_i/r_i, p_i/r_i)$

and $(q_j/r_j, p_j/r_j)$ being different if $i \neq j$. Suppose that $|\theta_1 q_i + \theta_2 p_i - r_i| < c(q_i p_i)^{-1}$ for $i = 1, \cdots, N$; then

$$N \leqq R^2\left[c\left(\frac{QP}{qp}\right)^2 + \frac{1}{2}\right]\left[\frac{4c}{pqr\theta} + 2\left(1 + \frac{c}{pqr}\right)\frac{\theta}{\theta_1\theta_2}\right] + 1,$$

where $P = \max p_i$, $p = \min p_i$, $Q = \max q_i$, $q = \min q_i$, $R = \max r_i$, $r = \min r_i$, $\theta = (\theta_1^2 + \theta_2^2)^{1/2}$. *J. Kubilius* (Vilnius)

J16-23 (27# 1415)

Karimov, B.

On linear Diophantine approximations. (Russian)

Dokl. Akad. Nauk SSSR **148** (1963), 504.

The author announces the following results without proof. If ω_1, ω_2, ω_3 are real numbers and there exist integers $x_0 \neq 0, y_0, z_0$ such that $x_0(\omega_1\omega_3 + \omega_2) + y_0\omega_1 + z_0 = 0$, then for any $\varepsilon > 0$ the system of inequalities (1) $|\omega_1 x - y| < \varepsilon x^{-1/2}$, $|\omega_2 x + \omega_3 y - z| < \varepsilon x^{-1/2}$ has infinitely many integral solutions x, y, z.

Suppose that θ is a real algebraic integer of degree 3, and let $\omega_1 = \theta$, ω_3 be any real number, $\omega_2 = \theta^2 - \omega_3\theta$, then there exists $\varepsilon > 0$ such that (1) has only finitely many integral solutions.

The system of inequalities

$$|\omega_1 x - y| \leqq cx^{-1/2}(\ln x)^{-\gamma},$$

$$|\omega_2 x + \omega_3 y - z| \leqq cx^{-1/2}(\ln x)^{-\gamma}$$

has finitely or infinitely many integral solutions x, y, z for almost all points $(\omega_1, \omega_2, \omega_3)$, $0 \leqq \omega_k \leqq 1$ $(k = 1, 2, 3)$, according as $\gamma > \frac{1}{2}$ or $0 < \gamma \leqq \frac{1}{2}$. *J. Kubilius* (Vilnius)

J16-24 (31# 5833)

Lindström, Bernt

On a combinatorial problem in number theory.

Canad. Math. Bull. **8** (1965), 477–490.

A standard result in elementary number theory [see, e.g., Cassels, *An introduction to Diophantine approximation*, p. 106, Cambridge, New York, 1957; MR **19**, 396] is the following. Let $L_j = \sum_{k=1}^n a_{jk} z_k$ be m linear forms with integral coefficients a_{jk} in $n > m$ variables z_k. Suppose all $|a_{jk}| \leqq a$. Then there exist values of the variables $z_1, z_2, \cdots, z_n$, not all 0, such that $L_j = 0$ and such that all $|z_j| < h$, where $h > 0$ and (1) $h^{n-m} = a^m n^m$. The author improves this theorem and generalizes results of L. Moser and B. Gordon by proving that there exists a constant $c < 4e$ such that (1) can be replaced by (2) $h^{n-m} = (ca)^m n^{m/2}$.

Let $A(m)$ be the total number of 1's in the binary representation of the first m positive integers. Put $G_2(m) = A(m)$ and when $h \geqq 3$, $G_h(m) = (A(m) - m)/\log_2 h$. The author (by a very clever combinatorial method) constructs m linear forms L_j, as above, with coefficients 0 or 1, in $n > G_h(m)$ variables, such that the equations $L_j = 0$, for $1 \leqq j \leqq m$, have no integral solutions z_k with $|z_k| \leqq h$, $1 \leqq k \leqq n$. This improves and generalizes earlier work of the author, E. Berlekamp, G. Clement, P. Erdős, W. H. Mills, A. Rényi, H. S. Shapiro, S. Söderberg, and the reviewer.

Since $A(m) \sim \frac{1}{2} m \log_2 m$, this result shows that in the first mentioned theorem, (2) cannot be replaced by any equation of the form $h^{n-m} = (c_1 a)^m n^{m/2 - \varepsilon}$ for any $\varepsilon > 0$ and any $c_1 > 0$. *D. G. Cantor* (Los Angeles, Calif.)

Citations: MR 19, 396h = J02-11.

J16-25 (32# 7503)

Lesca, J.

Sur un résultat de Jarník.

Acta Arith. **11** (1965/66), 359–364.

Let $\theta = \{\theta_{ij}\}$ $(i = 1, \cdots, n; j = 1, \cdots, m)$ be a system of real numbers. For $t \geqq 1$ let

$$\psi_\theta(t) = \min_{0 < x \leqq t} \max_{1 \leqq i \leqq n} \left|\sum_{1 \leqq j \leqq m} \theta_{ij} x_j + x_{m+i}\right|,$$

where $x = \max(|x_1|, \cdots, |x_{m+n}|)$ and x_k $(k = 1, \cdots, m+n)$ are integers. Let $\varphi(t)$ be a positive non-increasing function of t. Jarník has proved [Math. Z. **72** (1959/60), 187–191; MR **22** #4701] the existence of algebraically independent systems such that $\psi_\theta(t) \leqq \varphi(t)$ for sufficiently large t in the following cases: (a) $m > 1$, $n \geqq 1$; (b) $m = 1$, $n > 1$ and $\lim_{t\to\infty} t\varphi(t) = \infty$. The projection of any coordinate axis of the set of such algebraically independent θ_{ij} in any given open set has the power of the continuum.

The author shows that this theorem is true in the case $m = 1$, $n > 1$, $\limsup_{t\to\infty} t\varphi(t) = \infty$. *J. Kubilius* (Vilnius)

Citations: MR 22# 4701 = J16-20.

J16-26 (32# 7504)

Lesca, J.

Existence de systèmes p-adiques admettant une approximation donnée.

Acta Arith. **11** (1965/66), 365–370.

The author extends Jarník's theorem [Math. Z. **72** (1959/60), 187–191; MR **22** #4701] on the existence of a system of linear forms with given approximation to p-adic numbers in case $m > 1$ [see the preceding review #7503]. *J. Kubilius* (Vilnius)

Citations: MR 22# 4701 = J16-20.

J16-27 (37# 1321)

Davenport, H.; Schmidt, W. M.

A theorem on linear forms.

Acta Arith. **14** (1967/68), 209–223.

The main result of this paper is the following theorem on linear forms. Let $m \geqq 1$, $n \geqq m+2$ be integers. Let $L(x)$, $P_1(x), \cdots, P_m(x)$ be independent linear forms in $x = (x_1, \cdots, x_n)$. Then there is a constant $c = c(L, P_1, \cdots, P_m)$ such that $|L(x)| \leqq c \max(|P_1(x)|, \cdots, |P_m(x)|)|x|^{-m-2}$ has an infinity of solutions in integer vectors x, where $|x| = \max(|x_1|, \cdots, |x_n|)$. Conversely, there exist independent linear forms $L, P_1, \cdots, P_m$ such that for all $\varepsilon > 0$ and every integer vector x,

$$|L(x)| \geqq c(\varepsilon)\max(|P_1(x)|, \cdots, |P_m(x)|)|x|^{-m-2-\varepsilon}.$$

The case $m = 1$, $n = 3$ of the first half of the theorem was essentially proved in a previous paper of the authors [same Acta **13** (1967/68), 169–176; MR **36** #2558] where they used it to settle the question of the approximation of real numbers by quadratic numbers. The above theorem does not suffice to prove the analogous result concerning the approximation of real numbers by algebraic numbers of degree at most k, although they do give an application concerning the approximation of real numbers by algebraic numbers. In proving the second half of the theorem they prove the following theorem in simultaneous diophantine approximations. Let $\psi(t) > 0$ be defined for $t = 1, 2, \cdots$. Given any $k > 1$ there are real numbers $\alpha_1, \cdots, \alpha_k$ such that $1, \alpha_1, \cdots, \alpha_k$ are linearly independent over the rational numbers and such that for all t sufficiently large there are integers $q, p_1, \cdots, p_k$ satisfying $1 \leqq q \leqq t$ and $|q\alpha_i - p_i| < \psi(t)$ for $i = 1, \cdots, k$, with one possible exception $i_0 = i_0(t)$. Moreover, these points are dense in k-space. The proofs, exploiting an analogy with continued fractions, are very clever. They are also elementary in that they require nothing more sophisticated than the Minkowski convex body theorem. *W. W. Adams* (Berkeley, Calif.)

Citations: MR 36# 2558 = J56-30.

Referred to in J16-31, J68-61.

J16-28 **(40# 1342)**

Jarník, Vojtěch

Un théorème d'existence pour les approximations diophantiennes.

Enseignement Math. (2) **15** (1969), 171–175.

The paper is devoted to a proof of Theorem A: Let $m \geqq 1$, $n \geqq 1$, $\varepsilon > 0$ be given and let $\phi(x)$ be a continuous positive function, decreasing for sufficiently large real x. Suppose, further, that $x^{1/k}\phi(x)$ $(1 \leqq k \leqq m)$, $x^{1+\varepsilon}\phi(x)$, $x^{(n-1)/m}\phi(x)$ are monotone for all sufficiently large x. Finally, let $\lambda(x)$ be a positive continuous function tending monotonely to 0 as $x \to \infty$. Then there are real numbers θ_{ij} $(1 \leqq i \leqq m, 1 \leqq j \leqq n)$ such that (i) there are infinitely many integral solutions $x_1, \cdots, x_n, y_1, \cdots, y_m$ of (*) $|\theta_{i1}x_1 + \cdots + \theta_{in}x_n - y_i| < \phi(x)$ $(1 \leqq i \leqq m)$, where $x = \max(|x_1|, \cdots, |x_n|)$, but (ii) there are only finitely many solutions when $\phi(x)$ in (*) is replaced by $\lambda(x)\phi(x)$.

It is stated that similar methods can be used to prove a similar result in which $\lambda(x)$ is replaced by a constant λ, $0 < \lambda < 1$, provided that $x\phi(x)$ tends monotonely to 0, $x^{(n-1)/m}\phi(x)$ is monotone and (**) $\int^\infty x^{n-1}(\phi(x))^m\,dx$ is convergent.

When (**) diverges, Theorem A follows from a metric theorem of A. V. Grošev [Dokl. Akad. Nauk SSSR **19** (1938), 151–152] and when (**) converges and $n=1$, it was proved by K. Černý [Acta Fac. Nat. Univ. Carol. Prague No. 188 (1948); MR **12**, 163; Czechoslovak Math. J. **2 (77)** (1952), 191–220; MR **15**, 857]. The author gives a quite simple reduction of the case $n > 1$ to $n = 1$ for convergent (**). *J. W. S. Cassels* (Cambridge, England)

Citations: MR 12, 163b = J12-16; MR 15, 857e = J12-25.

J16-29 **(40# 1344)**

Schmidt, Wolfgang M.

Badly approximable systems of linear forms.

J. Number Theory **1** (1969), 139–154.

For $i = 1, \cdots, m$, let $\mathbf{L}_i(\mathbf{x}) = \alpha_{i1}x_1 + \cdots + \alpha_{in}x_n$ be a linear form in n variables with real coefficients. Call the system of linear forms $\mathbf{L}_1, \mathbf{L}_2, \cdots, \mathbf{L}_m$ badly approximable in case there is a constant $c > 0$ depending only on $\mathbf{L}_1, \cdots, \mathbf{L}_m$ such that

$$(*)\ (\max\{|x_1|, \cdots, |x_n|\})^n(\max\{\|\mathbf{L}_1(\mathbf{x})\|, \cdots, \|\mathbf{L}_m(\mathbf{x})\|\})^m > c$$

for every integer point $\mathbf{x} \neq \mathbf{0}$, where $\|\cdots\|$ denotes the distance from the nearest integer. Let $N = N(m, n)$ be the set of all $m \times n$ real matrices $A = (\alpha_{ij})$ for which the associated forms $\mathbf{L}_1, \cdots, \mathbf{L}_m$ are badly approximable.

O. Perron [Math. Ann. **83** (1921), 77–84] showed $N(m, n)$ non-empty by constructing elements with algebraic entries, and A. Hinčin [Math. Z. **24** (1926), 706–714] showed that $N(m, n)$ has Lebesgue measure zero in mn-dimensional space. The author generalizes results of Jarnik ($N(1, 1)$ has Hausdorff dimension 1 [V. Jarnik, Prace Mat.-Fiz. **36** (1928), 91–106]), of H. Davenport [Mathematika **1** (1954), 51–72; MR **16**, 223] and J. W. S. Cassels [Proc. London Math. Soc. (3) **5** (1955), 435–448; MR **17**, 715] ($N(1, m)$ and $N(m, 1)$ have the power of the continuum) to the following theorem: $N(m, n)$ has the power of the continuum, and, in fact, its Hausdorff dimension is mn.

The proof consists of a very ingenious inductive argument which is a development of a method of J. W. S. Cassels [Mathematika **3** (1956), 109–110; MR **18**, 875] using K. Mahler's transference theorem on the successive minima of polar convex bodies [Časopis Pěst. Mat. Fys. **68** (1939), 93–102]. *L. C. Eggan* (Normal, Ill.)

Citations: MR 16, 223a = J12-26; MR 17, 715a = J12-31; MR 18, 875d = J12-32.

J16-30 **(40# 2609)**

Goldsmith, Donald L.

On a question related to diophantine approximation.

Acta Arith. **16** (1969/70), 57–70.

Theorem: In $(k+1)$-space, there are continuum-many lattices having no point except the origin in r tubes $|x_j - \lambda_q^{(j)}x_0| < \rho\min(1, |x_0|^{-1/k})$ and s layers $|x_0 + \mu_t^{(1)}x_1 + \cdots + \mu_t^{(k)}x_k| < \rho(1 + \max_j|x_j|)^{-k}$ $(j = 1, \cdots, k)$, where $\lambda_q^{(j)}$, $\mu_t^{(j)}$ $(j = 1, \cdots, k; q = 1, \cdots, r; t = 1, \cdots, s)$ are arbitrarily given real numbers and $\rho > 0$ only depends on the $\lambda_q^{(j)}$, $\mu_t^{(j)}$. The special cases $r = 0$, or $s = 0$, were dealt with by H. Davenport [*Studies in mathematical analysis and related topics*, pp. 77–81, Stanford Univ. Press, Stanford, Calif., 1962; MR **26** #3671]. The question of finding a lattice that avoids domains surrounding subspaces in R^{k+1} of intermediate dimension remains open. The result is generalized to the case of a denumerably infinite set of tubes and layers of the type considered. The proof makes use of the theory of polar reciprocal bodies; it proceeds by constructing two nested sequences of tubes in the space of lattices.

C. G. Lekkerkerker (Amsterdam)

Citations: MR 26# 3671 = J12-40.

J16-31 **(42# 7602)**

Davenport, H.; Schmidt, W. M.

Supplement to a theorem on linear forms.

Number Theory (Colloq., János Bolyai Math. Soc., Debrecen, 1968), pp. 15–25. *North-Holland, Amsterdam,* 1970.

The result of this paper extends a previous theorem of the authors [Acta Arith. **14** (1967/68), 209–223; MR **37** #1321]. They prove: If $L(x)$, $P_1(x)$, $\cdots$, $P_m(x)$ are independent linear forms in $x = (x_1, \cdots, x_n)$, where $n \geqq m+2$, then there is a constant $c = c(L, P_1, \cdots, P_m)$ such that $|L(x)| \leqq c|x|^{-2}\max(|P_1(x)|, \cdots, |P_m(x)|)^{m+1}$ has an infinity of solutions in integer vectors x, where

$$|x| = \max(|x_1|, \cdots, |x_n|).$$

The proof follows the proof of the earlier result closely but requires many changes in detail.

W. W. Adams (College Park, Md.)

Citations: MR 37# 1321 = J16-27.

J20 INHOMOGENEOUS LINEAR FORMS

See also Section J32.

See also reviews J02-11, J16-13, J32-5, J32-16, J56-17, J64-10, J64-14, J64-23, K15-30, K15-41, M15-14, M30-35, M50-2, P52-8, Z10-20, Z10-30, Z10-31.

J20-1 **(7, 273j)**

Jogin, I. I. Zur Theorie der Diophantischen Approximationen. Uchenye Zapiski Moskov. Gos. Univ. Matematika **73**, 37–40 (1944). (Russian. German summary)

[In the original, the author's name was transliterated Schogin. The Russian spelling is Žogin.] The following theorem is proved. Let θ be irrational and let $F(t) > 0$ be an increasing function for $t \geqq 1$ such that $F(2t)/F(t)$ is bounded as $t \to \infty$. There exists a constant $\Gamma > 0$, such that $1 \leqq x < \Gamma F(t)$, $|x\theta - y - \alpha| < 1/t$ is soluble in integral x, y for all $t \geqq 2$ and all real α if and only if there exists a constant $\gamma > 0$ such that $|q\theta - p| \geqq \gamma/F(q)$ for all integral $p, q \geqq 1$. [The reviewer remarks that this is a special case of an analogous theorem in n dimensions, which can be proved, for example, by means of his "Übertragungssatz"; see Nederl. Akad. Wetensch., Proc. **41**, 634–637 (1938); Časopis Pěst. Mat. Fys. **68**, 93–102 (1939); these Rev. **1**, 202]. *K. Mahler.*

Citations: MR 1, 202c = H05-2.

Referred to in J08-5.

J20-2 (7, 274a)

Jogin, I. I. Über eine Frage der Theorie der Diophantischen Approximationen. Uchenye Zapiski Moskov. Gos. Univ. Matematika **73**, 41–44 (1944). (Russian. German summary)

A proof that the inequality $|x\theta-y-\alpha|<5^{-\frac{1}{2}}|x|$ has an infinity of integral solutions if θ is irrational and α an arbitrary real number. [The author adds that $5^{-\frac{1}{2}}$ is the best possible constant; this is not correct unless x is restricted to positive integers. Compare Koksma, Diophantische Approximationen, Ergebnisse der Math. **4**, no. 4, Springer, Berlin, 1936, chapter 6, §§ 2–3.] *K. Mahler* (Manchester).

J20-3 (9, 10e)

Khintchine, A. Deux théorèmes liés au problème de Tchebycheff. Bull. Acad. Sci. URSS. Sér. Math. [Izvestia Akad. Nauk SSSR] **11**, 105–110 (1947). (Russian. French summary)

The author proves the following two theorems. (1) Let θ be irrational, and let β and C be positive constants. Then there exists a real number α such that $\theta x-y-\alpha\neq 0$ for all integers x, y, and that furthermore

(a) $$|\theta x-y-\alpha|<1/t, \quad |x|<Ct^{\beta}$$

is soluble in integers x, y for every real $t\geqq 1$ if and only if there is a positive constant γ such that $|q\theta-p|<\gamma q^{-1/\beta}$ for an infinity of pairs of integers $p, q>0$. (2) Denote by ξ_q the distance from ξ to the nearest fraction of denominator q; thus $0\leqq\xi_q\leqq\frac{1}{2}q^{-1}$. The conditions (a) are soluble in integers x, y for every $t\geqq 1$ if and only if there is a constant Γ such that $\alpha_q<\Gamma\theta_q^{1/(1+\beta)}$ for every integer $q\geqq 1$. *K. Mahler.*

Referred to in J20-6.

J20-4 (9, 10f)

Jarník, Vojtěch. On linear inhomogeneous Diophantine approximations. Rozpravy II. Třídy České Akad. **51**, no. 29, 21 pp. (1941). (Czech)

Let $\theta_{ij}, \alpha_i, \beta_j$ $(1\leqq i\leqq r,\ 1\leqq j\leqq s)$ be given real numbers, and let a, b denote integers. Define, for $t\geqq 1$, $\psi_1(t; \alpha_1, \cdots, \alpha_r)$ as the minimum for $0<\max|a_j|\leqq t$ and $1\leqq j\leqq s$ of

$$\max_{1\leqq i\leqq r}|\theta_{i1}a_1+\cdots+\theta_{is}a_s+a_{i+s}+\alpha_i|;$$

$\psi_2(t; \beta_1, \cdots, \beta_s)$ as the minimum for $0<\max|b_{s+i}|\leqq t$ and $1\leqq i\leqq r$ of

$$\max_{1\leqq j\leqq s}|\theta_{1j}b_{s+1}+\cdots+\theta_{rj}b_{s+r}-b_j-\beta_j|,$$

and put $\psi_1(t)=\psi_1(t; 0, \cdots, 0)$, $\psi_2(t)=\psi_2(t; 0, \cdots, 0)$. The function ψ_1' is defined in a similar way to ψ_1 except that the condition $0<\max|a_j|$ is omitted. Further, let $\varphi(t)$, $\sigma(t)$ be any increasing continuous functions of t for $t\geqq 0$, for which $\varphi(0)=\sigma(0)=0$ and such that, for some $\eta>0$, $t^{-\eta}\varphi(t)$ is increasing and tends to ∞ as $t\rightarrow\infty$; let $\rho(t)$ be the inverse function to $\varphi(t)$.

The author proves the following results, where the limits are taken as $t\rightarrow\infty$. (I) $\limsup t^{s/r}\psi_1(t)\leqq 1$, $\limsup t^{r/s}\psi_2(t)\leqq 1$. (II) If $\liminf t^{r/s}\psi_2(t)>0$, then $\liminf t^{s/r}\psi_1(t)>0$. (III) If $A>0$ and $\limsup\varphi(t)\psi_2(t)>A$, then

$$\liminf \rho(t)\sup\psi_1(t; \alpha_1, \cdots, \alpha_r)\leqq\Delta,$$

where sup denotes the least upper bound for $0\leqq\alpha_i<1$, and $\Delta=\frac{3}{2}(r+s)!(r+s)\max[1, \{(r+s)!(r+s)/\Delta\}^{1/\eta}]$.

(IV) If $\limsup\varphi(t)\psi_2(t)<\infty$ then there exists a system $(\alpha_1, \cdots, \alpha_r)$ such that

$\liminf\rho(t)\psi_1(t; \alpha_1, \cdots, \alpha_r)\geqq\liminf\rho(t)\psi_1'(t; \alpha_1, \cdots, \alpha_r)>0.$

(V) If the series (A) $\sum_{x=1}^{\infty}x^{s-1}\sigma^{-r}(x)$ converges, then, for almost all systems $(\alpha_1, \cdots, \alpha_r)$, (B) $\liminf\sigma(t)\psi_1(t; \alpha_1, \cdots, \alpha_r)=\liminf\sigma(t)\psi_1'(t; \alpha_1, \cdots, \alpha_r)=\infty$. (VI) If $\lim t^{-s/r}\sigma(t)=\infty$, and $\liminf\psi_2(t)t^{r/s}>0$, then (a) if the series (A) converges (B) is satisfied for almost all $(\alpha_1, \cdots, \alpha_r)$, and (b) if (A) diverges then for almost all $(\alpha_1, \cdots, \alpha_r)$

$\liminf\sigma(t)\psi_1(t; \alpha_1, \cdots, \alpha_r)=\liminf\sigma(t)\psi_1'(t; \alpha_1, \cdots, \alpha_r)=0.$

The proof of (I), (II) and (III) follows the method of two earlier papers [Mahler, Časopis Pěst. Mat. Fys. **68**, 93–102 (1939); Jarník, ibid., 103–111 (1939)]. *R. A. Rankin.*

J20-5 (9, 227a)

Hinčin, A. Ya. On a limiting case of Kronecker's approximation theorem. Doklady Akad. Nauk SSSR (N.S.) **56**, 563–565 (1947). (Russian)

Theorem. If θ_i, α_i $(1\leqq i\leqq n)$ are real numbers, then the system of equations (1) $\theta_i x-y_i-\alpha_i=0$ $(1\leqq i\leqq n)$ has integral solutions x, y_i if and only if there is a constant $\Gamma>0$ such that for any integers a_i, b, there is a further integer c satisfying

(2) $$\left|\sum_{i=1}^{n}a_i\alpha_i+c\right|<\Gamma\left|\sum_{i=1}^{n}a_i\theta_i+b\right|.$$

Proof. The condition is necessary since, by (1),

$$\sum_{i=1}^{n}a_i\alpha_i+c=x\left(\sum_{i=1}^{n}a_i\theta_i+b\right)$$

if $c=xb+\sum_{i=1}^{n}a_iy_i$. To show that it is sufficient, assume first that $n=1$, $\theta_1=\theta$ is irrational, $\alpha_1=\alpha$, and write θ_q, α_q for the shortest distance of θ and α from the nearest fractions of denominator $q\geqq 1$. Then, by hypothesis, $\alpha_q<\Gamma\theta_q$ for all integers $q\geqq 1$. Let p_n/q_n be the nth convergent of the continued fraction for θ, and let r_n be the integer with $\alpha_{q_n}=|\alpha-r_n/q_n|$. Put $x_n=r_nq_{n+1}-r_{n+1}q_n$, $y_n=r_np_{n+1}-r_{n+1}p_n$. Then

$$|x_n|=q_nq_{n+1}\left|\frac{r_n}{q_n}-\frac{r_{n+1}}{q_{n+1}}\right|\leqq q_nq_{n+1}(\alpha_{q_n}+\alpha_{q_{n+1}})$$
$$\leqq\Gamma q_nq_{n+1}(\theta_{q_n}+\theta_{q_{n+1}})=\Gamma,$$

since

$$\theta_{q_n}+\theta_{q_{n+1}}=\left|\frac{p_n}{q_n}-\frac{p_{n+1}}{q_{n+1}}\right|=\frac{1}{q_nq_{n+1}}.$$

Further,

$$|x_n\theta-y_n-\alpha|=\left|x_n\left(\theta-\frac{p_n}{q_n}\right)-\left(\alpha-\frac{r_n}{q_n}\right)\right|\leqq\Gamma\theta_{q_n}+\alpha_{q_n}\leqq 2\Gamma\theta_{q_n}.$$

Since x_n is bounded, x_n is a fixed integer x for an infinity of indices n, hence y_n is a fixed integer y, and so finally $x\theta-y-\alpha=0$ since $2\Gamma\theta_{q_n}\rightarrow 0$. Next let $n\geqq 2$, and assume again that θ_1 is irrational. By (2), there exists for given integers p, q a third integer r such that $|q\alpha_1-r|<\Gamma|q\theta_1-p|$, and so $\alpha_1=x\theta_1-y_1$ with certain integral x, y_1, by what has just been proved. Choose, for $2\leqq i\leqq n$, integers k_i and l_i such that $k_i\theta_1=l_i+\theta_i+\gamma_i$ with arbitrarily small $|\gamma_i|$. By (2), there exist integers m_i such that

$$|\delta_i|=|k_i\alpha_1-m_i-\alpha_i|\leqq\Gamma|k_i\theta_1-l_i-\theta_i|=\Gamma|\gamma_i|.$$

Now $\alpha_i=x\theta_i+(xl_i-y_1k_i-m_i)+(x\gamma_i-\delta_i)$. Here $xl_i-y_1k_i-m_i$ is an integer, and $|x\gamma_i-\delta_i|$ is arbitrarily small; hence the constant $\alpha_i-x\theta_i$ differs arbitrarily little from an integer, i.e., is itself an integer, $-y_i$ say, whence the assertion. The assertion is nearly obvious if all θ's are rational. *K. Mahler* (Manchester).

Referred to in J20-8.

J20-6 **(9, 227b)**

Hinčin, A. Ya. On some general theorems of the theory of linear Diophantine approximations. Doklady Akad. Nauk SSSR (N.S.) 56, 679–681 (1947). (Russian)

Let θ_{ij} $(1\leqq i\leqq s,\ 1\leqq j\leqq r)$ be rs real numbers, and let $L_j(x, y)=\sum_{i=1}^{s}\theta_{ij}x_i-y_j$ $(1\leqq j\leqq r)$. Assume that the system of inequalities (1) $L_j(x, y)=0$ $(1\leqq j\leqq r)$, $\sum_{i=1}^{s}x_i^2>0$ has no solutions in integers x_i, y_j. The system of equations (2) $L_j(x, y)-\alpha_j=0$ $(1\leqq j\leqq r)$ is called nondegenerate if it has no solution in integers. It is said to allow total approximate solutions of order $\beta>0$ if there is a $\gamma>0$ such that the inequalities

$$(3)\qquad |L_j(x, y)-\alpha_j|<1/t,\quad |x_i|<\gamma t^\beta,\qquad 1\leqq i\leqq s,\ 1\leqq j\leqq r$$

have integral solutions x_i, y_j for every $t\geqq 1$; it is said to allow partial approximate solutions of order β if, for some $\gamma>0$, (3) has integral solutions for an infinity of arbitrarily large t. Generalizing an earlier result of his [Bull. Acad. Sci. URSS. Sér. Math. [Izvestia Akad. Nauk SSSR] 11, 105–110 (1947); these Rev. 9, 10], the author proves the following theorem. There exist real numbers α_j for which (2) is nondegenerate and allows total approximate solutions of order β, if and only if (1) allows partial approximate solutions of order β.

The difficulty lies in the proof of sufficiency, which proceeds as follows. Choose an infinite sequence of positive λ_n with $\lambda_{n+1}>2\lambda_n$ such that the inequalities

$$\xi_j^{(n)}=|L_j(u^{(n)}, -v^{(n)})|<\lambda_n^{-1},\quad |u_i^{(n)}|<\gamma\lambda_n^\beta,\quad \sum_{i=1}^{s}u_i^{(n)2}>0$$

have integral solutions u_i, v_j, but that the stronger inequalities

$$|L_j(x, y)|<2/\lambda_{n+1},\quad |x_i|<2\gamma'\lambda_n^\beta,\quad \sum_{i=1}^{s}x_i^2>0,$$

where $\gamma'=\gamma\sum_0^\infty 2^{-n\beta}$, have no solution in integers. Put

$$\alpha_j=\sum_{n=1}^{\infty}\xi_j^{(n)},\quad k_i^{(n)}=\sum_{q=1}^{n-1}u_i^{(q)},\quad l_j^{(n)}=\sum_{q=1}^{n-1}v_j^{(q)}.$$

Then the series α_j evidently converge, and

$$|k_i^{(n)}|<\gamma\sum_{q=1}^{n-1}\lambda_q^\beta<\gamma'\lambda_{n-1}^\beta,$$

$$R_j^{(n)}=|L_j(k^{(n)}, -l^{(n)})-\alpha_j|\leqq\sum_{q=1}^{n-1}|\xi_j^{(q)}-\alpha_j|\leqq\sum_{q=n}^{\infty}|\xi_j^{(q)}|<2/\lambda_n.$$

Determine now n, as a function of $t\geqq\lambda_1$, by $\lambda_{n-1}\leqq 2t<\lambda_n$, so that

$$R^{(n)}<1/t,\quad |k_i^{(n)}|<\gamma'\lambda_{n-1}^\beta\leqq 2^\beta\gamma' t^\beta,$$

as asserted. The equations (2) are nondegenerate, since if $\alpha_j=L_j(a, -b)$, then

$$R_j^{(n)}=|L_j(k^{(n)}-a, b-l^{(n)})|<2/\lambda_n,\quad |k_i^{(n)}-a_i|<2\gamma'\lambda_{n-1}^\beta,$$

so that finally $k_i^{(n)}=a_i$, $u_i^{(n)}=k_i^{(n+1)}-k_i^{(n)}=0$, contrary to hypothesis. *K. Mahler* (Manchester).

Citations: MR 9, 10e = J20-3.

J20-7 **(9, 412i)**

Hinčin, A. Ya. On the theory of linear Diophantine approximations. Doklady Akad. Nauk SSSR (N.S.) 59, 865–867 (1948). (Russian)

Let θ_{ij} and α_j $(1\leqq i\leqq m,\ 1\leqq j\leqq n)$ be real numbers, and put

$$S_j=\sum_{i=1}^{m}\theta_{ij}x_i-y_j,\quad T_i=\sum_{j=1}^{n}\theta_{ij}u_j-v_i,$$

$$u=\max_{1\leqq j\leqq n}|u_j|,\quad T=T(u_j, v_i)=\max_{1\leqq i\leqq m}|T_i|,$$

$$A=A(u_j)=|\sum_{j=1}^{n}\alpha_j u_j-w|,$$

where w is the integer nearest to $\sum_{j=1}^{n}\alpha_j u_j$, hence $0\leqq A\leqq\frac{1}{2}$. Denote by $\varphi(t)$ a positive monotonic increasing function of the positive variable t, and by $\psi(t)$ the function inverse to $t\varphi(t)$. Then the following theorem is proved. There exist positive constants c_1 and c_2 such that the system of inequalities

$$|S_j-\alpha_j|<c_1/t,\quad |x_i|<c_2\varphi(t),\qquad 1\leqq i\leqq m,\ 1\leqq j\leqq n,$$

is solvable for every $t>0$ in integers x_i, y_j, if and only if there exists a positive constant Γ such that, for every system of integers u_j, v_i,

$$A\begin{cases}=0, & uT=0,\\ <\Gamma u/\psi(u/T), & uT>0.\end{cases}$$

K. Mahler (Manchester).

J20-8 **(9, 569d)**

Hinčin, A. Ya. A quantitative formulation of the approximation theory of Kronecker. Izvestiya Akad. Nauk SSSR. Ser. Mat. 12, 113–122 (1948). (Russian)

Let θ_{ij} and α_j $(1\leqq i\leqq m,\ 1\leqq j\leqq n)$ be real constants, and

$$S_j=\sum_{i=1}^{m}\theta_{ij}x_i-y_j,\quad T_i=\sum_{j=1}^{n}\theta_{ij}a_j+a_{n+i},$$

where x_i $(1\leqq i\leqq m)$, y_j $(1\leqq j\leqq n)$, a_k $(1\leqq k\leqq m+n)$ are sets of integer variables. Let A_a be the (nonnegative) distance of $\sum_{j=1}^{n}\alpha_j a_j$ from the nearest integer, let $a=\max_{j=1}^{n}|a_j|$ and $T_a=\max_{i=1}^{n}|T_i|$. Let $\phi(t)$ be any positive continuous and nondecreasing function of the real variable t, and let c_1 and c_2 be constants. The author is concerned with finding conditions under which the inequalities (1) $|S_j-\alpha_j|<c_1t^{-1}$ $(1\leqq j\leqq n)$ and (2) $|x_i|<c_2\phi(t)$ $(1\leqq i\leqq m)$ are soluble in integers x_i, y_j for every $t>0$. The classical theorem of Kronecker [see Koksma, Diophantische Approximationen, Ergebnisse der Math., v. 4, no. 4, Springer, Berlin, 1936, chap. 7] states a necessary and sufficient condition for (1) alone to be soluble, viz. that $A_a=0$ for every set of integers a_k for which $T_a=0$. In this paper a necessary and sufficient condition is found for both (1) and (2) to be soluble for some c_1, c_2 and every $t>0$, viz. that (i) $A_a=0$ for every set of integers a_k for which $aT_a=0$ and (ii) there exists a positive constant Γ such that $A_a\leqq\Gamma a/(\psi(a/T_a))$ for all other sets a_k, where $\psi(t)$ is the inverse function to $t\phi(t)$.

Kronecker's theorem is the limiting case $\phi(t)\to\infty$ of the main theorem. There is another limiting case of interest, when $\phi(t)$ is constant, $\psi(t)=t$; this case gives the following new theorem. For the equations $S_j=\alpha_j$ $(1\leqq j\leqq n)$ to be soluble in integers, it is necessary and sufficient that $A_a\leqq\Gamma T_a$ for some constant Γ and every set of integers a_k. This theorem was previously proved by the author only in the case $m=1$ [Doklady Akad. Nauk SSSR (N.S.) 56, 563–565 (1947); these Rev. 9, 227].

The proof of the main theorem starts from the identity

$$(3)\qquad \sum_{j=1}^{n}a_j(S_j-\alpha_j)=\sum_{i=1}^{m}x_iT_i-\sum_{j=1}^{n}a_j\alpha_j-g,$$

where $g=\sum_{i=1}^{m}a_{n+i}x_i+\sum_{j=1}^{n}a_jy_j$ is an integer. From this the "necessity" half follows trivially. The proof of sufficiency is still elementary, but too long for adequate summary here. The idea is to construct $m+n$ independent sets of integers a_{lk} $(1\leqq l\leqq m+n,\ 1\leqq k\leqq m+n)$ such that $|a_{lj}|\leqq\lambda^m K_l$, $|T_i(a_{lk})|\leqq\lambda^{-n}K_l$ $(1\leqq j\leqq n,\ 1\leqq i\leqq m,\ 1\leqq l\leqq m+n)$ with $\lambda=(t\phi(t))^{1/(m+n)}$ and $K_1K_2\cdots K_{m+n}\leqq(m+n)!$. That this is possible was proved by Mahler [Nederl. Akad. Wetensch., Proc. 41, 634–637 (1938)]. These sets a_{lk} then give $m+n$ equations of the form (3); the right-hand sides can be made simultaneously small by a single choice of integers x_i, y_j which makes each "g" equal minus the nearest integer to the corresponding $\sum_{j=1}^{n}a_j\alpha_j$; the equations being solved for the variables $S_j-\alpha_j$ can then be shown to lead to the inequalities (1). The equations defining the x_i, being also solved explicitly, show that the same choice of x_i satisfies (2).

F. J. Dyson (Princeton, N. J.).

Citations: MR 9, 227a = J20-5.

Referred to in J20-35.

J20-9 (10, 18d)

Hinčin, A. Ya. Regular systems of linear equations and a general problem of Čebyšev. Izvestiya Akad. Nauk SSSR. Ser. Mat. **12**, 249–258 (1948). (Russian)

Let $L_j=\sum_{i=1}^m \theta_{ij}x_i-y_j$ $(1\leqq j\leqq n)$, $x=\max|x_i|$, where θ_{ij} are mn real constants and x_i and y_j are $m+n$ integer variables. The set of forms L_j is called a "Čebyšev set" if to every set of real numbers α_j there corresponds a Γ such that the set of inequalities

$$\text{(I)}\qquad |L_j-\alpha_j|<\Gamma x^{-m/n}\qquad (1\leqq j\leqq n)$$

is soluble in x_i and y_j for arbitrarily large values of x. The set is called "regular" if for some positive δ and some arbitrarily large values of t the inequalities

$$\text{(II)}\qquad |L_j|\leqq t^{-1}\quad (1\leqq j\leqq n),\; 0<x\leqq \delta t^{n/m},$$

are not soluble. Result: a set is a Čebyšev set if and only if it is regular. The proof of "if" is simple, using Mordell's device of replacing α_j by $u\alpha_j$ in (I), where u is a new integer variable; the modified (I) will have solutions in x_i, y_j, u by Minkowski's classical theorem on convex regions, with u not exceeding a bound depending on Γ only; the case $u=0$ is excluded by the condition of regularity, and the case $u\neq 0$ leads back to the unmodified (I) with a new value of Γ. The proof of "only if" is longer, but equally easy apart from one interesting twist; it is proved first that a set is a Čebyšev set only if the transposed set of forms (with an obvious definition) is regular; then the complete proof of equivalence follows the cycle (Čebyšev→transposed regular →transposed Čebyšev→regular→Čebyšev). These results had been proved previously by the author in the case $n=1$ only [Acta Arith. **2**, 161–172 (1937)]. *F. J. Dyson.*

J20-10 (10, 250a)

Tornehave, Hans. On a generalization of Kronecker's theorem. Danske Vid. Selsk. Mat.-Fys. Medd. **24**, no. 11, 21 pp. (1948).

A function H is called a phase-modulated oscillation if there exists a real c (mean motion of H) and a real almost periodic function $g(t)$ such that $H(t)=e^{i(ct+g(t))}$. The following is a generalization of Kronecker's theorem. If $H_1, \cdots, H_m$ have rationally independent mean motions, then to any real sequence $v_1, \cdots, v_m$ and $\epsilon>0$ there exists a t such that, for $\nu=1, \cdots, m$, $|H_\nu(t)-e^{iv_\nu}|<\epsilon$. The t satisfying this inequality are relatively dense. Another theorem asserts that any complex-valued almost periodic function f, bounded away from zero, can be written as $f=r\cdot H$, where r is positive almost periodic and H is a phase-modulated oscillation. Its mean motion is called the mean motion of f. A consequence of these two theorems is the following generalization of a result obtained by Jessen [Math. Ann. **111**, 355–363 (1935)]. If $f_1, f_2, \cdots$ is a sequence of almost periodic functions bounded away from zero and α_i are complex numbers such that $\sum_{\nu=1}^\infty|\alpha_\nu|\sup_t|f_\nu(t)|<\infty$ and $\sum_{\nu=1}^\infty\alpha_\nu f_\nu=0$, then a finite subset of the mean motions of f_ν is rationally dependent. Finally this theorem can be extended even to relations of the form $Q_2(f_1, f_2, f_3)=k$, where Q_2 denotes a homogeneous quadratic form of its three arguments, and k a constant. *František Wolf* (Berkeley, Calif.).

J20-11 (12, 245b)

Davenport, H. On a theorem of Khintchine. Proc. London Math. Soc. (2) **52**, 65–80 (1950).

Khintchine [Rend. Circ. Mat. Palermo **50**, 170–195 (1926)] proved the existence of a positive absolute constant δ, such that for any real number α there exists at least one real number β with the property that $|\alpha x+\beta-y|>\delta/x$ for all integers $x>0, y$. Fukasawa [Jap. J. Math. **4**, 41–48 (1927)] gave a construction for β and a numerical estimate for δ, namely $\delta\geqq 1/457$ [see the bibliography in the reviewer's book, Diophantische Approximationen, Springer, Berlin, 1936, chapter VI]. The object of this paper is to give another arithmetical construction for β by means of the expansion of α in a semi-regular continued fraction by the method of the nearest integer. The author also deduces a better estimate for δ and he remarks that his pupil Prasad, by improving the method, has found a still better estimate, namely $\delta\geqq 3/32$. *J. F. Koksma* (Amsterdam).

Referred to in J20-12.

J20-12 (13, 116a)

Prasad, A. V. On a theorem of Khintchine. Proc. London Math. Soc. (2) **53**, 310–330 (1951).

The author proves: For any real number α there exists a real number β, such that for all integers $x>0$, y:

$$|\alpha x-y+\beta|>\delta/x,\quad \text{where}\quad \delta=3/32.$$

(Khintchine had proved that such an absolute constant $\delta>0$ exists; Fukasawa had proved that $\delta=1/457$ is such a constant; Davenport [same Proc. (2) **52**, 65–80 (1950); these Rev. **12**, 245; see this review for references to Khintchine and Fukasawa] had proved the same result with $\delta=1/73.9$.) The author expands α as a regular continued fraction, α being considered irrational and then he constructs β in a way which is similar to Davenport's process [loc. cit.]. The case that α is rational is treated by means of approximation of α by irrationals. *J. F. Koksma* (Amsterdam).

Citations: MR 12, 245b = J20-11.

Referred to in J20-19.

J20-13 (12, 801c)

Hewitt, Edwin, and Zuckerman, H. S. A group-theoretic method in approximation theory. Ann. of Math. (2) **52**, 557–567 (1950).

The main result of this paper is Theorem 1: Let S be any group, and let Σ be any group of characters of S (i.e. homomorphisms of S into the multiplicative group T of complex numbers of absolute value 1). Then, if χ is any character of Σ, if $\sigma_1, \sigma_2, \cdots, \sigma_m$ are arbitrary elements of Σ, and if ϵ is any positive real number, there exists an element $s\epsilon S$ such that

$$|\chi(\sigma_j)-\sigma_j(s)|<\epsilon\qquad (j=1, 2, \cdots, m).$$

The proof makes use of Zorn's lemma, of the Weierstrass-Stone approximation theorem, and, very essentially of a theorem of T. Tannaka [Tôhoku Math. J. **45**, 1–12 (1938)] and M. Krein [C. R. (Doklady) Acad. Sci. URSS (N.S.) **30**, 9–12 (1941); these Rev. **2**, 316] which asserts the following. Let $\mathfrak{T}$ be the set of all finite linear combinations (with complex coefficients) of the (irreducible continuous) representation coefficients of a topological group G. If H is an algebra-homomorphism of $\mathfrak{T}$ into the complex number system, such that $H(f\bar{f})\geqq 0$ for any $f\epsilon\mathfrak{T}$, then H is continuous in the uniform topology for $\mathfrak{T}$ and can therefore be extended continuously over the uniform closure of $\mathfrak{T}$ in the algebra of all continuous functions on G. Theorem 2 is a consequence of Theorem 1: Every character χ of a locally compact Abelian group G is the pointwise limit of continuous characters, in the sense that, for every $\epsilon>0$ and every finite subset $g_1, \cdots, g_m$ of elements of G, there is a continuous character τ of G such that

$$|\chi(g_j)-\tau(g_j)|<\epsilon\qquad (j=1, 2, \cdots, m).$$

These theorems are first applied to derive some classical approximation theorems, such as Kronecker's (Theorems 3–5). Here, and often in the sequel, a Hamel basis construction is also used. Next, it is proved (Theorem 6) that a real linear topological space X, when considered as an Abelian group (with respect to vector-addition), has only the continuous characters $\varphi(x)=e^{2\pi i f(x)}$, where $f(x)$ is a (uniquely determined) continuous linear functional on X. From this result and from Theorem 1 one gets Theorem 7: Let $y_1, y_2, \cdots, y_m$ be rationally independent continuous linear functionals on the real linear topological space X, and let $\alpha_1, \alpha_2, \cdots, \alpha_m$ be arbitrary real numbers. Then, for any $\epsilon>0$, it is possible to find an $x \varepsilon X$ such that

$$|\alpha_j - y_j(x)| < \epsilon \pmod 1 \quad (j=1, 2, \cdots, m).$$

In the particular case $X=C(0, 1)$, this theorem may be formulated as follows (Theorem 11): Let $g_1, g_2, \cdots, g_m$ be rationally independent, real-valued functions of bounded variation of the interval $[0, 1]$. If $\alpha_1, \alpha_2, \cdots, \alpha_m$ are any real numbers and $\epsilon>0$, then there exists a continuous real-valued function $f(x)$ on $[0, 1]$ such that

$$\left|\alpha_j - \int_0^1 f(x)dg_j(x)\right| < \epsilon \pmod 1 \qquad (j=1, 2, \cdots, m).$$

As further applications, some theorems are derived on approximation properties of p-adic numbers. The various examples, classical and new, illustrate the range of the present method. Some possible ways of generalization are pointed out. *B. Sz.-Nagy* (Szeged).

J20-14 (12, 807b)

Jarník, Vojtěch. Sur les approximations diophantiques linéaires non homogènes. Acad. Tchèque Sci. Bull. Int. Cl. Sci. Math. Nat. **47** (1946), 145–160 (1950).

Let $\|x\|$ denote the minimum of $|n+x|$ $(x=0, \pm 1, \pm 2, \cdots)$ and let Θ_{ij} be real numbers $(1 \leqq i \leqq r, 1 \leqq j \leqq s)$. For real α_i put

$$\psi_1(t; \alpha_1, \cdots, \alpha_r) = \min_{|a_j| \leqq t} \max_i \|\sum_j \Theta_{ij} a_j + \alpha_i\|,$$

$$\psi_2(t) = \min_{|b_i| \leqq t} \max_j \|\sum_i \Theta_{ij} b_i\|,$$

where a_j, b_i are integers,

$$(a_1, \cdots, a_s) \neq (0, \cdots, 0); \quad (b_1, \cdots, b_r) \neq (0, \cdots, 0).$$

Put $\psi_1(t)=\psi_1(t; 0, \cdots, 0)$. The author considers how the behavior of $\psi_1(t; \alpha_1, \cdots, \alpha_r)$ and $\psi_1(t)$ is determined by that of $\psi_2(t)$, the principal results being: (I) $\liminf_{t\to\infty} \psi_2(t)t^{r/s}>0$ implies $\liminf_{t\to\infty} \psi_1(t)t^{s/r}>0$. (This is a particular case of a result of Dyson [Proc. London Math. Soc. (2) **49**, 409–420 (1947); these Rev. **9**, 271].) (II) Let $\phi(t)$ be a continuous function with $\phi(0)=0$ and $\phi(t)t^{-\eta}$ increasing monotonely to ∞ for some $\eta=0$. Then $\limsup_{t\to\infty} \phi(t)\psi_2(t)>A>0$ implies $\liminf_{t\to\infty} \sup_{0\leqq\alpha_j<1} \rho(t)\psi_1(t; \alpha_1, \cdots, \alpha_r) \leqq \Delta$ where $\rho(t)$ is the inverse function of $\phi(t)$ and Δ is a given function of A, r, s, η only. (III) If $\limsup_{t\to\infty} \phi(t)\psi_2(t)<\infty$ then there is an $(\alpha_1, \cdots, \alpha_r)$ such that $\liminf_{t\to\infty} \rho(t)\psi_1(t; \alpha_1, \cdots, \alpha_r)>0$. (IV) Let $\sigma(t)$ be continuous and monotonely increasing, $\sigma(0)=0$. Suppose that $\sum_{x=1}^{\infty} x^{s-1}/\sigma^r(x)$ converges. Then $\liminf_{t\to\infty} \sigma(t)\psi_1(t; \alpha_1, \cdots, \alpha_r)=\infty$ for almost all $(\alpha_1, \cdots, \alpha_r)$. (V) Suppose that $\sum_{x=1}^{\infty} x^{s-1}/\sigma^r(x)$ diverges and that further $\lim_{t\to\infty} \sigma(t)t^{-s/r}=\infty$, $\liminf_{t\to\infty} t^{r/s}\psi_2(t)>0$. Then

$$\liminf_{t\to\infty} \sigma(t)\psi_1(t; \alpha_1, \cdots, \alpha_r)=0$$

for almost all $(\alpha_1, \cdots, \alpha_r)$. The author has previously obtained somewhat similar results in which the rôles of lim sup and lim inf are interchanged [Časopis Pěst. Mat. Fys. **68**, 103–111 (1939)]. *J. W. S. Cassels.*

Citations: MR 9, 271e = J16-4.

J20-15 (12, 807d)

Chabauty, Claude, et Lutz, Élisabeth. Approximations diophantiennes linéaires réelles. II. Problème non homogène. C. R. Acad. Sci. Paris **231**, 938–939 (1950).

En utilisant les théorèmes de la note précédente [même tome, 887–888 (1950); ces Rev. **12**, 483] les auteurs montrent deux théorèmes sur la solution approximative d'un système linéaire non homogène, qui généralisent et précisent des théorèmes connus de Blichfeldt et de Khintchine, à savoir les théorèmes qui sont cités dans le livre du rapporteur [Diophantische Approximationen, Springer, Berlin, 1936, p. 78, théorème 6; p. 82, théorème 3; p. 86, théorème 6]. Comme l'autrice fait remarquer dans un mémoir suivant il faut lire au lemme 3: $\rho_{h+1} \geqq (2q^{\frac{1}{2}}+1)q^{\frac{1}{2}}\rho_h$; quatres lignes avant le théorème 1: $\psi(z) \geqq \{24q^2 t(z)\}^{-1}$; au théorème 2: $f(p, n)=(p^{\frac{1}{2}}(p^{\frac{1}{2}}+1))^{-1}(4n)^{-n/p} \geqq (3p)^{-1}(4n)^{-n/p}$. *J. F. Koksma* (Amsterdam).

Citations: MR 12, 483d = J16-11.

Referred to in J32-31.

J20-16 (13, 630d)

Šneĭdmyuller, V. I. On the structure of two-dimensional Diophantine approximations. Doklady Akad. Nauk SSSR (N.S.) **80**, 713–716 (1951). (Russian)

The author extends a recent result of Postnikov [Doklady Akad. Nauk SSSR (N.S.) **76**, 493–496 (1951); these Rev. **12**, 595] to non-homogeneous Diophantine approximation. Let θ be real, and α be real and non-integral. Let p_i, q_i be integral solutions of $q_i|\theta q_i - p_i - \alpha|<c$, where

$$0<q=q_1<q_2<\cdots<q_t=Q.$$

Then $(Q/q)^2>(t-1)\mu/2c$, where μ is defined as follows. Let α' be the fractional part of α, and let Q_{n+1} be the first denominator of a convergent to the continued fraction for α' which is $>Q-q$. Put $d=Q_n+Q_{n+1}$; then

$$\mu=\min[\alpha', 1-\alpha', d^{-1}].$$

It seems to the reviewer that owing to the presence of the factor μ, which is obviously less than $1/t$, the result is of doubtful value. A similar non-homogeneous extension of Postnikov's two-dimensional result is also given. *H. Davenport* (London).

Citations: MR 12, 595f = J24-13.

J20-17 (13, 825d)

Kanagasabapathy, P. Note on Diophantine approximation. Proc. Cambridge Philos. Soc. **48**, 365–366 (1952).

A well-known theorem of Minkowski states: If θ is any irrational number and if α is a real number, not of the form $x-\theta y$ (x and y integers), then there are infinitely many integers x, y, with $y\neq 0$, such that $|(x-\theta y-\alpha)y|<\frac{1}{4}$. Moreover, it is known that the constant $\frac{1}{4}$ on the right is the best possible [cf. Koksma, Diophantische Approximationen, Springer, Berlin, 1936, p. 77]. The author shows that the constant $\frac{1}{4}$ still is the best possible if only one solution (x, y) of the Diophantine inequality is demanded instead of an infinity. *J. Popken* (Utrecht).

Referred to in J56-17.

J20-18 (14, 956c)

Cole, A. J. A problem of Diophantine approximation. Nederl. Akad. Wetensch. Proc. Ser. A. **56** = Indagationes Math. **15**, 144–157 (1953).

It is shown that

$$0.309\cdots \leqq \overline{\text{bd}} \liminf x|\theta x+\alpha-y| \leqq 0.409\cdots,$$

where x, y, are rational integers, θ is irrational and α not of the form $\theta m+n$ (m, n integers). That the lim inf $\leqq 5^{-\frac{1}{2}}$ is an old theorem of Khintchine [Math. Ann. **111**, 631–637 (1935)]. In a footnote it is shown that an example of Poitou and Descombes [C. R. Acad. Sci. Paris **234**, 581–583, 1522–1524 (1952); these Rev. **13**, 825, 921] allows $0.309\cdots$ to be replaced by $0.35\cdots$. *J. W. S. Cassels.*

Citations: MR 13, 825f = J04-11; MR 13, 921c = J04-12.

J20-19 (15, 293b)

Godwin, H. J. On a theorem of Khintchine. Proc. London Math. Soc. (3) **3**, 211–221 (1953).

Khintchine [Rend. Circ. Mat. Palermo **50**, 170–195 (1926)] proved the existence of a constant k_1 such that, if α denotes any real number, another real number β can be chosen so that $|x(\alpha x-y+\beta)|\geqq k_1$ for all integers $x>0$, y. Prasad [Proc. London Math. Soc. (2) **53**, 310–330 (1951); these Rev. **13**, 116] gave an estimate for k_1. Using his argument, the author, by means of simple continued fractions and distinguishing several cases, proved that for any such k_1, $k_1<0.2114$, whereas $k_1=68/483=0.14078$ still is admissable. *J. F. Koksma* (Amsterdam).

Citations: MR 13, 116a = J20-12.

J20-20 (15, 293d)

Szüsz, P. Verschärfung eines Hardy-Littlewoodschen Satzes. Acta Math. Acad. Sci. Hungar. **4**, 115–118 (1953). (Russian summary)

Let ω be irrational, and let $\alpha_1, \cdots, \alpha_k$ be arbitrary numbers with $0\leqq\alpha_\kappa<1$. Put $\{x\}=x-[x]$. It is shown that to each $\epsilon>0$ there corresponds a positive integer $m=m(\epsilon, k, \omega)$ such that for every positive integer ν there is an integer n with $\nu\leqq n\leqq\nu+m$ for which $|\{n^\kappa\omega\}-\alpha_\kappa|<\epsilon$ for $\kappa=1, 2, \cdots, k$. This strengthens a well-known theorem of Hardy and Littlewood [Acta Math. **37**, 155–191, 193–239 (1914)]. The proof is by induction on k. *W. J. LeVeque.*

Referred to in J20-21.

J20-21 (16, 341e)

Szüsz, Petér. Sharpening of a theorem of Hardy and Littlewood. Magyar Tud. Akad. Mat. Fiz. Oszt. Közleményei **4**, 205–208 (1954). (Hungarian)

Hungarian version of Acta Math. Acad. Sci. Hungar. **4**, 115–118 (1953); these Rev. **15**, 293.

Citations: MR 15, 293d = J20-20.

J20-22 (15, 687d)

Cassels, J. W. S. Über

$$\varliminf_{x\to+\infty} x|\vartheta x+\alpha-y|.$$

Math. Ann. **127**, 288–304 (1954).

Improving earlier theorems, the author shows that if α is not of the form $m\vartheta+n$ (m, n integers), then the limit of the title is at most 4/11, except when ϑ and α have the forms

$$\vartheta=(A\omega+B)/(C\omega+D),$$
$$\alpha=\Delta(-3\omega-7+14E+14F\omega)/14|C\omega+D|,$$

where $A, \cdots, F$ are integers with $\Delta=AD-BC=\pm1$, and $\omega=7^{1/2}$. In this exceptional case, the aforementioned limit is $27/28\sqrt{7}$. The proof (which involves a considerable amount of computation) depends on the following lemma. Let p_n/q_n be the convergents of the regular continued fraction expansion of ϑ, where $0<\vartheta<1$, and put $\epsilon_n=q_n\vartheta-p_n$. Then to each n there corresponds a pair of integers P_n and Q_n such that if $\alpha_n=Q_n\vartheta+\alpha-P_n$, then either $0\leqq Q_n<q_n$ and $\alpha_n(\alpha_n+\epsilon_{n-1})<0$, or $q_n\leqq Q_n<q_n+q_{n-1}$ and $\alpha_n(\alpha_n-\epsilon_n)<0$. It is asserted that the same method can be used to improve Khintchine's theorem [Math. Ann. **111**, 631–637 (1935)], that $F(\vartheta)=\max_\alpha \liminf_{x\to\infty} x|\vartheta x+\alpha-y|\leqq\frac{1}{3}$ for almost all ϑ, giving $F(\vartheta)<\frac{1}{4}$ for almost all ϑ. *W. J. LeVeque* (Ann Arbor, Mich.).

Referred to in J20-24, J20-27, J20-29, J20-31, J20-32.

J20-23 (17, 715b)

Barnes, E. S. On linear inhomogeneous Diophantine approximation. J. London Math. Soc. **31** (1956), 73–79.

Let ϕ be irrational and α such that $\phi x+y+\alpha\neq0$ for any integral x, y. It is then proved that for any k with $0\leqq k\leqq\frac{1}{4}$ there exist continuum-many values of ϕ to each of which correspond continuum-many values of α, such that

$$\liminf_{|x|\to\infty} |x(\phi x+y+\alpha)|=k.$$

It is further proved that the theorem remains true, if in the final statement one replaces $|x|\to\infty$ by $x\to\infty$. These theorems include as special cases results of Cassels (unpublished) and S. Fukasawa [Jap. J. Math. **3** (1926), 91–106]. They are proved by the aid of an algorithm used in former cases by the author [Acta Math. **92** (1954), 235–264; MR **16**, 802] and by the author and H. P. F. Swinnerton-Dyer [ibid. **92** (1954), 199–234; MR **16**, 802]. *J. F. Koksma* (Amsterdam).

Citations: MR 16, 802c = J32-48; MR 16, 802d = J32-49.

Referred to in J32-54.

J20-24 (17, 948g)

Descombes, Roger. Sur un problème d'approximation diophantienne. I, II. C. R. Acad. Sci. Paris **242** (1956), 1669–1672, 1782–1784.

Let ξ be a real irrational number, and η a real number, and put $c(\xi, \eta)=\limsup (v|v\xi-u-\eta|)^{-1}$, the limit being over all pairs of integers $v\neq0$, u. J. W. S. Cassels [Math. Ann. **127** (1954), 288–304; MR **15**, 687] used an algorithm similar to a continued fraction to evaluate the smallest value of $c(\xi, \eta)$ and to show that it is an isolated value. In the present notes the author announces the results of further investigations based on Cassel's method. He defines equivalence of pairs (ξ, η), points out that equivalent pairs give the same value of c, and lists the values of $c(\xi, \eta)$ between the minimum and the smallest limit point, together with the corresponding ξ and η. The smallest limit is $366795/(773868-28547\sqrt{510})=2.839\cdots$. *W. J. LeVeque* (Ann Arbor, Mich.).

Citations: MR 15, 687d = J20-22.

J20-25 (19, 124d)

Perron, Oskar. Ein neuartiges diophantisches Problem. Math. Z. **67** (1957), 176–180.

Let a, b denote two different real numbers and let

$$f(x)=|a-x|\,|b-x|,$$

where x is a rational integral variable. In considering the minimum of $f(x)$ as x varies it is sufficient to take $0<a<1$, $a<b$. Then for the particular range $c=b-a\leqq\sqrt{5}$, the expected inequality $f(x)\leqq1$ is shown to be soluble, with strict inequality except in the case $a=\frac{1}{2}(3-\sqrt{5})$, $c=\sqrt{5}$. The same inequality is established later for the extended range $\sqrt{5}<c\leqq\sqrt{8}$. [For another proof, see C. S. Davis, Quart. J. Math., Oxford Ser. (2) **1** (1950), 241–242; MR **12**, 393.] If now a, b denote complex numbers and $c=|b-a|$, the inequality $f(x)\leqq1$, where x is a gaussian integer, is stated to be soluble for $c<\sqrt{3}$. *J. H. H. Chalk.*

Citations: MR 12, 393b = J28-13.

Referred to in J20-26.

J20-26 **(20# 5767)**

van der Waerden, B. L. Ein diophantisches Problem von O. Perron. Arch. Math. **9** (1958), 54–58.

Let $f(x)=|a-x||b-x|$, where a, b are real numbers and x is a rational integral variable. Put $a=d-\frac{1}{2}c$, $b=d+\frac{1}{2}c$, and let $M(c, d)=\min f(x)$ over all x and let $G(c)=\max M(c, d)$ over all real d. Perron [Math. Z. **67** (1957), 176–180; MR **19**, 124] has shown that $G(c)\leqq 1$ for $c^2\leqq 8$, equality holding when $c^2=5$, 8. The problem of determining $G(c)$ is now solved; the graph of $G(c)$ as a function of c^2 being piecewise linear. In particular, $G(c)>1$, except when $c^2\leqq 8$ or $12\leqq c^2\leqq 13$.

J. H. H. Chalk (Hamilton, Ont.)

Citations: MR 19, 124d = J20-25.

J20-27 **(19, 253b)**

Descombes, Roger. Sur la répartition des sommets d'une ligne polygonale régulière non fermée. Ann. Sci. Ecole Norm. Sup. (3) **73** (1956), 283–355.

When ξ, ξ', η, η' are real and ξ, ξ' are irrational numbers, the number couples (ξ, η), (ξ', η') are defined to be equivalent if integers A, B, C, D exist so that $AD-BC=\pm 1$,

$$\xi'=\frac{A\xi+B}{C\xi+D},\ \eta'=\frac{AD-BC}{C\xi+D}\eta+\frac{E\xi+F}{C\xi+D}\quad (C\xi+D>0).$$

Khintchine [Math. Ann. **111** (1935), 631–637] determined the best possible constant γ_1 such that for each couple (ξ, η) and a given ε, $\varepsilon>0$, an infinite number of integer couples (u, v) exists with $v>0$ so that $v|v\xi-u-\eta|<\gamma_1^{-1}+\varepsilon$. If (ξ, η) is not equivalent to a certain number couple (ξ_1, η_1), Cassels [ibid. **127** (1954), 288–304; MR **15**, 687 determined the best possible constant γ_2 so that infinitely many integer couples (u, v) with $v>0$ exist so that $v|v\xi-u-\eta|<\gamma_2^{-1}+\varepsilon$. The present paper gives sequences $\gamma_2, \gamma_3, \cdots$ and (ξ_2, η_2), (ξ_3, η_3), $\cdots$ with the property that if (ξ, η) is not equivalent to any of the couples (ξ_1, η_1), $\cdots$, (ξ_{r-1}, η_{r-1}), $r\geqq 2$, then γ_r is the best possible constant so that infinitely many integers couples (u, v) with $v>0$ exist for which $v|v\xi-u-\eta|<\gamma_r^{-1}+\varepsilon$. For $r>5$ the numbers γ_r, ξ_r, η_r are given by recurrence formulas.

The author shows that if (ξ, η) is not equivalent to any pair (ξ_r, η_r) $(r=1, 2, \cdots)$, then an infinity of integer pairs (u, v) with $v>0$ exists so that $v|v\xi-u-\eta|<\gamma^{-1}+\varepsilon$, γ being the limit of the sequence γ_r. There is a continuous infinity of number couples (ξ, η) for which γ is the best possible constant. The proofs depend on an algorithm defined by the use of the continued fraction development of ξ close to the one used by Cassels. They are too detailed to be described here. *D. Derry* (Vancouver, B.C.).

Citations: MR 15, 687d = J20-22.

Referred to in J20-28, J20-29, J20-32.

J20-28 **(27# 4795)**

Ennola, Veikko

Extension of a result of J. W. S. Cassels on inhomogeneous Diophantine approximation.

Ann. Univ. Turku. Ser. A I No. 67 (1963), 24 *pp.*

The author has partially rediscovered the results of R. Descombes [Ann. Sci. École Norm. Sup. (3) **73** (1956), 283–355; MR **19**, 253].

J. W. S. Cassels (Cambridge, England)

Citations: MR 19, 253b = J20-27.

J20-29 **(20# 1670)**

Sós, Vera T. On the theory of Diophantine approximations. II. Inhomogeneous problems. Acta Math. Acad. Sci. Hungar. **9** (1958), 229–241.

The author continues the discussion of her algorithm for 1-dimensional inhomogeneous diophantine approximation [Part I, same Acta **8** (1957), 461–472; MR **20** #34], which, as she remarks, is essentially that of the reviewer [Math. Ann. **127** (1954), 288–304; MR **15**, 687]. For a pair of numbers α, β, with α irrational and β not of the shape $u+v\alpha$ (u, v integers), the algorithm gives a sequence of pairs of integers q_k, q_k' such that the lower limit of $\lambda(x)=x\|\alpha x+\beta\|$ for positive integral x is the lower limit when x is confined to be a q_k or a q_k'. (Here $\|\gamma\|$ denotes the distance of γ from the nearest integer.) [R. Descombes, Ann. Sci. Ecole Norm. Sup. (3) **75** (1956), 283–355; MR **19**, 253.] The author shows that

$$(1)\qquad \mu_k=\min\{\lambda(q_k), \lambda(q_k')\}<\tfrac{2}{3},$$

$$(2)\qquad \min\{\mu_k, \mu_{k+1}\}<\tfrac{1}{2},$$

for every k, the constants $\frac{2}{3}$ and $\frac{1}{2}$ being the best possible. She shows further, that for any $c<5^{-\frac{1}{2}}$ and any integer l there exist α, β such that

$$(3)\qquad \mu_k>c$$

for l consecutive values of k. The results (1) and (2) are an analogue to classical results of Borel about homogeneous approximation; but (3) is a contrast since the lower limit of $\lambda(x)$ is at most $27/28(7)^{\frac{1}{2}}$ [Cassels, loc. cit.].

J. W. S. Cassels (Cambridge, England)

Citations: MR 15, 687d = J20-22; MR 19, 253b = J20-27; MR 20# 34 = K10-19.

Referred to in J20-32.

J20-30 **(20# 5395)**

Theiler, G. A direct demonstration of Kronecker's theorem in the theory of almost periodic functions and some of its consequences. Lucrarile Inst. Petrol Gaze Bucureşti **4** (1958), 275–291. (Romanian. Russian and English summaries)

Expository article. *R. P. Boas, Jr.* (Evanston, Ill.)

J20-31 **(21# 4145)**

Chalk, J. H. H.; and Erdös, P. On the distribution of primitive lattice points in the plane. Canad. Math. Bull. **2** (1959), 91–96.

Let θ be a given irrational number and α be any real number; then it is well known that there exists an absolute constant λ such that

$$x|y-\theta x+\alpha| < \lambda$$

is satisfied by infinitely many integers x, y with $x>0$ [for the latest results on this classical problem see J. W. S. Cassels, Math. Ann. **127** (1954), 288–304; MR **15**, 687]. The authors show, under the same conditions for θ and α, that there exists an absolute constant λ such that the inequality

$$x|y-\theta x+\alpha| < \lambda(\log x/\log\log x)^2$$

is satisfied by infinitely many coprime integers x, y with $x>0$. The proof uses the continued fraction expansion of θ.

J. Popken (Amsterdam)

Citations: MR 15, 687d = J20-22.

J20-32 **(22# 3725)**

Sós, Vera T. On a problem of S. Hartman about normal forms. Colloq. Math. **7** (1959/60), 155–160.

If ξ is a real number denote as usual by $\|\xi\|$ the difference between ξ and the nearest integer (taken positively). A pair of real numbers (α, β) is called normal [respectively, positively normal, negatively normal] if there exist positive constants t_0, c depending only on (α, β), such that for any $t>t_0$ there exists an integer x for which $\|x\alpha-\beta\|\leqq t^{-1}$ and $|x|<ct$ [respectively, $0<x<ct$, $0<-x<ct$]. The author shows that there exist pairs (α, β) which are normal but neither positively nor negatively normal, thereby answering a question of S. Hartman [same Colloq. **6**

(1958), 334]. The proof uses the algorithm developed by the author [Acta Math. Acad. Sci. Hungar. **9** (1958), 229–241; MR **20** #1670], which is substantially the same as that employed by the reviewer [Math. Ann. **127** (1954), 288–304; MR **15**, 687] and by R. Descombes [Ann. Sci. École Norm. Sup. (3) **73** (1956), 283–355; MR **19**, 253].

J. W. S. Cassels (Cambridge, England)

Citations: MR 15, 687d = J20-22; MR 19, 253b = J20-27; MR 20# 1670 = J20-29.

J20-33 **(23# A3117)**

Skolem, Th.

A simple proof of a theorem concerning diophantine approximations.

Norske Vid. Selsk. Forh. Trondheim **32** (1959), 139–144 (1960).

If 1, $\xi_1, \xi_2, \cdots, \xi_n$ are real numbers, linearly independent over the rational field, Q is the unit cube $0 \leqq x_r < 1$ $(r = 1, 2, \cdots, n)$ in n-dimensional Euclidean space and R is any n-dimensional cube in Q with sides of length ε, then a famous theorem of Kronecker asserts that there exists a point $P(x_1, \cdots, x_n)$ in R and an integer t such that

(1) $$x_r \equiv t\xi_r \pmod 1,\ r = 1, 2, \cdots, n.$$

The proof is by induction on n and introduces a "box" argument to show that, for any given $\varepsilon > 0$, there is a positive integer k such that $k\xi_r \equiv \varepsilon_r \pmod 1$, $r = 1, 2, \cdots, n$, where $\varepsilon_1^2 + \cdots + \varepsilon_n^2 < \varepsilon^2$. It is completed by considering the system of straight lines $x_r = \varepsilon_r t + c_r$, $r = 1, 2, \cdots, n$, where the c_r are arbitrary integers, and noting that their points of intersection with the space $x_1 = 0$ are densely distributed in $x_1 = 0$.

For given $\xi_1, \xi_2, \cdots, \xi_n, \varepsilon$, an upper bound for the least integer t is determined. *J. H. H. Chalk* (Toronto)

J20-34 **(24# A1251)**

Descombes, Roger

Problèmes d'approximation diophantienne.

Enseignement Math. (2) **6** (1960), 18–26.

Let $\|\xi\|$ denote the distance of the irrational number ξ from the nearest integer. This paper contains a short historical account of work done on the calculation of $\liminf q\|q\xi\|$, $\liminf_{r\neq 0} |r|\ \|r\xi - \eta\|$ and $\liminf_{r>0} r\|r\xi - \eta\|$ ($q > 0$ and r are integers, η real).

H. Halberstam (London)

Referred to in Z10-17.

J20-35 **(30# 3063)**

Ostrowski, A. M.

On n-dimensional additive moduli and Diophantine approximations.

Acta Arith. **9** (1964), 391–416.

The main theorem of this paper (Theorem 4) is a quantitative form of Kronecker's theorem, which appears to be both less general and weaker than Hinčin's [Izv. Akad. Nauk SSSR Ser. Mat. **12** (1948), 113–122; MR **9**, 569], which is reproduced with a proof by Birch in the reviewer's tract [*An introduction to Diophantine approximation*, Tracts in Mathematics and Mathematical Physics, No. 45, Chapter V, Theorem 17, Cambridge Univ. Press, New York, 1957; MR **19**, 396]. For his proof the author states that he had to "take up the geometric investigation of n-dimensional lattices ab ovo. This study of metric relations in an n-dimensional lattice turned out to be quite rewarding—this is a more or less new chapter in the affine geometry." In fact this chapter is not quite so new as the author implies, and although his viewpoint is somewhat original, he appears to have rediscovered in a somewhat different form several ideas which are already in the literature. *J. W. S. Cassels* (Cambridge, England)

Citations: MR 9, 569d = J20-8; MR 19, 396h = J02-11.

J20-36 **(31# 4764)**

Wills, Jörg M.

Widerlegung einer Aussage von E. Borel über diophantische Approximationen.

Math. Z. **89** (1965), 411–413.

E. Borel [Leçons sur les séries divergentes, pp. 166–167, Gauthier-Villars, Paris, 1928] made a statement which can be reduced to the following: Let $\gamma_1, \cdots, \gamma_N$ be N irrationals; then the inequalities $|k\gamma_i - n_i - \frac{1}{2}| \leqq \frac{1}{4}$ have solutions in integers $k, n_1, \cdots, n_N$. For the proof he appealed to Kronecker's theorem, which is not applicable since $1, \gamma_1, \cdots, \gamma_N$ need not be linearly independent over the rationals. The author proves that Borel's statement is wrong if $N \geqq 3$ but correct if $N \leqq 2$. If $1, \gamma_1, \gamma_2$ are linearly dependent over the rationals, the distribution of points $(k\gamma_1 - [k\gamma_1], k\gamma_2 - [k\gamma_2])$, k an integer, is completely described in I. Niven, *Diophantine approximations* [Interscience, New York, 1963; MR **26** #6120].

R. P. Bambah (Columbus, Ohio)

Citations: MR 26# 6120 = J02-15.

J20-37 **(33# 1284)**

Goršenin, S. D.

Approximations by the fractional parts of the values of a polynomial of a special form. (Russian. English summary)

Vestnik Moskov. Univ. Ser. I Mat. Meh. **1964**, *no.* 6, 41–47.

Let A be a given real number and let k, r and n be positive integers satisfying $n > 24$, $k \geqq 2(n+1)r$, $1 \leqq r \leqq \sqrt{(n/24)}$. Let p be a sufficiently large prime and suppose that a is prime to p. Write $f(z) = ap^{-k}z^n$ for any integer z. The author proves that there exists a positive integer $z < p^{2k/r}$ and an integer v, such that $|f(z) - v - A| < p^{-k/(12r^2)}$.

R. A. Rankin (Glasgow)

J20-38 **(34# 141)**

Sudan, Gabriel

Sur le problème du rayon réfléchi.

Rev. Roumaine Math. Pures Appl. **10** (1965), 723–733.

The author considers the path of a particle which moves inside of a cube (or square). The path consists of line segments and satisfies the usual law of reflection. Results due to König and Szücz [Rend. Circ. Mat. Palermo **36** (1913), 79–90] are obtained in an elementary way without using the theorem of Kronecker [S.-B. Preuss. Akad. Wissensch. **1884**, 1179–1193] on simultaneous approximation of real numbers, modulo one. *Donald C. Benson* (Davis, Calif.)

J20-39 **(35# 1550)**

Mahler, K.

A remark on Kronecker's theorem.

Enseignement Math. (2) **12** (1966), 183–189.

The author briefly reviews some basic results in the geometry of numbers relating to polar reciprocal convex bodies and their successive minima with respect to a lattice. He then shows how these can be immediately applied to give a simple proof of Kronecker's theorems on inhomogeneous simultaneous approximation.

E. S. Barnes (Adelaide)

J20-40 **(37# 2697)**

Wills, Jörg M.

Zwei Sätze über inhomogene diophantische Approximation von Irrationalzahlen.

Monatsh. Math. **71** (1967), 263–269.

The following theorems are proved. Theorem 1: For each natural n there is a system of real irrational numbers $\alpha_1\alpha_2\cdots\alpha_n$ and real numbers $\beta_1, \beta_2, \cdots, \beta_n$ such that the system of inequalities $\|q\alpha_\nu - p_\nu - \beta_\nu\| \leqq \frac{1}{2}(n-1)/n$ ($\nu = 1, 2, \cdots, n$) has no solution in natural numbers $p_1, \cdots, p_n, q$. ($\|z\|$ is the distance of z from the nearest integer.) Further, for each $C > \frac{1}{2}(n-1)/n$ the system (*) $\|q\alpha_\nu - p_\nu - \beta_\nu\| < C$ ($\nu = 1, 2, \cdots, n$) has a solution. Theorem 2: Two irrational numbers ζ and τ are said to be essentially different if $\varphi \pm \tau$ are irrational. There is a system of pairwise essentially different irrationals $\alpha_1\alpha_2\cdots\alpha_n$ and real numbers $\beta_1\beta_2\cdots\beta_n$, such that the system of inequalities (**) $\|q\alpha_\nu - p_\nu - \beta_\nu\| \leqq \frac{1}{2}(n-1)/(n+1)$ ($\nu = 1, 2, \cdots, n$) has no solution in natural numbers p_ν and q.

Of course, $1, \alpha_1, \cdots, \alpha_n$ are not linearly independent, for (*), or (**) would contradict a classical theorem of Kronecker. The essential point in the paper is the determination of the minimum of Δ for which the system $\|q\alpha_\nu - p_\nu - \beta_\nu\| < \Delta$ ($\nu = 1, 2, \cdots, n$) has a solution for all $\alpha_1, \cdots, \alpha_n$ and $\beta_1, \beta_2, \cdots, \beta_n$ in the case when nothing is assumed about the linear independence of $1, \alpha_1, \cdots, \alpha_n$.

P. Szüsz (Stony Brook, N.Y.)

J20-41 **(44# 1763)**

Moskvin, D. A.

Certain topologico-algebraic structures that are connected with a ring of integers. (Russian)

Kazan. Gos. Univ. Učen. Zap. **127** (1967), *kn.* 3, 99–108.

The author studies and classifies all possible topologies τ on the additive group of the ring Z of rational integers that admit bicompact closure. To each such topology there corresponds a certain additive subgroup G_τ of the additive group of real numbers R_1 that contains elements of arbitrarily high order mod 1. Moreover a pseudobasis of 0 in the topology τ consists of the sets

$$V(\lambda, \varepsilon) = \{x : x \in Z, \|\lambda x\| < \varepsilon\},$$

where $\lambda \in G_\tau$, $\varepsilon > 0$ and $\|x\|$ is the distance to the nearest integer. To a countable G_τ corresponds a metrizable topology, and to a rational G_τ a ring topology on Z.

The author interprets linear Diophantine approximations in metric-topological terms. He considers the character groups of bicompact closures σ_τ of the group Z with respect to the topology τ. In particular, the character group σ_∞ that corresponds to $G_\tau = R_1$ is the rotation group of the circumference endowed with the discrete topology.

E. Novoselov (RŽMat **1969** #3 A125)

J20-42 **(44# 1764)**

Moskvin, D. A.

Compactifications of an additive group of integers and the Kronecker theorem on joint approximations. (Russian)

Kazan. Gos. Univ. Učen. Zap. **128** (1968), *kn.* 2, 83–85.

The author presents a brief algebraic-topological proof of the familiar Kronecker theorem in the theory of Diophantine approximations. He establishes the connection between this theorem and the properties of compactifications of the additive group of integers that he investigated previously in more detail [see #1763 above].

E. Novoselov (RŽMat **1969** #4 A74)

J20-43 **(44# 6613)**

Aggarwal, Satish K.

Inhomogeneous approximation in the field of formal power series.

J. Indian Math. Soc. (*N.S.*) **32** (1968), *suppl. I*, 403–419 (1970).

The author proves the following analogue of a theorem of Hinčin [see, e.g., J. W. S. Cassells, *An introduction to Diophantine approximation*, p. 85, Cambridge Univ. Press, New York, 1957; MR **19**, 396; Russian translation, Izdat. Inostran. Lit., Moscow, 1961; MR **22** #10976] in the field of formal power series over a finite field. Let K be a finite field, $K[t]$ the ring of polynomials over K, $K\{t\}$ the field of formal power series over K with valuation $|f(t)| = b^s$ (fixed $b > 1$) when $f(t) = \sum_{i=s}^{\infty} a_i t^i$ with $a_s \neq 0$, and with $\|f(t)\| = |\sum_{i=-1}^{-\infty} a_i t^i|$; if $L_1(U), L_2(U), \cdots, L_m(U)$ are m linear forms over $K\{t\}$ in n variables, then there exist $c_1, c_2, \cdots, c_m$ in $K\{t\}$ such that $(\max_{1\leq i\leq m}\|L_i(U)+c_i\|)^m \times (\max_{1\leq j\leq n}|u_j|)^n \geqq b^{-m-2n}$ for all non-zero vectors $U = (u_1, u_2, \cdots, u_n)$ in $(K[t])^n$. (Note that the exponents are different from those in Hinčin's theorem.) The author also shows the constant can be improved to b^{-2} in the case where $m = n = 1$, and proves an analogue of a theorem of Kronecker [cf., J. F. Koksma, *Diophantische Approximationen*, Kapitel 7, Ergeb. Math. Grenzgeb., Band 4, Heft 4, Springer, Berlin, 1936]. No new techniques are used.

L. C. Eggan (Normal, Ill.)

Citations: MR 19, 396h = J02-11; MR 22# 10976 = J02-12.

J24 FREQUENCY OR LOCALIZATION OF SOLUTIONS OF DIOPHANTINE INEQUALITIES

See also Section K05.

See also reviews J02-18, J02-20, J02-27, J12-13, J12-16, J12-25, J16-23, J20-20, J64-14, J64-23, J84-4, J84-46, K02-6, K05-77, K10-21, K50-48.

J24-1 **(1, 202d)**

Mahler, Kurt. Note on the sequence $\sqrt{n}$ (mod 1). Nieuw Arch. Wiskde **20**, 176–178 (1940).

Let ξ be any real irrational number, η any real number, ϵ any positive number, c a suitable absolute constant. Theorem 1: the inequality

$$|\xi - x^{\frac{1}{2}} - y| < \frac{1+\epsilon}{2\sqrt{5}\cdot x}$$

has an infinity of integral solutions x, y with $x > 0$. Theorem 1a: the inequality

$$|\xi - (x+\eta)^{\frac{1}{2}} - y| < c/x$$

has an infinity of similar solutions. These are deduced from a theorem of Khintchine [Math. Ann. **111**, 631–637 (1935)]. On the other hand, if ξ is rational, there exists $c = c(\xi) > 0$ such that, for all integers x, y ($x > 0$), either

$$|\xi - x^{\frac{1}{2}} - y| \geqq c/x^{\frac{1}{2}} \quad \text{or} \quad \xi - x^{\frac{1}{2}} - y = 0.$$

H. Davenport (Manchester).

Referred to in J24-8.

J24-2 **(1, 202e)**

Koksma, J. F. Über die asymptotische Verteilung gewisser Zahlfolgen modulo eins. Nieuw Arch. Wiskde **20**, 179–183 (1940).

The author proves the following modification of Theorem 1 of the preceding paper. Let ξ be a real irrational number, but not an algebraic integer of degree 2. For any $\epsilon>0$, at least one of the two inequalities

$$|\xi-x^{\frac{1}{2}}-y|<(1+\epsilon)/8x, \quad |\xi+x^{\frac{1}{2}}-y|<(1+\epsilon)/8x$$

has an infinity of integral solutions with $x>0$. Using Vinogradov's work, an inequality is also given for $|\xi-x^{1/q}-y|$ which has an infinity of solutions. *H. Davenport.*

Referred to in J24-8.

J24-3 **(1, 202f)**

Koksma, J. F. Ueber die asymptotische Verteilung eines beliebigen Systems (f_ν) von n reellen Funktionen f_ν der m ganzzahligen Veränderlichen $x_1, x_2, \cdots, x_m$ modulo Eins. Nederl. Akad. Wetensch., Proc. **43**, 211–214 (1940).

The author proves a general theorem on distributions mod 1; his result in the simplest case of one dimension is the following one: Let $f(1), f(2), \cdots$ be an arbitrary sequence of real numbers, $\varphi(1), \varphi(2), \cdots$ a sequence of positive numbers such that $\sum_{x=1}^{\infty}\varphi(x)$ converges. Then for "nearly all" real α the Diophantine inequalities

$$-\varphi(x)\leqq f(x)-y-\alpha\leqq\varphi(x)$$

have only a finite number of integral solutions $x\geqq1$, y.
K. Mahler (Manchester).

J24-4 **(2, 253b)**

Koksma, J. F. Ueber die Diskrepanz (mod 1) **und die ganzzahligen Lösungen gewisser Ungleichungen.** Nederl. Akad. Wetensch., Proc. **44**, 75–80 (1941).

Suppose $N\geqq1$; let U be a system of N real numbers $f(1), f(2), \cdots, f(N)$, and α, β a pair of real numbers satisfying $\alpha\leqq\beta\leqq\alpha+1$. Let $A(\alpha,\beta)$ be the number of solutions of the Diophantine inequality $\alpha\leqq f(x)<\beta$ (mod 1), and write

(1) $$R=R(\alpha,\beta)=A(\alpha,\beta)-(\beta-\alpha)N.$$

The upper limit D of the numbers $|R|/N$ for all pairs (α,β), $\alpha\leqq\beta\leqq\alpha+1$, is called (after van der Corput) the discrepancy (Diskrepanz) (mod 1) of the system U. Let U^* be the system of the N^2 differences $f(x)-f(z)$ $(x, z=1, 2, \cdots, N)$, D^* the discrepancy (mod 1) of U^*. van der Corput and Pisot [references given in the author's paper] have shown that

(2) $$D\leqq2^{\frac{1}{2}+(1/4\epsilon)}\cdot D^{*\frac{1}{2}-\epsilon}, \qquad \epsilon>0.$$

If $D(N)$ is the discrepancy (mod 1) of the first N terms of the system composed of an infinite series of real numbers $f(1), f(2), \cdots$, and if $H(N)$ tends with increasing N monotonically to zero and satisfies $H(N)>D(N)$, then it follows at once from the definition of D that the Diophantine inequality

(3) $$\alpha-H(x)<f(x)<\alpha+H(x) \text{ (mod 1)}$$

possesses infinitely many distinct integral solutions x. If $D^*(N)\to0$, the van der Corput-Pisot inequality (2) implies that, for every α, $H(x)$ in (2) may be taken approximately of the order $(D^*(x))^{\frac{1}{2}}$. Stated roughly, the author shows that, for almost all α, if $H(x)$ is taken approximately of order $D^*(N)$, then (3) has an infinity of solutions.

The author's result is deduced from the following identity [first stated in terms of sums by Vinogradoff, later in its present form by van der Corput and Pisot; Nederl. Akad. Wetensch., Proc. **42**, 476–486, 554–565, 713–722 (1939); cf. these Rev. **1**, 66]:

(4) $$\int_0^1 R^2(\alpha-t,\alpha+t)d\alpha=\int_0^{2t}R^*(-\alpha,\alpha)d\alpha, \quad 0\leqq t\leqq\tfrac{1}{4},$$

where $R(\alpha,\beta)$, $R^*(\alpha,\beta)$ are the error terms defined by (1) in terms of the systems U, U^*, respectively. The author gives a simple direct proof of (4). *D. C. Spencer.*

Citations: MR 1, 66b = K05-1; MR 1, 66c = K05-2.

J24-5 **(3, 71c)**

Duffin, R. J. and Schaeffer, A. C. Khintchine's problem in metric Diophantine approximation. Duke Math. J. **8**, 243–255 (1941).

Let $\{\alpha_q\}$ be a sequence of positive numbers satisfying: (i) $\sum_{q=1}^{\infty}\alpha_q=\infty$; (ii) $q\alpha_q$ is a decreasing function of q. Khintchine [Math. Ann. **92**, 115–125 (1924); Math. Z. **24**, 706–714 (1926)] has proved under these circumstances that for almost all x there are infinitely many rational numbers p/q for which (1) $|x-p/q|<\alpha_q/q$. The authors suppose only that $\alpha_q\geqq0$, $\sum\alpha_q=\infty$, replace condition (ii) by the weaker hypothesis (iii) that there is a number $c>0$ such that $\sum_{\nu=1}^{n}\alpha_\nu\phi(\nu)/\nu>c\sum_{\nu=1}^{n}\alpha_\nu$ for arbitrarily many n, where $\phi(\nu)$ is Euler's function, and prove that, for almost all x, (1) has an infinity of solutions in relatively prime p and q. An example is constructed which shows that (i) alone is not sufficient to secure the conclusions of Khintchine's theorem. *D. C. Spencer* (Cambridge, Mass.).

Referred to in J24-22, J24-27, J24-31.

J24-6 **(5, 256d; 5, 256e)**

Koksma, J. F. Contribution à la théorie métrique des approximations diophantiques non-linéaires. I. Nederl. Akad. Wetensch., Proc. **45**, 176–183 (1942).

Koksma, J. F. Contribution à la théorie métrique des approximations diophantiques non-linéaires. II. Nederl. Akad. Wetensch., Proc. **45**, 263–268 (1942).

A well-known theorem of Khintchine states that, if n is a positive integer, and if $\omega(x)$ is a positive function of the positive integer x such that $x[\omega(x)]^n$ tends monotonically to zero as $x\to\infty$, then the n simultaneous Diophantine inequalities

$$|\theta_\nu x-y_\nu|<\omega(x), \qquad \nu=1, 2, \cdots, n,$$

where $P(\theta_1, \theta_2, \cdots, \theta_n)$ is a given point of the space R_n, have an infinite number of integral solutions $\{x\geqq1, y_1, y_2, \cdots, y_n\}$ for almost all points P in R_n, provided that the series $\sum_{x=1}^{\infty}[\omega(x)]^n$ is divergent.

The object of the papers is to generalize Khintchine's result to the case in which x can take (for each ν) an assigned sequence s_ν of integral values $f_\nu(1), f_\nu(2), \cdots$ instead of all possible integral values. The Diophantine inequalities considered become

$$|\theta_\nu f_\nu(x)-y_\nu|<\omega_\nu(x), \qquad \nu=1, 2, \cdots, n,$$

under the hypotheses: $\omega_\nu(x)$ positive and monotonic decreasing; $x\prod_1^n\omega_\nu(x)\to0$ for $x\to\infty$; $\sum_1^\infty\prod_1^n\omega_\nu(x)$ divergent (it is also shown that these hypotheses can be transformed so as to become less stringent). Then Khintchine's result holds provided that the sequences s_ν satisfy certain conditions which would be too long to state here, but which are satisfied for very wide types of sequences. *R. Salem.*

Citations: MR 5, 256e = J24-6.

Referred to in H15-11, J24-6.

J24-7 **(7, 370c)**

Koksma, J. F. Sur la théorie métrique des approximations diophantiques. Nederl. Akad. Wetensch., Proc. **48**, 249–265 = Indagationes Math. **7**, 54–70 (1945).

The paper gives a new extension of the well-known theorem of Khintchine on the metric theory of Diophantine approximation. In particular, the following theorem is proved. Let $f(n)$ be a positive function of the integer n such that $f(n+1)-f(n)$ is larger than a positive constant independent of n. Let $g(n)$ be a positive function of n such that $f(n)g(n)$ is monotonic nondecreasing for large n. Let α be a given real number and let $\varphi(n)$ be positive and monotonic nonincreasing. Then for almost all $\theta \geqq 1$ the inequality

$$-\varphi(n) \leqq g(n)\theta^{f(n)} - p - \alpha \leqq \varphi(n)$$

has only a finite number of integral solutions in p, n ($n \geqq 1$) if $\sum \varphi(n) < \infty$, whereas both inequalities

$$0 < g(n)\theta^{f(n)} - p - \alpha < \varphi(n),$$
$$-\varphi(n) < g(n)\theta^{f(n)} - p - \alpha < 0$$

have an infinite number of solutions if $\sum \varphi(n) = \infty$.

Another result is the following: α and $\varphi(n)$ having the same meaning and ω being not greater than 1, if $\sum \varphi([n^{1/\omega}]) = \infty$, then for almost all $\theta \geqq 1$ both inequalities

$$\alpha < \theta^{n^\omega} - p < \alpha + \varphi(n), \quad \alpha - \varphi(n) < \theta^{n^\omega} - p < \alpha$$

have an infinite number of solutions in p, n ($n \geqq 1$). The paper gives other more general theorems of the same type, which there is not space to state here. *R. Salem.*

J24-8 (10, 183a)

Hlawka, Edmund. Über Folgen von Quadratwurzeln komplexer Zahlen. Österreich. Akad. Wiss. Math.-Nat. Kl. S.-B. IIa. **156**, 255–262 (1948).

This paper gives an extension to complex numbers of previous investigations of Mahler [Nieuw Arch. Wiskunde (2) **20**, 176–178 (1940); these Rev. **1**, 202] and Koksma [Nieuw Arch. Wiskunde (2) **20**, 179–183 (1940); these Rev. **1**, 202] on the approximability of real numbers by square roots of rational integers. Let $k(i)$ be the field of rational complex numbers, and let ξ be a complex number which is neither in $k(i)$ nor an integer in any quadratic field over $k(i)$. Then it is shown that for each $\epsilon > 0$ there are infinitely many pairs $x, z \epsilon k(i)$ such that $|\xi + ez^{\frac12} + x| < (1+\epsilon)/4|z|$, where $e = \pm 1$ or $\pm i$ is a unit depending only on ξ, and where $\Re z^{\frac12} \geqq 0$, $\Im z^{\frac12} \geqq 0$. It is not shown that 4 is the best possible constant. A similar theorem is proved when z is restricted to be a complex integer, where now ξ is any number not in $k(i)$. When ξ is in $k(i)$, there is a constant $C(\xi) > 0$ such that for each $z \neq 0$ with $\Re z^{\frac12} \geqq 0$, $\Im z^{\frac12} \geqq 0$ and for each $x \epsilon k(i)$, either $|\xi - z^{\frac12} + x| > C/|z|^{\frac12}$ or $\xi - z^{\frac12} + x = 0$. *W. J. LeVeque.*

Citations: MR 1, 202d = J24-1; MR 1, 202e = J24-2.

J24-9 (10, 354j)

Teghem, J. Sur un système d'inéquations diophantiennes. Acad. Roy. Belgique. Bull. Cl. Sci. (5) **34**, 593–603 (1948).

The following theorem is proved. Let $\delta < 1$, let c_1 and c_2 be positive constants, and let $\theta_1, \theta_2, \cdots$ be a set of real numbers with $\theta_1 > \frac{4}{3}$ and $\theta_\nu - \theta_{\nu-1} \geqq c_1 \nu^{-1}$ ($\nu = 2, 3, \cdots$). Let n be a function of x going to infinity with x, and let $\phi(n)$ be a positive increasing number-theoretic function such that $\lim (\log \phi(n)/\log n) > 0$ and $\liminf (\log \phi(n)/\log \theta_n) \geqq 3$ as $n \to \infty$. Let $c_2 \exp(-\phi(\nu)) \leqq \theta_\nu - [\theta_\nu] \leqq 1 - c_2 \exp(-\phi(\nu))$ for $\nu = 1, 2, \cdots$. Then the system of inequalities

$$0 < x^{\theta_\nu} < \exp(-\log^\delta x) \pmod 1,$$

($\nu = 1, 2, \cdots, n$) has infinitely many integral solutions x if $\limsup (\log \phi(n) + \log n)/\log \log x < 1 - \delta$ as $x \to \infty$. This generalizes an earlier theorem of J. F. Koksma [thesis, Groningen, 1930] in which $\phi(n) = 2^{c_3 n}$ and $\theta_\nu = c_3 \nu + \omega_\nu$, where $c_3 > 0$ and $\omega_1, \omega_2, \cdots$ are bounded and such that $c_3 + \omega_\nu - \omega_{\nu-1} \geqq \alpha > 0$ for some α. The proof makes use of an estimate of Weyl sums obtained by the author in a paper which has apparently not yet appeared.

W. J. LeVeque (Cambridge, Mass.).

Referred to in L15-12.

J24-10 (10, 682e)

Gel'fond, A. O. On some general cases of the distribution of the fractional parts of functions. Doklady Akad. Nauk SSSR (N.S.) **64**, 437–440 (1949). (Russian)

Let L be an increasing (in the wide sense) sequence of real numbers $y_1, y_2, \cdots$ and M a sequence of points $\tau_k = (\beta_{1,k}, \beta_{2,k}, \cdots, \beta_{\nu,k})$ in ν-dimensional space everywhere dense in the unit cube $0 \leqq \beta_{i,k} \leqq 1$ ($1 \leqq i \leqq \nu$, $k = 1, 2, \cdots$). The fractional parts $\{f_i(y)\}$ of the ν functions $f_i(y)$ are said to be (φ, M) everywhere densely distributed if the inequalities

$$0 \leqq \{f_i(y)\} - \beta_{i,k} \leqq \varphi(y) \qquad i = 1, 2, \cdots, \nu,$$

where $\varphi(y)$ is any given monotonically decreasing function satisfying $0 \leqq \varphi(y) \leqq 1$ and $\lim_{y\to\infty} \varphi(y) = 0$, possess for every point τ_k of M an infinity of solutions in values of y belonging to L. The author proves the following two theorems. (I) Let $f_1(t_1, y), f_2(t_2, y), \cdots, f_\nu(t_\nu, y)$ satisfy the following three conditions. (1) The functions $f_i(t_i, y)$ are defined for all $y \epsilon L$ in the segment $a_i \leqq t_i \leqq b_i$ ($1 \leqq i \leqq \nu$). (2) For all i, k, $\partial f_i/\partial t_j = f_i'(t_i, y_k)$ is continuous and increasing in the segment $[a_i, b_i]$ and $f_i'(a_i, y_k) \geqq 1$. (3) For fixed t_i in the segment $[a_i, b_i]$, $f_i'(a_i, y_k)$ tends monotonically to infinity with k so that $f_i(t_i, y_k) \to \infty$ as $k \to \infty$ ($1 \leqq i \leqq \nu$). Then, for any fixed sequence M, arbitrary $\epsilon > 0$ and c_i, d_i such that $a_i \leqq c_i < c_i + \epsilon = d_i \leqq b_i$, and for any given function $\varphi(y)$ of the type defined above, there exists at least one point $(\alpha_1, \alpha_2, \cdots, \alpha_\nu)$ with $c_i \leqq \alpha_i \leqq d_i$ ($1 \leqq i \leqq \nu$) such that the ν functions $f_i(\alpha_i, y)$ are (φ, M) everywhere densely distributed.

(II) Suppose that, in addition to (1), (2) and (3) of the first theorem,

$$\lim_{k\to\infty} \frac{f_i'(a_i, y_{k+1})}{f_i'(b_i, y_k)} \varphi(y_k) = \infty, \qquad 1 \leqq i \leqq \nu.$$

Also let M be an arbitrary infinite sequence of points τ_k in the ν-dimensional unit cube consisting, possibly, of only a finite number of distinct points. Also let $\varphi(y)$ be any given function such that $0 \leqq \varphi(y) \leqq 1$ ($\varphi(y) = 1$ is allowed). Then, in any segment $[c_i, d_i]$ inside $[a_i, b_i]$ there is at least one point such that

$$\lim_{k\to\infty} [\{f_i(\alpha_i, y_k)\} - \beta_{i,k}]/\varphi(y_k) = 0.$$

Examples, for $\nu = 1$, of functions $f(t, y)$ satisfying the conditions of the theorems are given and it is shown that (II) can be used to prove Fabry's theorem on gap series.

R. A. Rankin (Cambridge, England).

Referred to in Q05-20.

J24-11 (12, 162c)

Cassels, J. W. S. Some metrical theorems of Diophantine approximation. II. J. London Math. Soc. **25**, 180–184 (1950).

[For part I cf. the preceding review*.] Generalising previous results of Khintchine and Koksma the author proves the following result: Let $f_n(\theta)$, $n = 1, 2, \cdots$, be an infinite sequence of differentiable functions defined in the interval $(0, 1)$. We assume that $f_n'(\theta) > 0$, $f_n'(\theta)$ is monotone increasing in θ for fixed n and $\lim_{n\to\infty} f_n'(\theta) = \infty$ for fixed θ. Further let $0 \leqq \varphi_n < 1$, $\psi_n = \sum_{k=1}^n \varphi_k$. The author remarks that if $\sum \varphi_n$ converges then (1) $\{f_n(\theta)\} < \varphi_n$ has only a finite number of solutions for almost all θ ($\{x\}$ denotes the fractional part of x). The proof is easy. His second result is very much deeper. Write

$$F_n(\theta) = (f_n'(\theta))^{-1} \sum_{k=1}^{n} f_n'(\theta), \quad E_n = \int_0^1 F_n(\theta) d\theta.$$

*MR **12**, 162B = K30-7.

Assume that $\sum \varphi_n$ diverges and that

$$\sum_{k=1}^{n} \varphi_k E_k = o(\psi_n^2), \quad \sum \varphi_n n/f_n'(0) = o(\psi_n^2).$$

Then the inequality (1) has infinitely many solutions for almost all θ. Interesting special cases are obtained by putting $f_k(\theta) = n_k\theta$. *P. Erdös* (Aberdeen).

Citations: MR 12, 162b = K30-7.

J24-12 (12, 162d)

Cassels, J. W. S. Some metrical theorems of Diophantine approximation. III. Proc. Cambridge Philos. Soc. **46**, 219–225 (1950).

[For part II cf. the preceding review.] Let $\psi(n)>0$ be any function of the positive integer n. The author first of all proves the following theorems. The inequality $0 \leqq n\theta - m - \alpha \leqq \psi(n)$ has an infinity of integer solutions $n>0$ and m for almost all or almost no sets of numbers θ, α according as $\sum\psi(n)$ diverges or converges ("almost all" is understood in the sense of two-dimensional measure). If $\psi(n)$ is monotonically decreasing, then the inequality $0 \leqq n\theta - m < \psi(n)$ has an infinity of integer solutions $n>0$ and m for almost all or almost no θ according as $\sum\psi(n)$ diverges or converges. The condition of monotonicity cannot be omitted. Let λ_n be an increasing sequence of integers. Denote by μ_n the number of fractions of the form k/λ_n, $0<k<\lambda_n$, which are not of the form j/λ_m, $m<n$. The author defines λ_n to be a $\sum$ sequence if $\liminf N^{-1}\sum_{n\leqq N}\mu_n\lambda_n^{-1}>0$. The author proves that if $\psi(n)$ is monotonically decreasing and λ_n is a $\sum$ sequence, then the inequality $0 \leqq \lambda_n\theta - m < \psi(n)$ has an infinity of integer solutions $n>0$ and m for almost all or for almost no θ according as $\sum\psi(n)$ diverges or converges. He also shows that not all sequences are $\sum$ sequences. *P. Erdös.*

Referred to in K40-41.

J24-13 (12, 595f)

Postnikov, A. G. On the structure of two-dimensional Diophantine approximations. Doklady Akad. Nauk SSSR (N.S.) **76**, 493–496 (1951). (Russian)

Let θ be real and p_i, q_i be integer solutions of

$$|\theta - p_i/q_i| < c/q_i^2,$$

where $0<q=q_1<q_2<\cdots<q_t=Q$. Then $Q/q>c^{-\frac{1}{2}[\frac{1}{2}(t-1)]}$. This improves a lemma of Khintchine: $Q/q>((t-1)/2c)^{\frac{1}{2}}$. [Rend. Circ. Mat. Palermo **50**, 170–195 (1926)]. The following two-dimensional analogue is proved. Let $\theta_1>0$, $\theta_2>0$, and let $q_i>0$, $p_i>0$, r_i be N integer solutions of $|q_i\theta_1+p_i\theta_i-r_i|<c/q_ip_i$. Put $P=\max p_i$, $p=\min p_i$, $Q=\max q_i$, $q=\min q_i$. Then

$$(QP/qp)^2 > (n+2m-2)/2nc,$$

where $n+m=N$ and n, m are respectively the number of points of the type $(q_i/r_i, p_i/r_i)$ on and inside their convex cover. *J. W. S. Cassels* (Cambridge, England).

Referred to in J20-16.

J24-14 (16, 1003c)

de Vries, Dirk. Metrische onderzoekingen van Diophantische benaderingsproblemen in het niet-lacunaire geval. [Metrical investigations of Diophantine approximation-problems in the non-lacunary case.] Thesis, Free University of Amsterdam, 1955. 93 pp.

A well known theorem of A. Hinčin [Math. Z. **24**, 706–714 (1926)] asserts that if $\psi(x)$ is positive and continuous and $x\psi(x)$ tends monotonically to zero as $x\to\infty$, then the inequality $|x\theta-y|<\psi(x)$ has infinitely many integral solutions x, y $(x>0)$ for almost all or almost no θ, according as $\sum_1^N \psi(x)$ diverges or converges. J. W. S. Cassels [Proc. Cambridge Philos. Soc. **46**, 209–218 (1950); MR **12**, 162] generalized this theorem, requiring only that $\psi(x)$ be a positive decreasing function of the positive integral variable x, and replacing the above inequality by $|f(x)\theta-y|<\psi(x)$, where $f(x)$ is a Σ-function, i.e., a positive increasing integral-valued function with the following property: if μ_x is the number of fractions $j/f(x)$ $(0<j<f(x))$ which cannot be simplified to the form $k/f(z)$ $(z<x)$, then

$$\liminf_{N\to\infty} N^{-1}\sum_{x=1}^{N}\frac{\mu_x}{f(x)}>0.$$

Up to now, the only non-lacunary functions known to be Σ-functions were the powers x^k, k a positive integer. The first part of this thesis is devoted to extending the list, so that it now includes (among others) polynomials with integral coefficients, $[P_1(x)]$, $[P_1(x)/P_2(x)]$, $[(P_3(x))^{1/n}]$, and $[\log^d x]$, where P_1, P_2, P_3 are polynomials with rational coefficients and $n|\deg P_3$. Second, it is shown that if $\psi(x)$ is positive and decreasing, and $x\psi(x)$ is monotone, and if $a>0$ and b are rational, then the inequality

$$0 \leqq (ax+b)^{1/n}\theta - y < \psi(x)$$

has infinitely many integral solutions for almost all or almost no θ, according as $\sum\psi(x)$ diverges or converges. Cassels' theorem is not applicable here, since $(ax+b)^{1/n}$ is not integral-valued. Finally, V. Jarník's work [Math. Z. **33**, 505–543 (1931)], on the Hausdorff dimension of the exceptional θ-set when $\sum\psi(x)$ converges, is extended; it is shown for example that if $P(x)$ is a polynomial of degree n with integral coefficients, and if $\alpha>n+1$ is constant, then $\dim V=(n+1)/\alpha$, where V is the set of numbers θ for which the inequality $|\theta-y/P(x)|<x^{-\alpha}$ has infinitely many solutions.

All of these results depend on evaluating limits of sums of the form $N^{-1}\sum_1^N \varphi(f(x))/f(x)$, where φ is the Euler function. If such a limit is positive, the inequality occurring above in the definition of Σ-function follows from the fact that $\mu_x \geqq \varphi(x)$. *W. J. LeVeque* (Ann Arbor, Mich.).

Citations: MR 12, 162b = K30-7.

J24-15 (17, 133f)

Szüsz, P. Bemerkungen zur Approximation einer reellen Zahl durch Brüche. Acta Math. Acad. Sci. Hungar. **6** (1955), 203–212. (Russian summary)

For irrational α and $0\leqq\delta\leqq1$, let $M(\alpha,\delta,n)$ be the number of positive integers $q\leqq n$ for which $|q\alpha-p|<q^{\delta-1}$ for suitable integer p. A proof due to Turán is given, showing that $M(\alpha,\delta,n)>n^\delta$ for $\delta>0$. It is also shown that for $\alpha=(1+5^{\frac{1}{2}})/2$ there are infinitely many n for which $M(\alpha,\delta,n)<K(\delta)n^\delta$ for some constant K. Moreover, for $\varepsilon>0$ there is an $N_0=N_0(\varepsilon,\delta)$ such that for every real α there is an n between N and $N^{1+\delta}$ for which $|n\alpha-m|<n^{-\delta}$ for suitable m, if $N>N_0$, but that there is no such n between N and $g(N)N^{1+\delta}$, for infinitely many N, if $g(x)\downarrow0$ as $x\to\infty$. *W. J. LeVeque* (Ann Arbor, Mich.).

Referred to in J24-20, J24-29.

J24-16 (17, 466c)

Kurzweil, J. On the metric theory of inhomogeneous diophantine approximations. Studia Math. **15** (1955), 84–112.

Let ϕ be the homomorphism mapping the additive group of reals onto a circle K of unit circumference, such that for x real, $x'=\phi(x)=\{(2\pi)^{-1}\cos 2\pi z, (2\pi)^{-1}\sin 2\pi x)\}$. Lebesgue measure μ is defined on K in an obvious way. If g and h are real, $g<h$, then the interval $I[g,h]$ is the set of points x', where $g\leqq x\leqq h$. Let B denote a non-empty set of non-increasing sequences $\{b_k\}$ of positive numbers such that $\sum b_k$ diverges. A number $x\in(0,1)$ belongs to the set $\alpha(B)$ if for every $\{b_k\}\in B$, almost every $\eta\in K$ belongs to an infinite number of intervals $I[kx-b_k, kx+b_k]$ $(k=1,2,$

$\cdots$). Let $\tilde{B}$ be the union of all sets B. It is shown that $\alpha(\tilde{B})$ coincides with the set Y of numbers $y \in (0, 1)$ for which there is a constant $c=c(y)>0$ such that $|y-p/q|> c/q^2$ for all sufficiently large q. This settles a question raised by Steinhaus (whether $\alpha(\tilde{B})$ contains all irrational numbers) since Y is of measure 0. The theorem is generalized (e.g., to the case of s linear forms in n variables), and the complementary theorem is proved that if B consists of a single sequence, then $\mu\alpha(B)=1$.

W. J. LeVeque (Ann Arbor, Mich.).

J24-17 **(20# 1671)**

Szüsz, P. Über die metrische Theorie der Diophantischen Approximation. Acta Math. Sci. Hungar. **9** (1958), 177–193.

Let $f(x)$ be a nonnegative function of the positive integer x. The author considers the set $\mathfrak{S}_f$ of pairs (α, β) such that the inequality

$$x\|\alpha x-\beta\|\leqq f(x)$$

has infinitely many positive integer solutions x, where $\|y\|$ denotes the distance of y from the nearest integer. He proves first that if $f(x)$ is monotone decreasing, then for fixed β almost all or almost no α are such that $(\alpha, \beta) \in \mathfrak{S}_f$ according as $\sum x^{-1}f(x)$ diverges or converges. For $\beta=0$ this result is due to Khintchine (=Hinčin) [Math. Ann. **92** (1924), 115–124], and the author's proof is a development of Khintchine's. The author remarks that although Khintchine's result generalizes to n dimensions, it is not clear whether his will. Secondly, the author shows that there is no comparable result when the rôles of α and β are interchanged: more precisely, if $f(x)$ tends to 0, however slowly, there exist α such that $(\alpha, \beta) \in \mathfrak{S}_f$ for almost no β.

J. W. S. Cassels (Cambridge, England)

Referred to in J24-30, J24-33.

J24-18 **(20# 2314)**

Leveque, William J. On the frequency of small fractional parts in certain real sequences. Trans. Amer. Math. Soc. **87** (1958), 237–261.

Let $X_1, X_2, \cdots$ be a sequence of independent random variables, each uniformly distributed on $[0, \frac{1}{2}]$. Then if $f(k)$ is an arbitrary function of the positive integer k, whose values lie in $[0, \frac{1}{2}]$, the equation $\Pr\{X_k<f(k)\}=2f(k)$ holds, and from the Borel-Cantelli lemmas [cf. W. Feller, Probability theory and its applications, Wiley, New York, 1950; MR **12**, 424], it follows that the probability that the inequality (1) $X_k<f(k)$ is satisfied for infinitely many k is zero or one, according as (2) $\sum_{k=1}^{\infty} f(k)<$ or $=\infty$. If one puts (3) U_k=distance $\langle kx\rangle$ between kx and its nearest integer ($k\geqq1$, x a random real variable), the sequence $U_1, U_2, \cdots$ is a sequence of dependent random variables, uniformly distributed on $[0, \frac{1}{2}]$. A well-known theorem of A. Khintchine [Math. Ann. **92** (1924), 115–125] shows that nevertheless the above statement concerning (1) remains true if one replaces X_k by U_k. The author investigates whether the U_k resemble the X_k also in their finer structure. He considers the case where (2) diverges and finds a result which is not quite what one might expect from the case of independent variables. If however in (3) the sequence $\langle kx\rangle$ is replaced by $\langle r_1r_2\cdots r_kx\rangle$, where $r_1<r_2<\cdots$ is a fixed increasing sequence of positive integers, the sequence $U_1, U_2, \cdots$ is much less strongly dependent and the results are in accordance with the situation in case of independency.

The results would take too much space to be quoted here in full. Moreover the author pointed out in a letter that he discovered an error in the last sentence of p. 246, which forces him to replace in Theorem 1 (the case (3)) the quantity T_n by

$$U_n=\mathrm{No}\{m\leqq n|\langle mx\rangle<g(m),\ (m, l_m)=1\},$$

where $l_m=l_m(x)$ is the integer nearest to mx. There will appear a further paper correcting and extending the above indicated results.

J. F. Koksma (Amsterdam)

Referred to in J24-24, J24-30, J24-35, J24-49.

J24-19 **(21# 1290)**

Erdös, P.; Szüsz, P.; and Turán, P. Remarks on the theory of diophantine approximation. Colloq. Math. **6** (1958), 119–126.

For fixed $A>0$, $c>1$ and integer $N\geqq 2$, denote by $S(N, A, c)$ the set of $\alpha \in [0, 1]$ such that for some integers x and y with $(x, y)=1$ and $N\leqq y<cN$, the inequality $|\alpha-x/y|<A/y^2$ holds. The question is considered whether $\lim_{N\to\infty}$ meas $S(N, A, c)=f(A, c)$ exists, and if so, what its nature is. It is shown that for $0<A<c/(1+c^2)$, $f(A, c)=12A\pi^{-2}\log c$, and that for $A>10$ and $c>10$ (for example), meas $S(N, A, c)$ is bounded away from 0 and 1 as $N\to\infty$, explicit bounds being given. From the lower bound it follows immediately that if $R(N, c)$ is the set of $\alpha \in [0, 1]$ for which the interval $N\leqq y\leqq cN$ contains the denominator q_ν of at least one convergent to the regular continued fraction expansion of α, then $\liminf_{N\to\infty}$ meas $R(N, c)\geqq 3\pi^{-2}\cdot(1-c^{-2})$.

W. J. LeVeque (Ann Arbor, Mich.)

Referred to in J24-26, J24-28, J24-47.

J24-20 **(21# 6358)**

Szüsz, P. Über die Approximation einer reellen Zahl durch Brüche. Acta Math. Acad. Sci. Hungar. **10** (1959), 69–75. (Russian summary, unbound insert)

Refining the results of an earlier paper [same Acta **6** (1955), 203–212; MR **17**, 133], the following theorems are proved. (I) If α is an arbitrary irrational number and $\varepsilon>0$, there is an $n_0=n_0(\alpha, \varepsilon)$ such that for each $n\geqq n_0$ there is a natural number x for which the relations $\|\alpha x\|<(5^{1/2}x)^{-1}$ and $n\leqq x<\varepsilon n^2$ hold. (II) Let $g(x)\to 0$ monotonically as $x\to\infty$, and let $f(x)$ be an increasing function. Then there is an irrational number α and an increasing function $f_1(x)$ with the following properties: (a) for infinitely many natural numbers x, $f_1(x)=f(x)$, and (b) for infinitely many positive integers N the inequality $\|\alpha x\|>f_1(x)/x$ holds for all integers x with $N\leqq x<g(N)N^2$.

W. J. LeVeque (Ann Arbor, Mich.)

Citations: MR 17, 133f = J24-15.

J24-21 **(22# 160)**

Gosselin, Richard P. On Diophantine approximation and trigonometric polynomials. Pacific J. Math. **9** (1959), 1071–1081; erratum, **10** (1960), 1479.

Le résultat principal concerne la divergence des polynômes trigonométriques d'interpolation. On désigne par $I_{n,u}(x; f)$ le polynôme trigonométrique d'ordre n prenant les mêmes valeurs que f aux points $u+2k\pi(2n+1)^{-1}$, et par ψ une fonction croissante définie sur $(0, \infty)$. Théorème: Il existe une fonction f telle que $\psi(|f|) \in L^1(0, 2\pi)$ et telle que la suite $I_{n,n}(x; f)$ diverge p.p. dans le carré $0\leqq x\leqq 2\pi$, $0\leqq u\leqq 2\pi$. Pour $\psi(x)=x^p$ $(1\leqq p<2)$, on retrouve un résultat de Marcinkiewicz et Zygmund [Fund. Math. **28** (1936), 131–166]. Un théorème auxiliaire concerne la mesure de l'ensemble des points $x \in [0, 1]$ qui satisfont la condition suivante (où $\gamma>0$ et m entier sont fixés): il existe une fraction irréductible p/q, telle que $\gamma m<q\leqq m$ et $|x-p/q|\leqq 1/\gamma m^2$ [resp. $2/\gamma^2m^2$].

J.-P. Kahane (Montpellier)

J24-22 **(22# 7971)**

Hartman, S.; Szüsz, P. On congruence classes of denominators of convergents. Acta Arith. **6** (1960), 179–184.

Verff. beweisen den folgenden Satz: (c_k) sei eine abnehmende Folge, $\sum_1^\infty c_k=\infty$. Sei $c_k'=c_k$ für $k\equiv b \pmod{a}$, $=0$ sonst. Dann ist für fast alle reellen Zahlen α für unendlich viele k die Ungleichung $|\alpha k-p|<c_k'$ mit passendem p, $(k,p)=1$, erfüllt. Der Beweis beruht auf einem Resultat von Duffin und Schaeffer [Duke Math. J. **8** (1941), 243–255; MR **3**, 71]. Einige weitere Anwendungen des Satzes werden gegeben. *H.-E. Richert* (Göttingen)

Citations: MR 3, 71c = J24-5.

Referred to in J24-30.

J24-23 (22# 9482)

Schmidt, Wolfgang. A metrical theorem in diophantine approximation. Canad. J. Math. **12** (1960), 619–631.

The author proves the following more precise version of an old theorem of Khintchine (Hinčin) [Math. Z. **24** (1926), 706–714]:

Let $\psi_1(q), \cdots, \psi_n(q)$ be n non-negative functions of the positive integer q and suppose that

$$(1)\qquad \psi(q) = \prod_{1\leq i\leq n} \psi_i(q)$$

is monotone decreasing. For real numbers $\theta_1, \cdots, \theta_n$ denote by $N(h;\theta_1,\cdots,\theta_n)$ the number of integral solutions of

$$(2)\qquad 0 \leq q\theta_i-p_i < \psi_i(q) \quad (1\leq i\leq n)$$

with $1\leq q\leq h$, and put $\Psi(h)=\sum_{1\leq q\leq h}\psi(q)$, $\Omega(h)=\sum_{1\leq q\leq h}q^{-1}\psi(q)$. Then

$$(3)\quad N(h;\theta_1,\cdots,\theta_n) = \Psi(h)+O(\Psi^{1/2}(h)\Omega^{1/2}(h)\log^{2+\varepsilon}\Psi(h))$$

for almost all sets $\theta_1, \cdots, \theta_n$.

Khintchine's theorem is, substantially, the statement that $N(h;\theta_1,\cdots,\theta_n)$ is bounded for almost all $\theta_1,\cdots,\theta_n$ as $h\to\infty$ if and only if $\Psi(h)$ is bounded.

The author also indicates the proof of a similar theorem in which $q\theta_i$ and $\psi_i(q)$ in (2) are replaced respectively by a linear form $\theta_{i1}q_1+\cdots+\theta_{im}q_m$ in integers $q_1,\cdots,q_m$, and by a non-negative function $\psi_i(q_1,\cdots,q_m)$ respectively. In this second theorem there is no analogue of the condition that $\psi(q)$ be monotone but, presumably, the explicit error term that is given in the analogue of (3) is worse than that in (3).

The proofs require considerable modification of pre-existing techniques.

J. W. S. Cassels (Cambridge, England)

Referred to in J24-30, J24-33, J24-35, J24-39, J24-49, J24-52, J24-59.

J24-24 (22# 12089)

LeVeque, W. J. On the frequency of small fractional parts in certain real sequences. II. Trans. Amer. Math. Soc. **94** (1960), 130–149.

This paper corrects and extends Theorem 1 of a previous paper under the same title [same Trans. **87** (1958), 237–261; MR **20** #2314]. The correction needed was already referred to in the review cited. The extension reads as follows. Let $\langle x\rangle$ be the distance between x and the integer nearest x and let $g(x)$ satisfy the following conditions: (a) $xg(x)$ is non-increasing and $0<xg(x)<\frac{1}{2}$, for $x\geq 0$; (b) $xg(x)\to 0$ as $x\to\infty$; (c) $\sum_1^\infty g(k)=\infty$. Then for almost all x, the number of solutions $m\leq n$ of the inequality $\langle mx\rangle<g(m)$ is asymptotic to $2\sum_1^n g(k)$. A conjecture is formulated which has since been proved by Erdös.

C. G. Lekkerkerker (Amsterdam)

Citations: MR 20# 2314 = J24-18.

Referred to in J24-26, J24-30, J24-35, J24-49.

J24-25 (22# 12090)

LeVeque, W. J. On the frequency of small fractional parts in certain real sequences. III. J. Reine Angew. Math. **202** (1959), 215–220.

Theorem: Let $f(n)$ be a non-negative non-increasing function of the natural numbers n such that $f(1)\leq\frac{1}{2}$ and such that $\sum f(n)$ diverges. Let a_k $(k=1,2,3,\cdots)$ be an increasing sequence of positive integers and suppose that

$$\sum_{k\leq n}\frac{f(k)}{a_k}\sum_{l\leq k}(a_k,a_l) = O\{(\sum_{k\leq n}f(k))^{2-\eta}\}$$

for some constant $\eta>0$, as $n\to\infty$. Then for every sequence α_k of real numbers the number of solutions of

$$(1)\qquad \|a_lx-\alpha_l\| \leq f(l) \qquad (l\leq k)$$

is $\{2+o(1)\}f(k)$ $(k\to\infty)$ for almost all real numbers x. Here $\|y\|$ denotes the distance of the real number y from the nearest integer.

This theorem thus gives a condition under which an analogue of the strong law of large numbers holds for the events (1). The analogue of the strong law of large numbers has been shown to hold in other cases, not all of them covered by this theorem [see #12089 and reference therein; see also #12091]. The proof of the theorem is particularly simple, consisting of little more than an estimate of variance and an application of the Borel-Cantelli lemmas. The cross-terms occurring in the variance are estimated by an application of trigonometric sums which may have wider application.

J. W. S. Cassels (Cambridge, England)

Referred to in J02-20, J24-28, J24-33, J24-45, J24-56, K50-48.

J24-26 (22# 12091)

Erdös, P. Some results on diophantine approximation. Acta Arith. **5**, 359–369 (1959).

Let $\varphi(n,\varepsilon,C)$ be the set of $\alpha\in(0,1)$ for which the inequality $|q\alpha-p|<\varepsilon q^{-1}$ has no integral solutions p and q with $n<q<Cn$ and $(p,q)=1$. The author, Szüsz and Turán conjectured [Colloq. Math. **6** (1958), 119–126; MR **21** #1290] that for every ε and C, $\lim_n \text{meas}[\varphi(n,\varepsilon,C)]$ exists. This question remains open, but it is shown here that for every ε and η there is a $C=C(\varepsilon,\eta)$ such that for every n, $\text{meas}[\varphi(n,\varepsilon,C)]<\eta$. The same kind of proof (depending in the end on Chebyshev's inequality from probability theory) also yields the following theorem: Let $l(n)>0$ be a nondecreasing function for which $\sum(l(n))^{-1}$ diverges, and denote by $N(l,\alpha,n)$ the number of integers m such that $1\leq m\leq n$ and $m\alpha-[m\alpha]<(l(m))^{-1}$. Then for almost all α, $N(l,\alpha,n)\sim\sum_1^n(l(m))^{-1}$ as $n\to\infty$. This extends a theorem of the reviewer [#12089] in which it was further required that $l(n)=o(n^{-1})$.

W. J. LeVeque (Ann Arbor, Mich.)

Citations: MR 21# 1290 = J24-19; MR 22# 12089 = J24-24.

Referred to in J24-30, J24-33, J24-35, J24-49.

J24-27 (24# A3131)

Gallagher, Patrick

Approximation by reduced fractions.

J. Math. Soc. Japan **13** (1961), 342–345.

Let $\{\delta(n)\}$ be a sequence of non-negative real numbers. The author shows that the set of real numbers x in $0\leq x<1$ for which (1) $|x-m/n|<\delta(n)$ has infinitely many integer solutions (m,n), $n>0$, with (2) $\gcd(m,n)=1$ has measure either 0 or 1. The corresponding result, but without the condition (2), was already known [cf. the reviewer, Proc. Cambridge Philos. Soc. **46** (1950), 209–218; MR **12**, 162]; the author uses considerable ingenuity to deal with the condition (2). The result is of interest in view of the conjecture that a necessary and sufficient condition for measure 1 is that the series $\sum\delta(n)$ diverge. Duffin and Schaeffer [Duke Math. J. **8** (1941), 243–255; MR **3**, 71] have given an example of a sequence $\delta(n)$ such that the measure without the condition (2) is 1, but with the condition (2) is 0.

J. W. S. Cassels (Cambridge, England)

Citations: MR 3, 71c = J24-5; MR 12, 162b = K30-7.

J24-28 **(25# 1142)**

Kesten, Harry

Some probabilistic theorems on Diophantine approximations.

Trans. Amer. Math. Soc. **103** (1962), 189–217.

Let $\langle \xi \rangle$ be the distance between ξ and the integer nearest ξ, let $x_1, \cdots, x_j$ be independent random variables, each uniformly distributed on [0, 1], and let $N(m, \gamma, j)$ be the number of indices $k \leqq m$ for which simultaneously $\langle kx_i \rangle \leqq \gamma$ for $i = 1, \cdots, j$. Theorem 1: $\lim_{m\to\infty} \text{Prob}\{N(m, \alpha/m, 1) \geqq 1\} = F(\alpha)$, where

$$F(\alpha) = 0, \quad \text{if } \alpha < 0,$$
$$= 12\alpha\pi^{-2}, \quad \text{if } 0 < \alpha \leqq \tfrac{1}{2},$$
$$= 12\pi^{-2}\left(\alpha - \int_{1/2}^{\alpha}\left(2 - \frac{1}{y} - \frac{1-y}{y}\log\frac{y}{1-y}\right)dy\right), \quad \text{if } \tfrac{1}{2} < \alpha \leqq 1,$$
$$= 1, \quad \text{if } 1 < \alpha.$$

This extends a result of Friedman and Niven [same Trans. **92** (1959), 25–34; MR **21** #4947].

Let $S(m, \alpha, c)$ be the set of $\xi \in [0, 1]$ such that there exist integers a, b for which $m \leqq b \leqq mc$, $(a, b) = 1$, and $|b\xi - a| \leqq \alpha/b$. Erdős, Szüsz and Turán [Colloq. Math. **6** (1958), 119–126; MR **21** #1290] evaluated $\lim_{m\to\infty} \text{meas}\{S(m, \alpha, c)\}$ for $\alpha < c/(1+c^2)$. In Theorem 2, this limit is evaluated for $c/(1+c^2) \leqq \alpha \leqq \max(\frac{1}{2}, 1/c)$; the result is complicated.

Theorem 3: If $\gamma \in [0, \frac{1}{2}]$ is fixed, and $m, p \to \infty$ in such fashion that $m(2\gamma)^p \to \lambda > 0$, then $N(m, \gamma, p)$ has asymptotically a Poisson distribution with mean λ:

$$\lim \text{Prob}\{N(m, \gamma, p) = k\} = e^{-\lambda}\frac{\lambda^k}{k!}, \quad k = 0, 1, \cdots.$$

The proof, which is not simple, is by the method of moments, using a result of the reviewer [J. Reine Angew. Math. **202** (1959), 215–220; MR **22** #12090] for the first case in an induction. *W. J. LeVeque* (Ann Arbor, Mich.)

Citations: MR 21# 1290 = J24-19; MR 21# 4947 = K10-21; MR 22# 12090 = J24-25.

J24-29 **(25# 3009)**

Sós, Vera T.

On a problem in the theory of simultaneous approximation.

Ann. Univ. Sci. Budapest. Eötvös Sect. Math. **3–4** (1960/61), 291–294.

The author proves the following extension of Dirichlet's Theorem: Let $\alpha_1, \cdots, \alpha_k$ be irrational and $0 < \theta < 1/k$. Then there exist integers $p_1, \cdots, p_k, q > 0$ such that

$$(*) \qquad |q\alpha_\nu - p| < 2^{k+1}/q, \quad \nu = 1, \cdots, k,$$

and

$$(**) \qquad N \leqq q \leqq N^{(1+\theta)/1-\theta(k-1)},$$

where N is arbitrary. It is not known if the exponent in (**) is best possible. The question of localisation of q was raised by Turán and was completely solved by Szüsz for $k = 1$, with 1 instead of 4 in (*) [Szüsz, Acta Math. Acad. Sci. Hungar. **6** (1955), 203–212; MR **17**, 133].

R. P. Bambah (Chandigarh)

Citations: MR 17, 133f = J24-15.

Referred to in J24-41.

J24-30 **(27# 3585)**

Szüsz, P.

Über die metrische Theorie der diophantischen Approximation. II.

Acta Arith. **8** (1962/63), 225–241.

Part I appeared in Acta Math. Sci. Hungar. **9** (1958), 177–193 [MR **20** #1671]. Let $f(k)$ be a monotonely decreasing positive function of the natural number k such that $f(k) \leqq \frac{1}{2}$ and $\sum k^{-1}f(k)$ diverges. The function f is also supposed to satisfy a rather elaborate condition to ensure that it does not decrease too quickly, but in a footnote the author states that he can now dispense with it. Then for fixed k' and r and almost all real α the number $N^*(n)$ of solutions of $|\alpha k - l| \leqq k^{-1}f(k)$, $(l, k) = 1$, $0 < k \leqq n$, with (P) $k \equiv k' \pmod r$, satisfies the asymptotic equality

$$N^*(n) \sim \frac{12}{\pi^2}\frac{r\varphi((r, k'))}{C(r)(r, k')}\sum_{k=1}^{n}\frac{f(k)}{k},$$

where $C(r) = r^2 \prod_{p|r}(1 - p^{-2})$. There is also a similar but considerably more elaborate expression for the number $N(n)$ when the condition $(l, k) = 1$ is omitted. These results give quantitative form to earlier qualitative results of the author and Hartman [Acta Arith. **6** (1960), 179–184; MR **22** #7971]. Similar asymptotic expressions when the condition (P) is omitted were obtained independently by LeVeque [Trans. Amer. Math. Soc. **87** (1958), 237–261; MR **20** #2314; ibid. **94** (1960), 130–149; MR **22** #12089], Erdős [Acta Arith. **5** (1959), 359–369; MR **22** #12091] and Schmidt [Canad. J. Math. **12** (1960), 619–631; MR **22** #9482]. The author's proof uses probabilistic methods and the theory of continued fractions. Consequently, it would be interesting to see if the result can be generalized to simultaneous approximation.

J. W. S. Cassels (Cambridge, England)

Citations: MR 20# 1671 = J24-17; MR 20# 2314 = J24-18; MR 22# 7971 = J24-22; MR 22# 9482 = J24-23; MR 22# 12089 = J24-24; MR 22# 12091 = J24-26.

J24-31 **(28# 1167)**

Gallagher, P.

Metric simultaneous diophantine approximation.

J. London Math. Soc. **37** (1962), 387–390.

Denote by I the cube $0 \leqq x_i < 1$ $(1 \leqq i \leqq r)$ and by $|A|$ the measure of a measurable set A in I. Let $\{A_n\}$ be a sequence of measurable subsets of I of decreasing measure such that, whenever the point $(y_1, \cdots, y_r)$ is in A_n, so also is the cube $0 \leqq x_i \leqq y_i$ $(1 \leqq i \leqq r)$. The author shows that for almost all points X of R^r there are infinitely many positive integers n and lattice vectors L with all components prime to n such that $nX - L \in A_n$, provided $\sum |A_n|$ diverges. This is an extension of results of Khintchine [Math. Ann. **92** (1924), 115–125], Duffin and Schaeffer [Duke Math. J. **8** (1941), 243–255; MR **3**, 71] and Cassels [Proc. Cambridge Philos. Soc. **46** (1950), 209–218; MR **12**, 162], and the author remarks that his proof is a modification of the work of Cassels. *A. C. Woods* (Columbus, Ohio)

Citations: MR 3, 71c = J24-5; MR 12, 162b = K30-7.

Referred to in J24-39.

J24-32 **(28# 2088)**

Sprindžuk, V. G.

On theorems of Khinchin and Kubilius. (Russian. Lithuanian and English summaries)

Litovsk. Mat. Sb. **2** (1962), *no.* 1, 147–152.

Für reelles x bedeute $\|x\|$ den Abstand von der nächstliegenden ganzen Zahl. Sei $n \geqq 1$ ganz, $m = n(n+3)/2$, $k = (n/m) - 1$, $\varphi(q)$ eine positive Zahlenfolge mit konvergenter Reihe $\sum_{q=1}^{\infty} \varphi^n(q)q^k$. Es wird bewiesen, daß für fast alle Zahlen-n-tupel $(\omega_1, \cdots, \omega_n)$ die Ungleichung $\max_{i,j=1,\cdots,n} (\|\omega_i q\|, \|\omega_i\omega_j q\|) < \varphi(q)$ höchstens eindlich-viele ganzzahlige Lösungen q hat. {In der englischen Zusammenfassung fehlt bei $\omega_i\omega_j$ der Faktor q.} Daraus wird mit Hilfe des Khintchineschen Übertragungsprinzips [Rend. Circ. Mat. Palermo **50** (1926), 170–195] die Aussage hergeleitet, daß für gegebenes $\delta > 0$ bei fast allen $(\omega_1, \cdots, \omega_n)$ jedes quadratische Polynom $\Phi(x_1, \cdots, x_n)$ mit ganzzahligen Koeffizienten der Ungleichung $|\Phi(\omega_1, \cdots, \omega_n)| > \|\Phi\|^{-m-\delta}$ genügt, sofern seine Höhe $\|\Phi\|$ oberhalb einer Schranke $h_0(\omega_1, \cdots, \omega_n; \delta)$ liegt.

B. Volkmann (Mainz)

Referred to in J12-60.

J24-33 (28# 3018)

Schmidt, Wolfgang M.

Metrical theorems on fractional parts of sequences.

Trans. Amer. Math. Soc. **110** (1964), 493–518.

Let $\{x\}$ be the fractional part of x and $\|x\|$ the distance from x to the nearest integer. Let $I_{j,q}$ $(j = 1, \cdots, n;$ $q = 1, 2, \cdots)$ be intervals in $[0, 1)$ of length $L_{j,q}$. (For present purposes, besides ordinary intervals, also sets of the form $[1-a, 1) \cup [0, b)$ will be called intervals of length $a+b$ if $a+b<1$.) The following metric theorem is the main result of the paper. Let $P_1, \cdots, P_n$ be n non-constant polynomials with integer coefficients and $\psi(h) = \sum_{q=1}^{h} \prod_{j=1}^{n} L_{j,q}$. Denote by $N(h; \alpha_1, \cdots, \alpha_n)$ the number of integers q, $1 \leqq q \leqq h$ with $\{\alpha_j P_j(q)\} \in I_{j,q}$. If the intervals are decreasing, i.e., $I_{j,1} \supset I_{j,2} \supset \cdots$, then, for each $\varepsilon > 0$,

$$N(h; \alpha_1, \cdots, \alpha_n) = \psi(h) + O(\psi(h)^{1/2+\varepsilon})$$

for almost all real n-tuples $\alpha_1, \cdots, \alpha_n$.

If $\theta_j \in \bigcap_{q=1}^{\infty} I_{j,q}$ one can look on the problem as one concerning the simultaneous approximation of $\theta_1, \cdots, \theta_n$ by $\alpha_1 P_1, \cdots, \alpha_n P_n$. The above theorem unifies and generalizes a large number of special cases which had been proved by Khintchine [Math. Ann. **92** (1924), 115–125], Cassels [Proc. Cambridge Philos. Soc. **46** (1950), 209–218; MR **12**, 162], Szüsz [Acta Math. Acad. Sci. Hungar. **9** (1958), 177–193; MR **20** #1671], LeVeque [J. Reine Angew. Math. **202** (1959), 215–220; MR **22** #12090], Erdős [Acta Arith. **5** (1959), 359–369; MR **22** #12091] and the author [Canad. J. Math. **12** (1960), 619–631; MR **22** #9482].

The proof is based on an application of Chebyshev's inequality, and careful analysis is needed for the estimate of the variance of an approximation to $N(h; \alpha_1, \cdots, \alpha_n)$.

In the special case $P_j(q) = q$, $I_{j,q} = I_j$, $q = 1, 2, \cdots, j = 1, \cdots, n$ the error term $\psi(h)^{1/2+\varepsilon}$ may be replaced by $(\log h)^{n+1+\varepsilon}$. This is derived from an estimate for

$$\sum_{1 \leqq q_i \leqq h} \left(q_1, \cdots, q_n \left\| \sum_{i=1}^{n} \alpha_i a_i(q) + \theta \right\|\right)^{-1}$$

for arbitrary fixed increasing sequences of integers $a_i(q)$ $q = 1, 2, \cdots$ and fixed θ.

There is also a discussion of the case where the restrictions $\{\alpha_j P_j(q)\} \in I_{j,q}$ are replaced by restrictions of the form $\{\sum_{j=1}^{n} \alpha_j P_j(q_j)\} \in I_{q_1 \cdots q_n}$ for intervals $I_{q_1 \cdots q_n}$.

H. Kesten (Ithaca, N.Y.)

Citations: MR 12, 162b = K30-7; MR 20# 1671 = J24-17; MR 22# 9482 = J24-23; MR 22# 12090 = J24-25; MR 22# 12091 = J24-26.

Referred to in J24-39, J24-54.

J24-34 (31# 3382)

Erdős, P.

Problems and results on diophantine approximations.

Compositio Math. **16**, 52–65 (1964).

This is a discussion of several unsolved problems in diophantine approximations, some of which are preceded by a brief survey of what is known. The following results are announced without proof. For any positive ε let $\{\varepsilon_q; q = 1, 2, \cdots\}$ be a sequence with $\varepsilon_q = 0$ or $\varepsilon_q = \varepsilon$. Then in order that for almost all real α there exist infinitely many rational numbers p/q such that $|\alpha - p/q| < \varepsilon_q/q^2$, it is necessary and sufficient that $\sum_{q=1}^{\infty} \varepsilon_q \varphi(q)/q^2$ diverge, where $\varphi(q)$ is the Euler φ function. Next, let $n_1, n_2, \cdots$ be a sequence of positive integers. Then in order that for almost all α, infinitely many of the n_i be denominators of the continued fraction expansion of α, it is necessary and sufficient that $\sum_{i=1}^{\infty} \varphi(n_i)/n_i^2$ diverge.

I. Niven (Eugene, Ore.)

Referred to in K05-56, K05-60, K05-62, K05-70, K10-36.

J24-35 (31# 3383)

Lang, Serge

Asymptotic approximations to quadratic irrationalities. I, II.

Amer. J. Math. **87** (1965), 481–487; *ibid.* **87** (1965), 488–496.

Let β be a real quadratic irrational. The theorem of the first part states that if $c \geqq 1$ is given, the number $\lambda(B)$ of integers q such that $0 < q\beta - p < c/q$ $(p \in \mathbf{Z})$, $|q| \leqq B$, satisfies the inequality $|\lambda(B) - c_1 \log B| \leqq c_2$, where c_1, c_2 are constants depending only on β, c ($\mathbf{Z}$ is the set of rational integers). The main theorem of the second part is too circumstantial to be quoted here. A particular case (Theorem 2) states that the number of solutions of $|q\beta - p| < \psi(|q|)$ $(p \in \mathbf{Z})$, $|q| \leqq B$, is

$$(*) \qquad c_3 \int_1^B \psi(t)\,dt + O(\omega(B) + \log B\omega(B)^{1/2}),$$

for some constant c_3, where $\psi(t)$ is a decreasing function such that $\omega(t) = t\psi(t)$ is increasing, and $\omega'(t) > t^{-2}$ for all sufficiently large t. These results are compared with the asymptotics of the "almost all real β" case [LeVeque, Trans. Amer. Math. Soc. **87** (1958), 237–361; MR **20** #2314; ibid. **94** (1960), 130–149; MR **22** #12089; Erdős, Acta Arith. **5** (1959), 359–369; MR **22** #12091; and Schmidt, Canad. J. Math. **12** (1960), 619–631; MR **22** #9482; of whom the author mentions only the last].

The proofs depend on a study of the automorphisms of the integral quadratic form satisfied by β and of their effect on the representation of comparatively small integers.

{In (*) the constant c_3 (called c_1 in the paper) is said to depend only on β. The reviewer believes it to be obvious that it is, in fact, 4, and that a stronger estimate than (*) is valid for all β with bounded partial quotients.}

J. W. S. Cassels (Cambridge, England)

Citations: MR 20# 2314 = J24-18; MR 22# 9482 = J24-23; MR 22# 12089 = J24-24; MR 22# 12091 = J24-26.

Referred to in J02-20, J24-40, J24-61.

J24-36 (31# 5836)

Davenport, H.

On a theorem of Mrs. Turán.

Proc. London Math. Soc. (3) **14a** (1965), 76–80.

The following theorem is proved. Let k be fixed and let ϑ satisfy $0 < \vartheta \leqq 1/k$. Then for any positive numbers D, δ, there exist irrational numbers $\alpha_1, \cdots, \alpha_k$ with the property that for a certain sequence $N_1 < N_2 < \cdots$ of positive integers, the simultaneous inequalities $\|\alpha_j x\| < Dx^{-\vartheta}$ $(j = 1, 2, \cdots, k)$ are not satisfied by any integer x in any of the intervals $N_i \leqq x \leqq N_i^{\varphi-\delta}$, where $\varphi = (1+\vartheta)/(1-(k-1)\vartheta)$ and $\|\xi\|$ denotes the distance from a real number ξ to the nearest integer.

J. E. Cigler (Groningen)

Referred to in J24-41.

J24-37 **(32# 91)**

Adams, W.; Lang, S.

Some computations in diophantine approximations.

J. Reine Angew. Math. **220** (1965), 163–173.

Let $w_1, \cdots, w_m$ be fixed real numbers linearly independent over the rationals, and let $\lambda(B, c)$ denote the number of solutions of the inequality $|q_1w_1 + \cdots + q_mw_m| \leqq c/\max|q_i|^{m-1}$ in integers $q_1, \cdots, q_m$ with $\max|q_i| \leqq B$. It is known that $\lambda(B, c)/\log B$ tends to a limit for fixed c as B tends to infinity for almost all sets of real numbers $(w_1, \cdots, w_m)$. This paper consists primarily of about nine pages of computer-output giving the values of this ratio for increasing B for certain values (e.g., $\log 2$ or $e+\pi$) of the w_m ($m=2$ or 3) and for certain values of c (chosen apparently for convenience of the computer). The authors state that "if one plots the graph of $\lambda(B, c)$ against B, one finds that it is a step function which fits a curve $c_1 \log B$ rather well", but apparently no statistical test of goodness of fit was applied, and there is no discussion of how the behaviour of the error term fits the estimates for the error which are known to hold for almost all $(w_1, \cdots, w_m)$. *J. W. S. Cassels* (Cambridge, England)

Referred to in J24-38.

J24-38 **(32# 4085)**

Adams, William W.

Asymptotic Diophantine approximations to *e*.

Proc. Nat. Acad. Sci. U.S.A. **55** (1966), 28–31.

Using the explicit continued fraction expansion, it is shown that the number $\lambda_1(B)$ of solutions of $|qe-p| < q^{-1}$, $1 \leqq q \leqq B$, in integers p, q, and the number $\lambda_2(B)$ of solutions in coprime integers, satisfy the estimates $\lambda_1(B) = \{2G(B)\}^{3/2}/3 + O(G(B))$, $\lambda_2(B) = 3G(B) + O(1)$, where $G(B)$ is the inverse function of $4^x\Gamma(x+\frac{3}{2})$. Since $G(x) \sim \log x/\log\log x$, this shows that e does not behave like "almost any" real number (in the sense of Lebesgue) in this respect, in spite of the numerical evidence adduced by the author and Lang [J. Reine Angew. Math. **220** (1965), 163–173; MR **32** #91].

J. W. S. Cassels (Cambridge, England)

Citations: MR 32# 91 = J24-37.

J24-39 **(32# 5593)**

Gallagher, P. X.

Metric simultaneous diophantine approximation. II.

Mathematika **12** (1965), 123–127.

The author proves the following interesting results which sharpen and extend several previous results on this subject. Denote by $U(a)$ the set $(y_1, \cdots, y_r)$ in r-dimensional space for which $0 \leqq y_i \leqq a$. Let V be a vector with integer coordinates in r-dimensional space; write $(V, n) = 1$ if the greatest common divisor of n and the components of V is 1. x will denote a point in r-dimensional space. Theorem 1: Let $r \geqq 2$. Let $0 \leqq a_n \leqq 1$ be any sequence of real numbers. For almost all x the equation $nx - V \in U(a_n)$, $(V, n) = 1$, has infinitely many solutions in n if and only if $\sum_{n=1}^{\infty} a_n^{\,r}$ diverges.

Put $S_N = \sum_{n=1}^{N} a_n^{\,r}$ and denote by $T_N(x)$ the number of solutions n ($1 \leqq n \leqq N$) and V of $nx - V \in U(a_n)$.

Theorem 2: Let $r \geqq 3$. For almost all x one has, for every $\varepsilon > 0$,

$$T_N(x) = S_N + O(S_N^{\lambda_r + \varepsilon}),$$

where $\lambda_r < 1$ is a constant which depends on r.

(See also W. M. Schmidt [Canad. J. Math. **12** (1960), 619–631; MR **22** #9482; Trans. Amer. Math. Soc. **110** (1964), 493–518; MR **28** #3018]. Several other relevant references are quoted in the paper.)

{Part I appeared in J. London Math. Soc. **37** (1962), 387–390 [MR **28** #1167].} *P. Erdős* (Budapest)

Citations: MR 22# 9482 = J24-23; MR 28# 1167 = J24-31; MR 28# 3018 = J24-33.

Referred to in J24-54.

J24-40 **(32# 5594)**

Lang, Serge

Asymptotic Diophantine approximations.

Proc. Nat. Acad. Sci. U.S.A. **55** (1966), 31–34.

Let $g(t)$ be a positive increasing function of the positive real variable t. An irrational number α is said to be of type g if for all sufficiently large integers B there is a solution in coprime integers p, q of the inequalities $0 < |q\alpha - p| < q^{-1}$, $B/g(B) \leqq q < B$. Let $\omega(t)$ be a function increasing to infinity and such that $\{\omega(t)\}^{1/2}g(t)/t$ is decreasing for all large t. Then it is shown that the number $\lambda(B)$ of solutions of $0 \leqq q\alpha - p < t^{-1}\omega(t)$, $1 \leqq q < B$, has the asymptotic estimate

$$\lambda(B) = \int_1^B \frac{\omega(t)}{t}\,dt + O\left\{\int_1^B \frac{\{\omega(t)\}^{1/2}g(t)}{t}\,dt\right\}.$$

The author recently gave less perfect results in the same direction for quadratic irrationalities [Amer. J. Math. **87** (1965), 488–496; MR **31** #3383]. The proof follows Behnke [Abh. Math. Sem. Univ. Hamburg **3** (1922), 252–267] and Ostrowski [ibid. **2** (1922), 77–98].

{Connoisseurs of the American use of English should note Lang's statement that he has "trivialized the proof" of Behnke and Ostrowski. This apparently means to repeat the argument in a slightly more general setting.}

J. W. S. Cassels (Cambridge, England)

Citations: MR 31# 3383 = J24-35.

Referred to in J02-20, J24-49, J24-51, N28-73.

J24-41 **(34# 140)**

Davenport, H.

On a theorem of Mrs. Turán. II.

Proc. London Math. Soc. (3) **16** (1966), 570–576.

A generalization and strengthening of a result of Vera T. Sós [Ann. Univ. Sci. Budapest. Eötvös Sect. Math. **3–4** (1960/61), 291–294; MR **25** #3009] is established. Let $L, L_1, L_2, \cdots, L_k$ be $k+1$ real linear forms in $x, x_1, \cdots, x_k$ of determinant 1. Assume that the cofactors of the coefficients of L in the coefficient matrix are not all in rational ratios, and that none of them is zero. Let θ be fixed, $0 < \theta < 1/k$. Then for any $\varepsilon > 0$ and any $N > N_0(\varepsilon)$, there exist integers $x, x_1, \cdots, x_k$ such that $|L_i| < \varepsilon L^{-\theta}$ ($i = 1, \cdots, k$) and $N \leqq L \leqq N^{\phi}$, where $\phi = (1+\theta)\{1-(k-1)\theta\}^{-1}$. This result is generalized from the situation of one form L versus k forms L_i to an arbitrary separation of the $k+1$ forms, with appropriate inequalities satisfied. This result, when applied to the forms $x, \alpha_1 x - x_1, \cdots, \alpha_k x - x_k$, gives the following theorem. Let $\alpha_1, \cdots, \alpha_k$ be real numbers, at least some irrational. Then with θ as before, for any $\varepsilon > 0$ and any $N > N_0(\varepsilon)$ there exists an integer x such that $\|x\alpha_i\| < \varepsilon x^{-\theta}$ ($i = 1, \cdots, k$), with $N \leqq x \leqq N^{\phi}$. Here $\|y\|$ means $\min|y-n|$ over all integers n. This conclusion holds, even for $\theta = 1/k$, for every positive integer N and all real $\alpha_1, \cdots, \alpha_k$ if ε is specified as $(2^{1+\theta}+1)^{1/k}$. Reference should also be made to a related paper [Proc. London Math. Soc. (3) **14a** (1965), 76–80; MR **31** #5836] by the author. *I. Niven* (Eugene, Ore.)

Citations: MR 25# 3009 = J24-29; MR 31# 5836 = J24-36.

J24-42 **(34# 2529)**

Schmidt, Wolfgang M.

Simultaneous approximation to a basis of a real number-field.

Amer. J. Math. **88** (1966), 517–527.

Let $\psi(q) > 0$ be a decreasing function of the positive integer q. Define $\Psi(h) = \sum_{q=1}^{h} \psi(q)^n$. Let $1, \beta_1, \cdots, \beta_n$ be a basis of a real algebraic number field. Let $N(h; \beta_1, \cdots, \beta_n)$ be the number of solutions in integers $q, p_1, \cdots, p_n$ of the inequalities $1 \leqq q \leqq h$, $0 < \beta_i q + p_i \leqq \psi(q)$ $(1 \leqq i \leqq n)$. The author proves that if $q\psi(q)^n \to \infty$ with q, then $N(h; \beta_1, \cdots, \beta_n) \sim \Psi(h)$ as $h \to \infty$.

It is known [see, e.g., J. W. S. Cassels, *An introduction to Diophantine approximation*, p. 79, Cambridge Univ. Press, New York, 1957; MR **19**, 396; Russian translation, Izdat. Inostr. Lit., Moscow, 1961; MR **22** #10976] that if $\psi(q) = cq^{-1/n}$, where $0 < c \leqq c_0(\beta_1, \cdots, \beta_n)$, then $N(h; \beta_1, \cdots, \beta_n) = 0$ for all h.

The author states a dual theorem on the number of solutions in integers $p, q_1, \cdots, q_n$ of the inequalities $1 \leqq q_i \leqq h$, $0 < \beta_1 q_1 + \cdots + \beta_n q_n + p \leqq \varphi(\max\{q_1, \cdots, q_n\})$, where $\varphi(q) > 0$ is decreasing and $q^n\varphi(q) \to \infty$ with q.

R. P. Bambah (Columbus, Ohio)

Citations: MR 19, 396h = J02-11; MR 22# 10976 = J02-12.

Referred to in J24-49, J24-57.

J24-43 **(34# 4239)**

Ennola, Veikko

On the distribution of fractional parts of sequences.

Ann. Univ. Turku. Ser. A I No. 91 (1966), 7 *pp.*

The author proves the theorem: Let α be an irrational number such that the partial quotients of its regular continued fraction expansion are bounded. Let $f(n)$ be an arbitrary positive decreasing function, defined for every natural number n, with values < 1. Let $N(P, \alpha)$ denote the number of integers n such that $0 < \{n\alpha\} \leqq f(n)$, $1 \leqq n \leqq P$. Then $N(P, \alpha) = \sum_{n=1}^{P} f(n) + O(f(P)^{-2}P^{2/3} \log P)$.

{See also #4240 below.} *J. E. Cigler* (Groningen)

J24-44 **(34# 4240)**

Ennola, Veikko

On the distribution of fractional parts of sequences. II.

Ann. Univ. Turku. Ser. A I No. 92 (1966), 9 *pp.*

The author proves some theorems on the number $N(P, \alpha)$ of integers n satisfying $0 < n\alpha - [n\alpha] \leqq f(n)$, $1 \leqq n \leqq P$, where α denotes an irrational number such that the partial quotients of its regular continued fraction expansion are bounded. A typical result is Theorem 5: $N(P, \alpha) = P^{1-\gamma}(1-\gamma)^{-1} + O(P^{2(1-\gamma)/3} \log P)$ if $f(n) = n^{-\gamma}$, $0 < \gamma < 1$.

{For Part I, see #4239 above.}

J. E. Cigler (Groningen)

Referred to in J24-46.

J24-45 **(34# 5755)**

Philipp, Walter

Some metrical theorems in number theory.

Pacific J. Math. **20** (1967), 109–127.

This paper is concerned with the metric number theory associated with transformations of the unit interval of types (A) $T: x \to \{ax\}$ ($a > 1$ an integer), (B) $T: x \to \{1/x\}$ and (C) $T: x \to \{\theta x\}$ ($\theta > 1$ is not an integer). (Here $\{\ \}$ denotes fractional part.) For each type an estimate for $(|E \cap T^{-n}F| - |E|\,|F|)|F|^{-1}$ is given, where E is an interval and F is measurable. ($|\ |$ denotes the unique T-invariant normalised measure equivalent to Lebesgue measure.) From this, an estimate of the discrepancy in the mixing of degree r formula for intervals is deduced. In particular, V. A. Rohlin's result [Izv. Akad. Nauk SSSR Ser. Mat. **25** (1961), 499–530; MR **26** #1423; translated in Amer. Math. Soc. Transl. (2) **39** (1964), 1–36; see MR **28** #3909] that T is mixing of all degrees follows. Together with a Borel-Cantelli lemma, these estimates facilitate the proof of the following theorem. Let $\{I_n\}$ be a sequence of sub-intervals of the unit interval. Let $A(N, x)$ denote the number of positive integers $n \leqq N$ such that $T^n x \in I_n$. Then $A(N, x) = \phi(N) + O(\phi^{1/2}(N) \log^{3/2+\varepsilon} \phi(N))$, $\varepsilon > 0$, for almost all $x \in [0, 1)$, where $\phi(N) = \sum_{n \leqq N} |I_n|$.

As the author points out, the novelty of this result is in the arbitrariness of the sequence $\{I_n\}$. Unlike W. J. LeVeque's similar theorems [J. Reine Angew. Math. **202** (1959), 215–220; MR **22** #12090], $|I_n|$ need not be decreasing. The theorem also generalises well-known results on the distribution of sequences arising from T (cf. J. Cigler and G. Helmberg [Jber. Deutsch. Math.-Verein. **64**, Abt. 1, 1–50 (1961); MR **23** #A2409]). The paper finishes with some theorems on continued fractions and a full discussion of the relationship between the author's results and those of Khintchine, LeVeque, de Vroedt, Lévy and others (for references, see the paper).

W. Parry (Brighton)

Citations: MR 22# 12090 = J24-25; MR 23# A2409 = K02-1; MR 26# 1423 = K50-13.

Referred to in J24-54, K50-41, K50-48, K55-41.

J24-46 **(35# 153)**

Ennola, Veikko

On the distribution of fractional parts of sequences. III.

Ann. Univ. Turku. Ser. A I No. 97 (1967), 6 *pp.*

For every real number β, let $\|\beta\|$ denote the difference, taken positively, between β and the nearest integer. Let $f(x)$ be an arbitrary real function, defined for every real $x \geqq 1$, such that for some η $(0 < \eta < 1)$ one has (1) $x^{-1/2} \leqq f(x) \leqq 1 - \eta$ for $x \geqq 2$. For any irrational number α, let $N(P, \alpha)$ denote the number of integers n such that $0 < \{n\alpha\} \leqq f(n)$, $1 \leqq n \leqq P$. Write $h(m) = \sum_{n=1}^{m} |f(n) - f(n+1)|$ $(m = 1, 2, \cdots)$. Theorem 1: Suppose there exist positive numbers $\gamma \geqq 1$ and c such that (2) $n^\gamma\|n\alpha\| \geqq c$ for every natural number n. Then

$$N(P, \alpha) = \sum_{n=1}^{P} f(n) + O(P^{\gamma - 1/2}(1 + h(P)) \log^2 P).$$

Theorem 2: For almost all α, $N(P, \alpha) = \sum_{n=1}^{P} f(n) + O(P^{1/2}(1 + h(P)) \log^{2+\varepsilon} P)$.

{Part II appeared in same Ann. No. 92 (1966) [MR **34** #4240].} *J. E. Cigler* (Groningen)

Citations: MR 34# 4240 = J24-44.

J24-47 **(35# 1549)**

Kesten, H.; Sós, V. T.

On two problems of Erdős, Szüsz and Turán concerning diophantine approximations.

Acta Arith. **12** (1966/67), 183–192.

Let $S(N, A, c)$ denote the set of real ξ $(0 < \xi < 1)$ for which there is a natural b with $\|b\xi\| < A/b$, $N \leqq b \leqq cN$. ($\|z\|$ denotes the distance of z from the nearest integer; $|S|$ will denote the Lebesgue measure of S.)

P. Erdős, the reviewer and P. Turán [Colloq. Math. **6** (1958), 119–126; MR **21** #1290] considered the problem of whether (*) $\lim_{N\to\infty}|S(N, A, c)| = f(A, c)$ exists. We proved that for $0 < A < c/(1+c^2)$ the limit (*) exists and $f(A, c) = (12A/\pi^2) \log c$.

A related problem is the following. Let $T(N, c)$ be the set of ξ in $(0, 1)$ for which $N < q_n \leqq cN$ for at least one continued fraction convergent p_n/q_n of ξ. Does (**) $\lim_{N\to\infty}|T(N, c)| = g(c)$ exist? The authors show the existence of the limit (**) and outline a proof of the existence of the limit (*). The value of $f(A, c)$ is not determined.

P. Szüsz (Stony Brook, N.Y.)

Citations: MR 21# 1290 = J24-19.

J24-48 **(35# 4166)**

Sprindžuk, V. G.

Asymptotic behavior of the number of solutions of certain Diophantine inequalities. (Russian)

Dokl. Akad. Nauk SSSR **173** (1967), 770–772.

The author proves a theorem to the general effect that if the real number α is well-behaved and if the decreasing function $\chi(t)$ is also well-behaved, then for any fixed real β the number of integer solutions q of $\|\alpha q+\beta\|<\chi(q)$ with $q \leqq Q$ is approximately $2\sum_1^Q \chi(q)$ provided that the latter diverges. More precisely, he supposes that for any $\varepsilon>0$ there are only finitely many integral solutions q of $\|\alpha q\|<q^{-1-\varepsilon}$ and that $\chi(t)$ is a continuously differentiable function of t for $t\geqq 1$ such that (1) $\chi'(t)<0$, (2) $|\chi'(t)|$ decreases monotonely to zero as $t\to\infty$, (3) $|\chi'(t)|\gg t^{-2}$, and (4) $|\chi'(t)|\ll|\chi'(t_1)|$ for $t<t_1\ll t$. Then for any fixed β the number of integer solutions q of $\|\alpha q+\beta\|<\chi(q)$ with $q<Q$ is

$$2\Psi(q)+O(\Psi(Q)\chi^{-1}(Q)|Q'(Q)|^{1/2}Q^{\varepsilon})+O(\chi(Q)|\chi'(Q)|^{-1/2}),$$

where $\Psi(Q)=\int_1^Q \chi(t)\,dt$.

{This article has appeared in English translation [Soviet Math. Dokl. **8** (1967), 484–486].}

J. W. S. Cassels (Cambridge, England)

J24-49 (36# 3730)

Adams, William W.

Simultaneous asymptotic diophantine approximations.

Mathematika **14** (1967), 173–180.

Let $(\theta_1,\cdots,\theta_k)$ be k real numbers and let $\psi(t)$ be a positive decreasing function of the positive variable t. Denote by $\lambda(N)$ for positive integral N the number of solutions in integers $p_1,\cdots,p_k,q$ of the inequalities $1\leqq q\leqq N$, $0\leqq q\theta_i-p_i<\psi(q)$ $(1\leqq i\leqq k)$. It is known [P. Erdős, Acta Arith. **5** (1959), 359–369; MR **22** #12091; W. J. LeVeque, Trans. Amer. Math. Soc. **87** (1958), 237–261; MR **20** #2314; ibid. **94** (1960), 130–149; MR **22** #12089; W. M. Schmidt, Canad. J. Math. **12** (1960), 619–631; MR **22** #9482] that $\lambda(N)\sim\int_1^N(\psi(t))^k\,dt$ for almost all k-tuples $(\theta_1,\cdots,\theta_n)$. In this paper the author establishes conditions on $(\theta_1,\cdots,\theta_n)$ which ensure that this asymptotic relation holds with an explicit estimation of the error. More precisely, if g is an increasing positive function, the author says that $(\theta_1,\cdots,\theta_k)$ is "of type $\leqq g$" provided that for all integers $p,q_1,\cdots,q_k$ with $Q=\max(|q_1|,\cdots,|q_k|)$ sufficiently large, the inequality $|q_1\theta_1+\cdots+q_k\theta_k|\geqq(Q^kg(Q))^{-1}$ holds. Suppose that $(\theta_1,\cdots,\theta_k)$ is of type $\leqq g$, that $\psi(t)$ is decreasing and that

$$F(t) = t\{\psi(t)\}^k\{g(t)\}^{-k(k+2)/(k+1)}$$

is increasing to infinity. Then

$$\lambda(N) = \int_1^N \psi(t)^k\,dt+O(\int_1^N \psi(t)^kF(t)^{-1/(k+1)}\,dt) = \{1+o(1)\}\int_1^N \psi(t)^k\,dt.$$

This generalizes results of S. Lang [Proc. Nat. Acad. Sci. U.S.A. **55** (1966), 31–34; MR **32** #5594; *Introduction to diophantine approximations*, Addison-Wesley, Reading, Mass., 1966; MR **35** #129] and is related to results of W. M. Schmidt [Amer. J. Math. **88** (1966), 517–527; MR **34** #2529]. *J. W. S. Cassels* (Cambridge, England)

Citations: MR 20# 2314 = J24-18; MR 22# 9482 = J24-23; MR 22# 12089 = J24-24; MR 22# 12091 = J24-26; MR 32# 5594 = J24-40; MR 34# 2529 = J24-42; MR 35# 129 = J02-20.

J24-50 (36# 3731)

Sprindžuk, V. G.

On the metric theory of linear Diophantine approximations. (Russian)

Dokl. Akad. Nauk SSSR **176** (1967), 43–45.

Let R_n be n-dimensional Euclidean space with unit cube E_n and denote the set of all $m\times n$ matrices $\omega=(\omega_{ij})$, $0\leqq\omega_{ij}<1$, by Ω_{mn}. The natural one-to-one correspondence between Ω_{mn} and E_{mn} induces a Lebesgue measure on Ω_{mn}. Using the Borel-Cantelli lemma relative to this measure, the author proves the following simultaneous inhomogeneous extension of a theorem due to A. Ja. Hinčin [*Continued fractions* (Russian), ONTI NKTP, Moscow, 1935; second edition, GITTL, Moscow, 1949; MR **13**, 444; English translations, Noordhoff, Groningen, 1963; MR **28** #5038; Univ. Chicago Press, Chicago, Ill., 1964; MR **28** #5037] on homogeneous approximations. Let $n\geqq 2$, a be an integral primitive vector in R_n and $\{a\omega_j\}$ be the fractional part of the inner product of a with the jth row ω_j of ω. If to each a there corresponds a measurable set $A(a)$ in E_m, then there are finitely [infinitely] many a for which (*) $(\{a\omega_1\},\cdots,\{a\omega_m\})\in A(a)$, for almost all ω, precisely when the sum of the measures of all $A(a)$ is finite [infinite]. The number of a for which $\max_{1\leqq j\leqq n}|a_j|\leqq h$ and (*) is true for some fixed ω is given by $\Phi(h)+O(\Phi^{1/2+\varepsilon}(h)\ln h)$, where $\Phi(h)$ is the sum of the measures of all $A(a)$ under consideration. Analogues in the geometry of numbers are given.

{This article has appeared in English translation [Soviet Math. Dokl. **8** (1967), 1051–1054].}

J. B. Roberts (Portland, Ore.)

Citations: MR 13, 444f = Z02-18; MR 28# 5037 = Z02-21; MR 28# 5038 = Z02-22.

J24-51 (36# 5082)

Adams, William W.

Asymptotic diophantine approximations and Hurwitz numbers.

Amer. J. Math. **89** (1967), 1083–1108.

Let α be a real irrational number. Let $\omega(t)$ be a positive monotone function such that $\omega(t)/t$ is decreasing. For real $B\geqq 1$, define $\lambda(B,\omega)$ to be the number of solutions in integers p,q of the inequalities $|q\alpha-p|<\omega(q)/q$, $1\leqq q\leqq B$. The author presents a counting scheme for the number $\lambda(B,\omega)$ in terms of the simple continued fraction expansion of α. Then he determines the asymptotic behaviour of $\lambda(B,\omega)$, under a mild restriction on ω, in the case that α is a Hurwitz number. (A Hurwitz number is an irrational number α whose continued fraction expansion is in the form $[a_0,\cdots,a_v,\phi_1(1),\cdots,\phi_k(1),\phi_1(2),\cdots,\phi_k(2),\cdots]$, where $v\geqq 0$ and where $\phi_1,\cdots,\phi_k$ $(k>0)$ are polynomials of degree $\geqq 0$; an example is $e=[2,1,1,2,1,1,4,1,1,6,\cdots]$.) The final results are as follows: Let ϕ_i have degree ρ_i and leading coefficient b_i $(i=1,\cdots,k)$. Let C be a positive constant. Put $\nu=\rho_1+\cdots+\rho_k$, $\mu=\max\rho_i$, $a=\prod b_i$, $b=\sum_{\rho_i=\mu}b_i$, $b'=\sum_{\rho_i=\mu}[b_iC]$. Suppose that $\nu>0$. Next, let $\tau=G(t)$ denote the inverse of the function $t=a^\tau\Gamma(\tau)^\nu$; one has $m=G(q_{km})+O(1)$, where q_n is the denominator of the nth convergent to α. Then the following assertions hold: if $\omega(t)=C/G(t)^\mu$, then $\lambda(B,\omega)=b'G(B)+O(1)$; if $\omega(t)\leqq C$ and $\omega(t)/t$ satisfies some smoothness condition, then $\lambda(B,\omega)\sim b\int_1^{G(B)}\{x^\mu\omega(G^{-1}(x))\}^{1/2}\,dx$; if $\omega(t)\to\infty$, $\omega(t)\leqq CG(t)^\mu$ and $\omega(t)/t$ satisfies some smoothness condition, then

$$\lambda(B,\omega)\sim b\int_1^{G(B)}\{x^\mu\omega(G^{-1}(x))\}^{1/2}\,dx + 2\nu\int_1^{G(B)}\omega(G^{-1}(x))\log x\,dx.$$

These assertions complement a result of S. Lang [Proc. Nat. Acad. Sci. U.S.A. **55** (1966), 31–34; MR **32** #5594] concerning the case of a sufficiently fast growing function $\omega(t)$. More specific results are obtained when $\omega(t)\equiv C$ and $\alpha=e$. *C. G. Lekkerkerker* (Amsterdam)

Citations: MR 32# 5594 = J24-40.

Referred to in J80-47.

J24-52 (36# 5083)

Adams, William W.

A lower bound in asymptotic diophantine approximations.

Duke Math. J. **35** (1968), 21–35.

Let α be a real irrational number. For real $B\geqq 1$, define $\lambda(B,\alpha)$ to be the number of solutions in integers p,q of the inequalities $|q\alpha-p|<1/q$, $1\leqq q\leqq B$. Then for almost all

α, $\lambda(B,\alpha)\sim 2\log B$ $(B\to\infty)$, by a general theorem of W. Schmidt [Canad. J. Math. **12** (1960), 619–631; MR **22** #9482]. The present author proves that for all irrationals α, $\lambda(b,\alpha)\geqq c_0\log B+O(1)$ $(B\to\infty)$, where $c_0=3/\log((9+\sqrt{77})/2)=1.373\cdots$. The proof makes use of the theory of continued fractions; the author gives an explicit formula for $\lambda(q_n,\alpha)$, q_n the denominator of the nth convergent to α, and estimates very carefully the quotient $\lambda(q_n,\alpha)/\log q_n$. The result is sharp in the sense that there are (uncountably many) irrationals α such that $\lambda(B,\alpha)\sim c_0\log B$. On the other hand, the quotient $\lambda(B,\alpha)/\log B$ is unbounded if the set $\{q_n^{1/n}\}$ is unbounded.

C. G. Lekkerkerker (Amsterdam)

Citations: MR 22# 9482 = J24-23.

J24-53 (36# 6354)

Adams, William W.

Asymptotic diophantine approximations and equivalent numbers.

Proc. Amer. Math. Soc. **19** (1968), 231–235.

Let α be an irrational real number. For $\beta\geqq 1$ define $\lambda(\beta,\alpha)$ to be the number of solutions in integers p,q of $|q\alpha-p|<1/q$, $0<q\leqq\beta$. The author proves some results about the difference between $\lambda(\beta,\alpha)$ and $\lambda(\beta,\alpha')$, where α' is equivalent to α. In particular, he proves that $\lambda(\beta,\alpha)\sim\lambda(\beta,\alpha')$, as $\beta\to\infty$, for almost all α. He also proves the existence of equivalent α,α' for which $\lambda(\beta,\alpha)$ is not asymptotic to $\lambda(\beta,\alpha')$. *R. P. Bambah* (Columbus, Ohio)

J24-54 (37# 162)

Ennola, Veikko

On metric diophantine approximation.

Ann. Univ. Turku. Ser. A I No. 113 (1967), 8 *pp.*

Let $\{a_j\}$ be an increasing sequence of positive integers, and for $i=1,2,\cdots,k$, let $\{\alpha_n^{(i)}\}$ be a sequence on the half open interval $[0,1)$ and f_i a function from the positive integers into the open interval $(0,1)$. For every point $\mathbf{x}=(x_1,x_2,\cdots,x_k)$ in k-dimensional Euclidean space, denote by $N(h,\mathbf{x})$ the number of positive integers n not greater than h which satisfy the simultaneous inequalities $\alpha_n^{(i)}\leqq a_nx_i\leqq\alpha_n^{(i)}+f_i(n)$ (mod 1) for $i=1,2,\cdots,k$. The author gives an interesting generalization of Theorem 2 of P. X. Gallagher [Mathematika **12** (1965), 123–127; MR **32** #5593] with a somewhat better error term as follows: If $k\geqq 3$ and if for each pair of i,j $(1\leqq i,j\leqq k)$, $\liminf f_i(n)/f_j(n)>0$ as $n\to\infty$, then for every $\varepsilon>0$ and almost all $\mathbf{x}$, $N(h,\mathbf{x})=F(h)+O(F(h)^{1-1/2k}\log^{3/2+\varepsilon}F(h))$, where $F(h)=\sum_{n=1}^h\prod_{i=1}^k f_i(n)$. Notice that as in Gallagher's paper, no monotonicity assumptions are needed. The author also shows that the result holds for $k=2$ provided $\prod_{i=1}^k f_i$ is monotonically decreasing and $a_n=Q(n)$, where $Q(n)$ is an integral polynomial taking positive values for $n\geqq 1$. Finally, for any k, if the latter hypothesis is changed by omitting the lim inf condition but requiring in addition that $Q(n)$ be irreducible of degree $\geqq 2$, then the error term becomes $O(F(h)^{1/2}\log^{3/2+\varepsilon}F(h))$. For a general result along these latter lines, see W. M. Schmidt [Trans. Amer. Math. Soc. **110** (1964), 493–518; MR **28** #3018].

The proofs use a generalized version of a "quantitative Borel-Cantelli lemma" of W. Philipp [Pacific J. Math. **20** (1967), 109–127; MR **34** #5755].

L. C. Eggan (Tacoma, Wash.)

Citations: MR 28# 3018 = J24-33; MR 32# 5593 = J24-39; MR 34# 5755 = J24-45.

J24-55 (37# 6248)

Szüsz, P.

On the metric theory of continued fractions. (Hungarian)

Magyar Tud. Akad. Mat. Fiz. Oszt. Közl. **14** (1964), 361–400.

The main result of the paper is the following: For r,k_1,k_2 (r fixed) integers $(0\leqq k_1\leqq r-1;\ 0\leqq k_2\leqq r-1)$, let $m_0(k_1,k_2,x)$ be non-negative, monotonic increasing, twice continuously differentiable functions satisfying $m_0(k_1,k_2,0)=0$, $\sum_{k_1,k_2}m_0(k_1,k_2,1)=1$,

$$|\{m_0'(k_1,k_2,x)\cdot(1+x)\}'|<K.$$

Let $S(\nu)$ be defined by $0\leqq S(\nu)<r$, $S(\nu)\equiv k_2-\nu k_1 \pmod r$ and let

$$m_{n+1}(k_1,k_2,x)=\sum_{\nu=1}^\infty\{m_n(S(\nu),k_1,\nu^{-1})-m_n(S(\nu),k_1,(x+\nu)^{-1})\}$$

for $n=0,1,\cdots$. Then, if $(k_1,k_2,1)$, we have

$$|m_n'(k_1,k_2,x)-C/(1+x)|<Kq^n,$$

where C, q depend only on r and $q<1$. Among many other results the author obtains as an application the following theorem: Let $f(k)$ be a monotonic decreasing positive function with $f(k)\leqq\frac12$, $\sum_1^\infty k^{-1}f(k)=+\infty$. Let $N^*(\alpha,f,k',n)$ denote the number of solutions of $|\alpha k-\nu|\leqq k^{-1}f(k)$, $k\equiv k' \pmod r$, $(k,\nu)=1$, $k\leqq n$. Then for almost all α in $(0,1)$ we have

$$N^*(\alpha,f,k',n)\sim(D\phi((k',r))/(k',r))\sum_{k=1}^n k^{-1}f(k)$$

as $n\to\infty$, where D depends on r only. {Since most of the interesting results of this long paper could never be reproduced in abstracts, the reviewer feels that a revised English version of the article would be highly desirable.}

A. Meir (Edmonton, Alta.)

J24-56 (39# 132)

Kátai, Imre

On the greatest common divisor of values of a polynomial. (Hungarian. English summary)

Mat. Lapok **19** (1968), 93–99.

Author's summary: "Let a_n be an increasing sequence of natural numbers, $A(x,D,l)=\sum_{a_n\equiv l(D):a_n\leqq x}1$, $A(x,1,1)=A(x)$, satisfying the following conditions:

$$\sum_{D\leqq x^\gamma}\max_{(l,D)=1}|A(x,D,l)-A(x)/\varphi(D)|\ll A(x)(\log x)^{-B}$$

for some $\gamma>0$, and for arbitrary B. Further, let $A(x)\gg x(\log x)^{-c}$ (γ, B, c are constants). Let $P(x)$ be an irreducible polynomial with integer coefficients and of degree at least 2.

"Generalizing a result of V. Ennola [Acta Math. Acad. Sci. Hungar. **19** (1968), 329–335; MR **38** #105] we prove the following assertion. Theorem 1: Under the above conditions for a_n and $P(x)$ we have

$$\sum_{a_n\leqq x}(1/P(a_n))\sum_{m<n}(P(a_m),P(a_n))\ll A(x),$$

where (a,b) denotes the greatest common divisor of a and b.

"Hence we obtain immediately the following result in the theory of diophantine approximation using a theorem of W. J. LeVeque [J. Reine Angew. Math. **202** (1959), 215–220; MR **22** #12090]. Let f denote a nonincreasing function on the positive integers for which $0\leqq f(k)\leqq\frac12$ for

$k=1, 2$, and $\sum_1^\infty f(k)=\infty$. For real t let $\|t\|$ be the positive distance between t and the integer nearest t. For an increasing sequence c_k of natural numbers and an arbitrary sequence α_k of real numbers, let $N_n(x; c_k, \alpha_k, f(k))$ denote the number of solutions of the inequality $\|c_k x-\alpha_k\|<f(k)$ $(k=1, \cdots, n)$. Theorem 2: Under the conditions stated for a_k and $P(x)$ in Theorem 1, for any real sequence α_k and for $c_k=P(a_k)$ we have $N_n(x; c_k, \alpha_k, f(k))=(1+o(1))2\times\sum_{k=1}^n f(k)$ for almost all x."

Citations: MR 22# 12090 = J24-25; MR 38# 105 = C05-39.

J24-57 **(39# 1409)**

Adams, William W.

Simultaneous asymptotic diophantine approximations to a basis of a real cubic number field.

J. Number Theory **1** (1969), 179–194.

Sei $1, \beta_1, \beta_2$ Basis eines reellen kubischen Zahlkörpers und sei $C>0$ konstant. Weiter sei $\lambda_B=\lambda_B(\beta_1, \beta_2, C)$ die Anzahl ganzzahliger Lösungstupel q, p_1, p_2 von $0<q\beta_i-p_i<Cq^{1/2}$ $(i=1, 2; 1\leqq q\leqq B)$. Dann ist $\lambda_B=O(1)$ oder es gibt eine Konstante $C'>0$ mit $\lambda_B\sim C'\log B$ $(B\to\infty)$. Der Satz knüpft an Ergebnisse von W. Schmidt [Amer. J. Math. **88** (1966), 517–527; MR **34** #2529] und S. Lang [*Introduction to diophantine approximations*, Addison-Wesley, Reading, Mass., 1966; MR **35** #129] an. Zum Beweis werden Eigenschaften der Einheiten kubischer Zahlkörper und der Gleichverteilungssatz benötigt. Ein eng verwandter Satz, in dem statt des obigen Systems die Ungleichung $0<q_1\beta_1+q_2\beta_2-p<C/q^2$ untersucht wird, wird angegeben. Zur naheliegenden Frage, ob und wie sich die Ergebnisse auf höhere Dimensionen übertragen, wird mit zahlengeometrischen Überlegungen eine Vermutung aufgestellt. Die Hindernisse, schon für die nächsthöhere Dimension, werden erklärt. *J. M. Wills* (Berlin)

Citations: MR 34# 2529 = J24-42; MR 35# 129 = J02-20.

Referred to in J24-61.

J24-58 **(39# 5483)**

Ennola, Veikko

On a theorem of Erdős and Turan.

Math. Scand. **22** (1968), 75–80.

Let $f(n)$ be a positive decreasing function defined on the positive integers with $f(1)<1$, and let $x_1, x_2, \cdots$ be a sequence of real numbers. Let $N(P)$ denote the number of n with $1\leqq n\leqq P$ and $0<\{x_n\}\leqq f(n)$, where $\{x\}$ is the fractional part of x. The author proves the following theorem: Let

$$\varphi_q(P) = \max\{|\sum_{n=1}^k \exp[2\pi iq(x_n-f(n))]|: k=1, 2, \cdots, P\},$$

$$\psi_q(P) = \max\{|\sum_{n=1}^k \exp(2\pi iqx_n)|: k=1, 2, \cdots, P\}.$$

Then, for any positive integer m, we have

$$|N(P)-\sum_{n=1}^P f(n)| < 32(P/m+\sum_{q=1}^{2m-2} (\varphi_q(P)+\psi_q(P))/q).$$

This theorem generalizes work of P. Erdős and P. Turán [Nederl. Akad. Wetensch. Proc. **51** (1948), 1146–1154; ibid. **51** (1948), 1262–1269; MR **10**, 372], where the case $f(n)=f$ constant is considered.

O. P. Stackelberg (Durham, N.C.)

Citations: MR 10, 372c = K05-10; MR 10, 372d = K05-10.

J24-59 **(40# 112)**

de Mathan, Bernard

Approximation diophantienne par les éléments de certaines suites.

Séminaire Delange-Pisot-Poitou: 1967/68, *Théorie des Nombres, Fasc.* 1, *Exposé* 12, 9 *pp.* *Secrétariat mathématique, Paris,* 1969.

This is an analogue for ultrametric compact spaces of a theorem of W. Schmidt [Canad. J. Math. **12** (1960), 619–631; MR **22** #9482], who considered inequalities of type $0\leqq\theta_i-p_i/q<\psi_i(q)/q$, $i=1, 2, \cdots, n$, and obtained an asymptotic formula for the number of solutions in integers $p_1, \cdots, p_n, q\ (>0)$ valid for almost all real points $(\theta_1, \cdots, \theta_n)$.

The theorem proved by the author states that if E is a compact ultrametric space (i.e., the distance function $d(x, y)$ satisfies $d(x, y)\leqq\max(d(x, z), d(y, z))$ for all points x, y, z) with probability measure μ whose support is E; if $(u_n)_{n=1,2,\ldots}$ is any sequence of points of E which is uniformly distributed in a certain sense (too long to be specified here) and $\{B_n\}_{n=1,2,\ldots}$ is any sequence of balls in E such that $u_n\in B_n$, $\{\mu(B_n)\}_{n=1,2,\ldots}$ is decreasing and $\sum_{n=1}^\infty \mu(B_n)$ is divergent; if $N(h, x)=\operatorname{card}\{n; 1\leqq n\leqq h, x\in B_n\}$ and $\phi(h)=\sum_{n=1}^h \mu(B_n)$, then for almost all x, $N(h, x)=\phi(h)+O((\phi(h))^{1/2}(\log\phi(h))^{(3+\varepsilon)/2})$, as $h\to\infty$, for every $\varepsilon>0$. Presumably the theorem has applications to p-adic fields. *I. Danicic* (Aberystwyth)

Citations: MR 22# 9482 = J24-23.

J24-60 **(42# 5941)**

Erdős, P.

On the distribution of the convergents of almost all real numbers.

J. Number Theory **2** (1970), 425–441.

Let $n_1<n_2<\cdots$ be an infinite sequence of integers. The author shows that the necessary and sufficient condition that for almost all α, the inequality $|\alpha-a_i/n_i|<\varepsilon/n_i^2$ with $(a_i, n_i)=1$ should have infinitely many solutions is that $\sum\phi(n_i)/n_i^2$ diverges (here $\phi(n)$ is Euler's function). He conjectures that when this series diverges, then for almost all α, the number of solutions of the inequality with $i\leqq N$ is asymptotic to $2\varepsilon\sum_{i=1}^N \phi(n_i)/n_i^2$ as $N\to\infty$.

For an irrational number α, let $p_i^{(\alpha)}/q_i^{(\alpha)}$, $i=1, 2, 3, \cdots$, denote the sequence of continued fraction convergents to α. As an easy consequence of the first theorem, the author proves that for almost all α infinitely many $q_i^{(\alpha)}$ lie in the sequence n_i if and only if $\sum_{i=1}^\infty \phi(n_i)/n_i^2$ diverges.

D. G. Cantor (Los Angeles, Calif.)

J24-61 **(44# 2708)**

Adams, William W.

Simultaneous asymptotic diophantine approximations to a basis of a real number field.

Nagoya Math. J. **42** (1971), 79–87.

Let $1, b_1, \cdots, b_n$ be a basis of a real algebraic number field. For a fixed positive real number C denote by $N(B)$ the number of solutions in rational integers $q, p_1, \cdots, p_n$ of the inequalities $0<qb_i-p_i<C/q^{1/n}$ $(1\leqq i\leqq n)$ and $1\leqq q\leqq B$, where B is a positive integer. The author proves that there exists a positive real number C' such that as $B\to\infty$ either $N(B)$ remains bounded or $N(B)\sim C'\log B$.

The result for $n=1, 2$ was already known, that for $n=1$ being due to S. Lang [Amer. J. Math. **87** (1965), 481–487;

ibid. **87** (1965), 488–496; MR **31** #3383] while that for $n=2$ is due to the author [J. Number Theory **1** (1969), 179–194; MR **39** #1409]. *A. C. Woods* (Columbus, Ohio)

Citations: MR 31# 3383 = J24-35; MR 39# 1409 = J24-57.

J24-62 (44# 6610)

de Mathan, Bernard
Un critère de non-eutaxie.
C. R. Acad. Sci. Paris Sér. A-B **273** (1971), A433–A436.

Let $u=(u_n)$ be a sequence in $\mathbf{R}/\mathbf{Z}=\mathbf{T}$. We say that (u_n) is eutaxic if, for all sequences of intervals I_n* of decreasing lengths $|I_n|$ with $\sum |I_n|=\infty$, we have that the set of all $x\in\mathbf{T}$ in an infinite number of the I_n has measure 1. The notion was introduced by J. Lesca ["Sur les approximations diophantiennes à une dimension", Doctoral Thesis, Univ. Grenoble, Grenoble, 1968] who proved, for example, that, for irrational α, $(n\alpha-[n\alpha])$ is eutaxic if and only if α has bounded partial quotients. The present author defines a function λ on the space $\mathbf{T}^{\mathbf{N}}$ of a sequence of elements of $\mathbf{T}$ by $\lambda(u)=\liminf N^{-1}\lambda(u,N)$ $(N\to\infty)$, where $\lambda(u)=$ number of k, $0\le k<N$ such that some u_n, $0\le n\le N$ satisfies $k/N\le u_n<(k+1)/N$. He shows that $\lambda(u)=0$ implies that u is not eutaxic. However, $\lambda(u)\ge\frac{1}{2}$ for almost all u in $\mathbf{T}^{\mathbf{N}}$. But if $u=(n\alpha-[n\alpha])$ and α has unbounded partial quotients then $\lambda(u)=0$ thus recovering half of Lesca's result. He also gives a theorem about sequences of real functions $\phi_n(\chi)$ satisfying certain conditions which allow him to conclude that for almost all χ, $\lambda((\phi_n(\chi)))>0$.
W. W. Adams (College Park, Md.)

[*centered at u_n]

J28 PRODUCT OF TWO HOMOGENEOUS LINEAR FORMS IN TWO VARIABLES; MINIMA OF BINARY QUADRATIC FORMS

The special form $q(q\xi-p)$ is also the object of study in **J04** and **J08**, and there is much overlap.

See also reviews E16-72, J02-11, J36-17, J36-18, J36-23, J52-17, J56-9, J56-10, J56-16, J56-21, J56-23, J56-27, R12-23, Z10-20.

J28-1 (1, 39f)

Mahler, K. A proof of Hurwitz's theorem. Mathematica, Zutphen. B. **8**, 57–61 (1939).

A simple proof of the following theorem: Let a, b, c, d be four real numbers of determinant $ad-bc=1$, ϵ a positive number. Then there are two integers u and v, such that

$$|(au+bv)(cu+dv)|\leqq 1/\sqrt{5},\quad |au+bv|<\epsilon;\ u^2+v^2>0.$$

If

$$a=\frac{1+\sqrt{5}}{2\sqrt[4]{5}},\quad b=\frac{1}{\sqrt[4]{5}},\quad c=\frac{1-\sqrt{5}}{2\sqrt[4]{5}},\quad d=\frac{1}{\sqrt[4]{5}},$$

then

$$|(au+bv)(cu+dv)|\geqq 1/\sqrt{5}$$

for all integers u and v that do not vanish simultaneously.
J. F. Koksma (Amsterdam).

Referred to in J28-12.

J28-2 (7, 51c)

Ollerenshaw, Kathleen. The minima of a pair of indefinite, harmonic, binary quadratic forms. Proc. Cambridge Philos. Soc. **41**, 77–96 (1945).

Let K_μ be the set of points (x, y) in the Euclidean plane for which $|xy|\leqq 1$, $|x^2-y^2|\leqq 2\mu$. A K_μ-admissible lattice Λ is defined to be a lattice such that the origin is its only point interior to K_μ. Let $d(\Lambda)$ denote the area of a fundamental parallelogram of Λ. The lower bound of $d(\Lambda)$ for all K_μ-admissible lattices is designated by $\Delta(K_\mu)$. A lattice for which $d(\Lambda)=\Delta(K_\mu)$ is known as a critical lattice. In this paper $\Delta(K_\mu)$ is determined for all positive μ and all critical lattices are constructed. Except for the use of a result of Mahler, the methods are elementary.

By means of an affine transformation of these results the author obtains the principal result of the paper, which is as follows. Let $f_1(x, y)$, $f_2(x, y)$ be a pair of real, binary, indefinite quadratic forms $a_1x^2+2b_1xy+c_1y^2$, $a_2x^2+2b_2xy+c_2y^2$ for which the harmonic relation $a_1c_2-2b_1b_2+c_1a_2=0$ holds. Then the minimum number k is determined, depending only on the discriminants d_1 and d_2 of the forms, for which integers x, y exist, not both zero, so that $|f_1(x, y)|\leqq k$, $|f_2(x, y)|\leqq k$. When the equality sign holds in at least one of the above relations the pair of forms is defined to be critical. A critical pair is constructed for each pair of discriminant values d_1 and d_2. *D. Derry* (Saskatoon, Sask.).

Referred to in J28-14.

J28-3 (8, 502f)

Oppenheim, A. Two lattice-point problems. Quart. J. Math., Oxford Ser. **18**, 17–24 (1947).

The author considers the class $E(f)$ of ellipses which have center O and, as a pair of conjugate diameters, the given pair of straight lines $f(x, y)\equiv ax^2+2hx+by^2=0$, where $h^2>ab$, $D=2(h^2-ab)^{\frac{1}{2}}$ and f has no rational factors. Let Δ_0 be the lower bound of the area of those ellipses of $E(f)$ which have a lattice point on the boundary and no lattice points inside (except O); Δ_1 is defined analogously, but now the ellipses must have at least two lattice points P, Q (not collinear with O) on the boundary and none inside. It is shown that Δ_0 equals the lower bound of $2\pi|f(x, y)|D^{-1}$ for integral x, y (not both zero), whence $\Delta_0\leqq 2\pi/5^{\frac{1}{2}}$. The problem of finding Δ_1 turns out to be equivalent to that of the lower bound $\mathfrak{M}(C_f)$ of $|(A+B)/2H|$, as $Ax^2+2Hxy+By^2$ runs through all forms unimodularly equivalent to f. Using continued fractions, the author shows that $\mathfrak{M}\leqq\frac{1}{2}$; $\mathfrak{M}=\frac{1}{2}$ only if either $f\sim c(x^2+2xy-2y^2)$ or $f\sim c(x^2+4xy-3y^2)$. Now it is readily proved that $\Delta_1=\pi/(1-\mathfrak{M}^2)^{\frac{1}{2}}$ and hence $\pi\leqq\Delta_1\leqq 2\pi/3^{\frac{1}{2}}$. *N. G. de Bruijn* (Delft).

J28-4 (9, 334c)

Hallum, Kathleen C., and Mahler, Kurt. On the minimum of a pair of positive definite Hermitian forms. Nieuw Arch. Wiskunde (2) **22**, 324–354 (1948).

Let $f(x, y)=ax\bar{x}+b\bar{x}y+\bar{b}x\bar{y}+cy\bar{y}$ be a positive definite Hermitian form of determinant $ac-b\bar{b}=1$. Two such forms f_1, f_2 have the joint invariant $j=a_1c_2+a_2c_1-b_1\bar{b}_2-\bar{b}_1b_2$, which necessarily satisfies $j\geqq 2$. The problem studied here is the determination of $m(j)$, defined as the least number m such that the simultaneous inequalities $f_1(x, y)\leqq m$, $f_2(x, y)\leqq m$ are soluble in integers x, y (not both zero) of the field $k(i)$, for every pair of forms of invariant j. The problem is approached through the consideration of critical pairs of forms, i.e., pairs for which the above inequalities are not soluble if $m<m(j)$. It is shown that each of the forms of a critical pair is equivalent to a form of a special type.

Finally, a rule is obtained for determining $m(j)$, which involves only a finite number of trials for any given value of j. The authors have found $m(j)$ explicitly for $2 \leqq j \leqq 6$, and the results [p. 353] show close analogy to Mahler's results for the corresponding problem with two quadratic forms.
H. Davenport (London).

J28-5 **(10, 19b)**

Macbeath, A. M. The minimum of an indefinite binary quadratic form. J. London Math. Soc. **22** (1947), 261–262 (1948).

A simple proof is given of the following theorem due to Korkine and Zolotareff [Math. Ann. **6**, 366–389 (1873)]. Let $f(x, y) = ax^2 + bxy + cy^2$, where a, b, c are real and $d = b^2 - 4ac > 0$. Then integers x, y exist such that $|f(x, y)| \leqq \sqrt{(\frac{1}{5}d)}$. The equality sign is necessary if and only if $f(x, y)$ is equivalent to a multiple of $x^2 + xy - y^2$.
H. S. A. Potter (Aberdeen).

J28-6 **(10, 182h)**

Pall, Gordon. The minimum of a real, indefinite, binary quadratic form. Math. Mag. **21**, 255 (1948).

A simple evaluation is given of the first two Markoff minima of such forms. *H. S. A. Potter* (Aberdeen).

Referred to in J28-13.

J28-7 **(10, 236d)**

Ollerenshaw, Kathleen. On the minima of indefinite quadratic forms. J. London Math. Soc. **23**, 148–153 (1948).

Let Λ be an admissible lattice for the star domain $|xy| \leqq 1$ for which the lower bound of the product $|\xi\eta|$ is 1 when (ξ, η) runs through all points of Λ other than the origin. The author locates all Λ which have minimum determinant $\sqrt{5}$ and shows the minimum determinant of all the remaining lattices Λ is $2\sqrt{2}$. By an affinity which maps the region $|xy| \leqq 1$ into itself a lattice Λ is found which has a point on the line $y = x$ within a certain neighborhood of the point $(1, 1)$. The results are obtained by studying the intersection points of the hyperbolas $xy = \pm 1$ and the lines of minimum distance from the origin which are parallel to the line $y = x$ and contain points of the above Λ. The first two of the Markoff minima for real indefinite binary quadratic forms [Dickson, Studies in the Theory of Numbers, University of Chicago Press, 1930, chap. 7] may be determined from the above results. *D. Derry* (Vancouver, B. C.).

Referred to in J28-13.

J28-8 **(10, 236e)**

Delone, B. On the work of A. A. Markov "On binary quadratic forms with positive determinant." Uspehi Matem. Nauk (N.S.) **3**, no. 5(27), 2–5 (1948). (Russian)

Cf. the following review.

J28-9 **(10, 236f)**

Markov, A. On binary quadratic forms with positive determinant. Uspehi Matem. Nauk (N.S.) **3**, no. 5(27), 6–51 (1948). (Russian)

The work in question was originally published in St. Petersburg in 1880.

J28-10 **(11, 501d)**

Prachar, K. Über höhere Minima quadratischer Formen. Monatsh. Math. **53**, 268–277 (1949).

The author defines M_k for $k = 1, 2, \cdots$ to be the least number such that the binary quadratic form $ax^2 + 2bxy + cy^2$, with real coefficients satisfying $ac - b^2 = 1$, assumes values not exceeding M_k for at least k pairs of integers (x, y) and $(-x, -y)$, excluding $(0, 0)$. He proves that if k is greater than a certain absolute constant, then $M_k = k^2(k^2 - \frac{1}{4})^{-\frac{1}{2}}$. He also evaluates M_2 $(= 8/\sqrt{15})$, and the constant analogous to M_2 for binary Hermitian forms with variables in certain particular quadratic fields. *H. Davenport* (London).

J28-11 **(11, 643b)**

Cassels, J. W. S. The Markoff chain. Ann. of Math. (2) **50**, 676–685 (1949).

This paper gives a simple proof of the theorem by A. Markoff [Math. Ann. **15**, 381–406 (1879); **17**, 379–399 (1880)] on the successive minima of indefinite binary quadratic forms. The theorem is formulated as follows. Let $f(x, y) = ax^2 + bxy + cy^2$ be a real binary form of positive discriminant $d = b^2 - 4ac$, and let 1 be the lower bound of $|f(x, y)|$ for integers x, y not 0, 0. Then if $d < 9$, f is equivalent to a form $\phi_P = (x + hy)^2 - ky^2$ $(h = 3/2 - Q/P,\ k = 9/4 - 1/P^2)$, where P is a solution of the Diophantine equation (i) $P^2 + P_1^2 + P_2^2 = 3PP_1P_2$ $(P \geqq P_1, P \geqq P_2)$, and Q is the least nonnegative solution of $\pm P_1 Q \equiv P_2 \pmod{P}$. The proof is a considerable simplification of Remak's [Math. Ann. **92**, 155–182 (1924)], avoiding continued fractions. The first three minima are obtained by a method modeled on one by Oppenheim [Quart. J. Math. Oxford Ser. (1) **3**, 10–14 (1932)]. Properties of the solutions of (i) are developed in part 2, and these are applied to complete the proof by induction, in part 3. *G. Pall* (Chicago, Ill.).

Referred to in J28-13, J36-17.

J28-12 **(12, 245d)**

Mullender, P. On a theorem of Korkine-Zolotareff. Mathematica, Zutphen B. **13**, 23–27 (1944). (Dutch)

It was shown by Korkine and Zolotareff [Math. Ann. **6**, 366–389 (1873)] that the inequality

$$|(\alpha x + \beta y)(\gamma x + \delta y)| \leqq |\Delta|/A$$

has a solution in integers x, y not both zero for all real α, β, γ, δ with $\Delta = \alpha\delta - \beta\gamma \neq 0$ if and only if $A \leqq \sqrt{5}$. Mahler [same journal **8**, 57–61 (1939); these Rev. **1**, 39] showed that indeed there are infinitely many such solutions if $A \leqq \sqrt{5}$; it is this theorem for which the present author gives a new proof, using a method devised by Heilbronn [cf. Hardy and Wright, An Introduction to the Theory of Numbers, Oxford, 1938, pp. 389–390] for the original Korkine-Zolotareff theorem. *W. J. LeVeque*

Citations: MR 1, 39f = J28-1.

J28-13 **(12, 393b)**

Davis, C. S. The minimum of an indefinite binary quadratic form. Quart. J. Math., Oxford Ser. (2) **1**, 241–242 (1950).

If m is the lower bound of $|f(x, y)|$ for integers x, y not both zero, where f is a real, binary quadratic form of positive discriminant d, and $k = d/m^2$, a simple proof is given that no forms exist with $0 < k < 5$, $5 < k < 8$, or $12 < k < 13$. Replacing f by an equivalent form $\pm m(x - \alpha_1 y)(x - \alpha_2 y)/(1 - \epsilon)$, the proof is based on a discussion of integers which may lie between the roots of $(x - \alpha_1)(x - \alpha_2) = 1 - \epsilon$ and $-1 + \epsilon$. Compare Pall [Math. Mag. **21**, 255 (1948); these Rev. **10**, 182], Cassels [Ann. of Math. (2) **50**, 676–685 (1949); these Rev. **11**, 643], and Ollerenshaw [J. London Math. Soc. **23**, 148–153 (1948); these Rev. **10**, 236] for other simple proofs.
G. Pall (Chicago, Ill.).

Citations: MR 10, 182h = J28-6; MR 10, 236d = J28-7; MR 11, 643b = J28-11.

Referred to in H15-72, J20-25.

J28-14 **(14, 624d)**

Ollerenshaw, Kathleen. On the region defined by $|xy| \leq 1$, $x^2+y^2 \leq t$. Proc. Cambridge Philos. Soc. **49**, 63–71 (1953).

Let K_t denote the region of points (x, y) for which $x^2+y^2 \leq t$, $|xy| \leq 1$, $2 \leq t$. The author determines $\Delta(K_t)$, the lower bound of the determinants of the lattices whose only point interior to K_t is the origin, for all values of t. The critical lattices, the lattices of determinant $\Delta(K_t)$ whose only point interior to K_t is the origin, are also given. Where $f_1 \equiv a_1\xi^2+2b_1\xi\eta+c_1\eta^2$, $f_2 \equiv a_2\xi^2+2b_2\xi\eta+c_2\eta^2$, $d_1=b_1^2-a_1c_1>0$, $d_2=b_2^2-a_2c_2<0$, $a_1c_2-2b_1b_2+a_2c_1=0$, the results of the paper are applied to determine the least k, in terms of d_1, d_2, so that integers ξ, η, not both zero, always exist so that $\max(|f_1|, |f_2|) \leq k$. The critical form pairs, i.e., the form pairs for which $\max(|f_1|, |f_2|)=k$ for at least one integer pair ξ, η, are also given. The results are complementary to the analogous results given by the author for pairs of indefinite, binary quadratic forms [same Proc. **41**, 77–96 (1945); these Rev. **7**, 51]. *D. Derry.*

Citations: MR 7, 51c = J28-2.

J28-15 **(16, 801e)**

Cohn, Harvey. Approach to Markoff's minimal forms through modular functions. Ann. of Math. (2) **61**, 1–12 (1955).

The author considers a subgroup G of genus 1 of the modular group and generated by two modular matrices A and B with a commutator $B^{-1}A^{-1}BA$ of the form $K=\begin{pmatrix}-1 & k\\ 0 & -1\end{pmatrix}$. Then if $C=(AB)^{-1}$ an identity of Fricke yields $S(A)^2+S(B)^2+S(C)^2=S(A)S(B)S(C)$, where S denotes the trace of the matrices. The integers $S(A)$, $S(B)$, $S(C)$ must be divisible by 3. Writing $S(A)=3a$, $S(B)=3b$, $S(C)=3c$, Markoff's equation $a^2+b^2+c^2=3abc$ is obtained. All integral solutions of this equation can be obtained from the smallest solution $a=b=c=1$ by means of the two operations which replace the triples A, B, C by BA, A^{-1}, B^{-1} and A, $A^{-1}B$, B^{-1}. The number k is shown to be ± 6. Markoff's forms (whose minima are greater in absolute value than one-third the square root of the discriminant) are obtained by associating to $A=\begin{pmatrix}a_{11} & a_{12}\\ a_{21} & a_{22}\end{pmatrix}$ the form $a_{21}x^2+(a_{22}-a_{11})xy-a_{12}y^2$, whose minimum and discriminant are $a=a_{12}$ and $S(A)^2-4=9a^2_{21}-4$ respectively. The author finds new forms by taking $k=-12$. The unimodular matrices A of G are not integral in this case but $2A$ is integral. The doubled forms have a minimum which exceeds one-sixth of the square root of the discriminant.

H. D. Kloosterman (Leiden).

J28-16 **(16, 1090a)**

Barnes, E. S. A problem of Oppenheim on quadratic forms. Proc. London Math. Soc. (3) **5**, 167–184 (1955).

Let $f(x, y)=ax^2+bxy+cy^2$ be an indefinite real quadratic form with real coefficients and discriminant D. If $M(f)=\max\{a^2+\frac{1}{4}b^2, c^2+\frac{1}{4}b^2\}$, M is defined to be the lower bound of $M(f)$ where f runs through a complete class of equivalent forms. The author determines a constant q with the property that $M<qD$ unless the class of the forms which defines M contains a multiple of one of the forms x^2-3y^2, $3x^2+27xy-19y^2$, $6x^2+60xy-(\sqrt{(957)}+11)y^2$. For $\epsilon>0$ infinitely many nonequivalent forms exist for which $M>(q-\epsilon)D$. The proof depends on the Frobenius theory of infinite chains of reduced equivalent forms and their associated continued-fraction developments. The assumption $M \geqq qD$ is shown to imply that the continued-fraction development must be one of three types which correspond to the three classes of exceptional forms. *D. Derry.*

J28-17 **(20# 5185)**

Varnavides, P. Antisymmetric Markoff forms. Nederl. Akad. Wetensch. Proc. Ser. A **61**=Indag. Math. **20** (1958), 463–469; erratum, **62** (1959), 328.

Markov forms are integral indefinite forms

$$f(x, y)=mx^2+pxy+qy^2,$$

with homogeneous minimum m and discriminant $d=9m^2-4$. Here m can be an arbitrary positive integer for which the diophantine equation $m^2+y^2+z^2=3myz$ is soluble. The present author uses Markov's continued fraction machinery and constructs two sets of forms of the above type, with $d=9m^2+4$; the numbers m are those for which $m^2+y^2-z^2=3myz$ has a solution with $z=1$ and $z=2$, respectively. *C. G. Lekkerkerker* (Amsterdam)

J28-18 **(20# 6409)**

Fáy, Árpád. On Markoff's numbers. Mat. Lapok **7** (1956), 262–270. (Hungarian. Russian and English summaries)

The author defines a set M of real numbers as follows: $z \in M$ if and only if there exist homogeneous linear forms $\xi(x, y)$, $\eta(x, y)$ with real coefficients for which $z=m/|\Delta|$, where Δ is the determinant of $\xi\eta$ and $m=\inf|\xi\eta|$ over all pairs of integers $(x, y)\neq(0, 0)$. Markoff determined all the points of the set M in $(1/3, \infty)$. The author gives new proofs for some of these results of Markoff and also shows that M is closed and has no points in the open interval $(12^{-1/2}, 13^{-1/2})$. {This last result was known and is due to Shibata [Koksma, Diophantische Approximationen, Springer, Berlin, 1936; p. 33].}

P. Erdös (Birmingham)

J28-19 **(21# 40)**

Vinogradov, A.; Delone, B.; and Fuks, D. Rational approximations to irrational numbers with bounded partial quotients. Dokl. Akad. Nauk SSSR (N.S.) **118** (1958), 862–865. (Russian)

For an indefinite quadratic form $f(x, y)$ with determinant 1 put $m(f)=\inf|f(x, y)|$, where x, y run through all integer values not both zero. For real numbers θ put

$$\mu(\theta) = \liminf y|y\theta-x|,$$

where x, y are integers and $y\to\infty$. The authors show that the set of values taken by $\mu(\theta)$ as θ varies (the spectrum of μ) is a subset of the set of values taken by $m(f)$ as f varies (the spectrum of m). This is, of course, an immediate consequence of Mahler's work [Proc. Roy. Soc. London Ser. A **187** (1946), 151–187; Nederl. Acad. Wetensch. Math. Proc. **49** (1946), 331–343, 444–454, 524–532, 622–631; MR **8**, 195, 12] on bodies with automorphs, but the authors present a proof more or less from first principles but along the obvious lines. It then appears that the authors intended to show that the two spectra are in fact identical, but clearly a fair-sized chunk of manuscript got lost between the bottom of page 864 and the top of page 865, and it is not clear to the reviewer how this is to be replaced. Finally, they deduce from a result of Marshall Hall, Jr. [Ann. of Math. (2) **48** (1947), 966–993; MR **9**, 226] that the spectrum for m contains a complete interval to the right of the origin.

{The reviewer takes the opportunity of remarking that he was told by Marshall Hall of this application of his result some years ago, and, indeed, this application provided the motivation for Hall's paper.}

J. W. S. Cassels (Cambridge, England)

Citations: MR 8, 12b = H15-8; MR 8, 12d = H15-6; MR 8, 195f = H15-7; MR 9, 226b = A54-7.

J28-20 **(21# 6361)**

Mahler, K. An arithmetic property of groups of linear transformations. Acta Arith. **5** (1959), 197–203.

The main result of this paper is the following theorem. Let

$$F = \left\{ s_k \middle| \; s_k(z) = \frac{\alpha_k z + \beta_k}{\gamma_k z + \delta_k}, \quad \alpha_k, \beta_k, \gamma_k, \delta_k \text{ real}, \; \alpha_k\delta_k - \beta_k\gamma_k = 1 \right\}$$

be a Fuchsian group with compact fundamental region. Let $f(u, v) = au^2 + 2buv + cv^2$ be a positive definite quadratic form. Then the values $f(\alpha_k, \gamma_k)$ lie everywhere dense in the positive real axis. This is in contrast to the situation for the modular group, where the values $f(\alpha_k, \gamma_k)$ are nowhere dense (and have a minimum

$$\mu_f = \min_{\alpha,\gamma \text{ integers}} f(\alpha, \gamma) \leqq (2/\sqrt{3}) \sqrt{(ac-b^2)}).$$

The method of proof is as follows. If g is any form, represent g by the point z in the upper half plane such that $g(z, 1) = 0$. If f is represented by z, the F-equivalent form $f_k(u, v) = f(\alpha_k u + \beta_k v, \gamma_k u + \delta_k v)$ is represented by the point $z_k = x_k + iy_k = s_k^{-1}(z)$, where $y_k = \sqrt{(ac-b^2)}/f(\alpha_k, \gamma_k)$. The y_k are proved dense in the positive real numbers by showing that there is an image $s_k^{-1}(z)$ between any pair of horizontal lines. This follows from a theorem of G. A. Hedlund which states that if the fundamental region of F is compact, then the images $s_k(H)$ of any horocycle H are everywhere dense in the upper half plane.

The author conjectures analogous results when f is indefinite or when F is a Kleinian polyhedral group and f is Hermitian. The reviewer has proved these conjectures.

L. Greenberg (Providence, R.I.)

J28-21 **(24# A2563)**

Surányi, János

Über einen Satz von G. Szekeres in der Geometrie der Zahlen.

Ann. Univ. Sci. Budapest. Eötvös Sect. Math. **3–4** (1960/61), 319–326.

Sind

$$x_i = \sum_{1}^{2} {}_j\, a_{ij}u_j \qquad (i = 1, 2)$$

homogene Linearformen mit der Determinante 1, so besagt der Satz von Szekeres, daß für jeden Gitterpunkt, der auf keiner der Achsen liegt, das Produkt $|x_1x_2| \geqq \frac{1}{2}(1 + 1/\sqrt{5})$ ist. Der Verfasser beweist dies mit dem von Delone entwickelten "Algorithmus der zerteilten Parallelogramme" [Izv. Akad. Nauk SSSR Ser. Mat. **11** (1947), 505–538; MR **9**, 334]. Merkwürdigerweise erwähnt er die Arbeiten von Delone nicht. *O.-H. Keller* (Halle)

Citations: MR 9, 334e = J32-9.

J28-22 **(29# 3443)**

Von Wolff, M. R.

Application of the domain of action method to $|xy| \leqq 1$.

Illinois J. Math. **8** (1964), 500–522.

Let $\mathscr{S}$ be a star domain in the affine plane, symmetric about O. A set of points $\mathscr{P}$ is said to provide a packing for $\mathscr{S}$ if the domains $\{\mathscr{S} + P\}$, where $P \in \mathscr{P}$, have the property that no domain $(\mathscr{S} + P_0)$ contains the center of another in its interior; such a point set $\mathscr{P}$ is called an admissible point set for $\mathscr{S}$. Let $A(t)$ denote the number of points of $\mathscr{P}$ in the square $|x| < t$, $|y| < t$. Then the density $\mathscr{D}(\mathscr{P})$ of $\mathscr{P}$ is defined as $\lim\sup_{t\to\infty} A(t)/4t^2$. A norm distance $N(X) = N(OX)$ is a real-valued function which is non-negative, continuous and homogeneous. A convex distance function or Minkowski distance, where M is a norm distance with the additional properties that $M(PQ) = 0$ implies $P = Q$, $M(PQ) \leqq M(PR) + M(RQ)$.

Let $\mathscr{P}$ be a point set in the plane, and let M be a Minkowski distance. The domain of action $D(P) = D(P, M, \mathscr{P})$ of a point P relative to M and $\mathscr{P}$ is the set of all points X in the plane for which $M(PX) \leqq M(QX)$, $Q \in \mathscr{P}$, $Q \neq P$.

Consider the domain $\mathscr{S}: |xy| \leqq 1$. Let

$$M(P_1, P_2) = \tfrac{1}{2}(|x_2 - x_1| + |y_2 - y_1|).$$

N. E. Smith [Ph.D. Dissertation, McGill Univ., Montreal, Que., 1951] proved that a critical lattice gives the closest packing of $\mathscr{S}: |xy| \leqq 1$; in other words, if $\mathscr{P}$ is any $\mathscr{S}$-admissible point set, $\mathscr{D}(\mathscr{P}) \leqq 1/\sqrt{5}$, M. Rahman [Ph.D. Dissertation, McGill Univ., Montreal, Que., 1957] indicated that one might get as sharp a result by using domain of action method; this is indeed the case as the author proves that if O is any point of an admissible point set for the region $\mathscr{S}: |xy| \leqq 1$, then $|D(O)| \geqq \sqrt{5}$.

M. S. Cheema (Tucson, Ariz.).

J28-23 **(29# 4722)**

Chowla, S.

On the inequality $|x^2 - y^2 - 2xyk| \geqq 2k$ (x, y, k odd).

Norske Vid. Selsk. Forh. (Trondheim) **34** (1961), 91.

In this note the author proves that if k is a positive integer, then the equation $u^2 - (k^2+1)v^2 = m$ has no solutions in positive integers u, v if $|m| < 2k$, $|m| \neq t^2$. This strengthens the inequality of the title, which is an unpublished result of Davenport.

R. J. Crittenden (Providence, R.I.)

J28-24 **(29# 4735)**

Williams, K. S.

A note on the quadratic form $ax^2 + 2bxy + cy^2$.

Math. Gaz. **48** (1964), 290–291.

The author gives a simple geometric proof of the following well-known result. Let Λ be a 2-dimensional lattice of determinant $d(\Lambda) \neq 0$. Suppose $a > 0$, $ac > b^2$. Then there is a point $(u, v) \neq (0, 0)$ of the lattice Λ such that

$$au^2 + 2buv + cv^2 \leqq \frac{2}{\sqrt{3}} \sqrt{((ac-b^2)d(\Lambda))}.$$

D. J. Lewis (Ann Arbor, Mich.)

J28-25 **(37# 4012)**

Jackson, T. H.

Small positive values of indefinite binary quadratic forms.

J. London Math. Soc. **43** (1968), 730–738.

Let $f(x, y)$ be an indefinite quadratic form of discriminant $d > 0$ which does not represent zero for integral $(x, y) \neq (0, 0)$. Let $P = P(f)$ be the lower bound of positive values of f for integral x, y; and let $N = N(f) = P(-f)$. For all $k > 0$ (and there is no loss of generality in taking $k \geqq 1$), the author determines $\inf d^{(k+1)/2}/P^kN$ over all forms f. For positive integral n, set $F_n(x, y) = nx^2 + nxy - y^2$; then $P(F_n) = n$, $N(f_n) = 1$ and the infimum is attained by F_1 when $1 \leqq k < 3/2$, by F_n when $k = 1 + n/2$, and by one of F_n, F_{n+1} when $1 + n/2 < k < 1 + (n+1)/2$. "Second minima" are also found when $1 \leqq k < 3/2$ and when $k = 1 + n/2$.

These results, which generalize some earlier particular results of the reviewer, are preliminary to the author's work on the small positive values of indefinite forms in three or more variables. *E. S. Barnes* (Adelaide)

J28-26 **(42# 3024)**

Hightower, Collin J.

The minima of indefinite binary quadratic forms.
J. Number Theory **2** (1970), 364–378.

Let $f(x, y)$ be an indefinite quadratic form, $D(f)$ its discriminant, $m(f)$ the infimum of $|f(x, y)|$ over all integers (x, y) not both zero, and $\mu(f)=m(f)D(f)^{-1/2}$. The classical result of A. A. Markov as improved by L. E. Dickson [*Studies in the theory of numbers*, reprint, Chelsea, New York, 1957] is that $\mu(f)$ has only a discrete set of values above $\frac{1}{3}$ $(5^{-1/2}, 8^{-1/2}, 5\cdot(221)^{-1/2}, \cdots, \to\frac{1}{3})$, each corresponding to a unique equivalence class of (proportional) forms. Thus an infinitude of open intervals $(5^{-1/2}, 8^{-1/2})$, etc. are devoid of values of $\mu(f)$. A result of O. Perron [S.-B. Heidelberg. Akad. Wiss. Natur. Kl. **12** (1921), Essay 4; ibid. **12** (1921), Essay 8] establishes $((13)^{-1/2}, (12)^{-1/2})$ as another such interval. In the other direction, M. Hall [Ann. of Math. (2) **48** (1947), 966–993; MR **9**, 226], established a $\delta_0(>0)$ below which no such open intervals occur. The author shows the following theorem: There is a sequence of open intervals (A_n, B_n), (C_n, D_n) devoid of values of $\mu(f)$ satisfying $.305\ldots=(3249/34933)^{1/2}=A_1<B_1<A_2<B_2<\cdots<(6+2\cdot 5^{1/2})^{-1/2}=.309\ldots<\cdots<C_2<D_2<C_1<D_1=(49/480)^{1/2}=-320$. These numbers are constructed from the convergents to $(1+5^{1/2})/2$.

Harvey Cohn (Tucson, Ariz.)

Citations: MR 9, 226b = A54-7.

J28-27 **(44# 5277)**

Cohn, Harvey

Representation of Markoff's binary quadratic forms by geodesics on a perforated torus.
Acta Arith. **18** (1971), 125–136.

Let $Q(x, y)=ax^2+bxy+cy^2$, $d=b^2-4ac>0$ be a quadratic form with integral coefficients and irrational roots ξ, η (i.e., $Q(\xi, 1)=Q(\eta, 1)=0$). Let $G(\eta, \xi)$ be the geodesics from η to ξ on the upper half plane $\{z=x+iy, y>0\}$ with the metric $ds=|dz|/\operatorname{Im} z$. The author considers the geodesics transferred to the torus, whose period parallelogram corresponds to the fundamental domain of Γ_6 a subgroup of the modular group Γ_1 of index 6. He obtains some characterization of the geodesic of a Markoff form as the geodesic which remains on a perforated torus with open disks excised according to the images on the torus of $\operatorname{Im} z>3/2$. There are also some remarks on a homology class of the geodesics for a Markoff form.

A. N. Andrianov (Leningrad)

J32 PRODUCT OF TWO INHOMOGENEOUS LINEAR FORMS IN TWO VARIABLES; INHOMOGENEOUS MINIMA OF BINARY QUADRATIC FORMS

See also Section J20.

See also reviews H15-55, J40-43, J48-3, J52-7, J56-1, J56-8, J56-10, J56-18, J56-24, J56-25, J56-29.

J32-1 **(3, 70d)**

Mordell, L. J. On the product of two non-homogeneous linear forms. J. London Math. Soc. 16. 86–88 (1941).

Let $a_1, b_1, c_1, a_2, b_2, c_2$ be real numbers and $a_1b_2-a_2b_1=1$. A well-known theorem of Minkowski states that the inequality

$$|(a_1x+b_1y+c_1)(a_2x+b_2y+c_2)|\leqq\tfrac{1}{4}$$

has an integral solution x, y. A new proof is given, suggested by geometric considerations. *C. L. Siegel.*

J32-2 **(4, 189f)**

Pall, Gordon. On the product of linear forms. Amer. Math. Monthly **50**, 173–175 (1943).

A new proof is given for the case $n=2$ of Minkowski's theorem on linear forms. If $F_1(x_1, \cdots, x_n), \cdots, F_n(x_1, \cdots, x_n)$ are n nonhomogeneous linear forms of determinant $D\neq 0$, with real coefficients, there are integral values for the variables $x_1, \cdots, x_n$ such that $|F_1\cdots F_n|=2^{-n}D$. The proof shows first that without loss of generality we may restrict ourselves to the canonical forms $F_1=x+\rho y-\kappa$, $F_2=-\sigma x+y-\lambda$, where $0<\sigma\leqq\rho<1$, and then that integers x, y may be chosen to satisfy simultaneously

(A) $\quad -\frac{1}{2}<x+\rho y-\kappa\leqq\frac{1}{2}$,

(B) $\quad |(x+\rho y-\kappa)(-\sigma x+y-\lambda)|\leqq(1+\rho\sigma)/4$.

The latter part of the proof is an algorithm on the succession of solutions of (A) each of which will also satisfy (B) for a certain range of values for λ and shows that these ranges for λ include all values. *M. Hall* (Washington, D. C.).

J32-3 **(8, 137f)**

Mahler, Kurt. A problem of Diophantine approximation in quaternions. Proc. London Math. Soc. (2) **48**, 435–466 (1945).

In the first chapter several inequalities in the field of quaternions are derived. Chapter II deals with Hermitian forms in quaternions and their reduction. By combining the results of both chapters the author proves his main theorem which forms an equivalent to a well-known theorem of Minkowski on the product of linear (inhomogeneous) forms. If $\alpha, \beta, \gamma, \delta, \rho, \sigma$ are constant quaternions such that $\alpha\bar\alpha\delta\bar\delta+\beta\bar\beta\gamma\bar\gamma-\alpha\bar\gamma\delta\bar\beta-\beta\bar\delta\gamma\bar\alpha=1$, then there are two integral quaternions x, y satisfying $|\alpha x+\beta y+\rho|\,|\gamma x+\delta y+\sigma|\leqq\frac{1}{2}$.

J. F. Koksma (Amsterdam).

J32-4 **(8, 444b)**

Davenport, H. Non-homogeneous binary quadratic forms. Nederl. Akad. Wetensch., Proc. **49**, 815–821 = Indagationes Math. **8**, 518–524 (1946).

If $\alpha, \beta, \gamma, \delta$ are real numbers with $\Delta=\alpha\delta-\beta\gamma\neq 0$, then the theorem of Minkowski on nonhomogeneous linear forms asserts that for any two real numbers λ, μ there exists a lattice point (x, y) such that

$$|(\alpha x+\beta y+\lambda)(\gamma x+\delta y+\mu)|\leqq\tfrac{1}{4}|\Delta|.$$

This theorem can be expressed as a property of the indefinite binary form $ax^2+bxy+cy^2=(\alpha x+\beta y)(\gamma x+\delta y)=f(x, y)$. The author deduces the following theorem which is a little better than the theorem quoted above. For any indefinite quadratic form $f(x, y)$ which does not represent zero for any lattice point other than $(0, 0)$, there exists a number $M<\frac{1}{4}$, such that for any two numbers x_0, y_0 there exist x, y with $x\equiv x_0 \pmod 1$, $y\equiv y_0 \pmod 1$, such that $|f(x, y)|\leqq Md^{\frac{1}{2}}$, $d=b^2-4ac=\Delta^2$. The author defines $M(f)$ to be the lower bound of all such numbers M and investigates its properties. First he deduces an estimate $M(f)$ in terms of any value of f which corresponds to coprime integral values of x, y and which satisfies $0<f_1<d^{\frac{1}{2}}$. Furthermore, he proves that, if $f(x, y)=x^2+2kxy-y^2$, where k is a positive integer, then $M(f)=\frac{1}{4}k(k^2+1)^{-\frac{1}{2}}$ which includes the assertion that the inequality $M(f)<\frac{1}{4}$ is the best possible general inequality for $M(f)$. Finally the author proves that for any form f of Markoff's series [see, for example, Bachmann, Die Arithmetik der quadratischen Formen, vol. 2, Teubner, Leipzig, 1923, chap. 4] we have $M(f) < \frac{1}{4}(5/9)^{\frac{1}{2}}$. *J. F. Koksma.*

Referred to in J32-6, J32-7, J32-12, J32-16, J32-17, J32-25, J32-26, J32-33, J32-41, J32-43, J32-69, J40-18, J48-3, R12-23.

J32-5 (8, 444c)

Khintchine, A. Sur le problème de Tchebycheff. Bull. Acad. Sci. URSS. Sér. Math. [Izvestia Akad. Nauk SSSR] **10**, 281–294 (1946). (Russian. French summary)

Let θ be an irrational real number, α a real number such that $\theta x - y - \alpha \neq 0$ for integral x, y. Denote by $\lambda(\theta, \alpha)$ the lower bound of the positive numbers C for which $|\theta x - y - \alpha| < C/|x|$ is soluble in integers x, y with arbitrarily large $|x|$; by λ the number $\lambda(\theta, 0)$; and by $\mu(\theta)$ the upper bound of $\lambda(\theta, \alpha)$ for all α satisfying the condition above. A classical theorem of Minkowski [for the literature, see Koksma, Diophantische Approximationen, Ergebnisse der Math., v. 4, no. 4, Springer, Berlin, 1936, chap. 6] states that $\mu(\theta) \leqq \frac{1}{4}$; the author improves this inequality to $\mu(\theta) \leqq \frac{1}{4}(1-4\lambda^2)^{\frac{1}{2}}$; he shows that the right-hand side cannot be replaced by a smaller analytic function of λ, and that the equality sign holds for

$$\theta = \frac{1}{k+}\frac{1}{k+}\frac{1}{k+}\cdots$$

when $k \geqq 1$ is an even integer, but not when it is an odd integer. [Compare also the paper of Davenport reviewed above.] *K. Mahler* (Manchester).

Referred to in J32-13, J32-16.

J32-6 (8, 565c)

Davenport, H. Non-homogeneous binary quadratic forms. II. Nederl. Akad. Wetensch., Proc. **50**, 378–389 = Indagationes Math. **9**, 236–247 (1947).

[For part I see the same Proc. **49**, 815–821 = Indagationes Math. **8**, 518–524 (1946); these Rev. **8**, 444.] The author investigates the particular case $\xi = x + \theta y$, $\xi' = x + \theta' y$, where $\theta = \frac{1}{2}(1+5^{\frac{1}{2}})$, $\theta' = \frac{1}{2}(1-5^{\frac{1}{2}})$, and asks whether there are integers x, y such that (1) $|(\xi - a)(\xi' - b)| \leqq M$. Theorem 1. The inequality (1) always holds for $M = \frac{1}{4}$; the sign of equality is needed if and only if (2) $a = \frac{1}{2}\tau + \xi_1$, $b = \frac{1}{2}\tau' + \xi_1'$, where τ is any unit and ξ_1 is any integer of the field $k(\theta)$ and τ', ξ_1' their conjugates. Theorem 2. If a and b are not of the form (2), then (1) holds for $M = \frac{1}{5}$; the sign of equality is needed if and only if (3) $a = 5^{-\frac{1}{2}}\tau + \xi_1$, $b = -5^{-\frac{1}{2}}\tau' + \xi_1'$. Theorem 3. If a and b are not of the form (3), nor of the form (2), then (1) holds for $M = 1/6.34$, where the number 6.34 could be improved slightly. Theorem 4 gives an answer to the question of whether there are infinitely many integers x, y such that (1) holds: if $M > \frac{1}{4}$, (1) has infinitely many solutions with $|\xi - a|$ arbitrarily small and also infinitely many solutions with $|\xi' - b|$ arbitrarily small. *J. F. Koksma.*

Citations: MR 8, 444b = J32-4.

Referred to in J32-7, J32-8, J32-14, J32-19, J32-41, R12-23.

J32-7 (9, 79d)

Davenport, H. Non-homogeneous binary quadratic forms. III. Nederl. Akad. Wetensch., Proc. **50**, 484–491 = Indagationes Math. **9**, 290–297 (1947).

In this paper the author treats the case

$$f(x, y) = 5x^2 - 11xy - 5y^2.$$

Parts I and II appeared in the same Proc. **49**, 815–821 (1946); **50**, 378–389 (1947); these Rev. **8**, 444, 565. *J. F. Koksma* (Amsterdam).

Citations: MR 8, 444b = J32-4; MR 8, 565c = J32-6.

Referred to in J32-8, J32-16, J32-41, R12-23.

J32-8 (9, 412j)

Davenport, H. Non-homogeneous binary quadratic forms. IV. Nederl. Akad. Wetensch., Proc. **50**, 741–749, 909–917 = Indagationes Math. **9**, 351–359, 420–428 (1947).

[For papers II and III of this series see the same Proc. **50**, 378–389, 484–491 = Indagationes Math. **9**, 236–247, 290–297 (1947); these Rev. **8**, 565; **9**, 79. For definitions, formulae and notation see the review of paper II, of which this paper is a continuation.] The paper deals with the minima $M(a, b)$ of $|(\xi - a)(\xi' - b)|$. Suppose that a, b are not of the form (2), nor of the form (4) $a = \tau/a_m + \xi_0$, $b = \tau'/a_m' + \xi_0'$ (or vice versa), where τ is a unit and ξ_0 an integer of $k(\theta)$, and where m is an odd positive integer and a_m is defined by $a_m = 2(\theta^{n+1} - 1)/(\theta^n + 1)$ ($n = 3m$); then the lower bound $M(a, b)$ (for integers ξ of $k(\theta)$) satisfies $M(a, b) \leqq \frac{1}{4}\theta^{-1}$. If a, b are of the form (2), we have $M(a, b) = \frac{1}{4}$; if a, b are of the form (4), we have $M(a, b) = 1/|a_m a_m'|$ and in both cases this minimum is attained for an infinity of integers ξ of $k(\theta)$. This theorem not only solves the question of the third minimum which was left open in paper II, but establishes the existence of an infinite sequence of minima which may be expressed in terms of the Fibonacci numbers, as the author points out. *J. F. Koksma* (Amsterdam).

Citations: MR 8, 565c = J32-6; MR 9, 79d = J32-7.

Referred to in J32-14, J32-41, R12-23.

J32-9 (9, 334e)

Delone, B. N. An algorithm for the "divided cells" of a lattice. Izvestiya Akad. Nauk SSSR. Ser. Mat. **11**, 505–538 (1947). (Russian)

"The subject of this paper is a new algorithm and its application to Minkowski's problem of the product of two nonhomogeneous linear forms in two variables" [author's abstract]. The first part of the paper gives a geometrical exposition of the theory of minima of indefinite homogeneous binary quadratic forms, originally due to Markov. In this theory a decisive part is played by the algorithm for the development of a real number in a canonical continued fraction.

The second part of the paper deals with the corresponding nonhomogeneous problem, which is, in geometrical language, to find the minimum of $|xy|$ for the points of a lattice Γ not containing the origin of the (x, y)-plane. A "divided cell" is defined as a fundamental cell of Γ with one vertex in each quadrant. From one divided cell another can in general be derived as follows: the two edges intersecting the x-axis are produced until they cut the y-axis and on each edge is chosen the pair of neighbouring lattice-points straddling the y-axis; the four points thus obtained are the vertices of a divided cell. Given a lattice without points on, or edges parallel to, the coordinate axes, the above algorithm can be applied repeatedly, forwards and backwards, and leads to a series of divided cells continuing indefinitely in both directions. Results: (I) every lattice has at least one divided cell; (II) given any one divided cell, every divided cell can be obtained from it by the algorithm.

To each step in the algorithm corresponds a pair of positive integers, giving the number of lattice-points passed over in the displacement of each of the two edges originally cutting the x-axis, and a sign (+ or −) signifying that the displacement of both is either clockwise or anti-clockwise. The complete algorithm thus defines a series of pairs of positive integers with appropriate signs. Result: (III) given such a series of pairs of integers and signs, there is, with certain exceptions, one essentially unique lattice to which it belongs by the algorithm.

It follows easily from (I) that, if $|xy|$ has a lower bound 1 for all points of Γ, then the determinant of Γ is greater than or equal to 4 (this is Minkowski's theorem). Further results for this type of lattice are as follows. (IV) Every point of Γ such that $|xy|<\frac{5}{8}$ is a vertex of a divided cell, the constant $\frac{5}{8}$ being best possible. (V) For every integer n, there exists a Γ_n of this type with determinant $4(1+n^{-2})^{\frac{1}{2}}$. The Γ_n for different n are equivalent neither to each other nor to the critical lattice Γ_∞ of the Minkowski theorem (which has determinant 4). Result (V) answers negatively the question whether the minimum determinants in the Minkowski theorem have a discrete structure similar to that of the Markov forms. The lattice Γ_n is obtained according to (III) by taking the defining series to be

$$\cdots, nn+, nn-, nn+, nn-, \cdots.$$

The corresponding product of two nonhomogeneous linear forms is $(1+n^2)x(x+1)-y(y+n-1)+\frac{1}{2}n$, which has minimum $\frac{1}{2}n$ and determinant $2(1+n^2)^{\frac{1}{2}}$.

Finally, it is shown that an obvious extension of the notion of "divided cell" to 3 dimensions is unprofitable, since a face-centered cubic lattice exists which possesses no divided cells. *F. J. Dyson* (Ithaca, N. Y.).

Referred to in J28-21, J32-10, J32-44, J32-48, J32-51, J32-54, J32-57, J32-58, J32-62.

J32-10 (12, 82d)

Delone, B. N. Algorithmus der zerteilten Parallelogramme. Sowjetwissenschaft **1948**, no. 2, 178–210 (1948).

Translated from Izvestiya Akad. Nauk SSSR. Ser. Mat. **11**, 505–538 (1947); these Rev. **9**, 334.

Citations: MR 9, 334e = J32-9.

Referred to in J32-62.

J32-11 (9, 413a)

Davenport, H., and Heilbronn, H. Asymmetric inequalities for nonhomogeneous linear forms. J. London Math. Soc. **22**, 53–61 (1947).

Let $\alpha, \beta, \gamma, \delta$ be real numbers with $\Delta=\alpha\delta-\gamma\beta\neq0$ and let ξ, η denote the linear forms $\xi=\alpha x+\beta y$, $\eta=\gamma x+\delta y$. Minkowski proved that for any real numbers λ, μ there exist integers x, y for which

$$(1) \qquad |(\xi+\lambda)(\eta+\mu)|\leqq\tfrac{1}{4}|\Delta|$$

and it is plain that this result is the best possible, as is shown by the case $\xi=x$, $\eta=y$, $\lambda=\mu=\frac{1}{2}$. Now the authors prove that if one replaces (1) by the inequality

$$(2) \qquad (\xi+\lambda)(\eta+\mu)<|\Delta|,$$

there always exist integers x, y for which not only (2) holds, but also the supplementary conditions $\xi+\lambda\geqq0$, $\eta+\mu\geqq0$. Moreover, they prove that for any $c<1$ there exist $\alpha, \beta, \gamma, \delta, \lambda, \mu$ such that the inequalities

$$(\xi+\lambda)(\eta+\mu)<c|\Delta|, \quad (\xi+\lambda)(\eta+\mu)\geqq0$$

cannot hold simultaneously for integers x, y.

J. F. Koksma (Amsterdam).

Referred to in J32-52, J40-6.

J32-12 (9, 500g)

Varnavides, P. Note on non-homogeneous quadratic forms. Quart. J. Math., Oxford Ser. **19**, 54–58 (1948).

Let $f(x,y)=ax^2+bxy+cy^2$ be an indefinite binary quadratic form with real coefficients and discriminant $d=b^2-4ac>0$. Let $\varphi(f)$ denote the lower bound of the numbers M for which the inequality $|f(x,y)|\leqq M$ always has a solution in real x, y satisfying $x\equiv x_0 \pmod 1$, $y\equiv y_0 \pmod 1$, where x_0, y_0 are arbitrarily given real numbers. Now the author improves some results of Davenport, who gave a method for estimating an upper bound for $\varphi(f)$ in general cases [Nederl. Akad. Wetensch., Proc. **49**, 815–821 = Indagationes Math. **8**, 518–524 (1946); these Rev. **8**, 444] and he applies this improvement in a particular case which had been considered by Heinhold [Math. Z. **44**, 659–688 (1939)]. Improvements of Heinhold's results are given. *J. F. Koksma.*

Citations: MR 8, 444b = J32-4.

Referred to in R12-24, R12-25.

J32-13 (9, 569e)

Godunov, S. K. On a problem of Minkowski. Doklady Akad. Nauk SSSR (N.S.) **59**, 1525–1528 (1948). (Russian)

Let u, v be two homogeneous linear forms with unit determinant in x, y; let λ be the minimum of $|uv|$ for integral nonzero x, y. The author proves that for any real numbers α, β there exist integral x, y such that

$$|(u+\alpha)(v+\beta)|\leqq\tfrac{1}{4}(1-4\lambda^2)^{\frac{1}{2}},$$

and exhibits various forms for which the constant either is, or is not, best possible. The proof is elementary, and modelled on the proof by Khintchine [Bull. Acad. Sci. URSS. Sér. Math. [Izvestia Akad. Nauk SSSR] **10**, 281–294 (1946); these Rev. **8**, 444] of the corresponding theorem for the minimum of $|x(\theta x-y-\alpha)|$. Standard methods of the geometry of numbers are used; after a preliminary transformation to bring one generating vector of the (u, v) lattice to an angle of 45° with the axes, the proof utilises only the shape of the region $|uv|\leqq C$ in the neighborhood of the origin. *F. J. Dyson* (Princeton, N. J.).

Citations: MR 8, 444c = J32-5.

J32-14 (10, 19c)

Varnavides, P. Non-homogeneous binary quadratic forms. I, II. Nederl. Akad. Wetensch., Proc. **51**, 396–404, 470–481 = Indagationes Math. **10**, 142–150, 164–175 (1948).

Minkowski zeigte: Sind $\xi=\alpha x+\beta y$, $\eta=\gamma x+\delta y$ zwei Linearformen mit der Determinante $\Delta=\alpha\delta-\beta\gamma\neq0$, a und b beliebige reelle Zahlen, so gibt es ganze Zahlen x und y, dass $|(\xi-a)(\eta-b)|\leqq\frac{1}{4}|\Delta|$ ist, und dabei lässt sich die Konstante auf der rechten Seite nicht verbessern. In der vorliegenden Arbeit werden nur spezielle Linearformen untersucht, indem ξ eine beliebige ganze Zahl aus dem quadratischen Körper $k(\sqrt{2})$ und ξ' ihre Konjugierte sein soll. Dann wird gezeigt: Für beliebige reelle Zahlen a, b gibt es eine solche Zahl ξ mit $|(\xi-a)(\xi'-b)|\leqq\frac{1}{2}$. Dabei gilt Gleichheit nur für $a=\pm2^{-\frac{1}{2}}+\xi_0$, $b=\mp2^{-\frac{1}{2}}+\xi_0'$ und beliebiges ganzzahliges ξ_0 aus $k(\sqrt{2})$; und weiter: Sind a, b von dieser Form, so besteht die Ungleichung $|(\xi-a)(\xi'-b)|\geqq\frac{1}{2}$ für alle ganzen ξ aus $k(\sqrt{2})$ und hier tritt die Gleichheit für unendlich viele Werte von ξ ein.

Sind hingegen a, b von der Form $a=(\omega/\alpha_n)+\xi_0$, $b=(\omega'/\alpha_n')+\xi_0'$ oder $a=(\omega/\alpha_n')+\xi_0$, $b=(\omega'/\alpha_n)+\xi_0'$, wobei ω eine Einheit von $k(\sqrt{2})$, ξ_0 eine ganze Zahl aus $k(\sqrt{2})$, n eine ungerade positive ganzrationale Zahl und $\alpha_n=2^{\frac{1}{2}}(\tau^{n+1}-1)/(\tau^n+1)$, $\tau=1+\sqrt{2}$ bedeute, dann gibt es eine ganze Zahl ξ mit $|(\xi-a)(\xi'-b)|\leqq1/(\alpha_n\alpha_n')$, und wenn a, b nicht von einer der angegebenen Spezialformen ist, ergibt sich: Untere Grenze $(|(\xi-a)(\xi'-b)|)\leqq1/(2\tau)=1/4.828\cdots$. Sind jedoch a, b von der zuletzt angegebenen speziellen Gestalt, dann gilt $|(\xi-a)(\xi'-b)|\geqq1/(\alpha_n\alpha_n')$ für alle ganzzahligen ξ aus $k(\sqrt{2})$ und die Gleichheit tritt hier für unendlich viele ganzzahlige ξ ein.

Diese Ergebnisse und die Beweise derselben sind analog früheren Ergebnissen von H. Davenport [Nederl. Akad. Wetensch., Proc. **50**, 378–389, 741–749, 909–917 = Indagationes Math. **9**, 236–247, 351–359, 420–428 (1947); diese Rev. **8**, 565; **9**, 412] die sich auf die ganzen Zahlen ξ aus $k(\sqrt{5})$ beziehen. *T. Schneider* (Göttingen).

Citations: MR 8, 565c = J32-6; MR 9, 412j = J32-8.

Referred to in J32-20, J32-41.

J32-15 **(10, 182i)**

Macbeath, A. M. Non-homogeneous linear forms. J. London Math. Soc. **23**, 141–147 (1948).

Two more (geometrical) proofs are given of Minkowski's theorem that, if $a, b, \cdots, f$ are real with $ad-bc=\pm 1$, then there are integers x, y such that

$$|(ax+by+e)(cx+dy+f)| \leqq \tfrac{1}{4}.$$

A new proof is also given of the following theorem, due to Chalk [Quart. J. Math., Oxford Ser. **18**, 215–227 (1947); these Rev. **9**, 413]: if L_i $(i=1, 2, \cdots, n)$ are n nonhomogeneous linear forms in n variables with real coefficients and unit determinant, then (i) integers (x) exist such that $L_i>0$ $(i=1, 2, \cdots, n)$ and $L_1L_2\cdots L_n \leqq 1$; (ii) integers (x) exist such that $L_1\geqq 0$, $L_i>0$ $(i=2, 3, \cdots, n)$ and $L_1L_2\cdots L_n<1$.

W. J. LeVeque (Cambridge, Mass.).

Citations: MR 9, 413b = J40-6.

J32-16 **(10, 183c)**

Cassels, J. W. S. The lattice properties of asymmetric hyperbolic regions. II. On a theorem of Davenport. Proc. Cambridge Philos. Soc. **44**, 145–154 (1948).

[For part I see the same vol., 1–7 (1948); these Rev. **9**, 335.] Let $R\geqq 0$, $S\geqq 0$. Put $\bar{\omega}(z)=16Sz+4p^2z^2$ for $z>0$, where $p=[(R+S)/z]$, $\bar{\omega}(0)=\bar{\omega}(0+0)=4(R+S)^2$. The only discontinuities of $\bar{\omega}$ are at the points $z=(R+S)/q$, $q=1, 2, \cdots$, where $\bar{\omega}(z)=\bar{\omega}(z-0)>\bar{\omega}(z+0)$. Let $f(x, y)$ be an indefinite quadratic form with the determinant Δ^2 $(\Delta>0)$. Let m, n be two coprime integers and put $f(m, n)=\Delta z$, supposing $f(m, n)\geqq 0$. Theorem: let $z\leqq 4R$, $\bar{\omega}(z)\geqq 1$; then to every point (x_0, y_0) there is a point $(x, y)\equiv(x_0, y_0)$ (mod 1) such that (1) $-\Delta S\leqq f(x, y)\leqq \Delta R$. If $\bar{\omega}(z)>1$, the equality signs in (1) are unnecessary except, perhaps, when either $z=4R$, $(x_0, y_0)\equiv(\frac{1}{2}m, \frac{1}{2}n)$ or $\bar{\omega}(z+0)\leqq 1$ or $RS=z=0$ [see also Davenport, Nederl. Akad. Wetensch., Proc. **49**, 815–821 (1946); **50**, 378–389, 484–491 (1947) = Indagationes Math. **8**, 518–524 (1946); **9**, 236–247, 290–297 (1947); these Rev. **8**, 444, 565; **9**, 79]. Applications. (I) The Euclidean algorithm is valid in the field $R(\sqrt{t})$ for $t=2, 3, 6, 7, 11, 5, 13, 17, 21, 29, 33, 37, 41$. (II) Let θ be irrational, let α be any number which cannot be represented in the form $m'-n'\theta$ (m', n' integers). Suppose that there is a constant z and an infinite sequence of pairs of coprime integers (m_k, n_k), $|n_k|\to\infty$, such that either $\varphi_k=n_k(m_k-n_k\theta)\to z-0$, $0<z\leqq 4R$, $\bar{\omega}(z)>1$, or $\varphi_k\to z+0$, $0<z<4R$, $\bar{\omega}(z+0)\geqq 1$, or $\varphi_k\to +0$, $\bar{\omega}(0)>1$, or $\varphi_k\to +0$, $\bar{\omega}(0)=1$, $S\geqq R$. Then the inequalities $-S<n(m-n\theta-\alpha)<R$, $n\neq 0$, have an infinite number of solutions in integral m, n. [See also A. Khintchine, Bull. Acad. Sci. URSS. Sér. Math. [Izvestia Akad. Nauk SSSR] **10**, 281–294 (1946); these Rev. **8**, 444.] [On p. 153, line 14, read $\bar{\omega}(z+0)$ instead of $\bar{\omega}(z)$.]

V. Jarník (Prague).

Citations: MR 8, 444b = J32-4; MR 8, 444c = J32-5; MR 9, 79d = J32-7; MR 9, 335a = J08-5.

J32-17 **(10, 183d)**

Cassels, J. W. S. Lattice properties of asymmetric hyperbolic regions. III. A further result. Proc. Cambridge Philos. Soc. **44**, 457–462 (1948).

[Cf. the preceding review.] Let $f(x, y)=ax^2+bxy+cy^2$ be an indefinite quadratic form with $a>0$. (I) If x_0, y_0 are real numbers, there is a point $(x, y)\equiv(x_0, y_0)$ (mod 1) such that (1) $\frac{1}{4}e\leqq f(x, y)\leqq\frac{1}{4}a$; here $e=\min(0, e_1)$, $e_1=\min f(r, 1)$, where r runs through all integers [a classical result of Minkowski gives $|f(x, y)|\leqq\frac{1}{4}(b^2-4ac)^{\frac{1}{2}}$]. The second equality sign in (1) is unnecessary except perhaps if $(x_0, y_0)\equiv(\frac{1}{2}, 0)$; the first one is unnecessary except perhaps if there is an integer r such that either $f(r, 1)=e_1<0$, $(x_0, y_0)\equiv(\frac{1}{2}r, \frac{1}{2})$ or $f(r, 1)=e_1=0$, $x_0\equiv ry_0$. (II) For special forms $f=k(x^2-xy)-y^2$ (k integer, $k>0$) (1) says that (2) $-\frac{1}{4}\leqq f(x, y)\leqq\frac{1}{4}k$; but if k is odd, (2) can be replaced by the sharper relation (3) $-\frac{1}{4}\leqq f(x, y)\leqq\frac{1}{4}k-\frac{1}{4}(k+4)^{-1}$. Both (2) and (3) are best possible results. E.g.: if k is even, there is no point $(x, y)\equiv(\frac{1}{2}, 0)$ for which $-\frac{1}{4}(k+4)<f(x, y)<\frac{1}{4}k$ and no point $(x, y)\equiv(\frac{1}{2}, \frac{1}{2})$ or $(0, \frac{1}{2})$ for which $-\frac{1}{4}<f(x, y)<\frac{1}{4}(2k-1)$. There is an analogous result concerning (3) (k odd). For other results concerning special forms f see Davenport [Nederl. Akad. Wetensch., Proc. **49**, 815–821 = Indagationes Math. **8**, 518–524 (1946); these Rev. **8**, 444]. [In (7) (p. 460) read $-\frac{1}{4}$ instead of $\frac{1}{4}$.]

V. Jarník (Prague).

Citations: MR 8, 444b = J32-4.

J32-18 **(10, 355b)**

Sawyer, D. B. The product of two non-homogeneous linear forms. J. London Math. Soc. **23**, 250–251 (1948).

Let $x=a\xi+b\eta+p$, $y=c\xi+d\eta+q$ define a nonhomogeneous lattice of determinant $\Delta=|ad-bc|$, where a, b, c, d, p, q are real. Minkowski showed that integral values of ξ, η exist such that (1) $|xy|\leqq\Delta/4$. The author finds another proof by considering lattice points in relation to the region $|xy|\leqq\Delta/4$. This short proof may also be used to determine those lattices for which equality in (1) is necessary.

L. Tornheim (Ann Arbor, Mich.).

Referred to in J32-24, J32-44.

J32-19 **(10, 682g)**

Varnavides, P. On the quadratic form x^2-7y^2. Proc. Roy. Soc. London. Ser. A. **197**, 256–268 (1949).

Es wird die Frage nach der Approximation zweier inhomogener Linearformen der Form $x+y\sqrt{7}-a$ und $x-y\sqrt{7}-b$ durch ganze rationale Zahlen x, y, wobei a, b beliebige reelle Grössen sind, durch folgenden Satz beantwortet. Für beliebige reelle a, b gibt es ganze Zahlen ξ aus $K(\sqrt{7})$, die die Ungleichung $|(\xi-a)(\xi'-b)|\leqq\frac{9}{14}$ befriedigen, wenn ξ' die Konjugierte von ξ ist. Sind a, b speziell von der Form $a=\omega/\alpha_0+\xi_0$, $b=\omega'/\alpha_0'+\xi_0'$, wo $\alpha_0=\frac{1}{9}(7+5\sqrt{7})$, $\alpha_0'=\frac{1}{9}(7-5\sqrt{7})$, ω irgendeine Einheit von $K(\sqrt{7})$ und ξ_0 eine ganze Zahl aus $K(\sqrt{7})$ bedeuten, so gibt es unendlich viele Lösungen obiger Ungleichung, für die sogar das Gleichheitszeichen eintritt, aber keine Lösungen mit dem Kleinerzeichen. Wenn aber a und b nicht diese spezielle Gestalt haben, gibt es eine ganze Zahl ξ von $K(\sqrt{7})$, die die schärfere Ungleichung

$$|(\xi-a)(\xi'-b)|\leqq\frac{1}{1.56}<\frac{9}{14}$$

erfüllt. Für den Körper $K(\sqrt{11})$ hat der Verf. in seiner Dissertation [London, 1948] bereits ähnliche Resultate gezeigt. Der Beweis gründet sich auf eine Davenportsche Methode [Nederl. Akad. Wetensch., Proc. **50**, 378–389 = Indagationes Math. **9**, 236–247 (1947); diese Rev. **8**, 565].

T. Schneider (Göttingen).

Citations: MR 8, 565c = J32-6.

Referred to in J32-25, J32-29, J32-33.

J32-20 **(11, 83c)**

Varnavides, P. Quadratic forms near to x^2-2y^2. Quart. J. Math., Oxford Ser. **20**, 124–128 (1949).

Heinhold [Math. Z. **44**, 659–688 (1939)] zeigte, dass die Ungleichung $|f(x+x_0, y+y_0)|\leqq\frac{1}{2}=\frac{1}{4}(d/2)^{\frac{1}{2}}$ für

$$f(x, y)=f_0(x, y)=x^2-2y^2$$

und irgendwelche reellen Zahlen x_0, y_0 in ganzzahligen x, y stets lösbar ist (d bedeutet die Diskriminante der quadratischen Form $f(x, y)$). Verf. [Nederl. Akad. Wetensch., Proc. **51**, 396–404 = Indagationes Math. **10**, 142–150 (1948); diese Rev. **10**, 19] bewies eine Verschärfung dieser Ungleichung auf $\frac{1}{4}=\frac{1}{8}(d/2)^{\frac{1}{2}}$, falls nur $x_0\equiv 0(1)$, $y\equiv\frac{1}{2}(1)$ nicht gilt. Nun wird gezeigt, dass eine solche Verschärfung auch möglich ist für Formen $f(x, y)$, die "nahe bei $f_0(x, y)$" liegen. Der Satz lautet: seien g und h beliebige reelle Zahlen mit $7/5<g<10/7$, $7/5<h<10/7$ und sei $f(x, y)=(x-gy)(x+hy)$ mit $d=(g+h)^2$.

Dann existieren für beliebige reelle x_0, y_0 ganze Zahlen x, y, dass $|f(x+x_0, y+y_0)|<d^{\frac{1}{2}}/5.898<0.96\cdot\frac{1}{4}(d/2)^{\frac{1}{2}}$ gilt, ausser wenn $g=h=\sqrt{2}$ und $x_0\equiv 0(1)$, $y\equiv\frac{1}{2}(1)$ erfüllt sind.
T. Schneider (Göttingen).

Citations: MR 10, 19c = J32-14.

J32-21 (11, 501c)

Steuerwald, Rudolf. Bemerkungen zu einer Arbeit von Herrn J. Heinhold. Math. Z. **52**, 394–400 (1949).

The author settles some of the undecided cases in Heinhold's tables [Math. Z. **44**, 659–688 (1939)], relating to the minimum of the quadratic form x^2-Dy^2 or $x^2+xy-\frac{1}{4}(D-1)y^2$ for $x\equiv x_0$ (mod 1), $y\equiv y_0$ (mod 1). *H. Davenport.*

J32-22 (11, 582f)

Davenport, H. Indefinite binary quadratic forms. Quart. J. Math., Oxford Ser. (2) **1**, 54–62 (1950).

Let $f(x, y)=ax^2+bxy+cy^2$ have real coefficients and discriminant $d=b^2-4ac>0$, and $f(x, y)\neq 0$ for integral $(x, y)\neq(0, 0)$. Then there exist real numbers ξ, η such that $|f(x+\xi, y+\eta)|>\kappa^2\sqrt{d}$ for all integral x, y, where κ is an absolute constant. If a, b, c are integers then ξ, η can be taken to be rational. In the proof ξ, η are given explicitly, based upon a "regular" chain of equivalent forms selected from a complete chain of reduced forms according to Hermite. The author states that he has shown in a paper in the course of publication that κ^2 may be taken as 1/128, but the present proof is based on more general ideas, and the method can be extended to cubic fields of negative discriminant. *L. Tornheim* (Ann Arbor, Mich.).

Referred to in R12-19, R12-21, R16-9, R58-26.

J32-23 (12, 82b)

Blaney, Hugh. Some asymmetric inequalities. Proc. Cambridge Philos. Soc. **46**, 359–376 (1950).

Let $\xi=\alpha x+\beta y$, $\eta=\gamma x+\delta y$ be two real linear forms of determinant Δ, and μ a number with $-1<\mu\leqq 1$. The object of this article is to determine real continuous functions $g(\mu)$ for which the inequalities, $-\mu g(\mu)\Delta<(\xi+\theta)(\eta+\phi)\leqq g(\mu)\Delta$, have integral solutions x, y for all real θ and ϕ and to determine the values of μ for which these inequalities are the best possible (i.e., values of μ such that these inequalities are no longer true in the strict sense for any integral x, y for some ξ, η, θ, ϕ). For $0<\mu\leqq 1$ the author shows that the inequalities have solutions if $g(\mu)=\frac{1}{4}\mu^{-\frac{1}{2}}$ and that they are best possible for $\mu=m/(m+2)$, $m=1, 2, 3, \cdots$. Similar results are obtained if $g(\mu)=[(1+\mu)(1+9\mu)]^{-\frac{1}{2}}$, $0\leqq\mu\leqq\frac{1}{3}$, in which case the inequalities are best possible for $\mu=1/(4m-1)$, $m=1, 2, 3, \cdots$, and $\mu=0$. For negative μ a function $g(\mu)$ is given for which the inequalities are best possible for only a finite number of values μ. Using geometric methods, Davenport [Acta Math. **80**, 65–95 (1948); these Rev. **10**, 101] determined the first of these three functions $g(\mu)$ without enumerating all the values of μ for which the inequalities are the best possible. The author makes use of Davenport's method in this case and an adaptation of it to develop the third $g(\mu)$. All his other proofs are arithmetic.
D. Derry (Vancouver, B. C.).

Citations: MR 10, 101i = J48-3.

J32-24 (12, 82c)

Sawyer, D. B. A note on the product of two non-homogeneous linear forms. J. London Math. Soc. **25**, 239–240 (1950).

The theorem of Davenport [Acta Math. **80**, 65–95 (1948); these Rev. **10**, 101] that integer values ξ, η exist such that $-\lambda\Delta/4\leqq xy<\Delta/4\lambda$, where $x=a\xi+b\eta+p$, $y=c\xi+d\eta+q$ with a, b, c, d, p, q real and $\Delta=ad-bc>0$, is demonstrated by a slight modification of a proof given by the author when $\lambda=1$ [J. London Math. Soc. **23**, 250–251 (1948); these Rev. **10**, 355]. This is generalized to a set of four inequalities of the form $|x^\theta y^{1-\theta}|<\lambda\Delta^{\frac{1}{2}}$ (with equality to be allowed in one of them), one for each quadrant, one of which is satisfied by a point (x, y). Here θ and λ depend on the quadrant, $0<\theta<1$, and the θ's and λ's satisfy an inequality. *L. Tornheim.*

Citations: MR 10, 101i = J48-3; MR 10, 355b = J32-18.

J32-25 (12, 320b)

Inkeri, K. On the Minkowski constant in the theory of binary quadratic forms. Ann. Acad. Sci. Fennicae. Ser. A. I. Math.-Phys. no. **66**, 35 pp. (1950).

Let $f(x, y)=ax^2+bxy+by^2$ be a real quadratic form with positive discriminant $d=b^2-4ac$. Let $\mathfrak{M}$ be the least number with the property that for any real x_0, y_0, real X, Y exist with $X\equiv x_0$ (mod 1), $Y\equiv y_0$ (mod 1), and $|f(X, Y)|\leqq\mathfrak{M}$. Minkowski proved $\mathfrak{M}<\frac{1}{4}\sqrt{d}$. The purpose of this paper is either to determine $\mathfrak{M}$ exactly or to give estimates for it for special classes of forms. Forms representing 0 are excluded. Let $D=d/4a^2$. It is assumed $\frac{1}{4}<D$, otherwise $f(x, y)$ is replaced by an equivalent form with $\frac{1}{4}<D$. Let r be a real number with $r\equiv -b/2a$ (mod 1). The principal results of the paper consist of a number of different bounds for $\mathfrak{M}$ in the case where $D<1$ which depend on r, a, and D. Typical of these is the following. For $r\geqq 0$, $D\geqq r^2+r$, $\mathfrak{M}\leqq C|a|$, where C is

$$\max\{\tfrac{1}{4}D, \tfrac{1}{4}[(2D-r^2)^{\frac{1}{2}}-r]^2, \tfrac{1}{4}[1-(D-r^2-r)^2/D]\}.$$

These results for forms with appropriate values of r are improvements of Davenport's result that $\mathfrak{M}\leqq\frac{1}{4}|a|$ [Nederl. Akad. Wetensch., Proc. **49**, 815–821 = Indagationes Math. **8**, 518–524 (1946); these Rev. **8**, 444]. Davenport's result is a consequence of one of the author's theorems which also yields a proof of the following result of Heinhold [Math. Z. **44**, 659–688 (1939)]. For a squarefree integer m let $f(x, y)=x^2-my^2$ or $x^2+xy+\frac{1}{4}(1-m)y^2$ according as $m\not\equiv 1$ (mod 4) or $m\equiv 1$ (mod 4). If p, p_1, q are positive integers with $m=q^2+p=(q+1)^2-p_1$ or $m=(2q+1)^2+4p=(2q+3)^2-4p_1$ according as $m\not\equiv 1$ (mod 4) or $m\equiv 1$ (mod 4), then

$$\mathfrak{M}\leqq\tfrac{1}{4}\max(p, p_1)=M_0.$$

For $m\equiv 1$ (mod 4) it is shown that $p=p_1$ implies $\mathfrak{M}=M_0$, and conversely. For the remaining m a number of criteria are given which imply that $\mathfrak{M}=M_0$. Criteria are also given which insure that $\mathfrak{M}<M_0$. The author applies his results to the forms defined by the squarefree integers m with $m\leqq 101$. In most cases where $\mathfrak{M}<M_0$ a smaller bound for $\mathfrak{M}$ is given than M_0. For $m=7$ and $m=69$ simple computations give the exact value of $\mathfrak{M}$ $(<M_0)$. The first of these values was given earlier by Varnivides [Proc. Roy. Soc. London. Ser. A. **197**, 256–268 (1949); these Rev. **10**, 682], the second is new. *D. Derry* (Vancouver, B. C.).

Citations: MR 8, 444b = J32-4; MR 10, 682g = J32-19.

Referred to in J32-43, J32-69.

J32-26 (12, 678a)

Varnavides, P. The Minkowski constants associated with quadratic forms near x^2-2y^2. Bull. Soc. Math. Grèce **25**, 153–163 (1951). (Greek)

Let $d=b^2-4ac>0$ be the discriminant of the quadratic form $f(x, y)=ax^2+bxy+cy^2$. Minkowski showed that if x_0, y_0 are arbitrary real numbers, then there exist integers x, y such that $|f(x+x_0, y+y_0)|\leqq\frac{1}{4}\sqrt{d}$, and the constant $\frac{1}{4}$ cannot be improved since it is attained for $f(x, y)=xy$, $x_0=y_0=\frac{1}{2}$. The author denotes by $M(f)=M(a, b, c)$ the lower bound of all values λ having the property that for arbitrary real x_0, y_0 there exist integers x, y such that $|f(x+x_0, y+y_0)|\leqq\lambda\sqrt{d}$ and calls $M(f)$ the constant of Minkowski associated with the form f. Heinhold [Math. Z. **44**, 659–686 (1930)] and Davenport [Nederl. Akad. Wetensch., Proc. **49**, 815–821; these Rev. **8**, 444] obtained $M(f_0)=1/4\sqrt{2}$ for the form $f_0(x, y)=x^2-2y^2$. In the present paper the author asserts that for certain forms f differing little from $f_0(x, y)$ the Minkowski constant $M(f)<1/4\sqrt{2}-10^{-7}$ and states that this result is a consequence of the following theorem: Let g, h be arbitrary real numbers satisfying $|g-\sqrt{2}|<\tau^{-14}$, $|h-\sqrt{2}|<\tau^{-14}$, where $\tau=1+\sqrt{2}$, and let $\xi=x+gy$, $\eta=x-hy$. Then, given real numbers x_0, y_0 there exist integers x, y such that $|(\xi-x_0)(\eta-y_0)|\leqq\frac{1}{2}-\tau^{-14}$ unless $g=h=\sqrt{2}$ and $x_0=u+(v+\frac{1}{2})\sqrt{2}$, $y_0=u-(v+\frac{1}{2})\sqrt{2}$, u and v being integers. Since g and h are nearly $\sqrt{2}$, the form $f(x, y)=(x+gy)(x-hy)$ is "near" x^2-2y^2. The proof is practically unreadable because of the large number of misprints, of which we note the following: p. 154, line 13, x^2-2y^2 instead of x^2-2y; p. 158, equation (29), $\xi-x$ instead of $E-x$; the sign $<$ should be replaced by $>$ in (30), (33), (36), (39), (44), and p. 159, line 15. The relations (46), (47), (50), (53), and (54) are so badly printed that they are meaningless.

T. M. Apostol (Pasadena, Calif.).

Citations: MR 8, 444b = J32-4.

J32-27 (13, 16b)

Mordell, L. J. The product of two non-homogeneous linear forms. IV. J. London Math. Soc. **26**, 93–95 (1951).

In a previous paper [same J. **18**, 218–221 (1943); these Rev. **6**, 38] the author stated a refinement of a result of Minkowski's as follows: Theorem A. Let $L=ax+by$, $M=cx+dy$ where a, b, c, d are real and $ad-bc=\Delta\neq 0$. If a/b is irrational and $\epsilon>0$ is arbitrary, then for any real numbers p and q infinitely many integers (x, y) exist such that

(1) $|L+p|\cdot|M+q|<\frac{1}{4}|\Delta|$, (3) $0<|L+p|<\epsilon$.

The author remarks that he overlooked the exceptional case when $L+p=0$ for integers x and y. In this case the author shows that under the conditions of Theorem A and for an arbitrary $\delta>0$ there exist integers x and y satisfying

(4) $|L+p|\cdot|M+q|<\Delta/\sqrt{5}+\delta$, (5) $0<|L+p|<\epsilon$.

The constant $1/\sqrt{5}$ is now best possible. The author mentions the corresponding problem for n homogeneous linear forms in n variables. *B. W. Jones* (Boulder, Colo.).

Citations: MR 6, 38a = J52-7.

Referred to in J40-17.

J32-28 (13, 536e)

Hametner, Herbert. Über die Approximation von indefiniten binären quadratischen Formen. Monatsh. Math. **55**, 300–322 (1951).

Let $f(x, y)$ be an indefinite binary quadratic form with real coefficients. The paper is concerned with the determination of numbers M with the property that the inequality $|f(x-x_0, y-y_0)|\leqq M$ is soluble in integers x, y for any real numbers x_0, y_0. The author's result is somewhat complicated, and depends on the choice of a certain set of integer pairs. The result is applied to the particular cases when $f(x, y)$ is the norm-form of the quadratic fields generated by $\sqrt{47}$, $\sqrt{71}$, $\sqrt{79}$, $\sqrt{53}$, $\sqrt{85}$; and except in the first case smaller values of M are obtained than those previously published by other workers. *H. Davenport* (London).

J32-29 (13, 627e)

Bambah, R. P. Non-homogeneous binary quadratic forms. I. Two theorems of Varnavides. Acta Math. **86**, 1–29 (1951).

Let $f(x, y)=x^2-7y^2$. Then given any two real numbers x_0, y_0, there exist x, y with $x\equiv x_0$, $y\equiv y_0$ (mod 1) such that $|f(x, y)|\leqq 9/14$, equality being required if and only if $x_0\equiv\frac{1}{2}$, $y_0\equiv\pm 5/14$ (mod 1); otherwise one can have $|f(x, y)|<1/1.56$. The first half of the proof is geometric and reduces the problem to a consideration of certain small regions in the plane. The second half involves properties of the integers of the field generated by $\sqrt{7}$. A similar theorem holds for $f(x, y)=x^2-11y^2$. The first minimum is 19/22, being required only if $x_0\equiv\frac{1}{2}$, $y_0\equiv\pm 7/22$ (mod 1) and the next minimum $<1/1.16$. These results are a slight improvement upon Varnavides' [Proc. Roy. Soc. London. Ser. A. **197**, 256–268 (1949); these Rev. **10**, 682], whose proofs were entirely arithmetic. *L. Tornheim* (Ann Arbor, Mich.).

Citations: MR 10, 682g = J32-19.

J32-30 (13, 628a)

Bambah, R. P. Non-homogeneous binary quadratic forms. II. The second minimum of $(x+x_0)^2-7(y+y_0)^2$. Acta Math. **86**, 31–56 (1951).

The theorem on $f(x, y)=x^2-7y^2$ of the preceding review is improved by showing that the second minimum is $\frac{1}{2}$ and is required if and only if $x_0\equiv\frac{1}{2}$, $y_0\equiv\frac{1}{2}$ (mod 1). Also this time the proof is entirely geometric. The second half depends on a lemma of J. W. S. Cassels: if S is an integral, unimodular, nonhomogeneous linear transformation of the plane with real characteristic roots $\neq\pm 1$, if Λ is the lattice of points with integer coordinates, and if R is a certain type of region, then $\bigcap_{n=-\infty}^{+\infty} S^n(R+\Lambda)$ consists only of a finite number of certain nonhomogeneous lattices.

L. Tornheim (Ann Arbor, Mich.).

J32-31 (13, 919a; 13, 919b)

Cassels, J. W. S. The inhomogeneous minimum of binary quadratic, ternary cubic and quaternary quartic forms. Proc. Cambridge Philos. Soc. **48**, 72–86 (1952).

Cassels, J. W. S. Addendum to the paper, The inhomogeneous minimum of binary quadratic, ternary cubic, and quaternary quartic forms. Proc. Cambridge Philos. Soc. **48**, 519–520 (1952).

Let (x) denote $(x^{(1)}, \cdots, x^{(n)})$ and let $f((x))$ be a real homogeneous form in (x) of degree n. The author's results are of the type: There exist real (x_0) such that $|f((x))|\geqq|\Delta|/\gamma$ for all $(x)\equiv(x_0)$ (mod 1), where Δ depends on f and γ is a positive constant depending only on n. For $n=2$, f indefinite, $f((x))=f(x, y)=ax^2+2bxy+cy^2$, $\Delta=\sqrt{(b^2-ac)}$, $\gamma\backsim 48$ (see "Addendum" for error on p. 76 and other errors). Further results for $n=2$ are: If f represents 0, there are indenumerably many such incongruent (x_0); if a, b, and c are rational then (x_0) may be chosen rational; with $\gamma=87$, there exists a transcendental (x_0); for almost all (x_0): min $|f((x))|=0$, $(x)\equiv(x_0)$ (mod 1), provided f is not a multiple of a product of two linear forms with rational coefficients; the last proviso is necessary. For $n=3$ the factorable case only is dealt with: $f((x))=L_1L_2L_3$, L's linear, L_1 real, L_2 and L_3 conjugate complex, $\Delta=\det(L_1, L_2, L_3)\neq 0$, $\gamma=420$. Also for $n=4$ only the factorable case is dealt with: $f((x))=L_1L_2L_3L_4$, L's linear, $L_2=\bar{L}_1$, $L_4=\bar{L}_3$, $\Delta=\det(L_1, \cdots, L_4)$, $\gamma=5300$. In these cases, if the L's are conjugate linear forms of an algebraic field, then a rational (x_0) exists. The γ's for $n=3$, 4 are considerable improvements of results of Davenport [Acta Math. **84**, 159–179 (1950); Trans. Amer. Math. Soc. **68**, 508–532 (1950); these Rev. **12**, 594], whose γ's are 8×10^{13} and 10^{132}, respectively. For $n=2$, the author's γ (see "Addendum") is less of an improvement of the $\gamma=128$ of Davenport [Proc. London. Math. Soc. (2) **53**, 65–82 (1951); these Rev. **13**, 15]. The author's methods differ from those of Davenport. They are related to those of Chabauty and Lutz [C. R. Acad. Sci. Paris **231**, 887–888, 938–939 (1950); these Rev. **12**, 483, 807].

R. Hull (Lafayette, Ind.).

Citations: MR 12, 483d = J16-11; MR 12, 594b = R16-9; MR 12, 807d = J20-15; MR 13, 15f = R12-19.

Referred to in H15-60, J32-56, J40-43.

J32-32 **(13, 919e)**

Yûjôbô, Zuiman. On a theorem of Minkowski and its proof of Perron. Proc. Japan Acad. **27**, 263–267 (1951).

The theorem of Minkowski meant is the following: For each couple of forms $L_1=\alpha x+\beta y-\sigma$, $L_2=\gamma x+\delta y-\tau$ ($\Delta=\alpha\delta-\beta\gamma\neq0$) there exists at least one lattice point (x, y) such that $|L_1L_2|\leqq\frac{1}{4}|\Delta|$. Using the method by which Perron proved the above theorem, the author proves that even an infinity of such lattice points with $|x|\to\infty$, $|y|\to\infty$ exist, assuming that $\delta\neq0$, γ/δ is irrational, and $L_2\neq0$ for all lattice points (x, y). The case $L_1=x$, $L_2=\theta x-y-\vartheta$ has already been treated in the same way by the reviewer [Math. Ann. **116**, 464–468 (1939); for further literature see this paper].

J. F. Koksma (Amsterdam).

J32-33 **(14, 143b)**

Varnavides, P. The Minkowski constant of the form x^2-11y^2. Bull. Soc. Math. Grèce **26**, 14–23 (1952). (Greek summary)

The Minkowski constant $m(f)$ of the quadratic form $f(x, y)=ax^2+bxy+cy^2$ with discriminant $d>0$ is the lower bound of numbers λ such that for arbitrary real x_0, y_0 there exist integers x, y such that $f(x+x_0, y+y_0)\leqq\lambda\sqrt{d}$. This constant was determined for x^2-2y^2 by Davenport [Nederl. Akad. Wetensch., Proc. **49**, 815–821 (1946); these Rev. **8**, 444] and for x^2-7y^2 by the author [Proc. Roy. Soc. London. Ser. A. **197**, 256–268 (1949); these Rev. **10**, 682]. The form x^2-11y^2 is treated in this paper with the result $m(f)=19/44\sqrt{11}$, an improvement on Perron's inequality $m(f)\leqq1/2\sqrt{11}$. The method is similar to that used in the 1949 paper but deals with the field $k(\sqrt{11})$ instead of $k(\sqrt{7})$.

T. M. Apostol (Pasadena, Calif.).

Citations: MR 8, 444b = J32-4; MR 10, 682g = J32-19.

J32-34 **(14, 252c)**

Chalk, J. H. H. The minimum of a non-homogeneous bilinear form. Quart. J. Math., Oxford Ser. (2) **3**, 119–129 (1952).

Consider values of the real polynomial

$$B=(\alpha x+\beta y+c_1)(\gamma z+\delta t+c_2),$$

where $\Delta=\alpha\delta-\beta\gamma\neq0$, for integral x, y, z, t such that $xt-yz=\pm1$. Then there is a solution of $|B|\leqq\frac{1}{4}|\Delta|$. Furthermore, if α/β, γ/δ are irrational and $(c_1, c_2)\neq(0, 0)$, then $|B|\leqq\frac{1}{8}|\Delta|$. The case $(c_1, c_2)=(0, 0)$ has been treated by Davenport and Heilbronn [same J. **18**, 107–121 (1947); these Rev. **9**, 79]. The critical lattices of $|(x+c)y|\leqq1$ for $c\neq0$ have determinant $2+\sqrt{2}$. This is a consequence of the following lemma: Let $L_1=\alpha x+\beta y$, $L_2=\gamma x+\delta y$, $\Delta=\alpha\delta-\beta\gamma\neq0$, and $c\neq0$. Then there exist relatively prime integers x, y such that $|(L_1+c)L_2|\leqq k|\Delta|$ where $k=1-1/\sqrt{2}$; here k is least possible. If α/β is rational, then $|(L_1+c)L_2|<m|\Delta|$ where $m=1/4$. *L. Tornheim.*

Citations: MR 9, 79c = J44-6.
Referred to in J32-45.

J32-35 **(14, 252d)**

Davenport, H. Note on a result of Chalk. Quart. J. Math., Oxford Ser. (2) **3**, 130–138 (1952).

In the notation of the paper above it is shown that $m=1/4$ is least possible by exhibiting suitable L_1, L_2. If the condition that x, y be relatively prime is replaced by $(x, y)\neq(0, 0)$, then m can be replaced by 1/4.1; also an example is given to prove that the least possible value of m exceeds 1/5.06.

L. Tornheim (Ann Arbor, Mich.).

Referred to in J32-37, J32-59, J32-66.

J32-36 **(14, 252e)**

Kanagasabapathy, P. On the product of two linear forms, one homogeneous and one non-homogeneous. Quart. J. Math., Oxford Ser. (2) **3**, 197–205 (1952).

The constants 4.1 and 5.06 in the preceding review are improved to 4.2575 and 4.28471, respectively.

L. Tornheim (Ann Arbor, Mich.).

Referred to in J32-45, J32-59, J32-66.

J32-37 **(16, 340b)**

Davenport, H. Corrigendum to 'Note on a result of Chalk'. Quart. J. Math., Oxford Ser. (2) **5**, 211 (1954).

See same J. **3**, 130–138 (1952); these Rev. **14**, 252.

Citations: MR 14, 252d = J32-35.

J32-38 **(14, 730a)**

Barnes, E. S., and Swinnerton-Dyer, H. P. F. The inhomogeneous minima of binary quadratic forms. I. Acta Math. **87**, 259–323 (1952).

Let $f(P)=f(x, y)=ax^2+bxy+cy^2$ be a real indefinite form, where P represents the real point (x, y). Write $(x_1, y_1)\equiv(x_2, y_2)$ if x_1-x_2 and y_1-y_2 are integers, and define $M(f, P)=\inf|f(Q)|$ over all $Q\equiv P$, and $M(f)=\sup M(f, P)$ over all P. $M(f)$ is called the inhomogeneous minimum of f; it is the inf of those numbers m such that for any P_0 there exists $P\equiv P_0$ with $|f(P)|\leqq m$. If $M(f)$ is itself such an m, it is said to be an attained minimum, and a theorem of Heinhold [Math. Z. **44**, 659–688 (1939), p. 660] guarantees the existence of a point P with $M(f, P)=M(f)$, and the further existence of a corresponding point $Q\equiv P$ with $|f(Q)|=M(f)$. The authors also consider $M_2(f)=\sup M(f, P)$ taken over all P for which $M(f, P)<M(f)$. If $M_2(f)<M(f)$, then $M(f)$ is called an isolated minimum, and $M_2(f)$ the second minimum. A criterion for the existence of isolated minima is given.

The evaluation of $M(f)$ is of significance in determining the existence of a Euclidean algorithm in the quadratic field $k(\sqrt{m})$, $m>0$; if $f_m(x, y)=\|x+\omega y\|$ is the norm of the general integer $x+\omega y$ of $k(\sqrt{m})$, then $k(\sqrt{m})$ is Euclidean if and only if $M(f_m, P)<1$ for all rational points P. Consequently, $k(\sqrt{m})$ is Euclidean if $M(f_m)<1$.

The authors evaluate $M(f_m)$ and $M_2(f_m)$ for many values of m, and tabulate these and the previously known results for $m\leqq101$. Rédei's statement [Math. Ann. **118**, 588–608 (1942), p. 601, footnote; these Rev. **6**, 38] that for every rational point P either $M(f_{61}, P)=41/39$ or $M(f_{61}, P)<1$, is disproved, and it is shown that indeed

$$41/39=M_2(f_{61})<M(f_{61}).$$

The authors also prove that $k(\sqrt{97})$ is not Euclidean; the contrary was stated by Rédei [ibid., p. 607]. An example is given of a form with unattained first minimum.

Let $\phi(\alpha)$ be the inf of those numbers $m(\alpha)$ with the following property: for every real λ there exists an integer x such that $|(x+\lambda)^2-\alpha|\leqq m(\alpha)$. The authors evaluate $\phi(\alpha)$ explicitly, and show (trivially) that $M(f, P)\leqq|a|\cdot\phi(Dy^2/4a^2)$, where $P=(x, y)$. If K is taken as the supposed value of $M(f)$, the above considerations enable them to find a set R^* of points P where $M(f, P)<K$. Those exceptional points P where $M(f, P)\geqq K$ cannot transform (mod 1) into any point of R^* under any automorph of f. In some cases this permits the exceptional points to be determined easily; in the other cases the methods used are too complicated to describe here. The problem is then resolved into the evaluation of $M(f, P)$ at the exceptional points P. This is ac-

complished (in the simpler cases) by a lemma stating that if $M(f, P_i) < K$ $(i=1, \cdots, N)$, where $P_1, \cdots, P_N$ are N incongruent points which are permuted (mod 1) by some automorph of f, then there exists a point $P(x, y) \equiv P_i$ for some i, with $|f(P)| < K$, and for which $|y|$ cannot exceed an explicitly given bound depending on f, T and K. An extensive bibliography is given at the end of the paper.

I. Reiner (Urbana, Ill.).

Referred to in J32-41, J32-49, J32-50, J32-53, R12-49.

J32-39 (14, 730b)

Inkeri, K. Non-homogeneous binary quadratic forms. Den 11te Skandinaviske Matematikerkongress, Trondheim, 1949, pp. 216–224. Johan Grundt Tanums Forlag, Oslo, 1952. 27.50 kr.

This paper gives a survey of results connected with Minkowski's theorem on the product of two non-homogeneous linear forms. In particular, attention is drawn to the form $f(x, y) = x^2+xy-3y^2$, which has the property that the lower bound of $|f(x+x_0, y+y_0)|$ for integral x, y is always $\leqq \frac{1}{3}$, although for certain x_0, y_0 the inequality

$$|f(x+x_0, y+y_0)| \leqq \tfrac{1}{3}$$

is insoluble. [For later results and a list of references, see the paper reviewed above.] *H. Davenport* (London).

J32-40 (14, 954a)

Cassels, J. W. S. Yet another proof of Minkowski's theorem on the product of two inhomogeneous linear forms. Proc. Cambridge Philos. Soc. **49**, 365–366 (1953).

1. If $\xi = \alpha x + \beta y$, $\eta = \gamma x + \delta y$ be homogeneous linear forms in x, y with real coefficients and determinant $\alpha\delta - \beta\gamma = \Delta \neq 0$, then for any real constants p, q there are integers x, y such that (1) $|\xi+p|\,|\eta+q| \leqq \frac{1}{4}|\Delta|$. 2. If $\alpha x + \beta y \neq 0$ for all integers $x, y \neq 0, 0$, then there is a solution of (1) with $|\xi+p| \leqq \epsilon$ for every positive ϵ. These theorems are deduced in a few lines from the following lemma. Let θ, ϕ, ψ, ω be four real numbers such that $|\theta\omega - \phi\psi| \leqq \frac{1}{2}|\Delta|$, $|\psi\omega| \leqq |\Delta|$, $\psi > 0$. Then there is an integer x such that $\lambda = \theta + \psi x$, $\mu = \phi + \omega x$ satisfy the inequalities $|\lambda\mu| \leqq \frac{1}{4}|\Delta|$, $|\lambda| \leqq \psi$. Proof. By writing $\theta + r\psi$, $\phi + r\omega$ for θ, ϕ with a suitable integer r, we may suppose $-\psi \leqq \theta < 0$ and by writing $-\phi$, $-\omega$, $-\mu$ for ϕ, ω, μ, if need be, we also may suppose $\phi \geqq 0$. Then by distinguishing the cases $\phi+\omega \leqq 0$ and $\phi + \omega \geqq 0$ it is easily verified that $x=0$ or $x=1$ will do. *J. F. Koksma.*

J32-41 (14, 956a)

Barnes, E. S., and Swinnerton-Dyer, H. P. F. The inhomogeneous minima of binary quadratic forms. II. Acta Math. **88**, 279–316 (1952).

Using the methods and notation of Part I [Acta Math. **87**, 259–323 (1952); these Rev. **14**, 730], the authors consider the problem of finding an enumerably infinite number of inhomogeneous minima for the norm-forms x^2-11y^2 and $x^2+xy-3y^2$. Results of this kind have been obtained previously by Davenport [Nederl. Akad. Wetensch., Proc. **49**, 815–821 (1946); **50**, 378–389, 484–491, 741–749, 909–917 (1947); these Rev. **8**, 444, 565; **9**, 79, 412] for the form x^2+xy-y^2, and by Varnavides [ibid. **51**, 396–404, 470–481 (1948); these Rev. **10**, 19] for x^2-2y^2. The authors' results for x^2-11y^2 are analogous to those of Davenport and Varnavides, and include furthermore a proof of the existence for each $\epsilon>0$ of an enumerably infinite set of points P for which $M' > M(f, P) > M' - \epsilon$, where $M' = \lim M_k(f)$. The behavior of the minima of $x^2+xy-3y^2$ is surprisingly different, and the authors show the existence of countably many isolated minima extending below M'. The basic concept in both proofs is that of the set of transforms of a given point by all powers of the fundamental automorph of the given form. The authors conjecture that for any form f, $M_1(f)$ is rational, isolated, and taken at rational points (and possibly also at irrational points in $k(\sqrt{d})$, that $M_2(f)$ exists and is taken at points in $k(\sqrt{d})$, and further that these results are the strongest possible ones true for all forms. The paper concludes with a brief indication of the extension of its methods to norm-forms in n variables for real and complex fields.

I. Reiner (Urbana, Ill.).

Citations: MR 8, 444b = J32-4; MR 8, 565c = J32-6; MR 9, 79d = J32-7; MR 9, 412j = J32-8; MR 10, 19c = J32-14; MR 14, 730a = J32-38.

Referred to in J32-49, J32-65.

J32-42 (14, 1065d)

Chalk, J. H. H. A theorem of Minkowski on the product of two linear forms. Proc. Cambridge Philos. Soc. **49**, 413–420 (1953).

Consider the grid $\Lambda(C)$, i.e., the set of all points $(x_1, x_2) = (\alpha u + \beta v + c_1, \gamma u + \delta v + c_2)$ $(u, v = 0, \pm 1, \pm 2, \cdots)$, which may be obtained from the lattice Λ of the points $(\alpha u + \beta v, \gamma u + \delta v)$ by a translation which brings O to the position $C = (c_1, c_2)$. Let $\Delta = |\alpha\delta - \beta\gamma| > 0$. Then three points $X^{(1)}, X^{(2)}, X^{(3)}$ of $\Lambda(C)$ with coordinates $x_1^{(i)}, x_2^{(i)}$ $(i=1, 2, 3)$ and determinant

$$\pm \begin{vmatrix} x_1^{(1)} & x_2^{(1)} & 1 \\ x_1^{(2)} & x_2^{(2)} & 1 \\ x_1^{(3)} & x_2^{(3)} & 1 \end{vmatrix} = \Delta$$

form a generating set for the grid, i.e., each point $X \in \Lambda(C)$ can be written in the form $X = X_3 + m(X_1 - X_3) + n(X_2 - X_3)$ $(m, n$ integers). The author shows: For each grid $\Lambda(C)$ with $\Delta > 0$ there is a set of generating points in the region $|x_1x_2| \leqq \frac{1}{2}\Delta$. Unless both α/β and γ/δ are rational, there is an infinity of such sets even in the region $|x_1x_2| < \frac{1}{2}\Delta$ and the proof shows that each of these sets has a point in the region $|x_1x_2| \leqq \frac{1}{4}\Delta$. Moreover, an infinity of (u, v) satisfy the inequalities $|x_1x_2| < \frac{1}{2}\Delta$, $|x_1| < \epsilon$, where ϵ denotes an arbitrary positive constant. These are extensions of well-known results by Minkowski [cf. the reviewer's Diophantische Approximationen, Springer, Berlin, 1936, Kap. VI and III20]. *J. F. Koksma* (Amsterdam).

Referred to in J36-26.

J32-43 (14, 1065e)

Inkeri, K. Über einige Verschärfungen eines Minkowskischen Satzes. Ann. Acad. Sci. Fennicae. Ser. A. I. Math.-Phys. no. **136**, 16 pp. (1952).

Let $f(x, y)$ be an indefinite quadratic form with real coefficients. Barnes [Quart. J. Math., Oxford Ser. (2) **1**, 199–210 (1950); these Rev. **13**, 16] proved that, if

$$\mu(f) = \max\{|f(1,0)|, |f(0,1)|, \min(|f(1,1)|, |f(1,-1)|)\}$$

then $\frac{1}{4}\mu(f)$ is an admissible bound in Minkowski's theorem; i.e., for any real x_0, y_0 there exist integers x, y such that $|f(x+x_0, y+y_0)| \leqq \frac{1}{4}\mu(f)$. The author gives another proof of this, and shows that there is a form $q(x, y)$, equivalent to $f(x, y)$, reduced (in the sense of Gauss), and with $\mu(q) \leqq \mu(f)$. The author discusses the relations between this bound and those, which have been shown to be admissible in Minkowski's theorem, by Heinhold [Math. Z. **44**, 659–688 (1939)], Davenport [Nederl. Akad. Wetensch., Proc. **49**, 815–821 (1946); these Rev. **8**, 444], and the author [Ann. Acad. Sci. Fennicae. Ser. A. I. Math.-Phys. no. **66** (1950); these Rev. **12**, 320]. *C. A. Rogers* (London).

Citations: MR 8, 444b = J32-4; MR 12, 320b = J32-25; MR 13, 16a = R12-23.

J32-44 (15, 106c)

Mordell, L. J. Note on Sawyer's paper "The product of two non-homogeneous linear forms". J. London Math. Soc. **28**, 510–512 (1953).

Sawyer [same J. **23**, 250–251 (1948); these Rev. **10**, 355] made the unproved assertion that for a given non-homogeneous lattice Λ admissible for the region $|xy| \leqq k$, there exists a quadrilateral with vertices in Λ each in a different quadrant and containing no other points of Λ. A simple geometrical proof is given that either this is the case or there is a triangle of the same type and containing the origin. Sawyer's proof is modified to take care of the second case. Reference is made to Delone [Izvestiya Akad. Nauk SSSR. Ser. Mat. **11**, 505–538 (1947); these Rev. **9**, 334] who proved there is a parallelogram satisfying the assertion of Sawyer. *L. Tornheim* (Ann Arbor, Mich.).

Citations: MR 9, 334e = J32-9; MR 10, 355b = J32-18.

J32-45 (16, 340a)

Chalk, J. H. H. On the primitive lattice points in the region $|(x+c)y| \leqslant 1$. Quart. J. Math., Oxford Ser. (2) **5**, 203–211 (1954).

Let $L_1=\alpha x+\beta y$, $L_2=\gamma x+\delta y$ be real forms with

$$\Delta=\alpha\delta-\beta\gamma\neq 0$$

and $c\neq 0$. Then there exist coprime integers x, y such that $|(L_1+c)L_2| \leqq (1-2^{-1/2})|\Delta|$, with equality necessary if and only if $L_1=\lambda_1 y$, $L_2=\lambda_2(x+2^{1/2}-1)y$, $c=-(1-2^{-1/2})\lambda_1$, apart from an integral unimodular substitution on x, y. Furthermore, this value of c is best possible even if α/β is assumed to be irrational. This corrects a previous statement of the author [same J. (2) **3**, 119–129 (1952); these Rev. **14**, 252]. If the condition $(x, y)=1$ is not required, see Kanagasabapathy, ibid. **3**, 197–205 (1952) [these Rev. **14**, 252]. *L. Tornheim* (Ann Arbor, Mich.).

Citations: MR 14, 252c = J32-34; MR 14, 252e = J32-36.

J32-46 (16, 340c)

Cassels, J. W. S. On the product of two inhomogeneous linear forms. J. Reine Angew. Math. **193**, 65–83 (1954).

Let $\xi=\alpha x+\beta y+p$, $\eta=\gamma x+\delta y+q$ be inhomogeneous linear forms in x, y with real coefficients and determinant

$$\Delta=\alpha\delta-\beta\gamma\neq 0.$$

The author investigates the minimum of $|\xi\eta|$ for integer values of x, y under the condition $\xi\eta\neq 0$. Calling the pairs of forms (ξ, η) and (ξ_1, η_1) equivalent if there are integers a, b, c, d, x_0, y_0 and non-zero constants λ, μ such that

$$\xi_1(ax+by+x_0, cx+dy+y_0)=\lambda\xi(x, y),$$
$$\eta_1(ax+by+x_0, cx+dy+y_0)=\mu\eta(x, y)$$

identically in x, y, then clearly $\mathfrak{M}=|\Delta|^{-1}\min_{\xi\eta\neq 0}|\xi\eta|$ is the same number for two equivalent pairs of forms. By a simple argument the author proves that, for each given constant $\epsilon>0$, there are indenumerably many inequivalent pairs of forms, not equivalent to homogeneous pairs or to pairs of the type

(1) $\xi=x,\quad \eta=y+d$ (d arbitrary)

such that $\mathfrak{M}\geqq\frac{1}{3}-\epsilon$.

Further, he investigates the pairs with $\mathfrak{M}\geqq\frac{1}{3}$ and succeeds in giving a complete classification of all such pairs, giving for each of them the exact value of $\mathfrak{M}$. His most interesting table is too long to be reproduced here. *J. F. Koksma.*

J32-47 (16, 574a)

Godwin, H. J. On the inhomogeneous minima of certain convergent sequences of binary quadratic forms. Quart. J. Math., Oxford Ser. (2) **5**, 28–46 (1954).

Let $f(x, y)$ be an indefinite binary quadratic form and $M(a, b)$ denote min $f(x, y)$ for all x, y congruent mod 1 to the given numbers a and b respectively. Let $M_1(f)$ be the least upper bound of $M(a, b)$ for all values of a and b. In this paper M_1 is evaluated for the forms

$$\left(x+\frac{q_{n+1}}{q_n}y\right)\left(x-\frac{q_{n-1}}{q_n}y\right),$$

where q_n is the denominator of the nth convergent in the continued fraction expansion of $\frac{1}{2}(1+\sqrt{5})$. Also, M_1 is evaluated for the forms $x^2-p_n^2q_n^{-2}y^2$, where p_n/q_n is the nth convergent in the continued fraction expansion of $\sqrt{2}$. The limiting forms are, in the respective cases, x^2+xy-y^2 and x^2-2y^2 and the limiting values of M_1 are easily determined. *B. W. Jones* (Boulder, Colo.).

J32-48 (16, 802c)

Barnes, E. S., and Swinnerton-Dyer, H. P. F. The inhomogeneous minima of binary quadratic forms. III. Acta Math. **92**, 199–234 (1954).

Delauney (Delone) showed that every inhomogeneous two-dimensional lattice ('grid') $\mathfrak{L}$ of determinant Δ with no points on the co-ordinate axes always has 4 points A, B, C, D, one in each quadrant, forming a parallelogram of determinant Δ [Izv. Akad. Nauk. SSSR. Ser. Mat. **11**, 505–538 (1947); MR **9**, 334]. The authors show that such parallelograms ('divided cells') may be arranged in a sequence with a simple algorithm relating successive cells. This they use to investigate the inhomogeneous critical determinant D_m of the region $-1<\xi\eta<m$ $(m\geqq 1)$. There are always critical lattices of the form

$$\xi=\alpha(x-\tfrac{1}{2})+\beta(y-\tfrac{1}{2}),\quad \eta=\gamma(x-\tfrac{1}{2})+\delta(y-\tfrac{1}{2})$$

(x, y run through the integers). As an application, D_m is evaluated in a neighbourhood of $m=2$ and is shown to have a very complicated structure as a function of m. An estimate of the second inhomogeneous minimum for $m=2$ is given which is better than Davenport's [Acta Math. **80**, 65–95 (1948); MR **10**, 101]. *J. W. S. Cassels.*

Citations: MR 9, 334e = J32-9; MR 10, 101i = J48-3.

Referred to in J08-13, J20-23, J32-51, J32-54, J32-57, J32-60, J48-8.

J32-49 (16, 802d)

Barnes, E. S. The inhomogeneous minima of binary quadratic forms. IV. Acta Math. **92**, 235–264 (1954).

The author continues the discussion of the 'algorithm of the divided cell' [see the preceding review]. Every such algorithm defines a sequence of pairs of integers h_n, k_n satisfying simple restrictions and conversely every such sequence of integer pairs defines an inhomogeneous lattice $\mathfrak{L}$. The sums h_n+k_n characterise the corresponding homogeneous lattice. It is claimed that this algorithm is more effective than that used in I, II of the series [Acta Math. **87**, 259–323; **88**, 279–316 (1952); MR **14**, 730, 956]. As an illustration $M(x^2-19y^2)=170/171$ (wrongly given in I) and $M(x^2-46y^2)=76877/48668$ are evaluated. See also the paper reviewed below. *J. W. S. Cassels.*

Citations: MR 14, 730a = J32-38; MR 14, 956a = J32-41.

Referred to in J20-23, J32-53, J32-54, J32-57.

J32-50 **(16, 802e)**

Godwin, H. J. On the inhomogeneous minima of certain norm-forms. J. London Math. Soc. **30**, 114–119 (1955).

Using the notation of Barnes and Swinnerton-Dyer [Acta Math. **87**, 259–323 (1952); MR **14**, 730], the author fills in the gaps in their tables by finding the value of the inhomogeneous minimum $M_1(f_m)$ and an upper bound for $M_2(f_m)$ for $m=46$, 57, 67, 71, 86 and 94. He also evaluates $M_1(f_{73})$, $M_2(f_{73})$ and finds an upper bound for $M_3(f_{73})$. His method is a modification of the method of Barnes and Swinnerton-Dyer to make it easier to apply to norm-forms of fields with a large fundamental unit. *H. S. A. Potter.*

Citations: MR 14, 730a = J32-38.

J32-51 **(18, 21e)**

Bambah, R. P. Divided cells. Res. Bull. Panjab Univ. no. **81** (1955), 173–174.

A proof on B. N. Delone's theorem that an inhomogeneous plane lattice contains a divided cell [Izv. Akad. Nauk SSSR. Ser. Mat **11** (1947), 505–538; MR **9**, 334; see also E. S. Barnes and H. P. F. Swinnerton-Dyer, Acta Math. **92** (1954), 199–234; MR **16**, 802]. *J. W. S. Cassels.*

Citations: MR 9, 334e = J32-9; MR 16, 802c = J32-48.

J32-52 **(19, 19b)**

Blaney, Hugh. On the Davenport-Heilbronn theorem. Monatsh. Math. **61** (1957), 1–36.

Let $a, b, c, d, \lambda_0, \mu_0$ be real numbers, with $ad-bc=\pm 1$, and let

$$Q(x, y)=(ax+by+\lambda_0)(cx+dy+\mu_0).$$

It follows from a result of Davenport and Heilbronn [J. London Math. Soc. **22** (1947), 53–61; MR **9**, 413] that there exist integers x, y such that $0<Q\leqq 1$. The author has already proved [Thesis, Univ. of London 1949] the existence of a constant ξ $(0<\xi<1)$ such that the inequality $\xi\leqq Q\leqq 1$ is soluble in integers x, y, and he now shows that ξ can be chosen as:

$$\xi=\tfrac{1}{2}(-11+\sqrt{126})=[0, \overline{8, 1, 8, 11}]=0.11248\cdots.$$

With this value, the strict inequality $\xi<Q<1$ is soluble unless Q is equivalent (in an appropriate sense) to one of the three forms

$$xy,\ \tfrac{1}{2}y(2x+y+1),\ \tfrac{1}{2}y(2x+(1+2\xi)y+1).$$

The above value for ξ is best possible in the stronger sense that for any small $\delta>0$ there exist infinitely many forms Q such that the inequality $\xi+\delta<Q<1$ is insoluble. The proof of the results just stated is difficult, and requires the consideration of many cases. The treatment is on different lines according as a/b and c/d are both irrational, or not. *H. Davenport* (London).

Citations: MR 9, 413a = J32-11.

J32-53 **(19, 125a)**

Inkeri, K.; and Ennola, V. The Minkowski constants for certain binary quadratic forms. Ann. Univ. Turku. Ser. A. **25** (1957), 19 pp.

If $f(x, y)=ax^2+bxy+cy^2$ is a real indefinite quadratic form with discriminant $d=b^2-4ac>0$ and, for any real numbers x_0, y_0, the lower bound of $|f(x+x_0, y+y_0)|$ taken over all integer sets x, y is denoted by $M(f; x_0, y_0)$, it is known by a theorem of Minkowski that $M(f; x_0, y_0)\leqq\frac{1}{4}\sqrt{d}$. The Minkowski constant $M(f)$ is defined to be the upper bound of $M(f; x_0, y_0)$ taken over all real sets x_0, y_0. By two slightly differing methods, the authors determine the values of $M(f_m)$ for the principal forms:

$$f_m(x, y)=\begin{cases} x^2-my^2 & (m=7, 11, 19), \\ x^2+xy-\frac{1}{4}(m-1)y^2 & (m=41). \end{cases}$$

Proofs for the cases $m=7$, 11 had been found previously by Inkeri (unpublished) and others, and for the cases $m=19$ (unpublished), 41 by Barnes and Swinnerton-Dyer [for details and full references see Acta Math. **87** (1952), 259–323; MR **14**, 730]. The case $m=19$ was considered later by Barnes [ibid. **92** (1954), 235–264; MR **16**, 802]. *J. H. H. Chalk* (Hamilton, Ont.).

Citations: MR 14, 730a = J32-38; MR 16, 802d = J32-49.

J32-54 **(19, 125b)**

Birch, B. J. A grid with no split parallelepiped. Proc. Cambridge Philos. Soc. **53** (1957), 536.

It has been shown by B. N. Delone [Izv. Akad. Nauk SSSR. Ser. Mat. **11** (1947), 505–538; MR **9**, 334] that every 2-dimensional grid (inhomogeneous lattice) contains a parallelogram with one vertex in each quadrant and of area equal to the determinant of the grid. This has been exploited by Delone himself (loc. cit.), by Barnes and Swinnerton-Dyer [Acta Math. **93** (1954), 199–234; MR **16**, 802] and Barnes [ibid. **92** (1954), 235–264; J. London Math. Soc. **31** (1956), 73–76; MR **16**, 802; **17**, 715]. The author gives an elegant counterexample to show that no generalization to 3 dimensions is possible.
J. W. S. Cassels (Cambridge, England).

Citations: MR 9, 334e = J32-9; MR 16, 802c = J32-48; MR 16, 802d = J32-49; MR 17, 715b = J20-23.

J32-55 **(19, 396f)**

Sawyer, D. B. The product of two linear forms whose values are restricted. J. London Math. Soc. **32** (1957), 213–217.

Let $L=\alpha u+\beta v+\rho$, $M=\gamma u+\delta v+\sigma$ be non-homogeneous linear forms in u, v with real coefficients such that $D=\alpha\delta-\beta\gamma\neq 0$. Then Minkowski's theorem states that $|LM|\leqq\frac{1}{4}|D|$ for some integers u, v. The author proves that integers u, v exist such that $|LM|\leqq(1/\sqrt{5})|D|$ and either $L>0$ or $M>0$, and further that integers u, v exist such that $|LM|\leqq\frac{1}{2}|D|$ and $L+M>0$. In his proof the author uses an idea of Rogers [same J. **29** (1954), 133–143; MR **15**, 941]; his (geometric) method also leads to a proof of Minkowski's theorem stated above. *J. F. Koksma.*

Citations: MR 15, 941c = H05-31.

J32-56 **(20# 3825)**

Ennola, Veikko. On the first inhomogeneous minimum of indefinite binary quadratic forms and Euclid's algorithm in real quadratic fields. Ann. Univ. Turku. Ser. AI **28** (1958), 58 pp.

Let $f(x, y)=ax^2+bzy+y^2$ be an indefinite quadratic form with discriminant $d=b^2-4ac$. In his chapter I the author shows that there exist real numbers x_0, y_0 such that

$$|f(x_1, y_1)|\geqq\kappa d^{\frac{1}{2}}$$

for all numbers $x_1\equiv x_0 \pmod 1$, $y_1\equiv y_0 \pmod 1$, where $\kappa=(16+6\sqrt{6})^{-1}\doteqdot(30.69)^{-1}$. This value of κ is a very considerable improvement on the best hitherto known [Cassels, Proc. Cambridge Philos. Soc. **48** (1952), 72–86, 519–520; MR **13**, 919]. The proof is a modification of the original method of Davenport [Proc. London Math. Soc. (2) **53** (1951), 65–82; MR **13**, 15], which involves developments in continued fractions to the nearest integer(!). It is remarkable that the author can give a simple prescription for finding x_0, y_0 in terms of Davenport's algorithm.

In the rest of his paper the author gives a systematic determination of all the real quadratic fields with a euclidean algorithm. In chapter II he shows that all the fields with a euclidean algorithm do in fact possess it, and in chapter III he shows that no other fields have it. It is very useful to have such a systematic treatment of

results otherwise scattered through the literature and there attacked by diverse methods.
J. W. S. Cassels (Cambridge, England)

Citations: MR 13, 15f = R12-19; MR 13, 919a = J32-31; MR 13, 919b = J32-31.

Referred to in H30-86, J32-57.

J32-57 **(21# 670)**

Pitman, Jane. The inhomogeneous minima of a sequence of symmetric Markov forms. Acta Arith. **5** (1958), 81–116 (1959).

Using the 'algorithm of the divided cell' as introduced by Delone [Izv. Akad. Nauk SSSR. Ser. Mat. **11** (1947), 505–538; MR **9**, 334] and elaborated by Barnes and Swinnerton-Dyer [Acta Math. **92** (1954), 199–234; MR **16**, 802] and Barnes [Acta Math. **92** (1954), 235–264; MR **16**, 802], the author determines the inhomogeneous minimum of a certain sub-class of the Markoff binary quadratic forms, and in some cases also the second minimum. The results show that the lower bound of the expression

$$(\text{inhomogeneous minimum})/(\text{determinant})^{\frac{1}{2}}$$

over all indefinite quadratic forms can be at most equal to 1/12, which is a new low. Together with the recent result of Ennola [Ann. Univ. Turku Ser. A I **28** (1958); MR **20** #3825] this gives a much narrower range in which this lower bound must lie. The paper contains an almost complete account of the algorithm of the divided cell. {Miss Pitman has since shown that the lower bound in question is definitely smaller than 1/12, but has not yet published her work.}
J. W. S. Cassels (Cambridge, England)

Citations: MR 9, 334e = J32-9; MR 16, 802c = J32-48; MR 16, 802d = J32-49; MR 20# 3825 = J32-56.

J32-58 **(21# 3402)**

Rédei, L. Neuer Beweis eines Satzes von Delone über ebene Punktgitter. J. London Math. Soc. **34** (1959), 205–207.

The author gives a neat proof of the theorem of Delone [Izv. Akad. Nauk SSSR. Ser. Mat. **11** (1947), 505–538; MR **9**, 334] that every 2-dimension lattice with no points on the coordinate axes contains a 'split parallelogram'.
J. W. S. Cassels (Cambridge, England)

Citations: MR 9, 334e = J32-9.

J32-59 **(22# 2594)**

Kanagasabapathy, P. On the product $(ax+by+c)(dx+ey)$. Bull. Calcutta Math. Soc. **51** (1959), 1–7.

Let a, b, c, d, e be real numbers satisfying $ad-bc=1$, $e\neq 0$, and suppose that neither of the forms $ax+by$ and $cx+dy$ represents zero for integers x, y not both zero. Using a lemma due to Davenport [Quart. J. Math. Oxford Ser. (2) **3** (1952), 130–138; MR **14**, 252], the author proves that there exist integers x, y not both zero such that

$$|(ax+by+e)(cx+dy)| < (4.2577)^{-1}.$$

This improves slightly an earlier result by the author [ibid. (2) **3** (1952), 197–205; MR **14**, 252].
I. Reiner (Urbana, Ill.)

Citations: MR 14, 252d = J32-35; MR 14, 252e = J32-36.

Referred to in J32-66.

J32-60 **(22# 2595)**

Mihaljinec, Mirko. Notes on the E. S. Barnes and H. P. F. Swinnerton-Dyer's paper: The inhomogeneous minima of binary quadratic forms. III. Glasnik Mat.-Fiz. Astr. Društvo Mat. Fiz. Hrvatske. Ser. II **14** (1959), 121–134. (Serbo-Croatian summary)

[The paper of the title is Acta Math. **92** (1954), 199–234; MR **16**, 802.] Let $\mathscr{L}$ denote a two-dimensional vector space of points (ξ, η) with determinant Δ. A lattice $\mathscr{L}$ is said to be R_m-admissible if for the region R_m: $-1<\xi\eta<m$, with $m\geqq 1$ and real, there is no point of $\mathscr{L}$ contained in R_m. The lattice $\mathscr{L}_2$ is a multiple of lattice $\mathscr{L}_1$, if there are real numbers $C\geqq 1$ and k such that the transformation $f(\xi, \eta)=(C\xi/k, Ck\eta)$ transforms the points of $\mathscr{L}_1$ onto points of $\mathscr{L}_2$.

The "minimal determinants" $D_m^{(k)}$, for $k=1, 2, 3, \cdots$, are defined inductively as follows: $D_m^{(k)}=\inf\Delta(\mathscr{L})$, where $\mathscr{L}$ runs over all R_m-admissible lattices which are not multiples of any R_m-admissible lattice with determinant $\leqq D_m^{(k-1)}$ and $\Delta(\mathscr{L})>D_m^{(k-1)}$ if $k\geqq 2$, and without restriction if $k=1$.

Barnes and Swinnerton-Dyer have determined $D_m^{(1)}$ for m between $1.9090\cdots$ and $2.1251\cdots$ and Davenport has shown that $D_2^{(1)}$ is isolated. The author determines the values of $D_m^{(k)}$ for $m=2001/1000$ and $k=1, 2, 3, 4$. The first three of these minima are isolated. The minima for $k=4$ and $k=5$ are equal. Associated results are also obtained.
B. W. Jones (Boulder, Colorado)

Citations: MR 16, 802c = J32-48.

J32-61 **(22# 7977)**

Pitman, Jane. Davenport's constant for indefinite binary quadratic forms. Acta Arith. **6** (1960), 37–46.

Let $f(x, y)$ be an indefinite binary quadratic form with real coefficients and discriminant $D=D(f)$ and write $\Delta=\Delta(f)=\sqrt{D(f)}$. Let $P=(x', y')$ be any real point and $M(f; P)=\inf[|f(x+x', y+y')|; x, y \text{ integral}]$. Define $M(f)=\sup_P M(f; P)$ over all real points P. Davenport has shown the existence of a constant k such that, for all f, $M(f)>k\Delta(f)$. Thus an absolute constant K may be defined by $K=\sup[k; M(f)>k\Delta(f)]$, where the supremum is taken over all forms f.

Ennola showed that $K\geqq 1/30.69$ and the author has shown that $K\leqq 1/12$. This paper is an account of the theoretical basis for computations on EDSAC 2 at the Cambridge Mathematical Laboratory which establish the fact that $K\leqq 1/12.921$.
B. W. Jones (Boulder, Colo.)

J32-62 **(24# A91)**

Surányi, János

Über zerteilte Parallelogramme.

Acta Sci. Math. Szeged **22** (1961), 85–90.

Let Γ be an inhomogeneous lattice in the plane having no points on the coordinate axes. Then a theorem of Delone [Izv. Akad. Nauk SSSR Ser. Mat. **11** (1947), 505–538; MR **9**, 334; Sowjetwissenschaft **1948**, no. 2, 178–210; MR **12**, 82] says that Γ contains four points P_1, P_2, P_3, P_4, one in each quadrant, such that the parallelogram with vertices P_i $(i=1, 2, 3, 4)$ contains no other points of Γ in its interior or on its boundary. The present author gives a new and simple proof of this theorem and derives from it three classical results concerning the product of two binary linear forms.
C. G. Lekkerkerker (Amsterdam)

Citations: MR 9, 334e = J32-9; MR 12, 82d = J32-10.

J32-63 **(25# 48)**

Niven, Ivan

Minkowski's theorem on nonhomogeneous approximation.

Proc. Amer. Math. Soc. **12** (1961), 992–993.

Let θ be irrational and let γ be any real number not of the form $\theta a+b$, where a and b represent integers. The author gives a short proof of Minkowski's classic result that there are infinitely many pairs of integers x, y satisfying $|x|\ |\theta x+y+\gamma|<\frac{1}{4}$.
J. Popken (Amstelveen)

J32-64 **(25# 3008)**

Eggan, L. C.; Maier, E. A.

A result in the geometry of numbers.

Michigan Math. J. **8** (1961), 161–166.

The authors evaluate $m(c)=\max_{\alpha,\beta}[\{\min_u|\alpha-u|\,|\beta-u|$, u a rational integer$\}$, α, β real with $|\alpha-\beta|=2c]$. They also prove that for α,β real, there exist integers u such that $|\beta-u|<1$, and $|(\beta-\mu)(\alpha-u)|\leqq\frac{1}{4}$ if $|\alpha-\beta|<\frac{1}{2}$, and $<|\alpha-\beta|/2$ if $|\alpha-\beta|\geqq\frac{1}{2}$. This implies a theorem of Minkowski [Theorem IIA, p. 48, *An introduction to Diophantine approximation*, by Cassels, Cambridge Univ. Press, New York, 1957; MR **19**, 396]. *R. P. Bambah* (Chandigarh)

Citations: MR 19, 396h = J02-11.

Referred to in J56-29.

J32-65 **(26# 6125)**

Godwin, H. J.

On a conjecture of Barnes and Swinnerton-Dyer.

Proc. Cambridge Philos. Soc. **59** (1963), 519–522.

Barnes and Swinnerton-Dyer [Acta Math. **88** (1952), 279–316; MR **14**, 956] offered several conjectures on the inhomogeneous minima of an indefinite binary quadratic form with integral coefficients; in particular, they conjectured that, while the first minimum is always isolated, the second minimum need not be. The author produces an example of this suggested behaviour by showing that the form x^2-23y^2 has second minimum $\mu=(20\sqrt{23}-31)/46$ and by giving a sequence of points with minima μ_n, where $\lim\mu_n=\mu$. *E. S. Barnes* (Adelaide)

Citations: MR 14, 956a = J32-41.

J32-66 **(37# 5155)**

Blanksby, P. E.

On the product of two linear forms, one homogeneous and one inhomogeneous.

J. Austral. Math. Soc. **8** (1968), 457–511.

Extending earlier work of H. Davenport [Quart. J. Math. Oxford Ser. (2) **3** (1952), 130–138; MR **14**, 252] and P. Kanagasabapathy [ibid. (2) **3** (1952), 197–205; MR **14**, 252; Bull. Calcutta Math. Soc. **51** (1959), 1–7; MR **22** #2594], the author proves the following hybrid of the two classical results of Hurwitz and Minkowski on indefinite quadratic forms: Suppose $X=\alpha x+\beta y$, $Y=\gamma x+\delta y$ have determinant Δ, and do not represent 0 nontrivially in integers. If η is a nonzero constant, then (*) $\inf|X(Y+\eta)|\leqq k\Delta$, where the infimum is taken over all integer pairs $x,y\neq0$, 0 and

$$k=\frac{(3/49)(366458018\phi-7320551)}{(82387308\theta+392361)\phi-(1645818\theta+7838)}=$$
$$0.234254343\ldots,$$

with $\phi=(147+\sqrt{21651})/6$, $\theta=(104250+2\sqrt{10})/9005$. There exist forms for which the equality holds in (*) and if $0\leqq k'<k$ there exist uncountably many pairs of forms X, Y and real number η such that $\inf|X(Y+\eta)|=k'\Delta$. *D. J. Lewis* (Ann Arbor, Mich.)

Citations: MR 14, 252d = J32-35; MR 14, 252e = J32-36; MR 22# 2594 = J32-59.

J32-67 **(39# 4099)**

Blanksby, P. E.

A restricted inhomogeneous minimum for forms.

J. Austral. Math. Soc. **9** (1969), 363–386.

An indefinite binary form not representing zero is equivalent to the form $f(x,y)=\pm\Delta(\theta x+y)(x+\varphi y)/k$, where θ and φ are irrational numbers with absolute value greater than 1, $\sqrt{\Delta}$ is the discriminant of f and $k=\theta\varphi-1$. Such a form is called I-reduced. The author writes $M^+(f;P)=\inf\Delta r|s/k|$, where $P=(x_0,y_0)$, $s=x+\varphi y+x_0+\varphi y_0$ and the infimum is taken over all integers x and y such that $r=x\theta+y+\theta x_0+y_0>0$. If the infimum is taken over $r<0$, we have $M^-(f;P)$. Similarly, $M^\pm(g;Q)$ are defined symmetrically for $Q=(y_0,x_0)$. Then the author writes $M^*(f)=\sup_P\max\{M^\pm(f;P),M^\pm(g;Q)\}$.

The author's main results are as follows: (A) If f does not represent zero, $M^*(f)\leqq27\Delta/28\sqrt{7}$. (B) Except for an equivalence class of forms for which equality holds in A, $M^*(f)\leqq359\Delta/45\sqrt{510}$. *Burton W. Jones* (Boulder, Colo.)

J32-68 **(41# 6777)**

Varnavides, P. L.

The nonhomogeneous minima of a class of binary quadratic forms.

J. Number Theory **2** (1970), 333–341.

The inhomogeneous minima of four sequences of indefinite binary quadratic forms are simply established. The result for the forms $2x^2-\frac{1}{2}[(2p+1)^2+1]y^2$ $(p=1,2,\cdots)$, namely, $(2p+1)/4$, is stated here for the first time. The remaining forms are norm forms which have been discussed by several writers [see, for example, J. Heinhold, Math. Z. **44** (1939), 659–688]. *E. S. Barnes* (Adelaide)

J32-69 **(41# 6789)**

Varnavides, P.

Non-homogeneous binary quadratic forms.

Ann. Acad. Sci. Fenn. Ser. A I No. 462 (1970), 28 *pp.*

Let $f(x,y)=ax^2+bxy+cy^2=[a,b,c]$ be an indefinite quadratic form with real coefficients. The nonhomogeneous minimum of f is $\mathscr{M}(f)=\sup\inf|f(x+x_0,y+y_0)|$, where the infimum is over all integers x, y and the supremum over all real numbers x_0, y_0. If $p\geqq0$, $q>0$ and $r>0$ are integers, it is shown that $\mathscr{M}(f)$ for the forms $[1,0,-(2p+1)^2-1]$, $[1,0,1-4q^2]$ and $[1,1,-r^2]$ has the values $\frac{1}{2}(2p+1)$, $\frac{1}{2}(2q-1)$ and $r/4$, respectively. Furthermore, for suitable choices of x_0 and y_0, these minima are achieved for an infinity of integers x and y. It is also shown that $\mathscr{M}([2,-4,-3])=\frac{3}{4}$. Alternative proofs of part of these results are included in a later paper by the same author [#6777 above]. (See also H. Davenport [Nederl. Akad. Wetensch. Proc. Ser. A **49** (1946), 815–821; MR **8**, 444] and K. Inkeri [Ann. Acad. Sci. Fenn. Ser. A I No. 66 (1950); MR **12**, 320].) *D. G. James* (University Park, Pa.)

Citations: MR 8, 444b = J32-4; MR 12, 320b = J32-25.

J36 PRODUCT OF *n* HOMOGENEOUS LINEAR FORMS IN *n* VARIABLES

See also reviews A60-11, H05-1, H15-4, H15-42, H15-66, J02-15, J16-2, J40-6, J40-42, J56-7, J64-17, J64-20, J64-22, J64-23, J64-24, J64-25, P28-3.

J36-1 **(2, 149a)**

Hofreiter, Nikolaus. Über das Produkt von Linearformen.

Monatsh. Math. Phys. **49**, 295–298 (1940).

If d is the absolutely least discriminant for all algebraic fields of degree n with r real and $2s$ imaginary conjugates $(r+2s=n)$, it is observed that there exist linear forms $L_1,\cdots,L_n$ of determinant 1 (r with real, and s pairs with conjugate complex, coefficients) in rational integer variables $x_1,\cdots,x_n$ such that $|L_1\cdots L_n|<c$ has only the solution x_i all zero, if $c^2\leqq1/d$. This follows from the fact that $N(\omega_1x_1+\cdots+\omega_nx_n)\geqq1$, being the norm of an algebraic integer. The special cases $n=2$, $r=2$; $n=2$, $r=0$; $n=3$,

$r=3$; and $n=3$, $r=1$ give the results known to be best possible, $c^2=1/5$, $1/3$, $1/49$, $1/23$ [cf. H. Davenport, Proc. London Math. Soc. (2) **44**, 412–431 (1938), and **45**, 98–125 (1939)]. If $n=4$ the determination of d [J. Mayer, Akad. Wiss. Wien, S.-B. **138**, 733–742 (1929)] gives $c^2=1/725$, $1/275$ and $1/117$, according as $r=4$, 2 or 0.

G. Pall (Princeton, N. J.).

J36-2 **(3, 70f)**

Davenport, H. Note on the product of three homogeneous linear forms. J. London Math. Soc. **16**, 98–101 (1941).

Let L_1, L_2, L_3 be three homogeneous linear functions in x, y, z with real coefficients and determinant 1. Some time ago, the author proved [Proc. London Math. Soc. (2) **44**, 412–431 (1938)] that the lower bound of $|L_1L_2L_3|$ for integral values of $x, y, z \neq 0, 0, 0$ is not greater than $1/7$ (best possible constant). In the present note the original laborious proof is replaced by a very short and simple one, using the reduction of binary quadratic forms and a new lemma on the minimum of cubic polynomials.

C. L. Siegel (Princeton, N. J.).

Referred to in H20-7, J36-7, J36-15, J36-32.

J36-3 **(3, 167f)**

Mordell, L. J. The product of homogeneous linear forms. J. London Math. Soc. **16**, 4–12 (1941).

$L_1, \cdots, L_n$ are n homogeneous linear forms in n variables with real coefficients and determinant 1. Let K_n be the lower bound of $|L_1L_2\cdots L_n|$, if the variables run over all lattice points except $0, \cdots, 0$. The following estimates are proved: $K_3 \leqq 4/27$, $K_4 \leqq 1/14.0\cdots$, $K_5 \leqq 132.4\cdots$. These are deduced from the theorem: Let there be a symmetric convex domain in the $(x_1, \cdots, x_{n-1})$ space with center at the origin and of volume $2^{n-1}V$, such that every point of it satisfies the inequality $|x_1\cdots x_{n-1}(x_1+\cdots+x_{n-1})| \leqq 1$; then $K_n \leqq V^{-n/(n-2)}$. The proof of this theorem depends upon Minkowski's theorem concerning lattice points in convex bodies. The inequality $K_3 \leqq 4/27$ is less sharp than the best possible result $K_3 \leqq 1/7$ proved by Davenport [J. London Math. Soc. **13**, 139–145 (1938)]. *C. L. Siegel.*

J36-4 **(3, 167g)**

Žilinskas, G. On the product of four homogeneous linear forms. J. London Math. Soc. **16**, 27–37 (1941).

The method of the paper by Davenport mentioned in the above review is applied to the proof of the estimate $K_4 \leqq 3/20\sqrt{5} = 1/14.9\cdots$, which is better than Mordell's result $K_4 \leqq 1/14.0\cdots$. *C. L. Siegel* (Princeton, N. J.).

J36-5 **(4, 131d)**

Mordell, L. J. The product of three homogeneous linear ternary forms. J. London Math. Soc. **17**, 107–115 (1942).

Let $L_r = a_rx_1 + b_rx_2 + c_rx_3$ $(r=1, 2, 3)$ be three homogeneous linear forms of the real variables x_1, x_2, x_3 of determinant d. Two types of forms are considered: (I) the coefficients of L_1, L_2, L_3 are all real, $d=1$; (II) the coefficients of L_1 are real and those of L_2, L_3 are conjugate complex numbers, $d=i$. The author proves that the inequality $|L_1L_2L_3| < c+\epsilon$ is solvable, for any $\epsilon>0$, in integers x_1, x_2, x_3 not all zero, where $c=1/7$ in case (I) and $c=(23)^{-\frac{1}{2}}$ in case (II); in both cases, c is the best possible value of the constant on the right-hand side, and the corresponding extreme cases for L_1, L_2, L_3 are also determined. These results were discovered earlier by H. Davenport [Proc. London Math. Soc. (2) **44**, 412–431 (1938)]. The present proof is much more direct, since solutions of the inequality are actually constructed by means of a "descente infinie," whereas Davenport's results are in the nature of existence theorems. The proof depends upon a new theorem concerning the minimum of a binary cubic form, which is still unpublished.

C. L. Siegel (Princeton, N. J.).

Referred to in H20-5, H20-7, J52-8.

J36-6 **(4, 189e)**

Estermann, T. A new proof of a theorem of Minkowski. J. London Math. Soc. **17**, 158–161 (1942).

The author gives a new arithmetical proof for the well-known theorem of Minkowski on linear forms.

P. Erdös (Philadelphia, Pa.).

J36-7 **(4, 212a)**

Davenport, H. On the product of three homogeneous linear forms. IV. Proc. Cambridge Philos. Soc. **39**, 1–21 (1943).

Part I of the series appeared in J. London Math. Soc. **13**, 139–145 (1938); part II in Proc. London Math. Soc. (2) **44**, 412–431 (1938); part III in the same Proc. **45**, 98–125 (1941). A paper in J. London Math. Soc. **16**, 98–101 (1941) [these Rev. **3**, 70] is also connected with the present one. These papers will be referred to as [1], $\cdots$, [4], respectively.

Let L_1, L_2, L_3 be three real linear forms in u, v, w of determinant 1, and let M denote the lower bound of $|L_1L_2L_3|$ for integral u, v, w not all zero. The author previously [2] proved that $M \leqq 1/7$. In the present paper he develops the method of a later proof of his [4] and so shows: "Either $M=1/7$ and L_1, L_2, L_3 are equivalent in some order to

$$\lambda_1(u+\theta v+\phi w),\quad \lambda_2(u+\phi v+\psi w),\quad \lambda_3(u+\psi v+\theta w),$$

where θ, ϕ, ψ are the roots of $t^3+t^2-2t-1=0$ and $|\lambda_1\lambda_2\lambda_3| = 1/7$, or $M=1/9$ and L_1, L_2, L_3 are equivalent in some order to

$$\lambda_1(u+\theta'v+\phi'w),\quad \lambda_2(u+\phi'v+\psi'w),\quad \lambda_3(u+\psi'v+\theta'w),$$

where θ', ϕ', ψ' are the roots of $t^3-3t-1=0$ and $|\lambda_1\lambda_2\lambda_3| = 1/9$, or $M<1/9.1$."

For the proof, it can be assumed that $M \geqq 1/9.1$ and that

$$L_i = L_i^*(u+\alpha_iv+\beta_iw),$$

where

$$|L_1^*L_2^*L_3^*| = \frac{M}{1-\epsilon},\quad \begin{vmatrix} 1 & \alpha_1 & \beta_1 \\ 1 & \alpha_2 & \beta_2 \\ 1 & \alpha_3 & \beta_3 \end{vmatrix} = \mp\frac{1-\epsilon}{M},$$

and where $\epsilon>0$ is arbitrarily small; then α_i, β_i satisfy the inequalities

$$\left|\prod_{i=1}^{3}(u+\alpha_iv+\beta_iw)\right| \geqq 1-\epsilon$$

for integral u, v, w not all zero. It may further be assumed that the positive definite quadratic form

$$\{(\alpha_1-\alpha_2)v+(\beta_1-\beta_2)w\}^2+\{(\alpha_2-\alpha_3)v+(\beta_2-\beta_3)w\}^2 + \{(\alpha_3-\alpha_1)v+(\beta_3-\beta_1)w\}^2 = 2(Av^2+Bvw+Cw^2)$$

is reduced: $|B| \leqq A \leqq C$; for sufficiently small ϵ, its determinant satisfies

$$4(AC-B^2) = 3((1-\epsilon)/M)^2 < 248.5.$$

These conditions restrict the triples of numbers (α_i), (β_i) to certain types, according to the integral parts of α_i, β_i, $\alpha_i+\beta_i$, $\alpha_i-\beta_i$; all but two of these cases can be excluded.

K. Mahler (Manchester).

Citations: MR 3, 70f = J36-2.

Referred to in H20-5.

J36-8 **(5, 201b)**

Mordell, L. J. The product of n homogeneous forms. Rec. Math. [Mat. Sbornik] N.S. **12(54)**, 273–276 (1943). (English. Russian summary)

Let $X_1, X_2, \cdots, X_n$ be n linear forms in $x_1, x_2, \cdots, x_n$ with real or complex coefficients; X is the matrix of the coefficients of the X_i and is of determinant unity. The contragredient forms are those associated with X'^{-1} as the X_i are with X. Moreover, $K(X)$ is the lower bound of $|X_1X_2\cdots X_n|$ for integer values of the x's not all zero and $k(X)$ is $K(X)$ with X_n replaced by $X_1+X_2+\cdots+X_{n-1}$. Finally, K and k are the upper bounds of $K(X)$ and $k(X)$, respectively, over all sets of forms X_i with $|X|=1$. The author proves the following theorem. If $n>2$ and none of the contragredient forms represents zero, then

$$K(X)\leqq k[K(X'^{-1})]^{1/(n-1)}.$$

Further $K(X)\leqq k^q$ and $K\leqq k^q$, where $q=(n-1)/(n-2)$.
B. W. Jones (Ithaca, N. Y.).

Referred to in H15-66.

J36-9 (8, 565d)

Davenport, H. The product of n homogeneous linear forms. Nederl. Akad. Wetensch., Proc. **49**, 822–828 = Indagationes Math. **8**, 525–531 (1946).

Let $L_1, \cdots, L_n$ be n homogeneous linear forms in n variables $u_1, \cdots, u_n$ with real coefficients and determinant 1. Then, improving results of Minkowski and Blichfeldt, the author proves that the lower bound M of $|L_1\cdots L_n|^{1/n}$ for integral values of $u_1, \cdots, u_n$ satisfies $M\leqq(n^{-1}\gamma_nk^{-1+1/n})^{\frac{1}{2}}$, where $k>1.4$ and $\gamma_n\leqq 2\pi^{-1}\{\Gamma(2+\frac{1}{2}n)\}^{n/2}$. [For literature see, for example, the reviewer's "Diophantische Approximationen," Ergebnisse der Math., vol. 4, no. 4, Springer, Berlin, 1936.] *J. F. Koksma* (Amsterdam).

Referred to in H05-19, J36-12.

J36-10 (10, 18e)

Chalk, J. H. H. On the positive values of linear forms. II. Quart. J. Math., Oxford Ser. **19**, 67–80 (1948).

[For part I cf. the same J., Oxford Ser. **18**, 215–227 (1947); these Rev. **9**, 413.] Let $L_1, \cdots, L_n$ be n homogeneous forms in $n\geqq 2$ variables $u_1, \cdots, u_n$, whose determinant Δ is not zero. The author investigates the question whether there are infinitely many sets of integers $u_1, \cdots, u_n$, such that $L_1>0, L_2>0, \cdots, L_n>0$ and $L_1L_2\cdots L_n\leqq|\Delta|$. He proves that the answer is positive for all sets of forms for which not all of the n sets S_i of $n-1$ equations $L_j=0$ ($j=1, \cdots, n$; $j\neq i$; $i=1, 2, \cdots, n$) are soluble in integers not all zero. The proof depends on a similar theorem for inhomogeneous forms which is deduced by the author from Kronecker's theorem. The author points out the relationship between his results and some theorems of K. Mahler [Proc. London Math. Soc. (2) **49**, 128–157 (1946); Nederl. Akad. Wetensch., Proc. **49**, 331–343, 444–454, 524–532, 622–631 = Indagationes Math. **8**, 200–212, 299–309, 343–351, 381–390 (1946); these Rev. **8**, 12]. *J. F. Koksma* (Amsterdam).

Citations: MR 8, 12a = H15-9; MR 8, 12b = H15-8; MR 8, 12d = H15-6; MR 9, 413b = J40-6.

Referred to in J40-14.

J36-11 (10, 512b)

Rogers, C. A. The product of n homogeneous linear forms. J. London Math. Soc. **24**, 31–39 (1949).

Let $x_1, \cdots, x_n$ be n real homogeneous linear forms in n integral variables $u_1, \cdots, u_n$ with determinant unity. Let M be the lower bound of $|x_1\cdots x_n|$ for all sets of integers $(u_1, \cdots, u_n)$ other than $(0, \cdots, 0)$. The author obtains an upper bound for M which is asymptotic to $(2\pi)^{\frac{1}{2}}(\frac{1}{2}+\log 2)n^{\frac{1}{2}}(e\sqrt{e})^{-n}$ for large n. This number is less than the corresponding bounds which have been obtained by Blichfeldt and the reviewer. However, in a note added before going to press the author observes that an upper bound for M, which is less than his, is contained implicitly in a paper by Blichfeldt [Monatsh. Math. Phys. **48**, 531–533 (1939); these Rev. **1**, 68]. The proof is based on a lemma which gives an upper bound for $\prod|z_r-z_s|/(\sum|z_t|)^{\frac{1}{2}m(m-1)}$ ($1\leqq r<s\leqq m$, $1\leqq t\leqq m$). The author has succeeded in sharpening this lemma, and he can now improve on Blichfeldt's result for large n. *R. A. Rankin.*

Citations: MR 1, 68c = R28-1.

Referred to in J36-13.

J36-12 (10, 512c)

Chabauty, Claude. Sur le minimum du produit de formes linéaires réelles. C. R. Acad. Sci. Paris **228**, 1361–1363 (1949).

By using the method of a previous note* [see the 5th preceding review] the author proves the following general theorem. Let S and T be two star sets in n-dimensional space, S being contained in T, with critical determinants $\Delta(S)$ and $\Delta(T)$. Let G be a critical lattice of T, and let λ be any number such that the points of G in the set λS have a linear dimension less than h. Then

$$\Delta(T)\geqq 2^{1-h}\lambda^{n+1-h}\Delta(S).$$

This provides a possible method of improving, in particular cases, on the obvious inequality $\Delta(T)\geqq\Delta(S)$. By taking S to be $|x_1|+\cdots+|x_n|<n$, and T to be $|x_1\cdots x_n|<1$, and h to be 2, and by modifying a method of Davenport [see Nederl. Akad. Wetensch., Proc. **49**, 822–828 = Indagationes Math. **8**, 525–531 (1946); these Rev. **8**, 565], the author deduces the following theorem. Let $L_i(x_1, \cdots, x_n)$, where $i=1, \cdots, n$, be real linear forms of determinant 1. Then, provided n is sufficiently large, there exist integers $x_1, \cdots, x_n$, not all zero, such that

$$\left|\prod_{i=1}^{n}L_i(x_1, \cdots, x_n)\right|^{1/n}<\frac{1}{4.1}.$$

It may be mentioned, however, that a more precise result has been obtained by Rogers [see the preceding review]. [Errata, all on page 1362: (1) On line 10, omit n^{-1} in both formulae. (2) On line 8 from below, for $1/4n$ read $\frac{1}{4}$. (3) In the theorem, the possibility that $x_1=\cdots=x_n=0$ must be excluded, the statement "$0<$" must be deleted from the displayed formulae, and n must be deleted in each denominator.] *H. Davenport* (London).

*MR 10, 511F = H15-30.

Citations: MR 8, 565d = J36-9.

J36-13 (11, 501e)

Rogers, C. A. The product of n real homogeneous linear forms. Acta Math. **82**, 185–208 (1950).

Let $x_1, \cdots, x_n$ be n real homogeneous linear forms in n integral variables $u_1, \cdots, u_n$, with determinant unity. The lower bound of $|x_1x_2\cdots x_n|$ for all $(u_1, \cdots, u_n)$ other than $(0, 0, \cdots, 0)$ depends upon the coefficients of the forms and has an upper bound M_n. Let $M=\limsup_{n\to\infty}(M_n)^{1/n}$. It is shown that $M\leqq\frac{1}{4}\pi e^{-\frac{1}{2}}=1/5.70626\cdots$. This estimate is an improvement on an earlier result of the author [J. London Math. Soc. **24**, 31–39 (1949); these Rev. **10**, 512] and also on a result $M\leqq(2\pi)^{-\frac{1}{2}}e^{-\frac{1}{2}}=1/5.30653\cdots$ which he shows to be implicit in a paper of Blichfeldt. The proof is based upon a lemma which gives a sharper estimate for the upper bound of $K_m=\prod|z_r-z_s|^{2/\{m(m-1)\}}/\{\sum|z_t|/m\}$ ($1\leqq r<s\leqq m$, $1\leqq t\leqq m$, z_t real) than that given in the earlier paper. It is for large m that this estimate is required, and what the author does is, in effect, to approximate to K_m by the expression

$$K=\exp\left\{\int_0^1 dx\int_0^1\log|\alpha(x)-\alpha(\xi)|\,d\xi\right\}\Big/\int_0^1|\alpha(x)|\,dx,$$

where $\alpha(x)$ is an increasing function satisfying certain conditions. The determination of the function $\alpha(x)$ which makes K a maximum is a problem in the calculus of variations which the author (who makes acknowledgment to L. A. Wigglesworth for help) solves; in fact, max $K=\frac{1}{2}\pi e^{-\frac{1}{2}}$.

R. A. Rankin (Cambridge, England).

Citations: MR 10, 512b = J36-11.

Referred to in H30-24, J36-27, J36-28.

J36-14 (12, 319f)

Müller, Oskar. Über das Minimum des Produktes dreier ternärer linearer Formen. Thesis, University of Zürich, 1947. 36 pp.

This thesis consists of a geometric proof of a theorem obtained arithmetically by Davenport [Proc. London Math. Soc. (2) **44**, 412–431 (1938)], namely: If ξ, η, ζ are three linear forms in three variables with real coefficients and determinant D, then there exist rational integral values of the variables not all zero such that $|\xi\eta\zeta| \leqq |D|/7$, the equality holding only when $\xi\eta\zeta$ is equivalent to a multiple of $x^3+y^3+z^3-x^2y-x^2z-2xy^2-2xz^2+3xyz+3y^2z-4yz^2$. Brief applications are given.

B. W. Jones (Boulder, Colo.).

J36-15 (12, 320a)

Godwin, H. J. On the product of five homogeneous linear forms. J. London Math. Soc. **25**, 331–339 (1950).

Let $L_1, \cdots, L_5$ be five homogeneous linear forms, with real coefficients, in five variables $u_1, \cdots, u_5$. The determinant of the forms is taken, without loss of generality, to be 1. The author proves that there exist integral values, not all zero, of $u_1, \cdots, u_5$ for which $|L_1 \cdots L_5| < (57.02)^{-1}$. The method is based on that devised by the reviewer for the product of three linear forms [same J. **16**, 98–101 (1941); these Rev. **3**, 70]. The essential lemma is that if $|(n-\alpha)(n-\beta)(n-\gamma)(n-\delta)(n-\epsilon)| \geqq 1$ for all integers n, then $\sum(\alpha-\beta)^2 > 35.71$. The proof of this depends, as one might expect, on a somewhat elaborate division into cases, with a variety of treatments for the various cases.

H. Davenport (London).

Citations: MR 3, 70f = J36-2.

J36-16 (12, 678c)

Chalk, J. H. H., and Rogers, C. A. On the product of three homogeneous linear forms. Proc. Cambridge Philos. Soc. **47**, 251–259 (1951).

Let Λ be a lattice of points $X=(x_1, x_2, x_3)$ of determinant Δ. Let $P(X)=|x_1x_2x_3|$. Davenport [Proc. London Math. Soc. (2) **44**, 412–431 (1938)] showed that there is a point A of Λ not the origin such that $P(A) \leqq \Delta/7$. This result is improved upon by showing that Λ has an infinity of sets of generators A, B, C for which $P(A)P(B)P(C) \leqq (\Delta/7)^3$ and, except for a stated case, $P(A)P(B)P(C) \leqq (\Delta/7.2)^3$.

L. Tornheim (Ann Arbor, Mich.).

J36-17 (13, 627d)

Barnes, E. S. The minimum of the product of two values of a quadratic form. I. Proc. London Math. Soc. (3) **1**, 257–283 (1951).

Let $Q(x,y)=ax^2+bxy+cy^2$ be an indefinite real binary quadratic form of discriminant $D=b^2-4ac>0$, and let $M=M(Q)$ be the lower bound of $|Q(x,y)\,Q(u,v)|$ for all integers x, y, u, v satisfying $xv-uy=1$. Set $\lambda=\lambda(Q)=D/M$. In this paper it is proved that $\lambda \geqq (3+13\sqrt{10})/6$ unless Q is equivalent under unimodular transformation to a multiple of some one of a sequence of forms $Q_{-1}, Q_0, Q_1, Q_3, Q_5, \cdots$ given by: $Q_{-1}=x^2-xy-y^2$, $Q_0=x^2-2.5y^2$, and

$$Q_n=(\xi_n+2\eta_n)^{-1}\{(\xi_n+2\eta_n)x^2-(\xi_n-3\eta_n)xy-(\xi_n+10\eta_n)y^2\}$$

for positive odd n, where ξ_n and η_n are the integers defined by $\xi_n+\eta_n\sqrt{10}=(3+\sqrt{10})^{n+1}$. Setting $\lambda_n=\lambda(Q_n)$, the author proves that $\lambda_{-1}=5$, $\lambda_0=20/3$, and

$$\lambda_n=\{(3\xi_n+7\eta_n)^2-4\}/(\xi_n+2\eta_n)(\xi_n+4\eta_n)$$

for positive odd n, whence it follows that $\lambda_{-1}, \lambda_0, \lambda_1, \lambda_3, \cdots$ strictly increases toward the limit $(3+13\sqrt{10})/6$. The proof consists of normalizing $Q(x,y)$ so that $m=1$, where m is the lower bound of $|Q(x,y)|$ for integers x, y not both zero, and then considering separately various ranges for D by means of certain inequalities. The continued fraction expansion for the positive root of $Q_n(x,1)=0$ is used to derive the formula for $\lambda(Q_n)$. The result derived in this paper is analogous to a famous theorem of Markoff [Math. Ann. **15**, 381–409 (1879); **17**, 379–399 (1880); for a simple proof, see Cassels, Ann. of Math. (2) **50**, 676–685 (1949); these Rev. **11**, 643] which states that $D/m^2 \geqq 9$ unless Q is equivalent to a multiple of some one of a certain sequence of forms.

I. Reiner (Urbana, Ill.).

Citations: MR 11, 643b = J28-11.

Referred to in J36-18, J36-23, J52-21.

J36-18 (13, 825a)

Barnes, E. S. The minimum of the product of two values of a quadratic form. II. Proc. London Math. Soc. (3) **1**, 385–414 (1951).

Let $Q(x,y)=ax^2+bxy+cy^2$ be an indefinite quadratic form of positive determinant $D=b^2-4ac$. Excluding hereafter the trivial case where Q is a zero form, let M and $-N$ be the numerically least positive and negative values respectively of $Q(x,y)Q(u,v)$, where x, y, u, v are integers such that $xv-uy=1$. Setting $\lambda(Q)=D/M$ and $\mu(Q)=D/N$, the existence of integers x, y, u, v with $xv-uy=1$ for which $-D/m \leqq Q(x,y)Q(u,v) \leqq D/l$ is equivalent to the condition that either $\lambda(Q) \geqq l$ or $\mu(Q) \geqq m$.

In a previous paper [same Proc. (3) **1**, 257–283 (1951); these Rev. **13**, 627] the author obtained a sequence of forms $f_{-1}, f_0, f_1, f_3, f_5, \cdots$ for which $\lambda_n=\lambda(f_n)$ strictly increases to $\lambda_\infty=(13\sqrt{10}+3)/6$, such that $\lambda(Q) \geqq \lambda_\infty$ or $\mu(Q) \geqq \lambda_\infty$ unless Q is equivalent to a multiple of some f_n. In this paper asymmetric bounds are obtained for $\lambda(Q)$ and $\mu(Q)$. A second sequence of forms $\psi_0, \psi_2, \psi_4, \cdots, \psi_\infty$ is found for which $\mu_n=\mu(\psi_n)$ strictly increases to $\mu_\infty=\mu(\psi_\infty)=2(\sqrt{10}+1)$, as well as a form f_∞ for which $\lambda(f_\infty)=\lambda_\infty$, and it is shown that $\lambda(Q)>\lambda_\infty$ or $\mu(Q)>\mu_\infty$ unless Q is equivalent to a multiple of one of the f_n or ψ_n (including f_∞ and ψ_∞). To prove this result, the author first shows that the condition "$\lambda(Q)>l$ or $\mu(Q)>m$" is equivalent (with one minor restriction) to the falsehood of at least one of an infinite set of inequalities which involve certain continued fractions associated with the sequence of reduced forms equivalent to Q. The major difficulty of the proof consists of showing that if all of these inequalities hold with $l=\lambda_\infty$ and $m=\mu_\infty$, then these continued fractions must have a very specific structure, and further that this structure can arise only from one of the f_n or ψ_n (including f_∞ and ψ_∞). The author also proves that the constants λ_∞ and μ_∞ are in a certain sense the best possible. All of the proofs in this paper are independent of the results of the previous paper, and so furnish another proof of those results.

I. Reiner (Urbana, Ill.).

Citations: MR 13, 627d = J36-17.

Referred to in J44-13.

J36-19 (13, 825b)

Barnes, E. S. The minimum of the product of two values of a quadratic form. III. Proc. London Math. Soc. (3) **1**, 415–434 (1951).

We shall use the notation of the preceding review. In this paper the author shows that $\lambda(Q) \geqq 4$ and $\mu(Q) \geqq 5$ for all Q. In particular, he proves that $\lambda(Q) \geqq \lambda_\infty=(100+13\sqrt{13})/27$ unless Q is equivalent to a multiple of some one of a sequence of forms whose λ's increase from 4 to λ_∞. It is also shown that $\mu(Q) \geqq \mu_\infty = 2(\sqrt{5}+1)$ unless Q is equivalent to a multiple of some one of another sequence of forms whose μ's increase from 5 to μ_∞. Furthermore, the author proves his results to be best possible. The same approach is used here as in the preceding paper; however, these results are somewhat more difficult to prove than the previous ones.

I. Reiner (Urbana, Ill.).

Referred to in J36-23.

J36-20 (13, 920a)

Ankeny, N. C., and Rogers, C. A. A condition for a real lattice to define a zeta function. Proc. Nat. Acad. Sci. U. S. A. **37**, 159–163 (1951).

Let x_i be real linear forms in u_j, i.e.

$$(1) \qquad x^{(i)} = \sum_{j=1}^{n} u_j w_j^{(i)} \quad (i=1, 2, \cdots, n);$$

$w_j^{(i)}$ is real; the determinant $\Delta = |w_j^{(i)}| > 0$; $P(X) = \prod_{i=1}^n x^{(i)}$. Let Λ denote the lattice consisting of all points in (1) where $u_1, \cdots, u_n$ take on all integer values. Suppose that there is no point X of Λ other than the origin 0 with $P(X)=0$, and suppose that $P(X)$ assumes only a finite number of different values α with $-\Delta \leqq \alpha \leqq \Delta$. Then it is shown, that the product $P(X)$ can be expressed in the form

$$P(X) = w \prod_{i=1}^{n} \left(\sum_{j=1}^{n} u_j w_j^{(i)} \right)$$

where $w_j^{(1)}$ $(j=1, \cdots, n)$ are algebraic integers in a field of degree n, and $w_j^{(i)}$ $(j=1, \cdots, n; i=2, \cdots, n)$ are their $(n-1)$ different algebraic conjugates. So one of the linear forms $\sum_{j=1}^n u_j w_j^{(1)}$ arises from a ring in an algebraic number field, and the other $(n-1)$ linear forms are the $(n-1)$ different conjugates of the first linear form. Hence, the product is essentially the norm of all numbers in an order of an algebraic number field, which clearly takes on only a finite number of values in any finite interval.

S. C. van Veen (Delft).

J36-21 (14, 142b)

Mordell, Louis Joel. The product of *n* homogeneous linear forms. Univ. Roma. Ist. Naz. Alta Mat. Rend. Mat. e Appl. (5) **10**, 12–23 (1951).

An expository article discussing many of the important results in the geometry of numbers; especially some of the recent results obtained by the English school. The central problem of the discussion is that of estimating the greatest lower bound for all integral $x_1, \cdots, x_n$ of $|L_1L_2 \cdots L_n|/D$, where $L_i = \sum_{j=1}^n l_{ij}x_j$ are linear forms with complex coefficients and D is the absolute value of the determinant $|l_{ij}|$.

E. G. Straus (Los Angeles, Calif.).

J36-22 (14, 729c)

Barnes, E. S. Isolated minima of the product of *n* linear forms. Proc. Cambridge Philos. Soc. **49**, 59–62 (1953).

Let $X_i = \sum_{j=1}^n a_{ij}u_j$ $(i=1, \cdots, n)$ with real a_{ij} and $\Delta = \det(a_{ij}) \neq 0$, and let $M(X)$ be the lower bound of $|X_1 \cdots X_n|$ over all integer sets $(u) \neq (0)$. Define γ_n as the upper bound of $M(X)/|\Delta|$ over all sets (X_i). It is known [see Davenport and Rogers, Philos. Trans. Roy. Soc. London. Ser. A. **242**, 311–344 (1950); these Rev. **12**, 394] that γ_n is finite, and that (for $n=2$ or 3) $M(X) = \gamma_n|\Delta|$ holds only when the set (X_i) is unimodularly equivalent to a multiple of a certain critical set (W_i) given by $W_i = \sum_{j=1}^n \omega_j^{(i)}u_j$, where $\omega_1, \cdots, \omega_n$ is an integral basis for the totally real algebraic number field of degree n and least discriminant, and where superscripts denote conjugates. Further, there exists a constant $\gamma_n^{(2)} < \gamma_n$ such that $M(X) \leqq \gamma_n^{(2)}|\Delta|$ for sets (X_i) not satisfying the above condition.

The author considers sets of forms analogous to the above critical set, and generalizes results of Davenport and Rogers [loc. cit.] to obtain the following theorem: Let $X_i = \sum_{j=1}^n \alpha_j^{(i)}u_j$ $(i=1, \cdots, n)$, where $\alpha_1, \cdots, \alpha_j$ are elements of a totally real algebraic field of degree n, and suppose that there exist integers (u) with $|X_1 \cdots X_n| = M(X)$ for which $X_1, \cdots, X_n$ have arbitrarily prescribed signs. An ϵ-neighborhood of the set (X_i) is gotten by varying each coefficient $\alpha_j^{(i)}$ by an arbitrary real number β_{ij} subject to $|\beta_{ij}| < \epsilon$. Then there exists a number $\delta > 0$ such that for each set (Y_i) in a sufficiently small ϵ-neighborhood of (X_i), either $(Y_i) = \text{const.} \times (X_i)$ or $M(Y_i) < (1-\delta)M(X_i)$.

I. Reiner (Urbana, Ill.).

Citations: MR 12, 394b = H15-42.

J36-23 (15, 936d)

Oppenheim, A. On indefinite binary quadratic forms. Acta Math. **91**, 43–50 (1954).

Let $Q(x, y)$ be an indefinite quadratic form of discriminant D^2. A theorem of Barnes states that there are integers u, v, x, y such that

$$(*) \qquad -D^2(2\sqrt{5}+2)^{-1} \leqq Q(x, y)Q(u, v) < 0, \quad xv - yu = \pm 1$$

except when Q is equivalent to one of a denumerable set of forms; but this is no longer true if $2\sqrt{5}+2$ is replaced by a smaller number [Proc. London Math. Soc. (3) **1**, 257–283, 385–414, 415–434 (1951); these Rev. **13**, 627, 825]. The author remarks that (*) means that Q is equivalent to a form $\alpha x^2 + \beta xy + \gamma y^2$ with

$$(**) \qquad -D^2(2\sqrt{5}+2)^{-1} \leqq \alpha\gamma < 0,$$

and that he has already proved that every Q is equivalent to a reduced form with not merely (**) but also

$$\frac{1}{|\alpha|} + \frac{1}{|\gamma|} > \frac{3+\sqrt{5}}{D}, \quad D - \beta < (2\sqrt{5}-4)(|\alpha|+|\gamma|);$$

with the same set of exceptions [ibid. (2) **44**, 323–335 (1938)]. He further proves that Q is always equivalent to a form with $\max(|\alpha|, |\gamma|) \leqq \frac{1}{2}D$; but there are indenumerably many forms which are never equivalent to a form with $\max(|\alpha|, |\gamma|) \leqq (\frac{1}{2}-\epsilon)D$ for any preassigned $\epsilon > 0$.

J. W. S. Cassels (Cambridge, England).

Citations: MR 13, 627d = J36-17; MR 13, 825b = J36-19.

J36-24 (17, 14f)

Cassels, J. W. S., and Swinnerton-Dyer, H. P. F. On the product of three homogeneous linear forms and indefinite ternary quadratic forms. Philos. Trans. Roy. Soc. London. Ser. A. **248**, 73–96 (1955).

The authors prove "local isolation" theorems in which the concept of "neighbourhood of a form f" is used. If $(x_1, \cdots, x_n)$ is an algebraic form of some specific type, then for any $\varepsilon > 0$ an ε-neighbourhood of f is defined as the set of all forms f^* of the same type whose coefficients lie within the usually defined ε-neighbourhood of the corresponding coefficients of f. Any set which contains some ε-neighbourhood of f is called a neighbourhood. Main result: Let $f(x, y, z) = L_1L_2L_3$ be the product of three homogeneous linear forms in x, y, z which represent zero only for $x=y=z=0$ and suppose that f has integer coefficients. Let (δ_1, δ_2) be any open interval. Then there is a neighbourhood of f, such that all forms in that neighbourhood which are not multiples of f itself, take some value in (δ_1, δ_2) for integer values of x, y, z. This theorem holds for neighbourhoods in the set of all ternary cubic forms; the authors however restrict themselves to the proof of the case of neighbourhoods in the set of all products of three linear forms. Further: Let L_1, L_2, L_3 be the forms of the above theorem and let L_2^*, L_3^* be any real linear forms such that the product $L_2^*L_3^*$ is not a numerical multiple of L_2L_3. Then the set of values taken by $L_1L_2^*L_3^*$ for integer values of x, y, z, is everywhere dense in $(-\infty, +\infty)$.

The authors further state several hypotheses on the product of three linear forms which at the moment neither can be proved nor denied, and deduce some theorems concerning the equivalence, etc. of these hypotheses. *J. F. Koksma* (Amsterdam).

Referred to in J36-34, J64-28, J68-30.

J36-25 (17, 350g)

Oliwa, Godfried. Eine Anwendung des Übertragungsprinzips von Hlawka. Anz. Österreich. Akad. Wiss. Math.-Nat. Kl. **1953**, 239–242.

Verf. folgert unmittelbar aus dem inhomogenen Linearformensatz mit der Schranke von Tschebotarev: Besitzt das System der Kongruenzen $Ay_i \equiv a_i \pmod{m_i}$ $(i=0, 1, 2, \cdots, n)$, $(A, a_0, \cdots, a_n, m_0, \cdots, m_n$ ganz) eine Lösung, dann ist für jedes System reeller Zahlen $\xi_1, \cdots, \xi_n$

$$|z_0||\xi_1 z_0 - z_1|\cdots|\xi_n z_0 - z_n| \leq 2^{-(n+1)/2} m_0 \cdots m_n$$

in ganzen Zahlen $z_0, z_1, \cdots, z_n$ mit $Az_i \equiv a_i(m_i)$ $(i=0, 1, \cdots, n)$ lösbar. [Bem. des Ref.: Die Formulierung des Satzes in der Note des Verf. ist richtig, wenn $a_0 \not\equiv 0 \pmod{m_0}$.] Verf. gibt noch eine Verallgemeinerung auf imaginärquadratische Zahlkörper.

E. Hlawka (Zbl **52**, 281).

J36-26 (17, 465f)

Chalk, J. H. H. On the product of n homogeneous linear forms. Proc. London Math. Soc. (3) **5** (1955), 449–473.

Let Λ be a lattice of determinant Δ in n-dimensional Euclidean space. The cases $n=3$, 4 are proved of the following conjecture: there exist n lattice points generating Λ and lying in the region $|x_1 \cdots x_n| \leq 2^{-n+1}\Delta$. Also, the situation when not all are interior points is determined. The case $n=2$ was proved by H. Minkowski [Math. Ann. **54** (1900), 91–124] and generalized by the author [Proc. Cambridge Philos. Soc. (3) **49** (1953), 413–420; MR **14**, 1065]. The proof uses the results of Remak and Dyson that there exists an ellipsoid $p_1x_1^2 + \cdots + p_nx_n^2 = 1$ having n independent points of Λ on its boundary and no point of Λ except 0 inside. For $n=4$ the proof is short but for $n=3$ it involves a lengthy examination of many possibilities to find which of the points on the ellipsoid and simple combinations of them are effective. *L. Tornheim.*

Citations: MR 14, 1065d = J32-42.

J36-27 (21# 722)

Mulholland, H. P. Inequalities between the geometric mean difference and the polar moments of a plane distribution. J. London Math. Soc. **33** (1958), 260–270.

Démonstration de l'inégalité suivante: pour tout nombre réel $k > 0$ et toute mesure $\mu \geqq 0$ du plan, de masse totale 1, on a

$$\exp\left[\iint \log|z-\zeta|\,d\mu(z)d\mu(\zeta)\right] \leqq (2/\sqrt{e})^{1/k}\left[\int |z|^k d\mu(z)\right]^{1/k}$$

l'égalité ayant lieu seulement lorsque μ est la mesure de densité $k|z|^{k-2}/4\pi m$ sur le disque $|z| \leqq (2m)^{1/k}$ $(m > 0)$.

La méthode repose sur les éléments de la théorie du potentiel logarithmique. Un résultat analogue avait été donné, à propos d'un problème de la théorie des nombres, par C. A. Rogers [Acta Math. **82** (1950), 185–208; MR **11**, 501] lorsque $k=1$ et que μ est portée par l'axe réel (la constante figurant au second membre est alors $\pi/2\sqrt{e}$). L'auteur annonce également des applications à la théorie des nombres. *J. Deny* (Strasbourg)

Citations: MR 11, 501e = J36-13.

Referred to in J36-28.

J36-28 (22# 4703)

Mulholland, H. P. On the product of n complex homogeneous linear forms. J. London Math. Soc. **35** (1960), 241–250.

Let $x_1, x_2, \cdots, x_n$ be n homogeneous linear forms in n variables with associated matrix A, and let $L(A) = \inf(x_1x_2\cdots x_n)^{1/n}$, for integral values of the variables not all zero. Let $b_\gamma(n)$ be the class of $n \times n$ matrices A, with $|\det A| = 1$, having γn rows complex (in pairs of conjugates) and the remaining $(1-\gamma)n$ rows real. Let $M_\gamma(n)$ be the supremum of $L(A)$ for $A \in b_\gamma(n)$. Also put $\mathscr{M}_c = \limsup M_\gamma(n)$ for $n \to \infty$ and $\gamma \to c$, where $0 \leqq c \leqq 1$. C. A. Rogers [Acta Math. **82** (1950), 185–208; MR **11**, 501] showed that $\mathscr{M}_0 \leqq \pi/4e^{3/2} = 1/5.70626\cdots$. The author proves that, for $0 \leqq c \leqq 1$,

$$\mathscr{M}_c \leqq (\tfrac{1}{2}\sqrt{\pi})^{2-3c}e^{-3/2} \leqq 2\pi^{-1/2}e^{-3/2} = 1/3.97179\cdots.$$

This improves on a result of Blichfeldt which yields $\mathscr{M}_c \leqq (\pi e)^{-1/2} = 1/2.92221\cdots$. The author also indicates how his result can be improved to give $\mathscr{M}_c \leqq 1/3.985900\cdots$. The proof depends upon estimations of the number

$$C_m(k) = \sup\left\{\prod_{\rho\neq\sigma} |z_\rho - z_\sigma|^{1/(m^2-m)} / \left(\sum |z_\rho|^k/m\right)^{1/k}\right\},$$

where $k > 0$ and $z_1, z_2, \cdots, z_m$ are complex variables not all zero. By using the theory of the logarithmic potential and an inequality for integrals, analogous to one established for real variables by Rogers, and proved earlier by the author [same J. **33** (1958), 260–270; MR **21** #722], it is shown that $C_m(k)$ decreases as m increases and that

$$(2e^{-1/2})^{1/k} m^{(1-\eta)/2m} \leqq C_m(k) \leqq (2e^{-1/2})^{m/(km-k)}\left(\frac{m^{m(m-1)}}{m-1}\right)^{1/\kappa},$$

where $\kappa = \min(k, 1)$ and $\eta = \eta(m, k) \to 0$ as $m \to \infty$. In particular, $C_m(k) \to (2e^{-1/2})^{1/k}$ as $m \to \infty$. The first estimate for $\mathscr{M}_c$ is obtained by taking $k=1$, and the second by taking $k=5/6$. *R. A. Rankin* (Glasgow)

Citations: MR 11, 501e = J36-13; MR 21# 722 = J36-27.

J36-29 (30# 1096)

Philipp, Walter

A note on a problem of Littlewood about Diophantine approximation.

Mathematika **11** (1964), 137–141.

Littlewood has raised the question whether or not $\inf u\,\|u\alpha\|\,\|u\beta\| = 0$ for all real numbers α and β, where the infimum is taken over all positive integers u and $\|y\|$ is the distance from y to the nearest integer. This is a very delicate question, and so far no answer has been forthcoming. Using standard transference principles one sees that this question is equivalent to determining whether or not (A) $\inf|xy|\,\|x\alpha + y\beta\| = 0$ for all real α, β such that $1, \alpha, \beta$ are linearly independent over the rationals, where the infimum is over all nonzero integers x, y. The author proves: To any $X > 1$ there correspond continuum many pairs α, β of real numbers such that $1, \alpha, \beta$ are linearly independent and $y|x\alpha + y\beta| > 1/4398X$ for all integers $y > 0$ and x such that $|x| \leqq X$. Thus, if (A) is true, 0 cannot be approached too fast. The proof uses very simple ideas from analytic geometry.

D. J. Lewis (Ann Arbor, Mich.)

J36-30 **(32# 2398)**

Godwin, H. J.

On the inhomogeneous minima of totally real cubic norm-forms.

J. London Math. Soc. **40** (1965), 623–627.

Let 1, θ, ω be an integral basis of a totally real cubic field K, $F(x, y, z) = N(x + y\theta + z\omega)$ the norm form of K, Δ the discriminant of K, and M the inhomogeneous minimum of F. It was proved by Davenport [Proc. Cambridge Philos. Soc. **43** (1947), 137–152; MR **8**, 444] that $M < (\sqrt{\Delta})/8$. The author obtains bounds for M better than Davenport's for $\Delta \leqq 107132$, and all cyclic fields. In particular, he shows that for cyclic fields with $\Delta \geqq 169$, $M \leqq (\sqrt{\Delta})/18 - 1/8$.

Morris Newman (Washington, D.C.)

Citations: MR 8, 444a = J40-5.

Referred to in R16-46.

J36-31 **(36# 2556)**

Karanikolov, Hristo

A second contribution on Minkowski's theorems for certain linear homogeneous forms. (Bulgarian. Russian and English summaries)

Godišnik Visš. Tehn. Učebn. Zaved. Mat. **2** (1965), *kn.* 3, 13–16 (1966).

The author proves the following theorem: Let a_{ij} $(i, j = 1, 2, \cdots, n;\ n \geqq 2)$ be arbitrary real numbers. If $\Delta = |\det a_{ij}| > 0$, then the inequality $\prod_{i=1}^{n} |\sum_{j=1}^{n} a_{ij}x_j| \leqq \Delta$ has infinitely many solutions in integers $x_1, x_2, \cdots, x_n$, $\sum_{j=1}^{n} x_j^2 > 0$.

The author cites the following theorems of Minkowski. Theorem I: There exist integers $x_j(\sum |x_j| > 0)$ such that $|\sum a_{1j}x_j| \leqq c_1$, $|\sum a_{ij}x_j| < c_i$ $(2 \leqq i \leqq n)$, where $c_1c_2 \cdots c_n \geqq |\det(a_{ij})| > 0$. Theorem II (Corollary): There exist integers $x_j(\sum |x_j| > 0)$ such that $\prod_{i=1}^{n} |\sum_{j=1}^{n} a_{ij}x_j| \leqq |\det(a_{ij})|$ if $|\det(a_{ij})| > 0$. The author's paper in same Godišnik **2** (1965), kn. 2, 1–6 (1967), is a contribution to Theorem II, as is the paper under review. *T. T. Tonkov* (Leningrad)

Citations: MR 39# 6827 = J36-33.

J36-32 **(37# 6246)**

Noordzij, P.

Über das Produkt von vier reellen, homogenen, linearen Formen.

Monatsh. Math. **71** (1967), 436–445.

If L_1, L_2, L_3, L_4 are four homogeneous linear forms with real coefficients and determinant 1, then there are integral values of the variables for which $0 < |L_1L_2L_3L_4| < 500^{-1/2}$. Furthermore, there is a system of four such linear forms for which every choice of integral values of the variables, not all zero, yields $725^{-1/2} \leqq |L_1L_2L_3L_4|$. These assertions are equivalent to the following theorems. Theorem 1: There exists a homogeneous lattice G with determinant $725^{1/2}$ which contains no point $X = (x_1, x_2, x_3, x_4)$ with $|x_1x_2x_3x_4| < 1$ except for the origin $O = (0, 0, 0, 0)$. Theorem 2: If a homogeneous lattice G with determinant D contains no point X with $|x_1x_2x_3x_4| < 1$ other than the origin O, then $D > 500^{1/2}$. The author proves these theorems by extending the method developed by H. Davenport [J. London Math. Soc. **16** (1941), 98–101; MR **3**, 70] to treat the analogous problem for three forms. After the author obtained his theorems he learned that somewhat weaker results had already been established in the 1964 Vienna dissertation of G. Böhm [see the bibliography of the article under review].

A. L. Whiteman (Los Angeles, Calif.)

Citations: MR 3, 70f = J36-2.

J36-33 **(39# 6827)**

Karanikolov, Hristo

The Minkowski theorems for certain linear homogeneous forms. (Bulgarian. Russian and French summaries)

Godišnik Visš. Tehn. Učebn. Zaved. Mat. **2** (1965), *kn.* 2, 1–6 (1967).

Author's summary: "We sharpen a theorem of Minkowski relating to a collection of homogeneous linear forms. Namely, we prove the theorem: if the real numbers a_{ij} $(i, j = 1, 2, \cdots, n)$ satisfy the condition $\Delta = |\det a_{ij}| > 0$, then the inequality $\prod_{i=1}^{n} |\sum_{j=1}^{n} a_{ij}x_j| \leqq \Delta$ has at least one solution $x_1, \cdots, x_n$ in integers such that the condition $x_1 \cdots x_n \neq 0$ is satisfied."

Referred to in J36-31.

J36-34 **(40# 1341)**

Chowla, S.; De Leon, M. J.

On a conjecture of Littlewood.

Norske Vid. Selsk. Forh. (Trondheim) **41** (1968), 45–47.

Let a, b be arbitrary irrational numbers. A well-known conjecture of Littlewood asserts that given any $e > 0$, there exist arbitrarily large integers z such that the inequality $|z(x - az)(y - bz)| < e$ is solvable in integers x, y. The only case other than the trivial ones where the theorem is known to hold is when a, b belong to the same cubic field [see J. W. S. Cassels and H. P. F. Swinnerton-Dyer, Philos. Trans. Roy. Soc. London Ser. A **248** (1955), 73–96; MR **17**, 14]. The authors observe that when $a = \sqrt{2}$ and $b = \sqrt{3}$, the conjecture is true provided $\liminf \|\frac{1}{4}(\sqrt{6})(\sqrt{2} + 1)^n\| = 0$, where $\|x\|$ is the absolute value of the difference between the real number x and the nearest integer. *A. C. Woods* (Columbus, Ohio)

Citations: MR 17, 14f = J36-24.

J40 PRODUCT OF n INHOMOGENEOUS LINEAR FORMS IN n VARIABLES

See also reviews D02-14, H15-35, H15-74, J32-15, J32-31, J32-41, J36-10, R16-9.

J40-1 **(2, 350d)**

Tschebotaröw, N. Beweis des Minkowski'schen Satzes über lineare inhomogene Formen. Vierteljschr. Naturforsch. Ges. Zürich **85** Beiblatt (Festschrift Rudolf Fueter), 27–30 (1940).

Let $y_k = b_k + \sum_{l=1}^{n} a_{kl}x_l$ $(k = 1, \cdots, n)$ be n nonhomogeneous linear functions with real coefficients, where the determinant $D = |a_{kl}| \neq 0$. An unproved hypothesis due to Minkowski asserts that the inequality $|y_1y_2 \cdots y_n| \leqq 2^{-n}|D|$ has an integral solution $x_1, \cdots, x_n$. The author considers the following simpler problem: To determine a number K_n depending only on n and not on the coefficients b_k and a_{kl}, such that the inequality $|y_1y_2 \cdots y_n| \leqq K_n|D|$ has an integral solution. He proves that the lower bound k_n of admissible numbers K_n is not greater than $2^{-n/2}$. The paper was originally published in Russian [N. G. Tschebotareff, Sci. Notes Kazan Univ. **94**, 14–16 (1934)]. *C. L. Siegel.*

J40-2 **(2, 350e)**

Mordell, L. J. Tschebotareff's theorem on the product of non-homogeneous linear forms. Vierteljschr. Naturforsch. Ges. Zürich **85** Beiblatt (Festschrift Rudolf Fueter), 47–50 (1940).

Proof of the inequality $k_n \leqq 2^{-n/2}\{1+(\sqrt{2}-1)^n\}^{-1}$ which is better than Tschebotareff's result $k_n \leqq 2^{-n/2}$ [cf. the preceding review]. *C. L. Siegel* (Princeton, N. J.).

Referred to in J40-27.

J40-3 **(8, 443h)**

Davenport, H. On a theorem of Tschebotareff. J. London Math. Soc. **21**, 28–34 (1946).

Let $L_1, \cdots, L_n$ be real linear forms in $x_1, \cdots, x_n$ with determinant 1. Then it is known that for a certain ω_n, depending on n only, the inequality $|(L_1+c_1)\cdots(L_n+c_n)| \leqq \omega_n$ has integral solutions $(x_1, \cdots, x_n)$ for all sets of real constants $c_1, \cdots, c_n$. Tschebotareff proved that $\omega_n \leqq (1+\epsilon)2^{-n/2}$ $(\epsilon>0)$, Mordell proved that $\omega_n \leqq \{(\sqrt{2})^n+(2-\sqrt{2})^n\}^{-1}$ and now the author deduces the improvement: $\omega_n \leqq \lambda_n 2^{-n/2}$, where $\lambda_n < 1$ and $\lambda_n \to (2e-1)^{-1}$ (cf. J. London Math. Soc. **24**, 316 (1949)) as $n \to \infty$. [Minkowski's conjecture asserts that $\omega_n = 2^{-n}$ and has been proved for $n=2$ [Minkowski] and $n=3$ [Remak].]

J. F. Koksma.

Referred to in J40-4, J40-11.

J40-4 **(11, 233b)**

Davenport, H. Corrigendum: On a theorem of Tschebotareff. J. London Math. Soc. **24**, 316 (1949).

The paper appeared in the same J. **21**, 28–34 (1946); cf. these Rev. **8**, 443; **10**, 855.

Citations: MR 8, 443h = J40-3.

Referred to in J40-11.

J40-5 **(8, 444a)**

Davenport, H. On the product of three non-homogeneous linear forms. Proc. Cambridge Philos. Soc. **43**, 137–152 (1947).

Let ξ, η, ζ be linear forms in u, v, w with real coefficients and determinant $\Delta \neq 0$. Minkowski's conjecture, proved by Remak [and afterward in a simpler way by Davenport, J. London Math. Soc. **14**, 47–51 (1939)], states that for any real numbers a, b, c there exist lattice points (u, v, w) such that $|(\xi-a)(\eta-b)(\zeta-c)| \leqq \frac{1}{8}|\Delta|$. By a modification of his method, the author proves that, if none of the forms represents zero for any lattice point $(u, v, w) \neq (0, 0, 0)$, there exists a number M, depending only on the coefficients of the forms and satisfying $M<\frac{1}{8}$, such that for any real a, b, c there exist lattice points (u, v, w) for which

$$|(\xi-a)(\eta-b)(\zeta-c)| \leqq M|\Delta|.$$

The author investigates the "true value" of M (i.e., the lower bound $\bar{M}$ of all such numbers M, which turns out to be a number M itself) for two special sets of linear forms (they are also the forms for which the minimum of $|\xi\eta\zeta|/|\Delta|$ has its greatest possible values): (a) $\xi=\theta u+\varphi v+\psi w$, $\eta=\varphi u+\psi v+\theta w$, $\zeta=\psi u+\theta v+\varphi w$, where θ, φ, ψ are the roots of $t^3+t^2-2t-1=0$, the determinant of these forms being 7; the author proves that $\bar{M}=1/49$; (b) $\xi=u+\theta' v+\varphi' w$, $\eta=u+\varphi' v+\psi' w$, $\zeta=u+\psi' v+\theta' w$, where θ', φ', ψ' are the roots of $t^3-3t-1=0$, the determinant of these forms being 9; the author proves $\bar{M}=1/27$. *J. F. Koksma.*

Referred to in J36-30, J40-9, J40-12, J40-41, R16-46.

J40-6 **(9, 413b)**

Chalk, J. H. H. On the positive values of linear forms. Quart. J. Math., Oxford Ser. **18**, 215–227 (1947).

Let $L_1, \cdots, L_n$ be n homogeneous linear forms in n variables $u_1, \cdots, u_n$ with determinant $\Delta \neq 0$. Then, if $c_1, \cdots, c_n$ are any real constants, there exist integral values of $u_1, \cdots, u_n$, such that simultaneously $L_1+c_1>0, \cdots, L_n+c_n>0$, $(L_1+c_1)(L_2+c_2)\cdots(L_n+c_n) \leqq |\Delta|$. (Putting $c_1=\cdots=c_n=0$ one sees that this theorem on inhomogeneous forms includes a nontrivial theorem on homogeneous forms.) The author further proves: unless the ratios of the coefficients in each of the forms $L_1, \cdots, L_n$ are all rational the result of the above theorem is valid with strict inequality. For $n=2$ these results were proved by B. Segre and by K. Mahler [Duke Math. J. **12**, 337–365, 367–371 (1945); these Rev. **6**, 258] in the homogeneous case and by Davenport and Heilbronn in the inhomogeneous case [cf. the preceding review].* *J. F. Koksma* (Amsterdam).

*MR 9, 413A = J32-11.

Citations: MR 6, 258a = J08-1; MR 6, 258b = J08-2; MR 9, 413a = J32-11.

Referred to in H05-29, J32-15, J36-10, J40-14.

J40-7 **(10, 19a)**

Dyson, F. J. On the product of four non-homogeneous linear forms. Ann. of Math. (2) **49**, 82–109 (1948).

A famous conjecture, attributed to Minkowski, asserts: Let $y_i=a_{i1}x_1+\cdots+a_{in}x_n$ $(i=1, \cdots, n)$ be n forms with real coefficients and the determinant $\Delta \neq 0$; let $\eta_1, \cdots, \eta_n$ be real numbers. Then there are n integers $x_1, \cdots, x_n$ such that $|y_1-\eta_1|\cdots|y_n-\eta_n| \leqq 2^{-n}|\Delta|$; the sign of equality is necessary if and only if $y_i=p_iv_i$, $\eta_i=p_i(m_i+\frac{1}{2})$, where $v_1, \cdots, v_n$ are unimodular forms in $x_1, \cdots, x_n$ with integer coefficients, $p_1, \cdots, p_n$ are real numbers whose product is Δ and $m_1, \cdots, m_n$ are integers. This conjecture has been proved for $n=2$ by Minkowski and for $n=3$ by Remak [Math. Z. **17**, 1–34 (1923); **18**, 173–200 (1923)] and later much more simply by Davenport [J. London Math. Soc. **14**, 47–51 (1939)]. An attempt to prove it for $n=4$ has been made by Hofreiter [Monatsh. Math. Physik **40**, 351–392, 393–406 (1933)]. Here a complete proof is given for $n=4$.

The result is an easy consequence of the following two theorems. (A) Let S be a four-dimensional space with Cartesian coordinates $y_1, \cdots, y_4$ and origin O. Let L be a set of points in S with the following properties: (i) O nonεL; (ii) every bounded set contains only a finite number of points of L; (iii) given any one A of the coordinate axes and any $\delta>0$, L contains a point the distance of which from A is less than δ. Then there are numbers $\lambda_i>0$ $(i=1, \cdots, 4)$ such that the ellipsoid $\lambda_1y_1^2+\cdots+\lambda_4y_4^2=1$ has no points of L inside it but has four points of L on its surface not lying in a three-dimensional subspace through O. (B) Let L be a lattice with determinant $\Delta \neq 0$ in four-dimensional space which contains five points O, P_1, P_2, P_3, P_4 not lying in a three-dimensional subspace and such that $OP_1=OP_2=OP_3=OP_4$ and $OP \geqq OP_1$ for every point $P \varepsilon L$ other than O. Then every hypersphere of radius $|\Delta|^{\frac{1}{4}}$ contains a point of L in its interior or on its surface; the supplement "or on its surface" is necessary if and only if L is a rectangular cubic lattice and the hypersphere has its centre at the centre of one of the cells of L. Theorem (A) is a purely geometrical theorem; its proof is based on a topological lemma which has been proved by the author in a separate paper [same Ann. **49**, 75–81 (1948); these Rev. **10**, 55]. The proof of (B) belongs to the geometry of numbers and uses the classical theorem on the minimum of a positive definite quaternary form. The author observes that he has not been able to find any reference to the conjecture in Minkowski's papers. [On p. 108, line 9 from below, read $\theta_{k,l}$ instead of $\varphi_{k,l}$.] *V. Jarník* (Prague).

Referred to in H30-67, H30-72, J40-32.

J40-8 **(10, 284f)**

Hlawka, Edmund. Inhomogene Linearformen in algebraischen Zahlkörpern. Akad. Wiss. Wien, S.-B. IIa. **155**, 63–73 (1947).

The author extends the method of Siegel [see H. Davenport, Acta Arith. **2**, 262–265 (1937)], relating to the product of n nonhomogeneous linear forms, to the case when the homogeneous parts of the linear forms are those arising from an "Ordnung" in an algebraic number-field, and its conjugates. The formulation of the main theorem is not altogether clear to the reviewer, and is incomplete as regards the reference to the determinant of the forms.

H. Davenport (London).

J40-9 (11, 12a)

Prasad, A. V. A nonhomogeneous inequality for integers in a special cubic field. I, II. Nederl. Akad. Wetensch., Proc. **52**, 240–250, 338–350 = Indagationes Math. **11**, 55–65, 112–124 (1949).

Let θ, ϕ, $\bar{\phi}$ be the real and complex roots of the equation $t^3-t-1=0$; then $k(\theta)$ is the cubic field of smallest discriminant. Let a be any real number and let b, $\bar{b}$ be any conjugate complex numbers. The following theorem is proved. There is an integer ξ of $k(\theta)$ with conjugates η, $\bar{\eta}$ in $k(\phi)$, $k(\bar{\phi})$ for which $|(\xi-a)(\eta-b)(\bar{\eta}-\bar{b})|<1/5.0001$, unless a, b, $\bar{b}$ are of the form

$$a=\frac{\sigma}{\theta^2+1}+\xi_1,\quad b=\frac{\tau}{\phi^2+1}+\eta_1,\quad \bar{b}=\frac{\bar{\tau}}{\bar{\phi}^2+1}+\bar{\eta}_1,$$

where ξ_1, η_1, $\bar{\eta}_1$ are conjugate integers and σ, τ, $\bar{\tau}$ are conjugate units of $k(\theta)$, $k(\phi)$, $k(\bar{\phi})$. If a, b, $\bar{b}$ are of this form then there is no integer ξ in $k(\theta)$ with conjugates η, $\bar{\eta}$ for which the above absolute value is smaller than $\frac{1}{5}$, but there are an infinite number for which it is equal to $\frac{1}{5}$. To each triple of conjugates ξ, η, $\bar{\eta}$ there correspond rational integers u, v, w such that

$$\xi=u+v\theta+w\theta^2,\quad \eta=u+v\phi+w\phi^2,\quad \bar{\eta}=u+v\bar{\phi}+w\bar{\phi}^2,$$

and conversely, so that this theorem can also be regarded as a statement about the product of three special complex nonhomogeneous linear forms. From this point of view one is more interested in $M(\xi,\eta,\bar{\eta})$, the lower bound of numbers λ such that there are rational integers u, v, w for which $|(\xi-a)(\eta-b)(\bar{\eta}-\bar{b})|<\lambda|\Delta|$, where Δ is the determinant of the forms. The above theorem shows that for these special forms $M(\xi,\eta,\bar{\eta})=1/5\sqrt{23}$. This paper is closely related to that by Davenport [Proc. Cambridge Philos. Soc. **43**, 137–152 (1947); these Rev. **8**, 444] on products of real forms in cubic fields.

W. J. LeVeque (Ann Arbor, Mich.).

Citations: MR 8, 444a = J40-5.

Referred to in J40-12.

J40-10 (11, 418c)

Schneider, Theodor. Eine Bemerkung zur Minkowskischen Vermutung über inhomogene Linearformen. Arch. Math. **2**, 87–89 (1950).

Let $\xi_1, \cdots, \xi_n$ be n homogeneous linear forms in n variables x_i, with determinant Δ. Let $\eta_1, \cdots, \eta_n$ be n real numbers. Then a conjecture of Minkowski asserts that integral x_i exist such that (1) $\prod_1^n|\xi_i-\eta_i|\leqq 2^{-n}|\Delta|$. The author proves (1) under the following additional hypothesis: there exist positive real t_i with $\prod_1^n t_i=|\Delta|$, such that no integral values of the x_i not all zero satisfy the inequalities $(|\xi_i|/t_i)<1$, $i=1, 2, \cdots, n$, and (2) $\sum_1^n(|\xi_i|/t_i)<\frac{1}{2}n$. The proof is short and elementary. A previous theorem of S. Kowner [Rec. Math. [Mat. Sbornik] (1) **32**, 528–541 (1925)] proved (1) on the stronger hypothesis obtained from the above by omitting condition (2).

F. J. Dyson (Birmingham).

J40-11 (11, 501f)

Chalk, J. H. H. On the product of non-homogeneous linear forms. J. London Math. Soc. **25**, 46–51 (1950).

Let $L_1, \cdots, L_n$ be arbitrary real linear forms in integral variables $u_1, \cdots, u_n$ with determinant 1, and let $c_1, \cdots, c_n$ be arbitrary real numbers. Let γ_n be the lower bound of the numbers γ such that there always exist integers $u_1, \cdots, u_n$ satisfying (*) $|(L_1+c_1)\cdots(L_n+c_n)|\leqq\gamma$. By a result of Davenport, which is a refinement of an earlier result of Tschebotareff, we know that $\gamma_n\leqq\gamma_n'$ where $\gamma_n'<(\sqrt{2})^{-n}$ and $\gamma_n'\sim(2e-1)^{-1}(\sqrt{2})^{-n}$ as $n\to\infty$ [H. Davenport, same J. **21**, 28–34 (1946); **24**, 316 (1949); these Rev. **8**, 443; **11**, 233, and references given there]. The author supposes that $\gamma>\gamma_n'$ and that none of the forms $L_1+c_1, \cdots, L_n+c_n$ assumes the value zero, and he proves that the inequality (*) has an infinity of sets of integral solutions if and only if the forms $L_1, \cdots, L_n$ are not expressible in the form

$$L_\nu=\lambda_\nu(a_{\nu1}u_1+\cdots+a_{\nu n}u_n),\qquad \nu=1, \cdots, n,$$

where the a's are integers and the λ's are real.

C. A. Rogers (Princeton, N. J.).

Citations: MR 8, 443h = J40-3; MR 11, 233b = J40-4.

J40-12 (12, 678d)

Clarke, L. E. On the product of three non-homogeneous linear forms. Proc. Cambridge Philos. Soc. **47**, 260–265 (1951).

Let θ, ϕ, ψ be the roots of t^3-4t+2. Let $\xi=u+v\theta+w\theta^2$, $\eta=u+v\phi+w\phi^2$, $\zeta=u+v\psi+w\psi^2$. Then corresponding to any real numbers a, b, c there exist rational integers u, v, w for which $|(\xi-a)(\eta-b)(\zeta-c)|\leqq\frac{1}{2}$, and for certain values a, b, c, always $|(\xi-a)(\eta-b)(\zeta-c)|\geqq\frac{1}{2}$ for all rational integers u, v, w. This result for the cubic field $R(\theta)$ is analogous to theorems of Davenport [Proc. Cambridge Philos. Soc. **43**, 137–152 (1947); these Rev. **8**, 444] for the cyclic cubic fields of discriminant 49 and 81 and of Prasad [Nederl. Akad. Wetensch., Proc. **52**, 240–250, 338–350 = Indagationes Math. **11**, 55–65, 112–124 (1949); these Rev. **11**, 12] for discriminant -23. The choice here is the totally real non-cyclic field with smallest discriminant, viz. 148. The result implies that $R(\theta)$ has a Euclidean algorithm.

L. Tornheim (Ann Arbor, Mich.).

Citations: MR 8, 444a = J40-5; MR 11, 12a = J40-9.

J40-13 (13, 104g)

Lekkerkerker, C. G. Lecture in the Actualiteiten series on the Minkowski-Hajós theorem. Math. Centrum Amsterdam. Rapport ZW 1951-008, 13 pp. (1951). (Dutch)

An exposition of the theorem of the title, following Rédei's simplification [Acta Univ. Szeged. Sect. Sci. Math. **13**, 21–35 (1949); these Rev. **10**, 683]. See also Rédei [Comment. Math. Helv. **23**, 272–282 (1949); these Rev. **11**, 318].

W. J. LeVeque (Manchester).

Citations: MR 10, 683c = J16-8; MR 11, 318d = J16-9.

J40-14 (13, 726a)

Cole, A. J. On the product of n linear forms. Quart. J. Math., Oxford Ser. (2) **3**, 56–62 (1952).

Let $L_1, L_2, \cdots, L_n$ be n homogeneous linear forms in n variables $u_1, u_2, \cdots, u_n$ with real coefficients and determinant $\Delta\neq0$ and let $c_1, c_2, \cdots, c_n$ be any n real numbers. Then the author proves that there exist integers $u_1, u_2, \cdots, u_n$, such that

(1) $L_1+c_1>0,\ L_2+c_2>0, \cdots, L_{n-1}+c_{n-1}>0,$

(2) $(L_1+c_1)(L_2+c_2)\cdots(L_{n-1}+c_{n-1})|L_n+c_n|\leqq\frac{1}{2}|\Delta|.$

He proves that this result is best possible and he deduces sufficient conditions for $L_1, \cdots, L_n$ such that (1) and (2) have an infinity of integer solutions $(u_1, \cdots, u_n)$. These theorems are analogous to former theorems of Chalk [same J. **18**, 215–227 (1947); **19**, 67–80 (1948); these Rev. **9**, 413; **10**, 18], who considered the inequalities $L_1+c_1>0, \cdots, L_n+c_n>0$ instead of (1), in which case the number $\frac{1}{2}|\Delta|$ in (2) must be replaced by $|\Delta|$.

J. F. Koksma.

Citations: MR 9, 413b = J40-6; MR 10, 18e = J36-10.

J40-15 **(13, 918b)**

Clarke, L. E. Non-homogeneous linear forms associated with algebraic fields. Quart. J. Math., Oxford Ser. (2) **2**, 308–315 (1951).

Let k be an algebraic number field of degree n, with r real and $2s$ complex conjugates, $n=r+2s$. Employing a basis $\omega_1^{(j)}, \cdots, \omega_n^{(j)}$ of the integers of the conjugate $k^{(j)}$ of k, $j=1, \cdots, n$, the author considers n linear forms $L_j=\sum_\nu \omega_\nu^{(j)}x_\nu$ $(j=1, \cdots, n)$, with the agreement that the forms are real for $j=1, \cdots, r$, and conjugate complex for $j=v$ and $j=v+s$, $v=r+1, \cdots, r+s$. The determinant of the n forms is $\sqrt{d}$, where d is the discriminant of k. Let $a_1, \cdots, a_n$ be any n complex numbers of which the first r are real and the others are conjugate complex in pairs, numbered as are the forms. The author's first theorem is that there exists a number μ_n, depending only on n, such that the inequality

$$(1) \qquad |(L_1-a_1)\cdots(L_n-a_n)| \leqq \mu_n |d|^{n/2}$$

is solvable in rational integers $x_1, \cdots, x_n$. If $s=0$, it has been known that the exponent $n/2$ may be replaced by $1/2$. The author gives an example showing that the exponent $n/2$ cannot be improved when $n=2$, $r=0$, $s=1$. He says that it is probable that a smaller exponent will suffice in all other cases. His second theorem states that the exponent $n/2$ may be replaced by $n/2-(n-2)/(n-1)$ when n is an odd prime. Theorem 3 is that for certain cubic fields the exponent $2/3$ suffices, while Theorem 4 is that there are infinitely many cubic fields for which this is best. Theorem 1 is proved by associating a positive definite quadratic form with the set of linear forms, and applying Minkowski's reduction theory to it. Refinements give the second theorem and special arguments give the third and fourth. *R. Hull.*

J40-16 **(13, 918c)**

Davenport, H. Linear forms associated with an algebraic number-field. Quart. J. Math., Oxford Ser. (2) **3**, 32–41 (1952).

The inequality (1) of the preceding review is improved to the extent of replacing the exponent $n/2$ by $n/2(n-s)$. The author employs inequalities due to Minkowski and to Mahler for which he gives proofs. His method is a modification of a method devised by Siegel for the case $s=0$. He discusses its scope in the last section. By a refinement like Clarke's, he obtains the further result that, if $s>1$, and n is a prime, the exponent can be replaced by $n/2(n-s)-(s-1)/(n-1)(n-s)$. By the same method the author also proves: for any ideal $\mathfrak{A}$ in k, any α (integral or not) in k, there exists an integer η in k such that $\eta\equiv\alpha \pmod{\mathfrak{A}}$ and $|N\eta| < C_n(N\mathfrak{A})|d|^{n/2(n-s)}$, where N denotes norm.

R. Hull (Lafayette, Ind.).

Referred to in J40-19, J40-25.

J40-17 **(13, 918d)**

Varnavides, P. On the product of three linear forms. Proc. London Math. Soc. (3) **2**, 234–244 (1952).

The author proves the following theorem: Let L_1, L_2, L_3 be linear forms in u_1, u_2, u_3 with determinant 1 and let c_2, c_3 be any given real numbers. Then for any $\epsilon>0$ we can satisfy

$$|L_1(L_2-c_2)(L_3-c_3)| < 1/9.1, \quad |L_1| < \epsilon,$$

in integers u_1, u_2, u_3 not all zero, except when c_2, c_3 are both zero and L_1, L_2, L_3 are multiples of P, Q, R (see below) in some order. In these cases, the product $|L_1L_2L_3|$ has the minimum value 1/7 or 1/9. Here P, Q, R are defined to be $u_1+\theta_iu_2+\theta_i^2u_3$, $i=1, 2, 3$ where θ_i are the roots of the equation $t^3+t^2-2t-1=0$ or of $t^3-3t-1=0$. The constant 9.1 is not the best possible constant but the excluded cases are essential. This result differs in character from that of Mordell for two forms L_1, L_2-c_2 [J. London Math. Soc. **26**. 93–95 (1951); these Rev. **13**, 16] in that the constant is not the same as for the homogeneous approximation.

B. W. Jones (Boulder, Colo.).

Citations: MR 13, 16b = J32-27.

J40-18 **(14, 358d)**

Cassels, J. W. S. The product of n inhomogeneous linear forms in n variables. J. London Math. Soc. **27**, 485–492 (1952).

Let $L_1, \cdots, L_n$ be n real linear forms in $\xi=(x_1, \cdots, x_n)$ of determinant $\Delta\neq0$. It has been conjectured that the minimum of $|L_1\cdots L_n|$ for $\xi\equiv\xi_0 \pmod 1$ is always at most $2^{-n}|\Delta|$; this has been proved for $n\leqq4$. Equality is required for the forms $L_j^*=p_jx_j$, with $p_j\neq0$, and $\xi_0\equiv(\frac12, \cdots, \frac12) \pmod 1$. The author proves that there are sets of forms $L_1^{(k)}, \cdots, L_n^{(k)}$, inequivalent to $L_1^*, \cdots, L_n^*$, for which the minimum of $|L_1^{(k)}\cdots L_n^{(k)}|$ for $\xi\equiv(\frac12, \cdots, \frac12) \pmod 1$ has $2^{-n}|\Delta|$ as a limit as $k\to\infty$. The result, but not the proof, extends that for $n=2$ by Davenport [Nederl. Akad. Wetensch., Proc. **49**, 815–821 = Indagationes Math. **8**, 518–524 (1946); these Rev. **8**, 444]. *L. Tornheim.*

Citations: MR 8, 444b = J32-4.

J40-19 **(15, 105f)**

Barnes, E. S. Note on non-homogeneous linear forms. Proc. Cambridge Philos. Soc. **49**, 360–362 (1953).

Let θ be an algebraic number of degree n, with conjugates $\theta^{(1)}, \cdots, \theta^{(n)}$ of which the first r are real and the last $2s$ are arranged in complex conjugate pairs $\theta^{(r+j)}$, $\theta^{(r+s+j)}$ $(j=1, \cdots, s)$. Define the linear forms

$$L_i(u_1, \cdots, u_n)=u_1\omega_1^{(i)}+\cdots+u_n\omega_n^{(i)} \quad (i=1, \cdots, n),$$

where $\omega_1, \cdots, \omega_n$ is a basis for the integers of $k(\theta)$. Let $c_1, \cdots, c_r$ be real, c_{r+j} and c_{r+s+j} complex conjugates $(j=1, \cdots, s)$, and call M_c the lower bound of $|(L_1-c_1)\cdots(L_n-c_n)|$ for integral $u_1, \cdots, u_n$. Davenport [Quart. J. Math., Oxford Ser. (2) **3**, 32–41 (1952); these Rev. **13**, 918] proved for any set of numbers $c_1, \cdots, c_n$ that $M_c<C_n|d|^{n/2(n-s)}$, where C_n depends only on n, and d is the discriminant of $k(\theta)$. In the present paper, the author uses a method of Tschebotareff to give a very simple proof of Davenport's result.

I. Reiner (Urbana, Ill.).

Citations: MR 13, 918c = J40-16.

J40-20 **(15, 106a)**

Davenport, H. On the product of n linear forms. Proc. Cambridge Philos. Soc. **49**, 190–193 (1953).

The author uses Barnes' methods [see the preceding review] to improve the above-stated result to

$$M_c<C_n|d|^{(2n-2)/(2n-2s-1)}$$

for $s>0$, which gives a better estimate except when $r=0$.

I. Reiner (Urbana, Ill.).

J40-21 **(15, 857b)**

Rogers, C. A. The product of n non-homogeneous linear forms. Proc. London Math. Soc. (3) **4**, 50–83 (1954).

Let x_i be n real linear forms in the n variables u_j and with determinant Δ, and let b_i be n real constants. Consider the inequalities

$$(1) \qquad |(x_1+b_1)\cdots(x_n+b_n)| < 1,$$

$$(2) \qquad (x_1+b_1)^2+\cdots+(x_n+b_n)^2<\epsilon^2,$$

$$(3) \qquad x_1+b_1>0, \cdots, x_n+b_n>0,$$

$$(4) \qquad (x_1+b_1)\cdots(x_{n-1}+b_{n-1})<\epsilon^{n-1}.$$

Theorems, usually giving necessary and sufficient conditions, are proved for the existence of an infinity of solutions such that (1), (2) hold; or (1); or (1), (4); or (1), (2), (3). The proofs depend on slightly more general geometric theorems. The conditions used are natural ones. One of the basic conditions imposed is that the number $\lambda_1 b_1 + \cdots + \lambda_r b_r$ is an integer for every set of real numbers $\lambda_1, \cdots, \lambda_r$ such that the form $\lambda_1 x_1 + \cdots + \lambda_r x_r$ has integral coefficients for the u_j. A geometric equivalent is shown to be that there is a point of a grid (non-homogeneous lattice) Γ in the smallest linear manifold M which is generated by points of the lattice Λ of Γ and which contains the points satisfying $x_1 = \cdots = x_r = 0$. *L. Tornheim* (Ann Arbor, Mich.).

J40-22 **(16, 340d)**

Samet, P. A. The product of non-homogeneous linear forms. I. Proc. Cambridge Philos. Soc. **50**, 372–379 (1954).

Let θ, ϕ, ψ be the three (real) roots of the cubic equation $t^3 - 4t + 2 = 0$ and let ξ, η, ζ be the linear forms $x + y\theta + z\theta^2$, $x + y\phi + z\phi^2$, $x + y\psi + z\psi^2$, respectively. For a given point $P(x_0, y_0, z_0)$, $M(P)$ is defined to be the minimum of $|\xi\eta\zeta|$ for all x, y, z for which $x \equiv x_0$, $y \equiv y_0$, $z \equiv z_0$ (mod 1). The author shows that $M(P) \leqq \frac{1}{2}$ which is the best possible refinement of the result obtained by applying Minkowski's theorem for inhomogeneous forms. It is shown that $M(P) = \frac{1}{2}$ only if $P \equiv (0, 0, \frac{1}{2})$ (mod 1). The principal result of the paper is that if $P \not\equiv (0, 0, \frac{1}{2})$ (mod 1), then $M(P) \leqq \frac{1}{4}$. This second minimum is attained only if either $P \equiv (0, \frac{1}{2}, 0)$ or $P \equiv (0, \frac{1}{2}, \frac{1}{2})$ (mod 1). *D. Derry* (Vancouver, B. C.).

Referred to in R16-46.

J40-23 **(16, 340e)**

Samet, P. A. The product of non-homogeneous linear forms. II. The minimum of a class of non-homogeneous linear forms. Proc. Cambridge Philos. Soc. **50**, 380–390 (1954).

Let θ, ϕ, ψ be the three (real) roots of the equation $t^3 + at^2 - 2t - a = 0$, where a is a large positive integer. Let ξ, η, ζ denote the linear forms $x + y\theta + z\theta^2$, $x + y\phi + z\phi^2$, $x + y\psi + z\psi^2$ respectively. For a given point $P(x_0, y_0, z_0)$, $M(P)$ is defined to be the minimum of $|\xi\eta\zeta|$ for all x, y, z for which $x \equiv x_0$, $y \equiv y_0$, $z \equiv z_0$ (mod 1). If $a \equiv 2$ (mod 4) it is shown that $M(P) \leqq a^2/8$ and that $M(P) = a^2/8$ implies $P \equiv (0, 0, \frac{1}{2})$ (mod 1). In the case for which $a \equiv 1$ or 3 mod 4 then $M(P) \leqq (a^2/8)(1 - a^{-1})^3$ and $M(P) = (a^2/8)(1 - a^{-1})^3$ implies $P \equiv \pm(a^{-1}, 0, \frac{1}{2} - \frac{1}{2}a^{-1})$ (mod 1). *D. Derry.*

J40-24 **(16, 451a)**

Swinnerton-Dyer, H. P. F. The inhomogeneous minima of complex cubic norm forms. Proc. Cambridge Philos. Soc. **50**, 209–219 (1954).

Let K_1, K_2, K_3 be conjugate cubic algebraic number fields (K_1 real, K_2, K_3 complex conjugate). Let $\omega_{11}, \omega_{12}, \omega_{13}$ be a basis for the integers of K_1, $\omega_{i1}, \omega_{i2}, \omega_{i3}$ the conjugate basis for the integers of K_i and $\xi_i = \omega_{i1}x_1 + \omega_{i2}x_2 + \omega_{i3}x_3$. Let $M(K; x_1', x_2', x_3') = \min |\xi_1\xi_2\xi_3|$ taken over all points

$$(x_1, x_2, x_3) \equiv (x_1', x_2', x_3') \pmod 1$$

and $M(K) = \max M(K; x_1', x_2', x_3')$ taken over all points (x_1', x_2', x_3') in the unit cube $0 \leqq x_j' < 1$. Let d be the discriminant of K_1, K_2 and K_3. The author shows that if $|d| < 1237$ then $M(K) < |d|^{2/3}/16\sqrt[3]{2}$, and that this result is best possible, in the sense that neither the exponent $\frac{2}{3}$ nor the numerical constant can be improved.
H. S. A. Potter (Aberdeen).

J40-25 **(17, 829e)**

Davenport, H.; and Swinnerton-Dyer, H. P. F. Products of inhomogeneous linear forms. Proc. London Math. Soc. (3) **5** (1955), 474–499.

Suppose $L_k = \sum_{j=1}^n a_{kj}x_j$, where $\Delta = \det(a_{kj}) \neq 0$, a_{kj} is real if $k = 1, \cdots, r$ and $j = 1, \cdots, n$, the remaining coefficients are complex and $a_{k+s,j} = \bar{a}_{kj}$ if $k = r+1, \cdots, r+s$ and $j = 1, \cdots, n$. Let $r + 2s = n$, $P(x_1, \cdots, x_n) = L_1L_2 \cdots L_n$, $M_H = \min|P(x_1, \cdots, x_n)|$ taken over integers $x_1, \cdots, x_n$, not all zero, $M_I(x') = \min|P(x_1 + x_1', \cdots, x_n + x_n')|$ taken over all integers $x_1, \cdots, x_n$, where $x_1', \cdots, x_n'$ are real numbers, and $M_I = \max M_I(x')$ taken over all sets of real numbers $x_1', \cdots, x_n'$.

If $r > 0$ and $s > 1$, the authors prove that the inequality

(1) $$M_H^\lambda M_I^{1-\lambda} < C|\Delta|$$

holds with

$$\lambda = \max\left(\frac{s-1}{n-1}, \frac{s}{2n-2s}\right).$$

This is an improvement on earlier results; for instance, Davenport [Quart. J. Math. Oxford Ser. (2) **3** (1952), 32–41; MR **13**, 918] first proved that (1) holds with $\lambda = s/n$. More refined results are also proved in the cases when $s = 2$ and $r = 0$ or 1. *H. S. A. Potter* (Aberdeen).

Citations: MR 13, 918c = J40-16.

J40-26 **(18, 22a)**

Birch, B. J.; and Swinnerton-Dyer, H. P. F. On the inhomogeneous minimum of the product of n linear forms. Mathematika **3** (1956), 25–39.

The authors develop new techniques to attack Minkowski's conjecture that the inhomogeneous minimum of the product of n linear forms L_j in n variables x_j of determinant $\Delta \neq 0$ is at most $2^{-n}|\Delta|$, with equality only when the set of forms is equivalent to $L_j = \lambda_j x_j$ for constants λ_j under an integral unimodular transformation. They first show that the conjecture is certainly true for all sets of forms sufficiently near to $L_j = \lambda_j x_j$ in the sense of Mahler's topology for lattices. Secondly, they show that the conjecture is generally true for n forms if it is true for all smaller numbers of forms and also for the very special sets of n forms whose product has an attained strictly positive homogeneous minimum. Finally they give a verification of the conjecture for $n = 3$ independent of the theory of quadratic forms used by earlier workers [R. Remak, Math. Z. **17** (1923), 1–34; **18** (1923), 173–200; H. Davenport, J. London Math. Soc. **14** (1939), 47–51].
J. W. S. Cassels (Cambridge, England).

J40-27 **(22# 1564)**

Mordell, L. J. Tschebotareff's theorem on the product of non-homogeneous linear forms. II. J. London Math. Soc. **35** (1960), 91–97.

[For part I see Vierteljschr. Naturf. Ges. Zürich **85** Beiblatt (1940), 47–50; MR **2**, 350.] Let $L_1, L_2, \cdots, L_n$ be n linear forms in the n variables $(u) = (u_1, u_2, \cdots, u_n)$ with real coefficients a_{rs} such that $\Delta = |\det a_{rs}| > 0$. Let $c = (c_1, c_2, \cdots, c_n)$ be any real numbers, let M denote the lower bound for all sets of integers (u) of $\prod_{r=1}^n |L_r + c_r|$, and let T_n be $2^{n/2}M/\Delta$. Minkowski conjectured that $T_n \leqq 2^{-n/2}$ and this has been proved for $n = 2, 3, 4$. Tschebotareff proved that $T_n \leqq 1$ and the author and A. C. Woods improved this result. Davenport showed that $T_n < \gamma_n^{-1}$, where $\gamma_n > 1$ and $\gamma_n \to 2e - 1$ as $n \to \infty$. In this paper the author proves

$$T_n \leqq [4 - 2(2 - (3/4) \cdot 2^{1/2})^n - 2^{-n/2}]^{-1}.$$

This improves Woods' result but is not as good as Davenport's for large n. *B. W. Jones* (Boulder, Colo.)

Citations: MR 2, 350e = J40-2.

Referred to in D02-14.

J40-28 **(24# A1259)**

Rieger, G. J.

Einige Bemerkungen über inhomogene Linearformen.

J. Reine Angew. Math. **203** (1960), 126–129.

Let $L_1, \cdots, L_n$ be homogeneous linear forms in real

variables $x_1, \cdots, x_n$ with determinant $D \neq 0$, where $L_1, \cdots, L_r$ have real coefficients and if $r<n$, $L_{r+s+j}=\overline{L_{r+j}}$ $(j=1, \cdots, s)$, $r+2s=n$. Let $\alpha_1, \cdots, \alpha_n$ be arbitrary numbers of the same type as the forms, i.e., $\alpha_1, \cdots, \alpha_r$ real and $\alpha_{r+s+j}=\overline{\alpha_{r+j}}$ $(j=1, \cdots, s)$. If $\mathbf{x}=(x_1, \cdots, x_n)$ is integral and $\boldsymbol{\alpha}=(\alpha_1, \cdots, \alpha_n)$, put $M_H=\inf_{\mathbf{x}\neq\mathbf{0}}|L_1\cdots L_n|$, $M_I(\boldsymbol{\alpha})=\inf_{\mathbf{x}}|(L_1+\alpha_1)\cdots(L_n+\alpha_n)|$, $M_P(\boldsymbol{\alpha})=\inf_{\mathbf{x}}|(L_1+\alpha_1)\cdots(L_n+\alpha_n)|$ where $L_k+\alpha_k>0$ $(k=1, \cdots, r)$, $M_S(\boldsymbol{\alpha})=\inf_{\mathbf{x}}|(L_1+\alpha_1)\cdots(L_n+\alpha_n)|$ where $R(L_k+\alpha_k)>0$ $(k=1, \cdots, r+s)$. Then, denoting by M_I, M_P, M_S the upper bounds of $M_I(\boldsymbol{\alpha})$, $M_P(\boldsymbol{\alpha})$, $M_S(\boldsymbol{\alpha})$ over all such $\boldsymbol{\alpha}$, respectively, we have $M_I \leqq M_P \leqq M_S$. Various unequalities connecting M_I, and M_P, with M_H and D are known; here, corresponding results are given for M_S. Thus, using Siegel's method, it follows that

$$\text{(i)}\quad M_S < c(n)|D|\left(\frac{|D|}{M_H}\right)^{s/(n-s)} \quad \text{if } M_H \neq 0,$$

$$\text{(ii)}\quad M_S < c'(n)|D|\left(\frac{|D|}{M_H}\right)^{(2s-1)/(2n-1-2s)} \quad \text{if } M_H \neq 0 \text{ and } s>0,$$

for suitable constants $c(n)$, $c'(n)$.

J. H. H. Chalk (Toronto)

J40-29 (24# A1886)

Bombieri, Enrico

Su un teorema di A. C. Woods sul prodotto di n forme lineari reali non omogenee. (English summary)

Boll. Un. Mat. Ital. (3) **16** (1961), 288–294.

Let M be the lattice formed by n linear forms $\xi_1, \xi_2, \cdots, \xi_n$ (with real coefficients) in $x_1, \cdots, x_2$; let D be the determinant of this lattice. Let S be the region $|x_i|<\lambda_i$, $1 \leqq i \leqq n$, and suppose that S contains no points of M (except the origin). Then set $\gamma=\sup_S(\lambda_1\lambda_2\cdots\lambda_n)/D$. A classical theorem of Minkowski enables us to conclude that $0 \leqq \gamma \leqq 1$. The author proves that, if $\gamma>\frac{1}{2}$, and if we set $\mu=2(2\gamma^{1/n}-(2\gamma-1)^{1/n})$, then, for any set of real numbers $\rho_1, \cdots, \rho_n$ (and any positive ε) there exists a point with integral coordinates so that $\prod_{i=1}^n|\xi_i-\rho_i| \leqq D/\mu^n+\varepsilon$. If $\gamma=1$, we get $\mu=2$, and therefore the author's result proves, for that case, the well-known conjecture usually attributed to Minkowski.

H. W. Brinkmann (Swarthmore, Pa.)

J40-30 (24# A1986)

Woods, A. C.

On a theorem of Tschebotareff.

Duke Math. J. **25** (1958), 631–637.

Es seien $L_1, L_2, \cdots, L_n$ reelle Linearformen in den Variablen $u_1, u_2, \cdots, u_n$ mit der Determinante 1. Ferner seien $c_1, c_2, \cdots, c_n$ beliebig gegebene reelle Zahlen. Minkowski vermutete, daß $M=\inf\prod_{i=1}^n|L_i{}^r+c_i| \leqq 2^{-n}$ ist, wobei die Variablen alle ganzen Zahlen durchlaufen. Tschebotareff zeigte: $M<2^{-n/2}$, was von Mordell und Davenport verbessert wurde. Es seien $(x_1, x_2, \cdots, x_n)$ kartesische Koordinaten eines Punktes X, K_1 und K_2 seien die Punktmengen: $\prod_{i=1}^n|x_i+1| \leqq 1$ bzw. $\prod_{i=1}^n|x_i-1| \leqq 1$ und es sei $K=K_1 \cup K_2$. Verfasser zeigt, daß die Vermutung von Minkowski mit der Vermutung äquivalent ist, daß die kritische Determinante $\Delta(K)=2^n$ ist. Dann beweist er, daß $M \leqq 2^{-n/2}(2-(2-\sqrt{2})^n)^{-1}$ ist.

N. Hofreiter (Zbl **84**, 46)

J40-31 (26# 2402)

Macbeath, A. M.

Factorization of matrices and Minkowski's conjecture.

Proc. Glasgow Math. Assoc. **5**, 86–89 (1961).

The author restates Minkowski's conjecture on the inhomogeneous minimum of the product of n linear forms as follows. For $\mathbf{y}=(y_1, \cdots, y_n)$, write $\prod(\mathbf{y})=|y_1y_2\cdots y_n|$; call M a Minkowski matrix if for any n-vector $\mathbf{a}$ there is an integer n-vector $\mathbf{x}$ such that $\prod(M\mathbf{x}-\mathbf{a}) \leqq 2^{-n}|\det M|$. Minkowski conjectured that every non-singular $n\times n$ matrix is a Minkowski matrix, and this has been proved for $n \leqq 4$. Call U unimodular if all the elements of both U and U^{-1} are integers, and call T unit-triangular if $t_{ii}=1$ and $t_{ij}=0$ for $i>j$. A matrix is a DOTU-matrix if it is a product of a diagonal, an orthogonal, a unit-triangular, and a unimodular matrix in that order. The author shows that every DOTU-matrix is a Minkowski matrix, and deduces that every rational non-singular matrix has a neighbourhood consisting entirely of Minkowski matrices.

B. J. Birch (Manchester)

J40-32 (27# 98)

Karanikolov, Hr. [Karanicoloff. Chr.]

On a theorem of Minkowski and Čebotarev. (Russian)

Uspehi Mat. Nauk **18** (1963), *no.* 3 (111), 163–166.

A well-known conjecture of Minkowski states that if $L_i(x_1, \cdots, x_n)$ are n real linear forms in n variables with determinant $\Delta \neq 0$ then for every real set of numbers c_i there are integer values of the variables such that

$$\prod_i |L_i(x_1, \cdots, x_n)-c_i| \leqq 2^{-n}|\Delta|.$$

This is known for $n \leqq 4$ [Dyson, Ann. of Math. (2) **49** (1948), 82–109; MR **10**, 19] and is trivial for all n for the special case when the forms have rational coefficients. In this note it is shown that if the conjecture holds for all sets of r forms in r variables than it also holds for n forms in n variables provided that $n-r$ of them have rational coefficients.

J. W. S. Cassels (Cambridge, England)

Citations: MR 10, 19a = J40-7.

J40-33 (28# 1178)

Bombieri, Enrico

Alcune osservazioni sul prodotto di n forme lineari reali non omogenee. (English summary)

Ann. Math. Pura Appl. (4) **61** (1963), 279–285.

Denote by $L_1, \cdots, L_n$ n real linear forms in n variables of determinant $D>0$. Further, let $c=(c_1, \cdots, c_n)$ be a point of R_n. Put $\operatorname{Inf}\prod_{1\leqq i\leqq n}|L_i+c_i|=m(c)D$, where the lower bound is extended over all integral values of the variables. It is a classical conjecture of Minkowski that $m(c) \leqq 2^{-n}$ for all c, but so far this has been established only for $n \leqq 4$. The author proves that for $n \geqq 5$ there exists an integer h in the interval $1 \leqq h \leqq (\sqrt{2})^{n+1}$ depending, perhaps, on the forms and c such that $m(hc) \leqq 2^{-n}$.

A. C. Woods (Columbus, Ohio)

Referred to in J40-37.

J40-34 (28# 3005)

Bombieri, E.

Sul teorema di Tschebotarev. (English summary)

Acta Arith. **8** (1962/63), 273–281.

Let Λ be an n-dimensional lattice with determinant $d(\Lambda)\neq 0$. Write $G(\mathbf{x})=|x_1\cdots x_n|$, and define $\mu(\Lambda)=\sup_{x_0}\inf_{x\equiv x_0(\Lambda)}G(\mathbf{x})$ and $\mathscr{D}_1=\sup_\Lambda \mu(\Lambda)/d(\Lambda)$. Minkowski's conjecture on the inhomogeneous minimum of the product of n linear forms asserts that $\mathscr{D}_1=2^{-n}$. Tchebotarev proved that $\mathscr{D}_1 \leqq 2^{-n/2}$, and this has been improved by various authors. The author improves the previous record by a factor of about 3 by showing that for large enough n, $\mathscr{D}_1 \leqq \gamma_n 2^{-n/2}$ with $\gamma_n<\frac{1}{3}(2e-1)^{-1}$; his main new idea is an improvement of a theorem of Siegel's (see the author [Boll. Un. Mat. Ital. (3) **17** (1962), 283–288; MR **26** #4972]).

B. J. Birch (Manchester)

Citations: MR 26# 4972 = H15-74.

Referred to in J40-36.

J40-35 **(33# 4014)**

Karanikolov, Chr. [Karanikolov, Hr.]

Historiques sur une hypothèse de H. Minkowski.

Mat. Vesnik **2** (**17**) (1965), 201–203.

From the author's introduction: "The conjecture of Minkowski in question is that if $L_k=\sum_{i=1}^n a_{ki}x_i+b_k$ ($k=1, 2, \cdots, n\geqq 2$) are given forms with a_{ki} and b_{ki} real numbers, subject only to the condition $\Delta=|\det a_{ki}|>0$ ($k, i=1, \cdots, n$), then the inequality $\prod_{k=1}^n |L_k|\leqq\Delta/2^n$ has at least one solution in integers $[x_1, x_2, \cdots, x_n]$."

J40-36 **(35# 1556)**

Gruber, Peter

Eine Erweiterung des Blichfeldtschen Satzes mit einer Anwendung auf inhomogene Linearformen.

Monatsh. Math. **71** (1967), 143–147.

Let Λ be an n-dimensional lattice with determinant $d(\Lambda)\neq 0$. Write $G(x)=|x_1x_2\cdots x_n|$ and define $\mu(\Lambda)=\sup_{x_0}\inf_{x\equiv x_0\,(\Lambda)} G(x)$. It was proved by E. Bombieri [Acta Arith. **8** (1962/63), 273–281; MR **28** #3005] that for large enough n, $\mu(\Lambda)\leqq d(\Lambda)(\beta_n 2^{n/2})^{-1}$ with $\beta_n>(3+10^{-4})(2e-1)$. The present author gives a simple proof of the slightly weaker result that $\mu(\Lambda)\leqq d(\Lambda)(\gamma_n 2^{n/2})^{-1}$ with $\gamma_n\to 3(2e-1)$ $(n\to\infty)$. His main tool is a theorem on measurable sets in R^n which is a generalization of Blichfeldt's theorem; this theorem replaces a similar auxiliary result of Bombieri and is proved in an elementary way.

C. G. Lekkerkerker (Amsterdam)

Citations: MR 28# 3005 = J40-34.

J40-37 **(36# 3729)**

Gruber, Peter

Über das Produkt inhomogener Linearformen.

Acta Arith. **13** (1967/68), 9–27.

Let $L_i=\sum_{j=1}^n a_{ij}u_j$, $1\leqq i\leqq n$, be n real homogeneous linear forms having determinant with absolute value $d>0$. Let $c_1, c_2, \cdots, c_n$ be n arbitrary real numbers. Minkowski conjectured that for suitable integral values of the variables u_j, one has $\prod_{i=1}^n |L_i+c_i|\leqq d2^{-n}$. Up to now, Minkowski's conjecture has been proved only for $n\leqq 4$.

The principal theorem of the present paper is the following sharpening of certain results of E. Bombieri on products of linear forms [Atti Accad. Naz. Lincei Rend. Cl. Sci. Fis. Mat. Natur. (8) **33** (1962), 45–48; MR **26** #3646; Ann. Mat. Pura Appl. (4) **61** (1963), 279–285; MR **28** #1178]. If $0<c\leqq 2^{-n/2}d$ one may select $h\leqq 2^{(1-n)/2}c^{-1}d$ so that $\prod_{i=1}^n |L_i+hc_i|<c$ for suitable integer values of the variables. If $c<n(n!)^{-2}2^{-n(n+3)/2}d$, this inequality may be obtained with $h\leqq 2^{1-n}c^{-1}d$.

A measure can be introduced in the space of systems of n linear forms in n variables in a natural way. The author shows that if Minkowski's conjecture is true in dimension n, it is true almost everywhere in dimension $n+1$. He also shows that the constant $\frac{1}{4}$ which occurs when $n=2$ may be replaced by $\frac{1}{8}$ almost everywhere.

J. B. Kelly (Tempe, Ariz.)

Citations: MR 26# 3646 = H15-75; MR 28# 1178 = J40-33.

Referred to in H30-86.

J40-38 **(38# 2086)**

Narzullaev, H. N.

On Minkowski's problem relative to a system of inhomogeneous linear forms. (Russian)

Dokl. Akad. Nauk SSSR **180** (1968), 1298–1299.

This is an announcement of two theorems bearing on the problem mentioned in the title. Let K, φ, Δ, E be real, unimodular $n\times n$ matrices such that K is diagonal, φ is orthogonal, $\Delta=[\varepsilon_{ij}]$ is triangular and E has integer entries. Further, let x and α be vectors and suppose x has integer components. Set $y=K\varphi\Delta Ex+\alpha$, $L=|y_1y_2\cdots y_n|$. Results: (1) $\min_x L\leqq(\sum_{i=1}^n \varepsilon_{ii}^2/n)^{n/2}/2^n$. (If $\varepsilon_{ii}=1$, $i=1, 2, \cdots, n$, this is Minkowski's result.) (2) If A is a unimodular $n\times n$ matrix and $n=2$ or 3, then there are K, φ, Δ, E of the type described above, with $\varepsilon_{ii}=1$, for which $A=K\varphi\Delta E$. {In the text this last equality sign is misprinted as $\neq$.} The proof of (2) is sketched; it is stated that the proof of (1) does not rest on Minkowski's theorem on central convex bodies in the geometry of numbers.

{This article has appeared in English translation [Soviet Math. Dokl. **9** (1968), 766–767].}

W. J. Firey (Corvallis, Ore.)

J40-39 **(39# 1478)**

Narzullaev, H. N.

On Minkowski's problem concerning a system of linear inhomogeneous forms. (Russian)

Mat. Zametki **5** (1969), 107–116.

A real unimodular n by n matrix A can be represented in the form $A=K\varphi\Delta E$, where K, φ, Δ, E are also unimodular and K is diagonal, φ is orthogonal, Δ is lower triangular, E is integral. If T is an arbitrary column vector and X an integral column vector, let $Y=AX+T$ and $L(A, T, X)=|y_1y_2\cdots y_n|$, where y_i are the entries of Y. Theorem 1 asserts that

$$\min_X L(A, T, X)\leqq(n^{-1}\textstyle\sum_{i=1}^n \varepsilon_{ii}^2)^{n/2}2^{-n},$$

where $\varepsilon_{11}, \cdots, \varepsilon_{nn}$ are the diagonal entries of Δ. The conjecture of Minkowski about nonhomogeneous linear forms would hold if one could choose the matrices K, φ, Δ, E so that $\varepsilon_{11}=\cdots=\varepsilon_{nn}=1$. The author proves that this is possible if $n=2$ or 3 by studying the underlying system of equations.

D. Ž. Djoković (Waterloo, Ont.)

J40-40 **(40# 5547)**

Karanikolov, Hristo

The Minkowski conjecture concerning the product of several linear inhomogeneous forms. (Bulgarian. Russian and English summaries)

Godišnik Visš. Tehn. Učebn. Zaved. Mat. **3** (1966), *kn.* 1, 21–28 (1967); *errata, ibid.* **3** (1966), *kn.* 1, 132 (1967).

In the present paper the author proves the following theorem: Suppose one is given the following nonhomogeneous linear forms $L_k=\sum a_{ki}x_i+b_k$, $k=1, 2, \cdots, n$, where a_{ki}, b_k are real numbers and $\det a_{ki}\neq 0$. Then there exist nonnegative numbers $A_1, \cdots, A_n$, all less than or equal to $|\det a_{ki}|$, such that for $\varepsilon\geqq 0$, the inequalities $L_k\leqq 1/2(A_k+\varepsilon)$ ($k=1, 2, \cdots, n$) possess at least one solution in integers $x_1, x_2, \cdots, x_n$.

R. Finkelstein (Bowling Green, Ohio)

J40-41 **(42# 7598)**

Gruber, Peter

Über einen Satz von Remak in der Geometrie der Zahlen.

J. Reine Angew. Math. **245** (1970), 107–118.

Let $L_1, \cdots, L_n$ be n real, homogeneous linear forms in $n\geqq 2$ real variables, of determinant 1. A famous conjecture in the geometry of numbers states that, for arbitrary real numbers $c_1, \cdots, c_n$, there exist integral values of the variables such that $\prod|L_i+c_i|\leqq 2^{-n}$. This conjecture is known to be true if $n=2$ or 3 and also if the matrix of coefficients A has the form $DOTU$, where D is a diagonal matrix, O an orthogonal matrix, T an upper triangular matrix and U an integral unimodular matrix. The present author shows that if $n=3$ or if $n\geqq 4$ and A has the form $DOTU$, the upper bound 2^{-n} may be replaced by $2^{-n}(1-\alpha(n)\lambda^{2/(n-1)})$, where $\alpha(n)$ is some positive constant depending on n and where λ is the homogeneous minimum of the product $\prod |L_i|$. The proof, which can be interpreted geometrically, involves some numerical work and requires the distinction of several cases. In the case $n=3$, use has to be made of an isolation theorem of H. Davenport [Proc. Cambridge Philos. Soc. **43** (1947), 137–152; MR **8**, 444]. *C. G. Lekkerkerker* (Amsterdam)

Citations: MR 8, 444a = J40-5.

J40-42 **(42# 7599)**

Gruber, P.

Über einige Resultate in der Geometrie der Zahlen. *Number Theory (Colloq., János Bolyai Math. Soc., Debrecen,* 1968), *pp.* 105–110. *North-Holland, Amsterdam,* 1970.

In this article several theorems concerning homogeneous and inhomogeneous linear forms, particularly the product of such forms, are reviewed. Some results of the author are included, but no proofs are given. *H. Groemer* (Tucson, Ariz.)

J40-43 **(43# 156)**

Aggarwal, Satish K.

Inhomogeneous minima of binary quadratic, ternary cubic and quaternary quartic forms in fields of formal power series. *Proc. Nat. Inst. Sci. India Part A* **35** (1969), 684–702.

The author proves analogues for power series fields over finite fields of theorems of H. Davenport [Proc. London Math. Soc. (2) **53** (1951), 65–82; MR **13**, 15; Acta Math. **84** (1950), 159–179; MR **12**, 594; Trans. Amer. Math. Soc. **68** (1950), 508–532; MR **12**, 594], but using the techniques of the reviewer [Proc. Cambridge Philos. Soc. **48** (1952), 72–86; MR **13**, 919; ibid. **48** (1952), 519–520; MR **13**, 919]. J. V. Armitage [Proc. London Math. Soc. (3) **7** (1957), 498–509; MR **20** #30] has proved analogues of Davenport's results in the binary and ternary cases using Davenport's techniques: but the author's results in these cases neither contain nor are contained in those of Armitage.

More precisely, let t be an indeterminate over the finite field K. Denote by $K\{t\}$ the completion of $K(t)$ with respect to the valuation which is trivial on K and for which $|t|=e>1$. Let $m\in K$ be such that $-m$ is not a square in K. For $n=2, 3, 4$ let $L_j(X)$ $(1\leqq j\leqq n)$ be linear forms in $X=(X_1, \cdots, X_n)$ with coefficients in $K\{t\}$ and of determinant $\Delta\neq 0$. Then there exist $x_1{}^0, \cdots, x_n{}^0\in K\{t\}$ such that for all $x_j\equiv x_j{}^0 \pmod{K[t]}$ we have, respectively,

$$|L_1(x)L_2(x)| \geqq |\Delta|/e^2, \quad (n=2)$$
$$|L_1(x)(L_2{}^2(x)+mL_3{}^2(x))| \geqq |\Delta|/e^3,$$
$$|(L_1{}^2(x)+mL_2{}^2(x))(L_3{}^2(x)+L_4{}^2(x))| \geqq |\Delta|/e^4.$$

The constants on the right hand sides are sharp. The details are carried out only in the case $n=4$. *J. W. S. Cassels* (Cambridge, England)

Citations: MR 12, 594b = R16-9; MR 12, 594c = R16-10; MR 13, 15f = R12-19; MR 13, 919a = J32-31; MR 13, 919b = J32-31; MR 20# 30 = R58-26.

J44 HOMOGENEOUS MINIMA OF QUADRATIC OR BILINEAR FORMS IN MORE THAN TWO VARIABLES

For real indefinite forms in sufficiently many variables, and which do not have positive minima, see **D76**.

See also reviews A04-4, D08-21, D76-1, D76-2, D76-8, D76-10, D76-11, D76-16, D76-23, E02-6, E12-17, E12-140, E20-74, H15-42, H25-1, H25-6, H25-7, H30-23, J36-24, J48-4, J52-21, J56-2.

J44-1 **(6, 57e)**

Mordell, L. J. Observation on the minimum of a positive quadratic form in eight variables. J. London Math. Soc. **19**, 3–6 (1944).

The author proves the following theorem which follows from work of Gauss and Hermite. If $M(f)$ is the minimum of a positive quadratic form in n variables with real coefficients and determinant Δ and $M(f)\leqq\gamma_n\Delta^{1/n}$, where γ_n is independent of the coefficients of f, then $\gamma_n\leqq\gamma_{n-1}^{(n-1)/(n-2)}$. *B. W. Jones* (Ithaca, N. Y.).

Referred to in J44-5.

J44-2 **(7, 368c)**

Remak, Robert. Ein Satz über die sukzessiven Minima bei definiten quadratischen Formen. Nederl. Akad. Wetensch., Proc. **44**, 1071–1076 (1941).

Let $\lambda_1, \cdots, \lambda_n$ be the successive minima associated with a convex body, centre at the origin, in n-dimensional space, as defined by Minkowski. For the special case when the body is given by $Q\leqq 1$, where Q is a positive definite quadratic form in n variables, of determinant D, the author proves that $\lambda_1{}^2\cdots\lambda_n{}^2\leqq\gamma_n{}^nD$. Here γ_n is the greatest minimum of any positive definite quadratic form of determinant 1. The result was already proved, in effect, by Minkowski [Geometrie der Zahlen, Leipzig, 1910, § 51]; the present proof differs from Minkowski's only in its notation. *H. Davenport* (London).

J44-3 **(8, 137g)**

Chaundy, T. W. The arithmetic minima of positive quadratic forms. I. Quart. J. Math., Oxford Ser. **17**, 166–192 (1946).

Let F be a positive definite quadratic form in n variables with given coefficients a_{rs}. If the variables x_r are restricted to integral values, not all zero, the corresponding values of F are necessarily positive, and will somewhere attain a minimum. Multiplying F by a suitable number, we can arrange that this minimum shall be 2. Then the determinant $\Delta=|a_{rs}|$ of F is a function of the coefficients. If we vary the coefficients, Δ will have an absolute minimum value Δ_n, depending only on n; e.g., $\Delta_1=2$ and $\Delta_2=3$, these being the determinants of the forms $2x_1{}^2$ and $2(x_1{}^2+x_1x_2+x_2{}^2)$. Using an inductive method, considerably simpler than that used by Gauss, Korkine and Zolotareff, Hofreiter, and Blichfeldt [Math. Z. **39**, 1–15 (1934)], he corroborates the known values $\Delta_3=\Delta_4=\Delta_5=4$, $\Delta_6=3$, $\Delta_7=2$, $\Delta_8=1$, and adds the new values $\Delta_9=1$, $\Delta_{10}=\frac{3}{4}$. The nonary form giving $\Delta_9=1$ is equivalent to the extreme form T_9 of Korkine and Zolotareff [Math. Ann. **6**, 366–389 (1873)], but the denary form giving $\Delta_{10}=\frac{3}{4}$ is entirely new.

Some readers have doubted the rigor of the author's argument because in varying the coefficients (as in the change from G_2 to G_2' on page 169) he admits forms which might be expected to attain values less than 2. The author agrees that this is a valid objection, but he believes the same results can be obtained by an improved method which he is now developing. *H. S. M. Coxeter* (Notre Dame, Ind.).

J44-4 **(8, 369g)**

Mahler, Kurt. On reduced positive definite quaternary quadratic forms. Nieuw Arch. Wiskunde (2) **22**, 207–212 (1946).

According to Minkowski's definition, a positive definite n-ary form $f(x)=\sum\sum a_{hk}x_hx_k$ is said to be reduced if $f(x)\geqq a_{hh}$ for all integers $x_1, \cdots, x_n$ such that $(x_h, x_{h+1}, \cdots, x_n)=1$ (this holding for each value of h) and also $a_{12}\geqq 0$, $a_{23}\geqq 0, \cdots, a_{n-1,n}\geqq 0$. Using a classical theorem of Korkine and Zolotareff [Math. Ann. **5**, 581–583 (1872)] on the extreme form $x_1^2+x_2^2+x_3^2+x_4^2+(x_1+x_2+x_3)x_4$, the author proves that every reduced quaternary form of determinant D satisfies $a_{11}a_{22}a_{33}a_{44}\leqq 4D$. *H. S. M. Coxeter* (Toronto, Ont.).

J44-5 (9, 10g)

Oppenheim, A. Remark on the minimum of quadratic forms. J. London Math. Soc. **21** (1946), 251–252 (1947).

The author proves again the result of Mordell [same J. **19**, 3–6 (1944); these Rev. **6**, 57] that $\gamma_n^{n-2}\leqq\gamma_{n-1}^{n-1}$, where γ_n is the upper bound of the minima of positive definite quadratic forms in n variables, of determinant 1. He presents the proof in such a way as not to presuppose the existence of γ_n. *H. Davenport* (Stanford University, Calif.).

Citations: MR 6, 57e = J44-1.

J44-6 (9, 79c)

Davenport, H., and Heilbronn, H. On the minimum of a bilinear form. Quart. J. Math., Oxford Ser. **18**, 107–121 (1947).

The authors study the minimum $M(B)$ of the absolute value of $B(x, y, z, t)=(\alpha x+\beta y)(\gamma z+\delta t)$, where $\alpha, \beta, \gamma, \delta$ are real constants and x, y, z, t are real variables subject to $xt-yz=\pm 1$, whereas they suppose $\Delta=\alpha\delta-\beta\gamma\neq 0$; α/β and γ/δ irrational. For all forms of this kind they prove

$$M(B)\leqq\frac{3-\sqrt{5}}{2\sqrt{5}}|\Delta|.$$

For all forms for which this minimum is not attained, they find $M(B)\leqq\frac{1}{4}(2-\sqrt{2})|\Delta|$ and for all forms where this second minimum is not attained, they find $M(B)\leqq\frac{1}{3}(\sqrt{2}-1)|\Delta|$. Whereas in the first two cases only the forms which are equivalent to a multiple of one particular form attain the minimum named above, in the third case the following unexpected theorem holds. For any $\delta>0$ there exists a set of forms, no one of which is equivalent to a multiple of another, for which $M(B)>\frac{1}{3}((\sqrt{2}-1)-\delta)|\Delta|$, and the set has the cardinal number of the continuum. This theorem shows a fundamental difference from the well-known similar classical results of Markoff on binary quadratic forms [Math. Ann. **15**, 381–407 (1879); **17**, 379–400 (1880)]. *J. F. Koksma* (Amsterdam).

Referred to in J32-34, J44-13, J44-14.

J44-7 (9, 135f)

Venkov, B. A. Sur le problème extrémale de Markoff pour les formes quadratiques ternaires indéfinies. Bull. Acad. Sci. URSS. Sér. Math. [Izvestia Akad. Nauk SSSR] **9**, 429–494 (1945). (Russian. French summary)

Consider an indefinite ternary quadratic form $f(x_1x_2x_3)=x'Ax$, $A=(a_{ij})$, $x'=[x_1x_2x_3]$, with real coefficients a_{ij} and of determinant $|A|=1$. The minimum $m(f)$ of a form f is the numerically smallest value attained by $f(x_1x_2x_3)$ for integral values of $x_1x_2x_3$ not all zero. The author constructs eleven forms $f_1, \cdots, f_{11}$ of determinant unity whose minima are equal, respectively, to $(\frac{2}{3})^{\frac{1}{3}}$, $(\frac{2}{5})^{\frac{1}{3}}$, $(\frac{1}{3})^{\frac{1}{3}}$, $(\frac{8}{25})^{\frac{1}{3}}$, $(\frac{7^3}{1200})^{\frac{1}{3}}$, $(\frac{2}{7})^{\frac{1}{3}}$, $(\frac{4}{15})^{\frac{1}{3}}$, $(\frac{5^3}{3\cdot 13^2})^{\frac{1}{3}}$, $(\frac{3^3}{112})^{\frac{1}{3}}$, $(\frac{2^5}{135})^{\frac{1}{3}}$, $(\frac{2}{9})^{\frac{1}{3}}$, in absolute value, and proves that unless a real indefinite ternary quadratic form f of determinant 1 is equivalent to one of the above eleven forms, we have $|m(f)|<(\frac{2}{9})^{\frac{1}{3}}$. Employing similar information regarding the minima of indefinite binary quadratic forms, Markoff [Math. Ann. **56**, 233–251 (1902); Mém. Acad. Imp. Sci. St. Pétersbourg (8) **23**, no. 7 (1909)] determined the first four minima. By similar methods Dickson [Studies in the Theory of Numbers, University of Chicago Press, 1930] recomputed these four minima. Fujiwara [Abh. Math. Sem. Hamburgischen Univ. **2**, 74–80 (1923)] used difficult geometric considerations to arrive at an incorrect value of $m(f_4)$. The above sequence of eleven minima and the sequence of associated forms represents the first significant extension of the results of Markoff. The method employed by the author is an extension of his geometrical ideas employed in the study of the integral automorphs of the indefinite ternary quadratic forms [Bull. Acad. Sci. URSS. Sér. Math. [Izvestia Akad. Nauk SSSR] **1**, 139–170 (1937)]. It makes use of information regarding the distribution with respect to the cone $f(x, y, z)=0$ of a set of nonequivalent (under the automorphisms of f) lattice points. *A. E. Ross* (Notre Dame, Ind.).

J44-8 (9, 334b)

Davenport, H. On a theorem of Markoff. J. London Math. Soc. **22**, 96–99 (1947).

A simpler proof is given of the following result due to Korkine and Markoff [Markoff, Math. Annalen **56**, 233–251 (1903)]: let $Q(x, y, z)$ be an indefinite ternary quadratic form with real coefficients and with determinant $D\neq 0$; then there exist integers x, y, z (not all zero) for which $|Q(x, y, z)|\leqq(\frac{2}{3}|D|)^{\frac{1}{3}}$. The sign of equality is necessary if and only if Q is equivalent to a multiple of $x^2+y^2-z^2+xz+yz$. *H. S. A. Potter* (Aberdeen).

J44-9 (10, 284d)

Mordell, L. J. The minimum of a definite ternary quadratic form. J. London Math. Soc. **23**, 175–178 (1948).

More than a century ago, Gauss proved that, of all positive definite ternary quadratic forms of minimum value 1 (for integral values of the variables, not all zero), those of minimum determinant are equivalent to $x^2+xy+y^2+yz+z^2$. Geometrically, this means that, of all sphere-packings (in ordinary space) where the centers of the spheres form a lattice, the densest is the cubic close-packing. The treatments by Gauss, Dirichlet, Hermite, Korkine and Zolotareff, and Minkowski are all quite complicated. The present paper contains a proof using nothing but elementary algebra. *H. S. M. Coxeter* (New York, N. Y.).

J44-10 (10, 593b)

Davenport, H. On indefinite ternary quadratic forms. Proc. London Math. Soc. (2) **51**, 145–160 (1949).

If $Q(x, y, z)$ denotes an indefinite ternary quadratic form with real coefficients and determinant $D\neq 0$, there exist integers x, y, z not all zero such that $|Q|\leqq(\frac{2}{3}|D|)^{\frac{1}{3}}$. Analogously to this result of Markov, the author proves that, if $D>0$, there exist integers x, y, z for which $0<Q\leqq(\frac{27}{4}D)^{\frac{1}{3}}$; if $D<0$ there exist integers x, y, z for which $0<Q\leqq(4|D|)^{\frac{1}{3}}$. Both theorems are best possible as is shown by the forms $x^2-\frac{1}{3}y^2-\frac{1}{3}z^2-\frac{2}{3}yz-\frac{2}{3}zx+\frac{2}{3}xy$, x^2+yz, respectively. *J. F. Koksma* (Amsterdam).

Referred to in J44-17, J44-18.

J44-11 (13, 443c)

Coxeter, H. S. M. Extreme forms. Canadian J. Math. **3**, 391–441 (1951).

Let $\sum a_{ij}x^ix^j$ be a positive definite quadratic form in n variables, with determinant Δ, and let M be the minimum of the form for integer values of the variables (not all zero). The form is said to be extreme if M^n/Δ is a local maximum for variation of the coefficients a_{ij}. Suppose that the linear transformation $\xi^i=\sum c^i{}_jx^j$ reduces the form to the unit form, and consider the lattice in Euclidean n-space generated by the vectors $\mathbf{t}_j=\sum c^i{}_j\mathbf{p}_i$, where the $\mathbf{p}_i$ are the unit vectors along the axes. This lattice determines the form, and a different set of generating vectors generates an equivalent form. A form is said to be connected if it cannot be expressed as the sum of two forms involving different sets of variables, and is said to be reflexible if the corresponding point-lattice is invariant under the reflections that reverse its n basic vectors in turn. The author proceeds to enumerate the classes of connected reflexible forms, to determine their minimal vectors of the associated point-lattices, and to identify the subgroups of the lattices which keep the origin fixed with well-known finite orthogonal groups (namely those with simplicial fundamental regions). The paper concludes with a table of extreme forms related to these groups. The many interesting details of this work cannot be summarized in a brief review. *J. A. Todd* (London).

Referred to in H30-57, J44-21, J48-18.

J44-12 **(13, 825c)**

Gastinger, Walter. Über die untere Grenze der positiven Werte reeller quadratischer Formen. Monatsh. Math. **56**, 49–60 (1952).

The author is concerned with finding numbers such that the inequality $0<P(x_1, \cdots, x_n)\leqq p_n|D|^{1/n}$ is soluble in integers $x_1, \cdots, x_n$ for every indefinite quadratic form P with real coefficients and non-zero determinant D. The best possible values of p_n are known when n is 2 or 3, and Blaney [J. London Math. Soc. **23**, 153–160 (1948); these Rev. **10**, 511] proved in a lemma that 2^{n-1} is always an admissible value for p_n. The author follows Blaney's method, and by a more detailed treatment of the inequalities he obtains smaller values for p_n for individual values of n, and the estimate $(4/3)^{(n-1)/2}$ for large n. He also points out that the example $P=x_1x_n+x_2x_{n-1}+\cdots$ shows that p_n cannot be taken less than 2 for even n, or less than $2^{1-1/n}$ for odd n.

H. Davenport (London).

Citations: MR 10, 511c = J48-4.

J44-13 **(14, 142a)**

Barnes, E. S. The minimum of a factorizable bilinear form. Acta Math. **86**, 323–336 (1951).

Let $B(x, y, z, t)=(\alpha x+\beta y)(\gamma z+\delta t)$ be a real bilinear form with $\Delta=\alpha\delta-\beta\gamma\neq 0$, and assume that it does not represent zero integrally. Let $M(B)$ be the greatest lower bound of $|B(x, y, z, t)|$ as x, y, z, t range over all integers such that $xt-yz=1$. Set $m(B)=M(B)/|\Delta|$. Davenport and Heilbronn [Quart. J. Math. **18**, 107–121 (1947); these Rev. **9**, 79] proved the following: (i) $m(B)<(2^{1/2}-1)/3$ unless B is equivalent to a multiple of one of three specific forms B_1, B_2, B_3 for which $m(B_1)>m(B_2)>m(B_3)=(2^{1/2}-1)/3$, and (ii) for any $\epsilon>0$, there are uncountably many forms for which $m(B)>m(B_3)-\epsilon$. In the present paper the author gives an alternate proof of this result, using the methods of his previous paper [Proc. London Math. Soc. (3) **1**, 385–414 (1951); these Rev. **13**, 825]. *I. Reiner* (Urbana, Ill.).

Citations: MR 9, 79c = J44-6; MR 13, 825a = J36-18.
Referred to in J44-14.

J44-14 **(14, 955d)**

Barnes, E. S. The minimum of a bilinear form. Acta Math. **88**, 253–277 (1952).

Let $B(x, y, z, t)=\alpha xy+\beta xt+\gamma yz+\delta yt$ be a real bilinear form. A form is said to be equivalent to B if it may be obtained from B by a substitution

$$\begin{pmatrix} x & z \\ y & t\end{pmatrix}=\begin{pmatrix} p & q \\ r & s\end{pmatrix}\begin{pmatrix} x' & z' \\ y' & t'\end{pmatrix} \quad\text{or}\quad \begin{pmatrix} x & z \\ y & t\end{pmatrix}=\begin{pmatrix} p & q \\ r & s\end{pmatrix}\begin{pmatrix} z' & x' \\ t' & y'\end{pmatrix},$$

where p, q, r, s are integers such that $ps-qr=\pm 1$. The quantities $\Delta(B)=\alpha\delta-\beta\gamma$ and $\theta(B)=|\beta-\gamma|$ are then equivalent invariants. With the bilinear form $B(x, y, z, t)$ is associated the quadratic form $B(x, y, x, y)$ of discriminant $D=\theta^2-4\Delta$, equivalent bilinear forms yielding equivalent quadratic forms. Define $\omega=\theta/\sqrt{|D|}$.

Let $M(B)$ be the greatest lower bound of $|B(x, y, z, t)|$ over integers x, y, z, t such that $xt-yz=\pm 1$. For symmetric forms, characterized by $\omega=0$, Schur [S.-B. Preuss. Akad. Wiss. **1913**, 212–231] has proved when $D>0$ that $M(B)\leqq D^{1/2}/2\sqrt{5}$, with equality for multiples of exactly one class of equivalent bilinear forms. An analogous result was obtained when $D>0$ for factorizable forms, characterized by $\omega=1$, by Davenport and Heilbronn [Quart. J. Math., Oxford Ser. **18**, 107–121 (1947); these Rev. **9**, 79; see also Barnes, Acta Math. **86**, 323–336 (1951); these Rev. **14**, 142]. In the present paper the author considers bilinear forms with arbitrary ω, and proves that $M(B)\leqq D^{1/2}/2\sqrt{5}$ for all ω when $D>0$, with equality if and only if B is a multiple of any form in a specific countable set of equivalence classes. When $D<0$, he shows that $M(B)\leqq(-D)^{1/2}/2\sqrt{3}$ for all ω, with equality under similar circumstances. He also solves the more difficult question of finding the best possible estimate for $M(B)$ as a function of ω and D for the case $D>0$, $0\leqq\omega\leqq\omega_0=1.24$ (approximately). A continuous function $\chi(\omega)$, whose graph is a zig-zag line, is defined for the indicated range, and a specific set of nine quadratic forms Q_i $(i=0, 1, \cdots, 8)$ is given, and the author then proves that $M(B)\leqq D^{1/2}\chi(\omega)/2$ when $D>0$. He shows further that for each ω $(0\leqq\omega\leqq\omega_0)$ there exists a form B for which equality holds and for which the quadratic form associated with B is equivalent to a multiple of some Q_i. This theorem includes as special cases the above-mentioned results of Schur, Davenport and Heilbronn.

In proving his results, the author first establishes a connection between $M(B)$ and the quadratic form Q associated with B, and shows easily that

$$M(B)=\tfrac{1}{2}|D|^{1/2}\inf|b\cdot|D|^{-1/2}-\omega|,$$

where b runs over all middle coefficients of forms equivalent to Q. The estimates for $M(B)$ independent of ω are simple consequences of the above formula, but in order to obtain the best possible results for given D and ω, some rather complicated reasoning is used which involves the chain of reduced forms equivalent to Q. The author remarks that his methods do not enable him to evaluate $\chi(\omega)$ for $\omega>\omega_0$.

I. Reiner (Urbana, Ill.).

Citations: MR 9, 79c = J44-6; MR 14, 142a = J44-13.

J44-15 **(14, 1065g)**

Rankin, R. A. On positive definite quadratic forms. J. London Math. Soc. **28**, 309–314 (1953).

Let $f(x_1, \cdots, x_n)$ be a positive quadratic form in n variables and of determinant 1. Let $M_r(f)$ denote the lower bound of any principal minor of order r of any form equivalent to f $(1\leqq r\leqq n-1)$. It is shown then that for given f the lower bound $M_r(f)$ is attained, that, if F is the form adjoint to f, we have $M_r(f)=M_{n-r}(F)$, and that $M_r(f)$ is bounded above for all forms of determinant 1 and attains its upper bound. If $\gamma_{n,r}$ denotes the maximum of $M_r(f)$ for all forms of determinant 1, then $\gamma_{n,m}\leqq\gamma_{r,m}(\gamma_{n,r})^{m/r}$ for $1\leqq m<r\leqq n-1$. The author further proves $\gamma_{4,2}=3/2$. From these results and the known values of $\gamma_{n,1}$ for $1\leqq n\leqq 8$ he obtains upper bounds for $\gamma_{n,r}$ for small values of n and r.

J. F. Koksma (Amsterdam).

J44-16 **(14, 1066a)**

Coxeter, H. S. M., and Todd, J. A. An extreme duodenary form. Canadian J. Math. **5**, 384–392 (1953).

A positive definite quadratic form of determinant Δ that assumes a minimum integral value M when its n arguments assume integral values not all zero is called extreme if Δ/M^n is a minimum for infinitesimal variations of the coefficients. The form which assumes the smallest minimum for a given n is called absolutely extreme. For $n=12$ a new extreme duodenary form K_{12} is described for which $\Delta(2/M)^{12}$

has a value 729/4096 that is less than the values 13, 4, 1, 1/4 for other known extreme duodenary forms, but it is not known whether K_{12} is absolutely extreme. Expressed as a sum of squares in variables $x_1, \cdots, x_6, y_1, \cdots, y_6$, the new form is

$$12K_{12}=\sum(6x_j+3y_j+2y_6)^2 +3(6x_6-y_6+2\sum y_j)^2+3\sum y_j^2+y_6^2,$$

where j is summed from 1 to 5. It is derived by considering a lattice of points whose coordinates are the Eisenstein integers $A_j\omega+B_j\omega^2$ in complex unitary six space. It is eutactic and perfect. *J. S. Frame.*

J44-17 (15, 291f)

Oppenheim, A. One-sided inequalities for quadratic forms. I. Ternary forms. Proc. London Math. Soc. (3) **3**, 328–337 (1953).

Let $f(x, y, z)$ be an indefinite ternary quadratic form with real coefficients in integral variables x, y, z with determinant $\Delta\neq 0$. Let P_1, P_2, P denote the lower bounds of the positive values of f, $-f$ and $|f|$ respectively, so that $P=\min(P_1, P_2)$. Assume $\Delta<0$ without loss of generality. Then Davenport has proved $P_1^3\leqq 4|\Delta|$, $4P_2^3\leqq 27|\Delta|$, equality in the respective cases implying equivalence to a positive multiple of x^2+yz, x^2+4yz [same Proc. (2) **51**, 145–160 (1949); these Rev. **10**, 593]. These results already had been asserted partly by the author [J. London Math. Soc. **6**, 222–226 (1931)] and he now proves Davenport's result and moreover that, if the above mentioned forms are excluded, we shall have $P_1^3\leqq(9/4)|\Delta|$, $P_2^3\leqq(343/64)|\Delta|$, equality implying equivalence to a positive multiple of $3x^2+4yz$, x^2+16yz, respectively. *J. F. Koksma* (Amsterdam).

Citations: MR 10, 593b = J44-10.

Referred to in J44-18, J44-22.

J44-18 (15, 607b)

Oppenheim, A. One-sided inequalities for quadratic forms. II. Quaternary forms. Proc. London Math. Soc. (3) **3**, 417–429 (1953).

[For part I see same Proc. (3) **3**, 328–337 (1953); these Rev. **15**, 291.] Let $f(x_1, \cdots, x_n)$ be an indefinite quadratic form in integral variables and nonzero determinant $\Delta(f)$ with signature s. Let $P_1(f)$ denote the lower bound of the positive values of f; let $P_2(f)=P_1(-f)$ and $J_1=P_1^n/|\Delta|$, $J_2=P_2^n/|\Delta|$. Davenport [same Proc. (2) **51**, 145–160 (1949); these Rev. **10**, 593] and the author (not published yet) deduced estimates for J_1 and J_2 in the ternary case. Now the author gives similar results in the quaternary case: If $n=4$, $s=0$, $\Delta(f)>0$, then either $J_1\leqq 256/81$, or else $J_1=16$ and f is $P_1(xy+zt)$, or $J_1=81/16$ and f is $\frac{1}{3}P_1(-x^2+4y^2+4zt)$, or $J_1=4$ and f is $P_1(xy+2zt)$ or $P_1(xy+z^2-t^2)$, or $J_1=16/5$ and f is $P_1(xy+z^2+zt-t^2)$ (the same for P_2). If $n=4$, $s=2$, $\Delta(f)<0$, then (a) either $J_1<2048/729$, or $J_1=16/3$ and f is $P_1(x^2+xy+y^2+zt)$ or $J_1=4$ and f is $P_1(x^2+y^2+zt)$ and (b) either $J_2<7.6$ or $J_2=256/27$ and f is $\frac{1}{2}P_2(x^2+xy+y^2+3zt)$. *J. F. Koksma* (Amsterdam).

Citations: MR 10, 593b = J44-10; MR 15, 291f = J44-17.

Referred to in J44-22, J44-46.

J44-19 (16, 18a)

Davenport, H., and Watson, G. L. The minimal points of a positive definite quadratic form. Mathematika **1**, 14–17 (1954).

Let $q(x_1, \cdots, x_n)=q(\mathbf{x})$ be a positive definite n-ary quadratic form with real coefficients, and let $\mathbf{x}^{(1)}, \cdots, \mathbf{x}^{(n)}$ be points at which $q(\mathbf{x})$ assumes its successive minima, so that for each k, $q(\mathbf{x}^{(k)})$ is the smallest value assumed by q for any $\mathbf{x}$ linearly independent of $\mathbf{x}^{(1)}, \cdots, \mathbf{x}^{(k-1)}$. Let N be the determinant of the coordinates of $\mathbf{x}^{(1)}, \cdots, \mathbf{x}^{(n)}$, and let $N(q)=\min|N|$ for all such sets $(\mathbf{x}^{(1)}, \cdots, \mathbf{x}^{(n)})$. Improving a result of the second author (apparently not published), it is shown that if n is large, there is a form q such that $N(q)>\pi^{-n/2}\Gamma(1+\frac{1}{2}n)$. *W. J. LeVeque.*

Referred to in J44-39, J44-50.

J44-20 (16, 1002d)

Kneser, Martin. Two remarks on extreme forms. Canad. J. Math. **7**, 145–149 (1955).

Let $f(x_1, \cdots, x_n)=\sum\sum a_{ij}x_ix_k$ be a positive definite quadratic form of determinant $D=\det(a_{ij})$, and let M be the minimum of f for integers $x_1, \cdots, x_n$ not all zero. Suppose the minimum is attained by s pairs of "minimal vectors" $\pm(m_{1k}, \cdots, m_{nk})$, so that $f(m_{1k}, \cdots, m_{nk})=M$ $(k=1, \cdots, s)$. The form is said to be extreme if M^n/D does not increase for any small variation of the coefficients a_{ij}. It is said to be perfect if it is determined by its minimal vectors. It is said to be eutactic if there exist s positive numbers ρ_k such that $\sum\rho_k(\sum m_{ik}x_i)^2$ is equal to the adjoint form. It was proved by Voronoï [J. Reine Angew. Math. **133**, 97–178 (1908)] that a form is extreme if and only if it is both perfect and eutactic. The author's first "remark" is a simpler proof for this theorem. The second is his announcement of a new extreme form in six variables:

$$\sum_1^3(x_j^2-x_jx_{j+3}+x_{j+3}^2)+\left(\sum_1^6 x_i\right)^2,$$

for which $M=2$, $D=2^{-6}\,3^3\,13$ [cf. the following review]. *H. S. M. Coxeter* (Toronto, Ont.).

Referred to in J44-32.

J44-21 (16, 1002e)

Barnes, E. S. Note on extreme forms. Canad. J. Math. **7**, 150–154 (1955).

The extreme forms [cf. the preceding review] were enumerated for $n\leqq 5$ by Korkine and Zolotareff [Math. Ann. **11**, 242–292 (1877)], who also found three such forms for $n=6$. A fourth was added by Coxeter [Canad. J. Math. **3**, 391–441 (1951), p. 439; MR **13**, 443]. A fifth was discovered simultaneously by M. Kneser and the author. In a sense, they were very nearly anticipated by N. Hofreiter [Monatsh. Math. Phys. **40**, 129–152 (1933)], whose incorrect form has determinant $2^{-8}\,3^3\,53$; we merely have to replace his 53 by 52 to obtain D for the new form.

The author also answers the reviewer's challenge [loc. cit., p. 393] to find a perfect form that is not extreme (and therefore not eutactic). One such form (in eleven variables) is

$$\sum_1^3(x_j^2-x_jx_{j+3}+x_{j+3}^2)+\left(\sum_1^{11} x_i\right)^2+\sum_7^{11} x_k^2.$$

H. S. M. Coxeter (Toronto, Ont.).

Citations: MR 13, 443c = J44-11.

J44-22 (16, 1002f)

Barnes, E. S. The non-negative values of quadratic forms. Proc. London Math. Soc. (3) **5**, 185–196 (1955).

Let $Q(x_1, x_2, \cdots, x_n)$ be an indefinite quadratic form in n integral variables, of determinant $d\neq 0$ and signature s. Define $k_{n,s}$, depending only on n and s, as the least constant for which the inequalities (*) $0\leqq Q(x_1, \cdots, x_n)\leqq(k_{n,s}|d|)^{1/n}$ always admit integral solutions $x_1, x_2, \cdots, x_n$, not all zero. For binary forms the results are classical and for $n\geqq 5$ it is conjectured that Q takes arbitrarily small values; hence, only the cases $n=3$ and $n=4$ are of interest. Oppenheim [same Proc. (3) **3**, 328–337, 417–429 (1953); MR **15**, 291, 607] solved the related problem with strict inequality on the left and with the exclusion of $i-1$ classes of zero forms $Q_{n,s}^{\nu}$ $(\nu=1, 2, \cdots, i-1)$. Oppenheim's constants $c_{n,s}^{(\nu+1)}$ are, therefore, upper bounds for the $k_{n,s}$. The author improves these bounds and obtains: $k_{3,1}=4/3$ (excluding the class equivalent to $-x^2+8(y^2+yz+z^2)$, the constant can be reduced to <1.332), $k_{3,-1}\leqq 4$; $k_{4,0}\leqq 64/81$, $k_{4,2}\leqq 32/27$, $k_{4,-2}\leqq 64/27$. A note adds the further improvements, due to Oppenheim, that $k_{4,0}=64/81$ and $k_{3,-1}=16/5$. All these results are obtained by comparatively simple arguments from the following key lemma (Theorem 3 of the paper): Let $\phi(x, y)$ be a non-zero, indefinite binary quadratic form of determinant $D>0$. Then ϕ represents, for integral x, y, values $p>0$, $-n<0$, satisfying $p^3n\leqq D^2/16$. If ϕ is not equivalent to a positive multiple of $-x^2+8y^2$, then the inequality may be replaced by $p^3n<D^2/16.07$. The proof of the lemma is based on the theory of continued fractions.

E. Grosswald (Philadelphia, Pa.).

Citations: MR 15, 291f = J44-17; MR 15, 607b = J44-18.

Referred to in J44-23, J48-12.

J44-23 (17, 128e)

Oppenheim, A., and Barnes, E. S. The non-negative values of a ternary quadratic form. J. London Math. Soc. 30 (1955), 429–439.

In connection with recent results of Barnes [Proc. London Math. Soc. (3) **5** (1955), 185–196; MR **16**, 1002] the following theorem is proved. Let $Q(x, y, z)$ be a quadratic form of determinant $d\neq 0$ and signature -1. Then there exist integers x, y, z, not all zero, satisfying

$$0\leqq Q(x, y, z)\leqq (16d/5)^{\frac{1}{3}}.$$

The second sign of equality is necessary if and only if Q is equivalent to a positive multiple of the special form $-x^2-xy-y^2+90z^2$. *J. F. Koksma* (Amsterdam).

Citations: MR 16, 1002f = J44-22.

J44-24 (18, 114f)

Rankin, R. A. On the minimal points of positive definite quadratic forms. Mathematika 3 (1956), 15–24.

The basic result proved is that if $\mathbf{y}_1, \cdots, \mathbf{y}_m$ are any m points on the surface of the unit sphere in n dimensional space, then

$$\sum \Delta^2(j_1, \cdots, j_n)\leqq (m/n)^n,$$

where the summation is over all choices of n distinct suffixes $j_1, \cdots, j_n$ from $1, \cdots, m$, and $\Delta(j_1, \cdots, j_n)$ denotes the determinant formed from the coordinates of $\mathbf{y}_{j_1} \cdots, \mathbf{y}_{j_n}$. From this are deduced results on the determinants of the sets of integer points at which a positive definite quadratic form in n variables attains some or all of its successive minima. In particular, if $n\geqq 4$ and $f(\mathbf{x})$ is a perfect form, there exist n integer points $\mathbf{x}_1, \cdots, \mathbf{x}_n$ at which f attains its minimum, such that

$$0<\{\det(\mathbf{x}_1, \cdots, \mathbf{x}_n)\}^2\leqq n^{-1}\gamma_n{}^n(1+1/n)^n,$$

where γ_n is Hermite's constant. Finally a result is obtained concerning the packing (supposed symmetrical about the centre) of spherical caps of angular radius α on a sphere, which is an improvement on an earlier result [Proc. Glasgow Math. Assoc. **2** (1955), 139–144; MR **17**, 523] when α is just less than $\frac{1}{4}\pi$. *H. Davenport.*

Referred to in J44-39.

J44-25 (19, 16c)

Dempster, A. P. The minimum of a definite ternary quadratic form. Canad. J. Math. **9** (1957), 232–234.

The author gives a very simple determination of the lattice in Euclidean space of three dimensions of smallest cell for given minimum distance between lattice points, using extremely elementary geometrical methods.

J. A. Todd (Cambridge, England).

J44-26 (19, 120g)

Barnes, E. S. On a theorem of Voronoi. Proc. Cambridge Philos. Soc. **53** (1957), 537–539.

Elementary algebraic proof of the fundamental theorem of Voronoi [J. Reine Angew. Math. **133** (1907), 97–178] that a positive definite quadratic form is extreme if, and only if, it is perfect and eutactic. *J. A. Todd.*

Referred to in J44-32.

J44-27 (19, 251d)

Barnes, E. S. The complete enumeration of extreme senary forms. Philos. Trans. Roy. Soc. London. Ser. A. **249** (1957), 461–506.

Let $\varphi=\sum a_{ij}x_ix_j$ be a positive definite quadratic form in n variables with determinant Δ, and let M be the minimum of the form for integral values of the variables. The form is said to be extreme if M^n/Δ is a local maximum for variation of the coefficients a_{ij}. If φ obtains its minimal value M for S values of the vector of variables x_i, it is said to be perfect if the form is completely defined by the values of these S minimal vectors. An extreme form is necessarily perfect.

The author makes an exhaustive study of the perfect forms in 6 variables following the methods of Voronoï [J. Reine Angew. Math. **133** (1907), 97–178]. Associated with a perfect form $\varphi=\sum a_{ij}x_ix_j$ with S minimal vectors $\lambda_k=\sum m_{ik}x_i$, a region R is defined in the $N=\frac{1}{2}n(n+1)$-dimensional space of the coefficients a_{ij} such that

$$\sum a_{ij}x_ix_j=\sum \rho_k\lambda_k^2 \quad (\rho_k\geqq 0;\ k=1, \cdots, S).$$

The region R is bounded by $(N-1)$-dimensional faces of the form $\sum p_{ij}a_{ij}=0$. Then for each such face, for a uniquely determined number ρ, $\varphi'(x)=\varphi(x)+\rho\sum p_{ij}x_ix_j$ is also a perfect form, called a neighbouring form of φ.

Starting with the perfect form

$$\varphi_0=\sum x_i^2+\sum_{i<j} x_ix_j,$$

the author finds all the inequivalent neighbouring forms, the neighbouring forms of these, and so on until he obtains a complete set of forms which includes equivalents of all the neighbours of these forms.

The following forms are obtained:

$$\varphi_0,\ \varphi_1=\varphi_0-x_1x_2,\ \varphi_2=\varphi_0-x_1x_2-x_1x_3,$$

$$\varphi_3=\varphi_0-\tfrac{1}{2}(x_1x_2+x_3x_4+x_5x_6),$$

$$\varphi_4=\varphi_0-\tfrac{1}{2}(x_1x_2+x_3x_4+x_3x_5+x_3x_6+x_4x_5+x_4x_6+x_5x_6),$$

$$\varphi_5=\varphi_0-\tfrac{1}{2}(x_1x_2+x_3x_4+x_3x_5+x_4x_5),$$

$$\varphi_6=\varphi_0-\tfrac{1}{2}(2x_1x_2+x_1x_3+x_1x_6+x_2x_5+x_4x_6+2x_5x_6).$$

Of these perfect forms φ_5 is not extreme, the others are extreme. *D. E. Littlewood* (Bangor).

Referred to in J44-32.

J44-28 (19, 251e)

Barnes, E. S. The perfect and extreme senary forms. Canad. J. Math. **9** (1957), 235–242.

The author describes the results (proved in detail in the paper reviewed above) on perfect and extreme senary forms. There are six classes of extreme forms, and one form which is perfect but not extreme. Among the classes of extreme forms is one, denoted here by ϕ_6, which has not been described before. As a representation of this form, we may take

$$\sum_1^6 x_i^2+\sum_{i<j} x_ix_j-\tfrac{1}{2}[2x_1x_2+x_1x_3+x_1x_6+x_2x_5+x_4x_6+2x_5x_6].$$

The lattice corresponding to this form has 21 minimal vectors, and its group, of order 672, is the direct product of

PGL(2, 7) with a group of order 2. The form appears as a particular case ($n=6$) of the form f_n, extreme for all $n \geqq 6$, corresponding to the lattice defined by $y_1, \cdots, y_{n+1}$ which takes integral values subject to

$$\sum y_i=0, \quad \sum iy_i \equiv 0 \pmod{n+1}.$$

The author incidentally constructs an example to disprove a conjecture by Coxeter that every eutactic form in n variables with at least $\frac{1}{2}n(n+1)$ minimal vectors is perfect. *J. A. Todd* (Cambridge, England).

J44-29 **(19, 635e)**

Watson, G. L. The minimum of an indefinite quadratic form with integral coefficients. J. London Math. Soc. 32 (1957), 503–507.

The object of this paper is to prove the theorem: Let f be a quadratic form in $k \leqq 3$ variables, which has integral coefficients and is primitive, non-singular, and indefinite; let d be the discriminant of f, and denote by $\min^+ f$ the least positive value assumed by f for integral values of the variables; then

$$\min{}^+ f = O(|d|^{1/(2k-2)+\varepsilon}),$$

where ε is any positive real number and the constant implied by the O-notation depends on ε but not on k. It follows immediately that the exponent of $|d|$ may be replaced by $1/k$ except for a finite number of values of d. The author states in fact that the only exceptions with $d \neq 1$ or -1 are three ternary forms. *B. W. Jones.*

J44-30 **(20# 3102)**

Watson, G. L. One-sided inequalities for integral quadratic forms. Quart. J. Math. Oxford Ser. (2) **9** (1958), 99–108.

Let f denote a quadratic form in k variables and $\min^+ f$ the lower bound of the strictly positive values of f for integral values of the variables. Let $d(f)$ be defined as $(-1)^{\frac{1}{2}k} \det A$, for k even; $\frac{1}{2}(-1)^{\frac{1}{2}(k-1)} \det A$, for k odd. Four exceptional forms are

$$f_1=4x_1x_2-x_3^2, \quad f_2=8x_1x_2-x_3^2, \quad f_3=16x_1x_2-x_3^2,$$

whose $\min^+ f$ are 3, 4, 7, respectively, and

$$f_4=3x_1x_2-x_3^2-3x_3x_4-3x_4^2,$$

whose $\min^+ f$ is 2.

The author proves the following theorem. Suppose f is an integral, primitive, non-singular quadratic form, which, if definite, is positive and of rank $k \leqq 8$. Suppose that f is not equivalent to f_1, f_3, or f_4. Then $(\min^+ f)^k/|d(f)| \leqq 1$, $\frac{1}{2}$, $\frac{1}{3}$, $\frac{1}{4}$ according as $s(f) \equiv 0$ or ± 1, ± 3, ± 2, 4 (mod 8), where $s(f)$ is the signature of f. The sign of equality is necessary for every pair k, s, but only if f is equivalent to f_2 or satisfies (4) $|d|=1, 2, 3, 4$ according as $\pm s \equiv 0$ or 1, 3, 2, 4 (mod 8).

These results are related to those of Davenport and others [see H. Davenport, Mathematika 3 (1956), 81–101; MR **19**, 19]. *B. W. Jones* (Boulder, Colo.)

J44-31 **(20# 6998)**

Barnes, E. S. The construction of perfect and extreme forms. I. Acta Arith. **5** (1958), 57–79 (1959).

Let $f(\mathbf{x})$ be a positive definite quadratic form of determinant D and order n; let M be the minimum of $f(\mathbf{x})$ for integral non-zero vectors $\mathbf{x}$. Then $f(\mathbf{x})$ assumes the value M for a finite number of integral vectors $\mathbf{x}=\pm\mathbf{m}_1, \cdots, \pm\mathbf{m}_s$. The form $f(\mathbf{x})$ is said to be perfect if the s relations

$$F(\mathbf{m}_k)=M \quad (k=1, \cdots, s)$$

uniquely determine the coefficients of f. Perfect forms include all extreme forms, that is, those for which $M/D^{1/n}$ is a local maximum. Most known perfect and extreme forms are listed by Coxeter.

The author presents in this paper a method which yields many perfect forms with little labor starting with a known perfect or extreme form and producing a new form either by extending the range of values or by increasing the dimension of the known form. To this end he defines the lattice of a form f to be the lattice Tx, over integral vectors x, where $f=x'T'Tx$, and denotes this lattice by $\Lambda(f)$. A form f' is called a refinement of f if $\Lambda(f')$ contains $\Lambda(f)$. If Λ and Λ' are both n-dimensional, his basic theorem is: Let f' be a refinement of f with the same minimum M. Then if f is perfect, so is f'; and if f is extreme, so is f'.

The case when Λ has lower dimension than Λ' will be taken up in part II of this paper.

B. W. Jones (Boulder, Colo.)

Referred to in J44-35.

J44-32 **(21# 3379)**

Vladimirov, V. S. On perfect forms in six variables. Mat. Sb. N.S. **44 (86)** (1958), 263–272. (Russian)

Voronoi proved that a positive definite quadratic form is extreme if and only if it is perfect and eutactic. [For definitions of these terms and simple proofs of Voronoi's theorem see M. Kneser, Canad. J. Math. **7** (1955), 145–149; MR **16**, 1002; or E. S. Barnes, Proc. Cambridge Philos. Soc. **53** (1957), 537–539; MR **19**, 120.] If $n \leqq 5$, all classes of perfect forms in n variables are known and all of these are in fact extreme. Using an algorithm devised by Voronoi, Barnes has found all classes of perfect forms in six variables [cf. Philos. Trans. Roy. Soc. London, Ser. A **249** (1957), 461–506; MR **19**, 251; and loc. cit.]. There are seven such classes, one of which is not extreme. The present author, working independently of Barnes (but a little later), obtains partial results in the same direction.

P. T. Bateman (Zbl **80**, 264)

Citations: MR 16, 1002d = J44-20; MR 19, 120g = J44-26; MR 19, 251d = J44-27.

J44-33 **(21# 5622)**

Barnes, E. S. Criteria for extreme forms. J. Austral. Math. Soc. **1** (1959/61), part 1, 17–20.

The positive definite quadratic form $f(x)=\sum a_{ij}x_ix_j$ $(a_{ij}=a_{ji})$ of determinant $\|a_{ij}\|=D$ and minimum M for integral $x \neq 0$ is said to be extreme if the ratio $\gamma_n(f)=M/D^{1/n}$ is a local maximum for small variations in the coefficients; it is absolutely extreme if $\gamma_n(f)=\gamma_n$ where $\gamma_n=\max \gamma_n(f)$ (taken over all forms in n variables). The minimal vectors of f are the integral solutions $x=\pm m_1, \pm m_2, \cdots, \pm m_s$ of $f(x)=M$. Let H be any subset of the minimal vectors, say $m_1, m_2, \cdots, m_t$ $(t \leq s)$. The author calls f H-perfect if f is uniquely determined by H and its minimum M; he calls f H-eutactic if the adjoint $F(x)=\sum A_{ij}x_ix_j$ is expressible as

$$F(x) \equiv \sum \rho_k(m_k'x)^2$$

with $\rho_k>0$ $(k=1, 2, \cdots, t)$. These definitions reduce to the accepted definitions of the terms perfect and eutactic if H is the set of all minimal vectors. Finally, let G denote the group of automorphs of f.

It is well known that f is extreme if and only if f is perfect and eutactic. The author proves, in a very simple way, the following two criteria for extreme forms. (1) f is extreme if and only if there exists a subset H of its minimal vectors such that f is H-perfect and H-eutactic. (2) If there exists a subset H of the minimal vectors of f such that f is H-perfect and G is transitive on H, then f is extreme.

These results have obvious important practical consequences in the calculations required to check that a form is extreme [see next review].

W. Moser (Winnipeg, Man.)

J44-34 **(21# 5623)**

Barnes, E. S.; and Wall, G. E. Some extreme forms defined in terms of Abelian groups. J. Austral. Math. Soc. **1** (1959/61), part 1, 47–63.

In this paper the authors describe N-variable forms ($N=2^n$) which are extreme and for which $\gamma_n(f)=(N/2)^{1/2}$. It is well known [see J. F. Koksma, *Diophantische Approximationen*, Springer, Berlin, 1936, Ch. II, § 6] that $1/2\pi e \leqq \liminf \gamma_n/n \leqq \limsup \gamma_n/n \leqq 1/\pi e$, but this is the first time that an infinite sequence of extreme forms with $\gamma_n(f)$ unbounded has been constructed.

Let V be the n-dimensional vector space over GF(2). In N-dimensional Euclidean space, consider integral vectors $x=(x_\alpha)$ with coordinates x_α indexed by the $N=2^n$ elements of V. If W is any subset of V, $[W]$ denotes the vector x defined by $x_\alpha=1$ if $\alpha\in W$, $x_\alpha=0$ if $\alpha\notin W$. Let $\lambda_0, \lambda_1, \cdots, \lambda_n$ be integral exponents satisfying $\lambda_0=0$, $\lambda_r-1\leqq\lambda_{r-1}\leqq\lambda_r$ for $1\leqq r\leqq n$. Denote by $\Lambda(\lambda)=\Lambda(\lambda_0, \lambda_1, \cdots, \lambda_n)$ the sublattice of Γ (the integer lattice) generated by all vectors $2^{\lambda_{n-r}}[C_r]$ where C_r runs through all cosets in V (considered as an additive group); r is the vector space dimension of C_r. The authors define $f_{(\lambda)}$ to be the N-dimensional form with lattice $\Lambda_{(\lambda)}$, so that the values assumed by $f_{(\lambda)}$ for integral values of its variables are those of $x^2=\sum_{\alpha\in V} x_\alpha^2$ for $x\in\Lambda(\lambda)$. In other words, if $\Lambda(\lambda)$ is specified by $\xi=Tx$ (x integral), then $f_{(\lambda)}(x)=x'T'Tx$.

The authors prove, using the results of the paper reviewed above, that "most" of the forms $f_{(\lambda)}$ are extreme, and give criteria for determining which are extreme. A simple computation shows that the extreme forms corresponding to the exponent set $\lambda_r=[r/2]$, $0\leqq r\leqq n$, have $\gamma_n(f)=(N/2)^{1/2}$. This yields a large number of extreme forms nearly all of which are new. A table at the end of the paper gives the exponent set (λ), $\log_2 M$, $\log_2 D$, s, and $\log_2\Delta$ (where $\Delta=(2/M)^N D$) of the extreme forms $f_{(\lambda)}$ for $N=4, 8, 16, 32$. Also, some further results which may be obtained in similar fashion are indicated, e.g., $\gamma_{15}\geqq 2^{7/8}$, $\Delta_{15}\leqq 2^{1/6}$. *W. Moser* (Winnipeg, Man.)

J44-35 **(21# 6360)**

Barnes, E. S. The construction of perfect and extreme forms. II. Acta Arith. **5** (1959), 205–222.

This article continues the work of a previous one [Acta Arith. **5** (1959), 57–79; MR **20** #6998] and describes another general method of constructing perfect and extreme forms. A quadratic form $g(x_1, x_2, \cdots, x_n, x_{n+1})$ is called an extension of f if f can be gotten from g by replacing x_{n+1} by 0. The object is to find an extension of a perfect form which is itself perfect.

The author denotes by Π_x the set of points $\mathbf{x}$ which, with the metric defined by the positive definite form $f(\mathbf{x})$, are at least as near to the origin as to any other point of the integral lattice. A point of Π_x lying on n linearly independent faces is called a vertex $\mathbf{v}$ of Π_x. The principal theorem is: Suppose that $f(\mathbf{x})$ is perfect with minimum M, and that, for some integral $t\geqq 1$, the point $t\boldsymbol{\lambda}$ is congruent to a vertex $\mathbf{v}$ of Π_x and satisfies

$$\mathscr{M}(t\boldsymbol{\lambda}) = \mathscr{M}(\mathbf{v}) < M.$$

Then the extension $g(\mathbf{x}, x_{n+1})$ defined by

$$g(\mathbf{x}, x_{n+1}) = f(x_1+\lambda_1 x_{n+1}, \cdots, x_n+\lambda_n x_{n+1})+cx_{n+1}^2$$

with $t^2c=M-\mathscr{M}(t\boldsymbol{\lambda})$, has $M_t(g)=M$; and g is perfect if its minimum is M.

Various applications to particular forms are given.
B. W. Jones (Mayaguez)

Citations: MR 20# 6998 = J44-31.

J44-36 **(27# 106)**

Newman, M.

Bounds for cofactors and arithmetic minima of quadratic forms.

J. London Math. Soc. **38** (1963), 215–217.

Let A be a real symmetric positive definite matrix of order n, and let d_r be the determinant of the minor of order $n-r$ obtained by deleting the first r rows and columns. Then, if γ_n is the Hermite constant, it is shown that, after a suitable congruence transformation, the numbers d_r satisfy $d_{n-r}\leqq\gamma_{r+1}(d_{n-r-1})^{r/(r+1)}$ $(1\leqq r\leqq n-1)$ and that $\gamma_{m+n}^{m+n}\geqq\gamma_m^m\gamma_n^n$. *R. A. Rankin* (Glasgow)

J44-37 **(28# 2092)**

Chalk, J. H. H.

Small positive values of indefinite quadratic forms.

Canad. Math. Bull. **6** (1963), 331–339.

Let Λ denote the lattice of points $X=(x_1, x_2, \cdots, x_n)$ with integral coordinates and let $Q(X)=\sum_{i,j=1}^n a_{ij}x_ix_j$ be any indefinite quadratic form in the integer variables $x_1, \cdots, x_n$ with real coefficients and non-zero determinant.

The author proves the following for $n\geqq 3$. Theorem: Let X_1 be any primitive point of Λ with $Q(X_1)>0$, and put $\theta=\theta(X_1)=Q(X_1)|d|^{-1/n}$. Then there is a basis $X_1, X_2, \cdots, X_n$ of Λ satisfying

$$0 < Q(X_i) \ll \theta^{\nu_n}|d|^{1/n}, \qquad \text{if } \theta < 1,$$
$$\ll \theta|d|^{1/n}, \qquad \text{if } \theta \geqq 1,$$

where $\nu_n=(1-n\cdot 2^{-n+1})(n-1)^{-2}>0$ and $\ll$ is the Vinogradov symbol implying a constant dependent only on n.
Burton W. Jones (Boulder, Colo.)

J44-38 **(28# 5039)**

Scott, P. R.

On perfect and extreme forms.

J. Austral. Math. Soc. **4** (1964), 56–77.

A positive quadratic form is said to be perfect if it is uniquely determined by the integral vectors for which it takes its least positive value. The author gives an extensive list of new perfect forms. These are direct sums of forms of type $x_1^2-x_1x_2+x_2^2-x_2x_3+\cdots+x_{r-1}^2-x_{r-1}x_r+x_r^2$, and certain linear sections of these forms. Conditions are obtained for such forms to be perfect; it turns out that only in a comparatively small number of cases are the forms not perfect.

The author then shows that certain of these forms are extreme, i.e., that M^n/D is a local minimum, where M is the minimum of the form for integral vectors, D the determinant of the form, and n the number of variables.
J. A. Todd (Cambridge, England)

J44-39 **(29# 2224)**

Rankin, R. A.

On the minimal points of perfect quadratic forms.

Math. Z. **84** (1964), 228–232.

A positive definite quadratic form $f(\mathbf{x})=\sum_{i,j=1}^n a_{ij}x_ix_j$ is said to be perfect if the set of integer points $\mathbf{x}$ for which $f(\mathbf{x})$ is minimal, say $f(\mathbf{x}_\nu)=M$ $(\nu=1, \cdots, p)$ with $\mathbf{x}_\nu\neq\mathbf{x}_\mu$ for $\nu\neq\mu$, is such that the equations $f(\mathbf{x}_\nu)=M$ completely determine the a_{ij}. For a perfect form we have $\frac{1}{2}n(n+1)\leqq p\leqq 2^n-1$. Form all possible n-by-n determinants Δ_μ having the $\mathbf{x}_\nu$'s as row vectors. Then $0\leqq\Delta_\mu^2\leqq\gamma_n^n$, where γ_n is Hermite's constant. Davenport and Watson [Mathematika **1** (1954), 14–17; MR **16**, 18] showed that this upper bound could not be improved appreciably. Let Δ^2 be the least positive value of the Δ_μ^2. The author [ibid. **3** (1956), 15–24; MR **18**, 114] showed $\Delta^2 \leqq (1/n)\{\gamma_n(1+1/n)\}^n < e\gamma_n^n/n$. He now shows that $\Delta = 1$ if $n \leqq 6$ and

$$\Delta^2 \leqq \frac{n(n!)}{(n+1)(2n)!}\{(n+1)\gamma_n\}^n \quad \text{for } n > 6.$$

For large n, this last bound represents a considerable improvement. The proof is essentially combinatorial.

D. J. Lewis (Ann Arbor, Mich.)

Citations: MR 16, 18a = J44-19; MR 18, 114f = J44-24.

Referred to in J44-47.

J44-40 **(32# 4091)**

Scott, P. R.

The construction of perfect and extreme forms.

Canad. J. Math. **18** (1966), 147–158.

The method of construction is as follows: Let ϕ_i be perfect forms in r_i variables ($1 \leqq i \leqq k$), each with minimum M for integral values of its variables (not all zero), and suppose that $\phi_i \geqq 2M$ if $\phi_i > M$. Let f be the direct sum of the ϕ_i. Congruence relations are now imposed on the variables of f to ensure that f does not assume the value M. It then follows that $\min f = 2M$, attained either when one summand takes the value $2M$ or when two summands each take the value M.

The author shows how the problem of determining whether the constructed form is perfect or extreme may be reduced to the case of two summands, and gives several examples of the construction. He obtains several new classes of extreme forms, as well as some known extreme forms in a new guise. *E. S. Barnes* (Adelaide)

J44-41 **(33# 5571)**

Watson, G. L.

On the minimum of a positive quadratic form in n ($\leqq 8$) variables. Verification of Blichfeldt's calculations.

Proc. Cambridge Philos. Soc. **62** (1966), 719.

The author announces that he has verified the calculations of H. F. Blichfeldt in Math. Z. **39** (1934/35), 1–15.

J44-42 **(35# 4161)**

Worley, R. T.

Minimum determinant of asymmetric quadratic forms.

J. Austral. Math. Soc. **7** (1967), 177–190.

From the author's introduction: "Let $f = f(x_1, x_2, \cdots, x_n)$ be an indefinite n-ary quadratic form of signature s and let $m_+(f)$, $m_-(f)$ denote the infimum of the non-negative values taken by f and $-f$, respectively, for integral $(x_1, x_2, \cdots, x_n) \neq (0, 0, \cdots, 0)$. Assume that $m_+(f) \neq 0$ and let $A(f) = m_-(f) \cdot m_+(f)^{-1} \geqq k$ for some integer k. Then Segré has shown that, for $n = 2$, f must have determinant $\det(f)$ satisfying $|\det(f)| \geqq \frac{1}{4}(m_+(f))^2(k^2+4k)$. In this paper the results of L. Tornheim [Duke Math. J. **22** (1955), 287–294; MR **16**, 1003] are used to extend Segré's result above, and this extension is used to find a bound such that there are only finitely many non-equivalent indefinite ternary quadratic forms f which attain the value $m_+(f) = 1$ and which have $d(f) = |\det(f)|$ less than this bound."

H. Gross (Zürich)

Citations: MR 16, 1003a = J08-13.

J44-43 **(35# 4162)**

Worley, R. T.

Asymmetric minima of indefinite ternary quadratic forms.

J. Austral. Math. Soc. **7** (1967), 191–238.

From the author's introduction: "Let $f = f(x_1, \cdots, x_n)$ be an indefinite n-ary quadratic form. The problem of asymmetric minima is to find, for a given signature s and for each $t \geqq 0$, the value $\phi_n^s(t)$, defined to be the infimum of the set of all positive α such that every normalized form f (that is, every form with $d(f) = |\det(f)| = 1$) takes a value in the closed interval $[-\alpha, t\alpha]$. The value $\phi_n^s(t)$ is thus a measure of the size of the least closed interval $I = [-a, b]$ containing the origin and with asymmetry $b/a = t$, such that every normalized form f takes a value in any open interval containing I. L. Tornheim [Duke Math. J. **22** (1955), 287–294; MR **16**, 1003] has shown how to calculate $\phi_2^0(t)$ for any given $t > 0$ in terms of infinite chains $[p_i]$, $-\infty < i < \infty$, of positive integers, and simple continued fractions associated with these chains. However, $\phi_2^0(t)$ appears to be an extremely complicated function. It comes therefore as a surprise to find that $t\phi_3^1(t)$ is a continuous piecewise linear function of t. This property of $\phi_3^1(t)$ is a consequence of a theorem which is the main result of the paper. A further consequence is that every normalized form f takes a value in the closed interval $I = [-\phi_3^1(t), t\phi_3^1(t)]$." *H. Gross* (Zürich)

Citations: MR 16, 1003a = J08-13.

J44-44 **(36# 5080)**

Watson, G. L.

Asymmetric inequalities for indefinite quadratic forms.

Proc. London Math. Soc. (3) **18** (1968), 95–113.

Let $f(x_1, \cdots, x_n)$, $n \geqq 2$, be an indefinite non-singular quadratic form with real coefficients and let s be the signature of f. Let $d(f) = (-4)^{[n/2]}\Delta(f)$ be the discriminant of f, where $\Delta(f)$ is the determinant of f. Denote by $P(f)$ the lower bound of the positive values assumed by f for integral vectors $(x_1, \cdots, x_n)$. The author is concerned with the inequalities $P^n(f) < c(n, s) \cdot |d(f)|$, where $c(n, s)$ depends at most on n and s; which lead, for example, to the estimation of class numbers with given n, d. When $n = 3$, it is shown that—with 15 exceptional cases—$P^3(f) < |d(f)|$ if $d(f) < 0$ and $3P^3(f) < d(f)$ if $d(f) > 0$. For zero forms with $n \geqq 4$, an inductive method from $n-2$ to n is given.

S. Konno (Osaka)

J44-45 **(37# 153)**

Prasad, Manoranjan

Note on an extreme form.

Pacific J. Math. **25** (1968), 167–176.

The author considers positive definite quadratic forms $f_n(x_1, x_2, \cdots, x_n)$. The definition of an extreme form is too complex to give here but, vaguely, a form is extreme if in a certain chain of forms it has the maximum number of representations of its minimum and if its adjoint possesses a certain property. All binary extreme forms are equivalent to the form $x^2 + xy + y^2$. The purpose of the paper is to find a positive definite form f_n which is extreme and for which each of the binary forms $f_2(x_i, x_j)$ is extreme.

Burton W. Jones (Boulder, Colo.)

J44-46 **(40# 4205)**

Jackson, T. H.

Small positive values of indefinite quadratic forms.

J. London Math. Soc. (2) **1** (1969), 643–659.

Let $f(x_1, \cdots, x_n)$ be an indefinite quadratic form with real coefficients, discriminant $d \neq 0$ and signature s; let $P = P(f)$ be the infimum of the positive values assumed by $f(\underset{\sim}{x})$ for integral $\underset{\sim}{x}$; define $\phi(f) = P^n/|d|$. Considerably extending known results for $n = 4$, due to A. Oppenheim [Proc. London Math. Soc. (3) **3** (1953), 417–429; MR **15**, 607], the author proves that if $s = 2$, $\phi(f) < \frac{1}{8}$ unless f is equivalent to a positive multiple of one of 6 specified forms; if $s = -2$, $\phi(f) < \frac{1}{4}$ apart from 6 exceptional forms; and if $s = 0$, $\phi(f) < 1/12$ apart from 20 exceptional forms. An upper bound, depending only on s (mod 8), is also given for $\phi(f)$, for all $n \geqq 5$ and all s, when f is not a multiple of an integral form (although it may be conjectured that then $P(f)=0$). *E. S. Barnes* (Adelaide)

Citations: MR 15, 607b = J44-18.

J44-47 **(41# 1645)**

Watson, G. L.

On the minimal points of perfect septenary quadratic forms.

Mathematika **16** (1969), 170–177.

It is proved that every perfect septenary quadratic form assumes its minimum value at a set of 7 points with integer coordinates whose determinant is 1. The reviewer proved [Math. Z. **84** (1964), 228–232; MR **29** #2224] that this was true with 7 replaced by $n\leqq 6$ and that, when $n=7$, the value of the determinant Δ could be taken to be either 1 or 2. It therefore suffices to exclude the case $\Delta=2$. For this purpose a set of six lemmas of a technical nature are proved, in which the author makes a detailed study of the subsets of the set of minimal vectors having a given number of odd components.

R. A. Rankin (Glasgow)

Citations: MR 29# 2224 = J44-39.

J44-48 **(42# 205)**

Mordell, L. J.

The minimum of a singular ternary quadratic form.

J. London Math. Soc. (2) **2** (1970), 393–394.

The author considers the singular ternary quadratic form $\phi(x, y, z)=ax^2+by^2+cz^2+2fyz+2gzx+2hxy$ with (zero determinant and) integral coefficients and shows that analogously to the binary form some integral triple $(x, y, z)\neq(0, 0, 0)$ exists for which $|\phi(x, y, z)|\leqq jd^{1/2}$, where $d=\text{g.c.d.}(bc-f^2, ca-g^2, ab-h^2)$ and $j=(4/3)^{1/2}$ in the semi-definite case and $(4/5)^{1/2}$ in the semi-indefinite case (a best possible result). *Harvey Cohn* (Tucson, Ariz.)

J44-49 **(43# 1925)**

Gruber, Boris

On the minimum of a positive definite quadratic form of three variables. **(Serbo-Croatian summary)**

Glasnik Mat. Ser. III **5 (25)** (1970), 1–18.

Let $f(X)=\sum_{i,k=1}^{3} a_{ik}x_ix_k$ be a positive-definite ternary quadratic form. Denote by Λ the lattice of points in 3-space with integral coordinates. According to H. Minkowski's reduction theory [*Gesammelte Abhandlungen*, Band II, pp. 53–100, Teubner, Leipzig, 1911], for given f there exists a basis $\{A_1, A_2, A_3\}$ of Λ such that (*) $f(p_1A_1+p_2A_2+p_3A_3)\geqq f(A_i)$ whenever i is an index with $1\leqq i\leqq 3$ and p_1, p_2, p_3 are integers with g.c.d. $(p_i, \cdots, p_3)=1$; furthermore, the inequalities (*) all hold if only a certain finite number of them hold. The inequalities (*) imply that $f(A_1)$ is the arithmetic minimum of f. The present author shows that f takes its arithmetic minimum at a given point $A_1\neq(0, 0, 0)$ of Λ if and only if there exist two further lattice points, A_2 and A_3, such that (i) $\{A_1, A_2, A_3\}$ is a basis of Λ and (ii) the following inequalities hold (for all signs): $f(A_1)\leqq f(A_2)\leqq f(A_3)$, $f(A_i\pm A_j)\geqq f(A_j)$ $(1\leqq i<j\leqq 3)$, and $f(A_1\pm A_2\pm A_3)\geqq f(A_3)$. The proof is preceded by several lemmas involving the values of the bilinear form $g(X, Y)=\sum_{i,k=1}^{3} a_{ik}x_iy_k$.

C. G. Lekkerkerker (Amsterdam)

J44-50 **(44# 6612)**

Watson, G. L.

On the minimum points of a positive quadratic form.

Mathematika **18** (1971), 60–70.

Let f_n denote a positive-definite real quadratic form in n variables and $\min f_n$ the minimum of $f_n(\mathbf{x})$ for integral $\mathbf{x}\neq\mathbf{0}$; a minimum point of f_n is an integral $\mathbf{x}$ with $f_n(\mathbf{x})=\min f_n$. The author continues his investigations into properties of the set of minimum points of f_n. He re-proves some results, attributed to Minkowski but first explicitly given by H. Davenport and the author in Mathematika **1** (1954), 14–17 [MR **16**, 18], that if Δ is the determinant of any set of n minimum points of f_n then (*) $\Delta\leqq 1$ $(n=1, 2, 3)$, $\Delta\leqq 2$ $(n=4, 5)$ and $\Delta\leqq 2^{n-4}$ $(n=6, 7, 8)$. These were originally derived from the inequality $\Delta\leqq\gamma_n^{n/2}$, where γ_n is Hermite's constant. Here they are proved by the use of no results for γ_n other than the crude inequality $\gamma_8^4<17$. The author also shows that equality in (*) for $n=4$, 6, 7 or 8 can occur only for a unique class of forms, and that strict inequality in (*) for $n=7$, 8 implies $\Delta\leqq 4$, 9 respectively. Such results may lead to simpler proofs of the precise values of γ_6, γ_7 and γ_8. *E. S. Barnes* (Adelaide)

Citations: MR 16, 18a = J44-19.

J48 INHOMOGENEOUS MINIMA OF QUADRATIC FORMS IN MORE THAN TWO VARIABLES

See also reviews D76-14, D76-18, D76-19, D76-28, H30-32, H30-33, H30-80, J12-44.

J48-1 **(2, 252a)**

Mahler, Kurt. **On a property of positive definite ternary quadratic forms.** J. London Math. Soc. **15**, 305–320 (1940).

Some time ago Davenport gave a simple proof of the following result of Remak [J. London Math. Soc. **14**, 47–51 (1939)]: Let $f(x, y, z)$ be a positive definite ternary quadratic form of determinant 1 which assumes its minimum in three linearly independent lattice points. Then given any three real numbers x_0, y_0, z_0, there are three integers x^1, y^1, z^1 such that

$$f(x_0+x^1, y_0+y^1, z_0+z^1)\leqq 3/4,$$

with equality if and only if $f(x, y, z)=x^2+y^2+z^2$ and $2x_0$, $2y_0$, $2z_0$ are odd integers. The author proves the following result which is shown to be more general than that of Davenport-Remak: Let

$$f(x, y, z)=a(x^2+y^2+z^2)+2bxy+2cxz+2dyz$$

be a positive definite ternary quadratic form of determinant 1 with coefficients satisfying the inequalities

$$1\leqq a\leqq 2^{1/3},\quad 0\leqq b\leqq \tfrac{1}{2}a.$$

Then given any three real numbers x_0, y_0, z_0 there are three real numbers x_1, y_1, z_1, such that $x_1\equiv x_0 \pmod 1$, $y_1\equiv y_0 \pmod 1$, $z_1\equiv z_0 \pmod 1$, $f(x_1, y_1, z_1)\leqq 3/4$, with equality if and only if $x_0\equiv y_0\equiv z_0\equiv\frac{1}{2} \pmod 1$, $f(x, y, z)=x^2+y^2+z^2$. The method of the proof is geometrical and has been used by the author in a previous paper for the study of Hermitian forms [J. London Math. Soc. **15**, 213–236 (1940); these Rev. **2**, 148].

P. Erdös (Philadelphia, Pa.).

Citations: MR 2, 148e = J56-1.

J48-2 **(8, 502g)**

Davenport, H. **Sur une extension d'un théorème de Minkowsky.** C. R. Acad. Sci. Paris **224**, 990–991 (1947).

(I) Let $Q(x, y, z)$ be an indefinite ternary quadratic form with the discriminant D. Let x_0, y_0, z_0 be arbitrary real numbers. Then there are three integers x, y, z such that

(1) $$|Q(x+x_0, y+y_0, z+z_0)|\leqq k|D|^{\frac{1}{3}},$$

where $k^3=.27$. (II) If (2) $Q=x^2+5y^2-z^2+5yz+zx$, $x_0=y_0=\frac{1}{2}$, $z_0=0$, it is impossible to satisfy the inequality (1) in the strict sense (i.e., with $<$ instead of $\leqq$). (III) There exists an absolute constant $k'<k$ so that k in (1) may be replaced by k' if we exclude forms $\alpha Q_0(x, y, z)$, where α is a real number and Q_0 is equivalent to (2). The proofs will appear

elsewhere. The following lemma is emphasized by the author. Let $\alpha, \beta, \gamma, \delta, c_1, c_2$ be real numbers, $\alpha\delta-\beta\gamma=\Delta\neq 0$, $\mu>0$, $\nu>0$, $\mu\nu\geqq\frac{1}{16}$. Then there exist integers x, y such that $-\nu|\Delta|\leqq(\alpha x+\beta y+c_1)(\gamma x+\delta y+c_2)\leqq\mu|\Delta|$. [If $\mu=\nu$, this is a well-known analogue of (1) for binary forms; see, e.g., Koksma, Diophantische Approximationen, Ergebnisse der Math., v. 4, no. 4, Springer, Berlin, 1936, pp. 18–20.]

V. Jarník (Prague).

J48-3 (10, 101i)

Davenport, H. Non-homogeneous ternary quadratic forms Acta Math. **80**, 65–95 (1948).

Let $\alpha, \beta, \gamma, \delta, c_1, c_2$ be real, with $\Delta=|\alpha\delta-\beta\gamma|\neq 0$. Then Minkowski [Geometrie der Zahlen, Teubner, Leipzig, 1910] showed that there exist integers x, y such that $|(\alpha x+\beta y+c_1)(\gamma x+\delta y+c_2)|\leqq\Delta/4$. This can be reformulated to assert that for any indefinite binary quadratic form $Q(x, y)$ and any real x_0, y_0, there exist integers x, y such that $|Q(x+x_0, y+y_0)|\leqq\Delta/4$. In the present paper this is generalized to indefinite ternary quadratic forms $Q(x, y, z)$ of determinant $D\neq 0$. It is shown that for any real x_0, y_0, z_0 there are integers x, y, z such that $|Q(x+x_0, y+y_0, z+z_0)|\leqq(27|D|/100)^{\frac{1}{3}}$, and that the constant is best possible. If, however, Q is restricted to null forms, it is shown that 27/100 can be replaced by $\frac{1}{4}$, which again is best possible. It is also shown that except for forms which are equivalent to a multiple of $x^2+5y^2-z^2+5yz+xz$, there are integers x, y, z such that $|Q(x+x_0, y+y_0, z+z_0)|\leqq(1-\delta)(27|D|/100)^{\frac{1}{3}}$ for some $\delta\geqq 10^{-40}$. This result has no analogue for binary quadratic forms [Nederl. Akad. Wetensch., Proc. **49**, 815–821 = Indagationes Math. **8**, 518–524 (1946); these Rev. **8**, 444].

In the course of the proof of the first theorem stated above, the following generalization of Minkowski's theorem is obtained: if μ, ν are positive numbers with $\mu\nu\geqq\frac{1}{16}$, then for any c_1, c_2 there are integers x, y such that

$$-\nu\Delta\leqq(\alpha x+\beta y+c_1)(\gamma x+\delta y+c_2)\leqq\mu\Delta.$$

W. J. LeVeque (Cambridge, Mass.).

Citations: MR 8, 444b = J32-4.

Referred to in J32-23, J32-24, J32-48, J48-8, J48-12, J48-19, J48-22, J52-21.

J48-4 (10, 511c)

Blaney, Hugh. Indefinite quadratic forms in n variables. J. London Math. Soc. **23**, 153–160 (1948).

Let $Q=Q(x_1, \cdots, x_n)$ denote an indefinite quadratic form in n variables with determinant $D\neq 0$. (I) For any $c\geqq 0$ a number $C=C(c, n)$ exists such that, for any Q, integers $x_1, \cdots, x_n$ with $(x_1, \cdots, x_n)=1$ exist, such that $c|D|^{1/n}<Q\leqq C|D|^{1/n}$. (II) For any $\gamma\geqq 0$ a number $\Gamma=\Gamma(\gamma, n)$ exists such that, for any Q and any real numbers $\alpha_1, \cdots, \alpha_n$, integers $x_1, \cdots, x_n$ exist, such that

$$\gamma|D|^{1/n}<Q(x_1+\alpha_1, \cdots, x_n+\alpha_n)\leqq\Gamma|D|^{1/n}.$$

Some applications to asymmetric approximations are given.

J. F. Koksma (Amsterdam).

Referred to in D76-2, H05-29, J44-12, J48-7, J48-11.

J48-5 (10, 511d)

Davenport, H. Note on indefinite ternary quadratic forms. J. London Math. Soc. **23**, 199–202 (1948).

The author proves the following theorem. Let $f(x, y, z)$ be an indefinite ternary form whose determinant D is negative. Denote the lower bound of the positive values of f for integral x, y, z by $L|D|^{\frac{1}{3}}$. Then for any real x_0, y_0, z_0 the lower bound of $|f(x_0+x, y_0+y, z_0+z)|$ for integral x, y, z does not exceed $\{2L^{\frac{1}{2}}-\frac{1}{4}L^2\}^{\frac{1}{3}}|D|^{\frac{1}{3}}$. From this theorem the author further deduces that there exist such forms f for which $f(x_0+x, y_0+y, z_0+z)$ takes arbitrarily small positive values, which is a contrast to a forthcoming result of the author on binary quadratic forms.

J. F. Koksma.

J48-6 (12, 806c)

Blaney, H. Indefinite ternary quadratic forms. Quart. J. Math., Oxford Ser. (2) **1**, 262–269 (1950).

Let $Q(x, y, z)$ be an indefinite ternary quadratic form with determinant $D<0$. Then for any real x_0, y_0, z_0, there exist integers x, y, z such that $0<Q(x+x_0, y+y_0, z+z_0)\leqq(-4D)^{\frac{1}{3}}$. This is true with strict inequality unless $Q(x+x_0, y+y_0, z+z_0)$ is equivalent to a positive multiple of either x^2+yz or $(x-y)(x+y+1)+2z^2$, in which cases it is not.

J. F. Koksma (Amsterdam).

J48-7 (14, 143a)

Rogers, C. A. Indefinite quadratic forms in n variables. J. London Math. Soc. **27**, 314–319 (1952).

Let $Q(u_1, \cdots, u_n)$ be an indefinite quadratic form in n variables with real coefficients and with non-zero determinant D. A theorem of Blaney [same J. **23**, 153–160 (1948); these Rev. **10**, 511] asserts that for any real numbers $\alpha_1, \cdots, \alpha_n$ integers $u_1, \cdots, u_n$ exist for which

$$(1)\qquad |Q(u_1+\alpha_1, \cdots, u_n+\alpha_n)|\leqq c_n|D|^{1/n},\quad c_n=2^{n-2}.$$

In order to improve on this result the author considers linear forms

$$x_i=a_i^{(1)}u_1+\cdots+a_i^{(n)}u_n\quad(i=1, \cdots, n),$$

such that

$$Q(u_1, \cdots, u_n)=x_1^2+\cdots+x_r^2-x_{r+1}^2-\cdots-x_n^2$$

and using a fundamental lemma of Mahler [Proc. Roy. Soc. London. Ser. A. **187**, 151–187 (1946), p. 156; these Rev. **8**, 195] reduces the problem to the case that (I) $D=1$, (II) the lattice Λ, consisting of the points $(x_1, \cdots, x_n)$ which arise from integral values of $u_1, \cdots, u_n$, contains n independent points on the boundary of the sphere $x_1^2+\cdots+x_n^2\leqq\{m(\Lambda)\}^2$, where $m(\Lambda)$ denotes the smallest distance between two different points of Λ. This leads to the result that in (1) one may take $c_n=\frac{1}{4}n^2\gamma_n$, where γ_n is Hermite's constant (for which we have $\limsup\gamma_n/n=1/\pi e$). In a note added in proof the author communicates an argument of Davenport showing that one may even take $c_n=\frac{1}{4}n\gamma_n$.

J. F. Koksma (Amsterdam).

Citations: MR 8, 195f = H15-7; MR 10, 511c = J48-4.

J48-8 (16, 802b)

Barnes, E. S. The inhomogeneous minimum of a ternary quadratic form. Acta Math. **92**, 13–33 (1954).

Let $Q(x, y, z)$ be an indefinite ternary quadratic form of determinant $D\neq 0$. It was proved by the reviewer [Acta Math. **80**, 65–95 (1948); MR **10**, 101] that there exist, for any real x_0, y_0, z_0, integers x, y, z such that

$$(*)\qquad |Q(x+x_0, y+y_0, z+z_0)|\leqq\left(\frac{27}{100}|D|\right)^{1/3}.$$

It was further proved that the constant 27/100 can be diminished slightly if Q is not equivalent to a multiple of Q_1, where

$$Q_1(x, y, z)=x^2+5y^2-z^2+5yz+zx.$$

In the present paper the author succeeds in proving more, namely that (*) can be improved to

$$|Q(x+x_0, y+y_0, z+z_0)|<\left(\frac{4}{15}|D|\right)^{1/3}$$

except in two special cases. These are the cases when $2x_0$, $2y_0$, $2z_0$ are odd integers and when in addition Q is equivalent to a multiple of either Q_1 or Q_2, where

$$Q_2(x, y, z)=2x^2-y^2+15z^2.$$

The proof makes use of the powerful technique developed by Barnes and Swinnerton-Dyer in the paper reviewed below* for investigating the best asymmetric inequalities satisfied by non-homogeneous indefinite binary quadratic forms. The details of the present paper are complicated and delicate, as might be expected. *H. Davenport* (London).

*I.e. MR **16**, 802C = J32-48.

Citations: MR 10, 101i = J48-3; MR 16, 802c = J32-48.
Referred to in J48-12.

J48-9 (17, 715c)

Mordell, L. J. The minimum of an inhomogeneous quadratic polynomial in n variables Math. Z. **63** (1956), 525–528.

Let $f(X)=\sum a_{rs}x_rx_s+2\sum a_rx_r$, with real coefficients, $a_{rs}=a_{sr}$, $a_{rr}>0$, det $(a_{rs})\neq 0$. Then there exist numbers $X=(x_r)$ with assigned residues modulo 1 such that $f(X)\leqq k$ if k is given by $\det\begin{Vmatrix}A & B\\ C & -k\end{Vmatrix}=0$, where $A=(a_{rs})$, B is the column vector $(\frac{1}{2}a_{rr}+a_r)$, and C the row vector $(a_s-\frac{1}{2}a_{ss})$ and if, further, the indices $1, \cdots, n$ can be divided into two sets $\lambda, \lambda', \cdots$ and $\mu, \mu', \cdots$ such that $a_{\lambda\lambda'}\leqq 0$ $(\lambda\neq\lambda')$ and $a_{\mu\mu'}\leqq 0$ $(\mu\neq\mu')$. This generalizes a result of Dirichlet for $n=2$. The proof is simple and uses algebraic identities. For $n=2$ the result is shown to be best possible; for $n>2$ it is pointed out that it cannot be.
L. Tornheim (Berkeley, Calif.).

Referred to in R16-50.

J48-10 (17, 1060c)

Barnes, E. S. The covering of space by spheres. Canad. J. Math. **8** (1956), 293–304.

Let $f(x)=f(x_1, \cdots, x_n)$ be a positive definite quadratic form of determinant D. Define the "inhomogeneous minimum"

$$m(f)=\sup_{\alpha}\inf_{x} f(x+\alpha),$$

where α and x run through all real vectors and all integer vectors respectively. The author, by analogy with the homogeneous case, says that $f(x)$ is extreme if the expression $D^{-1/n}m(f)$ is a (local) minimum, and absolutely extreme if it is a global minimum. For $n=3$ he shows that the only extreme forms are the multiples of

$$x_1^2+x_2^2+x_3^2(x_1-x_2)^2+(x_2-x_3)^2+(x_3-x_1)^2$$

and forms equivalent to them. He shows that these are absolutely extreme. It follows that

$$m(f)\geqq(125D/1024)^{\frac{1}{3}}$$

whenever $n=3$, a result due to Bambah [Proc. Nat. Inst. Sci. India **20** (1954), 25–52; MR **15**, 780]. The ingenious proof uses a type of reduction of quadratic forms developed by Voronoï [J. Reine Angew. Math. **133** (1908), 97–178; **134** (1908), 198–287; **136** (1909), 67–181]. It is claimed that this method is simpler than that of Bambah and opens up the possibility of attacking $n>3$.
J. W. S. Cassels (Cambridge, England).

Citations: MR 15, 780c = H30-32.
Referred to in H30-36, H30-73, H30-76, H30-80, J48-18.

J48-11 (18, 634a)

Foster, D. M. E. Indefinite quadratic polynomials in n variables. Mathematika **3** (1956), 111–116.

This paper is a sharpening of a result of H. Blaney [J. London Math. Soc. **23** (1948), 153–160; MR **10**, 511] which is as follows: If $q(x_1, \cdots, x_n)$ is an indefinite quadratic form in n variables with real coefficients and determinant $\Delta_n\neq 0$, for any $\gamma\geqq 0$, there is a number $\Gamma=\Gamma(\gamma, n)$ such that the inequalities

(1) $\gamma|\Delta_n|^{1/n}<q(x_1+\alpha_1, \cdots, x_n+\alpha_n)<\Gamma|\Delta_n|^{1/n}$

are solvable in integers $x_1, \cdots, x_n$ for any real $\alpha_1, \cdots, \alpha_n$.

The author shows that the same conclusion may be drawn for $\Gamma(\gamma, n)=\gamma+C_n\gamma^{1/2^n}+C_n'$, where C_n, C_n' are suitable positive numbers depending only on n.

B. W. Jones (Boulder, Colo.).

Citations: MR 10, 511c = J48-4.
Referred to in D76-18, T35-37.

J48-12 (19, 944d)

Barnes, E. S. The inhomogeneous minimum of a ternary quadratic form. II. Acta Math. **96** (1956), 67–97.

Let $Q(x, y, z)$ be an indefinite ternary quadratic form with real coefficients and determinant $D\neq 0$. For any real x_0, y_0, z_0 let $M(Q; x_0, y_0, z_0)$ denote the lower bound of $|Q(x+x_0, y+y_0, z+z_0)|$ for integers x, y, z. It was proved by Davenport [Acta Math. **80** (1948), 65–95; MR **10**, 101] that $M(Q; x_0, y_0, z_0)\leqq(27|D|/100)^{1/3}$, with equality for a particular form Q_1 when $x_0=y_0=z_0=\frac{1}{2}$, and it was further proved that the inequality is "isolated". This result was much improved by Barnes [Acta Math. **92** (1954), 13–33; MR **16**, 802] and is further improved in the present paper. It is now shown that $M(Q; x_0, y_0, z_0)\leqq(|D|/4)^{1/3}$ except when Q is equivalent to a multiple of either Q_1 or Q_2, where Q_2 is a particular form for which $M(Q_2; \frac{1}{2}, \frac{1}{2}, \frac{1}{2})=(4|D|/15)^{1/3}$. Even for Q_1 or Q_2 the inequality remains valid except when $2x_0$, $2y_0$, $2z_0$ are all odd integers. The constant $\frac{1}{4}$ has a certain relevance in the problem, since it is known that $M(Q; x_0, y_0, z_0)\leqq(|D|/4)^{\frac{1}{3}}$ if Q represents 0 for integers x, y, z not all 0. The work of the present paper is difficult and elaborate; it is based on a new inequality for a positive value of $Q(x, y, z)$ when $D<0$ [Proc. London Math. Soc. (3) **5** (1955), 185–196; MR **16**, 1002], and on the technique developed by Barnes and Swinnerton-Dyer [Acta Math. **92** (1954), 199–234; MR **16**, 802] for investigating the values of inhomogeneous indefinite binary quadratic forms. *H. Davenport.*

Citations: MR 10, 101i = J48-3; MR 16, 802b = J48-8; MR 16, 1002f = J44-22.

J48-13 (20# 5167)

Birch, B. J. The inhomogeneous minimum of quadratic forms of signature zero. Acta Arith. **4** (1958), 85–98.

Minkowski's theorem on the product of two non-homogeneous linear forms can be stated as follows. Let $Q(x, y)$ be an indefinite binary quadratic form of determinant D, with real coefficients. Let

$$M_I(Q)=\sup\inf Q(x^*+x, y^*+y),$$

where "inf" is over integers x, y and "sup" is over real x^*, y^*. Then $M_I(Q)\leqq|\frac{1}{4}D|^{\frac{1}{2}}$. The present paper contains a generalization of this: if Q_{2n} is a quadratic form in $2n$ variables of signature zero, with real coefficients, then $M_I(Q_{2n})\leqq|\frac{1}{4}D|^{1/(2n)}$. Further, the sign of equality is needed if and only if Q is equivalent to a multiple of the form $x_1x_2+x_3x_4+\cdots+x_{2n-3}x_{2n-2}+2x_{2n-1}x_{2n}$. The result is deduced from auxiliary theorems. Thus it is shown that a more precise result holds if $n\geqq 3$ and Q_{2n} does not represent zero properly; also that $M_I(Q_{2n})=0$ if $n\geqq 2$ and Q_{2n} is not a multiple of a rational form and assumes values arbitrarity near to 0 (including 0). The detailed work of the paper is difficult, though technically elementary, and uses a variety of methods.

H. Davenport (Cambridge, England)

Referred to in J48-17, J48-19.

J48-14 (22# 37)

Watson, G. L. The inhomogeneous minimum of an indefinite quadratic form. Proc. Cambridge Philos. Soc. **55** (1959), 368–370.

The author deals with the indefinite n-ary quadratic

form $f(x_1, x_2, \cdots, x_n)$ and defines $M_I(f)$ as

$$\sup_{x_1^*,\cdots,x_n^*} \{ \inf_{x_i \equiv x_i^* \pmod 1} |f(x_1, \cdots, x_n)| \},$$

where x_i^* range independently over all real numbers. He proves the theorem: Suppose that the quadratic form $f=f(x_1, \cdots, x_n)$ has integral coefficients and is indefinite, non-degenerate and primitive. Let ε be any positive constant, and write

$$\theta_n = \frac{3n-4}{4(n-1)^2}.$$

Then we have

$$M_I(f) \ll |d(f)|^{\theta_n+\varepsilon},$$

where $d(f)$ is the discriminant of f and the constant implied by the notation (that of Vinogradov) depends only on n and ε.

His proof is modeled on one of Birch and is related to results of Davenport, Ridout, Birch and others.

B. W. Jones (Boulder, Colo.)

J48-15 (23# A1610)

Barnes, E. S.

The inhomogeneous minima of indefinite quadratic forms.

J. Austral. Math. Soc. **2** (1961/62), 9–10.

Let $f(\mathbf{X})$ be a real indefinite n-ary quadratic form with determinant $d \neq 0$. Define, for any vector $\boldsymbol{\alpha}$, $m(f; \boldsymbol{\alpha}) = \inf f(\mathbf{X})$ over all $\mathbf{X} \equiv \boldsymbol{\alpha} \pmod 1$, $m(f) = \sup_{\boldsymbol{\alpha}} m(f; \boldsymbol{\alpha})$, and $c_n = \inf_f m(f)/|d|^{1/n}$. It is known that $c_2 > 0$ [see, e.g., J. W. S. Cassels, *Introduction to the geometry of numbers*, Springer, Berlin, 1959, p. 306]. The author shows that $c_n = 0$ for all $n \geqq 3$, by proving the sharper result: for any $n \geqq 3$, and any signature (r, s), $r \geqq 1$, $s \geqq 1$, $r+s=n$, there exists a quadratic form f of signature (r, s) with determinant ± 1 and $m(f)=0$. {Reviewer's correction: In the left-hand term in line 13, p. 10, read $-(ab)^{1/2} - \frac{1}{4}a$ for $-(ab)^{1/2} + \frac{1}{4}a$.}

R. P. Bambah (Chandigarh)

J48-16 (24# A2561)

Barnes, E. S.

The positive values of inhomogeneous ternary quadratic forms.

J. Austral. Math. Soc. **2** (1961/62), 127–132.

Let $f(x, y, z)$ be an indefinite ternary quadratic form of signature (2, 1) and determinant $d \neq 0$. It is shown that, given any triple (x_0, y_0, z_0) of real numbers, there exists a triple (x, y, z) such that $x-x_0$, $y-y_0$, $z-z_0$ are integers, and such that $0 < f(x, y, z) \leqq (4|d|)^{1/3}$. The equality sign is necessary for only a few possible choices of f and the triple (x_0, y_0, z_0); these possibilities are given explicitly.

I. Reiner (Urbana, Ill.)

Referred to in J48-20, J48-22.

J48-17 (27# 3597)

Chalk, J. H. H.

Integral bases for quadratic forms.

Canad. J. Math. **15** (1963), 412–421.

Let $Q_{n,s}(\mathbf{x}) = \sum a_{ij}x_i x_j$ be a real indefinite quadratic form in n variables with determinant $D = \det(a_{ij}) \neq 0$ and signature s. Define the inhomogeneous minimum of Q as usual by $M_I(Q) = \sup_{\mathbf{x}_0} \inf_{\mathbf{x}} |Q(\mathbf{x}+\mathbf{x}_0)|$, where $\mathbf{x}$ runs over integer points and $\mathbf{x}_0$ runs over real points. There is a best-possible constant $C_{n,s}$ such that $M_I(Q_{n,s}) \leqq C_{n,s}|D|^{1/n}$ for all forms Q of the right type; the value of $C_{n,s}$ is known for $n \leqq 3$ and for $s=0$. The author defines another minimum M_B by $M_B(Q) = \inf\{\max_{r=1,\cdots,n} |Q(\mathbf{x}_r)|\}$, where the infimum is taken over all bases $\mathbf{x}_1, \cdots, \mathbf{x}_n$ of the lattice of integral points. There is a best-possible constant $C''_{n,s}$ such that $M_B(Q_{n,s}) \leqq C''_{n,s}|D|^{1/n}$; the author conjectures that $C''_{n,s} = 4^{1/n}C_{n,s}$. This was known for $n \leqq 3$, and the author checks it for $s=0$ by proving $C''_{2m,0}=1$. The author also proves that $M_B(Q)=0$ for any form Q in at least 3 variables that represents arbitrarily small non-zero values. The author's reduction theory is similar to that of the reviewer [Acta Arith. **4** (1958), 85–98; MR **20** #5167].

B. J. Birch (Manchester)

Citations: MR 20# 5167 = J48-13.

J48-18 (33# 7307)

Dickson, T. J.

An extreme covering of 4-space by spheres.

J. Austral. Math. Soc. **6** (1966), 179–192.

Any positive definite quadratic form $f=f(x)=f(x_1, \cdots, x_n)$ has an "inhomogeneous minimum" $m=m(f)= \sup_\lambda \inf_x f(x+\lambda)$, where λ and x run through all real vectors and all integral vectors, respectively. This is regarded as a function of the coefficients of f. Another such function is their determinant D. By analogy with the homogeneous case [see, e.g., the reviewer, Canad. J. Math. **3** (1951), 391–441, 395; MR **13**, 443], f is said to be "extreme" if m^n/D is a local minimum, and "absolutely extreme" if m^n/D is a global minimum. Bleicher [ibid. **14** (1962), 632–650; MR **26** #6862] proved that the form $H_n = (n+1)\sum_1^n x_i^2 - (\sum_1^n x_i)^2 = n\sum_1^n x_i^2 - 2\sum_{i<j} x_i x_j$, which is "reciprocal" to $\sum x^2 + (\sum x)^2$, is extreme for all n. The corresponding lattice consists of the vertices of Hinton's honeycomb [see the reviewer, J. Math. Pures Appl. (9) **41** (1962), 137–156, 154; MR **25** #4417]. Barnes [Canad. J. Math. **8** (1956), 293–304; MR **17**, 1060] proved that every binary or ternary extreme form is equivalent to H_2 or H_3, respectively, in agreement with the known fact that H_2 and H_3 are absolutely extreme. Delone and Ryškov [Dokl. Akad. Nauk SSSR **152** (1963), 523–524; MR **31** #126] proved that H_4 is absolutely extreme, but left open the possibility of other extreme quaternary forms. The author provides one instance of such a form, $3\sum_1^4 x_i^2 - (\sum_1^4 x_i)^2 + 2(1+\alpha)x_1x_2 + 2(2-\alpha)x_3x_4$, where α is the smaller root of the equation $x^2 - 5x + 3 = 0$.

H. S. M. Coxeter (Toronto, Ont.)

Citations: MR 13, 443c = J44-11; MR 17, 1060c = J48-10; MR 26# 6862 = H30-54; MR 31# 126 = H30-65.

J48-19 (34# 7461)

Dumir, Vishwa Chander

Inhomogeneous minimum of indefinite quaternary quadratic forms.

Proc. Cambridge Philos. Soc. **63** (1967), 277–290.

Let $Q(x_1, \cdots, x_n)$ be an indefinite quadratic form with signature (r, s) and determinant $D \neq 0$. It is known that there exists a constant $C_{r,s}$ such that, given any real $c_1, \cdots, c_n$, there exist integral $x_1, \cdots, x_n$ satisfying (*) $|Q(x_1+c_1, \cdots, x_n+c_n)| \leqq (C_{r,s}|D|)^{1/n}$. Denote by $K_{r,s}$, for each (r, s), the lower bound of constants $C_{r,s}$ for which this inequality holds. For small $n=r+s$, the known values of $K_{r,s}$ are $K_{1,1}=1/4$ (a classical result of Minkowski), $K_{1,2}=K_{2,1}=27/100$ [H. Davenport, Acta Math. **80** (1948), 65–95; MR **10**, 101] and $K_{2,2}=1/4$ [B. J. Birch, Acta Arith. **4** (1958), 85–98; MR **20** #5167]. The author here completes the case $n=4$ by proving that $K_{1,3}=K_{3,1}=1/3$. The result (*) then holds with $C_{1,3}=C_{3,1}=1/3$, and the equality sign is needed if and only if Q is equivalent to a multiple of $x_1^2+x_2^2+3x_3^2-x_4^2$ with $c_i \equiv 1/2 \pmod 1$ $(i=1, \cdots, 4)$.

E. S. Barnes (Adelaide)

Citations: MR 10, 101i = J48-3; MR 20# 5167 = J48-13.

J48-20 **(35# 134)**

Dumir, Vishwa Chander

Asymmetric inequalities for non-homogeneous ternary quadratic forms.

Proc. Cambridge Philos. Soc. **63** (1967), 291–303.

Let $Q(x, y, z)$ be an indefinite ternary quadratic form of determinant $D>0$. The author proves that, given any real x_0, y_0, z_0, there exist integers x, y, z satisfying $0<Q(x+x_0, y+y_0, z+z_0)\leqq(8D)^{1/3}$. The equality sign is needed for just one class of forms.

This result complements one of the same form obtained by the reviewer [J. Austral. Math. Soc. **2** (1961/62), 127–132; MR **24** #A2561] for indefinite ternary forms with negative determinant. *E. S. Barnes* (Adelaide)

Citations: MR 24# A2561 = J48-16.

Referred to in J48-22.

J48-21 **(37# 4011a; 37# 4011b)**

Dumir, Vishwa Chander

Positive values of inhomogeneous quaternary quadratic forms. I.

J. Austral. Math. Soc. **8** (1968), 87–101.

Dumir, Vishwa Chander

Positive values of inhomogeneous quaternary quadratic forms. II.

J. Austral. Math. Soc. **8** (1968), 287–303.

For positive integers r, s, let $\Gamma_{r,s}$ be the least positive number with the following property: let $Q(x_1, \cdots, x_n)$ be an indefinite quadratic form with real coefficients, determinant $D\neq 0$, and signature $(r, s)(r+s=n)$, and let $c_1, \cdots, c_n$ be any real numbers; then there exists an integral $x_1, \cdots, x_n$ satisfying $0<Q(x_1+c_1, \cdots, x_n+c_n)\leqq(\Gamma_{r,s}|D|)^{1/n}$. It is known that $\Gamma_{1,1}=4$, $\Gamma_{2,1}=4$, $\Gamma_{1,2}=8$. Here the values $\Gamma_{3,1}=16/3$ and $\Gamma_{2,2}=16$ are established, along with the set of forms for which the equality sign is needed. The methods used are purely arithmetical, relying on known results for homogeneous and inhomogeneous forms in 2, 3 and 4 variables. *E. S. Barnes* (Adelaide)

J48-22 **(39# 5473)**

Dumir, Vishwa Chander

Asymmetric inequality for non-homogeneous ternary quadratic forms.

J. Number Theory **1** (1969), 326–345.

A function f is determined with the following property: If $Q(x, y, z)$ is an indefinite ternary quadratic form of determinant $D<0$, t is a non-negative real number and x_0, y_0, z_0 are real numbers, then there exist integers x, y, z satisfying

$$-t(f(t)|D|)^{1/3}<Q(x+x_0, y+y_0, z+z_0)\leqq(f(t)|D|)^{1/3}.$$

$f(t)$ is specified as the reciprocal of a cubic polynomial in each of six intervals; e.g., $f(t)=8/(1+t)^3$ for $t\geqq 7$ (with the case $t=\infty$ included as a natural limit). The result is best possible for only eight values of t, namely, 0, 1/7, 3/5, 1, 9/7, 3, 7 and ∞. The previously known (best possible) values of f were $f(0)$ [the reviewer, J. Austral. Math. Soc. **2** (1961/62), 127–132; MR **24** #A2561], $f(1)$ [H. Davenport, Acta Math. **80** (1948), 65–95; MR **10**, 101] and $f(\infty)$ [the author, Proc. Cambridge Philos. Soc. **63** (1967), 291–303; MR **35** #134]. *E. S. Barnes* (Adelaide)

Citations: MR 10, 101i = J48-3; MR 24# A2561 = J48-16; MR 35# 134 = J48-20.

J52 MINIMA OF HIGHER DEGREE FORMS AND OTHER FUNCTIONS

See also Sections D76, E76, J36, J40.

See also reviews D76-7, D76-12, H15-46, H20-3, H20-4, H20-5, J36-5.

J52-1 **(3, 70e)**

Mordell, L. J. On the minimum of a binary cubic form. J. London Math. Soc. **16**, 83–85 (1941).

Let $f(x, y)$ be an arbitrary binary cubic form with real coefficients and the discriminant d. There exist integers $x, y\neq 0, 0$ such that $|f(x, y)|\leqq k^{\frac{1}{4}}$, where $k=d/49$ for $d>0$ and $k=-d/23$ for $d<0$ (best possible constants). The proof of this theorem and of some other related results is to appear in the Proc. London Math. Soc. *C. L. Siegel.*

J52-2 **(3, 70g)**

Davenport, H. On a conjecture of Mordell concerning binary cubic forms. Proc. Cambridge Philos. Soc. **37**, 325–330 (1941).

The author here proves the following: Let ϵ be any positive number. There exists a binary cubic form $F(x, y)$ of determinant D and with real coefficients such that none of the roots of $F(x, 1)=0$ is equivalent to any root of $\varphi(x)=0$ with the property that $F(x, y)>\alpha(\epsilon)$ for all integers $x, y\neq 0, 0$, where $\varphi(x)$ is x^3+x^2-2x-1 or x^3-x-1 according as $D<0$ or >0 and

$$\alpha(\epsilon)=\left(\frac{|D|}{49+\epsilon}\right)^{\frac{1}{4}} \quad \text{or} \quad \left(\frac{D}{23+\epsilon}\right)^{\frac{1}{4}}$$

in the respective cases. This bears on a result of Mordell's which is soon to appear in the Proc. London Math. Soc., namely: $F(x, y)\leqq\alpha(0)$ for all real binary cubics. This paper shows that Mordell's upper limit cannot be bettered even by excluding forms allied with $\varphi(x)$. *B. W. Jones.*

J52-3 **(5, 172d)**

Mordell, L. J. On numbers represented by binary cubic forms. Proc. London Math. Soc. (2) **48**, 198–228 (1943).

Let $f(x, y)=ax^3+bx^2y+cxy^2+dy^3$ be a binary cubic form with real coefficients a, b, c, d and discriminant $D=27a^2d^2-18abcd-b^2c^2+4ac^3+4db^3$; the variables x, y are integral and not both zero. Two forms are called equivalent ($\sim$) if they can be transformed into each other by linear substitutions with integral coefficients. The following theorems are proved. (1) If $D<0$, then $|f(x, y)|\leqq(-D/49)^{\frac{1}{4}}$ is solvable; the equality sign is necessary when and only when $D=-49e^4$ and $f(x, y)\sim e(x^3+x^2y-2xy^2-y^3)$, where e is an arbitrary constant. (2) If $D>0$, then $|f(x, y)|\leqq(D/23)^{\frac{1}{4}}$ is solvable; the equality is necessary when and only when $D=23e^4$ and $f(x, y)\sim e(x^3-xy^2-y^3)$. (3) If $D=0$, then $|f(x, y)|<\epsilon$ is solvable, for arbitrary $\epsilon>0$. In the last case $f(x, y)\sim a(x+py)^2(x+qy)$ with real a, p, q; therefore the third theorem is an immediate consequence of Minkowski's theorem on linear forms. In the two other cases $f(x, y)$ can be transformed by a real linear substitution into $f_1(x, y)=x^3+x^2y-2xy^2-y^3$ or $f_2(x, y)=x^3-xy^2-y^3$, respectively. Let a lattice L be given by $x=\alpha\xi+\beta\eta$, $y=\gamma\xi+\delta\eta$, with arbitrary real $\alpha, \beta, \gamma, \delta$ satisfying $\alpha\delta-\beta\gamma=1$, where ξ, η run through all integer values. It is proved that a point of L different from (0, 0) lies in each of the regions R defined by $|f_1(x, y)|\leqq 1$, $|f_2(x, y)|\leqq 1$; a prominent part is played by the linear substitutions which transform R into itself;

moreover, Minkowski's theorem is applied to a particular set of parallelograms. *C. L. Siegel* (Princeton, N. J.).

Referred to in H15-3, H15-46, H20-5, H20-7, J52-5.

J52-4 **(5, 254f)**

Davenport, H. The minimum of a binary cubic form. J. London Math. Soc. **18**, 168–176 (1943).

The object of the paper is to give simple proofs of two theorems of Mordell. These theorems may be stated as follows, where $f(x, y)$ is a binary cubic form with real coefficients and discriminant D. Theorem: there exist integers x, y not both zero for which $|f(x, y)| \leqq (-D/23)^{\frac{1}{4}}$ or $(D/49)^{\frac{1}{4}}$ according as $D<0$ or $D>0$. The strict inequality holds except when f is equivalent to $(-D/23)^{\frac{1}{4}}(x^3-xy^2-y^3)$ or $(D/49)^{\frac{1}{4}}(x^3+x^2y-2xy^2-y^3)$. *B. W. Jones.*

Referred to in H20-5, H20-7.

J52-5 **(6, 37i)**

Mordell, L. J. The minimum of a binary cubic form. I. J. London Math. Soc. **18**, 201–210 (1943).

In a recent paper [Proc. London Math. Soc. (2) **48**, 198–228 (1943); these Rev. **5**, 172], the author proved the following theorem. Let $x=\alpha\xi+\beta\eta$, $y=\gamma\xi+\delta\eta$, where ξ, η assume all integral values, and $\alpha, \beta, \gamma, \delta$ are real numbers such that $\alpha\delta-\beta\gamma=1$, be any lattice L of determinant unity. Then a point P of L other than the origin O lies in the region R defined by $|x^3-xy^2-y^3|\leqq 1$; the equality sign is required for, and only for, the two lattices $x=\xi$, $y=\eta$ and $(3\theta^2-1)x=-\xi-(\theta+3)\eta$, $(3\theta^2-1)y=-3\theta\xi+\eta$, where θ is the real root of $t^3=t+1$. Now it is proved that P may be chosen in the finite part of R cut off by the parts of the lines $y=1$, $3\theta^2x-y=3\theta-1$ lying in the first quadrant, and their images in O. The proof uses the same geometric ideas, but it is simpler and involves less calculation than the former one. The author remarks that a further simplification is obtained by taking R to be the region $|x^3+y^3|\leqq 1$ instead of $|x^3-xy^2-y^3|\leqq 1$; this will be the subject of a further paper [cf. the two following reviews]. *C. L. Siegel.*

Citations: MR 5, 172d = J52-3.

Referred to in H20-5, H20-7.

J52-6 **(6, 37j)**

Mordell, L. J. The minimum of a binary cubic form. II. J. London Math. Soc. **18**, 210–217 (1943).

[Cf. the preceding and the following reviews.] A new proof of Minkowski's theorem on the product of two nonhomogeneous linear forms $f(x, y)$, $g(x, y)$, with determinant 1, is given by applying Minkowski's theorem on convex bodies to the three-dimensional region

$$\lambda|zf(x/z, y/z)|+\lambda^{-1}|zg(x/z, y/z)|\leqq 1, \qquad |z|\leqq 2,$$

λ any positive constant. *C. L. Siegel* (Princeton, N. J.).

J52-7 **(6, 38a)**

Mordell, L. J. The product of two non-homogeneous linear forms. III. J. London Math. Soc. **18**, 218–221 (1943).

[Cf. the two preceding reviews.] A simplified proof and a refinement of the following theorem: a point of L other than the origin lies in the region $|x^3-x^2y-2xy^2+y^3|\leqq 1$; the equality sign is required for, and only for, the two lattices $x=\xi$, $y=\eta$ and $(\theta-\phi)x=\theta\phi^2\xi+\psi\eta$, $(\theta-\phi)y=-\theta^2\xi-\theta\phi^2\eta$, where θ, ϕ, ψ are the three roots of $t^3+t^2=2t+1$ and $\theta>\phi>\psi$ [cf. the preceding review]. *C. L. Siegel.*

Referred to in H20-5, J32-27.

J52-8 **(6, 57c)**

Davenport, H. On the minimum of a ternary cubic form. J. London Math. Soc. **19**, 13–18 (1944).

The author proves the following results related to work of Mordell [J. London Math. Soc. **17**, 107–115 (1942); these Rev. **4**, 131]. Let m be the minimum of $f=x^3+y^3+z^3$, where x, y, z are linear forms in the integers u, v, w (not all zero) with real coefficients whose determinant is 1. Then

$$m\leqq\frac{8}{\Gamma^3(4/3)\{2+27\sqrt{3}/(2\pi)\}}=1.1897\cdots.$$

There is a particular form given for which m is .8151. *B. W. Jones* (Ithaca, N. Y.).

Citations: MR 4, 131d = J36-5.

Referred to in J52-12, J52-25.

J52-9 **(6, 57d)**

Mordell, L. J. On the minimum of a ternary cubic form. J. London Math. Soc. **19**, 6–12 (1944).

If f in the above review is replaced by $F=x^3+y^3+z^3+Kxyz$, the author derives the first upper bounds for the minimum of F which are lower than those obtained by Minkowski. For $K=0$, Mordell's result is not as good as that of Davenport above. *B. W. Jones* (Ithaca, N. Y.).

Referred to in H15-66.

J52-10 **(7, 418c)**

Davenport, H. The reduction of a binary cubic form. I. J. London Math. Soc. **20**, 14–22 (1945).

Let $f(x, y)=ax^3+bx^2y+cxy^2+dy^3$, where the coefficients are real and the discriminant D is positive. A simple arithmetical proof is given of Mordell's theorem [same J. **18**, 201–217 (1943); these Rev. **6**, 37] that there exist integers x, y not both zero for which $|f(x, y)|\leqq(D/49)^{1/4}$, and this with strict inequality unless $f(x, y)$ is equivalent to

$$(D/49)^{1/4}(x^3+x^2y-2xy^2-y^3).$$

The author also proves that, if $f(x, y)$ is reduced and $D=49$, then at least one of the products $f(1, 0)f(0, 1)$, $f(1, 0)f(1, 1)$, $f(1, 0)f(1, -1)$, $f(0, 1)f(1, 1)$, $f(0, 1)f(1, -1)$ does not exceed 1 numerically. One of them is numerically less than 1 except when $\pm f(x, y)=x^3+x^2y-2xy^2-y^3$ or

$$\pm f(x, y)=x^3+2x^2y-xy^2-y^3.$$

If $f(x, y)=0$ for integers x, y not both zero the above results are truisms. In this case the author proves that there is no upper bound in terms of D for a nonzero value of $f(x, y)$. *H. S. A. Potter* (Aberdeen).

Citations: MR 6, 37h = D32-15.

Referred to in J52-14.

J52-11 **(7, 418d)**

Davenport, H. The reduction of a binary cubic form. II. J. London Math. Soc. **20**, 139–147 (1945).

[Cf. the preceding review.] The author defines a method of reduction for the binary cubic form with real coefficients $f(x, y)=ax^3+bx^2y+cxy^2+dy^3$ and negative discriminant $D=18abcd+b^2c^2-4ac^3-4db^3-27a^2d^2$. He proves that, if $f(x, y)$ is a reduced binary cubic form of discriminant -23, then one at least of $f(1, 0)$, $f(0, 1)$, $f(1, -1)$, $f(1, -2)$ does not exceed 1 numerically. Moreover, one of them is numerically less than 1 except when $f(x, y)=x^3+x^2y+2xy^2+y^3$ (in which case all four values are ± 1). *H. S. A. Potter.*

J52-12 **(8, 443g)**

Davenport, H. On the minimum of $x^3+y^3+z^3$. J. London Math. Soc. **21**, 82–86 (1946).

Let $m(k, n)$ (k, n are positive integers) denote the lower bound of the numbers λ satisfying the following conditions: if $x_1, \cdots, x_n$ are linear forms in $u_1, \cdots, u_n$ with real coefficients and determinant 1, then there exist integral values of $u_1, \cdots, u_n$ for which $0<|x_1^k+\cdots+x_n^k|\leqq\lambda$. Using a theorem of Blichfeldt [Trans. Amer. Math. Soc. **15**, 227–235 (1914)] in the geometry of numbers the author proves that $m(3, 3) \leqq 1.1571 \cdots$. This is a better inequality than that of the author's recent note [J. London Math. Soc. **19**, 13–18 (1944); these Rev. **6**, 57]. The author shows that the same method is applicable to the investigation of the estimate of the number $m(k, n)$ in the general case, where n is a positive and k an odd positive integer.

V. Knichal (Prague).

Citations: MR 6, 57c = J52-8.
Referred to in J52-25.

J52-13 (11, 11i)

Ankeny, N. C. A note on the minimum of a binary form. Bull. Amer. Math. Soc. **55**, 615–618 (1949).

Let D be the absolute value of the discriminant of a binary n-ic form f with real coefficients, and let $L(f)$ denote the lower bound of $|f|$ for integral values of its variables not both zero. If s denotes the number of pairs of conjugate imaginary roots, it is known, for $n=2$ and 3, that $L(f) \leqq cD$, where c is a constant depending only on s and n. Thus, $c=\frac{1}{3}$ if $s=1$, $n=2$; $c=1/23$ if $s=1$, $n=3$. By demonstrating the existence of forms with D near zero but L bounded away from zero, it is shown here that no such inequality holds if $n>3$. The proof develops properties of continued fractions such as the following. (i) If p_n/q_n is the nth convergent to the continued fraction $\alpha=\{a_0, a_1, \cdots\}$ (α irrational, $\alpha>1$), then if $q_n \leqq y < q_{n+1}$, $|x-\alpha y| > 1/(3a_{n+1}y)$. (ii) If $\alpha>1$ and $\beta>1$ are irrationals whose continued fraction expansions have their partial quotients less than N, then $|(x-\alpha y)(x+\beta y)| > 1/(3N)$ for x, y not both zero.

G. Pall (Chicago, Ill.).

J52-14 (11, 331c)

Chalk, J. H. H. Reduced binary cubic forms. J. London Math. Soc. **24**, 280–284 (1949).

Let $f(x, y)=ax^3+bx^2y+cxy^2+dy^3$ have distinct real linear factors. Its discriminant $D>0$ and its quadratic covariant $Q(x,y)=Ax^2+Bxy+Cy^2$ has discriminant $-3D$. Assume that f is reduced, i.e., that Q is. Then $A \leqq D$. Davenport [same J. **20**, 14–22 (1945); these Rev. **7**, 418] has shown that $\min\{|f(1, 0)|, |f(0, 1)|, |f(1, 1)|, |f(1, -1)|\} \leqq (D/49)^{\frac{1}{4}}$ with equality only when $f(x, y)$ is a multiple of

$$x^3+x^2y-2xy^2-y^3$$

or $x^3+2x^2y-xy^2-y^3$. The author improves this by replacing D by A^2. He also finds all reduced f for which $A^2=D$.

L. Tornheim (Ann Arbor, Mich.).

Citations: MR 7, 418c = J52-10.

J52-15 (11, 582g)

Mordell, L. J. The minimum of a binary cubic form. Acta Univ. Szeged. Sect. Sci. Math. **13**, 69–76 (1949).

Lecture at the University of Szeged.

J52-16 (12, 678b)

Davis, C. S. The minimum of a binary quartic form. I. Acta Math. **84**, 263–298 (1951).

Every binary real quartic form without real roots can be mapped into a uniquely determined form $x^4+6mx^2y^2+y^4$, $-\frac{1}{3}<m\leqq\frac{1}{3}$, by an appropriate affine transformation. In the present paper the critical lattices of the region of points (x, y) for which $x^4+6mx^2y^2+y^4\leqq 1$ (i.e., lattices of minimum determinant with no points other than the origin interior to the region) are determined. For $-\frac{1}{6}<m<\frac{1}{3}$ there are two critical lattices, the one generated by the vectors (α, α), (β, δ), where (α, α), (β, δ) and $(\alpha+\beta, \alpha+\delta)$ are on the boundary of the region and the reflection of this lattice by the y-axis. The determination of these lattices is reduced to a minimum problem which is solved with the use of the invariants of the quartic form. For $-\frac{1}{3}<m\leqq-\frac{1}{6}$ there are two lattices generated by P, Q such that P, Q, $P+Q$, $P-Q$ are on the boundary of the region and each lattice is obtained from the other by an affine mapping of the region into itself. By use of the method developed by Mordell [Proc. London Math. Soc. (2) **48**, 339–390 (1945); these Rev. **6**, 257] for the solution of the corresponding problem for the cubic forms the author shows these two lattices are the only critical lattices of the region. Let $f(x, y)$ be a real quartic form without real roots, $\mathfrak{D}$ its discriminant, and m the uniquely defined number such that the region of points (x, y) for which $f(x, y)\leqq 1$ is mapped into the region defined by $x^4+6mx^2y^2+y^4\leqq 1$ by an affinity. The ultimate result of the paper is the construction of the best possible function $k(m)$ so that, for all forms $f(x, y)$ with given m and $\mathfrak{D}$, integers x, y exist, not both of which are zero, so that $f(x, y)\leqq k(m)\mathfrak{D}^{\frac{1}{6}}$. This result is equivalent to the determination of the determinant of the critical lattices of the region of points x, y for which $x^4+6mx^2y^2+y^4\leqq 1$.

D. Derry.

Citations: MR 6, 257d = H15-3.
Referred to in J52-17, J52-23.

J52-17 (13, 115a)

Davis, C. S. The minimum of a binary quartic form. II. Acta Math. **85**, 183–202 (1951).

Let $f(x, y)$ be a real homogeneous binary quartic form with at least one real root. Let k be a number with the property that the region of points (x, y) for which $|f(x, y)|\leqq(k+\epsilon)\Delta^2$ always contains a point of every lattice of determinant Δ other than the origin for all positive ϵ. Let k^* denote the greatest lower bound of the numbers k. The object of this paper is to determine k^* for all $f(x, y)$ or to give (upper) estimates for it. If $f(x, y)$ has multiple roots, k^* is completely determined except for the case in which all the roots are real and two of them are distinct in which case an estimate is given. For the case where $f(x, y)$ has double roots and imaginary roots the form is shown to have a chain of successive minima which are linked with the Markoff minima of indefinite binary quadratic forms.

If $f(x, y)$ has four distinct real roots the results are first developed for the form $f_m(x, y)=x^4+6mx^2y^2+y^4$, $-1\leqq m<-\frac{1}{3}$, and then extended to the general forms by use of invariants. Where $-\frac{1}{2}<m<-\frac{1}{3}$, a region, $f_{m'}(x, y)\leqq 1$, with appropriately chosen m' from the range $-\frac{1}{3}<m'<\frac{1}{6}$ is inscribed within the region $|f_m(x, y)|\leqq 1$. As $k^*(m')$ (the number k^* for the form $f_{m'}(x, y)$) is known from part I of this paper [Acta Math. **84**, 263–298 (1951); these Rev. **12**, 678] an estimate for $k^*(m)$ is given by $k^*(m)<k^*(m')$. For the range $-1\leqq m\leqq-\frac{1}{2}$ the author proves $k^*(m)\leqq 1$ and $k^*(m)=1$ if and only if $|f_m(x, y)|\geqq 1$ for all integral (x, y) other than $(0, 0)$. This latter condition is shown to be satisfied for a set of values m of positive measure and also not to be satisfied for a set of values m of positive measure.

For the case in which $f(x, y)$ has both real and imaginary roots, all of which are distinct, the known estimates of k^* are listed and some of them are improved by inscribing a rectangle within the region $|x^4+6mx^2y^2-y^4|\leqq 1$.

D. Derry (Vancouver, B. C.).

Citations: MR 12, 678b = J52-16.

J52-18 (12, 806d)

Davenport, H. Note on a binary quartic form. Quart. J. Math., Oxford Ser. (2) **1**, 253–261 (1950).

Let x and y be linear forms in two variables u and v with real coefficients and with determinant 1. Then there exist integers u and v, not both zero, such that

$$(1) \qquad |x^2(x^2-y^2)|\leqq 1/(1+2\sqrt{5}).$$

This constant is the least possible, in the sense that there exist special linear forms x, y for which (1) has no solution with strict inequality.

J. F. Koksma (Amsterdam).

J52-19 **(13, 114d)**

Bambah, R. P. Non-homogeneous binary cubic forms. Proc. Cambridge Philos. Soc. **47**, 457–460 (1951).

The author has proved the following theorem in his thesis [Cambridge University, 1950]: Let

$$f(x, y) = ax^3 + bx^2y + cxy^2 + dy^3$$

be a cubic binary form with real coefficients and positive discriminant $D = 18abcd + b^2c^2 - 4ac^3 - 4db^3 - 27a^2d^2$. Then, if f is reduced in the sense of Hermite and x_0, y_0 are any two real numbers, there exist integers x, y, such that

$$|f(x+x_0, y+y_0)| \leqq \max(|f(\tfrac{1}{2}, 0)|, |f(0, \tfrac{1}{2})|, |f(\tfrac{1}{2}, \tfrac{1}{2})|, |f(\tfrac{1}{2}, -\tfrac{1}{2})|)$$

and there exist forms f for which the sign "=" is necessary. The present note gives a summary of the proof. In an addendum the author states that the theorem remains true, if the form is not restricted to be reduced.

J. F. Koksma (Amsterdam).

Referred to in H15-61, H15-81, J52-20.

J52-20 **(13, 919c)**

Chalk, J. H. H. The minimum of a non-homogeneous binary cubic form. Proc. Cambridge Philos. Soc. **48**, 392–401 (1952).

Let $f(x, y)$ be a binary cubic form with real coefficients and negative discriminant. For any real numbers x_0, y_0 there exist x, y satisfying $(x, y) \equiv (x_0, y_0) \pmod 1$ such that $8|f(x, y)| \leqq \max\{|f(1, 0)|, |f(0, 1)|, |f(1, 1)|, |f(1, -1)|\}$. It is proved geometrically by showing that, for each of certain lattices, the region $|x|(x^2+y^2) \leqq 1$ and its homothetic images at all lattice points cover the plane. Bambah has proved the similar theorem for forms of positive discriminant [same Proc. **47**, 457–460 (1951); these Rev. **13**, 114].

L. Tornheim (Ann Arbor, Mich.).

Citations: MR 13, 114d = J52-19.

Referred to in H15-81, J52-22.

J52-21 **(14, 358e)**

Barnes, E. S. On indefinite ternary quadratic forms. Proc. London Math. Soc. (3) **2**, 218–233 (1952).

Let $Q(x, y, z)$ be a real indefinite ternary quadratic form of determinant $d \neq 0$, and let $M(Q)$ be the lower bound of $|Q(x_1, y_1, z_1) \cdot Q(x_2, y_2, z_2) \cdot Q(x_3, y_3, z_3)|$ for integral variables satisfying

$$\begin{vmatrix} x_1 & x_2 & x_3 \\ y_1 & y_2 & y_3 \\ z_1 & z_2 & z_3 \end{vmatrix} = \pm 1.$$

It is proved that $M(Q) < |d|/2.2$ except when Q is equivalent (under unimodular transformation) to a multiple of Q_1, Q_2, or Q_3 (three explicitly given forms for which $M(Q)/|d| = 2/3$, 12/25 and 12/25, respectively.) It is also shown that if Q assumes values arbitrarily close to 0, then $M(Q) = 0$. In proving these theorems the author uses his previous results on quadratic forms [same Proc. (3) **1**, 257–283 (1951); these Rev. **13**, 627], and also a lemma due to H. Davenport [Acta Math. **80**, 65–95 (1948); these Rev. **10**, 101].

I. Reiner (Urbana, Ill.).

Citations: MR 10, 101i = J48-3; MR 13, 627d = J36-17.

J52-22 **(14, 540d)**

Mordell, L. J. The minima of some non-homogeneous functions of two variables. Duke Math. J. **19**, 519–527 (1952).

The following theorem is proved for real functions $f(x, y)$ satisfying certain general geometrical conditions. For every real pair x_0, y_0 there exist x, y such that $x \equiv x_0$, $y \equiv y_0 \pmod 1$ and $|f(x, y)| \leqq k \max(|f(1, 0)|, |f(0, 1)|, |f(1, 1)|, |f(1, -1)|)$, where k depends only on $f(x, y)$. Examples, where

$$X = px + qy, \quad Y = rx + sy \quad (px - qr \neq 0),$$

are:

$$f(x, y) = Ye^{|X|^n} \quad (1 < n \leqq 2), \quad k = \tfrac{1}{2};$$
$$f(x, y) = X^3 + Y^3 + |t|(X+Y), \quad k = \tfrac{1}{2};$$
$$f(x, y) = X|X|^{n-1} + Y|Y|^{n-1}, \quad k = 2^{-n} \quad (n > 1).$$

This last case for $n = 3$ is a result due to Chalk [Proc. Cambridge Philos. Soc. **48**, 392–401 (1952); these Rev. **13**, 919]. The proof is geometrical and is related to that of Chalk.

L. Tornheim (Ann Arbor, Mich.).

Citations: MR 13, 919c = J52-20.

Referred to in H15-61, H15-81.

J52-23 **(14, 623e)**

Černý, Karel. On the minimum of binary biquadratic forms. Čehoslovack. Mat. Ž. **2**(77), 1–56 (1952). (Russian. English summary)

The author investigates the minimum of a binary quartic form $f(x, y) = a_0x^4 + \cdots + a_4y^4$ for integers $x, y \neq 0, 0$, when $a_0, \cdots, a_4$ are real and the roots of the equation $f(x, 1) = 0$ are all complex. His results coincide with those found by C. S. Davis [Acta Math. **84**, 263–298 (1951); these Rev. **12**, 678]. (The author remarks in a footnote that he saw the review of this paper only when his own was in course of printing.) Černý's first main result corresponds to Davis's Theorem 2′ and his second main result to Davis's Theorem 4. The parameter M of Černý is $6m$ in Davis's notation. The treatment in both papers is based on techniques developed by Mordell.

H. Davenport (London).

Citations: MR 12, 678b = J52-16.

J52-24 **(17, 826d)**

Mordell, L. J. Some Diophantine inequalities. Mathematika 2 (1955), 145–149.

A very general theorem is proved, of which the following is a more or less typical consequence: Let $f(x) = |x_1 \cdots x_n| g(x)$, where

$$g(x) = \sum_{r=1}^{n} \sum_{s=1}^{n} a_{rs}x_rx_s + \sum_{r=1}^{n} a_rx_r + a_0.$$

Suppose that $g(x) \geqslant 0$ for all points x with coordinates numerically smaller than 1, and that $a_{rr} \geqq 0$ for $r = 1, \cdots, n$. Then there is a point x congruent to any preassigned point (mod 1), such that

$$f(x) \leqq \frac{1}{2^{n+2}} \sum_{r=1}^{n} a_{rr} + \frac{1}{2^n} a_0.$$

The latter theorem generalizes a result (not yet published) of Birch and Swinnerton-Deyer.

W. J. LeVeque.

J52-25 **(39# 2706)**

Spohn, W. G.

On the lattice constant for $|x^3 + y^3 + z^3| \leq 1$.

Math. Comp. **23** (1969), 141–149.

Let x, y, z be linear forms in u, v, w with real coefficients and determinant D, and such that $|x^3 + y^3 + z^3| > 1$ except for $u = v = w = 0$. Let Δ be the g.l.b. of such D: then it was shown by H. Davenport [J. London Math. Soc. **19** (1944), 13–18; MR **6**, 57; ibid. **21** (1946), 82–86; MR **8**, 443] that $0.864 \leq \Delta \leq 1.226$. In the present paper it is conjectured that $\Delta = 0.948754\ldots$. This conjecture is based on the construction of lattices for which $|x^3 + y^3 + z^3| \geq 1$ is satisfied for all points with norm (i.e., $\max(|u|, |v|, |w|)$) less than some given number (250 ultimately). To facilitate the construction the values of $x^3 + y^3 + z^3$ at 9 particular lattice points are taken as coordinates, and the condition that $x^3 + y^3 + z^3 = 1$ for some other given lattice point gives a surface in 9-dimensional space. It is found that this surface is approximately linear, and linear programming techniques are used to minimize the determinant of the lattice, subject to the conditions that $|x^3+y^3+z^3|\geqq 1$ for various other lattice points. If another lattice point fails to satisfy this condition, the procedure is repeated with a different basic set of 9 points. The values of Δ obtained at successive stages are remarkably close together, but the author's claim of almost geometrical convergence does not seem to be borne out by his results. There is also no guarantee that some lattice point with a large norm will not have considerable effect on the preceding work.

H. J. Godwin (London)

Citations: MR 6, 57c = J52-8; MR 8, 443g = J52-12.

J56 THE ABOVE PROBLEMS FOR APPROXIMATION BY ALGEBRAIC NUMBERS

See also reviews A56-57, H05-1, H15-63, J02-8, J16-27, J28-10, J32-14, J32-19, J64-5, Z30-20.

J56-1 **(2, 148e)**

Mahler, Kurt. On the product of two complex linear polynomials in two variables. J. London Math. Soc. **15**, 213–236 (1940).

Let α, β, γ, δ be four complex numbers such that $\alpha\delta-\beta\gamma=1$ and ξ, η two other complex numbers. Consider the product $P(x, y)=(\alpha x+\beta y+\xi)(\gamma x+\delta y+\eta)$, where x and y are variable integers in the Gaussian field $K(\sqrt{-1})$. Some time ago E. Hlawka [Monatsh. Math. Phys. **46**, 324–334 (1938)] proved that the minimum of the absolute value of $P(x, y)$ is not greater than $\frac{1}{2}$; he determined also the limiting case, when the sign of equality is necessary. The author obtains a new proof using the theory of reduction of positive binary Hermitian forms and the solution of a certain simple geometrical extremum problem. Two analogous theorems for the quadratic fields $K(\sqrt{-2})$ and $K(\sqrt{-3})$ are proved by means of the same method.

C. L. Siegel.

Referred to in J48-1.

J56-2 **(2, 149b)**

Hofreiter, Nikolaus. Diophantische Approximationen komplexer Zahlen. Monatsh. Math. Phys. **49**, 299–302 (1940).

Set $\lambda=1$ if $m\not\equiv 3$, $\lambda=2$ if $m\equiv 3 \pmod 4$. If $\alpha_1, \cdots, \alpha_{n-1}$ are complex numbers, linearly independent with respect to the field $k(i\sqrt{m})$, there exist infinitely many integers $q_1, \cdots, q_n$ in this field, such that

$$|L|=|\alpha_1 q_1+\cdots+\alpha_{n-1}q_{n-1}-q_n|<\frac{K_n}{|q_1\cdots q_{n-1}|}.$$

Here $\lambda K_2=\sqrt{(m/2)}$, $K_3{}^2=8m^{3/2}/(27\lambda^3\sqrt{3})$, $K_4=m/(4\lambda^2)$, $K_n{}^2<(n+1)!(2\sqrt{m}/(n\pi\lambda))^n$. The proof uses the inequality

$$n(|q_1|^2\cdots|q_{n-1}|^2|L|^2)^{1/n}\leqq\frac{|q_1|^2}{t}+\cdots+\frac{|q_{n-1}|^2}{t}+|L|^2t^{n-1},$$

where $q_\nu=y_\nu+(\lambda-1+i\sqrt{m})z_\nu/\lambda$, and thus reduces the problem to one on minima of quadratic forms.

G. Pall.

J56-3 **(3, 71a)**

Koksma, J. F. und Meulenbeld, B. Diophantische Approximationen homogener Linearformen in imaginären quadratischen Zahlkörpern. Nederl. Akad. Wetensch., Proc. **44**, 426–434 (1941).

In a recent paper [Nederl. Akad. Wetensch., Proc. **44**, 62–74 (1941); see these Rev. **2**, 253] the authors (transforming Blichfeldt's method) obtained a new result concerning the approximation of linear forms. Here they apply their method to obtain the following analogous result in the field $k(i\sqrt{m})$, $m\geqq 1$ (m quadrat frei). Suppose that n is a positive integer, that $\lambda=1$ for $m\not\equiv 3 \pmod 4$, $\lambda=2$ for $m\equiv 3 \pmod 4$, and let

$$\rho_{n,m}=\frac{2^{n+1}n^n}{(n+1)^n}\left[\left\{\frac{(2n+1)m^{(n+1)/2}}{(n+1)\pi^{n+1}\lambda^{n+1}}\right\}\Big/\left\{1+\frac{(n-1)^{2n+2}}{n(n+1)^{2n+1}}+\frac{2^{2n+3}n^{2n+1}(2n+1)}{(n+1)^{2n+1}}\cdot\sum_{\mu=2n+3}^{\infty}\frac{1}{\mu}\left(\frac{n-1}{2n}\right)^{\mu}\right\}\right]^{\frac{1}{2}}.$$

Then to any system of complex numbers $(\theta_1, \theta_2, \cdots, \theta_n)$ and real $t>2$, there are "integers" $(P_1, P_2, \cdots, P_n, Q)$ of $k(i\sqrt{m})$, $1\leqq P=\max(|P_1|, |P_2|, \cdots, |P_n|)\leqq 2(\rho_{n,m}t)^{1/n}$, for which the linear form $L=\theta_1P_1+\theta_2P_2+\cdots+\theta_nP_n-Q$ satisfies $|L|\leqq\rho_{n,m}/P^n$, $|L|\leqq 2/t$.

D. C. Spencer (Cambridge, Mass.).

Citations: MR 2, 253a = J12-1.

Referred to in J56-6.

J56-4 **(3, 71b)**

Koksma, J. F. und Meulenbeld, B. Simultane Approximationen in imaginären quadratischen Zahlkörpern. Nederl. Akad. Wetensch., Proc. **44**, 310–323 (1941).

The authors carry over to complex numbers the fundamental method of Blichfeldt [Trans. Amer. Math. Soc. **15**, 227–235 (1914)], and obtain the following theorem on simultaneous approximation in the complex quadratic field $k(i\sqrt{m})$, where m is a positive (quadrat frei) integer. Suppose that n is a positive integer, that $\lambda=1$ for $m\not\equiv 3 \pmod 4$, $\lambda=2$ for $m\equiv 3 \pmod 4$, and let

$$\gamma_{n,m}=\frac{2n}{\pi^{\frac{1}{2}}(n+1)}\left[\left\{\frac{4(2n+1)m^{(n+1)/2}}{(n+1)\pi\lambda^{n+1}}\right\}\Big/\left\{1+\frac{(n-1)^{2n+2}}{n(n+1)^{2n+1}}+\frac{2^{2n+3}n^{2n+1}(2n+1)}{(n+1)^{2n+1}}\cdot\sum_{\mu=2n+3}^{\infty}\frac{1}{\mu}\left(\frac{n-1}{2n}\right)^{\mu}\right\}\right]^{1/2n}$$

Then to any system of complex numbers $(\theta_1, \theta_2, \cdots, \theta_n)$, and real $t>2$, there is at least one system of "integers" of $k(i\sqrt{m})$, $(Q, P_1, P_2, \cdots, P_n)$, say, where $1\leqq|Q|\leqq 2t^n\gamma^n_{n,m}$, for which

$$\left|\theta_\nu-\frac{P_\nu}{Q}\right|\leqq\frac{\gamma_{n,m}}{|Q|^{1+1/n}},\quad\left|\theta_\nu-\frac{P_\nu}{Q}\right|\leqq\frac{2}{t|Q|},\quad \nu=1, 2, \cdots, n.$$

D. C. Spencer (Cambridge, Mass.).

J56-5 **(3, 273b)**

Hlawka, Edmund. Über komplexe homogene Linearformen. Monatsh. Math. Phys. **49**, 321–326 (1941).

This paper gives a new proof of a theorem of Minkowski which may be stated as follows: Let α, β, γ, δ be any complex numbers for which $\alpha\delta-\beta\gamma=1$. Then there exist complex integers x, y, not both zero, such that $\alpha x+\beta y$ and $\gamma x+\delta y$ simultaneously do not exceed $\rho=((3^{\frac{1}{2}}+1)/6^{\frac{1}{2}})^{\frac{1}{2}}=1.05610\cdots$ in absolute value. The author uses the two Hermitian forms

$$|\alpha x+\beta y|^2\pm|\gamma x+\delta y|^2$$

and, denoting these by $F_1(x, y)$ and $F_2(x, y)$, shows the above theorem is equivalent to the assertion of the existence of x and y such that

$$f(x, y)=F_1(x, y)+|F_2(x, y)|\leqq 2\rho^2.$$

By supposing that $f(0, 1)>2\rho^2$ it is shown that $f(-1, 1)\leqq 2\rho^2$.

D. H. Lehmer (Berkeley, Calif.).

J56-6 (9, 335b)

Mullender, Pieter. Toepassing van de Meetkunde der Getallen op Ongelijkheden in $K(1)$ en $K(i\sqrt{m})$. [Application of the Geometry of Numbers to Inequalities in $K(1)$ and $K(i\sqrt{m})$]. Thesis, Free University of Amsterdam, 1945. x+85 pp.

Using Blichfeldt's principle in the geometry of numbers [Trans. Amer. Math. Soc. **15**, 227–235 (1914)] the reviewer and Meulenbeld have deduced a number of theorems on the Diophantine approximations of systems of linear forms [Nederl. Akad. Wetensch., Proc. **44**, 62–74, 426–434 (1941); **45**, 256–262, 354–359, 471–478, 578–584 (1942); these Rev. **2**, 253; **3**, 71; **5**, 256]. Simplifying their method and generalizing their results the author gives a systematic survey, adding various new theorems especially on systems of forms in certain complex algebraic fields. He also generalizes a theorem of van der Corput and Schaake [Acta Arith. **2**, 152–160 (1936)]. The first part of the author's paper reviewed below gives an English survey of this thesis.

J. F. Koksma (Amsterdam).

Citations: MR 2, 253a = J12-1; MR 3, 71a = J56-3; MR 5, 256f = J16-2; MR 5, 256g = J16-2; MR 5, 256h = J16-2; MR 5, 256i = J16-2.

Referred to in H10-13.

J56-7 (9, 335c)

Mullender, P. Homogeneous approximations. Nederl. Akad. Wetensch., Proc. **50**, 173–185 = Indagationes Math. **9**, 136–148 (1947).

Let $L_k = a_{k1}z_1 + \cdots + a_{kn}z_n$ $(k=1, \cdots, n)$ be n linear forms of determinant $\Delta \neq 0$. Let either $\omega = 1$ ("real case") or $\omega = im^{\frac{1}{2}}$, where $m>0$ is an integer without quadratic divisors ("complex case"). A point $[z_1, \cdots, z_n]$ is called a lattice point if $z_1, \cdots, z_n$ are integers of the field $K(\omega)$. In the real case it is supposed that the eventual nonreal forms L_k appear in pairs of conjugate complex forms. (I) The first part contains a summary (without proofs) of the author's dissertation [cf. the preceding review]. Let $p>0$, $q>0$ be integers, $p+q=n$, $\sigma>0$;

$$\Lambda_{p,q,\sigma} = q\cdot\sigma^{-1}\cdot(p+q)^{(p+q)/\sigma}(2p)^{-p/\sigma}(2q)^{-q/\sigma}$$
$$\times\int_0^{2q/(p+q)}(1-x)^{p/\sigma}x^{q/\sigma-1}dx$$
$$+p\cdot\sigma^{-1}\int_{(p+q)/(2p)}^{1}(1-x)^{p/\sigma}x^{-1}dx$$

for $p \geqq q$, $\Lambda_{p,q,\sigma} = \Lambda_{q,p,\sigma}$ for $p \leqq q$; $\zeta^{(\omega)} = 1$ for $\omega = 1$, $\zeta^{(\omega)} = \frac{1}{4}\pi\lambda m^{-\frac{1}{2}}$ for $\omega = im^{\frac{1}{2}}$; $\lambda = 1$ for $m \not\equiv 3 \pmod 4$, otherwise $\lambda = 2$; $\gamma_{p,q}^{(im^{\frac{1}{2}})} = (\zeta^{(\omega)})^{\frac{1}{2}(p+q)}2^{p+q}(\Lambda_{p,q,\frac{1}{2}})^{\frac{1}{2}}$, $\gamma_{p,q}^{(1)} = 2^{p+q}\Lambda_{p,q,1}$; $\sigma' = \sigma$ for $\omega = 1$, $\sigma' = \frac{1}{2}\sigma$ for $\omega = im^{\frac{1}{2}}$;

$$\rho_{p,q,\sigma}^{(\omega)} = (\zeta^{(\omega)})^{\sigma'(p+q)}2^{p+q}\left(\frac{(\Gamma(1+1/\sigma'))^{p+q}\Lambda_{p,q,\sigma'}}{\Gamma(1+p/\sigma')\Gamma(1+q/\sigma')}\right)^{\sigma'}.$$

Theorem: let $\sigma \geqq 1$,

$$P = |L_1|^\sigma + \cdots + |L_p|^\sigma, \quad Q = |L_{p+1}|^\sigma + \cdots + |L_{p+q}|^\sigma;$$

then to every $t>0$ there is at least one lattice point $[z_1, \cdots, z_n] \neq [0, \cdots, 0]$ such that

$$Q \leqq 2/t, \quad P \leqq 2(t^q|\Delta|^\sigma/\rho_{p,q,\sigma}^{(\omega)})^{1/p}, \quad P^pQ^q \leqq |\Delta|^\sigma/\rho_{p,q,\sigma}^{(\omega)};$$

and, to every $\epsilon>0$, there is an infinity of lattice points satisfying $Q<\epsilon$, $P^pQ^q \leqq |\Delta|^\sigma/\rho_{p,q,\sigma}^{(\omega)}$. There is a completely analogous theorem for $P' = \max(|L_1|, \cdots, |L_p|)$, $Q' = \max(|L_{p+1}|, \cdots, |L_{p+q}|)$, only with $|\Delta|$, $\gamma_{p,q}$ instead of $|\Delta|^\sigma$, $\rho_{p,q,\sigma}^{(\omega)}$. Further, extending a result of van der Corput and Schaake to the complex case, the author proves: if $\sigma \geqq 2$, $\omega = im^{\frac{1}{2}}$, there is a lattice point $[z_1, \cdots, z_n] \neq [0, \cdots, 0]$ such that $|L_1|^\sigma + \cdots + |L_n|^\sigma \leqq |\Delta|^{\sigma/n}/B$, if

$$0 < B < \frac{2(\zeta^{(\omega)})^{\frac{1}{2}\sigma}}{(1+2n/\sigma)^{\sigma/(2n)}}\frac{(\Gamma(1+2/\sigma))^{\sigma/2}}{(\Gamma(1+2n/\sigma))^{\sigma/(2n)}}.$$

(II) The second part contains the proof of the following theorem. Let $n=3$; we call $A>0$ an allowed value in the inequality $|L_1L_2L_3| \leqq |\Delta|/A$ if this inequality is satisfied by an infinity of lattice points, $[z_1\, z_2, z_3]$, whatever the coefficients a_{ik} may be (with the obvious condition in the "real case"). Let $C^{(\omega)}$ be the upper bound of all allowed values. Let $C_{2,1}^{(\omega)}$ have an analogous meaning with regard to the inequality $|\alpha_1x_1 + \alpha_2x_2 - x_3| < 1/(AX^2)$, where $X = \max(|x_1|, |x_2|) \geqq 1$. Then $C_{2,1}^{(\omega)} \leqq C^{(\omega)}$. It is easily seen that, if the factorization in prime numbers in $K(\omega)$ is unambiguous, we have $C^{(\omega)} \leqq (D_3^{(\omega)})^{\frac{1}{2}}$, where $D_3^{(\omega)}$ denotes the minimum of the absolute values of the discriminants of all algebraic fields of degree 3 relative to $K(\omega)$, e.g., $D_3^{(1)} = 23$ [Furtwängler], and so $C_{2,1}^{(1)} \leqq 23^{\frac{1}{2}}$. *V. Jarník* (Prague).

Referred to in J12-18.

J56-8 (9, 569f)

Perron, Oskar. Ein Analogon zu einem Satz von Minkowski. S.-B. Math.-Nat. Abt. Bayer. Akad. Wiss. **1945/46**, 159–165 (1947).

The author makes the following conjecture. Let α, β, γ, δ, ρ, σ be complex numbers such that $\alpha\delta - \beta\gamma = 1$; then there exist integers x, y in the field $K(i\sqrt{D})$ $(D>0$ and "quadratfrei") such that

$$|\alpha x + \beta y - \rho|\cdot|\gamma x + \delta y - \sigma| \leqq \begin{cases}\frac{1}{4}(1+D), & D \not\equiv 3 \pmod 4,\\ \frac{1}{16}(1+D)^2/D, & D \equiv 3 \pmod 4;\end{cases}$$

he proves that the right hand member cannot be replaced by a smaller one, states that he possesses a proof of the conjecture for $D = 1, 2, 3$ and gives the proof for $D=1$.

J. F. Koksma (Amsterdam).

Referred to in J56-10, J56-18.

J56-9 (10, 183b)

Hlawka, Edmund. Eine asymptotische Formel für Potenzsummen komplexer Linearformen. Monatsh. Math. **52**, 248–254 (1948).

Let $L_1(x, y)$ and $L_2(x, y)$ be linear forms with complex coefficients, whose determinant has absolute value unity. A theorem of Perron [Math. Ann. **103**, 533–544 (1930)] states that there exist integers x, y, not both zero, in the field $k(i)$ which satisfy $|L_1L_2| \leqq 3^{-\frac{1}{2}}$, this being the best possible constant. The author deduces from Perron's proof that the integers can be so chosen that $|L_1|$ and $|L_2|$ are also absolutely bounded. Now let $M(\alpha)$ denote the lower bound of the numbers λ for which the inequality $|L_1|^\alpha + |L_2|^\alpha < \lambda$ is always soluble in integers x, y, not both zero, of $k(i)$. The author deduces estimates for $M(\alpha)$ for $0<\alpha<1$, analogous to those given by Mahler [Proc. Cambridge Philos. Soc. **40**, 107–116, 116–120 (1944); J. London Math. Soc. **18**, 233–238 (1943); these Rev. **6**, 119] for real linear forms. *H. Davenport* (London).

Citations: MR 6, 119b = H20-4; MR 6, 119c = H20-4; MR 6, 119d = H20-4.

J56-10 (10, 593a)

Perron, Oskar. Diophantische Ungleichungen in imaginären quadratischen Körpern. Mat. Tidsskr. B. **1949**, 1–17 (1949).

The paper extends known results on the product of two homogeneous or nonhomogeneous linear forms to the fields of the title. Theorem 1. If α, β, γ, δ are complex numbers with $|\alpha\delta-\beta\gamma|=1$, there are infinitely many pairs of integers $x, y\epsilon k(i\sqrt{D})$ for which $|(\alpha x+\beta y)(\gamma x+\delta y)|\leqq K$, where $K=(6D)^{\frac{1}{2}}/\pi$ for $D\not\equiv 3 \pmod 4$, $K=(6D)^{\frac{1}{2}}/(2\pi)$ for $D\equiv 3 \pmod 4$. Theorem 2. If $\alpha, \beta, \gamma, \delta, \rho, \sigma$ are complex numbers with $|\alpha\delta-\beta\gamma|=1$, there are integers $x, y\epsilon k(i\sqrt{D})$ for which $|(\alpha x+\beta y-\rho)(\gamma x+\delta y-\sigma)|\leqq L$, where

$$L=(\tfrac{1}{4}(1+D))^h D^{1-h}(\Delta+(6D)^{\frac{1}{2}}/h\pi).$$

Here $h=1$ or 2 according as $D\not\equiv 3$ or $D\equiv 3 \pmod 4$, and Δ is the smallest number such that each class of ideals in $k(i\sqrt{D})$ contains an ideal with norm not exceeding Δ. These theorems are not best possible. In the case of theorem 1 the author has shown that K can be replaced by $1/\sqrt{3}$ for $D=1$ and $1/13^{\frac{1}{2}}$ for $D=3$ [Math. Z. **35**, 563–578 (1932)] and by $1/\sqrt{2}$ for $D=2$ [Math. Z. **37**, 749–767 (1933)], but by no smaller constants. Concerning theorem 2, the author had proved earlier [S.-B. Math.-Nat. Abt. Bayer. Akad. Wiss. **1945/46**, 159–165 (1947); these Rev. **9**, 569] that the best possible constant for $D=1$ is $\frac{1}{2}$; he proves in the present paper that the L above can be replaced by $\frac{3}{4}$ and $\frac{1}{3}$ in the cases $D=2$, $D=3$, but by no smaller numbers. The proofs given depend on Minkowski's theorem on lattice points in convex regions and the existence of integers of the field in question inside certain Cassini ovals in the complex plane.

W. J. LeVeque (Ann Arbor, Mich.).

Citations: MR 9, 569f = J56-8.

Referred to in J56-18.

J56-11 (12, 162a)

Descombes, Roger, et Poitou, Georges. Sur l'approximation dans $R(i\sqrt{11})$. C. R. Acad. Sci. Paris **231**, 264–266 (1950).

The authors call $C=\inf_x \limsup_{p,q}(|q(p-qx)|)^{-1}$, where p and q range over the integral elements of a quadratic field $R(i\sqrt{m})$, and where x ranges over all complex numbers, the Hurwitz constant of $R(i\sqrt{m})$. For $m=1, 2, 3$, and 7, $C=3^{\frac{1}{2}}, 2^{\frac{1}{2}}, 13^{\frac{1}{2}}$, and $8^{\frac{1}{2}}$, respectively. [For references, see Koksma, Diophantische Approximationen, Springer, Berlin, 1936, pp. 51–52; Hofreiter, Monatsh. Math. Phys. **45**, 175–190 (1937).] For $m=11$, it has been known that $C\leqq 5^{\frac{1}{2}}=1.495\cdots$. The authors find that $C=5^{\frac{1}{2}}/2=1.118\cdots$, but sketch their methods of proof very briefly. An extension of the theory of continued fractions is required, different from that of Hurwitz which does not yield "sequences of best approximation" for certain complex numbers x_0. The proofs also require the Euclidean algorithm which exists in the five cases $m=1, 2, 3, 7$, and 11.

R. Hull.

Referred to in J56-12, J56-22.

J56-12 (12, 594d)

Poitou, Georges, et Descombes, Roger. Sur l'approximation dans le corps des racines cubiques de l'unité. C. R. Acad. Sci. Paris **232**, 292–294 (1951).

In a previous note [same C. R. **231**, 264–266 (1950); these Rev. **12**, 162] the authors have sketched briefly a theory of continued fractions applicable in each of the five Euclidean imaginary quadratic fields $R(im^{\frac{1}{2}})$, $m=1, 2, 3, 7$, and 11, and their successful application of the theory, in the case $m=11$, to determining the inf C of $C(x)$, as x ranges over all complex numbers, for the "constants of approximation": $C(x)=\limsup_{p,q}|q(p-qx)|^{-1}$, as p and q range over the integers of $R(im^{\frac{1}{2}})$. In the present note, they sketch some applications of the theory in the case $m=3$. This case is unique among the five in having two properties in common with the rational case which simplify the study of $C(x)$. They report that they are able to establish (1) Perron's value $C=13^{\frac{1}{2}}=1.8988\ldots$, for the "first" constant of approximation, i.e., Hurwitz' constant, of $K=R(i3^{\frac{1}{2}})$; (2) that C is isolated among the values of $C(x)$ for K and corresponds to certain values of x which are quadratic over K; (3) that the second value of $C(x)$ for K is 2, which is also isolated and also corresponds to certain relative quadratic values of x; (4) the third value of $C(x)$ for K is $(32(3)^{\frac{1}{2}}/13)^{\frac{1}{2}}=2.06487\ldots$, which is also isolated, etc.; (5) every other value of $C(x)$ for K exceeds 2.070068; (6) $((28+16(3)^{\frac{1}{2}})/13)^{\frac{1}{2}}=2.0701693\ldots$, is a point of accumulation of values of $C(x)$ for K.

R. Hull.

Citations: MR 12, 162a = J56-11.

J56-13 (13, 323a)

Cassels, J. W. S., Ledermann, W., and Mahler, K. Farey section in $k(i)$ and $k(\rho)$. Philos. Trans. Roy. Soc. London. Ser. A. **243**, 585–626 (1951).

Let R be the ring of rational integers and k the field of rational numbers. The first type of Farey section in $k(i)$ is $H_N(i)$ the set of all fractions α/β with α, β in $R(i)$ having norms $N(\alpha), N(\beta)\leqq N$. The second type $H_N^*(i)$ does not require $N(\alpha)\leqq N$. For a given N, $R(\alpha, \beta)$ is the set of all complex numbers z for which α/β is the best approximation among the fractions in $H_N(i)$, i.e. for which $|\beta z-\alpha|\leqq|\beta' z-\alpha'|$ for all α'/β' in $H_N(i)$ (similarly, $R^*(\alpha, \beta)$ is defined). Next, α/β, α'/β' are defined to be adjacent if $R(\alpha, \beta)$ and $R(\alpha', \beta')$ have a common point. A median of α/β and α'/β' is a number $(\alpha+\epsilon\alpha')/(\beta+\epsilon\beta')$, where ϵ is a unit, i.e. ± 1 or $\pm i$. The following assertions are proved. Reduced fractions α/β, α'/β' are adjacent if and only if $|\alpha\beta'-\alpha'\beta|=1$ or $2^{\frac{1}{2}}$ and one of their medians is not in $H_N(i)$. Fractions of $H_{N+1}(i)$ not in $H_N(i)$ are medians of adjacent fractions in $H_N(i)$. Similar definitions and results occur for $H_N^*(i)$ and this Farey section is used to prove the known result that if $\alpha, \beta, \gamma, \delta$ are complex numbers, $\alpha\delta-\beta\gamma=1$, then there are ξ, η in $R(i)$, not both 0, such that $|\alpha\xi+\beta\eta|\leqq k^{\frac{1}{2}}$, $|\gamma\xi+\delta\eta|\leqq k^{\frac{1}{2}}$, where $k=2^{\frac{1}{2}}/(3-3^{\frac{1}{2}})$. The regions $R(\alpha, \beta)$ in the complex plane have boundaries which are arcs of circles. Where three or more such regions meet, the set of angles subtended may have only one of three sets of values. Furthermore $R(\alpha, \beta)$ is a star domain about α/β for N sufficiently large, while $R^*(\alpha, \beta)$ is always a star domain. Corresponding results (except those about star domains) are mentioned and proofs indicated for $k(\rho)$, where $\rho^2+\rho+1=0$. Diagrams of $R(\alpha, \beta)$ are given for $H_{10}(i)$, $H_b(i)$, and $H_{19}(\rho)$ and of $R^*(\alpha, \beta)$ for $H^*_{25}(i)$ and $H^*_{19}(\rho)$.

L. Tornheim.

Referred to in J56-14, J56-19, J56-22, J56-31.

J56-14 (13, 538e)

Mahler, K. Farey sections in the fields of Gauss and Eisenstein. Proceedings of the International Congress of Mathematicians, Cambridge, Mass., 1950, vol. 1, pp. 281–285. Amer. Math. Soc., Providence, R. I., 1952.

A clear review of a theory of Farey sections in $k(i)$ and $k(p)$, which meanwhile appeared with full proofs, in a joint paper by Cassels, Ledermann and the author [Philos. Trans. Roy. Soc. London. Ser. A. **243**, 585–626 (1951); these Rev. **13**, 323].

J. F. Koksma (Amsterdam).

Citations: MR 13, 323a = J56-13.

J56-15 (13, 921b)

Hofreiter, Nikolaus. Über die Approximation von komplexen Zahlen durch Zahlen des Körpers $K(i)$. Monatsh. Math. **56**, 61–74 (1952).

The author proves: For each complex number ξ at least one couple of integers p, $q\neq 0$ in $K(i)$ exists such that $(p, q)=1$, $|\xi-p/q|\leqq\sqrt{(2-\sqrt{3})}/|q|^2$; the constant $\sqrt{(2-\sqrt{3})}$ is best possible as the example $\xi=\frac{1}{2}(1+i\sqrt{3})$ shows. The analogous result and more general theorems in the real case have been proved by Prasad [J. London Math. Soc. **23**, 169–171 (1948); these Rev. **10**, 513].

J. F. Koksma (Amsterdam).

Citations: MR 10, 513d = J04-6.

J56-16 **(14, 358c)**

Cassels, J. W. S. Über einen Perronschen Satz. Arch. Math. **3**, 10–14 (1952).

Let $f(x, y)=ax^2+2bxy+cy^2$, where a, b, and c are complex numbers and $\Delta=b^2-4ac\neq 0$, and let $R(i)$ denote the ring of Gaussian integers. The author proves: There exists a universal constant $A_0>3/4$ such that $|f(x, y)|\leqq|\Delta/A_0|^{1/2}$ has a solution $x, y \in R(i)$, $(x, y)\neq(0, 0)$, unless f is equivalent to a multiple of x^2+xy+y^2. The related Perron theorem is: There exist infinitely many pairs $(x, y)\neq(0, 0)$, $x, y \in R(i)$, for which $|f(x, y)|\leqq|4\Delta/3|^{1/2}$. The author gives an independent proof of his theorem, which, however, he says can be obtained from Perron's theorem by employing the Heine-Borel theorem and the like. *R. Hull* (Lafayette, Ind.).

Referred to in J56-31.

J56-17 **(14, 359b)**

Schmetterer, L. Notiz zu einem Satz über Diophantische Approximationen. Monatsh. Math. **56**, 253–255 (1952).

It is shown that if θ non-$\in k(i)$ and β is not real, the inequality $|y\theta-x-\beta|<|h|/2|y|$ has no solution in Gaussian integers x, y with $y\neq 0$ if $0<|h|<1$. Moreover, 2 is the best possible constant, in the sense that the theorem becomes false if 2 is replaced by a larger number. This is an analogue of a theorem of Kanagasabapathy [Proc. Cambridge Philos. Soc. **48**, 365–366 (1952); these Rev. **13**, 825] in the real case. *W. J. LeVeque* (Ann Arbor, Mich.).

Citations: MR 13, 825d = J20-17.

J56-18 **(14, 623c)**

Schmetterer, Leopold. Über das Produkt zweier komplexer inhomogener Linearformen. Monatsh. Math. **56**, 339–343 (1952).

The author extends known results of Minkowski, Hlawka, Mahler, and Perron for $k(1)$, $k(i)$, $k(i\sqrt{2})$ to $k(i\sqrt{7})$ in proving the following theorem. Let α, β, γ, δ, ξ, η denote complex numbers such that $|\alpha\delta-\beta\gamma|=1$. Then integers x, y in $k(i\sqrt{7})$ exist such that $|\alpha x+\beta y-\xi||\gamma x+\delta y-\eta|\leqslant 4/7$; the constant $4/7$ is the best possible. The method used is that introduced by Perron [Mat. Tidsskr. B. **1949**, 1–17; S.-B. Math.-Nat. Abt. Bayer. Akad. Wiss. **1945/46**, 159–165; these Rev. **10**, 593; **9**, 569]. *J. F. Koksma.*

Citations: MR 9, 569f = J56-8; MR 10, 593a = J56-10.

J56-19 **(14, 850a)**

LeVeque, W. J. Geometric properties of Farey sections in $k(i)$. Nederl. Akad. Wetensch. Proc. Ser. A. **55** = Indagationes Math. **14**, 415–426 (1952).

By use of a diagram of spheres [Speiser, J. Reine Angew. Math. **167**, 88–97 (1932)] the boundaries of the Farey sections in the Gaussian field as defined by Cassels, Ledermann, and Mahler [Philos. Trans. Roy. Soc. London. Ser. A. **243**, 585–626 (1951); these Rev. **13**, 323] are given a simple construction; they are the projection of the center of a sphere which moves so as to be tangent to two spheres and the complex plane. This enables the author to simplify, clarify, and extend the earlier results and proofs. *L. Tornheim* (Ann Arbor, Mich.).

Citations: MR 13, 323a = J56-13.

Referred to in J56-22, J56-31.

J56-20 **(14, 850b)**

LeVeque, W. J. Continued fractions and approximations in $k(i)$. I, II. Nederl. Akad. Wetensch. Proc. Ser. A. **55** = Indagationes Math. **14**, 526–535, 536–545 (1952).

An algorithm and some properties for a regular continued fraction expansion for complex numbers with partial quotients which are Gaussian integers are obtained by using results of the paper reviewed above. The convergents form a subset of the best approximations as defined by the Farey sections of the complex plane. Unfortunately the successive partial quotients sometimes depend not only on the preceding complete quotient but also on the preceding convergent; there is a discussion of when this occurs. This difficulty prevents introducing the notion of equivalent numbers. Finally, several probability theorems are obtained on distributions involving the partial quotients and the complete quotients of the continued fractions and the approximating numbers in the Gaussian field. *L. Tornheim* (Ann Arbor, Mich.).

Referred to in J56-22, J56-31.

J56-21 **(14, 954e)**

Oppenheim, A. One-sided inequalities for hermitian quadratic forms. Monatsh. Math. **57**, 1–5 (1953).

Let $\phi(x, y)=ax\bar{x}+\beta x\bar{y}+\bar{\beta}\bar{x}y+cy\bar{y}$ be a binary Hermitian form, and let $\Delta=ac-\beta\bar{\beta}$ be its determinant. Let x, y be integers in $k(\sqrt{m})$. The form is indefinite if $m>0$, $\Delta\neq 0$ and also if $m<0$, $\Delta<0$. The author proves that the inequality

$$0<\phi(x, y)\leqq C|\Delta|^{1/2}$$

is soluble in integers x, y of $k(\sqrt{m})$, where $C=|m|^{1/2}$ if $m\equiv 1 \pmod 4$ and $C=2|m|^{1/2}$ otherwise, m being squarefree. If certain forms are excluded, the inequality is soluble with C replaced by $2^{-1/2}C$, and if certain other forms are excluded C can be replaced by $\frac{2}{3}C$. It is of interest that the results obtained are more complete than those known for $|\phi(x, y)|$. The proof is based on results for quaternary quadratic forms of signature zero, in papers by the author which are not yet published. There are some misprints, e.g., a factor a is missing from the first term on the right of (3). *H. Davenport* (London).

J56-22 **(16, 574c)**

Poitou, Georges. Sur l'approximation des nombres complexes par les nombres des corps imaginaires quadratiques dénués d'idéaux non principaux, particulièrement lorsque vaut l'algorithme d'Euclide. Ann. Sci. Ecole Norm. Sup. (3) **70**, 199–265 (1953).

Let m be a positive integer, and let x be a complex number not in the quadratic field $K=R(\sqrt{-m})$ of discriminant $-D$. Let $C(x)$ designate the limit superior of $|q(qx-p)|^{-1}$ for all integers p, q of K with $q\neq 0$. The set of numbers $C(x)$ is called the Markov spectrum of K, and the inf C_D of the spectrum is called the Hurwitz constant of K. The present paper is an elaborate investigation of the spectra and Hurwitz constants for certain small values of D. The results obtained are as follows: (a) For $D=3$, the numbers $13^{1/4}$, 2, $(32\cdot 3^{1/2}/13)^{1/2}$ are isolated points of the spectrum; every other value is greater than 2.070068. The number

$$((28+16\cdot 3^{1/2})/13)^{1/2}=2.0701\cdots$$

is a point of accumulation in the spectrum. (b) $C_{11}=\frac{1}{2}\sqrt{5}$. [This had been announced earlier; cf. Descombes and Poitou, C. R. Acad. Sci. Paris **231**, 264–266 (1950); MR **12**, 162.] It is an isolated point, the next value being larger than 1.21. (c) $C_{19}=1$. (d) $C_{20}\leqq 1$. (e) $C_{43}\leqq\sqrt{(5/11)}$.

The basic tools in the investigation are the notions of a sequence of best approximations and of a best regular sequence. If x non-$\in K$, a fraction p_0/q_0 in K is called a convergent (réduite) to x if there is no simpler fraction closer to x, i.e., if min $(|qx-p|)=|q_0x-p_0|$, the minimum being taken over all integers $p, q \in K$ such that $|q|\leqq|q_0|$. Choosing one representative from each set of equivalent convergents, and arranging these representatives by the size of $|q|$, we obtain the sequence of best approximations to x. Consequtive convergents p/q and p'/q' to x are said to be adjacent if $d=pq'-p'q$ is a unit of K; while such convergents are always adjacent in the rational case, they need not be in other fields. This was implicit in the work of J. W. S. Cassels, W. Ledermann and K. Mahler [Philos. Trans. Royal Soc. London. Ser. A. **243**, 585–626 (1951); MR **13**, 323] in the case $D=4$. As a preliminary step in the present work, the author finds all the permissible values of d for $D=3, 4, 7, 8, 11, 19$, in which cases the integers of the fields form Euclidean rings. The existence of a Euclidean algorithm in these cases also makes it possible to assert that, given any irreducible fraction p_0/q_0 in K, there is a p/q adjacent to p_0/q_0 such that $|qx-p|<|q_0x-p_0|$. If this fact is applied repeatedly, and at each stage the simplest fraction is chosen, a best regular sequence of approximations results, and this in turn yields a regular continued fraction expansion for x, the partial quotients being integers of K. In the case $D=3$, the only possible values of d are the units of the field; it is in part this apparently fortuitous circumstance that makes possible the relatively precise results listed above.

The author was apparently unaware of the work of Cassels, Ledermann and Mahler cited above, as well as that of the reviewer [Nederl. Akad. Wetensch. Proc. Ser. A. **55**, 415–426, 526–535, 536–545 (1952); MR **14**, 850] where the continued-fraction expansion just mentioned was considered in the Gaussian field. In addition to these papers and the copious references provided by Poitou, the interested reader should examine the recent paper by R. Mönkemeyer [Math. Nachr. **11**, 321–344 (1954); MR **16**, 223], in which similar ideas occur. *W. J. LeVeque.*

Citations: **MR** 12, 162a = J56-11; **MR** 13, 323a = J56-13; **MR** 14, 850a = J56-19; **MR** 14, 850b = J56-20; **MR** 16, 223b = A60-14.

Referred to in J56-31.

J56-23 (17, 17f)

Chalk, J. H. H. Rational approximations in the complex plane. J. London Math. Soc. **30**, 327–343 (1955).

Let $f(u, v)=C(u-\Omega v)(u-\Omega' v)$ be a form of discriminant $D\neq 0$, the coefficients C, Ω, Ω' are arbitrary complex numbers. Let $H(u, v)=auu^*+buv^*+b^*u^*v+cvv^*$ (* denotes complex conjugate) be an indefinite hermitian form (a and c real, $\Delta=bb^*-ac>0$). Perron [Math. Z. **35**, 563–578 (1932)] showed the existence of infinitely many solutions of $|f(u, v)|\leq(\frac{1}{3}|D|)^{\frac{1}{2}}$ in gaussian integers u, v. Oppenheim [Proc. Nat. Acad. Sci. U.S.A. **15**, 724–727 (1929)] obtained a result on real indefinite quartic forms which can be specialized to the fact that $|H(u, v)|\leq(\frac{2}{3}\Delta)^{\frac{1}{2}}$ has a solution. Subsequently, it was shown that there are infinitely many solutions in this case also.

The present author obtains these results by modification of a method of Ford [Trans. Amer. Math. Soc. **27**, 146–154 (1925)]; a central rôle is played by the Picard group [cf. Ford, ibid. **19**, 1–42 (1918)]. The author's explicit results are too long to be quoted here.

N. G. de Bruijn (Amsterdam).

Referred to in J56-25.

J56-24 (17, 1059e)

Rogers, K. Indefinite binary hermitian forms. Proc. London Math. Soc. (3) **6** (1956), 205–223.

The author considers indefinite binary hermitian forms: $f(x, y)=ax\bar{x}+b\bar{x}y+\bar{b}x\bar{y}+cy\bar{y}$ with non-zero determinant $-d=ac-b\bar{b}<0$. He defines $M(f; x_0, y_0)$ to be the greatest lower bound of $|f(x+x_0, y+y_0)|$ taken over all Gaussian integers x, y; $M(f)$ is the sup $M(f; x_0, y_0)$ taken over all complex number pairs x_0, y_0. The author's main result is the following, under the definitions and notations above: $M(f)<2d^{\frac{1}{2}}/5$ except when $\pm f$ is equivalent to one of the following forms:

(i) $f_1=d^{\frac{1}{2}}(e^{i\pi/4}\bar{x}y+e^{-i\pi/4}x\bar{y})$,

(ii) $f_2=(d/3)^{\frac{1}{2}}(x\bar{x}-3y\bar{y})$,

(iii) $f_3^{(1)}=d^{\frac{1}{2}}(x\bar{x}-y\bar{y})$,
$f_3^{(2)}=d^{\frac{1}{2}}(x\bar{y}+\bar{x}y)$,

(iv) $f_4=(d/21)^{\frac{1}{2}}(x\bar{x}-21y\bar{y})$,

(v) $f_5^{(1)}=(2d/3)^{\frac{1}{2}}(x\bar{x}+(1+i)\bar{x}y/2+(1-i)x\bar{y}/2-y\bar{y})$,
$f_5^{(2)}=(d/6)^{\frac{1}{2}}(x\bar{x}-6y\bar{y})$.

In each case the value of $M(f)$ is given and all points x_0, y_0 for which it is attained are given. If $|f(x, y)|$ assumes arbitrarily small non-zero values for Gaussian integers x, y, then $M(f)=0$.

Some of the details are omitted and reference made to the author's Ph. D. thesis [Cambridge, 1954].

B. W. Jones (Boulder, Colo.).

J56-25 (17, 1061a)

Chalk, J. H. H. Rational approximations in the complex plane. II. J. London Math. Soc. **31** (1956), 216–221.

[For part I see same J. **30** (1955), 327–343; MR **17**, 17.] The author proves the theorem of Hlawka, (accounting for extreme values), [Monatsh. Math. **46** (1938), 324–334], that for any complex numbers u_0, v_0 there exists gaussian integers u, v satisfying

$$|f(u+u_0, v+v_0)|\leq\tfrac{1}{2}|\Delta|,$$

where $f(u, v)=(\alpha u+\beta v)(\gamma u+\delta v)$, and $\alpha, \beta, \gamma, \delta$ are complex with nonvanishing determinant Δ. The theorem is referred to the Ford-like configuration of spheres $S(\Omega, R)$ of radius R lying in the upper half ξ, η, ζ space, tangent to plane $\zeta=0$, and with center directly above $\Omega=\xi+i\eta$. Thus if $\varrho e^{i\theta}$ ($\varrho^2\geqq 2$), u_0' are given complex constants, then for some gaussian integer u' the sphere $S(\varrho e^{i\theta}(u'+u_0'), \frac{1}{2}\varrho^2)$ will intersect a preassigned circle orthogonal to the ζ-plane. *Harvey Cohn* (Washington, D.C.).

Citations: **MR** 17, 17f = J56-23.

J56-26 (18, 287b)

Poitou, G. Approximations diophantiennes et groupe modulaire. Séminaire A. Châtelet et P. Dubreil de la Faculté des Sciences de Paris, 1953/1954. Algèbre et théorie des nombres. 2e tirage multigraphié, pp. 7-01 – 7-06. Secrétariat mathématique, 11 rue Pierre Curie, Paris, 1956.

This paper [also published in Séminaire d'algèbre et de théorie des nombres dirigé par A. Châtelet et P. Dubreil, 1953/1954, Fac. Sci. Paris, 1954; MR **16**, 1082] is an exposition of the applications of a geometric method of Speiser [see Koksma, Diophantische Approximationen, Springer, Berlin, 1936, p. 43]. The author shows among other things that this method easily leads to lower bounds for the Hurwitz constants of the quadratic fields $R(\sqrt{-m})$ ($m=1, 2, 3, 7, 11, 19$), which do not differ much from the known exact values of these constants.

C. G. Lekkerkerker (Amsterdam).

Citations: **MR** 16, 1082g = Z10-1.

J56-27 (18, 287d)

Mordell, L. J. Diophantine inequalities in complex quadratic fields. Publ. Math. Debrecen **4** (1956), 242–255.

The paper deals with the problem of finding the best possible constant $k=k(D)$ such that, if a, b, c, d are complex numbers and $|ad-bc|=1$, the inequality

(1) $$|(ax+by)(cx+dy)|\leqq k$$

can be satisfied by complex numbers x, y of a prescribed form, e.g. by integers in $K(i\sqrt{D})$. Such integers are of the form

$$m+n\left(\frac{D'+i\sqrt{D'}}{2}\right)=m+n\omega,$$

where $-D'$ is the discriminant of the field $K(i\sqrt{D})$ and where m, n are rational integers. The general problem considered is to investigate (1) for $x=u_1+v_1\omega$, $y=u^2+v_2\omega$, where u and v denote real numbers with assigned residues (mod 1). Several applications.

J. F. Koksma (Amsterdam).

J56-28 (21# 1291)

Poitou, G. Sur les fractions continues arithmétiques. Bull. Soc. Math. Belg. **9** (1957), 3–7.

A discussion of continued fraction expansions (in the rational and imaginary quadratic number fields) whose convergents are also best approximations.

W. J. LeVeque (Ann Arbor, Mich.)

J56-29 (27# 3583)

Eggan, L. C.; Maier, E. A.

On complex approximation.

Pacific J. Math. **13** (1963), 497–502.

It is proved that, given any complex numbers β and γ, there exists a Gaussian integer u such that $|\beta-u|<2$, and that $|\beta-u|\,|\gamma-u|<27/32$ in case $|\beta-\gamma|<\sqrt{(11/8)}$, $|\beta-u|\,|\gamma-u|<|\beta-\gamma|/\sqrt{2}$ in case $|\beta-\gamma|\geqq\sqrt{(11/8)}$. This is proved by an analysis of inequalities satisfied by numbers in certain regions on the complex plane. Related questions for real numbers were treated earlier by the authors [Michigan Math. J. **8** (1961), 161–166; MR **25** #3008]. The result is used to establish that there are infinitely many pairs of relatively prime Gaussian integers x, y such that $|x(x\theta-y-\alpha)|<\frac{1}{2}$ under the following hypotheses: θ is any irrational complex number; α is any complex number such that there are no Gaussian integers m, n satisfying $\alpha=m\theta+n$. {Further applications of the authors' result are to be found in Chapter 5 of the reviewer's book [*Diophantine approximations*, Interscience, New York, 1963; MR **26** #6120].}

I. Niven (Eugene, Ore.)

Citations: MR 25# 3008 = J32-64; MR 26# 6120 = J02-15.

J56-30 (36# 2558)

Davenport, H.; Schmidt, Wolfgang M.

Approximation to real numbers by quadratic irrationals.

Acta Arith. **13** (1967/68), 169–176.

In analogy with the well-known result that for any real irrational number ξ there are infinitely many rationals p/q satisfying $|\xi-p/q|<1/q^2$, the authors prove the following theorem: For any real ξ which is not rational or quadratic irrational, there is a constant C such that the inequality $|\xi-\alpha|<CH(\alpha)^{-3}$ has infinitely many solutions in rationals or real quadratic irrationals α (where $H(\alpha)$ denotes the height of α—the maximum of the absolute values of the coefficients of the minimal polynomial of α with relatively prime integer coefficients). The exponent -3 is seen to be best possible by looking at the norm of cubic irrationals in the usual way. A constant C is explicitly given.

W. W. Adams (Berkeley, Calif.)

Referred to in J16-27, J68-61.

J56-31 (39# 6831)

Schmidt, Asmus L.

Farey triangles and Farey quadrangles in the complex plane.

Math. Scand. **21** (1967), 241–295 (1969).

Farey sequences have been generalized by J. W. S. Cassels, W. Ledermann and K. Mahler [Philos. Trans. Roy. Soc. London Ser. A **243** (1951), 585–626; MR **13**, 323] to so-called Farey sections in the complex plane for the two quadratic fields $Q(i)$ and $Q(i\sqrt{3})$. The author gives a generalization of Farey fractions to imaginary quadratic fields along different lines, and applies his generalization to the approximation spectra of $Q(i\sqrt{m})$ in the cases $m=1, 2, 3, 7$. The approximation spectrum of a quadratic field $Q(i\sqrt{m})$ is the set of all approximation constants $C(\xi)=\lim\sup|q(q\xi-p)|^{-1}$, the lim sup being taken over all algebraic integers p, q in the field, except $q=0$. For $Q(i)$, the first minimum of the spectrum, $3^{1/2}$, was determined by L. R. Ford [Trans. Amer. Math. Soc. **27** (1925), 146–154] and O. Perron [Math. Ann. **103** (1930), 533–544; ibid. **105** (1931), 160–164] and shown to be isolated by J. W. S. Cassels [Arch. Math. **3** (1952), 10–14; MR **14**, 358]. The author establishes a definite lower bound for the second minimum. For $Q(i\sqrt{2})$ the first minimum, $2^{1/2}$, was found by O. Perron [Math. Z. **37** (1933), 749–767] and the author shows the second isolated minimum to be $3^{1/2}$. For $Q(i\sqrt{7})$ the first minimum, $8^{1/4}$, was found by N. Hofreiter [Monatsh. Math. Phys. **45** (1937), 175–190], and the author establishes again the second isolated minimum $3^{1/2}$. For $Q(i\sqrt{3})$ the first minimum was found by O. Perron [Bayer. Akad. Wiss. Math.-Natur. Kl. S.-B. **3** (1931), 129–154], the second and third minima by G. Poitou [Ann. Sci. École Norm. Sup. (3) **70** (1953), 199–265; MR **16**, 574]; the author gives a simple proof of the first and second minima.

The author introduces the concepts of Farey triangles and Farey quadrangles in $Q(i\sqrt{m})$, which give a simple characterization of equivalence classes of complex numbers in each of the fields studied. For example, a matrix $\begin{pmatrix} p_1 & p_2 & p_3 \\ q_1 & q_2 & q_3 \end{pmatrix}$ is a Farey matrix in $Q(i\sqrt{m})$ if each p_j and each q_j is an integer in the field and if $|p_jq_k-q_jp_k|=1$ for j, k any distinct pair from 1, 2, 3. Such a matrix gives a Farey triangle, namely, the convex hull of the three points p_j/q_j, $j=1, 2, 3$, in the complex plane. Similarly, a Farey quadrangle is obtained from an appropriately defined 2-by-4 Farey matrix. The author develops the basic properties of Farey triangles and quadrangles, and then works out the applications cited above. In addition to the excellent bibliography given in the paper, reference should also be made to work of W. J. LeVeque [Nederl. Akad. Wetensch. Proc. Ser. A **55** (1952), 415–426, 526–535, 536–545; MR **14**, 850]. *I. Niven* (Eugene, Ore.)

Citations: MR 13, 323a = J56-13; MR 14, 358c = J56-16; MR 14, 850a = J56-19; MR 14, 850b = J56-20; MR 16, 574c = J56-22.

Referred to in J56-32.

J56-32 (40# 7206)

Schmidt, Asmus L.

Farey simplices in the space of quaternions.

Math. Scand. **24** (1969), 31–65.

Extending the ideas of Farey triangles and Farey quadrangles first introduced by the author [Math. Scand. **21** (1967), 241–295 (1969); MR **39** #6831] to study approximation constants in $\mathbf{Q}(im^{1/2})$, $m=1, 2, 3, 7$, the author introduces Farey simplicies to study Diophantine approximations in the skew-field of quaternions. A quaternion ξ (viewed as a point in Euclidean four-space) is irrational in case at least one of its coordinates is irrational, and is an integer (in the sense of A. Hurwitz [*Vorlesungen über die Zahlentheorie der Quaternionen*, Springer, Berlin, 1919]) in case each of its coordinates is either an integer or half of an odd integer. For ξ an irrational quaternion, define the approximation constant $C(\xi)$ by $C(\xi)=\limsup(|q|\,|\xi q-p|)^{-1}$, where the lim sup is over all integral quaternions q, p ($q\neq 0$) and absolute value denotes Euclidean distance. The author gives an independent proof of a theorem of A. Speiser [J. Reine Angew. Math. **167** (1932), 88–97] which shows that $C(\xi)\geqq(5/2)^{1/2}$ for every irrational quaternion ξ.

More important, however, the author shows that the only approximation constant $C(\xi)>(2.51)^{1/2}$ is $C(\xi)=(5/2)^{1/2}$. (Speiser's paper did not even show $(5/2)^{1/2}$ was the best possible.) With equivalence defined as usual in terms of an integral (quaternion) unimodular transformation, the author also shows that the set $C^{-1}((5/2)^{1/2})$ consists of two distinct equivalence classes represented by $(\frac{1}{2}, (1+5^{1/2})/4, (1-5^{1/2})/4, 0)$ and $(\frac{1}{2}, (1-5^{1/2})/4, (1+5^{1/2})/4, 0)$. This set also has a simple characterization in terms of Farey simplicies. The determination of the second minimum appears to be even more involved and is still open.

L. C. Eggan (Normal, Ill.)

Citations: MR 39# 6831 = J56-31.

J60 METRIC THEOREMS CONCERNING THE ABOVE PROBLEMS

For the metric theory of continued fractions, see **K50**.

See also Section J24.

See reviews H15-69, J02-11, J02-18, J02-20, J02-27, J04-9, J04-13, J04-15, J04-21, J04-24, J04-25, J04-36, J04-41, J04-49, J04-51, J04-53, J04-60, J04-66, J04-69, J12-16, J12-25, J12-40, J12-41, J12-46, J12-62, J12-64, J12-66, J16-25, J16-28, J16-29, J2

J64 THE ABOVE PROBLEMS FOR NONARCHIMEDEAN VALUATIONS

See also reviews A54-44, H05-4, J02-14, J16-26, J20-43, J24-59, J40-43, K05-77, K40-14, Q20-49, R58-44, T55-14.

J64-1 **(1, 136d)**

Jarník, Vojtěch. Über einen p-adischen Übertragungssatz. Monatsh. Math. Phys. **48**, 277–287 (1939).

Let $\alpha_1, \cdots, \alpha_m$ be m p-adic integers, $\beta_1(\alpha_1, \cdots, \alpha_m)$ and $\beta_2(\alpha_1, \cdots, \alpha_m)$ the upper bounds of all exponents γ and δ such that the problems

$$(x_1, \cdots, x_m, p)=1, \quad \max(|x_0|, \cdots, |x_m|)\leqq\xi,$$
$$|x_0+\alpha_1x_1+\cdots+\alpha_mx_m|_p\leqq\xi^{-\gamma},$$

or

$$(y_0, p)=1, \quad \max(|y_0|, \cdots, |y_m|)\leqq\eta,$$
$$|\alpha_iy_0-y_i|_p\leqq\eta^{-\delta}, \qquad i=1, 2, \cdots, m,$$

have integral solutions x_i or y_i for an infinity of arbitrarily large ξ or η. It is easily proved that $\beta_1\geqq m+1$, $\beta_2\geqq(m+1)/m$, and that $\beta_1=\beta_2$ for $m=1$. K. Mahler has shown [Časopis Pěst. Mat. Fys. **68**, 85–92 (1939)] that

$$\beta_2\geqq\frac{m\beta_1}{1+(m-1)\beta_1}.$$

Author shows now that this is the best possible result: To every $\gamma\geqq m+1$ there exist m p-adic integers $\alpha_1, \cdots, \alpha_m$ such that

$$\beta_1=\gamma, \quad \beta_2=\frac{m\gamma}{1+(m-1)\gamma}.$$

The proof depends on the following lemmas: (1) To a given $\gamma\geqq 2$ there is a p-adic integer α for which $\beta_1(\alpha)=\gamma$. (2) Let $\beta_1(\alpha_1)=\gamma\geqq 2$. Then for all p-adic integers $\alpha_2, \cdots, \alpha_m$

$$\beta_1(\alpha_1, \cdots, \alpha_m)\geqq\max(m+1, \gamma),$$
$$\beta_2(\alpha_1, \cdots, \alpha_m)\geqq\max\left(\frac{m+1}{m}, \frac{m\gamma}{1+(m-1)\gamma}\right),$$

and for "nearly all" these integers

$$\beta_1(\alpha_1, \cdots, \alpha_m)=\max(m+1, \gamma),$$
$$\beta_2(\alpha_1, \cdots, \alpha_m)=\max\left(\frac{m+1}{m}, \frac{m\gamma}{1+(m-1)\gamma}\right).$$

For the method, compare author's paper [Prace Mat.-Fiz. **43**, 151–166 (1936)]; there the analogous problem in the real case, that is, for Khintchine's "Uebertragungssatz," is solved, with identical result. *K. Mahler* (Manchester).

J64-2 **(1, 295b)**

Mahler, Kurt. On a geometrical representation of p-adic numbers. Ann. of Math. (2) **41**, 8–56 (1940).

A geometric representation of a p-adic integer by means of a sequence $z_n=x_n+iy_n$ is obtained in the following manner. For any r(esidue) c(lass) $\zeta \bmod P^n$ there exist integer matrices $T_n=(p_n, p_n'; q_n, q_n')$ such that (a) $(1, \zeta)T_n\equiv(0, 0) \bmod P^n$; (b) the determinant of T_n is P^n; (c) the elements q_n, q_n' of the second row are relatively prime. The last condition ensures that the r.c. mod P^n represented by T_n is unique; also, if the r.c. of T_{n+i} is contained in the r.c. of T_n, we have $T_{n+i}=T_n\Omega$, where Ω is an integer matrix. In particular, if two matrices represent the same r.c., the factor Ω will belong to the modular group Γ; and if T_n is a sequence representing the r.c. of a fixed number ζ, then the quotients $T_n^{-1}T_{n+1}$ will have determinant P. Choosing a number λ in the classical fundamental domain F of Γ, we normalize the matrices of the sequence by asking that $T_n(z_n)=\lambda$ define numbers z_n in the fundamental domain F; and this sequence of complex numbers is the geometric representation of ζ. Now let $\Phi(X, Y)=2(X-\lambda Y)(X-\bar\lambda Y)/|\lambda-\bar\lambda|$ and consider the transforms $\Phi((X, Y)T_n)$; computing the coefficient of X^2 in two ways we obtain the equation $\Phi(p_n, q_n)=P^n/y_n$. The numbers y_n (and z_n) therefore provide information about the "size" of the pairs p_n, q_n which according to the above condition (a) solve the congruence $u+v\zeta\equiv 0 \bmod P^n$. For instance, since z_n is in F, $y_n\geqq 3^{1/2}/2$, we get: For all n this congruence has solutions p_n, q_n with $0<\Phi(p_n, q_n)\leqq P^n2\cdot 3^{-1/2}$ (shorter: $u+v\zeta$ satisfies $0<\Phi\leqq P^n2\cdot 3^{-1/2}$). The matrices T_n furnish the best solutions, that is, $\Phi(p, q)<\Phi(p_n, q_n)$ is impossible, equality holds only in specified cases (a "Lagrange" type theorem). It is shown that y_n tends towards infinity exactly for rational ζ; if ζ is irrational, $y_n\leqq P^{1/2}$ infinitely often, and it may almost always be greater than $P^{1/2}-\epsilon$. Consequence: if ζ is irrational, then for an infinity of indices $u+v\zeta$ does not satisfy $0<\Phi<P^nP^{-1/2}$; but for every $\epsilon>0$ there exists a ζ such that for all large indices $0<\Phi\leqslant(P^{-1/2}+\epsilon)P^n$ is satisfied

("Khintchine"). On the same basis: If ζ is an irrational and θ an arbitrary p-adic integer, then $u+v\zeta+\theta$ satisfies infinitely often $\Phi<P^n(P+1)/4P^{1/2}$ ("Tchebycheff").

A group of results is obtained (by force) for individual primes P. For $P=2$, 3 one out of every three, for $P=5$ one of every two subsequent $y_n \geqq 1/c_P$ (this number is $7^{1/2}/2$, $2^{1/2}$, 1, respectively). This implies: For every ζ, and at least one of three subsequent indices, $u+v\zeta$ satisfies $0<\phi\leqq c_P P^n$; and for $\epsilon>0$ there exists a ζ, such that for large indices $0<\phi\leqq(c_P-\epsilon)P^n$ is impossible ("Hurwitz-Borel"). The same theorem holds, with $c_P=2/3^{1/2}$, for all primes $P\equiv 1 \bmod 6$.

M. A. Zorn (Los Angeles, Calif.).

Referred to in J64-3.

J64-3 (1, 295a)

Davenport, H. Note on linear fractional substitutions with large determinant. Ann. of Math. (2) **41**, 59–62 (1940).

If the integer N is sufficiently large and if P is prime to N, then for any two intervals I, I' of length $N^{(5/6)+\epsilon}$ there exist integers α and β', respectively contained in I and I', such that $\alpha\beta'\equiv P(\bmod N)$. The proof of this lemma is "an exercise in the use of exponential sums." The number of solutions is first represented in the form $1/N^2$ times a sum of roots of unity. This expression is then decomposed into four parts, the first of which is asymptotically $\varphi(N)N^{-(1/3)+\epsilon}$ (trivial). The lemma will certainly be true if the three remaining parts are of lower order; this is proved on the basis of the following O-relations (letting $e(t)=\exp(2\pi it/N)$): (a) the Kloosterman sums $\sum e(au+bu^{-1})$ are $O(N^{(2/3)+\epsilon}(a, N))$, where $u=1, \cdots, N-1$; (b) for $x=1, \cdots, N-1$, $x'=\min(x, N-x)$, the sums (y in any interval) $\sum e(xy)$ are $O(N/x')$.

If $z_0(\Re z_0=r, |z_0|^2=m)$ is a complex number not equal to zero, then it may be approximated by fixed points $f(\Re f=(\alpha-\beta')/2\beta, \ |f|^2=-\alpha'/\beta)$ of linear substitutions $Z=(\alpha z+\alpha')/(\beta z+\beta')$, with integral coefficients of determinant P, such that (a) both $\Re f-\Re z_0$ and $|z_0|^2-|f|^2$ are $O(P^{-(1/12)+\epsilon})$, (b) the signs of these two differences may be preassigned arbitrarily.

The construction is such that β is chosen first, in dependence on P; afterwards the existence of the numbers α and β' is shown. With $k>m$ fixed, β is taken such that $(\beta, P)=1$, $P=k\beta^2+O(\beta)$; the conditions to be fulfilled by α and β' are: (a) $\alpha-\beta'=2r\beta+O(\beta^{(5/6)+\epsilon})$, (b) $\alpha\beta'=(k-m)\beta^2+O(\beta^{(11/6)+\epsilon})$, (c) $\alpha\beta'\equiv P(\bmod \beta)$. The author provides explicitly two intervals I, I' of length $\beta^{(5/6)+\epsilon}$ such that the first two conditions are satisfied whenever α, β' are taken from I, I'; the lemma guarantees that one of these pairs satisfies also the last condition.

M. A. Zorn (Los Angeles, Calif.).

Citations: MR 1, 295b = J64-2.

J64-4 (2, 251c)

Tornheim, Leonard. Linear forms in function fields. Bull. Amer. Math. Soc. **47**, 126–127 (1941).

Let $F(z)$ be the field of rational functions in z over F. Suppose that V is the valuation belonging to the pole of z. Then define the degree of an element r of $F(z)$ as $-V(r)$. The author proves the following analogue of Minkowski's theorem: "Let $L_i=\sum_{j=1}^n a_{ij}x_j$, $i=1, \cdots, n$, be n linear expressions with coefficients a_{ij} in $F(z)$ and with the determinant $|a_{ij}|$ of degree d. Then for any set of n integers $c_1, \cdots, c_n$ which satisfy the condition $\sum_{i=1}^n c_i>d-n$ there exists a set of values for $x_1, \cdots, x_n$ in $F[z]$ and not all zero such that each L_i has degree at most c_i." For the proof it suffices to consider the case when all c_i are equal. Furthermore, the elements a_{ij} may be taken as polynomials in z. Due to the validity of the Euclidean algorithm in $F[z]$ the matrix (a_{ij}) can be transformed into triangular form. The theorem is then readily proved by considering vector spaces of n-tuples of polynomials of bounded degree.

O. F. G. Schilling (Chicago, Ill.).

J64-5 (7, 245d)

Chabauty, Claude. Approximation par des nombres formés avec un nombre fini de facteurs premiers et arithmétique des suites récurrentes. C. R. Acad. Sci. Paris **219**, 17–19 (1944).

The author sketches a proof of the following theorem. Let K be an algebraic field of degree n, X, Y integers in K, $\overline{|X, Y|}$ the maximum of the absolute values of X, Y and their conjugates, $\theta\neq 0$ any algebraic number, $W(\xi)$ any valuation in $K(\theta)$ and c a positive number. Then

$$0<W(X/Y-\theta)<\overline{|X, Y|}^{-c}$$

has only a finite number of solutions X, Y for which the prime ideal factors of XY belong to a finite set. From this result, applications to recurrent sequences are made [cf. K. Mahler, Nederl. Akad. Wetensch., Proc. **38**, 50–60 (1935); **39**, 633–640, 729–737 (1936)].

K. Mahler.

J64-6 (7, 369e)

Jarník, Voitech. Sur les approximations diophantiques des nombres p-adiques. Revista Ci., Lima **47**, 489–505 (1945).

Let s be a positive integer and $\lambda(n)$ a positive function of the integer $n\geqq 1$ such that $n^{s+1}\lambda^s(n)$ decreases steadily to zero as n tends to infinity. Denote by $H(\lambda)$ and $h(\lambda)$ the sets of all systems of s p-adic integers $\{\mathfrak{a}_1, \mathfrak{a}_2, \cdots, \mathfrak{a}_s\}$ for which the conditions

$$(p, n)=1; \quad |n\mathfrak{a}_i-y_i|_p<\lambda(\max(|n|, |y_1|, \cdots, |y_s|)), \qquad i=1, \cdots, s,$$

have an infinite or a finite number of solutions in rational integers $n, y_1, \cdots, y_s$, respectively. Define the external measures $\mu H(\lambda)$ and $\mu h(\lambda)$ in analogy to the real case; then $\mu H(\lambda)=0$ if $\sum_1^\infty n^s\lambda^s(n)$ converges, $\mu h(\lambda)=0$ if $\sum_1^\infty n^s\lambda^s(n)$ diverges. This result corresponds to a well-known theorem of Khintchine for the real case [Math. Z. **24**, 706–714 (1926); see also V. Jarník, Math. Z. **33**, 505–543 (1931); A. Grošev, Bull. Acad. Sci. URSS. Sér. Math. [Izvestia Akad. Nauk SSSR] **1937**, 427–443 (1937)]. For the theory of measure in the p-adic field, see also the Amsterdam thesis of H. Turkstra [Metrische Bijdragen tot de Theorie der Diophantische Approximaties in het Lichaam der P-adische Getallen, Groningen, 1936]. *K. Mahler* (Manchester).

Referred to in J64-13.

J64-7 (7, 506i)

Monna, A. F. Généralisation P-adique d'un théorème de Minkowski sur les formes linéaires. Nederl. Akad. Wetensch., Proc. **49**, 162–166 = Indagationes Math. **8**, 59–63 (1946).

In the space R_n of all points $x=(x_1, \cdots, x_n)$ with P-adic coordinates, let $\|x\|=\max_{1\leqq i\leqq n}|x_i|_P$ be the distance of x from the origin. Denote by

$$y_i=\sum_{j=1}^n a_{ij}x_j, \qquad i=1, 2, \cdots, n,$$

a system of n linear forms with P-adic coefficients of non-vanishing determinant Δ and by

$$x_i=\Delta^{-1}\sum_{j=1}^n A_{ij}y_j, \qquad i=1, 2, \cdots, n,$$

the inverse forms. Assume that S is a set of points x in R_n without finite points of accumulation; let $\varphi(\lambda)$ be the number of its elements such that

$$\|x\| \leqq P^{\lambda}$$

and let $\lim \sup_{\lambda\to\infty} P^{n\lambda}/\varphi(\lambda) = a$. The author proves that, if

$$m(I_0) \equiv |\Delta|_P^{-n} \prod_{i=1}^{n} \max_{1\leqq j\leqq n} P^{c_j}|A_{ij}|_P > a,$$

then there exist two different elements a, b of S such that $x = a - b \neq 0$ satisfies the inequalities

$$\text{(P)} \qquad \left|\sum_{j=1}^{n} a_{ij}x_j\right|_P \leqq P^{c_i}, \qquad j = 1, 2, \cdots, n.$$

Here $m(I_0)$ is the P-adic measure of the parallelepiped defined by (P). *K. Mahler* (Manchester).

J64-8 (9, 79h)

Lock, Didericus Jacobus. Metrisch-Diophantische Onderzoekingen in $K(P)$ en $K^{(n)}(P)$. [Metric-Diophantine Investigations in $K(P)$ and $K^{(n)}(P)$]. Thesis, Free University of Amsterdam, 1947. vii+100 pp.

The aim of this thesis is to develop metrical theorems on Diophantine approximation in the field of P-adic numbers $K(P)$ and in metricised product fields $K^{(n)}(P)$. In close agreement with the thesis of H. Turkstra [Free University of Amsterdam, 1936], Lebesgue and Hausdorff measures are developed. In contradistinction to Turkstra, the author starts from the axioms of Carathéodory [chapter II]. Chapter III gives an outline of the Diophantine theory; chapter IV contains the proof of the p-adic analogue of Khintchine's theorem on the simultaneous approximation of the number system $\theta, \theta^2, \cdots, \theta^n$. This proof comes from Turkstra, who mentioned this result in his thesis. Chapter V contains the complete transcription of Khintchine's theorem on the approximation of the system $(\theta_1, \theta_2, \cdots, \theta_n)$, whereas chapter VI transcribes the refinement of this theorem in the case $n=1$ by Jarník with use of Hausdorff measure. The final chapter VI contains the p-adic analogue of investigations of Mahler and the reviewer on the classification of transcendental numbers. Some results on the extension fields of $K(P)$, familiar in modern algebra, which had to be used here, have been systematically included in the introductory chapter I. For literature see the reviewer's Diophantische Approximationen [Ergebnisse der Math., v. 4, no. 4, Springer, Berlin, 1936], and Monatsh. Math. Phys. **48**, 176–189 (1939); these Rev. **1**, 137. *J. F. Koksma.*

Citations: MR 1, 137a = J84-1.

Referred to in J64-13, J84-8, J84-13, K15-21.

J64-9 (10, 549a)

Monna, A. F. Sur les espaces linéaires normés. VI. Nederl. Akad. Wetensch., Proc. **52**, 151–160 = Indagationes Math. **11**, 40–49 (1949).

[Part V appeared in the same Proc. **51**, 197–210 = Indagationes Math. **10**, 68–81 (1948); these Rev. **9**, 517.] Let E be a complete totally non-Archimedean normed space over a valued field K satisfying the following conditions. (A) The set of norms $\|\xi\|$ forms a sequence $\{C_n\}$, $-\infty < n < +\infty$; (B) for every n, the set $\|\eta\| \leqq C_n$ modulo the set $\|\eta\| < C_n$ is a vector space over the residue field $\bar{K}$ of K in a natural way; let it be assumed that the dimension is 1. If $\xi_n \varepsilon E$ is such that $\|\xi_n\| = C_n$, then it is proved that every element of E is uniquely of the form $\sum_{i=0}^{\infty} a_{n-i}\xi_{n-i}$, where the a_{n-i} are taken from a system of representatives of $\bar{K}$ in K. Conversely if (A) is satisfied and this expansion holds, then (B) is also true. Again assuming (A) and (B), the author proves that E is locally compact if and only if $\bar{K}$ is finite. There is an application to Diophantine approximation in p-adic fields. *I. S. Cohen* (Cambridge, Mass.).

J64-10 (13, 116h)

Lutz, Élisabeth. Sur les approximations diophantiennes linéares P-adiques. I. Théorèmes généraux. C. R. Acad. Sci. Paris **232**, 587–589 (1951).

Soient P un nombre premier, Q le corps des nombres P-adiques, $|z|_P$ pour $z\varepsilon Q$ la valeur P-adique de z (avec $|P|_P = P^{-1}$). Soit E l'ensemble des entiers P-adiques. Considérons le système des formes

$$\text{(1)} \qquad L_j(x) = x_j + \sum_{1\leqq i\leqq q} a_{ij}x_{p+i} \quad (1 \leqq j \leqq p;\ p+q=n;\ a_{ij}\varepsilon E),$$

où $(x) = (x_1, \cdots, x_n)$ désigne un point à coordonnées entières et posons $H(x) = \max|x_h|$, $h = 1, 2, \cdots, n$. Alors pour tout entier rationnel $\lambda > 0$ le système

$$|L_j(x)|_P \leqq P^{-\lambda} \ (j = 1, 2, \cdots, p), \quad 0 < H(x) \leqq P^{\lambda p n^{-1}},$$

a une solution (x) et de plus le système

$$|L_j(x)|_P \leqq P^{-\lambda} \ (j = 1, 2, \cdots, p),$$
$$P^{\lambda p}(n!)^{-1} \leqq H(x^{(1)}) \cdots H(x^{(n)}) \leqq P^{\lambda p}$$

a n solutions linéairement indépendantes

$$(x^k) = (x_1^{(k)}, \cdots, x_n^{(k)}).$$

Si le système (1) est non-annulable, alors le système des inégalités $|L_j(x)|^p \cdot H(x)^n \leqq 1$ $(j = 1, \cdots, p)$ a une infinité de solutions $(x_1, \cdots, x_n)$ primitives. Si pour un λ et un $(x^{(0)}) = (x_1^{(0)}, \cdots, x_n^{(0)})$ avec $x_h^{(0)} \varepsilon E$, il n'y a pas de solution $(x) = (x_1, \cdots, x_n)$ à coordonnées entières au système

$$|L_j(x)|_P \leqq P^{-\lambda} \ (j = 1, \cdots, p), \quad 0 < H(x) \leqq c_1 P^{\lambda p n^{-1}},$$

ni au système

$$|L_j(x + x^{(0)})|_P \leqq P^{-\lambda} \ (j = 1, \cdots, p), \quad 0 \leqq H(x) \leqq c_2 P^{\lambda p n^{-1}},$$

on a $c_1^{n-1}c_2 \leqq \frac{1}{2}n$.

Ces énoncés qui sont étroitement liés aux résultats de K. Mahler [Jber. Deutsch. Math. Verein. **44**, 250–255 (1934)] sont des cas spéciaux des résultats annoncés par l'auteur sur des inégalités $f(x) \leqq c$ où $f(x)$ désigne une forme hyperconvexe, c'est-à-dire que $f \geqq 0$ pour $x\varepsilon Q^n$ pendant de plus $f(tx) = |t|_P f(x)$ pour tout $t\varepsilon Q$ et $f(x+y) \leqq \max(f(x), f(y))$. En outre des généralisations sont annoncées au cas où l'on considère simultanément plusieurs nombres premiers $P_1, \cdots, P_l$, comme l'a fait K. Mahler dans le cas des formes linéaires. *J. F. Koksma* (Amsterdam).

Referred to in J64-14.

J64-11 (13, 117a)

Lutz, Élisabeth. Sur les approximations diophantiennes linéaires P-adiques. II. Existence de systèmes remarquables. C. R. Acad. Sci. Paris **232**, 667–669 (1951).

En utilisant la notation de la note analysée ci-dessus, l'auteur considère des systèmes (1); elle démontre l'existence de systèmes (1) spéciaux comme elle l'a fait pour le cas réel dans une note antérieure [Chabauty et Lutz, même C. R. **231**, 887–888 (1950); ces Rev. **12**, 483]. Quelques uns de ses résultats sont liés à ceux de K. Mahler [Mathematica, Zutphen. B. **7**, 2–6 (1938), p. 5]. *J. F. Koksma.*

Citations: MR 12, 483d = J16-11.

Referred to in J64-14.

J64-12 (13, 117c)

Lutz, Élisabeth. Sur les approximations diophantiennes linéaires P-adiques. III. Problème non homogène. C. R. Acad. Sci. Paris **232**, 784–786 (1951).

Un théorème P-adique analogue au théorème de Kronecker et quelques théorèmes qui s'y rattachent sont annoncés. L'auteur utilise la relation entre systèmes associés de formes linéaires [cf. les deux analyses ci-dessus et l'analyse ci-dessous]. *J. F. Koksma* (Amsterdam).

Referred to in J64-14.

J64-13 **(13, 117d)**

Lutz, Élisabeth. Sur les approximations diophantiennes linéaires P-adiques. IV. Résultats métriques. C. R. Acad. Sci. Paris **232**, 1389–1392 (1951).

Des résultats métriques sur l'approximation des formes

$$b_j+x_j+\sum_{1\leqq i\leqq q} a_{ij}x_{p+i} \quad (1\leqq j\leqq p;\ p+q=n)$$

qui dans le cas homogène ($b_j=0$) généralisent des résultats de Jarník [Revista Ci., Lima **47**, 489–505 (1945); ces Rev. **7**, 369]. La méthode est analogue à celle de Cassels dans le cas réel [Proc. Cambridge Philos. Soc. **46**, 209–218 (1950); ces Rev. **12**, 162]. Pour la théorie métrique P-adique cf. aussi les thèses de H. Turkstra [Amsterdam, 1936] et de D. Lock [Université libre à Amsterdam, 1947; ces Rev. **9**, 79]. *J. F. Koksma* (Amsterdam).

Citations: MR 7, 369e = J64-6; MR 9, 79h = J64-8; MR 12, 162b = K30-7.

Referred to in J64-14.

J64-14 **(16, 1003d)**

Lutz, Elisabeth. Sur les approximations diophantiennes linéaires P-adiques. Actualités Sci. Ind., no. 1224. Hermann & Cie, Paris, 1955. 106 pp. 1200 francs.

This thesis constitutes an extensive investigation of approximation questions concerning a system Λ of p linearly independent linear forms $L_j(x)$ with P-adic coefficients, in n rational integral variables $(x_1, \cdots, x_n)=x$. (Λ is said to be of signature (p, n).) Certain of the results proved here were announced earlier [C. R. Acad. Sci. Paris **232**, 587–589, 667–669, 784–786, 1389–1392 (1951); MR **13**, 116, 117]. The treatment is self-contained, except for general topological and measure-theoretic background

Define P: a rational prime; Q: the rational numbers; Q_P: the P-adic completion of Q; $Q_P{}^n$: the n-dimensional vectors with components in Q_P; $E_P{}^n$: the vectors of $Q_P{}^n$ with P-adic integral components; Z^n: the vectors of $Q_P{}^n$ whose components are rational integers. If Λ is of signature (p, n), it is equivalent to a canonical system L, in which $L_j(x)=x_j+\sum_{i=1}^q a_{ij}x_{p+i}$ $(j=1, \cdots, p)$, where $p+q=n$. Λ is said to be free if the a_{ij} are algebraically independent. The system M, in which $M_i(y)=y_{p+i}-\sum_{j=1}^p a_{ij}y_j$ for $i=1, \cdots, q$, is called the associate of L. For $x \in Z^n$, put

$$\Lambda(x)=\max_{1\leqq j\leqq p}(|\Lambda_j(x)|_P) \quad\text{and}\quad H(x)=\max_{1\leqq i\leqq n}(|x_i|).$$

x is said to be primitive if $\gcd(x_1, \cdots, x_n)=1$. Define the rational integer $\delta(\Lambda)$ by the relation $P^{-\delta}=\max_D(|\det D|_P)$, where D ranges over the minors of rank p in the matrix of coefficients of Λ. $\rho(\Lambda)$ has a similar but more complicated definition. Finally, if $\varphi(\lambda)$ has non-negative value for positive rational integral λ, and $x^{(0)} \in E_P{}^n$, the notation $\Lambda_{x^{(0)}}\dashv\dashv\varphi(\lambda)$ means that there is a positive constant c such that for all sufficiently large λ there is a solution $x \in Z^n$ of the inequalities $\Lambda(x+x^{(0)})\leqq P^{-\lambda}$ and $H(x)c\leqq\varphi(\lambda)$. In case $x^{(0)}=0$, the restriction $x\neq 0$ is also imposed, and we write $\Lambda\dashv\dashv\varphi(\lambda)$.

The following are among the principal theorems: I. Let $\lambda_1, \cdots, \lambda_p$ be rational integers $\geqq\rho(\Lambda)$, and let c be a positive constant. Then there exists $x \in Z^n$, primitive except for a power of P, such that $|\Lambda_j(x)|_P\leqq P^{-\lambda_j}$ $(j=1, \cdots, p)$ and $0<H(x)\leqq c$, if $c^n\geqq P^{\lambda_0-\delta(\Lambda)}$, where $\lambda_0=\sum_1^p\lambda_j$. II. If $\Lambda(x)$ is never 0 for $x \in Z^n$, $x\neq 0$, the inequality $H^n(x)\Lambda^p(x)\leqq P^{-\delta}$ has infinitely many primitive solutions x. III. If $\varphi(\lambda)$ is as described above, and $1\leqq p\leqq\frac{1}{2}n$, there exists a free canonical system L of signature (p, n) such that $L\dashv\dashv\varphi(\lambda)$, and these systems are everywhere dense in the space $Q_P{}^{pq}$ of coefficients $(a_{11}, \cdots, a_{qp})$. IV. Given a canonical system L of signature (p, n) $(1\leqq p<n)$ and an $x^{(0)} \in E_P{}^n$, there exists a function $\epsilon(\lambda)$ tending to zero with λ^{-1} such that $L_{x^{(0)}}\dashv\dashv\epsilon(\lambda)P^\lambda$, if and only if $\sum_1^p L_j(x^{(0)})y_j$ is a rational integer for every $y \in Z^n$ such that $M(y)=0$. V. For every system L having integral P-adic coefficients, there is an $x^{(0)} \in E_P{}^n$ such that the inequality $L(x+x^{(0)})\leqq(nP^3H(x))^{-n/p}$ has no solution $x\neq 0$. VI. Let $f(l)$ be a sufficiently regular function, defined and positive for $l>0$. Let the forms of the system L have integral coefficients in Q_P. Then the number of solutions x of the inequality $L(x)\leqq f(H(x))$ is finite or infinite for almost all (Haar measure) $a=(a_{11}, \cdots, a_{qp}) \in E_P{}^{pq}$, according as the series $\sum_1^\infty h^{n-1}f^p(h)$ converges or diverges.

W. J. LeVeque (Ann Arbor, Mich.).

Citations: MR 13, 116h = J64-10; MR 13, 117a = J64-11; MR 13, 117c = J64-12; MR 13, 117d = J64-13.

Referred to in J02-14.

J64-15 **(18, 22b)**

Postnikov, A. G. Properties of solutions of Diophantine inequalities in the field of formal power series. Dokl. Akad. Nauk SSSR (N.S.) **106** (1956), 21–22. (Russian)

The author states without proof some results analogous to those of the geometry of numbers about the ring of formal power series over a field. Write $\omega(x)=\sum_{-l}^\infty a_u x^u$, where possibly $l<0$, and put $\|\omega(x)\|=l$ if $a_{-l}\neq 0$. For any $\omega_1(x)$, $\omega_2(x)$ and any integers m, n there are polynomials $P(1/x)$, $Q(1/x)$, $R(1/x)$ in $1/x$ such that

$$\|P(1/x)\|\leqq m,\ \|Q(1/x)\|\leqq n,$$

$$\|P(1/x)\omega_1(x)+Q(1/x)\omega_2(x)-R(1/x)\|\leqq -m-n-1.$$

It is possible to choose P, Q, R so that $\|P(1/x)\|=m$. Under certain restrictive conditions on the coefficients of $\omega_1(x)$, $\omega_2(x)$ there exist recurrence relations between the P, Q, R obtained for differing values of m and n.

J. W. S. Cassels (Cambridge, England).

Referred to in J64-16.

J64-16 **(18, 875e)**

Postnikov, A. G. Properties of solutions of Diophantine inequalities in the field of formal power series. Mat. Sb. N.S. **40(82)** (1956), 295–302. (Russian)

The author proves the results already announced and some similar but simpler ones [Dokl. Akad. Nauk SSSR (N.S.) **106** (1956), 21–22; MR **18**, 22].

J. W. S. Cassels (Cambridge, England).

Citations: MR 18, 22b = J64-15.

J64-17 **(20# 38)**

Armitage, J. V. The product of n linear forms in a field of series. Mathematika **4** (1957), 132–137.

Let K be a field, t an indeterminate, $K\{t\}$ the field of all formal series $\phi=\sum_{k=-\infty}^m a_k t^k$ where $a_k\in K$, and let $|\phi|$ denote the valuation on $K\{t\}$ defined by $|0|=0$, $|\phi|=e^m$ if $a_m\neq 0$. Let $L_i=\sum_{j=1}^n \alpha_{ij}u_j$ $(i=1, 2, \cdots, n)$, where $\alpha_{ij}\in K\{t\}$, be n linear forms of determinant $D\neq 0$. By K. Mahler [Ann. of Math. (2), **42** (1941), 488–522; MR **2**, 350], there exist polynomials $u_1, \cdots, u_n\in K[t]$ not all zero such that

$$|L_1L_2\cdots L_n|\leqq e^{-(n-1)}|D|.$$

In the opposite direction, the author proves that this result is best-possible if K has the characteristic 0, or if K is a finite field of at least $n-1$ elements. If, however, K is a finite field of $q\leqq n-2$ elements, then the stronger inequality

$$|L_1L_2\cdots L_n|\leqq e^{-n+2-\gamma}|D|$$

can be satisfied; here γ is the integer defined by $\gamma\geqq(n-1)/q>\gamma-1$. *K. Mahler* (Manchester)

Citations: MR 2, 350c = H05-4.

J64-18 **(21# 3398)**

Fenna, D. Simultaneous Diophantine approximation to series. J. London Math. Soc. **34** (1959), 173–176.

Let k denote an arbitrary nontrivial field, z an indeterminate, and K the field of formal Laurent series

$$x = \alpha_d z^d + \alpha_{d-1} z^{d-1} + \cdots$$

with coefficients in k. For a fixed real number $\kappa > 1$, define a valuation on K by $|0| = 0$ and $|x| = \kappa^d$ if α_d is the leading nonzero coefficient in $x \neq 0$. Let $c(n)$ be the supremum of the numbers c such that for all $t_1, \cdots, t_n \in K$ which are not all rational functions of z, there are infinitely many sets of polynomials $b_0, b_1, \cdots, b_n \in k[z]$ satisfying

$$|b_0(b_0 t_i - b_i)^n| \leqq \kappa^{-c} \quad (i = 1, 2, \cdots, n).$$

It is shown that $c(n) = n$.

W. J. LeVeque (Ann Arbor, Mich.)

J64-19 **(21# 3408)**

Eichler, Martin. Ein Satz über Linearformen in Polynombereichen. Arch. Math. **10** (1959), 81–84.

Let k be an arbitrary field, x an indeterminate and $k_\infty(x)$ the field of formal power series in x^{-1} over k. Let $d(a)$ denote the degree of a function a in $k_\infty(x)$ or in $k[x]$ (thus, e.g., $d(x^{-3}) = -3$). The author proves the following theorem: Let $M = (m_{ij})$ be any $n \times n$ non-singular matrix over $k_\infty(x)$ and let $\tilde{M} = (\tilde{m}_{ij})$ be the contragredient of M: $\tilde{M} = {}^t M^{-1}$. For any integers $\gamma_1, \cdots, \gamma_n$, let $l(M, \gamma)$ and $l(\tilde{M}, \tilde{\gamma})$ denote the dimensions of the vector spaces over k formed by all $(x_1, \cdots, x_n)$ and $(\tilde{x}_1, \cdots, \tilde{x}_n)$, respectively, satisfying

$$d\Big(\sum_i x_i m_{ij}\Big) \leqq \gamma_j, \qquad x_i \in k[x],$$

$$d\Big(\sum_i \tilde{x}_i \tilde{m}_{ij}\Big) \leqq \tilde{\gamma}_j = -2-\gamma_j, \quad \tilde{x}_i \in k[x], \quad i, j = 1, \cdots, n.$$

Then $l(M, \gamma) = l(\tilde{M}, \tilde{\gamma}) - d(|M|) + \sum_i \gamma_i + n$. It is also noticed that the theorem of Riemann-Roch is a consequence of this equality.

K. Iwasawa (Cambridge, Mass.)

J64-20 **(27# 4794)**

Davenport H.; Lewis, D. J.

An analogue of a problem of Littlewood.

Michigan Math. J. **10** (1963), 157–160.

Let K be an infinite field and let t be an indeterminate. Let $\mathfrak{P}_K$ denote the field of formal power series $M = m_h t^h + m_{h-1} t^{h-1} + m_{h-2} t^{h-2} + \cdots$ with coefficients in K, and let N be any non-zero polynomial in t with coefficients in K. The authors show that, for given Θ and Φ in $\mathfrak{P}_K$, the expression

$$|N|_K \|N\Theta\|_K \|N\Phi\|_K$$

need not take arbitrarily small values. Here $|M_k| = e^h$, if $m_h \neq 0$, and $\|M\|_K = |M_1|_K$, where M_1 is the part of the series for M comprising only the negative powers of t. It is an open question whether or not the corresponding proposition for the field of real numbers (instead of $\mathfrak{P}_K$) holds true.

C. G. Lekkerkerker (Amsterdam)

Referred to in J64-22, J64-25, J64-28.

J64-21 **(28# 1168)**

Bumby, Richard T.

An elementary example in p-adic diophantine approximation.

Proc. Amer. Math. Soc. **15** (1964), 22–25.

Let $g(t) > 0$ for $t > 1$, and let $c > 0$ be a constant. The author is interested in problems of the following type. "Do there exist integers x, y ($x \neq 0, y \neq 0, x \neq y$) such that $g(\max(|x|, |y|))\, |x|\, |y|\, |x|_2\, |y|_3\, |x-y|_5 < c$?" ($|x|_p$ denotes the p-adic value of x.) He proves that if $g(t) = \log t$ and $c > \frac{2}{5}(\log 2)(\log 3)/(\log 6)$, this problem has infinitely many integral solutions x, y such that $(x, y) = 1$.

K. Mahler (Canberra)

J64-22 **(29# 2218)**

Baker, A.

On an analogue of Littlewood's Diophantine approximation problem.

Michigan Math. J. **11** (1964), 247–250.

Littlewood has asked whether for each pair of real numbers θ and ϕ and each $\varepsilon > 0$ there exist integers $n \neq 0, m, q$, such that $|n| \cdot |n\theta - m| \cdot |n\phi - q| < \varepsilon$. This question remains unanswered. However, Davenport and the reviewer [same J. **10** (1963), 157–160; MR **27** #4794] obtained a negative answer for the analogous question concerning formal power series fields over infinite fields. In this paper the author gives specific examples of θ and ϕ for which a negative answer results. Specifically, he shows that for all polynomials $u(t) \neq 0$, $v(t)$, $w(t)$ with real coefficients, one has

$$|u(t)|_K |u(t)\theta - v(t)|_K |u(t)\phi - w(t)|_K \geqq e^{-5},$$

where $|a_m t^m + a_{m+1} t^{m+1} + \cdots|_K = e^m$ if $a_m \neq 0$.

D. J. Lewis (Ann Arbor, Mich.)

Citations: MR 27# 4794 = J64-20.

Referred to in J64-25, J64-28.

J64-23 **(32# 5592)**

Cantor, David G.

On the elementary theory of diophantine approximation over the ring of adeles. I.

Illinois J. Math. **9** (1965), 677–700.

Let K be an algebraic number field. Let T be any set of normalized valuations $|\ |_v$ of K including the normalized archimedean valuations of K. Denote by K_v the completion of K with regard to $|\ |_v$. The author defines the T-adele ring $A^T = A_K{}^T$ and deduces a number of approximation theorems which, when specialized to the case of the rational number field and no or one p-adic valuation, reduce to known results. The elements of A^T, or T-adeles, are the sets $a = (a_v)_{v \in T}$ such that $a_v \in K_v$ for all $v \in T$ and $|a_v|_v \leqq 1$ for almost all (= all but finitely many) $v \in T$; a basis for the neighbourhoods of an element $a \in A^T$ is formed by the parallelotopes $P(i, a) = \{x \in A^T : x - a \leqq i\}$, where i is an invertible T-adele, or T-idele, and where $x - a \leqq i$ means $|x-a|_v \leqq |i|_v$ for all $v \in T$. The role of integers in A^T is played by the adeles (k), $k \in K$, having the property that $|k|_v \leqq 1$ for each normalized valuation $v \notin T$. These adeles form a discrete ring K^T, and the quotient A^T/K^T is compact. A fundamental domain D^T for A^T/K^T and a Haar measure μ^T on A^T can be written down explicitly. The following topics are discussed.

(1) Minkowski's theorem on a system of homogeneous linear forms. Existence of infinitely many solutions $x, y \in K^T$ to

$$0 < \prod_v |\alpha x + \beta y|_v \leqq \frac{c(\alpha, \beta)}{\prod_v \max(|x|_v, |y|_v)},$$

where α, β are T-adeles with not both α_v and $\beta_v = 0$ for any $v \in T$, $c(\alpha, \beta)$ is a positive constant and $|a|_v$ is an abbreviation for $|a_v|_v$.

(2) Kronecker's theorem. The proof of this theorem is mainly based upon the following fact: If $\alpha = (\alpha_1, \cdots, \alpha_n) \in (A^T)^n$, $(A^T)^n$ the n-dimensional vector space over A^T, if S is an (additive) subgroup of $(A^T)^n$ and $\chi(\alpha) = 1$ for each continuous character χ on $(A^T)^n$ with $\chi(S) = 1$, then α is in the closure of S. This fact itself is a consequence of the duality theorem for locally compact abelian groups. A simple corollary of Kronecker's theorem is the strong approximation theorem in A^T.

(3) Uniform distribution mod K^T, Weyl's criterion, Van der Corput's fundamental inequality. The concepts and results discussed all concern functions on K^T (with values in A^T) rather than sequences of elements, as in the usual theory of uniform distribution (cf. J. Cigler and G. Helmberg [Jber. Deutsch. Math. Verein. **64**, Abt. 1, 1–50 (1961); MR **23** #A2409]); the lack of a linear order in K^T makes it necessary to take limits in the sense of nets.

(4) A metrical theorem. The following generalization of a theorem of Hinčin is proved. Suppose that T is not the entire set of normalized valuations of K. Let $(\varepsilon_q)_{q\in K^T}$ be a system of T-ideles with $\sum_q \prod_v |\varepsilon_q|_v < \infty$. Then for fixed $\alpha \in A^T$ and almost all $\theta \in A^T$, there exist only finitely many solutions $q \in K^T$ to (*) $q\theta - \alpha \leqq \varepsilon_q$ (mod K^T). If, however, $\sum_q \prod_v |\varepsilon_q|_v = \infty$, then for almost all pairs $(\theta, \alpha) \in (A^T)^2$ there exist infinitely many solutions to (*). Similar results hold for certain classes of sequences (ε_{qn}); these results are related to a refinement of Hinčin's result given by J. W. S. Cassels [Proc. Cambridge Philos. Soc. **46** (1950), 209–218; MR **12**, 162].

C. G. Lekkerkerker (Amsterdam)

Citations: MR 12, 162b = K30-7; MR 23# A2409 = K02-1.

J64-24 (35# 1548)

Cusick, T. W.

Analogue of a theorem of Khintchine in fields of formal power series.

Proc. Cambridge Philos. Soc. **62** (1966), 637–642.

Let k be a field and let $k[t]$ be the ring of polynomials in one variable t over k. Further, let $k\{t\}$ be the field of formal power series $\lambda = a_m t^m + a_{m-1}t^{m-1} + \cdots + a_0 + a_{-1}t^{-1} + \cdots$ with coefficients in k. Put $|\lambda| = e^m$ provided $a_m \neq 0$ and put $\|\lambda\| = e^{-h}$, where $a_{-1} = a_{-2} = \cdots = a_{-h+1}$, $a_{-h} \neq 0$. The author shows that if $L_j(X)$ are any n linear forms in m variables $x_1, x_2, \cdots, x_m$ with coefficients in $k\{t\}$, then there exist $\alpha_1, \alpha_2, \cdots, \alpha_n$ in $k\{t\}$ such that

$$(\max_{1\leqq j\leqq n} \|L_j(X) - \alpha_j\|)^n (\max |x_i|)^m \geqq e^{-n-m}$$

for all $x_1, x_2, \cdots, x_m$ not all zero in $k[t]$, and this result is best possible.

This theorem is an extension of work of S. K. Aggarwal [Notices Amer. Math. Soc. **12** (1965), 620, Abstract 65T-357].

A. C. Woods (Columbus, Ohio)

J64-25 (35# 6625)

Cusick, T. W.

Littlewood's Diophantine approximation problem for series.

Proc. Amer. Math. Soc. **18** (1967), 920–924.

One of the famous unsolved problems in Diophantine approximations is the following problem of Littlewood: For each pair of real numbers θ, ϕ and each positive real number ε, do there exist integers $q > 0$, a, b such that $q|\phi\theta - a|\,|q\phi - b| < \varepsilon$? There is a natural analogy for formal power series fields. Let K be a field of characteristic 0, t a transcendental over K; $K[t]$ denotes the ring of polynomials in t over K, $K(t)$ the set of rational functions in t and $K\{t\}$ the completion of $K(t)$ under the valuation having $|t| = e$. H. Davenport and the reviewer [Michigan Math. J. **10** (1963), 157–160; MR **27** #4794] have shown that there exist ϕ, ψ in $K\{t\}$ such that $|Q|\,|Q\theta - A|\,|Q\phi - B| \geqq e^{-2}$ for all $Q \neq 0$, A, $B \in K[t]$. Thus the answer to the analogy to Littlewood's problem in power series fields is no. Later, A. Baker [ibid. **11** (1964), 247–250; MR **29** #2218] showed that $|Q|\,|Qe^{1/t} - A|\,|Qe^{2/t} - B| \geqq e^{-5}$ for all $Q \neq 0$, A, $B \in K[t]$. The author now proves the following: Let $n \geqq 2$ and let $f(x)$ be any polynomial of degree $\geqq n$ over K; then there exists a continuum of θ in $K\{t\}$ such that

$$|Q|\,|Q\theta - A_1| \cdots |Q\theta^{n-1} - A_{n-1}|\,|Qf(\theta) - A_n| \geqq e^{-n},$$

for all $Q \neq 0$, $A_1, \cdots, A_n$ in $K[t]$. In particular, there exists a continuum of $\theta \in K\{t\}$ such that

$$|Q|\,|Q\theta - A_1| \cdots |Q\theta^n - A_n| \geqq e^{-n}.$$

{See also a paper by R. T. Bumby [Proc. Amer. Math. Soc. **18** (1967), 1125–1127; MR **36** #116] on this same subject.}

D. J. Lewis (Ann Arbor, Mich.)

Citations: MR 27# 4794 = J64-20; MR 29# 2218 = J64-22; MR 36# 116 = J64-26.

Referred to in J64-26, J64-28.

J64-26 (36# 116)

Bumby, R. T.

On the analog of Littlewood's problem in power series fields.

Proc. Amer. Math. Soc. **18** (1967), 1125–1127.

See the review of T. W. Cusick's paper [same Proc. **18** (1967), 920–924; MR **35** #6625] for background and definitions. The author proves that there exists an infinite sequence $\theta_1, \theta_2, \theta_3, \cdots$ in $K\{t\}$ such that for any $m \geqq 1$, $|Q| \cdot |Q\theta_1 - A_1| \cdots |Q\theta_m - A_m| \geqq e^{-m}$ for all $Q \neq 0, A_1, \cdots, A_m$ in $K[t]$. Furthermore, if $\theta_1, \cdots, \theta_m$ satisfies this relation then the sequence $\{\theta_i\}$ can be chosen with $\theta_1, \cdots, \theta_m$ as the first m elements. The author's proofs make abundantly clear that the problem is essentially the determination of the dimension of certain linear spaces.

D. J. Lewis (Ann Arbor, Mich.)

Citations: MR 35# 6625 = J64-25.

Referred to in J64-25, J64-28.

J64-27 (41# 177)

Aggarwal, Satish K.

Homogeneous approximation in the field of formal power series.

Proc. Nat. Inst. Sci. India Part A **34** (1968), 242–249.

Let $K(t)$ be a transcendental extension of the field K and denote by $K\{t\}$ the completion of $K(t)$ with respect to the valuation $|a/b| = e^{\deg a - \deg b}$, $e > 1$ (the base of natural logarithms), $a, b \in K[t]$. Let $P_n = K\{t\}^n$ be the ultrametric space with distance defined by $|X| = \max(|x_1|, \cdots, |x_n|)$, $X \in P_n$. For $x \in K\{t\}$, the "fractional part" of x has absolute value $\|x\| = \inf|x - a|$, $a \in K[t]$. The author proves the analogues for $K\{t\}$ of well known theorems of Diophantine approximation in **R**. The main results are as follows. (I) Let m, n be positive integers and suppose $\mathrm{Card}(K) \geqq m + n - 1$. Then there exist $\mu = \mu(K, m, n) > 0$, and m linear forms $L_i(U) = L_i(u_1, \cdots, u_n)$ $(1 \leqq i \leqq m)$, in n variables $u_1, \cdots, u_n$, with coefficients in $K\{t\}$, such that for all $U \neq 0$ in P_n, $|U|^n(\max_{1\leqq i\leqq m}\|L_i(U)\|)^m \geqq \mu$. (II) Let $L_i(U)$ $(1 \leqq i \leqq m)$ be a system of linearly independent forms with $1 \leqq m < n$ and suppose that there does not exist any $U \neq 0$ in P_n such that $L_i(U) = 0$, $1 \leqq i \leqq m$. Then the inequality $(\max_{1\leqq i\leqq m}|L_i(U)|)^m |U|^{n-m} \leqq \exp(n - m - \delta)$ has infinitely many solutions in P_n, where δ is an integer depending on the given set of forms.

J. V. Armitage (London)

J64-28 (42# 1768a; 42# 1768b)

Armitage, J. V.

An analogue of a problem of Littlewood.

Mathematika **16** (1969), 101–105.

Armitage, J. V.

Corrigendum and addendum: "An analogue of a problem of Littlewood".

Mathematika **17** (1970), 173–178.

Littlewood posed the following still unsolved problem: For each pair of real numbers θ, ϕ and each $\varepsilon>0$, does there exist an integer n such that $n\|n\theta\|\|n\phi\|<\varepsilon$? Here $\|\alpha\|$ is the distance from α to the nearest integer. There is an obvious analog of this problem for the field $k\{t\}$ of formal power series in t over a field k with the polynomials playing the role of the integers and distance being measured by $|a_nt^n+a_{n+1}t^{n+1}+\cdots|=q^n$ when $a_n\neq 0$ and where $q>1$. H. Davenport and the reviewer [Michigan Math. J. **10** (1963), 157–160; MR **27** #4794] showed that when characteristic $k=0$, there exist series θ, ϕ for which the answer to the series analog of Littlewood's problem is no. Later A. Baker [ibid. **11** (1964), 247–250; MR **29** #2218] showed one could choose $\theta(t)=\exp(1/t)$ and $\phi(t)=\exp(2/t)$. R. Bumby [Proc. Amer. Math. Soc. **18** (1967), 1125–1127; MR **36** #116] and T. W. Cusick [ibid. **18** (1967), 920–924; MR **35** #6625] also discussed extensions to more series. The author now proves that the answer to the series analog for Littlewood's problem is no when the characteristic of $k>3$. The proof is along the lines of that given by Baker, and the θ, ϕ used lie in the "real" cubic field defined by $y^3=(1+t^{-1})$. It is interesting to note that J. W. S. Cassels and H. P. F. Swinnerton-Dyer [Philos. Trans. Roy. Soc. London Ser. A **248** (1955), 73–96; MR **17**, 14] showed that in case θ, ϕ are elements of a real cubic field, then the answer to Littlewood's problem is yes. The first of the two papers contained some errors and was unnecessarily complicated. The second paper contains a revised, corrected and much simpler argument.

D. J. Lewis (Ann Arbor, Mich.)

Citations: MR 17, 14f = J36-24; MR 27# 4794 = J64-20; MR 29# 2218 = J64-22; MR 35# 6625 = J64-25; MR 36# 116 = J64-26.

J68 APPROXIMABILITY OF ALGEBRAIC NUMBERS (THUE-SIEGEL THEOREM, ETC.)

See also reviews D02-15, D24-66, D40-52, D48-45, J02-3, J02-11, J02-14, J02-23, J12-46, J12-57, J16-23, J40-15, J40-19, J40-22, J40-24, J76-30, J76-36, J76-67, J80-44, N32-14, Q05-34, Q10-8, Q15-3, Q99-9, R14-105, R58-4, Z10-4, Z10-47.

J68-1 (1, 67a; 1, 67b)

Hasse, H. Simultane Approximation algebraischer Zahlen durch algebraische Zahlen. Nachr. Ges. Wiss. Göttingen. Fachgruppe I, N. F. **1**, 209–212 (1939).

Hasse, H. Simultane Approximation algebraischer Zahlen durch algebraische Zahlen. Monatsh. Math. Phys. **48**, 205–225 (1939).

The theorem of Thue-Siegel [Math. Z. **10**, 173–213 (1921)] for the approximation of an algebraic number α by the numbers ξ of a fixed algebraic number field K can be expressed as follows. Let $r>1$ be the relative degree of α with regard to K. Let $f(x)=0$ be the irreducible equation for ξ in the rational number field, written with integral, relatively prime coefficients, and denote by $H(\xi)$ the maximum of the absolute values of these coefficients. If we have $|\alpha-\xi|\leqq C/H(\xi)^e$ for an infinite sequence of primitive numbers ξ of K, where the positive constants C and e are independent of ξ, then

$$e\leqq \min_{s=0,1,\cdots,r-1}\{s+r/(s+1)\}.$$

Siegel's paper: Über einige Anwendungen Diophantischer Approximationen [Abh. Preuss. Akad. Wiss. Phys. Math. Kl. **1929**, 1–70] contains implicitly a generalization of this theorem for the approximation of a system of algebraic numbers. The author formulates this generalization and gives a new proof for it in the following form, which is sharper than the result contained in Siegel's paper: Let $\alpha_1, \alpha_2, \cdots, \alpha_g$ be a fixed system of algebraic numbers and K an algebraic number field of degree k. Denote by r the relative degree of $K(\alpha_1, \alpha_2, \cdots, \alpha_g)$ over K and denote by d the smallest integer such that the $(g+d)!/g!d!$ power products of $\alpha_1, \alpha_2, \cdots, \alpha_g$ of degrees $0, 1, \cdots, d$ are linearly dependent with regard to K. Let $\xi_i^{(\kappa)}$ $(\kappa=1, 2, \cdots, k)$ be the conjugates of ξ_i $(i=1, 2, \cdots, g)$, $N(1, \xi_1, \cdots, \xi_g)$ the norm of the ideal $(1, \xi_1, \cdots, \xi_g)$ in K, and

$$H=H(\xi_1, \xi_2, \cdots, \xi_g)$$
$$=\{\prod_{\kappa=0}^{k-1}\max(1, |\xi_1^{(\kappa)}|, \cdots, |\xi_g^{(\kappa)}|)\}/N(1, \xi_1, \cdots, \xi_g).$$

If we have

$$|\alpha_i-\xi_i|\leqq C/H^e, \qquad i=1, 2, \cdots, g,$$

for an infinite sequence of different systems $\xi_1, \xi_2, \cdots, \xi_g$ of K, where $C>0$ and $e>0$ are independent of $\xi_1, \xi_2, \cdots, \xi_g$, then

$$e\leqq \min_{\delta=0,1,\cdots,d-1}\{\delta+(C_{g+\delta,g})^{-1}r\}.$$

For $g=1$, one obtains the theorem of Thue-Siegel and a new proof for this theorem. Moreover it follows that the theorem of Thue-Siegel is also true for the approximation by imprimitive numbers ξ and that it is true for $r=1$. It may be mentioned that the latter fact was also proved by the reviewer [J. Reine Angew. Math. **160**, 70–99 (1929)] as a special case of a theorem on the approximation of an algebraic number by algebraic numbers of a fixed degree.

A. Brauer (Princeton, N. J.).

J68-2 (2, 253c)

Parry, C. J. The p-adic generalization of the Thue-Siegel theorem. J. London Math. Soc. **15**, 293–305 (1940).

Let ξ be an algebraic number of degree n and s a positive integer. By a result of the reviewer [Math. Z. **10**, 173–213 (1921)], the inequality $|qr^{-1}-\xi|<r^{-(s+\epsilon+n(s+1)^{-1})}$ has for any positive ϵ only a finite number of solutions in integers q and $r>0$; a corresponding statement holds in the case when approximation is made by algebraic numbers of a fixed degree instead of the rational number q/r. Another generalization of Thue's theorem was found by Mahler [Math. Ann. **107**, 691–730 (1933) and **108**, 37–55 (1933)] who considered approximations to p-adic numbers by rational numbers and obtained an estimation of the number of representations of an arbitrary integer by binary forms. The author combines these results and extends them to the case of approximation by algebraic numbers for the different possible valuations. The proofs were to have been published in Acta Arithmetica; the present note contains only an abstract of the results.

C. L. Siegel.

J68-3 (2, 350a)

Gelfond, A. On the simultaneous approximations of algebraic numbers by rational fractions. Bull. Acad. Sci. URSS. Sér. Math. [Izvestia Akad. Nauk SSSR] **5**, 99–104 (1941). (Russian. English summary)

The author establishes an exact connection between the simultaneous approximation of powers of an algebraic number by fractions with equal denominators and the structure of linear forms of powers of the number, provided that the form is sufficiently rapidly convergent to zero. Relations between linear forms involving arbitrary numbers and simultaneous approximation of these numbers were investigated by Khintchine (Khintchinesches Übertragungsprinzip) [Rend. Circ. Mat. Palermo **50**, 170–195 (1926)], but the relation for a certain class of forms is here determined in a very precise way. *D. C. Spencer.*

J68-4 **(4, 266d)**

Pisot, C. Ein Kriterium für die algebraischen Zahlen. Math. Z. **48**, 293–323 (1942).

The main results of the paper are as follows. (I) Every algebraic number ξ of degree s can be approximated by a sequence of rational fractions u_n/v_n possessing the following properties: there exists a number $\alpha>1$, a positive ϵ and constants C_1, C_2, C_3, such that

(1) $|u_{n+1}-\alpha u_n|<C_1/|u_n|^\epsilon,\quad |v_{n+1}-\alpha v_n|<C_2/|v_n|^\epsilon$

and

(2) $$|u_n-\xi v_n|<C_3/|v_n|^\epsilon.$$

Here $\epsilon\leqq(s-1)^{-1}$, but can be taken arbitrarily close to $(s-1)^{-1}$ by a suitable choice of α. (II) Every number ξ which can be approximated by a sequence of rational fractions u_n/v_n possessing the properties (1) is algebraic of degree $s\leqq(\epsilon+1)/\epsilon$, and the approximating fractions u_n/v_n possess the property (2). It is also shown that the converse proposition (II) can be improved by imposing on the approximating fractions conditions less restrictive than (1), that is, by supposing only

$$\sum_{k=n}^{2n}|u_{k+1}-\alpha u_k|^2\leqq 1/4\alpha^2,\quad \sum_{k=n}^{2n}|v_{k+1}-\alpha v_k|^2\leqq 1/4\alpha^2$$

for every n larger than a fixed m. The paper is closely related to a previous article of the author [Ann. Scoula Norm. Super. Pisa (2) **7**, 205–248 (1938). *R. Salem.*

Referred to in K25-9, R06-19.

J68-5 **(9, 412h)**

Dyson, F. J. The approximation to algebraic numbers by rationals. Acta Math. **79**, 225–240 (1947).

C. L. Siegel [Math. Z. **10**, 173–213 (1921)] zeigte: Ist ξ algebraisch n-ten Grades, so hat die Ungleichung $|\xi-p/q|<q^{-\mu}$ für $\mu\geqq 2n^{\frac{1}{2}}$ höchstens endlich viele Lösungen in ganzen rationalen Zahlen p, q. Dyson verschärft nun den Exponenten auf $\mu>(2n)^{\frac{1}{2}}$. Der Beweis stützt sich auf einen Hilfssatz über Polynome in 2 Veränderlichen.

T. Schneider (Göttingen).

Referred to in A56-19, J02-3, J68-6, J68-7, J68-9, J68-14, J68-48.

J68-6 **(10, 354f)**

Gel'fond, A. O., and Linnik, Yu. V. On Thue's method in the problem of effectiveness in quadratic fields. Doklady Akad. Nauk SSSR (N.S.) **61**, 773–776 (1948). (Russian)

Thue's method in the theory of Diophantine equations leads to upper bounds for the number of solutions, but in general not to bounds for these solutions themselves. The implications of this fact in the problem of obtaining all imaginary quadratic fields of class number 1 are discussed. Further, a generalized form of the Thue-Siegel theorem is given without proof; it gives the same exponent as Dyson's recent improvement of this theorem [Acta Math. **79**, 225–240 (1947); these Rev. **9**, 412]. *K. Mahler.*

Citations: MR 9, 412h = J68-5.

J68-7 **(10, 592h)**

Schneider, Theodor. Über eine Dysonsche Verschärfung des Siegel-Thueschen Satzes. Arch. Math. **1**, 288–295 (1949).

A simplified proof is given of a theorem due to the reviewer [Acta Math. **79**, 225–240 (1947); these Rev. **9**, 412], that for any algebraic number θ of degree n the inequality $|\theta-p/q|<q^{-\mu}$ has at most a finite number of solutions for every $\mu>(2n)^{\frac{1}{2}}$. The proof follows the methods used in an earlier paper of the author [J. Reine Angew. Math. **175**, 182–192 (1936)]. The complicated lemma on which the reviewer's proof was based turns out to be entirely unnecessary. Since the constant $(2n)^{\frac{1}{2}}$ is presumablv far from the best possible, it is of interest to note that each of the two proofs makes use of some information which the other ignores. The question arises, whether a better result might be obtained by combining the two methods.

F. J. Dyson (Birmingham).

Citations: MR 9, 412h = J68-5.

Referred to in J02-3, J68-9.

J68-8 **(11, 159e)**

Mahler, K. On a theorem of Liouville in fields of positive characteristic. Canadian J. Math. **1**, 397–400 (1949).

Let k be an arbitrary field, x an indeterminate, $k\langle x\rangle$ the field of formal power series $z=a_fx^f+a_{f-1}x^{f-1}+\cdots$ $(a_i\epsilon k)$, and put $|z|=e^f$ for $a_f\neq 0$. It is proved that if $z\epsilon k\langle x\rangle$ and is algebraic of degree $n\geqq 2$ over $k(x)$, then there exists a constant $c>0$ such that $|z-a/b|\geqq c|b|^{-n}$ for all $a, b\epsilon k[x]$, $b\neq 0$. While this result can be improved for fields of characteristic 0, it is pointed out that for k of characteristic $p>0$ no improvement is possible. It is shown that for the (algebraic) number $z=\sum_0^\infty x^{-p^i}$ of degree p over $k(x)$ there exists a sequence of pairs in $k[x]$ such that $|z-a_n/b_n|=|b_n|^{-p}$, $\lim|b_n|=\infty$. *L. Carlitz* (Durham, N. C.).

Referred to in J68-58.

J68-9 **(11, 583a)**

Mahler, K. On Dyson's improvement of the Thue-Siegel theorem. Nederl. Akad. Wetensch., Proc. **52**, 1175–1184 = Indagationes Math. **11**, 449–458 (1949).

F. J. Dyson verschärfte [Acta Math. **79**, 225–240 (1947); diese Rev. **9**, 412] den Thue-Siegelschen Satz, indem er zeigte: Wenn zu irgendeiner algebraischen Zahl ξ vom Grade $n\geqq 2$ bei positivem μ die Ungleichung $|\xi-p/q|<q^{-\mu}$ mit ganzen Zahlen p, q und $q\geqq 1$ unendlich viele solcher Lösungen p/q besitzt, so ist $\mu\leqq(2n)^{\frac{1}{2}}$. In dieser Arbeit wird ein vereinfachter, besonders durchsichtiger Beweis des Haupthilfssatzes der Dysonschen Arbeit gegeben. Am Schluss wird ein weiterer ähnlicher Hilfssatz mit Anwendung auf Kettenbrüche algebraischer Zahlen angekündigt. [Ein Beweis des Referenten für den genannten Dysonschen Satz ist fälschlich mit Mathematische Nachrichten **2**, 288–295 (1949), statt wie es richtig heissen muss Archiv der Mathematik **1**, 288–295 (1949) zitiert; vgl. diese Rev. **10**, 592.]

T. Schneider (Göttingen).

Citations: MR 9, 412h = J68-5; MR 10, 592h = J68-7.

J68-10 **(12, 320d)**

Parry, C. J. The $\mathfrak{p}$-adic generalisation of the Thue-Siegel theorem. Acta Math. **83**, 1–100 (1950).

In der umfangreichen Abhandlung wird im wesentlichen ein Satz von Mahler verallgemeinert und aus dem neuen Ergebnis werden einige Folgerungen gezogen. Der Mahlersche Satz [Math. Ann. **107**, 691–730 (1933); **108**, 37–55 (1933)] gibt eine Aussage über die Approximation algebraischer Zahlen durch rationale bei archimedischer, wie auch nicht-archimedischer Bewertung, und ist in diesem Sinne als Übertragung des bekannten Thue-Siegelschen Satzes [Siegel, Math. Z. **10**, 173–213 (1921)] aufzufassen. Er lautet: Seien $P_1, \cdots, P_\sigma$ (mit $\sigma\geqq 0$) verschiedene natürliche Primzahlen und $\xi_0, \cdots, \xi_\sigma$ reelle P_1-adische, $\cdots, P_\sigma$-adische Wurzeln eines irreduziblen Polynoms $f(x)$ vom Grade $m\geqq 3$ mit ganzrationalen Koeffizienten. Sei $\alpha\equiv\min_{s=1,\cdots,m-1}(m/(s+1)+s)$ und β eine Zahl mit $\alpha<\beta\leqq m$, ferner c eine positive Konstante. Dann ist die Anzahl der

Lösungen der Ungleichung

$$\min\left(1, \left|\frac{p}{q}-\xi_0\right|\right)\prod_{k=1}^{\sigma}\min(1, |p-\xi_k q|_{P_k})\leqq c\max(|p|, |q|)^{-\beta}$$

in Paaren relativ primer ganzrationaler Zahlen p, q nicht grösser als $c_0 2^{\beta(1+\epsilon_0)\sigma/(\beta-\alpha)}$, wobei ϵ_0 eine positive Zahl und c_0 eine Konstante, die nur von ϵ_0, β, und $f(x)$, nicht aber von $P_1, \cdots, P_\sigma$ abhängt, ist ($|p-\xi_k q|_{P_k}$ berechne die P_k-adische Bewertung von $(p-\xi_k q)$). Der Verfasser verallgemeinert diesen Satz der Approximation durch rationale Zahlen nun auf die Annäherung algebraischer Zahlen durch algebraische Zahlen bei archimedischer und nicht-archimedischer Bewertung. Sein Hauptergebnis lautet: "Let $f(x)$ be a polynomial of degree $m\geqq 2$ with integral coefficients from $\mathfrak{K}$ and a nonzero discriminant. Let $\mathfrak{q}_1, \mathfrak{q}_2, \cdots, \mathfrak{q}_\rho$, where $0\leqq\rho\leqq r_1+r_2$, be ρ of the r_1+r_2 infinite prime ideals corresponding to the r_1 real and r_2 pairs of conjugate imaginary fields conjugate to $\mathfrak{K}$, and let $\mathfrak{r}_1, \mathfrak{r}_2, \cdots, \mathfrak{r}_\sigma$, where $\sigma\geqq 0$, be σ different finite prime ideals of $\mathfrak{K}$. Let $h_{k\delta}$ $(k=1, 2, \cdots, \sigma;\ \delta=1, 2, \cdots, G(\mathfrak{r}_k))$ be a natural number not greater than h^2. Let $\xi_{j\gamma}$ $(j=1, 2, \cdots, \rho;\ \gamma=1, G(\mathfrak{q}_j))$ be a real or complex root of $f(x)$ and $\eta_{k\delta\tau}$ $(k=1, 2, \cdots, \sigma;\ \delta=1, 2, \cdots, G(\mathfrak{r}_k);\ \tau=1, 2, \cdots, h_{k\delta})$ an $\mathfrak{r}_k$-adic root of $f(x)$, and let t be the total number of these roots. Let c and ϵ_0 be two positive numbers and α and β two numbers such that $\alpha=\min_{s=1,\cdots,m-1}(m/(s+1)+s)$ and $\beta>\alpha$. Then the number of different algebraic numbers λ of degree h (or any divisor of h) over $\mathfrak{K}$, lying in the perfect $\mathfrak{r}_1$-adic, $\mathfrak{r}_2$-adic, $\cdots$, $\mathfrak{r}_\sigma$-adic extensions of $\mathfrak{K}$ and satisfying the inequality

$$\prod_{j=1}^{\rho}\prod_{\gamma=1}^{G(\mathfrak{q}_j)}\min(1, |\lambda-\xi_{j\gamma}|_{\mathfrak{q}_{j\gamma}})\prod_{k=1}^{\sigma}\prod_{\delta=1}^{G(\mathfrak{r}_k)}\prod_{\tau=1}^{h_{k\delta}}\min(1, |\lambda-\eta_{k\delta\tau}|_{\mathfrak{r}_k})\leqq c\Lambda^{-h\beta}$$

is not greater than $k_0 2^{\beta(1+\epsilon_0)t/(\beta-\alpha)}$, where k_0 is a constant depending only on ϵ_0, β, c, $\mathfrak{K}$, $f(x)$, and h, and not on the number and choice of the roots to which approximation is made or on the corresponding ideals." *T. Schneider.*

Referred to in J68-65.

J68-11 (12, 483c)

Schneider, Theodor. Zur Annäherung der algebraischen Zahlen durch rationale. J. Reine Angew. Math. **188**, 115–128 (1950).

A detailed proof is given for the following theorem: Let the pairs of integers $p_\nu\neq 0$, q_ν $(\nu=1, 2, \cdots; 0<q_1\leqq q_2\leqq\cdots)$ be of the form $p_\nu=p_\nu'p_\nu''$ and $q_\nu=q_\nu'q_\nu''$ where p_ν'' and q_ν'' have bounded prime factors. Let $\zeta\neq 0$ be an algebraic number, and let $|\zeta-p_\nu/q_\nu|\leqq q_\nu^{-\mu}$ $(\nu=1, 2, \cdots)$, $\lim_{\nu\to\infty} q_\nu=\infty$, $\limsup_{\nu\to\infty}\log p_\nu'/\log p_\nu=\alpha$, $\limsup_{\nu\to\infty}\log q_\nu'/\log q_\nu=\beta$. If $\mu>\alpha+\beta$, then $\limsup_{\nu\to\infty}\log q_{\nu+1}/\log q_\nu=\infty$. [Compare the reviewer's note, Acta Arith. **3**, 89–93 (1938), lemma 3.] The proof is based on a paper of the author [same J. **175**, 182–192 (1936)], and on a simple idea in a paper of the reviewer [Akad. Wetensch. Amsterdam, Proc. **39**, 633–640, 729–737, (1936)]. In the last section, this theorem is generalized, and some applications to transcendency are indicated.

K. Mahler (Manchester).

Referred to in J76-30.

J68-12 (15, 107e)

Djerasimović, Božidar. L'intervalle de l'erreur

$$\left|\sqrt[n]{c}-\frac{p}{q}\right|.$$

Bull. Soc. Math. Phys. Serbie **5**, no. 1–2, 53–56 (1953). (Serbo-Croatian. French summary)

J68-13 (15, 406f)

Cugiani, Marco. Sui punti esclusi dalle coperture dell'insieme razionale. Boll. Un. Mat. Ital. (3) **8**, 294–300 (1953).

Let p_N/q_N be the Nth of the irreducible fractions, $0\leqq p/q\leqq 1$ arranged according to $p+q$ and subsidiarily according to p/q. For any $\epsilon>0$ the author defines a set $\mathcal{J}(\epsilon)$ consisting substantially of the points distant less than 2^{-N_ϵ} from some p_N/q_N; the author's definition is slightly different but much more elaborate. An immediate application of Liouville's theorem shows that to almost all α and in particular for all irrational algebraic α there is an $\epsilon_0=\epsilon_0(\alpha)>0$ such that α non-ε $\mathcal{J}(\epsilon)$ for all $\epsilon<\epsilon_0$. The $\mathcal{J}'(\epsilon)$ depending on other orderings of the p/q are also considered. The problem is attributed to W. R. Transue. *J. W. S. Cassels.*

J68-14 (17, 242d)

Roth, K. F. Rational approximations to algebraic numbers. Mathematika **2** (1955), 1–20; corrigendum, 168.

The author proves Siegel's famous conjecture that if α is an algebraic number of degree $n\geqq 2$ and if $|\alpha-h/q|<q^{-k}$ for infinitely many pairs of rational integers h, q with $q>0$, then $k\leqq 2$. The history of the problem is well known: Liouville observed [C.R. Acad. Sci. Paris **18** (1844), 883–885, 910–911; J. Math. Pures Appl. **16** (1851), 133–142] that $k\leqq n$; Thue proved [J. Reine Angew. Math. **135** (1909), 284–305] that $k\leqq\frac{1}{2}n+1$; Siegel proved [Math. Z. **10** (1921), 173–213] that $k\leqq s+n(s+1)^{-1}$ $(s=1, 2, \cdots, n-1)$, so that certainly $k\leqq 2n^{\frac{1}{2}}$; Dyson proved [Acta Math. **79** (1947), 225–240; MR **9**, 412] that $k\leqq(2n)^{\frac{1}{2}}$. The present result is of course the best possible. The author's method consists in using, together with techniques which have been familiar to the expert for some time, the following new lemma which, although similar in appearance to lemmas used in previous attacks, provides the push needed to secure the long sought result: Let R be a polynomial in m variables $X_1, \cdots, X_m$ with rational integral coefficients all of absolute value $\leqq b$; let δ be a real number with $0<\delta<m^{-1}$, let the degree of R in X_i be $\leqq r_i$ and suppose that $r_m>10\delta^{-1}$, $r_i>r_{i+1}\delta^{-1}$ $(1\leqq i\leqq m-1)$; if h_i, q_i are relatively prime rational integers with $q_i>0$, $\delta r_1\log q_1\geqq b$, $\delta\log q_1>m(2m+1)$, $r_i\log q_i\geqq r_1\log q_1$ $(2\leqq i\leqq m)$, then there exist rational integers $j_1, \cdots, j_m$ with $0\leqq j_i\leqq r_i$ $(1\leqq i\leqq m)$ and $\sum_{i=1}^m j_i/r_i<10^m\delta^{(\frac{1}{2})^m}$ such that $\partial^{j_1+\cdots+j_m}R(h_1/q_1, \cdots, h_m/q_m)/\partial X_1^{j_1}\cdots\partial X_m^{j_m}\neq 0$. The author states that his method yields also a generalization of the above theorem to a theorem on approximation to α by algebraic numbers of fixed degree. In the corrigendum he points out that this is not correct, but that a different generalisation does obtain. *E. R. Kolchin.*

Citations: MR 9, 412h = J68-5.

Referred to in D40-46, D40-48, D60-72, G30-26, J02-14, J68-15, J68-16, J68-17, J68-20, J68-22, J68-23, J68-27, J68-29, J68-46, J68-48, J68-50, J68-52, J68-54, J68-56, J68-61, J68-65, J68-68, K45-9.

J68-15 (17, 1060d)

Davenport, H.; and Roth, K. F. Rational approximations to algebraic numbers. Mathematika **2** (1955), 160–167.

Let α be an irrational algebraic number, ζ be real and >0. Roth proved previously [Mathematika **2** (1955), 1–20, 168; MR **17**, 242] Siegel's famous conjecture that $|\alpha-h/g|<q^{-2-\zeta}$ has only a finite number N of solutions in relatively prime rational integers h, g with $g>0$. In the present paper there is obtained (theorem 1 and corollaries) an upper bound for N depending on ζ, the degree of α, and the maximum of the absolute values of the coefficients of the defining equation of α. Similarly, there is obtained (theorem 2) an upper bound for the number of diophantine solutions of $f(x, y)=g(x, y)$, where f is an irreducible form with rational integral coefficients, g is a polynomial with rational coefficients, and $\deg f-\deg g>2$. Finally, it is shown (theorem 3) that if p_k/q_k is the kth convergent to the continued fraction for a real algebraic number β then $\log\log q_k < c(\beta)k/\log k)^{\frac{1}{2}}$, where $c(\beta)$ is independent of k. *E. R. Kolchin* (Paris).

Citations: MR 17, 242d = J68-14.

Referred to in D40-35, D40-48, D60-64, J68-39.

J68-16 (18, 22f)

Roth, K. F. Rational approximations to algebraic numbers. Colloque sur la Théorie des Nombres, Bruxelles, 1955, pp. 119–126. Georges Thone, Liège; Masson and Cie, Paris, 1956.

The author gives a simplified an readable outline of his proof of Siegel's conjecture on the approximation of algebraic numbers by rationals. The necessary bibliography can be found in his earlier work [Mathematika 2 (1955), 1–20, 168; MR 17, 242]. *Harvey Cohn.*

Citations: MR 17, 242d = J68-14.

J68-17 (18, 875h)

Cugiani, Marco. Sugli insiemi numerici del tipo $p^n - q^n a$. Ist. Lombardo Sci. Lett. Rend. Cl. Sci. Mat. Nat. **90** (1956), 209–220.

In a previous paper [Boll. Un. Mat. Ital. (3) **10** (1955), 489–497; MR **17**, 829] the author studied the set of points whose abscissas are of the form $p^2 - q^2\alpha$, where α is a fixed real number and p, q range over the integers. He showed that for almost all α this set has 0 for an accumulation point. The present paper deals with the set $I_n(\alpha)$ of points of the form $p^n - q^n\alpha$ where n is an integer $\geqq 3$. In this case $I_n(\alpha)$ has no finite accumulation point for almost all α. In particular this is true when α is any algebraic number, a fact which follows from the recent theorem of Roth [Mathematika **2** (1955), 1–20, 168; MR **17**, 242]. Other results for even values of $n \geqq 4$ show that the derivative of $I_n(\alpha)$ can be either of the two half-axes or a proper subset thereof, depending on the choice of α. *D. H. Lehmer* (Berkeley Calif.).

Citations: MR 17, 242d = J68-14; MR 17, 829g = J04-25.

J68-18 (20# 32)

Ridout, D. Rational approximations to algebraic numbers. Mathematika **4** (1957), 125–131.

If α is any real algebraic number and if $k > 2$, then the Thue-Siegel-Roth theorem asserts that there are at most finitely many rational numbers p/q $(q > 0)$ satisfying the inequality $|\alpha - p/q| < q^{-k}$. The author shows that the lower bound for k can be reduced if certain conditions are imposed on p and q. His result reads:

Let α be any algebraic number other than 0; let $P_1, \cdots, P_s, Q_1, \cdots, Q_t$ be distinct primes and let μ, ν, c be real numbers satisfying $0 \leqq \mu \leqq 1$, $0 \leqq \nu \leqq 1$, $c > 0$. Let p, q be restricted to integers of the form $p = p^* P_1^{\rho_1} \cdots P_s^{\rho_s}$, $q = q^* Q_1^{\sigma_1} \cdots Q_t^{\sigma_t}$, where $\rho_1, \cdots, \rho_s, \sigma_1, \cdots, \sigma_t$ are non-negative integers and p^*, q^* are integers satisfying $0 < |p^*| \leqq cp^\mu$, $0 \leqq |q^*| \leqq cq^\nu$. Then, if $k > \mu + \nu$, there are at most a finite number of solutions of the inequality $0 < |\alpha - p/q| < q^{-k}$.

There is a similar, but weaker, result in Th. Schneider's new book on transcendental numbers, "Einführung in die transzendenten Zahlen", Springer, Berlin-Göttingen-Heidelberg, 1957 [MR **19**, 252], p. 13, Satz 6. *J. Popken* (Amsterdam)

Citations: MR 19, 252f = J02-9.

Referred to in D60-72, J02-14, J68-26, J68-39, J68-44, J68-63, J76-36, J84-36, Q05-34.

J68-19 (20# 33)

Mahler, K. On the fractional parts of the powers of a rational number. II. Mathematika **4** (1957), 122–124.

The author shows: Let α be any positive algebraic number; let u and v be relatively prime integers satisfying $u > v \geqq 2$ and let $\varepsilon > 0$. Then the inequality

$$|\alpha(u/v)^n - \text{nearest integer}| < e^{-\varepsilon n}$$

is satisfied by at most a finite number of positive integers n.

The proof uses Ridout's extension of the Thue-Siegel-Roth theorem [see the preceding review].

The special case $\alpha = 1$, $u = 3$, $v = 2$, $\varepsilon = \log \frac{4}{3}$ is of interest for the solution of Waring's Problem. From this and from well-known results about $g(k)$ the author deduces: Except possibly for a finite number of values of k, we have

$$g(k) = 2^k + [(3/2)^k] - 2.$$

J. Popken (Amsterdam)

Referred to in J68-26, P08-40.

J68-20 (20# 3851)

Ridout, D. The p-adic generalization of the Thue-Siegel-Roth theorem. Mathematika **5** (1958), 40–48.

Let $f(x)$ be a polynomial of degree $n \geqq 2$ with rational coefficients, and let $\zeta, \zeta_1, \cdots, \zeta_t$ be a real root, a p_1-adic root, $\cdots$, a p_t-adic root of $f(x) = 0$, respectively; here $p_1, \cdots, p_t$ are distinct primes. Assume κ is a constant such that the inequality

$$\min\left(1, \left|\zeta - \frac{h}{q}\right|\right) \prod_{\tau=1}^{t} \min(1, |h - q\zeta_\tau|_{p_\tau}) \leqq (\max(|h|, q))^{-\kappa}$$

has infinitely many solutions in integers h, q, where $(h, q) = 1$ and $q > 0$. Using Siegel's method of the proof of the Thue-Siegel theorem, the reviewer showed in 1932 [Math. Ann. **107** (1933), 691–730] that $\kappa < 2\sqrt{n}$. The new method due to K. F. Roth [Mathematika **2** (1955), 1–20, 168; MR **17**, 242] enables the author to show that κ cannot be larger than 2, a result that is best possible. *K. Mahler* (Manchester)

Citations: MR 17, 242d = J68-14.

Referred to in B40-52, G30-26, J02-14, J68-47, J68-51.

J68-21 (21# 42)

Bombieri, Enrico. Sull'approssimazione di numeri algebrici mediante numeri algebrici. Boll. Un. Mat. Ital. (3) **13** (1958), 351–354. (English summary)

Elementary deduction of a lower bound for $|\xi_1 - \xi_2|$, where ξ_1, ξ_2 are nonconjugate algebraic numbers. A less precise result was given by A. Brauer [Jber. Deutsch. Math Verein **38** (1929), 47; for a more complete presentation, see J. Reine Angew. Math. **160** (1929), 70–99, especially pp. 75–78]. *C. G. Lekkerkerker* (Amsterdam)

Referred to in J68-24, J68-37.

J68-22 (22# 1560)

Bergström, Harald. The fundamental theorem of Roth. Nordisk Mat. Tidskr. **7** (1959), 57–72, 96. (Swedish. English summary)

An exposition of the original theorem of Roth [Mathematika **2** (1955), 1–20; corrigendum, 168; MR **17**, 242], somewhat simplified by the systematic use of vectors. *W. J. LeVeque* (Ann Arbor, Mich.)

Citations: MR 17, 242d = J68-14.

J68-23 (22# 3726)

Cugiani, Marco. Sulla approssimabilità dei numeri algebrici mediante numeri razionali. Ann. Mat. Pura Appl. (4) **48** (1959), 135–145.

The author uses methods of Roth [Mathematika **2** (1955), 1–20; corrigendum, 168; MR **17**, 242] and Schneider [*Einführung in die transzendenten Zahlen*, Springer, Berlin, 1957; MR **19**, 252; especially the theorem on page 13] to prove the following theorem. Let α be a fixed algebraic number of degree $g > 1$ and let b be a fixed natural number; suppose that there is a sequence of pairs (p_n, q_n) of coprime integers, where $q_n = q_n' q_n''$ and

q_n'' is a power of b, such that the following two conditions are satisfied: (a) there exist constants η, ω, with $0 \leqq \eta \leqq 1$, $0 \leqq \omega$, such that

$$\frac{\log q_n''}{\log q_n} \geqq 1 - \eta - \frac{\omega}{(\log \log \log q_n)^{1/2}};$$

(b) there exists an ε such that

$$\left| \alpha - \frac{p_n}{q_n} \right| < q_n^{-1-\eta-f(q_n)},$$

where

$$f(q_n) = [(4+2\eta)(2g+1)+\omega+\varepsilon](\log \log \log q_n)^{-1/2}.$$

Then

$$\limsup_{n\to\infty} \frac{\log q_{n+1}}{\log q_n} = +\infty.$$

J. W. S. Cassels (Cambridge, England)

Citations: MR 17, 242d = J68-14; MR 19, 252f = J02-9.

J68-24 **(22# 4705)**

Mahler, K. On a theorem by E. Bombieri. Nederl. Akad. Wetensch. Proc. Ser. A **63** = Indag. Math. **22** (1960), 245–253.

Here α and β denote roots of real or complex polynomials f and g of heights $H(f)$ and $H(g)$, degrees m and n, and non-zero resultant $R(f, g)$. Then

$$|\alpha-\beta| \geqq |R(f, g)|/H(f)^n H(g)^m C_{n,m}.$$

If α and β are distinct roots of the same polynomial f, and if $D(f)$ is the non-zero root-discriminant, then

$$|\alpha-\beta| \geqq |D(f)|/H(f)^{2m-1}C_m.$$

The constants C are not complicated, e.g., $C_m/(m+1)^{4m}4^{m^2} = 1$ if f has only real zeros and $=(m+1)^{m^2}$ otherwise. This improves results of A. Brauer [Jber. Deutsch. Math. Verein. **38** (1929), 47] and Bombieri [Boll. Un. Mat. Ital. (3) **13** (1958), 351–354; MR **21** #42] in regard to strength of inequality and extension to non-algebraic numbers, by use of the resultant and discriminant.

H. Cohn (Tucson, Ariz.)

Citations: MR 21# 42 = J68-21.

Referred to in J68-25, J68-37.

J68-25 **(23# A885)**

Mahler, K.

On a theorem by E. Bombieri.

Nederl. Akad. Wetensch. Proc. Ser. A **64** = *Indag. Math.* **23** (1961), 141.

The author indicates that the following remarks ought to be appended to an earlier paper [same Proc. **63** (1960), 245–253; MR **22** #4705]: "The exponent of a_0 in the formula (15) on p. 250 has been given incorrectly as $m-1$; it should be $m-2$. The same change must be made in the formulae (17) and (18) on p. 251, and in the formula on the 3rd line of p. 253. As a consequence, in lines 5, 8 and 21 of p. 253, the expression $|D(f)|$ is to be replaced by $|a_0 D(f)|$. Thus the final result in Theorem 2 is that

$$\Delta(f) \geqq \{\Gamma(f)H(f)^{2m-1}\}^{-1}|a_0 D(f)|.$$

For polynomials with integral coefficients this gives a slight improvement. I am indebted to Dr. Rainer Güting for this correction."

Citations: MR 22# 4705 = J68-24.

J68-26 **(22# 8000)**

Uchiyama, Saburô. On the Thue-Siegel-Roth theorem. I, II. Proc. Japan Acad. **35** (1959), 413–416, 525–529.

In part I, the T-S-R theorem is refined in the case of an imaginary quadratic field K, as follows: If α is a nonzero algebraic number, and $\kappa > 1$ is fixed, then the inequality $|\alpha-\xi| < (M(\xi))^{-\kappa}$ has only finitely many nonrational solutions ξ in K. Here $M(\xi)$ is the absolute value of the coefficient of x^2 in the polynomial with coprime rational integral coefficients which defines ξ. A corollary is that for $\nu > 2$, the inequality $0 < |\alpha - p/q| < |q|^{-\nu}$ has only finitely many integral solutions p, q in K.

In part II, statements are given of generalizations to the case of algebraic approximants of related theorems by Ridout [Mathematika **4** (1957), 125–131; MR **20** #32] and Mahler [ibid. 122–124; MR **20** #33]. Proof of the following theorem is also given: Let $\alpha \neq 0$ be algebraic, and let K be an algebraic number field. Let $\mathfrak{p}_1, \cdots, \mathfrak{p}_s$ be a finite set of prime ideals with distinct rational primes $p(\mathfrak{p}_1), \cdots, p(\mathfrak{p}_s)$ in an arbitrary finite extension L of $K(\alpha)$. Then for each $\kappa > 2$ the inequality

$$\prod_{k=1}^{s} |\alpha-\xi|_{\mathfrak{p}_k} < (H(\xi))^{-\kappa}$$

has only finitely many solutions ξ in K. Here $H(\xi)$ is the maximum of the absolute values of the coefficients in the polynomial with coprime rational integral coefficients which defines ξ.

W. J. LeVeque (Ann Arbor, Mich.)

Citations: MR 20# 32 = J68-18; MR 20# 33 = J68-19.

J68-27 **(22# 8001)**

Uchiyama, Saburô. On the Thue-Siegel-Roth theorem. III. Proc. Japan Acad. **36** (1960), 1–2.

Let k be a field of characteristic 0, t an indeterminate, and $k\langle t\rangle$ the field of all formal series $\alpha = \sum_{-\infty}^{f} c_n t^n$, where the c_n lie in k and f is finite. If $c_f \neq 0$, put $|\alpha| = e^f$; further put $|0| = 0$; then $|\alpha|$ is a valuation on $k\langle t\rangle$. $k\langle t\rangle$ contains the rational function field $k(t)$ as a subfield. In analogy to K. F. Roth's theorem [Mathematika **2** (1955), 1–20; corrigendum, 168; MR **17**, 242], the author notes that if $\alpha \in k\langle t\rangle$ is algebraic over $k(t)$, the inequality

$$0 < |\alpha - p/q| < q^{-\kappa} \quad (\kappa > 2)$$

has only finitely many solutions in polynomials p, $q \neq 0$. The same result was already proved by D. Fenna in his Manchester Ph.D. thesis of 1956.

K. Mahler (Manchester)

Citations: MR 17, 242d = J68-14.

J68-28 **(22# 8002)**

Cugiani, Marco. Sull'approssimabilità di un numero algebrico mediante numeri algebrici di un corpo assegnato. Boll. Un. Mat. Ital. (3) **14** (1959), 151–162. (English summary)

The following theorem is proved. Let K be an algebraic number field and let α be an algebraic number of degree $g \geqq 2$ over K. Let $\varepsilon > 0$ and let $f(q)$ denote the function

$$f(q) = (8g+\varepsilon)(\log \log \log q)^{-1/2}.$$

If $\{\xi_i\}$ is a sequence of primitive numbers of K such that $|\alpha - \xi_i| < q_i^{-\{2+f(q_i)\}}$, where q_i denotes the height of ξ_i, then we have

$$\limsup_{i\to\infty} \frac{\log q_{i+1}}{\log q_i} = +\infty.$$

The author's proof makes extensive use of the method of Roth-LeVeque, as exposed in W. J. LeVeque, *Topics in number theory* [Addison-Wesley, Reading, Mass., 1956; MR **18**, 283; Vol. 2, Ch. 4].

C. G. Lekkerkerker (Amsterdam)

Citations: MR 18, 283b = Z01-38.

Referred to in J68-47.

J68-29 **(22# 10977)**

Roth, K. F. Rational approximations to algebraic numbers. Proc. Internat. Congress Math. 1958, pp. 203–210. Cambridge Univ. Press, New York, 1960.

This is a report of the one-hour lecture which the author delivered at the Edinburgh Congress in 1958. He outlines a proof of his celebrated theorem on the approximation of algebraic numbers by rationals [Mathematika **2** (1955), 1–20; corrigendum, 168; MR **17**, 242] and lists some generalizations and extensions of it. Inter alia, he raises the problem of obtaining an analogous theorem concerning simultaneous approximations to two given algebraic numbers. *C. G. Lekkerkerker* (Amsterdam)

Citations: MR 17, 242d = J68-14.

J68-30 **(23# A111)**

Peck, L. G.

Simultaneous rational approximations to algebraic numbers.

Bull. Amer. Math. Soc. **67** (1961), 197–201.

Let $\beta_0, \cdots, \beta_n$ be elements of a real (but not necessarily totally real) field of degree $n+1$ which are linearly independent over the rationals. The author shows that there exists a constant $C=C(\beta_0, \cdots, \beta_n)$ such that there exist infinitely many integral solutions $q_0>0, q_1, \cdots, q_n$ of the set of inequalities:

$$|q_0\beta_j - q_j\beta_0| < Cq_0^{-1/n}(\log q_0)^{-1/(n-1)} \quad (1 \leqq j < n),$$
$$|q_0\beta_n - \beta_n q_0| < Cq_0^{-1/n}.$$

This is in striking contrast with the well-known fact that the system of inequalities $|q_0\beta_j - q_j\beta_0| < Dq_0^{-1/n}$ has only finitely many integral solutions if $D=D(\beta_0, \cdots, \beta_n)$ is chosen small enough (whereas there are infinitely many integral solutions with $D=|\beta_0|$, by Minkowski's convex body theorem). The author's result sharpens some of the conclusions of Cassels and Swinnerton-Dyer [Philos. Trans. Roy. Soc. London Ser. A **248** (1955), 73–96; MR **17**, 14] but does not shed any more light on the conjecture of Littlewood which they considered.

J. W. S. Cassels (Cambridge, England)

Citations: MR 17, 14f = J36-24.

J68-31 **(23# A856)**

Lang, Serge

On a theorem of Mahler.

Mathematika **7** (1960), 139–140.

A very short proof of the following theorem. Let K be of characteristic 0, α algebraic over K, Γ a finitely generated subgroup of K^*, and let χ and c be positive constants. The elements $\beta\in\Gamma$ such that $|\alpha-\beta|\leqq cH(\beta)^{-\chi}$ have bounded height. The theorem is deduced from Roth's theorem for K, but follows equally from Siegel's weaker theorem. *K. Mahler* (Manchester)

J68-32 **(23# A892)**

Fenna, D.

Rational approximation with series.

J. Austral. Math. Soc. **2** (1961/62), 107–119.

Let $\mathfrak{k}$ be a field of characteristic 0, z an indeterminate, $\mathfrak{K}$ the field of formal Laurent series $x=\sum_{-\infty}^{d}\alpha_n z^n$, where $\alpha_n\in\mathfrak{k}$. If $\alpha_d\neq 0$, put deg $x=d$. In his Manchester thesis of several years ago, published only now, the author obtains the following analogue to Roth's theorem: If $x\in\mathfrak{K}$ is algebraic over $\mathfrak{k}[z]$ and is not in $\mathfrak{k}(z)$, and if $\nu>2$, there are at most finitely many u and $v\neq 0$ in $\mathfrak{k}[z]$ such that $\deg(x-u/v)<-\nu \deg v$. The same result has also been obtained by several other mathematicians, e.g., by S. Lang. *K. Mahler* (Manchester)

J68-33 **(23# A1600)**

Hitzig, I.

On the approximation by rational numbers to an irrational algebraic number. (Romanian. Russian and French summaries)

Gaz. Mat. Fiz. Ser. A **12 (65)** (1960), 292–298.

Let $x_0>y_0$ (>0) be two rational numbers. Using geometric considerations, the author indicates an algorithm for the approximation of the irrational number $(x_0^2+y_0^2)^{1/2}$ by rationals. *E. Grosswald* (Philadelphia, Pa.)

J68-34 **(23# A1601)**

Skolem, Th.

A new version of some considerations of A. Thue.

Math. Scand. **8** (1960), 71–80.

In 1908, A. Thue obtained his famous theorem on the rational approximations of algebraic numbers, first in the special case of irrationalities $(a/b)^{1/r}$, where $r\geqq 3$. For the proof he essentially constructed the convergents $f_n(x)/g_n(x)$ of the continued fraction of $(1+x)^{1/r}$; thus $f_n(x)$ and $g_n(x)$ are polynomials such that $r_n(x)=f_n(x)-g_n(x)(1+x)^{1/r}$ vanishes to a high order at $x=0$. On putting $1+x=a/b$, rational approximations of $(a/b)^{1/r}$ are obtained, and these allow showing that for every $\varepsilon>0$ there can be at most finitely many pairs of integers u, v such that

$$|u(a/b)^{1/r}-v| < u^{-(\frac{1}{2}r+\varepsilon)}.$$

In the present note, the author gives a shorter construction of polynomials $f_n(x)$, $g_n(x)$ by means of recursive formulae and a very short proof that $f_{n+1}(x)g_n(x)-f_n(x)g_{n+1}(x)=c_nx^{2n+1}$, where $c_n\neq 0$. This leads to a simplification of Thue's proof.—See also K. Mahler, Math. Ann. **105** (1931), 267–276. *K. Mahler* (Manchester)

J68-35 **(23# A3116)**

Rauzy, Gérard

Approximation diophantienne des nombres algébriques.

Faculté des Sciences de l'Université de Paris. Mathématiques approfondies. Théorie des nombres. 1960/61.

Secrétariat mathématique, Paris, 1961. 60 *pp.* 20.00 *NF;* \$4.10.

The author gives a very lucid account of the work of Thue-Siegel (Chapter 2), Roth (Chapter 3), Mahler's p-adic generalization of Roth's theorem (Chapter 4). There is an application to the problem of the fractional part of $(3/2)^n$ (the latter application helped Mahler to complete the solution of Waring's problem for large exponents). *S. Chowla* (Boulder, Colo.)

Referred to in Z10-20.

J68-36 **(24# A1882)**

Uchiyama, Saburô

Rational approximations to algebraic functions.

J. Fac. Sci. Hokkaido Univ. Ser. I **15**, 173–192 (1961).

Let K be any field of characteristic χ, τ a primary irreducible polynomial in $K[t]$, deg τ its degree, $c>1$ a constant, $|x|$ and $|x|_\tau$ the valuations of $K(t)$ determined by $|t|=c$ and $|\tau|_\tau=c^{-\deg\tau}$, respectively, and $K\langle t^{-1}\rangle$ and $K\langle\tau\rangle$ the corresponding completions of $K(t)$. The author first proves the analogues of the theorem of Liouville for $K\langle t^{-1}\rangle$ and $K\langle\tau\rangle$ and shows that these are best possible if $\chi>0$; and next, for $\chi=0$, proves the analogues of Roth's theorem and shows that also this is best. He then obtains analogues for a theorem of Ridout and for one on the simultaneous approximation of roots of the same algebraic equation over $K[t]$ in several fields $K\langle t^{-1}\rangle$, $K\langle\tau_1\rangle$, $\cdots$,

$K\langle\tau_r\rangle$. These theorems can again be applied to the study of binary equations and of the powers of rational functions. Work similar to that of the author, with the same results, has already been done by several other mathematicians, in particular by S. Lang. For a detailed treatment of the analogous problems over the rational field see, e.g., K. Mahler, *Lectures on diophantine approximations*, Part I [Univ. Notre Dame Press, Notre Dame, Ind., 1961].

K. Mahler (Manchester)

J68-37 **(24# A2559)**

Güting, R.

Approximation of algebraic numbers by algebraic numbers.

Michigan Math. J. **8** (1961), 149–159.

If $P(x)=\sum_{j=0}^{n} a_j x^j$ where $a_n\neq 0$, denote by $h(P)$ the maximum and by $s(P)$ the sum of the absolute values of the a_j. A theorem by E. Bombieri [Boll. Un. Mat. Ital. (3) **13** (1958), 351–354; MR **21** #42] states that if α_1 and α_2 are zeros of two distinct irreducible polynomials $P_1(x)$ and $P_2(x)$ of degrees n_1 and n_2, then

$$|\alpha_1-\alpha_2| > (4n_1n_2)^{-3n_1n_2}h(P_1)^{-n_2}h(P_2)^{-n_1},$$

and this was generalised by the reviewer [Nederl. Akad. Wetensch. Proc. Ser. A **63** (1960), 245–253; MR **22** #4705] to zeros of two polynomials with arbitrary complex polynomials and also to a pair of zeros of one such polynomial; the right-hand side has then as a factor the absolute value of either the resultant or the discriminant of the polynomials. In the present paper the author gives a very much simpler proof for better inequalities. His main tool is the inequality $\prod_{j=1}^{n}\max(1,|\alpha_j|)\leq s(P)/|a_n|$, where the α_j are the zeros of $P(x)$ [Mahler, Mathematika **7** (1960), 98–100; MR **23** #A1779]. If $R(P_1,P_2)=R$ is the resultant of $P_1(x)$ and $P_2(x)$ and α_1 and α_2 are again zeros of $P_1(x)$ and $P_2(x)$, respectively, he shows that

$$|\alpha_1-\alpha_2| > 2^{-\max(n_1,n_2)+1}|R|s(P_1)^{-n_2}s(P_2)^{-n_1},$$

with a similar inequality involving the discriminant when α_1,α_2 are zeros of the same polynomial. More generally, he gives a lower bound for the absolute value of a complex polynomial

$$A(x_1,\cdots,x_m)=\sum_{i_1=0}^{N_1}\cdots\sum_{i_m=0}^{N_m} a_{i_1\cdots i_m}x_1^{i_1}\cdots x_m^{i_m}$$

when $x_1=\alpha_1,\cdots,x_m=\alpha_m$ are zeros of given complex polynomials $P_1(x),\cdots,P_m(x)$. His results become of particular interest, for applications to transcendental numbers, when all polynomials have integral coefficients. Now, if $P_1(x),\cdots,P_m(x)$, of degrees $n_1,\cdots,n_m$, have integral coefficients, and if $A(\alpha_1,\cdots,\alpha_m)\neq 0$, he shows that

$$|A(\alpha_1,\cdots,\alpha_m)| \geq s\{s(P_1)^{N_1/n_1}\cdots s(P_m)^{N_m/n_m}\}^{-q},$$

where q is the degree of the number field generated by $\alpha_1,\cdots,\alpha_m$, and $s=\sum_{i_1=0}^{N_1}\cdots\sum_{i_m=0}^{N_m}|a_{i_1\cdots i_m}|$.

K. Mahler (Manchester)

Citations: MR 21# 42 = J68-21; MR 22# 4705 = J68-24; MR 23# A1779 = J76-38.

Referred to in C05-33, J76-53.

J68-38 **(25# 47)**

Poitou, G.

Le théorème de Thue-Siegel-Roth.

Enseignement Math. (2) **7** (1961), 281–285 (1962).

Some historical remarks.

W. J. LeVeque (Ann Arbor, Mich.)

Referred to in Z10-17.

J68-39 **(26# 2393)**

Fraenkel, Aviezri S.

On a theorem of D. Ridout in the theory of Diophantine approximations.

Trans. Amer. Math. Soc. **105** (1962), 84–101.

The theorem mentioned in the title is the following extension of the Thue-Siegel-Roth Theorem. Let $\alpha\neq 0$ be an algebraic number. Let $\{P_1,\cdots,P_s\}$ and $\{Q_1,\cdots,Q_t\}$ be sets of primes. Let P denote the class of positive integers p' generated by $P_1,\cdots,P_s$, Q the class of positive integers q' generated by $Q_1,\cdots,Q_t$. Let μ,ν,c be real numbers satisfying $0\leq\mu\leq 1$, $0\leq\nu\leq 1$, $c>1$. Let p,q be integers restricted to the form

$$(1)\qquad p=p^*p',\quad q=q^*q',\quad p'\in P,\quad q'\in Q,$$

such that

$$(2)\qquad |p^*|<c|p|^{\mu},\quad 0<q^*<cq^{\nu}.$$

Then the inequality

$$(3)\qquad 0<|\alpha-p/q|<q^{-(\mu+\nu+\zeta)}$$

has only a finite number of solutions in p,q for every $\zeta>0$ [Ridout, Mathematika **4** (1957), 125–131; MR **20** #32].

The paper reviewed here consists of three parts. In the first the author proves a metrical analogue of Ridout's theorem. Here α need no longer be restricted to algebraic numbers; however, we retain the restrictions (1) and (2) put upon the integers p,q. It is shown that for any $\zeta>0$ the inequality (3) has infinitely many solutions (p,q) of the form (1) subject to (2) for almost no real numbers α. In fact the author proves a more general result. Instead of the classes P and Q, considered above, he takes classes C_1 and C_2 of positive integers $p'\in C_1$ and $q'\in C_2$ such that for every $\varepsilon>0$ both series $\sum(p')^{-\varepsilon}$ and $\sum(q')^{-\varepsilon}$ converge. Further, instead of (3), he considers its r-dimensional analogue, namely the set of simultaneous inequalities $|\alpha_i-p_i/q|<q^{-(\mu+\nu/r+\zeta)}$ $(i=1,\cdots,r)$. It is shown that this set of inequalities has infinitely many solutions for almost no real vectors $(\alpha_1,\cdots,\alpha_r)$ if the integers p,q are restricted to the form $p_i=p_i^*p'$, $q=q^*q'$, $p_i'\in C_1$, $q'\in C_2$, subject to $|p_i^*|<c|p_i|^{\mu}$, $0<q^*<cq^{\nu}$. Moreover, the author proves a second theorem of the same kind.

In the second part the author investigates whether Ridout's exponent $-(u+\nu+\zeta)$ in (3) is best possible. Although this problem is not solved completely, he obtains several results in this direction. Here again he generalizes the problem to the case of the approximation of a real vector $(\alpha_1,\cdots,\alpha_r)$ by vectors of the form $(p_1/q,\cdots,p_r/q)$. We state one of his results (Theorem IV) in the one-dimensional case: Let α be a real number. Let $0\leq\mu\leq 1$, $0\leq\nu\leq 1$, such that either $\mu+\nu\geq 1$, or $\mu=0$, or $\nu=0$ is satisfied. Let $K>0$ (with the provision $K=1$ if $\mu=\nu=1$). Let $S_1=\{P_1,\cdots,P_s\}$ and $S_2=\{Q_1,\cdots,Q_t\}$ be sets of primes. Then there exists a constant $c>1$ depending on K,α,μ,ν and the primes in S_1 and S_2, such that $|\alpha-p/q|<Kq^{-(\mu+\nu)}$ has infinitely many solutions in integers (p,q) of the form (1) subject to (2). The proof uses the pigeon-hole principle.

In the final part the author gives an upper bound for the number of solutions of (3), similar to the well-known estimation of Davenport and Roth in the case of Roth's Theorem [ibid. **2** (1955), 160–167; MR **17**, 1060].

J. Popken (Berkeley, Calif.)

Citations: MR 17, 1060d = J68-15; MR 20# 32 = J68-18.

J68-40 **(26# 6122)**

Roth, K. F.

Rational approximations to irrational numbers.

An Inaugural Lecture delivered at University College, London, 22 January, 1962.

Published for the College by H. K. Lewis & Co. Ltd., London, 1962. 13 *pp.* 3*s.* 6*d.*

An expository article on the author's well-known theorem and its background, with allusions to some problems remaining open. *W. J. LeVeque* (Ann Arbor, Mich.)

Referred to in J68-41.

J68-41 **(35# 1552)**

Roth, K. F.

Rational approximations to irrational numbers. (Polish)

Wiadom. Mat. (2) **9**, 199–208 (1966/67).

Polish translation of *Rational approximations to irrational numbers* [Lewis, London, 1962; MR **26** #6122].

Citations: MR 26# 6122 = J68-40.

J68-42 **(28# 5029)**

Baker, A.

Rational approximations to certain algebraic numbers.

Proc. London Math. Soc. (3) **14** (1964), 385–398.

Given any irrational number α and any real $k>2$, by Roth's theorem there exists a constant $c=c(\alpha, k)$ such that $|\alpha-p/q|>cq^{-k}$ for all rational numbers p/q. One purpose of this paper is to exhibit a class of algebraic numbers for which c can be made explicit. This class of numbers has the form $\alpha=(u_1a^{1/n}+u_2b^{1/n})/(u_3a^{1/n}+u_4b^{1/n})$, where the integers $a>b$, $n\geqq 3$, u_1, u_2, u_3, u_4 are chosen so that α is irrational, and where $a>(a-b)^\rho(3n)^{2\rho-2}$ with $\rho=\frac{5}{2}(2k-1)(k-2)^{-1}$. For these α an explicit value of c, not set forth in detail here, is obtained as a function of $a, b, u_1, u_2, u_3, u_4, n, k$. In proving this result the author establishes that with a satisfying the same inequalities, $|ax^n-by^n|\geqq C|x^{n-k}|$ for all integers x and y for a given explicit $C=C(a, b, n, k)$. This generalizes a theorem of C. L. Siegel [Math. Ann. **114** (1937), 57–68]. Furthermore, bounds on $|x|$ and $|y|$ are established for integer solutions x and y of $ax^n-by^n=f(x, y)$ in case f is a polynomial of degree at most $n-3$ with rational coefficients under the assumption $a>(a-b)^{25/2}(3n)^{23}$. The proofs employ the hypergeometric function to get a sequence of rational approximations to $(a/b)^{1/n}$ almost as good as the convergents of the continued fraction expansion but with denominators increasing comparatively slowly. *I. Niven* (Eugene, Ore.)

Referred to in J68-45, J80-18.

J68-43 **(29# 1182)**

Mahler, K.

On the approximation of algebraic numbers by algebraic integers.

J. Austral. Math. Soc. **3** (1963), 408–434.

Let K be an algebraic number field of degree n over the rationals, and let $\kappa\in K$ have conjugates $\kappa^{(1)}, \cdots, \kappa^{(n)}$. Let $h(\kappa)$ be the smallest positive rational integer such that the polynomial $h(\kappa)(x-\kappa^{(1)})\cdots(x-\kappa^{(n)})=b_0x^n+\cdots+b_n$ has rational integral coefficients. Put

$$q(\kappa) = \max(|b_0|, \cdots, |b_n|).$$

For arbitrary algebraic numbers $\xi_1, \cdots, \xi_n$, put

$$\Delta(\kappa) = \prod_{j=1}^{n} \min(1, |\kappa^{(j)}-\xi_j|).$$

Let $\Sigma=\{\kappa_l\}$ be an infinite sequence of distinct elements of K, and put $h(\kappa_l)=h_l$, etc. Suppose that there are constants c', c'', σ and τ, such that $c'>0$, $c''>0$, $\sigma>0$, $0\leqq\tau\leqq 1$, for which $\Delta_l\leqq c'q_l^{-\sigma}$ and $h_l\leqq c''q_l^\tau$ for all l. The main result of the paper is that under these circumstances, $\sigma\leqq 1+\tau$. The condition $h_l\leqq c''q_l^\tau$ holds automatically for $c''=\tau=1$, and this case gives the reviewer's extension of Roth's theorem [the reviewer, *Topics in number theory*, Vol. 2, Addison-Wesley, Reading, Mass., 1956; MR **18**, 283]. The other extreme, $c''=1$, $\tau=0$, gives information about the approximation of algebraic numbers by algebraic integers from K. Several other conclusions are drawn, among them the following new type of result on inhomogeneous approximation: If α is a real quadratic irrationality, if β is a real algebraic number, and if $\varepsilon>0$, then there are only finitely many pairs of rational integers $x\neq 0$, y such that $|x\alpha-y-\beta|<|x|^{-(1+\varepsilon)}$. *W. J. LeVeque* (Ann Arbor. Mich.)

Citations: MR 18, 283b = Z01-38.

J68-44 **(30# 1095)**

Fraenkel, Aviezri S.

Distance to the nearest integer and algebraic independence of certain real numbers.

Proc. Amer. Math. Soc. **16** (1965), 154–160.

Let $\|x\|$ denote the distance from x to the nearest integer. Let $P_1, \cdots, P_s, Q_1, \cdots, Q_t$ be primes, and suppose that $c\geqq 1$, $0\leqq\mu\leqq 1$, $a>0$, $\delta>0$. Let p and q be of the form $p=p^*p'$, $p'=P_1^{\rho_1}\cdots P_s^{\rho_s}$, $q=Q_1^{\sigma_1}\cdots Q_t^{\sigma_t}$, where $\rho_1, \cdots, \rho_s, \sigma_1, \cdots, \sigma_t$ are non-negative integers and p^* is any integer such that $0<p^*\leqq cp^\mu$. Theorem 1 asserts that for α algebraic, the inequality $0<\|\alpha p/q\|<q^{-(\mu+\delta)}$ has only finitely many solutions p, q for which $p\geqq aq$. Moreover, for almost all real r-tuples $(\alpha_1, \cdots, \alpha_r)$ the inequalities $0<\|\alpha_ip/q\|<p^{-(\mu/r+\delta)}$, $i=1, \cdots, r$, have only finitely many simultaneous solutions with $p\geqq aq$. (Special conditions if $\mu=1$.) The remaining theorems are concerned with the same kind of approximation problems, and with the algebraic and set-theoretic properties of various r-tuples $(\alpha_1, \cdots, \alpha_r)$ for which the above system has infinitely many solutions. The proofs depend heavily on Ridout's extension [Mathematika **4** (1957), 125–131; MR **20** #32] of Roth's theorem, and on Schmidt's work [Bull. Amer. Math. Soc. **68** (1962), 475–478; MR **25** #3903] on the algebraic independence of a set of numbers having certain Diophantine properties. *W. J. LeVeque* (Ann Arbor, Mich.)

Citations: MR 20# 32 = J68-18; MR 25# 3903 = J88-24.

J68-45 **(30# 1977)**

Baker, A.

Rational approximations to $\sqrt[3]{2}$ and other algebraic numbers.

Quart. J. Math. Oxford Ser. (2) **15** (1964), 375–383.

Let m and n be integers such that $n\geqq 3$, $1\leqq m<n$. Let a and b be positive integers for which $\frac{7}{8}a\leqq b<a$ and $n|(a-b)$. Suppose also that $\lambda=4b(a-b)^{-2}\mu_n^{-1}>1$, where $\mu_n=\prod_{p|n}p^{1/(p-1)}$. Finally, put $\alpha=(a/b)^{m/n}$. Then it is shown that $|\alpha-p/q|>c/q^\kappa$ for all integers p, q with $q>0$, where $\lambda^{\kappa-1}=2\mu_n(a+b)$ and $c^{-1}=2^{\kappa+2}(a+b)$. The condition $n|(a-b)$ can always be satisfied by replacing a and b by na and nb, with corresponding increases in the values of κ and c^{-1}. It follows as a corollary that always $|\sqrt[3]{2}-p/q|>10^{-6}q^{-2.955}$, and similar inequalities are given for the cube roots of 17, 19, 20, 37 and 43. From the result for $\sqrt[3]{2}$, it is easily deduced that if $x^3-2y^3=n$, then $\max(|x|, |y|)<(3\cdot 10^5|n|)^{23}$. The proofs, as in the author's earlier paper [Proc. London Math. Soc. (3) **14** (1964), 385–398; MR **28** #5029], depend on identities involving hypergeometric functions. *W. J. LeVeque* (Ann Arbor, Mich.)

Citations: MR 28# 5029 = J68-42.

Referred to in D24-66.

J68-46 **(31# 2206)**

Schmidt, Wolfgang M.

Über simultane Approximation algebraischer Zahlen durch Rationale.

Acta Math. **114** (1965), 159–206.

Let $\xi_1, \cdots, \xi_n$ be $n \geqq 2$ algebraic numbers such that $1, \xi_1, \cdots, \xi_n$ are linearly independent over the rational field. It has been conjectured that, for any constant $\varepsilon > 0$, both the inequality $\max(|\xi_1 - p_1/q|, \cdots, |\xi_n - p_n/q|) < q^{-(1+(1/n)+\varepsilon)}$ and the inequality $|q_1\xi_1 + \cdots + q_n\xi_n - p| < (\max(|q_1|, \cdots, |q_n|))^{-(n+\varepsilon)}$ have only finitely many solutions in integers $p_1, \cdots, p_n, q > 0$, and integers $q_1, \cdots, q_n, p$, respectively. In the lowest case $n = 2$, the author establishes now a result which is only slightly weaker, and in some directions is even stronger, than the conjecture.

Let $\|\alpha\|$ denote the distance of the real number α from the nearest integer. Also, for $x > 0$, put $L_0(x) = x$, $L_1(x) = \max(1, \log x)$, and $L_{k+1}(x) = L_1(L_k(x))$ $(k = 1, 2, 3, \cdots)$. A set $\mathfrak{Q} = \{q_1, q_2, q_3, \cdots\}$, where $0 < q_1 < q_2 < q_3 < \cdots$, is said to have the property E if $\mathfrak{Q}$ either is a finite set, or if otherwise

$$\limsup_{h \to \infty} \frac{L_k(q_{h+1})}{L_k(q_h)} = \infty$$

for all k. The main theorems of the author state that both the set of all integers $q > 0$ satisfying $\|q\xi_1\|\,\|q\xi_2\| q^{1+\varepsilon} < 1$ and the set of all integers $q > 0$ for which there are integers q_1, q_2 satisfying $\|q_1\xi_1 + q_2\xi_2\|\,|q_1|^{1+\varepsilon}|q_2|^{1+\varepsilon} < 1$, $|q_1| > 0$, $|q_2| > 0$, $q = \max(|q_1|, |q_2|)$, have this property E.

The rather involved proofs of these theorems are based on Roth's ideas [Mathematika **2** (1955), 1–20; MR **17**, 242] or rather on a generalisation of these ideas. As an application, the author shows that if $q_1, q_2, q_3, \cdots$ are the denominators of the successive convergents of a real quadratic irrational, and if c is any constant $> 1 + \sqrt{3}$, then the series $\sum_{h=1}^{\infty} 1/q_{[c^h]}$ is transcendental.

K. Mahler (Canberra)

Citations: MR 17, 242d = J68-14.

Referred to in D48-65, J68-56.

J68-47 **(31# 3384)**

Rodriquez, Gaetano

Approssimabilità di irrazionali p-adici mediante numeri razionali.

Ist. Lombardo Accad. Sci. Lett. Rend. A **98** (1964), 691–708.

Let $F(x)$ be a polynomial of degree $n \geqq 2$ with rational integer coefficients, irreducible in the field of rationals, and let p_τ $(\tau = 1, \cdots, t)$ be distinct primes. Further, let ζ and ζ_τ be a real and a p_τ-adic zero of $F(x)$, and let h_i/q_i $(i = 1, 2, \cdots)$ be an appropriate sequence of rational numbers with $(h_i, q_i) = 1$ and $\max(|h_i|, |q_i|) = H_i \to \infty$ $(i \to \infty)$. Put $U_i = \min(1, |\zeta - h_i/q_i|) \prod_\tau \min(1, |q_i\zeta_\tau - h_i|_{p_\tau})$, $f(x) = (8n + \varepsilon)(\log\log\log x)^{-1/2}$, where ε denotes any fixed positive constant.

Ridout [Mathematika **5** (1958), 40–48; MR **20** #3851] proved the following p-adic generalization of the Thue-Siegel-Roth theorem: If for a constant k and all i, $U_i \leqq H_i^{-k}$, then $k \leqq 2$. Cugiani [Boll. Un. Mat. Ital. **14** (1959), 151–162; MR **22** #8002] proved an extension of the theorem of Roth from which it follows, in particular, that if $|\zeta - h_i/q_i| \leqq H_i^{-(2+f(H_i))}$ (for all i), then $\limsup \log H_i/\log H_{i-1} = \infty$. The present author proves the same conclusion assuming that for all i, $U_i \leqq H_i^{-(2+f(H_i))}$.

E. Fogels (Riga)

Citations: MR 20# 3851 = J68-20; MR 22# 8002 = J68-28.

Referred to in J68-51.

J68-48 **(32# 4086)**

Hyyrö, Seppo

Über rationale Näherungswerte algebraischer Zahlen.

Ann. Acad. Sci. Fenn. Ser. A I No. 376 (1965), 15 *pp.*

In 1908, A. Thue proved his classical theorem [J. Reine Angew. Math. **135** (1909), 284–305]: If θ is real and algebraic of degree $n \geqq 3$, and if $\varepsilon > 0$ is constant, then the inequality (1): $|\theta - (p/q)| < q^{-(n/2)-1-\varepsilon}$ has at most finitely many solutions in integers $p, q > 0$. The proof consists in the effective construction of a constant $c_1(\theta, \varepsilon)$ and of a function $c_2(\theta, \varepsilon, q_1)$ with the following property: Either $q \leqq c_1$ for all the solutions p/q of (1); or there exist also solutions p/q with $q > c_1$. In the latter case, let $q = q_1$ be the smallest possible value of $q > c_1$ for such a solution; then $q \leqq c_2$ for all solutions p/q of (1). The author carries this a little further, using Dyson's method [Acta Math. **79** (1947), 224–240; MR **9**, 412] rather than Thue's. Let K be a number field of degree $n \geqq 3$, ρ a primitive and θ an irrational real element of K; further, let $C > 0$ be a constant and let κ and λ be constants such that $\kappa > 2$, $\lambda > 2$, $\kappa\lambda > 2n$. Then positive expressions $g(\rho)$ and $G(\rho, \theta, q_1)$ can be effectively constructed such that if there exist integers p_1, q_1 with $|\rho - (p/q)| < q_1^{-\kappa}$, $q_1 > g(\rho)$, then $q \leqq G(\rho, \theta, q_1)$ for all integers p, q with $|\kappa - (p/q)| < q^{-\lambda}$. A similar, but more complicated theorem could also be deduced from the proof of Roth's theorem [Mathematika **2** (1955), 1–20; corrigendum, 168; MR **17**, 242].

K. Mahler (Canberra)

Citations: MR 9, 412h = J68-5; MR 17, 242d = J68-14.

Referred to in J68-50.

J68-49 **(32# 5597)**

Ćetković, Simon

Existence of a set of transcendental numbers by which it is possible to approximate algebraic numbers with arbitrary precision. (Serbo-Croatian. French summary)

Bull. Soc. Math. Phys. Serbie **11** (1959), 93–97.

The author shows that for each interval (d, e), there is a sequence of transcendental numbers $\{t_n\}$ for which the inequality $g(n) < |a - t_n| < h(n)$ has infinitely many solutions n for each fixed algebraic number $a \in (d, e)$ (g and h being arbitrary real functions with $0 \leqq g(n) < h(n)$ for each n). The sequence is constructed recursively.

J68-50 **(34# 5752)**

Hyyrö, Seppo

Über Approximation algebraischer Zahlen durch rationale.

Ann. Univ. Turku. Ser. A I No. 84 (1965), 12 *pp.*

The following observation may serve as background for the paper under review. Let ϑ be an irrational algebraic number, $C > 0$, $\lambda > 2$. Then by K. F. Roth's theorem [Mathematica **2** (1955), 1–20; corrigendum, ibid. **2** (1955), 168; MR **17**, 242] there exists a positive number $S = S(\vartheta, \lambda, C)$ such that $|\vartheta - (u/v)| < Cv^{-\lambda}$ has no diophantine solutions with $v \geqq S$. However, as yet it is not possible to determine S effectively for given ϑ, λ and C.

The author's main theorem and most explicit result (Satz 4, p. 6) reads: Let K be a number field of degree $n \geqq 3$, ρ a primitive integral number for this field and ϑ an arbitrary irrational integral number of K. Further, let C, λ and κ be positive constants such that $n/2 < \kappa < n$, $(\lambda - 1) \times (\kappa - n/2) > n/2$. Then positive numbers $g_0 = g_0(\rho)$ and $S = S(\rho, \vartheta, \kappa, \lambda, C, q)$ can be effectively constructed such that the two diophantine inequalities $|\rho - (p/q)| < q^{-\kappa}$, $q > g_0$, $|\vartheta - (u/v)| < Cv^{-\lambda}$, $v \geqq S$, cannot be satisfied simultaneously.

Therefore, if one "good" approximation p/q for a primitive number ρ of K is known, then it is possible to solve the above problem for a particular value of λ and every irrational number ϑ of K.

The proof is based on the classical method used by A. Thue to derive his approximation theorem [J. Reine Angew. Math. **135** (1909), 284–305]. The present paper is the first of a series of three publications. The later papers of the author [Ann. Acad. Sci. Fenn. Ser. A I No. 376 (1965); MR **32** #4086; ibid. No. 394 (1967)] give sharper results but use, instead of Thue's idea, the more complicated later methods of Dyson, Roth and others.

J. Popken (Amsterdam)

Citations: MR 17, 242d = J68-14; MR 32# 4086 = J68-48.

J68-51 (35# 2832)

Rodriquez, Gaetano

Approssimabilità di irrazionali p-adici mediante numeri razionali. II.

Boll. Un. Mat. Ital. (3) **20** (1965), 232–244.

Let $p_1, \cdots, p_t$ be distinct primes and let $F(x)$ be a polynomial of degree $n \geqq 2$ with rational integral coefficients, irreducible over the field of rationals, which has a real zero ξ and a p_τ-adic zero ξ_τ, for $\tau = 1, \cdots, t$. Further, let $\{h_i/q_i\}$ be an infinite sequence of distinct rational numbers, with $(h_i, q_i) = 1$ and $H_i = \max(|h_i|, |q_i|) > e^e$. Put $U_i = \min(1, |\xi - h_i/q_i|) \cdot \prod_\tau |q_i \xi_\tau - h_i|_{p_\tau}$.

In a previous paper [Ist. Lombardo Accad. Sci. Lett. Rend. A **98** (1964), 691–708; MR **31** #3384] the author gave a quantitative refinement of a result of D. Ridout [Mathematika **5** (1958), 40–48; MR **20** #3851] by proving the following: If, for all i, $U_i \leqq H_i - (2 + f(H_i))$, where $f(x) = (8n + \varepsilon)(\log\log\log x)^{-1/2}$, then

$$\limsup_{i \to \infty} \log H_{i+1}/\log H_i = \infty.$$

He now derives the same conclusion taking $f(x) = 5(\sqrt{\log(4n)})(\log\log\log x)^{-1/2}$.

C. G. Lekkerkerker (Amsterdam)

Citations: MR 20# 3851 = J68-20; MR 31# 3384 = J68-47.

J68-52 (35# 4167)

Baker, A. [Baker, Alan]

Simultaneous rational approximations to certain algebraic numbers.

Proc. Cambridge Philos. Soc. **63** (1967), 693–702.

A well-known conjecture states that if $1, \alpha_1, \alpha_2, \cdots, \alpha_k$ are algebraic numbers which are linearly independent over the field of rational numbers, and if $\kappa > 1 + 1/k$, then there are only finitely many sets of integers $p_1, p_2, \cdots, p_k, q$ $(q > 0)$ such that (*) $|\alpha_j - p_j/q| < 1/q^\kappa$ $(j = 1, 2, \cdots, k)$. K. F. Roth [Mathematika **2** (1955), 1–20; corrigendum, 168; MR **17**, 242] established the case $k = 1$, and this certainly implies that for any positive integer k, (*) has only finitely many solutions for $\kappa > 2$. (The author has added in proof that W. M. Schmidt has established the conjecture for $k = 2$.)

The author's main result determines, for any $\kappa > 1 + 1/k$, a set $\alpha_1, \cdots, \alpha_k$ which generates an algebraic number field of degree exceeding $k + 1$, such that (*) has only finitely many solutions. This is a corollary of the main theorem, which also admits several other applications. In particular, a bound is calculated for the size of the solutions of certain Diophantine equations in three unknowns, and an inequality is derived which gives an estimate for the accuracy of approximations to α_j by algebraic numbers of bounded degree.

L. C. Eggan (Tacoma, Wash.)

Citations: MR 17, 242d = J68-14.

Referred to in J68-62.

J68-53 (35# 4169)

Hyyrö, Seppo

Zur Approximation algebraischer Zahlen.

Ann. Acad. Sci. Fenn. Ser. A I No. 394 (1967), 18 *pp.*

Nach dem bekannten Thue-Siegel-Rothschen Satz besitzt die Ungleichung (*) $|\vartheta - p/q| < C/q^\lambda$ für jedes algebraische ϑ, $\lambda > 2$ und positive C höchstens endlich viele Lösungen. Der Beweis des Thue-Sigel-Rothschen Satzes gibt aber keine Schranke für die Anzahl der möglichen Lösungen von (*) und auch keine Schranke G mit der Eigenschaft, dass (*) keine Lösung mit $q < G$ besitzt. In der vorliegenden Arbeit wird eine solche Schranke angegeben, allerdings unter etwas komplizierten Voraussetzungen über den algebraischen Zahlkörper, dem ϑ angehört.

P. Szüsz (Stony Brook, N.Y.)

J68-54 (36# 112)

Ramachandra, K.

Approximation of algebraic numbers.

Nachr. Akad. Wiss. Göttingen Math.-Phys. Kl. II **1966**, 45–52.

In the present paper the author proves the following theorem: Let α be an algebraic number of degree $g \geqq 2$ and let β be a real number which exceeds $\min(mh^{m-1}(g/m!)^{1/m})$, $m = 1, 2, \cdots$. Then there exist only finitely many solutions of the inequality $|\alpha - \zeta| < (H(\zeta))^{-\beta}$ in algebraic numbers ζ of degree h, where $H(\zeta)$ denotes the maximum of the absolute values of the coefficients of the irreducible equation with coprime rational integer coefficients satisfied by ζ. An essential tool in obtaining this result is K. F. Roth's lemma [Mathematika **2** (1955), 1–20; MR **17**, 242] as generalized by W. J. LeVeque [*Topics in number theory*, Vol. II, Chapter 4, Addison-Wesley, Reading, Mass., 1956; MR **18**, 283].

R. Finkelstein (Tempe, Ariz.)

Citations: MR 17, 242d = J68-14; MR 18, 283b = Z01-38.

Referred to in D56-33.

J68-55 (36# 2563)

Stepanov, S. A.

The approximation of an algebraic number by algebraic numbers of a special form. (Russian. English summary)

Vestnik Moskov. Univ. Ser. I Mat. Meh. **22** (1967), *no.* 6, 78–86.

In 1956, W. J. LeVeque published a generalisation of Roth's theorem to algebraic number fields [*Topics in number theory*, Vol. 2, Chapter 4, Addison-Wesley, Reading, Mass., 1956; MR **18**, 283]. This result is further extended by the author, who proves the following theorem. Let K be an algebraic number field of degree n over Q; let $\omega_1, \cdots, \omega_n$ be a fixed basis of the integers in K; let α be algebraic of degree $\nu \geqq 2$ over K; and let b be a fixed positive integer. Denote by $\gamma \geqq 1$, $c > 1$, $\kappa > 0$, and $\eta > 0$ four real constants. Assume there exist infinitely many sets of $n+1$ rational integers $p_1, \cdots, p_n, q > 0$ that are relatively prime and have the following properties: (1) $|\alpha - (p_1\omega_1 + \cdots + p_n\omega_n)/q| < h^{-\kappa n}$, where $h = \max(|p_1|, \cdots, |p_n|, q)$; (2) $q = b^s q'$, where $s \geqq 0$ and $q' \geqq 1$ are rational integers such that $\limsup(\log q'/\log q) = \eta$; (3) $q^\gamma \leqq h \leqq cq^\gamma$. Then $\kappa \leqq 2(1 - 1/n) - (1/\gamma)(1 - \eta - 2/n)$. This theorem can be further generalised.

K. Mahler (Canberra)

Citations: MR 18, 283b = Z01-38.

J68-56 (36# 6357)

Schmidt, Wolfgang M.

On simultaneous approximations of two algebraic numbers by rationals.

Acta Math. **119** (1967), 27–50.

The author proves several theorems which extend Roth's

theorem to simultaneous approximations. The proofs are complicated and depend heavily on an earlier paper of the author in which weaker results were proved [same Acta **114** (1965), 159–206; MR **31** #2206]. In fact the relationship between the results of the earlier paper and the present one is similar to the relationship between the result of an earlier paper of T. Schneider [J. Reine Angew. Math. **175** (1936), 182–192] and Roth's theorem [K. Roth, Mathematika **2** (1955), 1–20; MR **17**, 242]. Both the present and the earlier paper of the author depend on a generalization of Roth's ideas.

In particular, the following theorems are among those proved ($\|\xi\|$ denotes the distance from ξ to the nearest integer): Theorem 1: Let α, β be algebraic and $1, \alpha, \beta$ linearly independent over the rationals. Then for every $\varepsilon > 0$, there are only finitely many positive integers q with $\|q^\alpha\| \, \|q^\beta\| q^{1+\varepsilon} < 1$. Theorem 2: With α, β, ε as in Theorem 1, there are only finitely many pairs of rational integers $q_1 \neq 0$, $q_2 \neq 0$ with $\|q_1{}^\alpha + q_2{}^\beta\| \, |q_1 q_2|^{1+\varepsilon} < 1$.

These and other results are derived from a theorem (Theorem 6) concerning the successive minima of a parallelepiped defined by a "proper" set of linear forms. The definition of "proper" is as follows: Let $L_j = \sum_{i=1}^{n+1} \alpha_{ji} X_i$, $j = 1, \cdots, n+1$, be $n+1$ linear forms ($n \geqq 1$). Let S be a subset of $\{1, \cdots, n+1\}$ and let A_{ij} denote the cofactor of α_{ij} in the matrix (α_{hk}) $(1 \leqq h, k \leqq n+1)$. Then $L_1, \cdots, L_{n+1}$; S are called proper if (i) the α_{hk} are algebraic and $\det(\alpha_{hk}) \neq 0$, (ii) for every $i \in S$ the non-zero elements among $A_{i1}, \cdots, A_{i,n+1}$ are linearly independent over the rationals, and (iii) for every k, $1 \leqq k \leqq n+1$, there is an $i \in S$ with $A_{ik} \neq 0$.

Applying this theorem and another one involving the relationship between the second minimum of certain parallelepipeds and their first minimum ("transference principle", Theorem 7), the author deduces a general theorem in the case $n = 2$ (Theorem 12), and from this follow Theorems 1 and 2, among others.

An application of Theorem 6 to the case $n = 3$ (here the theorem yields a result on the "third minimum"), results in the following Theorem 5: Let α, β, γ be algebraic, $1, \alpha, \beta, \gamma$ linearly independent over the rationals, and $\varepsilon > 0$. Then there are only finitely many triples of non-zero integers q_1, q_2, q_3 satisfying $\|\alpha q_1 + \beta q_2 + \gamma q_3\| \, |q_1 q_2 q_3|^{5/3+\varepsilon} < 1$.

The author conjectures that here $5/3$ may be replaced by 1, but states that he is unable to prove this, and further cannot prove any result in this direction for more than three algebraic numbers. *S. L. Segal* (Rochester, N.Y.)

Citations: MR 17, 242d = J68-14; MR 31# 2206 = J68-46.

Referred to in D56-30, J68-65, M55-60.

J68-57 (36# 6592)

Güting, R.

Polynomials with multiple zeros.

Mathematika **14** (1967), 181–196.

For $P(x) = x^p + a_{p-1}x^p + \cdots + a_0 = \prod_{j=1}^p (x - \alpha_j)$, let $S(P) = 1 + |a_{p-1}| + \cdots + |a_0|$ and $D(P) = \prod_{j<k} (\alpha_j - \alpha_k)^2$. The resultant of the two polynomials P and Q is denoted by $\operatorname{Res}(P, Q) = Q(\alpha_1)Q(\alpha_2) \cdots Q(\alpha_p)$.

Using some lemmas, including two due to K. Mahler [Mathematika **7** (1960), 98–100; MR **23** #A1779], the author proves a number of theorems, including the following: (I) If Q has no multiple zeros, then its first derivative Q' satisfies the inequality $|Q'(\alpha)| \geqq [\frac{1}{2}q(q-1)]^{(1-q)/2} \times D(Q)^{1/2} \mathscr{S}(Q)^{2-q} \max(1, |\alpha|^{q-2})$, where $q = \deg Q$. (II) If P has integral coefficients and a kth order non-real zero α, then $|P^{(k)}(\alpha)| \geqq \lambda |\operatorname{Im} \alpha|^{k/2} S(P)^{(3k-p)/2k} \max(1, |\alpha|^{p-3k})$, where $\lambda = k!(p/k)^{k-p} 2^{p(2k-p)/k}$. (III) If P, R are polynomials of degrees p, r and if α is a kth order zero of P, then $|R(\alpha)| \geqq 2^{-pr/k} |\operatorname{Res}(P, R)^{1/k} S(P)^{r/k} S(R)^{(p-k)/k} \max(1, |\alpha|^r)$.
M. Marden (Milwaukee, Wis.)

Citations: MR 23# A1779 = J76-38.

Referred to in J84-43.

J68-58 (37# 2698)

Armitage, J. V.

The Thue-Siegel-Roth theorem in characteristic p.

J. Algebra **9** (1968), 183–189.

Let K be an algebraic number field and b any element of a finite extension of K, and let k be any real number > 2. Roth's theorem asserts that the elements c of K with $|b - c| < (H(c))^{-k}$ have bounded height $H(c)$. S. Lang has given an exposition of this theorem [see *Diophantine geometry*, Interscience, New York, 1962; MR **26** #119] for a larger class of fields K characterized by certain axioms which are also valid in function fields in one variable over fields of characteristic zero. On the other hand, in the case of characteristic p there is an example of a function field K by K. Mahler [Canad. J. Math. **1** (1949), 397–400; MR **11**, 159] for which the above statement becomes false, if one takes a suitable b from a cyclic extension of degree p over K. The author finds a set of axioms, e.g., satisfied by all function fields in one variable over finite fields of characteristic p, for which Roth's theorem can be proved, provided the elements b are excluded from cyclic extensions of degree a power of p over K. The proof is a slight modification of that by Lang. *O. H. Körner* (Marburg)

Citations: MR 11, 159e = J68-8; MR 26# 119 = G02-9.

J68-59 (37# 6249)

Davenport, H.

A note on Thue's theorem.

Mathematika **15** (1968), 76–87.

Liouville's classical theorem states that if θ is a real algebraic number of degree $r \geqq 2$, then there is a computable constant $c = c(\theta) > 0$ such that the inequality (*) $|\theta - p/q| < c/q^\kappa$ has for all $\kappa \geqq r$, no rational solutions p/q $(q > 0)$. Thue, Siegel and ultimately Roth successively improved the exponent κ, but the constant was no longer claimed computable. However, it had been stated that Thue's method enables one to compute a number $q_0 = q_0(\theta, \kappa)$ such that (*) has at most one solution with $c = 1$ and $q > q_0$. S. Hyyrö [Ann. Acad. Sci. Fenn. Ser. A I No. 355 (1964); MR **34** #5750] noted that there is no justification for this statement. The author shows that Thue's method does yield a computable q_0 provided Thue's restrictions on κ are strengthened somewhat. The author defines a function $\kappa = \kappa(r)$ and shows that for this value of κ there is a computable number $q_0 = q_0(\theta, \kappa)$ such that (*) has at most one solution with $c = 1$ and $q > q_0$. In particular, $\kappa(3) = 1 + \sqrt{3}$, $\kappa(r) = 1 + \sqrt{(\frac{1}{4}r^2 + r)}$ for every even $r \geqq 4$, and a more complicated function if r is odd, depending on whether $r \equiv 1$ or $3 \pmod 4$.

As the author notes, Schinzel, in his review of S. Hyyrö's paper [Zbl. Math. **137** (1967), 257] by using a remark of Thue [J. Reine Angew. Math. **135** (1909), 284–305] and Theorem I of Chapter I of A. O. Gel'fond's *Transcendental and algebraic numbers* (Russian) [GITTL, Moscow, 1952; MR **15**, 292; English translation, Dover, New York, 1960; MR **22** #2598], has shown that for $r \geqq 5$ and $\kappa > 3\sqrt{(r/2)}$ one can compute a constant $q_0 = q_0(\theta, r)$ such that the inequality (*) has at most one solution with $c = \frac{1}{2}$ and $q > q_0$. For $r \geqq 9$, this result is superior to that mentioned above. *L. C. Eggan* (Normal, Ill.)

Citations: MR 15, 292e = J02-6; MR 22# 2598 = J02-7; MR 34# 5750 = D48-55.

Referred to in J68-66.

J68-60 **(38# 5758)**

Kolchin, E. R.

Some problems in differential algebra.

Proc. Internat. Congr. Math. (*Moscow*, 1966), *pp.* 269–276. *Izdat.* "*Mir*", *Moscow*, 1968.

This paper first gives a brief, lucid description of the basic notions and results of differential algebra. This serves as an introduction to a detailed description of some of the most interesting and challenging problems in differential algebra. The only background required for reading this paper is some familiarity with the elementary notions of algebraic geometry (e.g., generic zero, specialization). The author reviews and brings up to date the following problems: (1) To determine when a singular zero of an irreducible differential polynomial Q lies in the general manifold of Q. (2) To find necessary and sufficient conditions for the extension of a differential specialization. A counterexample is given to show that the analog of the theorem, on extension of specialization, of algebraic geometry does not hold for extensions of differential specializations. (3) The author has previously shown [Bull. Amer. Math. Soc. **70** (1964), 570–573; MR **29** #5816] that the size of a manifold M of a system of differential polynomials can be described by a polynomial in one variable $P(M)$ with rational integer coefficients. The significance of these coefficients, their invariance under birational (differential birational) transformations, and their relation to some classical theorems and problems on the order of a system of differential equations, is discussed. (4) Given a linear algebraic group G and a differential field F (e.g., rational functions of x with complex coefficients), does there exist a Picard-Vessiot extension K of F such that G is the group of automorphisms of K over F. Also, given $K \supset F$, to find the group G. (5) A functional analog of Liouville's theorem on approximation of algebraic numbers by rational numbers, was given by the author [Proc. Amer. Math. Soc. **10** (1959), 238–244; MR **21** #6366]. The possibility and difficulty of extending these results, to obtain an analog of the Thue-Siegel-Roth theorem, is discussed.

Lawrence Goldman (New York)

Citations: MR 21# 6366 = Q10-8.

J68-61 **(40# 91)**

Davenport, H.; Schmidt, Wolfgang M.

Approximation to real numbers by algebraic integers.

Acta Arith. **15** (1968/69), 393–416.

Let ξ be a real number. Let S be a set of real algebraic numbers. Let $\lambda>0$ be real. We consider the problem of whether there are an infinite number of $\alpha\in S$ such that $0<|\xi-\alpha|\ll H(\alpha)^{-\lambda}$.

Let $n\geqq 1$ be an integer and let $S=$ all numbers of degree $\leqq n$. Assume ξ is not an algebraic number of degree $\leqq n$. Then, of course, if $n=1$ and $\lambda=2$, there is always an infinite number of solutions and the exponent "2" is best possible. The authors [same Acta **13** (1967/68), 169–176; MR **36** #2558] have shown that if $n=2$ and $\lambda=3$, there is an infinite number of solutions and again the exponent is best possible (for a related result see another paper of the authors [ibid. **14** (1967/68), 209–223; MR **37** #1321]). In general, E. Wirsing has shown [J. Reine Angew. Math. **206** (1960), 67–77; MR **26** #79] the same result for any n and $\lambda=(n+3)/2$. In the present paper we let S be all algebraic integers of degree $\leqq n$ $(n\geqq 2)$. Assume ξ is not an algebraic number of degree $\leqq(n-1)/2$ (or $\leqq 2$ if $n=3$). Then again there is an infinite number of solutions in the following cases: (1) $n=3$ and $\lambda=(3+\sqrt{5})/2$, (2) $n=4$ and $\lambda=3$, (3) $n\geqq 3$ and $\lambda=[(n+1)/2]$. It is not known whether these exponents are best possible.

To prove this result the authors consider the convex body $|x_0\xi^{n-1}+\cdots+x_{n-1}|<R^{-n+1}$, $|x_0|,\cdots,|x_{n-2}|<R$ and show that large values of R can be found such that the nth successive minimum of this set is not too big. This is accomplished by showing there are large values of R such that the first minimum of the polar body $|y_0|<R^{n-1}$, $|y_0\xi^m-y_m|<R^{-1}$ $(m=1,\cdots,n-1)$ is not too small. That is, it is shown that there are arbitrarily large values of X such that the inequalities $|x_0|<X$, $|x_0\xi^m-x_m|<cX^{-1/(\lambda-1)}$ $(1\leqq m\leqq n-1)$, where $c=c(n,\xi)$ is sufficiently small, have no solutions in integers $x_0,\cdots,x_{n-1}$ not all zero.

It is also interesting to note that E. Wirsing has shown that if ξ is algebraic and $S=$ all real algebraic numbers of degree $\leqq n$ and $\lambda>2n$, then there is only a finite number of solutions. This generalizes the famous Roth theorem [K. F. Roth, Mathematika **2** (1955), 1–20; corrigendum, ibid. **2** (1955), 168; MR **17**, 242] where $n=1$.

W. W. Adams (College Park, Md.)

Citations: MR 17, 242d = J68-14; MR 26# 79 = J84-20; MR 36# 2558 = J56-30; MR 37# 1321 = J16-27.

J68-62 **(40# 2612)**

Osgood, Charles F.

The simultaneous diophantine approximation of certain kth roots.

Proc. Cambridge Philos. Soc. **67** (1970), 75–86.

Let $1\leqq k_1<k$ be relatively prime integers. Let $0=s_1<s_2<\cdots<s_n$ be $n\geqq 2$ integers. Let $0<\varepsilon<1/2(n-1)$ be a real number. Let N be a large integer which depends explicitly on the s_i and k, k_1. Let C be any number, $0\leqq C\leqq 1$. Explicit functions $\phi(N,\varepsilon)$ and $\Lambda(N)$ are determined such that, setting $\alpha_j=C(N+s_j)^{k_1k^{-1}}$, we have for all integers $q, p_1,\cdots,p_n$ with $q>\phi(N,\varepsilon)$,

$$\max|\alpha_j-P_j/q| \geqq n^{-1}(2q)^{-\{1+(1+\varepsilon)/\Lambda(N)\}} \quad (1\leqq j\leqq n).$$

($\Lambda(N)$ tends to $n-1$ as $N\to\infty$.) An application to Diophantine equations is given. For similar results, see A. Baker [same Proc. **63** (1967), 693–702; MR **35** #4167] and N. I. Fel'dman [Mat. Zametki **2** (1967), 245–256; MR **36** #2560].

W. W. Adams (College Park, Md.)

Citations: MR 35# 4167 = J68-52; MR 36# 2560 = J80-25.

Referred to in J68-66.

J68-63 **(41# 156)**

Wallisser, Rolf

Zur Approximation algebraischer Zahlen durch arithmetisch charakterisierte algebraische Zahlen.

Arch. Math. (*Basel*) **20** (1969), 384–391.

In 1957 D. Ridout published a generalization of Roth's theorem about the rational approximations to an algebraic number [Mathematika **4** (1957), 125–131; MR **20** #32]. In the present paper the author extends Ridout's theorem to the approximation of an algebraic number by numbers of a given algebraic number field which satisfy certain arithmetical conditions. His result reads: "Let α be an algebraic number of degree n, K an algebraic number field of degree N. Further, let $(\xi_j)_{j\in\mathbf{N}}$ be a sequence of numbers taken from K with minimal polynomials $M(\xi_j)=a_{0j}x^r+a_{1j}x^{r-1}+\cdots+a_{rj}$, $1\leqq r\leqq N$, and with heights $q_j=\max_{0\leqq\nu\leqq r}|a_{\nu j}|$. Let $a_{0j}=k_{0j}{}^r$, $k_{0j}{}^{r-\nu}/a_{\nu j}$, $1\leqq\nu\leqq r$, $k_{0j}=k_j{}^*\prod_{i=1}^s k_i{}^{\sigma_i(j)}$, $a_{rj}=h_j{}^*\prod_{l=1}^t h_1{}^{\tau_1(j)}$, where $k_1,\cdots,k_s$, $h_1,\cdots,h_t$ are primes, $|k_j{}^*|$, $|h_j{}^*|$, $\sigma_1,\cdots,\sigma_s$, $\tau_1,\cdots,\tau_t$ nonnegative integers. Put $\tau=\liminf_{j\to\infty}\log|a_{0j}|/\log q_j$, $\sigma=\liminf_{j\to\infty}\log|a_{rj}|/\log q_j$, $\nu=\limsup_{j\to\infty}\log|k_j{}^*|/\log|k_{0j}|$, $\mu=\limsup_{j\to\infty}\log|h_j{}^*|/\log|a_{rj}|$, and let $K>1+\nu\tau-\sigma(1-\mu)$; then the inequality $|\alpha-\xi|<q^{-K}$ has only a finite number of solutions in the sequence $(\xi_j)_{j\in\mathbf{N}}$."

In the special case treated by Ridout one has $K=Q$, $N=1$, $\tau=\sigma=1$. As an application the author obtains: The

set $\{x \mid x=\sum_{\nu=0}^{\infty} \varepsilon_\nu((3-\sqrt{2})/9)^{3^\nu}$, $\varepsilon_\nu=0$ or 1, ε_ν not a Null sequence$\}$ forms a continuum of transcendental numbers.
J. Popken (Amstelveen)

Citations: MR 20# 32 = J68-18.

J68-64 (42# 3026)

Fel'dman, N. I.

Estimation of an incomplete linear form in certain algebraic numbers. (Russian)

Mat. Zametki **7** (1970), 569–580.

Let K be an imaginary quadratic field and let $b_1, \cdots, b_m$ be distinct non-zero numbers of K satisfying certain conditions. Put $\omega_k=b_k^{\nu}$, where ν is a positive rational number less than 1, and let μ be a real number with $\mu>m-1$. The author gives the following result. There exists a positive constant X_0 such that the inequality $|x_1\omega_1+\cdots+x_m\omega_m|>X^{-\mu}$ holds for all integers $x_1, \cdots, x_m$ of K satisfying $X=\max(|x_1|, \cdots, |x_m|)\geqq X_0$, and that X_0 can be effectively determined by $b_1, \cdots, b_m, \nu$, and μ. The proof is analogous to that of a theorem in the author's previous article [same Zametki **2** (1967), 245–246; MR **36** #2560].

{This article has appeared in English translation [Math. Notes **7** (1970), 343–349].}
E. Inaba (Tokyo)

Citations: MR 36# 2560 = J80-25.

J68-65 (42# 3028)

Schmidt, Wolfgang M.

Simultaneous approximation to algebraic numbers by rationals.

Acta Math. **125** (1970), 189–201.

If ξ is real, denote by $\|\xi\|$ its distance from the nearest integer; if ω is algebraic, let $H(\omega)$ be the maximum of the absolute values of the coefficients of the defining equation with rational integral coefficients for ω. Denote by α, $\alpha_1, \cdots, \alpha_n$ finitely many real algebraic numbers, where $\alpha_1, \cdots, \alpha_n$, 1 are linearly independent over Q. Further, let ε be any positive constant. The author establishes the following important theorems which are almost best possible. (1) The inequality $\|q\alpha_1\|\cdots\|q\alpha_n\|<|q|^{-1-\varepsilon}$ has only finitely many solutions in integers $q\neq 0$. Hence the simultaneous inequalities $|\alpha_h-(p_h/q)|<|q|^{-1-(1/n)-\varepsilon}$, $h=1, 2, \cdots, n$, have only finitely many solutions in integers $p_1, \cdots, p_n, q\neq 0$. (2) The inequality $\|q_1\alpha_1+\cdots+q_n\alpha_n\|<|q_1\cdots q_n|^{-1-\varepsilon}$ has only finitely many solutions in integers $q_1\neq 0, \cdots, q_n\neq 0$. Hence there are only finitely many sets of integers $p, q_1, \cdots, q_n$ not all zero satisfying

$$|p+q_1\alpha_1+\cdots+q_n\alpha_n| < \max(|q_1|, \cdots, |q_n|)^{-n-\varepsilon}.$$

(3) For every integer $k\geqq 1$ there are only finitely many algebraic real numbers ω of degree $\leqq k$ such that $|\alpha-\omega|^{-k-1-\varepsilon}$. For $n=1$, (1) and (2) reduce to Roth's theorem [K. F. Roth, Mathematika **2** (1955), 1–20; corrigendum, ibid. **2** (1955), 168; MR **17**, 242], and so does (3) for $k=1$. When $n=2$, (1) and (2) were earlier proved by the author [Acta Math. **119** (1967), 27–50; MR **36** #6357]. A weaker form of (3) with $-2k-\varepsilon$ instead of $-k-1-\varepsilon$ was already obtained by E. Wirsing ("Approximation to algebraic numbers by algebraic numbers of bounded degree", Report on the Number Theory Inst. (Stony Brook, N.Y., 1969), to appear). Theorems (1) and (2) are essentially equivalent to one another by the number geometrical theorem on the successive minima of polar convex bodies [the reviewer, Časopis Pěst. Mat.-Fys. **68** (1939), 93–102], and (3) can be deduced from (2). The proofs of either (1) or (2) are quite deep and make use of the theory of the successive minima of compound convex bodies [the reviewer, Proc. London Math. Soc. (3) **5** (1955), 358–379; MR **17**, 589] However, the main basis of the proof is a theorem by the author [loc. cit.] on the next to last successive minimum of a parallelotope the proof of which generalises that of Roth. For a more detailed discussion of the proofs of this paper see Chapters 4 and 5 of the author's "Lectures on Diophantine approximation" [lecture notes, Math. Dept., Univ. of Colorado, Boulder, Colo., 1970].

{In the opinion of the reviewer, it should be possible to generalise the methods of this paper so as to include both Archimedean and finitely many non-Archimedean valuations of the rational field. This might allow the proof of the finiteness of the number of integral solutions $(x_1, \cdots, x_n)$ of $x_1+\cdots+x_n=0$, where, for $h=1, \cdots, n$, $x_h\neq 0$ is allowed to have no prime factors except finitely many given ones. The application of Schmidt's theorems to the theory of decomposable forms is immediate [see, e.g., C. J. Parry, Acta Math. **83** (1950), 1–100; MR **12**, 320].}
K. Mahler (Columbus, Ohio)

Citations: MR 12, 320d = J68-10; MR 17, 242d = J68-14; MR 17, 589a = H05-46; MR 36# 6357 = J68-56.

J68-66 (43# 1927)

Osgood, Charles F.

The diophantine approximation of roots of positive integers.

J. Res. Nat. Bur. Standards Sect. B **74B** (1970), 241–244.

From his effective result concerning simultaneous diophantine approximations of certain kth roots [Proc. Cambridge Philos. Soc. **67** (1970), 75–86; MR **40** #2612], the author deduces the following theorem. Suppose that $k\geqq 150$ and m are fixed positive integers; then $|\sqrt[k]{(m)}-p/q|<q^{-(7/8)k}$ can hold for at most one pair of relatively prime integers p and q with $q\geqq 2^9(\sqrt[k]{(m)}+1)^6$. H. Davenport [Mathematika **15** (1968), 76–87; MR **37** #6249] and A. Schinzel in his review in Zbl **137**, 257 of the paper of S. Hyyrö in Ann. Acad. Sci. Fenn. Ser. A I No. 355 (1964) [MR **34** #5750] obtained similar results for real algebraic numbers but the lower bound for q was not as good, nor was it given explicitly.
W. W. Adams (College Park, Md.)

Citations: MR 34# 5750 = D48-55; MR 37# 6249 = J68-59; MR 40# 2612 = J68-62.

J68-67 (43# 3267)

Sprindžuk, V. G.

An effective estimate of rational approximations for algebraic numbers. (Russian)

Dokl. Akad. Nauk BSSR **14** (1970), 681–684.

Let α be an algebraic integer of degree $n\geqq 4$, and x and y rational integers, not both zero. Further, let S be the product of a finite number of prime ideals of the field $Q(\alpha)$. The author proves the following inequality by an effective method: (*) $|x-\alpha y|\prod_{\mathfrak{P}\in S}|x-\alpha y|>cX^{-n+1}\exp(\ln X)^{0.5-\varepsilon}$, where c is an effectively computable constant and $X=\max(x, y)$. The inequality (*) always holds if α is non-exclusive. α is exclusive if there exists an enumeration of its conjugates $\alpha^{(1)}, \cdots, \alpha^{(n)}$ such that

$$(\alpha^{(1)}-\alpha^{(i)})(\alpha^{(2)}-\alpha^{(j)})/((\alpha^{(2)}-\alpha^{(i)})(\alpha^{(1)}-\alpha^{(j)})) = (1-\zeta_j)/(1-\zeta_i),$$

for $i\neq j$, $3\leqq i, j\leqq n$, where ζ_i and ζ_j are roots of unity. The author proves that "almost all" α are non-exclusive.
R. Finkelstein (Bowling Green, Ohio)

Referred to in N32-67.

J68-68 (44# 6609)

Fel'dman, N. I.

An effective power sharpening of a theorem of Liouville. (Russian)

Izv. Akad. Nauk SSSR Ser. Mat. **35** (1971), 973–990.

The author proves the following theorem: Let α be an algebraic number of degree $n\geqq 3$; then there exist effective positive constants a and C, depending only on α, such that $|\alpha-p/q|>Cq^{a-n}$ for all rational integers p, q with $q>0$. This sharpens the famous Thue-Siegel-Roth theorem [see K. F. Roth, Mathematika **2** (1955), 1–20; corrigendum, ibid. **2** (1955), 168; MR **17**, 242], which is merely an existence theorem, as well as subsequent effective results by A. Baker and by the author. Furthermore, the author establishes an effective bound for the absolute values of the solutions of the Diophantine equation $f(x, y)=m$, where f is an irreducible form of degree $\geqq 3$.

B. Volkmann (Stuttgart)

Citations: MR 17, 242d = J68-14.

J68-69 **(44# 6611)**

Popken, J.
A contribution to the Thue-Siegel-Roth problem.
Number Theory (Colloq., János Bolyai Math. Soc., Debrecen, 1968), *pp.* 181–190. *North-Holland, Amsterdam,* 1970.

The late author strengthens a principal auxiliary result in the proof of the Thue-Siegel-Roth theorem (see Theorem II, p. 106 in J. W. S. Cassels' *An introduction to Diophantine approximation* [Cambridge Univ. Press, New York, 1957; MR **19**, 396; Russian translation, Izdat. Inostran. Lit., Moscow, 1961; MR **22** #10976]). Let α be an algebraic number of degree n and height a, and let m and $r_1, \cdots, r_m$ be given positive integers. One is interested in polynomials $R(x_1, \cdots, x_m)$ with rational integral coefficients, of degree $\leqq r_j$ in x_j $(j=1, \cdots, m)$, and having a positive height b such that $\log b<(r_1+\cdots+r_m)\log(4a+4)$.

Let Ω denote the collection of all m-tuples $(i_1, \cdots, i_m)$ of integers $0\leqq i_j\leqq r_j$ $(j=1, \cdots, m)$ having $|\Omega|=(r_1+1)\cdots(r_m+1)$ elements and let B denote a subset of Ω. Provided (*) $|B|/|\Omega|\leqq 1/(2n)$, it follows from the usual proof that there exists a polynomial R as above satisfying $(\partial/\partial x_1)^{i_1}\cdots(\partial/\partial x_m)^{i_m}R(\alpha, \cdots, \alpha)=0$ for each $(i_1, \cdots, i_m)$ in B. Let $\varepsilon>0$; in the usual proof one needs a set $B(\varepsilon)\subset\Omega$ defined by the inequality (1): $\sum_{j=1}^m (i_j/r_j)\leqq(m/2)(1-\varepsilon)$ and one shows that (*) holds for $B=B(\varepsilon)$ as soon as $m\geqq 8(n/\varepsilon)^2$, that is, $\varepsilon\geqq[(8/m)n^2]^{1/2}$.

In the present paper, the author shows that simpler proofs and sharper results can be obtained by using certain elementary results from probability theory. Namely, (*) can be written as $P(B)\leqq 1/(2n)$ provided one turns Ω into a probability space with a uniform distribution P on Ω. Further, (1) can be written as $S_m\leqq-\varepsilon m$ where $S_m=X_1+\cdots+X_m$ and $X_j=(2i_j-r_j)/r_j$. Here X_j (when regarded as a function on Ω) is a random variable with $|X_j|=1$, $E(X_j)=0$ and $\operatorname{Var}(X_j)=\frac{1}{3}(1+2/r_j)\leqq 1$. Moreover, the random variables $X_1, \cdots, X_m$ are independent. Hence $E(S_m)=0$, $\operatorname{Var}(S_m)\leqq m$ so that by Čebyšev's inequality $P(B(\varepsilon))\leqq P(|S_m|\geqq\varepsilon m)\leqq m/(\varepsilon m)^2$ (which is $\leqq 1/(2n)$ for $\varepsilon\geqq[(2/m)n]^{1/2}$). Using a version of the Bernstein inequality, the author further shows that $P(B(\varepsilon))\leqq P(|S_m|\geqq\varepsilon m)\leqq 2\cdot\exp(-\frac{1}{2}m\varepsilon^2)$ (which is $\leqq 1/(2n)$ for $\varepsilon\geqq[2/m\log(4n)]^{1/2}$).

J. H. B. Kemperman (Rochester, N.Y.)

Citations: MR 19, 396h = J02-11; MR 22# 10976 = J02-12.

J72 IRRATIONALITY; LINEAR INDEPENDENCE OVER A FIELD

For expansions of e, π, Euler's constant etc., see inter alia **A54, M10, Z30**.

See also reviews A68-3, G05-71, J80-45, J88-2, K55-26, Q05-81, R04-2, R04-9, R04-10, R04-11, Z01-37, Z02-32, Z15-9, Z15-32, Z15-33, Z15-67.

J72-1 **(1, 134b)**

Vijayaraghavan, T. On the irrationality of a certain decimal. Proc. Indian Acad. Sci., Sect. A. **10**, 341 (1939).

The decimal .2357111317 $\cdots$, where the sequence of digits is that of the primes p in ascending order, is irrational. For if this were a recurring decimal with k digits in the recurring part, and l digits in the nonrecurring part, then for any s the number of primes with s digits would not exceed $K=sk+l$ (the author has $k+l$), and $\sum(1/p)\leqq\sum_s K/10^{s-1}<\infty$. Compare proofs in Hardy and Wright, The Theory of Numbers, p. 112.

G. Pall (Montreal, Que.).

J72-2 **(2, 149e)**

Popken, J. On Lambert's proof for the irrationality of π. Nederl. Akad. Wetensch., Proc. **43**, 712–714 (1940).

The author condenses Lambert's demonstration for the irrationality of π into a new and very short proof whose simplicity is surprising, but his initial formula must also appear very surprising to a reader who is not at all acquainted with Lambert's proof.

G. Pólya.

Referred to in J72-6, J72-9, J72-18.

J72-3 **(3, 161a)**

Goodstein, R. L. Recurring digits in irrational decimals. Math. Gaz. **25**, 273–278 (1941).

J72-4 **(3, 267f)**

Chatterjee, B. C. On irrationality. Bull. Calcutta Math. Soc. **32**, 69–71 (1940).

The author proves the following interesting result. Let a_i denote the ith digit after the decimal point in the decimal expression for a number A. If for some constant n, and any prime number p of s digits, the digits of p will occur in proper order (but not necessarily contiguous) among the first $[4.19s+n]$ digits a_i, then A is an irrational number.

P. Franklin (Cambridge, Mass.).

J72-5 **(5, 143f)**

Lototsky, A. V. Sur l'irrationnalité d'un produit infini. Rec. Math. [Mat. Sbornik] N.S. **12(54)**, 262–271 (1943). (French. Russian summary)

A method developed by A. Gelfond [Rec. Math. [Mat. Sbornik] **40**, 42–47 (1933)] is used by the author to prove that the entire function $\varphi(z)=\prod_1^\infty[1+za^{-n}]$, $a>1$ positive integer, cannot take on a complex rational value for a complex rational value of z distinct from zero and $-a^n$. Use is made of the classical power series

$$\varphi(z)=1+\sum_1^\infty\{\prod_1^n(a^k-1)\}^{-1}z^n,$$

but the main tool is interpolation at the points $\cdots, 0, a^{k-1}\alpha, -a^k, \cdots$, where $k=1, 2, 3, \cdots$ and α is a given complex rational number. It is shown that the corresponding interpolation series converges to $\varphi(z)$ everywhere. Estimates of the coefficients of this series and a discussion of their arithmetical nature shows that the coefficients must be zero from a certain point on if $\varphi(\alpha)$ is assumed complex rational and this leads to the contradiction that $\varphi(z)$ is a polynomial. There are many disturbing misprints, in particular, in all remainder terms the symbol "O" should be replaced by "o."

E. Hille (New Haven, Conn.).

J72-6 **(8, 317f)**

Popken, J. On the irrationality of π. Euclides, Groningen **17**, 217–227 (1941). (Dutch)

The author's simplified version of Lambert's proof was given in Nederl. Akad. Wetensch., Proc. **43**, 712–714 (1940);

these Rev. **2**, 149. The present paper also contains a historical introduction. *R. P. Boas, Jr.* (Providence, R. I.).

Citations: MR 2, 149e = J72-2.

J72-7 (8, 443c)

Ricci, Giovanni. Sull'irrazionalità del rapporto della circonferenza al diametro. Atti Secondo Congresso Un. Mat. Ital., Bologna, 1940, pp. 147–151. Edizioni Cremonense, Rome, 1942.

Using a well-known method of Hermite the author gives simple proofs for the following two theorems. (I) If a, b and c denote rational integers, such that $(a+bi)/c\neq 0$, then it follows that $e^{(a+bi)/c}$ does not have the form $(A+Bi)/C$, where A, B and C also are rational integers. (II) For every positive integer m the number $e^{im^{\frac{1}{2}}}$ is not real. On account of $e^{2iz}=(1+i\tan z)/(1-i\tan z)$ it follows from (I) that for every rational $p/q\neq 0$ the number $\tan p/q$ is irrational; hence π is irrational. As a corollary of (II) we obtain, on account of $e^{i\pi}=-1$, that π^2 is irrational. *J. Popken.*

J72-8 (9, 10d)

Niven, Ivan. A simple proof that π is irrational. Bull. Amer. Math. Soc. **53**, 509 (1947).

Of all known proofs for the irrationality of π the author's seems to be the simplest. It proceeds as follows. Let $\pi=a/b$ be rational, where a and b are positive integers. Put $f(x)=b^nx^n(\pi-x)^n/n!$ and $F(x)=f(x)-f^{(2)}(x)+f^{(4)}(x)-\cdots+(-1)^nf^{(2n)}(x)$, the positive integer n being specified later. Then the relation (1) $\int_0^\pi f(x)\sin x\,dx=F(0)+F(\pi)$ can be proved. From this we deduce a contradiction; for the left-hand side represents a positive number, tending to zero if n increases indefinitely, and the right-hand side is an integer, as is easily shown. *J. Popken* (Utrecht).

Referred to in J72-12, J72-13, J72-15, J72-19, J72-21, Z01-28.

J72-9 (9, 135e)

Popken, J. On the irrationality of the tangent of a rational number. Norsk Mat. Tidsskr. **26**, 66–70 (1944). (Dutch)

The author proves the following theorem: If r designates a rational number, not 0, and ϵ an arbitrarily chosen positive constant, then we have for nearly all pairs of integers q, p (q positive) the inequality $|\tan r-p/q|>q^{-2-\epsilon}$. [For the method used see Popken, Nederl. Akad. Wetensch., Proc. **43**, 712–714 (1940); these Rev. **2**, 149.] The author does not make use of continued fractions, in distinction to Lambert in his classic investigations. *J. F. Koksma.*

Citations: MR 2, 149e = J72-2.

J72-10 (9, 334a)

Cugiani, Marco. Nuova osservazione sopra un vecchio teorema di Liouville. Boll. Un. Mat. Ital. (3) **2**, 125–128 (1947).

Before the transcendentality of e was demonstrated by Hermite, it was shown by Liouville, using elementary means, that equations of the form $ae^2+be+c=0$, $ae^4+be^2+c=0$, a, b, c rational integers, were impossible. Ricci [same Boll. (2) **13**, 89–92 (1934)] returned to the elementary attack of Liouville and showed that certain polynomial relations in the Gaussian field were impossible. The author continues in this direction, demonstrating that various polynomial relations in the field $K(i\sqrt{3})$ are impossible.

R. Bellman (Stanford University, Calif.).

J72-11 (9, 500f)

Chowla, S. On series of the Lambert type which assume irrational values for rational values of the argument. Proc. Nat. Inst. Sci. India **13**, 171–173 (1947).

The author shows that the function

$$g(x)=x/(1-x)-x^3/(1-x^3)+x^5/(1-x^5)-\cdots$$

has an irrational value when x has the form $1/t$, where $t\geqq 5$ is a positive integer. To this end he considers the sum $S=\sum_{n=1}^\infty r(n)/t^n$, where $r(n)$ is the number of representations of n as a sum of two squares. Using the "decimal" scale with basis t he shows by means of well-known arithmetical theorems that S is irrational for an integer $t\geqq 5$. The result stated above then follows on account of the relation $4g(x)=\sum_{n=1}^\infty r(n)x^n$. *J. Popken* (Utrecht).

Referred to in J72-17, J72-32, J76-65.

J72-12 (10, 354g)

Koksma, J. F. On Niven's proof that π is irrational. Nieuw Arch. Wiskunde (2) **23**, 39 (1949).

A modification of an argument of Niven [Bull. Amer. Math. Soc. **53**, 509 (1947); these Rev. **9**, 10] is used to prove the irrationality of e^s for any rational number $s\neq 0$, and thus also the irrationality of $\log r$ for positive rational values $r\neq 1$. The procedure is to assume $e^s=a/b$ for $s=p/q$, where a, b, p, q are positive integers. Then for positive integral n define $P_n(x)$ to be $bp^{2n+1}x^n(1-x)^n/n!$ so that $I_n=\int_0^1 P_n(x)dx$ is positive and tends to zero as n increases indefinitely. A contradiction is obtained by repeated partial integration of I_n, which shows that I_n is an integer, because $P_n(x)$ and all its derivatives are integers at $x=0$ and $x=1$. Minor correction: the last equation in the paper should have e^s instead of e^x. *I. Niven* (Eugene, Ore.).

Citations: MR 9, 10d = J72-8.

J72-13 (10, 354h)

Rado, R. An arithmetical property of the exponential function. J. London Math. Soc. **23**, 267–271 (1948).

For $j=1, 2, \cdots, n$, let $f_j(x)$ be a set of real functions, not all identically zero, of the real variable x satisfying the differential equations $f_j'=\sum_{k=1}^n c_{jk}f_k$ with determinant $|c_{jk}|\neq 0$. It is proved that for every rational number α, with one possible exception, at least one of the numbers $f_j(\alpha)$ is irrational. Next let $P_j(x)$ be a set of n polynomials with complex coefficients, $P_j(x)$ having degree less than an integer m_j. Let w_j be a set of nonzero complex numbers such that $\prod_{j=1}^n(x-w_j)^{m_j}$ has rational coefficients, the degree of this product being say r. Define $F(x)=\sum_{j=1}^n P_j(x)\exp(w_jx)$. If $R\{F(x)\}$ is not identically zero, then for every rational number α, with one possible exception, at least one of the numbers $F^{(q)}(\alpha)$ is irrational, $q=0, 1, 2, \cdots, r-1$. This second result is a corollary of the first, which is obtained by generalizing a method of the reviewer [Bull. Amer. Math. Soc. **53**, 509 (1947); these Rev. **9**, 10]. Two special cases of the second result are the irrationality of e^b and $\tan b$ for any rational number $b\neq 0$. *I. Niven.*

Citations: MR 9, 10d = J72-8.

Referred to in J76-24.

J72-14 (10, 432b)

Arnold, B. H., and Eves, Howard. A simple proof that, for odd $p>1$, $\arccos 1/p$ and π are incommensurable. Amer. Math. Monthly **56**, 20–21 (1949).

A short proof is given for the theorem stated in the title. The authors also remark that this result is a corollary of the following well-known theorem. Let m and n ($n>2$) be two relatively prime integers; then $\cos 2\pi m/n$ is an algebraic number of degree $\varphi(n)/2$, $\varphi(n)$ denoting Euler's function. *J. Popken* (Utrecht).

J72-15 (10, 432c)

Butlewski, Z. A proof that e^m is irrational. Colloquium Math. **1**, 197–198 (1948).

Niven [Bull. Amer. Math. Soc. **53**, 509 (1947); these Rev. **9**, 10] has given a simple proof for the irrationality of π. Using Niven's method the author proves that e^m is irrational for any positive integer m. The reviewer remarks that this proof resembles Hermite's proof for the same theorem [Oeuvres, v. 3, pp. 153–155]. A still more elementary proof was also given by Hermite [Oeuvres, v. 3, pp. 127–130]. *J. Popken* (Utrecht).

Citations: MR 9, 10d = J72-8.

J72-16 **(10, 432e)**

Popken, J. On the irrationality of π. Math. Centrum Amsterdam. Rapport **ZW 1948–014**, 5 pp. (1948). (Dutch)

Given two functions

$$R_0(x)=\sum_0^\infty a_{0m}x^{-2m}$$

and

$$R_1(x)=\sum_0^\infty a_{1m}x^{-2m-1},$$

with $a_{10}\neq 0$, the author recursively defines

$$R_\nu(x)=\sum_0^\infty a_{\nu m}x^{-2m-\nu}$$

by the equations

$$R_{\nu-2}(x)=\frac{a_{\nu-2,0}}{a_{\nu-1,0}}xR_{\nu-1}(x)+R_\nu(x), \quad \nu=2, 3, \cdots.$$

Applying this algorithm to $R_0(x)=\cos x^{-1}$, $R_1(x)=\sin x^{-1}$, he proves Lambert's theorem, that $\cot y$ is irrational for rational $y\neq 0$. This implies immediately that π is irrational.

Application is also made to the problem of location of the real zeros of a polynomial $B(x)=b_0x^n+b_1x^{n-1}+\cdots$. Let D_ν be the minor determinant formed from the first ν rows and columns of the matrix

$$\begin{bmatrix} b_1 & b_0 & 0 & 0 & 0 & 0 & \cdots \\ b_3 & b_2 & b_1 & b_0 & 0 & 0 & \cdots \\ b_5 & b_4 & b_3 & b_2 & b_1 & b_0 & \cdots \\ \cdots & & \cdots & & \cdots & & \end{bmatrix},$$

where $b_\nu=0$ for $\nu>n$. Then by applying the algorithm with slight modifications to the polynomials

$$R_0(x)=b_0x^n+b_2x^{n-2}+\cdots$$

and

$$R_1(x)=b_1x^{n-1}+b_3x^{n-3}+\cdots,$$

the author deduces Hurwitz's theorem, that if $b_0>0$ and $D_\nu>0$ for $\nu=1, 2, \cdots, n$, then $B(x)$ has no positive zeros. *W. J. LeVeque* (Ann Arbor, Mich.).

J72-17 **(10, 594c)**

Erdös, P. On arithmetical properties of Lambert series. J. Indian Math. Soc. (N.S.) **12**, 63–66 (1948).

Let $f(x)=x(1-x)^{-1}+x^2(1-x^2)^{-1}+x^3(1-x^3)^{-1}+\cdots$ and $g(x)=x(1-x)^{-1}-x^3(1-x^3)^{-1}+x^5(1-x^5)^{-1}-\cdots$. Chowla has proved that $g(1/t)$ is irrational for any integer $t\geqq 5$ [Proc. Nat. Inst. Sci. India **13**, 171–173 (1047); these Rev. **9**, 500]. In this paper the author extends Chowla's result by showing that both $f(1/t)$ and $g(1/t)$ are irrational for any integer t with $|t|>1$. To this end he uses the identities $f(x)=\sum_{n=1}^\infty d(n)x^n$ and $4g(x)=\sum_{n=1}^\infty r(n)x^n$, where $d(n)$ and $r(n)$ denote, respectively, the number of divisors of n and the number of representations of n as a sum of two squares. He then shows that the "decimal" representations of $\sum_{n=1}^\infty t^{-n}d(n)$ and $\sum_{n=1}^\infty t^{-n}r(n)$ in the scale of t will contain at least h consecutive zeros, h being an arbitrary positive integer. Since these representations are not finite it follows that these numbers are irrational. *J. Popken* (Utrecht).

Citations: MR 9, 500f = J72-11.

Referred to in J72-32, J72-39, J72-62, J76-65.

J72-18 **(10, 682c)**

Popken, J. Remark on my paper "On Lambert's proof for the irrationality of π." Nederl. Akad. Wetensch., Proc. **52**, 504 = Indagationes Math. **11**, 164 (1949).

The author acknowledges that the method of his paper [same Proc. **43**, 712–714 (1940); these Rev. **2**, 149] is nearly the same as that of Hermite [Cours de la faculté des sciences . . . , 4th ed., Hermann, Paris, 1891, pp. 74–75].

Citations: MR 2, 149e = J72-2.

J72-19 **(11, 418a)**

Wachs, Sylvain. Contribution à l'étude de l'irrationalité de certains nombres. Bull. Sci. Math. (2) **73**, 77–95 (1949).

In part I of this contribution the author gives a condition for the irrationality of the sum of an infinite series of rational terms. [This theorem is well-known; see Koksma, Diophantische Approximationen, Springer, Berlin, 1936, p. 54.] In part II he starts from the integral

$$(1) \qquad \int_0^\pi x^n(\pi-x)^n \sin x dx,$$

introduced by Niven to prove the irrationality of π [Bull. Amer. Math. Soc. **53**, 509 (1947); these Rev. **9**, 10]. By considering a larger class of similar integrals the author intends to obtain more general results. He gives applications by showing in this manner the irrationality of such numbers as π^2, $\log A$ $(A\neq 1)$, e^A, where A denotes a positive integer. [The paper contains some misprints and other mistakes; the most serious one at the end of § 6, where the quantity M introduced depends on n. In view of various papers giving generalizations of Niven's method it is perhaps of interest to remark that there exists a close connection between this method and the classical proofs for the irrationality of π and π^2 of Lambert, Hermite and others. Take for instance the integral used by Jordan [Cours d'Analyse, v. 2, 3rd ed., Gauthier-Villars, Paris, 1913, p. 106] for this purpose, i.e.,

$$\int_{-1}^1 (1-t^2)^n \cos zt dt.$$

Putting $z=\pi/2$, $1+t=2x/\pi$, we obtain, apart from the factor $(2/\pi)^{2n+1}$, the integral (1).] *J. Popken.*

Citations: MR 9, 10d = J72-8.

Referred to in J72-21.

J72-20 **(12, 243f)**

Skolem, Th. An arithmetical property of the function

$$\sum_{n=0}^\infty \frac{x^n}{\prod_{i=0}^\lambda p_i^{\kappa_i(n)}},$$

where the p_i are natural primes and the $\kappa_i(n)$ polynomials with integral coefficients. Norske Vid. Selsk. Forh., Trondheim **22**, no. 39, 183–187 (1950).

Mit einem kurzen elementaren Beweis wird gezeigt, dass die Funktion $f(x)=\sum_{n=0}^\infty\{x^n/\prod_{i=0}^\lambda p_i^{\kappa_i(n)}\}$, wobei $p_1, \cdots, p_\lambda$ Primzahlen und $\kappa_i(n)\equiv a_in^2+b_in$ eine in n quadratische Form mit nichtnegativen, ganzrationalen a_i $(i=1, \cdots, \lambda)$ und positivem a_1, sowie ganzrationalen b_i sei, für $x=0$ und $x=r=\pm 1, \cdots, \pm l$, wobei r die zu $P=\prod_{i=1}^\lambda p_i^{2a_i}$ relativ primen Reste durchlaufe, im Körper der rationalen Zahlen linear unabhängige Werte annimmt; es ist also

$$c_0 f(0)+\sum_{r=-l}^{+l} c_r f(r)=0$$

mit rationalen $c_0, c_{-l}, \cdots, c_l$ nur, wenn $c_0^2+\sum_{r=-l}^{l} c_r^2=0$. Damit wird in der Beantwortung der Frage nach den linear unabhängigen Werten $f(0), f(-l), \cdots, f(l)$ mit rationalen Koeffizienten eine Lücke geschlossen, da diese Tatsache bereits gezeigt ist [Skolem, in einer unveröffentlichten Arbeit] für jeden Grad der Polynome $\kappa_i(n)$, der >2 ist, und da diese Funktionswerte linear abhängig sind im Körper der rationalen Zahlen, falls alle Polynome $\kappa_i(n)$ vom ersten Grade sind. *T. Schneider* (Göttingen).

J72-21 (12, 318k)

Iwamoto, Yosikazu. A proof that π^2 is irrational. J. Osaka Inst. Sci. Tech. Part I. 1, 147–148 (1949).

The argument follows the reviewer's formulation [Bull. Amer. Math. Soc. **53**, 509 (1947); these Rev. **9**, 10] of the classical proof of the irrationality of π. A similar extension to π^2 has been done by Wachs [Bull. Sci. Math. (2) **73**, 77–95 (1949); these Rev. **11**, 418]. *I. Niven.*

Citations: MR 9, 10d = J72-8; MR 11, 418a = J72-19.

J72-22 (14, 541e)

Spiegel, M. R. On a class of irrational numbers. Amer. Math. Monthly **60**, 27–28 (1953).

Let $a_1, a_2, \cdots$ be an infinite sequence of integers, infinitely many of which are different from 0. Put

$$\phi=\sum_{n=1}^{\infty}\frac{a_n}{r^n(n!)^b}.$$

Assume that for all sufficiently large n, $|a_n|<Cn^\alpha$ where C and α are constants, $\alpha<b$. The author proves that ϕ is irrational. *P. Erdös* (Los Angeles, Calif.).

J72-23 (14, 851e)

Skolem, Th. Some theorems on irrationality and linear independence. Den 11te Skandinaviske Matematikerkongress, Trondheim, 1949, pp. 77–98. Johan Grundt Tanums Forlag, Oslo, 1952. 27.50 kr.

Of the many results obtained by the author we cite the following. (1) Let $f(x)=\sum_{n=0}^{\infty}[nx]/n!$ and let $x_1, x_2, \cdots, x_l$ be different positive numbers, such that 1 is not linearly dependent on them. Then the numbers $1, f(x_1), \cdots, f(x_l)$ are linearly independent. (Here "linear dependence" is always meant with respect to the field of rational numbers.) (2) Let $g, h_1, h_2, \cdots, h_m$ be integers >1 and let $h_1, h_2, \cdots, h_m$ be different. Then the m numbers $\sum_{n=0}^{\infty}g^{-h_r^n}$ $(r=1, 2, \cdots, m)$ are linearly independent. (3) Let g and h denote integers >1 and let $f(0), f(1), f(2), \cdots$ be integers, such that $0<f(0)\leqq f(1)\leqq f(2)\leqq\cdots$ and such that

$$\limsup \frac{1}{n}\log\frac{\log f(n)}{\log g}<\log h.$$

Then the number $\sum_{n=0}^{\infty}f(n)g^{-h^n}$ is transcendental.

The methods used are similar to the "Reihenmethode" in Koksma, "Diophantische Approximationen" [Springer, Berlin, 1936, p. 54].

By another method the author derives: (4) Let h and k be positive integers, $h>1$, let

$$f(x)=\sum_{n=0}^{\infty}h^{-\binom{n+1}{2}}k^{-n}x^n.$$

Moreover, let $k_1, k_2, \cdots, k_l$ be rational numbers $\neq 0$, such that none of the quotients k_i/k_j $(i\neq j)$ is equal to an integral power of h. Then $f(0)$ and the numbers $f^{(s)}(k_r)$ $(r=1, 2, \cdots, l;\ s=0, 1, \cdots)$ are linearly independent. (The reviewer remarks that this last result is closely related to a theorem of Tschakaloff [Math. Ann. **84**, 100–114 (1921)].) *J. Popken* (Utrecht).

J72-24 (14, 957e)

Churchhouse, R. F. A criterion for irrationality. Canadian J. Math. **5**, 253–260 (1953).

Using a criterion of Legendre for the irrationality of numbers represented by continued fractions the author proves that the continued fraction $K(x^{\psi(n)}/1)$ assumes irrational values for $x=r/s$, where r and s are positive integers with $(r, s)=1$ and such that $r<s^\gamma$. Here $\psi(n)$ is a strictly increasing positive integral-valued function of n and

$$\gamma=\liminf\frac{1}{\psi(n)}\sum_{r=0}^{n-1}(-1)^r\psi(n-r).$$

The result is applied to a number of special cases. *W. J. Thron* (St. Louis, Mo.).

J72-25 (15, 107d)

Pennisi, L. L. Elementary proof that e is irrational. Amer. Math. Monthly **60**, 474 (1953).

J72-26 (15, 781a)

Oppenheim, A. Criteria for irrationality of certain classes of numbers. Amer. Math. Monthly **61**, 235–241 (1954).

The author considers infinite series of the form

$$x=a_0+\frac{a_1}{b_1}+\frac{a_2}{b_1b_2}+\frac{a_3}{b_1b_2b_3}+\cdots,$$

where the a_i and the b_i are integers. The author obtains various criteria for the irrationality of x. Here are some of his theorems. Let $b_i\geqq 2$, $0\leqq a_i\leqq b_i-1$; then x is irrational if there exists an irrational α and an infinite sequence i_n so that $a_{i_n}/b_{i_n}\to\alpha$. The number x is irrational if we assume $b_i\geqq 1$; further, for every g there is a b_n satisfying $g|b_n$, and finally there is an infinite sequence i_n so that $0<x_{i_n}<1$ (mod 1) where $x_k=a_k/b_k+a_{k+1}/b_k\cdot b_{k+1}+\cdots$. *P. Erdös.*

Referred to in J72-29, J72-43, K55-26.

J72-27 (16, 224f)

Breusch, Robert. A proof of the irrationality of π. Amer. Math. Monthly **61**, 631–632 (1954).

J72-28 (16, 452e)

Beatty, S. Elementary proof that e is not quadratically algebraic. Amer. Math. Monthly **62**, 32–33 (1955).

J72-29 (16, 908g)

Diananda, P. H., and Oppenheim, A. Criteria for irrationality of certain classes of numbers. II. Amer. Math. Monthly **62**, 222–225 (1955).

Continuation of a paper by the second author [same Monthly **61**, 235–241 (1954); MR **15**, 781]. Two new criteria for the irrationality of numbers of the form

$$a_0+\sum_{i=1}^{\infty}a_i/b_1b_2\cdots b_i,$$

where the a_i and the b_i are integers, are given. *J. Popken* (Utrecht).

Citations: MR 15, 781a = J72-26.

Referred to in J72-43, K55-26.

J72-30 **(16, 1004b)**

Schneider, Theodor. Über die Irrationalität von π. S.-B. Math.-Nat. Kl. Bayer. Akad. Wiss. **1954**, 99–101 (1955).

By use of a cleverly designed interpolation series for the exponential function, the author gives a brief proof that, unless $b=0$, e^b and b cannot both belong to the field $R(i)$, the rational numbers with i adjoined. The irrationality of π is seen to be a consequence by taking $b=i\pi$. *I. Niven.*

J72-31 **(17, 466b)**

Turowicz, A. Sur une propriété des nombres irrationnels. Ann. Polon. Math. **2** (1955), 103–105.

It is shown that player A can always force the series $\sum_1^\infty a_n$ to converge to an irrational number, if he and B alternately choose the successive terms in such a way that $\{a_n\}$ is a decreasing sequence of positive real numbers. The proof depends on the Cantor expansion $\sum_1^\infty c_n/n!$ (c_n integers, $0\leqq c_n<n$) of a real number.

W. J. LeVeque (Ann Arbor, Mich.).

Referred to in J72-34.

J72-32 **(19, 252e)**

Erdös, P. On the irrationality of certain series. Nederl. Akad. Wetensch. Proc. Ser. A. **60**=Indag. Math. **19** (1957), 212–219.

Extending previous results of the reviewer [Proc. Nat. Inst. Sci. India **13** (1947), 171–173; MR **9**, 500] the author proved [J. Indian Math. Soc. **12** (1948), 63–66; MR **10**, 594] that, for every integer $t>1$, the series $\sum_1^\infty d(n)/t^n$ and $\sum_1^\infty r(n)/t^n$ are irrational, where $d(n)$ denotes the number of divisors of n and $r(n)$ the number of solutions of $n=x^2+y^2$. He remarked that he could not prove the irrationality of the series $\sum_1^\infty \sigma(n)/t^n$ where $\sigma(n)$ is the sum of the divisors of n. In this paper he proves that the series $\sum_1^\infty 1/t^{\sigma(n)}$ is irrational. Among many interesting problems raised by the author we mention: "I do not know if a series $\sum_1^\infty 1/t^{n_k}$ with $\limsup (n_k/k)=\infty$ can be an algebraic number." Here $n_1<n_2<n_3<\cdots$ is a sequence of increasing positive integers. *S. Chowla.*

Citations: MR 9, 500f = J72-11; MR 10, 594c = J72-17.

J72-33 **(19, 1159d)**

Yasinovyĭ, È. A. Irrationality of certain values of trigonometric functions. Mat. v Škole 1958, no. 3, 3–6. (Russian)

Simple elementary proofs of the irrationality of certain numbers; in particular, $\cos \pi/n$; $n>3$ and $\sin \pi/n$; $n\neq 1, 2, 6$.

J72-34 **(20# 5186)**

Šalát, Tibor. Zu einer Eigenschaft der Irrationalzahlen. Mat.-Fyz. Časopis. Slovensk. Akad. Vied **7** (1957), 128–137. (Slovak. Russian and German summaries)

Two players alternately choose positive real numbers $a_1, a_3, a_5, \cdots$ and $a_2, a_4, \cdots$, in such a way that for every n, $a_{n+1}<a_n$. It is shown that either player can force the following outcome: $\prod_1^\infty (1+a_n)$ converges to an irrational number. This is in analogy with a result due to A. Turowicz [Ann. Polon. Math. **2** (1955), 103–105; MR **17**, 466] concerning $\sum_1^\infty a_n$. *W. J. LeVeque* (Göttingen)

Citations: MR 17, 466b = J72-31.

J72-35 **(20# 5187)**

Erdös, Paul. Sur certaines séries à valeur irrationnelle. Enseignement Math. (2) **4** (1958), 93–100.

A. Oppenheim asked if

$$\sum_{n=1}^\infty \frac{p_n{}^k}{n!} \quad (k=1, 2, 3, \cdots)$$

is irrational. Here p_n is the nth prime.

The author proves this for $k=1$, and states that the proof is more complicated when $k>1$. When $k=1$ he uses $p_{n+1}-p_n=o(p_n)$. He also raises the problem: Is $\sum_1^\infty p_n/2^n$ irrational? *S. Chowla* (Boulder, Colo.)

J72-36 **(23# A1605)**

Inkeri, K.

The irrationality of π^2.

Nordisk. Mat. Tidskr. **8** (1960), 11–16, 63.

There exists a very simple method of Hermite [*Cours de la Faculté des Sciences*, 4th ed., Hermann, Paris, 1891, pp. 74–75] to prove the irrationality of π. This proof is fairly unknown because it is not reproduced in Hermite's *Oeuvres* [Gauthier-Villars, Paris, 1905–1917]. The author of the paper reviewed here uses this method to give a simple proof for the irrationality of π^2. He starts from the identity ($x\neq 0$)

$$P_h(x^{-1}) \sin x + Q_h(x^{-1}) \cos x =$$
$$2^h \sum_{n=0}^\infty (-1)^n(n+1)(n+2)\cdots(n+h)x^{2n}/(2n+2h+1)!$$

($h=0, 1, \cdots$), where $P_h(x)$ and $Q_h(x)$ are polynomials with integral coefficients of degree at most $2h+1$. The first polynomials are odd functions, the second ones are even. Denoting the right-hand side of the identity by $R_h(x)$, one obtains for $x=\pi$, clearly, $-Q_h(\pi^{-1})=R_h(\pi)$. If we now assume $\pi^2=a/b$, where a and b are positive integers, then it follows that $a^hR_h(\pi)$ must be integral for all h. On the other hand one obtains easily, from the series, $0<a^hR_h(\pi)<(2a)^h/h!$ for $h\geqq 1$. This gives a contradiction for sufficiently large h.

In an alternative approach the author replaces the series for $R_h(x)$ given above by an integral. The reviewer remarks that the resulting formula, too, has been derived first by Hermite [*Oeuvres, III*, p. 146] exactly for the purpose of proving the irrationality of π^2.

J. Popken (Amsterdam)

J72-37 **(23# A2378)**

Sudan, Gabriel

Sur l'irrationnalité de certaines expressions.

Rend. Mat. e Appl. (5) **19** (1960), 416–429.

Various conditions on the integers a_i and b_i are known, under which it can be asserted that the value of a continued fraction

$$\frac{a_1|}{|b_1}+\frac{a_2|}{|b_2}+\frac{a_3|}{|b_3}+\cdots$$

is irrational. (For example, Stern showed in 1832 that it suffices to have $b_i\geqq a_i>0$.) Many such theorems have been collected and proved in a uniform way in the present paper. *W. J. LeVeque* (Ann Arbor, Mich.)

J72-38 **(25# 3011)**

Briggs, W. E.

The irrationality of γ or of sets of similar constants.

Norske Vid. Selsk. Forh. (Trondheim) **34** (1961), 25–28.

Let k and r be positive integers with $0<r\leqq k$, let $x>0$. Further, let $\sum^* n^{-1}$ denote the sum of all terms of the form n^{-1} if n runs through all positive integers $\equiv r \pmod k$ and $\leqq x$. Then a constant $\delta_{r,k}$ may be introduced by

$$\sum{}^* n^{-1} = \log x/k+\delta_{r,k}+o(1) \text{ as } x\to\infty.$$

Similarly, a constant λ_k may be defined by the summation of n^{-1} if n runs through the positive integers relatively prime to k and $\leqq x$.

The author investigates relations existing between the Euler constant γ and the constant $\delta_{r,k}$ and γ_k. He further gives some theorems from which the following one is a sample: "If γ is rational, then $\delta_{a,a}$ is irrational for $a>1$".
J. Popken (Berkeley, Calif.)

J72-39 (27# 105)

Golomb, Solomon W.
On the sum of the reciprocals of the Fermat numbers and related irrationalities.
Canad. J. Math. **15** (1963), 475–478.

It is shown that $F(z)=\sum_{n=0}^{\infty} z^{2^n}/(1+z^{2^n})$ and $G(z)=\sum_{n=0}^{\infty} z^{2^n}/(1-z^{2^n})$ take irrational values whenever $z=1/t$, $t=2, 3, 4, \cdots$. In particular $F(1/2)$ is irrational: the sum of the reciprocals of the Fermat numbers $2^{2^n}+1$ is irrational.

The kernel of the proof is that in the expansion of $F(1/t)$ as a number written to the base t there occur arbitrarily long finite runs of zeros. As the author remarks, this approach is similar to P. Erdős's well-known method of proving that the Lambert series $\sum_{n=0}^{\infty} z^n/(1-z^n)$ takes irrational values for $z=1/t$, $t=1, 2, \cdots$ [J. Indian Math. Soc. (N.S.) **12** (1948), 63–66; MR **10**, 594]. The result for $G(z)$ follows from the identity $F(z)+G(z)=2z/(1-z)$.
J. Popken (Berkeley, Calif.)

Citations: MR 10, 594c = J72-17.
Referred to in J76-65.

J72-40 (27# 1419)

Haradze, A. K.
Application of Hermite's method to the proof of the irrationality of the roots of certain special functions. (Russian. Georgian summary)
Tbiliss. Gos. Univ. Trudy Ser. Meh.-Mat. Nauk **84** (1962), 45–52.

Hermite's method of proving the irrationality of π rests on the evaluation of the integral $\int_0^1 (1-t^2)^n \cos \frac{1}{2}\pi t\, dt$ and then letting n go to infinity. The author applies this method to prove the irrationality of the smallest positive zero of certain special functions connected with the Bessel functions. *S. Knapowski* (New Orleans, La.)

J72-41 (29# 2222)

Froda, Alexandre
Critères paramétriques d'irrationalité.
Math. Scand. **12** (1963), 199–208.

Let $\alpha=\lim(y_r/x_r)$, where x_r and y_r are positive integers, be the limit of an increasing sequence of rational numbers; let $q_1, q_2, q_3, \cdots$ be positive numbers, and let $p_{r+1}=q_{r+1}/q_r<x_{r+1}/x_r$ for all r, $\lim(x_r/q_r)=\infty$. Put

$$\xi_r = 1-\frac{x_r}{q_r}+\left[\frac{x_r}{q_r}\right], \quad \eta_r = \frac{y_r}{q_r}-\left[\frac{x_r}{q_r}\right] \quad (r = 1, 2, 3, \cdots).$$

Assume that for sufficiently large r

$$\frac{y_{r+1}-p_{r+1}y_r}{x_{r+1}-p_{r+1}x_r} > \frac{y_{r+2}-p_{r+2}y_{r+1}}{x_{r+2}-p_{r+2}x_{r+1}}$$

and that there exists an increasing sequence of suffixes $r_1, r_2, r_3, \cdots$ such that $\xi_{r_1}\geqq\xi_{r_2}\geqq\xi_{r_3}\geqq\cdots$, $\eta_{r_1}\geqq\eta_{r_2}\geqq\eta_{r_3}\geqq\cdots$. Then α is irrational. A similar criterion holds for the limits of decreasing sequences. When $q_r=1$ for all r, a result due to V. Brun is obtained [Arch. Math. Naturvid. **31** (1910), no. 3]. *K. Mahler* (Canberra)

Referred to in J72-42, J72-45, J72-46, J72-49, J72-52.

J72-42 (29# 3431)

Froda, Alexandre
Sur l'irrationalité du nombre 2e.
Atti Accad. Naz. Lincei Rend. Cl. Sci. Fis. Mat. Natur. (8) **35** (1963), 472–478.

Put

$$\varepsilon_r = \sum_{s=1}^{r}\frac{1}{s!}, \quad x_r = 1, \quad y_r = 2^{\varepsilon_r}, \quad q_r = \frac{2^{\varepsilon_r}}{r}, \quad p_r = \frac{q_{r+1}}{q_r},$$

$$\alpha_r = \frac{y_r}{x_r}, \quad \alpha = \lim\alpha_r = 2^{e-1}, \quad \xi_v = 1-\left[\frac{x_v}{q_v}\right]+\frac{x_v}{q_v},$$

$$\eta_v = \frac{y_v}{q_v} - \left[\frac{y_v}{q_v}\right].$$

Further, assume that $\alpha=u/v$, where $u\geqq 1$, $v\geqq 1$, $(u, v)=1$, is rational. The following four properties are shown to hold.

$$\frac{y_{v+1}}{q_{v+1}}-\frac{y_v}{q_v} \geqq \frac{y_{v+2}}{q_{v+2}}-\frac{y_{v+1}}{q_{v+1}};$$

$$0 < \frac{x_{v+1}}{q_{v+1}}-\frac{x_v}{q_v} < \frac{x_{v+2}}{q_{v+2}}-\frac{x_{v+1}}{q_{v+1}} \quad \text{for sufficiently large } \nu;$$

$\eta_v=0$ for all ν; $\xi_{r_v}\geqq\xi_{r_{v+1}}$ for an infinite sequence of suffixes r_v. In a previous paper [Math. Scand. **12** (1963), 199–208; MR **29** #2222] it was shown that these properties imply the irrationality of α, hence of 2^e, if the sequence $\{\alpha_v\}$ consists of rational elements. This is not so in the present case, but, according to the author, this condition is superfluous. *K. Mahler* (Canberra)

Citations: MR 29# 2222 = J72-41.
Referred to in J72-45, J72-46, J72-49, J72-50.

J72-43 (30# 3867)

Šalát, T.
On irrationality criteria for real numbers. (Slovak. Russian and German summaries)
Acta Fac. Natur. Univ. Comenian. **7**, 649–662 (1963).

The author derives a pair of irrationality criteria for numbers of the form (1) $\sum_{i=1}^{\infty} a_i/(b_1b_2\cdots b_i)$, where the a's are integers and the b's natural numbers. Writing $\{z\}$ for $z-[z]$, $x_n=\sum_{i=n}^{\infty} a_i/(b_nb_{n+1}\cdots b_i)$, $y_n=\{a_n/b_n\}+\{x_{n+1}/b_n\}$, it is shown that a number representable by a convergent (Cantor) series of type (1) (and every positive real number is so representable provided all the b's are >1 and $0<a_i<b_i-1$ for each i) is irrational if and only if the set of elements in one or the other of the sequences $\{x_n\}$ $(n=1, 2, \cdots)$, $\{y_n\}$ $(n=1, 2, \cdots)$ is infinite. From this result, the author derives a number of known irrationality criteria in terms of the sequence a_i/b_i $(i=1, 2, 3, \cdots)$ [A. Oppenheim, Amer. Math. Monthly **61** (1954), 235–241; MR **15**, 781; P. H. Diananda and A. Oppenheim, ibid. **62** (1955), 222–225; MR **16**, 908], as well as the following: Let a, b be integers with $a>1$, $b>a/(a-1)$, and let $\{\varepsilon_i\}$ $(i=1, 2, \cdots)$ be an infinite sequence of integers satisfying $0<\varepsilon_i\leqq i$, $\varepsilon_i=i$ infinitely often; then

$$x = \sum_{i=1}^{\infty}\frac{\varepsilon_i}{(a+b)(2a+b)\cdots(ia+b)}$$

is irrational. *H. Halberstam* (Nottingham)

Citations: MR 15, 781a = J72-26; MR 16, 908g = J72-29.

J72-44 (31# 124)

Erdős, P.; Straus, E. G.
On the irrationality of certain Ahmes series.
J. Indian Math. Soc. (N.S.) **27** (1963), 129–133 (1964).

Let $\{n_k\}$ be an increasing sequence of positive integers; let $\alpha=\sum_{k=1}^{\infty} 1/n_k$; and let N_k and N_k^* denote the least common multiple and the product of $n_1, n_2, \cdots, n_k$, respectively. The authors prove the following three theorems. (I) Let lim sup $n_k^2/n_{k+1} \leqq 1$, lim sup $N_k/n_{k+1} < \infty$; then α is rational if and only if $n_{k+1}=n_k^2-n_k+1$ for $k \geqq k_0$. (II) Let lim sup $n_k^2/n_{k+1} < \infty$, lim sup $N_k^*/n_{k+1} < \infty$. If α is rational, n_k^2/n_{k+1} has only finitely many limiting values, and the smallest is $\leqq 1$. (III) Let

$$\limsup \frac{n_k^2}{n_{k+1}} \leqq 1, \quad \limsup \frac{N_k}{n_{k+1}}\left(\frac{n_{k+1}^2}{n_{k+2}}-1\right) \leqq 0;$$

then α is rational if and only if the same condition as in (I) is satisfied. *K. Mahler* (Canberra)

Referred to in J72-54, J72-64, J76-65.

J72-45 (31# 2208)

Froda, Alexandre

Sur l'irrationalité des nombres réels, définis comme limites.

Rev. Roumaine Math. Pures Appl. **9** (1964), 565–575.

The author establishes a further pair of tests for irrationality for $\alpha=\lim \alpha_r$ or $\alpha=\lim \alpha_r^*$, where the α_r form an increasing, and the α_r^* a decreasing, sequence of real numbers, respectively. See his earlier papers [Math. Scand. **12** (1963), 199–208; MR **29** #2222; Atti Accad. Naz. Lincei Rend. Cl. Sci. Fis. Mat. Natur. (8) **35** (1963), 472–478, statement 6; MR **29** #3431]. *K. Mahler* (Canberra)

Citations: MR 29# 2222 = J72-41; MR 29# 3431 = J72-42.

J72-46 (31# 2209)

Froda, Alexandre

Sur des familles de critères d'irrationalité.

Math. Z. **89** (1965), 126–136.

Let $\{\alpha_r\}$ be an increasing sequence of positive numbers, $\alpha=\lim \alpha_r$. Let $\alpha_r=y_r/x_r$, where $x_r>0$ and $y_r>0$. Finally let $\{q_r\}$ be a sequence of positive numbers such that $q_1=1$, $x_r/q_r \to \infty$, and that $m_r=x_r/q_r$ is an integer that increases with r. Put $p_{r+1}=q_{r+1}/q_r$, and assume that always $q_{r+1}/q_r<x_{r+1}/x_r$, and write

$$\eta_r=\frac{y_r}{q_r}-\left[\frac{y_r}{q_r}\right].$$

Assume that (A) there is an increasing sequence S_0 of integers r for which the products $m_r(\alpha-\alpha_r)$ are strictly increasing, and (B) for the r in S_0, the numbers η_r are decreasing. Then α is irrational. A single condition (A) or (B) does not imply the irrationality of α. An analogous result holds when $\alpha=\lim \alpha_r^*$ is the limit of a decreasing sequence $\{\alpha_r^*\}$. (See the earlier papers [Math. Scand. **12** (1963), 199–208; MR **29** #2222; Atti Accad. Naz. Lincei Rend. Cl. Sci. Fiz. Mat. Natur. (8) **35** (1963), 472–478; MR **29** #3431; and #2208 above].) *K. Mahler* (Canberra)

Citations: MR 29# 2222 = J72-41; MR 29# 3431 = J72-42.

J72-47 (32# 1158)

Straus, E. G.

Rational dependence in finite sets of numbers.

Acta Arith. **11** (1965), 203–204.

Mikusiński and Schinzel [same Acta **9** (1964), 91–95; MR **29** #1205] proved that, in a finite set of points on the real line such that every distance except the maximal one occurs more than once, all distances are commensurable. In the present article there is given a generalization of this theorem. Let $x_1, x_2, \cdots, x_n$ be real numbers and let m be the dimension of the vector space spanned by $\{x_i-x_j|i,j=1,\cdots,n\}$ over the rationals. Let m' be the dimension of the rational vector space spanned only by those x_i-x_j for which $x_i-x_j \neq x_k-x_l$ whenever $(i,j) \neq (k,l)$. Then $m'=m$. *E. Inaba* (Tokyo)

Citations: MR 29# 1205 = C10-32.

Referred to in C10-45.

J72-48 (32# 1170)

Inkeri, K.

The irrationality of values of some transcendental functions. (Finnish. English summary)

Arkhimedes **1965**, *no.* 1, 15–21.

Author's summary: "By using an appropriate modification of the methods of Hermite [*Cours de la Faculté des Sciences*, quatrième édition, Hermann, Paris, 1891] and Niven [*Irrational numbers*, The Mathematical Association of America, New York, 1956; MR **18**, 195] it is proved in an elementary way that z^{-1} tg z and cos z are irrational, if z^2 is rational different from zero. This theorem contains, as special cases, the results of Niven [loc. cit., pp. 16–23] concerning trigonometric, hyperbolic, and exponential functions."

Citations: MR 18, 195c = Z01-37.

J72-49 (32# 5599)

Froda, Alexandre

La constante d'Euler est irrationnelle. (Italian summary)

Atti Accad. Naz. Lincei Rend. Cl. Sci. Fis. Mat. Natur. (8) **38** (1965), 338–344.

This proof that Euler's constant is irrational depends on the unsupported assertion that an irrationality criterion of the author [Math. Scand. **12** (1963), 199–208; MR **29** #2222] continues to hold when certain numbers, which were required to be rational, are permitted to take irrational values. As he has already resolved another long-unsolved problem by a similar technique [Atti Accad. Naz. Lincei Rend. Cl. Sci. Fis. Mat. Natur. (8) **35** (1963), 472–478; MR **29** #3431; cf. the last sentence of the review], one would like to see a detailed proof of the extended criterion. *J. W. S. Cassels* (Cambridge, England)

A proof of the extended criterion has been published by the author in C. R. Acad. Sci. Paris **261** (1965), 3012–3015 [MR **33** #5565].

Citations: MR 29# 2222 = J72-41; MR 29# 3431 = J72-42.

Referred to in J72-50.

J72-50 (33# 5565)

Froda, Alexandre

Nouveaux critères paramétriques d'irrationalité.

C. R. Acad. Sci. Paris **261** (1965), 3012–3015.

Suppose that $\alpha>0$ is the limit of an increasing [or decreasing] sequence $\{\alpha_n\}$. The author proves the irrationality of α under a number of conditions imposed on the α_n. His criterion is too complicated to be quoted here, but it is deduced in an elementary way from the classical criterion given, for example, in Koksma's *Diophantische Approximationen* [p. 53, Springer, Berlin, 1936]. The author claims that his criterion is capable of proving the irrationality of 2^e and Euler's constant, and refers to his papers in Atti Accad. Naz. Lincei Rend. Cl. Sci. Fis. Mat. Natur. (8) **35** (1963), 472–478 [MR **29** #3431] and ibid. (8) **38** (1965), 338–344 [MR **32** #5599]. *S. Knapowski* (Coral Gables, Fla.)

Citations: MR 29# 3431 = J72-42; MR 32# 5599 = J72-49.

J72-51 (34# 7455)

Estermann, T.

A theorem implying the irrationality of π^2.

J. London Math. Soc. **41** (1966), 415–416.

The paper contains an elementary proof of the following theorem: There is no number z such that z^2 and $z \coth z$ are rational. The author points out several consequences of this theorem and in particular the corollary that π^2 cannot be a rational number (take $z = \pi i/2$). *M. Cugiani* (Milan)

J72-52 **(36# 1395)**

Froda, Alexandre

Extension effective de la condition d'irrationalité de Viggo Brun.

Rev. Roumaine Math. Pures Appl. **10** (1965), 923–929.

Viggo Brun established a sufficient condition C_0 for a real number $\alpha > 0$ to be irrational [Arch. Math. Naturvid. **31** (1910), no. 3].

The author proved in a previous paper [Math. Scand. **12** (1963), 199–208; MR **29** #2222] that the condition C_0 can be substituted by another condition C_1.

In this paper he proves that the condition C_1 is an effective extension of the condition C_0. In particular, he shows that by condition C_1 the irrationality of e can be proved by a process that cannot be applied when we assume condition C_0 only. *M. Cugiani* (Milan)

Citations: MR 29# 2222 = J72-41.

J72-53 **(36# 5087)**

Sprindžuk, V. G.

Irrationality of the values of certain transcendental functions. (Russian)

Izv. Akad. Nauk SSSR Ser. Mat. **32** (1968), 93–107.

The author considers a class E^* of functions $\varphi(z) = \sum_{\nu=1}^{\infty} c_\nu z^\nu/\nu!$ satisfying the following conditions: all c_ν lie in the same algebraic number field K; all the conjugates of c_ν with respect to K are $O((\nu!)^{\alpha+\varepsilon})$ for a fixed $\alpha < 1$ and arbitrary $\varepsilon > 0$; there exist two sequences β_n and q_{nm} such that $0 \leq \beta_n = o(\sqrt{n})$, $q_{nm} < c\ (n, \varepsilon)\,(m!)^{\beta_n+\varepsilon}$, q_{nm} are positive integers and $q_{nm}c_{\nu_1}c_{\nu_2}\cdots c_{\nu_n}$ algebraic integers whenever $\nu_1+\nu_2+\cdots+\nu_n \leq m$. The class E^* contains E-functions of C. L. Siegel [*Transcendental numbers*. Ann. of Math. Studies No. 16. Princeton Univ. Press, Princeton, N.J., 1949; MR **11**, 330; German translation. Bibliographisches Institut, Mannheim, 1967; MR **35** #133] and also hypergeometric functions $\sum_{n=0}^{\infty} ([\alpha_1, n]\cdots[\alpha_l, n]\ [\beta_1, n]\cdots[\beta_m, n])z^{nt}$ ($[\alpha, n] = \alpha(\alpha+1)\cdots(\alpha+n-1)$) for algebraic values of α_i, $\beta_j \neq 0, -1, -2, \cdots$. The following theorem is proved. If a transcendental E^* function $\varphi(z)$ satisfies a differential equation $p(z)y' + q(z)y + r(z) = 0$, where p, q, r are polynomials over K, then for any positive integers d_0 and t there exists only a finite number $N(d_0, t)$ of algebraic z of degree $\leq t$ such that $\varphi(z)$ is algebraic of degree $\leq d_0$. An estimation for $N(t, d_0)$ and two specific applications are given.

A. Schinzel (Warsaw)

Citations: MR 11, 330c = J76-18; MR 35# 133 = J76-19.

J72-54 **(37# 164)**

Oppenheim, A. [Oppenheim, Alexander]

The irrationality of certain infinite products.

J. London Math. Soc. **43** (1968), 115–118.

The author proves the following theorem which generalises one by P. Erdős and E. G. Strauss [J. Indian Math. Soc. (N.S.) **27** (1963), 129–133; MR **31** #124]. Let $\{a_i\}$ be a sequence of positive integers such that $\limsup a_i^2/a_{i+1} < \infty$, $\limsup a_i = \infty$. Then $\prod_i \{1 + (1/a_i)\}$ converges. The value of this product is rational if and only if there exists a second sequence $\{c_i\}$ of positive integers such that $c_i a_{i+1} - c_{i+1} = (a_i+1)(c_{i-1}a_i - c_i) > 0$ for $i \geq i_0$; $0 < c_i < a_i$; $c_i = o(a_i)$; $\limsup c_i/c_{i-1} = \limsup a_i^2/a_{i-1}$. A further theorem for the more general products $\prod_i \{1 \mp (h_i/a_i)\}$, where the h_i are bounded positive integers, is stated.

K. Mahler (Canberra)

Citations: MR 31# 124 = J72-44.

Referred to in J72-64.

J72-55 **(38# 1225)**

Abian, Alexander

An example of a nonmeasurable set.

Boll. Un. Mat. Ital. (4) **1** (1968), 366–368.

A non-empty set S of real numbers is said to be linearly independent if no finite linear combination of distinct elements of S with rational coefficients, not all zero, is equal to zero. In this note the author proves the existence of a linearly independent set of real numbers which is Lebesgue nonmeasurable and which has at least one point in common with every nondenumerable closed subset of the real line. The construction of the set requires Zorn's lemma and the well-ordering theorem.

B. K. Lahiri (Kalyani)

J72-56 **(38# 3234)**

Schwarz, Wolfgang

Remarks on the irrationality and transcendence of certain series.

Math. Scand. **20** (1967), 269–274.

Let k, $b < t$ be positive integers. Let $G_k(bt^{-1}) = \sum_{n=0}^{\infty} b^{k^n}(t^{k^n} - b^{k^n})^{-1}$.

Using the standard proof of the irrationality of e, the irrationality of $G_k(bt^{-1})$ for $k \geq 2$, $t \geq 2$, and $0 < b < t^{1-1/k}$ is proved. Also, for $t \geq 2$, $G_2(t^{-1})$ is proved not to be a quadratic irrationality. Further, for $k > 2$, $t \geq 2$, and $0 < b < t^{1-5/2k}$, the number $G_k(bt^{-1})$ is proved to be transcendental by using a version of Roth's theorem. Another result of this nature and P-adic generalizations are also given.

W. W. Adams (Berkeley, Calif.)

Referred to in J76-65.

J72-57 **(39# 140)**

Bundschuh, Peter

Arithmetische Untersuchungen unendlicher Produkte.

Invent. Math. **6** (1969), 275–295.

Let K be an imaginary quadratic field, and let q be a fixed integer of K with norm > 1. The following two theorems are the principal results of this paper. Theorem 1: Let $f(z)$ be an integral function that satisfies the functional equation $f(z) = (1 - (z/q))f(z/q)$ with $f(0) = 1$, and let a be a complex number such that $a \neq 0$ and $a \neq q^n$ for all positive integers n. Then a and $f(a)$ cannot simultaneously belong to K. Theorem 2: Let $f(z)$ and a satisfy the hypotheses of Theorem 1 and assume in addition that a is a number of K. If $\varepsilon > 0$ is given, then the inequality

$$|f(a) - (P/Q)| \leq |Q|^{-(7/2+\varepsilon)}$$

is satisfied by only a finite number of fractions P/Q, where P and Q are integers of K and $Q \neq 0$. Among the important consequences of these two theorems are the following corollaries. (1) Let k be a positive integer and m a rational integer not divisible by k if m is negative. Then the product $\prod_{n=1}^{\infty}(1 \pm q^{-kn-m})$ does not belong to K. (2) The numbers $\prod_{n=1}^{\infty}(1+q^{-kn-m})$ and $\prod_{n=1}^{\infty}(1-q^{-kn-m})$ are irrational when q is a rational integer such that $|q| \geq 2$ and k, m are selected as in (1). (3) For every given $\varepsilon > 0$, there is a constant $c > 0$ depending only on ε such that the inequality $|\prod_{n=1}^{\infty}(1 \pm 2^{-n}) - P/Q| \geq c(\varepsilon)|Q|^{-(7/2+\varepsilon)}$ is satisfied for all pairs of rational integers (P, Q) with $Q \neq 0$. The methods are based on the theory of entire functions and involve some interesting interpolation series.

A. L. Whiteman (Los Angeles, Calif.)

J72-58 **(40# 1347)**

Pólya, G.

Entiers algébriques polygones et polyèdres réguliers.

Enseignement Math. (2) **15** (1969), 237–243.

Using elementary facts about algebraic integers, the author shows that certain trigonometric functions which depend on rational integers, such as $\tan(2\pi/n)$, are irrational except for a specified finite number of cases (in the above example, $n=1, 2, 4, 8$ if $n>0$). These results are employed to prove some statements about polygons whose vertices have integral coordinates and about the five regular polyhedra. The following two theorems are typical. (1) If the vertices of an n sided polygon in the plane have integral coordinates and equal angles, then $n=4$ or $n=8$. (2) If α, β, γ, δ, ε are the inner dihedral angles of the five regular polyhedra, then there is essentially only one dependence of the form $n_1\alpha+n_2\beta+n_3\gamma+n_4\delta+n_5\varepsilon=0$ with rational integers n_i. *H. Groemer* (Tucson, Ariz.)

J72-59 **(40# 5549)**

Mąkowski, A.

Angles of a parallelogram with vertices in lattice points.

Elem. Math. **24** (1969), 114–115.

The author proves the following theorem: Let four lattice points be the vertices of a parallelogram and α its angle $<\pi/2$. Then α/π is irrational or $\alpha=2\pi/n$, where $n=4$ or 8.

The main tool used for the proof is D. H. Lehmer's well-known theorem about the degree of the algebraic integer $2\sin 2k/n$, where k and n are positive integers. {Reviewer's remark: the proof can be abbreviated considerably by using instead a similar theorem about the tangent [cf. I. Niven, *Irrational numbers*, Corollary 3.12, p. 41, Wiley, New York, 1956; MR **18**, 195].} *J. Popken* (Amstelveen)

Citations: MR 18, 195c = Z01-37.

J72-60 **(41# 140)**

Lange, L. J.

A simple irrationality proof for *n*th roots of positive integers.

Math. Mag. **42** (1969), 242–244.

Using the well-ordering property, the author gives a simple proof that for positive integers n and a, any rational root of $x^n=a$ is an integer. *I. Niven* (Eugene, Ore.)

J72-61 **(41# 5305)**

Ptáček, František

Irrationality of certain continued fractions. (Czech. Russian, English, French and German summaries)

Sb. Prací Vys. Školy Doprav. v Žilině a Výzkum. Ústav. Doprav. v Praze No. 25 (1969), 85–100.

Let a, b, n be positive integers satisfying $a<b$, $n\geqq 3$, and write $x=a/b$. Define $r_1=1/n$, $r_{2\nu}=(n\nu-1)/((2\nu-1)\cdot 2n)$, $r_{2\nu+1}=(n\nu+1)/((2\nu+1)\cdot 2n)$ $(\nu=1, 2, 3, \cdots)$. The author proves that if $a\leqq n$, then the continued fraction

$$1+\frac{r_1x^n|}{|1}+\frac{r_2x^n|}{|1}+\frac{r_3x^n|}{|1}+\cdots$$

is an irrational number, and equals $(1+x^n)^{1/n}$. He deals also with the case $n=2$, and derives several other results of this type. The proofs are based on several classical theorems cited from O. Perron [*Die Lehre von den Kettenbrüchen. Band II: Analytisch-funktionentheoretische Kettenbrüche*, Teubner, Berlin, 1913; second edition, 1929; third edition, Stuttgart, 1957; MR **19**, 25]. *H. Halberstam* (Nottingham)

J72-62 **(41# 6787)**

Erdős, P.

On the irrationality of certain series.

Math. Student **36** (1968), 222–226 (1969).

In a previous paper [J. Indian Math. Soc. **12** (1948), 63–66; MR **10**, 594] the author proved that for every integer $t\geqq 2$ the series $\sum_{n=1}^{\infty}(t^n-1)^{-1}=\sum_{n=1}^{\infty}d(n)t^{-n}$ is irrational. In this paper he proves that if $(n_i, n_j)=1$ and $\sum_{i=1}^{\infty}1/n_i<\infty$, then for every integer $t\geqq 2$ the number $\sum_{i=1}^{\infty}(t^{n_i}-1)^{-1}$ is irrational. He asserts that the condition $(n_i, n_j)=1$ is superfluous. The proof is accomplished by showing that the t-ary expansion of the series is infinite and contains arbitrarily long strings of zeros. He also lists other series that he expects but cannot prove are irrational: $\sum_{p\text{ prime}}(t^p-1)^{-1}$, $\sum_{i=1}^{\infty}(t^{n!}-1)^{-1}$, $\sum_{n=1}^{\infty}(2^n-1)^{-1}t^{-n}$ and $\sum_{i=1}^{\infty}(t^{n_i}-1)^{-1}$, where $n_{k+1}-n_k\to\infty$ $(k\to\infty)$. *W. W. Adams* (College Park, Md.)

Citations: MR 10, 594c = J72-17.

J72-63 **(42# 1770)**

Călin, Dan; Kiss, Vladimir

Irrationality of the sum of some remarkable series. (Romanian)

Gaz. Mat. Ser. A **75** (1970), 161–164.

Mit Hilfe der klassischen Methode von Fourier (1815) zum Beweis der Irrationalität der Zahl e zeigen die Autoren die Irrationalität von Reihen folgender Typen: $\sum_{n=1}^{\infty}u_n(E_n!)^{-1}$, $\sum_{n=1}^{\infty}u_n(\prod_{i=1}^{n}E_i)^{-1}$, wo entweder $u_n=1$ oder $u_n=(-1)^{n+1}$ für alle natürlichen n gilt. $\{E_n\}_{n=1,2,\ldots}$ ist eine echt monoton wachsende Folge natürlicher Zahlen. *P. Bundschuh* (Freiburg)

J72-64 **(42# 7600)**

Oppenheim, A. [Oppenheim, Alexander]

The irrationality or rationality of certain infinite series.

Studies in Pure Mathematics (Presented to Richard Rado), pp. 195–201. *Academic Press, London,* 1971.

The author proves several theorems on the irrationality of certain classes of series with (positive, or positive and negative) terms. The simplest theorem is as follows. Let $\{a_i\}$, $\{n_i\}$, and $\{e_i\}$ be three sequences of positive integers, and let N_i denote the least common multiple of $n_1, n_2, \cdots, n_i$. Assume that $\lim\sup a_{i+1}n_i^2n_{i+1}^{-1}\leqq 1$, and that both $a_1a_2\cdots a_{i+1}N_iN_{i+1}^{-1}$ and e_i are bounded. Then $\sum_{i=1}^{\infty}a_1a_2\cdots a_ie_in_i^{-1}$ is rational if and only if $n_{i+1}-1=a_{i+1}n_i(n_i-1)$ and $e_i=\text{const}$ for all sufficiently large i. For earlier work see P. Erdős and E. G. Straus [J. Indian Math. Soc. (N.S.) **27** (1963), 129–133 (1964); MR **31** #124] and the author [J. London Math. Soc. **43** (1968), 115–118; MR **37** #164]. *K. Mahler* (Columbus, Ohio)

Citations: MR 31# 124 = J72-44; MR 37# 164 = J72-54.

J72-65 **(43# 7413)**

Erdős, P.; Straus, E. G.

Some number theoretic results.

Pacific J. Math. **36** (1971), 635–646.

The paper deals with two separate topics. In the first part it is shown that the maximal number $f(p)$ of a set A of residues (mod p) chosen so that the sums of different numbers of distinct elements of A are distinct satisfies $(4p)^{1/3}+o(p^{1/3})<f(p)<(288p)^{1/3}+o(p^{1/3})$. It is conjectured that $f(p)=(4p)^{1/3}+o(p^{1/3})$.

In the second part the irrationality of certain infinite series is established. It is shown that the series $\sum_{n=1}^{\infty}d(n)/(a_1a_2\cdots a_n)$ (where $d(n)$ denotes the number of divisors of n) is irrational whenever the integers a_n satisfy $2\leqq a_1\leqq a_2\leqq\cdots\leqq a_n\leqq\cdots$. It is conjectured that it is sufficient to assume that $a_n\to\infty$.

Also shown to be irrational are the series

$$\sum_{n=1}^{\infty} \phi(n)/(a_1a_2\cdots a_n) \quad \text{and} \quad \sum_{n=1}^{\infty} \sigma(n)/(a_1a_2\cdots a_n)$$

(where $\sigma(n)$ denotes the sum of the divisors of n) under the same monotonicity assumption as above and under the further hypothesis that $a_n \geqq n^{11/12}$ for all large n.

(The last equation in the proof of Lemma 2.14 does not appear to be correct. However, if one lets the "c" of Lemma 2.17 be the reciprocal of the "c" of Lemma 2.14, then the proof still goes through.)

J. B. Kelly (Tempe, Ariz.)

J76 TRANSCENDENCE PROOFS

See also Section J88.

See also reviews C05-29, C05-33, J02-3, J02-6, J02-9, J02-14, J02-17, J02-18, J02-21, J02-22, J02-23, J02-26, J02-27, J68-11, J68-46, J68-63, J72-23, J72-32, J72-56, J80-26, J88-5, J88-12, J88-13, J88-17, J88-20, J88-22, J88-35, J88-37, J88-49, K20-22, K20-23, K25-30, Q05-12, Q05-13, Q05-29, Q05-43, Q05-63, Q05-69, Q05-89, Q10-5, Q10-8, T99-7, U99-1, Z01-37, Z02-32, Z10-2, Z10-20, Z10-31, Z15-60, Z30-20.

J76-1 **(1, 71a)**

Niven, Ivan. The transcendence of π. Amer. Math. Monthly **46**, 469–471 (1939).

This paper repeats the classical proof as to be found in Perron's Irrationalzahlen, page 178 or Landau's Zahlentheorie II, page 93, but without avoiding integrals.

A trifling slip: On page 471, second formula from below, the factor τ^{p+1} is omitted. *P. Scherk.*

J76-2 **(2, 149d)**

Veldkamp, G. R. Ein Transzendenz-Satz für p-adische Zahlen. J. London Math. Soc. **15**, 183–192 (1940).

The well-known theorem of Gelfond and Schneider [C. R. (Doklady) Acad. Sci. URSS (N.S.) **2**, 1–6 (1934); J. Reine Angew. Math. **172**, 65–69 (1934)] concerning the transcendence of ω^θ for irrational algebraic θ and algebraic $\omega\neq 0, 1$ has an analogue in the field of p-adic numbers: Let θ and ω be p-adic algebraic integers, θ irrational, $\omega\neq 1$ and $\omega\equiv 1 \pmod p$ and define the p-adic power ω^θ by the convergent series

$$\omega^\theta = e^{\theta \log \omega} = \sum_{k=0}^{\infty} \frac{(\theta \log \omega)^k}{k!}, \quad \log \omega = -\sum_{k=1}^{\infty} \frac{(1-\omega)^k}{k};$$

then ω^θ is transcendent. The proof uses the method of Th. Schneider and a lemma from a paper of K. Mahler [Compositio Math. **2**, 259–275 (1935)], where another proof of this theorem based on Gelfond's idea had already been given. *C. L. Siegel* (Princeton, N. J.).

Referred to in J76-31, Q05-29.

J76-3 **(3, 263f)**

Wade, L. I. Certain quantities transcendental over $GF(p^n, x)$. Duke Math. J. **8**, 701–720 (1941).

Carlitz [Duke Math. J. **1**, 137–168 (1935)] has introduced several important quantities into the analysis of the field which is constituted by the power series in x^{-1/q^k} with coefficients from a finite field $GF(p^n)=GF(q)$. Let

$$[k]=x^{q^k}-x, \qquad L_k=[k][k-1]\cdots[1],$$
$$F_k=[k][k-1]^q\cdots[1]^{q^k}, \qquad \xi=\lim [1]^{q^k/(q-1)}/L_k,$$
$$\psi(t)=\sum(-1)^k t^{q^k}.$$

The function $\psi(t)$ shares with the "linear" polynomials $f=\sum A_i t^{q^i}$ the property $\psi(u+v)=\psi(u)+\psi(v)$; it has a multiplication theorem of the form $\psi(Mt)=f_M(\psi(t))$, where M is an "integer," that is, a polynomial form in x. Furthermore, $\psi(\xi)=0$ and $\psi(t+M\xi)=\psi(t)$.

The principal result of the present paper is the theorem that the values of the ψ-function (which is akin to the exponential function) are transcendental if the argument is an algebraic "number" not equal to 0. More specifically, it is shown: (1) Let $\zeta=\psi(1)=\sum(-1)^k/F_k$, or even $\sum B_k/F_k$ with certain restrictions on the degrees of the B_k, or finally $\zeta=\sum 1/[k]$. If f is a "linear" polynomial not equal to 0, then $f(\zeta)\neq 0$. (2) Let γ_i be the conjugates of an "algebraic integer" γ; then the product of the values $\psi(\gamma_i)$ and a fortiori $\psi(\gamma)$ do not vanish. (3) Under the same conditions for the γ_i, with f a "linear" polynomial whose coefficients are only slightly restricted, the product of the quantities $f(\psi(\gamma_i))$ and a fortiori $f(\psi(\gamma))$ cannot be zero.

In order to show that the respective quantities in (1)–(3) do not vanish, these formal power series in x are skillfully and laboriously decomposed into appropriate parts. The following elementary principles are then applied to these decompositions: (4) If $I\neq 0$ is integral, and Q is a sum of terms of negative degrees, $I+Q$ is not zero. (5) If $I\not\equiv 0 \pmod F$ then $I\neq 0$ (this principle occurs in several variants). (6) If D is not zero and is of smaller degree than P, it is not divisible by P.

The general transcendency theorem calls of course for arbitrary polynomials f and possibly fractional arguments γ. Generality is achieved on the basis of: (7) the (known) fact that every polynomial divides a "linear" polynomial; (8) the multiplication theorem for the ψ-function; (9) a special lemma (7.1 of the paper) which takes care of the slight restrictions on f mentioned under (3). *M. A. Zorn.*

Referred to in J76-6, J76-8, J76-26, J76-74.

J76-4 **(3, 266b)**

Schneider, Theodor. Zur Theorie der Abelschen Funktionen und Integrale. J. Reine Angew. Math. **183**, 110–128 (1941).

Suppose that $F(x, y)=0$ is an absolutely irreducible polynomial of genus q with algebraic numbers as coefficients. Let

$$\mathfrak{u}=\int_c^{(x,y)} \Re(x, y)dx, \quad \Re(x, y)=\{R_1(x, y), \cdots, R_q(x, y)\},$$

be a basis of the integrals of first kind on $F(x, y)=0$, so that the $R_i(x, y)$ have algebraic coefficients. Letting

$$\mathfrak{U}=\sum_{l=1}^{q} \int_{c_l}^{(x_l, y_l)} \Re(x, y)dx,$$

the author considers the Abelian functions $A(\mathfrak{U})$ which are rational symmetric functions of the q couples (x_i, y_i) with algebraic numbers as coefficients. The two main results of the paper are the following: "Theorem I. If $p+1$ functions $A(\mathfrak{U})$ possess q common periods which are linearly independent in the field of all algebraic numbers, then the functions are algebraically dependent. Theorem II. If the integrand of an integral of the first or second kind is an algebraic function with algebraic numbers as coefficients, then not all periods of the integral are algebraic." The first result is obviously a nontrivial refinement of a well-known fact on Abelian functions. For the proof the author has to study the coefficients of the local inversion problem. First it is shown that the R_i may be selected so as to have algebraic coefficients for their developments relative to x_i-c_i. Moreover, these coefficients can be selected to have bounded denominators and "uniform" upper bounds for all their conjugates. Then comparison of coefficients gives estimates for the local inversion. At various places it is necessary to select the lower limits c_i appropriately so as to stay away from poles, etc. After these preparations the author succeeds in proving both theorems by the method of Cauchy's integral theorem in q space. The latter was used previously by the author and Gelfond in connection with solving Hilbert's problem in the transcendency of α^β, α and β general algebraic numbers. *O. F. G. Schilling* (Chicago, Ill.).

Referred to in J76-51.

J76-5 (4, 191f)

Hille, Einar. Gelfond's solution of Hilbert's seventh problem. Amer. Math. Monthly **49**, 654–661 (1942).

The author gives a clear exposition of the following well-known theorem of Gelfond: α^β is transcendental if α and β are algebraic, β is irrational and $\alpha\neq 0$ or 1. *P. Erdös.*

J76-6 (5, 89a)

Wade, L. I. Certain quantities transcendental over $GF(p^n, x)$. **II.** Duke Math. J. **10**, 587–594 (1943).

Let $GF(q)$ denote a fixed finite field of order $q=p^n$, $GF[q, x]$ the ring of polynomials in the indeterminate x over $GF(q)$ and $GF(q, x)$ the quotient field of $GF[q, x]$. For a positive integer k and indeterminate t, $[k]=x^{q^k}-x$, $F_k=[k][k-1]^q\cdots[1]^{q^{k-1}}$, $L_k=[k]\cdots[1]$, $L_0=F_0=1$;

$$\Psi(t)=\sum_{j=0}^{\infty}(-1)^j(t^{q^j}/F_j).$$

It is shown that $\sum_{j=0}^{\infty}(1/L_j^\gamma)$, where γ is any positive rational integer, is transcendental over $GF(q, x)$. This yields a new proof of the transcendence of ξ, where $\xi\neq 0$ belongs to a super-field over $GF(q, x)$ and has the property that $\Psi(E\xi)=0$ for every element E of $GF[q, x]$. The author had used other methods to prove the transcendence of ξ and other quantities in an earlier paper of like title [Duke Math. J. **8**, 701–720 (1941); these Rev. **3**, 263].

J. L. Dorroh (Baton Rouge, La.).

Citations: MR 3, 263f = J76-3.

Referred to in J76-8, J76-74.

J76-7 (6, 119f)

Webber, G. Cuthbert. Transcendence of certain continued fractions. Bull. Amer. Math. Soc. **50**, 736–740 (1944).

The following results on special continued fractions of Hurwitz's type [A. Hurwitz, Mathematische Werke, vol. 2, Basel, 1933, pp. 276–302] are proved.

(1) Let $a_1, a_2, \cdots, a_{k-1}$ be ∓ 1; let all numbers $f_1, f_2, \cdots$ be $+1$ or all be -1; let $b_1, b_2, \cdots, b_{k-1}$ be arbitrary positive integers; and let $g_n=g_0+dn$, where $g_0\geqq 0$ and $d>0$ are integers. Then

$$g_0+\frac{a_1|}{|b_1}+\cdots+\frac{a_{k-1}|}{|b_{k-1}}+\frac{f_1|}{|g_1}+\frac{a_1|}{|b_1}+\cdots+\frac{a_{k-1}|}{|b_{k-1}}+\frac{f_2|}{|g_2}+\cdots$$

is a transcendental number. (2) If a, c, d are positive integers, and either c or d is even, and if further $b=c^2a+c^2d/2$, $e=c^2d$, then

$$\frac{1|}{|a}+\frac{1|}{|b}+\frac{1|}{|a+d}+\frac{1|}{|b+e}+\frac{1|}{|a+2d}+\frac{1|}{|b+2e}+\cdots$$

is transcendental. (3) If u^2 is a positive integer, then

$$\frac{1|}{|1}+\frac{1|}{|3u^2}+\frac{1|}{|5}+\frac{1|}{|7u^2}+\frac{1|}{|9}+\frac{1|}{|11u^2}+\frac{1|}{|13}+\cdots$$

and

$$\frac{1|}{|1}+\frac{1|}{|3u^2-2}+\frac{1|}{|1}+\frac{1|}{|3}+\frac{1|}{|1}+\frac{1|}{|7u^2-2}+\frac{1|}{|1}+\frac{1|}{|7}+\frac{1|}{|1}+\frac{1|}{|11u^2-2}+\frac{1|}{|1}+\frac{1|}{|11}+\cdots$$

are transcendental. For the proof of these theorems, the author transforms the classical continued fractions of $(e^2-e^{-2})/(e^2+e^{-2})$ and $\mathfrak{F}_{\lambda-1}(2)/\mathfrak{F}_\lambda(2)$ and applies the results of Lindemann and C. L. Siegel [Abh. Preuss. Akad. Wiss., Phys.-Math. Kl. **1929**, no. 1] on the transcendency of these functions. *K. Mahler* (Manchester).

J76-8 (6, 144h)

Wade, L. I. Two types of function field transcendental numbers. Duke Math. J. **11**, 755–758 (1944).

Continuation of ideas and problems developed in two earlier papers [same J. **8**, 701–720 (1941); **10**, 587–594 (1943); these Rev. **3**, 263; **5**, 89]. All three papers utilize certain expressions introduced into the theory by Carlitz. For methods and terms, such as the use of "transcendental" in this connection, compare the former papers and references.

"Let p denote a fixed prime, $GF(p^n)$ the finite field of order p^n, $n=1, 2, \cdots$, and Γ the algebraic closure of $GF(p)$. For e indeterminates $x_1, \cdots, x_e$, $\Gamma(x_1, \cdots, x_e)$ will denote the field of rational functions in $x_1, \cdots, x_e$ with coefficients from Γ. If $E\neq 0$ and $G\neq 0$ are two polynomials in $x_1, \cdots, x_e$ (with coefficients from Γ, or, what is the same thing, from some $GF(p^n)$), define deg $E/G=$deg $E-$deg G, where deg is an abbreviation for degree. If we write $-$deg $0=\infty$, then $-$deg defines an exponential valuation of $\Gamma(x_1, \cdots, x_e)$. Denote by Φ the corresponding completion of $\Gamma(x_1, \cdots, x_e)$. Here we shall consider the transcendence over $\Gamma(x_1, \cdots, x_e)$, or equivalence over $GF(p;\, x_1, \cdots, x_e)$ of two types of elements of Φ defined by infinite series." (I) The real numbers $\sum g^{-q^k}$, $k=0, 1, 2, \cdots$; $g>1$, $q>1$ rational integers, are known to be transcendental. For the corresponding series in Φ, $\sum G^{-q^k}$, G a polynomial in $\Gamma(x_1, \cdots, x_e)$ of deg $G>0$ and $q>1$ a rational integer, the following result is proved. The series is algebraic if $q>1$ is of the form p^s, transcendental otherwise. (II) The character of the real numbers $\sum g^{-k^q}$, $k=0, 1, 2, \cdots$; $g>1$, $q>1$ rational integers, is not known. The corresponding series in Φ, $\sum G^{-k^q}$, is always transcendental. *A. J. Kempner* (Boulder, Colo.).

Citations: MR 3, 263f = J76-3; MR 5, 89a = J76-6.

Referred to in J76-26, J76-32.

J76-9 (7, 52d)

Puri, Amritsagar. Transcendence of decimals. Math. Student **12**, 88–90 (1945).

Lacuna theorems on the algebraic or transcendental character of numbers represented by series. (I) If the sequence $q_{n+1}-rq_n$ is unbounded,

$$x=a_0+a_{q_1}/k^{q_1}+a_{q_2}/k^{q_2}+\cdots$$

cannot satisfy an algebraic equation with integral rational coefficients of degree less than $r+1$. Here a_i, q_i are integers, $a_0\geqq 0$, $a_{q_n}>0$, $q_n<q_{n+1}$ and k is an integer not less than 2. In particular, if the sequence q_{n+1}/q_n is unbounded, x is transcendental. (II) If (a) the positive integer u_n is divisible by the least common multiple of $u_1, u_2, \cdots, u_{n-1}$ and (b)

$2^{k_n} \leqq u_n \leqq 2^{k_n+1}$, then $\sum_1^\infty a_n/u_n$ (a_i positive integers) is transcendental if the sequence k_{n+1}/k_n is unbounded, and does not satisfy an algebraic equation with rational coefficients of degree less than $r+1$ if the sequence $k_{n+1}-rk_n$ is unbounded. The proofs are elementary. [In formula (3.4), read ξ for ξ_s; in theorem 3 read: "if the sequence $p_{n+1}-rp_n$ is unbounded."] *A. J. Kempner* (Boulder, Colo.).

J76-10 (7, 106a)

Brun, Viggo. Quadrature of the circle. Norsk Mat. Tidsskr. **23**, 13 pp. (1941). (Norwegian)

A historical lecture. A dozen assorted formulas involving π are quoted at the end.

The paper is identical with that quoted in these Rev. **3**, 97.

Referred to in A08-11.

J76-11 (8, 5g)

Cabannes, Henri. Application des fractions continues à la formation de nombres transcendants. Revue Sci. (Rev. Rose Illus.) **82**, 365–367 (1944).

Let $x_1=[\alpha_1; \alpha_2, \alpha_3, \cdots]$, the α_i positive integers; $p_n/q_n=[\alpha_1; \alpha_2, \alpha_3, \cdots, \alpha_n]$, $x_n=[\alpha_n; \alpha_{n+1}, \alpha_{n+2}, \cdots]$. Let $f_1(x)=a_0x^r+\cdots+a_r$, $f(x)$ irreducible in $R(1)$, $f_1(x_1)=0$. For $f_n(x)=x^rf_{n-1}(\alpha_{n-1}+1/x)$ ($n=2, 3, \cdots$) x_n is a simple root of $f_n(x)=0$. Several theorems are proved. (I) There exists an N such that for $n>N$ all roots of $f_n(x)=0$ except x_n have their real part $\Re(x)>-1$. (II) By the use of his inequality governing the approximation of algebraic numbers by rational numbers, Liouville was the first to construct sets of transcendental numbers. By an independent method, which is, however, related to the type of argument employed by Liouville, the author proves: if the positive integers $\alpha_1, \alpha_2, \alpha_3, \cdots$ of a number $[\alpha_1; \alpha_2, \alpha_3, \cdots]$ form an increasing sequence such that $\log \alpha_n/\log(\alpha_1\alpha_2\cdots\alpha_{n-1})$ increases indefinitely with n, the number is transcendental. (III) If the $\alpha_1, \alpha_2, \cdots$ of a number $[\alpha_1; \alpha_2, \alpha_3, \cdots]$ are such that $(\log \alpha_n)^{1/n}$ increases indefinitely with n, the number is transcendental. (IV) If $\log(1+\alpha_n)/\log\{(1+\alpha_1)\cdots(1+\alpha_{n-1})\}$ increases indefinitely with n, the number $[\alpha_1; \alpha_2, \alpha_3, \cdots]$ is transcendental. *A. J. Kempner* (Boulder, Colo.).

J76-12 (8, 370a)

Cohn, Harvey. Note on almost-algebraic numbers. Bull. Amer. Math. Soc. **52**, 1042–1045 (1946).

Consider a class of power series $\sigma(x)=x^{e_0}+a_1x^{e_1}+a_2x^{e_2}+\cdots$, where $a_1=r_1/S_1$, $a_2=r_2/S_2, \cdots$ are rational coefficients different from zero and $e_0, e_1, e_2, \cdots$ denote rapidly increasing integers. For any positive integer h let g_h be the maximum of the numbers $1, |a_1|, \cdots, |a_h|$ and let d_h be the least common multiple of the denominators $S_1, \cdots, S_h$. Then it is shown that $\sigma(x)$ has a transcendental value for every algebraic number $x\neq 0$ within its circle of convergence if the following three conditions are fulfilled:

$$\lim_{h\to\infty} e_{h+1}/e_h=\infty, \quad \lim_{h\to\infty} e_{h+1}/\log g_h=\infty, \quad \lim_{h\to\infty} e_{h+1}/\log d_h=\infty.$$

J. Popken (Utrecht).

Referred to in J76-68.

J76-13 (9, 79g)

Mahler, K. On the generating functions of integers with a missing digit. K'o Hsüeh (Science) **29**, 265–267 (1947). (Chinese)

Let N be the set of positive integers n which contain no digit equal to zero. The author proposed the problem of whether or not $\sigma=\sum 1/n$ for $n\varepsilon N$ is a transcendental number. In this paper the author considers an analogous problem for $f(z)=\sum z^n$, $n\varepsilon N$. He studies the analytic behavior of the function $f(z)$ and proves that $f(z)$ is a transcendental number if z is algebraic and $0<|z|<1$. [Cf. Math. Ann. **101**, 342–366 (1929).] *L.-K. Hua* (Princeton, N. J.).

Referred to in J76-14.

J76-14 (13, 213b)

Mahler, K. On the generating function of the integers with a missing digit. J. Indian Math. Soc. (N.S.) Part A. **15**, 33–40 (1951).

Translated from a paper which appeared in Chinese in K'o Hsüeh (Science) **29**, 265–267 (1947); these Rev. **9**, 79.

Citations: MR 9, 79g = J76-13.

J76-15 (10, 594a)

Carlitz, L. Hurwitz series: Eisenstein criterion. Duke Math. J. **16**, 303–308 (1949).

By an H-series of type k, $k\geqq 0$, is meant a formal power series of the form $f(x)=\sum_{m=0}^\infty a_mx^m/\{m, k\}$, where the coefficients a_m are rational integers and

$$\{m, k\}=m!(m+1)!\cdots(m+k)!/1!2!\cdots k!,$$

$\{m, 0\}=m!$. For fixed k, the totality of such series is denoted by $\mathfrak{H}_k$. The order of $f(x)$ is the integer r such that $a_0=\cdots=a_{r-1}=0$, $a_r\neq 0$. It is proved that $\mathfrak{H}_k$ is a domain of integrity; that $\mathfrak{H}_0\subset\mathfrak{H}_k\subset\mathfrak{H}_{k+1}$; and that the quotient of series of type k is expressible in terms of series of type $k+r$. The major result of the paper states that if the series $w(x)=\sum_0^\infty\gamma_mx^m$ with rational coefficients γ_m satisfies the equation $\sum_{i=0}^t A_i(x)w^i(x)=0$, where the coefficients $A_i(x)\varepsilon\mathfrak{H}_k$, then there exist integers $c\geqq 1$, $r\geqq k$ such that $cw(cx)\varepsilon\mathfrak{H}_r$. The function-theoretic arguments appearing in the usual proof of the Eisenstein criterion are here replaced by a simple lemma. As an application it is shown that for an arbitrary prime p the series $\sum_{m=0}^\infty p^{-m^2}x^m$ is transcendental relative to any $\mathfrak{H}_r$. *A. L. Whiteman.*

J76-16 (10, 594b)

Rudin, Walter. A theorem on Hurwitz series. Duke Math. J. **16**, 309–311 (1949).

[For the definition of an H-series of type k, of the set $\mathfrak{H}_k$ and of the symbol $\{m, k\}$, see the preceding review.] The author proves the transcendence of a class of H-series of type $k+1$ relative to $\mathfrak{H}_k$. As a special case of his theorem, he deduces that $\sum_0^\infty x^m/\{m, k+1\}$ is transcendental relative to $\mathfrak{H}_k$. In particular, considering the simplest case ($k=0$), he concludes that $\sum_0^\infty x^m/m!(m+1)!$ is transcendental relative to the set $\mathfrak{H}_0$ of series $\sum_0^\infty a_mx^m/m!$.

A. L. Whiteman (Los Angeles, Calif.).

J76-17 (11, 160a)

Schneider, Théodor. Ein Satz über ganzwertige Funktionen als Prinzip für Transzendenzbeweise. Math. Ann. **121**, 131–140 (1949).

The author proves a general theorem concerning the algebraic dependence of a set of functions which are, in an extended sense, integral valued. The following special case will illustrate the nature of the results. Let $\{\zeta_k\}$ be a sequence of complex numbers, not necessarily distinct. Let $r_m=\sup_{k\leq m}|\zeta_k|$ and set $\alpha=\liminf(\log m)/\log r_m$. For any ζ, let $L_m(\zeta)$ be the multiplicity with which ζ occurs in $\{\zeta_1, \zeta_2, \cdots, \zeta_m\}$ and assume that $L_m(\zeta)\leqq m/\log m$ for all m and ζ. Let I be the set of rational integers. Let f_i, $i=1, 2, \cdots, n$, be an entire function of order μ_i, and suppose that $\sum_1^n\mu_i<(n-1)\alpha$. If $f_i^{(\lambda)}(\zeta)\varepsilon I$ for each i, each m, each ζ in $\{\zeta_k\}$, and each $\lambda<L_m(\zeta)$, then $f_1, \cdots, f_n$ are algebraically dependent. This theorem is a consequence of the general theorem in which the functions f_i are allowed to be mero-

morphic, and in which I is replaced by an arbitrary field K of algebraic numbers. In this form, it may be applied to prove many of the known theorems on transcendence. For example, to obtain a solution of Hilbert's problem, choose $f_1(z)=a^z$, $f_2(z)=z$; then the independence of f_1 and f_2 proves that a^b is transcendental whenever a and b are algebraic and $a\neq 0, 1, b$ irrational. *R. C. Buck* (Madison, Wis.).

Referred to in J02-21, J76-31, Q05-64.

J76-18 **(11, 330c)**

Siegel, Carl Ludwig. Transcendental Numbers. Annals of Mathematics Studies, no. 16. Princeton University Press, Princeton, N. J., 1949. viii+102 pp. $2.00.

"This booklet reproduces with slight changes a course of lectures delivered in Princeton during the spring term 1946. It would be misleading to call it a theory of transcendental numbers, our knowledge concerning transcendental numbers being narrowly restricted. The text deals with a few special transcendency problems of some interest, but it is more than a mere collection of scattered examples, since it involves a method which might be useful in the search of more general results."

The first chapter begins with simple proofs of the irrationality of e and π. Their transcendency, and the general theorem of Lindemann-Weierstrass, is proved in a way somewhat similar to that in the paper of K. Mahler [J. Reine Angew. Math. **166**, 118–136 (1931)], using Hermite's polynomials $P_1(x), \cdots, P_m(x)$ of given degrees for which $P_1(x)e^{\rho_1 x}+\cdots+P_m(x)e^{\rho_m x}\neq 0$ has a zero of maximal order at $x=0$.

The second chapter deals with systems of differential equations

$$(1)\qquad y_k'=\sum_{l=1}^{m}Q_{kl}(x)y_l,\qquad k=1,\cdots,m,$$

where the Q's are rational functions with algebraic coefficients. Assume the matrix $Q=(Q_{kl})_{k,l=1,\cdots,m}$ consists of r square boxes $Q_t=(Q_{kl,t})_{k,l=1,\cdots,m_t}$, $1\leqq t\leqq r$, along the diagonal of Q, the elements of Q outside the boxes being zero, while $\sum_{t=1}^{r}m_t=m$. Thus (1) splits into r separate systems

$$(2)\qquad y'_{k,t}=\sum_{l=1}^{m_t}Q_{kl,t}(x)y_{l,t},\qquad k=1,\cdots,m_t;\ 1\leqq t\leqq r.$$

Then there is a matrix $Y=(y_{kl}(x))_{k,l=1,\cdots,m}$, the columns of which form m systems of independent solutions of (1), such that Y splits into r boxes $Y_t=(y_{kl,t}(x))_{k,l=1,\cdots,m_t}$, $1\leqq t\leqq r$, of the same orders as the boxes of Q, the columns of Y_t forming now m_t systems of independent solutions of (2). The boxes Y_t are called independent if

$$\sum_{t=1}^{r}\sum_{k=1}^{m_t}\sum_{l=1}^{m_t}P^*_{k,t}(x)c_{l,t}y_{kl,t}(x)\neq 0$$

for arbitrary polynomials $P^*_{k,t}(x)$ and arbitrary constants $c_{l,t}$, unless all products $P^*_{k,t}(x)c_{l,t}$ vanish simultaneously. A series $f(x)=\sum_{n=0}^{\infty}c_nx^n/n!$ is said to be of type E if (a) all c_n lie in the same algebraic field K of finite degree over the rational field; (b) if $\epsilon>0$, then all c_n and their conjugates with respect to K are $O(n^{\epsilon n})$; (c) moreover, there exists a sequence $q_0, q_1, \cdots$ of positive integers such that c_kq_n $(0\leqq k\leqq n)$ is integral in K and that $q_n=O(n^{\epsilon n})$. We also say that the functions $E_1(x), \cdots, E_m(x)$ of type E form a normal system if (A) $y_1=E_1(x), \cdots, y_m=E_m(x)$ is a solution of a system of differential equations (1) as defined above, and (B) if (1) has a solution matrix Y consisting of independent boxes Y_t. The following lemma is then proved. Let $E_1(x), \cdots, E_m(x)$ form a normal system; let further the number $\alpha\neq 0$ and the Taylor coefficients of $E_1(x), \cdots, E_m(x)$ all lie in the same field K of finite degree h over the rational field. If α is different from the poles of all functions $Q_{kl}(x)$ $(k, l=1, \cdots, m)$, then at least $\frac{1}{2}m/h$ of the numbers $E_1(\alpha), \cdots, E_m(\alpha)$ are linearly independent over K. From this result, one easily deduces the following theorem. Let the E-functions $y=E_1(x), \cdots, y=E_m(x)$ satisfy a system of differential equations (1) as defined above, and let, for $\nu=1, 2, \cdots$, the products $E_1(x)^{\nu_1}\cdots E_m(x)^{\nu_m}$ $(\nu_1+\cdots+\nu_m\leqq\nu)$ form a normal system. Let $\alpha\neq 0$ be an algebraic number different from the poles of the functions $Q_{kl}(x)$ $(k, l=1, \cdots, m)$. Then the numbers $E_1(\alpha), \cdots, E_m(\alpha)$ are algebraically independent. As a special case, this theorem contains the author's earlier transcendency theorem about Bessel functions [Abh. Preuss. Akad. Wiss., Phys.-Math. Kl. **1929**, 1–70 (1930)], and, in fact, the proof uses the same ideas as this earlier paper.

Chapter 3 contains the two proofs of Schneider and Gelfond of the transcendency of a^b for irrational algebraic b and algebraic $a\neq 0, \neq 1$. Of particular interest is the very short form given to Gelfond's proof. In the last chapter, the following theorem of Schneider is proved. Let $\mathfrak{p}_1=(\xi_1, \eta_1)$ and $\mathfrak{p}_2=(\xi_2, \eta_2)$ be two points on the Riemann surface $\mathfrak{R}$ of a curve $\varphi(\xi, \eta)=0$ of genus 1 with algebraic coefficients. Let ξ_1, η_1 and ξ_2, η_2 be algebraic numbers. Let $w(\mathfrak{p})$ be an indefinite elliptic integral of the second kind which is regular at $\mathfrak{p}_1$ and $\mathfrak{p}_2$ and is not a rational function. Then

$$w(\mathfrak{p}_2)-w(\mathfrak{p}_1)=\int_{\mathfrak{p}_1}^{\mathfrak{p}_2}dw$$

is transcendental unless $\mathfrak{p}_1$ and $\mathfrak{p}_2$ coincide and the path of integration on $\mathfrak{R}$ is homologous to 0. This result contains, e.g., the transcendency of the perimeter of an ellipse with algebraic axes. The booklet ends with a rather short bibliography. *K. Mahler* (Manchester).

Referred to in J02-6, J02-9, J72-53, J76-19, J76-39, J76-54, J76-72, J80-15, J80-19, J80-26, J88-9, J88-10, J88-12, J88-15, J88-16, J88-30, J88-37.

J76-19 **(35# 133)**

Siegel, Carl Ludwig

Transzendente Zahlen.

Übersetzung aus dem Englischen von B. Fuchssteiner und D. Laugwitz. B. I. Hochschultaschenbücher, Band 137*.

Bibliographisches Institut, Mannheim, 1967. 86 *pp. DM* 4.80.

German translation of *Transcendental numbers* [Ann. of Math. Studies, No. 16, Princeton Univ. Press, Princeton, N.J., 1949; MR **11**, 330].

Citations: MR 11, 330c = J76-18.

Referred to in J72-53, J80-26.

J76-20 **(11, 331a)**

Dietrich, Verne E., and Rosenthal, Arthur. Transcendence of factorial series with periodic coefficients. Bull. Amer. Math. Soc. **55**, 954–956 (1949).

The authors obtain the following result from Lindemann's general theorem. If the coefficients a_n in the series $\varphi(z)=\sum_{n=0}^{\infty}a_nz^n/n!$ are algebraic and form a periodic sequence from some a_n on, then $\varphi(z)$ is a transcendental number for every algebraic value of $z\neq 0$, except in the trivial case that $\varphi(z)$ is a polynomial in z. [This theorem is contained in a more general result of B. McMillan, J. Math. Physics **18**, 28–33 (1939); see theorem II, p. 30, and remark 2, p. 33. However, McMillan's proof is not quite correct and his results need supplementing.] *J. Popken* (Utrecht).

Referred to in J76-21.

J76-21 **(12, 318j)**

Dietrich, Verne E., and Rosenthal, Arthur. A remark about our note "Transcendence of factorial series with periodic coefficients." Proc. Amer. Math. Soc. **1**, 825 (1950).

Cf. Bull. Amer. Math. Soc. **55**, 954–956 (1949); these Rev. **11**, 331.

Citations: MR 11, 331a = J76-20.

J76-22 **(11, 501a)**

Duarte, F. J. Monograph on the numbers π and e. Historical and bibliographical notes. Estados Unidos de Venezuela. Bol. Acad. Ci. Fís. Mat. Nat. 11 (1948), no. 34–35, 1–252 (1949). (Spanish)

This monograph probably contains more information on π and e than has ever before been collected in one place. There are historical details; a large collection of numerical and geometrical approximations, formulas involving π and e, and decimal values of functions of these numbers; no less than five (classical) proofs of the transcendence of e and π; and even mnemonics, paradoxes and jokes. Some of the formulas and numerical values seem to be new.

R. P. Boas, Jr. (Providence, R. I.).

J76-23 **(13, 16f)**

LeVeque, William J. Note on the transcendence of certain series. Proc. Amer. Math. Soc. 2, 401–403 (1951).

The series studied by the author have the form

$$\sum_{n=0}^{\infty}(n+k)^{n+m}x^n/n!,$$

where k and m are integers. It is shown that the sum of such a series is a transcendental number for every algebraic value of x with $0<|x|<e^{-1}$ in each of the following two cases: (a) $k=0$ or 1 and $m\geqq 0$, (b) $k\neq 0$, 1 and $m\geqq -1$. (0^0 is interpreted as 1.) The proof depends on the well-known theorem, that e^α is transcendental for every algebraic value of $\alpha\neq 0$.

J. Popken (Utrecht).

Referred to in J76-34, J76-42.

J76-24 **(13, 114a)**

van der Sluis, A. An arithmetical theorem on systems of linear differential equations. Nederl. Akad. Wetensch. Proc. Ser. A. **54** = Indagationes Math. **13**, 252–255 (1951).

Let $f_j(z)$, $j=1, 2, \cdots, n$, be entire functions, not all identically zero, which satisfy $f_j'=\sum_{k=1}^{n}c_{jk}f_k$, the coefficients being algebraic numbers with determinant $|c_{jk}|\neq 0$. Then for every algebraic number α, with one possible exception, at least one of the $f_j(\alpha)$ is transcendental. This is an extension of a theorem of R. Rado [J. London Math. Soc. **23**, 267–271 (1948); these Rev. **10**, 354].

I. Niven.

Citations: MR 10, 354h = J72-13.

J76-25 **(13, 444c)**

Drach, Jules. Sur la transcendance du nombre π. Bull. Sci. Math. (2) **75**, 135–145 (1951).

This is the text of a lecture given by Drach in 1892 for his degree of agrégé. The proof of trancendency and of the general theorem of Lindemann is essentially that of Weierstrass.

K. Mahler (Manchester).

J76-26 **(13, 538g)**

Spencer, S. M., Jr. Transcendental numbers over certain function fields. Duke Math. J. **19**, 93–105 (1952).

Let F denote a field of arbitrary characteristic and let $x_1, \cdots, x_w$ be indeterminates. The first part of the paper contains a proof of an analog of a theorem of G. Faber [Math. Ann. **58**, 545–557 (1904)]; the analog asserts that certain entire functions $f(t)$ with coefficients in $F(x_1, \cdots, x_w)$ have the property that $f(\alpha)$ is transcendental for all algebraic $\alpha\neq 0$. In the second part of the paper F is assumed to have characteristic p; a number of theorems are proved which extend certain results due to L. I. Wade [same J. **8**, 701–720 (1941); **11**, 755–758 (1944); these Rev. **3**, 263; **6**, 144]. As a typical example we quote: the series $\sum_0^\infty G_k^{-q^k}$, where $G_k\varepsilon GF[p^n, x_1, \cdots, x_w]$, $G_k|G_{k+1}$, is transcendental for $q\neq p^s$, $q>1$.

L. Carlitz (Durham, N. C.).

Citations: MR 3, 263f = J76-3; MR 6, 144h = J76-8.

Referred to in J76-32.

J76-27 **(13, 929a)**

Gel'fond, A. O. Linear differential equations of infinite order with constant coefficients and asymptotic periods of entire functions. Trudy Mat. Inst. Steklov., v. 38, pp. 42–67. Izdat. Akad. Nauk SSSR, Moscow, 1951. (Russian) 20 rubles.

The author considers in the complex domain equations of the form $L[F]\equiv\sum_{n=0}^{\infty}a_nF^{(n)}(z)=\Phi(z)$, where $\phi(t)=\sum a_nt^n$ is an entire function of exponential type σ, and obtains several novel results. If α is a root of $\phi(t)=0$ of multiplicity $s+1$ (or more), then $z^se^{\alpha z}$ is a solution of $L[F]=0$, and the author shows first that any solution $F(z)$ of the homogeneous equation, regular in a circle $|z|\leqq\sigma_0$, $\sigma_0>\sigma$, is representable by a uniformly convergent series of these functions in some neighborhood of the origin. In fact,

$$(2\pi i)^{-2}\int_{|\xi|=\sigma_1+2\rho}\frac{F(\xi)}{\xi-z}\int_{|t|=R}\frac{dt}{\phi(t)}\int_0^\infty\frac{\phi(x/(\xi-z))-\phi(t)}{x/(\xi-z)-t}e^{-x}dxd\xi,$$

$\sigma_0>\sigma_1>\sigma$, is a linear combination of functions $z^se^{\alpha z}$, and the difference between this and $F(z)$ can be estimated. If $F(z)$ is an entire function, something can be said about the rapidity of convergence of the approximating sums.

The author next considers the inhomogeneous equation $L[F]=\Phi(z)$, with the object of showing that there is a solution of roughly the same growth as $\Phi(z)$ when $\Phi(z)$ is an entire function of greater than exponential type. His result is expressed in terms of a regularized maximum modulus $\bar{M}(r)$, defined for $r=|z|>1$ by

$$\log\bar{M}(r)=r\max_{1\leqq t\leqq r}t^{-1}\log M(t).$$

where $M(t)$ is the maximum modulus. Then if $\Phi(z)$ is of greater than exponential type, $\epsilon>0$ and $\theta>1$, the equation $L[F]=\Phi(z)$ has a solution $F_0(z)$ such that

$$|F_0(z)|<c(\epsilon,\theta)[\bar{M}(\theta\rho)]^{(1+\epsilon)/\log\theta},\quad |z|\leqq\rho.$$

A function F with period ω satisfies the equation

$$L[F]=F(z+\omega)-F(z)=0,$$

and the corresponding $\phi(z)$ is $e^{\omega z}-1$. Hence a generalization of the fact that a nonconstant entire function cannot have two independent periods is the theorem, which the author proves, that if $F(z)$ satisfies $L_1[F]=L_2[F]=0$, with $\phi_1(t)$ and $\phi_2(t)$ having no zeros in common, except perhaps the origin, then $F(z)$ must be a polynomial, of degree not exceeding one less than the multiplicity of the zero at the origin which has smaller multiplicity.

Finally the author generalizes Whittaker's definition of an asymptotic period [Interpolatory function theory, Cambridge, 1935] by requiring that $F(z+\omega)-F(z)$ is of lower order than $F(z)$ in a generalized sense. He calls $F(z)$ of greater order than $F_1(z)$ if there are positive increasing functions u and w, such that $u(2x)/u(x)\to 1$ and for some $\theta>1$

$$\limsup u[\log\bar{M}(r)]/w(r)>\limsup u[\log\bar{M}_1(\theta r)]/w(r),$$

$\bar{M}$ being defined as above. (If $u(x)$ and $w(x)$ are both $\log x$ we have the usual definition of order and asymptotic period.) Then if an entire function has any asymptotic periods in the generalized sense, they lie on a single straight line, they form a set of measure zero (these properties were proved by

Whittaker with the old definition), and the ratio of two asymptotic periods is either a rational or a transcendental number.

R. P. Boas, Jr. (Evanston, Ill.).

Referred to in J76-28, Q05-21.

J76-28 **(14, 739f)**

Gel'fond, A. O. Linear differential equations of infinite order with constant coefficients and asymptotic periods of entire functions. Amer. Math. Soc. Translation no. **84**, 31 pp. (1953).

Translated from Trudy Mat. Inst. Steklov. **38**, 42–67 (1951); these Rev. **13**, 929.

Citations: MR 13, 929a = J76-27.

J76-29 **(14, 541f)**

Redheffer, R. M. Power series and algebraic numbers. Amer. Math. Monthly **60**, 25–27 (1953).

The author proves the following theorem: Let $f(z)=\sum a_k z^k$ be a rational function with algebraic coefficients; assume further that $f(z)$ is not a polynomial and that $f(z)$ is regular at 0. Let m be the multiplicity of that root of the denominator of $f(z)$ which has maximum multiplicity. Then, apart from a set of at most m values, the value of the function $g(z)=\sum a_k z^k/k!$ is transcendental when the value of z is algebraic. Some related questions are discussed.

P. Erdös (Los Angeles, Calif.).

J76-30 **(15, 106h)**

Kasch, Friedrich. Zur Annäherung algebraischer Zahlen durch arithmetisch charakterisierte rationale Zahlen. Math. Nachr. **10**, 85–98 (1953).

In the well-known theorem of Th. Schneider [J. Reine Angew. Math. **188**, 115–128 (1950); these Rev. **12**, 483] concerning the rate of growth of the denominators in a sequence of rational numbers p_i/q_i which provide sufficiently good approximations to a given algebraic number, it is assumed that these denominators are essentially products of powers of a fixed set of primes. In the present work, an analogous result is obtained for the case that the q's are essentially factorials. Suppose that $(p_i, q_i)=1$, $q_i\leqq q_{i+1}$, and that $q_i=q_i'q_i''$, where $q_i''f_i=l_i!$ for some integers f_i and l_i, and

$$\limsup_{i\to\infty}\frac{\log q_i'}{\log q_i}=\omega_1,\quad \limsup_{i\to\infty}\frac{\log f_i}{\log q_i}=\omega_2.$$

Then if for some algebraic number α, the inequality $|\alpha-p_i/q_i|<q_i^{-\mu}$ holds for all i, where

$$\mu>1+\omega_1+\omega_2((1-\omega_1)(2+\omega_2)+\omega_1^2),$$

then $\limsup_{i\to\infty}(\log q_{i+1})/(\log q_i)=\infty$. Using this theorem, the author is able to show the transcendence of the values of certain series, of which a special case is $\sum_0^\infty x^n/([c^n])!$, where $c>1$ and $x\neq 0$ is rational. *W. J. LeVeque.*

Citations: MR 12, 483c = J68-11.

J76-31 **(15, 604f)**

Günther, Alfred. Über transzendente $\mathfrak{p}$-adische Zahlen. I. Ein Satz über algebraische Abhängigkeit $\mathfrak{p}$-adischer Funktionen als Prinzip für Transzendenzbeweise $\mathfrak{p}$-adischer Zahlen. J. Reine Angew. Math. **192**, 155–166 (1953).

A $\mathfrak{p}$-adic analog of a theorem of T. Schneider [Math. Ann. **121**, 131–140 (1949); these Rev. **11**, 160] concerning the algebraic dependence of a set of integral-valued functions is established. From this the author obtains the two following $\mathfrak{p}$-adic generalizations of two theorems of Mahler [J. Reine Angew. Math. **169**, 61–66 (1932); Compositio Math. **2**, 259–275 (1935)] and a theorem of Veldkamp [J. London Math. Soc. **15**, 183–192 (1940); these Rev. **2**, 149]: Let p be the rational prime number of which $\mathfrak{p}$ is a factor and put $p^{-1/(p-1)}=q$. 1. Let α be a $\mathfrak{p}$-adic number such that $0<|\alpha|_\mathfrak{p}<q$; then one at least of the two numbers α and e^α is transcendental. 2. Let $\alpha\neq 0$, 1 be such that $|\alpha-1|_\mathfrak{p}<q$ and β an irrational $\mathfrak{p}$-adic integer; then one at least of the numbers α, β, α^β is transcendental. It is further shown that for a $\mathfrak{p}$-adic number α with $0<|\alpha|_\mathfrak{p}<q$ one at least of the numbers α and $\sin\alpha$ (or $\cos\alpha$) is transcendental.

O. Taussky-Todd (Washington, D. C.).

Citations: MR 2, 149d = J76-2; MR 11, 160a = J76-17.

Referred to in J80-10, Q05-30.

J76-32 **(16, 907b)**

Boughon, Pierre, Nathan, Jacqueline, et Samuel, Pierre. Une classe de séries formelles transcendantes. Acad. Roy. Belg. Bull. Cl. Sci. (5) **41**, 93–96 (1955).

Let k denote a field of arbitrary characteristic and consider formal power series in an indeterminate T with coefficients in k. The object of the paper is to find sufficient conditions that a power series be transcendental relative to $k(T)$. Let $s=\sum_1^\infty a_i T^{d_i}$ $(a_i\in k, a_i\neq 0)$. Theorem 1. A sufficient condition that the series s not be algebraic of degree n relative to $k(T)$ is that there exist an $\epsilon>0$ and such that $d_{i+1}/d_i>n+\epsilon$ for infinitely many i. Theorem 2. A sufficient condition that s be transcendental is that the ratio d_{i+1}/d_i be unbounded. For example, $\sum T^{n!}$ is transcendental over $k(T)$ for any k. The series $(1-T)^{-1}+\sum T^a$, $a=2^{n^2}$, is transcendental over $k(T)$, $k=\mathrm{GF}(2)$; similarly for $(1-T)^{-1}+\sum T^b$, $b=3^{n^2}$, when $k=\mathrm{GF}(3)$. [For analogous results see, e.g. L. I. Wade, Duke Math. J. **11**, 755–758 (1944); MR **6**, 144; and S. M. Spencer, ibid. **19**, 93–105 (1952); MR **13**, 538.]

L. Carlitz (Durham, N. C.).

Citations: MR 6, 144h = J76-8; MR 13, 538g = J76-26.

J76-33 **(17, 117h)**

Popken, J. Un aperçu historique sur les nombres transcendants. Bull. Soc. Math. Belg. **1954,** 71–82 (1955).

J76-34 **(18, 566d)**

Popken, J. Un théorème sur les nombres transcendants. Bull. Soc. Math. Belg. **7** (1955), 124–130.

Let $F(z)$ be a polynomial with algebraic coefficients, which vanishes at zero but not identically. Let r be a rational number, and for $n=1, 2, \cdots$ put

$$F_n=\begin{cases}\dfrac{1}{n+r}\left(\dfrac{d^n}{dz^n}e^{(n+r)F(z)}\right)_{z=0} & \text{if } n+r\neq 0,\\ 0 & \text{if } n+r=0.\end{cases}$$

Then the series

$$f(z)=\sum_{n=1}^\infty F_n\frac{z^n}{n!}$$

defines an analytic function f. It is shown that for each algebraic number z different from the zeros of F, the function f is regular, and $f^{(k)}(z)$ is transcendental for $k=0, 1, 2, \cdots$. This theorem generalizes a result of the reviewer [Proc. Amer. Math. Soc. **2** (1951), 401–403; MR 13, 16] dealing with the special case $F(z)=z$.

W. J. LeVeque (Ann Arbor, Mich.).

Citations: MR 13, 16f = J76-23.

Referred to in J76-35, J76-42.

J76-35 **(19, 122f)**

Popken, J. Some theorems concerning transcendental numbers. Colloque sur la Théorie des Nombres, Bruxelles, 1955, pp. 107–110. Georges Thone, Liège; Masson and Cie, Paris, 1956.

Some theorems are stated, the proofs of which appear in full in Bull. Soc. Math. Belg. **7** (1955), 124–130 [MR **18**,

566]. Main theorem: Let $F(z)$ denote a polynomial with algebraic coefficients $\not\equiv 0$ such that $F(0)=0$; moreover let r denote an arbitrary rational constant. Put for $n=1, 2, \cdots$

$$F_n = \begin{cases} \frac{1}{n+r}\left(\frac{d^n}{dz^n} e^{(n+r)F(z)}\right)_{z=0} & \text{if } n+r\neq 0, \\ 0 & \text{if } n+r=0. \end{cases}$$

Then $f(z)=\sum_{n=1}^{\infty} F_n z^n/n!$ generates a many-valued analytic function. For every algebraic z, distinct from the zeros of $F(z)$, the function $f(z)$ is regular, and $f(z)$ and all its derivatives take there transcendental values.

J. F. Koksma (Amsterdam).

Citations: MR 18, 566d = J76-34.

J76-36 (21# 2640)

Fel'dman, N. I. Transcendence of numbers of certain classes. Uspehi. Mat. Nauk **14** (1959), no. 1 (85), 237–244. (Russian)

The author proves the following theorem and uses it to prove the transcendency of decimal fractions and series: "If α and $k>1$ are real constants, and if the inequality $|\alpha - p/q| < q^{-k}$ has infinitely many solutions in integers p, q such that $(p, q)=1$, $q=q'q''$, where q' and q'' are positive integers such that $\lim (\log q'/\log q)=0$ and q'' has only bounded prime factors, then α is transcendental." A slightly more general result was already published by D. Ridout [Mathematika **4** (1957), 125–131; MR **20** #32].

K. Mahler (Manchester)

Citations: MR 20# 32 = J68-18.

J76-37 (23# A112)

İçen, Orhan Ş.

Über die Funktionswerte der p-adischen Reihe $\sum_{\nu=0}^{\infty} p^{a\binom{\nu}{2}} z^{\nu}$.

J. Reine Angew. Math. **202** (1959), 100–106.

Let p be a prime, a a positive integer, z a p-adic variable, and

$$f(z) = \sum_{\nu=0}^{\infty} p^{a\binom{\nu}{2}} z^{\nu},$$

so that, identically in z, $zf(p^a z)=f(z)-1$. It suffices thus to investigate $f(z)$ when $p^{-a} < |z|_p \leqq 1$. The author proves: "If t is any positive integer, there can be at most $k(t)=32t^4$ distinct rational numbers z satisfying $p^{-a}<|z|_p \leqq 1$ such that $f(z)$ is algebraic at most of degree t." Thus there are rational $z\neq 0$ for which $f(z)$ either is transcendental, or is algebraic of arbitrarily high degree. The proof is based on the property of $f(z)$ not being an algebraic function of z.

K. Mahler (Manchester)

J76-38 (23# A1779)

Mahler, K.

An application of Jensen's formula to polynomials.

Mathematika **7** (1960), 98–100.

Let $f(x)=a_0x^n+a_1x^{n-1}+\cdots+a_n$ ($a_0\neq 0$, $a_n\neq 0$) be a polynomial with complex coefficients, and write $S(f)=|a_0|+|a_1|+\cdots+|a_n|$. It is obvious that

$$\frac{1}{2\pi}\int_0^{2\pi} \log |f(e^{i\theta})|\, d\theta \leqq \log S(f).$$

Making use of Jensen's integral formula, the author shows that there is a similar lower estimate for the integral, namely,

$$\frac{1}{2\pi}\int_0^{2\pi} \log |f(e^{i\theta})|\, d\theta \geqq \log S(f) - n \log 2.$$

From these relations he deduces inequalities of N. I. Fel'dman [Izv. Akad. Nauk SSSR Ser. Mat. **15** (1951), 53–74; MR **12**, 595] and A. O. Gel'fond [*Transcendental and algebraic numbers* (Russian), GITTL, Moscow, 1952; pp. 22–24, 168–173; MR **15**, 292]. The first of these states that, for any zeros $\xi_1, \cdots, \xi_m$ of f,

$$|a_0\xi_1\cdots\xi_m| \leqq S(f).$$

The second states that, if $f=f_1f_2\cdots f_s$, where the f_i are polynomials, then

$$S(f_1)S(f_2)\cdots S(f_s) \leqq 2^n S(f).$$

L. Mirsky (Sheffield)

Referred to in C05-29, J68-37, J68-57.

J76-39 (24# A715)

Schmidt, Hermann

Eine Anwendung der Siegelschen Transzendenzsätze für Bessel-Funktionen.

Math. Z. **77** (1961), 309–313.

Let the sequence of algebraic numbers A_ν ($\nu=0, 1, \cdots$) satisfy for $\nu \geqq \nu_0$ ($\geqq 0$) a linear difference equation $A_{\nu+m}+\beta_1 A_{\nu+m-1}+\cdots+\beta_m A_\nu=0$ of order m with constant algebraic coefficients β_μ ($\mu=1, 2, \cdots, m$). The author shows first that, if the power series $\sum_{\nu=0}^{\infty} (A_\nu/\nu!)z^\nu$ does not degenerate into a polynomial, then it represents an entire function $F(z)$ with the property that for at most a finite number of values of z both values z and $F(z)$ can be algebraic. He further derives the same property for the power series

$$\sum_{\nu=0}^{\infty} \frac{(-1)^\nu A_\nu}{\nu!(\alpha+1)\cdots(\alpha+\nu)} z^\nu,$$

where α denotes a rational number different from $-1, -2, \cdots$, and such that 2α is not an odd integer. Moreover, the author gives certain algebraic refinements of the above two theorems.

The result for the first power series is obtained (by means of standard formulae from the theory of difference equations) from the Lindemann-Weierstrass theorem. The second result follows in the same manner from a similar theorem of Siegel concerning Bessel functions [*Transcendental numbers*, Princeton Univ. Press, Princeton, N.J., 1949; p. 73; MR **11**, 330].

As the author remarks the first result is also contained in a theorem by the reviewer [Nederl. Akad. Wetensch. Proc. Ser. A **53** (1950), 1645–1656; MR **12**, 600; compare also T. Itihara and K. Ôishi, Tôhoku Math. J. **37** (1933), p. 209, Th. 1].

J. Popken (Amstelveen)

Citations: MR 11, 330c = J76-18; MR 12, 600g = Q10-5.

J76-40 (25# 2036)

Mahler, K.

On some inequalities for polynomials in several variables.

J. London Math. Soc. **37** (1962), 341–344.

Let

$$f(z_1, \cdots, z_n) = \sum_{k_1=0}^{m_1} \cdots \sum_{k_n=0}^{m_n} a_{k_1\cdots k_n} z_1^{k_1}\cdots z_n^{k_n}$$

be a polynomial with complex coefficients in the n indeterminants $z_1, \cdots, z_n$; and write

$$L(f) = \sum_{k_1=0}^{m_1}\cdots\sum_{k_n=0}^{m_n} |a_{k_1\cdots k_n}|, \quad H(f) = \max |a_{k_1\cdots k_n}|.$$

Further, define $M(f)$ as

$$\exp \int_0^1 dt_1 \cdots \int_0^1 dt_n \cdot \log |f(e^{2\pi i t_1}, \cdots, e^{2\pi i t_n})|$$

if f does not vanish identically, and as 0 otherwise. The author establishes a number of inequalities connecting

$L(f)$, $H(f)$, $M(f)$, and the coefficients of f. He shows, for example, that

$$|a_{k_1\cdots k_n}| \leq \binom{m_1}{k_1}\cdots\binom{m_n}{k_n}M(f)$$

and $M(f) \leqq L(f) \leqq 2^{m_1+\cdots+m_n}M(f)$. Again, let $f=f_1\cdots f_s$. It is then shown that

$$\prod_{k=1}^{s} L(f_k) \leq 2^{m_1+\cdots+m_n}L(f)$$

and an upper bound is obtained for $H(f_1)\cdots H(f_s)$ in terms of $H(f)$. This last result has been established previously in a different way by A. O. Gelfond [*Transcendental and algebraic numbers* (Russian), pp. 168–173, GITTL, Moscow, 1952; MR **15**, 292]. *L. Mirsky* (Sheffield)

Citations: MR 15, 292e = J02-6.

J76-41 (26# 2394)

Baker, A.

Continued fractions of transcendental numbers.

Mathematika **9** (1962), 1–8.

Let the real number ξ have the non-periodic continued fraction

$$\xi = a_0+\frac{1|}{|a_1}+\frac{1|}{|a_2}+\cdots$$

with approximants p_n/q_n. E. Maillet [*Introduction à la théorie des nombres transcendants*, Ch. VII, Paris, 1906] proved that ξ must be transcendental if there exist infinitely many n such that

(1) $$a_n = a_{n+1} = \cdots = a_{n+\lambda(n)-1},$$

where $\lambda(n)$ is greater than a certain function of q_n. The author sharpens this result considerably as follows: For the non-periodic continued fraction ξ, let the relation (1) hold for $n=n_1, n_2, \cdots$, such that

(2) $$(\log \lambda_i)(\log n_i)^{1/2}/n_i \to \infty \text{ as } i\to\infty,$$

where $\lambda_i=\lambda(n_i)$. Then ξ is transcendental. The author generalises this result as follows: According to (1) a single partial quotient is repeated $\lambda(n)-1$ times. It is now shown: Let $k_i<Cn_i$ for some constant C, let the block of consecutive partial quotients $(a_{n_i}, a_{n_i+1}, \cdots, a_{n_i+k_i-1})$ be repeated λ_i-1 times. Then the condition (2) implies that ξ is transcendental.

If ξ has bounded partial quotients a_n, then the condition (2) can be replaced by the much weaker: $\limsup(\lambda_i/n_i)>B$, where B is a constant depending on the upper bound given for the a_n.

Finally the author proves a similar result for so-called quasi-periodic continued fractions.

The proofs are based upon well-known theorems of Liouville type and K. F. Roth type for the approximation of an algebraic number by numbers from a quadratic field. *J. Popken* (Berkeley, Calif.)

J76-42 (27# 2484)

Juberg, R. K.

Transcendental values of a class of functions.

Nederl. Akad. Wetensch. Proc. Ser. A **66**=*Indag. Math.* **25** (1963), 387–393.

The author shows that for certain classes of multi-valued analytic functions each function $f(z)$ and all its derivatives assume only transcendental values for almost all algebraic numbers z ("almost all" means here that a finite set of numbers must be excluded).

In the following let α be an algebraic irrational and let $\beta=r\alpha+s$, where r and s are arbitrary rationals. Further set $p^{(0)}=1$, $p^{(n)}=p(p-1)\cdots(p-n+1)$. Then the author's most general result reads: "Let $F(z)$ be a polynomial with algebraic coefficients of degree at least one and such that $F(0)=0$. Set, for $n=1, 2, 3, \cdots$,

$$F_n = (n\alpha+\beta)^{-1}[(d/dz)^n(1+F(z))^{n\alpha+\beta}]_{z=0} \text{ if } n\alpha+\beta \neq 0,$$
$$= 0 \qquad \text{if } n\alpha+\beta = 0.$$

Then the series $\sum_{n=1}^{\infty} F_n(z^n/n!)$ generates an analytic function $f(z)$. For every algebraic number z, distinct from the zeros of $F(z)$, the function $f(z)$ is regular and $f(z)$ together with all its derivatives has only transcendental values."

For the special case $F(z)=z$ this leads to functions $f_k(z)$ generated by Taylor series

$$\sum_{n=0}^{\infty} (n\alpha+\beta')^{(n+k)}(z^n/n!) \quad (k = 0, 1, \cdots),$$

where β' has the form $R\alpha+S$ (R and S denoting rationals).

A second result of the author deals with functions $f_\lambda(z)$ generated by the series

$$\sum_{n=\lambda}^{\infty} (\alpha n+\beta)^{(n-\lambda)}(z^n/n!) \quad (\lambda = 1, 2, 3, \cdots).$$

The proofs of these theorems use from the theory of transcendental numbers only the Gelfond-Schneider theorem and the Lindemann-Weierstrass theorem. See for closely related results the paper of the reviewer [Bull. Soc. Math. Belg. **7** (1955), 124–130; MR **18**, 566], and also W. J. LeVeque [Proc. Amer. Math. Soc. **2** (1951), 401–403; MR **13**, 16]. *J. Popken* (Amstelveen)

Citations: MR 13, 16f = J76-23; MR 18, 566d = J76-34.

J76-43 (27# 3587)

Roth, H.

On a theorem of Maillet.

Tôhoku Math. J. (2) **15** (1963), 114–115.

According to K. Roth's theorem, a real number ρ is transcendental if there are infinitely many pairs of integers p, q such that $0<|\rho-p/q|<q^{-\tau}$ for some fixed $\tau>2$. The author observes that this is the case if and only if $\limsup(\log b_{n+1}/\log q_n) \geqq \tau-2$, where b_n and p_n/q_n are the nth partial quotient and convergent, respectively, in the continued fraction expansion of ρ.

W. J. LeVeque (Ann Arbor, Mich.)

J76-44 (27# 5729)

Platonov, M. L.

Examples of sets of transcendental numbers of the form ξ^η. (Russian)

Izv. Vysš. Učebn. Zaved. Matematika **1962**, *no.* 6 (31), 91–100.

The author quotes his earlier result [Trudy Irkutsk. Univ. **15** (1957), 82–98] and deduces from it a number of simpler criteria for transcendency of numbers of the form ξ^η. A particular example is the following. If $\xi=\sum_{n=1}^{\infty} 2^{-n^{6n}}$ and η is an irrational algebraic number, then ξ^η is transcendental. *S. Knapowski* (New Orleans, La.)

J76-45 (28# 95)

Lang, Serge

Transcendental points on group varieties.

Topology **1** (1962), 313–318.

Let exp be the exponential function on a group variety G defined over the field of algebraic numbers K, $\mathfrak{g}$ the Lie algebra of G at the origin and α a non-zero element of $\mathfrak{g}$, rational over K. Then the author proves that $\exp(\alpha)$ is transcendental over K, provided that $\exp(t\alpha)$ is not an algebraic function of t.

When G is a linear group, G has a matrix representation over K. Then α can be identified with a matrix M and $\exp(M)=\sum M^n/n!$. When M is rational over K, it can be conjugated to a matrix consisting of blocks of the type $aI+N$, where I is the identity matrix, N a nilpotent matrix and a an algebraic number. Since $\exp(aI+N) = e^a I \times$ (rational matrix), the above result follows from the corresponding result for ordinary exponential function in this case.

In general, G contains a maximal linear subgroup L such that $A = G/L$ is an Abelian variety. Let π be the canonical map: $G \to A$ and π_* the induced homomorphism on Lie algebras. If $\pi_* \alpha \neq 0$, then it is enough to prove that $\pi \circ \exp_G(\alpha) = \exp_A \circ \pi_*(\alpha)$ is transcendental. Hence the general case is reduced to the case of an Abelian variety A^n. Let Θ be a standard map of C^n onto A by means of theta functions θ_i. After making a suitable change of coordinates in C^n, we may assume that the $\partial/\partial z_i$ are defined over K, which form a basis of the Lie algebra of invariant derivations on A. In this case, $\exp(t\alpha) = \Theta(t\alpha)$ and α is a non-zero complex vector with n coordinates. Let k be an algebraic number field of finite degree over which A is defined. If $\exp(\alpha)$ is rational over k, so are the $\exp(t\alpha)$ for $t = 1, 2, \cdots$. Replacing K by a suitable k, and assuming that $\exp(\alpha)$ is rational over k, we can find m distinct integers t for which the $\exp(t\alpha)$ are rational over k. Let δ be the order of the θ_i and take $m > 10 \cdot \delta \cdot [k : \mathbf{Q}]$. We may assume that θ_0 does not vanish at these m distinct points. θ_i and $f_i = \theta_i/\theta_0$ induce functions $\bar{\theta}_i$ and $\bar{f}_i$ of one variable t obtained from $\Theta(t\alpha)$, and the ring $k[\bar{f}_1, \cdots, \bar{f}_N]$ is stable under d/dt. Since $\Theta(t\alpha)$ is not an algebraic function of t, $\dim_k k(t, \bar{f}_1, \cdots, \bar{f}_N) \geqq 2$. On the other hand, it can be proved, under these situations, that $m \leqq 10 \cdot \delta \cdot [k : \mathbf{Q}]$. This completes the proof.

T. Matsusaka (Waltham, Mass.)

Referred to in J02-22, J76-51, J76-52, J76-55.

J76-46 (28# 1170)

Kolberg, O.

A class of power series with transcendental sums for algebraic values of the variable.

Årbok Univ. Bergen Mat.-Natur. Ser. **1962**, *no.* 18, 6 *pp.*

Let k be a positive integer and r any rational number except $-1, -2, \cdots, -k+1$. Let $P(x)$ be a polynomial, not identically zero, with algebraic coefficients. If β is any non-zero algebraic number satisfying $|\beta| < e^{-1}$, then $\sum_1^\infty (n+r)^{n-k} P(n) \beta^n/n!$ is transcendental. The proof employs the Lindemann theorem that e^β is transcendental.

I. Niven (Eugene, Ore.)

J76-47 (29# 70)

Lang, Serge

Transzendente Zahlen.

Bonn. Math. Schr. No. 21 (1963), viii+22 *pp.*

The main text contains an elegant and elementary exposition of a theorem and proof due in substance to T. Schneider [*Einführung in die transzendenten Zahlen*, Satz 13, Springer, Berlin, 1957; MR **19**, 252], concerning the algebraic dependence of sets of functions satisfying suitable analytic and arithmetic conditions. In the foreword, various unsolved problems are discussed, one of them being to find circumstances under which the following assertion is true: Let G be an algebraic group variety of dimension n, defined over an algebraic number field. Let T be its tangent space at the unit element, with its structure as algebraic vector space. Let $\alpha \in T$ be a non-vanishing tangent vector, and let the curve $\exp(t\alpha)$ be of dimension n over the field of rational functions in t. Then the transcendence degree of the point $(\alpha, \exp(\alpha))$ is at least n.

W. J. LeVeque (Ann Arbor, Mich.)

Citations: MR 19, 252f = J02-9.

J76-48 (29# 402)

Jager, H.

A multidimensional generalization of the Padé table. I, II, III.

Nederl. Akad. Wetensch. Proc. Ser. A **67** = *Indag. Math.* **26** (1964), 193–198, 199–211, 212–225.

This series of papers is based on some unpublished work of K. Mahler, and concerns the development of an idea of Hermite ["Sur la fonction exponentielle", *Œuvres de Charles Hermite*, Tome III, pp. 150–181, Gauthier-Villars, Paris, 1912]. The author's starting point is the observation that for any m positive integers ρ_i ($i = 1, 2, \cdots, m$) there exists a non-trivial set of polynomials $a_i(x)$ (where $a_i(x)$ is of degree $\rho_i - 1$ at most) such that $r(x) = \sum_{i=1}^m a_i(x) f_i(x)$ is $O(x^{\sigma-1})$ as $x \to 0$, where $\sigma = \sum_{i=1}^m \rho_i$. Also there exists a set of non-trivial polynomials $\bar{a}_i(x)$ (where $\bar{a}_i(x)$ is of degree $\sigma - \rho_i$ at most) such that $\bar{r}_{kl}(x) = \bar{a}_l(x) f_k(x) - \bar{a}_k(x) f_l(x)$ for all $k, l = 1, 2, \cdots, m$ is $O(x^{\sigma+1})$ as $x \to 0$. Hermite considered the case in which $f_i(x) = \exp(w_i x)$; the author here systematically generalizes the theory and extends the investigations to cover the cases in which $f_i(x)$ is a formal power series. The study of the Padé table is included in the case $m = 2$ with $f_i(x) = 1$.

C. W. Clenshaw (Teddington)

Referred to in J76-49, J76-56, J76-59, J76-64.

J76-49 (29# 2576)

Jager, H.

A multidimensional generalization of the Padé table. IV, V, VI.

Nederl. Akad. Wetensch. Proc. Ser. A **67** = *Indag. Math.* **26** (1964), 227–239, 240–244, 245–249.

Each of these three papers develops in some detail a special case of the more general function system considered in the first three papers of the series [same Proc. **67** (1964), 193–198, 199–211, 212–225; MR **29** #402]. These special cases are (i) the binomial function system, (ii) the exponential function system, and (iii) the logarithmic function system.

C. W. Clenshaw (Teddington)

Citations: MR 29# 402 = J76-48.

Referred to in J76-56, J76-59, J76-64.

J76-50 (30# 1098)

Mignosi, Giuseppe

Sul teorema di A. Gelford [Gel'fond] relativo al settimo problema di Hilbert.

Atti Accad. Sci. Lett. Arti Palermo Parte I (4) **23** (1962/63), 95–127 (1964).

The author gives a detailed and carefully worked out version of Gelfond's classical proof of 1929 of the transcendency of β^η, where β and η are algebraic numbers, β is distinct from 0 and 1, and η is irrational.

K. Mahler (Canberra)

J76-51 (32# 7506)

Lang, Serge

Algebraic values of meromorphic functions.

Topology **3** (1965), 183–191.

In his previous paper [Topology **1** (1962), 313–318; MR **28** #95] in order to produce a transcendental number from an algebraic group variety defined over an algebraic number field K, the author proved the following theorem. Let the g_i, $1 \leqq i \leqq N$, be meromorphic functions of one variable t of order $\leqq \rho$ such that d/dt leaves the ring $K[g_1, \cdots, g_N] = \mathfrak{o}$ invariant and $\dim_K \mathfrak{o} \geqq 2$. Let S be a set of points in the complex plane, not containing the poles of the g_i, such that $g_i(P)$ is in K whenever P is in S. Then S is finite and the number of points in S is bounded by $2\rho[K:Q]$. In this paper, this theorem is generalized to the case of several variables. That is, the g_i may be replaced by meromorphic functions on C^n; d/dt is then replaced by n partial derivatives; assume that $\dim_K \mathfrak{o} \geqq n+1$; S is now a set of points in C^n such that g_i is defined and $g_i(P)$ is in K for all i and P in S; furthermore, S is assumed to become $S_1 \times \cdots \times S_n$ after a suitable change of coordinates, where each S_i is a set of m distinct complex numbers. Then the conclusion is that m is bounded by $b\rho[K:Q]$, where b is a constant which depends only on n. As an application, a theorem of Schneider on periods of theta functions is shown to follow directly from this generalized theorem [cf. T. Schneider, J. Reine Angew. Math. **183** (1941), 110–128; MR **3**, 266].

T. Matsusaka (Waltham, Mass.)

Citations: MR 3, 266b = J76-4; MR 28# 95 = J76-45.
Referred to in J02-22, J76-55, J76-58, Q05-64.

J76-52 (32# 7507)

Rauzy, Gérard

Points transcendants sur les variétés de groupe.

Séminaire Bourbaki, 1963/64, *Fasc.* 3, *Exposé* 276, 8 *pp. Secrétariat mathématique, Paris*, 1964.

This is an expository paper on the works of S. Lang [Topology **1** (1962), 313–318; MR **28** #95; and #7506 above].

T. Matsusaka (Waltham, Mass.)

Citations: MR 28# 95 = J76-45.
Referred to in Z10-31.

J76-53 (32# 7508)

Mahler, K.

Arithmetic properties of lacunary power series with integral coefficients.

J. Austral. Math. Soc. **5** (1965), 56–64.

The author calls a power series $f(z)=\sum_0^\infty f_h z^h$ admissible if (1) it has integral coefficients, (2) there are two infinite sequences of integers, $\{r_n\}$ and $\{s_n\}$, such that $0=s_0\leqq r_1< s_1\leqq r_2<s_2\leqq \cdots$ and $\lim_{n\to\infty} s_n/r_n=\infty$, for which $f_h=0$ if $r_n<h<s_n$ but $f_{r_n}\neq 0$, $f_{s_n}\neq 0$ $(n=1,2,3,\cdots)$, and (3) f has positive radius of convergence, R_f. It is shown that if α is algebraic and $|\alpha|<R_f$, where f is admissible, then $f(\alpha)$ is algebraic if and only if, for some $N=N(\alpha)$, $P_n(\alpha)=0$ for all $n\geqq N$, where $P_n(z)=\sum_{s_n}^{r_{n+1}} f_h z^h$. The proof depends on a lower bound given by R. Güting [Michigan Math. J. **8** (1961), 149–159; MR **24** #A2559] for the value of a polynomial at an algebraic number not a zero of the polynomial.

Among further results, the following theorem is proved. Suppose that $0<R\leqq 1$, let Σ be the set of all algebraic numbers satisfying $|\alpha|<R$, and let S be any subset of Σ which contains 0 and, with any element α, all the conjugates of α which lie in Σ. Then there is an admissible power series $f(z)$ such that $R_f=R$ and $S_f=S$, where S_f is the set of all algebraic β for which $|\beta|<R_f$ and $f(\beta)$ is algebraic.

W. J. LeVeque (Ann Arbor, Mich.)

Citations: MR 24# A2559 = J68-37.

J76-54 (33# 4012)

Peršikova, T. V.

On the transcendence of values of certain E-functions. (Russian. English summary)

Vestnik Moskov. Univ. Ser. I Mat. Meh. **21** (1966), *no.* 2, 55–61.

Let t be a positive integer, and let $f(z)=\sum_{n=0}^\infty C_n z^{nt}/n!$ be a Siegel E-function [C. L. Siegel, *Transcendental numbers*, Princeton Univ. Press, Princeton, N.J., 1949; MR **11**, 330] satisfying a linear differential equation $P_m y^{(m)}+\cdots+P_0 y=Q$, where $P_m=z^q, P_{m-1},\cdots,P_1,P_0,Q$ are polynomials of degree at most q. Assume that $f(z)$ has the following property: If $\beta_1,\cdots,\beta_k$ are algebraic numbers $\neq 0$ such that $\beta_h{}^t\neq\beta_i{}^t$ for $h\neq i$, then the numbers $1, f^{(j)}(\beta_i)$ $(i=1,2,\cdots,k; j=0,1,\cdots,\min(m-1,s-1))$ are linearly independent over the field of algebraic numbers. Denote by $\{\gamma_n\}$ a recursive sequence of algebraic numbers, $\gamma_{n+k}=a_{k-1}\gamma_{n+k-1}+\cdots+a_0\gamma_n$, where $a_0\neq 0,\cdots,a_{k-1}$ are algebraic numbers. Then, the function $F(z)=\sum_{n=0}^\infty \gamma_n C_n z^{nt}/n!$ is for $s\leqq m$ transcendental for all algebraic $z=\alpha\neq 0$, but for $s>m$ is transcendental for all except at most $(s-1)\max(q,1)$ algebraic $z=\alpha\neq 0$. The author applies this result to a number of examples.

K. Mahler (Canberra)

Citations: MR 11, 330c = J76-18.

J76-55 (33# 5564)

Adams, William W.

Transcendental numbers in the P-adic domain.

Amer. J. Math. **88** (1966), 279–308.

S. Lang, in several papers [Topology **1** (1962), 313–318; MR **28** #95; ibid. **3** (1965), 183–191; MR **32** #7506; see also the appendix of his book *Algebra*, pp. 493–499, Addison-Wesley, Reading, Mass., 1965; #5416 above], generalised the basic work by Th. Schneider on the solutions of algebraic differential equations [*Einführung in die transzendenten Zahlen*, Springer, Berlin, 1957; MR **19**, 252; French translation, Gauthier-Villars, Paris, 1959; MR **21** #5620]. While this work was concerned with analytic functions of complex variables, the author establishes here an analogous theory for analytic functions of a P-adic variable. From these results he deduces new proofs of the transcendency of the P-adic functions e^α and α^β, and also a measure of transcendency of α^β. He further shows that if $\alpha\neq 0$ and β lie in an algebraic extension of the P-adic field, α is not a root of unity, and β is of degree $r\geqq 4$ over Q, and if further $|\alpha-1|_P<1$, $|\beta^k\log\alpha|_P<P^{-1/(P-1)}$ for $k=1,2,\cdots,r$, then the transcendency degree of the r numbers $\alpha^\beta, \alpha^{\beta^2},\cdots,\alpha^{\beta^{r-1}}$ is at least 2.

K. Mahler (Canberra)

Citations: MR 19, 252f = J02-9; MR 21# 5620 = J02-10; MR 28# 95 = J76-45; MR 32# 7506 = J76-51.

J76-56 (34# 2527a; 34# 2527b; 34# 2527c)

Coates, John

On the algebraic approximation of functions. I.

Nederl. Akad. Wetensch. Proc. Ser. A **69** = *Indag. Math.* **28** (1966), 421–434.

Coates, John

On the algebraic approximation of functions. II.

Nederl. Akad. Wetensch. Proc. Ser. A **69** = *Indag. Math.* **28** (1966), 435–448.

Coates, John

On the algebraic approximation of functions. III.

Nederl. Akad. Wetensch. Proc. Ser. A **69** = *Indag. Math.* **28** (1966), 449–461.

The field F of "constants", the Euclidean domain ω of "polynomials", and the domain of integrity Ω of "functions" are to satisfy $F\subset\omega\subset\Omega$ and to have the following properties. The ring ω has a valuation $|a|$ such that $|0|=-\infty$; $|a|\geqq 0$ is an integer if $a\neq 0$; $|a+b|\leqq \max(|a|,|b|)$; $|ab|=|a|+|b|$. If a and $b\neq 0$ are in ω, there exist c and d in ω such that $a=bc+d$, and either $d=0$ or $|d|<|b|$. Thus unique prime factorisation holds in ω. F is the field consisting of 0 and of all $\alpha\in\omega$ for which $|\alpha|=0$.

Denote by $\Pi=p_1,p_2,p_3,\cdots$ an infinite sequence of (equal or distinct) primes in ω such that $|p_\lambda|=1$ for all λ, and put $\psi_0=1$, $\psi_\lambda=p_1p_2\cdots p_\lambda$ if $\lambda\geqq 1$. Assume that for every $f\in\Omega$ and for every p_λ there are a unique $\varphi_\lambda\in F$ and a unique $f_\lambda\in\Omega$ such that $f=\varphi_\lambda+p_\lambda f_\lambda$. Then, for any $n\geqq 1$, every $f\in\Omega$ has a unique representation $f=\sum_{\lambda=0}^{n-1}\varphi^{(\lambda)}\psi_\lambda+\psi_n f^{(n)}$, where all $\varphi^{(\lambda)}$ are in F, and $f^{(n)}$ is in

Ω. In particular, if $f \in \omega$, then $f^{(n)}=0$ for $n>|f|$, so that now the development terminates after finitely many terms. Denote by $(\psi_\lambda{}^n)$ the principal ideal in Ω generated by $\psi_\lambda{}^n$. If $f \in \Omega$, $f \neq 0$, and if n is a positive integer, put $\|f\|_n = \lambda$ if $f \in (\psi_\lambda{}^n)$ but $f \notin (\psi_\lambda{}^{n+1})$, and write $\|f\|$ instead of $\|f\|_1$.

Now let $m \geqq 2$, and let $\mathbf{f}=(f_1, \cdots, f_m)$ be a vector of m functions and $\boldsymbol{\rho}=(\rho_1, \cdots, \rho_m)$ a vector of m non-negative integers; further, put $\sigma=\rho_1+\cdots+\rho_m$. There exist then m "Latin" polynomials $a_1, \cdots, a_m$ and m "German" polynomials $\mathfrak{a}_1, \cdots, \mathfrak{a}_m$ such that $r=\sum_{k=1}^m a_k f_k$, $|a_k| \leqq \rho_k - 1$, $\|r\| \geqq \sigma - 1$ $(k=1, 2, \cdots, m)$, and $\mathfrak{r}_{jk}=\mathfrak{a}_k f_j - \mathfrak{a}_j f_k$, $|\mathfrak{a}_k| \leqq \sigma - \rho_k$, $\|\mathfrak{r}_{jk}\| \geqq \sigma+1$ $(j, k=1, 2, \cdots, m)$. For each $\boldsymbol{\rho}$ select once for all a fixed set each of Latin and German polynomials and denote them by $a_k(\rho_1, \cdots, \rho_m)$ and $\mathfrak{a}_k(\rho_1, \cdots, \rho_m)$. Next let δ_{hk} be 1 for $h=k$ and 0 for $h \neq k$; put $a_{hk}(\rho_1, \cdots, \rho_m)=a_k(\rho_1+\delta_{h1}, \cdots, \rho_m+\delta_{hm})$,

$$\mathfrak{a}_{hk}(\rho_1, \cdots, \rho_m) = \mathfrak{a}_k(\rho_1-\delta_{h1}, \cdots, \rho_m-\delta_{hm}).$$

The vector $\mathbf{f}$ is called normal at $\boldsymbol{\rho}$ if no prime in Π divides all its components and if the Latin polynomials can be chosen so that $|a_{hh}(\rho_1, \cdots, \rho_m)|=\rho_h$ for all h. Then also $|\mathfrak{a}_{hh}(\rho_1, \cdots, \rho_m)|=\sigma-\rho_h$ for all h, and we can define the matrices $A(\rho_1, \cdots, \rho_m)$ and $\mathfrak{A}(\rho_1, \cdots, \rho_m)$ consisting of the elements $\alpha_h{}^{-1} a_{hk}(\rho_1, \cdots, \rho_m)$ and $\beta_h{}^{-1}\mathfrak{a}_{hk}(\rho_1, \cdots, \rho_m)$, where α_h and β_h are the highest coefficients of $a_{hh}(\rho_1, \cdots, \rho_m)$ and of $\mathfrak{a}_{hh}(\rho_1, \cdots, \rho_m)$, respectively.

One shows that if $\mathbf{f}$ is normal at $\boldsymbol{\rho}$, then $\det A=\psi_\sigma$ and $\det \mathfrak{A}=\psi_\sigma{}^{m-1}$; moreover, $A\mathfrak{A}^T=\psi_\sigma I$, where I is the unit matrix.

Further results are entirely analogous to those in the papers by H. Jager [Nederl. Akad. Wetensch. Proc. Ser. A **67** (1964), 193–198, 199–211, 212–225; MR **29** #402; ibid. **67** (1964), 227–239, 240–244, 245–249; MR **29** #2576] and they generalise those by the reviewer (unpublished paper of 1934, to appear in Compositio Math.).

Of particular interest is the author's notion of a normality zigzag. This is an infinite system of parameter vectors $\Sigma=\{\boldsymbol{\rho}^{(n)}\}$ such that $\boldsymbol{\rho}^{(0)}$ is the zero vector; $\rho_k{}^{(n+1)}=\rho_k{}^{(n)}+\delta_{h_n k}$, where the suffix h_n depends only on n; and $\mathbf{f}$ is normal at all $\boldsymbol{\rho}^{(n)}$. The author shows that if $\mathbf{f}$ is normal at $\boldsymbol{\rho}$, then $\boldsymbol{\rho}$ belongs to a normality zigzag Σ of $\mathbf{f}$.

The paper concludes with explicit examples. In particular, the author determines the Latin and the German polynomials that belong to the vector $\mathbf{f}=(\alpha_1{}^z, \cdots, \alpha_m{}^z)$ when F is the complex field, ω is the ring $F[z]$, and Π is the set of prime polynomials $z, z-1, z-2, \cdots$.

K. Mahler (Canberra)

Citations: MR 29# 402 = J76-48; MR 29# 2576 = J76-49.

Referred to in J76-57, J76-64.

J76-57 (35# 1547)

Coates, John

On the algebraic approximation of functions. IV.

Nederl. Akad. Wetensch. Proc. Ser. A **70** = *Indag. Math.* **29** (1967), 205–212.

The author continues the investigations of the first three parts of this paper [same Proc. **69** (1966), 421–434; MR **34** #2527a; ibid. **69** (1966), 435–448; MR **34** #2527b; ibid. **69** (1966), 449–461; MR **27** #2527c] and obtains the following unrelated results. (1) Let G be a connected open set in the complex z-plane, S any finite or countable subset of G, and $f_1(z), \cdots, f_m(z)$ finitely many functions analytic in G which are linearly independent over the rational function field. Then there exist infinitely many sequences $\Pi=\{z_1, z_2, z_3, \cdots\}$ of distinct points of G, suitable subsequences of which converge to every point of S, such that the functions form a perfect system relative to Π. The proof is non-effective. (2) Let G be as in (1), let $\Pi=\{z_1, z_2, z_3, \cdots\}$ be a sequence of distinct points in G, and let $f_1(z), \cdots, f_m(z)$ be analytic in G and perfect relative to Π; denote further by $p_1(z), \cdots, p_m(z)$ polynomials of the exact maximal degree $\rho-1$, and let H be the highest coefficient of one of these polynomials which has the exact degree $\rho-1$. Under these very general assumptions, the author establishes a lower bound for

$$\max\{|p_1(z)f_1(z)+\cdots+p_m(z)f_m(z)|: z = z_1, \cdots, z_{m\rho}\}.$$

This lower bound involves the German polynomials $\mathfrak{A}_{hk}(z|\rho\cdots\rho)$ (defined in Part II, § 11), hence is effective only when these polynomials are known. An interesting case is given for $\Pi=\{0, 1, 2, \cdots\}$ by the functions $f_1(z)=\alpha_1{}^z, \cdots, f_m(z)=\alpha_m{}^z$, where $\alpha_1, \cdots, \alpha_m$ are distinct complex numbers not zero. Put $\Delta=\max_k|\alpha_k|$, $\omega=\min_k|\alpha_k|$, $\delta=\min_{k\neq l}|\alpha_k-\alpha_l|$. Then

$$\max\{|p_1(z)\alpha_1{}^z+\cdots+p_m(z)\alpha_m{}^z| : z = 0, 1, \cdots, m\rho-1\} \geqq H\min(1, \Delta^{m\rho-1})(\rho-1)!\,\omega^{\rho-1}\delta^{(m-1)\rho}/(6\Delta)^{m\rho-1}.$$

For a related result see T. Dancs and P. Turán [Publ. Math. Debrecen **11** (1964), 257–272; MR **30** #3217].

K. Mahler (Canberra)

Citations: MR 30# 3217 = M50-30; MR 34# 2527a = J76-56; MR 34# 2527b = J76-56; MR 34# 2527c = J76-56.

J76-58 (34# 2530)

Lang, Serge

Algebraic values of meromorphic functions. II.

Topology **5** (1966), 363–370.

This is a continuation of Part I [Topology **3** (1965), 183–191; MR **32** #7506]. In order to state the results, some definitions have to be made. Let α be an algebraic number and d a positive integer such that $d\alpha$ is an algebraic integer. d is then called a denominator for α. Then size $(\alpha) \leqq B$ if B is real positive, and if there is a denominator d for α such that $\log d \leqq B$ and $\log|\alpha'| \leqq B$, where the α' denote the conjugates of α. Next, let λ_1, λ_2 be two positive functions defined on some set. If there is a constant $c>0$ such that $\lambda_1 \leqq c\lambda_2$, one writes $\lambda_1 \ll \lambda_2$. Finally, let $S=\bigcup_n S_n$ be a set of complex numbers such that $S_n \subset S_{n+1}$ and f a meromorphic function defined on S with values in a number field K. Then f is said to be of arithmetic order ρ on S if size $(f(z)) \ll n^\rho$ and if there is an entire function h that has no zero on S, such that hf is entire and $\log|1/h(z)| \ll n^\rho$ for $z \in S_n$ and $n \to \infty$.

With S and K as above, the main theorem of the paper is as follows. Let f, g be two meromorphic functions defined on S with values in K, with the order and arithmetic order $\ll \rho$. If $\lambda \geqq 2$ and $\operatorname{card}(S_n) \gg \ll n^{\lambda\rho}$ as $n \to \infty$, then f and g are algebraically dependent over K. Some applications of this are discussed also. When (β_1, β_2), (z_1, z_2, z_3) are two sets of linearly independent complex numbers over Q, then not all of the $e^{\beta_i z_j}$ are algebraic. When β is irrational, there are at most two multiplicatively independent algebraic numbers $\alpha \neq 0$ such that α^β is algebraic. Finally, the main theorem is applied to a linear group G [Abelian variety A] over an algebraic number field and to a 1-parameter subgroup (complex analytic) $\varphi: C \to G$ $[A]$. It is shown that if the 1-parameter subgroup φ has algebraic dimension $\geqq 2$ in G $[A]$, and if Γ is a subgroup of C with at least three [seven] linearly independent elements over Q, then $\varphi(\Gamma)$ cannot be contained in the group of algebraic points of G $[A]$.

T. Matsusaka (Waltham, Mass.)

Citations: MR 32# 7506 = J76-51.

J76-59 (34# 4206)

Baker, A.

A note on the Padé table.

Nederl. Akad. Wetensch. Proc. Ser. A **69** = *Indag. Math.* **28** (1966), 596–601.

The Padé table considered here is a multidimensional generalisation of Padé's classical table, introduced by K. Mahler about 1935 (in an unpublished paper). Let $\rho_1, \cdots, \rho_n$ be a set of positive integers. A system $f_1(x), \cdots, f_n(x)$ of formal power series in x over the complex numbers is said to be normal with respect to $\rho_1, \cdots, \rho_n$ if for every set of polynomials $a_1(x), \cdots, a_n(x)$, not all identically zero, and with degrees at most $\rho_1 - 1, \cdots, \rho_n - 1$, respectively, the power series $a_1(x)f_1(x) + \cdots + a_n(x)f_n(x)$ cannot have a zero at $x = 0$ of order greater than $\rho_1 + \cdots + \rho_n - 1$.

Extending results of H. Jager [Nederl. Akad. Wetensch. Proc. Ser. A **67** (1964), 193–198, 199–211, 212–225; MR **29** #402; ibid. **67** (1964), 227–239, 240–244, 245–249; MR **29** #2576], the author proves the following theorem: "Let k, l denote positive integers. For every integer r with $0 \leqq r \leqq k$, let $\rho(r, s)$ $(s = 0, 1, \cdots, 1)$ denote integers satisfying $1 \leqq \rho(r, 0) \leqq \rho(r, 1) \leqq \cdots \leqq \rho(r, 1)$. Further, let $\omega_0, \omega_1, \cdots, w_k$ denote complex numbers, no two of which differ by an integer. Then the system $(1-x)^{\omega_r}(\log(1-x))^{1-s}$ $(0 \leqq r \leqq k,\ 0 \leqq s \leqq 1)$ is normal with respect to the set $\rho(r, s)$."

This rather deep result is obtained by extending a method used by Mahler in his paper on the logarithms of algebraic numbers [Philos. Trans. Roy. Soc. London Ser. A **245** (1953), 371–398; MR **14**, 624]. The author remarks that the fundamental identities derived in his paper can also be used to give new proofs for several classical results from the theory of transcendental numbers, such as Lindemann's theorem and the theorem that e^{π} is transcendental. *J. Popken* (Amsterdam)

Citations: MR 14, 624g = J80-8; MR 29# 402 = J76-48; MR 29# 2576 = J76-49.

J76-60 (35# 130)

Mahler, K.; Szekeres, G.

On the approximation of real numbers by roots of integers.

Acta Arith. **12** (1966/67), 315–320.

Let A be the set of all numbers $\alpha > 1$ for which none of the powers $\alpha, \alpha^2, \alpha^3, \cdots$ is an integer. For each $n > 0$, let g_n be a nearest integer to α^n, so that $|\alpha^n - g_n| \leqq \frac{1}{2}$, and put $P(\alpha) = \liminf_{n\to\infty} |\alpha^n - g_n|^{1/n}$. It is shown that (a) if $P(\alpha) = 0$, then α is transcendental, (b) in every neighborhood of every number $x > 1$ there exist uncountably many $\alpha \in A$ for which $P(\alpha) = 0$, and (c) for almost all $\alpha \in A$, $P(\alpha) = 1$, so that there are transcendental numbers with this property. *W. J. LeVeque* (Ann Arbor, Mich.)

J76-61 (37# 165)

Ramachandra, K.

Contributions to the theory of transcendental numbers. I, II.

Acta Arith. **14** (1967/68), 65–72; *ibid.* **14** (1967/68), 73–88.

If α is an algebraic number with the conjugates $\alpha^{(j)}$, and $d(\alpha)$ is the least positive integer such that $\alpha d(\alpha)$ is an algebraic integer, put size $\alpha = d(\alpha) + \max_j |\alpha^{(j)}|$. The main theorem of Part I uses the following notation: (1) $F_t(z) = H_t(z)/G_t(z)$ $(t = 1, \cdots, s)$ are $s \geqq 2$ meromorphic functions algebraically independent over the complex field C. For each t, $H_t(z)$ and $G_t(z)$ are entire functions without common zeros, at most of order ρ, and

$$M^{(t)}(R) = (1 + \max_{|z|=R} |H_t(z)|)(1 + \max_{|z|=R} |G_t(z)|).$$

(2) $S = \{(a_\mu, n_\mu)\}$ is an infinite sequence of pairs of one complex number a_μ and one positive integer n_μ, where all the a_μ are distinct, and $n_1 \leqq n_2 \leqq n_3 \leqq \cdots$. It is assumed that for every positive integer Q the expressions $D(Q) = \max_{a_\mu \leqq Q} |a_\mu|$ and $N(Q) = \sum_{a_\mu \leqq Q} 1$ are finite and satisfy $\liminf_{Q\to\infty} (\log N(Q))/(\log D(Q)) > \rho$. (3) $S_1 = \{(a_{\mu_r}, n_{\mu_r})\}$ is an infinite subsequence of S, $N_1(Q) = \sum_{a_{\mu_r} \leqq Q} 1$, and the following property holds. If $p(x_1, \cdots, x_s)$ is any polynomial with coefficients in C, and $P(z) = p(F_1(z), \cdots, F_s(z))$, then, for every Q, the property: $P(a_{\mu_r}) = 0$ for all a_{μ_r} satisfying $n_{\mu_r} \leqq Q$, implies the property: $P(a_\mu) = 0$ for all a_μ satisfying $n_\mu \leqq Q$. (4) Let $K(Q)$ be the finite extension of the rational number field generated by the function values $F_t(a_\mu)$, where $1 \leqq t \leqq s$ and $n_\mu \leqq Q$, and let, for every Q, $K(Q)$ be an algebraic number field, of degree $h(Q)$ say. (5) For each t and Q put $M_1^{(t)}(Q) = 1 + \max_{n_\mu \leqq Q}$ size $F_t(a_\mu)$; $M_2^{(t)}(Q) = 1 + \max_{n_\mu \leqq Q} |G_t(a_\mu)|^{-1}$. The main theorem asserts: If $r_1, \cdots, r_s$ are positive integers such that $r_1 \cdots r_s \sim h(q)(h(q)+1)N_1(q)$ as the integer $q \to \infty$, then there exists for each q a $Q \geqq q$ such that, for every $R > 0$,

$$(\textstyle\prod_{t=1}^s (M_1^{(t)}(Q))^{r_t})^{8h(Q)} (\prod_{t=1}^s (M_2^{(t)}(Q))^{r_t})$$
$$\times (\textstyle\prod_{t=1}^s (M^{(t)}(R))^{r_t})(8D(Q)/R)^{N(Q-1)} \geqq 1.$$

This result is specialised in Part 2. Let β be called a pseudo-algebraic point of $f(z)$ if either $f(\beta)$ is an algebraic number, or β is a pole of $f(z)$. Further denote by $\dim(F_1(z), \cdots, F_s(z))$ the dimension over the rational field of the vector space generated by the common pseudo-algebraic points of $F_1(z), \cdots, F_s(z)$. The principal result of the author is as follows. Let $F_1(z), \cdots, F_s(z)$, where $s \geqq 2$, be meromorphic functions algebraically independent over C, each function having an algebraic addition formula, and let ρ^* and ρ_* be the largest, and the smallest of the orders of these functions, respectively. Then $\dim(F_1(z), \cdots, F_s(z)) \leqq \rho^* + (\rho_* - \theta)/(s-1)$. Here θ denotes 1 or 0 according as the functions have, or have not, a common period. This principal result has, e.g., the following consequences. (1) Let $a \neq 0$, $b \neq 0$, and $c \neq 0$ be complex numbers such that $\log a$, $\log b$, and $\log c$ are linearly independent over the rational field $\mathbf{Q}$, and let $\omega_1, \omega_2, \cdots, \omega_n$ be $n \geqq 2$ complex numbers which are likewise linearly independent over $\mathbf{Q}$. Then at least $n-1$ of the values $a^{\omega_1}, b^{\omega_1}, c^{\omega_1}, \cdots, a^{\omega_n}, b^{\omega_n}, c^{\omega_n}$ are transcendental. (2) If $j(\omega)$, $\Delta(\omega_1, \omega_2)$, and $\wp(z; \omega_1, \omega_2)$ have their usual meaning from the theory of elliptic functions, and if a and b are positive algebraic numbers such that $(\log a)/(\log b)$ is irrational and $a < b < a^{-1}$, then at least one of the two function values $j(\log a/2\pi i)$, $\Delta(2\pi i, \log a)^{-1}(\wp''(\log b; 2\pi i, \log a))^3$ is transcendental. *K. Mahler* (Canberra)

Referred to in J76-67.

J76-62 (37# 1323)

Sprindžuk, V. G.

The finiteness of the number of rational and algebraic points on certain transcendental curves. (Russian)

Dokl. Akad. Nauk SSSR **177** (1967), 524–527.

Der Autor betrachtet, in Verallgemeinerung gewisser von C. L. Siegel untersuchter Funktionen, sogenannte E^*-Funktionen. Das sind Funktionen $\varphi(z) = \sum_{\nu=0}^{\infty} c_\nu z^\nu/\nu!$ mit folgenden Eigenschaften: (a) die Zahlen c_ν sind algebraisch und gehören zu einem Körper K endlichen Grades über dem Körper Q der rationalen Zahlen; (b) es existiert zu beliebigem $\varepsilon > 0$ eine Zahl α, $0 \leqq \alpha < 1$, für welche $\overline{|c_\nu|} < c(\varepsilon)(\nu!)^{\alpha+\varepsilon}$ ist ($\overline{|x|}$ bezeichnet für $x \in K$ den größten absoluten Betrag aller zu x in K konjugierten Werte); (c) es existiert eine Zahlenfolge β_n, $0 \leqq \beta_n = O(\sqrt{n})$ für $n \to \infty$, für welche bei beliebigem $\varepsilon > 0$, $q_{nm} < c(n, \varepsilon)(m!)^{\beta_n + \varepsilon}$ ist (q_{nm} bezeichnet für festes n die kleinste natürliche Zahl q, für welche alle Zahlen $qc_{\nu_1}c_{\nu_2}\cdots c_{\nu_n}$, $\nu_1 + \nu_2 + \cdots + \nu_n \leqq m$, ganz sind). Es wird u.a. gezeigt, daß für E^*-Funktionen $\varphi_1(z), \cdots, \varphi_m(z)$, welche das Gleichungssystem $y_l' = \sum_{j=1}^m Q_{jl}y_j + Q_l$ $(l = 1, 2, \cdots, m)$ mit rationalen Funktionen Q_{jl}, Q_l befriedigen und welche algebraisch unabhängig über dem Körper der rationalen Funktionen $C(z)$ sind, die folgende Behauptung gilt: Für eine beliebige natürliche Zahl d und einen beliebigen Körper algebraischer Zahlen K_1 von endlichem Grad über Q existieren nur endlich viele $z \in K_1$, für welche die Funktionen $\varphi_1(z), \cdots, \varphi_m(z)$ gleichzeitig algebraische Werte vom Grade $\leqq d$ über Q annehmen; insbesondere liegen auf der Kurve $\Gamma = (\varphi_1(z), \cdots, \varphi_m(z))$ für $z \in K_1$ nicht mehr als endlich viele Punkte aus ${K_1}^m$.

Darüber hinaus wird der Beweis kurz skizziert, daß die vorstehende Behauptung auch für ein beliebiges System von Funktionen $\varphi_1(z), \cdots, \varphi_m(z)$ gilt, wenn diese Funktionen das System $y_l' = \sum_{j=1}^m Q_{jl}y_j + Q_l$ $(l = 1, 2, \cdots, m)$ mit Polynomen Q_{jl}, Q_l befriedigen und algebraisch unabhängig über $C(z)$ sind.

Schließlich wird noch gezeigt, daß gewisse hypergeometrische Funktionen zur Klasse der E^*-Funktionen gehören.

{This article has appeared in English translation [Soviet Math. Dokl. **8** (1967), 1452–1455].}

O. Krötenheerdt (Halle)

J76-63 (38# 1060)

Šmelev, A. A.

A certain class of transcendental numbers. (Russian)

Mat. Zametki **4** (1968), 341–348.

Let a, α, β be algebraic numbers such that $a(a-1) \neq 0$, $\alpha \neq 0$, and β is irrational. The author, who was not aware of A. Baker's more general result [Mathematika **14** (1967), 220–228; MR **36** #3732], proves that $e^{\alpha}a^{\beta}$ is transcendental. His method is slightly different from that of Baker.

K. Mahler (Columbus, Ohio)

Citations: MR 36# 3732 = J80-26.
Referred to in J80-35.

J76-64 (39# 458)

Mahler, K.

Perfect systems.

Compositio Math. **19**, 95–166 (1968).

This is a hitherto unpublished manuscript prepared in 1934–5, on which are based a series of recent papers by H. Jager [Nederl. Akad. Wetensch. Proc. Ser. A **67** (1964), 193–198, 199–211, 212–225; MR **29** #402; ibid. **67** (1964), 227–239, 240–241, 245–249; MR **29** #2576] and J. Coates [ibid. **69** (1966), 421–434; MR **34** #2527a; ibid. **69** (1966), 435–448; MR **34** #2527b; ibid. **69** (1966), 449–461; MR **34** #2527c]. The antecedent result is an idea of C. Hermite [C. R. Acad. Sci. Paris **77** (1873), 18–24; ibid. **77** (1873), 74–79; ibid. **77** (1873), 226–233; ibid. **77** (1873), 285–293; reprinted in *Œuvres de Charles Hermite*, Tome III, pp. 150–181, Gauthier-Villars, Paris, 1912] dealing with the best formal rational approximations of exponential functions. Let $\langle z_n \rangle$ be an infinite sequence of points, distinct or not, in a domain G of the complex z-plane, and let $\psi_0(z) = 1$, $\psi_n(z) = \prod_{j=1}^n (z - z_j)$, $n = 1, 2, \cdots$. For a function $f(z)$ holomorphic in G, let $O(f)$ denote the integer n such that $f(z)/\psi_n(z)$ is holomorphic in G but $f(z)/\psi_{n+1}(z)$ is not. Now consider a finite system $f_1(z), f_2(z), \cdots, f_m(z)$ of functions holomorphic in G, such that at least one has a non-zero value at each point z_n, and also consider a finite vector $\boldsymbol{\rho} = (\rho_1, \rho_2, \cdots, \rho_m)$ of non-negative integers, with $\sigma = \sum_{k=1}^m \rho_k$. By elementary linear algebra, there exist polynomials $a_1(z), a_2(z), \cdots, a_m(z)$ of respective degrees at most $\rho_k - 1$, $k = 1, \cdots, m$, such that if $r(z) = \sum_{k=1}^m a_k(z)f_k(z)$, then $O(r) \geqq \sigma - 1$. Also there exist polynomials $b_1(z), \cdots, b_m(z)$ of respective degrees at most $\sigma - \rho_k$, $k = 1, \cdots, m$, such that if $r_{kl}(z) = b_l(z)f_k(z) - b_k(z)f_l(z)$, then $O(r_{kl}) \geqq \sigma + 1$, $k, l = 1, \cdots, m$.

The set $\{f_k(z)\}$ is called "perfect" if for every choice of $\boldsymbol{\rho}$, there exists a set $\{a_k = a_k(z; \boldsymbol{\rho})\}$ in which the polynomial a_k is of exact degree $\rho_k - 1$ for $k = 1, \cdots, m$. (A simple example of a perfect set is $\{\exp(\lambda_k z), k = 1, \cdots, m\}$, where the λ_k's are distinct.) It is shown that in the case of a perfect set, the respective exact degrees of the polynomials b_k are $\sigma - \rho_k$, $k = 1, \cdots, m$; and that the polynomials $\{a_k(z; \boldsymbol{\rho})\}$ and $\{b_k\}$ are uniquely determined up to a multiplicative constant. To study how these sets depend on the vector $\boldsymbol{\rho}$, matrices $A(z) = [a_k(z; \boldsymbol{\rho}_h)]$ and $B(z) = [b_k(z; \boldsymbol{\rho}_h')]$ are set up, where $\boldsymbol{\rho}_h = (\rho_1, \rho_2, \cdots, \rho_h + 1, \rho_{h+1}, \cdots, \rho_m)$ and in $\boldsymbol{\rho}_h'$, $\rho_h + 1$ is replaced by $\rho_h - 1$. It is shown that when these matrices are suitably normed, $\det A(z) = \psi_0(z)$, $\det B(z) = [\psi_0(z)]^{m-1}$, and $A(z)B^T(z) = \psi_\sigma(z)I$. This small sample of the results is intended merely to convey the general flavor of the work. The paper develops a wealth of identities and non-obvious properties of the polynomials $\{a_k(z; \boldsymbol{\rho})\}$, $\{b_k\}$ and the function $r(z)$ and $r_{kl}(z)$.

J. H. Curtiss (Coral Gables, Fla.)

Citations: MR 29# 402 = J76-48; MR 29# 2576 = J76-49; MR 34# 2527a = J76-56; MR 34# 2527b = J76-56; MR 34# 2527c = J76-56.

J76-65 (40# 2611)

Mahler, K.

Remarks on a paper by W. Schwarz.

J. Number Theory **1** (1969), 512–521.

From the author's text: "Recently, in a short paper, W. Schwarz [Math. Scand. **20** (1967), 269–274; MR **38** #3234] established a number of results on irrationality and transcendency of values of the function $G_k(z) = \sum_{h=0}^\infty z^{k^h}/(1 - z^{k^h})$ at certain rational points z; here k denotes a fixed integer $\geqq 2$. Schwarz also considered the analogous problem for p-adic numbers. He gives in his paper references to earlier work by the reviewer [Proc. Nat. Inst. Sci. India **13** (1947), 171–173; MR **9**, 500], P. Erdős [J. Indian Math. Soc. (N.S.) **12** (1948), 63–66; MR **10**, 594], P. Erdős and E. G. Straus [ibid. (N.S.) **27** (1963), 129–133 (1964); MR **31** #124] and S. W. Golomb [Canad. J. Math. **15** (1963), 475–478; MR **27** #105]. Surprisingly, Schwarz does not mention 3 papers of mine [Math. Ann. **101** (1929), 342–366; ibid. **103** (1930), 573–587; Math. Z. **32** (1930), 545–585] of almost 40 years ago in which the problem of the transcendency of functions like $G_k(z)$ was solved for all algebraic values of z, and very general theorems were proved. In this note I shall therefore give a short account of my old work and make suggestions for further investigations."

We shall quote one such suggestion from the conclusion of the paper. Let $F(z) = j(\log z/2\pi i) - z^{-1}$, where $j(w)$ is Weber's modular function of level 1. "However, a decision as to whether $F(z_0)$ is transcendental for algebraic z_0, where $0 < |z_0| < 1$, does not seem to be easy."

S. Chowla (University Park, Pa.)

Citations: MR 9, 500f = J72-11; MR 10, 594c = J72-17; MR 27# 105 = J72-39; MR 31# 124 = J72-44; MR 38# 3234 = J72-56.

J76-66 (41# 1654)

Wallisser, Rolf

Zur Transzendenz der Werte der Exponentialfunktion.

Monatsh. Math. **73** (1969), 449–460.

The author gives a new and simple proof of the following theorem: "Let the inequality

$$|\alpha - p/q| < \exp(-(\log q)^2 \log\log q)$$

have an infinity of solutions in coprime integers p and q ($q>0$). Then $\exp\alpha$ is transcendental." This result is originally due to K. Mahler [Philos. Trans. Roy. Soc. London Ser. A **245** (1953), 371–398; MR **14**, 624] and N. I. Fel'dman [Izv. Akad. Nauk SSSR Ser. Mat. **15** (1951), 53–74; MR **12**, 595].

Further, the author gives a proof of the p-adic analogue of the preceding theorem: "Let $|\alpha|_p < \exp(-c)\cdot p^{-1/(p-1)}$, $c > 1+\log 2$, and let the inequality

$$|\alpha - P/Q|_p < \exp(-(\log H)^2 \log\log H),$$

$H = \max(|P|, |Q|)$ have an infinity of solution in coprime integers P and Q. Then $\exp\alpha$ is transcendental."

J. Popken (Amstetveen)

Citations: MR 12, 595e = J80-4; MR 14, 624g = J80-8.

J76-67 (41# 5302)

Ramachandra, K.

Lectures on transcendental numbers.

The Ramanujan Institute Lecture Notes, 1.

The Ramanujan Institute, Madras, 1969. iii+73 *pp.* $2.00.

Im ersten Kapitel dieser hektographierten Vorlesungsausarbeitung werden einige später benötigte Definitionen sowie einfache Sätze über algebraische Zahlen und Zahlkörper angegeben.

Das zweite beginnt mit einem Transzendenzbeweis für die Zahl e. Daran schließen sich fünf Formulierungen des Lindemann-Weierstraßschen Satzes an, deren Äquivalenz nachgewiesen wird. Diejenige Formulierung wird bewiesen, die es gestattet, die Ideen des vorgeführten Beweises für Hermites Satz in vollem Umfang zu übernehmen und durch Hinzufügung eines einfachen Gedankens zu einem Beweis des allgemeineren Satzes zu gelangen.

Das dritte Kapitel ist dem Gel'fond-Schneiderschen Satz gewidmet. Zunächst wird dieser Satz gefolgert aus einer auf ganze Funktionen spezialisierten Form des Hauptsatzes zweier neuerer Arbeiten des Verfassers [Acta Arith. **14** (1967/68), 65–72; ibid. **14** (1967/68), 73–88; MR **37** #165]. Die hier bewiesene Form dieses Hauptsatzes liefert als Korollar z.B. auch die Transzendenz mindestens einer der Zahlen 2^π, 2^{π^2}, 2^{π^3}. Sodann wird der Beweis eines von T. Schneider [*Einführung in die transzendenten Zahlen*, Satz 13, Springer, Berlin, 1957; MR **19**, 252] herrührenden Satzes gegeben, der sich auf die algebraische Abhängigkeit zweier meromorpher Funktionen bezieht, die geeigneten analytischen und arithmetischen Bedingungen genügen. Aus diesem Satz wird der Gel'fond-Schneidersche Satz nochmals abgeleitet; ferner wird die bekannte Tatsache über die Modulfunktion $j(\tau)$ gezeigt: Ist α algebraisch, aber nicht imaginär-quadratisch, so ist $j(\alpha)$ transzendent.

Das letzte Kapitel beschäftigt sich mit der Approximation algebraischer Irrationalitäten durch rationale Zahlen. Zunächst wird das elementare Liouvillesche Ergebnis gezeigt und eine kurze Übersicht über die historische Entwicklung dieses Problems gegeben. Der weitaus größte Teil dieses Kapitels ist einer detaillierten Darstellung des Beweises für das abschließende Resultat von K. F. Roth gewidmet.

Jedem Kapitel ist ein kleines Literaturverzeichnis angefügt; eine Liste der Druckfehler, die allerdings etwa doppelt so lang sein müßte, beschließt die Schrift.

P. Bundschuh (Freiburg)

Citations: MR 19, 252f = J02-9; MR 37# 165 = J76-61.

J76-68 (41# 6785)

Baron, Gerd; Braune, Erhard

Zur Transzendenz von Lückenreihen mit ganzalgebraischen Koeffizienten und algebraischem Argument.

Compositio Math. **22** (1970), 1–6.

The authors show that the lacunary series $\sigma(\Theta) = \sum_1^\infty a_\nu \Theta^{e_\nu}$ is transcendental for algebraic $0<\Theta<1$ provided the coefficients a_ν are algebraic integers whose conjugates do not spread further than the interval $(D^{-\nu}, D^\nu)$ (D constant) and whose gaps are large compared with the degrees g_ν of a_ν, i.e., as $h\to\infty$, $\lim(e_h/e_{h+1})(\prod_{\nu=1}^h g_\nu)^2 = 0$. The result generalizes work of the reviewer [Bull. Amer. Math. Soc. **52** (1946), 1042–1045; MR **8**, 370] by using estimates of Liouville's type [see Th. Schneider, *Einführung in die transzendenten Zahlen*, pp. 7–11, Springer, Berlin, 1957; MR **19**, 252; French translation, Gauthier-Villars, Paris, 1959; MR **21** #5620]. The authors examine the conditions further, particularly in view of analytic noncontinuability.

Harvey Cohn (Tucson, Ariz.)

Citations: MR 8, 370a = J76-12; MR 19, 252f = J02-9; MR 21# 5620 = J02-10.

J76-69 (41# 6788)

Fel'dman, N. I.; Šidlovskiĭ, A. B.

The methods of A. O. Gel'fond in the theory of transcendental numbers. (Russian)

Uspehi Mat. Nauk **25** (1970), *no.* 1 (151), 201–202.

A short statement of some of the main results by Gel'fond in the theory of transcendental numbers.

K. Mahler (Columbus, Ohio)

J76-70 (42# 4499)

Waldschmidt, Michel

Solution d'un problème de Schneider sur les nombres transcendants.

C. R. Acad. Sci. Paris Sér. A-B **271** (1970), A697–A700.

The author generalizes some well-known theorems of A. O. Gel'fond [English translation, *Transcendental and algebraic numbers*, Chapter III, Theorem II and Lemma VII, Dover, New York, 1960; MR **22** #2598]. As special cases of his result he shows for any non-zero rational r, (i) either e^{e^r} or $e^{e^{2r}}$ is transcendental, (ii) if α is an algebraic number not 0 or 1 then either $\alpha^{(\log\alpha)^r}$ or $\alpha^{(\log\alpha)^{2r}}$ is transcendental. His main theorem is as follows: Let y_1, y_2 and x_1, x_2 be complex numbers, each pair linearly independent over the rational numbers; suppose there exist N_0, $\tau>0$ such that for all $N>N_0$, n_1, n_2 with $|n_1|, |n_2| \leqq N$, one has $|n_1+n_2(x_1/x_2)| > e^{-\tau N^2\log N}$; and suppose the field $\mathbf{Q}(x_1, x_2, y_1, y_2, e^{x_1y_1}, e^{x_2y_1})$ has transcendence degree $\leqq 1$ over $\mathbf{Q}$; then either $e^{x_1y_2}$ or $e^{x_2y_2}$ is transcendental. The proof depends on the author's generalization of Gel'fond's fundamental lemma: Let α be a complex number, let σ_1, σ_2 be strictly increasing functions tending to ∞ such that $\sigma_2(x) \leqq \sigma_1(x)$ and $\sigma_i(x+1) \leqq (\sqrt{2})\sigma_i(x)$ ($i=1, 2$, $x>0$); suppose for all $N>N_0$ there is a polynomial P_N with rational integer coefficients of height H_N and degree d_N such that $|P_N(\alpha)| < \exp(-12\sigma_1(N)\sigma_2(N))$, $\log H_N \leqq \sigma_1(N)$, $d_N \leqq \sigma_2(N)$; then α is algebraic.

W. W. Adams (College Park, Md.)

Citations: MR 22# 2598 = J02-7.

J76-71 (43# 159)

Baker, A. [Baker, Alan]

On the quasi-periods of the Weierstrass ζ-function.

Nachr. Akad. Wiss. Göttingen Math.-Phys. Kl. II **1969**, 145–157.

The author proves the following remarkable theorem: Let ζ_ν ($\nu=1, 2$) be Weierstrass' ζ-functions and $\wp_\nu = -\zeta_\nu'$ the associated Weierstrass' $\wp$-functions; assume the corresponding pairs of invariants g_2, g_3 to be algebraic and denote by (ω_v, ω_v') any pairs of fundamental periods of $\wp_v$ and by $\eta_v = 2\zeta_v(\omega_v/2)$ the quasi-periods of ζ_v. Then any linear combination of $\omega_1, \omega_2, \mu_1, \mu_2$ with algebraic coefficients is either zero or transcendental. This theorem generalizes well known famous results of Siegel and Schneider, who considered only one $\wp$-function, respectively, one pair of associated functions $(\wp, \zeta)$, with algebraic invariants. In a former paper [Symposia Mathematica, Vol. IV (INDAM, Rome, 1968/69), pp. 155–174, Academic Press, London, 1970] the author developed a new method to obtain the result above just for the periods of a pair of $\wp$-functions. In the present paper he uses essentially the same arguments and succeeds in including the quasi-periods of ζ-functions by an appropriate choice of the auxiliary function in paragraph 3. *H. Klingen* (Freiburg)

Referred to in J76-76.

J76-72 **(43# 579)**

Sprindžuk, V. G.

On the theory of Siegel's hypergeometric functions. (Russian)

Dokl. Akad. Nauk BSSR **13** (1969), 389–391.

The author proves that if λ is an algebraic number and $\mathbf{K} = \mathbf{Q}(\lambda)$ is a normal extension of the field of rational numbers $\mathbf{Q}$ with degree $g = [\mathbf{K}:\mathbf{Q}] > 1$, then the smallest positive integer Q_n for which $Q_n n!/((\lambda+1)(\lambda+2)\cdots(\lambda+n))$ is an algebraic integer increases at least as fast as $(n!)^{g-1-\varepsilon}$, $\varepsilon > 0$. Hence it follows, e.g., that a power series with the general term $a_k = [\alpha_1, n]\cdots[\alpha_l, n]/[\beta_1, n]\cdots[\beta_m, n]$ for $k = nt$, $t = m - l \geqq 1$ and $n = 0, 1, \cdots$, and $a_k = 0$ for $nt < k < (n+1)t$, $t = m - l \geqq 1$ and $n = 0, 1, \cdots$, where $[x, 0] = 1$, $[x, n] = (x+n-1)[x, n-1]$, $\alpha_i, \beta_j \neq 0, -1, -2, \cdots$, and $\alpha_i \neq \beta_j$ for $i = 1, 2, \cdots, l$, $j = 1, 2, \cdots, m$, i.e., a Siegel hypergeometric function [cf. C. L. Siegel, *Transcendental numbers*, Ann. of Math. Studies, No. 16, Princeton Univ. Press, Princeton, N.J., 1949; MR **11**, 330], is not an E-function if some α_i or β_j is an irrational algebraic number that generates a normal extension of **Q**.

J. Ławrynowicz (Łódź)

Citations: MR 11, 330c = J76-18.

J76-73 **(43# 1928)**

Van der Poorten, A. J.

On the arithmetic nature of definite integrals of rational functions.

Proc. Amer. Math. Soc. **29** (1971), 451–456.

The author proves the following results. Theorem: Let $P(x)$ and $Q(z)$ denote polynomials with algebraic coefficients and with no common polynomial factor, denote by $\alpha_1, \alpha_2, \cdots, \alpha_n$ the distinct zeros of $Q(z)$ and by $r_1, r_2, \cdots, r_n$ the residues respectively at the poles of the rational function $P(z)/Q(z)$; further, let Γ be some contour in the complex plane for which the definite integral $\int_\Gamma (P(z)/Q(z))\,dz$ exists, and suppose that Γ is either closed or has endpoints that are algebraic or infinite; then the definite integral is algebraic if and only if

$$\int_\Gamma \left(\sum_{k=1}^n r_k/(z-\alpha_k)\right) dz = 0.$$

Corollary: If, in addition, $\deg P(z) < \deg Q(z)$ and the zeros of $Q(z)$ are distinct, then $\int_\Gamma P(z)/Q(z)\,dz$ is either transcendental or zero.

D. G. Cantor (Los Angeles, Calif.)

J76-74 **(43# 1947)**

Geysel, J. M.

Transcendence properties of the Carlitz-Besselfunctions.

Math. Centrum Amsterdam Afd. Zuivere Wisk. **1971**, ZW-2, 19 *pp.* (*loose abstract*)

Let F_q denote the finite field of order $q = p^n$, p a prime and $n \geqq 1$. Define the non-archimedean valuation $|E| = q^{\deg E}$ $(E \in F_q[x], E \neq 0)$, $|0| = 0$. The completion of $F_q(x)$ with respect to $|\ |$ is denoted by $F_q((x^{-1}))$ and the algebraic closure of $F_q((x^{-1}))$ by Φ. The valuation $|\ |$ can be extended to Φ in a unique way.

The reviewer [Duke Math. J. **1** (1935), 137–168] introduced the function $\psi(t) = \sum_{k=0}^{\infty} (-1)^k t^{q^k}/F_k$, where $F_k = \prod_{j=0}^{k-1} (x^{q^k} - x^{q^j})$, $F_0 = 1$; $\psi(t)$ furnishes an explicit example of an entire function in an algebraically closed field with a non-archimedean valuation. The function $\psi(t)$ can also be written as an infinite product $\psi(t) = t\prod_E (1 - t/E\xi)$, where E runs through all non-zero elements of $F_q[x]$, $\xi = \lim_{k=\infty} (x^q - x)^{q^k/(q-1)}/L_k$, and $L_k = \prod_{j=1}^{k} (x^{q^j} - x)$, $L_0 = 1$. Moreover the function $\lambda(t) = \sum_{k=0}^{\infty} t^{q^k}/L_k$ satisfies the condition $\psi(\lambda(t)) = \lambda(\psi(t)) = t$.

An element $\alpha \in \Phi$ that is a zero of a polynomial with coefficients in $F_q(x)$ is said to be algebraic. L. I. Wade [ibid. **8** (1941), 701–720; MR **3**, 263; ibid. **10** (1943), 587–594; MR **5**, 89] showed that, for any algebraic $\alpha \neq 0$, $\psi(\alpha)$ is transcendental over $F_q(x)$ and also that ξ is transcendental. He also proved [ibid. **13** (1946), 79–85; MR **7**, 370] the following analogue of the Gel'fond-Schneider theorem. If $\alpha \neq 0$ and $\beta \notin F_q(x)$, then at least one of the quantities α, β and $\psi(\beta\lambda(\alpha))$ is transcendental.

The reviewer [ibid. **27** (1960), 139–158] introduced the function $J_m(t) = \sum_{k=0}^{\infty} (-1)^k (t^{q^{m+k}})/(F_{m+k} F_k^{q^m})$ as an analogue of the Bessel function. Define the operator Δ by means of $\Delta f(t) = f(xt) - xf(t)$; thus, for example, $\Delta\psi(t) = -\psi^q(t)$.

The principal result of the present paper can now be stated. Theorem: Suppose that $\alpha \neq 0$ and $\beta \notin F_q(x)$ and let m be an arbitrary non-negative integer; then at least one element of the set $\{\alpha, \beta, J_m(\alpha), \Delta J_m(\alpha), J_m(\alpha\beta), \Delta J_m(\alpha\beta)\}$ is transcendental over $F_q(x)$. *L. Carlitz* (Durham, N.C.)

Citations: MR 3, 263f = J76-3; MR 5, 89a = J76-6; MR 7, 370d = T99-6.

J76-75 **(43# 4768)**

Baker, A. [Baker, Alan]

On the periods of the Weierstrass $\wp$-function.

Symposia Mathematica, Vol. IV (INDAM, Rome, 1968/69), pp. 155–174. *Academic Press, London,* 1970.

Let $\wp(z)$ denote a Weierstrass $\wp$ function with g_2, g_3 its invariants in the usual equation $(\wp'(z))^2 = 4(\wp(z))^3 - g_2\wp(z) - g_3$. T. Schneider proved that if g_2 and g_3 are algebraic then any period of $\wp$ is transcendental [Math. Ann. **113** (1936), 1–13]. Suppose now that $\wp_1$ and $\wp_2$ are $\wp$ functions, both with algebraic invariants. Suppose ω_1 is a period of $\wp_1$ and ω_2 a period of $\wp_2$. The author considerably extends Schneider's result by showing that if α_1 and α_2 are algebraic numbers then $\alpha_1\omega_1 + \alpha_2\omega_2$ is either 0 or transcendental. *D. G. Cantor* (Los Angeles, Calif.)

Referred to in Q05-87.

J76-76 **(44# 3964)**

Coates, J.

The transcendence of linear forms in $\omega_1, \omega_2, \eta_1, \eta_2, 2\pi i$.

Amer. J. Math. **93** (1971), 385–397.

Let ζ denote a Weierstrass ζ-function and $\mathfrak{P} = -\zeta'$, the associated $\mathfrak{P}$-function. Let g_2 and g_3 be the usual invariants appearing in the algebraic equation $(\mathfrak{P}')^2 = 4\mathfrak{P}^3 - g_2\mathfrak{P} - g_3$. Denote by ω_1 and ω_2 a fundamental set of periods for $\mathfrak{P}$ with $\mathrm{Im}(\omega_1/\omega_2) > 0$, and set $\eta_j = 2\zeta(\omega_j/2)$, $j = 1, 2$. Using a method of A. Baker [Nachr. Akad. Wiss. Göttingen Math.-Phys. Kl. II **1969**, 145–157; MR **43** #159], the author proves that any nonvanishing linear form in $\omega_1, \omega_2, \eta_1, \eta_2, 2\pi i$, with algebraic coefficients, is transcendental whenever g_2, g_3 are algebraic numbers. This generalizes a result of Baker [op. cit.].
I. Kra (Stony Brook, N.Y.)

Citations: MR 43# 159 = J76-71.

J80 MEASURE OF TRANSCENDENCE OR IRRATIONALITY

See also reviews G05-71, J02-3, J02-9, J02-21, J02-22, J02-23, J68-64, J76-55, J88-5, J88-6, M50-40, P60-25, Q05-21, Q05-87, R14-105, Z10-67.

J80-1 (1, 295c)

Gelfond, A. Sur l'approximation du rapport des logarithmes de deux nombres algébriques au moyen de nombres algébriques. Bull. Acad. Sci. URSS. Sér. Math. [Izvestia Akad. Nauk SSSR] **1939**, 509–518 (1939). (Russian. French summary)

It is proved that, if α, β, θ are algebraic numbers with $\alpha \neq 0, 1$, $\beta \neq 0, 1$, $\log\alpha/\log\beta$ irrational, and $H_0\theta^n + H_1\theta^{n-1} + \cdots + H_n = 0$, H_i rational integers, $\max|H_i| \leqq H$, then

$$\left|\frac{\log\alpha}{\log\beta} - \theta\right| > e^{-(\log H)^{3+\epsilon}}$$

for any $\epsilon > 0$ and all sufficiently large H. This is a sharpened form of the theorem that β^θ cannot be algebraic if β, θ are algebraic with $\beta \neq 0, 1$, and θ irrational [A. Gelfond, Bull. Acad. Sci. URSS. Sér. Math. **1934**, 623–630 (1934); T. Schneider, J. Reine Angew. Math. **172**, 65–69 (1934)], and the proof is a development of the author's proof of the original theorem. It is deduced that (apart from trivial cases of exception not specified by the author) an equation of the form

$$\alpha^x + \sum_{k=1}^{m_1} P_k(x, y)\alpha_k^x = \beta^y + \sum_{l=1}^{m_2} Q_l(x, y)\beta_l^y,$$

where $\alpha, \alpha_k, \beta, \beta_l$ are algebraic, $|\alpha| > |\alpha_k|$, $|\beta| > |\beta_l|$, and P_k, Q_l are polynomials with integral algebraic coefficients, can have at most a finite number of solutions in rational integers x, y. *A. E. Ingham* (Berkeley, Calif.).

Referred to in J02-3, J80-3, J80-29, J80-33, L25-5, N28-1.

J80-2 (8, 317g)

Mordoukhay-Boltovskoy, D. Sur les conditions pour qu'un nombre s'exprime au moyen d'équations transcendantes d'un type général. C. R. (Doklady) Acad. Sci. URSS (N.S.) **52**, 483–486 (1946).

In this note the author intends to prove the following theorem. Let $a_1, \cdots, a_n$ be algebraic numbers and let $\Omega(u_1, \cdots, u_n, v)$ be a polynomial in $u_1, \cdots, u_n, v$ with algebraic coefficients. If ξ is an irrational number satisfying the equation $\Omega(a_1^\xi, \cdots, a_n^\xi, \xi) = 0$, then there exist two positive numbers λ, k, such that, for every rational p/q (integral p, q with $q > 0$), $|\xi - p/q| > e^{-\lambda q^k \log q}$. The proof contains many printer's errors and in one place there seems to be an essential difficulty. The author also gives some indications for the proof of a more general theorem. *J. Popken.*

J80-3 (11, 232a; 11, 232b)

Fel'dman, N. I. The approximation of certain transcendental numbers. Doklady Akad. Nauk SSSR (N.S.) **66**, 565–567 (1949). (Russian)

Fel'dman, N. I. On the measure of transcendency of the logarithms of algebraic numbers and elliptic constants. Uspehi Matem. Nauk (N.S.) **4**, no. 1(29), 190 (1949). (Russian)

[In the second paper the author's initials are incorrectly given as N. M.] Let $P(x)$ denote a polynomial with integer coefficients, of order n and of height H. The following theorems are stated. (1) There exists a positive constant c_1 such that for any polynomial $P(x)$,

$$|P(\pi)| \geqq \exp(-c_1 n \cdot \max[n \log^2 n; \log H \cdot \log\log H]).$$

(2) Let $\alpha \neq 1$ denote an algebraic number. There exists a positive constant c_2 depending only on α and on the branch of $\log\alpha$ which has been chosen, such that

$$|P(\log\alpha)| \geqq \exp(-c_2 n^2(1 + \log n) \max[n \log^2 n; \log H \cdot \log\log H]).$$

(3) Let $\wp(z)$ denote the elliptic function of Weierstrass with algebraic invariants g_2 and g_3, and let ω denote a period of $\wp(z)$. There exist a constant c_3 depending only on the values of g_2 and g_3 and on the period ω, such that

$$|P(\omega)| \geqq \exp(-c_3 n^4 \max[n \log^5 n; \log H \cdot (\log\log H)^4]).$$

(4) Let us suppose again that the invariants g_2 and g_3 of $\wp(z)$ are algebraic numbers, and let us choose β in such a manner that $\wp(\beta)$ shall be an algebraic number. It follows that

$$|P(\beta)| \geqq \exp(-\max[\exp\{n^{2+\kappa} + n^{(2+\kappa)(\frac{1}{2}+\epsilon)+1}\}; \log H \cdot \exp\{n(\log\log H)^{\frac{1}{2}+\epsilon}\}]),$$

where κ and ϵ are positive numbers, and it is supposed that

$$\max(\log H, \exp m^{2+\kappa}) \geqq C,$$

where the constant C depends on $\epsilon, \kappa, \beta, g_2$ and g_3. Theorems 3 and 4 imply that ω and β are transcendental numbers, which was first proved by T. Schneider [Math. Ann. **113**, 1–13 (1936)]. It is stated that these results are proved by means of the methods developed by A. O. Gelfond [Bull. Acad. Sci. URSS. Cl. Sci. Math. Nat. [Izvestiya Akad. Nauk SSSR] **1934**, 623–634; Bull. Acad. Sci. URSS. Sér. Math. [Izvestiya Akad. Nauk SSSR] **1939**, 509–518; these Rev. **1**, 295; C. R. (Doklady) Acad. Sci. URSS (N.S.) **7** (**1935 II**), 177–182]. *A. Rényi* (Budapest).

Citations: MR 1, 295c = J80-1.

J80-4 (12, 595e)

Fel'dman, N. I. The approximation of certain transcendental numbers. I. Approximation of logarithms of algebraic numbers. Izvestiya Akad. Nauk SSSR. Ser. Mat. **15**, 53–74 (1951). (Russian)

The author proves a number of important results on the approximation of the logarithm of an algebraic number by means of algebraic numbers of both variable degree and variable height; for instance: There exists a positive constant γ such that

$$|P(\pi)| > \exp\{-\gamma n(n \log n + 1 + \log H) \times \log(n \log n + 2 + \log H)\}$$

for every polynomial $P(z) \not\equiv 0$ of degree n with rational integral coefficients of absolute value not greater than H. The proofs are given with full details. They depend on the study of polynomials in z and e^z that, together with their derivatives up to a high order, vanish at a great number of points of an arithmetic progression, and are based on the ideas used by Gel'fond in his classical proof of the transcendency of α^β [Bull. Acad. Sci. URSS Cl. Sci. Math. Nat. [Izvestiya Akad. Nauk SSSR] **1934**, 623–634].

K. Mahler (Manchester).

Referred to in J76-66, J80-7, J80-9, J80-13, J84-6.

J80-5 **(12, 679a)**

Gel'fond, A. O., and Fel'dman, N. I. On the measure of relative transcendentality of certain numbers. Izvestiya Akad. Nauk SSSR. Ser. Mat. **14**, 493–500 (1950). (Russian)

The authors improve an earlier result due to Gel'fond [Doklady Akad. Nauk SSSR (N.S.) **64**, 277–280 (1949); these Rev. **10**, 682]. Let α be a root of an irreducible equation of degree 3 and $a\neq 0, 1$ an algebraic number. Let $P(x, y)$ be any polynomial in x and y having integral coefficients each of modulus not exceeding H and of degrees n_1 and n_2 in x and y. Then, for any $\epsilon>0$,

$$|P(a^\alpha, a^{\alpha^2})|\geqq e^{-e^{\sigma^{4+\epsilon}}}$$

provided that $\sigma=\max(n_1+n_2, \log H)>\sigma_0$. The proof makes use of ideas similar to those introduced in the paper referred to above. *R. A. Rankin* (Cambridge, England).

Citations: MR 10, 682d = J88-3.

J80-6 **(13, 117e)**

Fel'dman, N. I. The approximation of certain transcendental numbers. II. The approximation of certain numbers connected with the Weierstrass function $\wp(z)$. Izvestiya Akad. Nauk SSSR. Ser. Mat. **15**, 153–176 (1951). (Russian)

The following results are proved: Denote by $\wp(z)$ the Weierstrass elliptic function, with algebraic invariants g_2 and g_3, further by ξ a variable algebraic number of degree n and height H, by $P(z)$ a variable polynomial $\neq 0$ with integral coefficients and likewise of degree n and height H. (1) If $\omega\neq 0$ is a period of $\wp(z)$, then

$$|\xi-\omega|>\exp\{-c_1n^4(n\log n+\log H+1)\times\log^4(n\log n+\log H+2)\},$$
$$|P(\omega)|>\exp\{-c_2n^4(n\log n+\log H+1)\times\log^4(n\log n+\log H+2)\},$$

where $c_1>0$ does not depend on ξ and $c_2>0$ not on $P(z)$. (2) Let α be such that $\wp(\alpha)$ is an algebraic number, and let $\kappa>0$. Then

$$|\xi-\alpha|>\exp\{-\exp(n^{2+\kappa}+\log\log H+c_3n(n^{2+\kappa}+\log\log H)^{\frac{1}{2}})\},$$
$$|P(\alpha)|>\exp\{-\exp(n^{2+\kappa}+\log\log H+c_4n(n^{2+\kappa}+\log\log H)^{\frac{1}{2}})\},$$

where $c_3>0$ and $c_4>0$ do not depend on ξ and $P(z)$, respectively. *K. Mahler* (Manchester).

J80-7 **(13, 213c)**

Fel'dman, N. I. On the joint approximation by algebraic numbers of the logarithms of several algebraic numbers. Doklady Akad. Nauk SSSR (N.S.) **75**, 777–778 (1950). (Russian)

This note announces a generalization to the simultaneous approximation of the logarithms of several algebraic numbers by algebraic numbers of the results in his earlier paper [Izvestiya Akad. Nauk SSSR. Ser. Mat. **15**, 53–74 (1951); these Rev. **12**, 595]. *J. W. S. Cassels.*

Citations: MR 12, 595e = J80-4.

J80-8 **(14, 624g)**

Mahler, K. On the approximation of logarithms of algebraic numbers. Philos. Trans. Roy. Soc. London. Ser. A. **245**, 371–398 (1953).

The author gives a new measure of transcendency for the logarithm of an algebraic number $\neq 0$, $\neq 1$ (Theorem 3). This important result is free of unknown numerical constants. One of its corollaries is that the logarithm of an algebraic number is not a U-number. This extends a well-known result of the author for the logarithm of a rational number [J. Reine Angew. Math. **166**, 137–150 (1932), p. 149].

The proof of Theorem 3 uses an idea due to Siegel and is based on a system of identities of the form

$$\sum_{k=0}^{m}A_{hk}(x)(\ln x)^k=R_h(x)\quad(h=0, 1, \cdots, m),$$

where the A's are polynomials of degree not greater than n with integral coefficients and, if $x\neq 1$, with a nonvanishing determinant, while the R's have at $x=1$ a zero of order at least $(m+1)n$.

The same method enables the author to study more closely the rational approximations to the logarithms of rational numbers. In particular, he derives a very simple measure of irrationality for the logarithm of a rational number (Theorem 5). As an application he obtains a number of results of the following type: $|2^\alpha-e^{\alpha_1}|\geqq|2^3-e^2|$ for all pairs of positive integers α, α_1 and with equality only in the case $\alpha=3$, $\alpha_1=2$. Finally he derives by means of these methods inequalities, such as $\ln f-[\ln f]>f^{-40\ln\ln f}$, $e^\alpha-[e^\alpha]>a^{-40\alpha}$, where both f and α are sufficiently large positive integers. *J. Popken* (Utrecht).

Referred to in J76-59, J76-66, J80-9, J80-13, J80-20.

J80-9 **(14, 957a)**

Mahler, K. On the approximation of π. Nederl. Akad. Wetensch. Proc. Ser. A. **56** = Indagationes Math. **15**, 30–42 (1953).

The author derives the remarkable inequality

$$|\pi-p/q|>q^{-42}$$

for all positive integers p, $q\geqq 2$. This measure of irrationality for π is stronger than any previous result in this direction. Furthermore, he gives the following measure of transcendency for π. Let ω be a real or complex algebraic number. Denote by R the rational field K if ω is real, and the Gaussian imaginary field $K(i)$ if ω is non-real. Further denote by ν the degree of ω over R, by

$$a_0z^\nu+a_1z^{\nu-1}+\cdots+a_\nu=0\quad(a_0\neq 0),$$

an equation for ω with integral coefficients in R which is irreducible over this field. Put

$$m=[20\cdot 2^{5(\nu-1)/2}],$$
$$a=\max(|a_0|, |a_1|, \cdots, |a_\nu|; (m+1)^{(m+1)/\nu}).$$

Then

$$|\pi-\omega|>e^{m+1}(m+1)^{-(m+1)}a^{-(m+1)\nu\log(m+1)}.$$

This measure of transcendency is compared with a similar result of Fel'dman [Izvestiya Akad. Nauk SSSR. Ser. Mat. **15**, 53–74 (1951); these Rev. **12**, 595]. The paper closes with some applications. The proofs depend on a basic result on a system of approximation forms of the type

$$\sum_{k=0}^{m}A_{hk}(x)(\log x)^k\quad(h=0, 1, \cdots, m+1)$$

where the A's are polynomials with integral coefficients, obtained in an earlier paper [Philos. Trans. Roy. Soc. London. Ser. A. **245**, 371–398 (1953); these Rev. **14**, 624]. *J. Popken* (Utrecht).

Citations: MR 12, 595e = J80-4; MR 14, 624g = J80-8.
Referred to in J80-13, J80-20, U99-2, Z02-32.

J80-10 **(16, 451f)**

Günther, Alfred. Über transzendente $\mathfrak{p}$-adische Zahlen. II. Zur approximation transzendenter $\mathfrak{p}$-adischer Zahlen durch rationale. J. Reine Angew. Math. **193**, 1–10 (1954).

This is a continuation of an earlier paper [same J. **192**, 155–156 (1953); MR **15**, 604]. The author establishes two theorems related to a theorem of A. Gelfond [Mat. Sb. N.S. **7**(**49**), 7–25 (1940); MR **1**, 292]. (1) Let α be an algebraic $\mathfrak{p}$-adic number with $0<|\alpha-1|_\mathfrak{p}<p^{-1/(p-1)}$, β a $\mathfrak{p}$-adic number with $0<|\beta-1|_\mathfrak{p}<p^{-1/(p-1)}$, and $\eta=\log\alpha/\log\beta$ an irrational algebraic $\mathfrak{p}$-adic integer. Then for every $\epsilon>0$ the inequality

$$(*)\qquad \left|\beta-\frac{q_1}{q_2}\right|_\mathfrak{p}<p^{-(\log q)^{2+\epsilon}/\epsilon}$$

has at most a finite number of solutions in relatively prime rational integers q_1, q_2 with $q=\max\{|q_1|,|q_2|\}$. (2) Let α be a $\mathfrak{p}$-adic number with $0<|\alpha-1|_\mathfrak{p}<p^{-1/(p-1)}$, β an algebraic $\mathfrak{p}$-adic number with $0<|\beta-1|_\mathfrak{p}<p^{-1/(p-1)}$, and $\eta=\log\alpha/\log\beta$ an irrational algebraic $\mathfrak{p}$-adic integer. Then for every $\epsilon>0$ the inequality

$$\left|\alpha-\frac{q_1}{q_2}\right|_\mathfrak{p} p^{-(\log q)^{2+\epsilon}/\epsilon}$$

has at most a finite number of solutions in relatively prime rational integers q_1, q_2 with $q=\max\{|q_1|,|q_2|\}$. From these theorems, the following corollaries are deduced. (1.1) If η is an irrational algebraic $\mathfrak{p}$-adic integer and β a $\mathfrak{p}$-adic number with $0<|\beta-1|_\mathfrak{p}<p^{-1/(p-1)}$, and if for every $\epsilon>0$ the inequality (*) has an infinite number of solutions in relatively prime rational integers q_1, q_2, with $q=\max\{|q_1|,|q_2|\}$, then β^η is transcendental. (2.1) If β is an algebraic $\mathfrak{p}$-adic number with $0<|\beta-1|_\mathfrak{p}<p^{-1/(p-1)}$ and if η is an irrational algebraic $\mathfrak{p}$-adic integer, then for every $\epsilon>0$ the inequality

$$\left|\beta^\eta-\frac{q_1}{q_2}\right|_\mathfrak{p}<p^{-(\log q)^{2+\epsilon}/\epsilon}$$

has at most a finite number of solutions in relatively prime rational integers q_1, q_2 with $q=\max\{|q_1|,|q_2|\}$.

M. Newman (Washington, D. C.).

Citations: MR 1, 292d = D60-1; MR 15, 604f = J76-31.

J80-11 **(17, 948e)**

Platonov, M. L. An estimate of the approximation of algebraic irrationalities by ratios of logarithms of increasing natural numbers. Irkutsk. Gos. Univ. Trudy. **8** (1953), no. 1, 53–62. (Russian)

Let α be an irrational algebraic number and let ε be positive. Then there is a $q_0>0$ such that for every pair of positive integers p and $q>q_0$ for which $\log p/\log q$ is irrational, the inequality

$$\left|\alpha-\frac{\log p}{\log q}\right|>\exp(-\log^2 q\log^{5+\varepsilon}\log q)$$

holds. The proof depends on a method described by A. O. Gelfond [Uspehi Mat. Nauk (N.S.) **4** (1949), no. 4(32), 19–49; MR **11**, 231]. *W. J. LeVeque* (Ann Arbor, Mich.).

Citations: MR 11, 231c = J02-3.

J80-12 **(20# 5895)**

Fel'dman, N. I. Joint approximations of the periods of an elliptic function by algebraic numbers. Izv. Akad. Nauk SSSR Ser. Mat. 22 (1958), 563–576. (Russian)

"This article introduces an estimate from below for the sum $|\omega-\xi|+|\omega_1-\xi_1|$, where ω and ω_1 are the periods of the function $\wp(z)$ with algebraic invariants and ξ, ξ_1 are algebraic numbers. The estimate depends on the degree and height of the numbers ξ, ξ_1 and on the degree of the field formed by the adjunction of ξ and ξ_1 to the rational field." *Author's summary*

J80-13 **(21# 5619)**

Fel'dman, N. I. On the measure of transcendence of the number π and of the logarithms of algebraic numbers. Dokl. Akad. Nauk SSSR **126** (1959), 1214–1215. (Russian)

The author states that the method of his paper [Izv. Akad. Nauk SSSR. Ser. Mat. **15** (1951), 53–74; MR **12**, 595] can be improved so as to give the following results. Let $\zeta=\log\alpha$ be the natural logarithm of an algebraic number $\neq 0$, $\neq 1$. There exists a positive constant γ_0 depending only on α, such that, if $H>\exp(n^4)$,

$$|a_0+a_1\zeta+\cdots+a_n\zeta^n|>H^{-\gamma_0 n^2(\log(n+1))^2}$$

for all integers $a_0, a_1, \cdots, a_n$ not all zero satisfying $\max_{0\le h\le n}|a_h|\le H$. For $\zeta=\pi$ the better estimate

$$|a_0+a_1\pi+\cdots+a_n\pi^n|>H^{-\gamma n\log(n+1)}$$

if $H>\exp(n^4)$ holds where γ is an absolute constant. Both estimates are much sharper than the so far best due to the reviewer [Philos. Trans. Roy. Soc. London, Ser. A **245** (1953), 371–398; Nederl Akad. Wetensch. Proc. Ser. A **56** (1953), 30–42; MR **14**, 624, 957].

K. Mahler (Manchester)

Citations: MR 12, 595e = J80-4; MR 14, 624g = J80-8; MR 14, 957a = J80-9.

J80-14 **(22# 5623a; 22# 5623b)**

Fel'dman, N. I. The measure of transcendency of the number π. Izv. Akad. Nauk SSSR. Ser. Mat. **24** (1960), 357–368. (Russian)

Fel'dman, N. I. Approximation by algebraic numbers to logarithms of algebraic numbers. Izv. Akad. Nauk SSSR. Ser. Mat. **24** (1960), 475–492. (Russian)

In these two important papers, which are closely connected, measures of transcendency for π and $\log\alpha$ (α algebraic) are found which are far better than any obtained before. In the case of π it is shown that there exists an absolute constant Λ_0 such that

$$|\pi-\xi|>H^{-\Lambda_0 n\log(n+2)}$$

for all algebraic numbers ξ, of degree n and height H, provided that $H>\exp(n^2\log^4(n+2))$. Next let $\alpha_1,\cdots,\alpha_m$ be finitely many algebraic numbers distinct from 0 and such that $\log\alpha_1,\cdots,\log\alpha_m$ are linearly independent over the rational field R. Let $\xi_1,\cdots,\xi_m$ be arbitrary algebraic numbers, of degrees $n_1,\cdots,n_m$ and heights $h_1,\cdots,h_m$ respectively; let n be the degree of the extension field $R(\alpha_1,\cdots,\alpha_m,\xi_1,\cdots,\xi_m)$, and let

$$H=\exp\left\{n\left(\frac{\log h_1}{n_1}+\cdots+\frac{\log h_m}{n_m}\right)\right\}.$$

Then a constant Λ depending only on $\alpha_1,\cdots,\alpha_m$ exists such that

$$\sum_{\mu=1}^m|\log\alpha_\mu-\xi_\mu|>H^{-\Lambda\{n\log(n+2)\}^{(m+1)/m}},$$

provided $H>\exp(n^4)$.

The proofs in both papers use similar ideas. They are based on Dirichlet's Schubfachprinzip and on two interpolation formulae that express the coefficients C_{kl} in

$$f(z)=\sum_{k=0}^{q_0-1}\sum_{l=0}^{q-1}C_{kl}z^ke^{lz}$$

either in terms of the values $f(2\pi xi/q)$, where $x=0, 1, \cdots, qq_0-1$, or in terms of the values $f^{(s)}(2\pi xi)$, where $x=0, 1, \cdots, q_0-1$; $s=0, 1, \cdots, q-1$. Both proofs depend on the fact that for all integers k and l, and for large s, the coefficients $B_{kl}^{(s\kappa)}$ in

$$\left(e^z \frac{d}{dz}\right)^s z^k e^{lz} = e^{(l+s)z} \sum_{\kappa=0}^{k} B_{kl}^{(s\kappa)} z^{k-\kappa}$$

have a greatest common divisor which is likewise large and for which a lower bound can be obtained by means of the prime number theorem. *K. Mahler* (Manchester)

Referred to in J80-16, J80-38, J80-42.

J80-15 (26# 6124)

Lang, Serge

A transcendence measure for *E*-functions.

Mathematika **9** (1962), 157–161.

Let $f_1(z), \cdots, f_s(z)$ be E-functions over an algebraic number field K [cf., e.g., C. L. Siegel, *Transcendental numbers*, p. 33, Princeton Univ. Press, Princeton, N.J., 1949; MR **11**, 330]. Let these functions satisfy a system of linear differential equations

$$f_i'(z) = \sum_{j=1}^{s} Q_{ij}^*(z) f_j(z) \qquad (i = 1, \cdots, s),$$

where the coefficients Q_{ij}^* are rational functions in $K(z)$. Let α be an algebraic number, distinct from zero and from the poles of the functions Q_{ij}^*. Then by a famous result of Šidlovskiĭ the algebraic independence of the functions $f_1, \cdots, f_s$ over $K(z)$ implies the algebraic independence of the numbers $f_1(\alpha), \cdots, f_s(\alpha)$ over the rationals [Šidlovskiĭ, Izv. Akad. Nauk SSSR Ser. Mat. **23** (1959), 35–66; MR **21** #1295; compare also Mahler's exposition, "On a theorem of Shidlovsky", Mimeographed Notes, Math. Centrum Amsterdam, Amsterdam, 1959].

The author derives in this paper the corresponding transcendence measure: Let all conditions of Šidlovskiĭ's theorem be satisfied, including the algebraic independence of $f_1, \cdots, f_s$. Let now $g(X_1, \cdots, X_s)$ be a polynomial in $X_1, \cdots, X_s$ of degree d and of height $G \neq 0$. Then

$$g(f_1(\alpha), \cdots, f_s(\alpha)) \geqq aG^{-bd^s},$$

where a is a positive number depending on the f_i, s, Q_{ij}^*, α, and d, while b depends only on s and on the degree $N = [K(\alpha) : \mathbf{Q}]$.

The proof uses a generalization to E-functions of the already classical methods employed in the special cases of exponential functions and of Bessel functions.

J. Popken (Berkeley, Calif.)

Citations: MR 11, 330c = J76-18; MR 21# 1295 = J88-15.

Referred to in J80-32.

J80-16 (27# 4798)

Fel'dman, N. I.

On a measure of transcendence of the number *e*. (Russian)

Uspehi Mat. Nauk **18** (1963), *no.* 3 (111), 207–213.

The author proves the following two equivalent results. (I) There is an absolute positive constant Λ_0 such that

$$|e-\xi| > \exp(-\Lambda_0 n^2 N \log^3 N), \qquad N = n + \log H,$$

for every algebraic number ξ of degree n and height H. (II) There is a second absolute positive constant Λ_1 such that

$$|P(e)| > \exp(-\Lambda_1 n^2 N \log^3 N), \qquad N = n + \log H,$$

for every polynomial $P(x)$ of degree n with rational integral coefficients of height H. Both inequalities are better than any so far obtained. The proof uses methods similar to those in the author's earlier paper [Izv. Akad. Nauk SSSR Ser. Mat. **24** (1960), 475–492; MR **22** #5623b].

K. Mahler (Canberra)

Citations: MR 22# 5623b = J80-14.

J80-17 (28# 2091)

Fel'dman, N. I.

Arithmetic properties of the solutions of a transcendental equation. (Russian. English summary)

Vestnik Moskov. Univ. Ser. I Mat. Meh. **1964**, *no.* 1, 13–20.

The following theorem is proved: Let α be an irrational algebraic number, $P(x, y)$ an irreducible polynomial whose coefficients are rational integers, and such that $P_x(x, y) \neq 0$, $P_y(x, y) \neq 0$, $P(1, 1) \neq 0$. If a transcendental number η satisfies the equation $P(z, z^\alpha) = 0$, then there exists a number $\Lambda_0 > 0$, dependent upon α and η only, such that for arbitrary rational integers $Q > 0$ and P the inequality (1) $|\eta - P/Q| > \exp(-\Lambda_0 \ln^2 Q \ln^{-1} \ln \ln Q)$ holds. Two more inequalities are stated without proof: (1) If in (1) P/Q is replaced by an algebraic number θ of degree n and height H, then $|\eta - \theta| > \exp(-\Lambda_1 \ln^2 H \ln^{-1} \ln \ln H)$, where $\Lambda_1 = \Lambda_1(n)$. (2) If ζ is a root of the equation $P(z, e^z) = 0$, and θ is an algebraic number of degree n and height H, then $|\zeta - \theta| > \exp(-\Lambda_2 \ln^2 H \ln^2 \ln H)$, where $\Lambda_2 = \Lambda_2(n)$. *J. W. Andrushkiw* (S. Orange, N.J.)

Referred to in J80-42, Z10-43.

J80-18 (30# 1976)

Baker, A.

Approximations to the logarithms of certain rational numbers.

Acta Arith. **10** (1964), 315–323.

In analogy to the author's recent results on the rational approximations of algebraic numbers [Proc. London Math. Soc. (3) **14** (1964), 385–398; MR **28** #5029], and by means of a similar technique, the following theorems are proved.

Theorem 1: Let $a, b, h = b - a$, and n be positive integers, and let κ be a real number greater than n. Put $\lambda = 50\kappa(\kappa+1)$, $\rho = (n+1)(\kappa+1)(\kappa-n)^{-1}$, $c = a^{-\lambda \log a}$, $\alpha = b/a$, and assume that $a > (32^{n/2}h)^\rho$. Then

$$|x_0 + x_1 \log \alpha + \cdots + x_n (\log \alpha)^n| > cX^{-\kappa}$$

for all integers $x_0, x_1, \cdots, x_n$ satisfying $X = \max(|x_0|, |x_1|, \cdots, |x_n|) > 0$. Corollary: Let ξ be a variable algebraic number of height H and degree at most n. Put $\delta = n(\kappa - n)$. Then $|\log \alpha - \xi| > H^{-n-1-\delta}$ if H is sufficiently large, and there exists an infinite sequence of ξ's such that $|\log \alpha - \xi| < H^{-n-1+\delta}$.

Theorem 2: Let $a > 0$, p, and $q > 0$ be integers. Put

$$\begin{aligned} \kappa(a) &= 12.5 && \text{if } a = 1, \\ &= 7 && \text{if } a = 2, \\ &= 2\frac{\log\{4\sqrt{2}\, a^2/(a+1)\}}{\log\{\sqrt{2}\, a^3/(a+1)^2\}} && \text{if } a \geqq 3, \\ c(a) &= 10^{-10^5} && \text{if } a = 1, \\ &= \{\sqrt{2}\, a\}^{-10^4} && \text{if } a \geqq 2 \end{aligned}$$

(thus $\kappa(a) \downarrow 2$ as $a \to \infty$). Then

$$\left|\log \frac{a+1}{a} - \frac{p}{q}\right| > c(a) q^{-\kappa(a)}.$$

Theorem 3: Let p and q be positive integers such that $\zeta = (e^{p/q} - 1)^{-1} > 1$. Then the fractional part of ζ is greater than $c(1)q^{-\kappa(1)}$. *K. Mahler* (Canberra)

Citations: MR 28# 5029 = J68-42.

J80-19 (31# 2204)

Baker, A.

On some Diophantine inequalities involving the exponential function.

Canad. J. Math. **17** (1965), 616–626.

Let $\theta_1 = e^{r_1}, \cdots, \theta_k = e^{r_k}$, where $r_1 \neq 0, \cdots, r_k \neq 0$ are distinct rational numbers. Denote by $c > 0$ a certain positive constant that depends only on $k, \theta_1, \cdots, \theta_k$, and put $\varepsilon(x) = c(\log\log x)^{-1/2}$. The author proves the following two results, which are equivalent by a simple transfer theorem. (I) For all sufficiently large positive integers n,

$$n^{1+\varepsilon(n)}\|n\theta_1\| \cdots \|n\theta_k\| \geqq 1.$$

(II) For all but finitely many sets of k integers $x_1 \neq 0, \cdots, x_k \neq 0$,

$$|x_1 \cdots x_k(x_1\theta_1 + \cdots + x_k\theta_k)| \geqq x^{1-\varepsilon(x)},$$

where $x = \max(|x_1|, \cdots, |x_k|)$.

The proof involves a slight generalisation of Siegel's method for the proof of the transcendency of the Bessel functions [C. L. Siegel, *Transcendental numbers*, Princeton Univ. Press, Princeton, N.J., 1949; MR **11**, 330].

K. Mahler (Canberra)

Citations: MR 11, 330c = J76-18.
Referred to in J80-21, K45-28.

J80-20 (34# 5754)

Mahler, K.

Applications of some formulae by Hermite to the approximation of exponentials and logarithms.

Math. Ann. **168** (1967), 200–227.

Let $w, w_1, \cdots, w_m, q, q_1, \cdots, q_m$ and λ be $2m+3$ integers satisfying $w \geqq 1, q \geqq 1, 0 < w_1 < w_2 < \cdots < w_m = \lambda, \lambda \geqq 2$, and (*) $w^m > qm^{\lambda-m}e^{10\lambda}$. Then $\max_{k=1,\cdots,m}|\exp(w_k/w) - q_k/q| > (qwm^{1+\lambda}\exp(6\lambda+5))^{-1}$. Now let the $2m+3$ integers satisfy (*) and $w_i \neq w_j$ for $i \neq j$, $\lambda = \max w_j$, $\lambda \geqq 2$. Then $\max_{k=1,\cdots,m}|\log q_k/q - w_k/w| > (qwm^{1+\lambda}\exp(7\lambda+7))^{-1}$. A different bound for this maximum is established in case (*) is not satisfied. Next, let p_j be the jth prime. Let $w, w_1, \cdots, w_m$ be $m+1$ positive integers with $m \geqq 10$. Then $\max_{k=1,\cdots,m}|\log p_k - w_k/w| > (2m)^{-m-5}$ if $1 \leqq w \leqq K^{-1}m \log 2m$, where $K = (10 + \log m)\log(2m \log m)$. Another inequality, not given here, is proved for a different set of values of w. These results are applied to show that $|e^g - a| > g^{-33g}$ for any sufficiently large positive integer g, where a is the closest integer to e^g. The author [Philos. Trans. Roy. Soc. London Ser. A **245** (1953), 371–398; MR **14**, 624] had established the analogous result with 40 instead of 33. Another application gives a lower bound for $\pi - p/q$ where p/q is a rational approximation to π, but the author [Nederl. Akad. Wetensch. Proc. Ser. A **56** (1953), 30–42; MR **14**, 957] obtained a better result earlier. The proofs use specially designed Hermite polynomials, and some estimates of J. P. Rosser and L. Schoenfeld [Illinois J. Math. **6** (1962), 64–94; MR **25** #1139].

I. Niven (Eugene, Ore.)

Citations: MR 14, 624g = J80-8; MR 14, 957a = J80-9; MR 25# 1139 = N04-37.

J80-21 (35# 132)

Fel'dman, N. I.

Estimates from below for certain linear forms. (Russian. English summary)

Vestnik Moskov. Univ. Ser. I Mat. Meh. **22** (1967), *no.* 2, 63–72.

The author gives a very ingenious proof of the following result. Let $\varphi_\lambda(z) = \sum_{\nu=0}^{\infty} z^\nu/((\lambda+1)(\lambda+2)\cdots(\lambda+\nu))$, and let $\alpha \neq 0, \lambda_1, \cdots, \lambda_n$ be rational numbers such that the λ_k are distinct from $-1, -2, -3, \cdots$ and such that $\lambda_h \not\equiv \lambda_k$ (mod 1) if $h \neq k$. Further, let $x_1, \cdots, x_n$ be n arbitrary integers not all zero, and let y be an integer. There exists a constant $c_0 > 0$ depending only on $\alpha, \lambda_1, \cdots, \lambda_n$ such that $|x_1\varphi_{\lambda_1}(\alpha) + \cdots + x_n\varphi_{\lambda_n}(\alpha) + y| \geqq X^{-1-(c_0/\log\log(X+2))}$, where $X = \prod_{k=1}^{n}\max(1, |x_k|)$. For this result compare, in particular, A. Baker [Canad. J. Math. **17** (1965), 616–626; MR **31** #2204].

K. Mahler (Canberra)

Citations: MR 31# 2204 = J80-19.

J80-22 (35# 2833)

Kappe, Luise-Charlotte

Zur Approximation von e^α.

Ann. Univ. Sci. Budapest. Eötvös Sect. Math. **9** (1966), 3–14.

The following theorem is proved. Let α be an algebraic number of degree s over the Gaussian field. Then the inequality $|e^\alpha - p/q| < |q|^{-(4s^2-2s+\varepsilon)}$ has at most a finite number of solutions; here p and q are integers in $\Gamma(i)$.

P. Szüsz (Stony Brook, N.Y.)

J80-23 (35# 5399a; 35# 5399b)

Šidlovskiĭ, A. B.

On estimates of the measure of transcendentality of the values of E-functions. (Russian)

Uspehi Mat. Nauk **22** (1967), *no.* 3 (135), 245–246.

Šidlovskiĭ, A. B.

Estimates of the degree of transcendence of the values of E-functions. (Russian)

Mat. Zametki **2** (1967), 33–44.

These two notes, of which only the second one contains proofs, deal with the same results.

Recently, the author [Dokl. Akad. Nauk SSSR **171** (1966), 810–813; MR **34** #4209] suggested a simple change in Siegel's approximation method which allows one to obtain better lower bounds for the rank of linear approximation forms of values of E-functions. By means of this method he now derives improved measures of linear and algebraic independence [cf. for weaker measures, S. Lang, *Introduction to transcendental numbers*, 96–99; #5397 above].

Let K be either the rational number field, or any imaginary quadratic number field. The Taylor coefficients and the arguments ξ of all E-functions mentioned are to lie in K. The main theorem is as follows and is almost best-possible. Let $y_1 = f_1(z), \cdots, y_m = f_m(z)$, where $m \geqq 2$, be E-functions linearly independent over $\mathbf{C}(z)$ which satisfy the system of homogeneous linear differential equations (H) $y_k' = \sum_{i=1}^{m} Q_{ki}y_i$ $(k = 1, 2, \cdots, m)$, where the Q_{ki} are in $\mathbf{C}(z)$. If $\xi \neq 0$ lies in K and is not a pole of any Q_{ki}, then to every $\varepsilon > 0$ there exists a constant $c = c(\varepsilon) > 0$ such that $|a_1f_1(\xi) + \cdots + a_mf_m(\xi)| > c\max(|a_1|, \cdots, |a_m|)^{-(m-1+\varepsilon)}$ for all rational integers $a_1, \cdots, a_m$ not all zero. If the E-functions $y_1 = f_1(z), \cdots, y_m = f_m(z)$, where $m \geqq 1$, instead satisfy the inhomogeneous system of linear differential equations (I) $y_k' = Q_{k0} + \sum_{i=1}^{m} Q_{ki}y_i$ $(k = 1, 2, \cdots, m)$, where the Q_{k0} and the Q_{ki} lie in $\mathbf{C}(z)$, it suffices to include the further E-function $y_0 = E_0(z) \equiv 1$. If now $\xi \neq 0$ lies in K and is not a pole of any Q_{k0} or Q_{ki}, one deduces that

$$|a_0 + a_1f_1(\xi) + \cdots + a_mf_m(\xi)| > c^* \max(|a_0|, |a_1|, \cdots, |a_m|)^{-(m+\varepsilon)}$$

for all rational integers $a_0, a_1, \cdots, a_m$ not all zero, provided that $1, f_1(z), \cdots, f_m(z)$ are linearly independent over $\mathbf{C}(z)$. Here c^* depends on ε, but not on the a_k. Next assume that the E-functions $y_1 = f_1(z), \cdots, y_m = f_m(z)$ satisfy (I) and are algebraically independent over $\mathbf{C}(z)$. Let $n > 0$ be any integer, and let $P(y_1, \cdots, y_m) = \sum p_{k_1\cdots k_m}y_1^{k_1}\cdots y_m^{k_m} \not\equiv 0$ be a polynomial with rational integral coefficients of total degree n; thus the summation extends over the $(m+n)!/(m!n!) = N$ say, systems of suffixes $k_1, \cdots, k_m$ for which $k_1 \geqq 0, \cdots, k_m \geqq 0, k_1 + \cdots + k_m \leqq n$.

Under the same hypothesis for ξ as in the last result, there exists for every $\varepsilon > 0$ a constant $C = C(\varepsilon) > 0$ independent of the coefficients of P such that

$$|P(f_1(\xi), \cdots, f_m(\xi))| > C \max(|p_{k_1\cdots k_m}|)^{-(N-1+\varepsilon)}.$$

This theorem can be generalised to the case when simultaneously the solutions in E-functions of more than one system of differential equations are considered. Several explicit examples are considered, and the author also mentions in the first note that even stronger lower estimates can be obtained if the upper bound $O(n^{\varepsilon n})$ in the definition of the E-functions is replaced by $O(C^n)$.

K. Mahler (Canberra)

Citations: MR 34# 4209 = J88-30; MR 35# 5397 = J02-22.

J80-24 (36# 1396)

Sprindžuk, V. G.

Concerning Baker's theorem on linear forms in logarithms. (Russian)

Dokl. Akad. Nauk BSSR **11** (1967), 767–769.

Durch Heranziehung p-adischer Methoden verschärft der Autor in gewissen Fällen ein Ergebnis von A. Baker [Mathematika **13** (1966), 204–216; vgl. auch ibid. **14** (1967), 102–107; ibid. **14** (1967), 220–228; MR **36** #3732]. Baker hatte für vorgegebene multiplikativ unabhängige algebraische Zahlen $\alpha_\nu \neq 0, 1$ die Linearform $|\beta_1 \log \alpha_1 + \cdots + \beta_{n-1} \log \alpha_{n-1} - \log \alpha_n|$ durch (*) $C(\alpha, d, \varepsilon) \exp\{-(\log H)^{n+1+\varepsilon}\}$ nach unten abgeschätzt; hierbei durchlaufen die β_ν die algebraischen Zahlen einer Höhe $\leqq H$ und vom Grade $\leqq d$.

Der Autor zeigt, daß man in (*) den Exponenten von $\log H$ durch jedes $\kappa > 3$ ersetzen kann, wenn die α_ν und β_ν gewisse weitere Bedingungen erfüllen (die multiplikative Unabhängigkeit der α_ν braucht nicht gefordert zu werden).

W. Schwarz (Freiburg)

Citations: MR 36# 3732 = J80-26.

Referred to in D40-49, D60-78.

J80-25 (36# 2560)

Fel'dman, N. I.

An estimate of the absolute value of a linear form in the logarithms of certain algebraic numbers. (Russian)

Mat. Zametki **2** (1967), 245–256.

The author establishes the following effective result. Let K be an imaginary quadratic field, m a positive integer, μ a real number greater than m, and $m_0 = \frac{1}{4}(m-1+\sqrt{(m^2+4m+1)})$. Let $a_1, \cdots, a_m$ be m distinct integers not zero in K. Put $a = \max_{i=1,2,\cdots,m}|a_i|$,

$$G_k = \text{lcm}_{i=1,2,\cdots,m; i\neq k}\{a_k, a_k - a_i\},$$

$$G = \text{lcm}_{i,j=1,2,\cdots,m; i\neq j}\{a_i, \overline{a_i - a_j}\},$$

where lcm denotes the least common multiple in K. Further, let $E_k = |a_k \prod_{i=1; i\neq k}^m (a_k - a_i)|$,

$$E = \max_{k=1,2,\cdots,m} \mathscr{E}_k, \quad E_0 = \min_{k=1,2,\cdots,m} \mathscr{E}_k,$$

$$B = 3a^m(m+1)^{m+1}(m_0+1)^{m_0+1}m_0^{-2m_0}(m-m_0)^{m_0-m},$$

$$F = \{(EE_0^{-1}a^{m+1})^\mu e^{(m^2+1)(\mu+1)} \times B^{m(\mu+1)}|G|^{2(\mu+1)} \prod_{k=1}^m |G_k|^{\mu+1}\}^{1/(\mu-m)}.$$

Finally, denote by q a positive rational integer, and by γ an integer in K, such that $q > F$,

$$0 < |\gamma| < (qF^{-1})^{(\mu-m)/\mu(m+1)}.$$

Then a constant $X_0 = X_0(a_1, \cdots, a_m, \gamma, q, \mu) > 0$ can be determined effectively such that

$$|x_0 + x_1 \log\{1-(a_1\gamma/q)\} + \cdots + x_m \log\{1-(a_m\gamma/q)\}| > X^{-\mu}$$

for all sets of $m+1$ integers $x_0, x_1, \cdots, x_m$ in K satisfying $X = \max(|x_0|, |x_1|, \cdots, |x_m|) \geqq X_0$.

The proof depends on an explicit construction of sets of $m+1$ linearly independent linear forms in $1, \log(1-a_1z), \cdots, \log(1-a_mz)$ with coefficients that are polynomials in z. The interest of this theorem lies in the fact that the constant μ can be chosen arbitrarily close to m.

K. Mahler (Canberra)

Referred to in J68-62, J68-64.

J80-26 (36# 3732)

Baker, A. [Baker, Alan]

Linear forms in the logarithms of algebraic numbers. I, II, III.

Mathematika **13** (1966), 204–216; *ibid.* **14** (1967), 102–107; *ibid.* **14** (1967), 220–228.

This is a very interesting sequence of papers. We quote from the introduction: "In 1934 Gel'fond and Schneider proved, independently, that the logarithms of an algebraic number to an algebraic base, other than 0 or 1, is either rational or transcendental and thereby solved the famous seventh problem of Hilbert. Among the many subsequent developments, Gel'fond obtained ... a positive lower bound for the absolute value of $\beta_1 \log \alpha_1 + \beta_2 \log \alpha_2$, where β_1, β_2 denote algebraic numbers, not both 0, and α_1, α_2 denote algebraic numbers not 0 or 1, with $\log \alpha_1/\log \alpha_2$ irrational. Of particular interest is the special case in which β_1, β_2 denote integers. In this case it is easy to obtain a trivial positive lower bound, and the existence of a non-trivial bound follows from the Thue-Siegel-Roth theorem. But Gel'fond's result improves substantially on the former, and, unlike the latter, it is derived by an effective method of proof. Gel'fond [*Transcendental and algebraic numbers* (Russian), GITTL, Moscow, 1952; MR **15**, 292; English translation, p. 177, Dover, New York, 1960; MR **22** #2598] remarked that an analogous theorem for linear forms in arbitrarily many logarithms of algebraic numbers would be of great value for the solution of some apparently very difficult problems of number theory. It is the object of this paper to establish such a result."

Define the height of an algebraic number to be the maximum of the absolute values of the relatively prime integer coefficients in its minimal defining polynomial.

The main result of Part I is Theorem 1.1: Let $\alpha_1, \cdots, \alpha_n$ $(n \geqq 2)$ denote algebraic numbers, not 0 or 1, such that (for any fixed branch of the logarithm) $\log \alpha_1, \cdots, \log \alpha_n$ and $2\pi i$ are linearly independent over the rationals Q; suppose $k > n+1$ and let d be any positive integer; then there is an effectively computable number $C = C(n, \alpha_1, \cdots, \alpha_n, k, d) > 0$ such that for all algebraic numbers $\beta_1, \cdots, \beta_n$, not all 0, with degree at most d, we have (*) $|\beta_1 \log \alpha_1 + \cdots + \beta_n \log \alpha_n| > Ce^{-(\log H)^k}$, where H denotes the maximum of the heights of $\beta_1, \cdots, \beta_n$. An immediate consequence is Corollary 1.1: If $\alpha_1, \cdots, \alpha_n$ denote non-zero algebraic numbers and $\log \alpha_1, \cdots, \log \alpha_n$, $2\pi i$ are linearly independent over Q, then $\log \alpha_1, \cdots, \log \alpha_n$ are linearly independent over the field A of all algebraic numbers. Starting with Corollary 1.1 for $n=1$, complete induction gives Corollary 1.2: If $\alpha_1, \cdots, \alpha_n$ denote positive real algebraic numbers other than 1 and $\beta_1, \cdots, \beta_n$ denote real algebraic numbers with $1, \beta_1, \cdots, \beta_n$ linearly independent over Q, then $\alpha_1^{\beta_1} \cdots \alpha_n^{\beta_n}$ is transcendental. It is remarked that, according to Gel'fond and Linnik, Theorem 1.1 suffices, at least in principle, to settle a conjecture by C. F. Gauss [*Disquisitiones arithmeticae*, § 303, G. Fleischer, Leipzig, 1801; German translation, Springer, Berlin, 1889; reprinting of German translation, pp. 351–352, § 303, Chelsea, New York, 1965; MR **32** #5488; English translation, Yale Univ. Press, New Haven, Conn., 1966; MR **33** #5545] that there are only nine imaginary quadratic fields with class number 1; H. Stark's different approach is mentioned. In order to prove Theorem 1.1, it is first shown that the following modified form of the theorem is sufficient. Theorem 1.1′: Under the hypothesis of Theorem 1.1 there is an effectively computable number C such that for all algebraic numbers $\beta_1, \cdots, \beta_{n-1}$ with degrees at most d, we have $|\beta_1 \log \alpha_1 + \cdots + \beta_{n-1} \log \alpha_{n-1} - \log \alpha_n| \geqq e^{-(\log H)^k}$, where H denotes any number not less than C and the height of $\beta_1, \cdots, \beta_{n-1}$. The proof itself then depends on the construction of a function ϕ of $n-1$ variables; ϕ is a direct generalization of a function of one variable used in Gel'fond's work (cf. $R(x)$ in C. L. Siegel's *Transcendental numbers* [p. 81, Ann. of Math. Studies No. 16, Princeton Univ. Press, Princeton, N.J., 1949; MR **11**, 330; German translation, Bibliograph. Inst., Mannheim, 1967; MR **35** #133]) and has the form $\phi(z_1, \cdots, z_{n-1}) = \sum_{\lambda_1=0}^{L} \cdots \sum_{\lambda_n=0}^{L} p(\lambda_1, \cdots, \lambda_n) \alpha_1^{\gamma_1 z_1} \cdots \alpha_{n-1}^{\gamma_{n-1} z_{n-1}}$ with integer coefficients and $\gamma_r = \lambda_r + \lambda_n \beta_r$ $(1 \leqq r < n)$.

In Part II it is shown that the number $2\pi i$ can be omitted in Corollary 1.1; this gives Corollary 2.1: Let $\alpha_1, \cdots, \alpha_n$ denote non-zero algebraic numbers; then $\log \alpha_1, \cdots, \log \alpha_n$ are linearly independent over Q if and only if they are linearly independent over A. Also Corollary 1.2 is sharpened to Corollary 2.2: If $\alpha_1, \cdots, \alpha_n$ denote algebraic numbers other than 0 or 1 and if $\beta_1, \cdots, \beta_n$ denote algebraic numbers with $1, \beta_1, \cdots, \beta_n$ linearly independent over Q, then $\alpha_1^{\beta_1} \cdots \alpha_n^{\beta_n}$ is transcendental. Both corollaries follow from Theorem 2.2: Let again $\alpha_1, \cdots, \alpha_n$ $(n \geqq 2)$ denote non-zero algebraic numbers such that $\log \alpha_1, \cdots, \log \alpha_n$ are linearly independent over Q; suppose $k > 2n+1$ and let d be any positive integer; then there is an effectively computable number $C > 0$ such that for all algebraic numbers $\beta_1, \cdots, \beta_n$, not all 0, with degree at most d, we have (*).

In Part III an inhomogenous analogue is considered. Theorem 3.1: Let $\alpha_1, \cdots, \alpha_n, \beta_0, \beta_1, \cdots, \beta_n$ denote non-zero algebraic numbers; suppose $k > n+1$, and let d $[H]$ denote the maximum of the degrees [heights] of $\beta_0, \cdots, \beta_n$; then $|\beta_0 + \beta_1 \log \alpha_1 + \cdots + \beta_n \log \alpha| > Ce^{-(\log H)^k}$ for some effectively computable number $C = C(n, \alpha_1, \cdots, \alpha_n, k, d) > 0$. The method of proof can be adapted to $\beta_0 = 0$ and yields Theorem 3.2: Let $\alpha_1, \cdots, \alpha_n, \beta_1, \cdots, \beta_n$ denote non-zero algebraic numbers; suppose that either $\log \alpha_1, \cdots, \log \alpha_n$ or $\beta_1, \cdots, \beta_n$ are linearly independent over Q; suppose that $k > n$ and let d $[H]$ denote the maximum of the degrees [heights] of $\beta_1, \cdots, \beta_n$; then (*) holds for some effectively computable number $C = C(n, \alpha_1, \cdots, \alpha_n, k, d) > 0$. Theorem 3.1 implies that $e^{\beta_0} \alpha_1^{\beta_1} \cdots \alpha_n^{\beta_n}$ is transcendental for any non-zero algebraic numbers $\alpha_1, \cdots, \alpha_n, \beta_0, \beta_1, \cdots, \beta_n$. Furthermore, $\pi + \log \alpha$ is transcendental for any algebraic number $\alpha \neq 0$ and also $e^{\alpha\pi + \beta}$ is transcendental for all algebraic numbers α, β with $\beta \neq 0$. It is conjectured that Theorems 3.1 and 3.2 essentially hold with $k = 1$. Further applications to binary forms and to Liouville's theorem of 1844 concerning the approximation of algebraic numbers by rational numbers are announced.

{See also #3746 below.} *G. J. Rieger* (Munich)

Citations: MR 11, 330c = J76-18; MR 15, 292e = J02-6; MR 22# 2598 = J02-7; MR 32# 5488 = Z25-20; MR 33# 5545 = Z25-21; MR 35# 133 = J76-19; MR 36# 3746 = J80-27.

Referred to in D24-71, D40-49, D40-50, D56-32, E16-76, J76-63, J80-24, J80-27, J80-28, J80-31, J80-33, J80-34, J80-36, J80-40, J80-41, M55-60, R14-79, R14-81, R14-82, R14-84, R14-105.

J80-27 (36# 3746)

Brumer, Armand

On the units of algebraic number fields.

Mathematika **14** (1967), 121–124.

The author shows that the method of A. Baker [Mathematika **13** (1966), 204–216; see #3732 above] extends to cover the p-adic analogue of Baker's theorem on the approximation to logarithms of algebraic numbers: Let $\alpha_1, \cdots, \alpha_n$ be p-adic units algebraic over the rationals whose p-adic logarithms are linearly independent over the rationals; then these numbers are linearly independent over the algebraic closure of the rationals in the p-adic field. The author then notes that, as observed by J. Ax [Illinois J. Math. **9** (1965), 584–589; MR **31** #5858], this implies: If E is an abelian extension of the rationals (or an imaginary quadratic) then the p-adic regulator of E does not vanish. *D. J. Lewis* (Ann Arbor, Mich.)

Citations: MR 31# 5858 = R22-27; MR 36# 3732 = J80-26.

Referred to in J80-26.

J80-28 (41# 5303)

Fel'dman, N. I.

A refinement of two effective inequalities of A. Baker. (Russian)

Mat. Zametki **6** (1969), 767–769.

In A. Baker's fundamental paper [Philos. Trans. Roy. Soc. London Ser. A **263** (1967/68), 173–191, Theorems 1 and 2; MR **37** #4005] appears the condition $k > n+1$. In another paper by Baker [Mathematika **14** (1967), 220–228; MR **36** #3732] this is relaxed to $k > n$. In the paper under review, this condition is, under certain circumstances involving the number of fundamental units in a certain field, modified, and in some cases relaxed.

{This article has appeared in English translation [Math. Notes **6** (1969), 925–926].} *S. Stein* (Davis, Calif.)

Citations: MR 36# 3732 = J80-26; MR 37# 4005 = D40-50.

J80-29 (36# 5086)

Schinzel, A.

On two theorems of Gelfond and some of their applications.

Acta Arith. **13** (1967/68), 177–236.

The two theorems referred to in the title are Gel'fond's ordinary and $\mathfrak{p}$-adic measure of irrationality of the ratio of two logarithms of algebraic numbers [see A. O. Gel'fond, Izv. Akad. Nauk SSSR (1939), 509–518; MR **1**, 295; *Transcendental and algebraic numbers* (Russian), GITTL, Moscow, 1952; MR **15**, 292; English translation, Dover, New York, 1960; MR **22** #2598]. For the author's purpose the theorems must be given without unspecified constants. He proves the following version of Gel'fond's theorem. Let α, β be algebraic numbers, let n, m be rational integers with $N = \max\{|n|, |m|\} > 0$. Then

$$\log|\alpha^n - \beta^m| - \max\{n \log|\alpha|, m \log|\beta|\} > -c_1(\log N + c_2)^3,$$

where c_1, c_2 are explicitly given constants depending only on α, β (one special case of the case where $|\alpha|, |\beta|$ are multiplicatively dependent is excluded). He gives a similar estimate in the $\mathfrak{p}$-adic case.

The author's first application of Gel'fond's theorems concerns the study of linear recurrences of the second order. That is, let u_n be a sequence of rational integers satisfying the formula $u_{n+1} = Pu_n - Qu_{n-1}$, where P, Q are rational integers and $PQ \neq 0$, $\Delta = P^2 - 4Q \neq 0$, $u_1^2 - Pu_1u_0 + Qu_0^2 \neq 0$. It is well known that $u_n = \Omega\omega^n + \Omega'\omega'^n$, where ω, ω' are the roots of $z^2 - Pz + Q$ and Ω, Ω' are chosen suitably. If $\Delta > 0$, u_n is easy to estimate. Here the case where $\Delta < 0$ is considered. For example, the author shows that if $\Delta < 0$, $P^2 \neq Q$, $2Q$, $3Q$ and ω/ω', Ω/Ω' are multiplicatively dependent, then for n sufficiently large (explicitly given) we have

$$|u_n| > |\Delta|^{-1/2}|Q|^{n/2} \exp(-c_3 \log^2 n),$$

where $c_3(P, Q, u_1, u_0)$ is given explicitly. As an application of this theorem the author proves that if d is an odd negative integer $\neq 1-2^k$, the diophantine equation $x^2-d=2^m$ has at most one solution for $m>80$, $x>0$. This result is generalized to the case where the "2" on the right side of the equation is replaced by certain primes. He further obtains lower bounds for the largest prime factor of u_n.

Next, the author turns to the expression $x^\nu \pm P_1^{n_1} \cdots P_k^{n_k}$ for $\nu=2, 3$ and $P_1, \cdots, P_k$ positive integers. He gives an estimate for its size and in certain cases for its greatest prime factor. He uses the latter result to show how to solve effectively all diophantine equations of the form $q_1^{y_1} \cdots q_i^{y_i} \pm r_1^{z_1} \cdots r_j^{z_j} = s^x$, where $q_1, \cdots, q_i, r_1, \cdots, r_j$ are distinct primes and s is a positive integer.

Finally, he gives estimates for the greatest prime factor of certain polynomials. For example, he proves that $q(Ax^\nu - E) \geqq c \log\log x$, for $\nu=2, 3$, A, E non-zero integers and c given explicitly ($q(n)$ = greatest prime factor of n). Or he shows $\liminf(\log q(f(x)))/\log|f(x)|$ $(x \to \infty)$ is bounded above by an explicitly given constant depending only on the degree of f, where f is any polynomial with rational integer coefficients.

Many references to the earlier literature are given in the paper. *W. W. Adams* (Berkeley, Calif.)

Citations: MR 1, 295c = J80-1; MR 15, 292e = J02-6; MR 22# 2598 = J02-7.

Referred to in B40-57.

J80-30 (37# 4025)

Fel'dman, N. I.

Estimation of a linear form in the logarithms of algebraic numbers. (Russian)

Mat. Sb. (N.S.) **76** **(118)** (1968), 304–319.

Suppose $\alpha_1, \cdots, \alpha_m$ are algebraic numbers, and $\ln \alpha_1, \cdots, \ln \alpha_m$ are fixed values of their logarithms which are linearly independent over the rational field. Let $\beta_0, \beta_1, \cdots, \beta_m$ be algebraic numbers, not all 0, of degrees $\leqq n$ and heights $\leqq H$. The author proves that for any $\varepsilon>0$, there is a constant $C_1 = C_1(\varepsilon, n, \ln \alpha_1, \cdots, \ln \alpha_m)$ such that

$$|\beta_0+\beta_1 \ln \alpha_1 + \cdots + \beta_m \ln \alpha_m| > C_1 \exp(-\ln H \ln^{m+\varepsilon} \ln H).$$

He also shows that if $\alpha \neq 0$ is algebraic, $y \neq 0$ an integer, and P/Q any rational number of the form $p_1^{x_1} \cdots p_m^{x_m}$, where $p_1, \cdots, p_m$ are fixed primes, then for any $\varepsilon>0$,

$$|e^{\alpha y} - P/Q| \geqq C_2 \exp(-\ln\ln Q \ln^{m+\varepsilon} \ln\ln Q),$$

where $C_2 = C_2(\varepsilon, \alpha, p_1, \cdots, p_m)$. Moreover, if the only solution of $\alpha^{y'} = p_1^{x_1'} \cdots p_m^{x_m'}$ is $x_1' = \cdots = x_m' = y' = 0$, then $|\alpha^y - P/Q| \geqq C_3 \exp(-\ln\ln Q \ln^{m+1+\varepsilon} \ln\ln Q)$, where $C_3 = C_3(\varepsilon, \alpha, p_1, \cdots, p_m)$. *B. Gordon* (Los Angeles, Calif.)

Referred to in J80-33, R46-36.

J80-31 (37# 4026)

Fel'dman, N. I.

A linear form in the logarithms of algebraic numbers. (Russian)

Uspehi Mat. Nauk **23** (1968), *no.* 3 (141), 185–186.

The author announces the following improvement of a recent result of A. Baker [Mathematika **13** (1966), 204–216; MR **36** #3732]. Let $\alpha_1, \alpha_2, \cdots, \alpha_m$, $m \geqq 2$, be given algebraic numbers whose logarithms are linearly independent over the rationals. Then for every integer $n \geqq 1$ and every $\varepsilon>0$ there exist constants $c=c(n, \alpha_1, \alpha_2, \cdots, \alpha_m)$, $C=C(\varepsilon, n, \alpha_1, \alpha_2, \cdots, \alpha_m)$ such that $|\beta_0+\beta_1 \log \alpha_1 + \cdots + \beta_m \log \alpha_m| > C \exp\{-(\log H^{-c})(\log\log H^{-c})^{(\varepsilon+m)}\}$ holds for every choice $\beta_0, \beta_1, \cdots, \beta_m$, not all zero, of algebraic numbers whose degrees and heights do not exceed n and H.

An application is mentioned to a Thue-Siegel-Roth type theorem with restricted approximants.

D. H. Lehmer (Berkeley, Calif.)

Citations: MR 36# 3732 = J80-26.

J80-32 (37# 5156)

Galočkin, A. I.

Estimate of the mutual transcendence measure of the values of E-functions. (Russian)

Mat. Zametki **3** (1968), 377–386.

The author generalizes theorems proved by C. L. Siegel in his classical memoir [Abh. Deutsch. Akad. Wiss. Berlin Kl. Phys.-Math. **1929**, no. 1; reprinted in *Gesammelte Abhandlungen*, Vol. I, pp. 209–266, Springer, Berlin, 1966; see MR **33** #5441] concerning E-functions [see reprint cited, p. 223]. A typical formula of Siegel gives a lower bound [reprint cited, p. 229] for $|g(J_0(\xi), J_0'(\xi))|$ (where ξ is a non-zero algebraic number, $J_0(x)$ is Bessel's function and $g(x, y)$ is a polynomial in x and y with integral coefficients (not all zero) and absolutely $\leqq G$) in terms of G, the dimension of g and the degree of ξ. See also a paper by S. Lang [Mathematika **9** (1962), 157–161; MR **26** #6124]. The author's theorem (pp. 378–379 of his paper) is too long to be quoted here. *S. Chowla* (University Park, Pa.)

Citations: MR 26# 6124 = J80-15; MR 33# 5441 = Z25-23.

J80-33 (38# 1059)

Fel'dman, N. I.

An improvement of the estimate of a linear form in the logarithms of algebraic numbers. (Russian)

Mat. Sb. (N.S.) **77** **(119)** (1968), 423–436.

In a previous work [same Sb. (N.S.) **76** **(118)** (1968), 304–319; MR **37** #4025] the author proved the following inequality:

$$|\beta_0+\beta_1 \ln \alpha_1 + \cdots + \beta_m \ln \alpha_m| > C_1 \exp[-\ln H \ln^{m+\varepsilon} \ln H].$$

Here $\varepsilon>0$; $\ln \alpha_1, \cdots, \ln \alpha_m$ are fixed values of logarithms of algebraic numbers, linearly independent over the field of rationals; $\beta_0, \beta_1, \cdots, \beta_m$ are arbitrary algebraic numbers not all 0 whose degrees do not exceed n and H, respectively; and C_1 is a constant depending on ε, n and the logarithms of the α's.

In the present paper the author sharpens the above inequality by proving the following theorem: There exists an effective absolute constant c such that for any linearly independent logarithms of algebraic numbers $\ln \alpha_1, \cdots, \ln \alpha_m$ (over the field of rationals), $m \geqq 2$, and arbitrary algebraic numbers $\gamma_0, \gamma_1, \cdots, \gamma_m$ ($|\gamma_0| + \cdots + |\gamma_m| > 0$), the following inequality holds:

$$|\gamma_0+\gamma_1 \ln \alpha_1 + \cdots + \gamma_m \ln \alpha_m| > \exp\{-n(n+\ln H_0)(4^{6m^2-2m+1} n^{1/2}[c+\ln h])^{12m^2+4m-3}\},$$

where n is the degree of the field $R(\alpha_1, \cdots, \alpha_m, \gamma_0, \cdots, \gamma_m)$, and $H_0 = \operatorname{Max}\{L(\gamma_0), L(\gamma_1), \cdots, L(\gamma_m)\}$,

$$h = \operatorname{Max}\{L(\alpha_1), \cdots, L(\alpha_m), e^{|\ln \alpha_1|}, \cdots, e^{|\ln \alpha_m|}\}.$$

Finally, $L(\zeta)$ denotes the sum of the moduli of the coefficients of the irreducible polynomial with relatively prime integer coefficients satisfied by ζ.

The proof of this theorem relies upon known methods of Gel'fond and Baker [cf. A. O. Gel'fond, Izv. Akad. Nauk SSSR Ser. Mat. **5/6**, 509–518 (1939); MR **1**, 295; A. Baker, Mathematika **13** (1966), 204–216; ibid. **14** (1967), 102–107; ibid. **14** (1967), 220–228; MR **36** #3732].

Finally, the author derives some simple corollaries of the above theorem. *R. Finkelstein* (Bowling Green, Ohio)

Citations: MR 1, 295c = J80-1; MR 36# 3732 = J80-26; MR 37# 4025 = J80-30.

Referred to in J80-46.

J80-34 **(38# 2092)**

Fel'dman, N. I.

A linear form in the logarithms of algebraic numbers. (Russian)

Dokl. Akad Nauk SSSR **182** (1968), 1278–1279.

Der Autor kündigt folgende weitgehende Verschärfung von Ergebnissen von A. O. Gel'fond [Uspehi Mat. Nauk **4** (1949), no. 5 (33), 14–48; MR **11**, 231] und A. Baker [Mathematika **13** (1966), 204–216; ibid. **14** (1967), 102–107; MR **36** #3732] an: Sind α_μ, β_0 und β_μ $(\mu=1, \cdots, m)$ Zahlen aus einem algebraischen Körper vom Grade n über dem rationalen Zahlkörper $\mathbf{Q}$ und sind $\log \alpha_1, \cdots, \log \alpha_m$ linear unabhängig über $\mathbf{Q}$, so ist $|\beta_0+\sum_{1\leqq\mu\leqq m}\beta_\mu\cdot\log\alpha_\mu| > \exp\{-(c_0+c_1)^{16m^2}\cdot\log H\}$, wobei H das Maximum der Höhen der β's bezeichnet; die Konstante c_1 wird explizit in Abhängigkeit von n, m und den α's gegeben; c_0 ist eine absolute, effektiv berechenbare Konstante. Eine analoge Ungleichung gilt im p-adischen.

{This article has appeared in English translation [Soviet Math. Dokl. **9** (1968), 1284–1285].}

W. Schwarz (Freiburg)

Citations: MR 11, 231d = J88-5; MR 36# 3732 = J80-26.

J80-35 **(39# 133)**

Šmelev, A. A.

The approximation of a certain class of transcendental numbers. (Russian)

Mat. Zametki **5** (1969), 117–128.

Let $a\neq 0$, $\neq 1$, $\alpha\neq 0$, $\beta\neq 0$, and $\xi\neq 0$ be algebraic numbers, n_0 a fixed positive integer, and ε a constant satisfying $0<\varepsilon<\frac{1}{2}$. Denote by N and H the degree and the height of ξ, respectively. The author shows that there exists a constant H_0 depending only on a, α, β, n_0, and ε such that $|e^\alpha a^\beta-\xi|>\exp(-(\log H)^{4+\varepsilon})$ if $N\leqq n_0$ and $H>H_0$. See also the earlier paper by the author [same Zametki **4** (1968), 341–348; MR **38** #1060].

K. Mahler (Columbus, Ohio)

Citations: MR 38# 1060 = J76-63.

J80-36 **(40# 94)**

Ramachandra, K.

A note on Baker's method.

J. Austral. Math. Soc. **10** (1969), 197–203.

Let $\alpha_1, \alpha_2, \cdots, \alpha_n$ $(n\geqq 2)$ be multiplicatively independent nonzero algebraic numbers and set $M(H)=\min|\beta_1\log\alpha_1+\cdots+\beta_n\log\alpha_n|$, where the minimum is taken over all algebraic numbers $\beta_1, \cdots, \beta_n$ not all equal to zero, of degrees not exceeding a fixed natural number d_0 and heights not exceeding an arbitrary natural number H. Then a result of A. Baker [Mathematika **13** (1966), 204–216; MR **36** #3732] states that $M(H)>Ae^{-(\log H)^{n+1+\varepsilon}}$ for every fixed $\varepsilon>0$ and an explicit constant A depending on $\alpha_1, \cdots, d_0$ and ε. In the present paper the author generalizes this result by proving the following theorem: Let $\alpha_1, \cdots, \alpha_{n+f}$ $(n\geqq 1, f\geqq 1)$ be multiplicatively independent nonzero algebraic numbers and set $M(H, \alpha_{n+i})=\min|\beta_1\log\alpha_1+\cdots+\beta_n\log\alpha_n-\log\alpha_{n+i}|$ $(i=1, \cdots, f)$, where the minimum is taken over all algebraic numbers $\beta_1, \cdots, \beta_n$ of degrees not exceeding a fixed natural number d_0 and heights not exceeding an arbitrary natural number H. Then we have $\sum_{i=1}^f M(H, \alpha_{n+i})>Ae^{-(\log H)^{(n+1)/f+1+\varepsilon}}$, with an explicit positive constant A depending only on the α_i, d_0 and ε.

Further, the author deduces the following corollary: Given the α_i as in the theorem and a positive ε, there exists an algebraic number α_{n+1} for which $M(H, \alpha_{n+1})$ exceeds $Ae^{-(\log H)^{1+\varepsilon}}$ for infinitely many H with an explicit positive constant A depending only on the α_i, d_0 and ε.

R. Finkelstein (Bowling Green, Ohio)

Citations: MR 36# 3732 = J80-26.

J80-37 **(40# 1345)**

Fel'dman, N. I.

An elliptic analog of an inequality of A. O. Gel'fond. (Russian)

Trudy Moskov. Mat. Obšč. **18** (1968), 65–76.

The author proves the following theorem. Let $\wp(z)$ be the Weierstrass elliptic function with the invariants g_2 and g_3 which are assumed to be algebraic numbers. Further, let α and β be two numbers such that $\wp(\alpha)$ and $\wp(\beta)$ are algebraic while $\zeta=\alpha/\beta$ is irrational. Then there exists a positive constant $\Lambda=\Lambda(g_2, g_3, \alpha, \beta)$ such that $|\zeta-(P/Q)|>\exp(-e^{\Lambda\sqrt{\log Q}})$ for all rational integers P and $Q>0$.

K. Mahler (Columbus, Ohio)

Referred to in J80-44.

J80-38 **(40# 1346)**

Haneke, Wolfgang

Über ein System simultaner diophantischer Approximationen.

Math. Z. **110** (1969), 378–384.

Seien $\alpha_1, \cdots, \alpha_r\neq 0$ feste reelle algebraische Zahlen mit $\sum_{\nu=1}^r x_\nu\log\alpha_\nu\neq 0$ für alle ganzen $(x_1, \cdots, x_r)\neq 0$. Seien $\xi_1, \cdots, \xi_r$ beliebige reelle algebraische Zahlen und seien G_ν bzw. H_ν Grad bzw. Höhe von ξ_ν. R sei der rationale Zahlkörper, G der Grad des Körpers $R(\alpha_1, \cdots, \alpha_r; \xi_1, \cdots, \xi_r)$ und es werde $H=\exp[G\sum_{\nu=1}^r(\log H_\nu)/G_\nu]$ gesetzt. Dann zeigt der Autor: Zu jedem $r\geqq 2$ und jedem $\varepsilon>0$ gilt für alle reellen $t\geqq 1$ mit einem nur von $\alpha_1, \cdots, \alpha_r$ und ε abhängigen c_1:

$$\sum_{\nu=1}^r |t\log\alpha_\nu-\xi_\nu| > e^{-c_1(G\log H)^{(1-1/r)^{-1}+\varepsilon}}$$

für $G<\log H$. Hilfsmittel sind Methoden aus A. O. Gel'fonds Buch [*Transcendental and algebraic numbers* (Russian), GITTL, Moscow, 1952; MR **15**, 292; English translation, Dover, New York, 1960; MR **22** #2598]. Das Ergebnis ist eine Verschärfung eines Resultats von N. I. Fel'dman [Izv. Akad. Nauk SSSR Ser. Mat. **24** (1960), 475–492; MR **22** #5623b]. *J. M. Wills* (Berlin)

Citations: MR 15, 292e = J02-6; MR 22# 2598 = J02-7; MR 22# 5623b = J80-14.

J80-39 **(40# 2610)**

Fel'dman, N. I.

A certain inequality for a linear form in the logarithms of algebraic numbers. (Russian)

Mat. Zametki **5** (1969), 681–689.

The main result of this paper is the following theorem: Let $\ln\alpha_1, \cdots, \ln\alpha_m$, $m\geqq 2$, be fixed logarithmic branches of the nonzero algebraic numbers $\alpha_1, \cdots, \alpha_m$ such that $\ln\alpha_1, \cdots, \ln\alpha_{m-1}$ are linearly independent over the field of rational numbers. Further, let $b_1, \cdots, b_{m-1}$ be rational integers, $H=\max|b_k|$ and δ, η positive constants, with $A=$height α_m. Then there exists an effectively computable constant γ, independent of A and H, such that if $|b_1\ln\alpha_1+\cdots+b_{m-1}\ln\alpha_{m-1}-\ln\alpha_m|<\exp(-\delta H)$, then $H<\max(\gamma, \ln^{m-1+\eta}A)$.

This generalizes a result of A. Baker [Philos. Trans. Roy. Soc. London Ser. A **263** (1967/68), 173–191; MR **37** #4005], who obtained an analogous result with -1 replaced by $+1$ in the exponent of $\ln A$.

{This article has appeared in English translation [Math. Notes **5** (1969), 408–412].}

R. Finkelstein (Bowling Green, Ohio)

Citations: MR 37# 4005 = D40-50.

J80-40 **(41# 3402)**

Baker, A.

Linear forms in the logarithms of algebraic numbers. IV.

Mathematika **15** (1968), 204–216.

The following theorem is proved: Let $\alpha_1, \cdots, \alpha_n$ be algebraic numbers different from zero with heights $\leqq A$ and degrees $\leqq d$. It is supposed that $d \geqq 4$ and $A \geqq 4$. Further, let δ be a real number satisfying $0<\delta<1$. Then if for some rational integers $b_1, \cdots, b_n$ with $|b_\nu| \leqq H$ ($\nu=1, 2, \cdots, n$) the inequality

$$0<|b_1 \log \alpha_1+\cdots+b_m \log \alpha_m|<e^{-\delta H}$$

holds, then we have $H<(4^{n^2}\delta^{-1} d^{2n} \log A)^{(2n+1)2}$.

The importance of the above theorem lies in the fact that it gives "calculable" values for the upper bound of the absolute values of possible solutions of certain systems of Diophantine equations.

{Parts I through III appeared in Mathematika **13** (1966), 204–216; ibid. **14** (1967), 102–107; ibid. **14** (1967), 220–228 [MR **36** #3732].} *P. Szüsz* (Stony Brook, N.Y.)

Citations: MR 36# 3732 = J80-26.

J80-41 **(41# 5304)**

Sprindžuk, V. G.

Estimates of linear forms with p-adic logarithms of algebraic numbers. (Russian)

Vescī Akad. Navuk BSSR Ser. Fīz.-Mat. Navuk **1968**, *no.* 4, 5–14.

The main result of this paper is the following theorem, which is the p-adic analogue of a theorem of A. Baker [Mathematika **13** (1966), 204–216; MR **36** #3732]: Let $\alpha_1, \cdots, \alpha_n$ be algebraic numbers different from 0, 1 and let $\chi>n+1$. Then for arbitrary algebraic numbers $\beta_1, \cdots, \beta_n$ whose degrees do not exceed d and which are not all 0, we have $|\beta_1 \log \alpha_1+\cdots+\beta_n \log \alpha_n|>p^{(-\ln H)^\chi+c}$, where p is a rational prime, H is the largest of the heights of $\beta_1, \cdots, \beta_n$ and $c=c(\alpha_1, \cdots, \alpha_n, d, \chi)$ is an effectively computable number.

In addition to the above result, the author proves some results which give a better bound, but with the restriction that the corresponding q-adic logarithms (q a prime number $\neq p$) are not smaller in the q-adic metric.

R. Finkelstein (Bowling Green, Ohio)

Citations: MR 36# 3732 = J80-26.

J80-42 **(41# 8345)**

Šmelev, A. A.

Approximation of the roots of certain transcendental equations. (Russian)

Mat. Zametki **7** (1970), 203–210.

Let $P(x_1, x_2)$ be a polynomial with integer coefficients, $\partial P/\partial x_1 \not\equiv 0$, $\partial P/\partial x_2 \not\equiv 0$. Suppose that no polynomial $P(x_1)$ with integer coefficients is a factor of $P(x_1, x_2)$.

Denote by ζ an irrational root of the equation $P(z, a^z)=0$. Then for any n there exists a constant $\Lambda=\Lambda(n, \zeta)$ such that for every algebraic number ξ of degree $\leqq n$ and of height H, the inequality $|\zeta-\xi|>\exp(-\Lambda \ln^3 H(\ln\ln H)^5)$ holds.

The proof is based on the method of A. O. Gel'fond [*Transcendental and algebraic numbers* (Russian), GITTL, Moscow, 1952; MR **15**, 292; English translation, Dover, New York, 1960; MR **22** #2598] and some ideas of N. I. Fel'dman [Izv. Akad. Nauk SSSR Ser. Mat. **24** (1960), 475–492; MR **22** #5623b; Vestnik Moskov. Univ. Ser. I Mat. Meh. **1964**, no. 1, 13–20; MR **28** #2091].

{This article appeared in English translation [Math. Notes **7** (1970), 122–126].} *J. Kubilius* (Vilnius)

Citations: MR 15, 292e = J02-6; MR 22# 2598 = J02-7; MR 22# 5623b = J80-14; MR 28# 2091 = J80-17.

J80-43 **(42# 3027)**

Galočkin, A. I.

Estimates from below of linear forms in the values of certain hypergeometric functions. (Russian)

Mat. Zametki **8** (1970), 19–28.

Let d be a positive integer, J either the rational field Q or the quadratic field $Q(\sqrt{-d})$. For arbitrary complex numbers $\xi_1, \cdots, \xi_m$ put

$$L(\xi_1, \cdots, \xi_m; H) = \min|h_1\xi_1+\cdots+h_m\xi_m|,$$

where the minimum is extended over all integers $h_1, \cdots, h_m$ in J which are not all 0 and satisfy $|h_i| \leqq H$ for $i=1, \cdots, m$. Next, let $\lambda_1, \cdots, \lambda_m$ be algebraic numbers distinct from $-1, -2, \cdots$; let a be a positive integer such that $a\lambda_1, \cdots, a\lambda_m$ are algebraic integers; let

$$\Lambda = \max(a, |\lambda_1|, \cdots, |\lambda_m|),$$

and let T be the differential operator $z\, d/dz$. Put $[\lambda, \nu]=1$ if $\nu=0$, and $[\lambda, \nu]=\lambda(\lambda+1)\cdots(\lambda+\nu-1)$ if $\nu=1, 2, \cdots$. Further, put $\psi(z)=\sum_{\nu=0}^\infty ([\lambda_1+1, \nu]\cdots[\lambda_m+1, \nu])^{-1}(z/m)^{m\nu}$, $\psi_1(z)=\psi(z)$, $\psi_{s+1}(z)=T\psi_s(z)$. Assume that $\alpha^m \neq 0$ lies in J and that H is sufficiently large. The author proves that if $\lambda_1, \cdots, \lambda_m$ are rational, then $L(1, \psi_1(\alpha), \cdots, \psi_m(\alpha), H)>H^{-m-(\gamma_1 m^2/\log\log H)}$, where γ_1 depends only on α and Λ. If further $\lambda_1, \cdots, \lambda_m$ are algebraic such that the polynomial $(x+\lambda_1)\cdots(x+\lambda_m)$ has coefficients in J and is $\neq 0$ for $x=0$, and if $m \geqq 2$, then

$$L(\psi_1(\alpha), \cdots, \psi_m(\alpha), H) > H^{-1-m-(\gamma_2 m^2/\log\log H)},$$

where again γ_2 depends only on α and Λ. The transfer theorem by A. Ja. Hinčin [Rend. Circ. Mat. Palermo **50** (1926), 170–195] implies in both cases an analogous estimate for simultaneous approximations.

{This article has appeared in English translation [Math. Notes **8** (1970), 478–484].} *K. Mahler* (Columbus, Ohio)

J80-44 **(43# 7410)**

Coates, J.

An application of the Thue-Siegel-Roth theorem to elliptic functions.

Proc. Cambridge Philos. Soc. **69** (1971), 157–161.

Let (*) $\alpha_j=e^{u_j}$ ($j=1, \cdots, n$) be algebraic, such that $u_1, u_2, \cdots, u_n, 2\pi i$ are linearly independent over the field Q of rationals; then, according to the Thue-Siegel-Roth theorem, to each $\delta>0$ there are only finitely many integers $b_1, \cdots, b_n$ satisfying $|\sum_{j=1}^n b_j u_j|<e^{-\delta H}$, with $H=\max_j|b_j|$. The main result of the present paper is an analogous statement, with the u_j defined not as solutions of (*), but of $\alpha_j=\wp(u_j)$, with algebraic α_j's. Here $\wp(z)$ is Weierstrass' elliptic function of primitive periods ω_1, ω_2, and corresponding to algebraic g_2, g_3, such that $u_1, u_2, \cdots, u_n, \omega_1, \omega_2$ are linearly independent over Q. To be precise, it is shown that under the stated conditions and for every $\delta>0$, (⁑) $|\sum_{j=1}^n b_j u_j|<e^{-\delta H^2}$ has only finitely many solutions in integers b_j, with $H=\max_j|b_j|$. The proof is not "effective". In fact, by methods somewhat similar to those of C. L. Siegel [Abh. Akad. Wiss. Berlin **1929**, no. 1; reprinted in *Gesammelte Abhandlungen*, Band I, pp. 209–266, Springer, Berlin, 1966; see MR **33** #5441], the problem is reduced to a Diophantine approximation problem in an algebraic number field, to which LeVeque's generalization of Roth's theorem applies directly. If (⁑) would have infinitely many solutions, then so would the last-named problem, contradicting the Roth-LeVeque theorem. The author remarks that the proof could be made effective, by use of a Baker-type result, instead of the Thue-Siegel-Roth-LeVeque theorem, but that the existing results of this type are not strong enough for a successful use in connection with (⁑). For the particular

case $n=2$, N. I. Fel'dman [Trudy Moskov. Mat. Obšč. **18** (1968), 65–76; MR **40** #1345; translated in Trans. Moscow Math. Soc. **18** (1968), 71–84; see MR **40** #2486] has proven an effective theorem, actually stronger than (*). {On the last 2 lines of p. 158 and first line on p. 159, read l instead of some of the 1's.} *E. Grosswald* (Philadelphia, Pa.)

Citations: MR 33# 5441 = Z25-23; MR 40# 1345 = J80-37.

J80-45 **(44# 162)**

Nurmagomedov, M. S.

The arithmetic properties of the values of a certain class of analytic functions. (Russian)

Mat. Sb. (N.S.) **85 (127)** (1971), 339–365.

In his classical paper [Abh. Preuss. Akad. Wiss. Phys.-Math. Kl. **1929**, no. 1] C. L. Siegel started the investigation of the transcendency of a very general class of entire functions which he called E-functions. This work has in recent years been much extended by A. B. Šidlovskiĭ and his students [see, e.g., Izv. Akad. Nauk SSSR Ser. Mat. **23** (1959), 35–66; MR **21** #1295]. Siegel already considered in that paper a more general class of power series, the G-series, but could obtain only very special results for these. This work on G-series is now greatly extended by the author. He considers power series with coefficients in either the rational field or an imaginary quadratic field, of finite radius of convergence, and such that the least common denominators of the first n coefficients increase only like the nth power of a constant; just as in Siegel's paper, a set of finitely many such series is to satisfy a homogeneous or inhomogeneous system of linear differential equations in which the coefficients are rational functions of z. The author can then obtain theorems on the linear independence, and measures of irrationality, for the values of these G-functions at either rational points or points in the imaginary quadratic field. The exact statements of his theorems are rather involved and depend on a large number of parameters. *K. Mahler* (Columbus, Ohio)

Citations: MR 21# 1295 = J88-15.

J80-46 **(44# 1631)**

Kaufman, R. M.

An estimate of a linear form of logarithms of algebraic numbers in a $\mathfrak{P}$-adic metric. (Russian. English summary)

Vestnik Moskov. Univ. Ser. I Mat. Meh. **26** (1971), *no.* 2, 3–10.

In an algebraic number field K consider a prime ideal $\mathfrak{P}$ that divides a prime number p. The $\mathfrak{P}$-adic norm of a nonzero number a in K is defined to be $|a|_{\mathfrak{P}}=N(\mathfrak{P})^{-v_{\mathfrak{P}}(a)}$, where $v_{\mathfrak{P}}(a)$ is the order of a at $\mathfrak{P}$. Let $\alpha_1, \cdots, \alpha_m$ be fixed algebraic numbers that are multiplicatively independent over the rational field Q and such that their heights are not greater than a given number h. Further, let $\beta_0, \beta_1, \cdots, \beta_m$ be arbitrary algebraic numbers with heights not exceeding a number $H>1$. If $|\alpha_i-1|_{\mathfrak{P}}\leqq N(\mathfrak{P})^{-1}$, $i=1, \cdots, m$, with respect to the $\mathfrak{P}$-adic norm in $K=Q(\alpha_1, \cdots, \beta_m)$, then the $\log \alpha_i$ are defined in the $\mathfrak{P}$-adic sense and the inequality $|\beta_0+\beta_1 \log \alpha_1+\cdots+\beta_m \log \alpha_m|_{\mathfrak{P}}>N(\mathfrak{P})^{-c\log H}$ holds with a constant c that depends only on p, h, m and $(K:Q)$. This is the $\mathfrak{P}$-adic analogue of the result for the case of the ordinary norm [N. I. Fel'dman, Mat. Sb. (N.S.) **77 (119)**, (1968), 423–436; MR **38** #1059]. *E. Inaba* (Tokyo)

Citations: MR 38# 1059 = J80-33.

J80-47 **(44# 3962)**

Bundschuh, Peter

Irrationalitätsmaße für e^a, $a\neq 0$ rational oder Liouville-Zahl.

Math. Ann. **192** (1971), 229–242.

The author gives very effective irrationality measures for the numbers appearing in the title. The simplest is that for all integers $p, q>0$, $|e-p/q|>1/q^2g(q)$ where $g(q)=(18 \log 4q)/(\log\log 4q)$. A similar result holds for $e^{1/t}$ ($t\neq 0$ integral) with $g(q)=(c \log q)/(\log\log q)$ again (c here and below is an explicit constant). He shows these results on the order of magnitude of $g(t)$ are best possible. They are also implicit in the reviewer's paper [Amer. J. Math. **89** (1967), 1083–1108; MR **36** #5082] without the precise effectiveness in c. (Of course the diophantine approximations to powers of e have a very large literature which I cannot reproduce here.) Considering now e^a for a any non-zero rational, the author obtains $g(t)=\exp((c \log q)/(\log\log q))$. He obtains the same type of results for $(1/\sqrt{a}) \tanh\sqrt{a}$, $a=4/t$ for $t>0$ integral or $a>0$ rational, with $g(q)$ as above for $e^{1/t}$ and e^a respectively. Finally for certain Liouville numbers α he obtains a similar result. *W. W. Adams* (College Park, Md.)

Citations: MR 36# 5082 = J24-51.

J80-48 **(44# 3963)**

Coates, J.

An application of the division theory of elliptic functions to diophantine approximation.

Invent. Math. **11** (1970), 167–182.

Let $\wp(z)$ be the Weierstrass elliptic function with the invariants g_2 and g_3 and a pair of independent periods ω_1 and ω_2 such that ω_1/ω_2 does not lie in an imaginary quadratic field. Denote by u a real or complex number not of the form $r_1\omega_1+r_2\omega_2$ with rational coefficients r_1 and r_2 but such that $\wp(u)$ is an algebraic number, by α an algebraic number distinct from 0 and 1, by K any algebraic number field of finite degree over Q, and by $\varepsilon>0$ a constant. The author proves that if β_1 and β_2 are elements of K, not both zero and of height at most H, then $|\beta_1 u+\beta_2 \log \alpha|>C\cdot\exp(-H^\varepsilon)$, where $C>0$ does not depend on H. The proof is based on an algebraic result on the division values $\wp(u/n)$ by Tate which is also proved in this paper. *K. Mahler* (Canberra)

J84 CLASSES OF TRANSCENDENTAL NUMBERS

See also reviews B28-13, J02-23, J04-49, J12-60, J24-32, J64-8, J80-8, K15-95, K20-28, Q10-2, Q99-5, Q99-8, Z01-38.

J84-1 **(1, 137a)**

Koksma, J. F. Über die Mahlersche Klasseneinteilung der transzendenten Zahlen und die Approximation komplexer Zahlen durch algebraische Zahlen. Monatsh. Math. Phys. **48**, 176–189 (1939).

Let $M(n, a)$, for integral $n\geqq 1$, $a\geqq 1$, be the set of all polynomials

$$f(x)=\sum_{h=0}^{n} a_h x^h \neq 0$$

of degree not greater than n with integral coefficients of absolute value not greater than a. Further, let θ be a real or complex transcendental number, $\omega(n, a)$ the minimum of $|f(\theta)|$ taken over all elements of $M(n, a)$, and $\omega^*(n, a)$ the minimum of $|\theta-\alpha|$, where α is the root of any one of the primitive irreducible polynomials belonging to $M(n, a)$. Put

$$\omega(n)=\limsup_{n\to\infty} \frac{\log(1/\omega(n, a))}{\log a},$$

$$\omega^*(n)=\limsup_{a\to\infty} \frac{\log(1/(a\omega^*(n, a)))}{\log a},$$

and

$$\omega = \limsup_{n\to\infty} \frac{\omega(n)}{n}, \qquad \omega_1 = \text{upper bound}_{n\geq 1} \frac{\omega(n)}{n};$$

$$\omega^* = \limsup_{n\to\infty} \frac{\omega^*(n)}{n}, \qquad \omega_1^* = \text{upper bound}_{n\geq 1} \frac{\omega^*(n)}{n},$$

and denote by μ (μ^*) the number ∞, if $\omega(n)$ ($\omega^*(n)$) is finite for all n, otherwise the smallest index for which this function is infinite. K. Mahler [J. Reine Angew. Math. **166**, 118–136 (1932)] called θ an S-, T- or U-number, according as $\omega<\infty$, $\mu=\infty$, $\omega=\mu=\infty$, or $\omega=\infty$, $\mu<\infty$; the author analogously defines S^*-, T^*-, and U^*-numbers by means of ω^* and μ^* instead of ω and μ. Then he shows that θ has the same letter in both classifications by proving that

$$\mu=\mu^* \quad\text{and}\quad \omega_1^*\leqq\omega_1\leqq\omega_1^*+2$$

(the numbers ω and ω_1 (ω^* and ω_1^*) are simultaneously finite or infinite). He further shows that for "nearly all" real (complex) numbers $\omega_1^*\leqq 1$ ($\omega_1^*\leqq 1/2$), and therefore $\omega_1\leqq 3$ ($\omega_1\leqq 5/2$); this forms a slight improvement on a result of Mahler [Math. Ann. **106**, 131–139 (1932)]. The proof of the surprising inequality $\omega_1\leqq\omega_1^*+2$ depends on the lemma: "Let r be any integer in $(1, n)$, $m^{(1)}, \cdots, m^{(r)}$ any r positive integers of sum not greater than n, and $a^{(1)}, \cdots, a^{(r)}$ any r positive integers of product not greater than $(4n)^n a$. If $\Omega(n, a)$ is the minimum of all expressions

$$\prod_{\rho=1}^{r} \omega^*(m^{(\rho)}, a^{(\rho)}),$$

then

$$\Omega(n, a)\leqq Ba^{2n-3}\omega(n, a),$$

where B is a positive constant depending only on θ and n." It would be of interest to replace the exponent $2n-3$ of a by a smaller number. *K. Mahler* (Manchester).

Referred to in J64-8, J84-2, J84-3, J84-6, J84-48.

J84-2 (2, 148d)

Mahler, Kurt. Über Polynome mit ganzen rationalen Koeffizienten. Mathematica, Zutphen. B. **8**, 173–182 (1940).

Verfasser betrachtet das Polynom $f(x)=a_0x^n+a_1x^{n-1}+\cdots+a_n$ ($n\geqq 3$, $a_0\neq 0$) und die zugehörige Binärform $F(x, y)=a_0x^n+a_1x^{n-1}y+\cdots+a_ny^n$. Durch Abschätzung der Resultante von F_x' und F_y' leitet Verfasser verschiedene Ungleichungen her, unter anderm eine Abschätzung nach unten von $|f(x)|$, die von Wichtigkeit ist bei der Bestimmung des Masses aller Mahlerschen S-Zahlen. [Für weitere Literatur vgl. diese Rev. **1**, 137.] *J. F. Koksma.*

Citations: MR 1, 137a = J84-1.

J84-3 (11, 82j)

Kubilyus, I. On the application of I. M. Vinogradov's method to the solution of a problem of the metric theory of numbers. Doklady Akad. Nauk SSSR (N.S.) **67**, 783–786 (1949). (Russian)

Let $s\geqq 1$ be a rational integer and c any positive constant. Then Mahler showed [Math. Ann. **106**, 131–139 (1932)] that, for almost all real numbers θ and for $\omega\geqq 3$ the inequality

$$\left|\sum_{\sigma=0}^{s} a_\sigma\theta^\sigma\right| < ca^{-(1+\omega)s}, \qquad a=\max_{0\leqq\sigma\leqq s}|a_\sigma|,$$

possesses only a finite number of solutions in rational integers $a_1, a_2, \cdots, a_s$. Koksma replaced the condition $\omega\geqq 3$ by $\omega\geqq 2$ [Monatsh. Math. Phys. **48**, 176–189 (1939); these Rev. **1**, 137] and it is conjectured that the result holds for any positive ω. That this is so for $s=1$ is almost trivial, and the purpose of the author's paper is to prove the conjecture for $s=2$. His proof is based on an estimate of Vinogradov for a trigonometric sum and upon the following lemma. Let $\varphi(q)$ be a positive decreasing function of the integer q for $q\geqq q_0>0$ such that the series $\sum_{q\geqq q_0} q^{\frac{1}{2}}\varphi(q)\tau(q)\log^{\frac{1}{2}} q$ is convergent, where $\tau(q)$ is the number of divisors of q. Then the system of inequalities $|\theta-p_1/q|<\varphi(q)$, $|\theta^2-p_2/q|<\varphi(q)$ possesses only a finite number of solutions for almost all θ. The main result then follows from this lemma by taking a particular function $\varphi(q)$ and applying a known result due to Khintchine and Mahler [Mahler, Rec. Math. [Mat. Sbornik] N.S. **1**(43), 961–962 (1936)]. *R. A. Rankin.*

Citations: MR 1, 137a = J84-1.

Referred to in J84-4, J84-6, J84-7, J84-15.

J84-4 (12, 679b)

Cassels, J. W. S. Some metrical theorems in Diophantine approximation. V. On a conjecture of Mahler. Proc. Cambridge Philos. Soc. **47**, 18–21 (1951).

Denote by $\|\xi\|$ the difference between ξ and the nearest integer. It is well known that if $\theta_1, \theta_2, \cdots, \theta_n$ are any n real numbers then (1) $\max_{j=1,2,\cdots,n}\|g\theta_j\|\leqq g^{-1/n}$ has infinitely many integer solutions $g>0$. In particular, (1) holds if $\theta_j=\alpha^j$. Khintchine showed that, if $g^{-1/n}$ is replaced by $\epsilon g^{-1/n}$, (1) has infinitely many solutions for almost all $\theta_1, \theta_2, \cdots, \theta_n$; he also showed that for almost all α, $\max_{j=1,2,\cdots,n}\|g\alpha^j\|<\epsilon g^{-1/n}$ has infinitely many solutions. On the other hand, it is well known that for any $\epsilon>0$ and almost all $\theta_1, \cdots, \theta_n$

$$\max_{j=1,2,\cdots,n}\|g\theta_j\|<g^{-1/n-\epsilon}$$

has only a finite number of solutions. Mahler conjectured [Math. Ann. **106**, 131–139 (1932)] that for any ϵ and almost all α, $\max_{j=1,2,\cdots,n}\|g\alpha^j\|<g^{-1/n-\epsilon}$ has only a finite number of solutions. For $n=1$ this is well known. Recently Kubilyus proved it for $n=2$ [Doklady Akad. Nauk SSSR (N.S.) **67**, 783–786 (1949); these Rev. **11**, 82]. For $n=2$ the author proves the following more general theorem: Let $\phi(g)$ and $\psi(g)$ satisfy $\sum_g\phi(g)\psi(g)<\infty$, $\psi(g)\geqq\max\{\phi(g), g^{-\frac{1}{2}}\log g\cdot d(g)\}$, where $d(g)$ is the number of divisors of g. Then for almost all α there are only a finite number of solutions of

$$\|g\alpha\|\leqq\phi(g), \qquad \|g\alpha^2\|\leqq\psi(g).$$

Mahler's conjecture follows by putting $\psi(g)=g^{-\frac{1}{2}+\frac{1}{2}\epsilon}$, $\phi(g)=g^{-\frac{1}{2}-\epsilon}$. Finally the author states the following conjecture which would imply Mahler's conjecture for general n. For all $\epsilon>0$ there is some $\delta>0$ depending only on ϵ and such that $\delta\to 0$ as $\epsilon\to 0$ with the following property: The number of solutions N of

$$0\leqq r<g, \qquad \|r^j/g\|\leqq r^{1-1/n-\epsilon} \qquad (j=2, \cdots, n)$$

satisfies $N=O(g^{1/n+\delta})$. *P. Erdös* (Aberdeen).

Citations: MR 11, 82j = J84-3.

J84-5 (14, 956e)

LeVeque, W. J. On Mahler's U-numbers. J. London Math. Soc. **28**, 220–229 (1953).

The complex number ξ is called a U-number if for some fixed integer $n\geqq 1$ and for all $\omega>0$ there exist infinitely many sets of $n+1$ integers $a_0, a_1, \cdots, a_n$ with $A=\max(|a_0|, \cdots, |a_n|)$ such that

$$0<|a_0+a_1\xi+\cdots a_n\xi^n|<A^{-\omega}.$$

The author calls ξ a U_m-number if this condition is fulfilled for $n=m$ but for no $n\leqq m-1$. He proves several theorems on U_m-numbers (U_1-numbers are identical with Liouville numbers), e.g., there exist U_m-numbers for every $m\geqq 1$. Also he proves that numbers satisfying a certain transcendency condition due to Th. Schneider are not U_m-numbers for too small values of m. [Cf. Mahler, Nederl. Akad. Wetensch., Proc. 39, 633–640, 729–737 (1936).] ***J. F. Koksma*** **(Amsterdam).**

Referred to in J84-21, J84-28.

J84-6 **(14, 956f)**

LeVeque, W. J. Note on S-numbers. Proc. Amer. Math. Soc. **4**, 189–190 (1953).

The transcendental number ξ is said to be an S-number if there is a constant $\gamma=\gamma(\xi)>0$ and a sequence of positive constants $\Gamma_m=\Gamma_m(\xi)$ $(m=1, 2, \cdots)$ such that for each polynomial $f(x)\equiv a_0+a_1x+\cdots+a_mx^m$ of arbitrary degree m and integer coefficients $a_0, \cdots, a_m$ such that

$$a=\max(|a_0|, \cdots, |a_m|)\geqq 1$$

the inequality $|f(\xi)|\geqq\Gamma_m a^{-\gamma m}$ holds. Now let γ_r denote the infimum of all numbers γ' such that almost all real ξ are S-numbers with $\gamma\leqq\gamma'$ and let γ_c denote the infimum of all γ' such that almost all complex numbers ξ are S-numbers with $\gamma\leqq\gamma'$. Mahler conjectured that $\gamma_r=1$, $\gamma_c=1/2$ and he proved $\gamma_r\leqq 4$, $\gamma_c\leqq 4$ [Math. Ann. **106**, 131–139 (1932)]; the reviewer proved $\gamma_r\leqq 3$, $\gamma_c\leqq 5/2$ [Monatsh. Math. Phys. **48**, 176–189 (1939); these Rev. **1**, 137]; Kubilyus proved the conjecture in the special case $m=2$ [Doklady Akad. Nauk SSSR (N.S.) **67**, 783–786 (1949); these Rev. **11**, 82]. In this paper the author, combining Mahler's original argument with a theorem of Fel'dman [Izvestiya Akad. Nauk SSSR. Ser. Mat. **15**, 53–74 (1951); these Rev. **12**, 595], proves $\gamma_r\leqq 2$, $\gamma_c\leqq 3/2$. *J. F. Koksma* (Amsterdam).

Citations: MR 1, 137a = J84-1; MR 11, 82j = J84-3; MR 12, 595e = J80-4.
Referred to in J84-8, J84-19.

J84-7 **(21# 1296)**

Kasch, Friedrich. Über eine metrische Eigenschaft der S-Zahlen. Math. Z. **70** (1958), 263–270.

Let ξ denote a given (real or complex) transcendental number; put

$$\omega_n(H, \xi) = \min P(\xi),$$
$$\bar{\omega}_n(\xi) = \limsup_{H\to\infty} \log(\omega_n(H, \xi))^{-1}(\log H)^{-1},$$

where the minimum is extended over all polynomials $P(x)$ of degree $\leqq n$ with integer coefficients and of height $\leqq H$ $(\geqq 1)$. Put $\vartheta_n(\xi)=\bar{\omega}_n(\xi)n^{-1}$. S-numbers are those numbers for which $\vartheta(\xi)=\sup\vartheta_n(\xi)$ $(n\to\infty)$ is finite. Mahler conjectured: $\vartheta_n(\xi)=1$ for almost all real ξ and $\vartheta_n(\xi)=\frac{1}{2}$ for almost all complex ξ $(n\geqq 1)$. Kubilyus [Dokl. Akad. Nauk SSSR **67**(1949), 703–706; MR **11**, 82] proved the conjecture for $n=2$ by means of Vinogradoff's methods. The author gives an elementary treatment of this case $n=2$, which also allows one to deduce the complex analogue, and discusses the possibility of a generalization for $n>2$. [Cf. also the following review.] *J. F. Koksma* (Amsterdam)

Citations: MR 11, 82j = J84-3.
Referred to in J84-10, J84-11, J84-12, J84-15.

J84-8 **(21# 1297)**

Kasch, Friedrich; und Volkmann, Bodo. Zur Mahlerschen Vermutung über S-Zahlen. Math. Ann. **136** (1958), 442–453.

For the notation cf. the preceding review. The authors prove (a) for almost all real ξ we have $1\leqq\vartheta_n(\xi)\leqq 2-2/n$ $(n\geqq 3)$; (b) for almost all complex ξ we have $1/2-1/2n\leqq\vartheta_n(\xi)\leqq 3/2-2/n$ $(n\geqq 3)$; these results contain the best estimates attained till now, viz. by LeVeque [Proc. Amer. Math. Soc. **4** (1953), 189–190; MR **14**, 956]. Using Hausdorff's notion of dimension, they refine their results still further. {It seems that the authors don't know the work of D. J. Lock [Thesis, Free Univ. of Amsterdam, 1947; MR **9**, 79].} *J. F. Koksma* (Amsterdam)

Citations: MR 9, 79h = J64-8; MR 14, 956f = J84-6.
Referred to in J84-9, J84-11, J84-16, J84-27.

J84-9 **(23# A2379)**

Volkmann, Bodo

Ein metrischer Beitrag über Mahlersche S-Zahlen. I. *J. Reine Angew. Math.* **203** (1960), 154–156.

Let ξ be a transcendental number. For any given positive integer n let $\vartheta_n(\xi)$ denote the least upper bound of all positive numbers σ such that for an infinity of polynomials $p_i(x)$ of degree $\leqq n$ with integral coefficients and of height $\|p_i\|$, the inequality $|p_i(\xi)|\leqq\|p_i\|^{-n\sigma}$ $(i=1, 2, \cdots)$ is satisfied (put $\vartheta_n(\xi)=0$ if there does not exist any such σ). S-numbers are transcendental numbers such that $\vartheta(\xi)=\sup_{n=1,2,\cdots}\vartheta_n(\xi)$ is positive and finite. Mahler conjectured that $\vartheta(\xi)=1$ for almost all real ξ and $\vartheta(\xi)=\frac{1}{2}$ for almost all complex ξ. The author proves here that for almost all complex ξ

$$\frac{1}{2}-\frac{1}{2n}\leqq\vartheta_n(\xi)\leqq 1-\frac{1}{2n} \qquad (n=2, 3, \cdots);$$

hence $\vartheta(\xi)\leqq 1$ for almost all complex ξ. He thus sharpens an earlier result of Kasch and himself [Math. Ann. **136** (1958), 442–453; MR **21** #1297], where the upper bound $3/2-2/n$ was obtained instead of $1-1/2n$.
J. Popken (Amsterdam)

Citations: MR 21# 1297 = J84-8.
Referred to in J84-12, J84-19, J84-27.

J84-10 **(23# A2380)**

Kasch, Friedrich

Ein metrischer Beitrag über Mahlersche S-Zahlen. II. *J. Reine Angew. Math.* **203** (1960), 157–159.

The author gives a still better upper bound for $\vartheta_n(\xi)$ than the one obtained by Volkmann [cf. the preceding review]. He obtains for almost all complex numbers:

$$\text{(A)} \qquad \frac{1}{2}-\frac{1}{2n}\leqq\vartheta_n(\xi)\leqq 1-\frac{1}{n} \qquad (n=2, 3, \cdots),$$

and moreover $\vartheta_4(\xi)\leqq 5/8$. With respect to the conjecture of Mahler the score at present is as follows. For almost complex numbers: $\vartheta_2(\xi)=1/4$ [Kasch, Math. Z. **70** (1958), 263–270; MR **21** #1296]; $1/3\leqq\vartheta_3(\xi)\leqq 1/2$ [Volkmann, Math. Ann. **139** (1959), 87–90]; $3/8\leqq\vartheta_4(\xi)\leqq 5/8$ [the present paper]; and formula (A) for $n\geqq 5$.
J. Popken (Amsterdam)

Citations: MR 21# 1296 = J84-7.
Referred to in J84-12, J84-27.

J84-11 **(23# A3711)**

Volkmann, Bodo

Zum kubischen Fall der Mahlerschen Vermutung. *Math. Ann.* **139**, 87–90 (1959).

Let ξ be a transcendental number; let $p_i(x)$ denote a polynomial of degree $n\geqq 1$ with integer coefficients and with height $\|p_i\|$ (=maximum of the absolute values of the coefficients). Let $\vartheta_n(\xi)$ be the least upper bound of those $\sigma>0$ for which sequences $p_i(x)$ $(i=1, 2, \cdots; \|p_i\|\to\infty)$ exist such that $|p_i(\xi)|<\|p_i\|^{-n\sigma}$ $(i=1, 2, \cdots)$ (if no such numbers $\sigma>0$ exist, put $\vartheta_n(\xi)=0$). Then Mahler's conjecture may be stated as follows: For almost all real ξ we have (1) $\vartheta_n(\xi)=1$ $(n\geqq 1)$. For almost all complex ξ we have (2) $\vartheta_n(\xi)=1/2-1/2n$ $(n\geqq 1)$. Kubilyus and Kasch proved the conjecture for $n=2$, and Kasch and the author found some inequalities for $\vartheta_n(\xi)$ concerning the general case $n\geqq 3$ [for references see F. Kasch, Math. Z. **70** (1958), 263–270; MR **21** #1296; F. Kasch and the author, Math. Ann. **136** (1958), 442–453; MR **21** #1297]. Here it is proved that $\frac{1}{3}\leqq\vartheta_3(\xi)\leqq\frac{1}{2}$ for almost all complex ξ. [Cf. also #A3712.]
J. F. Koksma (Amsterdam)

Citations: MR 21# 1296 = J84-7; MR 21# 1297 = J84-8.

J84-12 **(23# A3712)**
Volkmann, Bodo
Zur Mahlerschen Vermutung im Komplexen.
Math. Ann. **140** (1960), 351–359.

For notation and references cf. #A3711 above. Improving known results, the author proves the following statements: For all complex ξ we have $\vartheta_2(\xi)=\frac{1}{4}$, $\vartheta_3(\xi)=\frac{1}{3}$, $\frac{3}{8}\leqq\vartheta_4(\xi)\leqq\frac{1}{2}$, $\frac{2}{5}\leqq\vartheta_5(\xi)\leqq\frac{3}{5}$. The first statement was already known by Kasch [Math. Z. **70** (1958), 263–270; MR **21** #1296], but the general simple method used by the author which does not make use of the discriminant of $p_i(x)$ leads to a simpler proof also for $\vartheta_2(\xi)$. [Cf. also the author, J. Reine Angew. Math. **203** (1960), 154–156; MR **23** #A2379; F. Kasch, ibid. **203** (1960), 157–159; MR **23** #A2380.]
J. F. Koksma (Amsterdam)

Citations: MR 21# 1296 = J84-7; MR 23# A2379 = J84-9; MR 23# A2380 = J84-10.

J84-13 **(23# A3713)**
Kasch, Friedrich; Volkmann, Bodo
Metrische Sätze über transzendente Zahlen in P-adischen Körpern.
Math. Z. **72** (1959/60), 367–378.

For notation and references cf. #A3711, A3712. Instead of real or complex numbers ξ, one may consider P-adic numbers ξ in the field K_P, where P denotes a fixed prime number $\geqq 2$. The number $\vartheta_n(\xi)$ can be defined as in the preceding reviews, for $n\geqq 1$, if the absolute value $|p_i(x)|$ is replaced by the P-adic value $|p_i(\xi)|_P$. By Turkstra [Thesis, Free University, Amsterdam, 1936] a measure in K_P was introduced and several metrical results concerning $\vartheta_n(\xi)$ were deduced. His investigations were continued and refined by Lock [Thesis, Free University, Amsterdam 1947; MR **9**, 79]. Using Turkstra's measure, the authors prove that, for almost all numbers $\xi\in K_P$,

$$\vartheta_1(\xi)=2,\quad \vartheta_2(\xi)=\frac{3}{2},\quad 1+\frac{1}{n}\leqq\vartheta_n(\xi)\leqq 2-\frac{1}{2n}\ (n\geqq 2).$$

These results (contain or) improve the best known results on $\vartheta_n(\xi)$ in the P-adic case. *J. F. Koksma* (Amsterdam)

Citations: MR 9, 79h = J64-8.
Referred to in J84-18, J84-32, J84-44.

J84-14 **(24# A1253)**
Davenport, H.
A note on binary cubic forms.
Mathematika 8 (1961), 58–62.

Let $D(f)$ be the discriminant of a binary cubic form $f(x,y)=ax^3+bx^2y+cxy^2+dy^3$ with integral coefficients. It is shown that

$$\sum |D(f)|^{-1/2} < CH^2 \quad \text{as } H\to\infty$$

for some constant C, where the summation is taken over all forms f which are irreducible over the rationals (so that in particular $D(f)\neq 0$) and such that

$$\max(|a|,|b|,|c|,|d|)\leqq H.$$

This sum is of interest in connection with a conjecture of Mahler; see the following review [#A1254].
W. J. LeVeque (Ann Arbor, Mich.)

Referred to in J84-18.

J84-15 **(24# A1254)**
Volkmann, B.
The real cubic case of Mahler's conjecture.
Mathematika 8 (1961), 55–57.

Using the result of Davenport described in the preceding review [#A1253], the author shows that the following (modification of a) conjecture of Mahler holds when $n=3$: Let $\vartheta_n(\xi)$ be the supremum of the set of all positive numbers σ for which there exist infinitely many polynomials $p_1(x), p_2(x),\cdots$ of degree n, with integral coefficients, satisfying $|p_i(\xi)|<\|p_i\|^{-n\sigma}$, $i=1,2,\cdots$, where $\|p_i\|$ is the maximum of the absolute values of the coefficients of p_i. Then for almost all real ξ, $\vartheta_n(\xi)=1$ for $n=1,2,\cdots$. Proofs for the case $n=2$ were given by J. F. Kubiljus [Dokl. Akad. Nauk SSSR **67** (1949), 783–786; MR **11**, 82] and F. Kasch [Math. Z. **70** (1958), 263–270; MR **21** #1296]. *W. J. LeVeque* (Ann Arbor, Mich.)

Citations: MR 11, 82j = J84-3; MR 21# 1296 = J84-7.
Referred to in J84-18.

J84-16 **(24# A1883)**
Schmidt, Wolfgang M.
Bounds for certain sums; a remark on a conjecture of Mahler.
Trans. Amer. Math. Soc. **101** (1961), 200–210.

For any transcendental number ζ let $\theta_n(\zeta)$ be the least upper bound of the set of real numbers σ such that there exists a sequence of different polynomials $P_i(x)$ of degree n with integral coefficients, satisfying $|P_i(x)|<H(P_i)^{-n\sigma}$, where $H(P)$ is the maximum of the absolute values of the coefficients of the polynomial $P(x)$. It is known that $\theta_n(\zeta)\geqq 1$ for real ζ. Mahler [Math. Ann. **106** (1932), 131–139] conjectured that $\theta_n(\zeta)=1$ for almost all real ζ. The best estimate was obtained by Kasch and Volkmann [ibid. **136** (1958), 442–453; MR **21** #1297], viz., $\theta_n(\zeta)\leqq 2-2/n$ $(n\geqq 2)$ almost everywhere. The author improves this estimate to $\theta_n(\zeta)\leqq 2-F/(3n)$ $(n\geqq 3)$.

The proof is based on the following theorem. Let $Q(x,y)=\sum_{l=0}^{n}Q_l(y)x^l$ be a polynomial in x, y of degree d with integral coefficients, m the degree of $Q_n(y)$, $H\geqq 1$ and $\rho>\frac{1}{3}$ be real numbers, $n\geqq 1$, $3m\geqq n+3$, $n\rho\geqq 1$, $m\rho\geqq 1$. If $Q(x,y)-k$ has no rational linear factor for $k\neq 0$, then

$$\sum |Q(x,y)|^{-\rho}\leqq \gamma_{d\rho}H^{2/3},$$

where the sum is taken over integers x, y, $|x|\leqq H$, $|y|\leqq H$, $Q(x,y)\neq 0$; the constant $\gamma_{d\rho}$ does not depend on the coefficients of $Q(x,y)$. *J. Kubilius* (Vilnius)

Citations: MR 21# 1297 = J84-8.

J84-17 **(24# A3134)**
Erdős, P.
Representations of real numbers as sums and products of Liouville numbers.
Michigan Math. J. **9** (1962), 59–60.

It is shown that every real number is the sum (or product) of two Liouville numbers. Both a constructive and a nonconstructive proof are given, the latter by a category argument. *W. J. LeVeque* (Ann Arbor, Mich.)

J84-18 **(25# 2035)**
Kasch, Friedrich; Volkmann, Bodo
Metrische Sätze über transzendente Zahlen in P-adischen Körpern. II.
Math. Z. **78** (1962), 171–174.

For notations and references cf. Part I, i.e., same Z. **72** (1959/60), 367–378 [MR **23** #A3713]. It is proved that the analogue of Mahler's conjecture as formulated in I for the P-adic case holds true for $n=3$: for almost all $\xi\in K_P$ one has $\vartheta_3(\xi)=4/3$. The real case $n=3$ was settled by Volkmann [Mathematika **8** (1961), 55–57; MR **24** #A1254]. Both proofs make use of an estimate due to Davenport concerning cubic forms [ibid. **8** (1961), 58–62; MR **24** #A1253]. *J. F. Koksma* (Amsterdam)

Citations: MR 23# A3713 = J84-13; MR 24# A1253 = J84-14; MR 24# A1254 = J84-15.

J84-19 **(25# 5036)**

Volkman, Bodo

Zur metrischen Theorie der S-Zahlen.

J. Reine Angew. Math. **209** (1962), 201–210.

This paper contains new results related to Mahler's conjecture about S-numbers. Let ξ be a transcendental number; let $\vartheta_n(\xi)$ $(n=1, 2, \cdots)$ denote the supremum of all positive numbers σ such that, for an infinity of polynomials $p_i(x)$ of degree $\leqq n$ with integral coefficients and height $\|p_i\|$, the inequality $|p_i(\xi)| \leqq \|p_i\|^{-n\sigma}$ $(i=1, 2, \cdots)$ holds (put $\vartheta_n(\xi)=0$ if there does not exist any such σ). Mahler conjectured $\vartheta_n(\xi)=1$ for almost all real ξ and $\vartheta_n(\xi)=\frac{1}{2}-\frac{1}{2}n$ for almost all complex ξ, and hence $\vartheta(\xi)=\sup_{n=1,2,\cdots}\vartheta_n(\xi)=1$ for almost all real ξ and $\vartheta(\xi)=\frac{1}{2}$ for almost all complex ξ. In recent years this problem has drawn much attention. The rest results obtained earlier for $\vartheta(\xi)$ were as follows. For almost all ξ: (in the real case) $1 \leqq \vartheta(\xi) \leqq 2$ [LeVeque, Proc. Amer. Math. Soc. **4** (1953), 189–190; MR **14**, 956]; (in the complex case) $\frac{1}{2} \leqq \vartheta(\xi) \leqq 1$ [the author, J. Reine Angew. Math. **203** (1960), 154–156; MR **23** #A2379]. In the present paper the author sharpens these inequalities to the following. For almost all ξ: (in the real case) $1 \leqq \vartheta(\xi) \leqq \frac{3}{2}$; (in the complex case) $\frac{1}{2} \leqq \vartheta(\xi) \leqq \frac{3}{4}$. *J. Popken* (Berkeley, Calif.)

Citations: MR 14, 956f = J84-6; MR 23# A2379 = J84-9.

Referred to in J84-26, J84-30, J84-34, J84-38.

J84-20 **(26# 79)**

Wirsing, Eduard

Approximation mit algebraischen Zahlen beschränkten Grades.

J. Reine Angew. Math. **206** (1960), 67–77.

Zur Beschreibung des Verhaltens einer komplexen oder reellen Zahl ξ bei Approximation durch algebraische Zahlen werden bei den Klasseneinteilungen nach Koksma bzw. nach Mahler Parameter $w_n^*(\xi)$ bzw. $w_n(\xi)$ $(n=1, 2, \cdots)$ eingeführt [siehe z.B. T. Schneider, *Einführung in die transzendenten Zahlen*, Springer, Berlin, 1957; MR **19**, 252]. Diese Folgen $w_n^*(\xi)$ und $w_n(\xi)$ sind bekanntlich beschränkt, wenn ξ algebraisch ist. Während für $w_n(\xi)$ auch die Umkehrung bekannt ist, fehlte bei $w_n^*(\xi)$ bisher ein solches Kriterium. Der Verfasser schliesst diese Lücke, indem er beweist, dass bei jedem n für alle ξ, die nicht algebraisch vom Grad $\leqq n$ sind, im Reellen $w_n^*(\xi) \geqq (n+1)/2$, im Komplexen $w_n^*(\xi) \geqq n/4$ gilt. Allgemeiner wird bewiesen, dass im reellen Fall für solche ξ stets $w_n^*(\xi) \geqq (w_n(\xi)+1)/2$ gilt. Ein weiterer Satz besagt, dass unter den gleichen Voraussetzungen auch $w_n^*(\xi) \geqq w_n(\xi)/(w_n(\xi)-n+1)$ ist, und ähnliche Ergebnisse werden im komplexen Fall bewiesen. *B. W. Volkmann* (Mainz)

Citations: MR 19, 252f = J02-9.

Referred to in J04-66, J68-61, J84-27, J84-29.

J84-21 **(26# 1284)**

Fel'dman, N. I.

Transcendental numbers with an approximation of given type. (Russian)

Uspehi Mat. Nauk **17** (1962), *no.* 5 (107), 145–151.

For a given transcendental α, let

$$\varphi(H, n, \alpha) = -\log\big(\min_{|a_i|\leqq H} |a_0\alpha^n + a_1\alpha^{n-1} + \cdots + a_n|\big)/\log H,$$

where a_i are rational integers not all zero. The author proves the following theorem. Let $\nu_1, \cdots, \nu_s$ $(s \geqq 2)$ be any different positive integers, $\nu_0 = \max_{1\leqq i\leqq s}\nu_i$, $\omega_1, \cdots, \omega_s$ be any given numbers $\geqq 2\nu_0$, and $\Omega = \max_{1\leqq i\leqq s}\omega_i$. There exists a transcendental number α such that

$$(\nu_0+3)(1+\nu_0\Omega+\nu_0) \geqq \limsup_{H=\infty} \varphi(H, \nu_i, \alpha) \geqq \omega_i \qquad (i = 1, 2, \cdots, s).$$

This is of interest in connection with results of LeVeque [J. London Math. Soc. **28** (1953), 220–229; MR **14**, 956]. *A. Schinzel* (Warsaw)

Citations: MR 14, 956e = J84-5.

J84-22 **(26# 6123)**

Roth, H.

Sur un théorème de Maillet.

Bull. Sci. Math. (2) **86** (1962), *1ère partie*, 61–63.

The author generalizes a result of Maillet on Liouville numbers [E. Maillet, *Théorie des nombres transcendants*, p. 50, Gauthier-Villars, Paris, 1906] as follows. Let ξ be a "generalized Liouville number" in a sense to be defined later. Let η be of the form $\eta=(a\xi+b)/(c\xi+d)$, $ad-bc\neq 0$, a, b, c, d denoting non-vanishing integers. Then η is again a generalized Liouville number. Here a generalized Liouville number is a real number ξ satisfying an infinity of inequalities

$$|\xi-(p_n/q_n)| \leqq q_n^{-\sigma_n} \qquad (n = 1, 2, \cdots),$$

where the p_n, $q_n>0$ are integers, pairwise coprime, and where the σ_n are real numbers such that $\limsup \sigma_n > 2$. The Thue-Siegel-Roth theorem implies that any generalized Liouville number is transcendental.

The proof uses the fractions $(ap_n+bq_n)/(cp_n+dq_n)$ as approximations to η. *J. Popken* (Berkeley, Calif.)

J84-23 **(27# 1417)**

Güting, R.

On Mahler's function θ_1.

Michigan Math. J. **10** (1963), 161–179.

Let n be a positive integer, let $\theta_n(\gamma)=w_n(\gamma)/n$ be the function of a complex variable γ introduced by Mahler in his well-known classification of transcendental numbers [Th. Schneider, *Einführung in die transzendenten Zahlen*, pp. 64–86, Springer, Berlin, 1957; MR **19**, 252]. It is known that $\theta_n(\gamma)>1$ for all real transcendental numbers γ. The author puts the following two questions forward: (1) Given any $c \geqq 1$, is it possible to find transcendental numbers γ such that $\theta_n(\gamma)=c$? (2) If the answer is affirmative, what is the fractional dimension of all real numbers γ such that $\theta_n(\gamma)=c$? These questions are of course far from a solution yet, but as a first step the author solves these in the linear case $n=1$. Using continued fractions he constructs for every $c>1$ a real number γ such that $\theta_1(\gamma)=c$. His main result is the answer to the question (2) for $n=1$: For any $c>1$ the fractional dimension of all real numbers γ with $\theta_1(\gamma)=c$ is equal to $2/(c+1)$.

In the general case he obtains only a partial result; he shows that for $c>1$ the fractional dimension of all real numbers between 0 and 1 such that $\theta_n(\gamma) \geqq c$ is at least $2/(nc+1)$ $(n=1, 2, 3, \cdots)$.

In the proofs the author uses methods on sets of fractional dimension in the theory of Diophantine approximations due to Jarnik [Mat. Sb. **36** (1929), 371–382] and Besicovitch [J. London Math. Soc. **9** (1934), 126–131]. *J. Popken* (Amstelveen)

Citations: MR 19, 252f = J02-9.

J84-24 **(27# 3586)**

Roth, H.

On Borel series.

Tôhoku Math. J. (2) **15** (1963), 103–113.

The convergence of the series $\sum_{p,q=1}^{\infty} A^{p+q}/|qx-p|$ is studied, where A is a constant with $|A|<1$. It is shown that the series converges for all irrational non-Liouville numbers, and that it converges for some Liouville numbers and diverges for others; examples of both kinds are given. *W. J. LeVeque* (Ann Arbor, Mich.)

J84-25 **(27# 5730)**

Sprindžuk, V. G.

On a classification of transcendental numbers. (Russian. Lithuanian and English summaries)

Litovsk. Mat. Sb. **2** (1962), *no.* 2, 215–219.

Let, for a given transcendental number ω, $w_n(\omega, H)=\min|P(\omega)|$, where P runs through all polynomials of degree $\leqq n$ and height $\leqq H$. The classification of transcendental numbers due to K. Mahler [J. Reine Angew. Math. **166** (1931), 118–136] is based on the behaviour of $w_n(\omega, H)$ when $H\to\infty$ and $n\leqq n_0$, as $n_0\to\infty$. The author takes at first $H\leqq H_0$ and $n\to\infty$, then $H_0\to\infty$, and divides accordingly all transcendental numbers into three classes, one of them containing almost all numbers. The proofs are only sketched. *A. Schinzel* (Warsaw)

J84-26 **(28# 61)**

Volkmann, Bodo

Zur metrischen Theorie der S-Zahlen. II.

J. Reine Angew. Math. **213** (1963/64), 58–65.

The author sharpens and simplifies the method already given in an earlier paper with the same title [same J. **209** (1962), 201–210; MR **25** #5036]. In this way he obtains in the present article further progress towards a proof of Mahler's conjecture about S-numbers. If $\vartheta(\xi)$ denotes Mahler's function, then the author proves: (a) For almost all complex ξ, $\frac{1}{2}\leqq\vartheta(\xi)\leqq\frac{2}{3}$, and (b) For almost all real ξ, $1\leqq\vartheta(\xi)\leqq\frac{4}{3}$. *J. Popken* (Amstelveen)

Citations: MR 25# 5036 = J84-19.

Referred to in J84-34, J84-38.

J84-27 **(28# 1165)**

Sprindžuk, V. G.

On some general problems of approximating numbers by algebraic numbers. (Russian. Lithuanian and German summaries)

Litovsk. Mat. Sb. **2** (1962), *no.* 1, 129–145.

Bei den Klasseneinteilungen der transzendenten Zahlen nach K. Mahler und nach J. F. Koksma definiert man für jedes komplexe ξ gewisse Parameter $\vartheta_n(\xi)$ bezw. $\vartheta_n^*(\xi)$ (siehe z.B. Th. Schneider [*Einfuhrung in die transzendenten Zahlen*, Springer, Berlin, 1957; MR **19**, 252] dort werden $w_n(\xi)=n\vartheta_n(\xi)$ und $w_n^*(\xi)=n\vartheta_n^*(\xi)$ verwendet). Verfasser beweist folgende Aussagen: (1) Es gibt Zahlen θ_n, θ_n^*, η_n und η_n^* $(n=1, 2, \cdots)$, derart, daß für fast alle reellen ξ die Gleichungen $\vartheta_n(\xi)=\theta_n$, $\vartheta_n^*(\xi)=\theta_n^*$, ebenso für fast alle komplexen ξ die Gleichungen $\vartheta_n(\xi)=\eta_n$, $\vartheta_n^*(\xi)=\eta_n^*$ gelten; (2) $\theta_n\leqq 2-(2/n)$ $(n\geqq 2)$; (3) $\eta_n\leqq 1-(3/2n)$ $(n\geqq 2)$; (4) $\theta_n^*\geqq(2/3)\theta_n-1/3$ $(n\geqq 3)$; (5) $\eta_n^*\geqq(2/3)\eta_n-(1/6)-(1/6n)$ $(n\geqq 4)$. Von diesen Ergebnissen war (2) im Wesentlichen schon bekannt [F. Kasch und der Referent, Math. Ann. **136** (1958), 442–453; MR **21** #1297] während (3) etwas schärfer ist als die besten vor der vorliegenden Arbeit veröffentlichten Resultate [der Referent, J. Reine Angew. Math. **203** (1960), 154–156; MR **23** #A2379; F. Kasch, ibid. **203** (1960), 157–159; MR **23** #A2380]. Ferner sind (4) und (5) weniger scharf als die Ungleichungen (6) bezw. (6') von E. Wirsing [ibid. **206** (1961), 67–77; MR **26** #79]. *B. Volkmann* (Mainz)

Citations: MR 19, 252f = J02-9; MR 21# 1297 = J84-8; MR 23# A2379 = J84-9; MR 23# A2380 = J84-10; MR 26# 79 = J84-20.

Referred to in J84-29, J84-30.

J84-28 **(28# 1171)**

Baker, A.

On Mahler's classification of transcendental numbers.

Acta Math. **111** (1964), 97–120.

Let ξ be a transcendental number, and let $H(f)$ and $n(f)$ be the height (maximum of absolute values of coefficients) and degree of the polynomial f over the rational integers. Put $\omega(n, H, \xi)=\min(|f(\xi)|)$, the minimum extending over all nonzero f with $H(f)\leqq H$ and $n(f)\leqq n$, and put $\omega(n, H, \xi)=H^{-n\rho(n,H,\xi)}$. Then Mahler [J. Reine Angew. Math. **166** (1931), 118–136; ibid. **166** (1932), 137–150] called ξ an S-, U-, or T-number, according as $\rho(n, H, \xi)$ is uniformly bounded for all n and H, or is such that $\limsup_H \rho(n, H, \xi)=\infty$ for suitable n, or neither of these, respectively. The author applies the reviewer's extension [*Topics in number theory*, Vol. 2, pp. 120–148, Addison-Wesley, Reading, Mass., 1956; MR **18**, 283] of Roth's method, originally devised to investigate the approximability of algebraic numbers by numbers from a fixed algebraic number field, to prove the following theorem: Let $\alpha_1, \alpha_2, \cdots$ be a sequence of distinct numbers in an algebraic number field, with heights $H(\alpha_1), H(\alpha_2), \cdots$, such that

$$(*)\quad |\xi-\alpha_j| < (H(\alpha_j))^{-\kappa}, \quad \limsup_{j\to\infty}\frac{\log H(\alpha_{j+1})}{\log H(\alpha_j)} < \infty.$$

Then there is a positive constant μ such that always $|f(\xi)|>X^{-\mu_n(f)}$, where $X=\max(2, H(f))$ and $\log\log\mu_n=\mu n^2$. In particular, if (*) holds, then ξ is not a U-number; this extends a result of the reviewer [J. London Math. Soc. **28** (1953), 220–229; MR **14**, 956]. Moreover, the theorem implies that the number $\xi=\sum_{n=1}^{\infty}2^{-(N+2)^n}$ (N a positive integer), which has unbounded partial quotients, is either a T-number or an S-number of type at least N, i.e., one for which $\rho(n, H, \xi)\geqq N$ for some n (in this case $n=1$) and, for each such n, for arbitrarily large H.

The author then turns to transcendental numbers with bounded partial quotients. He considers a quasi-periodic continued fraction

$$\xi = [a_0, a_1, \cdots, a_{n_0-1}, \underbrace{a_{n_0}, \cdots, a_{n_0+k_0-1}}_{\longleftarrow\lambda_0\longrightarrow}, \underbrace{a_{n_1}, \cdots, a_{n_1+k_1-1}}_{\longleftarrow\lambda_1\longrightarrow}, \cdots],$$

where the λ's indicate the number of times the distinct blocks of partial quotients are repeated. Suppose that $a_i<A$, $k_i\leqq K$ for all i, and put $C=4A^K$. Let $L=\limsup\lambda_i/\lambda_{i-1}$, and $l=\liminf\lambda_i/\lambda_{i-1}$. It is shown that if $L=\infty$ and $l>1$, then ξ is a U-number of degree 2 (i.e., $\rho(2, H, \xi)$ is unbounded), and that if $L<\infty$ and $l>C$, then ξ is either a T-number or an S-number of type $\geqq l/C$. The first conclusion is proved directly; the second follows from the first theorem. *W. J. LeVeque* (Boulder, Colo.)

The author remarks that, beginning on p. 119 of the paper, the misprint $X=2A^k$ in place of $X=2^{A^k}$ appears worthy of attention.

Citations: MR 14, 956e = J84-5; MR 18, 283b = Z01-38.

Referred to in J84-41.

J84-29 **(28# 2089)**

Sprindžuk, V. G.

On algebraic approximations in the field of power series. (Russian. English summary)

Vestnik Leningrad. Univ. Ser. Mat. Meh. Astronom. **18** (1963), *no.* 3, 130–134.

Sei K ein Körper der Charakteristik 0, $K\langle x\rangle$ der Körper der formalen Laurent-Reihen der Form $\xi(x)=\sum_{j=-\infty}^{m}\alpha_j x^j$ ($\alpha_m\neq 0$) über K. Es wird mit Hilfe der Bewertung $|\xi(x)|=\lambda^{-m}$ ($\lambda>1$, fest) für beliebige Polynome $P(z)$ in $K[x][z]$ die Höhe $\|P\|$ als Maximum der Werte der Koeffizienten $a_i(x)$ definiert, und damit lassen sich Parameter $w_n(\xi)$ sowie $w_n{}^*(\xi)$ einführen, die den bei der Mahlerschen und Koksmaschen Klasseneinteilung der transzendenten Zahlen auftretenden [vgl. der Verfasser, Litovsk. Mat. Sb. **2** (1962), no. 1, 129–145; MR **28** #1165] analog sind. In diesem Sinne werden zwei Ungleichungen von E. Wirsing [J. Reine Angew. Math. **206** (1960), 67–77; MR **26** #79] übertragen; die im vorliegenden Rahmen im Wesentlichen besagen, daß für jedes bezüglich des rationalen Funktionenkörpers $K(x)$ transzendente Element $\xi\in K\langle x\rangle$ sowohl $w_n{}^*(\xi)\geqq w_n(\xi)-n+1$ als auch $w_n{}^*(\xi)\geqq\frac{1}{2}(w_n(\xi)+1)$ gilt. *B. Volkmann* (Mainz)

Citations: MR 26# 79 = J84-20; MR 28# 1165 = J84-27.

Referred to in J84-33.

J84-30 (28# 2090a; 28# 2090b)

Sprindžuk, V. G.
On Mahler's conjecture. (Russian)
Dokl. Akad. Nauk SSSR **154** (1964), 783–786.

Sprindžuk, V. G.
More on Mahler's conjecture. (Russian)
Dokl. Akad. Nauk SSSR **155** (1964), 54–56.

Let ω be a transcendental number, and let $w_n(\omega)$ be the least upper bound of those w for which there exist infinitely many polynomials P over the integers, of degrees not exceeding n, for which $|P(\omega)|<h(P)^{-w}$, where $h(P)$ is the maximum of the absolute values of the coefficients of P. Put

$$\frac{1}{n}w_n(\omega) = \theta_n(\omega) \text{ for } \omega \text{ real},$$
$$= \eta_n(\omega) \text{ for } \omega \text{ complex}.$$

It is well known that $\eta_n(\omega)\geqq\frac{1}{2}(1-1/n)$ and $\theta_n(\omega)\geqq 1$, and Mahler conjectured [Math. Ann. **106** (1932), 131–139] that $\sup_n\eta_n(\omega)=\frac{1}{2}$ and $\sup_n\theta_n(\omega)=1$ for almost all complex and almost all real numbers ω, respectively. Partial results on this conjecture have been obtained by a number of authors; see B. Volkmann [J. Reine Angew. Math. **209** (1962), 201–210; MR **25** #5036] for literature. In the present papers, the author proves these two conjectures, in the order indicated. The proofs depend on inequalities proved earlier by him [Litovsk. Mat. Sb. **2** (1962), no. 1, 129–145; MR **28** #1165; ibid. **2** (1962), no. 2, 221–226], of which the complex case is the following:

Let P have degree n, height h, and distinct zeros $\kappa_1,\cdots,\kappa_n$. Let ω be a complex number, and suppose that $|\omega-\kappa_1|=\min|\omega-\kappa_i|$. Then if $|P(\omega)|<h^{-w}$,

$$|\omega-\kappa_1|^2 < c(n, \operatorname{Im}\omega)h^{-(5+4w-n)/3}|D(P)|^{-1/6},$$

where $D(P)$ is the discriminant of P.
W. J. LeVeque (Boulder, Colo.)

Citations: MR 25# 5036 = J84-19; MR 28# 1165 = J84-27.

Referred to in J84-34, J84-38, J84-44.

J84-31 (29# 2223)

Fraenkel, Aviezri S.
Transcendental numbers and a conjecture of Erdős and Mahler.
J. London Math. Soc. **39** (1964), 405–416.

Das Hauptanliegen dieser Arbeit ist es, aus Nebenbedingungen für die Näherungsbrüche (im Sinne der Kettenbruchlehre) reeller Zahlen auf die arithmetische Natur und die Verteilung dieser Zahlen zu schliessen [vgl. auch Erdős und Mahler, dasselbe J. **14** (1939), 12–18]. Neben anderem wird gezeigt, dass es eine überall dichte Menge von Liouville-Zahlen gibt derart, dass für jede dieser Zahlen sowohl der Zähler als auch der Nenner von unendlich vielen Näherungsbrüchen gewisse Bedingungen der Teilbarkeit und der Grössenordnung erfüllen. Bei den Beweisen werden der Approximationssatz von Kronecker und der Satz von Linnik über die kleinste Primzahl in einer primen Restklasse herangezogen.
G. J. Rieger (Munich)

J84-32 (29# 5784)

Sprindžuk, V. G.
On the measure of the set of S-numbers in a p-adic field. (Russian)
Dokl. Akad. Nauk SSSR **151** (1963), 1292.

Aus der Dissertation des Verfassers werden hier folgende Ergebnisse (ohne Beweis) angegeben, die sich auf das Analogon zur Mahlerschen Vermutung im p-adischen Fall [siehe F. Kasch und den Referent, Math. Z. **72** (1959/60), 367–368; MR **23** #A3713] beziehen: Für fast alle $\xi\in K_p$ ist $\vartheta_n(\xi)\leqq\frac{5}{4}(1+1/(2n))$ ($n=3$, 4, 5, 6, 7), $\vartheta_n(\xi)\leqq\frac{4}{3}n$ ($n\geqq 8$), $\vartheta_3(\xi)=\frac{4}{3}$. *B. Volkmann* (Stuttgart)

Citations: MR 23# A3713 = J84-13.

J84-33 (30# 1978)

Sprindžuk, V. G.
Metric theorems on algebraic approximation in the field of power series. (Russian. Lithuanian and English summaries)
Litovsk. Mat. Sb. **2** (1962), *no.* 2, 207–213.

{Die Arbeit berührt sich teilweise mit einer später im Vestnik Leningrad. Univ. Ser. Mat. Meh. Astronom. **18** (1963), no. 3, 130–134 [MR **28** #2089] erschienenen Untersuchung des Verfassers.} Sei K ein Körper der Charakteristik p mit p^f Elementen und $K\langle x\rangle$ der Körper der formalen Laurent-Reihen der Form $\xi(x)=\sum_{j=-\infty}^{m}\alpha_j x^j$ ($\alpha_m\neq 0$) über K. In $K\langle x\rangle$ wird die Bewertung $|\xi(x)|=p^{-fm}$ und außerdem in naheliegender Weise ein Maß μ eingeführt sowie für jedes Polynom $P(z)=\sum_{\nu=0}^{n}a_\nu(x)z^\nu$ mit $a_\nu(x)\in K[x]$ die Höhe $\overline{|P|}=\max(|a_0(x)|,\cdots,|a_n(x)|)$ definiert. Mit diesen Begriffen lassen sich für jedes $\xi\in K\langle x\rangle$ Analoga zu den Parametern $w_n(\xi)$ erklären, die bei der Mahlerschen Klasseneinteilung der transzendenten Zahlen auftreten (vergleiche oben angegebene Arbeit). In diesem Sinne beweist der Verfasser, daß fast alle $\xi\in K\langle x\rangle$ den Ungleichungen $n\leqq w_n(\xi)<\frac{4}{3}n$ ($n=1,2,\cdots$) genügen und daß für jedes n die Funktion $w_n(\xi)$ fast überall den gleichen Wert annimmt. *B. Volkmann* (Stuttgart)

Citations: MR 28# 2089 = J84-29.

J84-34 (30# 1979)

Sprindžuk, V. G.
On a conjecture of K. Mahler concerning the measure of the set of S-numbers. (Russian. Lithuanian and English summaries)
Litovsk. Mat. Sb. **2** (1962), *no.* 2, 221–226.

Für die bei der Mahlerschen Klasseneinteilung der transzendenten Zahlen auftretenden Parameter $w_n(\xi)$ wird im wesentlichen bewiesen, daß für fast alle reellen bezw. für fast alle komplexen ξ die Ungleichung $\sup_n w_n(\xi)/n\leqq\frac{4}{3}$ bezw. $\sup_n w_n(\xi)/n\leqq\frac{2}{3}$ ist. (In der Einleitung muß es auf S. 221, Z.16/17 statt "$\leqq$" offensichtlich "$\geqq$" heißen). Als Hilfsmittel wird eine vom Referenten [J. Reine Angew. Math. **209** (1962), 201–210; MR **25** #5036] eingeführte Klasseneinteilung der Polynome gegebenen Grades mit ganzrationalen Koeffizienten verwendet, die auf die Größenordnung der Abstände zwischen den Wurzeln jedes Polynoms Bezug nimmt. Ein entsprechendes Ergebnis ist ohne Kenntnis der vorliegenden Arbeit auch vom Referenten [ibid. **213** (1963/64), 58–65; MR **28** #61] bewiesen worden. Inzwischen sind jedoch beide Arbeiten durch den vom Verfasser erbrachten Beweis der Mahlerschen Vermutung [Dokl. Akad. Nauk SSSR **154** (1964), 783–786; MR **28** #2090a; ibid. **155** (1964), 54–56; MR **28** #2090b] überholt. *B. Volkmann* (Stuttgart)

Citations: MR 25# 5036 = J84-19; MR 28# 61 = J84-26; MR 28# 2090a = J84-30; MR 28# 2090b = J84-30.

J84-35 (30# 1980)

Schmidt, Wolfgang M.

Metrische Sätze über simultane Approximation abhängiger Grössen.

Monatsh. Math. **68** (1964), 154–166.

Ein reelles Zahlen-n-tupel $(\alpha_1, \cdots, \alpha_n)$ heisse extrem, falls für jedes $\varepsilon > 0$ die Ungleichung $|\alpha_1 q_1 + \cdots + \alpha_n q_n - p| < (\max|q_i|)^{-n-\varepsilon}$ höchstens endlichviele ganzzahlige Lösungen $(p, q_1, \cdots, q_n)$ besitzt. Dann besagt die Mahlersche Vermutung über die transzendenten Zahlen, daß für fast alle reellen x das n-tupel $(x, x^2, \cdots, x^n)$ extrem ist, und in dieser Formulierung erscheint sie als Spezialfall einer entsprechenden Frage bezüglich allgemeinerer algebraischer Kurven bezw. Flächen. Der Verfasser beweist die beiden folgenden Spezialfälle dieses allgemeineren Problems: (1) Bei fast allen s ist das Paar $(x_1(s), x_2(s))$ extrem, wenn die Funktionen $x_1(s)$ und $x_2(s)$ dreimal stetig differenzierbar sind und die zugehörige Kurve eine fast überall von Null verschiedene Krümmung besitzt. (2) Bei fast allen x_n $(n > 1)$ ist das n-tupel $(x_1, \cdots, x_n)$ extrem, wenn die x_i Gleichungen der Form $x_i = a_i x_n + b_i$ $(i = 1, \cdots, n-1)$ mit extremen $(a_1, \cdots, a_{n-1})$ oder $(b_1, \cdots, b_{n-1})$ genügen. Der Beweis von (1) verwendet wesentlich ein früheres Ergebnis des Verfassers [dieselben Monatsh. **68** (1964), 59–74; MR **28** #5042] über die Anzahl der Gitterpunkte in der "Nähe" gewisser Flächen. *B. Volkmann* (Stuttgart)

Citations: MR 28# 5042 = D48-44.

J84-36 (31# 2207)

Fraenkel, A. S.; Borosh, I.

Fractional dimension of a set of transcendental numbers.

Proc. London Math. Soc. (3) **15** (1965), 458–470.

The authors generalize the following classical result of Jarník [Mat. Sb. **36** (1929), 371–382]: "For $\lambda > 2$, denote by E_λ the set of all real α satisfying $|\alpha - p/q| < q^{-\lambda}$ for an infinity of integers p, q. Then the Hausdorff dimension, $\dim(E_\lambda)$, of this set E_λ equals $2/\lambda$." Observe for the following that Jarník's result implies that the set of all Liouville numbers has Hausdorff dimension zero. A class C of integers p' will be called restricted if C contains 1 and if $\sum_{p' \in C} (p')^{-\varepsilon} < \infty$ for every $\varepsilon > 0$. For instance, $C = \{P_1^{m_1} \cdots P_s^{m_s}\}$ is a restricted class if $\{P_1, \cdots, P_s\}$ is a finite set of fixed primes and if $m_1, \cdots, m_s$ run through all non-negative integers. Now let C_1, C_2 be two restricted classes of integers p' and q'. Let $c > 1$, $0 \leqq \mu \leqq 1$, $0 \leqq \nu \leqq 1$. Denote by $S(C_1, C_2, \mu, \nu)$ the set of all ordered pairs of integers (p, q) of the form $p = p^* p'$, $q = q^* q' > 0$, where $p' \in C_1$, $q' \in C_2$ and p^*, q^* are integers satisfying $|p|^\mu \leqq |p^*| < C|p|^\mu, q^\nu \leqq q^* < cq^\nu, (p, q) = (p^*, p') = (q^*, q') = 1$. The authors prove the following generalization of Jarník's theorem. "Let $0 \leqq \mu \leqq 1$, $0 \leqq \nu \leqq 1$. Let C_1, C_2 be any restricted classes such that (i) if $\mu < 1$, C_1 contains all the powers of a prime P, and if $\nu < 1$, C_2 contains all the powers of a prime $Q \neq P$, and (ii) if $\mu < 1$, $\nu < 1$, then $(1-\nu) \log P (1-\mu) \log Q$ is irrational. Let $\lambda > \mu + \nu$, and denote by E_λ the set of all real α satisfying $|\alpha - p/q| < q^{-\lambda}$ for an infinity of (p, q) in $S(C_1, C_2, \mu, \nu)$. Then $\dim(E_\lambda) = (\mu + \nu)/\lambda$."

If in the above theorem the restricted classes C_1 and C_2 are of the form $C_1 = \{P_1^{m_1} \cdots P_s^{m_s}\}$, $C_2 = \{Q_1^{n_1} \cdots Q_t^{n_t}\}$, then the set E_λ will be denoted by $E_\lambda(R)$. In this case, by a well-known result of Ridout [Mathematika **4** (1957), 125–131; MR **20** #32], all numbers of $E_\lambda(R)$, except perhaps zero, are transcendental. Moreover, if $\mu + \nu > 0$, then it follows from the theorem given above that "almost all" numbers of $E_\lambda(R)$ are non-Liouville transcendental numbers. "Almost all" means here that the exceptional set has lower dimension, in this case, zero.

The authors state in a footnote that the conditions on p^* and q^* in their theorem can be relaxed to $|p^*| < |p|^\mu$, $0 < q^* < q^\nu$. *J. Popken* (Amstelveen)

Citations: MR 20# 32 = J68-18.

Referred to in J84-37.

J84-37 (32# 5598)

Fraenkel, A. S.; Borosh, I.

Corrigendum: Fractional dimension of a set of transcendental numbers.

Proc. London Math. Soc. (3) **16** (1966), 192.

The original article appeared in same Proc. **15** (1965), 458–470 [MR **31** #2207]. The authors close a gap in the proof of Lemma 5.

Citations: MR 31# 2207 = J84-36.

J84-38 (31# 4762)

Sprindžuk, V. G.

A proof of Mahler's conjecture on the measure of the set of S-numbers. (Russian)

Izv. Akad. Nauk SSSR Ser. Mat. **29** (1965), 379–436.

Die inzwischen berühmt gewordene, auf K. Mahler [Math. Ann. **106** (1932), 131–139] zurückgehende Vermutung über die transzendenten Zahlen ist vor kurzem von dem Verfasser in zwei Noten [Dokl. Akad. Nauk SSSR **154** (1964), 783–786; MR **28** #2090a; ibid. **155** (1964), 54–56; MR **28** #2090b] bewiesen worden. Während die dortige Darstellung knapp gehalten ist und zahlreiche Hilfssätze verwendet werden, die in der Literatur verstreut liegen, wird in der vorliegenden Arbeit ein vollständiger, sehr ausführlich geschriebener Beweis erbracht. Folgende Beweisschritte erscheinen dem Referenten als die wesentlichsten.

(1) Es wird gezeigt, dass stets $\tilde{w}_n(\omega) = w_n(\omega)$ ist, wenn das Symbol ~ bedeutet, daß zur Approximation nur irreduzible Polynome mit ganzzahligen Koeffizienten zugelassen werden. Dieser Teil ist zwar grundsätzlich nicht neu, enthält jedoch einen neuen Beweis (mit Hilfe von Lagrangeschen Interpolationspolynomen) der Tatsache, dass für zwei Polynome P_1, P_2 stets $h(P_1 P_2) \gg h(P_1) h(P_2)$ ist, d. h. dass der Quotient beider Größen oberhalb einer Konstanten liegt, die nicht von den Höhen h, sondern nur vom Grad der Polynome abhängt.

(2) Es wird gezeigt, dass man sich auf die Betrachtung von Polynomen der Form $P(x) = a_n x^n + \cdots + a_0$ mit $h(P) = a_n$ beschränken kann.

(3) Es wird gezeigt, dass es genügt, zur Approximation an einer gegebenen Stelle ω solche Polynome P zu betrachten, die "in der Nähe" von ω wenigstens zwei Nullstellen besitzen.

(4) Es wird bewiesen, dass für jedes natürliche n Werte θ_n und η_n existieren, die von der Funktion $\theta_n(\omega)$ für fast alle reellen bzw. für fast alle komplexen ω angenommen werden.

(5) Bei gegebenem, geeignet gewähltem Exponenten $w_0 > 0$ werden jedem Polynom $P(x)$ die eindeutig bestimmten, einfachzusammenhängenden Gebiete $\sigma_i(P)$ (um seine Nullstellen herum) zugeordnet, in denen $|P(\omega)| < h(P)^{-w_0}$ ist. Ein solches Gebiet wird unwesentlich genannt, wenn mehr als die Hälfte seiner Fläche von einem entsprechenden Gebiet überdeckt wird, das zu einem Polynom $Q \neq P$ mit gleichem Grad und gleicher Höhe gehört. Es

wird bewiesen, dass es zum Beweis der Mahlerschen Vermutung genügt, nur solche ω zu betrachten, die (bei gegebenem Grad) in höchstens endlich vielen unwesentlichen Gebieten liegen.

(6) Es wird eine vom Referenten [J. Reine Angew. Math. **209** (1962), 201–210; MR **25** #5036; ibid. **213** (1963/64), 58–65; MR **28** #61] eingeführte Klasseneinteilung der Polynome verwendet, die auf die Größenordnung der Abstände ihrer Nullstellen Bezug nimmt. Unter diesen Klassen von Polynomen werden solche erster Art und zweiter Art unterschieden, je nachdem ob die Nullstellen in der Nähe der betrachteten Zahl ω "weit auseinander" oder "nahe beisammen" liegen. Für beide Arten von Klassen werden verschiedene Abschätzungen für den Abstand von ω zur nächstgelegenen Nullstelle hergeleitet.

(7) Zur Gewinnung von oberen Schranken für die Anzahl der Polynome in einer gegebenen Klasse wird ein Überdeckungsargument verwendet. Dabei spielt eine Hilfsbetrachtung eine besondere Rolle, die besagt, dass die geometrische Form eines Gebietes $\sigma_i(P)$ "nicht allzu flach" sein kann.

(8) Der eigentliche Beweis der Mahlerschen Vermutung wird durch vollständige Induktion erbracht: Der Verfasser gibt zunächst Beweise für die bereits bekannten Gleichungen $\theta_2=1$, $\eta_2=\frac{1}{4}$ und $\eta_3=\frac{1}{3}$ an und weist dann (mit der Schreibweise $w_n=n\eta_n$ bzw. $w_n=n\theta_n$) induktiv nach, dass im Komplexen bzw. im Reellen die Ungleichungen $w_n\leqq\frac{1}{2}+w_{n-1}$ $(n=3, 4, \cdots)$ und $w_n\leqq 1+w_{n-1}$ $(n=2, 3, \cdots)$ gelten, aus denen die eigentliche Behauptung unmittelbar folgt.

Die Arbeit enthält zwar einige Unstimmigkeiten; jedoch kann der Leser sie ohne viel Mühe beheben. {Im Beweis von Hilfssatz 18 ist eine Lücke. Die Konstante 4 in Hilfssatz 9 muss, soweit der Referent feststellen konnte, durch 7 ersetzt werden. Auf Seite 418, Zeile 8 muß die Gleichung $\delta=2/m+\varepsilon$ etwa durch $\delta=3/m+\varepsilon$ ersetzt werden, um auf Seite 419, Zeile 2 auch im Fall $w_{n-1}=\frac{1}{2}(n-2)$ die Ungleichheit zu gewährleisten. Entsprechendes gilt im reellen Fall; vergleiche Seite 427, Zeile 7 und 8.} *B. Volkmann* (Stuttgart)

Citations: MR 25# 5036 = J84-19; MR 28# 61 = J84-26; MR 28# 2090a = J84-30; MR 28# 2090b = J84-30.

Referred to in J84-40, J84-44.

J84-39 (32# 2378)

Berţi, Ştefan N.

On a set of transcendental numbers. (Romanian. Russian and French summaries)

Acad. R. P. Romîne Fil. Cluj Stud. Cerc. Mat. **14** (1963). 35–41.

Let A be a rational integer, $B\neq 1$ a natural integer and $\{a_n\}$ a sequence of natural integers, $0\leqq a_n\leqq B-1$, with infinitely many $a_n\neq 0$. If $\lambda_{A,B,\{a_n\}}=A+\sum_{n=1}^{\infty}a_nB^{-n!}$, then the set $\mathscr{L}_{A,B}=\bigcup_{\{a_n\}}\lambda_{A,B,\{a_n\}}$ consists of transcendental numbers that generalize those of Liouville (which correspond to $A=0$, $B=10$). Furthermore, set $\mathscr{L}_B=\bigcup_{A=-\infty}^{\infty}\mathscr{L}_{A,B}$, $\mathscr{L}=\bigcup_{B=2}^{\infty}\mathscr{L}_B$ and denote by $\mathscr{S}_{A,B}$ the set of maximal open intervals having an empty intersection with $\mathscr{L}_{A,B}$. Then the following results are established (among others): $\mathscr{L}_{A,B}\subset(A, A+1)$; $\mathscr{L}_{A,B}$, $\mathscr{L}_B$ and $\mathscr{L}$ have the cardinality of the continuum, but measure zero; $(a, b)\in\mathscr{S}_{A,B}$ if and only if it is of the form

$$\left(A+\sum_{n=1}^{i}a_nB^{-n!}+(B-1)\sum_{A=i+1}^{\infty}B^{-n!},\right.$$
$$\left.A+\sum_{n=1}^{i}a_nB^{-n!}+B^{-i!}\right)$$

or $(A+\sum_{n=1}^{\infty}(B-1)B^{-n!}, A+1)$. Finally, every point of the real axis is a condensation point for $\mathscr{L}$.

E. Grosswald (Philadelphia, Pa.)

J84-40 (34# 5756)

Sprindžuk, V. G.

A metric theorem on the smallest integral polynomials of several variables. (Russian)

Dokl. Akad. Nauk BSSR **11** (1967). 5–6.

Es bezeichne $\mathfrak{P}_{n_1,\cdots,n_m}$ die Menge der Polynome P in m Veränderlichen x_i vom Grade n_i in x_i $(i=1, \cdots, m)$. Mit h_P werde die Höhe des Polynoms P (h_P = Maximum der Absolutbeträge der Koeffizienten von P) bezeichnet. Für einen Punkt $\omega=(\omega_1, \cdots, \omega_m)$ des m-dimensionalen euklidischen Raumes R^m definiere man $w_{n_1,\cdots,n_m}(\omega)$ als Supremum aller $w>0$, für die die Ungleichung $|P(\omega_1, \cdots, \omega_m)|<h_P^{-w}$ unendlich viele Lösungen in Polynomen $P\in\mathfrak{P}_{n_1,\cdots,n_m}$ besitzt. Der Autor skizziert, an seinen früheren Beweis der Mahlerschen Vermutung über S-Zahlen [Izv. Akad. Nauk SSSR Ser. Mat. **29** (1965). 379–436; MR **31** #4762] anknüpfend, einen Beweis für folgenden Satz: Zu jedem vorgegebenen m-Tupel $(n_1, \cdots, n_m)$ natürlicher Zahlen existieren Zahlen $a_{n_1,\cdots,n_m}$ so daß für fast alle Punkte $\omega\in R^m$ die Beziehung $w_{n_1,\cdots,n_m}(\omega)=a_{n_1,\cdots,n_m}$ gilt.

W. Schwarz (Freiburg)

Citations: MR 31# 4762 = J84-38.

J84-41 (35# 1554)

Baker, A. [Baker, Alan]

On Mahler's classification of transcendental numbers. II. Simultaneous Diophantine approximation.

Acta Arith. **12** (1966/67). 281–288.

H. Davenport [Mathematika **1** (1954). 51–72; MR **16**, 223] has shown that there are uncountably many pairs of irrational numbers θ, ϕ such that all rational fractions p/q, r/q $(q>0)$ satisfy $\max(|\theta-p/q|, |\phi-r/q|)>cq^{-3/2}$, where c is a positive constant. It is shown here that there are U-numbers θ, ϕ for which this inequality always holds, and that for every $\Theta>1$, there are transcendental numbers θ, ϕ, each of which is a T-number or an S-number of type exceeding Θ, with this property. Similar one-variable theorems were obtained earlier by the author [Acta Math. **111** (1964). 97–120; MR **28** #1171; errata, MR **28**, p. 1247].

W. J. LeVeque (Ann Arbor, Mich.)

Citations: MR 16, 223a = J12-26; MR 28# 1171 = J84-28.

J84-42 (37# 361)

Mycielski, Jan

Algebraic independence and measure.

Fund. Math. **61** (1967). 165–169.

This paper extends a previous result of the author [Fund. Math. **55** (1964), 139–147; MR **30** #3855] by replacing the supposition "first category" by "measure zero". Let $A=R^{m_0}$ and let $\langle A, R_i\rangle$, $i<\omega$, be a relational structure that is closed under identification of variables and such that for $i<\omega$, $R_i\subsetneqq A^{r_i}$ implies R_i has m_0r_i-dimensional measure zero. Then there exists a perfect set $P\subseteq A$ which is independent in $\langle A, R_i\rangle$, that is, $f\colon P\to A$, $i<\omega$, $(x_1, \cdots, x_{r_i})\in R_i$ imply $(f(x_1), \cdots, f(x_{r_i}))\in R_i$. It follows that for each set X of measure zero of irrational numbers there is a perfect set $P\subseteq R$ that generates a field disjoint with X. In particular, there exists a subfield of R of power $\mathfrak{c}$ that contains no Liouville numbers. This solves a problem of P. Erdős [Riveon Lematematika **9** (1955), 45–48; MR **17**, 460]. It follows also that there exists a perfect subset of R whose set of distances is disjoint with a given set of measure zero, and that each perfect subset of R contains a perfect set of algebraically independent numbers.

J. C. Oxtoby (Bryn Mawr, Pa.)

Citations: MR 17, 460d = B28-13.

J84-43 **(38# 2097)**

Güting, Rainer

Zur Berechnung der Mahlerschen Funktionen w_n.

J. Reine Angew. Math. **232** (1968), 122–135.

Let γ be a real or complex number, and $w_n(\gamma)$ the function arising in the Mahler classification of γ. The author obtains upper and lower estimates for the value of $w_n(\gamma)$ under various assumptions, and in certain cases an exact evaluation. If γ is real, $0<\gamma<1$, $w_n(\gamma)$ is estimated in terms of lim sup p_{i+1}/p_i, where $\sum 2^{-p_i}$ is the dyadic expansion of γ^n, and similar formulas are given in terms of the regular continued fraction expansion of γ^n. These extend a previous result of the author for the case $n=1$ [Math. Z. **90** (1965), 382–387; MR **32** #2376]. A sufficient condition is given for γ to be a T-number. The proofs are based on earlier work of the author on polynomials with multiple zeros [Mathematika **14** (1967), 181–196; MR **36** #6592].

D. Rearick (Boulder, Colo.)

Citations: MR 32# 2376 = J04-49; MR 36# 6592 = J68-57.

J84-44 **(39# 6832)**

Sprindžuk, V. G. [Спринджук, В. Г.]

Mahler's problem in metric number theory [Проблема Малера в метрической теории чисел].

Izdat. "Nauka i Tehnika", Minsk, 1967. 181 *pp.* 0.75 *r.*

Das Buch stellt im Teil 1 eine gut lesbare Version des vom Verfasser früher [Dokl. Akad. Nauk SSSR **154** (1964), 783–786; MR **28** #2090a; ibid. **155** (1964), 54–56; MR **28** #2090b] erbrachten Beweises der Mahlerschen Vermutung über die transzendenten Zahlen dar. Die hier gegebene Beweisanordnung stimmt im wesentlichen mit der überein, die bereits in den Izv. Akad. Nauk SSSR Ser. Mat. **29** (1965), 379–436 [MR **31** #4762] veröffentlicht wurde. An Vorkenntnissen werden in diesem Teil nur die Anfangsgründe der Algebra, der komplexen Funktionentheorie und der Maßtheorie vorausgesetzt.

In Teil II wird das analoge Problem für nichtarchimedisch bewertete Körper gelöst. Die benötigten Tatsachen über diskrete Bewertungen und über das Haarsche Maß in lokalkompakten Körpern K werden zunächst (mit Literaturhinweisen) zusammengestellt, ausgehend von der eindeutigen Darstellung jedes Elementes $\omega\in K$ als Potenzreihe $\sum_{s=l}^{\infty}\varepsilon_s\pi^s$, wobei π ein erzeugendes Element des Ringes der ω mit $|\omega|<1$ ist und die Koeffizienten ε_s aus einem festen Repräsentantensystem T des zugehörigen Restklassenkörpers gewählt sind. So kann K als unendliches kartesisches Produkt $T\times T\times\cdots$ aufgefaßt werden, und das benötigte Maß läßt sich bequem in der üblichen Weise mit Hilfe von Zylindermengen einführen.

Bei der detaillierten Durchführung des Problems werden als die beiden typischen Fälle die p-adischen Körper Q_p und Potenzreihenkörper über endlichen Körpern betrachtet. Im ersteren Fall wird als Analogon zur Mahlerschen Vermutung der folgende Satz bewiesen: Für jedes $\omega\in Q_p$ sei $w_n(\omega)$ das Infimum der Menge der $w>0$, für die die Ungleichung $|F(\omega)|_p<h(F)^{-w}$ (h=Höhe) durch unendlich viele Polynome F vom Grad $\leqq n$ mit ganzrationalen Koeffizienten erfüllt wird. Für fast alle $\omega\in Q_p$ ist dann $w_n(\omega)=n+1$ $(n=1, 2, \cdots)$. Damit wird eine in dieser Form zuerst von F. Kasch und dem Referenten [Math. Z. **72** (1959/60), 367–378; MR **23** #A3713] ausgesprochene Vermutung bestätigt.

Im Fall des Körpers $K\langle x\rangle$ aller Potenzreihen $\omega=\omega(x)=\sum_{s=l}^{\infty}\alpha_s x^{-s}$ mit Koeffizienten aus einem endlichen Körper K wird die Bewertung $|\omega|=q^{-l}$ verwendet, wobei q die Kardinalzahl von K ist. Bei der Definition des Parameters $w_n(\omega)$ werden nunmehr in der Ungleichung $|F(\omega)|<h(F)^{-w}$ Polynome F vom Grad $\leqq n$ betrachtet, deren Koeffizienten ihrerseits Polynome aus $K[x]$ sind. Es wird bewiesen, daß (im Sinne des Haarschen Maßes auf $K\langle x\rangle$) fast überall $w_n(\omega)=n$ $(n=1, 2, \cdots)$ ist. Dieser zweite Teil lehnt sich im Beweisaufbau weitgehend an den Teil I an, erfordert aber in mehreren Abschnitten eigene Methoden.

Der Anhang behandelt zusätzliche Ergebnisse, die mit dem Hauptthema des Buches in mehr oder weniger engem Zusammenhang stehen und zum Teil schon anderweitig veröffentlicht sind. Auch eine Reihe von ungelösten Problemen werden hier formuliert.

{Eine vom Referenten besorgte englische Übersetzung des Buches wurde von der Amer. Math. Soc. veröffentlicht [see #6833 below].}

B. Volkmann (Stuttgart)

Citations: MR 23# A3713 = J84-13; MR 28# 2090a = J84-30; MR 28# 2090b = J84-30; MR 31# 4762 = J84-38.

Referred to in J12-60, J84-46.

J84-45 **(39# 6833)**

Sprindžuk, V. G.

Mahler's problem in metric number theory.

Translated from the Russian by B. Volkmann. Translations of Mathematical Monographs, Vol. 25.

American Mathematical Society, Providence, R.I., 1969. vii+192 *pp.* $12.70.

The Russian original is reviewed above [#6832].

Referred to in J12-60, J84-46.

J84-46 **(40# 4210)**

Sprindžuk, V. G.

On the metric theory of "nonlinear" diophantine approximations. (Russian)

Dokl. Akad. Nauk BSSR **13** (1969), 298–301.

Suppose S is a sequence of n dimensional integral vectors, P is a real polynomial in n variables, and λ an arbitrary non-negative real function defined over S. Suitably defining a measure in the class of all polynomials of fixed degree $\geqq 2$ and writing $\|x\|$ for the distance from x to the nearest integer, the author proves that, for almost all P, the number of a for which $\|P(a)\|\leqq\lambda(a)$ is finite if $\sum_{a\in S}\lambda(a)$ converges and is infinite otherwise. In the latter case, one can give an asymptotic estimate for the number of a. The work is based on two theorems, only the second of which is proved here, about the number of integral a satisfying $((a,\omega_1), \cdots, (a,\omega_n))\in A(a)$ mod 1, where $\omega_j=(\omega_{1j}, \cdots, \omega_{nj})$, $0\leqq\omega_{ij}\leqq 1$, (a,ω_j) is the inner product of a and ω_j, and $A(a)$ is a measurable (suitably defined) subset of the set of $1\times m$ matrices ω. The first theorem asserts the number of such a is finite or infinite depending on the convergence or divergence of the series $\sum_{a\in S}|A(a)|$ and the second theorem gives, in the case of divergence, an asymptotic estimate for the number of such a with bounded coordinates lying in a set S whose elements are pairwise linearly independent. The first theorem and a less sharp form of the second theorem are given on p. 162 of the English translation of the author's book *Mahler's problem in metric number theory* [Amer. Math. Soc., Providence, R.I., 1969; MR **39** #6833; Russian original, Izdat. "Nauka i Tehnika", Minsk, 1967; MR **39** #6832].

J. B. Roberts (Halifax, N.S.)

Citations: MR 39# 6832 = J84-44; MR 39# 6833 = J84-45.

J84-47 **(43# 157)**

Slesoraĭtene, R. [Sliesoraitienė, R.]

The Mahler-Sprindžuk theorem. (Russian. Lithuanian and English summaries)

Litovsk. Mat. Sb. **10** (1970), 367–374.

Let $\varepsilon > 0$; let $1 \leqq m \leqq 4$; let Σ_m denote the set of all polynomials $S(x) = a_0x^3 + a_1x^2 + a_2x + a_3$ with rational integral coefficients of which exactly m are distinct from 0; and let $h = \max_{0 \leqq j \leqq 3}|a_j|$. The author proves that for almost all real x, in the sense of Lebesgue, there are at most finitely many polynomials $S(x)$ in Σ_m for which $|S(x)| < h^{-m+1-\varepsilon}$. His proof depends on a special treatment of each of the separate possible cases. *K. Mahler* (Columbus, Ohio)

Referred to in J84-50.

J84-48 **(43# 4769)**

Schmidt, Wolfgang M.

***T*-numbers do exist.**

Symposia Mathematica, Vol. IV (INDAM, Rome, 1968/69), pp. 3–26. Academic Press, London, 1970.

Almost 40 years ago, K. Mahler [Math. Z. **31** (1930), 729–732; J. Reine Angew. Math. **166** (1931/32), 118–136] introduced his classification of the complex numbers into A-, S-, T- and U-numbers and showed that two numbers from different classes are algebraically independent. He also showed that the A-numbers are exactly the algebraic numbers, that almost all numbers are S-numbers, and that Liouville numbers are U-numbers. His work left open the question of the existence of T-numbers, and despite many efforts in the intervening years, this question has remained unsettled until the present paper.

The proof is an intricate inductive construction. As a warm-up the author first gives a (similar, but simpler) proof of the fact that if α belongs to the class of real irrational numbers that are not Liouville numbers, then there is a β in the same class such that α/β is a Liouville number.

Concerning the main theorem, the author first notes that it suffices to prove the existence of real T^*-numbers, as in J. F. Koksma's classification [Monatsh. Math. Phys. **48** (1939), 176–189; MR **1**, 137]. He shows in fact that if $A_1, A_2, \cdots$ are either real numbers or $+\infty$, with $A_1 > 8$ and $A_i \geqq 3i^2(A_{i-1}+1)$ for $i > 1$, then there exist continuum many numbers ξ with $\omega_i^*(\xi) = A_i$, $i = 1, 2, \cdots$. This implies the uncountability of the T^*-numbers, when all the A_i are finite, and of the U^*-numbers of degree n, when $A_{n-1} < \infty = A_n = A_{n+1} = \cdots$. The construction leading to this more general theorem depends on a theorem of E. Wirsing (unpublished) that if α is algebraic, there are only finitely many algebraic numbers β of degree n for which $|\alpha - \beta| < H(\beta)^{-2n-\varepsilon}$. This result has since been superseded by the author's remarkable work on simultaneous approximation to several algebraic numbers by rational numbers, which implies that the exponent $2n$ above can be replaced by $n+1$. *W. J. LeVeque* (Claremont, Calif.)

Citations: MR 1, 137a = J84-1.

J84-49 **(43# 4770)**

Slesoraĭtene, R. [Sliesoraitienė, R.]

An analogue of the Mahler-Sprindžuk theorem for certain third degree polynomials in two variables. I. (Russian. Lithuanian and English summaries)

Litovsk. Mat. Sb. **10** (1970), 545–564.

Let $P(x, y) = a_{111}x^3 + a_{112}x^2y + a_{122}xy^2 + a_{222}y^3 + a_{113}x^2 + a_{123}xy + a_{233}y^2 + a_{133}x + a_{233}y + a_{333}$ be a polynomial with integral coefficients of which n $(1 \leqq n \leqq 10)$ are not zero. Denote by $D(x)$ the discriminant of $P(x, y)$ relative to y, and by ε a positive constant. Further assume that $D(x)$ is of the exact degree 4 in x and that $a_{222} = 0$, but $a_{111}a_{112}a_{122} \neq 0$. Finally let at least one of the numbers $|a_{111}|$ and $|a_{333}|$ be distinct from both 0 and $h = \max_{h,i,j}|a_{hij}|$. Then, for almost all pairs of real numbers x, y satisfying $D(x) \neq 0$, the inequality $|P(x, y)| < h^{-n+1-\varepsilon}$ holds for only finitely many such polynomials P. The proof depends on a large number of separate cases. For the similar case of quadratic polynomials compare the earlier paper of the author [same Sb. **9** (1969), 627–634; MR **41** #8371].
K. Mahler (Columbus, Ohio)

Citations: MR 41# 8371 = J12-60.

Referred to in J84-50.

J84-50 **(44# 160)**

Slesoraĭtene, R. [Sliesoraitienė, R.]

The Mahler-Sprindžuk theorem for third degree polynomials in two variables. II. (Russian. Lithuanian and English summaries)

Litovsk. Mat. Sb. **10** (1970), 791–813.

Let $P(x, y) = a_{111}x^3 + a_{112}x^2y + a_{122}xy^2 + a_{222}y^3 + a_{113}x^2 + a_{123}xy + a_{223}y^2 + a_{133}x + a_{233}y + a_{333}$ be an arbitrary cubic polynomial with integral coefficients, where $h = \max_{i,j,k}|a_{ijk}| \geqq 1$. Assume that exactly n of the coefficients a_{ijk} are distinct from zero, $1 \leqq n \leqq 10$, and denote by ε any positive constant. The author proves that for almost all pairs of real numbers x, y each of the two sets of conditions $|P(x, y)| < h^{-n+1-\varepsilon}$, $a_{111}a_{222}a_{333} = 0$, and $|P(x, y)| < h^{-9-\varepsilon}$, $a_{111}a_{222}a_{333} \neq 0$ is satisfied by at most finitely many polynomials P. [Compare the author's previous papers, same Sb. **10** (1970), 367–374; MR **43** #157; ibid. **9** (1969), 627–634; MR **41** #8371; ibid. **10** (1970), 545–564; MR **43** #4770.] *K. Mahler* (Columbus, Ohio)

Citations: MR 41# 8371 = J12-60; MR 43# 157 = J84-47; MR 43# 4770 = J84-49.

J88 ALGEBRAIC INDEPENDENCE

See also reviews J02-6, J02-21, J02-22, J02-24, J76-4, J76-17, J76-47, J84-42, Q05-61.

J88-1 **(9, 413c)**

Skolem, Th. A proof of the algebraic independence of certain values of the exponential function. Norske Vid. Selsk. Forh., Trondhjem **19**, no. 12, 40–43 (1947).

In this note the author gives a simple and short proof of the following result, equivalent to the well-known theorem of F. Lindemann. The numbers $e^{\alpha_1}, e^{\alpha_2}, \cdots, e^{\alpha_n}$ are algebraically independent if $\alpha_1, \alpha_2, \cdots, \alpha_n$ are algebraic numbers, linearly independent relative to the field of rational numbers. The author's proof is a natural extension of the ordinary proof for the transcendence of e and π, well-known from textbooks, where a suitably chosen prime number plays an important part. *J. Popken* (Utrecht).

Referred to in J88-2.

J88-2 **(10, 18b)**

Skolem, Th. A proof of the algebraic independence of e and $e^{\sqrt{-d}}$, d positive integer, with another proof of the irrationality of $\log x$ and $\operatorname{arctg} x$ for rational x. Norsk Mat. Tidsskr. **28**, 97–104 (1946).

The author has given a new and simple proof for the theorem of Lindemann that the numbers $e^{\alpha_1}, \cdots, e^{\alpha_n}$ are algebraically independent if $\alpha_1, \cdots, \alpha_n$ are linearly independent algebraic numbers [Norske Vid. Selsk. Forh., Trondhjem **19**, no. 12, 40–43 (1947); these Rev. **9**, 413]. In the present paper he treats the special case $n = 2$, α_1 rational and α_2 belonging to an imaginary quadratic number field, and then his proof is almost identical with the usual proof for the transcendence of e [see, e.g., E. Landau, Vorlesungen über Zahlentheorie, v. 3, Hirzel, Leipzig, 1932, p. 92].

The second section of the paper contains proofs for the irrationality of $\log x$ and $\tan^{-1} x$ for rational values of x (if naturally the trivial cases $x=0$, 1 or $x=0$, respectively, are excluded). The starting point is the formula of Hermite,

$$\int_{-1}^{1}(1-x^2)^n \cos zx dx = n! z^{-2n-1}(P \sin z + Q \cos z),$$

where P and Q are polynomials in z with integral coefficients.

J. Popken (Utrecht).

Citations: MR 9, 413c = J88-1.

J88-3 **(10, 682d)**

Gel'fond, A. O. On the algebraic independence of algebraic powers of algebraic numbers. Doklady Akad. Nauk SSSR (N.S.) **64**, 277–280 (1949). (Russian)

Let α be a root of an irreducible algebraic equation of degree 3 and $a\neq 0, 1$ any algebraic number. The author shows that, for any fixed interpretation of $\log a$, the numbers $a^\alpha = e^{\alpha \log a}$ and $a^{\alpha^2} = e^{\alpha^2 \log a}$ are algebraically independent in the field of the rational numbers. The proof, which is not given in full detail, makes use of results said to have been proved by the author in another paper inaccessible to the reviewer. It is stated that the method can be applied without appreciable change to prove that if $\omega_1, \omega_2, \cdots, \omega_\nu$ (ω_1 rational) form a basis for the ring of algebraic integers of an algebraic field K, then no relation $P(a^{\omega_i}, a^{\omega_j})=0$ ($i\neq j\neq 1$) is possible where $P(x, y)$ is a polynomial in x, y with rational coefficients; an improvement of this result, which is to appear elsewhere, is announced. [Note. The reviewer suspects that in lemma 3 the case $k_1=k_2=k_3=0$ is meant to be excluded. Otherwise the lemma becomes trivial since B_ν is then a multiple of $f^{(s)}(0)$ which is zero by hypothesis.] *R. A. Rankin* (Cambridge, England).

Referred to in J02-6, J80-5, J88-4, J88-5.

J88-4 **(11, 83b)**

Gel'fond, A. O. On the algebraic independence of transcendental numbers of certain classes. Doklady Akad. Nauk SSSR (N.S.) **67**, 13–14 (1949). (Russian)

Three general theorems are stated which, it is claimed, can be deduced by a variation of the methods used in an earlier paper [same Doklady (N.S.) **64**, 277–280 (1949); these Rev. **10**, 682]. These theorems are too complicated to state here, but the following consequences may be mentioned. (I) If a and α are algebraic numbers and $a\neq 0, 1$, and if the degree of α exceeds 2, then it is not possible to express each of the four numbers a^α, a^{α^2}, a^{α^3}, a^{α^4} algebraically in terms of one of them. When α is a cubic irrational it follows that a^α and a^{α^2} are algebraically independent in the field of the rational numbers, a result already proved in the paper referred to. (II) If a is as before and ν is rational and not zero, then it is not possible to express each of the four numbers a^{e^ν}, $a^{e^{2\nu}}$, $a^{e^{3\nu}}$, $a^{e^{4\nu}}$ algebraically in terms of one of them and, in particular, at least one of them is transcendental. A similar result holds for the first three of the four numbers. *R. A. Rankin* (Cambridge, England).

Citations: MR 10, 682d = J88-3.

J88-5 **(11, 231d)**

Gel'fond, A. O. On the algebraic independence of transcendental numbers of certain classes. Uspehi Matem. Nauk (N.S.) **4**, no. 5(33), 14–48 (1949). (Russian)

By means of a method similar to one used by him already previously [Doklady Akad. Nauk SSSR (N.S.) **64**, 277–280 (1949); these Rev. **10**, 682], the author proves the following three theorems. (1) Let η_0, η_1, η_2, α_1, α_2 be real or complex numbers such that η_0, η_1, η_2, as well as 1, α_1, α_2, are linearly independent over the rational field. Let there be two positive constants τ, x' such that $|x_0\eta_0+x_1\eta_1+x_2\eta_2|>\exp(-\tau x \log x)$ for all rational integers x_0, x_1, x_2 satisfying

$$x=\max(|x_0|, |x_1|, |x_2|)>x'.$$

Then at least two of the 11 numbers

(a) $\alpha_1, \alpha_2, \exp(\eta_i\alpha_k), \quad i=0, 1, 2; k=0, 1, 2; \alpha_0=1,$

are algebraically independent over the rational field.

(2) Let η_0, η_1, α_1, α_2 be real or complex numbers such that $\eta_0\neq 0$, η_1/η_0 is irrational, and 1, α_1, α_2 are linearly independent over the rational field. Let there be two positive constants τ, x' such that $|x_0+x_1\eta_1/\eta_0|>\exp(-\tau x^2 \log x)$ for all rational integers x_0, x_1 satisfying $x=\max(|x_0|, |x_1|)>x'$. Then at least two of the 10 numbers

$$\eta_0, \eta_1, \alpha_1, \alpha_2, \exp(\eta_i\alpha_k), \qquad i=0, 1; k=0, 1, 2; \alpha_0=1,$$

are algebraically independent over the rational field.

(3) Let a, b, α, β be algebraic real or complex numbers such that a is different from 0 and 1, and that b and $(\log \alpha)/(\log \beta)$ are irrational. Then there exists to every $\epsilon>0$ a number H_0 with the following properties. If $P(x)\not\equiv 0$ is a polynomial of degree at most s and with rational integral coefficients of absolute value at most H, then

$$|P(a^b)|>\exp\left\{-\frac{s^3}{\log^3 s}(s+\log H)\log^{2+\epsilon}(s+\log H)\right\},$$

$$\left|P\left\{\frac{\log \alpha}{\log \beta}\right\}\right|>\exp\{-s^2(s+\log H)^{2+\epsilon}\},$$

if $H>H_0$.

On specializing the numbers η_i and α_k, (1) and (2) give, e.g., the following corollaries. (4) If α and $a\neq 0$, $\neq 1$ are algebraic and α is of higher than the third degree, then at least two of the numbers a^α, a^{α^2}, a^{α^3}, a^{α^4} are algebraically independent. If α is of the third degree, then already a^α and a^{α^2} are algebraically independent. [See the paper cited at the beginning of this review.] (5) If $a\neq 0$, $\neq 1$ is algebraic, and $\eta\neq 0$ is rational, then at least two of the numbers $a^{\exp(k\nu)}$, $k=1, 2, 3, 4$, are algebraically independent. (6) If $\nu\neq 0$ is a rational number, then at least one of the numbers $\exp(e^{k\nu})$, $k=1, 2, 3$, is algebraically independent of e, thus transcendental; and at least one of the numbers $\exp(\pi^{k\nu+1})$, $k=1, 2, 3$, is transcendental. Theorem 3 is also of interest because, e.g., the Thue-Siegel theorem on algebraic numbers implies only the weaker inequality $|P\{(\log \alpha)/(\log \beta)\}|>e^{-\epsilon H}$.

The proofs of (1)–(3) are rather similar; they are based on a number of lemmas which are of interest in themselves. For instance, (1) is proved as follows. Assume that every two of the 11 numbers (a) are algebraically dependent. Then there exist two transcendental numbers ω and ω_1, where $\omega_1^\nu+c_1(\omega)\omega_1^{\nu-1}+\cdots+c_\nu(\omega)=0$ and $c_1(\omega)$, $c_2(\omega)$, $\cdots$, $c_\nu(\omega)$ are polynomials in ω with integral coefficients, such that the numbers (a) are of the form S_i/T_i ($i=1, \cdots, 11$), the S_i and T_i being polynomials in ω of arbitrary degree and in ω_1 of degree at most $\nu-1$ with integral coefficients. Denote by N a sufficiently large positive integer, and put $T=T_1T_2\cdots T_{11}$, and $p=[N^3\log^{-\frac{1}{2}} N]+1$, $p_1=[N^3/\log^{\frac{1}{2}} N]$, $\lambda=3+\tau$, $\mu=[11\lambda N^3 \log^{-\frac{1}{2}} N]$. In the function

$$f(z)=\sum_{k_0=0}^{p-1}\sum_{k_1=0}^{N}\sum_{k_2=0}^{N}\sum_{k_3=0}^{N} A_{k_0k_1k_2k_3}z^{k_0}\exp\{(k_1\eta_1+k_2\eta_2+k_3\eta_3)z\}$$

let the A's be of the form

$$A_{k_0\cdots k_3}=\sum_{k_4=0}^{p_1} C_{k_0\cdots k_4}\omega^{k_4},$$

where the C's are integers not all zero. Then the numbers

$$f_{k_0k_1k_2}=T^\mu f(k_0+k_1\alpha_1+k_2\alpha_2),$$

where $0\leqq k_i\leqq\lambda N^2\log^{-\frac{1}{2}} N$, $i=0, 1, 2$, are polynomials in ω

and ω_1 with integral coefficients, and at least one of them is not zero. Denote by $\lambda_2, \lambda_3, \cdots$ positive constants independent of N. By means of Dirichlet's principle, integral values of the C's may be chosen such that

$$|C_{k_0\cdots k_4}| < \exp(\lambda_3 N^2 \log^{-\frac{1}{2}} N),$$

$f_{k_0k_1k_2}=0$, i.e., $f(k_0+k_1\alpha_1+k_2\alpha_2)=0$, for

$$0 \leqq k_i \leqq [\lambda_2 N^2 \log^{-\frac{1}{2}} N] = q,$$

$i=0, 1, 2$. If Γ is the circle $|\zeta|=2N^{9/2}$, therefore,

$$f(z)=\frac{1}{2\pi i}\int_\Gamma \prod_{k_0=0}^{q}\prod_{k_1=0}^{q}\prod_{k_2=0}^{q}\left\{\frac{z-k_0-k_1\alpha_1-k_2\alpha_2}{\zeta-k_0-k_1\alpha_1-k_2\alpha_2}\right\}\frac{f(\zeta)d\zeta}{\zeta-z},$$

and so

$$|f(z)| < \exp(-\lambda_2 N^6 \log^{-\frac{1}{2}} N), \qquad |z| \leqq N^3,$$

whence

$$|f_{k_0k_1k_2}| < \exp(-\tfrac{1}{2}\lambda_2 N^6 \log^{-\frac{1}{2}} N),$$
$$0 \leqq k_i \leqq \lambda N^2 \log^{-\frac{1}{2}} N;\ i=0, 1, 2.$$

Select now k_0, k_1, k_2 according to these inequalities such that $f_{k_0k_1k_2} \neq 0$. Then $f_{k_0k_1k_2}=P(\omega, \omega_1)$ is a polynomial in ω and ω_1 with integral coefficients. Let H be its height (i.e., the maximum of the absolute values of the coefficients), and n its degree in ω; it may be assumed of degree $\nu-1$ in ω_1. Then one shows that $n+\log H < \lambda_4 N^3 \log^{-\frac{1}{2}} N$. Denote by $\omega_1, \omega_2, \cdots, \omega_\nu$ the conjugates of ω_1 over the field generated by ω. The norm $P_0(\omega)=\prod_{k=1}^{\nu} P(\omega, \omega_1)$ does not vanish and is a polynomial in ω with integral coefficients, say of degree n_0 and height H_0. One shows that

$$|P_0(\omega)| < \exp(-\lambda_5 N^6 \log^{-\frac{1}{2}} N),$$
$$\max(n_0, \log H_0) < \lambda_6 N^3 \log^{-\frac{1}{2}} N.$$

By a new test of transcendency given in the paper, ω must then be algebraic contrary to hypothesis. *K. Mahler.*

Citations: MR 10, 682d = J88-3.
Referred to in J02-6, J80-34, J88-6, Z10-51.

J88-6 (13, 727f)

Gel'fond, A. O. On the algebraic independence of transcendental numbers of certain classes. Amer. Math. Soc. Translation no. 66, 46 pp. (1952).

Translated from Uspehi Matem. Nauk (N.S.) **4**, no. 5(33), 14–48 (1949); these Rev. **11**, 231.

Citations: MR 11, 231d = J88-5.
Referred to in Z10-51.

J88-7 (13, 921d)

Steinberg, R., and Redheffer, R. M. Analytic proof of the Lindemann theorem. Pacific J. Math. **2**, 231–242 (1952).

The authors give a proof of the Lindemann-Weierstrass theorem: Let $\alpha, \beta, \cdots, \sigma$ be different algebraic numbers, let $a, b, \cdots, s$ be arbitrary algebraic numbers; then

$$ae^\alpha+be^\beta+\cdots+se^\sigma=0$$

is possible only for $a=b=\cdots=s=0$. The paper is mainly expository. The method used is that in Hilbert's proof for the transcendence of e and π [Gesammelte Abhandlungen, Bd. 1, Springer, Berlin, 1932, pp. 1–4]. *J. Popken.*

J88-8 (14, 957d)

Čudakov, N. G. On algebraic independence of values of the exponential function. Ukrain. Mat. Žurnal **3**, 211–217 (1951). (Russian)

A new variant of the Hilbert-Hurwitz proof of Lindemann's theorem, using the language of modern algebra.
K. Mahler (Manchester).

J88-9 (16, 117e)

Šidlovskiĭ, A. B. On transcendentality and algebraic independence of the values of entire functions of certain classes. Doklady Akad. Nauk SSSR (N.S.) **96**, 697–700 (1954). (Russian)

C. L. Siegel has introduced the notion of E-functions, and has proved the following theorem [Transcendental numbers, Princeton, 1949, p. 52; these Rev. **11**, 330]: Let the E-functions $E_1, \cdots, E_n$ be solutions of a system of m homogeneous linear differential equations of first order, whose coefficients Q_{kl} are rational functions with algebraic numerical coefficients, and suppose that the power-products $E_1^{k_1}\cdots E_m^{k_m}$ $(k_1+\cdots+k_m \leqq N)$ form a normal system, for all positive integers N. If α is any algebraic number different from the zeros and poles of any of the Q_{kl}, then the m numbers $E_1(\alpha), \cdots, E_m(\alpha)$ are algebraically independent. In the present paper, the notion of an irreducible system of E-functions is introduced: $E_1, \cdots, E_m$ are said to form an irreducible system if they are solutions of a system A of linear homogeneous differential equations of the type just specified, none is identically zero, and the relation $\sum_1^m P_k(z)y_k=0$, where the P_k are arbitrary polynomials in z and the y_k are any solutions of A, is possible only in case $P_ky_k=0$ $(1 \leqq k \leqq m)$ identically. It is announced that Siegel's theorem has been generalized by dropping the condition of homogeneity and replacing normality by irreducibility. Various applications are mentioned, showing the algebraic independence of values of certain functions satisfying third and fourth order differential equations, for algebraic argument. No proofs are given. *W. J. LeVeque*

Citations: MR 11, 330c = J76-18.
Referred to in J88-10, J88-11, J88-12.

J88-10 (16, 907a)

Šidlovskiĭ, A. B. On a criterion of algebraic independence of the values of a class of entire functions. Dokl. Akad. Nauk SSSR (N.S.) **100**, 221–224 (1955). (Russian)

The following theorem is announced: Let the E-functions $f_1(z), \cdots, f_m(z)$ be arbitrary solutions of the system of m linear differential equations

$$y_k'=Q_{k,0}(z)+\sum_{i=1}^{m} Q_{k,i}(z)y_i \quad (k=1, \cdots, m)$$

whose coefficients $Q_{k,i}(z)$ are rational functions of z with algebraic numerical coefficients, and let α be an algebraic number distinct from 0 and the poles of all the $Q_{k,i}(z)$. Then the numbers $f_1(\alpha), \cdots, f_m(\alpha)$ are algebraically independent over the field of rational functions of z. This further extends the author's earlier generalization [same Dokl. (N.S.) **96**, 697–700 (1954); MR **16**, 117] of Siegel's work [Transcendental numbers, Princeton, 1949, p. 52; MR **11**, 330] on E-functions, in that $f_1(z), \cdots, f_m(z)$ are no longer required to form an irreducible system. Various applications are indicated. No proofs are given. *W. J. LeVeque.*

Citations: MR 11, 330c = J76-18; MR 16, 117e = J88-9.
Referred to in J88-11, J88-12, J88-52.

J88-11 (17, 466a)

Šidlovskiĭ, A. B. On transcendental numbers of certain classes. Dokl. Akad. Nauk SSSR (N.S.) **103** (1955), 977–980. (Russian)

This is a continuation of the author's recent work on E-functions [same Dokl. (N.S.) **96** (1954), 697–700; **100** (1955), 221–224; MR **16**, 117, 907]. The new results are similar to those announced earlier. *W. J. LeVeque.*

Citations: MR 16, 117e = J88-9; MR 16, 907a = J88-10.
Referred to in J88-12.

J88-12 **(17, 947c)**

Šidlovskiĭ, A. B. On transcendentality of the values of a class of entire functions satisfying linear differential equations. Dokl. Akad. Nauk SSSR (N.S.) **105** (1955), 35–37. (Russian)

This is one of a series of papers by the author [same Dokl. (N.S.) **96** (1954), 697–700; **100** (1955), 221–224; **103** (1955), 977–980; MR **16**, 117, 907; **17**, 466] extending and generalizing Siegel's work [Transcendental numbers, Princeton, 1949, p. 52; MR **11**, 330] on E-functions. The following theorems are announced: I. Let the E-function $f(z)$ be a solution of the linear differential equation $P_m(z)y^{(m)}+\cdots+P_1(z)y'+P_0(z)y=Q(z)$, whose coefficients $P_i(z)$ and $Q(z)$ are polynomials in z with algebraic numerical coefficients, and let α be an algebraic number different from 0 and the zeros of $P_i(z)$. Then the l numbers $f^{(i)}(\alpha)$, $i=0, 1, \cdots, l-1$, $1\leq l\leq m$, are algebraically independent if and only if the functions $f^{(i)}(z)$, $i=0, \cdots, l-1$ are algebraically independent over the field of rational functions of z. II. If a function f as in I is transcendental, and if $\alpha\neq 0$ is algebraic and not a zero of $P_m(z)$, then the numbers $f^{(i)}(\alpha)$, $i=0, 1, \cdots$ are transcendental and all zeros and all A-values of f and all its derivatives, except 0 and the zeros of $P_m(z)$ are transcendental for arbitrary algebraic A.

These theorems are very much easier to apply than Siegel's theorem, since Siegel's condition of normality has disappeared. It follows from II, for example, that every hypergeometric E-function [cf. Siegel, op. cit., p. 54] and each of its derivatives assumes a transcendental value for algebraic argument different from 0. *W. J. LeVeque.*

Citations: MR 11, 330c = J76-18; MR 16, 117e = J88-9; MR 16, 907a = J88-10; MR 17, 466a = J88-11.

Referred to in J88-13, J88-14.

J88-13 **(17, 1057d)**

Šidlovskiĭ, A. B. On a new criterion of the transcendentality and algebraic independence of values assumed by a class of entire functions. Dokl. Akad. Nauk SSSR (N.S.) **106** (1956), 399–400. (Russian)

Continuing the authors' earlier work on E-functions [for references and definitions see same Dokl. (N.S.) **105** (1955), 35–37; MR **17**, 947], the following theorems are announced. I. Let the set of E-functions $f_1(z), \cdots, f_m(z)$ be solutions of a system of m linear differential equations of first order

$$y_k'=Q_{k0}(z)+\sum_{i=1}^{m} Q_{ki}(z)y_i \ (k=1, \cdots, m)$$

whose coefficients $Q_{ki}(z)$ are rational functions of z with algebraic numerical coefficients, and let α be an arbitrary algebraic number different from the zeros and poles of all $Q_{ki}(z)$. Then in order that the l numbers $f_1(\alpha), \cdots, f_l(\alpha)$, $1\leq l\leq m$, be algebraically independent, it is necessary and sufficient that the functions $f_1(z), \cdots, f_l(z)$ be algebraically independent over the field $R(z)$ of rational functions of z. In particular, the numbers $f_i(\alpha)$, $i=1, \cdots, m$, are transcendental if and only if the corresponding functions $f_i(z)$ are transcendental. II. Under the same hypotheses, if the functions $f_1(z), \cdots, f_l(z)$, $2\leq l\leq m$ are algebraically independent over $R(z)$, then the $l-1$ numbers $f_i(\alpha)/f_s(\alpha)$, $i=1, \cdots, l$, $i\neq s$, are algebraically independent, for $s=1, \cdots, l$.

These theorems contain all the author's previous results. *W. J. LeVeque* (Ann Arbor, Mich.).

Citations: MR 17, 947c = J88-12.

J88-14 **(18, 382f)**

Šidlovskiĭ, A. B. On algebraic independence of transcendental numbers of a certain class. Dokl. Akad. Nauk SSSR (N.S.) **108** (1956), 400–403. (Russian)

The theorem announced by the author in a previous paper [same Dokl. (N.S.) **105** (1955), 35–37; MR **17**, 947] is applied to prove the algebraic independence of each of various sets of hypergeometric E-functions, such as

$$\psi_k(z)=\sum_{n=0}^{\infty}\frac{1}{(n!)^k}\left(\frac{z}{k}\right)^{kn} \ (k=1, 2, \cdots, r),$$

or of the values of such functions at $z=\alpha$, a nonzero algebraic number. *W. J. LeVeque.*

Citations: MR 17, 947c = J88-12.

J88-15 **(21# 1295)**

Šidlovskiĭ, A. B. A criterion for algebraic independence of the values of a class of entire functions. Izv. Akad. Nauk SSSR. Ser. Mat. **23** (1959), 35–66. (Russian)

In several earlier notes the author had announced without proof generalisations of Siegel's results on the transcendency of E-functions satisfying linear differential equations [C. Siegel, *Transcendental numbers*, Princeton Univ. Press, Princeton, N.J., 1949; MR **11**, 330]. The present paper contains the detailed proofs for the following two main results of the author. (I) Let $f_1(z), \cdots, f_m(z)$ be m E-functions such that

$$y_k' = Q_{k0}(z)+\sum_{i=1}^{m} Q_{ki}(z)y_i \quad (k = 1, 2, \cdots, m),$$

where the Q's are rational functions, and let $\alpha\neq 0$ be an algebraic number distinct from the poles of these rational functions. Then the numbers $f_1(\alpha), \cdots, f_m(\alpha)$ are algebraically independent over the field of rational numbers if and only if the functions $f_1(z), \cdots, f_m(z)$ are algebraically independent over the field of rational functions of z. (II) Let $f_1(z), \cdots, f_m(z)$ be m E-functions satisfying the homogeneous differential equations

$$\text{(D)} \qquad y_k' = \sum_{i=1}^{m} Q_{ki}(z)y_i \quad (k = 1, 2, \cdots, m),$$

where the Q's are rational functions, and let $\alpha\neq 0$ be an algebraic number distinct from the poles of these rational functions. Then the numbers $f_1(\alpha), \cdots, f_m(\alpha)$ do not satisfy any homogeneous algebraic equation with rational integral coefficients if and only if the functions $f_1(z), \cdots, f_m(z)$ do not satisfy any homogeneous algebraic equations with coefficients that are polynomials in z.

The proofs of these important results are based on Siegel's method. The essentially new part consists in the following fundamental lemma of the author. (III) Let $f_1(z), \cdots, f_m(z)$ be m fixed integral functions that satisfy a system of differential equations (D) where the Q's are rational functions. For any positive integer n denote by $P_{11}(z), \cdots, P_{1m}(z)$ m polynomials, of degree at most $2n-1$, not all identically zero, such that in the power series for

$$R_1(z) = \sum_{i=1}^{m} P_{1i}(z)f_i(z) = \sum_{\nu=0}^{\infty} \alpha_\nu \frac{x^\nu}{\nu!}$$

all coefficients α_ν with $0\leq\nu\leq 2mn-n-1$ vanish. Further, denote by $T(z)$ the least common denominator of the Q's, and define further linear forms

$$R_k(z) = \sum_{i=1}^{m} P_{ki}(z)f_i(z) \quad (k = 2, 3, \cdots)$$

in $f_1(z), \cdots, f_m(z)$ with polynomial coefficients $P_{ki}(z)$ by

$$R_k(z) = T(z)\frac{d}{dz}R_{k-1}(z)$$

where $f_1'(z), \cdots, f_m'(z)$ are to be replaced by their linear expressions in $f_1(z), \cdots, f_m(z)$ from (D). There exist positive integers n_0, p, and q independent of n such that, for

$n \geqq n_0$, the determinant

$$\Delta(z) = |P_{ki}(z)|_{k,i=1,2,\cdots,m}$$

does not vanish identically and is of the form $\Delta(z) = z^{(2m-1)n-m-p-1}\Delta_1(z)$ where $\Delta_1(z) \not\equiv 0$ is a polynomial in z of degree not exceeding $n+p+\frac{1}{2}qm(m-1)-1$.

K. Mahler (Notre Dame, Ind.)

Citations: MR 11, 330c = J76-18.

Referred to in J02-22, J80-15, J80-45, J88-16, J88-17, J88-20, J88-21, J88-22, J88-25, J88-28, J88-31, J88-33, J88-36, J88-43, J88-49, J88-52.

J88-16 (21# 2641)

Yakabe, Iwao. Note on arithmetic properties of Kummer's function. Mem. Fac. Sci. Kyushu Univ. Ser. A **13** (1959), 49–52.

Let $J_{\kappa,\lambda}(x) = \exp(-x/2) \cdot {}_1F_1(\kappa; \lambda; x)$, where ${}_1F_1(\kappa; \lambda; x)$ denotes the confluent hypergeometric function

$$1 + \kappa x/(\lambda \cdot 1!) + \kappa(\kappa+1)x^2/(\lambda(\lambda+1)\cdot 2!) + \cdots.$$

Let κ and λ be rational numbers, such that κ and $\kappa - \lambda$ are not integers. The author applies Siegel's method of E-functions in order to prove that for any algebraic $\alpha \neq 0$ the numbers $J_{\kappa,\lambda}(\alpha)$ and $J_{\kappa,\lambda}'(\alpha)$ are algebraically independent over the field of algebraic numbers [cf. e.g., C. L. Siegel, *Transcendental numbers*, Princeton Univ. Press, 1949; MR **11**, 330; Chapter II]. It follows, e.g., that at least one of the two values ${}_1F_1(\kappa; \lambda; \alpha)$, ${}_1F_1'(\kappa; \lambda; \alpha)$ is transcendental. There is no reference to the results of Šidlovskiĭ on E-functions [cf. A. B. Šidlovskiĭ, Izv. Akad. Nauk SSSR. Ser. Mat. **23** (1959), 53–66; MR **21** #1295; and earlier publications]. *J. Popken* (Amsterdam)

Citations: MR 11, 330c = J76-18; MR 21# 1295 = J88-15.

Referred to in J88-25.

J88-17 (21# 6357)

Šidlovskiĭ, A. B. Transcendentality and algebraic independence of the values of certain functions. Trudy Moskov. Mat. Obšč. **8** (1959), 283–320. (Russian)

Recently [Izv. Akad. Nauk SSSR. Ser. Mat. **23** (1959), 35–66; MR **21** #1295] the author published a detailed proof of his important theorem on the algebraic independence of the values of E-functions which satisfy a system of differential equations

$$(1) \qquad y_k' = Q_{k0}(z) + \sum_{l=1}^{m} Q_{kl}(z) y_l \quad (k = 1, 2, \cdots, m)$$

with rational functions $Q_{kl}(z)$ as coefficients. He now applies this theorem to prove a number of new results on the transcendency of the values of certain generalised hypergeometric functions. The simplest of his theorems is as follows. Let λ_0 be a non-negative integer; let $\lambda_1, \cdots, \lambda_m$ be non-integral rational numbers such that their differences $\lambda_{j_1} - \lambda_{j_2}$ $(1 \leqq j_1 < j_2 \leqq m)$ are not integers; let $\beta_1, \cdots, \beta_n$ be algebraic numbers linearly independent over the rational field; let $\alpha_1, \cdots, \alpha_n$ be distinct algebraic numbers not zero; and let

$$\varphi_\lambda(z) = \sum_{k=0}^{\infty} \frac{z^k}{(\lambda+1)(\lambda+2)\cdots(\lambda+k)}.$$

Then the $(m+1)n$ numbers

$$\varphi_{\lambda_0}(\beta_i), \quad \varphi_{\lambda_j}(\alpha_i) \quad (j = 1, 2, \cdots, m; \quad i = 1, 2, \cdots, n)$$

are algebraically independent over the rational field. The proof consists of the following three steps. (i) The functions $\varphi_{\lambda_0}(\beta_i z)$ and $\varphi_{\lambda_j}(\alpha_i z)$ are E-functions. (ii) These functions satisfy a system of linear differential equations of the form (1). (iii) These functions are algebraically independent over the field of rational functions of z.

K. Mahler (Manchester)

Citations: MR 21# 1295 = J88-15.

Referred to in J88-18, J88-22.

J88-18 (27# 1418)

Šidlovskiĭ, A. B.

On transcendentality and algebraic independence of the values of some functions.

Amer. Math. Soc. Transl. (2) **27** (1963), 191–230.

The original Russian article appeared in Trudy Moscov. Mat. Obšč. **8** (1959), 283–320 [MR **21** #6357].

Citations: MR 21# 6357 = J88-17.

J88-19 (23# A1604)

Kneser, Hellmuth

Eine kontinuumsmächtige, algebraisch unabhängige Menge reeller Zahlen.

Bull. Soc. Math. Belg. **12** (1960), 23–27.

J. von Neumann [Math. Ann. **99** (1928), 134–141] constructed a set of real numbers with the power of the continuum and with the property that m arbitrarily chosen different numbers of this set are always algebraically independent with respect to the field of rational numbers. In the paper reviewed here the author gives a more general, and at the same time more simple, construction of such a set. As a special case, let

$$f(x) = \sum_{n=1}^{\infty} 2^{-[n^{n+x}]} \quad (0 \leqq x < 1);$$

if $x_1, x_2, \cdots, x_m$ are arbitrarily chosen numbers from the interval $[0, 1)$, then $f(x_1), f(x_2), \cdots, f(x_m)$ are algebraically independent. *J. Popken* (Amstelveen)

Referred to in J88-24.

J88-20 (24# A89)

Šidlovskiĭ, A. B.

On the transcendence and algebraic independence of values of the E-functions related to an algebraic equation over the rational function field. (Russian. English summary)

Vestnik Moskov. Univ. Ser. I Mat. Meh. **1960**, *no.* 5, 19–28.

The author gives further applications of his basic paper [Izv. Akad. Nauk SSSR Ser. Mat. **23** (1959), 35–66; MR **21** #1295]. His two main results are as follows: (1) Assume the $m(\geqq 2)$ E-functions $f_1(z), \cdots, f_m(z)$ satisfy a system of differential equations $y_k' = \sum_{i=1}^m Q_{ki} y_i$ $(k = 1, \cdots, m)$ with rational functions Q_{ki} as coefficients, but do not satisfy any homogeneous algebraic equation with polynomial coefficients. Then, for algebraic α, each number $f_i(\alpha)$ is transcendental if the corresponding function $f_i(z)$ is transcendental, except for at most finitely many algebraic α. (2) Assume the $m(\geqq 2)$ E-functions $f_1(z), \cdots, f_m(z)$ satisfy a system of differential equations $y_k' = Q_{k0} + \sum_{i=1}^m Q_{ki} y_i$ $(k = 1, \cdots, m)$ with rational functions Q_{ki} as coefficients. Let exactly $m-1$ of the functions $f_1(z), \cdots, f_m(z)$ be algebraically independent over the rational function field; thus these functions are connected by an algebraic equation $P(f_m(z), f_{m-1}(z), \cdots, f_1(z)) = 0$, where $P \neq 0$ is an irreducible polynomial in $f_m(z), f_{m-1}(z), \cdots, f_1(z)$ with coefficients that are polynomials in z. On ordering the terms of P lexicographically, let the highest term be free exactly of $f_l(z), f_{l-1}(z), \cdots, f_1(z)$ where $1 \leqq l \leqq m-1$. If the algebraic number $\alpha \neq 0$ is not a zero or pole of any Q_{ki} or a zero of the coefficient of the highest term of P, then $f_1(\alpha), \cdots, f_l(\alpha)$ are algebraically independent over the rationals. *K. Mahler* (Manchester)

Citations: MR 21# 1295 = J88-15.
Referred to in J88-22, J88-25.

J88-21 (24# A1885)

Šidlovskiĭ, A. B.

A generalization of Lindemann's theorem. (Russian)
Dokl. Akad. Nauk SSSR **138** (1961), 1301–1304.

From his basic theorem [Izv. Akad. Nauk SSSR Ser. Mat. **23** (1959), 35–66; MR **21** #1295] the author deduces the following generalisation of Lindemann's theorem. Let $P(z)$, $Q(z)$, $R(z)$ be polynomials, and let $f(z)$ be an E-function satisfying

$$\text{(I)} \qquad P(z)y' + Q(z)y = R(z).$$

If $\alpha_1, \cdots, \alpha_m$ are algebraic numbers, the values $f(\alpha_1), \cdots, f(\alpha_m)$ are algebraically independent over the rationals if either of the following conditions are satisfied: (A) $\alpha_1, \cdots, \alpha_m$ are linearly independent over the rationals; or (B) the differential equation (I) has no solution which is a rational function, and $\alpha_1, \cdots, \alpha_m$ are distinct from 0, from one another, and from the zeros of $P(z)$.
K. Mahler (Manchester)

Citations: MR 21# 1295 = J88-15.
Referred to in J88-22.

J88-22 (25# 49)

Šidlovskiĭ, A. B.

Transcendence and algebraic independence of the values of entire functions of certain classes. (Russian)
Moskov. Gos. Univ. Uč. Zap. No. 186 (1959), 11–70.

This important paper was submitted by the author in June, 1954, but appeared only in 1959. Hence several of its results have already been published elsewhere by the author; see, in particular, Trudy Moskov. Mat. Obšč. **8** (1959), 283–320 [MR **21** #6357] and for the basic theorem the detailed proof in Izv. Akad. Nauk SSSR Ser. Mat. **23** (1959), 35–66 [MR **21** #1295]. The paper contains a large number of applications of this basic theorem to the transcendency and algebraic independence of values at algebraic points of particular classes of E-functions satisfying linear differential equations with rational coefficients. For the proofs it is necessary to show the algebraic independence of the functions over the rational function field in z. This is done in full detail, and in a similar way as for the Bessel functions by C. L. Siegel in Abh. Preuss. Akad. Wiss. Phys.-Math. Kl. **1929**, 1–70. The main functions considered are of the form (λ, μ rational)

$$\sum_{n=0}^{\infty} \frac{z^n}{(\lambda+1)\cdots(\lambda+n)}, \quad \sum_{n=0}^{\infty} \frac{(-z^2/4)^n}{n!(\lambda+1)\cdots(\lambda+n)},$$

$$\sum_{n=0}^{\infty} \frac{(-z^2/4)^n}{(\lambda+1)\cdots(\lambda+n)(\mu+1)\cdots(\mu+n)},$$

as well as functions derived from these by the operators $e^z z^{-\lambda} \int^z \cdots t^{\lambda-1} e^{-t} dt$ and $z^{-\lambda} \int^z \cdots t^{\lambda-1} dt$. By way of example, if $\alpha \neq 0$ is algebraic, then $\int_0^z (e^t - 1) dt/t$ and e^α are algebraically independent. See also the author's papers in Vestnik Moskov. Univ. Ser. I Mat. Meh. **1960**, no. 5, 19–28 [MR **24** #A89] and Dokl. Akad. Nauk SSSR **138** (1961), 1301–1304 [MR **24** #A1885].
K. Mahler (Manchester)

Citations: MR 21# 1295 = J88-15; MR 21# 6357 = J88-17; MR 24# A89 = J88-20; MR 24# A1885 = J88-21.
Referred to in J88-23, J88-26, J88-42, J88-44.

J88-23 (25# 3010)

Šidlovskiĭ, A. B.

On the transcendence and algebraic independence of the values of certain E-functions. (Russian. English summary)
Vestnik Moskov. Univ. Ser. I Mat. Meh. **1961**, *no.* 5, 44–59.

Let

$$\omega_{s,\mu}(z) = \sum_0^\infty \frac{z^n}{n!(n+\mu)^s}, \quad \gamma_{s,\mu}(z) = \sum_0^\infty \frac{(-1)^n z^{2n}}{(2n)!(2n+\mu)^s},$$

$$\delta_{s,\mu}(z) = \sum_1^\infty \frac{(-1)^{n-1} z^{2n-1}}{(2n-1)!(2n-1+\mu)^s},$$

where μ is a fractional rational number or zero. Then, for algebraic numbers $\alpha \neq 0$, the $m+1$ numbers $\omega_{0\mu}(\alpha)$, $\omega_{1\mu}(\alpha), \cdots, \omega_{m\mu}(\alpha)$, the $m+1$ numbers $\gamma_{0\mu}(\alpha), \gamma_{1\mu}(\alpha), \cdots, \gamma_{m\mu}(\alpha)$, and the $m+1$ numbers $\delta_{0\mu}(\alpha), \delta_{1\mu}(\alpha), \cdots, \delta_{m\mu}(\alpha)$ are algebraically independent. In particular, $\sin \alpha$ and $\int_0^\alpha (\sin t)/t\, dt$, and $\cos \alpha$ and $\int_0^\alpha (1-\cos t)/t\, dt$ are algebraically independent. Similar and more general results hold also for the series in which $n!$ has been replaced by $(n!)^k$, where k is a positive integer. For the proofs compare the remarks in the author's paper in Moskov. Gos. Univ. Uč. Zap. No. 186 (1959), 11–70 [MR **25** #49].
K. Mahler (Canberra)

Citations: MR 25# 49 = J88-22.
Referred to in J88-33.

J88-24 (25# 3903)

Schmidt, Wolfgang M.

Simultaneous approximation and algebraic independence of numbers.
Bull. Amer. Math. Soc. **68** (1962), 475–478.

Let $\|\xi\|$ denote the distance of a real number ξ to the nearest integer. The author proves the following theorem: Let $\xi_1, \cdots, \xi_n$ be an n-tuple of reals such that for every $d > 0$ there exists an integer q such that

$$0 < \|\xi_k q\| < (\|\xi_1 q\| \cdots \|\xi_{k-1} q\|)^d q^{1-nd} \quad (k = 1, \cdots, n).$$

Then $\xi_1, \cdots, \xi_n$ are algebraically independent over the rationals. Clearly this is a generalization of a well-known result of Liouville. As a corollary he obtains that the set of real numbers $\xi(x) = \sum_{t=1}^\infty 2^{-[t^{t+x}]}$, for $0 \leq x < 1$, consists of algebraically independent numbers, a result due to H. Kneser [Bull. Soc. Math. Belg. **12** (1960), 23–27; MR **23** #A1604] (cf. also a paper of F. Kuiper and the reviewer [Nederl. Akad. Wetensch. Proc. Ser. A **65** (1962), 385–390]). However the method applied by the author is quite different. In this way he gives a new contribution to von Neumann's problem to give effectively a set of algebraically independent numbers with the power of the continuum.

The author derives the above result from a still more general theorem. *J. Popken* (Berkeley, Calif.)

Citations: MR 23# A1604 = J88-19; MR 26# 2422 = J88-27.
Referred to in J68-44.

J88-25 (26# 1285)

Oleĭnikov, V. A.

The transcendence and algebraic independence of the values of certain E-functions. (Russian. English summary)
Vestnik Moskov. Univ. Ser. I Mat. Meh. **1962**, *no.* 6, 34–38.

Let

$$K_{\alpha,\beta}(z) = \sum_{n=0}^\infty \frac{\beta(\beta+1)\cdots(\beta+n-1)}{n!\alpha(\alpha+1)\cdots(\alpha+n-1)} z^n, \quad \alpha, \beta \neq 0, -1, -2.$$

The author proves the following theorem. If α, β are rationals, $\beta\neq 1, 2, \cdots$ and $\beta-\alpha\neq 0, 1, 2, \cdots$, then the numbers $K_{\alpha,\beta}(z)$ and $K_{\alpha,\beta}'(z)$ are algebraically independent for every algebraic value of $z\neq 0$. This is an improvement of the result of Yakabe [Mem. Fac. Sci. Kyushu Univ. Ser. A **13** (1959), 49–52; MR **21** #2641]. The proof is based on certain theorems of Šidlovskiĭ [Izv. Akad. Nauk SSSR Ser. Mat. **23** (1959), 35–66; MR **21** #1295; Vestnik Moskov. Univ. Ser. I Mat. Meh. **1960**, no. 5, 19–28; MR **24** #A89].
A. Schinzel (Warsaw)

Citations: MR 21# 1295 = J88-15; MR 21# 2641 = J88-16; MR 24# A89 = J88-20.

J88-26 (26# 2395)

Šidlovskiĭ, A. B.

Transcendency and algebraic independence of values of E-functions related by an arbitrary number of algebraic equations over the field of rational functions. (Russian)

Izv. Akad. Nauk SSSR Ser. Mat. **26** (1962), 877–910.

The author continues with his general theory of Siegel E-functions and their transcendency [Moskov. Gos. Univ. Uč. Zap. **186** (1959), 11–70; MR **25** #49]. The present paper contains detailed proofs of several theorems announced and applied by him already in a number of earlier papers. His most striking results are as follows. (1) Let $f_1(z), \cdots, f_m(z)$ be E-functions that satisfy a system of differential equations

$$y_k' = \sum_{i=1}^{m} Q_{ki}(z)y_i+Q_{k0}(z) \qquad (k = 1, 2, \cdots, m),$$

where the Q's are rational functions. Let $\alpha\neq 0$ be an algebraic number distinct from the poles of all the Q's. Then, for $1\leqq l\leqq m$, the values $f_1(\alpha), \cdots, f_l(\alpha)$ are algebraically independent over the field of rational numbers if and only if the functions $f_1(z), \cdots, f_l(z)$ are algebraically independent over the field of rational functions of z. (2) Let $f_1(z), \cdots, f_m(z)$ be E-functions that satisfy a homogeneous system of differential equations

$$y_k' = \sum_{i=1}^{m} Q_{ki}(z)y_i \qquad (k = 1, 2, \cdots, m),$$

where again the Q's are rational functions. Let α be as in (1). Then the values $f_1(\alpha), \cdots, f_l(\alpha)$, where $1\leqq l\leqq m$, are not connected by a homogeneous algebraic equation with rational coefficients if and only if the functions $f_1(z), \cdots, f_l(z)$ are not connected by a homogeneous algebraic equation with coefficients that are rational functions of z.
K. Mahler (Manchester)

Citations: MR 25# 49 = J88-22.

Referred to in J88-31, J88-34, J88-37, J88-42, J88-51.

J88-27 (26# 2422)

Kuiper, F.; Popken, J.

On the so-called von Neumann numbers.

Nederl. Akad. Wetensch. Proc. Ser. A **65** = *Indag. Math.* **24** (1962), 385–390.

In this paper, the authors generalize the construction of von Neumann numbers. The problem is to construct a set of real numbers with the power of the continuum with the property that any n members of the set are algebraically independent over the rationals.

Let F be a field with valuation $|\ |$ which is complete in $|\ |$. Let R be the ring of elements of F with the property that $a\in R$ if and only if $|a|\geqq 1$ for any $a\neq 0$. Theorem: Let F be of characteristic 0, and suppose that R contains the identity of F. Let

$$A_\nu = \sum_{m=0}^{\infty} \frac{a_{\nu m}}{\beta_0\beta_1\cdots\beta_m} \qquad (\nu = 1, 2, \cdots, m),$$

where $a_{\nu m}$, β_m denote non-zero elements of R with $|\beta_m|\rightarrow\infty$, as $m\rightarrow\infty$. Suppose, moreover, that as $m\rightarrow\infty$ $|a_{1m}|=o(|a_{2m}|)$, $|a_{2m}|=o(|a_{3m}|)$, $\cdots$, $|a_{n-1,m}|=o(a_{n,m})$, $|a_{n,m}|=O(|b_{m-1}|)$ where $b_m=\beta_0\beta_1\cdots\beta_m$, $\log|b_{m-1}|=o(\log|b_m|)$. Then $A_1, A_2, \cdots, A_n$ are algebraically independent over R.
R. Ayoub (University Park, Pa.)

Referred to in J88-24.

J88-28 (33# 7302)

Šidlovskiĭ, A. B.

Transcendence and algebraic independence of values of E-functions satisfying linear nonhomogeneous differential equations of the second order. (Russian)

Dokl. Akad. Nauk SSSR **169** (1966), 42–45.

Let $y_k'=A_{k0}+\sum_{i=1}^m A_{ki}y_i$ $(k=1, \cdots, m;\ m\geqq 2)$ be a system of differential equations with coefficients A_{ki} in a field L of analytic functions of z which is closed under differentiation. Let $y_1^0, \cdots, y_m^0$ be a solution of this system such that $y_1^0, \cdots, y_{m-1}^0$ are algebraically independent over L, but such that $P(y_1^0, \cdots, y_m^0)=0$, where $P\not\equiv 0$ is a polynomial in $L[x_1, \cdots, x_m]$. Let Q be the highest homogeneous part of P. Then the homogeneous system of differential equations $y_k'=\sum_{i=1}^m A_{ki}y_i$ $(k=1, \cdots, m)$ has a solution $y_1^*, \cdots, y_m^*$ satisfying $Q(y_1^*, \cdots, y_m^*)=0$. An analogous lemma holds for the solutions of differential equation $y^{(m)}+A_{m-1}y^{(m-1)}+\cdots+A_0y=B$ $(m\geqq 2)$ with coefficients in L. From these lemmas and from his general theorem [Izv. Akad. Nauk SSSR Ser. Mat. **23** (1959), 35–66; MR **21** #1295], the author deduces the following result. Let λ, μ, ν be rational numbers which are not negative integers and for which $\nu-\lambda$ and $\nu-\mu$ are not non-negative integers; further, let $\alpha\neq 0$ be an algebraic number. If

$$A_{\lambda\mu\nu}(z) = \sum_{n=0}^{\infty} \frac{(\nu+1)\cdots(\nu+n)}{(\lambda+1)\cdots(\lambda+n)(\mu+1)\cdots(u+n)}z^n,$$

then the two numbers $A_{\lambda\mu\nu}(\alpha)$ and $A_{\lambda\mu\nu}'(\alpha)$ are algebraically independent.

{This article has appeared in English translation [Soviet Math. Dokl. **7** (1966), 879–883].}

See also #7303 below. *K. Mahler* (Canberra)

Citations: MR 21# 1295 = J88-15.

Referred to in J88-44.

J88-29 (33# 7303)

Oleĭnikov, V. A.

Algebraic independence of values of E-functions satisfying linear nonhomogeneous differential equations of the third order. (Russian)

Dokl. Akad. Nauk SSSR **169** (1966), 32–34.

Let $\beta_1, \beta_2, \beta_3$ be rational numbers distinct from -1, $-2, \cdots$, and α_1, α_2 rational numbers distinct from $0, -1$, $-2, \cdots$; further, let $\gamma\neq 0$ be an algebraic number. Put $(\beta, n)=(\beta+1)\cdots(\beta+n)$, $(\beta_1, \beta_2, \beta_3, n)=(\beta_1, n)(\beta_2, n)(\beta_3, n)$, and $K_{\beta_1\beta_2\beta_3}(z)=\sum_{n=0}^\infty (z/3)^{3n}/(\beta_1, \beta_2, \beta_3, n)$; $K_{\beta_1\beta_2\beta_3\alpha_1}(z)=\sum_{n=0}^\infty(\alpha_1-1, n)(z/2)^{2n}/(\beta_1, \beta_2, \beta_3, n)$, $K_{\beta_1\beta_2\beta_3\alpha_1\alpha_2}(z)=\sum_{n=0}^\infty(\alpha_1-1, n)(\alpha_2-1, n)z^n/(\beta_1, \beta_2, \beta_3, n)$. From the lemma of the paper by Šidlovskiĭ reviewed above [#7302], the author deduces the following three results.

$K_{\beta_1\beta_2\beta_3}(\gamma)$, $K_{\beta_1\beta_2\beta_3}'(\gamma)$, and $K_{\beta_1\beta_2\beta_3}''(\gamma)$ are algebraically independent if none of $|2\beta_1-\beta_2-\beta_3|$, $|2\beta_2-\beta_1-\beta_3|$, $|2\beta_3-\beta_1-\beta_2|$ is a positive integer.

$K_{\beta_1\beta_2\beta_3\alpha_1}(\gamma)$, $K_{\beta_1\beta_2\beta_3\alpha_1}'(\gamma)$, and $K_{\beta_1\beta_2\beta_3\alpha_1}''(\gamma)$ are algebraically independent if simultaneously none of $2(\alpha_1-\beta_i)$, where $i=1, 2, 3$, and none of $\alpha_1+\beta_i-\beta_j-\beta_k+\frac{1}{2}$, where i, j, k are distinct suffixes 1, 2, 3, is an integer distinct from $+1$.

$K_{\beta_1\beta_2\beta_3\alpha_1\alpha_2}(\gamma)$, $K_{\beta_1\beta_2\beta_3\alpha_1\alpha_2}'(\gamma)$, and $K_{\beta_1\beta_2\beta_3\alpha_1\alpha_2}''(\gamma)$ are algebraically independent if simultaneously none of the differences $\alpha_i-\beta_j$ is an integer, and none of the differences $\alpha_1+\alpha_2-\beta_i-\beta_j$, where i and j are different suffixes 1, 2, 3, is an integer distinct from $+1$.

{This article has appeared in English translation [Soviet Math. Dokl. **7** (1966), 869–871].} *K. Mahler* (Canberra)

Referred to in J88-36.

J88-30 (34# 4209)

Šidlovskiĭ, A. B.

On a general theorem on the algebraic independence of values of E-functions. (Russian)

Dokl. Akad. Nauk SSSR **171** (1966), 810–813.

The author sketches a variation of Siegel's approximation method in the theory of the transcendency of E-functions [C. Siegel, *Transcendental numbers*, Ann. of Math. Studies, No. 16, Princeton Univ. Press, Princeton, N.J., 1949; MR **11**, 330] which allows an improved lower estimate for the rank of the function values. In this way he obtains, in particular, the following result. Let $f_1(z), \cdots, f_m(z)$ be E-functions with coefficients in an imaginary quadratic field K satisfying a system of differential equations $y_k' = Q_{k,0} + \sum_{i=0}^m Q_{k,i} y_i$ ($k = 1, 2, \cdots, m$), where the Q's are rational functions, say with the least common denominator $T(z)$. Let α be any number in K such that $\alpha T(\alpha) \neq 0$. Then $f_1(\alpha), \cdots, f_m(\alpha)$, 1 are linearly independent over K if and only if the functions $f_1(z), \cdots, f_m(z)$, 1 are linearly independent over the field of rational functions of z. This result implies an analogous one for algebraic relations of higher degree in the functions, and the function values, respectively.

{This article has appeared in English translation [Soviet Math. Dokl. **7** (1966), 1569–1572].}

K. Mahler (Canberra)

Citations: MR 11, 330c = J76-18.

Referred to in J80-23.

J88-31 (35# 1555)

Belogrivov, I. I.

Transcendentality and algebraic independence of values of certain E-functions. (Russian. English summary)

Vestnik Moskov. Univ. Ser. I Mat. Meh. **22** (1967), *no.* 2, 55–62.

Let $\lambda \neq -1, -2, -3, \cdots$ be a rational number, and let

$$K_\lambda(z) = \sum_{n=0}^{\infty} \frac{(-z^2/4)^n}{n!(\lambda+1)(\lambda+2)\cdots(\lambda+n)},$$

$$U_\lambda(z) = \sum_{n=0}^{\infty} \{(\lambda+1)^{-1} + (\lambda+2)^{-1} + \cdots + (\lambda+n)^{-1}\} \times \frac{(-z^2/4)^n}{n!(\lambda+1)(\lambda+2)\cdots(\lambda+n)}.$$

Thus $K_0(z) = J_0(z)$. The author applies the deep results by A. B. Šidlovskiĭ [Izv. Akad. Nauk SSSR Ser. Mat. **23** (1959), 35–66; MR **21** #1295; ibid. **26** (1962), 877–910; MR **26** #2395] to establish the following important results. (I) Let $\alpha_1, \cdots, \alpha_n$ be algebraic numbers $\neq 0$ such that $\alpha_h^2 \neq \alpha_k^2$ if $h \neq k$. Then any three of the five sets of numbers $\{J_0(\alpha_h)\}$, $\{J'(\alpha_h)\}$, $\{U_0(\alpha_h)\}$, $\{U_0'(\alpha_h)\}$, $\{U_0''(\alpha_h)\}$, where $h = 1, 2, \cdots, n$, consist of algebraically independent numbers. (II) Let $\alpha \neq 0$ be an algebraic number, and let λ be a rational number such that 2λ is not an integer. Then any four of the six numbers $K_\lambda(\alpha)$, $K_\lambda'(\alpha)$, $U_\lambda(\alpha)$, $U_\lambda'(\alpha)$, $U_\lambda''(\alpha)$, $U_\lambda'''(\alpha)$ are algebraically independent. Analogous results are established with $K_\lambda(z)$ and $U_\lambda(z)$ replaced by the functions

$$\varphi_\lambda(z) = \sum_{n=0}^{\infty} \frac{z^n}{(\lambda+1)(\lambda+2)\cdots(\lambda+n)},$$

$$V_\lambda(z) = \sum_{n=0}^{\infty} \{(\lambda+1)^{-1} + (\lambda+2)^{-1} + \cdots + (\lambda+n)^{-1}\} \times \frac{z^n}{(\lambda+1)(\lambda+2)\cdots(\lambda+n)}.$$

K. Mahler (Canberra)

Citations: MR 21# 1295 = J88-15; MR 26# 2395 = J88-26.

J88-32 (35# 2739)

Melzak, Z. A.

An informal arithmetical approach to computability and computation. III.

Canad. Math. Bull. **9** (1966), 593–609.

The author continues his study of what he has called Q-machines. [For Parts I and II, see same Bull. **4** (1961), 279–293; MR **27** #1364; ibid. **7** (1964), 183–200; MR **31** #4727.]

In Part II, the author defined a machine which he called $Q_5{}^0$. He now introduces the notion of a $\bar{Q}_5{}^0$ computable real number x. Roughly, x must be a limit of a sequence p_n/q_n of rational numbers which are obtained in a prescribed way from a prescribed type of (improper) $Q_5{}^0$ program. The span of such a program is the number of sequences appearing in its description. The span of a real irrational $\bar{Q}_5{}^0$ computable number is the smallest span of a program which computes it. Among the results proved are the following. (1) An irrational number is algebraic if and only if it is computable by a program of span 2. (2) Let a_1 and a_2 be two real irrational $\bar{Q}_5{}^0$ computable numbers of spans s_1 and s_2, respectively. If $|s_1 - s_2| < 2$, then a_1 and a_2 are algebraically independent.

The author turns to a consideration of the machine $Q_2{}^0$, also defined in Part II. He proceeds to arithmetize the description of a program for this machine and defines a Gödel number for each such program. He then outlines proofs that (1) there is a program which will recognize whether or not a nonnegative integer is the Gödel number of a program, and (2) there is no program which can determine whether or not a nonnegative integer is the Gödel number of a program which terminates when started with empty registers (unsolvability of the halting problem).

As stated in the title, some of the considerations are presented informally. *J. A. Schatz* (Albuquerque, N.M.)

J88-33 (35# 5398)

Belogrivov, I. I.

The transcendentality and algebraic independence of the values of certain hypergeometric E-functions. (Russian)

Dokl. Akad. Nauk SSSR **174** (1967), 267–270.

Put $[\lambda, 0] = 0$, $[\lambda, n] = \lambda(\lambda+1)\cdots(\lambda+n-1)$ and define $A_{m,s}(z) = \sum_{n=0}^{\infty} \prod_{i=1}^{s} [\lambda_i + 1, n]^{-m_i} (z/m)^{nm}$,

$$A_{m,s,\mu}(z) = 1 + \sum_{n=1}^{\infty} \prod_{j=1}^{s} [\lambda_j + 1, n]^{-m_j} \times \prod_{j=i}^{s} (\lambda_j + n)^{-1} (\lambda_i + n)^{q_i - 1 - \mu} (z/m)^{nm}$$

(where the parameters are suitably defined). The author states several theorems concerning the algebraic independence of the m numbers $A_{m,s}(\alpha), A'_{m,s}(\alpha), \cdots, A_{m,s}^{(m-1)}(\alpha)$ and of the l numbers $A_{m,s,\mu}(\alpha)$, $1 \leqq \mu \leqq l$, where α is an arbitrary non-zero algebraic number and $l \geqq m = m_1 + \cdots + m_s$.

These results are said to depend on the main theorem of A. B. Šidlovskiĭ [Izv. Akad. Nauk SSSR Ser. Mat. **23** (1959), 35–66; MR **21** #1295] and the author's generalization, also stated, of a lemma of Šidlovskiĭ [Vestnik Moskov. Univ. Ser. I Mat. Meh. **1961**, no. 5, 44–59; MR **25** #3010]. They contain as special cases some previous results due to Šidlovskiĭ, Siegel and Oleĭnikov. No proofs are given.

{This article has appeared in English translation [Soviet Math. Dokl. **8** (1967), 610–613].}

J. B. Roberts (Portland, Ore.)

Citations: MR 21# 1295 = J88-15; MR 25# 3010 = J88-23.

J88-34 (36# 2561)

Maler, K. [Mahler, K.]

A lemma of A. B. Šidlovskiĭ. (Russian)

Mat. Zametki **2** (1967), 25–32.

In den Untersuchungen von A. V. Šidlovskiĭ [vgl. Izv. Akad. Nauk SSSR Ser. Mat. **26** (1962), 877–910; MR **26** #2395] über die algebraische Unabhängigkeit von Werten Siegelscher E-Funktionen spielt folgendes Lemma eine wesentliche Rolle: Sei R_0 der Körper der rationalen, bei $z=0$ regulären komplexen Funktionen, Q eine n-reihige Matrix über R_0, $w=(w_1, \cdots, w_n)$ ein Lösungsvektor des Systems (*) $w'=Qw$ von Differentialgleichungen. Dabei seien die w_h algebraische Funktionen, die bei $z=0$ höchstens einen Pol haben. Wenn dann $\mu<m$ die Maximalzahl linear unabhängiger Funktionen w_h ist, gibt es eine Untermenge $(w_{i_1}, \cdots, w_{i_\mu})$ von w, die ein zu (*) analoges System von Gleichungen (mit μ statt n) erfüllt.

Für dieses Lemma gibt der Verfasser einen neuen, einfacheren Beweis, der im wesentlichen Hilfsmittel aus der linearen Algebra verwendet. *B. Volkmann* (Stuttgart)

Citations: MR 26# 2395 = J88-26.

J88-35 (36# 2562)

Šidlovskiĭ, A. B.

Transcendence of the values of E-functions. (Russian)

Proc. Fourth All-Union Math. Congr. (Leningrad, 1961) *(Russian), Vol. II, pp.* 147–158. *Izdat. "Nauka", Leningrad,* **1964.**

Author's summary: "This is a survey article on the development of the well-known Hermite-Lindemann-Siegel method in the theory of transcendental numbers, which permits one to establish the transcendentality and algebraic independence of the values of the E-functions for algebraic values of the independent variable when the E-functions are the solutions of differential equations with polynomial coefficients." (RŽMat **1965** #2 A180)

J88-36 (36# 5085)

Oleĭnikov, V. A.

The transcendence and algebraic independence of the values of certain entire functions. (Russian)

Izv. Akad. Nauk SSSR Ser. Mat. **32** (1968), 63–92.

Let $P(z, y, y', \cdots, y^{(n)})$ be a polynomial in $y, y', \cdots, y^{(n)}$ over the field of rational functions in z. The author calls the equation $P(z, y', \cdots, y^{(n)})=0$ differentially irreducible if no analytic solution $y(z)\not\equiv 0$ of this equation satisfies a similar equation of order less than n. He proves a criterion of differential irreducibility and applies it to the study of algebraic independence of E functions satisfying linear differential equations of order 3 with polynomial coefficients. In virtue of the fundamental theorem of A. B. Šidlovskiĭ [same Izv. **23** (1959), 35–66; MR **21** #1295], this leads to three theorems concerning algebraic independence of the values of hypergeometric E-functions, announced in an earlier paper of the author [Dokl. Akad. Nauk SSSR **169** (1966), 32–34; MR **33** #7303] and quoted in its review. *A. Schinzel* (Warsaw)

Citations: MR 21# 1295 = J88-15; MR 33# 7303 = J88-29.

J88-37 (37# 1322)

Mahler, K.

Applications of a theorem by A. B. Shidlovski.

Proc. Roy. Soc. Ser. A **305** (1968), 149–173.

Author's summary: "Šidlovskiĭ's deep theorem on Siegel E-functions satisfying systems of linear differential equations is applied in this paper to the study of the arithmetical properties of the partial derivatives $C_k(z)=(1/k!)\,\partial^k J_\nu(z)/\partial\nu^k|_{\nu=0}$ of the Bessel function $J_0(z)$. As a by-product, expressions involving Euler's constant γ and the constant $\zeta(3)$ are obtained for which transcendency is established."

Šidlovskiĭ's theorem [A. B. Šidlovskiĭ, Izv. Akad. Nauk SSSR Ser. Mat. **26** (1962), 877–910; MR **26** #2395] (for an account of the theorem in English, see S. Lang [*Introduction to transcendental numbers*, Addison-Wesley, Reading, Mass., 1966; MR **35** #5397]) states the following. Let $\mathbf{C}(z)$ denote the rational functions with coefficients in the complex numbers $\mathbf{C}$. Let $w_1=f_1(z), \cdots, w_m=f_m(z)$ denote E-functions (see C. L. Siegel [*Transcendental numbers*, Ann. of Math. Studies, No. 16, Princeton Univ. Press, Princeton, N.J., 1949; MR **11**, 330]) satisfying a system of differential equations $w_h'=q_{h0}+\sum_{k=1}^m q_{hk}w_k$ $(h=1, \cdots, m)$, where $q_{hk}\in\mathbf{C}(z)$. Let α be a non-zero algebraic number which is not a pole of any of the q_{hk}. Then the transcendence degree of $f_1(z), \cdots, f_m(z)$ over $\mathbf{C}(z)$ equals the transcendence degree of $f_1(\alpha), \cdots, f_m(\alpha)$ over the rational numbers $\mathbf{Q}$.

The author's purpose is to verify the hypothesis of Šidlovskiĭ's theorem in certain special cases. Let $J_\nu(z)$ denote the Bessel function of the first kind. Define $K_\nu(z)$ by the formula $J_\nu(z)=(\frac{1}{2}z)^\nu(\Gamma(\nu+1))^{-1}K_\nu(z)$. Define $A_k(z)$ in the same way $C_k(z)$ was defined above with K_ν replacing J_ν. Then the author shows, for example, that $A_0, A_1, A_2, A_3, A_0', A_2'$ are algebraically independent over $\mathbf{C}(z)$. From this, then, he deduces that if α is a non-zero algebraic number, $A_0(\alpha), A_1(\alpha), A_2(\alpha), A_3(\alpha), A_0'(\alpha), A_2'(\alpha)$ are algebraically independent over $\mathbf{Q}$. In particular, they are all transcendental. He gives this last result also for three other sets of six numbers involving the A's.

Now the author turns to the $C_k(z)$. Note that $C_0(z)=J_0(z)$ and $C_1(z)=Y_0(z)$, a Bessel function of the second kind. The author writes the A's as linear combinations of the C's with explicit coefficients which involve in particular γ and $\zeta(3)$. The above result then gives the algebraic independence of certain expressions involving the C_k. For example, if α is a non-zero algebraic number, then $(\pi Y_0(\alpha)/2J_0(\alpha))-\gamma-\log(\frac{1}{2}\alpha)$ is transcendental. In particular, $(\pi Y_0(2)/2J_0(2))-\gamma$ is transcendental. The expressions involving $\zeta(3)$ are more complicated. It must be noted that these results give no information about γ and $\zeta(3)$ themselves. However, this is the first time any results of an arithmetical nature have been found that involve these constants non-trivially. The paper ends with a discussion of possible generalizations of the work.

W. W. Adams (Berkeley, Calif.)

Citations: MR 11, 330c = J76-18; MR 26# 2395 = J88-26; MR 35# 5397 = J02-22.

J88-38 (38# 3247)

Brown, W. S.

Rational exponential expressions and a conjecture concerning π and e.

Amer. Math. Monthly **76** (1969), 28–34.

The rational exponential (REX) expressions are defined to be those which are obtained by addition, subtraction, multiplication, division and substitution, starting with the integers, the constants π and i, indeterminates and the exponential function. Using a natural definition of simplification for REX expressions, the author shows that the decision problem for the equivalence of simplified REX expressions is soluble, and provides a simplification algorithm, on the assumption that the following hypothesis holds: If $p_1, p_2, \cdots, p_n$ are non-zero REX expressions such that the set $\{p_1, p_2, \cdots, p_n, i\pi\}$ is linearly independent over the rationals, then the set $\{\exp p_1, \exp p_2, \cdots, \exp p_n, z, \pi\}$ is algebraically independent over the rationals.

This hypothesis implies, for instance, that e and π are algebraically independent, yet it is not known even that $e+\pi$ is irrational. *R. L. Goodstein* (Leicester)

J88-39 **(38# 4418)**

Oleĭnikov, V. A.

The algebraic independence of the values of E-functions. (Russian)

Mat. Sb. (*N.S.*) **78** (**120**) (1969), 301–306.

The author defines the E-function

$$(1)\quad f(z) = \sum_{n=0}^{\infty} \frac{1}{n!(\lambda+1)\cdots(\lambda+n)(\mu+1)\cdots(\mu+n)}\left(\frac{z}{3}\right)^{3n}$$

with rational $\lambda, \mu \neq -1, -2, -3, \cdots$; it satisfies the linear differential equation of the third order (2) $y'''+3(\lambda+\mu+1)y''/z+(3\lambda+1)(3\mu+1)y'/z^2-y=0$. We quote the author's first theorem (Theorem 2 of the paper). The functions $f(z), f'(z), f''(z)$ are algebraically dependent over the field of rational functions of z if and only if we have simultaneously $\lambda=\pm\frac{1}{3}+n_1$; $\mu=\mp\frac{1}{3}+n_2$, where n_1 and n_2 are arbitrary rational integers.

S. Chowla (University Park, Pa.)

J88-40 **(38# 4419)**

Šidlovskiĭ, A. B.

Algebraic independence of the values of certain hypergeometric E-functions. (Russian)

Trudy Moskov. Mat. Obšč. **18** (1968), 55–64.

The author defines

$$A_{\lambda,k}(z) = \textstyle\sum_{n=0}^{\infty}[(\lambda+1)\cdots(\lambda+n)]^{-k}(z/k)^{kn},$$

$$A_{\lambda,k,s}(z) = 1+\textstyle\sum_{n=1}^{\infty}[(\lambda+1)\cdots(\lambda+n-1)]^{-k}(\lambda+n)^{-s}(z/k)^{kn}$$

($\lambda\neq -1, -2, \cdots$; $k\geqq 1$, $s=1, 2, \cdots, m_k$, $m_k\geqq k$). The author proves three theorems, of which we quote the first: Let k and m be natural numbers, $m\geqq k$, λ a rational number ($\lambda\neq -1, -2, \cdots$), $\alpha\neq 0$ an arbitrary algebraic number. Then (1) the numbers $A_{\lambda,k,s}(\alpha)$ ($s=1, 2, \cdots, m$) are algebraically independent, and (2) the numbers $A_{\lambda,k}(\alpha)$, $A_{\lambda,k}'(\alpha)$, $\cdots$, $A_{\lambda,k}^{(k-1)}(\alpha)$ are algebraically independent.

S. Chowla (University Park, Pa.)

J88-41 **(38# 5722)**

Šmelev, A. A.

The algebraic independence of certain numbers. (Russian)

Mat. Zametki **4** (1968), 525–532.

Let Q be the rational number field, ω a real or complex transcendental number, Q_0 the extension field $Q_0=Q(\omega)$, and Q_1 a simple algebraic extension of Q_0. The author proves the following theorem: Let the four numbers $1, \alpha_1, \alpha_2, \alpha_3$ and the two numbers η_1, η_2 both be linearly independent over Q, and let τ and $x_0=x_0(\alpha_1, \alpha_2, \alpha_3, \eta_1, \eta_2)$ be positive constants as follows. For every $x>x_0$, $|x_1+x_2(\eta_2/\eta_1)|>\exp(-\tau x^2\log x)$ for all pairs of integers x_1, x_2 not both zero satisfying $|x_1|+|x_2|\leqq x$. Then the field obtained by adjoining the 11 numbers $\alpha_i, \eta_j, e^{\alpha_i\eta_j}$ ($i=1, 2, 3$; $j=1, 2$) to Q is not a field Q_1. The proof is based on ideas by A. O. Gel′fond [*Transcendental and algebraic numbers* (Russian), GITTL, Moscow, 1952; MR **15**, 292; English translation, Dover, New York, 1960; MR **22** #2598]. The theorem implies that if $\alpha_1, \alpha_2, \alpha_3, \beta_1, \beta_2$ are algebraic numbers such that neither $\log\alpha_1, \log\alpha_2, \log\alpha_3$, nor $1, \beta_1, \beta_2$ are linearly dependent over Q, at least two of the six numbers $\alpha_i^{\beta_j}$ ($i=1, 2, 3$; $j=1, 2$) are algebraically independent over Q.

K. Mahler (Columbus, Ohio)

Citations: MR 15, 292e = J02-6; MR 22# 2598 = J02-7.

J88-42 **(39# 5481)**

Šidlovskiĭ, A. B.

Transcendentality and algebraic independence of values of E-functions. (Russian)

Proc. Internat. Congr. Math. (*Moscow*, 1966), *pp.* 299–307. *Izdat. "Mir", Moscow*, 1968.

In this expository lecture the author describes his important work on the transcendency of E-functions satisfying systems of linear differential equations, and he reports on the related work by his students (see, in particular, his papers [Moskov. Gos. Univ. Učen. Zap. No. 186 (1959), 11–70; MR **25** #49; Izv. Akad. Nauk SSSR Ser. Mat. **26** (1962), 877–910; MR **26** #2395]).

{An English version has appeared in Amer. Math. Soc. Transl. Ser. 2, Vol. 70, Providence, R.I., 1968 [see MR **37** #1213].}

K. Mahler (Columbus, Ohio)

Citations: MR 25# 49 = J88-22; MR 26# 2395 = J88-26.

J88-43 **(40# 93)**

Nesterenko, Ju. V.

The algebraic independence of the values of E-functions which satisfy linear inhomogeneous differential equations. (Russian)

Mat. Zametki **5** (1969), 587–598.

Let F be a field of analytic functions of z; let (Q_i) $y_k'=q_{k0}^{(i)}+\sum_{j=1}^{m_i} q_{kj}^{(i)}y_j$ ($k=1, 2, \cdots, m_i$; $i=1, 2, \cdots, r$) be a set of r inhomogeneous systems of linear differential equations with coefficients $q_{kj}^{(i)}$ in F, and let (Q_i^*) $y_k'=\sum_{j=1}^{m_i} q_{kj}^{(i)}y_j$ ($k=1, 2, \cdots, m_i$; $i=1, 2, \cdots, r$) be the corresponding set of homogeneous systems. Assume that the $\sum_{i=1}^{r} m_i$ functions of every non-zero solution of the systems Q_i^* are algebraically independent over F. If (1) $f_{i1}, \cdots, f_{im_i}$ ($i=1, 2, \cdots, r$) is a solution of the systems Q_i where, for no i, all of $f_{i1}, \cdots, f_{im_i}$ lie in F, then the $\sum_{i=1}^{r} m_i$ functions (1) are algebraically independent over F. On combining this theorem with results of A. B. Šidlovskiĭ [Izv. Akad. Nauk SSSR Ser. Mat. **23** (1959), 35–66; MR **21** #1295], shorter proofs of the algebraic independence of the values at algebraic points of solutions of special linear differential equations can be deduced.

K. Mahler (Columbus, Ohio)

Citations: MR 21# 1295 = J88-15.

J88-44 **(40# 96)**

Šmelev, A. A.

Algebraic independence of values of certain E-functions. (Russian)

Izv. Vysš. Učebn. Zaved. Matematika **1969**, *no.* 4 (83), 103–112.

A. B. Šidlovskiĭ [Moskov. Gos. Univ. Učen. Zap. No. 186 (1959), 11–70; MR **25** #49] proved the following generalization of an earlier theorem of C. L. Siegel. Let α be a non-zero algebraic number, k an integer, λ and μ rational numbers, other than negative integers, such that $\lambda-\mu\neq(2k+1)/2$, and

$$K_{\lambda\mu}(z) = \textstyle\sum_{n=0}^{\infty}((-1)^n/((\lambda+1)\cdots(\lambda+u)(\mu+1)\cdots(\mu+n)))(z/2)^{2n}.$$

Then the numbers $K_{\lambda\mu}(\alpha)$ and $K_{\lambda\mu}'(\alpha)$ are algebraically independent over the rationals. In the present paper the following further generalization is proved along the lines of Šidlovskiĭ's paper. Suppose λ_i, μ_i, τ_j ($1\leqq i\leqq m$, $1\leqq j\leqq l$) are rational numbers, other than negative integers, $\beta_i=\lambda_i-\mu_i$, and that none of $\beta_{i_1}\pm\beta_{i_2}$, $\tau_{s_1}\pm\tau_{s_2}$, $\beta_i\pm\tau_j$ are integral. Then, if $\alpha_1, \cdots, \alpha_n$ are non-zero algebraic numbers with distinct squares, the $2(l+m)n$ numbers $K_{\lambda_i\mu_i}(\alpha_j)$, $K'_{\lambda_i\mu_i}(\alpha_j)$, $K_{0\tau_s}(\alpha_j)$, $K'_{0\tau_s}(\alpha_j)$ are algebraically independent over the rationals. The author points out that it is not known if the simplified methods for attacking such problems given in 1966 by A. B. Šidlovskiĭ [Dokl. Akad. Nauk SSSR **169** (1966), 42–45; MR **33** #7302] may be used to derive these new results. *J. B. Roberts* (Halifax, N.S.)

Citations: MR 25# 49 = J88-22; MR 33# 7302 = J88-28.

J88-45 (41# 5307)

Spira, Robert

A lemma in transcendental number theory.

Trans. Amer. Math. Soc. **146** (1969), 457–464.

This paper concerns a lemma of A. O. Gel'fond, which plays an important part in the theory of algebraic independence of numbers [see Gel'fond, *Transcendental and algebraic numbers*, pp. 140–141, Lemma III, Dover, New York, 1960; MR **22** #2598]. As was, however, pointed out by N. I. Fel'dman in a letter to the editor of the Trans. Amer. Math. Soc., the paper contains a mistake, which cannot easily be repaired, viz. on p. 462, where a false integral representation for $B_{i,n}$ is given. Although there might be some justification in the author's objections against Gel'fond's formulation and proof of the lemma, the reviewers, with Fel'dman, think the criticism of the author on this beautiful work of Gel'fond is far too severe. But the author's idea that Gel'fond's lemma can be sharpened is quite correct. This appears from a result by the second reviewer, which improves the author's assertion and will appear in Nederl. Akad. Wetensch. Proc. Ser. A.

J. Popken (Amstelveen) and *R. Tijdeman* (Amsterdam)

Citations: MR 22# 2598 = J02-7.

J88-46 (43# 160)

Šidlovskiĭ, A. B.

A certain theorem of C. Siegel. (Russian. English summary)

Vestnik Moskov. Univ. Ser. I Mat. Meh. **24** (1969), *no.* 6, 39–42.

Put $\mathscr{K}_\lambda(z)=\sum_{n=0}^\infty((-1)^n/(n!(\lambda+1)\cdots(\lambda+n)))(z/2)^{2n}$. Denote by λ_0 a non-negative integer, by $\lambda_1, \cdots, \lambda_m$ rational numbers $\not\equiv 0, \not\equiv \frac{1}{2} \pmod 1$ such that none of the differences $\lambda_i-\lambda_j$ $(i\neq j)$ are integers, by μ_0 half an odd integer, and by $\mu_1, \cdots, \mu_m$ rational numbers such that all $\lambda_k+\mu_k$ are integers. Further, let $\beta_1, \cdots, \beta_m$ be algebraic numbers which are linearly independent over the rational field, and let $\alpha_1\neq 0, \cdots, \alpha_m\neq 0$ be algebraic numbers such that $\alpha_i^2\neq\alpha_j^2$ $(i\neq j)$. Then $3n(m+1)$ function values $K_{\lambda_0}(\alpha_i)$, $K_{\lambda_0}'(\alpha_i)$, $K_{\lambda_k}(\alpha_i)$, $K_{\lambda_k}'(\alpha_i)$, $K_{\mu_0}(\beta_i)$, $K_{\mu_0}(\alpha_i)$ $(1\leqq i\leqq n; 1\leqq k\leqq m)$ are algebraically independent over the rational field. This generalises one of C. L. Siegel's results [Abh. Preuss. Akad. Wiss. Phys.-Math. Kl. **1929**, no. 1]. The proof is based on the general theory of the author.

K. Mahler (Columbus, Ohio)

J88-47 (43# 194)

Oleĭnikov, V. A.

The differential irreducibility of a linear inhomogeneous equation. (Russian)

Dokl. Akad. Nauk SSSR **194** (1970), 1017–1020.

Let k be a subfield of the field of meromorphic functions closed under differentiation. Let $L(y)=y^{(n)}+p_1y^{(n-1)}+\cdots+p_ny$ $(p_i\in k, i=0, \cdots, n)$, and consider the equation $L(y)=p_0$. This equation is said to be differentially [homogeneously] reducible if it has a solution y_0 transcendental over k and satisfying equation $R(y, y', \cdots, y^{(m)})=0$, where $m<n$ and R is a polynomial [a homogeneous polynomial with respect to $y, y', \cdots, y^{(m)}$] with coefficients in k. The main theorem says that the equation $L(y)=p_0$ $(p_0\neq 0)$ is differentially reducible if and only if the equation $p_0(p_0^{-1}L(y))'=0$ is homogeneously reducible. The result is then applied to obtain some theorems on algebraic independence of certain complex numbers.

{This article has appeared in English translation [Soviet Math. Dokl. **11** (1970), 1337–1341].}

Andrzej Białynicki-Birula (Warsaw)

J88-48 (43# 3215)

Ax, James

On Schanuel's conjectures.

Ann. of Math. (2) **93** (1971), 252–268.

The following conjecture of S. Schanuel [see S. Lang, *Introduction to transcendental numbers*, pp. 30–31, Addison-Wesley, Reading, Mass., 1966; MR **35** #5397] is proved: Let $y_1, \cdots, y_n$ in $t\mathbf{C}[[t]]$ be linearly independent over $\mathbf{Q}$; then $\dim_{\mathbf{C}(t)}\mathbf{C}(t, y_1, \cdots, y_n, \exp y_1, \cdots, \exp y_n)\geqq n$. This theorem is a power-series analogue for a number-theoretic conjecture of Schanuel concerning the exponential function that embodies all its known transcendency properties as well as some long-standing transcendency conjectures.

A technical device important in the proof of this theorem is the fact that if A and B, $A\subset B$, are commutative rings and D is a derivation of B such that $D(A)\subset A$, then D acts canonically on the module of Kähler differentials $\Omega^1_{B/A}$.

The author proves several statements related to his main result. He also produces a counter-example to the following question posed by Z. I. Borevič and I. R. Šafarevič [*The theory of numbers* (Russian), p. 397, Izdat. "Nauka", Moscow, 1964; MR **30** #1080; English translation, p. 300, Academic Press, New York, 1966; MR **33** #4001; German translation, Birkhäuser, Basel, 1966; MR **33** #4000; French translation, Gauthier-Villars, Paris, 1967; MR **34** #5734]: Let C be a field, $\mathbf{Q}\subset C$, $n\geqq 2$ and $y_1, \cdots, y_n$ in $tC[[t]]$ with

$$\operatorname{rank}_C(y_1, \cdots, y_n)+\operatorname{rank}_C(\exp y_1, \cdots, \exp y_n)\ \leqq\ n;$$

do there exist $i\neq j$ such that $y_i=y_j$?

The counter-example is

$$\{y_1, \cdots, y_n\} = \{a\log(1-t)+b\log(1+t)|a, b \text{ in } \mathbf{N}, a+b<N$$

where $N\in\mathbf{N}$, $N\geqq 3\}$. In particular, when $N=3$, $n=6$ and we get a counter-example for the lowest value of n for which the question had not been answered. The author shows how, with some strengthening of the hypotheses, one may get $y_i=y_j$ and suggests there should be further work in this direction. *J. L. Johnson* (New Brunswick, N.J.)

Citations: MR 30# 1080 = Z02-44; MR 33# 4000 = Z02-45; MR 33# 4001 = Z02-46; MR 34# 5734 = Z02-47; MR 35# 5397 = J02-22.

J88-49 (43# 3216)

Belogrivov, I. I.

The transcendence and algebraic independence of the values of certain hypergeometric E-functions. (Russian)

Mat. Sb. (N.S.) **82 (124)** (1970), 387–408.

From the author's summary: "We study the arithmetical behavior of the values of the functions

$$A_{m,s}(z) = \sum_{n=0}^\infty[\lambda_1+1, n]^{-m_1}[\lambda_2+1, n]^{-m_2}\cdots [\lambda_s+1, n]^{-m_s}(z/m)^{mn}$$

and

$$A_{m,s,\mu}(z) = 1+\sum_{n=1}^\infty[\lambda_1+1, n]^{-m_1}\cdots [\lambda_{i-1}+1, n]^{-m_{i-1}}[\lambda_i+1, n-1]^{-m_i}\cdots [\lambda_s+1, n-1]^{-m_s}[\lambda_i+n]^{q^{i-1}-\mu}(z/m)^{mn},$$

where $\lambda_1, \lambda_2, \cdots, \lambda_s$ are rational numbers different from

$-1, -2, \cdots; \mu = q_{i-1}+1, q_{i-1}+2, \cdots, q_i, i=1, 2, \cdots, s$; $s \geqq 1$; $[\lambda, 0]=1$, $[\lambda, n]=\lambda(\lambda+1)\cdots(\lambda+n-1)$, $n \geqq 1$; furthermore, $m_1, m_2, \cdots, m_s$ are arbitrary non-negative rational integers, $m_0=0$, $m=m_1+m_2+\cdots+m_s$, $m \geqq 1$; $q_i=m_1+m_2+\cdots+m_i$, $i=1, 2, \cdots, s-1$, $q_0=0$, $q=q_s=m_1+m_2+\cdots+m_{s-1}+t^s$, and $t_s \geqq m_s$ is an integer. The function $A_{m,s}$ is shown to be the solution of a linear differential equation of order m with polynomial coefficients. The system of functions $A_{m,s,\mu}(z)$, $\mu=1, 2, \cdots, q$, is a solution of a system of q linear differential equations whose coefficients are rational functions in z. Applying a general theorem of A. B. Šidlovskiĭ [Izv. Akad. Nauk SSSR Ser. Mat. **23** (1959), 35–66; MR **21** #1295] on the transcendence and algebraic independence of E-functions, we prove six theorems on the mutual algebraic independence of the values of the functions in each of the systems $A_{m,s}(z)$, $A'_{m,s}(z), \cdots, A_{m,s}^{(m-1)}(z)$ and $A_{m,s,\mu}(z)$, $\mu=1, 2, \cdots, q$, for arbitrary algebraic points $\alpha \neq 0$ and different rational values of the parameters $\lambda_1, \lambda_2, \cdots, \lambda_s$, $s \geqq 1$, and arbitrary values of $m_1, m_2, \cdots, m_s$."

{This article has appeared in English translation [Math. USSR-Sb. **11** (1970), 355–376].}

B. Volkmann (Stuttgart)

Citations: MR 21# 1295 = J88-15.

J88-50 **(43# 6163)**

Waldschmidt, Michel

Amélioration d'un théorème de Lang sur l'indépendance algébrique d'exponentielles.

C. R. Acad. Sci. Paris Sér. A-B **272** (1971), A413–A415.

This is an announcement of some results on the algebraic independence of values of the exponential function. The main result is the following: Let $x_1, \cdots, x_n$ $[y_1, \cdots, y_m]$ be complex numbers linearly independent over the rational numbers, and assume that either $m \geqq 3$ and $n \geqq 6$, or $m \geqq 4$ and $n \geqq 4$; then at least two of the mn numbers $e^{x_i y_j}$ $(1 \leqq i \leqq n, 1 \leqq j \leqq m)$ are algebraically independent. Further, some Diophantine approximation type conditions (similar to S. Lang's "transcendence types" [*Introduction to transcendental numbers*, p. 51, Addison-Wesley, Reading, Mass., 1966; MR **35** #5397]) are introduced on a field and similar types of results are obtained. The above results generalize results of Lang [op. cit., Chapters II, V]. This paper overlaps considerably with recent work done independently by D. Brownawell at Pennsylvania State University.

W. W. Adams (College Park, Md.)

Citations: MR 35# 5397 = J02-22.

J88-51 **(44# 5279)**

Belogrivov, I. I.

The transcendence and algebraic independence of the values of Kummer functions. (Russian)

Sibirsk. Mat. Ž. **12** (1971), 961–982.

Let $\nu, \mu \neq 0, -1, -2, \cdots$ be rational numbers, and let

$$K_{\nu,\mu}(z) = \sum_{n=0}^{\infty} \frac{\mu(\mu+1)\cdots(\mu+n-1)}{\nu(\nu+1)\cdots(\nu+n-1)} \frac{z^n}{n!}$$

and $B_{\nu,\mu}(z)=e^{-z^2/2}K_{\nu,\mu}(z)$. These functions are Siegel E-functions, and hence A. B. Šidlovskiĭ's general theory [Izv. Akad. Nauk SSSR Ser. Mat. **26** (1962), 877–910; MR **26** #2395] can be applied to the values of such functions at algebraic points provided the algebraic independence of the functions over $\mathbf{C}(z)$ can be established. The author proves four new results of this kind. The simplest one is as follows. Let $\alpha_1^2, \cdots, \alpha_n^2$ be algebraic numbers distinct from one another and from 0. Let ν_j, μ_j $(1 \leqq j \leqq m)$ be rational numbers such that $\nu_j \neq 0, -1, -2, \cdots$; $\mu_j \neq 0, \mp 1, \mp 2, \cdots$; $\mu_j-\nu_j \neq 0, \mp 1, \mp 2, \cdots$ $(1 \leqq j \leqq m)$; $(2\mu_j-\nu_j)-(2\mu_k-\nu_k) \mp \nu_j \mp \nu_k \neq 0, \mp 2, \mp 4, \cdots$ $(1 \leqq j < k \leqq m)$. Then the $2mn$ function values $B_{\nu_j,\mu_j}(\alpha_i)$, $B'_{\nu_j,\mu_j}(\alpha_i)$ $(1 \leqq i \leqq n; 1 \leqq j \leqq m)$ are algebraically independent over $\mathbf{Q}$. The proofs depend on several general lemmas on the algebraic independence over $\mathbf{C}(z)$ of solutions of systems of linear differential equations which have much interest in themselves.

{This article has appeared in English translation [Siberian Math. J. **12** (1971), 690–705].} *K. Mahler* (Canberra)

Citations: MR 26# 2395 = J88-26.

J88-52 **(44# 6614)**

Galočkin, A. I.

The algebraic independence of the values of the E-functions at certain transcendental points. (Russian. English summary)

Vestnik Moskov. Univ. Ser. I Mat. Meh. **25** (1970), no. 5, 58–63.

The author proves the following generalization of a known theorem due to A. B. Šidlovskiĭ [Dokl. Akad. Nauk SSSR **100** (1955), 221–224; MR **16**, 907; Izv. Akad. Nauk SSSR Ser. Mat. **23** (1959), 35–66; MR **21** #1295]: Let $f_1(z), f_2(z), \cdots, f_s(z)$ be E-functions that are solutions of the system of linear differential equations $y_i'=Q_{i0}(z)+\sum_{j=1}^{s} Q_{ij}(z)y_j$, $i=1, 2, \cdots, s$ with coefficients Q_{ij} in $C(z)$, let the functions $f_1, f_2, \cdots, f_s$ be algebraically independent over $C(z)$ and let α be any complex number (that is not a pole of any Q_{ij}) such that for any positive ε, κ the inequality $|\alpha-\Theta|<e^{-h^\varepsilon}$ is satisfied for infinitely many algebraic numbers Θ of height h and degree $\leqq \kappa$ (i.e., α is an algebraic number or a transcendental number which admits of sufficiently good approximation by algebraic numbers); then $\alpha, f_1(\alpha), f_2(\alpha), \cdots, f_s(\alpha)$ are algebraically independent. *B. Novák* (Prague)

Citations: MR 16, 907a = J88-10; MR 21# 1295 = J88-15.

J99 NONE OF THE ABOVE, BUT IN THIS CHAPTER

See also reviews M15-12, P52-8, Z10-1, Z15-63.

J99-1 **(1, 295d)**

Segal, B. Approximation of complex numbers by a sum of powers of integers with a given complex exponent. Rec. Math. [Mat. Sbornik] N.S. **5** (47), 147–184 (1939). (Russian. English summary)

". . . The main result of the present paper may be stated in the form of the following theorem: Let $a+bi$ be a given complex number, where $a>1$ and $b \neq 0$, $n=[a]+2$, $a'=n-a-1$, $r \geqq r_0$ an integer, where

$$r_0=[2n/\rho_n]+1=(2n/\rho_n)+\epsilon', \qquad 0<\epsilon' \leqq 1;$$

ρ_n depends only on n, and $0<\rho_n \leqq 1/n$. Further, let $N=N_1+N_2i$ be any complex number, $X=|N|^{1/a}$, $\Delta_1=X^{-\alpha}$, $\alpha=a'\epsilon'/r_0$, I_N the number of representations of N in the form

$$N_1+N_2i=h_1+h_2i+\sum_{\lambda=1}^{r} x_\lambda^{a+bi},$$

where $x_1, \cdots, x_r$ are integers satisfying the condition $0<x_\lambda<X$, and h_1, h_2 satisfy the conditions $-\Delta_1 \leqq h_1, h_2 \leqq \Delta_1$. Then we have the asymptotic formula

$$I_N=LX^{r-2a-2\alpha}(1+O(X^{-\alpha})),$$

where L exceeds a positive number depending only on a, b and r." The value of ρ_n is connected with the estimation of trigonometrical sums of the form

$$\sum_{A \leqq X \leqq B} \exp(2\pi i f(x)).$$

We can put

$$r_0=\begin{cases} n\cdot 2^{n+1}+1 & \text{if } 1<a<14, \\ [8n^4 \log n]+1 & \text{if } \quad a\geqq 14. \end{cases}$$

A less exact result was published by the author in a note in C. R. (Doklady) Acad. Sci. URSS (N.S.) **19**, 667–670 (1938). *K. Mahler* (Manchester).

J99-2 (5, 143g)

Hua, Loo-Keng. On Diophantine approximation. C. R. (Doklady) Acad. Sci. URSS (N.S.) **32**, 395–396 (1941).

A brief announcement of results to be published later.
H. S. Zuckerman (Seattle, Wash.).

J99-3 (10, 354i)

Fomin, A. M. On a class of nonlinear Diophantine approximations. Doklady Akad. Nauk SSSR (N.S.) **63**, 7–10 (1948). (Russian)

Let $\phi(x)$ be a positive real function of the positive real variable x, with $x\phi''(x)$ a nondecreasing function of x, and $\phi''(x)$ a monotone decreasing function tending to zero as $x\to\infty$. Let r be a given integer, N any sufficiently large number, and μ the solution of the equation $N=r\phi(\mu)$. Then there exist integers $x_1, \cdots, x_r$, satisfying the inequalities

$$|x_i-\mu|<6r^{3/2}(\phi'(\mu)/\phi''(\mu))^{1/2}, \qquad i=1, \cdots, r,$$
$$0<\phi(x_1)+\cdots+\phi(x_r)-N<c_r(\phi'(\mu)/\phi''(\mu))^{2^{1-r}}\phi''(\mu),$$

with c_r a constant depending only on r. The author states the above theorem, with two others of a similar but more general character. Proofs are only briefly sketched.
F. J. Dyson (Princeton, N. J.).

J99-4 (23# A861)

Supnick, Fred

On a quadratic Diophantine inequality.

Proc. Amer. Math. Soc. **12** (1961), 164–173.

If S_p denotes the solid n-sphere obtained by moving the circumsphere S of a solid unit n-cube C to a center p ($p\in C$), the problem arises of identifying those vertices of C that are points of S_p. With appropriate selection of a cartesian coordinate system, this problem becomes the algebraic one of finding the lattice points $(x_1, x_2, \cdots, x_n)$, $x_i=0$ or 1 $(i=1, 2, \cdots, n)$, that satisfy the quadratic inequality $\sum_{i=1}^{m}(x_i-p_i)^2\leqq n/4$, where $(p_1, p_2, \cdots, p_n)$ are the coordinates of p; moreover, the author seeks to obtain those points by a minimum number of operations. A constructive process for achieving this is found.
L. M. Blumenthal (Columbia, Mo.)

J99-5 (32# 1168)

Schaefer, Paul

The density of certain classes of rationals.

Amer. Math. Monthly **72** (1965), 894–895.

A positive rational number a/b is a P-rational if and only if the pair of integers a and b has the property P. Possible P properties are that a^2+b^2 be the cube of an integer or that a^3+b^3 be the square of an integer. The author gives conditions under which P-rationals are dense in an interval of the non-negative reals.

J99-6 (34# 1271)

Samušenok, I. N.

Arithmetic characteristic of a class of irrational numbers. (Russian)

Proc. Second Sci. Conf. Math. Dept. Ped. Inst. Volga Region, No. I (Russian), pp. 9–13. *Kuĭbyšev. Gos. Ped. Inst., Kuybyshev,* 1962.

J99-7 (35# 4165)

Schmidt, Wolfgang M.

On heights of algebraic subspaces and diophantine approximations.

Ann. of Math. (2) **85** (1967), 430–472.

The author generalizes diophantine approximation to the approximation of linear subspaces in n-dimensional euclidean space E_n by linear subspaces defined over a fixed algebraic number-field K (possibly of another dimension). The program is to formulate precisely and solve the following problem. Let K be a fixed real algebraic number-field of finite degree over the rationals. Let $0<d<n$, $0<e<n$, and suppose that A is a d-dimensional subspace of real euclidean space E_n (A need only be defined over the reals). To find a subspace B of E_n of dimension e defined over K which is given by equations with "small" coefficients (B has small "height") and which is "close" to A.

The precise formulation of the definitions and theorems is too complicated to give here, but is quite natural. There are analogues of much of diophantine approximation including a duality principle (which deals in subspaces of complementary dimension) and analogues of the transference principles of Perron and Hinčin (=Khintchine) relating approximation by subspaces of different dimensions. There are also counter-examples to show that certain results cannot be greatly improved (for which, as one might expect, A is defined over a finite algebraic extension of K). *J. W. S. Cassels* (Cambridge, England)

Referred to in H99-7.

J99-8 (42# 3025)

Diviš, Bohuslav; Novák, Břetislav

A remark on the theory of diophantine approximations.

Comment. Math. Univ. Carolinae **11** (1970), 589–592.

Preliminary announcement of results.

K. DISTRIBUTION MODULO 1; METRIC THEORY OF ALGORITHMS

For the distribution (mod m) of sequences of integers, see **B48**. For Schnirelman sums of real sets, see **B08** as cross-referenced in **L99**.

K02 BOOKS AND SURVEYS

See also reviews J02-11, J02-15, J02-20, K05-47, K15-76, K15-77, K25-10, K35-20, K40-30, K40-40, K45-11, K45-15, K45-24, K45-25, K50-19, P02-4, Q15-7, Q15-20, Z01-37, Z02-61.

K02-1 **(23# A2409)**

Cigler, Johann; Helmberg, Gilbert

Neuere Entwicklungen der Theorie der Gleichverteilung.

Jber. Deutsch. Math. Verein. **64**, *Abt.* 1, 1–50 (1961).

This article presents a readable and complete account of work in the theory of distribution modulo 1 of real sequences. The authors put special emphasis on developments since 1935, since Koksma's *Diophantische Approximationen* [J. Springer, Berlin, 1936] provides complete coverage of earlier work.

W. J. LeVeque (Ann Arbor, Mich.)

Referred to in J24-45, J64-23, K05-64, K40-35, K40-65.

K02-2 **(26# 6146)**

Postnikov, A. G.

Arithmetic modeling of random processes. **(Russian)**

Trudy Mat. Inst. Steklov. **57** (1960), 84 *pp.*

This monograph deals with the construction by number-theoretical methods of infinite sequences of numbers having prescribed statistical properties, and summarizes the work done by the author, I. I. Pjateckiĭ-Šapiro, N. M. Korobov and others. It consists of 27 sections and a bibliography containing 61 references.

Let $\varepsilon=\{\varepsilon_n\}$ $(n=1, 2, \cdots)$ be an infinite sequence consisting of the numbers $0, 1, \cdots, g-1$, where $g\geqq 2$ is a fixed integer. From this sequence we form for each $s\geqq 1$ the sequences of s-tuples $(\varepsilon_1, \varepsilon_2, \cdots, \varepsilon_s)$, $(\varepsilon_2, \varepsilon_3, \cdots, \varepsilon_{s+1})$, $\cdots$, $(\varepsilon_P, \varepsilon_{P+1}, \cdots, \varepsilon_{P+s-1})$, $\cdots$. Let Δ be a fixed s-tuple $(s=1, 2, \cdots)$ consisting of the numbers $0, 1, \cdots, g-1$, and let $N_P(\varepsilon, \Delta)$ denote the number of s-tuples $(\varepsilon_k, \varepsilon_{k+1}, \cdots, \varepsilon_{k+s-1})$ $(k=1, 2, \cdots, P)$ which are identical with Δ. The sequence $\varepsilon=\{\varepsilon_n\}$ is called a normal sequence if for any s and any s-tuple Δ one has $\lim_{P\to+\infty} N_P(\varepsilon, \Delta)/P= 1/g^s$. Put $\alpha=\sum_{n=1}^{\infty} \varepsilon_n/g^n$. The sequence (αg^k) $(k=1, 2, \cdots)$, where (x) denotes the fractional part of the real number x, is uniformly distributed in the interval $(0, 1)$ if and only if the sequence $\{\varepsilon_n\}$ is normal. It has been shown by Pjateckiĭ-Šapiro [Izv. Akad. Nauk SSSR Ser. Mat. **15** (1951), 47–52; MR **13**, 213] that if there exists a positive constant C such that $\limsup_{P\to+\infty} N_P(\varepsilon, \Delta)/P\leqq C/g^s$ for any s and any s-tuple Δ, then $\{\varepsilon_n\}$ is normal. This theorem can be used to prove the normality of certain sequences (e.g., that of D. G. Champernowne [J. London Math. Soc. **8** (1933), 254–260]). A sequence $\{\varepsilon_n\}$ consisting of the numbers 0 and 1 is called a normal Bernoulli sequence with parameter p $(0<p<1)$ if for any s-tuple Δ consisting of j ones and $s-j$ zeros one has $\lim_{P\to+\infty} N_P(\varepsilon, \Delta)/P= p^j(1-p)^{s-j}$. The theorem of Pjateckiĭ-Šapiro mentioned above can be generalized as follows. If there exists a constant C such that for any s-tuple Δ

$$\limsup_{P\to+\infty} N_P(\varepsilon, \Delta)/P \leqq Cp^j(1-p)^{s-j},$$

then the sequence $\{\varepsilon_n\}$ is a normal Bernoulli sequence with parameter p. This criterion can be used for constructing effectively normal Bernoulli sequences. The notion of a normal Bernoulli sequence is equivalent to the notion of admissible numbers introduced by A. H. Copeland [Amer. J. Math. **50** (1928), 535–552]. A further generalization is that of a Markov sequence. In §§ 18–22 the connection of the theory of normal sequences and its generalizations with ergodic theory is discussed in detail. Finally the book deals with continued fractions. Let $\Delta=(\delta_1, \cdots, \delta_s)$ be an arbitrary s-tuple of natural numbers and let Δ' denote the set of all real numbers α $(0<\alpha<1)$ the continued fraction of which has $\delta_1, \delta_2, \cdots, \delta_s$ as its first s digits. A continued fraction

$$\cfrac{1}{c_1+\cfrac{1}{c_2+\cfrac{1}{c_3+\cdots}}}$$

is called normal if the sequence $c=\{c_n\}$ is such that for any s-tuple of natural numbers Δ $\lim_{P\to+\infty} N_P(c, \Delta)/P= \int_{\Delta'} 1/(1+x) \log 2dx$. The analogue of Pjateckiĭ-Šapiro's theorem is valid again and can be applied for the construction of normal continued fractions.

A. Rényi (Budapest)

Citations: MR 13, 213d = K15-14.
Referred to in K45-24.

K02-3 **(30# 3874)**

Koksma, J. F.

The theory of asymptotic distribution modulo one.

Compositio Math. **16**, 1–22 (1964).

Diese Arbeit bildet die Einleitung zu einer Reihe von Arbeiten, die alle in Comp. Math. **16** erschienen sind. (Siehe auch den Verfasser und L. Kuipers [*Asymptotic distribution modulo* 1, Noordhoff, Groningen, 1965; MR **31** #113].) Sie gibt eine sehr gute Darstellung der Fragestellungen und Resultate der Theorie der asymptotischen Verteilung mod 1.

J. E. Cigler (Groningen)

Citations: MR 31# 113 = K02-4.

K02-4 **(31# 113)**

Koksma, J. F.; Kuipers, L. (Editors)

Asymptotic distribution modulo 1.

Papers presented at the NUFFIC International Summer Session in Science, sponsored by NATO, held in Breukelen (The Netherlands), August 1–11, 1962.

P. Noordhoff N. V., Groningen, 1965. iv+203 *pp.* *Dfl.* 10.50.

This volume contains papers of J. F. Koksma, J. P.

Bertrandias, P. Erdős, S. Hartman, I. Niven, W. Phillipp, Ch. Pisot and R. Salem, I. J. Schoenberg, N. B. Slater, C. de Vroedt, two papers each by G. Helmberg, E. Hlawka, J. H. B. Kemperman, B. Volkmann, and three papers by J. Cigler. These papers were published in Compositio Math. **16** (1964), fasc. 1–2, and were reviewed from that journal.

Referred to in K02-3.

K02-5 **(33# 254)**

Billingsley, Patrick

Ergodic theory and information.

John Wiley & Sons, Inc., New York-London-Sydney, 1965. xiii+195 *pp.* $8.50.

This book, the first in the new series "Tracts on Probability and Mathematical Statistics", originated in a series of lectures given by the author to one of the London Mathematical Society's conferences. It is a charming, well written and informative monograph. The exposition is simple enough to be read without pen and paper at hand, and in the space of 180 pages the author covers a remarkably wide selection of topics, sacrificing at times deep technicalities for a cohesive account of ergodic theory and its applications and ramifications. These will perhaps interest the specialist in spite of the disclaimer of the author that the book is intended for readers to whom the subject is new. In fact, the one main concession to the non-specialist is Chapter 3 on elementary conditional probability and conditional expectation.

Chapter 1 treats the ergodic theorem, giving two proofs based on the maximal ergodic theorem. There are numerous examples including an analysis of Gauss' problem on continued fractions, and an application to Diophantine approximation. Chapter 2 treats the Kolmogorov-Sinai application of Shannon's entropy to the isomorphism problem of ergodic theory. Chapter 4 treats the convergence of entropy, including the Shannon-McMillan-Breiman theorem and an application to dimension theory, while Chapter 5 applies the preceding material to the principal theorems of coding theory.

D. A. Darling (Ann Arbor, Mich.)

Referred to in K20-16.

K02-6 **(38# 1718)**

Billingsley, Patrick

Convergence of probability measures.

John Wiley & Sons, Inc., New York-London-Sydney, 1968. xii+253 *pp.* $12.50.

This book is a welcome addition to the series of books being written on specialized topics in probability and statistics. The subject matter is of great current interest and the exposition is lucid and elegant.

The author's preface is an accurate summary of this book and hence it suffices to quote from it. The author says: "This book is about weak-convergence methods in metric spaces, with applications sufficient to show their power and utility.

"The introduction motivates the definitions and indicates how the theory will yield solutions to problems arising outside it. Chapter 1 sets out the basic general theorems, which are then specialized in Chapter 2 to the space of continuous functions on the unit interval and in Chapter 3 to the space of functions with discontinuities of the first kind. The results of the first three chapters are used in Chapter 4 to derive a variety of limit theorems for dependent sequences of random variables.

"Although standard measure-theoretic probability and metric-space topology are assumed, no general (non-metric) topology is used, and the few results required from functional analysis are proved in the text or in an appendix.

"Mastering the impulse to hoard the examples and applications till the last, thereby obliging the reader to persevere to the end, I have instead spread them evenly through the book to illustrate the theory as it emerges in stages."

M. M. Siddiqui (Ft. Collins, Colo.)

K05 DISTRIBUTION (mod 1): GENERAL THEORY

See also reviews B08-65, B08-95, G05-65, J02-11, J20-41, J64-23, K02-3, K10-14, K15-14, K15-23, K15-78, K20-16, K35-12, K35-18, K35-24, K35-33, K45-1, L02-5, M55-76, Q05-20, Q15-2, Q15-21, Q20-17, Q99-2, Z10-9, Z10-10, Z10-30, Z10-31, Z10-34, Z10-47.

K05-1 **(1, 66b)**

van der Corput, J. G. et Pisot, Ch. Sur la discrépance modulo un. II. Nederl. Akad. Wetensch., Proc. **42**, 554–565 (1939).

Let for real α and λ the number of those elements u_ν of a finite system U of real numbers $u_1, u_2, \cdots, u_n$, for which

$$0 \leqq u_\nu - \alpha - [u_\nu - \alpha] < \lambda - [\lambda],$$

be denoted by $N(\alpha, \alpha+\lambda)$ and let

$$N(\alpha, \alpha+\lambda) = (\lambda - [\lambda] + D(\alpha, \alpha+\lambda))n.$$

Then the upper bound $D(U)$ of $|D(\alpha, \alpha+\lambda)|$ for all α and λ is called the discrepancy of U. In the first part of the paper [same vol., 476–486] the authors proved

$$(1) \qquad D(U) < 2^{(7/2)+(1/4\epsilon)}(D(U^*))^{\frac{1}{2}-\epsilon},$$

for any $\epsilon > 0$, where U^* is the system of numbers $u_\nu - u_\rho$ $(\nu = 1, \cdots, n; \rho = 1, \cdots, n)$. In this second part of the paper the authors deduce from (1): Let

$$\varphi(y) = \alpha\frac{y^k}{k!} + \alpha_1 y^{k-1} + \cdots + \alpha_k$$

be a polynomial of degree $k \geqq 1$ and real coefficients α, and let

$$\left|\alpha - \frac{a}{q}\right| \leqq \frac{\tau}{q^2},$$

where a/q is an irreducible fraction with $q > 0$ and where $\tau \geqq 1$. Then the discrepancy $D(Y)$ of the system $\varphi(1), \varphi(2), \cdots, \varphi(X)$ (where X is an integer not less than 3) satisfies for any $\epsilon > 0$ the inequality

$$D(Y) < Cx^\omega \xi^{(1-\epsilon)2^{1-k}},$$

where $x = \log X$, $\xi = (\tau + qX^{-1})(q^{-1} + X^{1-k})$ and where C is a constant and ω depends only on k and ϵ. Since, as the authors prove too:

$$\left| n^{-1}\sum_{\nu=1}^{n} e^{2\pi i u_\nu} \right| \leqq 2\pi D(U),$$

this theorem gives non-trivial approximations for the Weyl-sums.

H. D. Kloosterman (Leiden).

Referred to in J24-4, K05-20, K40-21.

K05-2 **(1, 66c)**

van der Corput, J. G. et Pisot, Ch. Sur la discrépance modulo un. III. Nederl. Akad. Wetensch., Proc. **42**, 713–722 (1939).

The authors prove: Let U be a system of n real numbers $u_1, u_2, \cdots, u_n$ and let V_l for any integer l with $0 \leqq l < n$ be the system of numbers $u_{l+1}-u_1, u_{l+2}-u_2, \cdots, u_n-u_{n-l}$. Then

$$D(U) \leqq \frac{2l}{n} + 2^{\alpha}(\omega)^{1/2},$$

where

$$\alpha = \frac{7}{2} + \left(\frac{|\log \omega|}{\log 2} \right)^{1/2},$$

$$\omega = \frac{1}{l+1}\left(1+\frac{1}{n}\right) + 2\frac{D(V_1)+\cdots+D(V_l)}{l+1} + \frac{2l}{n}.$$

Here $D(U), D(V_1), \cdots, D(V_l)$ are the discrepancies of the systems $U, V_1, \cdots, V_l$, respectively [for the definition of this notion, cf. the reference on the second part of the paper under the same title]. Furthermore, the authors prove: If U is the system of numbers $f(1), f(2), \cdots, f(n)$, where $f(y)$ is a real function whose kth difference $\Delta^k f(y)$ is not less than $r > 0$ for $y = 1, 2, \cdots, n-k$ (where k is an integer less than n), then

$$D(U) < C\left\{ (\rho^{-2}r)^{-1/(K-1)} + (rn^k)^{-2/K} + (\rho^{-1}rn)^{-2/K} \log \frac{1}{\rho} \right\}.$$

Here C is a constant and

$$\rho = \frac{1}{n-k}(\Delta^{k-1}f(n-k+1) - \Delta^{k-1}f(1)), \qquad K = 2^k.$$

H. D. Kloosterman (Leiden).

Referred to in J24-4, K40-21.

K05-3 (3, 2f)

Steinhaus, H. Sur les fonctions indépendantes. VI. Studia Math. **9**, 121–132 (1940). (French. Ukrainian summary)

Let $f(t)$, defined for $0 \leqq t < \infty$, be measurable with respect to relative measure [Kac and Steinhaus, Studia Math. **7**, 1–15 (1938)], and let $\{f(t)\} = f(t) - [f(t)]$, where $[x]$ denotes the greatest integer in x. One says that $f(t)$ is uniformly distributed (mod 1) if for every $\lambda (0 \leqq \lambda \leqq 1)$ the relative measure of the set of t's for which $\{f(t)\} < \lambda$ is equal to λ. The author proves several theorems showing the connection between statistical independence and uniform distribution (mod 1). For instance, if for every pair h, k of integers $(h^2+k^2>0)$ the function $hf(t)+kg(t)$ is uniformly distributed (mod 1), then $\{f(t)\}$ and $\{g(t)\}$ are statistically independent. As an application it is shown that for every pair a_1, a_2 of different real numbers the functions $\sin 2\pi(t+a_1)^2$ and $\sin 2\pi(t+a_2)^2$ are statistically independent. This last result (which gives an answer to a question of Kampé de Fériet) is included as a particular case in a more general investigation by Agnew and the reviewer [Bull. Amer. Math. Soc. **47**, 148–154 (1941); cf. these Rev. **2**, 229]. *M. Kac* (Ithaca, N. Y.).

Referred to in K40-26.

K05-4 (7, 146a)

Teghem, Jean. Sur l'application de la théorie des sommes de Weyl à des problèmes d'inégalités diophantiennes. Bull. Soc. Roy. Sci. Liége **11**, 4–6 (1942).

The author gives, without proofs, a number of results improving considerably on earlier ones by J. F. Koksma [Over Stelsels Diophantische Ongelijkheden, Groningen thesis, 1930]. *K. Mahler* (Manchester).

Referred to in L15-12.

K05-5 (7, 370a; 7, 370b)

Koksma, J. F. A general theorem from the theory of uniform distribution modulo 1. Mathematica, Zutphen. B. **11**, 7–11 (1942). (Dutch)

Koksma, J. F. Some integrals in the theory of uniform distribution modulo 1. Mathematica, Zutphen. B. **11**, 49–52 (1942). (Dutch)

If $f(k)$ is a real function of the integral variable k, let just $N_{\alpha,\beta}(N)$ of the numbers (1) $f(k)-[f(k)]$, $k=1, \cdots, N$, lie in the interval $\alpha \leqq u < \beta$. Put $R_{\alpha,\beta}(N) = N_{\alpha,\beta}(N) - (\beta-\alpha)N$ and denote by $D(N)$ the upper bound of $R_{\alpha,\beta}(N)/N$ for $0 \leqq \alpha \leqq \beta \leqq 1$. In another paper the author will show that, if $w(t)$ is of period 1 and of total variation T in $0 \leqq t \leqq 1$, then

$$\left| \sum_{k=1}^{N} w(f(k)) - N\int_0^1 w(t)dt \right| \leqq TND(N).$$

From this result he deduces, in the first note, that

$$\int_0^1 R_{0,t}(N)dw(t) = N\int_0^1 w(t)dt - \sum_{k=1}^{N} w(f(k)).$$

In the second note he obtains some further identities, for example,

$$\int_0^1 R_{0,t}^2(N)dt = \tfrac{1}{3}N^2 + N\sum_{k=1}^{N} f^2(k) + \sum_{k=1}^{N} f(k) - 2\sum_{k=1}^{n}\sum_{h=1}^{k} \max(f(h), f(k)).$$

K. Mahler (Manchester).

Referred to in K05-35.

K05-6 (7, 376l)

van Aardenne-Ehrenfest, T. Proof of the impossibility of a just distribution of an infinite sequence of points over an interval. Nederl. Akad. Wetensch., Proc. **48**, 266–271 = Indagationes Math. **7**, 71–76 (1945).

Let $\{a_n\}$ be an infinite sequence in an interval I; α, β are subintervals of I, $A_n(\alpha)$ is the number of points $a_m \subset \alpha$ with $m \leqq n$. The sequence $\{a_n\}$ is called "just" if, for some constant C, $|A_n(\alpha) - A_n(\beta)| \leqq C$ whenever $|\alpha| = |\beta|$. The result is as indicated in the title, slightly extended. *H. D. Ursell* (Leeds).

Referred to in K05-13, K40-48.

K05-7 (8, 317h)

van der Corput, J. G. Rhythmic system. I. Nederl. Akad. Wetensch., Proc. **49**, 708–721 = Indagationes Math. **8**, 416–429 (1946).

The author gives essentially simplified proofs of theorems contained in his paper in Acta Math. **59**, 209–328 (1932). [P. 714 = 422, line 3 from below: add "absolutely" before "rhythmic."] *V. Jarník* (Prague).

K05-8 (9, 135d)

Drewes, Aart. Diophantische Benaderingsproblemen. [Diophantine Approximation Problems]. Thesis, Free University of Amsterdam, 1945. vi+72 pp.

Chapter I contains a general survey on the metrical theories of Diophantine approximation and a summary of the author's results. Chapter II gives a generalization for the many-dimensional case of a theorem of the reviewer [Nederl. Akad. Wetensch., Proc. **39**, 225–240 (1936)]. The method used is a transcendental one and is based on a theorem of van der Corput and Koksma [see the reviewer's book, Diophantische Approximationen, Ergebnisse der

Math., v. 4, no. 4, Springer, Berlin, 1936, Satz 4]. Chapter III uses the same method and a theorem of Vinogradov and deals with the uniform distribution of systems of polynomials $\{f_\nu(x)\}$ ($\nu=1, \cdots, n$; $x=1, 2, \cdots$). Chapters IV to VI deal with the distribution (mod 1) of slowly increasing functions, e.g., $(\theta x)^{1/k}$ ($k>1$) and chapter V contains similar research on functions $f(x)$ whose nth differential quotient is a slowly increasing function. This chapter is remarkable for the elementary method used, although the results are less sharp than those developed by other authors with use of transcendental methods. Most of the theorems proved by the author are very general; in most cases an upper bound for the remainder in the formula of uniform distribution is given. It would take too much space to quote particular results. *J. F. Koksma* (Amsterdam).

K05-9 (10, 292b)

Gál, István Sándor, et Koksma, Jurjen Ferdinand. Sur l'ordre de grandeur des fonctions sommables. C. R. Acad. Sci. Paris **227**, 1321–1323 (1948).

A theorem is stated without proof which is asserted to be a generalization of the following Tauberian theorem [cf. Kaczmarz and Steinhaus, Theorie der Orthogonalreihen, Warsaw and Lwów, 1935, p. 8]. Let S be a measurable set, and let $f_n(x)$, $n=1, 2, \cdots$, be in $L^p(S)$, $p\geqq 1$. Let $F(N, x)=\sum_1^N f_n(x)$, and let ϕ be a positive increasing function with $\sum\phi(N)^{-p}<\infty$. Then if $\int_S|F(N, x)|^p dx\leqq\psi(N)$, the relation $F(N, x)=o(\psi(N)^{1/p}\phi(N))$ holds for almost all $x\varepsilon S$. The generalization is too long to be given here. *W. J. LeVeque* (Cambridge, Mass.).

K05-10 (10, 372c; 10, 372d)

Erdös, P., and Turán, P. On a problem in the theory of uniform distribution. I. Nederl. Akad. Wetensch., Proc. **51**, 1146–1154 = Indagationes Math. **10**, 370–378 (1948).

Erdös, P., and Turán, P. On a problem in the theory of uniform distribution. II. Nederl. Akad. Wetensch., Proc. **51**, 1262–1269 = Indagationes Math. **10**, 406–413 (1948).

In a forthcoming paper the authors prove the following theorem. If $f(z)=a_0+\cdots+a_nz^n$ satisfies $|f(z)|\leqq M$ on $|z|=1$, and has roots $z_1, \cdots, z_n$, then the number of roots with $\alpha\leqq\arg z_\nu\leqq\beta$ differs from $n(\beta-\alpha)/2\pi$ by less than $16\{n\log(M|a_0a_n|^{-\frac{1}{2}})\}^{\frac{1}{2}}$. It is pointed out that a similar result cannot hold in terms of $M(\theta)$, an upper bound for $|f(z)|$ on $|z|=\theta$, where θ is fixed and $0<\theta<1$. The main theorem of the present paper is that such a theorem does hold if it is further postulated that all the roots are outside $|z|=1$. If $M(\theta)=|a_0a_n|^{\frac{1}{2}}\exp(n/g(n, \theta))$, where $n\geqq g(n, \theta)\geqq 2$, it is proved that the number of roots with $\alpha\leqq\arg z_\nu\leqq\beta$ differs from $n(\beta-\alpha)/2\pi$ by less than $C(\log 4\theta^{-1})(n/\log g(n, \theta))$, where C is a numerical constant. The proof uses first a method of Schur to reduce the general case to that when all the roots are on $|z|=1$, and the proof of that case is based on the following theorem in Diophantine approximation. Let $\varphi_1, \cdots, \varphi_n$ be real, and suppose $|\sum_{\nu=1}^n e^{ik\varphi_\nu}|\leqq\psi(k)$ for $k=1, \cdots, m$. Then the number of the φ_ν with $\alpha\leqq\varphi_\nu\leqq\beta$ (mod 2π) differs from $n(\beta-\alpha)/2\pi$ by less than $C(n/(m+1)+\sum_{k=1}^m\psi(k)/k)$, with a numerical constant C. This is a finite analogue of Weyl's criterion for equal distribution. The paper contains a number of remarks, in addition to the main theorem. *H. Davenport* (London).

Referred to in J24-58, K05-16, K05-19.

K05-11 (10, 433c)

Koksma, J. F. On a definite integral in the theory of uniform distribution. Nieuw Arch. Wiskunde (2) **23**, 40–54 (1949).

Let $f(1), f(2), \cdots$ be an infinite real sequence. Let $\nu_{\alpha,\beta}(N)$ be the number of those among the first N elements of the sequence which belong (mod 1) to the interval $\alpha\leqq u<\beta$. Writing $\nu_{\alpha,\beta}(N)=(\beta-\alpha)N+R_{\alpha,\beta}(N)$, the term $R_{\alpha,\beta}(N)$ is called the remainder, which, in case of a uniformly distributed sequence, satisfies $R_{\alpha,\beta}(N)=o(N)$ for any α, β satisfying $\beta-\alpha\leqq 1$. The purpose of the paper is to consider sequences of the form $f(x, \theta)$ ($x=1, 2, \cdots$) when θ is a parameter and to give the following result about the order of magnitude of $R_{\alpha,\beta}(N, \theta)$ in mean. Let $f(x, \theta)$ denote a real function which is defined for $x=1, 2, \cdots, N$ ($N\geqq 2$), $a\leqq\theta\leqq b$, and which for fixed x has a derivative $f_\theta'(x, \theta)$ which is a positive and nondecreasing function of θ on the segment $a\leqq\theta\leqq b$. Define $L_i=L_i(N)$ ($i=1, 2, 3$) by

$$L_1=\sum_{x=1}^N\frac{1}{f_\theta'(x, a)},\quad L_2=\sum_{x=1}^N\frac{x}{f_\theta'(x, a)},$$

$$L_3=\sum_{x=2}^N\sum_{y=1}^{x-1}\int_a^b\frac{f_\theta'(y, w)}{f_\theta'(x, w)}dw.$$

Let α, β denote a pair of real numbers such that $\tau=\beta-\alpha\leqq 1$ and let $R(N)=R_{\alpha,\beta}(N;\theta)$ denote the remainder with respect to the uniform distribution (mod 1) of the numbers $f(1, \theta), f(2, \theta), \cdots, f(N, \theta)$ corresponding to the interval $\alpha\leqq u<\beta$. Then we have

$$\int_a^b R^2(N)d\theta=(b-a)\tau N-(b-a)\tau^2N$$
$$\pm 16\eta\tau^2NL_1\pm 24\eta\tau L_2\pm 8\eta\tau L^3\ (0\leqq\eta\leqq 1).$$

Various applications are given. *R. Salem.*

K05-12 (11, 336d)

van Aardenne-Ehrenfest, T. On the impossibility of a just distribution. Nederl. Akad. Wetensch., Proc. **52**, 734–739 = Indagationes Math. **11**, 264–269 (1949).

Let $A=\{a_n\}$ be a sequence of points in a unit interval I, α a subinterval of I, $|\alpha|$ its length and $A_n(\alpha)$ the number of the $a_m\subset\alpha$ with $m\leqq n$. Let $F_n(A)$ be the upper bound of $|A_n(\alpha)-n|\alpha||$ for α varying in I. The sequence A is said to be "just" if $F_n(A)$ is bounded. The author shows that $F_n(A)$ is never $o(\log\log n/\log\log\log n)$, so that no just distribution exists, and remarks that where $F_n(A)$ has been calculated it has been not $o(\log n)$. *H. D. Ursell.*

Referred to in K05-21, K05-62, K40-48.

K05-13 (11, 423i)

de Bruijn, N. G., and Erdös, P. Sequences of points on a circle. Nederl. Akad. Wetensch., Proc. **52**, 14–17 = Indagationes Math. **11**, 46–49 (1949).

Let $a=\{a_n\}$ be a sequence of points on a circle of radius $1/(2\pi)$. The first n points divide the circle into n arcs: the greatest and least values of the length of r consecutive arcs are written $M_n^r(a)$, $m_n^r(a)$. Defining

$$\Lambda_r(a)=\limsup nM_n^r(a),\qquad \Lambda_r=\inf\Lambda_r(a),$$
$$\lambda_r(a)=\liminf nm_n^r(a),\qquad \lambda_r=\sup\lambda_r(a),$$
$$\mu_r(a)=\limsup M_n^r(a)/m_n^r(a),\qquad \mu_r=\inf\mu_r(a),$$

the authors evaluate $(\Lambda_1, \lambda_1, \mu_1)$, obtain bounds for $(\Lambda_r, \lambda_r, \mu_r)$ and conjecture that $r(\Lambda_r-1)$, $r(\mu_r-1)$, $r(1-\lambda_r)$ all become infinite with r. The last conjecture would imply van Aardenne-Ehrenfest's theorem [same Proc. **48**, 266–271

= Indagationes Math. **7**, 71–76 (1945); these Rev. **7**, 376].
H. D. Ursell (Leeds).

Citations: MR 7, 376l = K05-6.
Referred to in K05-26, K05-32.

K05-14 (11, 647h)

Rényi, Alfréd. On the measure of equidistribution of point sets. Acta Univ. Szeged. Sect. Sci. Math. **13**, 77–92 (1949).

Let E be a set on $(0, 1)$, of measure $|E|$, and characteristic function $F(x)$; $F(x)$ is extended so as to be periodic with period 1. Let $G(t) = \int_0^1 F(x)F(x+t)dx$, $m(E) = \min G(t)$, $\mu(E) = m(E)/|E|^2$, so that $0 \leqq \mu(E) < 1$ if $0 < |E| < 1$. The author calls $\mu(E)$ the measure of equidistribution of E; it actually measures the equidistribution of the set of distances between points of E. The main purpose of the paper is to construct a set E with assigned (arbitrarily small) positive measure and $\mu(E)$ arbitrarily close to 1. Such a set always intersects all its translates in a set whose measure is bounded from zero. The author then introduces a measure of k-fold equidistribution and proves a weaker theorem for $k > 1$ which, however, implies a generalization of this intersection property. Finally he makes a connection with Singer's "difference bases" in number theory and proves a theorem about them. *R. P. Boas, Jr.* (Evanston, Ill.).

K05-15 (12, 86c)

Koksma, J. F. An arithmetical property of some summable functions. Nederl. Akad. Wetensch., Proc. **53**, 959–972 = Indagationes Math. **12**, 354–367 (1950).

Let $g(x) \varepsilon L^2$ have period one and be such that, if S_m denotes the mth partial sum of its Fourier series, one has $\sum m^{-1} \int_0^1 (g - S_m)^2 dx < \infty$. Let $u_n(x)$, $n = 1, 2, \cdots$, be a sequence of functions defined for $0 \leqq x \leqq 1$. The author proves that the relation

$$\lim_{n \to \infty} [g(u_1(x)) + \cdots + g(u_n(x))]/n = \int_0^1 g(t)dt$$

holds almost everywhere in x for a wide class of sequences $u_n(x)$, among which one can quote in particular $u_n(x) = nx$, $u_n(x) = \lambda_n x$ $(\lambda_{n+1} - \lambda_n \geqq \delta > 0)$, $u_n(x) = (1+x)^n$.
R. Salem (Cambridge, Mass.).

Referred to in K20-6.

K05-16 (12, 394c)

Koksma, J. F. Some theorems on Diophantine inequalities. Math. Centrum Amsterdam, Scriptum no. **5**, i+51 pp. (1950).

The author proves the following theorem. Let Q denote the m-dimensional parallelepiped $a_\mu \leqq x_\mu < b_\mu$ $(\mu = 1, \cdots, m)$, where $a_1, \cdots, a_m, b_1, \cdots, b_m$ are integers. Let $\alpha_1, \cdots, \alpha_n, \beta_1, \cdots, \beta_n$ be $2n$ real numbers satisfying $\alpha_\nu < \beta_\nu \leqq \alpha_\nu + 1$ $(\nu = 1, \cdots, n)$, and let $f_1(x), \cdots, f_n(x)$ be n real functions defined for all lattice points $(x) = (x_1, \cdots, x_m)$ of Q. Let $\lambda_1, \cdots, \lambda_n$ be n positive numbers $\geqq 1$ and let

$$M_\nu = \lambda_\nu \log \{e \min (n, \lambda_\nu)\},$$

$\nu = 1, \cdots, n$. Let $N(Q)$ denote the number of lattice points $(x) \varepsilon Q$ and let $T(Q)$ be defined by $T(Q)N(Q)$

$$= \sum_{(h)}{}^* \Big| \sum_{(x) \varepsilon Q} \exp \{2\pi i(h_1 f_1(x) + \cdots + h_n f_n(x)\} \Big| \prod_{\nu=1}^n p_{h_\nu, \nu},$$

where

$$p_{h_\nu, \nu} = \min \{\beta_\nu - \alpha_\nu + 75/\lambda_\nu,\ 1 - (\beta_\nu - \alpha_\nu) + 75/\lambda_\nu,\ 30/|h_\nu|\}$$

for $h_\nu \neq 0$, $\nu = 1, \cdots, n$, $p_{0, \nu} = \beta_\nu - \alpha_\nu + 75/\lambda_\nu$, $\nu = 1, \cdots, n$, and $\sum_{(h)}^*$ is extended over all nonzero lattice points $(h) = (h_1, \cdots, h_n)$ in n-dimensional space such that $|h_\nu| \leqq M_\nu$, $\nu = 1, 2, \cdots, n$. Then the number $N^*(Q)$ of solutions of the Diophantine system $\alpha_\nu \leqq f_\nu(x) < \beta_\nu \pmod 1$, $\nu = 1, \cdots, n$, satisfies the inequality

$$|N^*(Q)/N(Q) - \prod_{\nu=1}^n (\beta_\nu - \alpha_\nu)|$$
$$\leqq \{\prod_{\nu=1}^n (\beta_\nu - \alpha_\nu + 75/\lambda_\nu) - \prod_{\nu=1}^n (\beta_\nu - \alpha_\nu)\} + T(Q).$$

This result is an improvement of a theorem said to have been obtained by the author and van der Corput in 1935 but never published [the one-dimensional case is given in the author's Diophantische Approximationen, Springer, Berlin, 1936, Kapitel IX, Satz 4]. The author also gives two corollaries of his main theorem, namely (1) an improvement of an unpublished theorem of van der Corput [quoted without proof, loc. cit., Kapitel X, Satz 2] and (2) a refinement of a theorem of Erdös and Turán [Nederl. Akad. Wetensch., Proc. **51**, 1146–1154, 1262–1269 = Indagationes Math. **10**, 370–378, 406–413 (1948); these Rev. **10**, 372, last theorem quoted in the review]. *P. T. Bateman* (Urbana, Ill.).

Citations: MR 10, 372c = K05-10; MR 10, 372d = K05-10.
Referred to in K05-19, K45-10.

K05-17 (13, 539e)

Cassels, J. W. S. A theorem of Vinogradoff on uniform distribution. Proc. Cambridge Philos. Soc. **46**, 642–644 (1950).

Let D denote the discrepancy (mod 1) of the N numbers $a_1, \cdots, a_N$ and E the discrepancy (mod 1) of the N^2 differences $a_n - a_m$ $(n, m = 1, 2, \cdots, N)$. Then, by a theorem of Vinogradoff, (1) $D^3 \leqq cE$, where c denotes an absolute constant. The author proves this inequality with $c = 12$ in a short and remarkable way by showing that a certain integral lies between the numbers $N^2D^3/24$ and $N^2E/2$. The reviewer remarks that van der Corput and Pisot [Nederl. Akad. Wetensch., Proc. **42**, 476–486 = Indagationes Math. **1**, 143–153 (1939)] have replaced (1) by a sharper inequality which contains the estimate $D^{2+\epsilon} \leqq c(\epsilon)E$ for every constant $\epsilon > 0$, $c(\epsilon)$ denoting a constant which depends on ϵ only.
J. F. Koksma (Amsterdam).

Referred to in K05-18, K05-20.

K05-18 (13, 630b)

Cassels, J. W. S. Corrigenda: A theorem of Vinogradoff on uniform distribution. Proc. Cambridge Philos. Soc. **48**, 368 (1952).

Cf. same Proc. **46**, 642–644 (1950); these Rev. **13**, 539.

Citations: MR 13, 539e = K05-17.
Referred to in K05-20.

K05-19 (15, 15c)

Szüsz, Péter. Über ein Problem der Gleichverteilung. Comptes Rendus du Premier Congrès des Mathématiciens Hongrois, 27 Août–2 Septembre 1950, pp. 461–472. Akadémiai Kiadó, Budapest, 1952. (Hungarian. Russian and German summaries)

The author proves the following theorem: Let

$$P_\nu = \{(x_\nu^{(1)}, x_\nu^{(2)}, \cdots, x_\nu^{(p)})\}$$

be a sequence of points in p-dimensional space. Assume that

$$\sum_{\nu=1}^n \exp [i(k_1 x_\nu^{(1)} + k_2 x_\nu^{(2)} + \cdots + k_p x_\nu^{(p)})] \leqq \psi(k_1, k_2, \cdots, k_p).$$

Let Y be a p-dimensional rectangle, i.e., $y = (y_1, y_2, \cdots, y_p)$ is in Y if $\alpha_i \leqq y_i \leqq \beta_i$ for $i = 1, 2, \cdots, p$. We further assume $0 \leqq \alpha_i < \beta_i \leqq 2\pi$, $i = 1, 2, \cdots, p$. Let $N(n, Y)$ denote the number of P_ν, $1\leqq\nu\leqq n$, whose coordinates reduced mod 2π are in Y, and finally let m be any integer. We then have

$$\left|N(n, Y)-\frac{M(Y)}{(2\pi)^p}n\right|$$
$$<k_p\left(\frac{n}{m+1}+\sum_{k_1=-m}^{m}{}' \sum_{k_2=-m}^{m}{}' \cdots \sum_{k_p=-m}^{m}{}' \frac{\psi(k_1, k_2, \cdots, k_p)}{(|k_1|+1)\cdots(|k_p|+1)}\right)$$

where $M(Y)=\prod_{i=1}^{p}(\beta_i-\alpha_i)$ is the volume of Y, k_p depends only on p and the dash in the summation sign indicates that the term $k_1=k_2=\cdots=k_p=0$ should be omitted.

For $p=1$ this result was proved by the reviewer and Turán [Nederl. Akad. Wetensch., Proc. **51**, 1146–1154, 1262–1269 (1948); these Rev. **10**, 372], and the author's proof is similar to ours. The author quotes a paper by Koksma [Math. Centrum Amsterdam, Scriptum no. **5** (1950); these Rev. **12**, 394] who discovered the same result simultaneously with the author. *P. Erdös.*

Citations: MR 10, 372c = K05-10; MR 10, 372d = K05-10; MR 12, 394c = K05-16.

K05-20 (15, 293a)

Cassels, J. W. S. A new inequality with application to the theory of diophantine approximation. Math. Ann. **126**, 108–118 (1953).

Let $a_1, a_2, \cdots, a_N$ be a set of real numbers. For $0\leqq\alpha<\beta\leqq 1$ denote by $F(\alpha, \beta)$ the number of solutions of $\alpha\leqq a_n<\beta$ (mod 1). Then

$$D=\max_{(\alpha,\beta)}\left|\beta-\alpha-\frac{1}{N}F(\alpha, \beta)\right|$$

is called the discrepancy of the set. Improving results of Vinogradoff [Izvestiya Akad. Nauk SSSR (6) **20**, 585–600 (1926)] and van der Corput and Pisot [cf. Nederl. Akad. Wetensch., Proc. **42**, 476–486, 554–565, 713–722 (1939) = Indagationes Math. **1**, 143–153, 184–195, 260–269 (1939); these Rev. **1**, 66; and also the author's earlier paper, Proc. Cambridge Philos. Soc. **46**, 642–644 (1950); **48**, 368 (1952); these Rev. **13**, 539, 630], the author proves the fundamental inequality $D\leqq AE^{1/2}(1+|\log E|)$, where A is an absolute constant and where E denotes the discrepancy of the set of the N^2 differences a_m-a_n. Several applications.

J. F. Koksma (Amsterdam).

Citations: MR 1, 66b = K05-1; MR 13, 539e = K05-17; MR 13, 630b = K05-18.

Referred to in K05-38.

K05-21 (16, 575c)

Roth, K. F. On irregularities of distribution. Mathematika **1**, 73–79 (1954).

If $P_1, \cdots, P_N$ are N points in $(0\leqq x\leqq 1, 0\leqq y\leqq 1)$ and $S(x', y')$ is the number of them in $(x<x', y<y')$, then $\int_0^1\int_0^1[S(x, y)-Nxy]^2dxdy>c\log N$. It follows that there exists (x', y'), $|S(x', y')-Nx'y'|>c'\sqrt{\log N}$, and also that

$$F_n(A)>c''\sqrt{\log n},$$

greatly improving the result $F_n>k\log\log n/\log\log\log n$ of van Aardenne-Ehrenfest [Nederl. Akad. Wetensch., Proc. **52**, 734–739 (1949); MR **11**, 336] on the impossibility of a "just" distribution. *H. D. Ursell* (Leeds).

Citations: MR 11, 336d = K05-12.

Referred to in K05-23, K05-56, K05-60, K05-67, K05-69, K05-72, K05-75.

K05-22 (16, 682h)

Koksma, J. F. Estimations de fonctions à l'aide d'intégrales de Lebesgue. Bull. Soc. Math. Belg. **6** (1953), 4–13 (1954).

The following general theorem is proved: Let $\{u_\nu(x)\}$ be a sequence of real functions in $L^{(p)}(a, b)$, $p>1$. Let $\eta(t)$ be a positive decreasing function of $t>0$, such that $\sum_{N=1}^{\infty}\eta(N)/N<\infty$. Let $C_1, C_2, C_3, \alpha, \beta, \gamma$ be non-negative constants with $\alpha>0, \gamma>1$. Put $F(M, N)=\sum_{M+1}^{M+N}u_\nu(x)$, and suppose that for every $M\geqq 1$, $N\geqq 1$ such that $M=2^n-1$ and $N|(M+1)$, the inequality

$$\int_a^b|F(M, N)|^pdx\leqq C_1N^\alpha G(M, N)+C_2(M+N)^{p-\gamma}N^\gamma\eta(N)$$

holds, where G is such that

$$\sum_{k=0}^{(M+1)/N}G(M+kN, N)\leqq C_3N^\beta G(M, M+1)$$

and $\sum_1^\infty G(2^n-1, 2^n)/2^n(p-\alpha-\beta)<\infty$. Then for almost all $x\,\varepsilon\,[a, b]$, $F(0, N)=o(N)$.

As an application, it is shown that if $g\,\varepsilon\,L^{(2)}(0, 1)$ is of period 1 and has Fourier coefficients c_k such that $\sum_{k=1}^{\infty}|c_k|^2\sum_{d|k}d^{-1}<\infty$, then

$$N^{-1}\sum_{n=1}^{N}g(nx)\to\int_0^1 g(t)\,dt$$

for almost all x. This improves an earlier result of the author [J. Indian Math. Soc. (N.S.) **15**, 87–96 (1952); MR **13**, 827].

W. J. LeVeque (Ann Arbor, Mich.).

Citations: MR 13, 827c = K10-11.

Referred to in K20-6.

K05-23 (18, 566a)

Davenport, H. Note on irregularities of distribution. Mathematika **3** (1956), 131–135.

Roth [Mathematika **1** (1954), 73–79; MR **16**, 575] has shown that there exists an absolute positive constant c with the following property: Let (x_n, y_n) $(1\leqq n\leqq N)$ be N points in the unit square $0\leqq x<1$, $0\leqq y<1$; and for $0\leqq\xi<1$, $0\leqq\eta<1$ let $S(\xi, \eta)$ denote the number of these in $0\leqq x_n<\xi$, $0\leqq y_n<\eta$. Then

$$\int_0^1\int_0^1(S(\xi, \eta)-N\xi\eta)^2d\xi d\eta>c\log N.$$

Davenport shows that c cannot be replaced by any function of N tending to infinity with N. The counterexample makes use of the fractional parts of a real number with bounded partial quotients. It is shown that the corresponding result of Roth in three dimensions would similarly be the best possible if a problem of Littlewood about the existence of a pair of irrationals with certain properties has a positive solution. *J. W. S. Cassels.*

Citations: MR 16, 575c = K05-21.

K05-24 (19, 638a)

Ostrowski, Alexander. Zum Schubfächerprinzip in einem linearen Intervall. Jber. Deutsch. Math. Verein. **60** (1957), Abt. 1, 33–39.

Let $a_1=0, a_2, a_3, \cdots$ be distinct numbers in the interval $0\leqq x<1$. The first n of them divide the unit interval into n subintervals; let u_n be the length of the shortest subinterval. Then for every n, the inequality $nu_n\leqq 1$ holds, and the constant 1 cannot be reduced, since $a_1, \cdots, a_n$ may be equally spaced. The author considers the problem of finding a constant $\omega<1$ such that $nu_n\leqq\omega$ for infinitely many (but not necessarily all) n, whatever the sequence $\{a_k\}$ may be. He shows that $\omega_0=(4-2\sqrt{2})^{-1}=0.853\cdots$ is an allowable value, and in fact that for all sufficiently large n either $nu_n\leqq\omega_0$ or $Nu_N\leqq\omega_0$, where $N=[n\sqrt{2}]$. On the other hand, a simple example shows that ω certainly cannot be taken smaller than $\frac{1}{2}$. It is proved in a later paper [reviewed below] that $\omega=(\log 4)^{-1}=0.721\cdots$ is an allowable value.

W. J. LeVeque (Ann Arbor, Mich.).

Referred to in K05-27.

K05-25 **(19, 638e)**

Ostrowski, Alexander. Eine Verschärfung des Schubfächerprinzips in einem linearen Intervall. Arch. Math. 8 (1957), 1–10.

If $a_1=0, a_2, a_3, \cdots, a_n$ is a sequence of points on the half open interval $\langle 0, 1)$ it is called an a-sequence. Let u_n denote the length of the minimal interval of the set of the subdivision of $\langle 0, 1)$ formed by the points $a_1=0, a_2, \cdots, a_n$. Then $u_1 \geqq u_2 \geqq u_3 \cdots, u_\nu \leqq \nu^{-1}$. It has been shown by the author [see the paper reviewed above] that $\liminf \nu u_\nu = \Omega \leqq 0.854$. In the present paper the author establishes the relation $\frac{1}{2} \leqq \Omega \leqq 1/\log 4 = 0.7213\cdots$. *R. L. Jeffery.*

Referred to in K05-26, K05-32.

K05-26 **(20# 35)**

Schönhage, Arnold. Zum Schubfächerprinzip im linearen Intervall. Arch. Math. 8 (1957), 327–329.

Consider the sequences $A=(a_0, a_1, \cdots)$ with $a_0=1$, $a_1=0$, $0 \leqq a_\nu \leqq 1$ and $a_\mu \neq a_\nu$ for $\mu \neq \nu$. The points $a_0, a_1, \cdots, a_n$ of such a sequence divide the interval $0 \leqq x \leqq 1$ into n subintervals; let u_n be the shortest subinterval, so that $0 < u_n \leqq 1/n$, and let $\omega = \sup_A\{\liminf nu_n\}$. A. Ostrowski showed [Arch. Math. 8 (1957), 1–10; MR 19, 638] that $\frac{1}{2} \leqq \omega \leqq 1/\log 4 = 0.7213\cdots$. It is shown here that $\omega = 1/\log 4$. It is pointed out in a footnote that this result has been obtained earlier by Erdös and de Bruijn [Proc. Nederl. Akad. Wetensch. Proc. Ser. A. 52, 46–49 (1949); MR 11, 423]. *W. J. LeVeque* (Göttingen)

Citations: MR 11, 423i = K05-13; MR 19, 638e = K05-25.

Referred to in K05-32.

K05-27 **(20# 36)**

Ostrowski, Alexander. Bemerkungen zu meiner Mitteilung: Eine Verschärfung des Schubfächerprinzips in einem linearen Intervall. Arch. Math. 8 (1957), 330.

Acknowledgment of the priority of Erdös and de Bruijn, as regards the evaluation of ω [see the preceding review]. Certain 'finite' results of the paper of the title [and also Jber. Deutsch. Math. Verein. 60 (1957), 33–39; MR 19, 638] were not obtained by those authors. *W. J. LeVeque* (Göttingen)

Citations: MR 19, 638a = K05-24.

K05-28 **(20# 37)**

Toulmin, G. H. Subdivision of an interval by a sequence of points. Arch. Math. 8 (1957), 158–161.

Another proof that $\omega = 1/\log 4$. [See the two preceding reviews.] It is also shown here that the bound is actually attained, for example by the sequence in which $a_n = \{\log_2(2n-1)\}$ for $n \geqq 1$, where $\{x\} = x - [x]$ and $\log_2$ denotes the logarithm to base 2. Furthermore, it is shown that $\limsup nv_n \geqq 1/\log 2$ for all sequences A, where v_n is the longest of the subintervals into which the points $a_0, a_1, \cdots, a_n$ divide the unit interval. This lower bound is also attained by the above sequence. *W. J. LeVeque* (Göttingen)

Referred to in K05-32.

K05-29 **(20# 848)**

Coles, W. J. On a theorem of van der Corput on uniform distribution. Proc. Cambridge Philos. Soc. 53 (1957), 781–789.

Van der Corput proved [Acta Math. 56 (1931), 373–456; 59 (1932), 209–328] that a sequence of points (α_n, β_n) $(1 \leqq n < \infty)$ in the plane is uniformly distributed mod 1 if and only if for all pairs of integers u, v excluding $u=v=0$ the one-dimensional sequence $u\alpha_n + v\beta_n$ $(1 \leqq n < \infty)$ is uniformly distributed mod 1. The author gives a quantitative form to this qualitative criterion. Denote by $F^{(N)}(x_0, x_1; y_0, y_1)$ the number of points (α_n, β_n) $(1 \leqq n \leqq N)$ satisfying

$$x_0 \leqq \alpha_n \leqq x_1 \pmod 1), \quad y_0 \leqq \beta_n \leqq y_1 \pmod 1).$$

Put

$$D^{(N)} = \max \frac{1}{N} |F^N(x_0, x_1; y_0, y_1) - N(x_1-x_0)(y_1-y_0)|.$$

$D^{(N)}$ is said to be the discrepancy of the sequence (α_n, β_n) $(1 \leqq n \leqq N)$. Denote by $D_{u,v}^{(N)}$ the (one-dimensional) discrepancy of $u\alpha_n + v\beta_n$ $(1 \leqq n \leqq N)$. The author proves that there is an absolute constant C such that for every $\varepsilon > 0$

$$D^{(N)} \leqq \varepsilon + C(D_{0,1}^{(N)} + D_{1,0}^{(N)} + \sum_{\substack{(u,v)=1 \\ u>0,\, v>0}} f_{uv}(\varepsilon) D_{uv}^{(N)}),$$

where $f_{u,v}(\varepsilon) = \min(|uv|^{-1}, |\varepsilon uv|^{-2})$. *P. Erdös* (Toronto, Ont.)

Referred to in K05-54.

K05-30 **(20# 3404)**

Kennedy, P. B. A note on uniformly distributed sequences. Quart. J. Math. Oxford Ser. (2) 7 (1956), 125–127.

For a divergent sequence $\{x_n\}$ a well-known Tauberian theorem of Hardy [Proc. London Math. Soc. (2) 8 (1910), 310–320; Divergent series, Clarendon Press, Oxford, 1949; MR 11, 25; p. 121] asserts that (*) $\limsup_{n\to\infty} n|\Delta x_n| = +\infty$, where $\Delta u_n = u_n - u_{n-1}$. And Littlewood showed [Proc. London Math. Soc. (2) 9 (1911), 434–448; Theorem C] that (*) cannot be improved for general divergent sequences which have $(C, 1)$-limits. The author shows that (*) cannot be improved even for uniformly distributed sequences. *S. Ikehara* (Tokyo)

K05-31 **(22# 5620)**

Schützer, W. On non-equidistributed sequences of numbers mod. 1. Bol. Soc. Mat. São Paulo 12 (1957), 1–9 (1960).

An unnecessarily complicated proof of the existence of a sequence of numbers in [0, 1] having a particular non-uniform limiting distribution. The desired distribution (constant densities on three subintervals together comprising [0, 1]) is erroneously described, as the integral of the density function is not 1. *W. J. LeVeque* (Ann Arbor, Mich.)

K05-32 **(22# 8245)**

Groemer, Helmut. Über den Minimalabstand der ersten N Glieder einer unendlichen Punktfolge. Monatsh. Math. 64 (1960), 330–334.

Let $\{P_m\}$ be a sequence of points on the unit interval [0, 1], and write $d_N = \min |P_i - P_k|$ $(1 \leqq i, k \leqq N;\ i \neq k)$. The estimate

$$(*) \qquad \liminf_{N\to\infty} Nd_N \leqq 1/\log 4$$

was established, in recent years, by a number of authors; cf. N. G. de Bruijn and P. Erdös [Nederl. Akad. Wetensch. Proc. 52 (1949), 14–17 = Indag. Math. 11, 46–49; MR 11, 423], A. Ostrowski [Arch. Math. 8 (1957), 1–10; MR 19, 638], G. H. Toulmin [ibid. 158–161; MR 20 #37], A. Schönhage [ibid. 327–329; MR 20 #35]. In the present paper a generalization of the problem for n-dimensional euclidean space R_n is considered. Let $d(P, Q)$ be a normalized Minkowski metric for R_n, i.e., $d(P, Q)$ is a real-valued function defined on $R_n \times R_n$ and satisfying the following conditions. (i) $d(P, Q) > 0$ for $P \neq Q$. (ii) $d(P, Q) = d(R, S)$ for $P - Q = R - S$. (iii) $d(tP, tQ) = |t| d(P, Q)$ for all real t.

(iv) $d(P, R) \leqq d(P, Q) + d(Q, R)$. (v) For some fixed point P_0, the set of points X such that $d(P_0, X) \leqq 1$ has volume 2^n. Further, let B be a bounded subset of R_n having Jordan volume 1. Finally, let $\{P_m\}$ be an arbitrary sequence of points in B, and write

$$d_N = \min d(P_i, P_k) \quad (1 \leq i, k \leq N; i \neq k).$$

It is then shown that

$$(**) \qquad \liminf_{N\to\infty} N d_N^n \leq \left(1 + n \int_0^1 \frac{(1-x)^n}{x+1}\, dx\right)^{-1}.$$

For $n = 1$, the expression on the right-hand side of (**) is equal to 1/log 4. Thus the result proved in the present paper contains (*) as a special case. It is known that the constant in (*) is best possible; whether the same is true for (**) remains an open question.

{Reviewer's remark. The factor $(1-x)^n$ on the right-hand side of (**) is printed incorrectly as $(x-1)^n$.}

L. Mirsky (Sheffield)

Citations: MR 11, 423i = K05-13; MR 19, 638e = K05-25; MR 20# 35 = K05-26; MR 20# 37 = K05-28.

K05-33 (23# A3731)

Chauvineau, Jean

Sur la répartition modulo 1 de certaines fonctions périodiques.

C. R. Acad. Sci. Paris **252** (1961), 4090–4092.

Let $u_1, u_2, \cdots$ be a sequence of real numbers, equidistributed mod 1. Let $F(x)$ be a continuous and monotonic function on the interval (0, 1), mapping that interval onto $(-\infty, \infty)$. The author shows that the sequence $F(u_n - [u_n])$, again considered mod 1, has a distribution function, and he investigates when this sequence is equidistributed. *N. G. de Bruijn* (Eindhoven)

Referred to in K05-68.

K05-34 (24# A108)

Hlawka, Edmund

Cremonatransformation von Folgen modulo 1.

Monatsh. Math. **65** (1961), 227–232.

An infinite sequence of points in s-dimensional space is said to have density function ρ (mod 1) if for every measurable set Q in the unit cube E^s, the limiting frequency of the points of the sequence, reduced (mod 1), which lie in Q is $\int_Q \rho(\xi) d\xi$. J. Schoenberg proved in the case $s = 1$ [Math. Z. **28** (1928), 171–199] that if $x_1, x_2, \cdots$ is uniformly distributed (mod 1) (i.e., has density function 1), and if $x_i = (\xi_{i1}, \cdots, \xi_{is})$, then the sequence $y_1, y_2, \cdots$, in which $y_i = (\xi_{i1}^{-1}, \cdots, \xi_{is}^{-1})$, has density function

$$\rho(\xi) = \prod_{i=1}^{s} \sum_{n=1}^{\infty} (n + \xi_i)^{-2},$$

(mod 1). This theorem is proved here for arbitrary $s \geqq 1$. It is also shown that if D_n is the discrepancy for the sequence $x_1, \cdots, x_n$, and D_n^* that for $y_1, \cdots, y_n$, then $D_n^* \leqq 2 \cdot 12^s D_n^{1/(s+1)}$. *W. J. LeVeque* (Ann Arbor, Mich.)

K05-35 (25# 3029)

Hlawka, Edmund

Funktionen von beschränkter Variation in der Theorie der Gleichverteilung.

Ann. Mat. Pura Appl. (4) **54** (1961), 325–333.

Let $\omega_n = \{x_1, x_2, \cdots, x_n\}$ be a finite set of points in the s-dimensional unit cube E^s, and let Q be an s-dimensional subinterval of E^s. Let $n'(Q)$ be the number of points from among $x_1, \cdots, x_n$ which belong to Q, and let $\mu(Q)$ be the volume of Q. The quantity $D_n = \sup_{Q \in E^s} |n'(Q)/n - \mu(Q)|$ is called the discrepancy of ω_n. Let w be a Riemann integrable function on E^s, and put $m(w, \omega_n) = n^{-1} \sum_{k=1}^n w(x_k)$. It is shown that if w is of bounded variation on E^s, then

$$\left| m(w, \omega_n) - \int_E w(x)\, dx \right| \leq V(w) D_n,$$

where V is a complicated function of the total variation of w on E^s and its lower-dimensional projections. This generalizes the 1-dimensional result due to J. F. Koksma [Mathematica (Zutphen) B **11** (1942), 7–11; MR **7**, 370].

W. J. LeVeque (Ann Arbor, Mich.)

Citations: MR 7, 370a = K05-5.

Referred to in K05-39, K45-10, K45-13.

K05-36 (25# 4273)

Cigler, Johann

Über eine Verallgemeinerung des Hauptsatzes der Theorie der Gleichverteilung.

J. Reine Angew. Math. **210** (1962), 141–147.

A sequence formed by real numbers $x_1, x_2, \cdots$ is distributed uniformly modulo 1 if the sequence $x_{h+1} - x_1, x_{h+2} - x_2, \cdots$ is distributed uniformly for each given positive integer h (the reviewer's main theorem in the theory of uniform distribution modulo 1). By means of a modification of a method of J. Bass [Bull. Soc. Math. France **87** (1959), 1–64; MR **23** #A476], the author gives a generalization of this theorem on strongly regular matrices $A = (a_{nk})$. [Strongly regular means that $a_{nk} \geqq 0$ and that, as $n \to \infty$, $\sum_{k=1}^{\infty} a_{nk} \to 1$, $a_{nk} \to 0$ and $\sum_{k=1}^{\infty} (a_{n,k+1} - a_{nk}) \to 0$.] A function $\gamma_f(h)$ of the integers h is called an A-correlation function of a bounded function $f(k)$ $(k = 1, 2, \cdots)$, if it is possible to find a monotonically increasing sequence m such that

$$\gamma_f(h) = \lim_m \sum_k a_{mk} f(k+h) \overline{f(k)}.$$

The author deduces a number of properties of these functions. (1) $\gamma_f(h)$ is positive definite. (2) There exists a positive Radon measure $\sigma_f = \sigma_{\gamma f}$ in [0, 1] with $\gamma_f(h) = \int_0^1 e^{2\pi i h x} d\sigma_\gamma(x)$ (h integer) and $M_h \gamma_f(h) = \sigma_f(0) \geqq 0$, where the left-hand side of the last relation is the limit of $\sum_h a_{nh} \gamma_f(h)$ as $n \to \infty$.

Application of these general results to the special functions $f_l(k) = e^{2\pi i l x_k}$ $(l = 0, 1, \cdots)$ leads not only to the main theorem formulated above, but also to some generalizations of this theorem. Finally, the author treats the corresponding results for sequences in compact groups and for the uniform distribution of functions which are defined on a locally compact group.

J. G. van der Corput (Berkeley, Calif.)

Citations: MR 23# A476 = K40-22.

Referred to in K05-37.

K05-37 (29# 6214)

Cigler, Johann

The fundamental theorem of van der Corput on uniform distribution and its generalizations.

Compositio Math. **16**, 29–34 (1964).

The contents of this paper were published earlier in a slightly different form [J. Reine Angew. Math. **210** (1962), 141–147; MR **25** #4273].

Citations: MR 25# 4273 = K05-36.

K05-38 (27# 109)

Hlawka, Edmund

Über die Diskrepanz mehrdimensionaler Folgen mod 1.

Math. Z. **77** (1961), 273–284.

Upper and lower bounds are given for the discrepancy (mod 1), $D_n(\omega)$, of a finite sequence $\omega = \{x_1, \cdots, x_n\}$ in s-dimensional Euclidean space in terms of the discrepancies (mod 1) of its projections $\omega(h) = \{hx_1, \cdots, hx_n\}$, where

$h=(h_1, \cdots, h_s)$ has integer coordinates. For instance, for each positive integer M,

$$D_n(\omega) < (31)^{s+1}[M^{-1} + \sum_{0<\|h\|\leq M}' D_n(\omega(h))/N(h)].$$

where the sum is extended over the lattice points h for which the components have a g.c.d. equal to 1. Further, $\|h\| = \max|h_i|$ and $N(h) = \prod_1^s \max(1, |h_i|)$.

Extending a result of Cassels [Math. Ann. **126** (1953), 108–118; MR **15**, 293], the author further shows that

$$D_n(\omega) < (16)^{4s} B^{1/2}(1+\log^s 1/B),$$

whenever $1 \leq q \leq n$. Here,

$$B = \{q^{-1}+h^{-1}\}\{1+2\sum_{j=1}^{q-1} D_{n-j}(x_{j+1}-x_1, \cdots, x_n-x_{n-j})\}.$$

J. H. B. Kemperman (Rochester, N.Y.)

Citations: MR 15, 293a = K05-20.

K05-39 (27# 3615)

Hlawka, Edmund

Geordnete Schätzfunktionen und Diskrepanz.

Math. Ann. **150** (1963), 259–267.

The discrepancy of sequences in $R=R^s$ $(s \leq 1)$ which has been investigated in recent papers by the author [Ann. Mat. Pura Appl. (4) **54** (1961), 325–333; MR **25** #3029; Monatsh. Math. **66** (1962), 140–151; MR **26** #888] is considered here in a slightly more general form. If $x=(\xi_1, \cdots, \xi_s) \subset R$, let $\bar{x} \subset E$ denote the reduced point $\bar{x} \equiv x \pmod 1$, where $E=E^s$ denotes the unit cube $0 \leq \xi_j < 1$ $(j=1, 2, \cdots, s)$. Let $Q=Q^s$ denote an arbitrary interval $\alpha_j \leq \xi_j < \beta_j \subset E$ and $\chi(Q\,;x)$ its characteristic function. Put, if ω_n denotes a given n-tuple of points $x_k \subset R$,

$$\nu = \nu(Q\,;\omega_n) = \frac{1}{n}\sum_{k=1}^{n} \chi(Q\,;\bar{x}_k);$$

then for a given non-negative Jordan integrable function $\rho(x)$ we call

$$(1) \qquad D = D(\omega_n\,;\rho) = \sup |\nu - \int_E \chi(Q\,;x)\rho(x)\,dx|$$

the discrepancy of ω_n with respect to the density $\rho(x)$. In (1) the supremum is extended over all $Q \subset E$ with $\alpha_j=0$. If the supremum is extended over all $Q \subset E$ we write D^* instead of D. For $\rho(x) \equiv 1$ we have the ordinary discrepancy (D^*) of ω_n. Now, using the methods of the theory of ordered estimators in mathematical statistics [cf. B. L. van der Waerden, *Mathematische Statistik*, § 17, Springer, Berlin, 1957; MR **18**, 771; errata, MR **18**, p. 1119], the author constructs, to a given finite sequence ω_n and a given density $\rho(x)$, other such sequences and densities, say $\tilde{\omega}_n$ and $\tilde{\rho}$, and gives estimates for $D(\tilde{\omega}_n, \tilde{\rho})$ in terms of $D(\omega_n, \rho)$. Several applications furnish approximation formulae for multiple integrals and an improvement of a theorem of Birnbaum and Pyke [Ann. Math. Statist. **29** (1958), 179–187; MR **20** #393; cf. also M. Dwass, ibid. **29** (1958), 188–191; MR **20** #2051; N. H. Kuiper, ibid. **30** (1959), 251–252; MR **21** #356].

J. F. Koksma (Amsterdam)

Citations: MR 25# 3029 = K05-35; MR 26# 888 = K45-10.

Referred to in K05-55.

K05-40 (27# 3618)

Davenport, H.; Erdős, P.; LeVeque, W. J.

On Weyl's criterion for uniform distribution.

Michigan Math. J. **10** (1963), 311–314.

Let $\{s_n(x)\}$ be a sequence of real-valued functions of the real parameter x, each $s_n(x)$ being bounded and integrable in $a \leq x \leq b$. Put

$$S(N, x) = \frac{1}{N}\sum_{n=1}^{N} e(ms_n(x)), \quad I(N) = \int_a^b |S(N, x)|^2\,dx.$$

The authors prove that if the series $\sum N^{-1}I(N)$ converges for each non-zero integer m, then the sequence $s_n(x)$ is uniformly distributed modulo 1 for almost all x in $a \leq x \leq b$. On the other hand, given any increasing function $\phi(M)$ which tends to infinity with M, there exists a sequence $s_n(x)$ which is not uniformly distributed modulo 1 for any x and which satisfies the inequality $\sum_1^M N^{-1}I(N) < \phi(M)$.

D. J. Lewis (Ann Arbor, Mich.)

Referred to in K20-9, K40-29, K40-49.

K05-41 (28# 5189)

Bateman, P. T.

Sequences of mass distributions on the unit circle which tend to a uniform distribution.

Amer. Math. Monthly **71** (1964), 165–172.

The principal result established in this paper is as follows. Let ϕ be a complex-valued function of a real variable which is continuous, has period 1, and is of bounded variation on (0, 1], so that the curve C: $z=\phi(t)$, $0 \leq t \leq 1$, is closed and rectifiable. Let N_1, N_2 be two mass distributions on C, each of total mass 1. For $0<\beta-\alpha\leq 1$ and $i=1, 2$, let $N_i(\alpha, \beta)$ denote the amount of mass from N_i lying on the part of C specified by the inequality $\alpha < t \leq \beta$. Write

$$T_i = \sup_{V \subseteq N_i} \left|\int_0^1 \phi(t)\,dV(0, t)\right|,$$

where the upper bound is taken over all mass distributions V such that $V(\alpha, \beta) \leq N_i(\alpha, \beta)$ whenever $0<\beta-\alpha\leq 1$. Put

$$\Delta = \sup_{\alpha<\beta\leq\alpha+1} |N_1(\alpha, \beta) - N_2(\alpha, \beta)|$$

and let $L=\sup_\lambda L(\lambda)$, where $L(\lambda)$ denotes the total variation on the unit interval of the function ϕ_λ given by

$$\phi_\lambda(t) = \max\{0, \Re(e^{-2\pi i\lambda}\phi(t))\}.$$

Then $|T_1-T_2| \leq L\Delta$.

A number of specializations of this theorem are discussed and, in particular, the following result which had been conjectured by the reviewer [same Monthly **69** (1962), 772–775; MR **27** #2473] is proved. Let $\{d_n\}$ be a sequence of positive integers tending to ∞; and suppose that, for any n, D_n is a set of d_n complex numbers of unit modulus. If $0<\beta-\alpha\leq 1$, denote by $N_n(\alpha, \beta)$ the number of $z \in D_n$ such that $2\pi\alpha < \arg z \leq 2\pi\beta$. Suppose that

$$\lim_{n\to\infty} \frac{N_n(\alpha, \beta)}{d_n} = \beta - \alpha$$

whenever $0<\beta-\alpha\leq 1$. Then, writing

$$T_n = \max_{D \subseteq D_n} |\sum_{z\in D} z|,$$

we have

$$\lim_{n\to\infty} \frac{T_n}{d_n} = \frac{1}{\pi}.$$

L. Mirsky (Sheffield)

Citations: MR 27# 2473 = C15-25.

K05-42 (29# 76)

Gerl, Peter

Einige metrische Sätze in der Theorie der Gleichverteilung mod 1.

J. Reine Angew. Math. **216** (1964), 50–66.

Generalizing methods and results of J. F. Koksma

[Composito Math. **2** (1935), 250–258] and W. J. LeVeque [Michigan Math. J. **1** (1953), 139–162; MR **14**, 1067], the author proves some metrical theorems on the uniform distribution mod 1 of sequences in r-dimensional space. The first part of the paper is concerned with the usual concept of uniform distribution and the second one with uniform distribution with respect to certain weighted means. There are some applications to sequences of matrices, among which the following is typical: The sequence $\begin{pmatrix} a & b \\ c & d \end{pmatrix}^n$ is uniformly distributed mod 1 for almost all a, b, c, d in the domain $a \geqq 0$, $b \geqq 0$, $a+d \geqq 1$, $ad-bc<0$. *J. E. Cigler* (Vienna)

Citations: MR 14, 1067d = K30-10.

K05-43 (29# 2225)

Gerl, Peter

Zur Gleichverteilung auf Flächen.

J. Reine Angew. Math. **216** (1964), 113–122.

The author studies the asymptotic behavior of sequences on r-dimensional surfaces Φ. A sequence of points $u(n) \in \Phi$ is said to be uniformly distributed with density ρ on a Jordan domain $Q \subseteq \Phi$ with area $\|Q\| < \infty$ if for every complex-valued Riemann integrable function f on Q the relation

$$\lim_{N\to\infty} \frac{1}{N} \sum_{n \leqq N} f(u(n)) = \frac{1}{\|Q\|} \int_Q f(u)\rho(u)\,du$$

holds. The author characterizes sequences $u(n)$ with density ρ on Q in terms of uniformly distributed sequences in the r-dimensional unit cube E_r and obtains generalizations of a theorem of H. Weyl. An application is made to the description of the structure of uniformly distributed sequences in the group G of rotations in 3-dimensional space: They are of the form $u(n) = (\varphi_1(n), \vartheta(n), \varphi_2(n))$ if $\varphi_1, \vartheta, \varphi_2$ denote the Eulerian angles of an element $g \in G$, where $\varphi_1(n) = 2\pi y_1(n)$, $\vartheta(n) = \arccos(1-2y_2(n))$, $\varphi_2(n) = 2\pi y_3(n)$, and $(y_1(n), y_2(n), y_3(n))$ is uniformly distributed in E_3. *J. E. Cigler* (Vienna)

K05-44 (30# 1115)

Bertrandias, Jean-Paul

Suites pseudo-aléatoires et critères d'équirépartition modulo un.

Compositio Math. **16**, 23–28 (1964).

Verfasser gibt eine Verallgemeinerung des van der Corput'schen Hauptsatzes der Theorie der Gleichverteilung mod 1 unter Verwendung positiv-definiter Funktionen. *J. E. Cigler* (Groningen)

Referred to in K20-31.

K05-45 (30# 3494)

Kemperman, J. H. B.

Probability methods in the theory of distributions modulo one.

Compositio Math. **16**, 106–137 (1964).

Various problems relating to complete A-distributions are treated. This notion is defined as follows. $(x_{n,k})$ is a double sequence of points in a compact metric space G. G^∞ is the countable-product space; $x_{n,k}{}^\infty$ is that point of G^∞ whose rth coordinate is $x_{n,k+r}$. A is a summation method determined by an appropriate matrix (a_{nk}). $\mu_{\infty,n}$ is the measure on G^∞ defined by $\mu_{\infty,n}(f) = \sum_k a_{nk} f(x_{n,k}{}^\infty)$ for $f \in C(G^\infty)$. If the sequence $\mu_{\infty,n}$ has a unique limit point μ_∞, then the latter is called the complete A-distribution of $(x_{n,k})$. *J. G. Wendel* (Ann Arbor, Mich.)

K05-46 (30# 3495)

Kemperman, J. H. B.

On the distribution of a sequence in a compact group.

Compositio Math. **16**, 138–157 (1964).

Using techniques from the paper reviewed above [#3494] the author generalizes the classical result of van der Corput: if $\{x_k\}$ is a sequence of real numbers such that for every n the sequence $\{x_{k+n} - x_k\}$ is uniformly distributed (mod 1), then so is the original sequence. *J. G. Wendel* (Ann Arbor, Mich.)

K05-47 (30# 4745)

Hlawka, Edmund

Discrepancy and uniform distribution of sequences.

Compositio Math. **16**, 83–91 (1964).

The author gives a survey of some new results in the quantitative theory of uniform distribution mod 1. Among the topics discussed are the Erdős-Turán-Koksma formula, a quantitative form of van der Corput's fundamental theorem, the connection between discrepancies of one- and n-dimensional sequences, and proposals for suitable definitions of the concept "discrepancy" in arbitrary compact spaces. *J. E. Cigler* (Groningen)

K05-48 (31# 3401)

Leveque, W. J. [LeVeque, W. J.]

An inequality connected with Weyl's criterion for uniform distribution.

Proc. Sympos. Pure Math., Vol. VIII, pp. 22–30. *Amer. Math. Soc., Providence, R.I.*, 1965.

Sei $\{x_k\}$ eine beliebige Folge reeller Zahlen, $\alpha \in [0,1]$, $N_\alpha(n)$ die Anzahl der Indices $k \leqq n$ mit $x_k - [x_k] \in [0, \alpha)$ und $D_n = \sup_{0 \leqq \alpha \leqq 1} |N_\alpha(n)/n - \alpha|$ die in der Theorie der Gleichverteilung übliche Diskrepanz. Der Verfasser beweist, dass stets

$$D_n \leqq \left\{ 6\pi^{-2} \sum_{m=1}^{\infty} \frac{1}{m^2 n^2} \left| \sum_{k=1}^{n} \exp(2\pi i m x_k) \right|^2 \right\}^{1/3}$$

gilt, wobei der Exponent $\frac{1}{3}$ nicht zu verschärfen ist. Während das wohlbekannte Kriterium von H. Weyl [Math. Ann. **77** (1916), 313–352] nur besagt, dass die Folge $\{x_k\}$ genau dann gleichverteilt ist, wenn für jedes ganze $m \neq 0$ die Summe $(1/n) \sum_{k=1}^n \exp(2\pi i m x_k)$ gegen Null konvergiert, wird hier ein quantitativer Zusammenhang zwischen dieser Summe und der Diskrepanz hergestellt. Als wichtigstes Hilfsmittel wird dabei die in der Wahrscheinlichkeitstheorie übliche Methode der charakteristischen Funktionen verwendet, und dem Beweis des erwähnten Satzes wird eine neue, elegante Herleitung des Weylschen Kriteriums vorangestellt, die auf dieser Methode beruht. *B. Volkmann* (Stuttgart)

Referred to in K05-64.

K05-49 (31# 4777)

Rimkevičiūtė, L.

Concerning uniform distributions of points on an interval. (Lithuanian. Russian summary)

Vilniaus Valst. Univ. Mokslo Darbai Mat. Fiz. **8** (1958), 49–56.

Let $\{x_k\}$ be a sequence of real numbers belonging to the interval $[0, 1]$, and let E be a subset of $[0, 1]$. Denote by $\nu_n(E)$ the number of the points $x_1, \cdots, x_n$ lying in E. If $\{x_k\}$ is uniformly distributed (u.d.) and E is Jordan measurable, then (1) $\nu_n(E)/n \to mE$ as $n \to \infty$, where mE is the measure of E. If E is Lebesgue measurable but not Jordan measurable, then (1) is not always true. If (1) is true for a u.d. sequence $\{x_k\}$ and Lebesgue measurable set

E, then $\{x_k\}$ is called regular with respect to E and E is called regular with respect to $\{x_k\}$.

The author proves some results concerning open sets. For every open set O, $\liminf \nu_n(O)/n \geqq mO$ as $n \to \infty$. For every u.d. sequence $\{x_k\}$ there exists an open set which is non-regular with respect to $\{x_k\}$. For every Jordan non-measurable open set O there exist both regular and non-regular sequences with respect to O. If open intervals $I_1, I_2, \cdots$ constitute an open set O, then in order that the u.d. sequence $\{x_k\}$ be regular with respect to O, it is necessary and sufficient that the series $\nu_n(I_1)/n + \nu_n(I_2)/n + \cdots = \nu_n(O)/n$ converge uniformly in n. *J. Kubilius* (Vilnius)

K05-50 (31# 5857)

Postnikova, L. P.

Fluctuations in the distribution of fractional parts. (Russian)

Dokl. Akad. Nauk SSSR **161** (1965), 1282–1284.

Let m be a given positive integer. Let $\alpha_1, \alpha_2, \cdots, \alpha_k$ be k numbers in $[0, 1)$. Suppose that every subinterval of $[0, 1)$ of length $1/2m$, and also every two subintervals the sum of whose lengths is $1/2m$, contain fewer than $f(k)$ points α_i. Then

$$\left|\sum_{j=1}^{k} e^{2\pi i m\alpha_j}\right| \leqq 2mf(k) - k.$$

{This article has appeared in English translation [Soviet Math. Dokl. **6** (1965), 597–600].}

J. E. Cigler (Groningen)

K05-51 (32# 5630)

Uchiyama, Saburô

A note on a theorem of J. N. Franklin.

Math. Comp. **20** (1966), 139–140.

In Math. Comp. **18** (1964), 560–568 [MR **30** #3077], J. N. Franklin proved a theorem concerning the equidistribution of a sequence of d-dimensional vectors, using Weyl's criterion for equidistribution modulo one and the individual ergodic theorem due to F. Riesz. In the present paper the author gives a proof of Franklin's theorem using only Weyl's criterion. *F. Supnick* (New York)

Citations: MR 30# 3077 = K30-25.

K05-52 (32# 7502)

Jajte, R.

On rigid divisions of the segment $\langle 0, 1\rangle$. (Polish)

Wiadom. Mat. (2) **8**, 135–137 (1965).

Let Ω be the set of all sequences of different numbers from the interval $\langle 0, 1\rangle$. For every $\xi = \{x_k\} \in \Omega$, the numbers $x_1, x_2, \cdots, x_n$ divide $\langle 0, 1\rangle$ into some subintervals. Denote by $\Delta_n(\xi)$ the length of the largest among these subintervals. It is proved that

$$\inf_{\xi\in\Omega} \limsup_{n\to\infty} n\Delta_n(\xi) = \frac{1}{\log 2}.$$

J. Kubilius (Vilnius)

K05-53 (34# 4238)

Judin, A. A.

Irregularity in the distribution of sequences. (Russian. Tajiki summary)

Izv. Akad. Nauk Tadžik. SSR Otdel. Fiz.-Tehn. i Him. Nauk **1966**, *no.* 2 (20), 7–18.

The author proves the following theorem. Let $E = \{x_1, \cdots, x_k\}$ be k points in the n-dimensional unit cube, $m = (m_1, \cdots, m_n)$ a lattice point with all $m_i \geqq 1$, and $\langle x_k, m\rangle$ the inner product of x_k and m. Then $|S| \leqq (\pi/2)^{n-1} \times (2m_1m_2\cdots m_n f(k) - k)$, where $S = \sum_{j=1}^{k} e^{2\pi i\langle x_j, m\rangle}$. $f(k)$ denotes the maximal number of points in E contained in p disjoint parallelepipeds, where $2^{n-1} \leqq p < \frac{1}{2}(3^n+1)$ and the sum of the volumes is $1/(2m_1\cdots m_n)$.

J. E. Cigler (Groningen)

K05-54 (35# 4171)

Ungar, Peter

The theorem of Coles on uniform distribution.

Comm. Pure Appl. Math. **20** (1967), 609–618.

Nach einem Satz von W. J. Coles [Proc. Cambridge Philos. Soc. **53** (1957), 781–789; MR **20** #848] ist die Diskrepanz einer zweidimensionalen reellen Zahlenfolge (α_n, β_n) stets kleiner als ein gewisser Ausdruck, in dem die Diskrepanzen $D_{uv}^{(N)}$ der zugehörigen Folgen $u\alpha_n + v\beta_n$ $(1 \leqq n \leqq N;$ $(u, v) \neq (0, 0);$ $u \geqq 0,$ $v \geqq 0$, ganz) auftreten. (Wegen Einzelheiten vgl. das angegebene Referat.) Der Verfasser gibt für diese Ungleichung einen neuen, methodisch einfacheren Beweis. Er ist insofern elementarer, als z. B. beim Beweis eines Hilfssatzes keine Fourier-Entwicklung mehr benötigt wird und bei einem anderen Hilfssatz ein Doppelintegral ohne Zerlegung des Integrationsbereiches berechnet wird. *B. Volkmann* (Stuttgart)

Citations: MR 20# 848 = K05-29.

K05-55 (35# 4961)

Hlawka, Edmund; Kuich, Werner

Geordnete Schätzfunktionen und Diskrepanz. II.

Österreich. Akad. Wiss. Math.-Natur. Kl. S.-B. II **174** (1965), 235–286.

Authors' summary: "Es sei X eine Zufallsvariable mit der Verteilungsfunktion $F(X)$. Eine Stichprobe der Grösse s mit Verteilungsfunktion $F(X)$ ist definiert als s-dimensionale Zufallsvariable $(X_1, \cdots, X_s)$ mit Verteilungsfunktion $\prod_{i=1}^{s} F(X_i)$. Wir setzen voraus, dass die Elemente oder Komponenten $X_1, \cdots, X_s$ der Stichprobe voneinander unabhängig sind, jedoch dieselbe Verteilungsfunktion $F(X)$ besitzen.

"Wir betrachten eine Realisation der Stichprobe $(X_1, \cdots, X_s)$, nämlich $(x_1, \cdots, x_s)$ und ordnen $x_1, \cdots, x_s$ der Grösse nach zu $x_{i_1} \leqq x_{i_2} \leqq \cdots \leqq x_{i_s}$ und setzen $x_{i_h} = z((x_1, \cdots, x_s), h)$ für $h = 1, \cdots, s$. Die zufällige Variable $z((X_1, \cdots, X_s), h)$ nennen wir die h-te geordnete Schätzfunktion.

"Es sei nun eine Folge von reellen Zahlen modulo 1 gegeben. Es wird im ersten Teil der Arbeit untersucht, wie gross der Fehler ist, wenn man die Glieder der Folge als Realisation der Stichprobe nimmt und die Verteilungsfunktion durch die Schätzfunktion der Verteilungsfunktion ersetzt. Das Supremum des Absolutbetrages der Differenz zwischen der Verteilungsfunktion und ihrer Schätzfunktion wird Diskrepanz gennant.

"In dieser Arbeit, die sich an die Arbeit des ersten Verfassers anschliesst [Math. Ann. **150** (1963), 259–267; MR **27** #3615], wird unter Benutzung von S. S. Wilks [Ann. Math. Statist. **13** (1942), 400–409; MR **4**, 165] und A. Wald [ibid. **14** (1943), 45–55; MR **4**, 222] die Verteilungsfunktion einer verallgemeinerten Spannweite, worunter wir die Differenz der $(k+i)$-ten und der k-ten geordneten Schätzfunktion verstehen wollen, hergeleitet. Dann wird die Diskrepanz, also das Supremum des Absolutbetrages der Differenz zwischen der Verteilungsfunktion der verallgemeinerten Spannweite und ihrer Schätzfunktion nach oben abgeschätzt. Desgleichen wird die Verteilungsfunktion von Waldschen Bereichen, die durch die Methode der sukzessiven Elimination hergestellt werden, hergeleitet und die Diskrepanz nach oben abgeschätzt.

"Im zweiten Teil der Arbeit werden zwei s-dimensionale n-gliedrige Folgen $\omega_n^{(1)}$ und $\omega_n^{(2)}$ betrachtet, die gesetzmässig mit zahlentheoretischen Mitteln konstruiert werden können. Es wird dann gezeigt, dass die Diskrepanz dieser Folgen in bezug auf die hergeleiteten Dichten der verallgemeinerten Spannweite und der Waldschen Bereiche kleiner als jede beliebig vorgegebene Zahl ε gemacht werden kann, soferne nur der Parameter n eine bestimmte Grösse, die von ε abhängt, überschreitet. Das bedeutet, dass die verallgemeinerten Spannweiten und Waldschen Bereiche dieser Folgen $\omega_n^{(1)}$ und $\omega_n^{(2)}$ für genügend grosse n ein zufälliges Verhalten zeigen (pseudorandom numbers).

"Zum Abschluss wird ein numerisches Beispiel für die untere Schranke der Zahl n gegeben, so dass die verallgemeinerten Spannweiten und Waldschen Bereiche der Folgen $\omega_n^{(1)}$ und $\omega_n^{(2)}$ ein zufälliges Verhalten zeigen."

{Part I, by the first author, appeared in Math. Ann. **150** (1963), 259–267 [MR **27** #3615].}

J. E. Cigler (Groningen)

Citations: MR 27# 3615 = K05-39.

K05-56 (35# 6639)

Kátai, Imre

An irregularity phenomenon in the theory of numbers. (Hungarian. English summary)

Magyar Tud. Akad. Mat. Fiz. Oszt. Közl. **17** (1967), 85–88.

P. Erdős raised the following question [Compositio Math. **16** (1964), 52–65; MR **31** #3382]: Let $z_k = e^{i\theta_k}$ $(k=1, 2, \cdots)$ be an infinite sequence of complex numbers. Put $A_k = \limsup_{n\to\infty} |z_1{}^k + z_2{}^k + \cdots + z_n{}^k|$. Then is it true that $\lim_{k\to\infty} A_k = \infty$? The paper contains a lower estimation for A_k under the assumption that the answer to the question of Erdős is negative. The following theorem is proved. Suppose that $A_k < \infty$ if $k > t$. Then $A_k > c(k/t)^{1/2}$, with t a numerical constant.

The proof is based on a discrepancy-estimation of K. F. Roth [Mathematika **1** (1954), 73–79; MR **16**, 575].

P. Szüsz (Stony Brook, N.Y.)

Citations: MR 16, 575c = K05-21; MR 31# 3382 = J24-34.

K05-57 (36# 2579)

Postnikov, A. G.

Dynamical systems in number theory. (Russian)

Proc. Fourth All-Union Math. Congr. (Leningrad, 1961) *(Russian), Vol. II, pp.* 124–131. *Izdat. "Nauka", Leningrad,* 1964.

The author indicates that an array of problems of the theory of distribution of fractional parts may be looked upon as problems in the behavior of the trajectories of certain special dynamical systems. Here are some examples: problems on the fractional part of the exponential function (essentially the main content of the work), problems on the fractional part of a linear function, problems on the distribution of the fractional part of a second degree polynomial (a system closely akin to the last one was considered by P. R. Halmos [Bull. Amer. Math. Soc. **55** (1949), 1015–1034, 1025; MR **11**, 373]). There are formulated fundamental theorems obtained both by ergodic methods and the method of trigonometric sums. *L. Postnikova* (RŽMat **1964** #9 A80)

K05-58 (36# 3733)

Carroll, F. W.

Some properties of sequences, with an application to noncontinuable power series.

Pacific J. Math. **24** (1968), 45–50.

For a sequence $s = \{s(n)\}$, $n \in Z$, the set of nonnegative integers, let s^N denote the set of N-tuples $\{s(n+1), \cdots, s(n+N)\}$. Let Δs denote the difference sequence $\{s(n+1) - s(n)\}$ and $\Delta^j s$ the higher differences. A bounded real sequence s is said to have property (PN) if for some subsequence S of Z and for every N-tuple $(h_1, \cdots, h_N)$ of integers not all zero, $\lim |\sum_{j=1}^N h_j s(n+j)| > 0$ as $n \to \infty$ over S. The sequence s is said to have property (QN) if (i) for a subsequence S of Z, $\Delta^j s(n)$ converges (mod 1) as $n \to \infty$ over S, $j = 2, \cdots, N$, and (ii) the set $\{(((s(n))), ((\Delta s(n)))), n \in S\}$ is nowhere dense in the plane where, for a real x, $((x)) = x - [x]$. Let $e(x) = \exp(2\pi i x)$. With these definitions, the following are the main results of the paper. If $s(n) = \phi(\psi(n))$, where ϕ is function with period one and at most a nowhere dense set of points of discontinuity and $\psi(n)$ has property (QN), then either s has property (PN) or ϕ coincides with a polynomial of degree at most $N-2$ on some subinterval of $[0, 1]$ (Theorem 2.1). If for some subsequence S_0 of Z and for every n-tuple $(h_1, \cdots, h_N)$ of integers not all zero, $\lim |\sum h_j s(n+j)| = \infty$ as $n \to \infty$ over S_0, then for a subsequence S of S_0 and for almost all real α, the sequence (αs^N) restricted to S is uniformly distributed (mod 1) over the N-cube (Theorem 3.1). The next theorem gives a criterion for the condition of the last theorem to hold. The final result is an application to noncontinuable power series—if (a_n) is a bounded sequence, $\lim_{N\to\infty} \inf_{k\geq 0} \sum_{n=k+1}^{k+N} |a_n| = \infty$ and if for every N, s satisfies the hypothesis of Theorem 3.1, then for almost all real α, the unit circle is a natural boundary for the power series $\sum |a_n| e(\alpha s(n)) z^n$. This result is to be contrasted with the theorem, due to R. L. Perry [J. London Math. Soc. **35** (1960), 172–176; MR **22** #1657], that for every sequence s, there is a sequence $(|a_n|)$ such that $\sum |a_n| e(s(n)) z^n$ has radius of convergence one but can be continued across a semicircle of the unit circle.

V. Ganapathy Iyer (Annamalainagar)

K05-59 (37# 1536)

de Bruijn, N. G.; Post, K. A.

A remark on uniformly distributed sequences and Riemann integrability.

Nederl. Akad. Wetensch. Proc. Ser. A **71** = *Indag. Math.* **30** (1968), 149–150.

The authors prove the following theorem. Let f be a real valued function defined on the unit interval I. If the limit $\lim_{N\to\infty} N^{-1} \sum_{j=1}^N f(x_j)$ exists for every uniformly distributed sequence $\{x_j\}$, then f is Riemann integrable on I (and $\int_0^1 f(x)\,dx = \lim_{N\to\infty} N^{-1} \sum_{j=1}^N f(x_j)$). This theorem is a converse of a theorem of H. Weyl [Math. Ann. **77** (1916), 313–352].

A. M. Bruckner (Santa Barbara, Calif.)

K05-60 (37# 4029)

Schmidt, Wolfgang M.

Irregularities of distribution.

Quart. J. Math. Oxford Ser. (2) **19** (1968), 181–191.

Let $\{x_n\}$ be an infinite sequence of real numbers in $[0, 1]$. For any α in this interval and any integer $n > 0$, let $Z(n, \alpha)$ denote the number of elements x_i with $1 \leqq i \leqq n$ lying in the subinterval $0 \leqq x_i \leqq \alpha$. Define $D(n, \alpha) = |Z(n, \alpha) - n\alpha|$. It is shown that the numbers α having $\limsup\{D(n, \alpha)/2^{-19}(\log n)^{1/2}\} > 1$ is a winning set in the sense of a modified Banach-Mazur game; we omit details here. Also for almost all real numbers α in the unit interval $\limsup\{D(n, \alpha)/2^{-19}(\log\log n)^{1/2}\} > 1$. It follows that there exist numbers α depending on the sequence $\{x_i\}$ such that $\limsup D(n, \alpha) = \infty$, thus answering a question of P. Erdős [Compositio Math. **16**, 52–65 (1964); MR **31** #3382]. The proofs are based on a generalization of an integral inequality due to K. F. Roth [Mathematika **1** (1954), 73–79; MR **16**, 575]. *I. Niven* (Eugene, Ore.)

Citations: MR 16, 575c = K05-21; MR 31# 3382 = J24-34.

Referred to in K05-63, K05-69, K05-75.

K05-61 (37# 5157)

Brown, J. L., Jr.; Duncan, R. L.

A generalized Weyl criterion.

Duke Math. J. **35** (1968), 699–705.

The authors give a necessary and sufficient condition for a sequence of real numbers (mod 1) to have a given continuous distribution function g.

More precisely, let (φ_ν) be a sequence of periodic complex valued functions in $L_2(0, 1)$ which is complete in $L_2(0, 1)$. Put $a_\nu = \int_0^1 g(X)\varphi_\nu(X)\,dX$ and $\Psi_\nu(X) = 1 + (1/a_\nu)\int_1^X \varphi_\nu(t)\,dt$ if $a_\nu \neq 0$ and $\Psi_\nu(X) = \int_1^X \varphi_\nu(t)\,dt$ if $a_\nu = 0$. Then g is the distribution function of the sequence of real numbers $(X_j)_1^\infty$ if and only if for each integer $\nu \geqq 0$, $\lim_{n\to\infty}(1/n)\sum_{j=1}^n \Psi_\nu(X_j) = 0$.

Special cases and examples are given.

M. Mendès France (Paris)

K05-62 (37# 5158)

Lesca, J.

Démonstration d'une conjecture de P. Erdös.

Acta Arith. **14** (1967/68), 425–427.

Let u_n, $n = 1, 2, 3, \cdots$, be an arbitrary sequence of real numbers $\in [0, 1)$. For each $\beta \in (0, 1)$ we regard the function $\pi([0, \beta), n)$, which counts the number of indices $i \in \{1, \cdots, n\}$ for which $u_i \in [0, \beta)$. Then the function $E(\beta, n) = \pi([0, \beta), n) - n\beta$ is studied. It is shown: For every sequence u_n, $n = 1, 2, 3, \cdots$, of numbers $\in [0, 1)$ and every subinterval θ of $[0, 1)$ there exists a subset $B \subset \theta$ of the order-type of Cantor's discontinuum, such that for each $\beta \in B$ the sequence $n \to E(\beta, n)$ is not bounded. This theorem confirms a conjecture of P. Erdős [Compositio Math. **16**, 52–65 (1964); MR **31** #3382]. The proof makes use of the fact that the function $(\beta, n) \to E(\beta, n)$, which maps $[0, 1) \times \{1, 2, 3, \cdots\}$ into the real numbers, is not bounded; this is a conclusion of a theorem of T. van Aardenne-Ehrenfest [Nederl. Akad. Wetensch. Proc. **52** (1949), 734–739; MR **11**, 336]. *E. Harzheim* (Cologne)

Citations: MR 11, 336d = K05-12; MR 31# 3382 = J24-34.

K05-63 (38# 3237)

Schmidt, Wolfgang M.

Irregularities of distribution. II.

Trans. Amer. Math. Soc. **136** (1969), 347–360.

Let $p_1, \cdots, p_N$ be points (not necessarily distinct) in the unit cube $0 \leqq x_j < 1$, $1 \leqq j \leqq n$, of Euclidean E^n with $n > 1$. Let P be the set of all points $p_i + g$, where $1 \leqq i \leqq N$ and g is a point with integer coordinates. Let A be a bounded and Jordan-measurable set of measure $\mu(A)$, and define $\Delta(A) = |N\mu(A) - \nu(A)|$, where $\nu(A)$ denotes the number of points of P in A, counting multiplicities. Given a point $\mathbf{u} = (u_1, \cdots, u_n)$, write $B(\mathbf{u})$ for the box consisting of points $\mathbf{x} = (x_1, \cdots, x_n)$ with $0 \leqq x_j < |u_j|$, $1 \leqq j \leqq n$. If τ is an orthogonal transformation of E^n and $\mathbf{v}$ is any point in E^n, write $B(\mathbf{u}; \mathbf{v}; \tau)$ for the box consisting of points $\tau\mathbf{x} + \mathbf{v}$, where $\mathbf{x} \in B(\mathbf{u})$. The diameter of this box is $(u_1^2 + \cdots + u_n^2)^{1/2}$. If $n = 2$ and δ is real such that $N\delta^2 > 1$, it is proved that there is a rectangle $B = B(\mathbf{u}; \mathbf{v}; \tau)$ of diameter δ satisfying $\Delta(B) \geqq c_1(\log(N\delta^2))^{1/2}$. If $n > 2$ and δ, ε are real such that $N\delta^n > \varepsilon > 0$, it is proved that there is a box $B = B(\mathbf{u}; \mathbf{v}; \tau)$ of diameter δ satisfying $\Delta(B) \geqq c_2(\varepsilon, n)(N\delta^n)^{1/6}$. The constants c_1 and c_2 are positive and independent of N. These results extend previous work of the author [Quart. J. Math. Oxford Ser. (2) **19** (1968), 181–191; MR **37** #4029]. The proofs employ certain integral inequalities established by the author.

I. Niven (Eugene, Ore.)

Citations: MR 37# 4029 = K05-60.

Referred to in K05-69, K05-75.

K05-64 (39# 156)

Kuipers, L.

Remark on the Weyl-Schoenberg criterion in the theory of asymptotic distribution of real numbers.

Nieuw Arch. Wisk. (3) **16** (1968), 197–202.

Given a sequence $\{u_k\}$ of real numbers with $0 < u_k < 1$ for all k. Let $A(x, N)$, $0 \leqq x < 1$, denote the number of the u_k with $1 \leqq k \leqq N$ and $u_k < x$. For any non-decreasing real-valued function g on $[0, 1]$, with $g(0) = 0$ and $g(1) = 1$, the sequence is said to have g as asymptotic distribution function if for all continuity points x in $[0, 1]$, $A(x, N)/N \to g(x)$ as $N \to \infty$. It is known (see J. Cigler and G. Helmberg [Jber. Deutsch. Math.-Verein. **64** (1961), 1–50; MR **23** #A2409]) that Weyl's criterion holds for general g: A sequence $\{u_k\}$ has g as distribution function if and only if $\lim_{N\to\infty} N^{-1}\sum_{k=1}^N \exp(2\pi imu_k) = \int_0^1 \exp(2\pi imx)\,dg(x)$ for all integers m.

In this paper the author gives a new proof for a special case of Weyl's criterion. He considers distributions g with at most finitely many discontinuities not at the points of the set $\{0, 1, u_1, \cdots, u_2\}$. The proof is based on the relation

$$\int_0^1 \left[\frac{A(x, N)}{N} - g(x)\right]^2 dx - \left(\int_0^1 \left[\frac{A(x, N)}{N} - g(x)\right] dx\right)^2 = \frac{1}{2\pi^2}\sum_{m=1}^{\infty}\frac{1}{m^2}\left|\frac{1}{N}\sum_{k=1}^{N}\exp(-2\pi imu_k) - \int_0^1 \exp(-2\pi imx)\,dg(x)\right|^2,$$

which, in the case $g(x) \equiv x$, was given and used by W. J. LeVeque [*Theory of numbers* (Proc. Sympos. Pure Math., Vol. VIII, Calif. Inst. Tech., Pasadena, Calif., 1963), pp. 22–30, Amer. Math. Soc., Providence, R.I., 1965; MR **31** #3401] to prove the classical Weyl criterion.

O. P. Stackelberg (Durham, N.C.)

Citations: MR 23# A2409 = K02-1; MR 31# 3401 = K05-48.

K05-65 (39# 2200)

Holewijn, P. J.

Note on Weyl's criterion and the uniform distribution of independent random variables.

Ann. Math. Statist. **40** (1969), 1124–1125.

Let $\{X_n\}$ be a sequence of independent r.v.'s with corresponding characteristic function sequence $\{f_n\}$. Theorem: $\{X_n\}$ is almost surely uniformly distributed mod 1 if and only if $[\sum_1^N f_n(2\pi h)]/N \to 0$ with $1/N$ for all natural numbers h. *J. G. Wendel* (Ann Arbor, Mich.)

K05-66 (39# 5482)

Artémiadis, Nicolas; Kuipers, Laurens

On the Weyl criterion in the theory of uniform distribution modulo 1.

Amer. Math. Monthly **76** (1969), 654–656.

Soit une suite réelle (u_n), où $n \geqq 1$, et soit $I = [0, 1]$; désignons par $\{u_n\}$ la partie fractionnaire de u_n, et par $A(x, N)$, où $x \in I$, N entier $\geqq 1$, le nombre des n tels que $1 \leqq n \leqq N$ et $0 \leqq \{u_n\} < x$; posons $f(x, N) = A(x, N)/N - x$. On rappelle que la suite (u_n) est dite équirépartie [e.r.] (mod 1) si on a (1) $\lim_{N\to\infty} f(x, N) = 0$ pour tout $x \in I$. L'objet de la note est de donner une démonstration nouvelle du critère de H. Weyl [Math. Ann. **77** (1916), 313–352]: pour que la suite (u_n) soit e.r. (mod 1), il faut et il suffit qu'on ait (2) $\lim_{N\to\infty} \sigma_N(h)=0$ pour tout entier $h \geqq 1$; on a posé ici $N\sigma_N(h)=\sum_{n=1}^{n=N} \exp 2i\pi h u_n$. Cette démonstration s'appuie sur la relation suivante, due à J. F. Koksma, mais non publiée semble-t-il: (3) $\int_0^1 f^2(x, N)\,dx - (\int_0^1 f(x, N)\,dx)^2 = \sum_{h=1}^{h=\infty} (2\pi^2 h^2)^{-1}|\sigma_N(h)|^2$. L'application dans (3) du théorème de la convergence bornée de Lebesgue quand $N\to\infty$ montre que (1) implique (2). D'autre part, si les conditions (2) sont satisfaites, le premier membre de (3) tend vers 0 quand $N\to\infty$; utilisant alors le théorème de Helly, on montre que la suite $(f(x, N))$, où $N \geqq 1$, contient une sous-suite $(f(x, N_k))$, où $k \geqq 1$, qui converge vers 0 sur I; or ce dernier résultat exige que la suite $(f(x, N))$ converge vers 0 sur I; ainsi (2) implique (1).

Les auteurs signalent que leur méthode s'applique, plus généralement, à l'étude des suites réelles qui admettent une fonction de répartition (mod 1) continue.

J. Chauvineau (Paris)

K05-67 (39# 6834)

Beer, Susanne

Über die Diskrepanz von Folgen in bewerteten Körpern. (English summary)

Manuscripta Math. **1** (1969), 201–209.

Let k be a field which is locally compact with respect to a valuation. This paper discusses the theory of uniform distribution of sequences of elements with value $\leqq 1$. It is shown, as a consequence of a theorem of K. F. Roth [Mathematika **1** (1954), 73–79; MR **16**, 575], that for archimedean valuations there are sequences with discrepancy $\geqq \gamma\sqrt{(N^{-1}\log N)}$, where γ is an absolute constant. For non-archimedean valuations examples are given of sequences with discrepancy N^{-1}.

D. J. Lewis (Ann Arbor, Mich.)

Citations: MR 16, 575c = K05-21.

K05-68 (39# 6835)

Chauvineau, Jean

Sur la répartition dans R et dans Q_p.

Acta Arith. **14** (1967/68), 225–313.

This paper consists of 8 chapters. In Chapter I the notion of (λ, λ')-distribution (mod 1) is introduced. Let (x_n) $(n=1, 2, 3, \cdots)$ be a sequence of real numbers. Let a and b be two real numbers with $0\leqq a<b\leqq 1$. Let (N, a, b) denote the number of indices i such that $1\leqq i\leqq N$ and $a\leqq x_i<b$ (mod 1). Then the following limiting values exist: $\rho_*(a, b)=\rho_*(x, a, b)=\lim\inf_{N\to\infty}(N, a, b)/((b-a)N)$, $\rho^*(a, b)=\rho^*(x, a, b)=\lim\sup_{N\to\infty}(N, a, b)/((b-a)N)$.

Definition 1: Let λ and λ' be two real numbers such that $0<\lambda\leqq 1\leqq\lambda'$. The sequence (x_n) is said to be (λ, λ')-distributed ((λ, λ')-d.) (mod 1) if, for every pair of real numbers a and b with $0\leqq a<b\leqq 1$, (1) $\rho_*(a, b)\geqq\lambda$ and $\rho^*(a, b)\leqq\lambda'$. If only the first of these two conditions is satisfied, the sequence (x_n) is said to be (λ, ∞)-distributed ((λ, ∞)-d.) (mod 1). If only the second of the two conditions is satisfied, the sequence (x_n) is said to be $(0, \lambda')$-distributed ($(0, \lambda')$-d.) (mod 1). The author derives the following theorems.

Criterion 1: A sequence (x_n) is (λ, ∞)-d. (mod 1) if and only if, for every function f continuous on the unit interval and assuming real, nonnegative values,

$$(2) \qquad \lim\inf_{N\to\infty} N^{-1}\sum_{n=1}^N f(\langle x_n\rangle) \geqq \lambda\int_0^1 f(t)\,dt.$$

Criterion 1′: A sequence (x_n) is $(0, \lambda')$-d. (mod 1) if and only if, for every function f continuous on the unit interval and assuming real, nonnegative values,

$$(2') \qquad \lim\sup_{N\to\infty} N^{-1}\sum_{n=1}^N f(\langle x_n\rangle) \leqq \lambda'\int_0^1 f(t)\,dt.$$

Here $\langle x_n\rangle$ denotes the fractional part of x_n. (2) and (2′) together form a criterion for the (λ, λ')-d. (mod 1). Now define $e(t)=e^{2\pi it}$ and $\sigma_N(x)=\sigma_N(x, h)=N^{-1}\sum_{n=1}^N e(hx_n)$, where $h\in\mathbf{Z}$ ($\mathbf{Z}$ is the set of rational integers).

Criterion 2: A sequence (x_n) is (λ, ∞)-d. (mod 1) if and only if, for every integer $k\geqq 2$ and for every $l=1, 2, \cdots, k$, (3) $\lim\inf_{N\to\infty}\sum_{h\in\mathbf{Z}}(k(\pi h)^{-1}\sin(\pi h/k))^2 e(-hl/k)\sigma_N(h)\geqq\lambda$.

Criterion 2′: A sequence (x_n) is $(0, \lambda')$-d. (mod 1) if and only if, for every integer $k\geqq 2$ and for every $l=1, 2, \cdots, k$, (3′) $\lim\sup_{N\to\infty}\sum_{h\in\mathbf{Z}}(k(\pi h)^{-1}\sin(\pi h/k))^2 e(-hl/k)\sigma_N(h)\leqq\lambda'$. (3) and (3′) together form a second criterion for the (λ, λ')-d. (mod 1).

Criterion 2bis: A sequence (x_n) is (λ, ∞)-d. (mod 1) if and only if, for every pair (δ, τ) of real numbers such that $0<\delta\leqq\frac{1}{2}$ and $0<\tau\leqq 1$,

$$(4) \quad \lim\inf_{N\to\infty}\sum_{h\in\mathbf{Z}}((\pi h\delta)^{-1}\sin\pi h\delta)^2 e(-h\tau)\sigma_N(h) \geqq \lambda.$$

Criterion 2′bis: A sequence (x_n) is $(0, \lambda')$-d. (mod 1) if and only if, for every pair (δ, τ) of real numbers such that $0<\delta\leqq\frac{1}{2}$ and $0<\tau\leqq 1$,

$$(4') \quad \lim\sup_{N\to\infty}\sum_{h\in\mathbf{Z}}((\pi h\delta)^{-1}\sin\pi h\delta)^2 e(-h\tau)\sigma_N(h) \leqq \lambda'.$$

(4) and (4′) together form an alternate form of the second criterion, (3) and (3′), for the (λ, λ')-d. (mod 1).

This shows that in the case $\lambda=\lambda'=1$ the preceding results reduce to the classic Weyl criterion.

Definition 2: Let λ and λ' be a pair of real numbers such that $0<\lambda\leqq 1\leqq\lambda'$. The sequence (x_n) is said to be strictly (λ, λ')-d. (mod 1) if $\inf_{0\leqq a<b\leqq 1}\rho_*(a, b)=\lambda$ and $\sup_{0\leqq a<b\leqq 1}\rho^*(a, b)=\lambda'$.

By elementary argument the author shows that the sequence $(\alpha\log n)$ $(n=1, 2, \cdots)$, where α is a real constant with $\alpha\neq 0$, is strictly $(\gamma/(e^\gamma-1), \gamma e^\gamma/(e^\gamma-1))$-d. (mod 1) $(\gamma=1/|\alpha|)$. Finally, the author derives an alternate sufficient condition for the (λ, λ')-d. (mod 1), and also a necessary and sufficient condition for the sequence (x_n) to be everywhere dense (mod 1) in the unit interval $[0, 1]$.

In Chapter II the author defines u.d. (mod 0), u.d. (mod ∞) and u.d. (mod 1) in the mean. (u.d. stands for uniformly distributed.)

Let m be a natural number. Obviously a sequence (x_n) of real numbers is u.d. (mod $1/m$) if and only if the sequence (mx_n) is u.d. (mod 1).

Definition 1: The sequence (x_n) is said to be u.d. (mod 0) if, for every pair of real numbers a and b such that $0\leqq a<b\leqq 1$, (1) $\lim_{m\to\infty}\rho_*(mx, a, b)=\lim_{m\to\infty}\rho^*(mx, a, b)=1$. U.d. (mod 1) implies u.d. (mod 0).

Sufficient condition 1: Let f be a real-valued function defined on $[0, \infty)$, let f be positive except on a certain segment $[0, t_0]$, strictly monotone, differentiable, let $f'(t)$ be monotone except on a certain segment $[0, t_1]$, let $\lim_{t\to\infty}f'(t)=0$ and $t|f'(t)|$ be bounded below by a positive number as $t\to\infty$. Then the sequence $(f(n))$ $(n=1, 2, \cdots)$ is u.d. (mod 0), (analog of a theorem of Fejér).

Example: $(\alpha\log n)$ $(n=1, 2, \cdots;\ \alpha$ real, $\neq 0)$ is u.d. (mod 0).

Definition 2: The sequence (x_n) is said to be u.d. (mod ∞) if and only if for every pair a and b of real numbers with $0\leqq a<b\leqq 1$,

$$(2) \quad \lim_{m\to\infty}\rho_*(x/m, a, b) = \lim_{m\to\infty}\rho^*(x/m, a, b) = 1.$$

The author proves a statement (again an analog of Fejér's theorem) similar to the above sufficient condition 1.

Example: (αn) $(n=1, 2, \cdots;\ \alpha$ real, $\neq 0)$ is u.d. (mod ∞).

Definition 3: The sequence (x_n) is said to be u.d. (mod 1) in the mean if, for every pair of real numbers a and b such that $0\leqq a<b\leqq 1$, $\lim_{N\to\infty}N^{-1}\sum_{M=1}^N (M, a, b)/M=b-a$. The author gives two criteria similar to the well known Weyl criteria, using continuous functions and exponential functions, respectively.

Definition 4: The sequence (x_n) is said to be almost u.d. (mod 1) in the mean if there exists a strictly increasing subsequence $\varphi(N)$ of the sequence $\{N\}$ of natural numbers such that, for every pair of real numbers a and b with $0 \leqq a < b \leqq 1$, $\lim_{N\to\infty} N^{-1} \sum_{M=1}^{N} (\zeta(M), a, b)/\zeta(M) = b-a$ [see I. I. Pjateckiĭ-Šapiro, Mat. Sb. (N.S.) **30** **(72)** (1952), 669–676; MR **15**, 106]. The example $x_n = \alpha \log n$ is dealt with elaborately.

Chapter III deals with uniformly distributed sequences defined by certain periodic functions. See the author's paper in C. R. Acad. Sci. Paris **252** (1961), 4090–4092 [MR **23** #A3731].

Chapter IV concerns continuous and periodic distributions mod 1 in **R**. Let χ_0 be a distribution function on [0, 1] and continuous on [0, 1]. Let α be a real number such that $0 < \alpha \leqq 1$. The sequence (x_n) is said to be α-continuously and periodically distributed (α-cpd) (mod 1) with respect to χ_0 say, continuous on [0, 1] and such that $\chi - \chi_0$ is the restriction to [0, 1] of a periodic function, defined on **R**, and a period α. The author derives some criteria and treats an example. (In relation (5) on page 254 the second term on the left should be multiplied by 2π.)

In Chapters V through VIII the author investigates mainly distributions in $\mathbf{Q}_p$, the field of p-adic numbers. In Chapter VI an extension of a result of J. F. Koksma [Compositio Math. **2** (1935), 250–258] is derived. This extension was announced already in the author's paper in C. R. Acad. Sci. Paris **260** (1965), 6252–6255 [MR **31** #3400].

Chapter VII introduces and characterises the notion of C-uniform distribution (mod 1), develops some criteria and gives many applications. See also Ju. E. Alenicyn [*Contemporary problems in theory anal. functions* (Internat. Conf., Erevan, 1965) (Russian), pp. 9–11, Izdat. "Nauka", Moscow, 1966; MR **35** #5597].

Chapter VIII deals with the distribution of sequences of p-adic integers in the ring $\mathbf{Z}_p$. The author extends the analogous results of I. Niven and S. Uchiyama [I. Niven, Trans. Amer. Math. Soc. **98** (1961), 52–61; MR **22** #10971; S. Uchiyama, Proc. Japan Acad. **37** (1961), 605–609; MR **26** #2413] in that he considers arbitrary distribution functions. *L. Kuipers* (Carbondale, Ill.)

Citations: MR 15, 106g = K35-6; MR 22# 10971 = B48-6; MR 23# A3731 = K05-33; MR 26# 2413 = P52-5; MR 31# 3400 = B48-13.
Referred to in B48-22.

K05-69 (39# 6837)

Schmidt, Wolfgang M.
Irregularities of distribution. III.
Pacific J. Math. **29** (1969), 225–234.

The author continues his earlier work [Quart. J. Math. Oxford Ser. (2) **19** (1968), 181–191; MR **37** #4029; Trans. Amer. Math. Soc. **136** (1969), 347–360; MR **38** #3237] on irregularities of distribution. No results from the earlier papers are used in the present paper. Consider any N points on the unit sphere $S = S^n$ of Euclidean space E^{n+1}. Let $\surd(A)$ denote the number of these points lying in any subset A of S. Let A be a measurable set with measure $\mu(A)$, where $\mu(S) = 1$, and write $\Delta(A) = N\mu(A) - \surd(A)$, the discrepancy being defined as $|\Delta(A)|$. Let $\omega(\mathbf{x}, \mathbf{y})$ be the spherical distance of any two points $\mathbf{x}, \mathbf{y}$ on S, let $H(\mathbf{x})$ be the halfsphere of points z such that $\omega(\mathbf{x}, \mathbf{z}) \leqq \pi/2$, and define $L(\mathbf{x}, \mathbf{y}) = H(\mathbf{x}) \cap H(\mathbf{y})$. If $n = 2$ it is proved that $\iint \Delta(L(\mathbf{x}, \mathbf{y}))^2\, d\mathbf{x} d\mathbf{y} \geqq c_1 \log N$ where both integrals are over S. It follows that there are slices $L = L(\mathbf{x}, \mathbf{y})$ with $|\Delta(L)| \geqq c_2(\log N)^{1/2}$, and that there are spherical triangles T with $|\Delta(T)| \geqq c_3(\log N)^{1/2}$. If $n > 2$ then $\iint \Delta(L(\mathbf{x}, \mathbf{y}))^2\, d\mathbf{x} d\mathbf{y} \geqq c_4(n) N^{1-(2/n)}$, and it follows that there are slices $L(x, y)$ with $|\Delta(L)| \geqq c_5(n) N^{(1/2)-(1/n)}$. The theorems are similar to results of K. F. Roth [Mathematika **1** (1954), 73–79; MR **16**, 575] in the n-cube, but in contrast to Roth the proofs are based on integral equations.
I. Niven (Eugene, Ore.)

Citations: MR 16, 575c = K05-21; MR 37# 4029 = K05-60; MR 38# 3237 = K05-63.

K05-70 (39# 6838)

Schmidt, Wolfgang M.
Irregularities of distribution. IV.
Invent. Math. **7** (1969), 55–82.

The author continues his earlier work [see #6837 above] on irregularities of distribution. Because it relates closely to the paper in the preceding review, the second main result of the present paper is reviewed first. Given any r satisfying $0 < r \leqq \pi/2$ and any point c on $S = S^n$, define $\mathbf{C} = \mathbf{C}(r, \mathbf{c})$ to be the set of points $\mathbf{x}$ on S^n such that $\omega(\mathbf{x}, \mathbf{c}) \leqq r$. Thus C is the spherical cap of radius r and center $\mathbf{c}$. Define $E(r, s) = \int \Delta(\mathbf{C}(r, \mathbf{c}))\Delta(C(s, \mathbf{c}))\, d\mathbf{c}$, where the integral is over the sphere S. Now if $n > 1$, $\varepsilon > 0$ and δ satisfy $0 < \delta \leqq \pi/2$ and $N\delta^n > \varepsilon$, then $\int_0^\delta r^{-1} E(r, r)\, dr > c_1(N\delta^n)^{1-\varepsilon-(1/n)}$. This is used to strengthen certain results of the paper in the preceding review. Also from this it follows that there is a cap $\mathbf{C}(r, \mathbf{c})$ with $r \leqq \delta$ such that

$$|\Delta(\mathbf{C}(r, \mathbf{c}))| > c_2(N\delta^n)^{1/2-\varepsilon-1/2n}(1+\log N)^{-1/2},$$

proving a conjecture of P. Erdős [Compositio Math. **16** (1964), 52–65, cf. p. 54; MR **31** #3382].

Next, in Euclidean space E^n let U^n be the unit cube of points $\mathbf{x} = (x_1, \cdots, x_n)$ with each coordinate satisfying $0 \leqq x_j < 1$. Consider any N points $\mathbf{p}_1, \cdots, \mathbf{p}_N$ (not necessarily distinct) in U^n, and let P be the set of all points $\mathbf{p}_j + \mathbf{g}$, where $1 \leqq j \leqq N$ and g is any integer point. For any bounded Jordan measurable set A of measure $\mu(A)$, let $\nu(A)$ denote the number of points of P, multiplicities counted, in A. Again define the discrepancy as $|\Delta(A)|$, where $\Delta(A) = N\mu(A) - \nu(A)$. Let $B(r, \mathbf{c})$ be the closed ball with radius $r > 0$ and center $\mathbf{c}$. Let $D(r\ s)$ denote $\int \Delta(B(r, \mathbf{c}))\Delta(B(s, \mathbf{c}))\, d\mathbf{c}$, where the integral is over U^n. For $n > 1$, $\varepsilon > 0$, let $\delta > 0$ satisfy $N\delta^n > \varepsilon$. It is proved that $\int_0^\delta r^{-1} D(r, r)\, dr > c_3(N\delta^n)^{1-\varepsilon-1/n}$. Several applications are made of this basic result of which we cite two, first that there exists a ball $B(r, \mathbf{c})$ with $r \leqq \delta$ such that $|\Delta(B(r, \mathbf{c}))| > c_4(N\delta^n)^{1/2-\varepsilon-1/2n}(1+\log N)^{-1/2}$. Second, the author gives a proof of another conjecture of P. Erdős [same reference as above]. Let $\nu^*(r, \mathbf{c})$ be the number of points of T in $B(r, \mathbf{c})$, where T is a completely arbitrary discrete set of points in E^n. Define $\Delta^*(r, c) = \mu(B(r, \mathbf{c})) - \nu^*(r, \mathbf{c})$. For $n > 1$, $\varepsilon > 0$ and $R > 1$ there is a ball $B(r, \mathbf{c})$ with $r \leqq R$, $|\mathbf{c}| \leqq R^{n+2}$ and $|\Delta^*(r, \mathbf{c})| > c_5 R^{(n-1)/2-\varepsilon}$. The proofs use integral equations involving $D(r, s)$, where earlier papers in the series used $D(r, r)$. *I. Niven* (Eugene, Ore.)

Citations: MR 31# 3382 = J24-34.
Referred to in K05-75.

K05-71 (40# 1349)

O'Neil, P. E.
A new criterion for uniform distribution.
Proc. Amer. Math. Soc. **24** (1970), 1–5.

Soient $\delta \geqq 0$, $\varepsilon > 0$; une séquence strictement croissante $(a_1, a_2, \cdots, a_N)$ d'éléments de $U = [0, 1]$ est dite une (δ, ε)-progression presque arithmétique [(δ, ε)-p.p.a.] s'il existe η, avec $0 < \eta \leqq \varepsilon$, tel qu'on ait (i) $0 \leqq a_1 \leqq \eta + \delta\eta$; (ii) $a_n + \eta - \delta\eta \leqq a_{n+1} \leqq a_n + \eta + \delta\eta$ pour $n = 1, 2, \cdots, N-1$; (iii) $1 - \eta - \delta\eta \leqq a_N \leqq 1$. En particulier, une $(0, \varepsilon)$-p.p.a. est une progression arithmétique de raison au plus égale à ε, et on notera que, si $\delta' \geqq \delta \geqq 0$, toute (δ, ε)-p.p.a. est une (δ', ε)-p.p.a. On a le lemme suivant: Soit $\varepsilon'' > 0$; il existe un couple (δ, ε) tel que, pour tout sous-intervalle fermé I de

U, de longueur l, et pour toute (δ, ε)-p.p.a., soit (a_n), de cardinal N, on ait $|R_N(I)-l|<\varepsilon''$, où $NR_N(I)$ désigne le nombre des n tels que $1\leqq n\leqq N$ et $a_n\in I$. L'auteur démontre alors le critère annoncé: pour qu'une suite $(x_n)_{n\geqq 1}$ d'éléments de U soit équirépartie dans U, il faut et il suffit que, pour tout triplet $(\delta, \varepsilon, \varepsilon')$, où $\delta>0$, $\varepsilon>0$, $0<\varepsilon'<1$, il existe un entier N_0 tel que, pour $N>N_0$, la séquence $(x_1, x_2, \cdots, x_N)$, privée d'au plus $[\varepsilon' N]$ éléments convenablement choisis, soit une réunion de (δ, ε)-p.p.a. disjointes. Finalement, ce théorème est appliqué à la démonstration de quelques résultats particuliers bien connus, concernant l'équirépartition (mod 1) des suites réelles. *J. Chauvineau* (Paris)

Referred to in K05-80.

K05-72 (40# 5550)

Halton, J. H.; Zaremba, S. K.

The extreme and L^2 discrepancies of some plane sets.

Monatsh. Math. **73** (1969), 316–328.

Let $C_N=\{(x_1, y_1), \cdots, (x_N, y_N)\}$ be a set of N points in the square $Q^2=\{(x, y)|0\leqq x<1, 0\leqq y<1\}$. For any point (x, y) in this square, let $\nu(x, y)$ denote the number of points of C_N with $0\leqq x_i<x$ and $0\leqq y_i<y$. Set $g(x, y)=N^{-1}\nu(x, y)-xy$. The authors call

$$D(C_N) = \sup_{(x,y)\in Q^2}|g(x, y)|$$

the extreme discrepancy of C_N, and

$$T(C_N) = (\textstyle\int_0^1\int_0^1 g^2(x, y)\,dxdy)^{1/2}$$

the L^2 discrepancy of C_N. K. F. Roth [Mathematika **1** (1954), 73–79; MR **16**, 575] proved that for any set $C_N\subset Q^2$, $T(C_N)>cN^{-1}(\log N)^{1/2}$, where c is an absolute constant. He also constructed a set R_N consisting of 2^N points in Q^2 for which $D(R_N)=O(N2^{-N})$. The authors modify Roth's set R_N and obtain a set S_N for which they compute the exact values of both discrepancies, with $T(S_N)=O(N^{1/2}2^{-N})$. They also obtain the R_N discrepancies as a by-product of their arguments, with $T(R_N)=O(N2^{-N})$. *O. P. Stackelberg* (Durham, N.C.)

Citations: MR 16, 575c = K05-21.

K05-73 (41# 159)

Lacaze, Bernard

Réunions finies de suites et génération de nombres dont la fonction de répartition est une interpolée linéaire.

C. R. Acad. Sci. Paris Sér. A-B **269** (1969), A1000–A1002.

(I.1) Definitions: Let $\{\alpha_n^1\}, \{\alpha_n^2\}, \cdots, \{\alpha_n^p\}$ $(n=1, 2, \cdots)$ be a set of p real nondecreasing sequences. It is assumed that there is always a p such that $\alpha_p^i>\alpha_q^i$ for all i and q. Let $F_i(N, x)$ be the number of elements among $\alpha_1^i, \alpha_2^i, \cdots, \alpha_N^i$ such that (α_n^i) mod $1\in[0, x[$. The existence of the function $l_i(x)=\lim_{N\to\infty}F_i(N, x)/N$ (distribution function mod 1) is assumed. Let $\{\beta_n\}$ be the "union" of the sequences $\{\alpha_n^i\}$, $i=1, 2, \cdots, p$. This "union" is constructed in the following way: (1) Every element β_n is an element α_k^i, and conversely, every element α_k^i is an element of the sequence $\{\beta_n\}$. (2) If $\beta_n=\alpha_k^i$, every element $\alpha_l^j<\alpha_k^i$ is an element β_m such that $m<n$. (3) If k elements α_n^i are identical, there exist $n_1, n_2, \cdots, n_k$ such that $\beta_{n_1}=\beta_{n_2}=\cdots=\beta_{n_k}=\alpha_n^i$. All sequences $\{\beta_n\}$ constructed in this way are considered to be identical and ordered. The distribution function mod 1 of $\{\beta_n\}$ is denoted by $l(x)$. It is assumed that there exist numbers $a_{ij}\neq 0$ such that $\lim_{n\to\infty}\alpha_n^i/\alpha_n^j=a_{ij}$.

(I.2) Lemma: Assumptions: There exists an increasing $f(n)$ such that $\lim_{n\to\infty}f(n+1)/f(n)=1$, $\lim_{n\to\infty}\alpha_n^i/f(n)=a_i$, $\lim_{n\to\infty}\alpha_n^j/f(n)=a_j$. a_i and a_j are finite and $\neq 0$, N_i and N_j are such that $N-\alpha_{N_i}^i=\inf_{n\in\mathbf{L}_i}(N-\alpha_n^i)$, $N-\alpha_{N_j}^j=\inf_{n\in\mathbf{L}_j}(N-\alpha_n^j)$. $\mathbf{L}_i(N)$ and $\mathbf{L}_j(N)$ are sets of indices n defined by $N-\alpha_n^i\geqq 0$ and $N-\alpha_n^j\geqq 0$. Assertion: $\lim_{N\to\infty}(f(N_i)/f(N_j))=(a_j/a_i)$.

Now suppose that there exists an increasing $f(n)$ such that the sequences $\{\alpha_n^1\}, \{\alpha_n^2\}, \cdots, \{\alpha_n^p\}$ satisfy the conditions of the lemma. Suppose that one can determine b_{ij} depending on f, a_i and a_j such that $\lim_{N\to\infty}(N_i/N_j)=b_{ij}$; then $l(x)=\sum_{i=1}^p(l_i(x)/\sum_{j=1}^p b_{ji})$, where $l(x)$ is the distribution function mod 1 of $\{\beta_n\}$.

(II) The author defines filtered sequences derived from the given $\{\alpha_n^i\}$, based on a certain subinterval of the unit-interval, and enumerates some properties. By taking the union of disjoint filtered sequences, the author arrives at the analogue of the sequence β_n defined in (I) and its distribution function mod 1. Special cases are $\{n\mu_i\}$ and $\{n^\theta\mu_i\}$. *L. Kuipers* (Carbondale, Ill.)

K05-74 (41# 1657)

Kuipers, Lauwerens; Stam, Aart Johannes

On a general form of the Weyl criterion in the theory of asymptotic distribution. I, II.

Proc. Japan Acad. **45** (1969), 530–535; *ibid.* **45** (1969), 536–540.

The authors establish a general form of the Weyl criterion, which contains among others the following well known cases: (i) The sequence (u_n) of real numbers is equidistributed (mod 1) if and only if for any $q\in\mathbf{Z}^*$, one has $\lim_{n\to\infty}(1/n)\sum_{k=1}^n\exp 2i\pi qu_k=0$. This is Weyl's classical criterion. (ii) The sequence (a_n) of integers is equidistributed (mod m) ($m\geqq 2$ is a given integer) if and only if for any $q\in\{1, 2, \cdots, m-1\}$, one has

$$\lim_{n\to\infty}(1/n)\textstyle\sum_{k=1}^n\exp(2i\pi/m)qa_k = 0.$$

This again is Weyl's classical criterion applied for equidistribution on the additive group $\mathbf{Z}/m\mathbf{Z}$. (The authors claim that I. Niven [Trans. Amer. Math. Soc. **98** (1961), 52–61; MR **22** #10971] and S. Uchiyama [Proc. Japan Acad. **37** (1961), 605–609; MR **25** #1145] discovered this criterion, but indeed it is a special case of H. Weyl's result [Math. Ann. **77** (1916), 313–352].) (iii) The real-valued Borel measurable function f is equidistributed (mod 1) if and only if for any $q\in\mathbf{Z}^*$, one has $\lim_{T\to\infty}(1/T)\int_0^T\exp 2i\pi qf(t)\,dt=0$, another one of Weyl's theorems.

The authors give other applications of their general formulae (sequences and functions with a given distribution function and sequences (mod 1) Borel distributed to a given distribution function). The last example could be itself generalized in order to obtain Weyl's criterion for a sequence to be M-distributed to the distribution function F (here $M=(a_i^j)$ is a summation matrix, that is to say, such that $a_i^j\geqq 0$, $\sum_{j=1}^\infty a_i^j=1$, $\lim_{i\to\infty}a_i^j=0, \cdots$). *M. Mendès France* (Talence)

Citations: MR 22# 10971 = B48-6; MR 25# 1145 = B48-8.

K05-75 (41# 8348)

Schmidt, Wolfgang M.

Irregularities of distribution. V.

Proc. Amer. Math. Soc. **25** (1970), 608–614.

The author continues his earlier work [Quart. J. Math. Oxford Ser. (2) **19** (1968), 181–191; MR **37** #4029; Trans. Amer. Math. Soc. **136** (1969), 347–360; MR **38** #3237; Pacific J. Math. **29** (1969), 225–234; MR **39** #6937; Invent. Math. **7** (1969), 55–82; MR **39** #6838] but now with points with weights. Let $S=S^n$ be the unit sphere in Euclidean space E^{n+1}. Let $p_1, p_2, \cdots$ be points on S, with non-negative weights $w_1, w_2, \cdots$, respectively. It is assumed that $\sum_{i=1}^\infty w_i=\sigma$ is positive and finite. For any $\alpha\geqq 1$ define

$\sigma(\alpha)=\sum_{i=1}^{\infty} w_i^{\alpha}$, also positive and finite, and then $\sigma=\sigma(1)$. For any subset A of S define $\nu(A)=\sum w_j$, where the sum is over those points p_j in A. Thus $\nu(S)=\sigma$. Let d_Sx be the canonical volume element on S, normalized so that $\int_S d_Sx=1$. Given a measurable set A of S, write $\mu(A)=\int_A d_Sx$ and $\Delta(A)=\nu(S)\mu(A)-\nu(A)$. Let $C(r,c)$ be a spherical cap on S with center c and radius $r\leqq\pi/2$. Write $\Delta(r,c)$ for $\Delta(C(r,c))$, and $E(r,s)=\int_S\Delta(r,c)\Delta(s,c)d_Sc$. The central theorem is that if $n>1$ and $\varepsilon>0$, then $\int_0^{\pi/2} r^{-1}E(r,r)\,dr\gg\sigma(2+n^{-1}+\varepsilon)\sigma(1)^{-n^{-1}-\varepsilon}$. It follows that there is always a spherical cap such that the sum of the weights of the points of the cap is not the expected value; a corollary to the basic theorem sets this forth in detail. The proofs are adaptations of a method of K. F. Roth [Mathematika **1** (1954), 73–79; MR **16**, 575]. *I. Niven* (Eugene, Ore.)

The reviewer wishes to delete the last sentence of his review.

Citations: MR 16, 575c = K05-21; MR 37# 4029 = K05-60; MR 38# 3237 = K05-63; MR 39# 6838 = K05-70.

K05-76 (42# 4500)

Berlekamp, E. R.; Graham, R. L.

Irregularities in the distributions of finite sequences.

J. Number Theory **2** (1970), 152–161.

Authors' summary: "Suppose $(x_1,x_2,\cdots,x_{s+d})$ is a sequence of numbers with $x_i\in[0,1)$ which has the property that for each $r\leqq s$ and for each $k<r$, the subinterval $[k/n,(k+1)/n)$ contains at least one point of the subsequence $(x_1,x_2,\cdots,x_{r+d})$. For fixed d, we wish to find the maximum $s=s(d)$ for which such a sequence exists. We show that $s(d)<4^{(d+2)^2}$ for all d and that $s(0)=17$." *G. J. Rieger* (Munich)

K05-77 (43# 161)

de Mathan, Bernard

Approximations diophantiennes dans un corps local.

Bull. Soc. Math. France Suppl. Mém. 21 (1970), 93 *pp.*

From the author's introduction: "Le but de ce travail est essentiellement d'obtenir quelques résultats d'arithmétique diophantienne en remplaçant Q, et ses complétés R et Q_p, par le corps des fractions rationnelles à une indéterminée sur un corps, le plus souvent fini, et les complétés d'un tel corps."

After an introductory section, the first chapter culminates in an analogue, for fields of formal Laurent series, of a sufficient condition for equidistribution mod 1 due to J. F. Koksma [Compositio Math. **2** (1935), 250–258]. Chapter 2 is devoted to the study of the distribution mod 1 of sequences $\{\mu\theta^n\}$ with μ, θ in such fields, connections being established with Cantor sets, sets of uniqueness and normality.

In the third chapter the author gives (for both real and power series fields) the first sufficient conditions known to the reviewer on a sequence of nonlinear functions $\phi_n(x)$ which guarantee that the number N_n of positive integers $k\leqq n$ for which $\{\phi_k(x)\}<\varepsilon_k$ ($\{\cdot\}$ is the fractional part) is asymptotic to $\sum_1^n\varepsilon_k$, for a.a. x, in some interval, whenever $\sum\varepsilon_n$ diverges. In the real case, the conditions are: $\phi_n\in C^1[a,b]$; $A_n=\inf_{[a,b]}\phi_n'(x)>0$ for $n>0$ and $\sum n^{-1}<\infty$; if $A_{mn}=\inf_{[a,b]}\phi_n'(x)/\phi_m'(x)$ then $\sum_{m=1}^{n-1}A_{mn}^{-1}=O(1)$; finally, there exists a positive decreasing sequence c_n such that $\sum c_n<\infty$, while for every $x,x'\in[a,b]$ and for every n, $|1/\phi_n'(x)-1/\phi_n'(x')|\leqq c_n|x-x'|$. Under these circumstances, if $S_n=\sum_1^n\varepsilon_k$, it is shown that $N_n=S_n+O(S_n^{1/2}\log^{3/2+\eta}S_n)$ for almost all $x\in[a,b]$.

In the first portion of the fourth chapter the author shows how the classical theory of regular continued fractions carries over to fields of formal power series. Finally, an analogue of Hinčin's theorem (which asserts that if $n\varepsilon_n\searrow 0$, the inequality $|qx-p|<\varepsilon_q$ has finitely many [infinitely many] solutions for a.a. x, if the series $\sum\varepsilon_n$ converges [diverges]) is proved for approximation in the ring of adèles of the field of rational functions in one variable over a finite field. *W. J. LeVeque* (Claremont, Calif.)

K05-78 (43# 3407)

Hlawka, Edmund

Discrepancy and Riemann integration.

Studies in Pure Mathematics (Presented to Richard Rado), pp. 121–129. *Academic Press, London*, 1971.

Suppose that E^s is the unit cube $0\leqq\xi_i<1$ and $\omega=\{x_i\}$ a sequence in E^s that is uniformly distributed with density ρ, i.e., $(1/N)\sum_{i=1}^N\chi(J,x_i)\rightarrow\int_{E^s}\chi(J,x)\rho\,dx$ for all intervals $J\subseteq E^s$, where ρ is a nonnegative-valued Riemann-integrable function defined on $\bar{E}^s$ with $\int_{E^s}\rho\,dx=1$. For each positive integer N, $D_N(\omega,\rho)$ is defined as $\sup_J|(1/N)\sum_{i=1}^N\chi(J,x_i)-\int_{E^s}\chi(J,x)\rho\,dx|$ and $D_N(f,\omega,\rho)$ as $|(1/N)\sum_{i=1}^N f(x_i)-\int_{E^s}f\rho\,dx|$, where f is a Riemann-integrable function of period 1 defined on $\bar{E}^s$.

The author proves results concerning the estimation of $D_N(f,\omega,\rho)$ with the help of $D_N(\omega,\rho)$, an example of which is the following inequality for $\rho=1$: $D_N(f,\omega,1)\leqq(1+2^{2s-1})\sum(f,[D_N^{-1/s}(\omega,1)]^{-1})$, where, for each $K\geqq 0$, $\sum(f,K)=\sup\{\sigma(f;p): p$ an interval partition of $\bar{E}^s$ with norm $\leqq K$ and $\sigma(f;p)$ the mean oscillation of f for $p\}$. *W. D. L. Appling* (Denton, Tex.)

K05-79 (44# 1632)

Brown, J. L., Jr.; Duncan, R. L.

Asymptotic distribution of real numbers modulo one.

Amer. Math. Monthly **78** (1971), 367–372.

Les auteurs démontrent des résultats connus sur la répartition (mod 1). *M. Mendès France* (Talence)

K05-80 (44# 1633)

Niederreiter, H.

Almost-arithmetic progressions and uniform distribution.

Trans. Amer. Math. Soc. **161** (1971), 283–292.

L'auteur utilise une définition due à P. E. O'Neil [Proc. Amer. Math. Soc. **24** (1970), 1–5; MR **40** #1349]: Soient $\delta\geqq 0$, $\varepsilon>0$; une séquence strictement croissante $(a_1,\cdots,a_N)$ d'éléments de $U=[0,1]$ est dite une (δ,ε)-progression presque arithmétique ((δ,ε)-p.p.a.) s'il existe η, avec $0<\eta\leqq\varepsilon$, tel qu'on ait: (i) $0\leqq a_1\leqq\eta+\delta\eta$; (ii) $a_n+\eta-\delta\eta\leqq a_{n+1}\leqq a_n+\eta+\delta\eta$ pour $n=1,\cdots,N-1$; (iii) $1-\eta-\delta\eta\leqq a_N\leqq 1$. Ayant indiqué que la discrépance $D_N(a)$ d'une séquence croissante $(a_1,\cdots,a_N)$ d'éléments de U reçoit la forme $\max_{1\leqq n\leqq N}\max(|nN^{-1}-a_n|,|(n-1)N^{-1}-a_n|)$ (cf. l'auteur, "Discrepancy and convex programming", Ann. Mat. Pura Appl., à paraître), l'auteur démontre, par voie élémentaire, le théorème suivant: Soient $0\leqq\delta<1$, $0<\eta\leqq\varepsilon$; si $(a_1,\cdots,a_N)$ est une (δ,ε)-p.p.a. définie par η, alors on a: (i) $D_N(a)\leqq N^{-1}+\delta(1+(1-\delta^2)^{1/2})^{-1}$ si $\delta>0$; (ii) $D_N(a)\leqq\min(\eta,N^{-1})$ si $\delta=0$. Il obtient ensuite, pour une séquence $(a_1,\cdots,a_N)$ d'éléments de U qui, privée de N_0 de ses termes, est une superposition de (δ_i,ε_i)-p.p.a. à N_i éléments, avec $0\leqq\delta_i\leqq\delta<1$, où $i=1,\cdots,k$, les majorations successives suivantes:

$$D_N(a)\leqq(k+N_0)N^{-1}+\sum_{i=1}^{i=k}N_i\delta_i(N(1+(1-\delta_i^2)^{1/2}))^{-1}\leqq(k+N_0)N^{-1}+(N-N_0)\delta(N(1+(1-\delta^2)^{1/2}))^{-1}.$$

Il en déduit enfin que, si une fonction réelle f d'une variable réelle vérifie les conditions classiques de Fejér, si f' désigne sa fonction dérivée et $\langle f\rangle$ sa fonction partie fractionnaire, alors $D_N(\langle f\rangle)=O(f(N)N^{-1}+(Nf'(N))^{-1})$ quand $N\to\infty$; le théorème de Fejér, qui affirme, sous ces conditions, l'équirépartition mod 1 de la suite $(f(n))_{n\geqq 1}$, en résulte aussitôt. *J. Chauvineau* (Paris)

Citations: MR 40# 1349 = K05-71.

K05-81 **(44# 3966)**

Hlawka, Edmund

Zur Definition der Diskrepanz.

Acta Arith. **18** (1971), 233–241.

Let ω be a finite collection of ν points in the s-dimensional unit cube I and let J be an s-dimensional interval. If $\nu^*(J)$ is the number of points of ω lying in J and $V(J)$ is the volume of J then $D(\omega)=\sup_J|\nu^*(J)/\nu-V(J)|$ is the discrepancy of ω. If, instead of allowing J to range over intervals, one allows it to run over arbitrary convex bodies or over spheres, one obtains quantities $D_C(\omega)$ and $D_K(\omega)$. In the present paper the author proves the existence of positive numbers $\gamma_1(s)$, $\gamma_2(s)$ such that $D(\omega)\leqq D_C(\omega)\leqq \gamma_1(s)\log^{-\gamma_2(s)}1/D_K(\omega)$. This result is proved via the inequality $D_C(\omega)\leqq 72^sD^{1/s}(\omega)$, which, in turn, is a consequence of the lemma: Let B be a set in I whose boundary δB is a finite union of the boundaries δC of convex sets C in I. Let I be split into subcubes of side length w. Then the number $N(\delta B)$ of subcubes having a common point with δB satisfies $N(\delta B)w^s\leqq 72^s\beta(B)w$, where $\beta(B)$ is the number of δC which cover δB.

J. B. Roberts (Portland, Ore.)

K10 DISTRIBUTION (mod 1): $\{k\alpha\}$, AND HIGHER DEGREE POLYNOMIALS IN k

Papers in this and adjacent sections are concerned with the number of solutions $n\leq N$ of inequalities such as $0<n\alpha-[n\alpha]\leq x$, where $x\in[0,1]$ is an arbitrary constant, independent of n. Results concerning the number of solutions $n\leq N$ of $0<n\alpha-[n\alpha]<f(n)$, say, where $f(n)\to 0$ as $n\to\infty$, are to be found in **J04** and **J24**.

See also reviews J02-15, J02-20, J24-18, J24-28, K05-1, K05-8, K05-22, K05-57, K30-27, K30-28, K40-60, K45-19, L05-1, L15-6, Q20-15, T55-32, Z10-47.

K10-1 **(1, 203b)**

Spencer, D. C. On a Hardy-Littlewood problem of diophantine approximation. Proc. Cambridge Philos. Soc. **35**, 527–547 (1939).

Let ω_1 and ω_2 be two positive numbers whose ratio $\theta=\omega_1/\omega_2$ is irrational and ξ a real number with $0\leqq\xi<\omega_1$. Let further, for real $r\geqq 1$,

$$P_r(x)=-\sum\frac{2\cos(2\pi nx-\frac{1}{2}\pi r)}{(2\pi n)^r}.$$

The author gives a systematic treatment of the sums

$$R_r=R_r(\xi,m)=\sum_{\nu=0}^{m}P_r\left(\frac{\xi+\nu\omega_2}{\omega_1}\right),$$

which, for positive integers r, have been considered (with different methods) by Hardy-Littlewood, Ostrowski, Behnke and Khintchine [references in the reviewer's "Diophantische Approximationen," Berlin, 1936 (Ergebnisse der Mathematik IV, 4)]. The method depends on a formulation of the problem in terms of a contour integral of the function

$$\frac{e^{-(\omega_1+\omega_2)z}}{(1-e^{-\omega_1 z})(1-e^{-\omega_2 z})}\frac{e^{\eta z}}{z^r},\qquad \eta \text{ real},$$

along the lines of the transcendental method developed by Hardy-Littlewood in their second memoir on the lattice points of a right-angled triangle [Abh. Math. Sem. Hansischen Univ. 1, 212–249 (1922)].

O-results: If θ is of type I hc $(0\leqq c<\infty)$, we have $R_r=O(m^{1-r/h})$ if $r<h$; $R_1=O(\log m)$ if $h=1$; $R_r=O(\log\log m)$ if $h=r>1$ and $R_r=O(1)$ if $r>h\geqq 1$. If θ is of type II hc $(h>0,\ 0\leqq c<\infty)$, we have $R_r=O(m\log^{-r/h}m)$. If $\epsilon>0$, then for almost all points (ω_1,ω_2) in the plane $R_1=O(\log^{1+\epsilon}m)$; $R_r=O(1)$ if $r>1$. All results hold uniformly in ξ. Ω-result: If $r\geqq 1$, θ of type I hc $(0<c\leqq\infty)$, then for an infinity of m $R_r>\text{const.}\cdot m^{1-r/h}$. *J. F. Koksma* (Amsterdam).

Referred to in P28-3.

K10-2 **(3, 70c)**

Linés Escardó, E. A theorem on the frequency of the points of a lattice which are on a strip, interior to another, both of known width. Revista Mat. Hisp.-Amer. (4) **1**, 75–81 (1941). (Spanish)

The author gives a not quite adequate and rather complicated geometrical proof of a special case of the following well-known fact: Let k be irrational, $\delta>0$; then the number of solutions of $b<y-xk<b+\delta$ in integers x, y with $0\leqq x\leqq X$ is asymptotic to δX. *P. Scherk.*

K10-3 **(4, 36e)**

Pillai, S. S. On algebraic irrationals. Proc. Indian Acad. Sci., Sect. A. **15**, 173–176 (1942).

One of the results of the paper reviewed below is given with slight alterations in the proof. *H. S. Zuckerman.*

K10-4 **(4, 36f)**

Pillai, S. S. On a problem in Diophantine approximation. Proc. Indian Acad. Sci., Sect. A. **15**, 177–189 (1942).

Questions concerning the order of $S(\theta,N)=\sum_{n=1}^{N}\{n\theta\}$, with $\{x\}=x-[x]-\frac{1}{2}$, have been discussed by Hardy and Littlewood [Proc. London Math. Soc. (2) **20**, 15–36 (1921)] and by others. In this paper a reduction formula is obtained. This expresses $S(\theta,N)$ by means of $S(\theta,N_1)$, where N_1 is decidedly smaller than N. By suitable applications of the reduction formula the author proves the results of Hardy and Littlewood and some further results. The proofs are all elementary in character, depending upon properties of the continued fraction expansion of θ, while Hardy and Littlewood, besides using the continued fraction, also make use of transcendental methods. The proofs of theorems VII and IX are not quite complete but the details are easily supplied if the condition $n>7$ is added to theorem VII. Most of the misprints are not serious but an inequality is inverted in theorem XVI. The inequality in the conclusion should read $S(\theta,N)<\delta$. *H. S. Zuckerman.*

K10-5 **(7, 278d)**

Tsuji, Masatsugu. On the uniform distribution of values of a function mod. 1. Proc. Imp. Acad. Tokyo **19**, 66–69 (1943).

Let $f(x)$ be a real continuous function defined for $0\leqq x<\infty$ and $(f(x))=f(x)-[f(x)]$, so that $0\leqq(f(x))<1$. Let $r>0$, $0\leqq\alpha<\beta\leqq 1$, and denote by $E(r,\alpha,\beta)$ the set of points x satisfying the conditions (1) $0\leqq x\leqq r$, (2) $\alpha\leqq(f(x))\leqq\beta$. Denote its measure by $mE(r,\alpha,\beta)$. The values of $f(x)$ are said to be distributed uniformly mod 1 if for any α, β we have $\lim r^{-1}\mathrm{m}E(r, \alpha, \beta) = \beta - \alpha$ as $r \to \infty$. The following theorem is established. If $f(x)$ is a positive continuous increasing convex function of $\log x$ such that $\lim f(x)/\log x = \infty$ as $x \to \infty$, then the values of $f(x)$ are distributed uniformly mod 1. The concept of uniform distribution mod 1 is also extended to a set of m functions of n variables by replacing the interval (1) by the sphere $x_1^2 + \cdots + x_n^2 \leqq r^2$ and (2) by a parallelepiped inside the unit cube of m-dimensional space. It is then shown that a set of m polynomials in n variables, having no constant (nontrivial) linear combination, have values which are distributed uniformly mod 1.

I. J. Schoenberg (Philadelphia, Pa.).

K10-6 (7, 433f)

Hardy, G. H., and Littlewood, J. E. Notes on the theory of series. XXIV. A curious power-series. Proc. Cambridge Philos. Soc. **42**, 85–90 (1946).

[Note XXIII appeared in the same Proc. **40**, 103–107 (1944); these Rev. **6**, 47.] Answering a question put by W. R. Dean, the authors prove first that, θ being any irrational number, if the radius of convergence of the series (S) $\sum x^n/\sin n\pi\theta$ is ρ, then the radius of convergence of the series (T) $\sum x^n/(\sin \pi\theta \sin 2\pi\theta \cdots \sin n\pi\theta)$ is $\rho/2$. The proof is based on the use of an algebraic identity. Since for almost all θ the radius ρ is equal to 1, it follows that, for almost all θ, the radius of convergence of (T) is $\frac{1}{2}$. From this the authors deduce that for almost all θ

$$(*) \qquad \lim_{n\to\infty} n^{-1} \sum_{m=1}^{n} \log |\operatorname{cosec} m\pi\theta| = \log 2.$$

They observe that if $(*)$ is true the result about the radii of convergence of (S) and (T) may be extended to the more general series $\sum a_n x^n/\sin n\pi\theta$ and

$$\sum a_n x^n/(\sin \pi\theta \sin 2\pi\theta \cdots \sin n\pi\theta).$$

The authors then give a direct proof of the equality $(*)$. Indeed, they prove more generally that, if $f(x)$ has period 1, is Riemann integrable in $(\delta, 1-\delta)$ for every $\delta > 0$, increases steadily to infinity when $x \to 0$ and $x \to 1$, and if

$$\int_0^1 f(x)\{\log^2 (1/x) + \log^2 (1/(1-x))\} dx < \infty,$$

then the following extension of Weyl's classical theorem is true for almost all θ:

$$\lim_{n\to\infty} n^{-1} \sum_{m=1}^{n} f(m\theta) = \int_0^1 f(x)dx.$$

Finally, they show that the last equality is true whenever $\int_0^1 f(x)dx < \infty$ if the partial quotients of the expansion of θ in a continued fraction are bounded.

R. Salem.

K10-7 (10, 513b)

Erdös, P. Some remarks on Diophantine approximations. J. Indian Math. Soc. (N.S.) **12**, 67–74 (1948).

Let $d(m)$ be the number of positive divisors of m, let $r_k(m)$ be the number of representations of m as a sum of k squares, and let $\{\alpha\}$ be the distance from α to the nearest integer. Simple proofs are given here of the following relations, which hold for almost all real α:

$$\sum_{m=1}^{n} d(m)e^{2\pi i m\alpha} = O(n^{\frac{1}{2}} \log n),$$

$$\sum_{m=1}^{n} r_2(m)e^{2\pi i m\alpha} = O(n^{\frac{1}{2}} \log n),$$

$$\sum_{m=1}^{n} r_4(m)e^{2\pi i m\alpha} = O(n \log^2 n).$$

More complicated proofs of these relations were given earlier by Chowla [Math. Z. **33**, 544–563 (1931)]. A conjecture of Spencer, that for almost all α

$$\sum_{m=1}^{n} 1/m\{m\alpha\} = (1+o(1)) \log^2 n,$$

is also proved, and the following theorem (among others) is stated without proof. Let $f(n)$ be an increasing function of n for which $f(n) > (2+\epsilon)n \log n$ and $\sum_{n=1}^{\infty} 1/f(n)$ converges. Then for almost all α and $n > n_0(\alpha)$, $\sum_{m=1}^{n} 1/\{m\alpha\} < f(n)$. [There are many misprints: the first inequality in the third paragraph, p. 67, should be $|\alpha - a/b| < 1/2b^2$; the right sides of (8) and (11) should be $O(n \log^{2+\epsilon} n)$ and $O(n \log^2 n)$ respectively; in footnotes 2, 3, p. 68, the years of publication are 1932 and 1931, respectively; the upper limit on the summation in the last displayed formula of the paper should be n, rather than ∞; p. 73, lines 7, 8, $(m\alpha)$ is to be replaced by $\{m\alpha\}$; in (31) there should be an equality sign before $o(\log^2 n)$.]

W. J. LeVeque (Ann Arbor, Mich.).

K10-8 (10, 682f)

Revuz, André. Sur la répartition des points $e^{\nu i\theta}$. C. R. Acad. Sci. Paris **228**, 1466–1467 (1949).

The note contains results concerning the distribution of the numbers $\omega, 2\omega, 3\omega, \cdots$ (mod 1), expressed in terms of the continued fraction for the irrational number ω. They are, I believe, all known in substance, though perhaps not in the form given here.

H. Davenport (London).

K10-9 (13, 16g)

Slater, N. B. The distribution of the integers N for which $\{\theta N\} < \phi$. Proc. Cambridge Philos. Soc. **46**, 525–534 (1950).

In Verschärfung des Weylschen Theorems, dass $\{\theta N\}$ für irrationales θ gleichverteilt ist (mod 1), wird folgendes untersucht: Wie gross sind die Lücken, die zwischen aufeinanderfolgenden ganzen Zahlen N liegen, falls N der Ungleichung $\{\theta N\} < \phi$ genügt, wobei $\{\theta N\}$ den positiven Rest von θN mod 1 bedeutet? Wie häufig treten die verschiedenen Lücken dabei auf? Diese Fragen werden unter Verwendung der Kettenbruchdarstellung von θ in der Weise beantwortet, dass die Gestalt der möglichen Lücken angegeben und für endliche N auch deren Anzahl mitgeteilt wird, während für beliebig grosse N ein asymptotischer Wert bewiesen wird. Die Untersuchung wird für rationales und irrationales θ getrennt geführt. Die Beweise sind völlig elementar und nicht tief liegend.

T. Schneider (Göttingen).

K10-10 (13, 119e)

Hartman, S. Sur une méthode d'estimation des moyennes de Weyl pour les fonctions périodiques et presque périodiques. Studia Math. **12**, 1–24 (1951).

Weyl's mean of the function f of period 1 is the limit, if it exists, $\lim_{N\to\infty} \sum_1^N f(\kappa\xi)/N$. Weyl has shown that for irrational ξ, it is equal to $M_f = \int_0^1 f(x)dx$. The author introduces the notation $G(f, \xi, N) = \sum_1^N f(\kappa\xi) - NM_f$ and investigates estimates of G under two types of hypotheses. The first concerns the character of the number ξ expressed by its "type" [J. Koksma, Diophantische Approximationen, Springer, Berlin, 1936], the second hypothesis concerns the character of the function given by an estimate of its Fourier coefficients. A sample theorem is the following. If the Fourier coefficients of f are $O(n^{-\kappa})$ and if ξ is a number of type $\mu < \kappa$, then $G(f, \xi, N) = O(1)$. The author concerns himself also with almost periodic functions.

František Wolf.

K10-11 **(13, 827c)**

Koksma, J. F. A Diophantine property of some summable functions. J. Indian Math. Soc. (N.S.) **15** (1951), 87–96 (1952).

Let $g(t)$ be a periodic function in L^2 of period 1. Let $\sum_{k=-\infty}^{\infty} c_k e^{2\pi i k t}$ be the Fourier expansion of $g(t)$. Assume that

$$(1) \qquad \sum_{k=1}^{\infty}\left(|c_k|^2 \sum_{d|k} \frac{\log^2 d}{d}\right) < \infty.$$

Then the author proves that for almost all x

$$(2) \qquad \lim_{N\to\infty} \frac{1}{N} \sum_{n=1}^{N} g(nx) = \int_0^1 g(t)dt.$$

The author remarks that (1) can be replaced by

$$\sum_{k=1}^{\infty} |c_k|^2 (\log\log k)^3 < \infty.$$

The interesting question whether (2) holds for every periodic $g(t)$ in L^2 is left open; in fact, the reviewer did not succeed in finding a periodic $g(t)$ in L which does not satisfy (2).

P. Erdös (Los Angeles, Calif.).

Referred to in K05-22.

K10-12 **(14, 1066e)**

Peck, L. G. On uniform distribution of algebraic numbers. Proc. Amer. Math. Soc. **4**, 440–443 (1953).

Let K be a real algebraic number field of degree $n+1$ over the rational field R, and let $1, \omega_1, \cdots, \omega_n$ be an R-base for K. If $f(x_1, \cdots, x_n)$ is Riemann integrable over the unit cube in n-space, and periodic with period 1 in all arguments, then Weyl's [Math. Ann. **77**, 313–352 (1916)] metrical extension of the Kronecker approximation theorem asserts that

$$\sum_{m=0}^{M-1} f(m\omega_1, \cdots, m\omega_n) = \mathfrak{M}(f) + o(M),$$

where $\mathfrak{M}(f)$ is the average of f over the unit cube. The author proves that $o(M)$ can be replaced by $O(1)$ if f has the form

$$f(x_1, \cdots, x_n) = \sum_{q_1, \cdots, q_n = -\infty}^{\infty} a(q_1, \cdots, q_n) \times \exp\{2i\pi(q_1 x_1 + \cdots + q_n x_n)\},$$

where $a(q_1, \cdots, q_n) = O\{(|q_1| + \cdots + |q_n|)^{-n-c}\}$, $c>0$.

The proof depends on the convergence of the series $\sum_{\alpha \varepsilon \mathfrak{a}, |\alpha|<c} |N(\alpha)|^{-1}(m(\alpha))^{-c}$, where $\mathfrak{a}$ is an integral ideal of K, and $m(\alpha) = \max(|\alpha^{(1)}|, \cdots, |\alpha^{(n)}|)$; here $\alpha^{(0)}, \alpha^{(1)}, \cdots, \alpha^{(n)}$ are the conjugates of $\alpha = \alpha^{(0)}$. In order to show the convergence the author applies the Dirichlet theory of units as well as the convergence of the Dedekind series $\zeta_K(s)$ $(s>1)$.

N. G. de Bruijn (Amsterdam).

K10-13 **(15, 57h)**

Hartman, S. Über die Abstände von Punkten $n\xi$ auf der Kreisperipherie. Ann. Soc. Polon. Math. **25** (1952), 110–114 (1953).

Let ξ be an irrational number, and let $P_1, \cdots, P_n$ be points on the circumference of a circle of unit circumference, obtained by proceeding by successive steps of arc-length ξ from a given point P_0. Now consider the points $P_0, P_1, \cdots, P_n$ in the order in which they lie on the circumference, and let m_n be the shortest arc between two neighbouring points and M_n the longest arc. The author proves that, for almost all ξ,

$$\liminf_{n\to\infty} nm_n = 0, \quad \limsup_{n\to\infty} nm_n = 1,$$

$$\liminf_{n\to\infty} nM_n = 1, \quad \limsup_{n\to\infty} nM_n = \infty,$$

as was conjectured by Steinhaus. The proofs are not difficult, and the results are shown to hold for any ξ which has unbounded partial quotients in its continued fraction development.

H. Davenport (London).

K10-14 **(15, 511d)**

LeVeque, W. J. On uniform distribution modulo a subdivision. Pacific J. Math. **3**, 757–771 (1953).

Let $\Delta = (0 = z_0 < z_1 < z_2 < \cdots)$ $(z_n \to \infty)$ be a subdivision of the interval $(0, \infty)$, and let $\phi(x)$ be defined by (i) $\phi(z_n) = n$, (ii) $\phi(x)$ linear in each interval $[z_i, z_{i+1}]$. A sequence of positive numbers x_n is called u.d. (=uniformly distributed) (mod Δ) if $\phi(x_n)$ is u.d. (mod 1). The author gives a number of sufficient conditions. He proves, moreover, for some sequences containing a parameter, that they are u.d. (mod Δ) for almost all values of the parameter. We quote here: (i) If $\delta_n = z_n - z_{n-1} \downarrow 0$, $\delta_n = O(x^{-1})$, then $\{k\theta\}$ is u.d. (mod Δ) for almost all θ; (ii) if $z_n = g(n)$, $g(x) \uparrow \infty$, $\{g'(x)/g(x)\} \downarrow 0$, $\{g'(x)/g(x)\} = O(x^{-\frac{1}{2}})$, then $\{\alpha^k\}$ is u.d. (mod Δ) for almost all $\alpha > 1$.

The reviewer remarks that in the proof of theorem 4 the author gives a false quotation of a principle proved in an earlier paper: the integral in (3) should read

$$\int_a^b \exp[hi(f_j(x) - f_k(x))]dx,$$

and the formula should be required for each integer h. The remainder of the proof of theorem 4 is easily revised in this respect.

N. G. de Bruijn (Amsterdam).

Referred to in K10-30, K10-48, K20-9, K35-34, K40-22.

K10-15 **(16, 224a)**

Szüsz, P. Über die Verteilung der Vielfachen einer komplexen Zahl nach dem Modul des Einheitsquadrats. Acta Math. Acad. Sci. Hungar. **5**, 35–39 (1954). (Russian summary)

Let $z = x+iy$; define J_z as the parallelogram determined by the vectors

$$\min(x, y)/\max(x, y), \quad \min(x, y) + i\max(x, y).$$

Put $z_v = (v\alpha) + i(v\beta)$, where $(v\alpha)$ is the fractional part of $v\alpha$. Let $N_z(n, y)$ denote the number of z_v, $1 \leqq v \leqq n$, falling into J_z. The author proves that for all g

$$|N_z(n, J_{z_g}) - n\min((g\alpha), (g\beta))| < c_1, \quad c_1 = c_1(\alpha, \beta, g),$$

i.e., c_1 is independent of n ($\min((g\alpha), (g\beta))$ is the area of J_z).

P. Erdös (Jerusalem).

K10-16 **(17, 589d)**

Szüsz, P. Lösung eines Problems von Herrn Hartman. Studia Math. **15** (1955), 43–55.

Let $N = N_{\alpha,\beta}(n, I)$ be the number of integers in 1, 2, $\cdots$, n such that the fractional parts $(h\alpha)$, $(h\beta)$ of $h\alpha$ and $h\beta$ are the coordinates of a point in a rectangular subinterval I of the unit square, α and β being numbers in the interval 0 to 1 such that 1, α, and β are linearly independent over the rationals. If $|I|$ is the area of the rectangle I, then N/n approaches $|I|$ as n goes to infinity, but the value of $N-n|I|$ may not be bounded. S. Hartman [Colloq. Math. **1** (1948), 239–240] has asked if $|N-(q\alpha)(q\beta)n|$ is bounded, where q is an integer and I is the rectangle with (0, 0) a corner and sides $(q\alpha)$ and $(q\beta)$, since Ostrowski [Jbr. Deutsch. Math. Verein. **39** (1930), Abt. 1, 34–46] has shown the corresponding result to be true in one dimension. By a construction using continued fractions the author shows this to be false with $q=1$ for any irrational β and any one of an uncountable set of α's corresponding to the given β. In a note added in proof he states that with $(q\alpha)\leqq(q\beta)$ the difference $|N-(q\alpha)n|$ is bounded if I is the parallelogram whose sides are the complex numbers $(q\alpha)/(q\beta)$ and $(q\alpha)+i(q\beta)$.

Marshall Hall, Jr. (Columbus, Ohio).

K10-17 (19, 17d)

Steinhaus, H. On golden and iron numbers. Zastos. Mat. 3 (1956), 51–65. (Polish. Russian and English summaries)

Durch die Beziehungen $z_n=nz$, $s_n=n(1-z)$, wo $z=\frac{1}{2}(5^{\frac{1}{2}}-1)$ ist, definiert zuerst der Verf. die Folge $\{z_n\}$ von „goldenen" und die Folge $\{s_n\}$ von „silbernen" Zahlen, und benutzend die Eigenschaft, dass diese Zahlen sehr gleichmässig mod 1 verteilt sind und dass deshalb auch $n^{-1}\sum_{k=1}^{n}f(z_k)$ für beliebige integrierbare Funktion $f(x)$ mit der Periode 1 sehr gut gegen $\int_0^1 f(x)dx$ konvergiert, zeigt er den Vorteil ihrer Benutzung zum statistischen Zwecke. Für eine gegebene natürliche Zahl N definiert er weiter die endliche Folge von „eisernen" Zahlen $\{c_n\}$ $(n=1, \cdots, N)$ als eine gewisse Permutation von Zahlen $1, \cdots, N$ so, dass die Folge von Zahlen $c_nz-[c_nz]$ wachsend sei ($[\mu]$ bedeutet dabei die grösste ganze Zahl $\leqq\mu$). Es wird der Voıteil dieser Folge bei ihrer Benutzung zu den Zwecken der Stichprobe und zur Qualitätskontrolle gezeigt. Es handelt sich nämlich um eine in gewissen Sinne universale Folge, die repräsentative Auswahlen (samples) aus beliebiger Anzahl $k\leqq N$ von Elementen gibt. Der Artikel ist durch den Beweis des Satzes beendet, dass für jede irrationale Zahl z und für jede natürliche Zahl n die Punkte $z, 2z, \cdots, nz$ die Zahlenachse mod 1 in Intervalle zerteilen, unter denen höchstens drei verschiedene Länge haben können. *V. Knichal* (Prag).

K10-18 (19, 646b)

Kac, M.; and Salem, R. On a series of cosecants. Nederl. Akad. Wetensch. Proc. Ser. A. **60**=Indag. Math. **19** (1957), 265–267.

Using elementary properties of Fourier coefficients of the function $|\sin mx/\sin x|$, the authors prove that if $c_k\geqq 0$ and $\sum c_k<+\infty$, then $\sum_1^\infty c_k|\csc kx|$ converges on a set of positive measure, or almost everywhere, if and only if $\sum_1^\infty c_k\log(1/c_k)<+\infty$. They state that the corresponding question for the series $\sum c_k|\csc kx|^\alpha$, $\alpha>1$, seems to be more difficult. *G. G. Lorentz* (Detroit, Mich.).

Referred to in K10-50, K15-51.

K10-19 (20# 34)

Sós, Vera T. On the theory of diophantine approximations. I. Acta Math. Acad. Sci. Hungar. **8** (1957), 461–472.

Denote by $\{\alpha\}$ the fractional part of the real number α. For real α and positive integer N put

$$C_\alpha(N)=-\frac{N}{2}+\sum_{1\leq n\leq N}\{n\alpha\}.$$

It was shown by A. Ostrowski [Abh. Math. Sem. Univ. Hamburg **1** (1921), 77–98] that $C_\alpha(N)$ is unbounded for every irrational α. In this paper the author answers a question raised by Ostrowski by showing that there exist real α such that $C_\alpha(N)$ is bounded below (or above). It is remarked that a modification of the proof would produce an α such that $C_\alpha(N)>-\varepsilon$ for all N, where ε is an arbitrarily small positive number: but that it is impossible for $C_\alpha(N)$ to be positive for every N.

The proof depends on an explicit formula for $C_\alpha(N)$ related to one given by Ostrowski, but in which in addition to the denominators of the convergents to α (the Hauptnenner), as in Ostrowski, the bye-denominators (Nebennenner) also occur.

The formula is motivated by a 'geometric' interpretation of the continued fraction algorithm, which is stated to generalize naturally to an algorithm appropriate to inhomogeneous problems.

J. W. S. Cassels (Cambridge, England)

Referred to in A58-10, J20-29, K50-47.

K10-20 (21# 3404)

Świerczkowski, S. On successive settings of an arc on the circumference of a circle. Fund. Math. **46** (1959), 187–189.

Let N be a positive integer and C a directed circle whose circumference is not an integer $\leqq N$. Let P be an arbitrary point on C and let p_x $(x=0, 1, \cdots, N)$ denote the point arrived at on going arc length x from P in the positive direction. Consider now the points p_x in the order in which they are situated on C, let p_{a_r} be the point which immediately follows p_0 and p_{a_k} the one immediately followed by p_0. Proving a conjecture of H. Steinhaus the author shows that if p_y follows immediately p_x then $y-x$ takes one of the values a_r, $-a_k$, a_r-a_k. It is stated that two other (unpublished) proofs are due to P. Szüsz and P. Erdös and V. S. Turán. *A. Dvoretzky* (Jerusalem)

Referred to in K10-34, K10-42.

K10-21 (21# 4947)

Friedman, Bernard; and Niven, Ivan. The average first recurrence time. Trans. Amer. Math. Soc. **92** (1959), 25–34.

Let $\alpha\in(0, 1)$, $\varepsilon>0$, and define

$$t(\alpha, \varepsilon) = \inf\{n:|n\alpha-[n\alpha+\tfrac{1}{2}]|\leqq\varepsilon, n>0\}.$$

The authors estimate certain mean values $\mu_j(\varepsilon)=\int_0{}^1 t^j d\alpha$, and prove in particular that for $\varepsilon\to 0$

$$\mu_1(\varepsilon) = \frac{6\log 2}{\pi^2}\varepsilon^{-1}+O(\log\varepsilon),$$

$$\mu_2(\varepsilon) = \frac{2\log 2+1}{\pi^2}\varepsilon^{-2}+O(\varepsilon^{-1}\log\varepsilon).$$

Similarly, for a point $(\alpha_1, \alpha_2, \cdots, \alpha_k)$ in the k-dimensional unit cube C_k, defining t as the smallest positive integer such that all inequalities $|n\alpha_j-[n\alpha_j+\frac{1}{2}]|\leqq\varepsilon$ hold, the authors show that there exist two positive constants c_1 and c_2 such that $c_1\varepsilon^{-k}<\mu_1<c_2\varepsilon^{-k}$, where μ_1 is the integral of t over C_k. *D. A. Darling* (Ann Arbor, Mich.)

Referred to in J24-28.

K10-22 (22# 4695)

Kesten, Harry. Uniform distribution mod 1. Ann. of Math. (2) **71** (1960), 445–471.

Let $f(x)$ be the characteristic function of the interval $0<a\leqq x\leqq b<1$, and $f(x+1)=f(x)$. The problem of studying the asymptotic distribution of the sum $\sum_{k=1}^N f(y+kx)$, when x and y are independent random variables uniformly distributed over [0, 1], was suggested by the reviewer [Research Problem No. 6, Bull. Amer. Math. Soc. **64** (1958), 60]. The author shows that for rational $b-a$ the quantity $(\log N)^{-1}(\sum_{k=1}^N f(y+kx)-N(b-a))$ has a Cauchy distribution asymptotically. The proof requires some detailed analysis. *R. D. Bellman* (Santa Monica, Calif.)

Referred to in K10-27.

K10-23 **(23# A3686)**

Flor, Peter

Ein Verteilungsproblem für arithmetische Folgen.

Abh. Math. Sem. Univ. Hamburg **25** (1961), 62–70.

Let $(e_n|n$ is an integer) be a bisequence with $e_n = \pm 1$. The principal result of the paper is the establishment of a criterion for the existence of real numbers a and b such that $e_n = \operatorname{sgn} \sin \pi(an+b)$ for all integers n. The author points out that such bisequences are closely related to the Sturmian bisequences of Morse and Hedlund.

W. H. Gottschalk (Philadelphia, Pa.)

K10-24 **(24# A3135)**

Luthar, Indar S.

On an application of uniform distribution of sequences.

Colloq. Math. **8** (1961), 89–93.

Let ζ be a point of the complex plane having irrational coordinates. The number λ_n of points z with integral coordinates that lie in the circle $|z-n\zeta| \leqq 1$ is either 2, 3, or 4. Let $f_k(N)$ ($k=2, 3, 4$) be the number of values of n for which $\lambda_n = k$ and $1 \leqq n \leqq N$. Then $\lim f_k(N)/N$ exists.

J. C. Oxtoby (Bryn Mawr, Pa.)

K10-25 **(25# 1143)**

Vinogradov, I. M.

On the distribution of the fractional parts of the values of a polynomial. (Russian)

Izv. Akad. Nauk SSSR Ser. Mat. **25** (1961), 749–754.

Let $n \geqq 4$ be an integer, let $f(x) = \alpha_n x^n + \cdots + \alpha_1 x$ be a polynomial with real coefficients, and let $p > 1$ be real. Let x run through an increasing sequence of positive integers, let p_1 be the numbers of such integers with $x \leqq p$, and let $p_1(A, B)$ be the number of the latter integers for which $A \leqq f(x) < B \pmod 1$, for $0 \leqq B - A \leqq 1$. Further, let $k \geqq 1$ be real, and put $h = \ln k$, $\delta = (5k)^{-1}$, $\lambda = 1 - (h + 1.35)/k$. It is shown that in the n-dimensional cube $0 < \alpha_n \leqq 1, \cdots, 0 < \alpha_1 \leqq 1$, there is a region Ω, of volume not exceeding

$$(2^{3h+h^2} n^h)^{n^3} p^{-\lambda n(n+1)/2},$$

which contains all those points $(\alpha_n, \cdots, \alpha_n)$ of this cube for which, for arbitrary A and B,

$$|p_1(A, B) - (B-A)p_1| \geqq 3p^{1-\delta}.$$

W. J. LeVeque (Ann Arbor, Mich.)

K10-26 **(25# 2056)**

Karacuba, A. A.

Distribution of fractional parts of polynomials of a special type. (Russian. English summary)

Vestnik Moskov. Univ. Ser. I Mat. Meh. **1962**, *no.* 3, 34–39.

Let n, k, P be natural numbers, $k \geqq n$, $n \geqq 80^2$, p a prime number, and $f(x) = a_1 x + \cdots + a_n x^n$ a polynomial with integral coefficients $a_1, \cdots, a_n$ and $(a_n, p) = 1$. Further, let γ be a real number of the interval $[0, 1]$, $\{\beta\} = \beta - [\beta]$ the fractional part of a real number β, $N_p(\gamma)$ the number of fractional parts of the polynomial $f(x)/p^k$ contained in the interval $(0, \gamma)$ when x takes all integral values from 1 to P. Theorem 1 (proved by the method of N. M. Korobov): If $S = \sum_{x=1}^{P} e^{2\pi i f(x)/p^k}$, where $P = [p^{\alpha k}] + 1$, $\sqrt{((\log n)/n)} \leqq \alpha \leqq 1$, then $|S| \leqq k(e^{c_1 n \log n})(P^{1-1/cn})$, where c and c_1 are constants. Theorem 2: The formula $N_p(\gamma) = \gamma P + O(P^{1-1/cn})$ is valid.

J. W. Andrushkiw (Newark, N.J.)

K10-27 **(26# 101)**

Kesten, H.

Uniform distribution mod 1. II.

Acta Arith. **7** (1961/62), 355–380.

Let $[a, b]$ be a proper subinterval of $[0, 1]$ and let $f(\xi) = 1$ or 0 according as $a \leqq \xi \leqq b \pmod 1$ or not. The author showed in Part I [Ann. of Math. (2) **71** (1960), 445–471; MR **22** #4695] that when $b-a$ is rational,

$$\lim_{N\to\infty} \operatorname{meas}\{x, y | (\log N)^{-1} \sum_{k=1}^{N} (f(y+kx) - (b-a)) < \alpha, \quad 0 \leqq x, y \leqq 1\} = \frac{1}{\pi}\int_{-\infty}^{\rho\alpha} \frac{dt}{1+t^2},$$

where ρ is an explicitly given constant depending on $b-a$. In the present Part II, the restriction that $b-a$ be rational is removed. For irrational $b-a$, the constant ρ is independent of $b-a$; it is again explicitly determined in terms of a certain triple Fourier series.

A list of errata to Part I is appended.

W. J. LeVeque (Ann Arbor, Mich.)

Citations: MR 22# 4695 = K10-22.

K10-28 **(26# 3693)**

Vinogradov, I. M.

On the distribution of systems of fractional parts of values of several polynomials. (Russian)

Izv. Akad. Nauk SSSR Ser. Mat. **26** (1962), 793–796.

Let the integer $n \geqq 4$ be fixed, let $\{x\}$ run through an increasing sequence of positive integers, let further p be a positive integer and let p_1 be the number of terms in $\{x\}$ not exceeding p. Denoting by $\Phi(p)$ the number of x-values up to p for which the fractional part of all $f_s(x) = \alpha_{n,s}x^n + \cdots + \alpha_{1,s}x$, $s = 1, 2, \cdots, l$, satisfy $A_s \leqq f_s(x) < B_s \pmod 1$, where $0 \leqq B_s - A_s \leqq 1$, the author proves, by using his method of trigonometric sums, that given $\delta \leqq \frac{1}{8}$ for points of the nl-dimensional cube $0 < \alpha_{\nu,\mu} \leqq 1$ ($\nu = 1, 2, \cdots, n$; $\mu = 1, 2, \cdots, l$), apart from a set with volume tending rapidly to zero as $p \to \infty$ and independent of the choice of numbers $A_1, B_1, \cdots, A_l, B_l$, one can state the inequality

$$|\Phi(p) - (B_1 - A_1)(B_2 - A_2) \cdots (B_l - A_l)p_1| < (3 + \tfrac{2}{3}\log p)^l p^{1-\delta}.$$

S. Knapowski (New Orleans, La.)

K10-29 **(29# 3436)**

Callahan, Francis P.

Density and uniform density.

Proc. Amer. Math. Soc. **15** (1964), 841–843.

An elementary proof is given that the multiples modulo 1 of any irrational number θ are uniformly dense in the unit interval. Whereas the standard elementary proof uses rational approximations of θ, the device here is to prove that any subinterval I of length $1/k$ gets a proportional share of the multiples of θ in the limiting sense, k being any positive integer. This is done by choosing an integer t so that $t\theta$ is close to $1/k \pmod 1$ and then observing that I and its translates $I + rt\theta$ taken modulo 1 with $1 \leqq r \leqq k-1$ approximately cover the unit interval. The proof generalizes to the n-dimensional case of uniform density, if density in the n-cube is assumed.

I. Niven (Eugene, Ore.)

K10-30 **(29# 4750)**

Davenport, H.; Erdős, P.

A theorem on uniform distribution. (Russian summary)

Magyar Tud. Akad. Mat. Kutató Int. Közl. **8** (1963), 3–11.

Let $z_1 < z_2 < \cdots$ be a sequence of positive real numbers such that $z_n \to \infty$ as $n \to \infty$, and let α be a positive irrational number. For given λ, $0 < \lambda < 1$, let $F(N)$ be the number of positive integers $k \leqq N$ for which $k\alpha$ falls in one of the intervals $(z_j, z_j + \lambda(z_{j+1} - z_j))$. If $F(N)/N \to \lambda$ as $N \to \infty$, for each λ, then the sequence $\{k\alpha\}$ is said to be uniformly distributed relative to the sequence $\{z_j\}$. We suppose that $z_{j+1} \sim z_j$ as $j \to \infty$, since this is easily seen to be necessary for uniform distribution. It follows from work of the reviewer [Pacific J. Math. **3** (1953), 757–771; MR **15**, 511] and Davenport and the reviewer [Michigan Math. J. **10** (1963), 315–319; MR **27** #3619] that $\{k\alpha\}$ is uniformly distributed relative to $\{z_j\}$ for almost all $\alpha > 0$ if $z_{j+1} - z_j$ is monotonic. In the present paper this monotonicity condition is considerably weakened. Namely, it is a consequence of the slightly more general theorem actually proved, that if the number of z_j not exceeding N is $O(N^{2-\delta})$, for some fixed $\delta > 0$, then $\{k\alpha\}$ is uniformly distributed relative to $\{z_j\}$ for almost all $\alpha > 0$.

W. J. LeVeque (Ann Arbor, Mich.)

Citations: MR 15, 511d = K10-14; MR 27# 3619 = K20-9.

Referred to in K10-48.

K10-31 (29# 5807)

Kesten, H.

The discrepancy of random sequences $\{kx\}$.

Acta Arith. **10** (1964/65), 183–213.

For $0 \leqq a \leqq b \leqq 1$, let

$$f(\xi; a, b) = 1 \quad \text{if } a \leqq \xi \leqq b,$$

$$= 0 \quad \text{if } 0 \leqq \xi < a \text{ or } b < \xi \leqq 1,$$

$$f(\xi + 1; a, b) = f(\xi; a, b),$$

and for $1 < b \leqq 1 + a$, put

$$f(\xi; a, b) = f(\xi; a, 1) + f(\xi; 0, b-1).$$

For each real x, let

$$D_N(x) = N^{-1} \sup_{0 \leqq a \leqq b \leqq 1+a} \left| \sum_{k=1}^{N} f(kx; a, b) - N(b-a) \right|.$$

The principal result of this deep paper is that for every fixed $\varepsilon > 0$,

$$\lim_{N \to \infty} \operatorname{mes}\left\{ x \in [0;1] : \left| \frac{N D_N(x)}{\log N \log\log N} - \frac{2}{\pi^2} \right| > \varepsilon \right\} = 0.$$

The proof depends on a fairly explicit, but rather complicated, connection between $D_N(x)$ and the continued fraction expansion of x.

W. J. LeVeque (Ann Arbor, Mich.)

K10-32 (30# 3873)

Hudaĭ-Verenov, M. G.

On an everywhere dense set. (Russian)

Izv. Akad. Nauk Turkmen. SSR Ser. Fiz.-Tehn. Him. Geol. Nauk **1962**, *no.* 3, 3–11.

It is proved that the points $(\sum_{s=1}^{n+1} k_s\alpha_{1s}, \cdots, \sum_{s=1}^{n+1} k_s\alpha_{ns})$, where the k_s are arbitrary integers, form a dense set in n-space if and only if $\sum_{i=0}^{n} m_i\Delta_i \neq 0$ for all integers m_i, where the Δ_i are determinants formed from the coefficients α_{is}. An application is made to the system of differential equations $y_i' = \sum_{k=1}^{m} \lambda_{ik}/(x - a_k)$ $(i = 1, \cdots, n)$.

W. A. Coppel (Madison, Wis.)

K10-33 (33# 101)

Gillet, André

Sur la répartition modulo un des multiples d'un nombre réel.

C. R. Acad. Sci. Paris **261** (1965), 4946–4947.

Let G_s ($s \in [a, b]$) be a family of real, positive, increasing, concave functions such that $G_s \in O(G_t)$ for $s < t$. For every real unbounded F, $g(F) = \inf\{s : F \in O(G_s)\}$ is defined. If a is an irrational number, M an interval in $[0, 1)$ of length m, and $M(a, N)$ the number of $na - [na]$ $(0 \leqq n < N)$ which belong to M, then $S(a, N) = \sup_M |M(a, N) - mN|$ is such a function F. The author states some theorems on $g(S(a, N))$ in terms of the continued fraction expansion of a.

J. E. Cigler (Groningen)

K10-34 (34# 2528)

Halton, John H.

The distribution of the sequence $\{n\xi\}$ $(n = 0, 1, 2, \cdots)$.

Proc. Cambridge Philos. Soc. **61** (1965), 665–670.

The author considers the distribution of the points $\{n\xi\}$ $(n = 1, 2, \cdots, N)$ in the interval $(0, 1)$, where $\{z\}$ denotes the fractional part of z. The structure of the system of intervals is described, which consists of 0, 1 and the points $\{n\xi\}$. As a corollary the author obtains a proof of the fact conjectured by Steinhaus and proved first by S. Świerczkowski [Fund. Math. **46** (1959), 187–189; MR **21** #3404], that the distance of two "neighbors" $\{k\xi\}$, $\{l\xi\}$ (this means that the interval $(\{k\xi\}, \{l\xi\})$ does not contain any point of the form $\{m\xi\}$) can take at most three different values.

P. Szüsz (Stony Brook, N.Y.)

Citations: MR 21# 3404 = K10-20.

Referred to in K10-42.

K10-35 (34# 7457)

Kruse, A. H.

Estimates of $\sum_{k=1}^{N} k^{-s}\langle kx\rangle^{-t}$.

Acta Arith. **12** (1966/67), 229–261.

Write $\langle x\rangle = \min\{|x - n| : n \text{ integer}\}$. The paper is concerned with estimating sums of the form $\sum_{k=1}^{n} k^{-s}\langle kx\rangle^{-t}$, where s and t are non-negative real numbers. The author improves on some of the previously obtained results [G. H. Hardy and J. E. Littlewood, Abh. Math. Sem. Univ. Hamburg **1** (1922), 212–249; S. Chowla, Math. Z. **33** (1931), 544–563; A. Z. Val′fiš, ibid. **33** (1931), 564–601] and supplies several new ones. The author's main tool is the theory of continued fractions.

S. Knapowski (Coral Gables, Fla.)

K10-36 (35# 155)

Kesten, Harry

On a conjecture of Erdős and Szüsz related to uniform distribution mod 1.

Acta Arith. **12** (1966/67), 193–212.

Let ξ be an irrational number. It is well known that the numbers $\{k\xi\}$ $(k = 1, 2, \cdots)$ are uniformly distributed in $(0, 1)$; here $\{z\}$ denotes the fractional part of z. That is, if $R(M, \xi, a, b) = \sum_{v \leqq M} 1 - (b-a)M$, $a \leqq \{v\xi\} \leqq b$, then (*) $R(M, \xi, a, b) = o(M)$. According to a result due to E. Hecke [Abh. Math. Sem. Univ. Hamburg **1** (1921), 54–76], for some intervals, instead of (*) the stronger relation (**) $R(M, \xi, a, b) = O(1)$ holds. Namely, if $(b-a) = \{j\xi\}$ for some j, then one has (**).

The author proves that the converse holds: (**) holds if and only if $(b-a)$ has the form $b - a = \{j\xi\}$. This confirms a conjecture of P. Erdős and the reviewer [see P. Erdős, Compositio Math. **16**, 52–65 (1964); MR **31** #3382].

P. Szüsz (Stony Brook, N.Y.)

Citations: MR 31# 3382 = J24-34.

K10-37 (35# 6629)

Gillet, André

Sur la répartition modulo 1 des multiples d'un nombre réel.

C. R. Acad. Sci. Paris Sér. A-B **264** (1967), A223–A225.

The author states without proof the following theorem. Let $S_a(N)$ denote $\sup|M(a, N)-mN|$, where the supremum is taken over all arcs M of the one-dimensional torus T of measure m, a is an irrational number and $M(a, N)$ denotes the number of integers n, $0 \leqq n < N$, satisfying $an \in M$. Then $\limsup S_a(N)/\log N \geqq 0.06$.

J. E. Cigler (Groningen)

K10-38 (36# 114)

Slater, Noel B.

Gaps and steps for the sequence $n\theta$ mod 1.

Proc. Cambridge Philos. Soc. **63** (1967), 1115–1123.

For a given θ $(0<\theta<1)$ and $r=0, 1, 2, 3, \cdots$ there are two related problems on the fractional parts $\{r\theta\}$, namely, (i) the gap problem: for any ϕ $(0<\phi<1)$ to determine the gaps between the successive r for which $\{r\theta\}<\phi$; and (ii) the step problem: for the set $\{1\theta\}, \{2\theta\}, \cdots, \{N\theta\}$ rearranged in ascending order, to determine the steps into which the interval $[0, 1]$ is thereby partitioned. For general ϕ (or N) the gaps (or steps) have three lengths, one being the sum of the other two (the three-step result has been called the "Steinhaus conjecture"). The author integrates results of K. Florek (1951), S. Świerczkowski (1958), J. Surányi (1958), V. T. Sós (1957, 1958), J. H. Halton (1965), with his own (1950, 1964) putting the problems together, choosing the simplest methods and adding some further results.

F. Supnick (New York)

K10-39 (36# 1394)

Auluck, F. C.; Ahluwalia, H. S.

On partitions into non-integral numbers and the nearest-neighbour spacing distribution function.

J. Math. Sci. **1** (1966), 9–11.

The authors are concerned with the nearest-neighbour spacings of numbers of the form $m+n\lambda \leqq u$, for u large, λ a given positive irrational and m, n non-negative integers. Distribution functions are reported for $\lambda=e-2$, $\sqrt{2}-1$; $u=800$, 1000, 1200.

H. Gupta (Allahabad)

K10-40 (36# 2565a; 36# 2565b)

Williams, K. S.

A sum of fractional parts.

Amer. Math. Monthly **74** (1967), 978–980.

Williams, Kenneth S.

A sum of fractional parts. II.

J. Natur. Sci. and Math. **6** (1966), 209–216.

For $f(\bar{x})$ an integral polynomial of degree d in n indeterminants which does not vanish identically modulo the odd prime p, the author considers the problem of estimating the sum (*) $\sum\{f(\bar{x})/p\}$, where the sum is over all integer points $\bar{x}=(x_1, x_2, \cdots, x_n)$ with $0 \leqq x_i < p$ and $\{a\}$ denotes the fractional part of a. The first paper shows that (*) equals $(p^n/2)+O(p^{n-1/2}\log p)$, the constant depending on n and d, and improves the error term for diagonal and quadratic polynomials to $O(p^{n-1})$. In the second paper, it is shown that for almost all homogeneous polynomials with $n \geqq 3$ (*) is $(p^n/2)+O(p^{n-1/2})$.

L. C. Eggan (Tacoma, Wash.)

Referred to in K10-45.

K10-41 (37# 163)

Verbickiĭ, I. L.

Estimate of the remainder term in the Kronecker-Weyl approximation theorem. (Russian)

Teor. Funkciĭ Funkcional. Anal. i Priložen. Vyp. 5 (1967), 84–98.

A well-known theorem of Kronecker-Weyl can be stated as follows: Given real numbers $\omega_1, \cdots, \omega_n$ (linearly independent over the rationals), $\eta_1, \cdots, \eta_n$ and $\varepsilon_1, \cdots, \varepsilon_n>0$, denote by $E_a(T)$ the set of solutions $t \in [a, a+T]$ of the inequalities $|t\omega_j-\eta_j|<\varepsilon_j$ (mod 1). Then $\operatorname{mes} E_a(T)=(2\varepsilon_1)\cdots(2\varepsilon_n)T+\varphi_a(T)$, with $\varphi_a(T)=o(T)$, $T\to\infty$. The author gives some estimations of the remainder term $\varphi_a(T)$ under restrictions on the $\omega_1, \cdots, \omega_n$. We state two typical theorems: If $|k_1\omega_1+\cdots+k_n\omega_n|>ct^{-m}$ for certain constants c, m and for all $|k_j|<t$, then $|\varphi_a(x)|<Ax^{1-(1/m)+\varepsilon}$. Conversely: Suppose $|k_1\omega_1+\cdots+k_n\omega_n|<ct^{-m}$, $t=\max_j(|k_j|)$, holds infinitely often; then for almost all $\varepsilon_1, \cdots, \varepsilon_n$ in the unit cube, the relation

$$\limsup[(\sup|\varphi_a(x)|)\cdot|x|^{-1+(n/m)+\varepsilon}] = \infty$$

is true for every $\varepsilon>0$. These results are connected with a series of papers by A. Ostrowski on diophantine approximations. The methods used are applications of two general theorems on almost periodic functions, of which one is proved in the paper, the other is a slight extension of a theorem of B. M. Levitan.

F. Schweiger (Vienna)

K10-42 (37# 4027)

Graham, R. L.; van Lint, J. H.

On the distribution of $n\theta$ modulo 1.

Canad. J. Math. **20** (1968), 1020–1024.

If θ is an irrational real number, let a_i $(i=0, 1, \cdots, n)$ be the fractional parts of $m\theta$ for $m=0, 1, \cdots, n$, arranged in ascending order, and $d_\theta(n)=\max(a_i-a_{i-1})$, $1 \leqq i \leqq n+1$, with $a_{n+1}=1$. The authors prove that

$$\sup_\theta \liminf_{n\to\infty} nd_\theta(n) = (1+\sqrt{2})/2$$

and $\inf_\theta \limsup_{n\to\infty} nd_\theta(n)=1+2(\sqrt{5})/5$, and that these limits are attained for $\theta=1+\sqrt{2}$ and $\theta=(1+\sqrt{5})/2$, respectively. They also prove that $\limsup_{n\to\infty} nd_\theta(n)$ is finite if and only if the partial quotients b_m of the simple continued fraction for θ are bounded.

These results are obtained as elementary deductions from the following theorem, probably first proved by S. Świerczkowski [Fund. Math. **46** (1959), 187–189; MR **21** #3404] in a different form: Let h_m/k_m be the mth convergent of the continued fraction for θ and $\eta_m=|k_m\theta-h_m|$. Then the numbers a_i-a_{i-1} $(i=0, \cdots, n)$ can have at most three different values, viz., η_{m+1}, $\eta_m-r\eta_{m+1}$ and $\eta_m-(r-1)\eta_{m+1}$, where the integers m, r (and k) are uniquely determined by $n=k_m+rk_{m+1}+k$, $1 \leqq r \leqq b_{m+2}$, $1 \leqq k \leqq k_{m+1}$. A proof is also given by J. H. Halton in Proc. Cambridge Philos. Soc. **61** (1965), 665–670, Theorem 2, Corollary 3 [MR **34** #2528], where further references are found.

I. Danicic (Aberystwyth)

Citations: MR 21# 3404 = K10-20; MR 34# 2528 = K10-34.

K10-43 (38# 123)

Veech, William A.

A Kronecker-Weyl theorem modulo 2.

Proc. Nat. Acad. Sci. U.S.A. **60** (1968), 1163–1164.

Let X, θ, I be, respectively, the reals mod 1, an irrational number in X, and a non-empty interval in X. For $n>0$, let $S_n=S_n(\theta, I)$ be the number of j, $1 \leqq j \leqq n$, such that $j\theta \in I$. Let $X_n=S_n$ (mod 2) and define $\mu_\theta(I)=\lim N^{-1}\sum_1^N X_k$ if the limit exists. Let $\{m_k/n_k\}$ be the sequence of principle convergents to θ, obtained from its continued fraction representation $[0; a_1, a_2, \cdots]$. Theorem 1: $\mu_\theta(I)$ exists for arbitrary I if and only if $\sup_j a_j<\infty$.

Let

$$f^{(n)}(x) = f(x)f(x+\theta)\cdots f(x+(n-1)\theta)$$

and $\alpha_n=\int_X f^{(n)}(x)\,d\nu(x)$, where $\nu=$ Lebesgue measure. Assume that (1) $\lim_{n\theta\to 0} \alpha_n^2 = 1$. Let $I=[t_1, t_2]$ and let $t=t_2-t_1$. The set t for which (1) holds is a subgroup, say $K(\theta)\subseteq X$. Let $b_1\cdots b_n$. be a sequence of integers such that $|b_j|\leq a_j+1$. Then (2) $m\theta+\sum_{j=1}^{\infty} b_j n_j\theta$ is in X. Let $K_0(\theta)$ be the set of t in (2) such that (a) b_j is eventually even and (b) $\lim b_j n_j \|n_j\theta\|=0$, where $\|\ \|$ is a group invariant metric, and let $K_1(\theta)$ be the subset of $K_0(\theta)$ whose elements also satisfy $\sum |b_j|n_j\|n_j\theta\|<\infty$. K_0, K_1 are subgroups of measure 0. Theorem 2: $K_1(\theta)\subseteq K(\theta)\subseteq K_0(\theta)$.

Consider the equations (3) $g(x)g(x+\theta)=f(x)$, (4) $g(x)g(x+\theta)=-f(x)$, where f is measurable and takes ± 1 as its only values. Theorem 3: If $t\in K_1(2\theta)$, then one of (3) and (4) has a measurable solution. Theorem 4: (1) For every I, at least one of $\mu_\theta(I)$ and $\mu_\theta(X-I)$ exists and has the value $\frac{1}{2}$. (2) If $t\notin K(\theta)$, then $\mu_\theta(I)=\frac{1}{2}=\mu_\theta(X-I)$. (3) If $t\neq m\theta$, and if the equation (3) has a solution such that $\int g(x)\,d\nu\neq 0$, then $\mu_\theta(I-x)$ does not exist for uncountably many x.

No proofs are given. *J. H. Neuwirth* (Storrs, Conn.)

Referred to in K10-44.

K10-44 (39# 1410)

Veech, William A.

Strict ergodicity in zero dimensional dynamical systems and the Kronecker-Weyl theorem mod 2.

Trans. Amer. Math. Soc. **140** (1969), 1–33.

This paper contains the proofs of theorems stated in Proc. Nat. Acad. Sci. U.S.A. **60** (1968), 1163–1164 [MR **38** #123]. The technical aspects of this paper preclude a review that does it justice.

Let $X=\{x|0\leq x<1\}$ be the reals mod 1 and let $\theta\in X$ be an irrational number. For each interval $I\subseteq X$ and for each $n>0$, let $S_n=S_n(\theta, I)$ be the number of j, $1\leq j\leq n$, such that $j\theta\in I$. The well-known Kronecker-Weyl theorem states that $\lim S_n/n=\nu(I)=$ the Lebesgue measure of I.

In this paper the author proves the following elegant result. Theorem: Let $x_n=0$ or 1, depending on whether S_n is even or odd. Let $\mu_\theta(I)=\lim N^{-1}\sum_1^N x_k$, if it exists. Then a necessary and sufficient condition that $\mu_\theta(I)$ exist is that θ have bounded partial quotients in its continued fraction expansion. His proof depends on an idea of H. Furstenberg [Amer. J. Math. **88** (1961), 573–610; MR **24** #A3263].

Let f be -1 on I and 1 on the complement, and consider the equation (a) $g(x+\theta)=f(x)g(x)$. Let $f^{(n)}(x)=f(x)f(x+\theta)\cdots f(x+(n-1)\theta)$, $n>0$, and $\alpha_n=\int f^{(n)}(x)\,d\nu(x)$. The author discusses the connection between the existence of $\mu_\theta(I)$ and the existence of solutions to (a). Now consider $\lim \alpha_n=1$ and $\lim \alpha_n^2=1$. He gives an abstract discussion of the relation between the existence of solutions of (a) and the existence of these limits. This is done within the framework of a certain class of dynamical systems and the problem of the strict ergodicity of these systems. *J. H. Neuwirth* (Storrs, Conn.)

Citations: MR 38# 123 = K10-43.

K10-45 (38# 2081)

Tietäväinen, Aimo

On the distribution of the residues of a polynomial.

Ann. Univ. Turku. Ser. A I No. 120 (1968), 4 *pp.*

From the author's introduction: "Let L denote the set of points $x=(x_1,\cdots,x_n)$ with integral coordinates in Euclidean n-space. For any odd prime p, let $C=C(p)$ be the set of points of L in the cube $0\leq x_i<p$ $(i=1,2,\cdots,n)$. Suppose that $f(x)$ is any polynomial of degree d in $x_1,\cdots,x_n$ with integral coefficients which does not vanish identically (mod p). For any real number a, let $\{a\}$ be the fractional part of a and $\|a\|$ the distance from a to the nearest integer. K. S. Williams [Amer. Math. Monthly **74** (1967), 978–980; MR **36** #2565a] showed that $\sum_{x\in C}\{f(x)/p\}=\frac{1}{2}p^n+O(p^{n-1/2}\log p)$, as $p\to\infty$, where the constant implied in the O-symbol depends only upon n and d. The purpose of this note is to prove the theorem: $\sum_{x\in C}\|f(x)/p\|=\frac{1}{4}p^n+O(p^{n-1/2})$, as $p\to\infty$, where the implied constant depends only upon n and d."

Citations: MR 36# 2565a = K10-40.

K10-46 (38# 3235)

Lesca, Jacques

Sur la répartition modulo 1 des suites $(n\alpha)$.

Séminaire Delange-Pisot-Poitou: 1966/67, Théorie des Nombres, Fasc. 1, *Exp.* 2, 9 *pp. Secrétariat mathématique, Paris,* 1968.

Sei T ein orientierter Kreis der Länge 1 mit einem Nullpunkt. Ein zusammenhängender Teil von T heiße Intervall I. Sei $(x_m)_{m\in N}$ eine Punktfolge aus T, I ein Intervall aus T, $n\in N$, $\Pi(I,n)$ die Zahl der Punkte $x_m\in I$ mit $m\leq n$, $|I|$ die Länge von I und $E(I,n)=\Pi(I,n)-n|I|$. Ist für eine Folge (x_n) und für jedes $I\subset T$, $E(I,n)=o(n)$, dann heißt die Folge gleichverteilt. Verfasser untersucht Folgen $(n\alpha)$ mit irrationalen α und zeigt den folgenden Satz: Sei α irrational, I ein Intervall mit den Enden $\beta<\gamma$ und sei $\gamma-\beta\in Z\alpha$, dann ist $E(I,n)$ für die Folge $(n\alpha)$ beschränkt. Der Satz ist die Verallgemeinerung eines Satzes von Hecke (1922). Verfasser vermutet, daß auch die Umkehrung des Satzes gilt, und gibt dazu ein Teilergebnis an. *J. M. Wills* (Berlin)

K10-47 (39# 6836)

Cigler, J.

On a theorem of H. Weyl.

Compositio Math. **21** (1969), 151–154.

Let $p(t)=a_0t^k+\cdots+a_k$ be a polynomial with at least one irrational a_i, $i<k$. It is well-known that the sequence $\{p(n)\}$ is uniformly distributed mod 1. The author gives a new proof of this fact which is purely functional analytic and depends on the mean ergodic theorem only, by proving a more general k-dimensional theorem. The proof also shows that $\{p(n)\}$ is well distributed mod 1.

O. P. Stackelberg (Durham, N.C.)

K10-48 (40# 92)

Schmidt, W. M.

Disproof of some conjectures on Diophantine approximations.

Studia Sci. Math. Hungar. **4** (1969), 137–144.

Let $x_0=0$, x_1, x_2, $\cdots$ be a strictly increasing sequence of real numbers with $x_n\to\infty$ and $x_{n+1}/x_n\to 1$ $(n\to\infty)$. For $0<\lambda\leq 1$, let $M(\lambda)$ be the union of all the intervals $[x_n, x_n+\lambda(x_{n+1}-x_n)]$, $(n=0,1,2,\cdots)$. Let $\alpha>0$ and set $F_\alpha(N,\lambda)$ equal to the number of positive integers $k\leq N$ such that $k\alpha\in M(\lambda)$. We say the sequence $k\alpha$ is uniformly distributed relative to the sequence x_n provided $F_\alpha(N,\lambda)/N\to\lambda$ $(N\to\infty)$ for all λ (if $x_n=n$ this is the usual concept of uniform distribution). This concept is due to W. J. LeVeque [Pacific J. Math. **3** (1953), 757–771; MR **15**, 511]. H. Davenport and P. Erdős [Magyar Tud. Akad. Mat. Kutató Int. Közl. **8** (1963), 3–11; MR **29** #4750] conjectured that the sequence $k\alpha$ is uniformly distributed relative to x_n for almost all $\alpha>0$. This was shown to be true by LeVeque [loc. cit.] and H. Davenport and W. J. LeVeque [Michigan Math. J. **10** (1963), 315–319; MR **27** #3619] if $x_{n+1}-x_n$ is monotonic and by Davenport and Erdős [loc. cit.] if $x_n\gg n^{(1/2)+\delta}$ for some $\delta>0$. The present author shows the general conjecture to be false, by constructing a sequence such that if $f(x)$ is defined by $f(x)=1$ if $x\in M(\frac{1}{2})$ and $f(x)=-1$ otherwise, then for almost $\alpha>0$, $\lim\sup|N^{-1}\sum_{n=1}^N f(n\alpha)|=1$ $(N\to\infty)$.

The author also considers the conjecture of H. T. Croft

that if S is any set of positive reals of infinite Lebesque measure, then for almost all $\alpha>0$, infinitely many of the numbers α, 2α, 3α, $\cdots$ lie in S. He constructs an open set S of infinite measure such that for almost all $\alpha>0$ only finitely many $k\alpha$ lie in S, and thus a measurable set S^* of infinite measure such that for all $\alpha>0$ only finitely many $k\alpha$ lie in S. But if $\mu_s(\rho)$ equals the measure of $S\cap(0,\rho)$ and $\limsup \mu_s(\rho)/\rho=C>0$ $(\rho\to\infty)$, then for almost all $\alpha>0$ infinitely many $k\alpha$ lie in S.

W. W. Adams (College Park, Md.)

Citations: MR 15, 511d = K10-14; MR 27# 3619 = K20-9; MR 29# 4750 = K10-30.

Referred to in K10-49.

K10-49 **(44# 2710)**

Schmidt, W. M.

Remark on my paper: "Disproof of some conjectures on diophantine approximations".

Studia Sci. Math. Hungar. **5** (1970), 479.

The author notes that the conjecture of Croft which he disproved [same Studia **4** (1969), 137–144; MR **40** #92] had already been disproved by two other authors.

Citations: MR 40# 92 = K10-48.

K10-50 **(41# 1652)**

Muromskiĭ, A. A.

Series of cosecants with monotone coefficients. (Russian. English summary)

Vestnik Moskov. Univ. Ser. I Mat. Meh. **24** (1969), *no.* 4, 52–60.

The author considers series of the type (*) $\sum_{k=1}^{\infty} c_k|\sin k\pi x|^{-\alpha}$ with $\alpha>0$ and $c_k\downarrow 0$. In particular, he looks for sufficient conditions that the series converge on the set M of numbers having continued fractions with bounded elements. Using some inequalities, he generalizes the estimates given by G. H. Hardy and J. E. Littlewood [Bull. Calcutta Math. Soc. **20** (1930), 251–266] for sums of the form $\sum_{k=1}^{N}|\sin k\pi x|^{-\alpha}$, $\alpha\geqq 1$. His main results follow then by application of Abel's summation formula. Using results of M. Kac and R. Salem [Nederl. Akad. Wetensch. Proc. Ser. A **60** (1957), 265–267; MR **19**, 646], diophantine approximation and techniques developed in an earlier paper [the author, Izv. Akad. Nauk SSSR Ser. Mat. **28** (1964), 53–62; MR **28** #4290], related results are derived. Let us mention two typical theorems: If the series $\sum_{k=1}^{\infty} k^{\alpha-1}c_k$, $\alpha\geqq 2$, is convergent, then (*) converges on the set M. If the series $\sum_{k=1}^{\infty} k^{\alpha-1}(\log k)^{2-\alpha}c_k$, $1\leqq\alpha<\frac{3}{2}$, converges, then (*) converges almost everywhere.

F. Schweiger (Salzburg)

Citations: MR 19, 646b = K10-18.

K15 DISTRIBUTION (mod 1): $\{g^k\alpha\}$, NORMAL NUMBERS, METRIC PROPERTIES OF RADIX EXPANSIONS

Metric properties of various generalizations of the radix expansion are covered in **K55**.

See also reviews A62-61, B20-7, K02-2, K20-5, K20-7, K20-10, K20-30, K20-35, K30-22, K35-4, K35-5, K35-8, K35-17, K35-18, K35-22, K40-33, K45-24, K50-33, K55-3, K55-36, L99-7, Q15-20, Z01-37, Z10-31, Z10-39.

K15-1 **(1, 4b)**

Pillai, S. S. On normal numbers. Proc. Indian Acad. Sci., Sect. A. **10**, 13–15 (1939).

The author considers, among others, the number $.123\cdots$ (in the scale r) formed by writing the positive integers (in the scale r) in succession. He gives a proof that these numbers are simply normal, that is, each digit from 0 to $r-1$ appears with the asymptotic frequency $1/r$. The proof of the stronger statement that these numbers are normal is inadequate. *H. S. Zuckerman* (Seattle, Wash.).

K15-2 **(2, 33c)**

Pillai, S. S. On normal numbers. Proc. Indian Acad. Sci., Sect. A. **12**, 179–184 (1940).

This paper is an addition to an earlier paper [Proc. Indian Acad. Sci., Sect. A. **10**, 13–15 (1939); these Rev. **1**, 4] in which some of the proofs were inadequate. The proof of Champernowne's theorem, that $.123456789101112\cdots$ is normal, is corrected and rewritten and it is shown that the usual definition of a number being normal can be replaced by the simpler definition used in the first paper. The author states that the argument can be elaborated to complete the proofs of the other theorems of his earlier paper.

H. S. Zuckerman (Seattle, Wash.).

K15-3 **(2, 33d)**

Vijayaraghavan, T. On decimals of irrational numbers. Proc. Indian Acad. Sci., Sect. A. **12**, 20 (1940).

The paper contains a very short and simple proof of the theorem: Let $a_0.a_1a_2\cdots$ represent a number θ in the scale of m, $\theta_n=.a_na_{n+1}\cdots$ $(n=1, 2, 3, \cdots)$, and L be the set of limit points of $\theta_1, \theta_2, \cdots$. If θ is irrational, then L contains an infinity of points. A corollary is given: Let $\theta=\sqrt[\alpha]{m}$ be irrational. Then the set of limit points of the fractional parts of the powers of θ contains an infinity of points.

The author does not point out the place of his problem within the framework of diophantine approximations. Hardy and Littlewood [Acta Math. **37**, 183 (1914)] say: "Perhaps the most interesting special sequence falling under the general type $(f(n)\theta)$ $[a=[a]+(a)]$ is that in which $f(n)=a^n$, where a is a positive integer. When θ is expressed as a decimal in the scale of a, the effect of multiplication by a is merely to displace the digits." Compare also Weyl [Math. Ann. **77**, 349 (1916)] and Koksma [Diophantische Approximationen, Ergebnisse der Math., vol. 4, 1935, p. 116]. The articles referred to give the theorem of the paper, even in much stronger form, for all irrational θ except an unspecified set of θ's of measure zero ("almost all" type of theorem). Vijayaraghavan's paper gives a much weaker theorem, but which holds for all irrational θ. *A. J. Kempner.*

Referred to in K40-46.

K15-4 **(3, 169e)**

Fortet, R. Sur une suite également répartie. Studia Math. **9**, 54–70 (1940). (French. Ukrainian summary)

This paper is concerned with the study of uniformly distributed sequences of the form

(S) $$x_n\equiv a^n x \pmod 1,$$

where $0\leqq x_n<1$; $n=0, 1, 2, \cdots$; $0\leqq x\leqq 1$, and $a>1$ is a fixed positive integer. Let $f(x)\varepsilon L$ $(0\leqq x\leqq 1)$ and put $\varphi_n(x)=(n+1)^{-1}\sum_{k=0}^{n} f(x_k)$. The author interprets Raikov's result that $\lim \varphi_n(x)$ exists almost everywhere (as $n\to\infty$) as a particular case of the strong law of large numbers for certain stationary chains [see A. Kolmogoroff, Rec. Math. [Mat. Sbornik] N.S. **2** (**44**), 367–368 (1937)]. Assuming that $f(x)$ satisfies Hölder's condition and employing the methods of Fréchet [Recherches théoriques modernes sur le calcul des probabilités, livre 2, p. 146. This is tome 1, fasc. 3 of Borel's Traité du calcul des probabilités, Paris, 1938] and Doeblin [Thèse, Univ. Paris, 1938 or Bull. Math. Soc. Roum. Sci. **39**, no. 1, 57–115; no. 2, 3–61 (1937); see, in particular, p. 98] the author proves the following two theorems. (I) The distribution function of

$$\left\{(n+1)\varphi_n(x)-(n+1)\int_0^1 f(x)dx\right\}(n+1)^{-\frac{1}{2}}$$

tends to a Gaussian distribution provided that the limit of

$$(n+1)^{-1}\int_0^1\left[\sum_{k=0}^n f(x_k)-(n+1)\int_0^1 f(x)dx\right]^2 dx$$

as $n\to\infty$ is different from 0. (II) Under the above condition the law of the iterated logarithm holds for the sum $\sum_{k=0}^n f(x_k)$. (I) is a generalization of an earlier result of the reviewer [Studia Math. **7**, 96–100 (1938); see also J. London Math. Soc. **13**, 131–134 (1938)]. The last part of the paper considers analogous questions for subsequences of (S).
M. Kac (Ithaca, N. Y.).

Referred to in K15-7, K15-13, Q15-20.

K15-5 (4, 16j)

Erdös, Paul. On the law of the iterated logarithm. Ann. of Math. (2) **43**, 419–436 (1942).

Let

$$\epsilon_1(t)/2+\epsilon_2(t)/2^2+\cdots$$

be the dyadic expansion of the real number t ($0\leqq t\leqq 1$) and put

$$f_n(t)=\sum_{k=1}^n \epsilon_k(t)-n/2.$$

The author refines the well-known law of the "iterated logarithm" as follows: If

$$\varphi(n)=\left(\frac{n}{2\ \mathrm{loglog}\ n}\right)^{\frac{1}{2}}(\mathrm{loglog}\ n+\tfrac{3}{4}\log_3 n+\tfrac{1}{2}\log_4 n$$
$$+\cdots+\tfrac{1}{2}\log_{k-1} n+(\tfrac{1}{2}+\delta)\log_k n),\quad k>3,$$

and $\delta>0$, then $f_n(t)>\varphi(n)$ almost everywhere can hold for a finite number of n's only. On the other hand if $\delta<0$ there is an infinite sequence of n's for which $f_n(t)>\varphi(n)$ almost everywhere. The methods employed are elementary, although by no means simple. *M. Kac* (Ithaca, N. Y.).

K15-6 (7, 369f)

Koksma, J. F. On decimals. Nieuw Arch. Wiskunde (2) **21**, 242–267 (1943).

Let θ be a real number, a an integer not less than 2 and let, as usual, $(\theta a^n)=\theta a^n-[\theta a^n]$ denote the fractional part of θa^n. Let $\nu_{\alpha,\beta}(N)$ be the number of numbers (θa^n), $1\leqq n\leqq N$, which belong to the interval (α,β) interior to the interval (0, 1). Let $\nu_{\alpha,\beta}(N)=(\beta-\alpha)N+R_{\alpha,\beta}(N)$. By a classical theorem of Weyl, $R_{\alpha,\beta}(N)=o(N)$ for almost all θ. The present paper gives a number of results on the order of magnitude of $R_{\alpha,\beta}(N)$ as well as on the order of magnitude of $R^*(N)=\sum_{n=1}^N((\theta a^n)-\frac{1}{2})$ and $Q(N)=\sum_{n=1}^N(\rho_n-n/N)^2$, where $\rho_1,\cdots,\rho_N$ denote the first N numbers (θa^n) arranged in nondecreasing order of magnitude. We quote the following results; there are many more general theorems which there is not space to state here. For almost all θ,

$$R^*(N)=o(N^{\frac{1}{2}}(\varphi(N))^{\frac{1}{2}}),$$
$$Q(N)=o(N^{\frac{1}{2}}(\varphi(N))^{\frac{1}{2}}),$$
$$R_{\alpha,\beta}(N)=o(N^{\frac{1}{2}}(\varphi(N))^{\frac{1}{2}}),$$

where $\varphi(N)$ is any positive, nondecreasing function of n such that $\sum 1/(n\varphi(n))<\infty$ and $\varphi(n+1)\leqq(1+K/n)\varphi(n)$, K being independent of N. *R. Salem.*

Referred to in K15-18.

K15-7 (7, 436f)

Kac, M. On the distribution of values of sums of the type $\sum f(2^k t)$. Ann. of Math. (2) **47**, 33–49 (1946).

(i) Let $f(t)$ be a function of period 1, satisfying Hölder's condition of order greater than $\frac{1}{2}$. Suppose, moreover, that

$$n^{-1}\int_0^n\left\{\sum_0^n f(2^k t)\right\}^2 dt\to\sigma^2\neq 0.$$

Then the measure of the set of t's for which

$$a<n^{-\frac{1}{2}}\sum_0^n f(2^k t)<b$$

approaches, as $n\to\infty$, the integral

$$(*)\qquad (2\pi)^{-\frac{1}{2}}\sigma^{-1}\int_a^b \exp(-u^2/2\sigma^2)du.$$

(ii) If $f(t)$ satisfies the same conditions, and if $l_{k+1}-l_k$ is increasing, the measure of the set of points where

$$a<n^{-\frac{1}{2}}\sum_1^n f(2^{l_k}t)<b$$

approaches, as $n\to\infty$, the integral (*), where $\sigma^2=\int_0^1 f^2 dt$. (iii) If $l_{n+1}-l_n$ is increasing and $f(t)$ satisfies Hölder's condition of order greater than $\frac{1}{2}$, the divergence of the series $\sum c_n^2$ implies the divergence almost everywhere of $\sum c_n f(2^{l_n}t)$. [For (i) and (ii) see also Kac, Studia Math. **7**, 96–100 (1938); Fortet, Studia Math. **9**, 54–70 (1940); these Rev. **3**, 169.] *A. Zygmund* (Philadelphia, Pa.).

Citations: MR 3, 169e = K15-4.
Referred to in K15-13, K15-62, K15-66, Q15-20.

K15-8 (8, 194b)

Copeland, Arthur H., and Erdös, Paul. Note on normal numbers. Bull. Amer. Math. Soc. **52**, 857–860 (1946).

D. G. Champernowne proved [J. London Math. Soc. **8**, 254–260 (1933)] that the infinite decimal 0.123456789101112 ··· is normal in the scale of ten. He conjectured that the infinite decimal 0.12357111317 ···, formed by the sequence of primes, has the same property. This note contains a proof of the following more general result, which includes the conjecture of Champernowne as a special case. If $a_1, a_2, \cdots$ is an increasing sequence of integers such that, for every $\theta<1$, the number of a's up to N exceeds N^θ provided N is sufficiently large, then the infinite decimal $0.a_1a_2a_3\cdots$ is normal in the scale in which these integers are expressed. The proof of this result is based on the concept of (ϵ, k) normality due to Besicovitch [Math. Z. **39**, 146–156 (1934)].
R. D. James (Vancouver, B. C.).

Referred to in K15-19.

K15-9 (11, 83a)

Korobov, N. M. On sums of fractional parts. Doklady Akad. Nauk SSSR (N.S.) **67**, 781–782 (1949). (Russian)

Let q be an integer greater than unity and let $\{\beta\}$ denote the fractional part of β. Further, let A be the set of numbers $\alpha\varepsilon(0, 1)$ for which the function $\{\alpha q^x\}$ is equally distributed. Then the following results are stated. (I) If $\epsilon(P)$ is any positive function of P which tends to zero as $P\to\infty$, then there exists at least one $\alpha\varepsilon A$ such that $\sum_{x=1}^P\{\alpha q^x\}-\frac{1}{2}P=\Omega(P\epsilon(P))$. (II) If $\varphi(P)$ is any increasing function of P which tends to infinity with P as slowly as we please, then there exists at least one $\alpha\varepsilon A$ such that $\sum_{x=1}^P\{\alpha q^x\}-\frac{1}{2}P=o(\varphi(P))$. Proofs are not given. It is stated that (I) is obtained from Weyl's conditions for equal distribution [Math. Ann. **77**, 313–352

(1916)] and that (II) and a further result which is not precisely stated, are based on applications of ideas introduced in a paper of the author which appears to be unpublished. *R. A. Rankin* (Cambridge, England).

K15-10 (11, 88e)

Eggleston, H. G. The fractional dimension of a set defined by decimal properties. Quart. J. Math., Oxford Ser. **20**, 31–36 (1949).

Any real number x in (0, 1) being expressed in the scale of N, let $P(x, i, r)$ be the number of times that the digit r occurs in the first i digits of this expression. The paper proves a conjecture of I. J. Good that the set of x for which $P(x, i, r) \sim p_r i$ (as $i \to \infty$, for each r) has fractional dimension α given by $N^{-\alpha} = \prod p_r^{p_r}$. *H. D. Ursell* (Leeds).

Referred to in K15-58, K15-64, K15-93.

K15-11 (12, 321a)

Korobov, N. M. Concerning some questions of uniform distribution. Izvestiya Akad. Nauk SSSR. Ser. Mat. **14**, 215–238 (1950). (Russian)

If, for each choice of the positive integer s and of the integers $m_1, \cdots, m_s$, not all 0, the function

$$m_1 f(x+1) + \cdots + m_s f(x+s)$$

is uniformly distributed, then $f(x)$ is said to be completely uniformly distributed. Polynomials are not completely uniformly distributed although some of them are uniformly distributed. Using estimates of exponential sums due to Vinogradov and his followers, the author establishes the existence of a class of completely uniformly distributed functions $f(x)$ such that for suitable positive λ_1 and λ_2, $f(x) = o(x^{\lambda_1 \log\log x})$, $f(x) \neq o(x^{\lambda_2 \log\log x})$. Now suppose that $b_k = \pm 1$, $\lambda > 3$, $w(k) \geqq k^\lambda$, and $1 + 1/k \leqq w(k+1)/w(k) \leqq k$ for all sufficiently large k; using the previous result, the author shows that $f(x) = \sum_{k=0}^{\infty} b_k e^{-w(k)} x^k$ is uniformly distributed. By taking $b_k = 1$ and $w(k) = k^{1+2/(1-\epsilon)}$, the author shows that there is a function of this kind for which $f(x) > \exp\{(\log x)^{3/2-\epsilon}\}$.

The author next derives a necessary and sufficient condition on α for the function $\alpha q_1 \cdots q_x$ to be uniformly distributed under the assumption that the q_x's are integers greater than 1 and tend to infinity. He also obtains two sufficient conditions on α which insure the uniform distribution of αq^x if $q \geqq 2$. Taking $q_x = x+1$, it follows from the first of these results that $\alpha \cdot x!$ is uniformly distributed if $\alpha = \sum_{k=1}^{\infty} [k^{1+\lambda}]/k!$ and $0 < \lambda < 1$; other specializations of these results are also given. While the uniform distribution of these functions for almost all values of α (in the sense of measure 0) has been known since Weyl's work [Math. Ann. **77**, 313–352 (1916)], not a single value of α has been known for which these functions actually are uniformly distributed. *L. Schoenfeld* (Urbana, Ill.).

Referred to in K15-15, K15-31, K15-52, K20-19, K35-8, K35-13.

K15-12 (12, 321d)

Korobov, N. M. Normal periodic systems and a question on the sums of fractional parts. Uspehi Matem. Nauk (N.S.) **5**, no. 3(37), 135–136 (1950). (Russian)

The author gives a brief indication of the proofs of the following results, the second of which depends on a result proved in the paper reviewed above. If $\epsilon(p) \to 0^+$ as $p \to \infty$, then there exists an α for which αq^x is uniformly distributed and such that $S(p) \neq o(p\epsilon(p))$, where

$$S(p) = \sum_{x=1}^{p} (\alpha q^x - [\alpha q^x]) - \tfrac{1}{2}p;$$

thus, the error term $o(p)$, which is a consequence of the uniform distribution of αq^x, cannot be improved for all α. If $\phi(p) \to \infty$ as $p \to \infty$, then there exists an α for which αq^x is uniformly distributed and such that $S(p) = o(\phi(p))$. *L. Schoenfeld* (Urbana, Ill.).

Referred to in K15-87.

K15-13 (12, 406e)

Maruyama, Gisirô. On an asymptotic property of a gap sequence. Kōdai Math. Sem. Rep. **1950**, 31–32 (1950).

The author gives a simple proof of the following theorem: If $f(t+1) = f(t)$, $\int_0^1 f(t)\,dt = 0$, and $f(t)$ satisfies a Lipschitz condition of order α, then

$$\limsup_{n\to\infty} \sum_1^n f(2^k t)/(2n \log\log n)^{\frac{1}{2}} = \sigma,$$

where $\sigma^2 = \lim_{n\to\infty} n^{-1} \int_0^1 (\sum_1^n f(2^k t))^2\,dt$. This theorem was first stated by Fortet [Studia Math. **9**, 54–70 (1940); these Rev. **3**, 169] who sketched a proof based on Doeblin's generalization of the law of the iterated logarithm to Markoff chains. The author deduces the theorem from Kolmogoroff's law of the iterated logarithm by applying an approximation procedure introduced by the reviewer [Ann. of Math. (2) **47**, 33–49 (1946); these Rev. **7**, 436]. *M. Kac.*

Citations: MR 3, 169e = K15-4; MR 7, 436f = K15-7.

K15-14 (13, 213d)

Šapiro-Pyateckiĭ, I. I. On the laws of distribution of the fractional parts of an exponential function. Izvestiya Akad. Nauk SSSR. Ser. Mat. **15**, 47–52 (1951). (Russian)

Let q be an integer greater than unity and α a real number. The fractional parts $\{\alpha q^k\}$ for $k = 0, 1, 2, \cdots$ determine a distribution function $\sigma(x)$. Conversely, if a distribution function $\sigma(x)$, is given it may be possible to find an α such that $\sigma(x)$ is the distribution function for the sequence $\{\alpha q^k\}$. It is proved that a necessary and sufficient condition for a distribution function $\sigma(x)$ to be that of a sequence $\{\alpha q^k\}$ $(k = 0, 1, 2, \cdots)$ is that, for every continuous function $f(x)$ of period 1,

$$\int_0^1 f(x)\,d\sigma = \int_0^1 f(qx)\,d\sigma.$$

The necessity of the condition is easily shown. To prove it sufficient use is made of the theory of dynamical systems and work of Krylov and Bogolyubov. A criterion for uniform distribution is also given. *R. A. Rankin.*

Referred to in K02-2, K15-26, K15-34, K15-55, K15-78, Q15-20.

K15-15 (13, 213e)

Korobov, N. M. Normal periodic systems and their applications to the estimation of sums of fractional parts. Izvestiya Akad. Nauk SSSR. Ser. Mat. **15**, 17–46 (1951). (Russian)

Throughout q denotes an integer greater than unity. There are two parts to the paper. In part I the author defines a normal periodic system $\rho_n(q)$ to be a sequence of numbers $\delta_1\delta_2 \cdots \delta_t$, where $t = q^n + n - 1$, δ_i is an integer in the range $0 \leqq \delta_i \leqq q-1$ $(i = 1, 2, \cdots, t)$ and the sequence has the property that no set of n successive numbers occurs more than once. The number of sets of n successive numbers in the sequence is q^n and is the maximum number of different sets possible. The existence of such systems was proved by Good [J. London Math. Soc. **21**, 167–169 (1946); these Rev. **8**, 430], and the author develops two general methods for constructing them. His first method (method A_1) is simple, particularly for $q = 2$, but does not give all normal periodic systems. For in the case $q = 2$ method A_1 gives not more than 2^n systems, while de Bruijn has shown that the total number is 2^r, where $r=2^{n-1}-n$ [Nederl. Akad. Wetensch., Proc. **49**, 758–764 = Indagationes Math. **8**, 461–467 (1946); these Rev. **8**, 247]. By a very complicated analysis involving the consideration of certain special systems of digits associated with a given $\rho_n(q)$, the author derives a second method (method A_2), by means of which every normal periodic system can be obtained. Every system $\rho_n(q)$ has the property that its last $n-1$ digits are the same as the first $n-1$, and by omission of these final $n-1$ digits we obtain an associated system $\rho'_n(q)$.

In part II these systems are applied to the problem of the behaviour of the sum of the fractional parts of the function $\alpha q^x (x=0, 1, 2, \cdots)$, where α is a real number. If these fractional parts are uniformly distributed in the interval (0, 1), write $\alpha\varepsilon L(q)$. Also write $E(\mu, P)=\sum_{x=1}^{P}\{\alpha q^x\}-\frac{1}{2}P$, where $\mu=1, 2, 3, \cdots$, so that, if $\alpha\varepsilon L(q)$ then $E(1, P)=o(P)$.

The following theorems are proved: I. Let $\epsilon(P)$ be any positive function which tends to zero as $P\to\infty$; then there exists an $\alpha\varepsilon L(q)$ such that $E(1, P)=\Omega(P\epsilon(P))$. II. Let $\varphi(P)$ be any positive function which tends to infinity as $P\to\infty$; then there exists an $\alpha\varepsilon L(q)$ such that $E(1, P)=O(\varphi(P))$, and for every $\alpha\varepsilon L(q)$ it is not possible to replace this estimate by $O(1)$. The author has shown that $\alpha\varepsilon L(q)$ implies $\alpha\varepsilon L(q^\mu)$ for $\mu=1, 2, 3, \cdots$ [Izvestiya Akad. Nauk SSSR. Ser. Mat. **14**, 215–238 (1950); these Rev. **12**, 321]. That a similar result does not in general hold for $E(\mu, P)$ is shown by III. Let $\epsilon(P)$ and $\varphi(P)$ be as in I and II. Then there exists an $\alpha\varepsilon L(q)$ such that $E(1, P)=o(\varphi(P))$, $E(2, P)=\Omega(P\epsilon(P))$. IV. On the other hand, for any $\varphi(P)$ as in II, there exists an $\alpha\varepsilon L(q)$ such that $E(\mu, P)=o(P)$ for $\mu=1, 2, 3, \cdots$. V. Also, if $E(1, P)$ is sometimes large, $E(2, P)$ can be small for all P. Thus, for any $\epsilon(P)$ and $\varphi(P)$ as in I and II, we can find an $\alpha\varepsilon L(q)$ such that $E(1, P)=\Omega(P\epsilon(P))$, $E(2, P)=o(\varphi(P))$. VI. On the other hand there exist an $\alpha\varepsilon L(q)$ such that $E(\mu, P)=\Omega(P\epsilon(P))$ for each $\mu=1, 2, 3, \cdots$. These results are compared with the known results for the fractional parts of αx $(x=1, 2, 3, \cdots)$. Thus for αx I is true, but II is not true if $\varphi(P)$ increases slower than $\log P$. On the other hand $\sum_{x=1}^{P}\{\alpha x\}-\frac{1}{2}P=O(\log P)$ for almost all α while Khintchine has shown that $E(1, P)=O(\{P\log\log P\}^{\frac{1}{2}})$ for almost all α [Fund. Math. **6**, 9–20 (1924)].

In the proofs of these results α is expressed as a "decimal" to the base q in the form $0\cdot\delta_1\delta_2\delta_3\cdots$ and, where necessary, the sequence of δ_i is chosen from the digits of systems $\rho_n'(q)$, the rate at which n increases being determined by the given functions $\varphi(P)$ and $\epsilon(P)$. *R. A. Rankin*

Citations: MR 12, 321a = K15-11.
Referred to in K15-22, K15-25, K15-87.

K15-16 (13, 438a)

Niven, Ivan, and Zuckerman, H. S. On the definition of normal numbers. Pacific J. Math. **1**, 103–109 (1951).

Let R be a real number with fractional part $0.\,x_1, x_2, \cdots$ when written to the scale r. Let $N(b,n)$ denote the number of occurrences of the digit b in the first n places. R is said to be simply normal if for each b, $0\leqq b<r$,

$$\lim_{n\to\infty} N(b, n)/n=1/r.$$

Borel defined $\mathcal{R}$ to be normal in the scale r if all the numbers $\mathcal{R}, r\mathcal{R}, r^2\mathcal{R}, \cdots$ are simply normal to all the scales $r, r^2, r^3, \cdots$. Borel also stated that a characteristic property of a normal number is the following: Denote by B a sequence of v digits b_i, $0\leqq b_i<r$, $i=1, 2, \cdots, v$. Let $N(B, n)$ denote the number of occurrences of the sequence B in the first n decimal places. Then for any B

$$(1)\qquad \lim_{n\to\infty} N(B, n)/n=1/r^v.$$

Hardy and Wright [An introduction to the theory of numbers, Oxford, 1938] state without proof that (1) is equivalent to the definition of normality. It is indeed easy to see that a normal number has property (1), but the converse is by no means obvious. The authors prove that (1) is indeed equivalent to normality. *P. Erdös* (Aberdeen).

Referred to in K15-27.

K15-17 (13, 535a)

Maxfield, J. E. Sums and products of normal numbers. Amer. Math. Monthly **59**, 98 (1952).

K15-18 (13, 566c)

Tsuchikura, Tamotsu. On the function $t-[t]-\frac{1}{2}$. Tôhoku Math. J. (2) **3**, 208–211 (1951).

Let $f(t)=t-[t]-\frac{1}{2}$ and a be an integer $\geqq 2$. Let $\epsilon_k(t)$ be the kth digit in the a-dic expansion of t, $\delta_k(t)=\epsilon_k(t)-(a-1)/2$. Then a simple computation shows that

$$\sum_{i=0}^{N} f(a^i t)=(a-1)^{-1}\sum_{k=1}^{N}\delta_k(t)+O(1).$$

Applying the law of the iterated logarithm we obtain $\limsup_{N\to\infty}(\sum_{i=0}^{N} f(a^i t))/N(\log\log N)^{1/2}=[(a+1)/b(a-1)]^{1/2}$ for almost all t. Using a simple corollary of a result of the reviewer and Erdös [Ann. of Math. (2) **48**, 1003–1013 (1947); these Rev. **9**, 292] we obtain

$$\liminf_{N\to\infty}\left|\sum_{i=0}^{N} f(a^i t)\right|\leqq(a-1)^{-1}(|f(t)|+\tfrac{1}{2})$$

for almost all t. These results improve upon crude results of Koksma [Nieuw Arch. Wiskunde (2) **21**, 242–267 (1943); these Rev. **7**, 369] obtained by non-probabilistic methods. A category theorem is also proved which supplements the strong law of large numbers applied to the case on hand. *K. L. Chung* (Ithaca, N. Y.).

Citations: MR 7, 369f = K15-6.
Referred to in K15-89.

K15-19 (13, 825g)

Davenport, H., and Erdös, P. Note on normal decimals. Canadian J. Math. **4**, 58–63 (1952).

A number $0.a_1a_2\cdots$ expressed as a decimal is said to be normal (Borel), if every fixed combination among the 10^k possible combinations of k digits occurs with the frequency $1/10^k$. Let $N(t)$ denote the number of times the said combination occurs among the first t digits; then the condition is that $\lim_{t\to\infty} N(t)/t=1/10^k$. A positive integer $g=a_1a_2\cdots a_t$ expressed in the decimal scale is said to be (ϵ, k) normal [Besicovitch, Math. Z. **39**, 146–156 (1934)], if any combination of $h\leqq k$ digits occurs in the sequence $a_1\cdots a_t$ a number of times which lies between $(1-\epsilon)10^{-h}t$ and $(1+\epsilon)10^{-h}t$.

The authors prove the following theorems 1 and 2 which contain known results of Champernowne [J. London Math. Soc. **8**, 254–260 (1933)] and Besicovitch [loc. cit.] and the first of which was conjectured by Copeland and Erdös [Bull. Amer. Math. Soc. **52**, 857–860 (1946); these Rev. **8**, 194]. Theorem 1. Let $f(x)$ be any polynomial in x, all of whose values for $x=1, 2, \cdots$ are positive integers. Then the decimal $0.f(1)f(2)f(3)\cdots$, where $f(n)$ is written in the scale of 10 is normal. Theorem 2. For any $\epsilon>0$ and $k\geqq 1$ (k an integer) almost all the numbers $f(1), f(2), \cdots$ are (ϵ, k) normal, i.e. the number of indices $n\leqq x$ for which $f(n)$ is not (ϵ, k) normal is $o(x)$ as $x\to\infty$ for fixed ϵ and k. The proof follows the method with which Weyl's criterion for uniform distribution is proved [Math. Ann. **77**, 313–352 (1916)].

J. F. Koksma (Amsterdam).

Citations: MR 8, 194b = K15-8.

K15-20 (13, 826d)

Maxfield, John E. A short proof of Pillai's theorem on normal numbers. Pacific J. Math. **2**, 23–24 (1952).

The author gives a simple proof of the following theorem: The necessary and sufficient condition that a number α be normal to the base r is that it be simply normal to the bases $r, r^2, \cdots$. *P. Erdös* (Los Angeles, Calif.).

K15-21 (14, 23c)

Eggleston, H. G. Sets of fractional dimensions which occur in some problems of number theory. Proc. London Math. Soc. (2) **54**, 42–93 (1952).

The sets mentioned in the title occur in certain problems of diophantine approximation. If the set of real numbers θ for which a certain law of diophantine approximation holds, is a null-set, one may ask for its dimension in the sense of Hausdorff. In this domain Besicovitch, Jarník, Knichal [references in the reviewer's Diophantische Approximationen, Springer, Berlin, 1936] and Lock [Dissertation, 1947; these Rev. **9**, 79] proved several results concerning the decimal expansion of real numbers. In the first part of this paper the author deduces a number of general theorems on Hausdorff dimension, using mainly a method of Besicovitch. In the second part applications to the problems of rational approximation to irrational numbers are given, whereas part 3 contains questions concerning the distribution of digits in decimals. In part 4 among other results is proved: If a strictly increasing sequence $M_1<M_2<\cdots$ of integers is such that $M_{i+1}/M_i\to\infty$ as $i\to\infty$, then the set S_α of θ for which $M_i\theta\to\alpha \pmod 1$ $(0\leqq\theta\leqq1)$ as $i\to\infty$ has dimension 1 for any given α $(0\leqq\alpha\leqq1)$. This is in striking contrast to the case that M_{i+1}/M_i is bounded, as then S_α is at most enumerable for any α $(0\leqq\alpha\leqq1)$.

J. F. Koksma (Amsterdam).

Citations: MR 9, 79h = J64-8.

Referred to in K15-69, K20-7.

K15-22 (14, 143d)

Korobov, N. M. Fractional parts of exponential functions. Trudy Mat. Inst. Steklov., v. 38, pp. 87–96. Izdat. Akad. Nauk SSSR, Moscow, 1951. (Russian) 20 rubles.

Let $\varphi(x)$ be any completely uniformly distributed function, in the sense of the author's earlier paper [Uspehi Matem. Nauk **4**, no. 1 (29), 189–190 (1949); these Rev. **11**, 231], for which the system of functions $\varphi(x+1)$, $\varphi(x+2), \cdots, \varphi(x+s)$ retains the property of uniform distribution under any linear transformation $x=\lambda y$. Define α_ν $(\nu=1, 2, \cdots, n)$ by

$$\alpha_\nu=\sum_{k=1}^{\infty}[q_\nu\{\varphi(kn+\nu)\}]/q_\nu^k,$$

where the q_ν are any given integers greater than unity. Then the author proves that the function $F(x)=\alpha_1q_1^x+\cdots+\alpha_nq_n^x$ is uniformly distributed. It is also proved that the function $\alpha f(x)a^x$ is uniformly distributed when a is an integer greater than unity, $f(x)$ a polynomial of positive degree with integral coefficients and α is defined as a rapidly converging power series in $1/a$ with coefficients of a special form. Finally, the author considers the exponential sum $S=\sum_{x=1}^{P}e^{2\pi i\alpha q^x}$, where q is an integer greater than unity and $0<\alpha<1$. Since $\int_0^1|S|^2d\alpha=P$, it follows that, corresponding to every $\epsilon>0$, we can find a sufficiently large $C=C(\epsilon)$ such that $|S|<C\sqrt{P}$ for every α in (0, 1) except possibly for a set of measure less than ϵ. By means of his theory of normal periodic systems [Izvestiya Akad. Nauk SSSR. Ser. Mat. **15**, 17–46 (1951); these Rev. **13**, 213] the author constructs a wide class of numbers α for which the inequality $|S|<(4\pi+1)q\sqrt{P}$ holds. *R. A. Rankin* (Birmingham).

Citations: MR 11, 231b = K35-5; MR 13, 213e = K15-15.

Referred to in K15-33, K25-14.

K15-23 (14, 143e)

Korobov, N. M., and Postnikov, A. G. Some general theorems on the uniform distribution of fractional parts. Doklady Akad. Nauk SSSR (N.S.) **84**, 217–220 (1952). (Russian)

It is proved that if for every positive integer h the function $F(x+h)-F(x)$ is uniformly distributed, then, for every pair of integers λ, μ, the function $F(\lambda x+\mu)$ is also uniformly distributed. This is a generalisation of a result of van der Corput [Acta Math. **56**, 373–456 (1931)]. Several applications of this result are given. It is also used to prove a general theorem on trigonometrical sums from which the following corollary is deduced. Let α be a real number for which $|\sum_{x=1}^{P}e^{2\pi im\alpha q^x}|<cm\sqrt{P}$. Here the notation is that of the paper reviewed above, m is a positive integer, and the methods of that paper can be used to construct numbers α satisfying this inequality. Then it follows that for any positive integer

$$\left|\sum_{x=1}^{P}e^{2\pi im\alpha q^{\lambda x}}\right|<c(m)\lambda\frac{P}{\sqrt{\log P}}.$$

R. A. Rankin (Birmingham).

K15-24 (14, 144a)

Korobov, N. M. Some many dimensional problems of the theory of Diophantine approximations. Doklady Akad. Nauk SSSR (N.S.) **84**, 13–16 (1952). (Russian)

In this paper the author constructs, with the aid of his theory of normal periodic systems, s real numbers $\alpha_1, \alpha_2, \cdots, \alpha_s$ such that the function $(m_1\alpha_1+m_2\alpha_2+\cdots+m_s\alpha_s)q^x$ is uniformly distributed; i.e., the system of functions $\alpha_1q^x, \cdots, \alpha_sq^x$ is uniformly distributed in s-dimensional space. Here q is an integer greater than unity and $m_1, m_2, \cdots, m_s$ are integers not all zero. Two further theorems concerning systems of functions uniformly distributed in s-dimensional space, which are analogous to the first two results mentioned in the second preceding review, are given; it is stated that they may be proved by similar methods. *R. A. Rankin.*

Referred to in K25-14, K35-8.

K15-25 (14, 144b)

Korobov, N. M. On normal periodic systems. Izvestiya Akad. Nauk SSSR. Ser. Mat. **16**, 211–216 (1952). (Russian)

The author generalises his theory of normal periodic systems [same Izvestiya **15**, 17–46 (1951); these Rev. **13**, 213] in the following way. Let n, q, τ, δ_ν be integers satisfying $n\geqq1$, $q\geqq2$, $\tau\geqq n$, $0\leqq\delta_\nu\leqq q-1$. Here the δ_ν may be regarded as digits in the scale of q. Denote by E_n any set of τ numbers each containing n digits δ_ν; the numbers of E_n need not be distinct. Then he says that a normal periodic system is possible in the set E_n if there exists a sequence of digits

$$\delta_1\delta_2\delta_3\cdots\delta_{n-1}\delta_n\delta_{n+1}\cdots\delta_\tau\delta_1\delta_2\cdots\delta_{n-1}$$

with the property that E_n coincides with the set of τ numbers of n digits which can be formed from successive batches of n digits of this sequence. When E_n is the set of q^n n-digited numbers in the scale of q, this definition coincides with that given in his earlier paper. A number of $n-1$ digits $\beta_1\beta_2\cdots\beta_{n-1}$ is said to occur in E_n if E_n contains a number of one of the forms $\beta_1\beta_2\cdots\beta_{n-1}\beta$, $\beta^1\beta_1\beta_2\cdots\beta_{n-1}$. The following theorem is proved. A normal periodic system is possible in a set E_n if and only if the following two conditions are both satisfied: (i) For every number of $n-1$ digits $\beta_1\beta_2\cdots\beta_{n-1}$ occurring in E_n the numbers of numbers of the form $\beta_1\beta_2\cdots\beta_{n-1}\beta$ and $\beta^1\beta_1\beta_2\cdots\beta_{n-1}$ contained in E_n are equal; (ii) for every partition of E_n into two non-null parts E_n' and E_n'' there can be found at least one number of $n-1$ digits which occurs in each of these parts. The proof of the necessity of these conditions is straightforward. In order to prove them sufficient a method is given which actually constructs a normal periodic system in the set E_n. Various examples of the theorem are given and the existence of normal periodic systems in the earlier sense follows as a simple corollary. The author also investigates the set E_n of numbers defined as follows: Let u_n $(n=0, 1, 2, \cdots)$ denote the Fibonacci numbers, and write $\delta_1\delta_2\cdots\delta_n$ for the number $\delta_1u_{n-1}+\delta_2u_{n-2}+\cdots+\delta_nu_0$ where $q=2$ and no two neighbouring digits δ_i, δ_{i+1} are both unity. Every positive integer can be represented uniquely in this way for a suitable n. Then E_n denotes the set of all n-digited numbers of this form. It is easily shown that normal periodic systems do not exist in E_n. However, if E_n is modified by the omission of those numbers which contain 1 as first and last digit, then it is proved that normal periodic systems exist in the new set so formed. *R. A. Rankin.*

Citations: MR 13, 213e = K15-15.

K15-26 (14, 359d)

Postnikov, A. G. On the distribution of the fractional parts of the exponential function. Doklady Akad. Nauk SSSR (N.S.) **86**, 473–476 (1952). (Russian)

Let q be an integer greater than unity, α a real number, Δ a subinterval of (0, 1) of length $|\Delta|$ and $N_P(\Delta)$ the number of fractional parts $\{\alpha q^x\}$ $(x=1, 2, \cdots, P)$ which lie in Δ. I. I. Šapiro-Pyateckiĭ [Izvestiya Akad. Nauk SSSR. Ser. Mat. **15**, 47–52 (1951); these Rev. **13**, 213] showed that $\{\alpha q^x\}$ is uniformly distributed if there exists a constant C such that $\limsup_{P\to\infty} N_P(\Delta)/P<C|\Delta|$ for every Δ. The author improves this result by showing that $\{\alpha q^x\}$ is uniformly distributed if, for some positive C and k,

$$\limsup_{P\to\infty}\frac{N_P(\Delta)}{P}<C|\Delta|\left(1+\log\frac{1}{|\Delta|}\right)^k.$$

This he does by applying Vinogradov's method of estimating trigonometric sums with polynomial exponents to the sums $\sum e^{2\pi i m\alpha q^x}$. The argument is carried through in detail in an exceptionally lucid fashion. *R. A. Rankin.*

Citations: MR 13, 213d = K15-14.
Referred to in K15-42, K15-55, K20-35.

K15-27 (14, 454c)

Cassels, J. W. S. On a paper of Niven and Zuckerman. Pacific J. Math. **2**, 555–557 (1952).

Let $0\leqq a_i<10$, a_i integer. Denote by

$$R_m(a_1, a_2, \cdots, a_r; b_1, b_2, \cdots, b_s)$$

the number of solutions of

$$b_n=a_1,\ b_{n+1}=a_2,\ \cdots,\ b_{n+r-1}=a_r,\qquad 0<n<n+r\leqq s,\ n\equiv m \pmod r.$$

Let $x_1, x_2, \cdots$ be an infinite sequence of integers $0\leqq x_i<10$. Niven and Zuckerman showed [same J. **1**, 103–109 (1951); these Rev. **13**, 438] that if

$$(1)\quad \lim_{N\to\infty}\frac{1}{N}\sum_{m=1}^{r} R_m(a_1, a_2, \cdots, a_r; x_1, x_2, \cdots, x_N)=10^{-r}$$

for all r and integers $a_1, a_2, \cdots, a_r$ $(0\leqq a_i<10)$ (i.e., if x_1, x_2 is normal), then

$$(2)\qquad \lim_{N\to\infty}\frac{1}{N}R_m(a_1, a_2, \cdots, a_r; x_1, \cdots, x_N)=r^{-1}10^{-r}$$

for all integers r, m, and a_i. The author gives a very simple and ingenious proof of this result. *P. Erdös.*

Citations: MR 13, 438a = K15-16.

K15-28 (14, 553e)

Izumi, Shin-ichi. Notes on Fourier analysis. XLIV. On the law of the iterated logarithm of some sequences of functions. J. Math. Tokyo **1**, 1–22 (1951).

Generalizing results of T. Tsuchikura, R. Fortet, and G. Maruyama, the author proves that if $f(x)$ is bounded, has period 1, mean value zero, and satisfies

$$\int_0^1 \max_{0\leqq\nu\leqq u}|f(x+\nu)-f(x)|\,dx=O\left[1\Big/\left(\log\left(\frac{1}{u}\right)\right)^\alpha\right]$$

with $\alpha>1$, then for almost all t

$$(1)\qquad \limsup_{N\to\infty}\frac{\sum_1^N f(2^nt)}{(N\log\log N)^{1/2}}=\sigma^2,$$

where

$$\sigma^2=\lim_{N\to\infty}\frac{1}{N}\int_0^1\left(\sum_1^N f(2^nt)\right)^2dt.$$

The case $\alpha=1$ leads, under more complicated conditions, to a result in which (1) is replaced by an inequality. Applications are made to the study of the "discrepancy" in uniform distribution. *R. Salem* (Paris).

K15-29 (14, 851b)

Maxfield, John E. Normal k-tuples. Pacific J. Math. **3**, 189–196 (1953).

A k-tuple β is a k-tuple $(\alpha_1, \alpha_2, \cdots, \alpha_k)$ of real numbers. The nth k-digit of the k-tuple to the base r is

$$b^{(n)}=(A_1^{(n)}, \cdots, A_k^{(n)})$$

where $A_s^{(n)}$ is the nth digit of the fractional part of α_s to the base r. A k-tuple is said to be simply normal to the base r if the number n_c of occurrences of the kth digit c in the first n k-digits of the fractional part of β has the property $\lim_{n\to\infty} n_c/n=1/r^k$ for each of the r^k possible values of c. A k-tuple β is said to be normal to the scale r if β, $r\beta$, $r^2\beta$, $\cdots$ are each simply normal to all the scales $r, r^2, \cdots$ where $r^s\beta=(r^s\alpha_1, \cdots, r^s\alpha_k)$. The author extends many of the theorems on normal numbers to normal k-tuples and proves several new theorems. Perhaps his most interesting theorem is the following. The k-tuple $\beta=(\alpha_1, \alpha_2, \cdots, \alpha_k)$ is normal if and only if $\sum_{i=1}^k h_i\alpha_i$ is a normal number for all integers $(h_1, h_2, \cdots, h_k)\neq(0, 0, \cdots, 0)$. *P. Erdös.*

Referred to in K40-33.

K15-30 (14, 852a)

Korobov, N. M. On some problems of Čebyšev type. Doklady Akad. Nauk SSSR (N.S.) **89**, 397–400 (1953). (Russian)

It is known that, for given irrational α, the inequalities

$$0<x<ct,\quad |\alpha x-y-\beta|<\frac{1}{t}\quad (c=c(\alpha)),$$

have solutions in integers x and y for arbitrary real β and any $t\geqq 1$ if and only if the partial quotients of the continued fractions for α are bounded. The author proves the following analogous theorem, where αx is replaced by αq^x, q being a fixed integer greater than unity. Theorem: In order that there shall exist a number $C>0$ such that for arbitrary β

and $t \geqq 1$ a pair of integers x, y can be found to satisfy

$$0 \leqq x < Ct, \quad |\alpha q^x - y - \beta| < \frac{1}{t}$$

it is necessary and sufficient that α shall be of bounded ratio. These last two words are defined as follows. Let δ_n be the nth digit after the decimal point in the "decimal" expansion of α in the scale of q. Then $\lambda = \lambda(n)$ is defined to be the least positive integer with the property that among the first λ n-digited numbers $\delta_{x+1}\delta_{x+2} \cdots \delta_{x+n}$ $(x = 0, 1, \cdots, \lambda - 1)$ that can be chosen from the decimal expansion of α each of the q^n possible n-digited numbers occurs at least once. It is supposed that the fractional parts of αq^x are everywhere dense over the interval (0, 1). Then α is said to be of bounded ratio if there exists a constant $c_1 = c_1(\alpha)$ such that $\lambda(n)/q^n < c_1$ for all $n \geqq 1$. By means of his theory of normal periodic systems the author proves that numbers of bounded ratio do exist and shows how an infinity of them can be constructed. Generalisations of these results to s dimensions are stated without proof. *R. A. Rankin* (Birmingham).

Referred to in K15-41.

K15-31 (14, 1067b)

Korobov, N. M. On a question of diophantine inequalities. Comptes Rendus du Premier Congrès des Mathématiciens Hongrois, 27 Août–2 Septembre 1950, pp. 259–262. Akadémiai Kiadó, Budapest, 1952. (Russian. Hungarian summary)

After a brief historical account of some problems on uniform distribution the author proves the following result. If, for some function $\varphi(x)$ and arbitrarily chosen integers $m_1, m_2, \cdots, m_s$, not all zero,

$$m_1\varphi(x+1) + m_2\varphi(x+2) + \cdots + m_s\varphi(x+s)$$

is uniformly distributed, then αq^x is uniformly distributed, where q is an integer greater than unity and

$$\alpha = \sum_{k=1}^{\infty} [\{\varphi(k)\}q]/q^k.$$

This is actually a particular case of earlier results due to the author [Uspehi Matem. Nauk (N.S.) **4**, no. 1(29), 189–190 (1949); Izvestiya Akad. Nauk SSSR. Ser. Mat. **14**, 215–238 (1950); these Rev. **11**, 231; **12**, 321]. A general theorem giving sufficient conditions for a function $\varphi(x)$ to possess the above property is stated (see references quoted).

R. A. Rankin (Birmingham).

Citations: MR 11, 231b = K35-5; MR 12, 321a = K15-11.

K15-32 (14, 1070a)

Volkmann, Bodo. Über Hausdorffsche Dimensionen von Mengen, die durch Zifferneigenschaften charakterisiert sind. I. Math. Z. **58**, 284–287 (1953).

If g, e_n are integers and $0 \leqq e_n < g$, the set of fractions $\sigma = \sum f_n g^{-n}$ for which $f_n \leqq e_n$ (all n) is considered.

H. D. Ursell (Leeds).

K15-33 (15, 15d)

Korobov, N. M. Unimprovable estimates of trigonometric sums with exponential functions. Doklady Akad. Nauk SSSR (N.S.) **89**, 597–600 (1953). (Russian)

In an earlier paper [Trudy Mat. Inst. Steklov. **38**, 87–96 (1951); these Rev. **14**, 143] the author showed that, in a certain sense, for most values of the real number α the inequality

$$|S| = \left| \sum_{x=1}^{P} \exp(2\pi i m \alpha q^x) \right| < cP^{\frac{1}{2}}$$

holds, where c depends only upon the positive integers m and $q > 1$. He now shows that this upper bound $cP^{\frac{1}{2}}$ can be replaced by an arbitrarily slowly increasing function $\phi(P)$ tending to infinity with P, for a certain class of values of α. In fact, if $\phi(P)$ is such a function, integers n_ν $(\nu = 0, 1, 2, \cdots)$ can be chosen sufficiently large such that for

$$\alpha = \sum_{\nu=0}^{\infty} \frac{1}{m_{\nu+1} q^{n_\nu}}$$

we have $|S| = o\{\phi(P)\}$. Here $\{m_\nu\}$ is a sequence of integers greater than unity and prime to q defined recursively by $m_{\nu+1} = m_\nu(q^{\tau_\nu} - 1)$ $(\nu = 1, 2, 3, \cdots)$ where $\{\tau_\nu\}$ is an arbitrary strictly increasing sequence of positive integers such that τ_ν is a multiple of the exponent of m_ν modulo q_ν. The integers n_ν are then chosen to satisfy $n_\nu > \phi_1(\nu m_\nu \tau_\nu)$ where ϕ_1 is the inverse function to ϕ. The result is best possible in the sense that $o\{\phi(P)\}$ cannot be replaced by $O(1)$ for every α of the class considered. *R. A. Rankin* (Birmingham).

Citations: MR 14, 143d = K15-22.

Referred to in K15-40, K15-96.

K15-34 (15, 306e)

Pyateckiĭ-Šapiro, I. I. Fractional parts and some questions of the theory of trigonometric series. Uspehi Matem. Nauk (N.S.) **8**, no. 3(55), 167–170 (1953). (Russian)

The author gives a summary, without proofs, of his results published elsewhere [see, in particular, Izvestiya Akad. Nauk SSSR. Ser. Mat. **15**, 47–52 (1951); Doklady Akad. Nauk SSSR (N.S.) **85**, 497–500 (1952); these Rev. **13**, 213; **14**, 161]. Another paper whose results he quotes and which has appeared in the Moskov. Gos. Univ. Učenye Zapiski **155**, Matematika **5** (1952) is still not available for review.

A. Zygmund (Cambridge, England).

Citations: MR 13, 213d = K15-14.

K15-35 (15, 513d)

Volkmann, Bodo. Über Hausdorffsche Dimensionen von Mengen, die durch Zifferneigenschaften charakterisiert sind. II. Math. Z. **59**, 247–254 (1953).

[For part I see Math. Z. **58**, 284–287 (1953); these Rev. **14**, 1070.] The digits $0, \cdots, (g-1)$ being divided into any m sets $G_1, \cdots$ of $w_1, \cdots$ members respectively $(\sum w_\mu = g)$, the fractions whose g-adic developments digits of these sets have asymptotic frequencies ζ_μ $(\sum \zeta_\mu = 1)$ form a set of dimension $\sum \zeta_\mu \log_g (w_\mu/\zeta_\mu)$. Sharper dimensional classification is also obtained for certain sets of dimension 0.

H. D. Ursell (Leeds).

K15-36 (15, 513e)

Volkmann, Bodo. Über Hausdorffsche Dimensionen von Mengen, die durch Zifferneigenschaften charakterisiert sind. III. Math. Z. **59**, 259–270 (1953).

Theorem 1 determines the dimension of the set of fractions whose g-adic development does not contain a given sequence $F \equiv f_1 \cdots f_i$ as $\log_g \gamma$, where γ is the greatest root of an algebraic equation $f(z; F, g) = 0$ of degree i. Th. 2 is a related result on linear graphs. Th. 6 compares $\gamma(F)$ with $\gamma(F')$: the result 6(c) is misstated. Th. 7 concerns the number of integers $k \leqq x$ whose g-adic development does not contain F: some changes are needed if F begins with 0.

H. D. Ursell (Leeds).

Referred to in K15-44.

K15-37 (15, 691f)

Volkmann, Bodo. Über Hausdorffsche Dimensionen von Mengen, die durch Zifferneigenschaften charakterisiert sind. IV. Math. Z. **59**, 425–433 (1954).

[For parts I–III see Math. Z. **58**, 284–287; **59**, 247–254, 259–270 (1953); these Rev. **14**, 1070; **15**, 513.] For g-adic representations $\rho = \sum e_i g^{-i}$, the possible digits e $(0 \leqq e < g)$

are assigned non-negative weights $\lambda(e)$ and are divided into exclusive classes $G_1, \cdots, G_m$. Define $S_\mu(\rho, n) = \sum \lambda(e_i)$ ($i \leqq n$, $e_i \varepsilon G_\mu$). The author determines the Hausdorff dimension of the set $E(\lambda; G; \zeta)$ of fractions ρ for which $S_\mu \sim \zeta_\mu n$ (all μ) as $n \to \infty$. *H. D. Ursell* (Leeds).

K15-38 (16, 223c)

Hanson, H. A. Some relations between various types of normality of numbers. Canadian J. Math. **6**, 477–485 (1954).

Let $a_1, a_2, \cdots$ be an increasing sequence of integers. It is shown how the question of (k, ϵ)-normality of almost all of the a_i can be reduced to the question of ϵ-normality. Let $x = .a_1a_2\cdots$ be formed by writing the digits of the a_i in order. If the a_i satisfy a (k, ϵ)-normality condition and a condition restricting the rate of increase of the a_i, it is proved that x is normal. The author defines a quasi-normal number as a real number such that every number formed by selecting those digits whose positions form an arithmetic progression, is simply normal. A normal number is quasi-normal. If x is normal in the scale B and if $y = \sum_{j=1}^{\infty} r_j s^{-j}$, $r_j = [B^j x] - s[B^j x/s]$, then y is quasi-normal in the scale s. However, y is normal in the scale s if and only if s divides B.
H. S. Zuckerman (Seattle, Wash.).

K15-39 (16, 267f)

Fine, N. J. On the asymptotic distribution of certain sums. Proc. Amer. Math. Soc. **5**, 243–252 (1954).

Let $((t)) = t - [t] - \frac{1}{2}$. The asymptotic distribution of $n^{-\frac{1}{2}} \sum_{k=0}^{n-1} ((2^k t - \beta))$. is known to be normal for $\beta \neq \frac{1}{2}$ and the author shows that the variance is given by the formula

$$\sigma^2 = \sum_{n=1}^{\infty} 2^{-n}((\beta + [2^n\beta] - 2^n\beta))^2.$$

Another typical result is the following: If $\gamma_n(t; 0, \beta)$ denotes the relative frequency with which $2^k t - [2^k t]$ falls within $(0, \beta)$, then the asymptotic distribution of $n^{\frac{1}{2}}(\gamma_n(t; 0; \beta) - \beta)$ is normal with variance

$$\tau^2 = \beta - \beta^2 + 2\sum_{n=1}^{\infty} 2^{-n}\{\min(\beta, \beta_n) - \beta\beta_n\}.$$

where $\beta_n = 2^n\beta - [2^n\beta]$. *M. Kac* (Ithaca, N. Y.).

Referred to in K15-62.

K15-40 (17, 588a)

Polosuev, A. M. A multidimensional case of unimprovable estimates of trigonometric sums with exponential functions. Dokl. Akad. Nauk SSSR (N.S.) **104** (1955), 186–189. (Russian)

The author generalizes a result of N. M. Korobov [same Dokl. (N.S.) **89** (1953), 597–600; MR **15**, 15] by showing that given any function $\phi(p)$ which tends to infinity as $p \to \infty$, no matter how slowly, real numbers $\alpha_1, \alpha_2, \cdots, \alpha_s$ can be constructed such that

$$\sum_{x=1}^{p} \exp\{2\pi i \sum_{l=1}^{s} m_l \alpha_l q_l^x\} = o\{\varphi(p)\},$$

for any integers $m_1, m_2, \cdots, m_s$ not all zero. Here $q_1, q_2, \cdots, q_s$ are fixed integers greater than unity. It is shown, further, that for no real numbers $\alpha_1, \alpha_2, \cdots, \alpha_s$ can the right-hand side of this result be replaced by $O(1)$.

R. A. Rankin (Glasgow).

Citations: MR 15, 15d = K15-33.
Referred to in K15-96.

K15-41 (17, 590a)

Korobov, N. M. Numbers with bounded quotient and their applications to questions of Diophantine approximation. Izv. Akad. Nauk SSSR. Ser. Mat. **19** (1955), 361–380. (Russian)

Let q be an arbitrary integer greater than unity, and let

$$\alpha = 0 \cdot \delta_1\delta_2\delta_3 \cdots$$

be the 'decimal' expansion of a real number α in the scale of q. Suppose that, for each $n \geqq 1$, there exists an integer $\lambda = \lambda(n)$ such that among the n-figure numbers

$$\delta_1\delta_2\cdots\delta_n,\ \delta_2\delta_3\cdots\delta_{n+1},\ \cdots,\ \delta_\lambda\delta_{\lambda+1}\cdots\delta_{\lambda+n-1}$$

each of the q^n different n-figure numbers occurs at least once. If there exists a number $C_1 = C_1(\alpha)$ such that, for all $n \geqq 1$, $\lambda(n)/q^n < C_1$, the number α is said to be a number with bounded ratio. In the theory of the distribution of the fractional parts of the numbers αq^x ($x = 1, 2, \cdots$), such numbers play a role similar to that played by numbers with bounded partial quotients in the theory of Diophantine approximation. Numbers α of the latter kind have the property that for them, and only for them, does there exist a constant $C = C(\alpha)$ such that the inequalities

$$0 < x < Ct,\ |\alpha x - y - \beta| < t,$$

possess integer solutions in x only for arbitrary β and $t \geqq 1$.

The author shows that a necessary and sufficient condition for the existence of a number $C > 0$ such that for arbitrary β and $t \geqq 1$ there exist integers x and y satisfying the inequalities

$$0 \leqq x < Ct,\ |\alpha q^x - y - \beta| < 1/t,$$

is that α is a number of bounded ratio, and he constructs an infinite class of numbers of bounded ratio. He also constructs an infinite class of numbers α for which

$$|\alpha q^x - y - \beta| < (1+\varepsilon)/x$$

has infinitely many solutions in x, y for each β, where ε is an arbitrary positive number. Some of these results were already proved in an earlier paper of the author [Dokl. Akad. Nauk SSSR (N.S.) **89** (1953), 397–400; MR **14**, 852] where the generalisations now mentioned were stated without proof.

The concept of bounded ratio is extended to systems of s numbers α_i expressed in scales of bases q_i ($i = 1, \cdots, s$) and similar results are proved. These are applied to questions concerning the distribution of fractional parts, numbers $\alpha_1, \alpha_2, \cdots, \alpha_s$ being constructed such that the system of functions $\alpha_i q_i^x$ ($x = 1, 2, \cdots$) are uniformly distributed in s-dimensional space. It is also shown that the number $N(v)$ of such points lying in a region v of the s-dimensional cube, for $x \leqq P$, satisfies

$$N(v) = vP + O(P^{1-1/(s+1)}).$$

R. A. Rankin (Glasgow).

Citations: MR 14, 852a = K15-30.
Referred to in K15-52, K15-59.

K15-42 (18, 803c)

Postnikov, A. G. Estimation of an exponential trigonometric sum. Izv. Akad. Nauk SSSR. Ser. Mat. **20** (1956), 661–666. (Russian)

Let g be an integer greater than unity and let

$$S = \sum_{x=0}^{P-1} e^{2\pi i \alpha g^x}$$

where $0 \leqq \alpha < 1$. It is shown that this interval can be divided into two subsets $\mathfrak{M}_1$ and $\mathfrak{M}_2$ such that, for large P,

$$|S| \leqq K(\varepsilon) \frac{P}{\log^{\frac{1}{4}-\epsilon} P}$$

when $\alpha \in \mathfrak{M}_2$, and

$$\text{meas } \mathfrak{M}_1 = O\{\exp(-K \log^3 P + O(\log P))\}.$$

Here K is a positive constant and $K(\varepsilon)$ depends only on ε, which is positive.

The proof uses an upper bound for $|S|^{2d}$ obtained by the author in a previous paper [Dokl. Akad. Nauk SSSR (N.S.) **86** (1952), 473–476; MR **14**, 359] together with a lemma which states that the number of numbers of f digits in the scale of r in which a given digit b occurs more than $\eta f/r$ times, where $r > \eta > 2$, is $O\{f^{\frac{1}{2}} r^f \exp(-\frac{1}{2}f\eta^2/r^2)\}$.

R. A. Rankin (Glasgow).

Citations: MR 14, 359d = K15-26.
Referred to in K15-59.

K15-43 (19, 1038b)

Long, Calvin T. Note on normal numbers. Pacific J. Math. **7** (1957), 1163–1165.

It is proved that a number α is normal to base r if and only if there is an infinite set M of positive integers m_k such that α is simply normal to each of the bases r^{m_k}. The condition does not suffice if M is finite.

H. S. Zuckerman (Seattle, Wash.).

K15-44 (19, 1161f)

Volkmann, Bodo. Über Hausdorffsche Dimensionen von Mengen die durch Ziffereigenschaften charakterisiert sind. V. Math. Z. **65** (1956), 389–413.

Given an integer $g \geqq 2$, every real number ρ, $0 < \rho \leqq 1$, possesses a unique g-adic representation $\rho = \sum_{j=1}^{\infty} e_j g^{-j}$ where $\{e_j\}$ is an integer sequence containing infinitely many non-zero members and such that $0 \leqq e_j < g$; let $\{e_j\}$ be called the digital sequence of ρ. Let each of $F_1, F_2, \cdots, F_m$ be a finite, ordered set of non-negative integers less than g. The author computes the Hausdorff dimension α of the set of those numbers ρ, $0 < \rho \leqq 1$, whose digital sequences do not contain any of $F_1, \cdots, F_m$ as a subset. He remarks that his proof of the special case $m=1$ in paper III of the series [Math. Z. **59** (1953), 259–270; MR **15**, 513] contains a fault which is now obviated by the methods of the present paper; these enable him also to find upper and lower bounds for the corresponding α-dimensional Hausdorff measure. The formula for α is given in a number of alternative forms, some referring to other special cases.

H. Halberstam (London).

Citations: MR 15, 513e = K15-36.
Referred to in K15-45, K15-64.

K15-45 (20# 7008)

Volkmann, Bodo. Über Hausdorffsche Dimensionen von Mengen, die durch Ziffereigenschaften charakterisiert sind. VI. Math. Z. **68** (1958), 439–449.

[For part V, see same Z. **65** (1956), 389–413; MR **19**, 1161.] To the g-adic expansion of a number ρ with $0 < \rho \leqq 1$ the author associates the points

$$p_i(\rho) = \left(\frac{A_0(\rho, i)}{i}, \cdots, \frac{A_{g-1}(\rho, i)}{i}\right) \quad (i = 1, 2, \cdots),$$

where $A_j(\rho, n)$ is the number of occurrences of the digit j among the first n digits in the expansion of ρ. The set of limit points of $\{p_i(\rho)\}$ is called $V(\rho)$.

The author obtains the following results on the set $V(\rho)$. 1. $(V\rho)$ is always a continuum contained in the set H in which the hyperplane $\sum x_j = 1$ intersects the unit cube $0 \leqq x_j \leqq 1$. 2. Conversely, every continuum in H is $V(\rho)$ for some ρ. 3. Let $G(V)$ be the set of ρ for which $V(\rho) = V$ and let $d(x) = -\sum x_j \log x_j / \log g$ (with $0 \log 0 = 0$). Then the Hausdorff dimension of $G(V)$ is given by $\dim G(V) = \min_{x \in V} d(x)$. 4. If T, S are arbitrary sets in H; $T \subseteq S$ and $G(T, S)$ the set of all ρ with $T \subseteq V(\rho) \subseteq S$; then $G(T, S)$ is empty if the minimal continuum T^* which contains T is not in S; $\dim G(T, S) = \min_{x \in T^*} d(x)$ if T is non-empty; and $\dim G(T, S) = \max_{x \in S} d(x)$ if T is empty.

E. G. Straus (Los Angeles, Calif.)

Citations: MR 19, 1161f = K15-44.
Referred to in K15-64, K15-69, K15-93.

K15-46 (21# 663)

Postnikov, A. G.; and Pyateckiĭ, I. I. On Bernoulli-normal sequences of symbols. Izv. Akad. Nauk SSSR. Ser. Mat. **21** (1957), 501–514. (Russian)

A family of transformations T of a metric space R into itself is called a dynamical system, and a measure μ on R, with $\mu(R) = 1$, is said to be invariant if, for all measurable sets E, $T^{-1}E$ is measurable and $\mu T^{-1}E = \mu E$. We are thus concerned with measure-preserving set-transformations in the sense of J. L. Doob [*Stochastic processes*, Wiley, New York, 1953; MR **15**, 445; Chapter X]. Under these circumstances the author proves the ergodic theorem in the form that, for almost all $p \in R$,

$$\lim_{n \to \infty} n^{-1} \sum_{r=1}^{n} \varphi(T^{r-1}p) = \int_R \varphi(p) d\mu;$$

when φ is absolutely integrable and R cannot be decomposed into the union of two disjoint invariant sets of positive μ-measure.

This result is applied to the following situation. In what follows α_i is either 0 or 1; 1 corresponds to the occurrence of an event which has probability p and 0 corresponds to its non-occurrence, which has probability $q = 1 - p$. An infinite set of events or tests corresponds to a binary decimal $\alpha = \cdot\alpha_1\alpha_2\alpha_3\cdots$. A measure μ is defined on the space $R = [0, 1]$ $(0 \leqq \alpha \leqq 1)$ as follows. If $A = \cdot\alpha_1\alpha_2\cdots\alpha_n$, then $\mu[A, A + 2^{-n}]$ is defined to be the probability of achieving the results $\alpha_1, \alpha_2, \cdots, \alpha_n$ in a succession of n tests. This measure is invariant for the set of transformations T^k $(k \geqq 0)$ defined by $T^k\alpha = \{\alpha 2^k\}$ (fractional part). If $\alpha = .\alpha_1\alpha_2\alpha_3\cdots$ and $\Delta = .\varepsilon_1\varepsilon_2\cdots\varepsilon_s$, then $N_P(\alpha, \Delta)$ denotes the number of times that Δ occurs as s successive digits in $\alpha_1\alpha_2\cdots\alpha_{P+s-1}$. The number α is said to be "normal according to Bernoulli" (because of the connexion with the Bernoulli or binomial distribution) if

$$\lim_{P \to \infty} \frac{N_P(\alpha, \Delta)}{P} = \mu\Delta$$

for all Δ and $s \geqq 1$.

It is shown that the probability of a sequence of events being normal according to Bernoulli is 1; i.e., the μ-measure of the set of numbers α normal according to Bernoulli is 1. It is also shown that, if

$$\limsup_{P \to \infty} \frac{N_P(\alpha, \Delta)}{P} < C\mu\Delta,$$

for some constant C and all Δ and s, then α is normal according to Bernoulli.

An example based on the work of D. G. Champernowne [J. London Math. Soc. **8** (1933), 254–260] is given.

R. A. Rankin (Glasgow)

K15-47 (21# 664)

Postnikov, A. G.; and Pyateckiĭ, I. I. A Markov-sequence of symbols and a normal continued fraction. Izv. Akad. Nauk SSSR. Ser. Mat. **21** (1957), 729–746. (Russian)

The ideas of the paper reviewed above are extended and applied to a stationary Markov process, the results being of a similar type. Two kinds of dynamical system are considered. The first is similar to that introduced in the previous paper; i.e., the transformations are of the form $\{\alpha\}\to\{2\alpha\}$. In the second system the transformations are of the form $\{\alpha\}\to\{1/\alpha\}$ and involve the expression of α as a continued fraction. The associated measure μ is defined by

$$\mu\Delta = \frac{-1}{\log 2}\int_\Delta \frac{dx}{1+x}$$

on sets of numbers α having given binary digits $\varepsilon_1, \varepsilon_2, \cdots, \varepsilon_s$ in certain given places $i_1, i_2, \cdots, i_s$ $(s\geqq 1)$.

R. A. Rankin (Glasgow)

Referred to in K15-78.

K15-48 (21# 665)

Postnikov, A. G. A test for a completely uniformly distributed sequence. Dokl. Akad. Nauk SSSR **120** (1958), 973–975. (Russian)

Consider an infinite sequence $\{\alpha_i\}$ of real numbers with $0\leqq\alpha_i\leqq 1$ for all i. Let I_s be an s-dimensional interval lying in the s-dimensional unit cube and let $N_k(I_s)$ be the number of points $(\alpha_i, \cdots, \alpha_{i+s-1})$, $1\leqq i\leqq k$, contained in I_s. If there is a constant c such that

$$\limsup_{k\to\infty} N_k(I_s)/k < c|I_s|$$

for all $s\geqq 1$ and all I_s, where $|I_s|$ is the volume of I_s, then $\{\alpha_i\}$ is completely uniformly distributed. This theorem and its proof are very similar to one concerning normal numbers published by the present author and I. I. Pjateckiĭ [see the article reviewed second above].

L. Schmetterer (Berkeley, Calif.)

K15-49 (21# 666)

Postnikov, A. G. A criterion for testing the uniform distribution of an exponential function in the complex domain. Vestnik Leningrad. Univ. **12** (1957), no. 13, 81–88. (Russian. English summary)

A criterion for the uniform distribution of the exponential function on the real domain, which was given by the author and I. I. Pyateckiĭ [see the article reviewed third above] is extended to the complex domain. Let $N_P(\Delta)$ be the number of fractional parts $\{(\mu_1+i\mu_2)(a+ib)^x\}$ $(0\leqq x<P)$ lying in an arbitrary square Δ of area $|\Delta|$ in the interior of the unit square, the sides of Δ being parallel to the axes. Then, if there exists a constant C such that, for all Δ,

$$\limsup_{P\to\infty}\frac{N_P(\Delta)}{P} < C|\Delta|,$$

it is shown that the fractional parts are uniformly distributed in the unit square. Here $\{\mu_1+i\mu_2\}$ is defined to be $\{\mu_1\}+i\{\mu_2\}$, and $a+ib$ is a Gaussian integer, which is not real and is different from $\pm i$. For given $a+ib$ an elaborate construction of a complex number $z=\mu_1+i\mu_2$ is given, for which $\{z(a+ib)^x\}$ is shown to be uniformly distributed by applying the above criterion.

R. A. Rankin (Glasgow)

Referred to in K30-21.

K15-50 (21# 1962)

Mineev, M. P. A Diophantine equation involving an exponential function and its application to the study of an ergodic sum. Izv. Akad. Nauk SSSR. Ser. Mat. **22** (1958), 585–598. (Russian)

Let $g\geqq 2$, $m_1, \cdots, m_k, n_1, \cdots, n_k$ be fixed positive integers. The author proves that there are $cp^k+O(p^{k-1})$ integral solutions $(x_1, \cdots, x_k, y_1, \cdots, y_k)$ of the conditions

$$m_1g^{x_1}+\cdots+m_kg^{x_k} = n_1g^{y_1}+\cdots+n_kg^{y_k},$$
$$0\leqq x_1, \cdots, x_k, y_1, \cdots, y_k\leqq p-1,$$

as $p\to\infty$; here $c\geqq 0$ does not depend on p. For $m_1=\cdots=m_k=n_1=\cdots=n_k=1$ this was found by A. G. Postnikov [Fetschrift anlässlich des 250. Geburtstages Leonhard Eulers, Berlin, 1957]. The following application is made. Let $f(t)$ be a real function of period 1 satisfying

$$\int_0^1 f(t)dt = 0,\quad \int_0^1 f(t)e^{2\pi int}dt = O(|n|^{-\beta})\quad (|n|\to\infty),$$

where $\beta>1/2$, such that $\sigma>0$, where

$$\sigma^2 = \int_0^1 f(t)^2dt+2\sum_{k=1}^{\infty}\int_0^1 f(t)f(g^kt)dt.$$

Then

$$\lim_{p\to\infty} p^{-1}\int_0^1\left(\sum_{k=0}^{p-1} f(g^kt)\right)^2 dt = \sigma^2,$$

and the author proves that

$$\lim_{p\to\infty}\ \operatorname*{meas}_{0\leqq\alpha<1} E\left\{\sum_{x=0}^{p-1} f(g^x\alpha) < \lambda\sqrt{p}\right\} = \frac{1}{\sigma(2\pi)^{1/2}}\int_{-\infty}^{\lambda} e^{-z^2/2\sigma^2}dz,$$

by first considering finite trigonometric sums.

K. Mahler (Manchester)

Referred to in K30-23.

K15-51 (21# 3393)

Kesten, Harry. On a series of cosecants. II. Nederl. Akad. Wetensch. Proc. Ser. A **62** = Indag. Math. **21** (1959), 110–119.

[For part I, see M. Kac and R. Salem, same Proc. **60** (1957), 265–267; MR **19**, 646.] Let x be a random variable, uniformly distributed on $[0, 1]$. It is shown that $\sum_0^\infty c_k \csc 2\pi 2^kx$ converges or diverges with probability 1, according as $\sum_0^\infty |c_k|$ converges or diverges. Also, if $a_n\to\infty$ with n, then $a_n^{-1}\sum_0^{n-1}\csc 2\pi 2^kx\to 0$ in probability, i.e.,

$$\lim_{n\to\infty}\operatorname{Prob}\left\{\left|a_n^{-1}\sum_0^{n-1}\csc 2\pi 2^kx\right| > \varepsilon\right\} = 0$$

for every $\varepsilon>0$. The latter theorem depends strongly on special properties of the sine function, since it is also shown that $n^{-1/2}\sum_0^{n-1}\varphi(2^kx)$ is asymptotically normally distributed, where $(\varphi(t))^{-1}$ is a certain linear approximation to $\sin 2\pi t$.

W. J. LeVeque (Ann Arbor, Mich.)

Citations: MR 19, 646b = K10-18.

K15-52 (21# 4946)

Starčenko, L. P. Construction of sequences jointly normal with a given one. Izv. Akad. Nauk SSSR. Ser. Mat. **22** (1958), 757–770; erratum, **23** (1959), 635–636. (Russian)

A sequence of digits $\varepsilon_1, \varepsilon_2, \varepsilon_3, \cdots$ $(0\leqq\varepsilon_n<g)$ is called normal if, for every Δ_s of s digits,

$$\lim_{P\to\infty}\frac{N_P(\Delta_s)}{P} = \frac{1}{g^s},$$

where $N_P(\Delta_s)$ is the number of times that Δ_s occurs among the first $P+s-1$ digits of the sequence. This is generalized to a system of l sequences of digits $\varepsilon_1^{(k)}, \varepsilon_2^{(k)}, \varepsilon_3^{(k)}, \cdots$ $(1\leqq k\leqq l,\ 0\leqq\varepsilon_n^{(k)}<g_k)$ which may be regarded as forming a matrix of l rows and infinitely many columns. Let Δ_s be any matrix of l rows and s columns, the kth row consisting of non-negative integers less than g_k $(1\leqq k\leqq l)$, and let $N_P(\Delta_s)$ be the number of times that Δ_s occurs as a submatrix of s consecutive columns in the first $P+s-1$ columns of the infinite matrix. Then the l sequences of digits are called jointly normal if, for every Δ_s,

$$\lim_{P\to\infty}\frac{N_P(\Delta_s)}{P} = \frac{1}{(g_1g_2\cdots g_l)^s}.$$

This definition differs slightly from that given by N. M.

Korobov [same Izv. **19** (1955), 361–380; MR **17**, 590]. A sufficient condition for joint normality is that

$$\limsup_{P\to\infty} \frac{N_P(\Delta_s)}{P} < \frac{c}{(g_1 g_2 \cdots g_l)^s}$$

for some positive constant c.

Suppose that $\varepsilon_1, \varepsilon_2, \varepsilon_3, \cdots$ is a given normal sequence with base $g \geqq 2$. The author constructs an infinity of sequences $\varepsilon_1^{(k)}, \varepsilon_2^{(k)}, \varepsilon_3^{(k)}, \cdots$ $(k=1, 2, 3, \cdots)$ with the same base g such that, for any positive l, the sequences $\{\varepsilon_n\}, \{\varepsilon_n^{(1)}\}, \cdots, \{\varepsilon_n^{(l-1)}\}$ are jointly normal. A conjecture of Korobov [ibid. **14** (1950), 215–238; MR **12**, 321] that every normal sequence can be expressed as a sequence of the first digits (in the scale of g) of some completely uniformly distributed sequence is answered in the affirmative. Further, let $\{\alpha_n\}$, $\{\beta_n\}$ be two sequences of numbers in the interval [0, 1] and write $Q_j^{(2s)}$ for the point with coordinates $(\alpha_j, \alpha_{j+1}, \cdots, \alpha_{j+s-1}, \beta_j, \beta_{j+1}, \cdots, \beta_{j+s-1})$ in the $2s$-dimensional unit cube. The two sequences are said to be jointly completely uniformly distributed if, for every positive integer s, the sequence of points $Q_1^{(2s)}, Q_2^{(2s)}, Q_3^{(2s)} \cdots$ is completely uniformly distributed in the $2s$-dimensional unit cube. For a given completely uniformly distributed sequence $\{\alpha_n\}$ the author constructs a sequence $\{\beta_n\}$ such that the two sequences are jointly completely uniformly distributed.

Finally the author constructs a sequence $\{\varepsilon_n^{(2)}\}$ jointly normal with a given sequence $\{\varepsilon_n^{(1)}\}$ without the restriction $g_1 = g_2$. *R. A. Rankin* (Glasgow)

Citations: MR 12, 321a = K15-11; MR 17, 590a = K15-41.

K15-53 (21# 5705)

Volkmann, Bodo. Die Dimensionsfunktion von Punktmengen. Math. Ann. **138** (1959), 145–154.

For any set M in Euclidean n-space R_n, the Hausdorff dimension of M at the point x is defined by

$$\dim(x, M) = \lim_{\varepsilon\to 0} \dim\{M \cap K_\varepsilon(x)\},$$

where $K_\varepsilon(x)$ is the closed sphere of centre x, radius ε, and $\dim\{E\}$ denotes the Hausdorff dimension of the set E in the sense of Besicovitch. The dimension function of the set M is the real function defined for all $x \in R_n$ by $f(x) = \dim(x, M)$.

The properties of dimension functions are studied and the following proved. (i) [Theorems 1 and 2] For any set M the dimension function is strongly upper semicontinuous at each point of R_n in the sense that $f(x_0) = \limsup_{x\to x_0} f(x)$, but there exist strongly upper semicontinuous functions which are not dimension functions of any set. (ii) [Theorems 3 and 4] Any real function $f(x)$ defined for $x \in R_n$ which is piecewise continuous and strongly upper semi-continuous with $0 \leqq f(x) \leqq n$ is the dimension function of a suitable set. (iii) [Theorem 5] If T is a locally bounded transformation of E_1 onto E_2 and M_1 is a subset of E_1 with $T(M_1) = M_2$, then for every $x \in E_1$, $\dim(x, M_1) = \dim(T(x), M_2)$. Various special types of dimension function are then discussed, examples being given with a basis in number theory. *S. J. Taylor* (Birmingham)

K15-54 (22# 4694)

Cassels, J. W. S. On a problem of Steinhaus about normal numbers. Colloq. Math. **7** (1959), 95–101.

The real number ξ is called normal with respect to the integer b $(b > 1)$ as base, if in the "decimal" expansion of ξ to the base b every digit occurs with the same asymptotic frequency. H. Steinhaus [*The new Scottish book*, Wrocław 1946–1958, Problem 144, p. 14] asked whether normality with respect to infinitely many b's implies normality with respect to all other b's. The author answers this question in the negative, constructing continuously many ξ's which are non-normal with respect to 3, but normal with respect to every integer which is not a power of 3. Actually he proves the following: Let $f(x)$ be defined, for $0 \leqq x < 1$, by $f(\sum_1^\infty \varepsilon_j 2^{-j}) = \sum_1^\infty \varepsilon_j 3^{-j}$ for all sequences ε_j which only take the values 0 or 1 (and $\varepsilon_j = 0$ infinitely often). Then for almost all x the number $f(x)$ is normal with respect to every integer which is not a power of 3. And $f(x)$ is obviously not normal with respect to 3. *N. G. de Bruijn* (Eindhoven)

Referred to in K15-63, K15-75.

K15-55 (22# 4696)

Pyateckiĭ-Šapiro, I. I. On the distribution of the fractional parts of the exponential function. Moskov. Gos. Ped. Inst. Uč. Zap. **108** (1957), 317–322. (Russian)

The results of the author [Izv. Akad. Nauk SSSR Ser. Mat. **15** (1951), 47–52; MR **13**, 213] and A. G. Postnikov [Dokl. Akad. Nauk SSSR **86** (1952), 473–476; MR **14**, 359] are strengthened. Let α be a real number, $q > 1$ be an integer, $P_\nu(\alpha, \Delta)$ be the number of fractional parts $\{\alpha q^k\}$ $(k = 0, 1, \cdots, \nu = 1)$ belonging to the interval $\Delta \subset (0, 1)$, $|\Delta|$ be the length of the interval Δ, and $\varphi(t)$ be a monotone function defined on the interval [0, 1]. If

$$\liminf_{t\to 0} \varphi(t) t^{\varepsilon - 1} = 0$$

for arbitrary $\varepsilon > 0$, and if there exists a constant C such that for any interval Δ

$$(1) \qquad \limsup_{\nu\to\infty} P_\nu(\alpha, \Delta) \leqq C\varphi(|\Delta|),$$

then the sequence $\{\alpha q^k\}$ is uniformly distributed. If however $\liminf_{t\to 0} \varphi(t) t^{\varepsilon-1} > 0$ for some $\varepsilon > 0$, one may always take α such that (1) is fulfilled but the fractional parts $\{\alpha q^k\}$ are non-uniformly distributed. *I. P. Kubilyus* (RŽMat **1959** #2307)

Citations: MR 13, 213d = K15-14; MR 14, 359d = K15-26.

Referred to in K15-61, K15-78, K30-23.

K15-56 (22# 7994)

Schmidt, Wolfgang. On normal numbers. Pacific J. Math. **10** (1960), 661–672.

Let r and s be two integers greater than 1. Under what circumstances does normality of a real number ξ, $0 < \xi < 1$, in the scale of r imply normality of ξ in the scale of s? (Normality means that, for all $k \geqq 1$, each combination of k digits occurs with the proper frequency.) Let the notation $r \sim s$ stand for the fact that r and s are powers of the same integer. It is fairly obvious that, in the case $r \sim s$, normality of ξ in the scales of r and s imply each other. (The paper contains a formal proof of this fact.) If $r \nsim s$, then this implication does not hold. In fact, the author shows that in this case there are c (power of the continuum) numbers ξ which are normal in the scale of r but not even simply normal in the scale of s. (Simple normality means that each single digit occurs with the proper frequency.) He derives this result from the following theorem. Let $1 < t < s$; then almost all numbers $\xi = \sum a_n t^{-n}$, $0 < \xi < 1$, written in the scale of t, have the property that the corresponding number $\eta = \sum a_n s^{-n}$ (which is obviously not simply normal in the scale of s) is normal in every scale r for which $r \nsim s$. One of the principal tools used in the proof of this theorem is the theory of uniform distribution modulo 1. *F. Herzog* (E. Lansing, Mich.)

Referred to in K15-75, K15-90, K55-36.

K15-57 **(24# A1240)**

Šalát, Tibor

Absolut konvergente Reihen und dyadische Entwicklungen. (Czech. Russian and German summaries)

Mat.-Fyz. Časopis Slovensk. Akad. Vied. **9** (1959), 3–14.

Let $A=\sum_{n=1}^{\infty} a_n$ be a fixed convergent series of positive terms with the additional property $a_n > \sum_{k=1}^{\infty} a_{n+k}$ $(n=1, 2, 3, \cdots)$, and let W denote the set of all real numbers x representable in the form $x=\sum_{n=1}^{\infty} \varepsilon_n a_n$ where $\varepsilon_n = \pm 1$ $(n=1, 2, 3, \cdots)$. Assume that the Lebesgue measure $\mu(W)$ of W is positive (as may be assured by a suitable initial choice of the series $\sum a_n$). Let $f(n, x)$ denote the number of $+1$'s among $\varepsilon_1, \varepsilon_2, \cdots, \varepsilon_n$ corresponding to a number $x \in W$. The author proves the following analogue of Borel's famous theorem about 'normal' numbers: For almost all $x \in W$, $\lim_{n\to\infty} f(n, x)/n = \frac{1}{2}$; more precisely, $f(n, x) = \frac{1}{2}n + O(\sqrt{(n \log\log n)})$. This improves a weaker result of the author [same Časopis **7** (1957), 193–206; MR **21** #4315].

H. Halberstam (London)

K15-58 **(24# A1750)**

Billingsley, Patrick

Hausdorff dimension in probability theory.

Illinois J. Math. **4** (1960), 187–209.

Besicovitch [J. London Math. Soc. **9** (1934), 126–131] and Knichal [Mém. Soc. Roy. Sci. Bohème **1933**, no. 14, 1–18] found the fractional dimensional number of the set of binary numbers for which the relative frequency of zeros has an assigned upper limit. The reviewer [Proc. Cambridge Philos. Soc. **37** (1941), 199–228; MR **3**, 75; esp. p. 200] conjectured the dimensional number for decimals of arbitrary radix, when the relative frequencies of the digits tend to assigned limits, and the result was proved by H. G. Eggleston [Quart. J. Math. Oxford Ser. **20** (1949), 31–36; MR **11**, 88]. A connection with entropy was pointed out by the reviewer and by Shannon [*Symposium on information theory*, Ministry of Supply, London, 1950; Good, IRE Trans. IT **1953**, 170; Shannon, ibid. **1953**, 173–174]. Now let $x=(x_1, x_2, x_3, \cdots)$ be an infinite message, i.e., a discrete-time stochastic process in which each x_i is a letter of an alphabet of finite or enumerable size. The reviewer pointed out [J. Roy. Statist. Soc. Ser. B **13** (1951), 61–62] that, with the help of generalized and 'highly generalized' decimals, the messages, x, can be put into one-to-one correspondence with the points of the unit interval in such a way that the probability that the message belongs to a set, S, is equal to the Lebesgue measure of the corresponding set on the unit interval. (Generalized decimals had been defined independently and earlier by C. J. Everett [Bull. Amer. Math. Soc. **52** (1946), 861–869; MR **8**, 259] for the case of a finite alphabet, and for other purposes.) Sets of messages of zero probability can be compared by means of the dimensional numbers of the corresponding sets on the unit interval, and the dimensional number was stated by the reviewer, without proof, for the case of the set of random sequences whose digits have the 'wrong' frequencies. The author now gives, for the case of a finite alphabet, an alternative, but possibly equivalent, definition of the dimensional number of a set of messages. His definition is more abstract since he does not invoke generalized or highly generalized decimals. He proves a result for Markov processes which includes as a special case the result stated by the reviewer. His result also generalizes a theorem due to J. R. Kinney [Proc. Amer. Math. Soc. **9** (1958), 603–608; MR **20** #6157].

The author also gives a short inductive proof of a theorem due to Whittle [J. Roy. Statist. Soc. Ser. B **17** (1955), 235–242; MR **17**, 982] for the joint distribution of the doublets in a Markov chain. Another proof, using graph theory, had been given by R. B. Dawson and the reviewer [Ann. Math. Statist. **28** (1957), 946–956; MR **20** #339] and by L. A. Goodman [ibid. **29** (1958), 476–490; MR **20** #1356].

I. J. Good (Teddington)

Citations: MR 3, 75b = K50-2; MR 8, 259c = A68-2; MR 11, 88e = K15-10.

Referred to in K15-64, K15-91, K55-35.

K15-59 **(24# A1901)**

Postnikov, A. G.

The frequency with which the fractional part of an exponential function lies in a given interval. (Russian)

Uspehi Mat. Nauk **16** (1961), *no.* 3 (99), 201–205.

N. M. Korobov [Izv. Akad. Nauk SSSR Ser. Mat. **19** (1955), 361–380; MR **17**, 590] constructed a set of real numbers α such that, for large P, $N_P(\mathfrak{M}) = P|\mathfrak{M}| + O(\sqrt{P})$, where $\mathfrak{M}$ is a subinterval of $[0, 1]$ of length $|\mathfrak{M}|$ and $N_P(\mathfrak{M})$ is the number of fractional parts $\{\alpha g^x\}$ $(x=1, 2, \cdots, P)$ that lie in M, g being an integer greater than unity. He raised the question as to whether the error term could be improved. By the use of normal periodic systems, in conjunction with an earlier result of his on trigonometric sums [ibid. **20** (1956), 661–666; MR **18**, 803], the author constructs a set of real numbers α for which

$$N_P(\mathfrak{M}) = P|\mathfrak{M}| + O\{P^{1/2}(\log P)^{-1/8} \log\log P\}.$$

R. A. Rankin (Glasgow)

Citations: MR 17, 590a = K15-41; MR 18, 803c = K15-42.

Referred to in K15-61.

K15-60 **(24# A2416)**

Wouk, Arthur

On digit distributions for random variables.

J. Soc. Indust. Appl. Math. **9** (1961), 597–603.

For any positive integer $\beta > 1$ and real x, consider the expansion $x = \varepsilon \sum_{k=-\infty}^{\infty} x_k \beta^k$ $(0 \leq x_k \leq \beta - 1,\ \varepsilon = \pm 1)$, where infinite runs of $\beta - 1$ are excluded. If x is a random variable (r.v.) with distribution function (d.f.) $F(x)$, then x_k, for fixed k, is also a r.v. with β possible states. The probability distribution induced on $0, \cdots, \beta-1$ from $F(x)$ by fixing k is studied; it is explicitly computed when x is rectangular or exponential. Theorem: If $F(x)$ is absolutely continuous, $\Pr(x_k = n) \to 1/\beta$ as $k \to -\infty$ $(n=0, 1, \cdots, \beta-1)$. An estimate of the rapidity of approach to this asymptotic uniformity is given. Counter-examples are provided to show that for asymptotic uniformity, continuity and absolute continuity are not necessary, while continuity is not sufficient. For certain classes of d.f.'s., it is proved that $A_n \geqq A_{n+1}$, where $A_n = \Pr(x_k = n \text{ or } \beta - n - 1)$. The relation of these results to the problem of normal numbers is discussed.

T. V. Narayana (Bethesda, Md.)

K15-61 **(25# 1144)**

Kulikova, M. F.

Construction of a number α whose fractional parts $\{\alpha g^\nu\}$ are rapidly and uniformly distributed. (Russian)

Dokl. Akad. Nauk SSSR **143** (1962), 782–784.

By means of a lemma attributed to Pjateckiĭ-Šapiro [Moskov. Gos. Ped. Inst. Uč. Zap. **108** (1957), 317–322; MR **22** #4696] the author constructs a real number α with the property that if $\mathfrak{M}$ is a subinterval of $[0, 1)$ of length $|\mathfrak{M}|$, then

$$N_P(\mathfrak{M}) = P|\mathfrak{M}| + O\{P^{1/2}(\log P)^{-1/4}(\log\log P)^{3/2}\}.$$

Here $N_P(\mathfrak{M})$ is the number of fractional parts $\{\alpha g^x\}$ $(1 \leq x \leq P)$ lying in $\mathfrak{M}$, g being an integer greater than unity. This is an improvement on an estimate of Postnikov [Uspehi Mat. Nauk **16** (1961), no. 3 (99), 201–205; MR **24** #A1901]. *R. A. Rankin* (Glasgow)

Citations: MR 22# 4696 = K15-55; MR 24# A1901 = K15-59.

K15-62 **(25# 2055)**

Ciesielski, Z.; Kesten, H.

A limit theorem for the fractional parts of the sequence $\{2^k t\}$.

Proc. Amer. Math. Soc. **13** (1962), 596–600.

Suppose that for $0 \leqq t \leqq 1$ we define

$$x_n(t) = n^{-1/2} \sum_{i=0}^{n-1} [f_t(2^i x) - t],$$

where $f_t(x) = 1$ for $0 < x < t$, $= 0$ for $t \leqq x \leqq 1$, and is periodic with period 1. For fixed n, $\{x_n(t)\}$ can be considered a stochastic process with parameter t and sample variable $x \in [0, 1]$; the basic measure is Lebesgue. It is known [M. Kac, Ann. of Math. (2) **47** (1946), 33–49; MR **7**, 436; N. J. Fine, Proc. Amer. Math. Soc. **5** (1954), 243–252; MR **16**, 267; erratum, MR **16**, 1337] that as $n \to \infty$, the finite-dimensional distributions of the process $\{x_n(t)\}$ converge to those of a Gaussian process $\{x(t)\}$ with means 0 and covariance $\rho(s, t) = \lim_{n\to\infty} \int_0^1 x_n(t)x_n(s)dx$. The present authors apply a general "invariance theorem" of Skorohod [Teor. Verojatnost. i Primenen. **1** (1956), 289–319; MR **18**, 943] to prove that the distributions of many functionals of $x_n(t)$ converge to the corresponding things for $x(t)$. In particular, this holds for the functional sup $|x_n(t)|$, which establishes a conjecture of M. Kac. *J. W. Lamperti* (Hanover, N.H.)

Citations: MR 7, 436f = K15-7; MR 16, 267f = K15-39.

K15-63 **(25# 3902)**

Schmidt, Wolfgang M.

Über die Normalität von Zahlen zu verschiedenen Basen.

Acta Arith. **7** (1961/62), 299–309.

If r is a positive integer and ξ a positive real number, then ξ is called normal to the base r if, for every power s of r, the development of ξ in the scale of s shows every digit with the same asymptotic frequency s^{-1}. It was shown by Cassels [Colloq. Math. **7** (1959), 95–101; MR **22** #4694] that there exist numbers ξ which are normal to the base 3 but non-normal with respect to any base that is not a power of 3. This result is generalized here as follows. Let the set of all integers > 1 be split into two disjoint classes R and S, such that for every $n > 1$ all powers of n belong to the same class as n itself. Then there exists a continuum of numbers ξ which all have the property that ξ is normal with respect to every $r \in R$ and non-normal with respect to every $s \in S$. *N. G. de Bruijn* (Eindhoven)

Citations: MR 22# 4694 = K15-54.

K15-64 **(25# 4064)**

Cigler, Johann

Hausdorffsche Dimensionen spezieller Punktmengen.

Math. Z. **76** (1961), 22–30.

Given any real number in [0, 1], for a fixed positive integer $g \geqq 2$ there is a g-adic expansion $x = \sum_{\nu=1}^{\infty} x_\nu g^{-\nu}$, $0 \leqq x_\nu \leqq g-1$, which is unique if sequences $\{x_\nu\}$ for which $x_\nu = g-1$ for all $\nu \geqq N$ are disallowed. Any property of the sequence $\{x_\nu\}$ determines a subset of [0, 1]. The author considers a number of properties, each defined in terms of asymptotic frequencies of finite sequences of digits in the g-adic expansion. He obtains in each case bounds for the Hausdorff dimension of the subset of [0, 1] defined by the property. Thus he is extending the investigations of Eggleston [Quart. J. Math. Oxford Ser. **20** (1949), 31–36; MR **11**, 88] and Volkmann [Math. Z. **65** (1956), 399–413; MR **19**, 1161; ibid. **68** (1958), 439–449; MR **20** #7008] in this direction. Investigations of this nature have recently appeared more interesting because of the possible applications to information theory [see Kinney, Proc. Amer. Math. Soc. **9** (1958), 603–608; MR **20** #6157; and Billingsley, Illinois J. Math. **4** (1960), 187–209; MR **24** #A1750]. The sets considered and the new results obtained are too complicated to summarise effectively. *S. J. Taylor* (London)

Citations: MR 11, 88e = K15-10; MR 19, 1161f = K15-44; MR 20# 7008 = K15-45; MR 24# A1750 = K15-58.

K15-65 **(26# 1988)**

Pathria, R. K.

A statistical study of randomness among the first 10,000 digits of π.

Math. Comp. **16** (1962), 188–197.

Referred to in K15-80.

K15-66 **(26# 2807)**

Takahashi, Shigeru

On the distribution of values of the type $\sum f(q^k t)$.

Tôhoku Math. J. (2) **14** (1962), 233–243.

The following theorem is proved. Suppose $f(t)$ has period 1, $f \in L^2$, $\int_0^1 f(t)dt = 0$, and, for some $\varepsilon > 0$,

$$\left(\int_0^1 |f(t) - S_n(t)|^2 \, dt\right)^{1/2} = O(\log^{-1-\varepsilon} n)$$

as $n \to \infty$, where $S_n(t)$ is the nth partial sum of the Fourier series of $f(t)$. If q is any real number > 1,

$$\sigma^2 = \lim_{n\to\infty} \left(\lim_{T\to\infty} \frac{1}{2T} \int_{-T}^{T} \left| \frac{1}{\sqrt{n}} \sum_{k=0}^{n-1} f(q^k t) \right|^2 dt \right)$$

exists and, when $\sigma > 0$ and ω is any real number,

$$\lim_{n\to\infty} \mu_R\left\{t; \frac{1}{\sigma\sqrt{n}} \sum_{k=0}^{n-1} f(q^k t) \leqq \omega \right\} = \frac{1}{\sqrt{(2\pi)}} \int_{-\infty}^{\omega} e^{-u^2/2} \, du,$$

where, for any measurable set A,

$$\mu_R\{A\} = \lim_{T\to\infty} |A \cap (-T, T)|/2T.$$

For $q = 2$ the result reduces to one of M. Kac [Ann. of Math. (2) **47** (1946), 33–49; MR **7**, 436] whose hypothesis, however, is a little stronger than the one in this paper. *H. Burkill* (Sheffield)

Citations: MR 7, 436f = K15-7.

K15-67 **(26# 4384)**

Rogers, C. A.; Taylor, S. J.

On the law of the iterated logarithm.

J. London Math. Soc. **37** (1962), 145–151.

Let S_n be the sum of the first n digits in the binary expansion of x, $0 \leqq x \leqq 1$, and let $\phi(t)$ be a real-valued function. If, for almost all x, the inequality $S_n > n/2 + \phi(n)$ is satisfied for only finitely many n, the function ϕ belongs to the upper class, otherwise to the lower class. Every function ϕ belongs to one of the classes, and under certain continuity and monotonicity conditions on ϕ, there are simple growth criteria to distinguish them [cf. P. Lévy, *Théorie de l'addition des variables aléatoires*, Gauthier-Villars, Paris, 1937].

If a partial ordering on the set of functions ϕ is introduced by saying that $\phi_1 > \phi_2$ if $\phi_1(t) - \phi_2(t)$ is monotone increasing for sufficiently large t, and tends to $+\infty$ when $t \to \infty$, the authors prove that the upper [lower] class has no smallest [largest] member with respect to this ordering, and that neither class admits a cofinal sequence $\{\phi_r\}$ with respect to this ordering.

D. A. Darling (Ann Arbor, Mich.)

K15-68 **(26# 4982)**

Usol'cev, L. P.

An analogue of the Fortet-Kac theorem. (Russian)

Dokl. Akad. Nauk SSSR **137** (1961), 1315–1318.

The following theorem is proved, which is analogous to a theorem of M. Kac [Ann. of Math. (2) **47** (1946), 33–49; MR **7**, 310]. Let $f(t)$ be a real function periodic with period 1 and continuous except for a finite number of points in the interval (0, 1) and satisfying a Lipschitz condition in each of its interval of continuity. It is supposed that $\int_0^1 f(t)dt=0$. Let $g\geqq 2$ be a fixed integer; let p run over primes such that $(g,p)=1$ and $h=h(p)$ be an integral-valued function such that $h(p)\to+\infty$ for $p\to+\infty$, and $h\leqq \log p/2\log g$. Let $N_p(x)$ denote the number of integers a for which $0\leqq a\leqq p-1$ and $\sum_{k=0}^{h-1} f(ag^k/p)<x\sqrt{h}$. Then the limit

$$\lim_{p\to+\infty}\frac{1}{p}\sum_{a=0}^{p-1}\frac{1}{h}\left(\sum_{k=0}^{h-1} f\left(\frac{ag^k}{p}\right)\right)^2=\sigma^2$$

exists and one has (a) in case $\sigma\neq 0$

$$\lim_{p\to+\infty}\frac{N_p(x)}{p}=\frac{1}{\sqrt{(2\pi)}\sigma}\int_{-\infty}^{x} e^{-u^2/2\sigma^2}\,du,$$

and (b) in case $\sigma=0$

$$\lim_{p\to+\infty}\frac{N_p(x)}{p}=0 \quad \text{for } x<0,$$

$$=1 \quad \text{for } x>0.$$

The proof uses the method of moments.

A. Rényi (Budapest)

K15-69 **(26# 5122)**

Cigler, Johann; Volkmann, Bodo

Über die Häufigkeit von Zahlenfolgen mit gegebener Verteilungsfunktion.

Abh. Math. Sem. Univ. Hamburg **26** (1963), 39–54.

The authors consider sequences $s=(x_1,x_2,\cdots)$ of real numbers satisfying $0<x_j\leqq 1$ such that for each $t\in[0,1]$, $\mu(t)=\lim_{n\to\infty}N(s,t,n)/n$ exists, where $N(s,t,n)$ is the number of x_j with $x_j\leqq t$, $j\leqq n$. Such sequences are said to have the distribution function $\mu(t)$. $M(\mu)$ denotes the set of sequences whose distribution function is μ, and $G(\mu)$ is the set of sequences whose distribution function coincides with μ at all points of continuity of the latter. Clearly, $M(\mu)\subset G(\mu)\subset T$, the space of all sequences $s=(x_1,x_2,\cdots)$ with $0<x_j\leqq 1$. In order to examine the size of the sets $M(\mu)$, $G(\mu)$, the authors set up a theory of fractional dimension measures in the space T. If C denotes a finite-dimensional cylinder in T, i.e., a set of the form $i_1\times i_2\times\cdots$, where i_j is a closed interval of length z_j contained in [0, 1] and only finitely many of the z_j are different from 1, then $\|C\|$ denotes $\prod_{j=1}^{\infty} z_j$. $\mathcal{D}(\eta,j,M)$ denotes any covering $\bigcup C_i$ of the set M by cylinder sets C_i each of which satisfies $z_{j+1}=z_{j+2}=\cdots=1$, $z_i\leqq\eta$, $i=1,2,\cdots,j$. Then if $0\leqq\alpha\leqq 1$, $M\subset T$, they define the α-measure of M by $\{M\}^\alpha=\lim_{\eta\to 0}\{\lim_{j\to\infty}[\inf_{\mathcal{D}(\eta,j,M)}\sum\|C_i\|^\alpha]\}$, and $\dim M=\sup\{\alpha|\{M\}^\alpha=\infty\}$. The case $\alpha=1$ corresponds to the measure introduced by Jessen [Acta Math. **63** (1934), 249–323] in the space T. The elementary properties of these measures $\{M\}^\alpha$ are discussed, and it is then proved that if the distribution function $\mu(t)$ is continuous in [0, 1] apart from a finite number of jumps of sum h and satisfies a uniform Lipschitz condition in every interval excluding the discontinuities, then the sets $M(\mu)$ and $G(\mu)$ belonging to the distribution function μ satisfy $\dim M(\mu)=\dim G(\mu)=1-h$. A connection is obtained between $\dim M(\mu)$ and the Hausdorff dimension of the set of real numbers with given frequencies for the g-adic expansion. These sets were previously discussed by Eggleston [Proc. London Math. Soc. (2) **54** (1951), 42–93; MR **14**, 23] and in a series of papers by Volkmann [Math. Z. **68** (1958), 439–449; MR **20** #7008].

S. J. Taylor (London)

Citations: MR 14, 23c = K15-21; MR 20# 7008 = K15-45.

K15-70 **(27# 254)**

Gierl, Anton

Über das Hausdorffsche Mass gewisser Punktmengen in der Ziffertheorie.

J. Reine Angew. Math. **202** (1959), 183–195.

The Hausdorff dimension of the set of real numbers whose g-adic expansion omits a certain set of digits $\{s_1,\cdots,s_k\}\subset\{0,1,\cdots,g-1\}$, where $g\geqq 3$, $1\leqq k<g-1$, is known to be $\alpha=\log(g-k)/\log g$. In this paper the outer α-dimensional measure of these sets is investigated. Certain upper bounds, too complicated to repeat here, are obtained. These bounds both sharpen and extend results of E. Best [Proc. London Math. Soc. (2) **47** (1942), 436–454; MR **5**, 1].

E. G. Straus (Los Angeles, Calif.)

K15-71 **(28# 4558)**

Stackelberg, Olaf P.

On the law of the iterated logarithm. I, II.

Nederl. Akad. Wetensch. Proc. Ser. A **67** = *Indag. Math.* **26** (1964), 48–55, 56–67.

Denote by $f_i(x)$ the ith Rademacher function. The author proves that for almost all x

$$(1)\qquad \limsup_{N=\infty}\frac{\left|\sum_{i=1}^{N}(1-((i-1)/N))f_i(x)\right|}{\sqrt{(\frac{2}{3}N\log\log N)}}=1.$$

The author in fact proves a more general theorem, which is too long to state here, from which he deduces (1).

P. Erdős (Hamilton, Ont.)

Referred to in K50-27.

K15-72 **(28# 5300)**

Mendès France, Michel

Nombres normaux et fonctions pseudo-aléatoires.

Ann. Inst. Fourier (Grenoble) **13** (1963), *fasc.* 2, 91–104.

A pseudo-random function (fonction pseudo-aléatoire) $f(t)$ is a complex integrable function of the real positive variable t such that $\lim_{T\to\infty}(1/T)\int_0^T\overline{f(t)}f(t+h)\,dt$ exists and is continuous for $h=0$. The author establishes relations between certain such functions and the theory of normal numbers. He uses, in particular, the multiple correlation

$$\Gamma(k_1,k_2,\cdots,k_p)=\lim_{T\to\infty}\frac{1}{T}\int_0^T f(t+k_1)f(t+k_2)\cdots f(t+k_p)\,dt.$$

A central theorem is the following: Let $f(t)=r_k(x)$ where $k=[t]$ and $r_k(x)=\exp(i\pi[x\cdot 2^k])$, the Rademacher function. A necessary and sufficient condition that the real number x be normal is that $\Gamma(k_1,\cdots,k_p)=0$ for all sequences of integers $k_1,\cdots,k_p$ $(k_1<k_2<\cdots<k_p)$. The well-known result of Borel that almost all real numbers are normal reappears in this context.

H. W. Brinkmann (Swarthmore, Pa.)

Referred to in K15-81.

K15-73 **(29# 4751)**

Mendès France, Michel

Représentation des nombres réels.

C. R. Acad. Sci. Paris **258** (1964), 4643–4645.

A toute suite réelle positive $f(n)$ on associe l'application $\alpha \to \sum_{n=1}^{\infty} 2^{-n} \exp i\pi[\alpha f(n)]$ où $[x]$ représente la partie entière de x. Soit $E(f)$ l'image de $[0, 1]$ par cette application (c'est $[0, 1]$ si $f(n) = 2^n$). Théorème 1 (cas particulier): Si $f(n) = o(2^n)$, $E(f)$ est de mesure nulle; la dimension de Hausdorff de $E(f)$ est majorée par $\limsup \log^+ f(n)/(n \log 2)$. Théorème 2 (sans démonstration): S'il existe deux suites d'entiers p_n et q_n tendant vers ∞ telles que $f(p_n+q_n) = f(q_n)$, $E(f)$ est un ensemble d'unicité pour les séries trigonométriques. *J.-P. Kahane* (Orsay)

K15-74 **(29# 392)**

Gál, S.; Gál, L.

The discrepancy of the sequence $\{(2^n x)\}$.

Nederl. Akad. Wetensch. Proc. Ser. A **67** = *Indag. Math.* **26** (1964), 129–143.

Given any sequence $\mathscr{S} = (s_1, s_2, \cdots)$ of real numbers, let $\mathscr{N}(N, a_1, a_2, \mathscr{S})$ denote the number of indices i not exceeding N such that $a_1 \leqq (s_i) < a_2$. Put

$$R(N, a_1, a_2, \mathscr{S}) = \mathscr{N}(N, a_1, a_2, \mathscr{S}) - (a_2 - a_1)N$$

and define the discrepancy $D(N)$ of the sequence $\mathscr{S}$ through

$$ND(N) = \text{lub } R(N, a_1, a_2, \mathscr{S}),$$

where the lub is taken over every pair of real numbers a_1, a_2 satisfying $0 \leqq a_1 \leqq a_2 \leqq 1$. This paper is devoted to proving that if $n_1 < n_2 < \cdots$ are non-negative integers and if $D(N, x)$ is the discrepancy of the sequence $(2^{n_1}x, 2^{n_2}x, \cdots)$, then there is a constant c such that for almost all real x,

$$\limsup_{N \to \infty} ND(N, x)/\sqrt{(N \log \log N)} \leqq c.$$

This is achieved by majorising $|\int_0^1 R^p\, dx|$ using combinatorial arguments concerning products of Walsh functions, and by estimating the measure of sets of the form $\{x: R(x) \geqq y\}$. It is suggested that $c = 1$.

E. J. Akutowicz (Bologna)

Referred to in K50-33.

K15-75 **(29# 1224)**

Schmidt, Wolfgang M.

Normalität bezüglich Matrizen.

J. Reine Angew. Math. **214/215** (1964), 227–260.

Let ξ be an n-dimensional vector and A an n-dimensional matrix, with real elements in both cases. Say that ξ is normal with respect to A if the sequence $\{A^k\xi; k = 1, 2, 3, \cdots\}$ is uniformly distributed modulo 1, i.e., their images modulo 1 are uniformly distributed in the unit cube. The matrix A is said to be good if almost every vector ξ is normal with respect to A. Let

$$P(x) = (x-\lambda_1)^{a_1} \cdots (x-\lambda_r)^{a_r}$$

be the characteristic polynomial of A, with distinct λ_i. Let C_n denote the n-dimensional vector space whose components are complex numbers, and $C^{(i)}$ the subspace of C_n consisting of those vectors ζ of C_n such that $(A-\lambda_i)^{a_i}\zeta = 0$. Then C is the direct sum of the $C^{(i)}$ taken over $i = 1, \cdots, r$. By $(A-\lambda_i)C^{(i)}$ is meant the set of all $(A-\lambda_i)\zeta$ with $\zeta \in C^{(i)}$. Let $C(A)$ denote the direct sum of those $C^{(i)}$ for which $|\lambda_i| > 1$ and of those $(A-\lambda_i)C^{(i)}$ for which $|\lambda_i| = 1$. Say that a subset M of C_n lies in a rational plane if there is a non-zero n-dimensional vector u with rational integer components such that the scalar product of u with every element of M is zero. It is proved that A is good if $C(A)$ lies in no rational plane. Conversely, if $C(A)$ lies in a rational plane, then almost no ξ is normal with respect to A. A corollary to this is that A is good if all its eigenvalues exceed 1 in absolute value. This corollary is a special case of a result of W. Philipp [Arch. Math. **12** (1961), 429–433; MR **27** #3614].

Say that the matrix A is almost integral if its elements are rational numbers and its eigenvalues are algebraic integers. Under the assumption that A is almost integral, it is established that A is good if and only if neither zero nor a root of unity is among its eigenvalues. This and other results of the paper are closely related to work of V. A. Rohlin [Izv. Akad. Nauk SSSR Ser. Mat. **13** (1949), 329–340; MR **11**, 40], J. Cigler [#1223 above], and to the previously cited work of W. Philipp.

Say that an almost integral A is almost ergodic if neither zero nor a root of unity is among its eigenvalues. Write $A \sim B$ in case $A^m = B^k$ for some natural numbers m and k. Write $A \to B$ in case every vector ξ which is normal with respect to A is also normal with respect to B. Under the assumption that A and B are almost ergodic, then $A \sim B$ implies $A \to B$, and the author conjectures that the converse of the implication holds. The converse does hold in the one-dimensional case by work of J. W. S. Cassels [Colloq. Math. **7** (1959), 95–101; MR **22** #4694] and the author [Pacific J. Math. **10** (1960), 661–672; MR **22** #7994]. The converse also holds for the special class of two-dimensional matrices defined in the usual way from complex numbers. This is implied by the following result: Assuming that A and B are almost ergodic, $A \not\sim B$, $AB = BA$, and that all the eigenvalues of B exceed 1 in absolute value, then $A \not\to B$. *I. Niven* (Eugene, Ore.)

Citations: MR 22# 4694 = K15-54; MR 22# 7994 = K15-56; MR 27# 3614 = K25-22; MR 29# 1223 = K40-33.

K15-76 **(30# 2120)**

Cigler, Johann

Some applications of the individual ergodic theorem to problems in number theory.

Compositio Math. **16**, 35–43 (1964).

An expository paper. *J. R. Blum* (Albuquerque, N.M.)

K15-77 **(30# 4749)**

Volkmann, Bodo

On non-normal numbers.

Compositio Math. **16**, 186–190 (1964).

This paper is expository in nature. The author deals with the Hausdorff (or fractional) dimension of certain sets of non-normal numbers to the base $g \geqq 2$. A list of papers dealing with this subject is given, and some unsolved problems are mentioned. *F. Herzog* (E. Lansing, Mich.)

K15-78 **(31# 1241)**

Cigler, Johann

Der individuelle Ergodensatz in der Theorie der Gleichverteilung mod 1.

J. Reine Angew. Math. **205** (1960/61), 91–100.

Let X be a compact separable space, and $C(X)$ the space of functions continuous on X with the norm $\|f\| = \sup_{x \in X} |f(x)|$. The author investigates a semigroup T^k, $k > 0$, of mappings of X into X. It is shown that the set E_T of ergodic measures coincides with the set of extremal points of the closed convex set of all T-invariant measures. Theorems are proved on the uniform distribution of the fractional parts of the sequences $\{a^n x\}$ for integral a (related work is due to I. I. Pjateckiĭ-Šapiro [Izv. Akad.

Nauk SSSR Ser. Mat. **15** (1951), 47–52; MR **13**, 213; Moskov. Gos. Ped. Inst. Učen. Zap. **108** (1957), 317–322; MR **22** #4696; and (with A. G. Postnikov) Izv. Akad. Nauk SSSR Ser. Mat. **21** (1957), 729–746; MR **21** #664]; see also A. M. Polosuev [Dokl. Akad. Nauk SSSR **104** (1955), 186–189; MR **20** #6411]). Let F be a set of non-negative functions on X whose norm-closure coincides with $C(X)$. If there exists $\mu \in E_T$ and $c \geqq 1$ such that for all $f \in F$,

$$\lim (n+1)^{-1} \sum_{k=0}^{n} f(T^k x) \leqq c \int_X f(x)\mu(dx),$$

then

$$\lim (n+1)^{-1} \sum_{k=0}^{n} f(T^k x) = \int_X f(x)\mu(dx).$$

In generalisation of a theorem of J. Ville [*Étude critique de la notion de collectif*, p. 5, Gauthier-Villars, Paris, 1939] and Pjateckiĭ-Šapiro [first loc. cit.] on the distribution function of fractional parts of the exponential function, there is the following theorem: Let A be an integral $k \times k$ matrix, with $|\det A| > 1$, and, for some positive integer l, $\|A^{-l}\| < 1$, where $\| \ \|$ indicates the Hilbert norm $\|A\| = \sup_{|x| \leq 1} |Ax|$. Then a necessary and sufficient condition for the continuous function $\sigma(x)$ to be the distribution function of a sequence $\{A^n x\}$ (mod 1) is that $\int_{X_k} f(x)\,d\sigma(x) = \int_{X_k} f(Ax)\,d\sigma(x)$, for all $f \in C(X_k)$, where X_k is the k-dimensional torus.

Conditions are given which relate the uniform distribution (or completely uniform distribution, or complete distribution in measure) of a sequence $\{x_n\}$ with certain derived sequences, formed by partial sums or by differences. *N. M. Akuliničev* (RŽMat **1961** #10 B459)

Citations: MR 13, 213d = K15-14; MR 20# 6411 = K25-15; MR 21# 664 = K15-47; MR 22# 4696 = K15-55.

K15-79 (31# 1242)

Gel'fond, A. O.

On zeros of analytic functions with given arithmetic coefficients and representations of numbers. (Russian)

Acta Arith. **11** (1965), 97–114.

The zeros of the general power series with bounded integral coefficients have no simple arithmetic properties; for example, the author proves that (I) if $\alpha_1, \alpha_2, \alpha_3, \cdots$ are any real numbers and m is a positive integer such that $0 < |\alpha_k| < 1$ for all k, and $\prod_{k=1}^{n} |\alpha_k|^{-1} < m+1$ for all n, then there exists a power series $f(z) = \sum_{k=0}^{\infty} t_k z^k$ with integral coefficients such that $|t_k| \leqq m$ and $f(\alpha_k) = 0$ for all k.

Let $\{x\} = x - [x]$ be the fractional part of x. The simplest of several results by the author is as follows. (II) Let $f(z) = \sum_0^\infty a_n z^n$ and $F(z) = \sum_0^\infty c_n z^n$, where $a_0 = 1$, be real power series that converge for $|z| < 1$. Put $= \delta_0 c_0$, $\delta_n = \{c_n - \sum_{k=0}^{n-1} \delta_k a_{n-k}\}$ for $n = 1, 2, 3, \cdots$. Then the power series

$$\sum_{1}^{\infty} p_n z^n = F(z) - f(z) \sum_{0}^{\infty} \delta_n z^n$$

has integral coefficients satisfying $|p_n| < |c_n| + |c_0 a_n| + \sum_1^{n-1} |a_k|$ for all n. The author applies this simple remark to the study of the distribution (mod 1) of sequences $q^n\alpha$, and, more generally, of recursive sequences, and he obtains results that are valid for almost all α, or for almost all values of the first terms of the sequence.

K. Mahler (Canberra)

K15-80 (31# 4108)

Stoneham, R. G.

A study of 60,000 digits of the transcendental "e".

Amer. Math. Monthly **72** (1965), 483–500.

Some standard statistical tests are carried out on the first 60,000 decimal digits of e, in the hope of finding evidence of departure from "normality", but, as one would expect, little such evidence is found. (A normal decimal is one for which every k-nome has 10^{-k} as its limiting relative frequency.) {The reviewer has for many years thought that, with this motivation, it would be better to use the binary expansions rather than the decimal expansions of numbers like e, π, and $\sqrt{2}$.} The author, in common with several previous authors [M. G. Kendall and B. Babington Smith, J. Roy. Statist. Soc. **101** (1938), 147–166; ibid. Suppl. **6** (1939), 51–61; G. E. Forsythe, *Monte Carlo Method*, Nat. Bur. Standards, Appl. Math. Ser. No. 12, U.S. Government Printing Office, Washington, D.C., 1951; MR **13**, 162; and R. K. Pathria, Math. Comp. **16** (1962), 188–197; MR **26** #1988] applies the "serial test" incorrectly. In fact, let $\psi^2 = 100N^{-1} \sum_{ij} (n_{ij} - N/100)^2$, where N is the number of dinomes in a sequence, and n_{ij} is the frequency of the dinome (i, j). The error is to assume that ψ^2 has asymptotically a χ^2 distribution. The expectation of ψ^2 is 99, and not 90, which is the number of degrees of freedom since $\sum_j n_{ij} = \sum_j n_{ji}$ (see the reviewer [Proc. Cambridge Philos. Soc. **49** (1953), 276–284; MR **15**, 727], and T. E. Hull and A. R. Dobell [SIAM Rev. **4** (1962), 230–254; MR **26** #5710]). The author assumes that the expectation is 90 and does not mention that this value is exceeded in all twelve of his blocks of 5000 digits.

I. J. Good (Oxford)

Citations: MR 26# 1988 = K15-65; MR 26# 5710 = K45-11.

K15-81 (32# 4087)

France, Michel Mendès [Mendès France, Michel]

A set of nonnormal numbers.

Pacific J. Math. **15** (1965), 1165–1170.

For any number x in the unit interval which is not of the form $r/2^k$, where r and k are integers, let $e_n(x)$ be the nth digit in the binary expansion, and let $r_n(x)$ be the Rademacher function $1 - 2e_n(x)$. Let E be the set of x in the unit interval such that for some real polynomial ϕ and for some positive n_0, $r_n(x) = \exp i\pi[\phi(n)]$ for all integers $n > n_0$, where $[x]$ denotes the greatest integer function. The author proves that the numbers of E are non-normal, and that the Hausdorff dimension of E is 0. This is in contrast to the known result [W. A. Beyer, same J. **12** (1962), 35–46; MR **25** #4063] that the Hausdorff dimension of the set of non-normal numbers is 1. The proofs make use of a result of the author [Ann. Inst. Fourier (Grenoble) **13** (1963), fasc. 2, 91–104; MR **28** #5300] giving an alternative definition of a normal number in terms of $r_n(x)$.

I. Niven (Eugene, Ore.)

Citations: MR 28# 5300 = K15-72.

K15-82 (33# 2619)

Long, Calvin T.

On real numbers having normality of order k.

Pacific J. Math. **18** (1966), 155–160.

A decimal expansion to base r is said to have normality of order k if every sequence of k digits occurs with asymptotic frequency r^{-k}. The author shows that the recurring decimal whose period consists of the numbers 0 to $(r^k - 1)$ written consecutively in order as k-digit numbers is normal of order k, and makes some fairly obvious remarks about the relation between the distribution of the fractional part of $r^n x$ ($n=0, 1, 2, 3, \cdots$) and the normality of x of order k. *J. W. S. Cassels* (Cambridge, England)

K15-83 **(33# 5598)**

Ducray, S.

Normal sequences.

J. Univ. Bombay (*N.S.*) **31** (1962/63), *parts* 3 *&* 5, *sect. A*, 1–4.

Der Verfasser setzt auseinander, wie der Satz von Borel, wonach fast alle reellen Zahlen normale m-adische Zifferndarstellungen ($m \geqq 2$) besitzen, in bekannter Weise als Spezialfall des starken Gesetzes der grossen Zahlen aufgefasst werden kann, wenn man den entsprechenden Wahrscheinlichkeitsraum auf das Einheitsintervall abbildet. Sodann betrachtet er eine etwas allgemeinere Klasse von Wahrscheinlichkeitsräumen, die sich durch eine ähnliche Methode eineindeutig auf das Einheitsintervall abbilden lassen. Dabei werden statt m "Ziffern" abzählbar viele, statt gleicher Wahrscheinlichkeiten $1/m$ beliebige Werte f_i für die Wahrscheinlichkeit jeder Ziffer i zugelassen. Es wird ohne Beweis angegeben, wie sich der Satz von Borel in diesem Fall durch Verwendung des starken Gesetzes der grossen Zahlen, des Zentralen Grenzwertsatzes und des Satzes vom iterierten Logarithmus verschärfen bzw. verallgemeinern lässt.

{Die Arbeit enthält einige begriffliche Ungenauigkeiten (z.B. wird der Unterschied zwischen Mass und äusserem Mass übersehen) und Druckfehler (es muss auf S. 1 statt $f_i<0$ richtig $f_i>0$, auf S. 2 statt $P(\bar{X}_i)$ richtig $P(\bar{X}_n)$ heißen; die Intervalle müssen in der Form [,) statt (,] geschrieben werden, usw.); aber der Leser kann sie leicht beheben.} *B. Volkmann* (Stuttgart)

K15-84 **(33# 5600)**

Korobov, N. M.

Distribution of fractional parts of exponential functions. (Russian. English summary)

Vestnik Moskov. Univ. Ser. I Mat. Meh. **21** (1966), *no.* 4, 42–46.

Let q be an integer $\geqq 2$ and let $\alpha=0.\alpha_1\alpha_2\cdots\alpha_k\cdots$ ($0\leqq\alpha_k\leqq q-1$) be a number represented in the q-adic system. Denote by $N_\gamma(P)$ the number of fractional parts of the function αq^x that fall into the interval $[0,\gamma)$ when x assumes the integral values from 1 through P. The author determines the set of values of q and α for which $N_\gamma(P)=\gamma P+O((\sqrt[\beta]{P})\ln^{4/3}P)$ for $P\to\infty$, $\gamma\in(0,1]$, $\beta>\frac{1}{3}$. *G. Biriuk* (Ann Arbor, Mich.)

Referred to in K15-87.

K15-85 **(34# 1270)**

Šalát, Tibor

A remark on normal numbers.

Rev. Roumaine Math. Pures Appl. **11** (1966), 53–56.

The author proves that the set of all simply normal numbers and the set of all absolutely normal numbers are of the first Baire category, by proving the following theorem. Let g, r and n be integers, $0\leqq r<g$, $g>1$, and for $0\leqq x<1$ let $N_n(r,x)$ denote the number of occurrences of r in the first n terms of the expansion of x to base g. If $L(r,x)$ denotes the set of limit points of the sequence whose nth term is $N(r,x)/n$, then $L(r,x)=[0,1]$ for all $x\in[0,1)$ with the exception of a set of the first category. *L. C. Eggan* (Tacoma, Wash.)

Referred to in K50-38.

K15-86 **(34# 7491)**

Grasselli, Jože

Normal numbers. (Slovenian)

Obzornik Mat. Fiz. **10** (1963), 6–12.

Informal discussion of normal numbers, using the Borel and Niven-Zuckerman definitions. *T. T. Tonkov* (Leningrad)

K15-87 **(36# 2564)**

Usol'cev, L. P.

A construction problem connected with the uniform distribution of the fractional parts of exponential functions. (Russian)

Izv. Vysš. Učebn. Zaved. Matematika **1967**, *no.* 12 (67), 75–83.

In a series of papers N. M. Korobov has investigated normal periodic sequences and their applications to questions of uniform distribution. See, for example, the author [Izv. Akad. Nauk SSSR Ser. Mat. **15** (1951), 17–46; MR **13**, 213; Uspehi Mat. Nauk (N.S.) **5** (1950), no. 3 (37), 135–136; MR **12**, 321; Vestnik Moskov. Univ. Ser. I Mat. Meh. **21** (1966), no. 4, 42–46; MR **33** #5600]. In the last of these it is shown how to construct x for which the number of $x, xg, \cdots, xg^{P-1}$, where g is a prime, having fractional parts in $[0,\gamma)$ equals $\gamma P+O(P^{1/3}\ln^{4/3}P)$ as $P\to\infty$. In the present paper this construction is extended to $g=p^\lambda$, where p is a prime and λ a natural number not divisible by p. In a footnote, due to the editors, a minor correction is given, and it is indicated how the results may be extended to permit λ to be divisible by p. *J. B. Roberts* (Portland, Ore.)

Citations: MR 12, 321d = K15-12; MR 13, 213e = K15-15; MR 33# 5600 = K15-84.

K15-88 **(36# 3735)**

Mendes France, Michel

Nombres normaux. Applications aux fonctions pseudo-aléatoires.

J. Analyse Math. **20** (1967), 1–56.

Let $g\geqq 2$ be an integer, $\varphi(n)$ a polynomial with real coefficients. $\{x\}$ denotes the fractional part, $[x]$ the integral part of x. Denote by $E(g)$ the set of real numbers which can be written in the form $\sum_{n=1}^\infty [g\{\varphi(n)\}]/g^n$, where $\varphi(n)$ runs through all real polynomials. The author proves that $E(g)$ has Hausdorff dimension 0 and that it contains no normal numbers. Several other results are proved about the set $E(g)$.

Let $\theta>2$. Let $C(\theta)$ be the set of real numbers of the form $(\theta-1)\sum_{n=1}^\infty \varepsilon_n/\theta^n$, $\varepsilon_n=0$ or 1. The author also studies the distribution mod 1 of the sequence $x\theta^n$ (mod 1) $x\in C(\theta)$. The results depend on whether θ is a Pisot-Vijayaraghavan number or not. Some interesting unsolved questions are raised.

In the last chapter the author investigates the so-called pseudo-random functions of J. Bass [C. R. Acad. Sci. Paris **247** (1958), 1163–1165; MR **21** #1646]. *P. Erdős* (Budapest)

Referred to in K20-31.

K15-89 **(37# 1337)**

de Vroedt, C.

A short proof of a theorem of Tsuchikura.

Nederl. Akad. Wetensch. Proc. Ser. A **71** = *Indag. Math.* **30** (1968), 232–233.

Regarding decimal expansions (with bases $\geqq 2$) of real numbers on the unit interval as a (dependent) stochastic process, using nothing more sublime than the law of the iterated logarithm together with ordinary Lebesgue

measure, the author greatly simplifies the proof of the first theorem in a number-theoretic paper by T. Tsuchikura [Tôhoku Math. J. (2) **3** (1951), 208–211; MR **13**, 566].
A. A. Mullin (Forest Hill, Md.)

Citations: MR 13, 566c = K15-18.

K15-90 **(37# 2713)**

Colebrook, C. M.; Kemperman, J. H. B.
On non-normal numbers.
Nederl. Akad. Wetensch. Proc. Ser. A **71** = *Indag. Math.* **30** (1968), 1–11.

Let $s \geqq 2$ be an integer and ν denote a probability measure on $[0, 1]$, invariant under the transformation $T: x \to sx \bmod 1$. A real number $x \in [0, 1]$ is said to be ν-normal to the base s if the sequence $\{T^i x\}$ has the asymptotic distribution ν. The case $\nu = \lambda$, the Lebesgue measure, is well known. The most interesting consequence of a more involved theorem proved in the paper reads as follows. The bases r and s are said to be equivalent ($r \sim s$) if there exist integers m, n and $t \geqq 2$ with $r = t^m$ and $s = t^n$ (otherwise, $r \nsim s$). Then there exists a number $z \in [0, 1]$ which is ν-normal to the base s and simultaneously λ-normal to each base $r \nsim s$. The proof runs in several steps and is related to a paper of W. Schmidt [Pacific J. Math. **10** (1960), 661–672; MR **22** #7994]. *F. Schweiger* (Vienna)

Citations: MR 22# 7994 = K15-56.

K15-91 **(38# 947)**

Dutta, M.; Sen, M.
On application of information-theoretic notions in the theory of numbers.
Proc. Nat. Inst. Sci. India Part A **33** (1967), 66–73.

It is observed that Shannon's definition of entropy can be applied to the real numbers on (0, 1) based on their representation in base r, and that normal numbers have the highest entropy. The authors are evidently unaware of the concept of Hausdorff dimension and other related concepts, as discussed by P. Billingsley [Illinois J. Math. **4** (1960), 187–209; MR **24** #A1750] or J. R. Kinney [Proc. Amer. Math. Soc. **9** (1958), 603–608; MR **20** #6157].
S. W. Golomb (Los Angeles, Calif.)

Citations: MR 24# A1750 = K15-58.

K15-92 **(38# 4437)**

Kátai, I.; Mogyoródi, J.
On the distribution of digits.
Publ. Math. Debrecen **15** (1968), 57–68.

Denote by $\alpha(n)$ the sum of the digits of the K-adical representation of n (K a fixed integer >1). The authors investigate the limit distributions of $\alpha(n)$ and $\alpha(p)$, p a prime. One of their theorems runs as follows: Let $M_x = \frac{1}{2}(K-1)\log x/\log K$, $D_x^2 = (1/12)(K^2-1)\log x/\log K$, $\phi(y) = (2\pi)^{-1/2}\int_{-\infty}^{y}\exp(-\frac{1}{2}t^2)\,dt$. If $M_x(y)$ denotes the numbers of primes $p \leqq x$ with $\alpha(p) < M_x + yD_x$, then $M_x(y)/\mathrm{li}\, x = \phi(y) + O((\log\log x)^{-1/3})$ uniformly in y as $x \to \infty$, provided that the following density hypothesis holds for the zeros of Riemann's zeta function: $N(\sigma, T) < cT^{2(1-\sigma)}\log^2 T$ ($\frac{1}{2} \leqq \sigma \leqq 1$, $1 \leqq T < \infty$, c a suitable constant).

The authors prove a similar result for $\alpha(n)$ without requiring the above density hypothesis.
W. G. H. Schaal (E. Lansing, Mich.)

K15-93 **(41# 5321)**

Colebrook, C. M.
The Hausdorff dimension of certain sets of nonnormal numbers.
Michigan Math. J. **17** (1970), 103–116.

The paper generalizes the concept of s-adic normality of a real number x by considering the set $I(s)$ of all probability measures on $[0, 1]$ which are invariant under the s-adic shift operator T_s. For a given $\nu \in I(s)$, a number x is called ν-normal to the base s if the sequence $s_n x$, $n = 1, 2, \cdots$, has the distribution ν. A similar property is termed ν-regularity.

The Hausdorff dimension of the set $G(\nu, s)$ of all ν-regular numbers is shown to equal $h(\nu) = \lim_{c\to\infty} h(\nu, c)$, where $h(\nu, c) = \sum_{a=0}^{s^c-1} \nu_a(\log \nu_a)/\log(s^c)$, with

$$\nu_a = \nu([as^{-c}, (a+1)s^{-c})).$$

This result sharpens a theorem of H. G. Eggleston [Quart. J. Math. Oxford Ser. **20** (1949), 31–36; MR **11**, 88].

Furthermore, the author extends work of the reviewer [Math. Z. **68** (1958), 439–449; MR **20** #7008] on numbers x without a distribution ν. In particular, he considers an arbitrary, connected set V within the set $I(s)$ with the weak* topology and discusses the set $G(V, s)$ of all $x \in [0, 1)$ for which the sequence of "approximating" distributions $\mu_{nx}(f) = (1/n)\sum_{k=0}^{n-1} f(T_s^k x)$, $f \in C[0, 1)$, has all $\mu \in V$ (and only these) as weak* accumulation points. It is proved that the set $G(V, s)$ is never empty and has Hausdorff dimension $\inf_{\nu\in V} h(\nu)$. *B. Volkmann* (Stuttgart)

Citations: MR 11, 88e = K15-10; MR 20# 7008 = K15-45.

K15-94 **(42# 207a; 42# 207b)**

Stoneham, R. G.
On (j, ε)-normality in the rational fractions.
Acta Arith. **16** (1969/70), 221–237.

Stoneham, R. G.
A general arithmetic construction of transcendental non-Liouville normal numbers from rational fractions.
Acta Arith. **16** (1969/70), 239–253.

Sei $\mathbf{N}$ die Menge der natürlichen Zahlen. $Z \in \mathbf{N}$, $m \in \mathbf{N}$, $(Z, m) = 1$; $g \in \mathbf{N}$; $g > 1$. Sei (*) $Z/m = \sum_{\nu=1}^{\infty} b_\nu/g^\nu$; $0 \leqq b_\nu \leqq g-1$; sei $B_j = b_i b_{i+1} \cdots b_{i+j-1}$ irgendein "Block von j Ziffern"; $N(B_j, X_\lambda)$ bezeichne die Anzahl, mit welcher B_j in $b_1 b_2 \cdots b_\lambda$ vorkommt. $\omega(m) = \mathrm{Ord}_m\, g$ bezeichnet die Periodenlänge von Z/m in (*). Definition 1: Wenn $\lim_{\lambda\to\infty} |N(B_j, X_\lambda)\cdot\lambda^{-1} - g^{-j}| = |N(B_{j,g})\omega^{-1}(m) - g^{-j}| < \varepsilon$, $j, \varepsilon > 0$ vorgegeben, dann heiße Z/m "(j, ε)-normal in bezug auf g" ((j, ε)-normal in the scale g). Definition 2: x ist normal in bezug auf g (normal in the scale g), wenn $\lim_{\lambda\to\infty} N(B_j, X_\lambda)\cdot\lambda^{-1} = g^{-j}$ für $j = 1, 2, 3, \cdots$. Definition 3: Eine Folge von reellen Zahlen $0 = x_0 < x_1 < \cdots < x_n < x_{n+1} = 1$ besitzt eine ε-Gleichverteilung (uniform ε-distribution) auf $[0, 1]$, wenn für hinreichend großes n ein $\varepsilon > 0$ existiert, welches von n abhängt, und ein δ derart, daß $\max(x_{i+1} - x_i) \leqq \delta < \frac{1}{2}$ für $i = 0, 1, \cdots, n$, so daß $D_n = \sup_{0\leqq\alpha<\beta\leqq 1} |N(J)n^{-1} - (\beta - \alpha)| < \varepsilon$ für alle $\beta - \alpha > \delta$; $\alpha, \beta \in [0, 1]$; $N(J)$ ist die Anzahl der x_k ($k \leqq n$) in $[\alpha, \beta]$. Mit diesen Begriffsbildungen werden in den beiden Arbeiten eine Reihe von Sätzen bewiesen, z.B. $2 \leqq m$, $g \in \mathbf{N}$, $(m, g) = 1$. Für jedes y; $2 \leqq y \in \mathbf{N}$; $(y, g) = 1$ sei $\omega(y)$ die Ordnung von $g \bmod y$; für $n = 1, 2, \cdots$ seien a_n, Z_n zwei Folgen natürlicher Zahlen. $a_0 = Z_0 = 0$; $a_n \to \infty$ für $n \to \infty$; $1 \leqq Z_n < m^n$; $(Z_n, m) = 1$ für $n \geqq 1$. Wenn $S(n, m) = \sum_{i=1}^{n} a_i\omega(m^i)$; $S(0, m) = 0$, dann ist $x(g, m) = \sum_{n=0}^{\infty} (Z_{n+1} - mZ_n)m^{-n-1}g^{-S(n,m)}$ normal zur Basis g^t für jedes $t = 1, 2, \cdots$.
H. J. Kanold (Braunschweig)

K15-95 **(42# 3016)**

Spears, Janina Lupkiewicz; Maxfield, John E.
Further examples of normal numbers.
Publ. Math. Debrecen **16** (1969), 119–127.

The class of $\mathscr{L}$-numbers (in scale r) is defined as follows:

let c be a digit (or group of s digits) and let λ be a number whose decimal in the scale r has $o(N)$ non-c digits among the first N digits (or non-c digit groups among the first sN digits). The numbers λ constitute the $\mathscr{L}$ numbers. It is proved that if α is normal in scale r, then so is $\alpha+u\lambda+v$ for arbitrary rational numbers u, v and $\lambda\in\mathscr{L}$. This gives a continuum of normal numbers associated with α. If L is the class of Liouville numbers, it is shown that $\mathscr{L}\cap L\neq\varnothing$, $\mathscr{L}\not\subset L$, $L\not\subset\mathscr{L}$. *I. Danicic* (Aberystwyth)

K15-96 **(43# 1931)**

Polosuev, A. M.

An unimprovable estimate of a multidimensional trigonometric sum with exponential functions. (Russian. English summary)

Vestnik Moskov. Univ. Ser. I Mat. Meh. **25** (1970), *no.* 1, 9–16.

Let m and $q>1$ be integers and let $\phi(n)$ be a given function tending to infinity with n. In 1953 N. M. Korobov [Dokl. Akad. Nauk SSSR **89** (1953), 597–600; MR **15**, 15] constructed a set of values of α for which the sum $\sum_{x=1}^{n}\exp\{2\pi im\alpha q^n\}=O(\phi(n))$, this estimate being "unimprovable" in the sense that $o(\phi(n))$ cannot be replaced by $O(1)$ no matter how slowly ϕ tends to infinity. In 1955 the author [ibid. **104** (1955), 186–189; MR **17**, 588] extended this result to multiple exponential sums. In the present paper he presents an improved derivation of his result. *D. H. Lehmer* (Berkeley, Calif.)

Citations: MR 15, 15d = K15-33; MR 17, 588a = K15-40.

K15-97 **(43# 6161)**

Volkmann, Bodo

Über extreme Anormalität bei Ziffernentwicklungen.

Math. Ann. **190** (1970/71), 149–153.

Let x be a real number in $[0, 1)$, $g\geqq 2$ an integer, and write x in the g-adic system: $x=\sum_{i=1}^{\infty}e_{gi}(x)/g^i$, where the $e_{gi}(x)$ are integers in $[0, g-1]$. Let $A_{g,j}(x, n)$ be the number of integers $i\leqq n$ such that $e_{gi}(x)=j$, and put $\mathfrak{p}(x; g, n)=(A_{g,0}(x, n)/n, \cdots, A_{g,g-1}(x, n)/n)$. Clearly the points $\mathfrak{p}(x; g, n)$ lie in the simplex

$$H_g=\{\zeta=(\zeta_0, \cdots, \zeta_{g-1})\in\mathbf{R}_g|0\leqq\zeta_j\leqq 1, \textstyle\sum_{j=0}^{g-1}\zeta_j=1\}.$$

If the sequence $\mathfrak{p}(x; g, n)$ $(n=1, 2, \cdots)$ is everywhere dense in H_g, x is called g-adically extreme. If this holds for every base g, x is called absolutely extreme. The following results are proved. (1) The set E of absolutely extreme numbers has the cardinality of the continuum. (2) E has Hausdorff dimension 0. (3) E is a winning set for the Banach-Mazur interval game. (4) E is of the second category. *B. Gordon* (Los Angeles, Calif.)

K15-98 **(43# 6165)**

Korobov, N. M.

Trigonometric sums with exponential functions, and the distribution of the digits in periodic fractions. (Russian)

Mat. Zametki **8** (1970), 641–652.

Let $m\geqq a\geqq 1$, $q\geqq 2$ be integers with $(q, m)=1$, and denote by τ the order of q (mod m). Let p be a prime such that $p^2|m$, and such that q has order less than τ (mod m/p). The author proves that if $p\nmid a$, then the sum (1) $\sum_{x=1}^{\tau}\exp(2\pi iaq^x/m)$ vanishes. Next let $m=p_1^{\alpha_1}\cdots p_s^{\alpha_s}$, $m_0=p_1\cdots p_s$, and suppose q has order τ_0 (mod m_0). Put $\mu=1$ if $m\equiv 1$ (mod 2), $q\equiv 3$ (mod 4), and $\mu=0$ otherwise. Let $q^{(\mu+1)\tau_0}-1=u_0p_1^{\beta_1}\cdots p_s^{\beta_s}$, where $(u_0, m_0)=1$. It is shown that if $a\not\equiv 0$ (mod $p^{\alpha_\nu-\beta_\nu}$) for at least one ν, then (1) vanishes. Next suppose m is odd, and let g be a common primitive root of all the numbers $p_\nu^{\alpha_\nu}$ $(\nu=1, \cdots, s)$. Denote by τ_ν and t_ν the order of q (mod $p_\nu^{\alpha_\nu}$) and (mod m/p^{α_ν}) respectively. Put $n_\nu=t_\nu\,\mathrm{ind}_g\,q$, and assume that $\tau_1, \cdots, \tau_s$ are pairwise coprime. Under these conditions it is shown that (1) is equal to

$$(\tau/\varphi(m))\textstyle\sum_{x=1,(x,m)=1}^{m}\exp(2\pi i(a_1m_1x^{n_1}+\cdots+a_sm_sx^{n_s})/m),$$

where $a_1m_1+\cdots+a_sm_s\equiv a$ (mod m). Now keeping the above notation, suppose that $\alpha_\nu>\beta_\nu$ for $\nu\leqq r$ and $\alpha_\nu\leqq\beta_\nu$ for $\nu>r$. Put $m'=p_1^{\beta_1}\cdots p_r^{\beta_r}p_{r+1}^{\alpha_{r+1}}\cdots p_s^{\alpha_s}$, and let τ' be the order of q (mod m'). Define $\tau_0=\tau'/2$ if $\mu=1$ and $m\equiv 0$ (mod 4), and $\tau_0=\tau'$ otherwise. Let $\omega=2(1-1/q)$ if $m\equiv 0$ (mod 4), $\tau_0\equiv 1$ (mod 2), $q\equiv 3$ (mod 4), and $\omega=1-1/q$ otherwise. In the q-adic system one has $a/m=0.\gamma_1\gamma_2,\cdots$, where the digits γ_x satisfy $0\leqq\gamma_x<q$, and $\gamma_{x+\tau}=\gamma_x$. Using the above theorems, the author proves that if $N_\tau(\gamma)$ is the number of digits $\gamma_x=\gamma$ $(1\leqq x\leqq\tau)$, then $|N_\tau(\gamma)-\tau/q|\leqq\omega\tau_0$. On the other hand if q, τ and γ are given, one can find a fraction a/m in lowest terms such that $|N_\tau(\gamma)-\tau/q|\geqq\omega\tau_0-1$. *B. Gordon* (Los Angeles, Calif.)

K20 DISTRIBUTION (mod 1): OTHER SEQUENCES OF TYPE $\{a_k\alpha\}$

See also Section J24.

See also reviews B12-23, B28-6, B48-16, J04-26, J24-18, K05-73, K15-11, K15-21, K35-1, K35-6, K35-9, K35-15, K35-16, K35-18, K35-24, K35-25, K35-27, K35-29, K35-30, K35-32, K40-72, Z10-38.

K20-1 **(7, 506e)**

Cotlar, M., and Levi, B. Exercises on the cosine function. Math. Notae **5**, 193–214 (1945). (Spanish)

The authors consider the distribution of the values of sequences of the form $\{\cos(2\pi n_kx+2\pi\alpha_k)\}$ or equivalently of sequences of fractional parts of $n_kx+\alpha_k$. Using measure-theoretic arguments they show that, if $\{n_k\}$ is strictly increasing, the sequence is, for almost all x, dense in (0, 1). [Stronger results are known; cf., for example, Koksma, Diophantische Approximationen, Ergebnisse der Math., v. 4, no. 4, Springer, Berlin, 1936, pp. 8, 94.] The authors also discuss the frequency of the n_k for which

$$|\cos 2\pi n_{k_1}x-\cos 2\pi n_{k_2}x|$$

exceeds a given positive number. *R. P. Boas, Jr.*

K20-2 **(12, 82g)**

Hartman, S. Une généralisation d'un théorème de M. Ostrowski sur la répartition des nombres mod **1.** Ann. Soc. Polon. Math. **22** (1949), 169–172 (1950).

L'auteur démontre la généralisation suivante d'un théorème d'Ostrowski [Jber. Deutsch. Math. Verein **39**, 34–46 (1930)]. Soit I un intervalle de longueur $\leqq 1$, fermé à gauche et ouvert à droite et réduit mod 1, c'est-à-dire que tout point z de I a été remplacé par $z-[z]$. Etant donnés n nombres réels $\alpha_1, \cdots, \alpha_n$ et n tels intervalles $I_1, \cdots, I_n$, le nombre $N(\alpha_1, \cdots, \alpha_n, x, I_1, \cdots, I_n)$ pour tout entier $x\geqq 1$ désignera le nombre des nombres entiers y, $1\leqq y\leqq x$, pour lesquels simultanément $\alpha_iy-[\alpha_iy]\varepsilon I_i$ $(i=1, 2, \cdots, n)$. Théorème. (A) Étant donnés deux systèmes de nombres réels $\alpha_1, \cdots, \alpha_n$ et $\xi_1, \cdots, \xi_n$ $(0<\xi_i\leqq 1)$, où l'un au moins des ξ_i est irrationnel, et un nombre naturel x, alors: (I) Il existe deux systèmes d'intervalles $I_1', \cdots, I_n'$ et $I_1'', \cdots, I_n''$ de longueur $|I_i'| = |I_i''| = \xi_i$ $(i=1, \cdots, n)$, pour lesquels

$$N(\alpha_1, \cdots, \alpha_n, x, I_1', \cdots, I_n') > x \cdot \xi_1 \xi_2 \cdots \xi_n,$$
$$N(\alpha_1, \cdots, \alpha_n, x, I_1'', \cdots, I_n'') \leqq x \cdot \xi_1 \xi_2 \cdots \xi_n.$$

(B) Si tous les nombres $\xi_1, \xi_2, \cdots, \xi_n$ sont rationnels, on a (I), où le signe $\leqq$ peut être remplacé par $<$, ou bien on a (II) $N(\alpha_1, \cdots, \alpha_n, x, I_1, \cdots, I_n) = x \cdot \xi_1 \xi_2 \cdots \xi_n$ quels que soient les intervalles $I_1, \cdots, I_n$ de longueur $\xi_1, \cdots, \xi_n$ respectivement.

J. F. Koksma (Amsterdam).

K20-3 (13, 539b)

Koksma, J. F. On a certain integral in the theory of uniform distribution. Nederl. Akad. Wetensch., Proc. Ser. A. **54** = Indagationes Math. **13**, 285–287 (1951).

Let $\lambda_1 < \lambda_2 < \cdots$ be a sequence of positive integers, and let x be a real number. Denote by $N(\alpha, \beta, x, N)$ the number of integers $n \leqq N$ for which $\lambda_n x$ satisfies $\alpha \leqq \lambda_n x < \beta$ (mod 1). Further put

$$A(N) = \sum_{1 \leqq m, n \leqq N} \frac{(\lambda_m, \lambda_n)}{[\lambda_m, \lambda_n]},$$

where (λ_m, λ_n) denotes the greatest common divisor and $[\lambda_m, \lambda_n]$ denotes the least common multiple of λ_m and λ_n. Define further

$$R(\alpha, \beta, x, N) = N(\alpha, \beta, x, N) - (\beta - \alpha) N.$$

The author proves in a very simple way that

$$(*) \qquad \int_0^1 [R(\alpha, \beta, x, N)]^2 dx < \tfrac{1}{3} A(N).$$

Equation (*) led the author more than ten years ago to the problem of estimating $A(N)$. This problem was solved by I. S. Gál [Nieuw Arch. Wiskunde (2) **23**, 13–38 (1949); these Rev. **10**, 355]. *P. Erdös* (Aberdeen).

Citations: MR 10, 355a = B28-6.

K20-4 (16, 804i)

Salem, R. Uniform distribution and capacity of sets. Comm. Sém. Math. Univ. Lund [Medd. Lunds Univ. Mat. Sem.] Tome Supplémentaire, 193–195 (1952).

Consider the infinite sequence $\{n_k x\}$, where x is a real number between 0 and 1 and the n_k are distinct positive integers. It has been shown by H. Weyl [Math. Ann. **77**, 313–352 (1916)] that the terms of this sequence are uniformly distributed modulo 1, in the sense that the ratio

$$\rho(k) = \frac{1}{k}(e^{2\pi i h n_1 x} + \cdots + e^{2\pi i h n_k x})$$

tends to zero as $k \to \infty$ for every positive integer h, for almost all values of x between 0 and 1 in the sense of Lebesgue measure. Under the additional assumption that $\max(n_1, n_2, \cdots, n_k) = O(k^p)$ for some $p \geqq 1$, the author shows that Lebesgue measure can be replaced by the smaller "measure" of capacity with respect to a generalized potential $\int r^{-\alpha} d\mu(x)$, usually referred to as α-capacity, for any $\alpha > 1 - p^{-1}$.

It is also noted that some restriction of the above type on the rate of growth of the n_k is necessary for these sharpened results. This is proved by the construction of a sequence $\{n_k\}$, for any two numbers $c > 1$ and $\alpha < 1$, such that $\max(n_1, \cdots, n_k) = O(c^k)$ and such that the exceptional set of x between 0 and 1 for which the terms of $\{n_k x\}$ are not uniformly distributed modulo 1 has positive α-capacity.

B. Lepson (Washington, D. C.).

K20-5 (16, 1016g)

Erdös, P., and Gál, I. S. On the law of the iterated logarithm. I, II. Nederl. Akad. Wetensch. Proc. Ser. A. **58** = Indag. Math. **17**, 65–76, 77–84 (1955).

Le but de cette Note est de démontrer le résultat suivant: Soit $n_1, n_2, \cdots, n_k, \cdots$ une suite croissante de nombres positifs satisfaisant à la condition $n_{k+1}/n_k \geqq q > 1$ $(k = 1, 2, \cdots)$. Dans ces conditions

$$\limsup \frac{|\sum_{k=1}^N \exp 2\pi i n_k x|}{(N \log\log N)^{1/2}} = 1.$$

L'inégalité $\limsup \leqq 1$ avait été démontrée par Salem et Zygmund [Bull. Sci. Math. (2) **74**, 209–224 (1950); MR **12**, 605] pour le cas des n_k entiers, dans les conditions plus générales où chaque exponentielle est multipliée par un coefficient a_k satisfaisant aux conditions de Kolmogoroff:

$$A_N = \sum_1^N a_k^2 \to \infty, \quad a_n = o\left(\frac{A_N}{\log\log A_N}\right)^{1/2},$$

l'inégalité s'écrivant dans ce cas:

$$\limsup \frac{|\sum_1^N a_k \exp 2\pi i n_k x|}{(A_N \log\log A_N)^{1/2}} \leqq 1.$$

Les auteurs de la présente Note affirment, sans donner de démonstration, pouvoir étendre leur théorème au cas des coefficients quelconques, qui est plus compliqué (c'est à dire transformer l'inégalité précédente en égalité: la premier formule de la p. 67 de leur note est sans doute le résultat d'une erreur de plume). *R. Salem* (Paris).

Referred to in K50-27.

K20-6 (18, 380a)

Koksma, J. F. Sur les suites $(\lambda_n x)$ et les fonctions $g(t) \epsilon L^{(2)}$. J. Math. Pures Appl. (9) **35** (1956), 289–296.

Let $0 < \lambda_1 < \lambda_2 < \cdots$ be a sequence of integers. The author investigates conditions which would insure that for every $g(t) \epsilon L^{(2)}$ of period 1

$$(1) \qquad \frac{1}{N} \sum_{k=1}^N g(\lambda_k x) \to \int_0^1 g(t) dt,$$

holds for almost all x.

The author proved [Nederl. Akad. Wetensch., Proc. **53** (1950), 959–972; MR **12**, 86] that if $\sum_{h=2}^\infty |c_h|^2 \log h < \infty$ then (1) holds, for almost all x where the c_h are the Fourier coefficients of $g(t)$. The reviewer proved [Trans. Amer. Math. Soc. **67** (1949), 51–56; MR **11**, 375] that if $\lambda_{k+1} > (1+c)\lambda_k$ then there exists a $g(t)$ for which (1) does not hold, in fact for almost all x

$$\limsup \frac{1}{N} \sum_{k=1}^N g(\lambda_k x) = \infty.$$

But if $\lambda_{k+1} > (1+c)\lambda_k$ and $\sum_{h=2}^\infty |c_h^2| (\log\log h)^2 < \infty$, then (1) holds for almost all x. The author proved [Bull. Soc. Math. Belg. **6** (1953), 4–13; MR **16**, 682], that if $\lambda_n = n$ and $\sum_{h=2}^\infty |c_h^2| \log\log h < \infty$ then (1) holds for almost all x.

In the present note the author sharpens his result of 1950 cited above for some sequences λ_k. His results are fairly complicated, we only state here two special cases: 1) If $\lambda_n = n^k$ and $\sum_{h=2}^\infty |c_h|^2 \sum_{d^k | h} \frac{1}{d} < \infty$, then (1) holds for almost all x; 2) if $(\lambda_m, \lambda_n) = 1$ and

$$\sum_{h=2}^\infty |c_h^2| \sum_{\lambda_n | h} \frac{1}{n} < \infty,$$

then (1) holds for almost all x. Clearly there are still

many interesting and unsolved problems here.

P. Erdös (Birmingham).

Citations: MR 12, 86c = K05-15; MR 16, 682h = K05-22.

K20-7 (19, 1050b)

Erdös, P.; and Taylor, S. J. On the set of points of convergence of a lacunary trigonometric series and the equidistribution properties of related sequences. Proc. London Math. Soc. (3) **7** (1957), 598–615.

This paper consists of three parts.

I. If $0\leqq\mu_k\leqq 2\pi (k=1, 2, \cdots)$, and (n_k) is an increasing sequence of integers such that $1<t_k=n_{k+1}/n_k\leqq K<\infty$, then $\sin(n_kx-\mu_k)\to 0$ as $k\to\infty$ for at most countable x. If $1<\rho\leqq t_k\leqq K<\infty$, then $\sin(n_kx-y)\to 0$ as $k\to\infty$ for at most enumerable pairs (x, y) $(0\leqq x\leqq 2\pi,\ 0\leqq y\leqq 2\pi)$; and hence there is an at most enumerable linear set Q such that $\sum\sin(n_kx-y)$ does not converge for any x unless $y\in Q$. In the case $t_k\to\infty$, the set of x such that $\sum\sin(n_kx-\mu_k)$ converges has dimension 1.

II. If t_k is an integer for sufficiently large k and $t_k\to\infty$, then the set of x such that $\sum\sin n_kx$ converges absolutely has the power of the continuum. The condition that t_k is an integer is necessary. Further if $\sum t_k^{-1}$ converges, then for any (μ_k) $\sum\sin(n_kx-\mu_k)$ converges absolutely for x in a set of power of the continuum. Suppose that $\lambda k^\rho\leqq t_k\mu k^\rho$ for fixed λ, μ and ρ. If $0<\rho\leqq 1$, then the set of x for which $\sum\sin(n_kx-\mu_k)$ converges absolutely has zero dimension; if $\rho>1$, it has $(1-1/\rho)$ dimension.

III. It is well known that, if (n_k) is an increasing sequence of integers, then the set E of x such that the sequence (n_kx) (mod 1) is not equidistributed in (0, 1) has zero Lebesgue measure, and that if n_k is a polynomial in k with integral coefficients, then E is enumerable. If (n_k) is an increasing sequence of integers such that $n_{k+1}-n_k<$constant, then E is not necessarily enumerable, but has dimension zero. If $n_k<Ck^\rho$, C and ρ being constants, then E has dimension $\leqq 1-1/\rho$ and if $t_k\geqq\rho>1$, then E has dimension 1. For the proof of some of these theorems, Eggleston's theorem is used [same Proc. (2) **54** (1952), 42–93; MR **14**, 23]. *S. Izumi.*

Citations: MR 14, 23c = K15-21.

K20-8 (22# 10978)

Mikolás, Miklós. On a problem of Hardy and Littlewood in the theory of diophantine approximations. Publ. Math. Debrecen **7** (1960), 158–180.

The Hardy-Littlewood problem referred to is the problem of obtaining estimates for sums of the type $\sum_{j=1}^N B_r(n_jx-[n_jx])$ and their quadratic integral means, where $B_r(x)$ is the Bernoulli polynomial of degree r and the n_j are distinct positive integers. The case $r=1$ has been treated by I. S. Gál [Nieuw Arch. Wisk. (2) **23** (1949), 13–38; MR **10**, 355], by making use of the known identity

$$(1)\qquad \int_0^1 (au-[au]-\tfrac{1}{2})(bu-[bu]-\tfrac{1}{2})\,du = \frac{(a, b)}{12[a, b]},$$

where (a, b) is the g.c.d. and $[a, b]$ the l.c.m. of a and b. In a previous paper [Acta Sci. Math. Szeged **13** (1949), 93–117; MR **11**, 645] the present author has generalized (1) and indeed proved

$$(2)\qquad \int_0^1 \zeta(1-s, au-[au])\zeta(1-s, bu-[bu])\,du = 2(\Gamma(s))^2\frac{\zeta(2s)}{(2\pi)^{2s}}\left(\frac{(a, b)}{[a, b]}\right)^s \quad (\Re(s)>\tfrac{1}{2}),$$

where $\zeta(s, u)$ is the Hurwitz zeta-function.

The object of the present paper is to further generalize (2) and to study the corresponding generalized Hardy-Littlewood problem. The first main result is the following. Let

$$f_1(u)*f_2(u)\,|\,(x) = \int_0^1 f_1(x-t)f_2(t)\,dt$$

and put $Z_s(u)=(\Gamma(s))^{-1}\zeta(1-s, u-[u])$; also let s_1, s_2 be arbitrary complex numbers, α, β arbitrary integers, $\Lambda=[|\alpha|, |\beta|]$. Then if $x\neq j/\Lambda$ $(j=0, \pm 1, \pm 2, \cdots)$ and $\Re(s_1)>0$, $\Re(s_2)>0$, $\Re(s_1)+\Re(s_2)>1$, it follows that

$$Z_{s_1}(\alpha u)*Z_{s_2}(\beta u)\,|\,(x)$$
$$= \frac{|\alpha|^{s_1}|\beta|^{s_2}}{\Lambda^{s_1+s_2}}Z_{s_1+s_2}((\text{sg }\alpha)\Lambda x) \qquad (\text{sg }\alpha = \text{sg }\beta),$$
$$= \frac{|\alpha|^{s_1}|\beta|^{s_2}}{\Lambda^{s_1+s_2}\sin\pi(s_1+s_2)}[\sin\pi s_1Z_{s_1+s_2}((\text{sg }\alpha)\Lambda x) + \sin\pi s_2Z_{s_1+s_2}((\text{sg }\beta)\Lambda x)]$$
$$(\text{sg }\alpha\neq\text{sg }\beta,\ s_1+s_2\neq\text{integer}).$$

The remaining results of the paper are concerned with sums of the type $\sum_{j,k=1}^N (n_j, n_k)/n_j^{\sigma_1}n_k^{\sigma_2}$ and are of a rather complicated nature. The following special case may, however, be cited. Put $\mathfrak{A}^\rho_{\sigma_1,\sigma_2}(N)=\sum_{j,k=1}^N (j, k)^\rho/j^{\sigma_1}k^{\sigma_2}$. Then

$$\mathfrak{A}^\rho_{\sigma_1,\sigma_2}(N) = \frac{\zeta(\sigma_1)\zeta(\sigma_2)}{\zeta(\sigma_1+\sigma_2)}\zeta(\sigma_1+\sigma_2-\rho)+O(N^Q\log N),$$

where $Q=\max(\rho+1-\sigma_1-\sigma_2,\ 1-\sigma_1)$.

L. Carlitz (Durham, N.C.)

Citations: MR 10, 355a = B28-6; MR 11, 645a = M30-11.

K20-9 (27# 3619)

Davenport, H.; LeVeque, W. J.
Uniform distribution relative to a fixed sequence.
Michigan Math. J. **10** (1963), 315–319.

Let $\Delta=\{z_n\}$ be a sequence of increasing positive real numbers such that z_n tends to infinity with n. The fractional part of a positive real number t, relative to Δ, is $\langle t\rangle_\Delta=(t-z_{n-1})/(z_n-z_{n-1})$, provided $z_{n-1}\leqq t<z_n$. A sequence $\{s_n\}$ is said to be uniformly distributed modulo Δ if for each α, $0<\alpha<1$, the proportion of $s_1,\cdots,s_N$ for which $\langle s_k\rangle_\Delta<\alpha$ has the limit α as N tends to infinity. LeVeque [Pacific J. Math. **3** (1953), 757–771; MR **15**, 511] studied the uniform distribution, modulo Δ, of the sequence $\{kx\}$. In the present paper, this study is continued and the authors show that if z_n-z_{n-1} decreases with n and if $a_{k+1}-a_k>Ca_k/k$ $(C>0)$, then for almost all x, $\{a_kx\}$ is uniformly distributed modulo Δ. This improves a result in the earlier paper. The proof uses the theorem discussed in the preceding review [#3618].

D. J. Lewis (Ann Arbor, Mich.)

Citations: MR 15, 511d = K10-14; MR 27# 3618 = K05-40.
Referred to in K10-30, K10-48.

K20-10 (28# 1442)

Kahane, J.-P.; Salem, R.
Distribution modulo 1 and sets of uniqueness.
Bull. Amer. Math. Soc. **70** (1964), 259–261.

The authors call a sequence $\{u_k\}$ of real numbers badly distributed modulo 1 if there exists at least one characteristic function X of an open interval Δ in (0, 1) which, extended to have period 1, satisfies

$$\limsup k^{-1}\{X(u_1)+\cdots+X(u_k)\} < \int_0^1 X(x)\,dx = |\Delta|.$$

Let E be a subset of (0, 1) such that for some $\{n_k\}$, $n_k\uparrow\infty$, and for every $x\in E$, the sequence $\{n_kx\}$ is badly distributed modulo 1. The authors prove that in this case E is a U^* set, that is, a set such that there is no non-trivial Fourier-Stieltjes series converging to zero on the complement of E. In this connection they are led to ask whether the set of all non-normal numbers is a U^* set.

R. P. Boas, Jr. (Evanston, Ill.)

K20-11 (29# 412)

Szüsz, P.

Über die absolute Konvergenz lakunaerer trigonometrischer Reihen. (Russian summary)

Acta Math. Acad. Sci. Hungar. **12** (1961), 215–220.

Suppose K is an arbitrary given positive number. The author shows that there exists a lacunary sequence of integers $\{n_k\}$, with $n_{k+1}/n_k > K$, with the property that if $\{a_k\}$ is any non-increasing sequence of positive numbers, then $\sum |a_k \sin \pi n_k x| < \infty$ for some $x \not\equiv 0 \pmod 1$ implies $\sum a_k < \infty$. If, however, one insists on "large gaps", the analogous conclusion is false. More particularly, the author shows that if $\{n_k\}$ is any sequence of integers with the property that $n_{k+1}/n_k \to \infty$, then there exists a non-increasing sequence $\{a_k\}$ so that $\sum |a_k \sin \pi n_k x| < \infty$ is a set of x of the power of the continuum, while $\sum a_k = \infty$.

E. M. Stein (Princeton, N.J.)

K20-12 (30# 4743)

Amice, Yvette

Un théorème de finitude.

Ann. Inst. Fourier (Grenoble) **14** (1964), *fasc.* 2, 527–531.

Es wird der folgende Satz bewiesen: Sei $S = \{\lambda_n\}$ eine monoton wachsende Folge ganzer Zahlen mit positiver äusserer Dichte und sei $l < 1$. Dann ist die Menge der Zahlen $x \in [0, 1]$, für welche ein Intervall $J(x)$ existiert, dessen Länge kleiner als l ist, so daß $\lambda_n x - [\lambda_n x] \in J(x)$ gilt $(n = 1, 2, 3, \cdots)$, endlich.

J. E. Cigler (Groningen)

Referred to in K20-20.

K20-13 (30# 4747)

Kátai, I.

Zur Gleichverteilung modulo Eins.

Ann. Univ. Sci. Budapest. Eötvös Sect. Math. **7** (1964), 73–77.

Let α be a real irrational number such that for all large integers q the distance between αq and the nearest integer exceeds $q^{-1-\delta}$, where δ denotes a constant > 0. The author proves that for almost all sequences of distinct natural numbers $a_\nu \leqq N \to \infty$, the numbers αa_ν are uniformly distributed mod 1, with the remainder term $\ll N^{1/2+\delta'}$ (δ' any constant $> \delta$).

The result can be obtained much more simply. By the well known Lemmas 2 and 3 of the paper,

$$\Delta =_{\text{def}} \left| \sum_{\substack{\{\nu\alpha\} \leqq \beta, \\ \nu \leqq n}} 1 - \beta n \right| < C\left(\frac{n}{m+1} + \sum_{k=1}^{m} \frac{1}{k} \cdot \frac{1}{2\|k\alpha\|}\right),$$

and since $\|k\alpha\| > k^{-1-\delta_0}$, it follows for $m = [\sqrt{n}] - 1$ that

$$\Delta < C\left(\frac{n}{m+1} + \frac{1}{2} \sum_{k=1}^{m} k^{\delta_0}\right) < Cn^{(1+\delta_0)/2}.$$

Referred to in K20-14.

K20-14 (34# 1296)

Kátai, I.

Korrektion zu: "Zur Gleichverteilung modulo Eins".

Ann. Univ. Sci. Budapest. Eötvös Sect. Math. **9** (1966), 94.

Correction to an article in same Ann. **7** (1964), 73–77 [MR **30** #4747]. The definition of $H_A(\alpha; \beta, N)$ (p. 73) should read: Sei $H_A(\alpha; \beta, N)$ die Anzahl der Ungleichungen $\{\alpha a_\nu\} \leqq \beta$, $\nu = 1, 2, \cdots, B_A(N)$, wo α eine irrationale Zahl ist.

Citations: MR 30# 4747 = K20-13.

K20-15 (34# 160)

Gel'fond, A. O.

The distribution of fractional quantities.

J. Math. Anal. Appl. **15** (1966), 65–82.

Let $\vartheta_1, \vartheta_2, \vartheta_3, \cdots$ be a sequence of real numbers $\vartheta_n > 1$ such that $\vartheta_{n+q} = \vartheta_n$ for some integer $q \geqq 1$. Define $x_1^{(\nu)}(\alpha) = \alpha$ and, generally, $x_n^{(\nu)}(\alpha) = \vartheta_{n+\nu-2} x_{n-1}^{(\nu)}(\alpha) - [\vartheta_{n+\nu-2} x_{n-1}^{(\nu)}(\alpha)]$ $(n = 2, 3, \cdots; \nu \geqq 1; \alpha \in (0, 1))$. Then the following theorem is proved: If f is Lebesgue integrable over $[0, 1]$, then for almost all α there exists the limit

$$\lim_{N \to \infty} N^{-1} \sum_{n \leqq N} f(x_n^{(\nu)}(\alpha)) = \int_0^1 f(t) \sigma_0(t)\, dt,$$

where $\sigma_0(t)$ is explicitly determined.

J. E. Cigler (Groningen)

K20-16 (35# 4369)

Furstenberg, Harry

Disjointness in ergodic theory, minimal sets, and a problem in Diophantine approximation.

Math. Systems Theory **1** (1967), 1–49.

The approach to ergodic theory in this remarkable paper is complementary to the one developed, mainly by the Russian school, associated with numerical and group invariants. In fact, the relationship investigated here between two measure-preserving transformations (processes) and between two continuous maps (flows) is disjointness, an extreme form of non-isomorphism. The concept seems rich enough to warrant quite a few papers, and these papers will no doubt be largely stimulated by the present one. An interesting aspect of the paper, apart from the new results it contains, is the entirely novel demonstration of a number of established theorems. The paper is divided into four parts: (I) Disjoint processes; (II) Disjoint flows; (III) Properties of minimal sets; (IV) A problem in diophantine approximation. Two processes X, Y are said to be disjoint ($X \perp Y$) if whenever they are homomorphic images (factors) of the same process Z, then there is a homomorphism of Z onto $X \times Y$ which, when composed with the projections of $X \times Y$ to X, Y, yields the given homomorphisms. (The commutativity in the diagram of this definition is essential, as a quick examination of a process X which is isomorphic to $X \times X$ will reveal.) An equivalent definition insists that the inverse images of the two Borel fields be independent. The disjointness of two flows is defined similarly (but of course there is no analogous second definition). Two processes (flows) are co-prime if they have no non-trivial common factor. Disjointness implies co-primeness. Definitions are given of Bernoulli processes $\mathscr{B}$, Pinsker processes $\mathscr{P}$ (with completely positive entropy), deterministic processes $\mathscr{D}$ (with zero entropy) without reference to entropy. A particularly interesting class is the class $\mathscr{W}$ of Weyl processes, which, in view of the author's structural theorem [Amer. J. Math. **85** (1963), 477–515; MR **28** #602], is a measure-theoretic analogue of the class of distal flows. {In this connection the reviewer is a little puzzled by the omission of the condition that $\mathscr{W}$ be closed under inverse limits, for it seems that such a definition would still yield the result "Mixing processes are disjoint from Weyl processes" and would provide yet another proof of L. M. Abramov's result [Izv. Akad. Nauk SSSR Ser. Mat. **26** (1962), 513–530; MR **26** #606; translated in Amer. Math. Soc. Transl. (2) **39** (1964), 37–56; see MR **28** #3909] that processes with quasi-discrete spectrum have zero entropy.} Disjointness relations are established between the various classes, but M. S. Pinsker's result [Dokl. Akad. Nauk SSSR **133** (1960), 1025–1026; translated as Soviet Math. Dokl. **1** (1960), 937–938; MR **27** #2603] $\mathscr{P} \perp \mathscr{D}$ is not proved. Two processes with positive entropy are not disjoint, in fact are not co-prime. This is a consequence of Sinaĭ's weak isomorphism theorem, but it is good to see a proof which does not depend upon such a deep result. Part I ends with a discussion of the relationship between disjointness and a problem of filtering.

In Part II analogues of weak mixing $\mathscr{W}$ and ergodicity $\mathscr{E}$ are defined for flows. Distal flows $\mathscr{D}$ are those such that $T^{m_n}x \to z$, $T^{m_n}y \to z$ imply $x = y$. Flows T with a dense set of periodic points and such that T^n $(n \neq 0)$ is ergodic, are denoted by $\mathscr{F}$. The main results: If two flows are disjoint, one must be minimal ($\mathscr{M}$); $\mathscr{F} \perp \mathscr{M}$; $\mathscr{W} \times \mathscr{M} \subset \mathscr{E}$; $\mathscr{W} \perp \mathscr{D} \cap \mathscr{M}$.

Part III is devoted to an analysis of the smallness of minimal sets for endomorphisms of compact abelian groups, with special emphasis on the circle group and the endomorphism $Tz = z^n$ $(n \neq 0)$. In fact this endomorphism is an $\mathscr{F}$ flow and as such every minimal set is "restricted" and therefore cannot be a basis for the group. If A is a T invariant (closed) subset of the circle, then the topological entropy of T restricted to A is the Hausdorff dimension of A multiplied by $\log n$. {A reference to P. Billingsley [e.g., *Ergodic theory and information*, Wiley, New York, 1965; MR **33** #254] and other authors would have been appropriate here.}

An example of a minimal set with positive topological entropy is given (cf. F. Hahn and Y. Katznelson [Trans. Amer. Math. Soc. **126** (1967), 335–360; MR **34** #7772] for a sharper result). The main result of the final Part IV says that if Σ is a non-lacunary (multiplicative) semi-group of integers and if α is irrational, then $\{n\alpha \bmod 1 : n \in \Sigma\}$ is dense in the unit interval. *W. Parry* (Brighton)

Citations: MR 33# 254 = K02-5.

K20-17 (37# 2702)

Mendès France, M.

Deux remarques concernant l'équirépartition des suites.

Acta Arith. **14** (1967/68), 163–167.

Let φ be a real function defined on the natural numbers N and tending to infinity. It is shown that there exists a sequence of integers $(\lambda_n) \in N^N$ such that (i) $\lambda_n = O(\varphi(n))$, $n \to \infty$, and (ii) the sequence $(x\lambda_n)$ is uniformly distributed modulo 1 if and only if x is irrational. If it is also demanded that (λ_n) be nondecreasing, then (ii) fails to hold, as shown by F. Dress [#2703 below]. For each nonnegative integer n, let $n = \sum_{p=0}^{\infty} e_p(n)2^p$, $e_p(n) \in \{0, 1\}$. Let θ be any real number greater than 1 and let $f_\theta(n) = \sum_{p=0}^{\infty} e_p(n)\theta^p$. A necessary and sufficient condition that θ be a Pisot number is that the sequence $(f_\theta(n))$ not be uniformly distributed modulo 1. *B. Garrison* (San Diego, Calif.)

Referred to in K20-29, K20-30.

K20-18 (37# 2703)

Dress, F.

Sur l'équirépartition de certaines suites $(x\lambda_n)$.

Acta Arith. **14** (1967/68), 169–175.

Let $(\lambda_n) \in N^N$ be a nondecreasing sequence of integers satisfying $\lambda_n = o(\log n)$. It is shown that there does not exist a real number x such that $(x\lambda_n)$ is uniformly distributed modulo 1. The importance of the condition that (λ_n) be nondecreasing is shown by comparing this result with a theorem of M. Mendès France [see #2702 above]. It is also shown that if $f(n)$ is a real function defined on the positive integers such that $f(n)/\log n$ goes to infinity with n, then there exists a nondecreasing sequence of integers $(\lambda_n) \in N^N$ satisfying (i) $\lambda_n = O(f(n))$, and (ii) the sequence $(x\lambda_n)$ is uniformly distributed modulo 1 for any irrational number x. *B. Garrison* (San Diego, Calif.)

K20-19 (37# 5160)

Šalát, Tibor

Zu einigen Fragen der Gleichverteilung (mod 1).

Czechoslovak Math. J. **18** (**93**) (1968), 476–488.

This paper is in two parts. Part I concerns the function $f_t(x) = tq_1q_2\cdots q_x$ with t real, and $\{q_k\}$ a sequence of integers satisfying $q_k > 1$. Suppose (1) $\lim q_k = \infty$. N. M. Korobov [Izv. Akad. Nauk SSSR Ser. Mat. **14** (1950), 215–238; MR **12**, 321] has shown that $f_t(x)$ is uniformly distributed mod 1 if and only if t has the form

$$t = a_0 + \sum_{k \geq 1} [\theta_k q_k]/(q_1 q_2 \cdots q_k),$$

where a_0 is an integer and $\theta_k = \{\phi(k)\}$, the fractional part of a uniformly distributed (mod 1) function $\phi(x)$. In the present work the author relaxes condition (1) to (2) $\sum_{k=1}^{N} 1/q_k = o(N)$ as $N \to \infty$.

Part II establishes a similar result for the function $F_t(x) = \varepsilon_x(t)/q_x$. Furthermore, the author studies the structure of the set $H^* = H^*(q_1, q_2, \cdots) = \{t = \sum_{k \geq 1} \varepsilon_k(t)/(q_1 q_2 \cdots q_k) \in [0, 1) : F_t(x)$ is uniformly distributed mod 1$\}$. For any sequence of integers $\{q_k\}$, $q_k > 1$, H^* is a homogeneous (and hence of Lebesgue measure 0 or 1) Borel $G_{\delta\sigma\delta}$-set of first Baire category in $[0, 1)$. The Lebesgue measure of H^* is 1 if and only if (2) is satisfied. If $V = [0, 1) - H^*$, and (2) is satisfied, the measure of V is 0. Furthermore, if $\sum_{n \geq 1} q_n/(q_1 q_2 \cdots q_n)^\varepsilon$ converges for all $\varepsilon > 0$, then the Hausdorff dimension of V is 1. For $q_k = k+1$, $k = 1, 2, \cdots$, the set $\{t = \sum_{k \geq 1} \varepsilon_k(t)/(k+1) \in [0, 1) : \varepsilon_k(t)/(k+1)$ is uniformly distributed mod 1$\}$ has Lebesgue measure 0 and Hausdorff dimension 1.

O. P. Stackelberg (Durham, N.C.)

Citations: MR 12, 321a = K15-11.

K20-20 (38# 4439)

Kaufman, R. [Kaufman, Robert P.]

Density of integer sequences.

Proc. Amer. Math. Soc. **19** (1968), 1501–1503.

An increasing sequence $N = \{n_k\}$ of positive integers satisfies condition D if it contains a sequence of "blocks" $B_j = [u_j, v_j] \cap N$, $1 \leq u_j < v_j$, with $v_j - u_j \to \infty$ and $1 + v_j - u_j \leq |B_j|$, where $|B_j|$ denotes the number of elements in B_j. Let $N^1, \cdots, N^r$ be r sequences satisfying condition D, and for each real x define $\Delta(x) = \{(n_1 x, \cdots, n_r x) | n_1 < n_2 < \cdots < n_r,\ n_s \in N^s \text{ for } 1 \leq s \leq r\}$. Then for all but a denumerable set of real numbers x, $\Delta(x)$ is dense (modulo 2π) in R^r. For $r = 1$ this has been proved in slightly sharper form by Y. Amice [Ann. Inst. Fourier (Grenoble) **14** (1964), fasc. 2, 527–531; MR **30** #4743] and J.-P. Kahane [ibid. **14** (1964), fasc. 2, 519–526; MR **30** #4746]. A result on probability distributions established by the author is used in the proof, along with Weyl's criterion.

I. Niven (Eugene, Ore.)

Citations: MR 30# 4743 = K20-12; MR 30# 4746 = K35-21.

K20-21 (39# 136)

Mendès France, M.

Nombres transcendants et ensembles normaux.

Acta Arith. **15** (1968/69), 189–192.

Let $\Lambda = (\lambda_n)$ be an infinite sequence of real numbers. A real number x is called Λ-normal if the sequence $x\Lambda = (x\lambda_n)$ is uniformly distributed modulo 1. The set of Λ-normal numbers is denoted by $B(\Lambda)$. A set E is called a normal elementary set if there exists a sequence Λ such that $E = B(\Lambda)$, and a denumerable intersection of normal elementary sets is called a normal set. It is shown that the complement of any real algebraic field of finite degree is a normal elementary set, hence that the set of real transcendental numbers is a normal set. Some unsolved problems are mentioned. *B. Garrison* (San Diego, Calif.)

Referred to in K20-23, K20-28, K20-32.

K20-22 (39# 5485)

Mendès France, Michel

Nombres transcendants et ensembles normaux.

Séminaire Delange-Pisot-Poitou: 1967/68, *Théorie des Nombres, Fasc.* 2, *Exp.* 16, 4 *pp. Secrétariat mathématique, Paris,* 1969.

A real number x is Λ-normal, where $\Lambda = (\lambda_n)$ is a real sequence, if $x\Lambda = (x\lambda_n)$ is uniformly distributed modulo 1. The set of Λ-normal real numbers is denoted by $B(\Lambda)$. A set E is simply normal if $E = B(\Lambda)$ for some Λ. A denumerable intersection of simply normal sets is called normal. The author proves the theorem: The complement of every field of real algebraic numbers is a simply normal set. A corollary is that the set of transcendental numbers is normal. The proof is based on a reformulation of a theorem of C. Pisot ["La répartition modulo un et les nombres algébriques", Ph.D. Thesis, Univ. Paris, Paris, 1938; see Ann. Scuola Norm. Sup. Pisa (2) **7** (1938), 205–248] giving a necessary and sufficient condition for a real number greater than 1 to be a PV-number (an algebraic integer greater than 1 with all conjugates in the open unit disc) and on a theorem of the author concerning Weyl means. The paper closes with five exercises and open problems.

J. B. Roberts (Halifax, N.S.)

K20-23 (39# 5486)

Meyer, Yves

Nombres algébriques et répartition modulo 1.

C. R. Acad. Sci. Paris Sér. A-B **268** (1969), A25–A27.

In this paper, the author proves a theorem which had already been discovered by the reviewer [Acta Arith. **15** (1968/69), 189–192; MR **39** #136], namely: Let **K** be a field of real algebraic numbers of finite degree ($\mathbf{Q} \subset \mathbf{K} \subset \mathbf{R}$). There exists a sequence of real numbers $\Lambda = (\lambda_1, \lambda_2, \cdots, \lambda_n, \cdots)$ such that the sequence $x\Lambda = (x\lambda_1, x\lambda_2, \cdots, x\lambda_n, \cdots)$ is equidistributed modulo 1 if and only if x is not in **K**.

The author's proof, however, seems more powerful than the reviewer's, since by improving his technique, the author has shown, in an as yet unpublished paper ["Nombres transcendants et équiré partition modulo 1", Acta Arith. (to appear)], that there exists a sequence Λ of real numbers such that $x\Lambda$ is equidistributed modulo 1 if and only if x is transcendental. *M. Mendès France* (Talence)

Citations: MR 39# 136 = K20-21.

K20-24 (40# 7208)

Colombeau, Jean-François

Ensembles normaux et ensembles normaux au sens large.

C. R. Acad. Sci. Paris Sér. A-B **269** (1969), A270–A272.

Soit $\Lambda = (\lambda_k)_{k \geqq 1}$ une suite réelle; un réel x est dit Λ-normal si la suite $x\Lambda$ est équirépartie (mod 1). Soit $B(\Lambda)$ l'ensemble des réels Λ-normaux; un ensemble réel E est dit normal s'il existe une suite réelle Λ telle que $E = B(\Lambda)$. On appelle ensemble normal au sens large toute intersection dénombrable d'ensembles normaux. La note est consacrée à la démonstration du théorème suivant: si E est normal au sens large et si $\mathbf{R} \backslash E$ est dénombrable, alors E est normal. Pour l'établir, l'auteur utilise une partition appropriée de **N** associée à toute suite réelle positive (α_n) telle que $\sum_1^\infty \alpha_n = 1$; considérant alors une famille dénombrable de suites réelles $\Lambda^n = (\lambda_k{}^n)_{k \geqq 1}$, où n parcourt **N**, et posant $E = \bigcap_{n \geqq 1} B(\Lambda^n)$, il montre que, si on prend $\alpha_n = (1-a)a^{n-1}$ et si on choisit convenablement a dans $]0, 1[$, on peut construire une suite réelle $\Lambda_a = (\lambda_k)_{k \geqq 1}$ telle que $E = B(\Lambda_a)$. Le résultat obtenu est à rapprocher du suivant [F. Dress et M. Mendès France, Séminaire Delange-Pisot-Poitou: 1968/69, *Théorie des nombres*, Fasc. 2, Exposé 17, Secrétariat mathématique, Paris, 1969; MR **41** #1655]: si E est normal au sens large et si $E \subset \mathbf{Z}$, alors E est normal; d'autre part, il apparaît comme une conséquence d'un théorème concernant l'équirépartition dans les groupes compacts, dont l'auteur donne l'énoncé. *J. Chauvineau* (Paris)

Citations: MR 41# 1655 = K20-25.

K20-25 (41# 1655)

Dress, François; Mendès France, Michel

Caractérisation des ensembles normaux dans Z.

Séminaire Delange-Pisot-Poitou: 1968/69, *Théorie des Nombres, Fasc.* 2, *Exp.* 17, 5 *pp. Secrétariat mathématique, Paris,* 1969.

Soit $u = (u_n)_{n \geq 1}$ une suite réelle; un réel x est dit u-normal si la suite xu est équirépartie (mod 1). Soit $B(u)$ l'ensemble des réels u-normaux; un ensemble réel E est dit normal s'il existe une suite réelle u telle que $E = B(u)$. On appelle ensemble normal au sens large toute intersection dénombrable d'ensembles normaux. Notons $D(k)$ l'ensemble des diviseurs de k, si $k \in \mathbf{Z}^*$, et posons $D(0) = \varnothing$. Les auteurs démontrent le théorème suivant: pour toute partie E de $\mathbf{Z}^*$, les conditions ci-dessous sont équivalentes: (1) E est normal au sens large; (2) E est normal; (3) on a $qE \subset E$ pour tout $q \in \mathbf{Z}^*$; (4) il existe une suite entière $(k_n)_{n \geq 1}$ telle que $E = \bigcap_{n \geq 1} (\mathbf{Z}^* \backslash D(k_n))$. La preuve de ce que (4) implique (2) utilise une suggestion non publiée de D. Cantor (1968) et s'appuie sur le lemme suivant: si v désigne une suite de $[0, 1[$ et si $B(v) \cap \mathbf{Z}^* \neq \varnothing$, alors $B(v) \subset \mathbf{Z}^*$. L'équivalence entre (2) et (3) entraîne que la réunion d'une famille de parties normales de $\mathbf{Z}^*$ est une partie normale de $\mathbf{Z}^*$. On notera que les ensembles $m\mathbf{Z}^*$, où m entier $\geqq 1$, sont normaux. *J. Chauvineau* (Paris)

Referred to in K20-24, K20-26, K20-27.

K20-26 (42# 209)

Dress, F.; Mendès France, M.

Caractérisation des ensembles normaux dans *Z*.

Acta Arith. **17** (1970), 115–120.

Cet article présente les résultats concernant la caractérisation des parties normales de $\mathbf{Z}^*$ précédemment exposés par les auteurs [Séminaire Delange-Pisot-Poitou: 1968/69, *Théorie des nombres*, Fasc. 2, Exp. 17, Secrétariat mathématique, Paris, 1969; MR **41** #1655], qui donnent ici une nouvelle version de leur démonstration principale. Les auteurs signalent d'ailleurs que, depuis la rédaction de leur étude, G. Rauzy a obtenu, en toute généralité, une caractérisation des ensembles normaux (Rauzy, Bull. Soc. Math. France, à paraître). Le rapporteur ajoute qu'il résulte de cette caractérisation complète que $\mathbf{Q}^*$ est normal (Rauzy, Acta Arith., à paraître).

J. Chauvineau (Paris)

Citations: MR 41# 1655 = K20-25.

Referred to in K20-27.

K20-27 (43# 4771)

Dress, François; Mendès France, Michel

Erratum: "Caractérisation des ensembles normaux dans Z".

Séminaire Delange-Pisot-Poitou: 1969/70, *Théorie des Nombres, Fasc.* 2, *Exp.* 24, 1 *p. Secrétariat mathématique, Paris,* 1970.

The authors note an error in the lemma on p. 17.03 of their

article in Séminaire Delange-Pisot-Poitou: 1968/69, *Théorie des nombres*, Fasc. 2, Exp. 17, Secrétariat mathématique, Paris, 1969 [MR **41** #1655]. There is a correct proof of the theorem in their article in Acta Arith. **17** (1970), 115–120 [MR **42** #209].

Citations: MR 41# 1655 = K20-25; MR 42# 209 = K20-26.

K20-28 (41# 8347)

Meyer, Yves

Nombres algébriques, nombres transcendants et équirépartition modulo 1.

Acta Arith. **16** (1969/70), 347–350.

Let $\Lambda=(\lambda_n)$ be a sequence of real numbers. Let $B(\Lambda)$ be the set of those real numbers x such that the sequence $x\Lambda=(x\lambda_n)$ is equidistributed (mod 1). If E is a set of real numbers, one says that E is a normal set if there exists a Λ such that $E=B(\Lambda)$. In this article, the author proves that the set T of real transcendental numbers is a normal set, a result which was partially discovered by the reviewer [same Acta **15** (1968/69), 189–192; MR **39** #136]. The author proves in fact that for every $\varepsilon>0$, one can find a sequence $\Lambda=(\lambda_n)$ such that $|\lambda_n-n|<\varepsilon$ for every $n\in\mathbf{N}$ and such that $B(\Lambda)=T$.

Since this paper appeared, quite a lot of work concerning normal sets has been done. One should note the author's book [*Nombres de Pisot, nombres de Salem et analyse harmonique*, Springer, New York, 1970] and an article of G. Rauzy giving a complete characterization of normal sets ("Caractérisation des ensembles normaux", to appear in Bull. Soc. Math. France).

M. Mendès France (Talence)

Citations: MR 39# 136 = K20-21.
Referred to in K20-32.

K20-29 (42# 1772)

Mendès France, Michel

La réunion des ensembles normaux. (English summary)

J. Number Theory **2** (1970), 345–351.

A set $E\subset R$ is normal if there exists a sequence $\{\lambda_n\}_{n\geq 0}\subset R$ such that the sequence $\{x\lambda_n\}_{n\geq 0}$ is uniformly distributed mod 1 if and only if $x\in E$. This class of sets was introduced by the author [Acta Arith. **14** (1968), 163–167; MR **37** #2702]. The main result here states that if $\{c_k\}_{k\geq 0}\subseteq R$, $C=$set of zeros of $\prod_{k=0}^{\infty}|\cos\pi xc_k|$ and $B=\bigcap_{q=1}^{\infty}(1/q)C$, then B is a "good" normal set (i.e., one corresponding to a sequence $\{\lambda_n\}$ of a special form). Conversely, every "good" normal set is of this form. As corollaries it is shown that $\bigcup_{s=1}^{\infty}\alpha_s Z^*$ and $(\bigcap_{t=1}^{\infty}(R-\beta_t Q))\cup\bigcup_{s=1}^{\infty}\alpha_s Z^*$ are normal whenever α_s,β_t are real numbers and $\sum\alpha_t^{-2}<\infty$. However, the reviewer had trouble understanding the proof of the crucial Lemma 2. A final remark states that G. Rauzy has recently characterized normal sets. This characterization allows him to show that a countable union of normal sets is not necessarily normal, but Q^* is normal.

H. Kesten (Ithaca, N.Y.)

Citations: MR 37# 2702 = K20-17.

K20-30 (42# 1773)

Dress, François

Intersections d'ensembles normaux. (English summary)

J. Number Theory **2** (1970), 352–362.

Normal sets of real numbers were introduced by M. Mendès France [Acta Arith. **14** (1968), 163–167; MR **37** #2702] (see #1772 above for a definition). Several special results indicated that finite and countable intersections of normal sets are again normal. The author now proves this result in general. This leads to a new proof of the normality of the set of transcendental numbers and also shows the normality of the set of reals which are normal (in the sense of Borel) to every integer base. It is also shown that some more special normality properties of sets are preserved under intersections.

H. Kesten (Ithaca, N.Y.)

Citations: MR 37# 2702 = K20-17.
Referred to in K20-32.

K20-31 (42# 7605)

Lesca, J.; Mendès France, M.

Ensembles normaux.

Acta Arith. **17** (1970), 273–282.

A toute suite réelle $\Lambda=(\lambda_n)_{n\geq 1}$ on associe l'ensemble $B(\Lambda)$ des réels Λ-normaux. Soit S l'ensemble des suites croissantes d'éléments de $\mathbf{N}$. L'étude successive de cinq hypothèses appropriées sur les applications de $\mathbf{N}$ dans $\mathbf{N}$ permet aux auteurs de démontrer le théorème 1 suivant: Si φ est une application de $\mathbf{N}$ dans $\mathbf{N}$ et si $B(\varphi(\Lambda))=B(\Lambda)$ pour toute suite $\Lambda\in S$, alors il existe un entier h tel que $\varphi(n)=n+h$ pour tout entier n suffisamment grand. Soit ε_Λ la fonction caractéristique de $\Lambda\in S$ dans $\mathbf{N}$; on munit S de la mesure μ obtenue en transportant, au moyen de l'injection ε de S dans $D=\{0,1\}^{\mathbf{N}}$ ainsi définie, la mesure de Haar normalisée de D sur S. Utilisant la notion de suite complexe pseudo-aléatoire [J. Bass, Bull. Soc. Math. France **87** (1959), 1–64; MR **23** #A476; J.-P. Bertrandias, Compositio Math. **16**, 23–28 (1964); MR **30** #1115], celle de suite normale à deux valeurs et divers lemmes concernant ces suites [le deuxième auteur, J. Analyse Math. **20** (1967), 1–56; MR **36** #3735], les auteurs démontrent le théorème 2 suivant: Si φ est une application polynômiale non constante de $\mathbf{Z}$ dans $\mathbf{Z}$, alors $B(\varphi(\Lambda))=\mathbf{R}\backslash\mathbf{Q}$ pour μ-presque toutes les suites $\Lambda\in S$. On en déduit que, pour toute telle fonction φ, on a $B(\varphi(\Lambda))=B(\Lambda)$ pour μ-presque toutes les suites $\Lambda\in S$, ce qui met en évidence la finesse du résultat fourni par le théorème 1. {Dans la bibliographie, l. 4 et 6, lire (1964) au lieu de (1965).}

J. Chauvineau (Paris)

Citations: MR 23# A476 = K40-22; MR 30# 1115 = K05-44; MR 36# 3735 = K15-88.

K20-32 (43# 6164)

Rauzy, Gérard

Caractérisation des ensembles normaux.

Bull. Soc. Math. France **98** (1970), 401–414.

Let $u=(u_n)$ be an infinite sequence of real numbers. Define $B(u)$ as the set of those real x's such that the sequence (xu_n) is equidistributed (mod 1). A set B of real numbers is said to be normal if there exists a sequence u such that $B=B(u)$.

The study of normal sets was originated by the reviewer [Acta Arith. **15** (1968/69), 189–192; MR **39** #136] and several authors have since tried to characterize them [F. Dress, J. Number Theory **2** (1970), 352–362; MR **42** #1773; Y. Meyer, *Nombres de Pisot, nombres de Salem et analyse harmonique*, Springer, New York, 1970; Acta Arith. **16** (1969/70), 347–350; MR **41** #8347; A. Zame, "Normal sets, the complements of which are countable", private communication].

In this article, the author gives a complete answer to the problem, namely, a set B is normal if and only if (i) $0\notin B$; (ii) for every positive integer q, $qB\subset B$; (iii) there exist continuous real functions $f_1,f_2,\cdots$ such that $\lim_{n\to\infty}f_n(x)=0$ exactly for those x's which are elements of B. The author proves that if, in addition, $B+1=B$ then there exists a sequence $u=(u_n)$ of integers such that $B=B(u)$.

He also proves that a countable union of normal sets need not be normal. The problem whether a finite union of normal sets is normal, remains to be solved. (The problem of the intersection of normal sets was settled by Dress [loc. cit.].) *M. Mendès France* (Talence)

Citations: MR 39# 136 = K20-21; MR 41# 8347 = K20-28; MR 42# 1773 = K20-30.

K20-33 **(44# 2712)**

Meijer, H. G.

On uniform distribution of integers and uniform distribution mod 1.

Nieuw Arch. Wisk. (3) **18** (1970), 271–278.

In 1961, S. Uchiyama stated the following theorem: "If $\{b_n\}$ is a uniformly distributed sequence of integers (in the sense of I. Niven), then the sequence $\{b_n x\}$ is uniformly distributed mod 1 for almost all real x." But his proof was not correct. It is known [L. Kuipers and Uchiyama, Proc. Japan Acad. **44** (1968), 608–619; MR **39** #148] that this theorem becomes true if one replaces "uniformly distributed mod 1 (in the sense of I. I. Pjateckiĭ-Šapiro)".

In this paper the author proves that the original assertion of Uchiyama is false. He constructs a uniformly distributed sequence $\{b_n\}$ of integers such that there exists a subset V of $[0, 1[$ having positive Lebesgue measure and such that for all x in V the sequence $\{b_n x\}$ is not uniformly distributed mod 1.

The construction is based on some results on continued fractions. *F. Dress* (Talence)

Citations: MR 39# 148 = B48-16.

K20-34 **(44# 6015)**

Saltykova, Z. N.

Certain questions that are connected with the simultaneous distribution of functionally dependent random variables. (Russian. English summary)

Teor. Verojatnost. i Primenen. **16** (1971), 541–548.

Denoting by $(A)_m$ the smallest positive residue of the number A modulo m, for m a positive integer, and by $[A]_m$ the greatest integer not exceeding $(A)_m$, the author studies the limiting distribution as $s\to\infty$ of $(N+1)$-dimensional random vectors of the form $Z_M(s)=([sf_0(\eta)]_{m_0}, [sf_1(\eta)]_{m_1}, \cdots, [sf_N(\eta)]_{m_N})$, where η is a random variable uniformly distributed on the closed interval $[0, 2\pi]$, and $M=(m_0, m_1, \cdots, m_N)$ is a fixed vector with positive integral components. The random vector $Z_M(s)$ has a discrete distribution concentrated on the set Λ of lattice points (points with integer co-ordinates) in the $(N+1)$-dimensional rectangle $\Delta=\{(x_0, x_1, \cdots, x_N): 0\leq x_i<m_i, i=0, 1, \cdots, N\}$. There are $\mu=\prod_{k=0}^N m_k$ such lattice points in Λ, and the author seeks conditions on $f_0, f_1, \cdots, f_N$ which guarantee that $\lim_{s\to\infty} P\{Z_M(s)=L\}=\mu^{-1}$ for each $L\in\Lambda$, i.e., that $Z_M(s)$ is asymptotically uniformly distributed on Λ. Theorem 1 deals with the special case when each $f_k(\eta)=\lambda_k\eta$, and asserts that, for almost all $(N+1)$-tuples $\lambda=(\lambda_0, \lambda_1, \cdots, \lambda_N)$, where $\lambda_0=1$ and the λ_k's are positive and linearly independent over the field of rational numbers, the bound $|P\{Z_M(s)=L\}-\mu^{-1}|<Cs^{-1}$ holds, where C depends on N, λ, M, and L, but not on s. Theorem 2 gives necessary and sufficient conditions that $P\{Z_M(s)=L\}\to\mu^{-1}$ as $s\to\infty$ when each f_k is an analytic function of a complex variable in an ε-neighborhood of the real line segment $[0, 2\pi]$ in the complex plane.

{This article has appeared in English translation [Theor. Probability Appl. **16** (1971), 533–538].}

S. A. Book (Dominguez Hills, Calif.)

K20-35 **(44# 6616)**

Moskvin, D. A.

The distribution of fractional parts of a sequence that is more general than the exponential function. (Russian)

Izv. Vysš. Učebn. Zaved. Matematika **1970**, *no.* 12 (103), 72–77.

The author proves the following theorem, generalising a result of A. G. Postnikov [Dokl. Akad. Nauk SSSR **86** (1952), 473–476; MR **14**, 359]: Let $p_1, p_2, \cdots, p_s\geq 2$ be fixed integers that are pairwise relatively prime, let $a_1<a_2<\cdots$ be the sequence of all integers expressible in the form $a_j=p_1^{x_1}\cdots p_s^{x_s}$, $x_i=0, 1, 2, \cdots$, and suppose that for some $C>0$, $\kappa\geq 0$ and any interval $\Delta\subset(0, 1)$ with positive length, some real number α satisfies the inequality $\limsup_{N\to\infty} N^{-1}\sum_{n=0}^{N-1}\chi_\Delta(\{\alpha a_n\})\leq C|\Delta|(1-\log|\Delta|)^\kappa$; then the sequence $\{\alpha a_n\}$ is uniformly distributed modulo 1.

B. Volkmann (Stuttgart)

Citations: MR 14, 359d = K15-26.

K25 DISTRIBUTION (mod 1): $\{\alpha^k\}$, PV-NUMBERS

Note that there is a separate section, R06, for papers concerning the structure of the set of PV-numbers, of Salem numbers, etc. Most of these papers have nothing to do with distribution (mod 1); the exceptions are cross-referenced explicitly below.

See also reviews J02-11, K05-8, K05-42, K05-57, K05-77, K10-14, K15-88, K20-17, K20-18, K30-23, K35-9, K35-22, K40-9, K40-42, K40-43, K40-46, K40-55, K40-56, K40-60, K55-16, K55-18, K55-21, Q20-32, R06-21, R06-38, Z10-20, Z10-40.

K25-1 **(1, 52a)**

Erdös, Paul. On a family of symmetric Bernoulli convolutions. Amer. J. Math. **61**, 974–976 (1939).

Let a family of distribution functions $\lambda(x;a)$ be defined by the formula

$$\int_{-\infty}^{+\infty} e^{inx}d_x\lambda(x;a)=\prod_{n=0}^{\infty}\cos(a^n u)$$

(that is, $\lambda(x;a)$ is a convolution of the infinitely many Bernoulli distributions $\beta(a^{-n}x)$). In addition to the known results of Wintner and Kershner [Amer. J. Math. **57**, 576–577 and 837 (1935)], it is proved that, if $\alpha>1$ is an algebraic integer of degree m and such that all of its conjugates $\alpha_2, \cdots, \alpha_m$ satisfy the inequality $|\alpha_i|<1$, $\lambda(x;\alpha^{-1})$ is purely singular. For example, α may be the "Fibonacci" number $\frac{1}{2}(5^{1/2}-1)$ or the positive root of the equation $a^3+a^2-1=0$.

M. Kac (Ithaca, N. Y.).

Referred to in K25-2.

K25-2 **(1, 139e)**

Erdös, Paul. On the smoothness properties of a family of Bernoulli convolutions. Amer. J. Math. **62**, 180–186 (1940).

Continuing a former investigation [these Rev. **1**, 52 (1940)], where further references are given, the author proves that for every integer m there exists a positive $\delta(m)$ such that the set of points a of the interval $1<a<1+\delta(m)$, for which

$$L(u;\sigma_a)=\prod\cos(u/a^n)=o(|u|^{-m})$$

does not hold, is of measure zero. In other words, for almost every a in the interval $(1, 1+\delta(m))$ the distribution function $\sigma_a(x)$ whose Fourier-Stieltjes transform is $L(u;\sigma_a)$ possesses a continuous derivative of order $m-1$. The proof is based on a quite deep diophantine analysis suggested partially by the work of Ch. Pisot [La répartition modulo un et les nombres algébriques, Ann. Scuola Norm. Super. Pisa (2) 7, 238]. *M. Kac* (Ithaca, N. Y.).

Citations: MR 1, 52a = K25-1.

K25-3 **(2, 33e)**

Vijayaraghavan, T. On the fractional parts of the powers of a number. I. J. London Math. Soc. **15**, 159–160 (1940).

This paper contains a proof due to the author and an alternative (somewhat simpler) proof due to A. Weil for the following theorem: If $\theta>1$ is rational, then there are infinitely many points of accumulation of the fractional parts of θ^n, $n=1, 2, \cdots$. *W. Feller* (Providence, R. I.).

Referred to in K25-6.

K25-4 **(3, 167h)**

Gelfond, A. On fractional parts of linear combinations of polynomials and exponential functions. Rec. Math. [Mat. Sbornik] N.S. **9** (**51**), 721–726 (1941). (Russian. English summary)

Let

$$f(n)=\sum_{k=1}^{p} P_k(n)\alpha_k^n,$$

where the $P_k(z)$ are polynomials in z, and where the α_k are constants, all distinct, $|\alpha_k|>1$, $k=1, 2, \cdots, p$. The author supposes that the α_k and coefficients of the polynomials are (in general) complex numbers, and defines d_n to be the distance of $f(n)$ from the nearest integer of the Gaussian corpus. He proves (among other things) that there is a constant λ (depending only on the moduli of the α_k, the moduli of the coefficients of the P_k and the degrees of the polynomials) such that, if $d_n<\lambda n^{-\frac{1}{2}}$ for $n\geqq n_0$, then all α_k are algebraic integers; moreover, there is then a constant $\gamma>1$ for which $d_n<\gamma^{-n}$, $n\geqq n_1$. *D. C. Spencer.*

K25-5 **(3, 274c)**

Vijayaraghavan, T. On the fractional parts of the powers of a number. II. Proc. Cambridge Philos. Soc. **37**, 349–357 (1941).

Let $G(\theta)$ denote the limit points of the fractional parts of θ^n $(n=1, 2, \cdots)$. Let S denote the set of all algebraic integers θ whose conjugates $\theta, \theta_1, \cdots, \theta_k$ satisfy $|\theta_i|<1$ $(i=1, \cdots, k)$. Since $\theta^n+\sum\theta_i^n$ is an integer for every n, $G(\theta)$ has at most the two points 0 and 1; if θ is irrational the point 1 must belong to $G(\theta)$. The set S contains algebraic integers of all degrees. However, there exist algebraic integers for which $G(\theta)$ consists of the entire interval (0, 1). The principal result is: if θ is a real algebraic number not less than 1, and if $G(\theta)$ consists of a finite number of elements, then θ is an algebraic integer, and if θ is irrational, θ belongs to S. More generally, if the limit points of the fractional parts of $u_n=d_1\xi_1^n+\cdots+d_l\xi_l^n$ $(n=1, 2, 3, \cdots)$ are finite in number, the d_j being nonzero constants and the ξ_j distinct algebraic numbers numerically greater than 1, then the ξ_j are algebraic integers, and any conjugates of any of the ξ_j not among them are numerically less than 1. The proof uses a theorem of Fatou and Hurwitz [Math. Ann. **77**, 510–512 (1916)] on recurring sequences of rational integers, and several deep-lying results from algebra and analysis. An extension to nonreal algebraic numbers is indicated at the end. *G. Pall* (Montreal, Que.).

Referred to in K25-6, K25-7, K25-9, K25-11, K40-46, Q20-11, R06-1.

K25-6 **(4, 266c)**

Rédei, L. Zu einem Approximationssatz von Koksma. Math. Z. **48**, 500–502 (1942).

The purpose of this paper is to prove (1) that, if α is an algebraic integer whose conjugates have all moduli less than 1, α^n tends to zero (mod 1) as $n\to\infty$; (2) that the fractional part of $(p/q)^n$ $(p>q>1$, p/q irreducible) has an infinite number of points of accumulation. Both results are known. [See Pisot, Ann. Scuola Norm. Super. Pisa (2) **7**, 205–248 (1938); Vijayaraghavan, J. London Math. Soc. **15**, 159–160 (1940); Proc. Cambridge Philos. Soc. **37**, 349–357 (1941); these Rev. **2**, 33; **3**, 274.] *R. Salem.*

Citations: MR 2, 33e = K25-3; MR 3, 274c = K25-5.

K25-7 **(5, 35f)**

Vijayaraghavan, T. On the fractional parts of the powers of a number. III. J. London Math. Soc. **17**, 137–138 (1942).

Let $\theta>1$ and assume that the fractional part of θ^n $(n=1, 2, \cdots)$ has only a finite number of limit points. It is conjectured that any such θ is algebraic [Proc. Cambridge Philos. Soc. **37**, 349–357 (1941); these Rev. **3**, 274] and it is proved that their number is denumerable. The limit points can be enclosed in one of a denumerable sequence of sets T of intervals of maximum length δ; let Δ be the minimum interval of the complement of T in (0, 1). It can be assumed that $\Delta/\delta>4m+4$, where $m=[\theta]$, and that θ^n belongs to T if $n\geqq q$. If now θ' has the same property, $[\theta']=m$ and $0<\theta-\theta'\leqq\delta q^{-1}(m+1)^{1-q}$, then $\theta^q-\theta'^q<\delta$. Let N be the least positive integer for which $\theta^{N+1}-\theta'^{N+1}>2\delta$, whence $N\geqq q$; then $\theta^{N+1}-\theta'^{N+1}>\Delta>(4m+4)\delta>(\theta+\theta')\cdot 2\delta \geqq(\theta+\theta')(\theta^N-\theta'^N)>\theta^{N+1}-\theta'^{N+1}$, a contradiction. Hence $\theta-\theta'>\delta_q^{-1}(m+1)^{1-q}$; and the number of all θ's is denumerable. *G. Pall* (Montreal, Que.).

Citations: MR 3, 274c = K25-5.

Referred to in K25-12, K40-46.

K25-8 **(8, 6c)**

Rosenblatt, Alfred. On the natural diatonic scales. I, II. Actas Acad. Ci. Lima **8**, 165–182, 183–196 (1945). (Spanish)

The tones in an octave of the natural diatonic scale correspond to the numbers 1, 9/8, 5/4, 4/3, 3/2, 5/3, 15/8, 2. It is easy to express these in terms of the numbers $\alpha=9/8$, $\beta=10/9$, $\gamma=16/15$, which correspond to a "major tone," a "minor tone" and a "diatonic semi-tone," respectively. The numbers corresponding to all the natural tones of all the natural ascending scales (in all octaves) form a multiplicative set G_1. It is shown that this set is uniformly dense in the sense of Weyl and various sets of generators of G_1 are given. The tones generated by those corresponding to α, β, γ in all the octaves are also considered; the numbers corresponding to these form a set G_2 which contains G_1 as a proper subset. The characterization of G_1 as a subset of G_2 is given in a simple manner. *H. W. Brinkmann.*

K25-9 **(8, 194c)**

Pisot, Charles. Répartition (mod 1) **des puissances successives des nombres réels.** Comment. Math. Helv. **19**, 153–160 (1946).

Let $\alpha>1$ and λ be real, let $(\lambda\alpha^n)$ be the fractional part of $\lambda\alpha^n$ $(n=0, 1, 2, \cdots)$ and let E be the set of limit points of $(\lambda\alpha^n)$. The paper consists partly in the exposition of the following known results, some of them due to the author [Ann. Scuola Norm. Super. Pisa (2) **7**, 205–248 (1938)], some due to Vijayaraghavan [Proc. Cambridge Philos. Soc. **37**, 349–357 (1941); these Rev. **3**, 274]. (I) The set of all numbers α and λ such that E consists of a finite number of points (briefly: E is finite) is denumerable. (II) If α is algebraic, necessary and sufficient conditions for E to be finite are that α is an algebraic integer whose conjugates have their moduli all less than 1, and that λ is an algebraic number of the field of α. In addition the following theorem is proved. (III) If E is finite and k denotes the number of irrational numbers in E and if the convergence of $(\lambda\alpha^n)$ towards its limit points is $o(n^{-k-1})$, then α is algebraic and therefore α and λ satisfy the conditions of (II). The proof of (III) is based on a method already used by the author in a previous paper on criteria for algebraic numbers [see these Rev. 4, 266]. *R. Salem* (Cambridge, Mass.).

Citations: MR 3, 274c = K25-5; MR 4, 266d = J68-4.
Referred to in K25-10, K25-25, K25-31, K40-46.

K25-10 (13, 116g)

Pisot, Charles. Quelques résultats d'approximation diophantienne. Algèbre et Théorie des Nombres. Colloques Internationaux du Centre National de la Recherche Scientifique, no. 24, pp. 57–58. Centre National de la Recherche Scientifique, Paris, 1950.

This is an expository article outlining results which have been published in greater detail in the author's previous papers [e.g., Comment. Math. Helv. **19**, 153–160 (1946); these Rev. **8**, 194]. *R. Salem* (Cambridge, Mass.).

Citations: MR 8, 194c = K25-9.

K25-11 (9, 271f)

Rédei, L. Über eine diophantische Approximation im bereich der algebraischen Zahlen. Math. Naturwiss. Anz. Ungar. Akad. Wiss. **61**, 460–470 (1942). (Hungarian. German summary)

The author proves the following theorem. Let α be a real algebraic number, $|\alpha|>1$. A necessary and sufficient condition that the sequence $\alpha^n-[\alpha^n]$ should converge is that α is an algebraic integer and all conjugates of α are less than 1 in absolute value. If this condition is satisfied, then $\lim_{n\to\infty}(\alpha^n-[\alpha^n])=0$. [The same result was proved by Vijayaraghavan, Proc. Cambridge Philos. Soc. **37**, 349–357 (1941); these Rev. **3**, 274.] *P. Erdös* (Syracuse, N. Y.).

Citations: MR 3, 274c = K25-5.

K25-12 (10, 433b)

Vijayaraghavan, T. On the fractional parts of powers of a number. IV. J. Indian Math. Soc. (N.S.) **12**, 33–39 (1948).

[For part III cf. J. London Math. Soc. **17**, 137–138 (1942); these Rev. **5**, 35.] Let θ be a real number greater than 1. Let $G(\theta)$ denote the set of limit points of the set of fractional parts of the numbers $\theta, \theta^2, \theta^3, \cdots, \theta^n, \cdots$. The author proves, as a consequence of a more general result, that if E denotes the set of the numbers θ for which $G(\theta)$ is not the entire interval $(0, 1)$, then E has the power of the continuum. (It is known that E has Lebesgue measure zero.) He proves also that the intersection of E and any interval (a, b) $(1<a<b)$ has the power of the continuum. *R. Salem* (Cambridge, Mass.).

Citations: MR 5, 35f = K25-7.

K25-13 (15, 293e)

LeVeque, W. J. The distribution modulo 1 of trigonometric sequences. Duke Math. J. **20**, 367–374 (1953).

In an earlier paper by the author [Michigan Math. J. **1**, 139–162 (1953); these Rev. **14**, 1067] there is contained the following result: If $X=(x_1, \cdots, x_r)$ is an r-dimensional continuous variable, if R is a set having positive r-dimensional Lebesgue measure, and if $\{f_n(X)\}$ is a sequence of real-valued functions, such that

$$(1)\quad \left|\int_R \exp\,(i(f_n(X)-f_m(X)))dX\right| \leq \frac{C}{\max\,(1, |n-m|^\epsilon)}$$

where $C>0$ and $\epsilon>0$ are constants, then $\{f_n(X)\}$ is uniformly distributed (mod 1) for almost all $X \in R$. Moreover, we have for each fixed integer $\beta\neq 0$

$$\begin{aligned}\sum_{n=1}^{N} \exp\,(2\pi i\beta f_n(X)) &= o(N^{1-\frac{1}{2}\epsilon+\theta}), && \text{if } 0<\epsilon<1,\\ &= o(N^{\frac{1}{2}}\log^{5/2+\theta} N), && \text{if } \epsilon=1,\\ &= o(N^{\frac{1}{2}}\log^{3/2+\theta} N), && \text{if } \epsilon>1,\end{aligned}$$

θ denoting an arbitrary positive constant. In the present paper some cases are considered in which (1) is fulfilled. If one wants to show that the sequence $\{z^n\}$ for complex $z=re^{ix}$ is uniformly distributed (mod 1), one is lead to the question whether for constant values of B and α the sequence $\{Br^n\cos(nx-\alpha)\}$ (for real x) is uniformly distributed (mod 1). Therefore, more generally, the author investigates (1) for $f_n(x)=u_n\phi(v_nx-\alpha)$, R being the segment $0\leq x\leq\omega$, where $\{u_n\}$ and $\{v_n\}$ are sequences of real numbers and where ϕ denotes a periodic function of period ω satisfying some conditions of regularity. Several applications: $\{z^n\}$ is uniformly distributed for almost all complex z with $|z|>1$; if $\{u_n\}$ is an increasing sequence of positive integers, then $\{u_n\cos(u_nx-\alpha)\}$ is uniformly distributed (mod 1) for almost all x in $(0, 2\pi)$; etc. *J. F. Koksma.*

Citations: MR 14, 1067d = K30-10.

K25-14 (15, 511a)

Polosuev, A. M. On a problem concerned with a uniform distribution of a system of functions. Dokl. Akad. Nauk SSSR **122** (1958), 346–348. (Russian)

Let $f_1(x), f_2(x), \cdots, f_s(x)$ be polynomials with integral coefficients and not identically zero. A method of Korobov [Izv. Akad. Nauk SSSR. Ser. Mat. **17** (1953), 389–400; MR **15**, 511] is generalized to show that, for certain λ_i, α_i $(1\leq i\leq s)$, the system of functions

$$\alpha_1\lambda_1{}^x f_1(x), \cdots, \alpha_s\lambda_s{}^x f_s(x)$$

is uniformly distributed in s-dimensional space. *R. A. Rankin* (Glasgow)

Citations: MR 14, 143d = K15-22; MR 14, 144a = K15-24.
Referred to in K25-15, K25-18.

K25-15 (20# 6411)

Korobov, N. M. Multidimensional problems of the distribution of fractional parts. Izvestiya Akad. Nauk SSSR. Ser. Mat. **17**, 389–400 (1953). (Russian)

In this paper the author develops a new method for studying problems of uniform distribution of fractional parts of functions. Results previously obtained by him by several different methods are deduced and improved.

Thus Theorem 1 states that if $f(x)$ is an arbitrary polynomial with integral coefficients, not identically zero, and if $\lambda>1$ is an algebraic integer satisfying a certain condition, then the function $\alpha\lambda^x f(x)$ is uniformly distributed, where

$$\alpha=\sum_{i=1}^{\infty}\frac{\varphi^{(i)}\gamma_i}{p_i(\lambda^{\tau_i}-1)}\left(\frac{1}{\lambda^{n_i}}-\frac{1}{\lambda^{n_{i+1}}}\right).$$

Here the numbers p_i, n_i, τ_i, $\varphi(i)$ and γ_i are rational numbers (the first four are integers) possessing certain properties and may be chosen in a variety of different ways. Theorem 2 asserts that under the same conditions there are numbers $\alpha_1, \alpha_2, \cdots, \alpha_s$ of a similar form such that the system of functions $\alpha_1\lambda^x f(x), \alpha_2\lambda^x f(x), \cdots, \alpha_s\lambda^x f(x)$ are uniformly distributed in s-dimensional space. Theorem 3 states that under the same conditions the functions $\alpha\lambda^x f_1(x), \alpha\lambda^x f_2(x), \cdots, \alpha\lambda^x f_s(x)$ are uniformly distributed in s-dimensional space, where the polynomials $f_1(x), f_2(x), \cdots, f_s(x)$ have integral coefficients and are linearly independent. These theorems generalise results obtained by the author in earlier papers

[Trudy Mat. Inst. Steklov. **38**, 87–96 (1951); Doklady Akad. Nauk SSSR (N.S.) **84**, 13–16 (1952); these Rev. **14**, 143, 144]. They depend upon the following fundamental lemma which states that for any integers $a \geqq 0$, $r \geqq 1$,

$$\left| \sum_{x=1}^{rp\tau} \exp\left\{ \frac{2\pi i \psi(x) f(a+x)}{p(\lambda^\tau - 1)} \right\} \right| < Cr\{N_p + \tau + p \log(a + r\tau)\}.$$

Here p is any prime, $\psi(x)$ is a function defined recursively as a linear combination of $\psi(x-1), \psi(x-2), \cdots, \psi(x-n)$ and is periodic with period $\tau \equiv 0 \pmod p$, and N is the number of roots of the congruence $\psi(x) \equiv 0 \pmod p$ for $1 \leqq x \leqq \tau$.

R. A. Rankin (Birmingham).

Citations: MR 15, 511a = K25-14.
Referred to in K15-78.

K25-16 (22# 6777)

Supnick, Fred; Cohen, H. J.; Keston, J. F. On the powers of a real number reduced modulo one. Trans. Amer. Math. Soc. **94** (1960), 244–257.

Let $\alpha > 1$ and put $v_k = \alpha^k - [\alpha^k]$ $(k = 1, 2, \cdots)$. In this paper, the distribution of the v_k is investigated, more precisely the decomposition of the set of positive integers into classes C_ν $(\nu = 1, 2, \cdots)$ such that j, k belong to the same class if and only if $v_j = v_k$. A class C_ν is called unitary or binary if it consists of one or two integers respectively. It is easily shown that there can be nonunitary classes only if α is an irrational algebraic integer whose minimal polynomial $M_\alpha(x)$ has a negative constant term. Let $L(\alpha)$ be the number of nonzero terms of $M_\alpha(x)$. The authors now derive seven theorems. Some of them give sufficient conditions in order that all classes be unitary. Further, there is a complete discussion of the cases $L(\alpha) = 2, 3$. It is also shown that there can be only finitely many binary classes and that in the case $L(\alpha) \geqq 3$ each class is unitary or binary. A set of α's is given for which $L(\alpha) > 3$ and not all classes are unitary. *C. G. Lekkerkerker* (Amsterdam)

Referred to in K25-17, K25-20, K25-21.

K25-17 (23# A123)

Ehlich, Hartmut

Die positiven Lösungen der Gleichung $y^a - [y^a] = y^b - [y^b] = y^c - [y^c]$.

Math. Z. **76** (1961), 1-4.

A short and elegant solution of a problem due to Vijaraghavan (not the reviewer), to the effect that the equation in the title has a solution for positive y and distinct positive integers a, b, c if and only if y^a, y^b, y^c are integers. This has been solved independently by F. Supnick, H. J. Cohen and J. F. Keston [Trans. Amer. Math. Soc. **94** (1960), 244–257; MR **22** #6777] and by E. C. Posner (to be published in Illinois J. Math.).

E. G. Straus (Los Angeles, Calif.)

Citations: MR 22# 6777 = K25-16.

K25-18 (24# A106)

Polosuev, A. M.

On a problem of uniform distribution of a system of functions. (Russian. English summary)

Vestnik Moskov. Univ. Ser. I Mat. Meh. **1960**, *no.* 2, 21–32.

Let there be given the (not necessarily distinct) numbers $\lambda_1 > 1, \cdots, \lambda_s > 1$. It is well-known that almost all vectors $(\alpha_1, \cdots, \alpha_s)$ have the property that (*) the sequence of points $(\alpha_1 f_1(x)\lambda_1^x, \cdots, \alpha_s f_s(x)\lambda_s^x)$ $(x = 1, 2, \cdots)$, is uniformly distributed (mod 1) in the s-dimensional cube, this for each choice of the integral coefficient polynomials $f_j(x) \not\equiv 0$ $(j = 1, \cdots, s)$.

Assuming that the given λ_j are either integers or Pisot numbers satisfying a certain mild restriction, the author succeeds in constructing explicitly a vector $(\alpha_1, \cdots, \alpha_s)$ satisfying (*). In fact, the α_j are defined in terms of an infinite series containing many free parameters.

This result, and also its method of proof, is a generalization of the case $\lambda_1 = \cdots = \lambda_s$ considered by N. M. Korobov [Izv. Akad. Nauk SSSR Ser. Mat. **17** (1953), 389–400; MR **15**, 511].

J. H. B. Kemperman (Rochester, N.Y.)

Citations: MR 15, 511a = K25-14.

K25-19 (24# A2574)

Kulikova, M. F.

A construction problem connected with the distribution of fractional parts of the exponential function. (Russian)

Dokl. Akad. Nauk SSSR **143** (1962), 522–524.

Let $\lambda > 1$ be fixed. The author constructs a real number θ such that the sequence of fractional parts $\{\theta\lambda^n\}$, $n = 1, 2, \cdots$, is uniformly distributed.

S. Knapowski (Poznań)

K25-20 (25# 1129)

Posner, Edward C.

Diophantine problems involving powers modulo one.

Illinois J. Math. **6** (1962), 251–263.

Supnick, Cohen and Keston [Trans. Amer. Math. Soc. **94** (1960), 244–257; MR **22** #6777] have proved the following two theorems. (i) If θ is a real number and the fractional parts of three different positive integral powers of θ are equal, then θ raised to some positive integral power is a rational integer. (ii) If the fractional parts of two different powers of θ are equal for infinitely many pairs of powers, then the same conclusion follows. The present author gives different proofs of these theorems, and then derives a generalization (somewhat too involved to be stated here) of the second theorem to complex numbers. Finally, he applies his results to prove the following two theorems. (1) Let $b_{k+2} = Pb_{k+1} - Qb_k$, P, Q rational numbers, $b_0 = 0$, be a linear recurring sequence of order two with the property that there is no positive integer q such that $b_{nq} = 0$, $n = 0, 1, 2, \cdots$. Then if K is sufficiently large, those b_k with $k > K$ are all different. (2) If the positive integer y is sufficiently large, the equation $2^{m+2} - 7y^2 = x^2$ has at most one solution in positive integers x, m.

A. L. Whiteman (Los Angeles, Calif.)

Citations: MR 22# 6777 = K25-16.

K25-21 (27# 3589)

Cohen, Herman J.; Supnick, Fred

Concerning real numbers whose powers have nonintegral differences.

Proc. Amer. Math. Soc. **14** (1963), 626–627.

Let β be any algebraic number having minimal polynomial of the form $x^n - b_1 x^{n-1} - \cdots - b_n$, where the b_i are non-negative rational numbers and $n > 1$. Let $\alpha = c_0 + c_1\beta + \cdots + c_m\beta^m$, where the c_i are non-negative rational numbers, $c_0 \geqq 1$, $c_m > 0$, and $m \geqq 1$. Then it is shown that $\alpha^r - \alpha^s$ is not a rational integer for all positive integers r, s with $r \neq s$. This extends earlier results of these authors and J. F. Keston [Trans. Amer. Math. Soc. **94** (1960), 244–257; MR **22** #6777].

W. J. LeVeque (Ann Arbor, Mich.)

Citations: MR 22# 6777 = K25-16.

K25-22 (27# 3614)

Philipp, Walter

Ein metrischer Satz über die Gleichverteilung mod 1.

Arch. Math. **12** (1961), 429–433.

In der vorliegenden Arbeit wird der folgende Satz

bewiesen: Es sei A eine beliebige reelle nichtsinguläre m-zeilige Matrix, deren Eigenwerte λ_i ($1 \leqq i \leqq m$) sämtlich den Betrag $|\lambda_i| > 1$ besitzen. Ferner sei $l_1 \leqq l_2 \leqq \cdots \leqq l_n \leqq \cdots$ eine Folge natürlicher Zahlen. Gibt es zwei positive Konstanten ε und c derart, daß l—als Funktion des Index betrachtet—wenigstens um c zunimmt, wenn der Index von n ab um $n/(\log n)^{1+\varepsilon}$ wächst, dann ist die Folge $\{A^{l_n}\mathfrak{x}\}$ gleichverteilt mod 1 für fast alle $\mathfrak{x}$. Ist überdies die Matrix A symmetrisch und sind alle ihre Eigenwerte positiv, so bleibt die Behauptung richtig, wenn die Zahlen $l_1, l_2, \cdots, l_n, \cdots$ beliebig reell sind. Dieses Ergebnis ist eine Verallgemeinerung eines Satzes von H. Weyl [Math. Ann. **77** (1916), 313–352] auf den m-dimensionalen Fall. *J. Cigler* (Zbl **100**, 277)

Referred to in K15-75, K25-23.

K25-23 (29# 5808)

Philipp, Walter

An n-dimensional analogue of a theorem of H. Weyl.

Compositio Math. **16**, 161–163 (1964).

The results of this lecture were published in Arch. Math. **12** (1961), 429–433 [MR **27** #3614].

Citations: MR 27# 3614 = K25-22.

K25-24 (29# 59)

Rauzy, Gérard

Répartition modulo 1 pour des suites partielles d'entiers. Développements en série de Taylor donnés sur des suites partielles.

C. R. Acad. Sci. Paris **258** (1964), 4881–4884.

Let $\mathcal{J}$ be a set of positive integers. The least upper bound of the set of real numbers $A \geqq 1$ satisfying $\mathcal{J} \supseteq \{n : x \leqq n < Ax\}$ for infinitely many x is called the frequency of $\mathcal{J}$. The main theorem is the following: Let $\vartheta > 1$ be a real algebraic number not belonging to the class T of algebraic integers all of whose conjugates have modulus $\leqq 1$. Then there exist two constants $A(\vartheta) > 1$, $\varepsilon(\vartheta) > 0$ such that, for every set $\mathcal{J}$ of frequency $A > A(\vartheta)$,

$$\lim_{n \in \mathcal{J} \to \infty} \|\lambda \vartheta^n\| \geqq \varepsilon(\vartheta)$$

for all real $\lambda \neq 0$ ($\|X\| = \min_n |X - n|$). *J. E. Cigler* (Vienna)

K25-25 (30# 4748)

Pisot, Ch.; Salem, R.

Distribution modulo 1 of the powers of real numbers larger than 1.

Compositio Math. **16**, 164–168 (1964).

Let τ denote an algebraic integer > 1 such that all the conjugates of τ have modulus $\leqq 1$. It is proved that if the modulus of some conjugate of τ is unity, then the set of powers τ^m ($m = 1, 2, \cdots$) is everywhere dense mod 1, yet the distribution is not uniform. Some unsolved problems are presented. Other results about numbers τ were proved in previous papers [Pisot, Ann. Scuola Norm. Sup. Pisa (2) **7** (1938), 205–248; Comment. Math. Helv. **19** (1946), 153–160; MR **8**, 194; Salem, Duke Math. J. **11** (1944), 103–108; MR **5**, 254; ibid. **12** (1945), 153–172; MR **6**, 206]. *E. Fogels* (*Riga*)

Citations: MR 5, 254a = R06-1; MR 6, 206b = R06-3; MR 8, 194c = K25-9.

K25-26 (31# 2235)

Cantor, David G.

On powers of real numbers (mod 1).

Proc. Amer. Math. Soc. **16** (1965), 791–793.

The author proves the following theorem. Let $\lambda > 0$ and $\theta > 1$ be real numbers, let d_n be the nearest integer to $\lambda\theta^n$, and set $\varepsilon_n = \lambda\theta^n - d_n$, $\delta_n = \sup_{m \geq 0} \sum_{i=m}^{m+n} |\varepsilon_i|$. Suppose that for some $n \geqq 1$, $\lambda^{1/n}\delta_n < [\theta - 1]/[4(\theta+1)\theta^2]$ and $\lambda\theta^n \geqq \delta_n(\theta + 1)$. Then θ is an algebraic integer whose conjugates have absolute value no greater than 1. This theorem generalizes a result of Pisot and amounts to a new arithmetic condition on the $|\varepsilon_n|$ which will guarantee that θ be algebraic. *G. R. MacLane* (Lafayette, Ind.)

K25-27 (34# 161)

Rauzy, Gérard

Suites partiellement recurrentes (Applications à la répartition modulo 1 et aux propriétés arithmétiques des fonctions analytiques).

Ann. Inst. Fourier (*Grenoble*) **16** (1966), *fasc.* 1, 159–234.

The frequency of a set J of natural numbers is defined as the l.u.b. of the set of real numbers a such that J contains all integers in $[n, an)$ for infinitely many n. The author generalizes some theorems which were known for the case $J = N = \{0, 1, 2, \cdots\}$. E.g., let $\vartheta > 1$ be a real algebraic number. There exists a set J of frequency ∞ satisfying $\lim_{n \in J} \|\lambda\vartheta^n\| = 0$ ($\|x\| = \inf |x - k|$, k an integer) if and only if ϑ belongs to T (the set of all algebraic integers > 1 all of whose conjugates are in $|z| \leqq 1$). *J. E. Cigler* (Groningen)

K25-28 (36# 115)

Zame, Alan

The distribution of sequences modulo 1.

Canad. J. Math. **19** (1967), 697–709.

A "distribution function" is taken to be any non-decreasing function mapping the unit interval I into itself and the end points onto the end points. If (x_n) is a sequence (reduced modulo 1) and d is a distribution function, then "(x_n) has d" as its distribution function, if, for every number $a \in I$, $\lim_{n\to\infty} n^{-1} \sum_{k=1}^{n} \chi(x_k) = d(a)$, where χ is the characteristic function on the interval $[0, a]$. If the limit fails to exist for some a, then the sequence (x_n) does not have a distribution function. H. Helson and J.-P. Kahane [J. Analyse Math. **15** (1965), 245–262; MR **31** #5856] established the existence of uncountably many x such that the sequence $(\theta^n x)$ does not have a distribution function modulo 1, where θ is some fixed number > 1.

In this paper the author exhibits a class of sequences of functions, characterized by certain growth properties, which can be shown to have the non-summability property for uncountably many values of the argument. This class of sequences includes lacunary (i.e., there exists a number $r > 1$ such that $h_{n+1}/h_n \geqq r$ for all n) and exponential sequences. A second question which the author considers is the determination of the distribution functions that can arise as the distribution functions modulo 1 of sequences generated in a natural way by sequences of functions. Under the assumption of a stronger growth property he shows that all possible distribution functions do arise. Thus, if d is an arbitrary distribution function and (n_k) is any sequence of real numbers satisfying $\lim_{k\to\infty}(n_{k+1} - n_k) = \infty$, then there exists a number θ such that the sequence (θ^{n_k}) has d as its distribution function. The author also generalizes this result. *F. Supnick* (New York)

Citations: MR 31# 5856 = K35-23.

K25-29 (37# 2694)

Mahler, K.

An unsolved problem on the powers of 3/2.

J. Austral. Math. Soc. **8** (1968), 313–321.

The study of the distribution modulo one of $(3/2)^n$, of

"ideal Waring problem" fame, leads to the consideration of the following question: Let α be an arbitrary positive number, set $g_n=[\alpha(3/2)^n]$ and $\alpha(3/2)^n=g_n+r_n$, $0\leqq r_n<1$. α is said to be a Z-number if $0\leqq r_n<\frac{1}{2}$ for all integers $n\geqq 0$. The main result of the present paper is that the (possibly empty) set of Z-numbers is at most countable. The proof starts with the remark that if α is a Z-number, then $g_{n+1}=\frac{1}{2}(3g_n+\varepsilon_n)$, $r_{n+1}=\frac{1}{2}(3r_n-\varepsilon_n)$, with $\varepsilon_n=0$ if $r_n<\frac{1}{3}$, $\varepsilon_n=1$ if $\frac{1}{3}\leqq r_n$ ($<\frac{1}{2}$ by the assumption on α). One then shows that $3r_0=\sum_{m=0}^{\infty}(\frac{2}{3})^m\varepsilon_m$ (the same series converges in the 2-adic sense to $-3g_0$). If the integer g_0 is given, then these formulae permit the recursive computation of all ε_n's, hence that of r_0, so that $\alpha=g_0+r_0$ is uniquely determined and there exists at most one Z-number in $[g_0, g_0+1]$ (actually in $[g_0, g_0+\frac{1}{2})$, because $r_0<\frac{1}{2}$). Assuming the existence of Z-numbers (extensive computer work has failed to lead to the discovery of even a single Z-number), some of their properties are established; in particular, for sufficiently large x, there are at most $x^{0.7}$ Z-numbers satisfying $0<\alpha\leqq x$. *E. Grosswald* (Philadelphia, Pa.)

K25-30 **(40# 1348)**

Boyd, David W.
Transcendental numbers with badly distributed powers. *Proc. Amer. Math. Soc.* **23** (1969), 424–427.

An old result of Pisot states that if $\theta>1$ and $\lambda\geqq 1$ are such that $\|\lambda\theta^n\|\leqq(2e\theta(\theta+1)(1+\operatorname{Log}\lambda))^{-1}$, $n=0, 1, 2, \cdots$, then θ is algebraic and $\lambda\in\mathbf{Q}(\theta)$ (in fact, θ is then a Pisot number or a Salem number).

In this article, it is shown that C. Pisot's result [Ann. Scuola Norm. Sup. Pisa (2) **7** (1938), 205–248] is, in a certain way, the best possible: a striking consequence of the author's Theorem I is that given any constant $c>2e(1+\operatorname{Log}2)=9.24\ldots$, there exists a real number $\lambda\geqq 1$ and a transcendental number θ as large as we wish such that (1) $\|\lambda\theta^n\|\leqq c(2e\theta(\theta+1)(1+\operatorname{Log}\lambda))^{-1}$, $n=0, 1, 2, \cdots$. Furthermore, the θ's which satisfy (1) form an uncountable set. *M. Mendès France* (Talence)

K25-31 **(44# 3965)**

Pathiaux, Martine
Répartition modulo 1 de la suite $\lambda\alpha^n$. *Séminaire Delange-Pisot-Poitou:* 1969/70, *Théorie des Nombres, Fasc.* 1, *Exp.* 13, 6 *pp.* *Secrétariat mathématique, Paris*, 1970.

Using different techniques from those of Pisot, the author establishes sufficient conditions for a real number α to be a PV number. For example, let $\alpha>1$ and denote by $((x))$ the residue (mod 1) of the real number x that lies in $[-\frac{1}{2}, +\frac{1}{2}[$. It is known that if there exists a $\lambda\neq 0$ such that $|((\lambda\alpha^n))|=o(1/\sqrt{n})$, then α is a PV number [R. Salem, *Algebraic numbers and Fourier analysis*, p. 12, exercise 2, Heath, Boston, Mass., 1963; MR **28** #1169]. The author shows that if there exists a $\lambda\neq 0$ such that for any integer $n\geqq 0$, $|((\lambda\alpha^n))|\leqq 1/(2\alpha(\alpha+1)\sqrt{(n+1)})$ holds, then α is a PV number.

In fact, it is shown that for any q, $0\leqq q\leqq\frac{1}{2}$, there exists a constant $\psi(q,\lambda)$ (explicitly calculated) such that the condition $|((\lambda\alpha^n))|\leqq\psi(q,\lambda)/(1+n)^q$; $n=0, 1, 2, \cdots$ ensures that α is a PV number. The case $q=0$ was already known [C. Pisot, Comment. Math. Helv. **19** (1946), 153–160; MR **8**, 194].

{There is one misprint on page 1, line 14: read

$$\lim_{N\to\infty}(1/N)\sum_{n=0}^{N} n\varepsilon_n^{\,2}=0.\}$$

F. Mendès France (Talence)

Citations: MR 8, 194c = K25-9; MR 28# 1169 = R06-24.

K30 DISTRIBUTION (mod 1): OTHER SPECIAL SEQUENCES

Many of the estimates for exponential sums in **L15**, **L20** and **L99** have implications for this section.

See also reviews A26-32, B36-125, K05-11, K05-80, K10-5, K15-22, K15-79, K25-13, K25-14, K35-5, K35-8, L20-2, L20-3, L20-5, L20-7, L20-11, M30-22, M30-25, M50-8, N36-30, N36-37, N36-54, N40-47, N60-7, N60-37, N60-74, N72-22.

K30-1 **(2, 40a)**

Vinogradow, I. M. A general property of prime numbers distribution. Rec. Math. [Mat. Sbornik] N.S. **7 (49)**, 365–372 (1940). (Russian. English summary)

Let $0<\alpha<1$, $N\geqq 2$. An estimate is obtained for the sum $\sum e^{2\pi ikp^\alpha}$, where p ranges over the primes not greater than N, and $k\ll N^{\alpha/3}$ is a positive constant. Let $0<\sigma<1$, $b>0$, $\epsilon>0$. It is proved that the number T of primes not greater than N such that $0\leqq bp^\alpha-[bp^\alpha]<\sigma$ is given by the formula $T=\sigma\pi(N)+O(N^{1+\epsilon}\Delta)$, $\Delta=(bN^{\alpha-1}+b^{-1}N^{-\alpha}+N^{-2\alpha/3})^{1/5}$. Otherwise stated, set $\beta=b^{-m}$, $m=1/\alpha$; T is the number of primes not greater than N in the intervals $\beta c^m\leqq p<\beta(c+\sigma)^m$, for integers c. *G. Pall* (Princeton, N. J.).

Referred to in L20-3.

K30-2 **(8, 6b)**

Vinogradow, I. M. A general distribution law for the fractional parts of values of a polynomial with the variable running over the primes. C. R. (Doklady) Acad. Sci. URSS (N.S.) **51**, 491–492 (1946).

Estimates for trigonometric sums such as

$$\sum_{k=1}^{K}\left|\sum\exp(2\pi ikf(p))\right|,$$

where K is a positive constant, $f(p)=a_np^n+a_{n-1}p^{n-1}+\cdots+a_1p$, a_i real, and the inner sum is over all primes $p\leqq P$, have previously been made to depend on the approximation of the leading coefficient a_n by rational numbers a/q. In this paper an estimate is given which depends on the approximation of any coefficient of the polynomial by rational numbers. The new estimate is then applied to the problem indicated in the title. The main result is the following asymptotic formula for the number T of fractional parts $\{f(p)\}$ satisfying $0\leqq\{f(p)\}\leqq\sigma$, $p\leqq P$:

$$T=\sigma\pi(P)+O(P^{1+\epsilon}(q^{-1}+P^{-1})^\rho),$$

where $a_r=a/q+z$; r is one of the numbers $1, 2, \cdots, n$; $|z|\leqq 1/q\tau$; $\tau=p^{r/2}$; $\rho=1/(8n^2\log 12n)$. No proofs are given. *R. D. James* (Vancouver, B. C.).

Referred to in K30-3.

K30-3 **(9, 11c)**

Vinogradow, I. M. A general law of the theory of primes. C. R. (Doklady) Acad. Sci. URSS (N.S.) **55**, 471–472 (1947).

In a previous note with a similar title [same C. R. (N.S.) **51**, 491–492 (1946); these Rev. **8**, 6], the author obtained an estimate for $\sum_{k=1}^{K}|\sum\exp(2\pi ikf(p))|$, where K is a positive constant, the inner sum is over all primes $p\leqq P$, and $f(p)$ is a polynomial with real coefficients. He then used the estimate to establish an asymptotic formula for the number T of fractional parts $\{f(p)\}$ of the polynomial satisfying $0\leqq\{f(p)\}\leqq\sigma$, $p\leqq P$. In this note the results are extended to the case where $f(p)$ is no longer a polynomial, but a function which may be approximated by a polynomial in a certain sense explained in the note. Two examples are given. One is the function a/p^m to which the stated results apply, and the other is the function $a \log x$, to which the results do not apply. No proofs are given.

R. D. James (Vancouver, B. C.).

Citations: MR 8, 6b = K30-2.

K30-4 (11, 14b)

Erdös, P., and Koksma, J. F. On the uniform distribution modulo 1 of lacunary sequences. Nederl. Akad. Wetensch., Proc. **52**, 264–273 = Indagationes Math. **11**, 79–88 (1949).

Let $0\leqq\alpha<\beta\leqq 1$. Denote by $u_1, u_2, u_3, \cdots$ a sequence of real numbers, by N' the number of those elements of the set

$$u=u_1-[u_1],\ u_2-[u_2],\ \cdots,\ u_N-[u_N]$$

satisfying $\alpha\leqq u<\beta$. The upper bound of $|N'/N-(\beta-\alpha)|$ for variable α, β and fixed $N\geqq 1$ is the discrepancy $D(N)$ of the sequence; it satisfies $D(N)=o(1)$ if and only if the sequence is uniformly distributed [Koksma, Diophantische Approximationen, Ergebnisse der Math., v. 4, no. 4, Springer, Berlin, 1936, chaps. 8, 9]. The authors prove a very general result for sequences $u_n=f(n,\theta)$ and nearly all θ; it contains the following special theorem. Let $a<b$, $\delta>0$. Let $u_n=f(n,\theta)$ denote a sequence of real functions defined for $a\leqq\theta\leqq b$ satisfying

$$f_\theta'(n+1,\theta)\geqq(1+\delta)f_\theta'(n,\theta)>0;$$
$$f_\theta''(n+1,\theta)\geqq(1+\delta)f_\theta''(n,\theta)\geqq 0$$

$(n=1, 2, \cdots)$ for all θ in $a\leqq\theta\leqq b$. Let $\omega(n)$ be an increasing function of n tending to infinity with n. Then

$$D(N)=o\{N^{-\frac{1}{2}}(\log N)^{\frac{3}{2}}(\log\log n)^{\frac{1}{2}}\omega(N)\}$$

for nearly all θ in $a\leqq\theta\leqq b$. *K. Mahler* (Manchester).

Referred to in K30-9.

K30-5 (11, 331f)

Erdös, P., and Koksma, J. F. On the uniform distribution modulo 1 of sequences $(f(n,\theta))$. Nederl. Akad. Wetensch., Proc. **52**, 851–854 = Indagationes Math. **11**, 299–302 (1949).

As a special case of a still more general theorem, the following result is proved. Let $f(n,\theta)$ be defined for $n=1, 2, \cdots$ and all θ with $\alpha\leqq\theta\leqq\beta$; let $(\partial/\partial\theta)f(n,\theta)$ be continuous in θ; and let $(\partial/\partial\theta)f(n_1,\theta)-(\partial/\partial\theta)f(n_2,\theta)$, for $n_1\neq n_2$, be monotonic (increasing or decreasing) in θ and of absolute value at least δ, where $\delta>0$ is independent of n_1, n_2, and θ. Then for $\epsilon>0$ and almost all θ the discrepancy $D(N,\theta)$ of the sequence $f(1,\theta), f(2,\theta), f(3,\theta), \cdots$ satisfies the inequality $ND(N,\theta)=O(N^{\frac{1}{2}}\log^{(5/2)+\epsilon}N)$.

K. Mahler (Manchester).

Referred to in K30-7, K30-8, K30-9.

K30-6 (11, 431b)

Erdös, P., and Turán, P. On the distribution of roots of polynomials. Ann. of Math. (2) **51**, 105–119 (1950).

The authors prove the following theorem on equidistribution of the arguments $\varphi_1, \cdots, \varphi_n$ of the roots of a polynomial $f(z)=a_0+a_1z+\cdots+a_nz^n$. If $N(\alpha,\beta)$ denotes the number of φ's in the interval $\alpha\leqq\varphi\leqq\beta$ $(0\leqq\alpha<\beta\leqq 2\pi)$, then $|N(\alpha,\beta)-(2\pi)^{-1}n(\beta-\alpha)|<16(n\log P)^{\frac{1}{2}}$, where

$$P=|a_0a_n|^{-\frac{1}{2}}(|a_0|+\cdots+|a_n|).$$

The authors then show that two known theorems are consequences: (1) a theorem of E. Schmidt on the maximal number of real roots [S.-B. Preuss. Akad. Wiss. Phys.-Math. Kl. **1932**, 321] and (2) a theorem of Szegö on the equidistribution of the roots of partial sums of a power series whose radius of convergence is 1 [S.-B. Berlin. Math. Ges. **21**, 59–64 (1922)]. *N. G. de Bruijn* (Delft).

K30-7 (12, 162b)

Cassels, J. W. S. Some metrical theorems in Diophantine approximation. I. Proc. Cambridge Philos. Soc. **46**, 209–218 (1950).

Let $f_{1,j}(\theta_j), \cdots, f_{n,j}(\theta_j)$, $j=1, 2, \cdots, w$, be w sequences of differentiable functions, defined for $a_j\leqq\theta_j\leqq b_j$. Assume that there is an absolute constant K so that for $m\neq n$ $|f'_{m,j}(\theta_j)-f'_{n,j}(\theta_j)|\geqq K>0$, also that $f'_{m,j}(\theta_j)-f'_{n,j}(\theta_j)$ is monotonic. Let $0\leqq\beta_j\leqq\gamma_j\leqq 1$; denote by $F_N(\beta,\gamma;\theta)$ the number of $n\leqq N$ so that simultaneously $\beta_j\leqq f_{n,j}\theta\leqq\gamma_j$, $j=1, 2, \cdots, w$. The author proves the following theorem. For almost all sets θ_j,

$$|F_N(\beta,\gamma;\theta)-N\prod_{j=1}^{w}(\gamma_j-\beta_j)|<N^{\frac{1}{2}}(\log N)^{w+\frac{1}{2}+\epsilon}$$

holds for all β, γ, and all $N>N_0$, where N_0 depends only on θ and ϵ. An interesting corollary results when one takes $w=1$ and $f_n(\theta)=n_k\theta$, where $n_1<n_2<\cdots$ is a sequence of integers. A similar theorem has been found almost simultaneously by Koksma and the reviewer [Nederl. Akad. Wetensch., Proc. **52**, 851–854 = Indagationes Math. **11**, 299–302 (1949); these Rev. **11**, 331]. *P. Erdös* (Aberdeen).

Citations: MR 11, 331f = K30-5.

Referred to in J24-11, J24-14, J24-27, J24-31, J24-33, J64-13, J64-23.

K30-8 (12, 162e)

Cassels, J. W. S. Some metrical theorems of Diophantine approximation. IV. Nederl. Akad. Wetensch., Proc. **53**, 176–187 = Indagationes Math. **12**, 14–25 (1950).

The author improves an earlier result [see the third preceding review*] which had also been obtained by Erdös and Koksma [same Proc. **52**, 851–854 = Indagationes Math. **11**, 299–302 (1949); these Rev. **11**, 331] and proves the following result. Let $\varphi(n)$ be a positive function of the positive integer n such that $\log n\log\log n\geqq\varphi(n)\geqq c>0$ (c independent of n) for all sufficiently large n, and that $\varphi(n)$ and $\varphi(n)^{-1}\log n\log\log n$ are finally monotonic nondecreasing. Let $f_n(\theta)$ $(n=1, 2, 3, \cdots)$ have positive monotonic nondecreasing continuous derivatives for $0\leqq\theta\leqq 1$. For $0\leqq\alpha\leqq\beta\leqq 1$ denote by $F_N(\alpha,\beta,\theta)$ the number of integers n for which $\alpha\leqq f_n(\theta)<\beta \pmod 1$, and put

$$\mathfrak{R}_N(\theta)=\max_{\alpha,\beta}|F_N(\alpha,\beta,\theta)-N(\beta-\alpha)|.$$

Then, if $f'_n(\theta)\geqq e^{\varphi(n)}f'_{n-1}(\theta)$ $(n=2, 3, \cdots)$, $f'_1(\theta)\geqq 1$, for all θ, there is an absolute positive constant A_0 such that

$$\mathfrak{R}_N(\theta)\leqq A_0N^{\frac{1}{2}}\varphi(N)^{-\frac{1}{2}}(\log N)^{\frac{1}{2}}\log\log N$$

for almost all θ and all sufficiently large N.

K. Mahler (Manchester).

*Now the review above.

Citations: MR 11, 331f = K30-5.

K30-9 (12, 163a)

LeVeque, Wm. J. Note on a theorem of Koksma. Proc. Amer. Math. Soc. **1**, 380–383 (1950).

The theorem mentioned in the title [Compositio Math. **2**, 250–258 (1935)] states that under certain general conditions for the functions $g(x,n)$, $a\leqq x\leqq b$, $n=1, 2, \cdots$, the sequence (1) $g(x,1), g(x,2), \cdots$ is uniformly distributed (mod 1) for almost all values of x, $a\leqq x\leqq b$. Using a lemma of Kac, Salem, and Zygmund [Trans. Amer. Math. Soc. **63**, 235–243 (1948); these Rev. **9**, 426], the author provides a new proof for a general, important case of that theorem. With his method the author proves further, that under his assumptions, for almost all x in $a \leqq x \leqq b$,

$$\sum_{n=1}^{N} e^{2\pi i m g(x,n)} = o(N^{\frac{1}{2}+\theta}),$$

$\theta > 0$, m an integer $\neq 0$. The reviewer remarks that this result is contained in theorems of Erdös and Koksma [Nederl. Akad. Wetensch., Proc. **52**, 264–273, 851–854 = Indagationes Math. **11**, 79–88, 299–302 (1949); these Rev. **11**, 14, 331] and Cassels [see the preceding review]; they proved among other things (with different methods), that (under general conditions) $ND(N) = o(N^{\frac{1}{2}} \log^{5/2+\theta} N)$, $\theta > 0$, for almost all x, $a \leqq x \leqq b$. Here $D(N)$ means the discrepancy in the uniform distribution of the sequence (1) [cf. Erdös and Koksma, loc. cit.]. *J. F. Koksma* (Amsterdam).

Citations: MR 11, 14b = K30-4; MR 11, 331f = K30-5.

K30-10 **(14, 1067d)**

LeVeque, W. J. On n-dimensional uniform distribution modulo 1. Michigan Math. J. 1 (1952), 139–162 (1953).

Extending the method used in an earlier paper [Proc. Amer. Math. Soc. **1**, 380–383 (1950); these Rev. **12**, 163] the author generalizes results of Erdös and Koksma [Nederl. Akad. Wetensch., Proc. **52**, 264–273, 851–854 = Indagationes Math. **11**, 79–88, 299–302 (1949); these Rev. **11**, 14, 331] on the discrepancy modulo 1 of sequences $f_1(x)$, $f_2(x)$, $\cdots (a \leqq x \leqq b)$, by considering s-tuples of functions of r continuous variables $x_1, x_2, \cdots, x_r$ $(a_\rho \leqq x_\rho \leqq b_\rho, \rho = 1, 2, \cdots, r)$ and n sequential variables. The results are too complicated to be stated here. *J. F. Koksma* (Amsterdam).

Referred to in K05-42, K25-13.

K30-11 **(15, 410e)**

Kuipers, L. Continuous and discrete distribution modulo 1. Nederl. Akad. Wetensch. Proc. Ser. A. **56** = Indagationes Math. **15**, 340–348 (1953).

Let $f(t)$ be a differentiable function, $0 < t < \infty$. The author proves several theorems about the uniform distribution mod 1 of the sequence $f(n)$. Among others he proves the following theorems: 1) Let $f(t)$ satisfy $|tf'(t)| \leqq M$, $0 \leqq t < \infty$. Then the sequence $f(1), f(2), \cdots$ is uniformly distributed mod 1. 2) Let $f(t)$ be twice differentiable, $f'(t)$ and $f''(t)$ are bounded and $f'(t)$ tends to an irrational number as t tends to infinity. Then the sequence $f(1), f(2), \cdots$ is uniformly distributed mod 1. 3) The sequence $n^{1/2} + \sin n$, $n = 1, 2, \cdots$, is uniformly distributed (mod 1). 4) The sequence $\cos(n + \log n)$, $n = 1, 2, \cdots$, is not uniformly distributed mod 1. *P. Erdös* (South Bend, Ind.).

Referred to in K40-21.

K30-12 **(15, 855f)**

Vinogradov, I. M. An elementary proof of a theorem from the theory of prime numbers. Izvestiya Akad. Nauk SSSR. Ser. Mat. **17**, 3–12 (1953). (Russian)

The following result is proved. Let $0 \leqq \sigma \leqq 1$, let $\pi_\sigma(N, q)$ be the number of primes p not exceeding N such that $p/q - [p/q] < \sigma$; then $\pi_\sigma(N, q) = \sigma\pi(N) + O(R)$ where

$$R = N^{1+\epsilon}\{(1/q + q/N)^{1/2} + N^{-1/6}\}$$

so that for large q and N/q, $\pi_\sigma(N, q)$ is asymptotically $\sigma\pi(N)$. The author remarks that a modification of the method enables him to replace the exponent $-1/6$ of N by $-1/5$ and that the method is applicable to a number of other problems including the corresponding question for primes in a given arithmetic progression.

The method of proof is different from that in Chapter XI of the author's book [Trudy Mat. Inst. Steklov. **23** (1947); these Rev. **10**, 599] which depended on the estimation of exponential sums; the final result in the book corresponds to the estimate with $-1/6$ replaced by $-1/5$ and has a few other refinements as well. The present proof depends on the use of the function $\psi_\sigma(x)$ defined to be $1 - \sigma$ if $x - [x] < \sigma$ and to be $-\sigma$ otherwise. By applying devices similar to those the author previously used for giving elementary estimates for sums of the form $\sum_x (f(x) - [f(x)])$, he now gives upper bounds for the absolute values of sums of the form

$$\textstyle\sum_x \psi_\sigma(\alpha x + \beta), \quad \sum_x \sum_y \psi_\sigma(xy/q + h/q), \quad \sum_x \sum_y \psi_\sigma(axy/q),$$
$$\textstyle\sum_x \sum_y \sum_m \psi_\sigma(axym/q), \quad \sum_p \psi_\sigma(ap/q).$$

From the estimate of the last sum, the result immediately follows. The methods used here are technically simpler than those used in the author's book. *L. Schoenfeld.*

Citations: MR 10, 599a = L02-2.

Referred to in N12-9.

K30-13 **(17, 242b)**

Yüh, Ming-I. A note on Diophantine inequality with prime unknown. Acta Math. Sinica 3 (1953), 218–224. (Chinese. English summary)

The author proves the following analogue of a theorem of I. M. Vinogradov [Trudy Mat. Inst. Steklov. **23** (1947), Ch. V; MR **10**, 599; **15**, 941]. "Let $\varepsilon > 0$ be arbitrarily small; let $0 < \varkappa < 1$; let P be a sufficiently large positive integer; let $f(x) = a_{n+1}x^{n+1} + \cdots + a_1 x$ be a polynomial with real coefficients; and let

$$a_r = \frac{a}{q} + \frac{\vartheta}{q\tau}, \ a \text{ and } q \text{ integers},$$

$$(a, q) = \lambda, \ |\vartheta| < \lambda, \ \tau = P^{r/2}, \ P^\varkappa \ll q \leqq \tau$$

for some suffix r with $2 \leqq r \leqq n$. Put

$$\varrho = \begin{cases} \dfrac{0.041}{n^2(\log n + 2)} & \text{if } q > P^{1/4}, \\[2ex] \dfrac{0.37\varkappa}{n^2\{\log(n^2/\varkappa) + 4\}} & \text{if } q \leqq P^{1/4}. \end{cases}$$

For every real number A there exist a prime p and an integer v such that $|f(p) - v - A| < 6P^{-\frac{1}{2}\varrho + \varepsilon}$, $0 < p < P$." A similar theorem is stated for functions f the derivatives of which satisfy certain inequalities. The proof uses Vinogradov's methods. *K. Mahler* (Manchester).

Citations: MR 10, 599a = L02-2; MR 15, 941b = L02-3.

K30-14 **(19, 1164c)**

Knapowski, S. Über ein Problem der Gleichverteilungstheorie. Colloq. Math. **5** (1957), 8–10.

Let $a_1, a_2, \cdots$ be an increasing sequence of positive integers. The author studies the distribution of the sequence $a_1^{-1}, 2a_1^{-1}, 3a_1^{-1}, \cdots, (a_1 - 1)a_1^{-1}, a_2^{-1}, 2a_2^{-1}, \cdots, (a_2 - 1)a_2^{-1}, a_3^{-1}, \cdots$, abbreviated by $u_1, u_2, \cdots$. If $0 \leqq \alpha < \beta \leqq 1$, and $A(n)$ is the number of u_i with $\alpha < u_i \leqq \beta$, $i < n$, then $A(n) = (\beta - \alpha)n + O(a_k)$, where a_k is the denominator involved in u_n. Unless $\alpha = \beta - 1 = 0$, the symbol O cannot be replaced by o. In the proof of the latter statement the author assumes that $a_k/k \to \infty$, but it is easy to see that this assumption is superfluous.

N. G. de Bruijn (Amsterdam).

K30-15 **(20# 2310)**

Delange, Hubert. On some arithmetical functions. Illinois J. Math. 2 (1958), 81–87.

The author proves the following statement of P. Erdös: Let $\omega(n)$ be the number of prime divisors of the positive integer n, and let λ be any irrational number. Then the numbers $\lambda\omega(n)$ are uniformly distributed modulo

1. This means that for $0 \leq t \leq 1$, the number of n's less than or equal to x and such that $\lambda\omega(n) - I[\lambda\omega(n)] \leq t$ is $tx + o(x)$ as x tends to $+\infty$ ($I[n]$ denotes the greatest integer $\leq n$). It is pointed out that this result can be deduced, if not simply, from a later result of Erdös concerning the number of integers $n \leq x$ for which $\omega(n) = k$ [Ann. of Math. (2) **49** (1948), 53–66; MR **9**, 333] and also that a very short proof can be based on a formula of Atle Selberg [J. Indian Math. Soc. (N.S.) **18** (1954), 83–87; MR **16**, 676], which, however, requires the properties of the Riemann zeta function in the critical strip.

The result of Erdös and the similar result for the total number of prime divisors of n are proved to be immediate consequences of a well-known theorem of H. Weyl [Math. Ann. **77** (1916), 313–352, Satz 1], a theorem of the author [Ann. Sci. Ecole Norm. Sup. (3) **73** (1956), 15–74, § 2.3; MR **18**, 720] and a Tauberian theorem [D. W. Widder, The Laplace transform, Princeton, 1941; Th. 17; MR **3**, 232] as formulated in the present paper. In fact, the uniform distribution modulo 1 of $\lambda f(n)$ for any irrational λ holds for all functions of a certain family ($\mathfrak{F}$) [Delange, Colloque sur la théorie des nombres, Bruxelles, 1955, Thone, Liège, 1956, pp. 147–161; MR **19**, 17] and also for n running through a certain infinite set of positive integers, distinct from the set of all positive integers.

S. Ikehara (Tokyo)

Citations: MR 9, 333b = N24-9; MR 16, 676a = N24-22; MR 18, 720a = N24-28; MR 19, 17b = N60-22.

Referred to in N36-37.

K30-16 (20# 2311)

Delange, Hubert. Sur certaines fonctions arithmétiques. C. R. Acad. Sci. Paris **246** (1958), 514–517.

Let $\omega_E(n)$ denote the number of distinct prime divisors of the positive number n which belong to a set E of prime numbers, and let $\Omega_E(n)$ be the total number of prime divisors of n belonging to E. The author has studied distribution properties of these functions [see review above and references therein] under the assumption (H): There is a constant α such that $\sum_{p \in E} p^{-s} + \alpha \log(s-1)$ is regular throughout the closed halfplane $Rs \geq 1$. This is now replaced by a more general and natural assumption (H_1): There is a number $\alpha \geq 0$ such that the number of numbers of E not exceeding x will be equal to $\alpha(x/\log x) + o(x/\log x)$ as x tends to $+\infty$, and if $\alpha = 0$, we have $\sum_{p \in E} p^{-1} = +\infty$.

Previous results on arithmetical functions in the papers above may be all deduced under the new assumption, and a very general theorem is stated which includes the former results and which is susceptible of other applications.

S. Ikehara (Tokyo)

Referred to in N36-30, N36-37.

K30-17 (20# 2312)

Delange, Hubert. Sur la distribution de certains entiers. C. R. Acad. Sci. Paris **246** (1958), 2205–2207

Continuation of paper reviewed above, indicating how other results may also be generalized under the new condition (H_1), including results of B. Hornfeck [Monatsh. Math. **60** (1956), 96–109; MR **18**, 18] and of E. Wirsing [Arch. Math. **7** (1956), 263–272; MR **18**, 642].

S. Ikehara (Tokyo)

Citations: MR 18, 18c = N24-26; MR 18, 642e = N76-8.

Referred to in N36-30, N36-37.

K30-18 (20# 2828)

Bass, Jean. Sur certaines classes de fonctions admettant une fonction d'autocorrélation continue. C. R. Acad. Sci. Paris **245** (1957), 1217–1219.

The author considers the class A of complex functions $u(t)$ of the real variable t, $0 \leq t < +\infty$, subjected to the following conditions: (a) $u(t)$ is integrable in every finite interval; (b) $\lim_{T\to\infty} (1/T) \int_0^T u(t)dt = 0$; (c) $\gamma(h) = \lim_{T\to\infty} (1/T) \int_0^T u(t+h)\bar{u}(t)dt$ exists and is continuous for all h; (d) $\gamma(0) \neq 0$; (e) $\lim_{h\to\infty} \gamma(h) = 0$. From a theorem of H. Weyl [Math. Ann. **77** (1916), 313–352] it is proved that if $\varphi(t)$ is a polynomial of degree $q \geq 2$, the coefficient of t^q being not commensurable with π, then the function $u(t) = \exp[i\varphi(n)]$ in $n \leq t < n+1$ (n=non-negative integer) belongs to the class A.

J. Kampé de Fériet (Lille)

Referred to in K30-19, K40-22.

K30-19 (20# 2829)

Bass, Jean; et Bertrandias, Jean-Paul. Moyennes de sommes trigonométriques et fonctions d'autocorrélation. C. R. Acad. Sci. Paris **245** (1957), 2457–2459.

This paper is based on a series of lemmas concerning the function $\gamma(h)$ [see #2828 above], the most important being the following one: if $u(t) = \exp[i\varphi(n)]$ in $n \leq t < n+1$, then $\gamma(h)$ is continuous and linear in every interval $n \leq h < n+1$. From these lemmas the authors deduce a demonstration of a theorem, very close to a known result of J. G. van der Corput [Acta Math. **56** (1931), 373–456]: If there exists α, $0 < \alpha < 1$, and $\theta > 0$ such that $|\varphi(n) - \theta n^\alpha| \to 0$ if $n \to \infty$, then

$$\lim_{N\to\infty} (1/N) \sum_{n=0}^{n=N} \exp[i\varphi(n)] = 0.$$

J. Kampé de Fériet (Lille)

Citations: MR 20# 2828 = K30-18.

Referred to in K40-22.

K30-20 (21# 662)

Delange, Hubert. On a theorem of Erdös. Scripta Math. **23** (1957), 109–116 (1958).

The author gives a new proof for the following theorem of the reviewer [Ann. of Math. (2) **47** (1946), 1–20; MR **7**, 416]. Let $f(n)$ be an additive function satisfying $f(p) \to 0$ as $p \to \infty$ and $\sum_p (f^2(p)/p) = \infty$ (p prime). Then the number of integers $n \leq x$ satisfying $f(n) - [f(n)] \leqslant t$ equals $tx + o(x)$. Several further theorems on the number of prime factors of integers are proved by the author. His principal tools are Weyl's theorem on uniform distributions and Brun's method; probability theory is not used.

P. Erdös (Birmingham)

Citations: MR 7, 416c = N60-7.

K30-21 (22# 35)

Polosuev, A. M. Uniform distribution of a system of functions representing a solution of linear first order difference equations. Dokl. Akad. Nauk SSSR **123** (1958), 405–406. (Russian)

A difference equation $\Phi(x+1) = A\Phi(x)$, where $\Phi(x)$ is an n-component vector $(\phi_1(x), \cdots, \phi_n(x))$ and A is an n by n constant matrix, is considered. Two theorems are given specifying conditions under which the points $\Phi(x)$ determined by a solution of the system in the Φ-space for $x = 1, 2, 3, \cdots$ are uniformly distributed, modulo 1. It is stated that both theorems can be proved by the method used by A. G. Postnikov in Vestnik Leningrad. Univ. **12** (1957), no. 13, 81–88 [MR **21** #666].

H. L. Turrittin (Minneapolis, Minn.)

Citations: MR 21# 666 = K15-49.

Referred to in K30-22.

K30-22 (24# A107a; 24# a107b)

Polosuev, A. M.

On the uniformity of the distribution of a system of functions forming the solution of a system of linear

finite-difference equations of first order. I. Criterion for uniformity of distribution. (Russian. English summary)
Vestnik Moskov. Univ. Ser. I Mat. Meh. **1960**, *no.* 5, 29–39.

Polosuev, A. M.
On the uniformity of the distribution of a system of functions forming the solution of a system of linear finite-difference equations of first order. II. Construction of a uniformly distributed system. (Russian. English summary)
Vestnik Moskov. Univ. Ser. I Mat. Meh. **1960**, *no.* 6, 20–25.

Some of the results proved here were announced in Dokl. Akad. Nauk SSSR **123** (1958), 405–406 [MR **22** #35]. Let A be an integral $s\times s$ matrix, $\det(A)\neq 0$, such that (i) no eigenvalue λ_i of A is a root of unity. Then [V. A. Rohlin, Izv. Akad. Nauk SSSR Ser. Mat. **13** (1949), 329–340; MR **11**, 40] the transformation $x\to Ax$ (mod 1) of the s-dimensional torus T unto itself is ergodic with respect to Lebesgue measure, so that the sequence $\{A^k\alpha\}$ is uniformly distributed (mod 1) for almost all $\alpha\in T$.

Let $\alpha\in T$ be fixed and suppose that there exists a constant $C>1$ such that

$$(*)\qquad \limsup p^{-1}N_p(\Delta) \leqq C|\Delta|.$$

Here, Δ denotes an arbitrary cube in T of volume $|\Delta|$ having its sides parallel to the coordinate axes. Further, $N_p(\Delta)$ denotes the number of $k=1,\cdots,p$ with $A^k\alpha\in\Delta$ (mod 1). The main theorem of Part I asserts that, under the assumption (ii) $|\lambda_i|\neq 1$, the sequence $\{\psi_k\}$ defined by $\psi_0=\alpha$, $\psi_{k+1}=A\psi_k+\gamma$ is uniformly distributed (mod 1). Here, $\gamma\in T$ is arbitrary but fixed. Clearly, this assertion is false. For, letting $\delta=(A-I)^{-1}\gamma$, one has $\psi_k=A^k(\alpha+\delta)-\delta$; now, choose γ such that $\alpha+\delta$ is rational.

The error seems to enter the proof in formula (12). The proof is correct in the special case $\gamma=0$. That (*) implies the uniform distribution (mod 1) of the sequence $\{A^k\alpha\}$ was recently shown by J. Cigler [J. Reine Angew. Math. **205** (1960), 91–100; p. 96] under the sole side condition (i).

In Part II, assuming (iii) $|\lambda_i|>1$ and using criterion (*), the author gives a more or less explicit construction of an $\alpha\in T$ such that $\{A^k\alpha\}$ is uniformly distributed (mod 1).

J. H. B. Kemperman (Rochester, N.Y.)

Citations: MR 22# 35 = K30-21.
Referred to in K30-23.

K30-23 (24# A3147)

Muhutdinov, R. H.
A Diophantine equation with a matrix exponential function. (Russian)
Dokl. Akad. Nauk SSSR **142** (1962), 36–37.

Let A be a nonsingular matrix with integral entries, no eigenvalue of which is a root of unity, and let $\psi(x)$ be a nonconstant polynomial with integral coefficients. The author claims to have an asymptotic estimate for the number of solutions in integers $x_1,\cdots,y_b$ from the large interval $[0,p-1]$ of the equation

$$\tilde m_1\psi(A^{x_1})+\cdots+\tilde m_a\psi(A^{x_a}) = \tilde n_1\psi(A^{y_1})+\cdots+\tilde n_b\psi(A^{y_b}),$$

where $\tilde m_1,\cdots,\tilde n_b$ are fixed integral vectors. The estimate is not given here.

If $\tilde\alpha=(\alpha_1,\cdots,\alpha_n)$, let $\{\tilde\alpha\}$ be the vector $(\{\alpha_1\},\cdots,\{\alpha_n\})$, where $\{\alpha_j\}$ is the fractional part of α_j: $\alpha_j=[\alpha_j]+\{\alpha_j\}$. Let π be the unit hypercube $0\leqq x_i\leqq 1$, $i=1,\cdots,n$, and let $\varepsilon(t)$ be any non-negative function with $\lim_{t\to\infty}\varepsilon(t)=0$. The above estimate can be used to show that if, for suitable $c>0$, and every hypercube $\Delta\subset\pi$,

$$\limsup_{p\to\infty}\frac{N_p(\Delta)}{p}\leqq c|\Delta|^{1-\varepsilon(|\Delta|)},$$

then the sequence $\{\tilde\alpha A^x\}$, $x=0,1,\cdots$, is uniformly distributed in π. Here $|\Delta|$ is the volume of Δ, and $N_p(\Delta)$ is the number of vectors $\{\tilde\alpha A^x\}$, $x=0,1,\cdots,p-1$, which lie in Δ.

Finally, two central limit theorems are stated for sums $p^{-1/2}\sum_{t=0}^{p-1}f(\tilde\alpha\psi(A^t))$, where f is a real (complex)-valued periodic function of n arguments.

The above theorems represent extensions or refinements of work by M. P. Mineev [Izv. Akad. Nauk SSSR Ser. Mat. **22** (1958), 585–598; MR **21** #1962], I. I. Pjateckiĭ-Šapiro [Moskov. Gos. Ped. Inst. Uč. Zap. **108** (1957), 317–322; MR **22** #4696], A. M. Polosuev [Vestnik Moskov. Univ. Ser. I Mat. Meh. **1960**, no. 5, 29–39; MR **24** #A107a] and J. Cigler [J. Reine Angew. Math. **205** (1960/61), 91–100].

W. J. LeVeque (Ann Arbor, Mich.)

Citations: MR 21# 1962 = K15-50; MR 22# 4696 = K15-55; MR 24# A107a = K30-22.

K30-24 (27# 5746)

Gabai, Hyman
On the discrepancy of certain sequences mod 1.
Nederl. Akad. Wetensch. Proc. Ser. A **66** = *Indag. Math.* **25** (1963), 603–605.

Let $P_1,\cdots,P_N$ be a sequence of N points in the unit square $0\leqq\xi\leqq 1$, $0\leqq\zeta\leqq 1$. Let $S(x,y)$ denote the number of points of the given sequence which are in the rectangle $0\leqq\xi<x$, $0\leqq\zeta<y$. The difference $\Delta(x,y)=S(x,y)-Nxy$ is called the error of the given sequence at the point (x,y) and the quantity $D=\sup|\Delta(x,y)|$, where the supremum is taken over all points in the unit square, is called the discrepancy of the sequence. The author considers the special sequence $J(2,n)$ of the 2^n points of the form $\left(\frac{t_1}{2}+\frac{t_2}{2^2}+\cdots+\frac{t_n}{2^n}, \frac{t_n}{2}+\frac{t_{n-1}}{2^2}+\cdots+\frac{t_1}{2^n}\right)$, $t_i=0$ or 1. He shows that in this case $n/3<D<n/3+5/2$. This gives a negative answer to a conjecture of Halton [Numer. Math. **2** (1960), 84–90; MR **22** #12688]. Proofs are omitted.

J. E. Cigler (Vienna)

K30-25 (30# 3077)

Franklin, Joel N.
Equidistribution of matrix-power residues modulo one.
Math. Comp. **18** (1964), 560–568.

The author considers the sequence $x^{(n)}$ of d-dimensional column vectors, defined by recurrence formulas $x^{(n+1)}=\{Ax^{(n)}+b\}$ $(n=0,1,2,\cdots)$, where A is a $d\times d$ matrix with integer entries and b is a d-dimensional column vector. For a vector y with real components y_i, $\{y\}$ is the vector with components $\{y_i\}$. Here $\{y_i\}$ is the fractional part of y_i.

The main result is the following theorem: If the matrix A is non-singular and has no eigenvalue which is a root of unity, then $x^{(n)}$ is equidistributed for almost all $x^{(0)}$ in the unit cube $0\leqq x_i^{(0)}<1$ $(i=1,\cdots,d)$. If $b=0$, the above condition for A is not only sufficient but also necessary.

The author also proves a corollary. If the sequence of real numbers x_n is defined by $x_n=a_1x_{n-1}+a_2x_{n-2}+\cdots+a_dx_{n-d}$ $(n>d)$, where $a_1,a_2,\cdots,a_d$ are integers, then for almost all real initial values $x_1,\cdots,x_d$, the sequence of

d-dimensional column vectors

$$x^{(n)} = \begin{pmatrix} x_{n+1} \\ \vdots \\ x_{n+d} \end{pmatrix} \quad (n = 0, 1, \cdots)$$

is equidistributed modulo 1 if and only if $z^d \neq a_1 z^{d-1} + a_2 z^{d-2} + \cdots + a_d$ for z equal to zero or to a root of unity.

The proof is founded on the Weyl criterion [Math. Ann. **77** (1916), 313–352] and an ergodic theorem of F. Riesz [Comment. Math. Helv. **17** (1945), 221–239; MR **7**, 255].

J. Kubilius (Vilnius)

Referred to in K05-51.

K30-26 (32# 1185)

Fomenko, O. M.

On fractional parts of values of functions of two variables. (Russian)

Prace Mat. **7** (1962), 59–61.

The author announces the following extension of a theorem of I. M. Vinogradov [*Selected works* (Russian), pp. 133–134, Izdat. Akad. Nauk SSSR, Moscow, 1952; MR **14**, 610]. Let Ω be a region in the xy-plane, bounded by a closed curve K. Suppose $z(x, y)$ is a function of class C^3 defined in Ω. Assume that K has at most two points on each of the curves $y = \text{const}$ or $z_x = \text{const}$. Let r, s, l, p be absolute constants, and A, B, U, W parameters depending on Ω and $z(x, y)$. (In the applications, $A, B, U, W \to \infty$ as Ω and z are appropriately varied.) Suppose $A \geqq B \geqq 1$, $U \geqq \max(A, 2)$, $W \geqq \max(B, 2)$, $r, s \geqq 1$, $l \geqq 0$, $0 \leqq p \leqq 1$. Assume further that $A^{-1} \leqq z_{xx} \leqq rA^{-1}$, $B^{-1} \leqq z_{yy} \leqq sB^{-1}$, $|z_{xy}| \leqq p(z_{xx}z_{yy})^{1/2}$, $|z_{xxx}| \leqq l(AU^{-1})$. Suppose that the number of sign changes of $z_{xxx}z_{xy} - z_{xx}z_{xxy}$ on any curve $z_x = \text{const}$ is $O(\log U)$. Finally, suppose that Ω is contained in a rectangle (whose sides are parallel to the coordinate axes) with base $\leqq U$ and height $\leqq W$. Let n be the number of lattice points (x, y) in Ω, and arrange the fractional parts $\theta_i = \{z(x, y)\}$ at these points in increasing order: $\theta_1 \leqq \theta_2 \leqq \cdots \leqq \theta_n$. Then for any $n_1 \leqq n$, $\sum_{i=1}^{n_1} \theta_i = n_1^2/2n + Rn_1 + O(U^2W^2T/n)$. Here

$$T = (AB)^{-1/2} \log^2 U + A^{1/2}U^{-1} + A^{-2/5}U^{-1/5},$$

while R is independent of n_1, and is of the order of $UWT^{1/2}/n$. The proof (not given) is said to follow that of Vinogradov, and depends on estimating the sum $S_m = \sum_\Omega \exp(2\pi i m f(x, y))$ extended over the lattice points of Ω (where $m \in \mathbf{Z}$). If $1 \leqq m \leqq B$, one gets

$$S_m = O[(AB)^{-1/2}UWm \log U + W\left(\frac{A}{m}\right)^{1/2} + UW(m^2A^{-2}U^{-1})^{1/5}],$$

while if $B \leqq m \leqq A$, $S_m = O(UWm^{1/2}A^{-1/2})$.

B. Gordon (Los Angeles, Calif.)

Citations: MR 14, 610i = Z25-9.

K30-27 (34# 4241)

Karimov, B.

On the distribution of the fractional parts of certain linear forms in a unit square. (Russian. Uzbek summary)

Izv. Akad. Nauk UzSSR Ser. Fiz.-Mat. Nauk **10** (1966), *no.* 1, 19–22.

On the basis of Weyl's criterion, it is shown that if ω_1, ω_2, ω_3 are real numbers such that ω_1, $\omega_2 + \omega_1\omega_3$, 1 are linearly dependent over the rationals, then the points $(\omega_1 t, \omega_2 t + \omega_3[\omega_1 t])$ are uniformly distributed modulo the unit square. The proof utilizes an analytic device introduced by F. G. Bulaevskaja-Maslova.

W. J. LeVeque (Ann Arbor, Mich.)

Referred to in K30-28.

K30-28 (34# 5753)

Karimov, B.

On the question of the number of fractional parts of certain linear forms in a rectangle. (Russian. Uzbek summary)

Izv. Akad. Nauk UzSSR Ser. Fiz.-Mat. Nauk **10** (1966), *no.* 2, 21–28.

The author generalizes the result of his earlier paper [same Izv. **10** (1966), no. 1, 19–22; MR **34** #4241]. Let ω_1, ω_2, ω_3 be real numbers. Denote by $N_P(\Delta)$ the number of fractional parts $(\{\omega_1 x\}, \{\omega_2 x + \omega_3[\omega_1 x]\})$ belonging to a rectangular region Δ of the unit square as x runs over the integers $0, 1, 2, \cdots, P$. If for any integers L_1, L_2, L_3, $L_1^2 + L_2^2 \neq 0$, the inequality $|\omega_1 L_1 + (\omega_2 + \omega_1\omega_3)L_2 - L_3| \geqq cr^{-s}$ holds, where $r = \max(|L_1|, |L_2|)$, $c > 0$, then $N_P(\Delta) = Pv(\Delta) + O(P^{1-1/(8s+12)})$. Here $v(\Delta)$ is the area of Δ. The constant in the symbol O depends on c, s, ω_3.

The proof is based on the method of trigonometric sums.

J. Kubilius (Vilnius)

Citations: MR 34# 4241 = K30-27.

K30-29 (35# 152)

Dudley, Underwood

Oscillating sequences modulo one.

Trans. Amer. Math. Soc. **126** (1967), 420–426.

The main result is the following theorem. Suppose that f increases without bound, f' decreases, and $f(x) = O(x^{1/2-\varepsilon})$ for some $\varepsilon > 0$, as $x \to \infty$. Then $\{f(n) \cos ny\}$ is uniformly distributed (mod 1) for almost all y.

J. E. Cigler (Groningen)

K30-30 (35# 154)

Gabai, Hyman

On the discrepancy of certain sequences mod 1.

Illinois J. Math. **11** (1967), 1–12.

For any set of N points in the unit square, and any point (x, y) in the square, let $S(x, y)$ denote the number of points of the set which are in the rectangle $0 \leqq \xi < x$, $0 \leqq \eta < y$. The difference $\Delta(x, y) = S(x, y) - Nxy$ is called the error at (x, y), and the quantity $D = \sup |\Delta(x, y)|$, where the supremum is taken over all points (x, y) of the unit square, is called the discrepancy of the set of points.

Among other results, the author determines the exact value of the error for the set of 2^n points of the form $(t_1/2 + t_2/2^2 + \cdots + t_n/2^n,\ t_n/2 + t_{n-1}/2^2 + \cdots + t_1/2^n)$, $t_i = 0, 1$, and shows that $n/3 < D < n/3 + 3$ for this set.

J. E. Cigler (Groningen)

K30-31 (37# 166)

Kaufman, Robert

Some irregular distributions modulo 1.

Proc. Amer. Math. Soc. **19** (1968), 592–594.

Let $r(x)$ denote the distance of the real number x to the nearest integer. For $0 < y < 1$ and $0 < \alpha < \frac{1}{2}$, put $B(n)$ = the number of integers $k \geqq 1$ for which $r(2^{-k}ny) > \alpha$. The author proves that for any $b \in (0, 1)$, $\sum_{n=1}^\infty b^{B(n)} = +\infty$, which contradicts the conjecture $B(n) \sim (1 - 2\alpha) \log n/\log 2$. The proof is based upon some properties of random Fourier-Stieltjes transforms.

O. P. Stackelberg (Durham, N.C.)

K30-32 (41# 1656)

Kuipers, L.

Remark on a paper by R. L. Duncan concerning the uniform distribution mod 1 **of the sequence of the logarithms of the Fibonacci numbers.**

Fibonacci Quart. **7** (1969), *no.* 5, 465–466, 473.

R. L. Duncan has proved [same Quart. **5** (1967), 137–140; MR **39** #1412] that if μ_n is the nth Fibonacci number, the sequence $\{\log \mu_n\}$ is uniformly distributed (mod 1). The author notes that this also follows from a theorem of Van der Corput since $\lim_{n\to\infty} \log \mu_{n+1} - \log \mu_n = \log \frac{1}{2}(1+\sqrt{5})$ is irrational. Similarly, he proves, using a theorem of C. L. Vanden Eynden ["The uniform distribution of sequences", Ph.D. Thesis, Univ. of Oregon, Eugene, Oregon, 1962], that the sequence of integral parts $[\log \mu_n]$ is uniformly distributed (mod m) for every integer $m > 1$.

I. Danicic (Aberystwyth)

Citations: MR 39# 1412 = B36-125.

K30-33 **(44# 5281)**

Thorp, Edward; Whitley, Robert
Poincaré's conjecture and the distribution of digits in tables.
Compositio Math. **23** (1971), 233–250.

Let $E_i(n)$ be the number of even digits and $O_i(n)$ the number of odd digits occurring in the ith column of the table of logarithms of the first n integers to the base 10. Poincaré conjectured that $\lim_{n\to\infty} E_i(n)/n$ and $\lim_{n\to\infty} O_i(n)/n$ exist and equal $\frac{1}{2}$. It is known that this conjecture is false. In the present paper the authors first determine the set of accumulation points for the sequences $\{E_i(n)/n\}$ and $\{O_i(n)/n\}$; for both sequences they find the same closed interval. Then they generalize their method to the following case. Let f be a strictly increasing function mapping $[0, \infty)$ onto itself with $\lim_{x\to\infty} f(x)/x = 0$ and let $N[f, i, l, n]$, denote the number of occurrences of the digit l in the ith decimal place of the numbers $\{f(k): 1 \leq k \leq n\}$. The authors give formulas for the closed interval which is the set of clusterpoints of the sequences $\{N[f, i, l, n]/n\}$.

Their results are related to a well-known theorem of Koksma about the distribution function (mod 1) of a sequence $\{f(n)\}$, where f is a slowly increasing real function.

H. G. Meijer (Delft)

K35 DISTRIBUTION (mod 1): WELL-DISTRIBUTED SEQUENCES, OTHER VARIATIONS

See also reviews B48-26, B56-13, K10-47, K15-11, K15-48, K15-78, K20-25, K40-23, K40-36, K40-39, K40-50, K40-51, K40-59, Q25-23, Z10-20, Z10-30.

K35-1 **(9, 570a)**

Ammann, André. Sur une application d'un théorème de calcul intégral à l'étude des répartitions module 1. C. R. Séances Soc. Phys. Hist. Nat. Genève **64**, 58–61 (1947).

L'auteur désigne d'une façon générale par δ un intervalle (fermé à gauche) pris sur le segment (0, 1), par c la longeur de δ, par $\bar{y}$ le reste, module 1, d'un nombre quelconque y, par $\pi^\delta(y)$ la fonction de période 1 qui vaut 1 si $\bar{y}$ est sur δ et 0 dans le cas contraire. Soit une suite de nombres $\{y_n\}$: l'auteur la dit "unifiante" si, en posant $f_n^\delta = \sum_{i=1}^n \pi^\delta(y_i)/n$, la suite $\{f_n^\delta\}$ admet le point d'accumulation c, et cela pour tout δ; une suite peut être unifiante sans être équirépartie. L'auteur a pu montrer, étant donnée une suite de fonctions $\{y_n(x)\}$, que la suite de nombres $y_n(x)$, où x a une valeur fixée quelconque, est pour presque-tout x une suite unifiante, pourvu que les fonctions $y_n(x)$ satisfassent à des conditions convenables: par exemple, si les $y_n(x)$ sont linéaires de la forme: $y_n(x) = t_n x$, il suffit que les t_n tendent vers l'infini; l'auteur indique des conditions suffisantes assez larges, applicables à les $y_n(x)$ non nécessairements linéaires.

R. Fortet (Caen).

K35-2 **(10, 17g)**

Ammann, André. Quelques Propriétés Concernant la Répartition des Suites de Nombres Module Un. Thesis, University of Geneva, 1947. 39 pp.

Soit $\bar{x}$ le reste mod (ω) de x. On sait qu'une suite réelle $\{x_i\}$ est dite équirépartie mod (ω) si, f_n étant le rapport à n du nombre de ceux des n premiers $\bar{x}_i$ qui appartiennent à une partie (α, β) de $(0, \omega)$, on a; $\lim_{n\to+\infty} f_n = (\beta - \alpha)/\omega$, et cela quels que soient α et β; ou encore (définition équivalente), si l'on a pour toute fonction $P(x)$ de période ω et intégrable-R, $\lim_{n\to\infty} F_n = [P]$, où $[P] = \omega^{-1}\int_0^\omega P(x)dx$ et $F_n = \sum_{i=1}^n P(x_i)/n$. L'auteur introduit les définitions nouvelles suivantes: (a) $\{x_i\}$ est dite "unifiante" mod (ω) si l'on a la propriété plus faible: (1) $\liminf F_n \leqq [P] \leqq \limsup F_n$, qui entraine, si $P(x)$ est bornée, qu'un des points d'accumulation de la suite $\{F_n\}$ est égal à $[P]$ [l'unifiance est réalisée si seulement (1) est satisfaite pour les $P(x)$ de la forme $a_0 + \sum_{k=1}^l (a_k \cos 2k\pi\omega^{-1}x + b_k \sin 2k\pi\omega^{-1}x)$ avec a_0, a_k, b_k rationnels]; (b) $\{x_i\}$ est "équirépartie totalement" si elle est équirépartie pour tout module entier (et par suite pour tout module rationnel); (c) $\{x_i\}$ est "unifiante totalement" si elle est unifiante pour tout module entier (et par suite pour tout module rationnel).

Appelant "normale" toute fonction $x(t)$ définie sur (0, 1) et pourvue d'une dérivée $x'(t) > 0$, continue et non-décroissante, et "suite normale" toute suite $\{x_i(t)\}$ de fonctions normales $x_i(t)$ telles que $\lim_{i\to\infty} x_i'(0) = +\infty$, l'auteur démontre que: (1) si une suite de fonctions $\{x_i(t)\}$ est normale, la suite numérique $\{x_i(t)\}$ (où t à une valeur déterminée quelconque) est: (α) unifiante pour tout module ω pour presque toutes les valeurs de t; (β) unifiante totalement pour presque toutes les valeurs de t; (2) en prenant $x_i(t) = a_i t$ où a_i est indépendant de t, si la suite $\{x_i(t)\}$ (qui n'est pas forcément normale) est équirépartie mod (1) presque-partout, elle est équirépartie totalement presque-partout; si elle n'est pas équirépartie mod (1) presque-partout, elle n'est équirépartie totalement que sur un ensemble de mesure nulle.

Ces résultats sout obtenus essentiellement à l'aide des lemmes suivants. (A) Étant donnée une suite infinie $\{F_n(t)\}$ de fonctions réelles sommables sur (0, 1), si, quel que soit α, on a: $\lim_{n\to\infty} \int_0^\alpha F_n(t)dt = 0$ on a presque-partout $\liminf F_n(t) \leqq 0 \leqq \limsup F_n(t)$, pourvu que les $F_n(t)$ admettent une majorante fixe sommable [Note du reviewer: il suffit d'une condition sensiblement moins stricte]. (B) Si la suite de fonctions $\{x_i(t)\}$ est normale, on a: $\lim_{i\to\infty} \int_0^1 P[x_i(t)]dt = [P]$ pour toute fonction $P(x)$ périodique de période 1 et sommable. L'auteur termine par des compléments et des exemples.

R. Fortet (Caen).

K35-3 **(10, 18a)**

Ammann, A. Sur les répartitions des suites de nombres réels. Résumé d'une thèse présentée à l'Université de Genève. Comment. Math. Helv. **21**, 327–331 (1948).

Cf. the preceding review.

K35-4 **(10, 235e)**

Korobov, N. M. On functions with uniformly distributed fractional parts. Doklady Akad. Nauk SSSR (N.S.) **62**, 21–22 (1948). (Russian)

A function $f(x)$ is called uniformly distributed if the sequence $\{f(k)\}$, $k = 1, 2, \cdots$, where $\{z\}$ denotes the fractional part of z, fills the interval (0, 1) with asymptotically uniform density. A function $f(x)$ is called completely uniformly distributed if, for every set of integers $n_1, n_2, \cdots, n_s$, the function $n_1 f(x+1) + \cdots + n_s f(x+s)$ is uniformly distributed. Theorem 1: there exist completely uniformly distributed functions; an example is $f(x)=\sum_{k=0}^{\infty}x^k \exp(-e^k)$. Theorem 3: let $a\geqq 2$ be an integer, $f(x)$ a completely uniformly distributed function, and $b=\sum_{k=1}^{\infty}a^{-k}[a\{f(k)\}]$; then the function ba^x is uniformly distributed. Proofs of these theorems are not given, but are said to be obtainable by following the method of Weyl [Math. Ann. **77**, 313–352 (1916)] and using recent results of Vinogradoff [C. R. (Doklady) Acad. Sci. URSS (N.S.) **43**, 47–48 (1944); these Rev. **6**, 170] and others on the order of magnitude of trigonometrical sums. *F. J. Dyson* (Princeton, N. J.).

Citations: MR 6, 170i = L15-11.

K35-5 (11, 231b)

Korobov, N. M. Some problems on the distribution of fractional parts. Uspehi Matem. Nauk (N.S.) **4**, no. 1(29), 189–190 (1949). (Russian)

The following theorems are stated. (I) Let $f(x)=\sum_{\nu=0}^{\infty}a_\nu x^\nu$, where $|a_\nu|=e^{-\omega(\nu)}$. If, for every sufficiently large ν, $\omega(\nu)>\nu^\lambda$ and $(1+\beta_1/\nu)\omega(\nu)\leqq\omega(\nu+1)\leqq\beta_2\nu\omega(\nu)$, where $\lambda>3$, $\beta_1>1$, $0<\beta_2<1$, then $f(x)$ is uniformly distributed. If a function $g(x)$ possesses the property that for every set of integers $m_1, \cdots, m_s$ not all zero the function

$$F_s(x)=m_1g(x+1)+\cdots+m_sg(x+s)$$

is uniformly distributed it is said to be completely uniformly distributed. The function $f(x)$ is such a function. (II) For every integer $q\geqq 2$ put $\alpha=\sum_{k=1}^{\infty}[\theta_kq]q^{-k}$, where $\theta_k=\{\varphi(k)\}$ is the fractional part of a completely uniformly distributed function $\varphi(k)$. Then αq^x is uniformly distributed. (III) Let $q_1, q_2, \cdots$ be a sequence of integers with $2\leqq q_1\leqq q_2\leqq\cdots$, $q_x\to\infty$ as $x\to\infty$, and put $f(x)=\alpha q_1q_2\cdots q_x$. Then a necessary and sufficient condition for $f(x)$ to be uniformly distributed is that α may be represented in the form $\alpha=\sum_{k=1}^{\infty}[\theta_kq_k]/(q_1q_2\cdots q_k)$, where $\theta_k=\{\varphi(k)\}$ is the fractional part of a uniformly distributed function $\varphi(k)$. *R. A. Rankin* (Cambridge, England).

Referred to in K15-22, K15-31.

K35-6 (15, 106g)

Šapiro-Pyateckiĭ, I. I. On a generalization of the notion of uniform distribution of fractional parts. Mat. Sbornik N.S. **30**(72), 669–676 (1952). (Russian)

Suppose $\theta_1, \theta_2, \cdots$ is a sequence of real numbers in the interval $[0, 1)$. If $0\leqq\alpha<\beta\leqq 1$, denote by $P(N;\alpha,\beta)$ the number of positive integers k such that $\alpha\leqq\theta_k\leqq\beta$ and $k\leqq N$. The author calls the sequence $\theta_1, \theta_2, \cdots$ almost uniformly distributed if there exists an increasing sequence of positive integers $N_1, N_2, \cdots$ such that $\lim_{m\to\infty}N_m^{-1}P(N_m;\alpha,\beta)=\beta-\alpha$ for every pair α, β. (Uniform distribution itself refers to the case in which $N_m=m$ for every m.) The author obtains two separate sufficient conditions on a sequence of positive integers $n_1, n_2, \cdots$ under either of which the sequence of fractional parts $\{\alpha n_1\}, \{\alpha n_2\}, \cdots$ is almost uniformly distributed for all but denumerably many real α. Condition I: There exist two sequences of positive integers $m_1, m_2, \cdots$ and $\nu_1, \nu_2, \cdots$ such that $n_k=m_{\nu_k}$ for every k, the sequence $\nu_1, \nu_2, \cdots$ has positive density, and the sequence of fractional parts $\{\alpha m_1\}, \{\alpha m_2\}, \cdots$ is uniformly distributed for all but denumerably many α. Condition II: There exists a positive integer r such that $\int_0^1|\sum_{k=1}^{s}\exp(2\pi ixn_k)|^{2r}dx=O(s^{2r}/n_s)$. Each of these two conditions contains the condition of positive density as a special case, the former when $m_k=k$ for every k and the latter when $r=1$. However, the author shows by an example that even the condition of positive density of the sequence $n_1, n_2, \cdots$ is not enough to ensure uniform distribution of the sequence $\{\alpha n_1\}, \{\alpha n_2\}, \cdots$ for all but denumerably many α. The proofs in this paper are not difficult, but they require fairly powerful tools from the theory of functions of a real variable. Needless to say, the basic idea on which Weyl's treatment of uniform distribution is based [Math. Ann. **77**, 313–352 (1916)] also plays a rôle. *P. T. Bateman* (Urbana, Ill.).

Referred to in B48-16, K05-68.

K35-7 (15, 511b)

Tsuji, Masatsugu. On the uniform distribution of numbers mod. 1. J. Math. Soc. Japan **4**, 313–322 (1952).

The author generalizes the concept of uniform distribution mod 1 of sequences as follows: Let $\{\lambda_n\}$ be a non-increasing sequence of positive numbers such that $\sum_1^\infty\lambda_n=\infty$. Let $\{x_n\}$ be a sequence of reals of fractional parts $\bar{x}_n=x_n-[x_n]$. Let I be a subinterval of $[0, 1]$, $\varphi(x)$ its characteristic function. If for every I

$$\lim_{n\to\infty}\frac{\lambda_1\varphi(\bar{x}_1)+\cdots+\lambda_n\varphi(\bar{x}_n)}{\lambda_1+\cdots+\lambda_n}=\text{length of } I,$$

then we say that the sequence $\{x_n\}$ is $\{\lambda_n\}$-uniformly distributed mod 1. The (ordinary) uniform distribution mod 1 is obtained when $\lambda_n=1$ for all n. Weyl's theorem [Math. Ann. **77**, 313–352 (1916)] generalizes as follows: $\{x_n\}$ is $\{\lambda_n\}$-unif. distrib. mod 1 if and only if for $k=1, 2, \cdots$ we have $\sum_{\nu=1}^{n}\lambda_\nu\exp(2\pi kx_\nu i)=o(\sum_1^n\lambda_\nu)$. Fejér's theorem [Pólya and Szegö, Aufgaben und Lehrsätze aus der Analysis, Bd. 1, Springer, Berlin, 1925, p. 72, Problem 174] is derived and generalized by a new method using Euler's summation formula. A new generalization of van der Corput's fundamental inequality [Acta Math. **56**, 373–456 (1931), p. 407] allows one to derive the following extension of van der Corput's theorem: Let the above sequence $\{\lambda_n\}$ be also such that λ_n/λ_{n+k} is a decreasing function of n for each $k=1, 2, \cdots$. If $\{g(n+h)-g(n)\}$ is $\{\lambda_n\}$-unif. distrib. mod 1 for each $h=1, 2, \cdots$, then also $\{g(n)\}$ is $\{\lambda_n\}$-unif. distrib. mod 1. As a very special case of applications of these theorems it is found that $\{\log n\}$ is $\{1/n\}$-unif. distrib. mod 1, illustrating the smoothing effect of the sequence $\{\lambda_n\}$ in this theory. *I. J. Schoenberg*.

K35-8 (18, 720d)

Korobov, N. M. On completely uniform distribution and conjunctly normal numbers. Izv. Akad. Nauk SSSR. Ser. Mat. **20** (1956), 649–660. (Russian)

For a certain strictly increasing sequence of positive integers $\{n_\nu\}$ $(\nu=1, 2, 3, \cdots)$ and an integer $q>1$, the function $\alpha(x)$ is defined by

$$\alpha(x)=\sum_{\nu=1}^{\infty}\left(\frac{1}{q^{n_\nu}}-\frac{1}{q^{n_{\nu+1}}}\right)\frac{x^\nu}{p^2_\nu},$$

where $\{p_\nu\}$ is a strictly increasing sequence of primes. It is proved that the fractional parts of the function $\varphi(x)=\alpha(x)q^x$ are completely uniformly distributed. It is also shown that

$$\sum_{x=1}^{P}e^{2\pi i\varphi(x)}=O(P^{\frac{1}{2}}\log P),$$

$$\sum_{x=1}^{P}e^{2\pi iF_s(x)}=O(P^{\frac{1}{2}}\log P,$$

where

$$F_s(x)=m_1\varphi(x+1)+m_2\varphi(x+2)+\cdots+m_s\varphi(x+s),$$

and $m_1, m_2, \cdots, m_s$ are arbitrary integers not all zero. These theorems extend and improve earlier results of the author [Izv. Akad. Nauk SSSR. Ser. Mat. **14** (1950), 215–238; MR 12, 321].

It is also proved that if $\varphi(x)$ is any completely uniformly distributed function, and $q_1, q_2, \cdots, q_s$ are integers greater than unity, and $\alpha_1, \alpha_2, \cdots, \alpha_s$ are defined by

$$\alpha_\nu=\sum_{k=1}^{\infty}\frac{[\{\varphi(sk+\nu)\}q_\nu]}{q_\nu^{\,k}}\qquad(1\leqq\nu\leqq s)$$

then the system of functions $\alpha_1 q_1{}^x, \cdots, \alpha_s q_s{}^x$ is uniformly distributed in s-dimensional space. Here brackets $[\cdots]$ and $\{\cdots\}$ denote integral and fractional parts respectively. This result was stated earlier by the author without proof [Dokl. Akad. Nauk SSSR (N.S.) **84** (1952), 13–16; **MR 14**, 144]. *R. A. Rankin* (Glasgow).

Citations: MR 12, 321a = K15-11; MR 14, 144a = K15-24.

K35-9 (20# 2313a; 20# 2313b)

Petersen, G. M. 'Almost convergence' and uniformly distributed sequences. Quart. J. Math. Oxford Ser. (2) **7** (1956), 188–191.

Keogh, F. R.; Lawton, B.; and Petersen, G. M. Well distributed sequences. Canad. J. Math. **10** (1958), 572–576.

Let $I=[a, b]$ be a subinterval of $[0, 1]$, $I(x)$ the characteristic function of I. A sequence $0 \leqq s_n \leqq 1$, $n=1, 2, \cdots$ is called well-distributed if $I(s_n)$ almost converges to $b-a$ for each I, i.e., if $(I(s_{n+1})+\cdots+I(s_{n+p}))/p \to b-a$ for $p\to\infty$, uniformly in n. Well-distributed sequences form a subclass of uniformly distributed ones. In the first paper it is shown that s_n is well-distributed if and only if $\exp(2\pi i k s_n)$ almost converges to zero for $n\to\infty$ and each $k=1, 2, \cdots$. Examples of sequences which are or are not well-distributed are given in the second paper. (1) t_k is well-distributed if s_k is and if $s_k - t_k \to 0$. (2) If θ is irrational, there is a sequence of positive integers n_k, $n_{k+1}/n_k > \lambda > 1$, for which the sequence of fractional parts $\{n_k\theta\}$ is well-distributed. (3) If r is rational, the sequence $\{r^k\theta\}$ is not well-distributed for any θ. {Reviewer remarks that also the sequence $\{f(k)\}$ is well-distributed, where $f(x)$ is a polynomial with an irrational leading coefficient.} *G. G. Lorentz* (Syracuse, N.Y.)

Referred to in K35-11, K35-15, K35-16, K35-18, K35-33, K40-23, K40-36.

K35-10 (21# 7190)

Starčenko, L. P. The construction of a completely uniformly distributed sequence. Dokl. Akad. Nauk SSSR **129** (1959), 519–521. (Russian)

The author gives an example of a completely uniformly distributed sequence. Let $\{\alpha\}$ be the fractional part of α. Put $n_k = [1+\exp(k^3)]$ $(k=1, 2, 3, \cdots)$; also denote by $p_1 = 2, p_2 = 3, \cdots$ the successive primes. Then the sequence is as follows: $\{\log 2\}$, $\{2 \log 2\}, \cdots, \{n_1 \log 2\}$; $\{\log 2\}$, $\{\log 3\}$, $\{2 \log 2\}$, $\{2 \log 3\}, \cdots, \{n_2 \log 2\}$, $\{n_2 \log 3\}; \cdots;$ $\{\log p_1\}$, $\{\log p_2\}, \cdots, \{\log p_r\}$, $\{2 \log p_1\}$, $\{2 \log p_2\}, \cdots,$ $\{2 \log p_r\}, \cdots, \{n_r \log p_1\}$, $\{n_r \log p_2\}, \cdots, \{n_r \log p_r\}; \cdots$. *K. Mahler* (Manchester)

K35-11 (22# 703)

Lawton, B. A note on well distributed sequences. Proc. Amer. Math. Soc. **10** (1959), 891–893.

The concept of a well distributed sequence was introduced by Petersen [Quart. J. Math. Oxford Ser. (2) **7** (1956), 188–191; MR **20** #2313a]. The author points out that the analogues of two further theorems about uniform distribution hold for well distribution.

J. W. S. Cassels (Cambridge, England)

Citations: MR 20# 2313a = K35-9.
Referred to in K40-23.

K35-12 (22# 12097)

Cigler, Johann. Asymptotische Verteilung reeller Zahlen mod 1. Monatsh. Math. **64** (1960), 201–225.

Let G be the additive group of reals (mod 1), $T=(c_{mn})$ a regular summation method. Let $\{x_k\}$ be a sequence of points in G such that T-lim $f(x_k)=\mu\cdot f$ exists for each $f\in C(G)$. Then μ is a bounded linear functional on $C(G)$; the corresponding regular Borel measure on G is called the T-distribution of $\{x_k\}$. Approximating f by a trigonometric polynomial, we have (Weyl) that $\{x_k\}$ has the T-distribution ν if and only if T-lim $e^{2\pi i q x_k} = \nu\cdot e^{2\pi i q x}$ $(q=0, \pm 1, \cdots)$.

It is observed that, in view of the boundedness of $\{f(x_k)\}$, $\{x_k\}$ has a $(C, 1)$-distribution ν if and only if it has a (C, k)-distribution ν $(k\geqq 1)$, and this is true if and only if it has an A-distribution ν. Further, $(C, 1)$ and (M, n^r) are equivalent if $r>-1$. Here, (M, p_n) is defined by $c_{mn}=p_m/(p_0+p_1+\cdots+p_n)$ if $n\leqq m$, $c_{mn}=0$ if $n>m$, where $p_n\geqq 0$, $p_0>0$, $\sum p_n=\infty$. Some typical results in this paper are the following.

(I) If $f(x)$ has a continuous derivative, $f(x)\to\infty$ monotonely, $f'(x)\to 0$, then $\{f(k) \pmod 1\}$ has the $(C, 1)$-distribution ν if and only if $\{f(g(k)) \pmod 1\}$ has the $(M, g'(n))$-distribution ν, where (say) $g(n)=n^s$ with $0<s\leqq 1$. Note that the latter is equivalent with a $(C, 1)$-distribution ν.

(II) If $f(x)$ has a continuous second derivative, $f(x)\to\infty$ monotonely, $f'(x)\to 0$ monotonely, then the only $(C, 1)$-distribution which $\{f(k) \pmod 1\}$ can have is the uniform distribution on $(0, 1)$. {Note of the reviewer: By other methods one can show that if $f(n)\to\infty$, $\Delta f(n)\to 0$ monotonely, then $\{f(k) \pmod 1\}$ has no $(C, 1)$-distribution unless $n\Delta f(n)\to\infty$, in which case (Féjèr) it has the uniform $(C, 1)$-distribution.}

(III) Let $f'(t)\to 0$ monotonely, $tf'(t)\to\infty$. Then the double sequence $\{f(aj+bk) \pmod 1\}$ has the uniform distribution on b as its $(M, 1, 1)$-distribution, the latter with respect to the sequence of rectangles $1\leqq j\leqq cn$, $1\leqq k\leqq dn$ (a, b, c, d constants).

Finally, the author determines under appropriate conditions the upper and lower (M, p_n)-distribution functions of a sequence, generalizing a result of J. F. Koksma [*Diophantische Approximationen*, Springer, Berlin, 1936; p. 88]; special attention is given to the case that p_n is an arithmetic function like $\tau(n)$, $\sigma^a(n)$ and $r(n)$. *J. H. B. Kemperman* (Madison, Wis.)

K35-13 (23# A2408)

Postnikova, L. P.
On a theorem of N. M. Korobov.
Dokl. Akad. Nauk SSSR **134** (1960), 42–43 (*Russian*); *translated as Soviet Math. Dokl.* **1** (1961), 1019–1020.

A sequence $\alpha_1, \alpha_2, \cdots$ of numbers in $[0, 1]$ is said to be completely uniformly distributed if for every positive integer t the sequence $(\alpha_1, \cdots, \alpha_t)$, $(\alpha_2, \cdots, \alpha_{t+1}), \cdots$ is uniformly distributed in the t-dimensional unit cube. A finite set of sequences $a_{11}, a_{12}, \cdots; a_{21}, a_{22}, \cdots; \cdots;$ $a_{s1}, a_{s2}, \cdots$, in which each element in the ith sequence is one of the numbers $0, 1, \cdots, g_i-1$ $(g_i\geqq 2)$ for $i=1, \cdots, s$, is said to be jointly normal if every $s\times s$ matrix which could possibly arise by taking s consecutive columns from the above array does in fact occur with its proper limiting frequency. (The adverb "jointly" is to be omitted if $s=1$.) N. M. Korobov showed [Izv. Akad. Nauk SSSR. Ser. Mat. **14** (1950), 215–238; MR **12**, 321] that if the sequence $\alpha_1, \alpha_2, \cdots$ is completely uniformly distributed, then the sequence $[\alpha_1 q], [\alpha_2 q], \cdots$ is normal. It is shown here that under the same hypothesis the sequences $[\alpha_1 q_1], [\alpha_2 q_1], \cdots$ and $[\{\alpha_1 q_1\} q_2], [\{\alpha_2 q_1\} q_2], \cdots$ are jointly normal. As an application, the following problem is solved: given a normal sequence, find a second sequence (with specified g_2) which is jointly normal with it.

W. J. LeVeque (Ann Arbor, Mich.)

Citations: MR 12, 321a = K15-11.

K35-14 **(23# A2410)**

Hlawka, Edmund

Erbliche Eigenschaften in der Theorie der Gleichverteilung.

Publ. Math. Debrecen **7** (1960), 181–186.

Let $\langle x_k \rangle$ be a sequence of real numbers $x_1, x_2, x_3, \cdots$. Let $y_k^{(h)} = x_{k+h} - x_k$. The author calls a property E of sequences of real numbers "hereditary" (erblich) if the following is true: If $\langle y_k^{(h)} \rangle$ has the property E for every positive integral h, then $\langle x_k \rangle$ itself and all its subsequences of the form $\langle x_{ks+r} \rangle$ with integral r and s ($r \geqq 0$, $s \geqq 1$) also have property E. It was known previously that the property of equidistribution (uniform distribution) modulo 1 is hereditary. The author shows several other properties of sequences of real numbers to be hereditary. These properties are refinements of equidistribution modulo 1, and several examples of such hereditary properties are given below. [The notation $e(x)$ stands for $\exp(2\pi i x)$.] (a) $\langle x_k \rangle$ is a g-g-sequence (uniformly equidistributed), that is, $\lim_n (1/n) \sum_{k=1}^{n} e(x_{k+p}) = 0$ uniformly with respect to p ($p = 0, 1, 2, \cdots$). [$p = 0$ alone means equidistribution modulo 1.] (b) $\langle x_k \rangle$ is a v-n-g-sequence (equidistributed of degree v, where v is a positive integer), that is, the sequence of vectors $\langle (x_k, x_{k+1}, \cdots, x_{k+v-1}) \rangle$ is equidistributed modulo 1 in the space of v dimensions. (c) $\langle x_k \rangle$ is an n-g-sequence (normally equidistributed), that is, it is a v-n-g-sequence for every positive integral v. Relations between these various hereditary properties are derived, and it is also shown that, in a certain sense, almost all sequences of real numbers are n-g-sequences.

F. Herzog (E. Lansing, Mich.)

K35-15 **(26# 1299)**

Dowidar, A. F.; Petersen, G. M.

The distribution of sequences and summability.

Canad. J. Math. **15** (1963), 1–10.

The paper is broadly in two sections. Section 1: The main result is the following generalisation of a theorem of Henstock's [J. London Math. Soc. **25** (1950), 27–33; MR **11**, 429]. Let A be a regular matrix and (s_n) a real sequence with $|s_n| \leqq B$. Let $(x(n))$, $n = 1, 2, \cdots$, denote the subsequence of the positive integers for which $s_{x(n)} \leqq x$ and let $g_m(x) = \sum_{n=1}^{\infty} a_{m,x(n)}$. If $g_m(x) \to g(x)$ as $m \to \infty$ for all x in $[-B, B]$, then for each continuous function $f(x)$ defined on $[-B, B]$, $\lim_m \sum_{n=1}^{\infty} a_{mn} f(s_n) = \int_{-B}^{B} f(x)\, dg(x)$. If $a_{mn} \geqq 0$, then the above result is true if $\lim_m g_m(x)$ exists for an everywhere dense set in $[-B, B]$. Section 2: A sequence (s_n), $0 < s_n \leqq 1$, is said to be well-distributed if for each sub-interval $I = [a, b]$ of $[0, 1]$, the sequence $I(s_n)$ is almost convergent to $b - a$, where $I(x)$ denotes the characteristic function of the interval $[a, b]$, i.e., if $\lim_p (1/p) \sum_{k=n+1}^{n+p} I(s_k) = b - a$, uniformly in n. Well-distributed sequences are necessarily uniformly distributed. That the sequence of fractional parts $\{n!\alpha\}$ is uniformly distributed for almost all α is known. The present authors prove that for almost all α, $0 < \alpha \leqq 1$, the sequence of fractional parts $\{n!\alpha\}$ is not well-distributed. Indeed, they prove the stronger result that if $n(k)$ is a subsequence of the positive integers such that $n(k)/n(k-1) \uparrow +\infty$, then the set of α, $0 < \alpha \leqq 1$, for which the sequence of fractional parts $\{n(k)\alpha\}$ is not well-distributed is of measure one. The authors prove also, in partial amendment to one of their earlier results [Keogh, Lawton and Petersen, Canad. J. Math. **10** (1958), 572–576; MR **20** #2313b], that the sequence of fractional parts $\{r^k\theta\}$ is not well-distributed for any θ when r is an integer.

M. S. Ramanujan (Madras)

Citations: MR 20# 2313b = K35-9.

Referred to in K35-18, K35-22, K35-27, K35-29.

K35-16 **(27# 3620)**

Petersen, G. M.; McGregor, M. T.

On the structure of well distributed sequences.

Nieuw Arch. Wisk. (3) **11** (1963), 64–67.

Let $\{s_n\}$ be a sequence of points in $[0, 1]$; it is said to be well distributed provided that

$$p^{-1} \sum_{k=n+1}^{n+p} I_{[a,b]}(s_k) \to b - a,$$

uniformly in n, for every interval $[a, b] \subset [0, 1]$, where $I_{[a,b]}$ is the characteristic function of $[a, b]$ [Petersen, Quart. J. Math. Oxford Ser. (2) **7** (1956), 188–191; MR **20** #2313a]. The authors prove the following negative result: If $f(x) \uparrow \infty$ and $f(k) - f(k-1) \to 0$, then for almost all α the sequence $\{f(k)\alpha\}$ is not well distributed.

R. P. Boas, Jr. (Evanston, Ill.)

Citations: MR 20# 2313a = K35-9.

Referred to in K35-18.

K35-17 **(29# 2235)**

Schnabl, Roman

Zur Theorie der homogenen Gleichverteilung modulo 1.

Österreich. Akad. Wiss. Math.-Natur. Kl. S.-B. II **172** (1963), 43–77.

P. Erdős and G. G. Lorentz [Acta Arith. **5** (1958), 35–44; MR **21** #37] called a sequence (x_n) homogeneously equidistributed mod 1 if for every positive integer d the sequence (x_{nd}/d) is uniformly distributed mod 1. The author generalizes this concept to the case of uniform distribution with respect to weighted means (M, λ_n) and constructs examples of such sequences. He also generalizes the concept of normal number to the case of weighted means and proves that there are numbers which are $(M, 1/n)$-normal but not normal in the usual sense. Some metrical theorems and a generalization of a theorem of Erdős and Lorentz conclude the paper.

J. E. Cigler (Vienna)

Citations: MR 21# 37 = B52-9.

K35-18 **(30# 1112)**

Petersen, G. M.; McGregor, M. T.

On the structure of well distributed sequences. II.

Nederl. Akad. Wetensch. Proc. Ser. A **67** = *Indag. Math.* **26** (1964), 477–487.

Part I appeared in Nieuw Arch. Wisk. (3) **11** (1963), 64–67 [MR **27** #3620]. Let s_n be real numbers on $[0, 1]$; let $e(x) = e^{2\pi i x}$. Then the s_n are uniformly distributed if and only if

$$\lim n^{-1} \sum_{k=1}^{n} e(hs_k) = 0$$

for each positive integer h, and well distributed if and only if $\lim p^{-1} \sum_{k=n+1}^{n+p} e(hs_k) = 0$ as $p \to \infty$, uniformly in n. Let $\{\theta\}$ denote the fractional part of θ. A proof [Keogh, Lawton and Petersen, Canad. J. Math. **10** (1958), 572–576; MR **20** #2313b] that $\{r^k\alpha\}$, where r is rational, is never well distributed, is stated to be incorrect; the present authors show that at any rate for almost all α the sequence $\{r^k\alpha\}$ is not well distributed. Several other results are proved. First, a sequence is uniformly distributed if and only if almost all its subsequences are uniformly distributed. Next, if $n(k)$ is a subsequence of the integers and $n(k)/n(k-1) \to \infty$, then $\{n(k)\alpha\}$ is, for almost all α, not well distributed. The same is true if $n(k)$ can be rearranged into $n'(k)$ so that $n'(k)/n'(k-1) \to \infty$. Finally, if $n(k)$ has a subsequence $n(k_i)$ of positive lower density such that $n(k_i)/n(k_{i-1}) \to \infty$, then $\{n(k)\alpha\}$ is, for almost all α, not well distributed; if $\{n(k)\alpha\}$ is well distributed for almost all α, then any subsequence $n(k_i)$ with $n(k_i)/n(k_{i-1}) \to \infty$ has density 0. Several corrections are given to Dowidar and Petersen [ibid. **15** (1963), 1–10; MR **26** #1299].

R. P. Boas, Jr. (Evanston, Ill.)

Citations: MR 20# 2313b = K35-9; MR 26# 1299 = K35-15; MR 27# 3620 = K35-16.
Referred to in K35-22, K35-25, K35-26.

K35-19 (30# 1113)

Helmberg, Gilbert; Paalman-de Miranda, Aida
Almost no sequence is well distributed.
Nederl. Akad. Wetensch. Proc. Ser. A **67** = *Indag. Math.* **26** (1964), 488–492.

According to Dowidar and Petersen (reference at the end of the preceding review [#1112]) almost no sequence, in a certain sense, is well-distributed in [0, 1]. The authors establish this in a different sense, namely, that of the usual product measure on the set of all sequences with elements between 0 and 1. In fact, they prove a more general result. Let X be a second-countable compact Hausdorff space, μ a normed Borel measure on X, X_∞ the product of countably many X's, and μ_∞ the completed product measure on X_∞. An element of X_∞ is μ-uniformly distributed [μ-well distributed] in X if for every continuous f we have

$$\lim_{N\to\infty} N^{-1} \sum_{n=1+h}^{N+h} f(x_n) = \int_X f(x)\, d\mu(x)$$

for $h=0$ [uniformly for $h=0, 1, 2, \cdots$]. It is completely μ-uniformly distributed if $T^n\xi$ is μ_∞-well distributed in X_∞, where T takes $(x_1, x_2, \cdots)$ into $(x_2, x_3, \cdots)$. If μ is not a point measure and $\xi \in X_\infty$ is completely μ-uniformly distributed in X, then ξ is not μ-well distributed in X. Hence μ_∞-almost no $\xi \in X_\infty$ is μ-well distributed in X, and no $\xi \in X_\infty$ is such that $(T^n\xi)$ is μ_∞-well distributed in X_∞.

R. P. Boas, Jr. (Evanston, Ill.)

K35-20 (30# 4744)

Cigler, Johann
Methods of summability and uniform distribution.
Compositio Math. **16**, 44–51 (1964).

Let A be a regular summability method. A sequence $\{x_k\}$ is said to have the A-distribution function $z(x)$ if for each continuous function $f(x)$ with period 1, the sequence $f(x_n)$ is A summable to the value $\int_0^1 f(x)\, dz(x)$. This means, if $A=(a_{nk})$, that $\sum_{k=1}^{\infty} a_{nk} f(x_k)$ tends to this integral as $n\to\infty$. For sequences which are uniformly distributed mod 1, one has $z(x)=x$ and $a_{nk}=1$ $(k\leqq n)$, $a_{nk}=0$ $(k>n)$. Another particular case is that of the weighted means, where $a_{nk}=a_k$ $(k\leqq n)$, $a_{nk}=0$ $(k>n)$. In this expository paper, the author compares the efficiency of different methods of summability with respect to questions of the asymptotic distribution of a sequence, with generalizations to sequences of powers of elements of a compact group with countable base.

J. G. van der Corput (Berkeley, Calif.)

K35-21 (30# 4746)

Kahane, Jean-Pierre
Sur les mauvaises répartitions modulo 1.
Ann. Inst. Fourier (*Grenoble*) **14** (1964), *fasc.* 2, 519–526.

Seien M, M' positive Toeplitz- (Summations-) Matrizen. M' heisse schwächer als M, $M'<M$, wenn es eine positive Toeplitz-matrix Q gibt, so dass $M'=QM$. Eine Folge reeller Zahlen heisse schlecht verteilt mod 1 bezüglich M, wenn sie für kein $M'<M$ gleich verteilt mod 1 ist. Sei $\lambda=(\lambda_j)$ eine monoton wachsende Folge ganzer Zahlen. Eine reelle Zahl $x>0$ heisse normal bezüglich λ, wenn die Folge $(\lambda_j x)$ gleichverteilt mod 1 ist; sie heisse anormal, wenn $(\lambda_j x)$ schlecht verteilt mod 1 ist (bezüglich der Matrix des arithmetischen Mittels). Der Verfasser beweist unter anderem die folgenden Sätze. Satz 1: Sei $\lambda_j=O(j)$, dann ist die Menge der anormalen Zahlen bezüglich λ höchstens abzählbar. Satz 2: λ besitze positive äussere Dichte und Δ sei ein nicht leeres Intervall. Dann ist die Menge $H_\Delta{}^\lambda$ der x, für die $\lambda_j x \notin \Delta$ mod 1 $(j=1, 2, 3, \cdots)$, endlich.

J. E. Cigler (Groningen)

Referred to in K20-20.

K35-22 (31# 4776)

Murdoch, B. H.
A note on well-distributed sequences.
Canad. J. Math. **17** (1965), 808–810.

A sequence $\{x_k\}$ is well distributed (mod 1) if the limit

$$\lim_{N\to\infty} N^{-1} \sum_{k=p+1}^{p+N} \chi_I((x_k)) = |I|$$

exists uniformly in $p\geqq 0$ for all subintervals I of (0, 1) with length $|I|$ and characteristic function $\chi_I(x)$; (x) represents the fractional part of x. The reviewer, in a paper with A. F. Dowidar [same J. **15** (1963), 1–10; MR **26** #1299] and in a paper with M. T. McGregor [Nederl. Akad. Wetensch. Proc. Ser. A **26** (1964), 477–487; MR **30** #1112], has discussed the distribution of $\{r^k\theta\}$ for r an integer and r rational. The following theorem, proved in this paper, greatly extends these previous results: Given a real number α; then $\{\alpha^k\theta\}_1{}^\infty$ is not well distributed (mod 1) for almost all real numbers θ. In the proof two cases are considered, α algebraic and α transcendental. Both cases depend on proving the uniform distribution of $\{\alpha^{k+1}\theta, \cdots, \alpha^{k+v}\theta\}_1{}^\infty$ in the unit v-dimensional cube.

As a corollary, an interesting generalization of the results for integers in the first reference cited is obtained for algebraic integers r of degree v and the distribution of $\{r^{k+1}\theta, \cdots, r^{k+v}\theta\}$ in the v-dimensional unit cube.

G. M. Petersen (Christchurch)

Citations: MR 26# 1299 = K35-15; MR 30# 1112 = K35-18.

K35-23 (31# 5856)

Helson, H.; Kahane, J.-P.
A Fourier method in diophantine problems.
J. Analyse Math. **15** (1965), 245–262.

Let Λ be a sequence of real numbers, $\{\lambda_n\}$, $\lambda_n\to\infty$. Let $E(\Lambda, M)$ denote the set of real numbers u such that the sequence $\{e(\lambda_n u)\}$, where $e(x)=e^{2\pi ix}$, is assigned a limit by the regular Toeplitz matrix M. The theorems in this paper are concerned with the complement, $CE(\Lambda, M)$, of $E(\Lambda, M)$. In the case in which M is the Cesàro matrix, members of $CE(\Lambda, M)$ determine sequences which are not uniformly distributed.

Theorem 1: Let K be a compact subset of the real line such that the norms $\sup_{t\in K} |P(t)|$ and $\sum_{j\geqq k} |a_j|$ are equivalent for the set of trigonometric polynomials (1) $P(t)=\sum_{j\geqq k} a_j e(\lambda_j t)$, k being some fixed positive integer. Then K intersects $CE(\Lambda, M)$ in a non-denumerable set, for each regular Toeplitz method M.

If a compact set K has the property of the theorem relative to a sequence Λ, K is said to be appropriate to Λ. If the summation in (1) is over the range 1 to ∞, and there is an increasing sequence of integers such that $1=m_1<m_2<\cdots$ and

$$\sum_{i=1}^{\infty} \sum_{j=m_i}^{m_{i+1}-1} a_j e(\lambda_j t)$$

is uniformly convergent on K, then the series is said to converge uniformly by blocks on K.

Theorem 2: Each of the following conditions is necessary and sufficient for a compact set K to be appropriate to a sequence Λ: (a) there is a positive integer k such that every bounded sequence $\{d_k, d_{k+1}, \cdots\}$ has the representation $d_j = \int e(\lambda_j t)\, d\mu(t)$ $(j = k, k+1, \cdots)$ for some measure μ carried on K; (b) for every sequence $\{d_1, d_2, \cdots\}$ of modulus 1, there is a measure μ on K such that

$$\limsup_{j\to\infty} \left| d_j - \int e(\lambda_j t)\, d\mu(t) \right| < 1;$$

(c) each trigonometric series which converges uniformly by blocks on K is absolutely convergent.

Theorem 3: Let Λ be a lacunary sequence and suppose K carries a positive measure ν whose Fourier-Stieltjes transform $F(x)$ is $O(|x|^{-\alpha})$ as $x \to \infty$ for some positive α. Then K is appropriate to Λ. Theorem 4: If K is compact and non-denumerable, then K is appropriate to some sequence Λ.

The paper concludes with some well-chosen examples and a discussion of the relation of these theorems to others on uniform distribution. *G. M. Petersen* (Christchurch)

Referred to in K25-28.

K35-24 (32# 2393)

Gerl, Peter

Konstruktion gleichverteilter Punktfolgen.

Monatsh. Math. **69** (1965), 306–317.

In this paper a study is made of the uniform distribution [well-distribution] of r-dimensional sequences. The author proves the existence of such sequences under certain conditions. In fact, he gives different explicit constructions for α_i so that the r-dimensional sequence $\{f_1(n)\alpha_1, \cdots, f_r(n)\alpha_r\}$ is uniformly distributed [well-distributed]. There are five theorems of which the first and third will be quoted.

Theorem 1: Let $\{f_1(n), \cdots, f_r(n)\}$ $(n = 1, 2, \cdots)$ be an r-dimensional sequence of natural numbers which satisfies the following: (1) $f_i(n) | f_i(n+1)$ for $i = 1, \cdots, r$ and $n = 1, 2, \cdots$, (2) $f_i(n+1)/f_i(n) \to \infty$ as $n \to \infty$ $(i = 1, \cdots, r)$. The number α_i is defined by $\alpha_i = [\alpha_i] + \sum_{n=1}^{\infty} (c_{ni}/f_i(n))$. Conclusion: The sequence $\{f_1(n)\alpha_1, \cdots, f_r(n)\alpha_r\}$ is r-dimensionally uniformly distributed [well-distributed] if and only if $c_{ni} = [\langle x_i(n)\rangle (f_i(n)/f_i(n-1))]$ ($[x]$ is the largest integer contained in x and $\langle x\rangle = x - [x]$) and $\{x_1(n), \cdots, x_r(n)\}$ is r-dimensionally uniformly distributed [well-distributed]. Conventionally, $f_i(0) = 1$.

Theorem 3: Let $f(n)$ be a sequence of natural numbers. Set $q_n(c, k) = f(n) + c \cdot n^k$ ($c \neq 0$, k a natural number). For an appropriate pair (c, k) the following relation is fulfilled: There is a sequence of natural numbers a_n with

$$q_n(c, k)a_n = q_{n+1}(c, k) - q_{n-1}(c, k).$$

The continued fraction expansion of $q_2(c, k)/q_1(c, k)$ is $q_2(c, k)/q_1(c, k) = [b_m, \cdots, b_1]$. Define α by

$$\alpha = [b_1, \cdots, b_m, a_1, a_2, \cdots];$$

then the sequence $f(n)\alpha$ is uniformly distributed [well-distributed]. *G. M. Petersen* (Christchurch)

Referred to in K35-30.

K35-25 (33# 113)

Petersen, G. M.; McGregor, M. T.

On the structure of well distributed sequences. III.

Nederl. Akad. Wetensch. Proc. Ser. A **69** = *Indag. Math.* **28** (1966), 42–48.

Let χ be the characteristic function of the interval $[a, b]$. A sequence σ_k is called almost convergent if

$$\lim_{p\to\infty} p^{-1} \sum_{k=n+1}^{n+p} \sigma_k$$

exists, uniformly in n. A sequence s_n is called well distributed if $\chi(s_n)$ is almost convergent to $b - a$ for each $[a, b] \subset [0, 1]$. It is called A-uniformly distributed if, for each continuous g of period 1, $g(s_n)$ is A-summable to $\int_0^1 g(x)\, dx$, where A is a regular matrix summability method. A Tauberian condition for A is a condition $\Delta s_k = O(|f(k)|)$, $\lim f(k) = 0$, which together with A-summability of s_n implies the convergence of s_n. Let $\{\theta\}$ denote the fractional part of θ. The authors prove that if s_n satisfies a Tauberian condition for A (or for almost convergence), then $\{s_n\alpha\}$ is not A-uniformly distributed (or well distributed) for any α, $0 < \alpha < 1$; in particular, $\Delta s_n = o(1)$ prevents $\{s_n\alpha\}$ from being well distributed. They conjecture that if $n_k/n_{k-1} \geqq r > 1$, then $\{n_k\alpha\}$ is not well distributed, and prove that if n_k are integers, and n_k/n_{k-1} is an integer greater than 1, then $\{n_k\alpha\}$ is, for almost all α, not well distributed.

{Part II appeared in same Proc. **67** (1964), 477–487 [MR **30** #1112].} *R. P. Boas, Jr.* (Evanston, Ill.)

Citations: MR 30# 1112 = K35-18.
Referred to in K35-26, K35-29.

K35-26 (35# 157)

Petersen, G. M.

On the structure of well distributed sequences. IV.

Nederl. Akad. Wetensch. Proc. Ser. A **70** = *Indag. Math.* **29** (1967), 128–131.

For notation and terminology, see two earlier papers by the author and M. T. McGregor [same Proc. **26** (1964), 477–487; MR **30** #1112; ibid. **28** (1966), 42–48; MR **33** #113]. In the latter paper, the authors conjectured that if n_k is a sequence of reals such that $n_k | n_{k-1} \geqq r > 1$ for $k = 2, 3, \cdots$, then, for almost all α, $0 < \alpha \leqq 1$, the sequence $\{n_k\alpha\}$ is not well distributed. The results in the present paper establish the above conclusion if $n_k | n_{k-1} > M > 2$. *M. S. Ramanujan* (Ann Arbor, Mich.)

Citations: MR 30# 1112 = K35-18; MR 33# 113 = K35-25.

Referred to in K35-29, K35-30.

K35-27 (33# 2618)

Gerl, Peter

Eine Bemerkung zur gleichmässigen Gleichverteilung mod 1.

Monatsh. Math. **70** (1966), 106–110.

In this paper, three theorems are proved. Theorem 1: If r_n $(n = 1, 2, \cdots)$ is a sequence of real numbers such that $r_{n+1}/r_n \to \infty$ as $n \to \infty$, then $\{r_n\alpha\}$ is not well distributed for almost all α. (Here, $\{r_n\alpha\} = r_n\alpha - [r_n\alpha]$, where $[x]$ is the largest integer contained in x.) Theorem 2: If r_n is any sequence of real numbers satisfying $r_{n+1}/r_n \to \infty$ as $n \to \infty$, there exists a (even uncountably many) fixed number α for which the sequence $r_n\alpha$ is well distributed. Theorem 3: If the sequence $f(n)$ is so selected that for some constant c, $\lim_{N\to\infty}(1/N) \sum'_{n\leqq N} 1 = 0$, where, for each N, the sum $\sum'$ is extended over all $n \leqq N$ such that $f(n+1) - f(n) \geqq c$, then $g(n)\alpha$ is not well distributed for any α. Here, we define $g(n) = a^{f(2k+1)+i}$ for $0 \leqq i \leqq f(2k+2) - f(2k+1)$, $n = \sum_{j=1}^{k} (f(2j) - f(2j-1)) + k + i + 1$, and $k = 0, 1, \cdots$. Theorem 2 indicates that Theorem 1 cannot be improved upon. Theorem 1 is an improvement of a result by A. F. Dowidar and the reviewer [Canad. J. Math. **15** (1963), 1–10; MR **26** #1299]. *G. M. Petersen* (Christchurch)

Citations: MR 26# 1299 = K35-15.
Referred to in K35-30.

K35-28 **(33# 5599)**

Knuth, Donald E.

Construction of a random sequence.

Nordisk Tidskr. Informations-Behandling **5** (1965), 246–250.

A real sequence $[X_1, X_2, X_3, \cdots] = [X_n]_{n=1}^{\infty}$ is defined to be k-distributed (in [0, 1]) if the limit of $\nu_N(u_1, v_1; \cdots; u_k, v_k)/N$, as $N \to \infty$, exists and equals $\prod_{i=1}^{k}(v_i - u_i)$, for all $0 \leqq u_i \leqq v_i \leqq 1$ $(i = 1, \cdots, k)$, where ν_N is the number of k-tuples $[X_{n+1}, \cdots, X_{n+k}]$ with $0 \leqq n < N$ for which every $u_i \leqq X_{n+i} < v_i$ $(i = 1, \cdots, k)$. For such a sequence, the vectors $[X_{n+1}, \cdots, X_{n+k}]$ $(n = 1, 2, 3, \cdots)$ are asymptotically uniformly distributed in the unit hypercube of k dimensions $(0 \leqq x_i \leqq 1 : i = 1, \cdots, k)$. The sequence is completely equidistributed if it is k-distributed for all $k > 0$. The author gives the first explicitly constructed example of a completely equidistributed sequence.

His sequence takes the form $[X_n]_{n=1}^{\infty} = [\mathscr{S}_k]_{k=1}^{\infty}$, where each $\mathscr{S}_k$ is a segment (finite sequence) of real numbers. Each $\mathscr{S}_k$ is a $k \cdot 2^{2k}$-fold repetition of a segment $\mathscr{A}_k$, and any $\mathscr{A}_k$ consists of 2^{k^2} real numbers Z_r. Each Z_r belongs to the set $\mathfrak{F}_k = \{\langle m \cdot 2^{-k}\rangle : m = 1, \cdots, 2^k\}$, where $\langle \cdots \rangle$ denotes the fractional part; $Z_1 = \cdots = Z_k = 0$; thereafter, each successive Z_r is chosen so that it corresponds in $\mathfrak{F}_k$ to the least integer m such that the k-tuple $[Z_{r-k+1}, \cdots, Z_r]$ has not previously occurred in $\mathscr{A}_k$.

The author proves the complete equidistribution of his sequence, using a result of M. H. Martin [Bull. Amer. Math. Soc. **40** (1934), 859–864] and L. R. Ford, Jr. ["A cyclic arrangement of n-tuples", Report P-1070, Rand Corporation, Santa Monica, Calif., 1957] that an endless repetition of the segment $\mathscr{A}_k$ is k-distributed on the finite set of rationals $\mathfrak{F}_k$.

J. H. Halton (Madison, Wis.)

K35-29 **(35# 2834)**

Petersen, G. M.

On the structure of well distributed sequences. V.

Nederl. Akad. Wetensch. Proc. Ser. A **70** = *Indag. Math.* **29** (1967), 229–233.

Part IV appeared in same Proc. **70** (1967), 128–131 [MR **35** #157]. The article under review continues work on problems previously discussed by A. F. Dowidar and the author [Canad. J. Math. **15** (1963), 1–10; MR **26** #1299] and the author and M. T. McGregor [Nederl. Akad. Wetensch. Proc. Ser. A **69** (1966), 42–48; MR **33** #113]. Theorem 1: If $\{n(k)\}$ is a sequence of real numbers with $n(k)/n(k-1) > M > 1$ then, for almost all α, $0 < \alpha \leqq 1$, the sequence of fractional parts of $n(k)\alpha$ is not well distributed. Theorem 2: If $n(k) \geqq n(k-1)$ and $\{n(k)\}$ contains a subsequence $\{n(k_i)\}$ of positive lower density such that $n(k_i)/n(k_{i-1}) > M > 1$ then, for almost all α, the sequence of fractional parts of $n(k)\alpha$ is not well distributed. Here "positive lower density" means that the number of elements of the subsequence with indices between $m+1$ and $m+p$ exceeds p/R for some positive integer R, all m and all $p > P$.

R. P. Boas, Jr. (Evanston, Ill.)

Citations: MR 26# 1299 = K35-15; MR 33# 113 = K35-25; MR 35# 157 = K35-26.

Referred to in K35-30, K35-32.

K35-30 **(35# 4172)**

Zame, Alan

On the measure of well-distributed sequences.

Proc. Amer. Math. Soc. **18** (1967), 575–579.

A sequence of numbers (x_n) is well distributed if for every subinterval I of (0, 1),

$$\lim_{N\to\infty}(1/N)\sum_{n=k}^{k+N} I(x_n\,[\text{modulo } 1]) = \|I\|$$

uniformly in k, where $I(x)$ denotes the characteristic function of the subinterval and $\|I\|$ its length. A sequence (r_n) is lacunary if $\lim\inf_{n\to\infty}(r_n/r_{n-1}) > 1$. $\mu(E)$ denotes the Lebesgue measure of a set E.

The following theorems are proved. Theorem 1: Let (r_n) be a lacunary sequence and let $\mathscr{W}$ be the set of x for which the sequence $(r_n x)$ is well distributed. Then $\mu(\mathscr{W}) = 0$. Theorem 3: Suppose $\lim(r_n/r_{n-1}) = r > 1$ (where r may be infinite). Then (i) if $r = \infty$, the Hausdorff dimension of $\mathscr{W}$ is 1, and (ii) if r is finite, $\mathscr{W}$ may be empty or (uncountably) infinite. Theorem 4: There exists an increasing sequence of positive integers (n_k), with $1 \leqq n_{k+1} - n_k \leqq M$ (for some constant M), such that the sequence $(n_k x)$ is not well distributed for any x. The first theorem answers a question of the reviewer [Nederl. Akad. Wetensch. Proc. Ser. A **70** (1967), 128–131; MR **35** #157] and the other two some of the questions raised by P. Gerl [Monatsh. Math. **69** (1965), 306–317; MR **32** #2393; ibid. **70** (1966), 106–110; MR **33** #2618]. Theorem 1 is also proved in a paper by the reviewer [Nederl. Akad. Wetensch. Proc. Ser. A **70** (1967), 229–233; MR **35** #2834].

The paper has a final selection in which these results are carried over to compact Abelian groups and compared with previous known results in the same field.

G. M. Petersen (Christchurch)

Citations: MR 32# 2393 = K35-24; MR 33# 2618 = K35-27; MR 35# 157 = K35-26; MR 35# 2834 = K35-29.

K35-31 **(37# 6250)**

Gerl, Peter

Einige mehrdimensionale Sätze zur gleichmässigen Gleichverteilung mod 1.

Nederl. Akad. Wetensch. Proc. Ser. A **71** = *Indag. Math.* **30** (1968), 151–161.

The main result of this paper states that for every sequence (A_n) of matrices satisfying $\lim(A_n A_{n+1}^{-1}) = 0$ the set of all x such that $(A_n x)$ is well-distributed mod 1 is a non-empty null-set.

J. E. Cigler (Groningen)

K35-32 **(37# 6639)**

Zame, Alan

Almost convergence and well-distributed sequences.

Canad. J. Math. **20** (1968), 1211–1214.

A sequence (x_n) of real numbers is said to be well distributed modulo 1 if for each subinterval $I = [a, b]$ of [0, 1] we have that $\lim_{n\to\infty} n^{-1}\sum_{k+1}^{k+n}\chi_I(x_m) = b - a$ uniformly in $k = 0, 1, 2, \cdots$, where χ_I is the characteristic function of I modulo 1. Let (n_i) be a strictly increasing sequence of positive integers and $\phi(m) = 1$ if $m \in (n_i)$ and $\phi(m) = 0$ otherwise. If there are a $\delta > 0$ and strictly increasing sequences of positive integers (k_i) and (p_i) such that $p_i^{-1}\sum_{k_i+1}^{k_i+p_i}\phi(m) \geqq \delta$, we shall write $\text{dens}(n_i) \geqq \delta$. The main result of this paper is the following Theorem 1: Suppose that (r_n) is a sequence of positive numbers such that there exist $r > 1$, $\delta > 0$, and a subsequence $(r_{n(i)})$ for which (i) $r_{n(i+1)}/r_{n(i)} \geqq r$ for all i and (ii) $\text{dens}(n(i)) \geqq \delta$. Then for almost all x, the sequence $(r_n x)$ is not well distributed modulo 1. This answers a question proposed by the reviewer [see Nederl. Akad. Wetensch. Proc. Ser. A **70** (1967), 229–233; MR **35** #2834] and would seem to be the last word for theorems of this kind on well distributed sequences. Further structural results will most likely have to concern different properties of the sequence.

G. M. Petersen (Christchurch)

Citations: MR 35# 2834 = K35-29.

K35-33 **(39# 1411)**

Cigler, J.

Some remarks on the distribution mod 1 **of tempered sequences.**

Nieuw Arch. Wisk. (3) **16** (1968), 194–196.

Soit l un entier $\geqq 0$; une suite réelle $(f(n))$ est dite tempérée d'ordre l si f est $(l+1)$ fois continûment différentiable sur $[1, \infty[$ et si $f^{(l+1)}(t) \downarrow 0$, $f^{(l)}(t) \to \infty$, $tf^{(l+1)}(t) \to \infty$ quand $t \to \infty$. On sait, d'après L. Fejér et J. G. van der Corput, que toute suite réelle tempérée est équirépartie [e.r.] (mod 1). L'auteur, se référant à la notion plus forte de suite bien répartie [b.r.] (mod 1) [G. M. Petersen, Quart. J. Math. Oxford Ser. (2) **7** (1956), 188–191; MR **20** #2313a], démontre le théorème suivant: Aucune suite réelle tempérée n'est b.r. (mod 1). À cet effet, il établit d'abord (lemme 3) que, si $(f(n))$ est une suite réelle tempérée, alors il existe une suite strictement croissante d'entiers (a_n) telle que, pour tout $h=0, 1, 2, \cdots$, on ait $\lim_{n\to\infty} f(a_n+h)=0$ (mod 1). Notant $\langle x\rangle$ la partie fractionnaire du réel x, posons $\omega_h=(\langle f(n+h)\rangle)$, où $n=1, 2, \cdots$; soit Ω l'ensemble des suites ω_h, où $h=0, 1, 2, \cdots$, et soit $\bar{\Omega}$ la fermeture de Ω dans $\prod_1^\infty X_n$, où $X_n=T$ et T désigne le tore unité de dimension 1. Il résulte du lemme 3 que, si la suite réelle $(f(n))$ est tempérée, alors la suite nulle appartient à $\bar{\Omega}$; si de plus la suite ω_0 était b.r. dans $[0, 1]$, alors [l'auteur, Compositio Math. **17**, 263–267 (1966); MR **35** #1753] la suite nulle serait e.r. dans $[0, 1]$; cette contradiction manifeste assure le résultat annoncé.

Le rapporteur signale une erreur de date dans la liste des références: p. 196, l. 2, lire 1965 au lieu de 1967.

J. Chauvineau (Paris)

Citations: MR 20# 2313a = K35-9; MR 35# 1753 = K40-51.

K35-34 **(41# 157)**

Burkard, Rainer E.

Zur gleichmäßigen Gleichverteilung von Polynomwerten modulo Δ.

Österreich. Akad. Wiss. Math.-Natur. Kl. S.-B. II **177** (1969), 395–406.

Let $0=z_0<z_1<z_2<\cdots$ with $\lim_{n\to\infty} z_n=\infty$ be a subdivision of the positive real numbers. Now consider the origin-symmetric set (lattice) $\Delta=(\cdots, -z_2, -z_1, 0, z_1, z_2, \cdots)$. Let ω be a sequence of real numbers x_n. Then either $x_n\in[z_{l-1}, z_l)$ or $x_n\in(-z_l, -z_{l-1}]$ for some l. Each of these intervals is of length $|z_l-z_{l-1}|=l_*$, say. {The author denotes this length by l. Why?} Let $[x_n]_\Delta=\min\{|z_{l-1}|, |z_l|\}$ and $\{x_n\}_\Delta=|x_n-[x_n]_\Delta|/l_*$. The sequence ω is said to be uniformly distributed mod Δ (abbreviated: u.d. mod Δ) if the sequence $(\{x_n\}_\Delta)$ is u.d. mod 1, according to W. J. LeVeque [Pacific J. Math. **3** (1953), 757–771; MR **15**, 511].

A class of sequences $(\omega_\sigma)_{\sigma\in S}=(x_{n,\sigma})$ $(n=1, 2, \cdots; \sigma\in S)$ is called well-distributed (abbreviated w.d.) mod Δ if to any $\varepsilon>0$ and all $[\alpha, \beta)$ $(0\leqq\alpha<\beta\leqq L)$, there exists an N_0 independent of σ such that $|N^{-1}A_\sigma(\alpha, \beta, N)-(\beta-\alpha)/L|<\varepsilon$ {l_1 is a typographical error?} for all $N\geqq N_0$ and all $\sigma\in S$. Here $A_\sigma(\alpha, \beta, N)$ is the number of elements $x_{i,\sigma}$ among $x_{1,\sigma}, \cdots, x_{N,\sigma}$ for which $\alpha\leqq x_{i,\sigma}<\beta$. {Reviewer's remark: The author should give here more explanation in order to connect this definition with the first one in the paper.}

A sequence $\omega=(x_n)$ is said to be w.d. mod Δ if the class of sequences $(x_{n+\sigma})$ $(\sigma=1, 2, \cdots)$ is w.d. mod Δ. In Section 3 the author considers a lattice of the type $\Delta=(-f(k), f(k))$, $k=0, 1, 2, \cdots$, where $f(x)=b_px^p+b_{p-1}x^{p-1}+\cdots+b_1x$ (p an integer >1), and $f(x)$ is monotonically increasing in $[0, \infty)$. Let $g(1), g(2), \cdots$ be a sequence of real numbers, where $g(x)=a_qx^q+a_{q-1}x^{q-1}+\cdots+a_0$ (q an integer $\geqq 1$, $a_q\neq 0$). Supposedly $q>p$. Under these assumptions the author shows that the sequence $(g(n))$ $(n=1, 2, \cdots)$ is u.d. mod Δ. Example: $g(x)=x$, $f(x)=\frac{1}{2}x^2+\frac{1}{2}x$.

In Section 4 the author proves: If Δ is a lattice whose interval lengths go to ∞, and if ω is a mod Δ u.d. monotone sequence, then the class of sequences $(\omega_\sigma)_{\sigma\in S}=(x_n+\sigma)$ is w.d. mod Δ if S is a bounded set of real numbers.

In the final Section 5 it is shown that the sequence $(g(n))$ $(n=1, 2, \cdots)$ generated by a polynomial $g(x)$ of degree q is not w.d. mod Δ $(\Delta=(-f(k), f(k)))$ if the degree of $f(k)$ is greater than q. *L. Kuipers* (Carbondale, Ill.)

Citations: MR 15, 511d = K10-14.

K40 DISTRIBUTION (mod 1): CONTINUOUS, p-ADIC AND ABSTRACT ANALOGUES

See also reviews B48-13, B48-22, J64-23, K05-36, K05-47, K05-66, K05-67, K05-68, K05-74, K05-77, K35-20, K35-30, K50-13, K50-46, K50-48, Q25-23, R06-29, T55-42, Z10-18, Z10-33, Z10-34, Z10-38, Z10-47.

K40-1 **(4, 2f)**

Halmos, Paul R. and Samelson, H. On monothetic groups. Proc. Nat. Acad. Sci. U.S.A. **28**, 254–258 (1942).

In a topological group G, let $D(G)$ be the set of all $x\varepsilon G$ such that the subgroup $\{x^n\}$ $(n=0, \pm 1, \pm 2, \cdots)$ generated by x is everywhere dense in G. The authors call G "monothetic" if $D(G)$ is nonvoid; G must then be Abelian, and, if locally compact and not isomorphic to the additive group of integers, must be compact. It is shown that the compact Abelian group G is "monothetic" if and only if: (a) G is "separable" (there exists an everywhere dense enumerable subset of G); (b) the group of the characters of finite order of G is isomorphic to a subgroup of the additive group of rational numbers mod 1. Furthermore, if G is Abelian, compact, connected and satisfies the second countability axiom (that is, has an enumerable character group), $D(G)$ has the Haar measure 1, that is, almost every element of G belongs to $D(G)$. In a totally disconnected group, the measure of $D(G)$ may take any value between 0 and 1; if the second countability axiom is not satisfied, it is asserted that $D(G)$ need not be measurable. All these results are direct consequences of the duality theory of Abelian groups.

A. Weil (Bethlehem, Pa.).

Referred to in K40-2.

K40-2 **(6, 146a)**

Eckmann, Beno. Über monothetische Gruppen. Comment. Math. Helv. **16**, 249–263 (1944).

The first part of the paper treats monothetic groups, that is, topological groups with a dense cyclic subgroup. The results are identical with results obtained by Halmos and Samelson [Proc. Nat. Acad. Sci. U. S. A. **28**, 254–258 (1942); these Rev. **4**, 2; evidently the author did not have access to this paper]; both papers make use of Pontrjagin's duality theory. In the second part the author defines equidistribution of a sequence x_i in a compact topological group G: let $N(M)$ be the number of those x_i's among the first N which are contained in the subset M of G; then the limit value of $N(M)/N$ has to equal the measure of M for every measurable M. Theorem: if the element a of the compact group G does not have 1 as eigenvalue in any nontrivial irreducible representation of G, then the powers of a are equidistributed. This is an extension of Weyl's theorem on

equidistribution. For finite groups a simpler proof is given.
H. Samelson (Syracuse, N. Y.).

Citations: MR 4, 2f = K40-1.
Referred to in K40-16, K40-52.

K40-3 (10, 235f)

Kuipers, Lauwerens. De Asymptotische Verdeling Modulo 1 van de Waarden van Meetbare Functies. [The Asymptotic Distribution Modulo 1 of the Values of Measurable Functions]. Thesis, Free University of Amsterdam, 1947. vi+112 pp.

Most of the known results about asymptotic distribution deal with the case of a discrete sequence $f(1), f(2), f(3), \cdots$. The purpose of the present paper is to establish analogous results for functions $f(t)$ with a continuous range of arguments. The investigation is based on the following definition. A measurable function $f(t)$ defined for $t \geqq 0$ is said to have a uniform distribution mod 1 if, denoting by $S_T(\alpha, \beta)$ (where $T \geqq 0$, $0 \leqq \alpha \leqq \beta \leqq 1$) the set of values t satisfying $0 \leqq t \leqq T$; $\alpha \leqq f(t) < \beta$ (mod 1), one has, for every interval (α, β), $\lim_{T\to\infty} T^{-1}$ meas $S_T = \beta - \alpha$ (modifications of this definition are obtained by replacing the intervals (α, β) by countable sums of intervals or more generally by arbitrary measurable sets).

In analogy with Weyl's well-known criterion for uniform distribution of sequences, the author shows that a measurable function $f(t)$ has a uniform distribution mod 1 if and only if, for every positive integer h, $\lim_{T\to\infty} T^{-1}\int_0^T e^{2\pi i h f(t)} dt = 0$. A similar result is proved for the multidimensional distribution of values of a system of functions $f_1(t), \cdots, f_n(t)$. Among the remaining results of the paper we note the following theorem. A differentiable function $f(t)$ $(t \geqq 0)$, whose derivative $f'(t)$ is a nonincreasing function satisfying $tf'(t) \to \infty$ for $t \to \infty$, has a uniform distribution mod 1. [This is a generalized analogue of a theorem due to Pólya and Szegö.] Further results are analogues of theorems of Koksma, van der Corput and Schoenberg. *W. Hurewicz.*

Referred to in K40-26.

K40-4 (11, 423j; 11, 423k)

Kuipers, L., and Meulenbeld, B. Asymptotic C-distribution. I. Nederl. Akad. Wetensch., Proc. **52**, 1151–1157 = Indagationes Math. **11**, 425–431 (1949).
Kuipers, L., and Meulenbeld, B. Asymptotic C-distribution. II. Nederl. Akad. Wetensch., Proc. **52**, 1158–1163 = Indagationes Math. **11**, 432–437 (1949).

A measurable function $f(t)$, $0 \leqq t < \infty$, is said to be C^{I} uniformly distributed if, for any $0 \leqq \alpha \leqq \beta \leqq 1$,

$$\lim_{T\to\infty} T^{-1}\int_0^T \theta(\alpha, \beta; f(t))dt = \beta - \alpha,$$

where $\theta(\alpha, \beta; f(t)) = 1$ if $\alpha \leqq f(t) < \beta$ (mod 1) and $\theta(\alpha, \beta; f(t)) = 0$ otherwise. This is clearly analogous to the definition of a uniformly distributed sequence. One obtains C^{II} uniform distribution if the interval (α, β) is replaced by any countable union of disjoint intervals, and C^{III} uniform distribution if it is replaced by any measurable set. The authors prove that there exists a function $f(t)$ which is C^{I} uniformly distributed but which is not C^{II} uniformly distributed, but that C^{II} and C^{III} are equivalent. They also give various necessary and sufficient conditions for a function $f(t)$ to be C^{I} or C^{II} uniformly distributed, which are applied to various classes of functions. *P. Erdös.*

Referred to in K40-6, K40-7, K40-10, K40-11, K40-15, K40-21.

K40-5 (11, 424a)

Kuipers, L. On real periodic functions and functions with periodic derivatives. Nederl. Akad. Wetensch., Proc. **53**, 226–232 = Indagationes Math. **12**, 34–40 (1950).

Let $t \geqq 0$, $f(t)$ differentiable and assume that $f'(t)$ has period p. The author proves that if $f(t+p) - f(t) = \tau$ is irrational then $f(t)$ is C^{I}-uniformly distributed (mod 1) [for the definition of C^{I}-uniform distribution see the preceding review]. If τ is rational $f(t)$ is C^{I}-uniformly distributed (mod 1) if and only if $\int_0^p e^{2\pi i h f(t)} dt = 0$ for any integer $h \neq 0$ with $h\tau$ an integer. The author also proves that if $\pi\alpha$ is irrational, $|\beta| < \alpha$, then $\alpha t + \beta \sin t$ is C^{II} uniformly distributed (mod 1). Finally he proves that if $f''(t)$ is periodic and $f'(t)$ is not periodic, then $f(t)$ is C^{I} uniformly distributed (mod 1). *P. Erdös.*

Referred to in K40-21, K40-26.

K40-6 (11, 648a)

Kuipers, L., and Meulenbeld, B. Some theorems in the theory of uniform distribution. Nederl. Akad. Wetensch., Proc. **53**, 305–308 = Indagationes Math. **12**, 53–56 (1950).

Generalising one of their previous results [same Proc. **52**, 1158–1163 = Indagationes Math. **11**, 432–437 (1949); these Rev. **11**, 423] the authors prove the following result. Let $f(t)$ be differentiable, $t \geqq 0$, and assume that $tf'(t)$ is bounded and that for $t > T^*$, we have $|tf'(t) - A| < B < \frac{1}{2}\pi^{-1}$, where A and B are fixed numbers; then $f(t)$ is not C^{I} uniformly distributed (mod 1). *P. Erdös* (Aberdeen).

Citations: MR 11, 423k = K40-4.
Referred to in K40-10, K40-21.

K40-7 (11, 648b)

Meulenbeld, B. On the uniform distribution of the values of functions of n variables. Nederl. Akad. Wetensch., Proc. **53**, 311–317 = Indagationes Math. **12**, 59–65 (1950).

The author extends the definition of C^{I} uniform distribution to functions of n variables and gives various conditions for a function in n variables to be C-uniformly distributed. For the definition of C^{I} uniform distribution see Kuipers and the author [same Proc. **52**, 1151–1157 = Indagationes Math. **11**, 425–431 (1949); these Rev. **11**, 423].
P. Erdös (Aberdeen).

Citations: MR 11, 423j = K40-4.
Referred to in K40-15.

K40-8 (12, 15e)

Kuipers, L., and Meulenbeld, B. New results in the theory of C-uniform distribution. Nederl. Akad. Wetensch., Proc. **53**, 822–827 = Indagationes Math. **12**, 266–271 (1950).

The authors continue their work on C-uniform distribution. Among others they prove, sharpening previous results, the following theorem. Let F be a sequence of intervals Q: $0 \leqq t < T$ with $T \to \infty$. If $f(t)$ is differentiable and satisfies $|tf'(t)| < k$ for $t > t_0 > 0$, then $f(t)$ is not C-uniformly distributed (mod t) in the intervals Q of F. *P. Erdös.*

Referred to in K40-23.

K40-9 (12, 395d)

Chabauty, Claude. Sur la répartition modulo un de certaines suites p-adiques. C. R. Acad. Sci. Paris **231**, 465–466 (1950).

Soit p un nombre premier ($\neq 2$ pour simplifier). Soit Q_p l'ensemble des nombres p-adiques. Tout $\alpha \varepsilon Q_p$ est de la forme $\sum_{h=r}^{+\infty} a_h p^h$, r entier rationnel, a_h pris dans l'ensemble $0, \pm 1, \cdots, \pm\frac{1}{2}(p-1)$. On pose $[\alpha] = \sum_{(h \leqq 0)} a_h p^h$ (c'est un nombre rationnel), et $\{\alpha\}=\sum_{(h>0)}a_h p^h$ (c'est un nombre p-adique de valeur absolue <1 qu'on appelera la reste modulo un p-adique de α). L'auteur énonce avec de brèves indications sur les démonstrations, plusieurs théorèmes analogues à ceux de Pisot et le rapporteur sur la répartition modulo un au sens ordinaire. Il montre que parmi certains ensembles de nombres algébriques absolus, $\theta \varepsilon Q_p$, définis par des inégalités sur les valeurs absolus p-adiques et ordinaires de θ et de ses conjugués absolus, l'un peut être caractérisé par la décroissance géométrique de $|\{\lambda\theta^n\}|_p$ (λ nombre convenable, n entier rationnel tendant vers $+\infty$), l'autre par la convergence de $\sum[\lambda\theta^n]^2$ (convergence au sens réel, [ ] ayant la définition p-adique donnée ci-dessus), et que ce dernier ensemble est fermé et admet des ensembles dérivées non vides de tout ordre entier (pour la topologie p-adique). *R. Salem* (Paris).

Referred to in K40-46.

K40-10 (12, 489b)

Kuipers, L., and Meulenbeld, B. Uniform distribution (mod 1) in sequences of intervals. Nederl. Akad. Wetensch., Proc. **53**, 1038–1048 = Indagationes Math. **12**, 382–392 (1950).

Let $0\leqq\alpha<\beta\leqq 1$, and $\theta(\alpha, \beta; f(t))$ equal 1 if $f(t)$ is in (α, β) mod 1 and zero otherwise. The authors previously defined [same Proc. **52**, 1151–1157, 1158–1163 (1949); **53**, 305–308 (1950) = Indagationes Math. **11**, 425–431, 432–437 (1949); **12**, 53–56 (1950); these Rev. **11**, 423, 648] a function to be C^I uniformly distributed if

$$\lim_{T\to\infty} T^{-1}\int_0^T \theta(\alpha, \beta; f(t))dt \to \beta-\alpha$$

for every α and β. The authors now define a function to be *C-u-d* (mod 1) in the sequence (S_k, T_k), $k=1, 2, \cdots$, $T_k-S_k\to\infty$, if $\lim\,(T_k-S_k)^{-1}\int_{S_k}^{T_k}\theta(\alpha, \beta; f(t))dt\to\beta-\alpha$ for every α and β. A function is said to be totally not *C-u-d* (mod 1) if there is no sequence $T_k\to\infty$ so that $f(t)$ is *C-u-d* (mod 1) in $(0, T_k)$. Various further concepts are defined and their interrelations are investigated by some examples and theorems. *P. Erdös* (Aberdeen).

Citations: MR 11, 423j = K40-4; MR 11, 423k = K40-4; MR 11, 648a = K40-6.

Referred to in K40-23.

K40-11 (12, 686b)

Kuiper, N. H. Distribution modulo 1 of some continuous functions. Nederl. Akad. Wetensch., Proc. **53**, 1390–1396 = Indagationes Math. **12**, 460–466 (1950).

Let $f(x)$ be increasing for sufficiently large x. Assume that $\lim_{x\to\infty}f'(x)/ax^{a-1}=k>0$, where $a>0$. Let φ be an unbounded set and let $\varphi(x)$ denote the Lebesgue measure of the intersection of φ with the interval $(0, x)$. The author defines $V(\varphi)=\lim_{x\to\infty}\varphi(x)/x$ if this limit exists. He proves that $V(\varphi)=V[f^{-1}(\varphi)]$. He applies this result to the theory of C-distributions investigated by Kuipers and Meulenbeld [same Proc. **52**, 1151–1157, 1158–1163; these Rev. **11**, 423]. One of his results is the following: Let $\lim_{x\to\infty}f(x)=\infty$, $f(x)$ continuous; if, whenever $\gamma>\beta>M$ and $f(\gamma)-f(\beta)>\frac{1}{4}$ then $\beta(f(\gamma)-f(\beta))/(\gamma-\beta)<L$ (M and L positive constants), then $f(x)$ does not possess a C^I distribution mod 1.

P. Erdös (Aberdeen).

Citations: MR 11, 423j = K40-4; MR 11, 423k = K40-4.

K40-12 (13, 119d)

Ryll-Nardzewski, C. Sur les suites et les fonctions également réparties. Studia Math. **12**, 143–144 (1951).

The paper is devoted to the proof of the following theorem. If f is a real measurable function and if either (a) the sequence $\{f(n+t)\}$ is equally distributed mod 1 for almost all t, i.e. the asymptotic frequency of the subsequence $\{n_k\}$ such that $f(n_k+t)-[f(n_k+t)]<\alpha$ is α, or (b) the sequence $\{f(nt)\}$ is equally distributed mod 1, then f is equally distributed mod 1, i.e.

$$\lim_{T\to\infty} T^{-1}\operatorname{meas}\{t\,|\,f(t)-[f(t)]<\alpha,\ t\varepsilon(0, T)\}=\alpha.$$

František Wolf (Berkeley, Calif.).

K40-13 (13, 206e)

Maak, Wilhelm. Der Kronecker-Weylsche Gleichverteilungssatz für beliebige Matrizengruppen. Abh. Math. Sem. Univ. Hamburg **17**, 91–94 (1951).

The author states the Kronecker-Weyl approximation theorem in the following way. Let $\{D^1(x)\}$ and $\{D^2(x)\}$ be two classes of almost-periodic functions on a group, each being a ring and closed with respect to translations and conjugations. If their intersection consists only of constants, then for any continuous function $f(D^1(x), D^2(y))$ in the underlying matrices constituting the classes one has

$$M_xM_yf(D^1(x), D^2(y))=M_xf(D^1(x), D^2(x)).$$

For Abelian groups this had been stated by Pontryagin.

S. Bochner (Princeton, N. J.).

K40-14 (14, 23b)

Carlitz, L. Diophantine approximation in fields of characteristic p. Trans. Amer. Math. Soc. **72**, 187–208 (1952).

Let the Galois field $GF(p^n)$ for fixed p and n be given and let $\Phi=GF[p^n, x]$ denote the field of all expressions

$$(1)\qquad \alpha=\sum_{i=-\infty}^{m}c_ix^i\quad (c_i\varepsilon GF(p^n)),$$

where x is an indeterminate. We may suppose c_m to be $\neq 0$; in that case $m\lesseqgtr 0$ is called the degree of α (deg α). If $m\geqq 0$ the part $\sum_0^m$ of (1) is called the integer part $[\alpha]$ of α; if $m<0$, we put $[\alpha]=0$; further the part $\sum_{-\infty}^{-1}$ is called the fractional part $\{\alpha\}$ of α (if $m<-1$, then $c_i=0$ for $m+1\leqq i\leqq -1$). The author investigates Diophantine approximations to the numbers $\alpha\varepsilon\Phi$. First, an analogue of Kronecker's theorem is proved which was previously shown by Mahler [Ann. of Math. **42**, 488–522 (1941); these Rev. **2**, 350]. Further, uniform distribution is defined: Let a sequence $\alpha_1, \alpha_2, \cdots$ in Φ and an arbitrary number $\beta\varepsilon\Phi$ be given. Let for fixed $N\geqq 1$ the number $N_k=N_k(N, \beta)$ of those α_i $(1\leqq i\leqq N)$ be considered such that deg $\{\alpha_i-\beta\}<-k$; we then say that the sequence $\alpha_1, \alpha_2, \cdots$ is uniformly distributed (mod 1) if $N_k/N\to p^{-nk}$, as $N\to\infty$ for any fixed couple $k\geqq 1$ and $\beta\varepsilon\Phi$.

Finally, the function $e(\alpha)$ is defined on Φ: Let θ define $GF(p^n)$; then for c_{-1} in (1) put

$$c_{-1}=\gamma_1\theta^{n-1}+\gamma_2\theta^{n-2}+\cdots+\gamma_n\quad (\gamma_i\varepsilon GF(p))$$

and $e(\alpha)=e^{2\pi i\gamma_1/p}$. Then $e(\alpha)=e(\beta)$ for $\alpha\equiv\beta$ (mod 1), i.e., for $\alpha=\beta+A$, where A is an integer in Φ, i.e., where A is a polynomial in $GF[p^n, x]$. The author then deduces the analogue to Weyl's well-known criterion and gives applications, mainly to polynomials $\varphi(u)$ of degree $k<p$ with coefficients in Φ for which u runs through the integers $A\varepsilon\Phi$.

J. F. Koksma (Amsterdam).

Citations: MR 2, 350c = H05-4.

Referred to in K40-71, K40-72, T55-31, T55-44.

K40-15 (14, 736b)

Kuipers, L., and Meulenbeld, B. On real functions of n variables. Nederl. Akad. Wetensch. Proc. Ser. A. **55** = Indagationes Math. **14**, 490–497 (1952).

The authors continue their work in the theory of C-uniform distribution (for the definition see a previous paper of Meulenbeld, same Proc. **53**, 311–317 = Indagationes Math. **12**, 59–65 (1950); these Rev. **11**, 648; and a paper by Kuipers and Meulenbeld, same Proc. **52**, 1151–1157, 1158–1163 = Indagationes Math. **11**, 425–431, 432–437 (1949); these Rev. **11**, 423). The authors prove several theorems about the uniform distribution and non-uniform distribution of functions. We here state only one of their theorems: Let there be given a sequence F of n-dimensional intervals Q: $0 \leqq S_u < t_u < T_u$, $u = 1, 2, \cdots, n$, where $T_n - \phi_n$ and the volume of Q tends to infinity as Q runs through F. Further, assume that $f(t_1, t_2, \cdots, t_n)$ is defined in all the Q's and has a partial derivative with respect to t_n. Assume finally that $\lim_{t_n \to \infty} \partial f/\partial t_n = c \neq 0$ uniformly in $t_1, t_2, \cdots, t_{n-1}$. Then $f(t)$ is uniformly distributed in the intervals Q of F.

P. Erdös (Los Angeles, Calif.).

Citations: MR 11, 423j = K40-4; MR 11, 423k = K40-4; MR 11, 648b = K40-7.

K40-16 (16, 575d)

Hlawka, Edmund. Über einen Satz von van der Corput. Arch. Math. 6, 115–120 (1955).

Let G be a compact group, with Haar measure so normalized that the measure of G is 1. Let $D_j(x)$, for $x \varepsilon G$ and $j = 0, 1, \cdots$, be the irreducible unitary representations of G, with notation chosen so that D_0 is the identity representation. A sequence $f = \{x_i\}$ of elements of G is said to be equidistributed on G if for each continuous function ϕ on G, $\lim_{N\to\infty} M_N(\phi, f) = M(\phi)$, where $M_N(\phi, f) = N^{-1}\sum_{i=1}^{n} \phi(x_i)$, $M(\phi) = \int_G \phi\, dx$. The known extension [see B. Eckmann, Comment. Math. Helv. 16, 249–263 (1943); MR 6, 146] of Weyl's criterion asserts that f is equidistributed on G if and only if $\lim_{N\to\infty} M_N(D_i, f) = 0$ for each $j > 0$. The author calls a sequence f weakly uniformly equidistributed if for each continuous function ϕ,

$$\lim_{N\to\infty} \limsup_{H\to\infty} \frac{1}{H} \sum_{r=0}^{H} | M_N(\phi, f_{rN}) - M(\phi) | = 0,$$

where $f_k = \{x_{i+k}\}$. It is easily seen that such a sequence is equidistributed on G.

J. G. van der Corput has shown [Acta Math. 56, 373–456 (1931)] that if $\{x_i\}$ is a real sequence such that $\lim_{i\to\infty} (x_{i+1} - x_i) = \alpha$, where α is irrational, then $\{x_i\}$ is equidistributed (mod 1). Generalizing this, the present author proves the following theorem: if $f = \{y_i\}$ is weakly uniformly equidistributed on G, and if $f' = \{x_i\}$ has the property that $\lim_{i\to\infty} y_{i+1}^{-1} x_{i+1} x_i^{-1} y_i = e$ (the unit element of G), then f' is also weakly uniformly equidistributed on G. The theorem has content, since it is shown that almost all sequences are weakly uniformly equidistributed on G.

W. J. LeVeque (Ann Arbor, Mich.).

• Citations: MR 6, 146a = K40-2.

K40-17 (17, 594c)

Hlawka, Edmund. Zur formalen Theorie der Gleichverteilung in kompakten Gruppen. Rend. Circ. Mat Palermo (2) 4 (1955), 33–47.

This paper unifies and develops certain aspects of the theory of uniform distribution on a compact group G, a theory first studied by Eckmann.

Let $A = (a_{ik})$ be an infinite matrix of the type used in theory of summability, i.e., $\lim_{i\to\infty} \sum_k a_{ik} = 1$, $\lim_{i\to\infty} a_{ik} = 0$. A sequence $\{x_i\}$ of elements of G is said to be A-uniformly distributed (A-gleichverteilt) if, for every continuous real-valued function $\varphi(z)$,

$$\lim_{i\to\infty} \sum_k a_{ik}\varphi(x_k) = \int \varphi(x)dx.$$

Here the integral is the invariant integral on G, normalized so that the measure of the whole group is unity. When A is the $(C-1)$ matrix ($a_{ik} = 1/i$ if $i \geqq k$, $a_{ik} = 0$ otherwise), the A is dropped and the sequence is simply said to be uniformly distributed. (This last is the case considered by Eckmann, and in the classical theory of uniform distribution, where G is a torus group.) The sequence is said to be A-g.g. distributed (gleichmässig gleichverteilt), if, uniformly for integers $h \geqq 0$,

$$\sum_k a_{ik}\varphi(x_{k+h}) \to \int \varphi(x)dx \text{ as } i \to \infty.$$

In addition to establishing certain basic results, some of which had been proved previously in a less general form, the author proves, among others, the following theorems. Theorem 4. Let $\{x_i\}$, $\{y_i\}$, be two sequences such that $\lim y_{i+1}^{-1} x_{i+1} x_i^{-1} y_i = e$, the identity. Then, if $\{x_i\}$ is g.g. distributed, so is $\{y_i\}$. Theorem 6. Let x_i be a sequence of elements of G. If each of the sequences $f^{(h)} = \{x_i^{-1} x_{i+h}\}$ is uniformly (g.g.) distributed, then every sequence $\{x_{ki+l}\}$ is uniformly (g.g.) distributed, where k, l are arbitrary natural numbers.

In the last part of the paper two independent proofs, one using ergodic theory, are given of the fact that almost all sequences are uniformly distributed. It is also shown that every dense sequence can be rearranged to form a uniformly distributed sequence, and that, if $\alpha > 0$, almost all sequences are uniformly distributed with respect to the Cesaro matrix $(C-\alpha)$.

There is a misprint on p. 39, l.16: "$D_j(x)$" should be "$D_j(a)$".

A. M. Macbeath (Dundee).

Referred to in K40-36.

K40-18 (18, 390f)

Hlawka, Edmund. Folgen auf kompakten Räumen. Abh. Math. Sem. Univ. Hamburg **20** (1956), 223–241.

The author generalizes ideas and theorems centering around the definition by which a real sequence $x_1, x_2, \cdots$ is called uniformly distributed mod 1 if

$$\lim_{n\to\infty} \lambda_n(f) = \lim_{n\to\infty} \frac{f(x_1) + \cdots + f(x_n)}{n} = \int_0^1 f(x)dx = \mu(f)$$

whenever f is a continuous function having period 1. Instead of requiring $x_1, x_2, \cdots$ to be real numbers reduced mod 1, he requires them to be elements of a compact space X having a countable basis. Instead of considering only arithmetic means, he introduces a real matrix $A = (a_{nk})$ for which $\sum_{k=1}^{\infty} |a_{nk}| \leqq \|A\| < \infty$ and $\sum_{k=1}^{\infty} a_{nk} \to 1$ as $n \to \infty$. Let $M(X)$ denote the weakly topologized space of all linear continuous functionals (or Radon measures) $\mu(f)$ defined over the vector space $C(X)$ of functions $f(x)$ which are real and continuous over X. Attention is then formed upon the linear functionals $\sum_{k=1}^{\infty} a_{nk} f_k(x)$ in $M(X)$ and their weak limit points. If μ is a functional in $M(X)$ and $n(1), n(2), \cdots$ is an increasing sequence of integers such that

$$\lim_{p\to\infty} \sum_{k=1}^{\infty} a_{n(p),k} f(x_k) = \mu(f) = \int_X f\, d\mu$$

for each f in $C(X)$, then μ is called an A-distribution (A-Mass) of the sequence $x_1, x_2, \cdots$. The paper contains 20 theorems involving existence, uniqueness, and other properties of these A-distributions. Several of these involve methods and results closely related to work of J. D. Hill [Ann. of Math. (2) **46** (1945), 556–562; Pacific J. Math. **1** (1951), 399–409; **4** (1954), 227–242; MR **7**, 153; **13**, 340; **15**, 950] on the Borel property of transformations of sequences of zeros and ones.

R. P. Agnew.

Referred to in K40-19.

K40-19 **(20# 5995)**

Hlawka, Edmund. Folgen auf kompakten Räumen. II. Math. Nachr. **18** (1958), 188-202.

For part I see Abh. Math. Sem. Univ. Hamburg **20** (1956), 223-241 [MR **18**, 390]. The present part II continues with 17 theorems of the sort described in the review of part I.

Citations: MR 18, 390f = K40-18.

K40-20 **(18, 680c)**

Rivkind, Ya. I. Limit theorem of probability theory on compact topological groups. Grodnenskiĭ Gos. Ped. Inst. Uč. Zap. **1** (1955), 51-58. (Russian)

It is well-known that, if n independent random real variables all have the same absolutely continuous distribution, then as $n\to\infty$ the distribution of the fractional part of their sum tends to the uniform distribution over [0, 1). The author obtains a result of this type in a much more general setting. Let Ω be a connected compact topological group. Let $\mu(E)$ be the normalized (two-sidedly) invariant measure on Ω and $P(E)$ be a probability measure on Ω. Let $x_1, x_2, \cdots$ be independent randomly variable elements in Ω all having the distribution determined by $P(E)$. The distribution of the product $x_1x_2\cdots x_n$ is determined by $P^{(n)}(E)$, where

$$P^{(k)}(E)=\int_\Omega P^{(k-1)}(Ex^{-1})P(dE_x)\ (k=1,2,\cdots)$$

and $P^{(1)}(E)=P(E)$. Supposing $P(E)$ to be of the form $\int_E p(x)\mu(dE_x)$, with $p(x)$ in the Lebesgue class $L_\Omega{}^2$, the author shows that $P^{(n)}(E)\to\mu(E)$ as $n\to\infty$. His proof employs a generalized characteristic function.

H. P. Mulholland (Birmingham).

K40-21 **(22# 7996)**

Hlawka, Edmund. Über C-Gleichverteilung. Ann. Mat. Pura Appl. (4) **49** (1960), 311-325.

Let $a(t, T)$ ($t\geqq 0$; $T\geqq 0$) be $\geqq 0$ with $\int_0{}^T a(t,T)\,dt=1$. If $\chi(x;\ \alpha,\beta)$ is the characteristic function of the interval $\alpha\leqq x<\beta$, with $0\leqq\alpha<\beta\leqq 1$, if $\{u\}=u-[u]$, and if finally $f(t)$ ($t\geqq 0$) is integrable, then let

$$D=D(A,f,T,\alpha,\beta)=\int_0^T a(t,T)\,\chi(\{f(t)\};\alpha,\beta)\,dt-(\beta-\alpha),$$

where A denotes the integral operator $\int_0{}^T a(t,T)$. The function $f(t)$ is said to have a uniform A-distribution if, for each α and β with $0\leqq\alpha<\beta\leqq 1$, $D\to 0$ as $T\to\infty$; this relation holds then uniformly in α and β. In the special case $A(t,T)=T^{-1}$ the distribution is called a C-distribution. These distributions have already been treated by L. Kuipers and B. Meulenbeld [Nederl. Akad. Wetensch. Proc. Ser. A **52** (1949), 1151-1157, 1158-1163; **53** (1950), 226-232, 305-308; **56** (1953), 340-348; MR **11**, 423, 424, 648; **15**, 410]. The supremum $D(A,f,T)$ of D for $0\leqq\alpha<\beta\leqq 1$ is called the A-discrepancy of f. A function f has a uniform A-distribution if and only if $D(A,f,T)\to 0$ as $T\to\infty$. For each integer $g\neq 0$, $D(A,gf,T)\leqq 3|g|D(A,f,T)$; therefore, if f has a uniform A-distribution, then also gf has. The author gives an upper bound for the absolute value of $D(A,f,T)-D(A,g,T)$ under the assumption that an upper bound for the absolute value of $f-g$ is known. Two functions with a constant difference have the same discrepancy. If $w(t)$ is a function of t ($0\leqq t\leqq 1$) with bounded variation V, then

$$\left|\int_0^T a(t,T)w(\{f(t)\})\,dt-\int_0^1 w(t)\,dt\right|\leqq VD(A,f,T).$$

Under general conditions the author obtains upper bounds for the absolute value of $D(A,f,T)$, with applications to uniform distribution, for instance: If there exists a $\sigma>0$ such that $g(t+\sigma)-g(t)$ tends, for $t\to\infty$, monotonically decreasing to zero, with $t|g(t+\sigma)-g(t)|\to\infty$, then $g(t)$ has a uniform C-distribution. Under the assumption that an upper bound is known for the absolute value of $D(C,\{g(t+h\sigma)-g(t)\},T-|h|\sigma)$, the author finds an upper bound for the absolute value of $D(C,\{g\},T)$. This result, which implies for instance that $g(t)$ has a uniform C-distribution if this is the case with $g(t+h\sigma)-g(t)$ for each positive integer h, is a generalisation of the corresponding result obtained by Van der Corput in the theory of discrete uniform distributions, where t traverses the sequence of the positive integers [Van der Corput and Ch. Pisot, Nederl. Akad. Wetensch. Proc. **42** (1939), 476-486, 554-565, 713-722; MR **1**, 66].

J. G. van der Corput (Berkeley, Calif.)

Citations: MR 1, 66b = K05-1; MR 1, 66c = K05-2; MR 11, 423j = K40-4; MR 11, 423k = K40-4; MR 11, 424a = K40-5; MR 11, 648a = K40-6; MR 15, 410e = K30-11.

K40-22 **(23# A476)**

Bass, Jean

Suites uniformément denses, moyennes trigonométriques, fonctions pseudo-aléatoires.

Bull. Soc. Math. France **87** (1959), 1-64.

Let $f(t)$ be a bounded complex function of the real variable t, which is 0 for $t<0$. Suppose that the mean

$$M(f)=\lim_{T\to\infty}T^{-1}\int_0^T f(t)\,dt$$

and the correlation function $\gamma(h)=M\{f(t)\bar{f}(t+h)\}$ exist. Then $f(t)$ is said to be a pseudo-random function in case $\gamma(h)$ is continuous and

$$M(f)=0,\qquad \gamma(0)\neq 0,\qquad \gamma(\infty)=0.$$

This concept, which was introduced by N. Wiener [Acta Math. **55** (1930), 117-258], has been studied by the author and colleagues [Bertrandias and Krée] in several recent papers [C. R. Acad. Sci. Paris **245** (1957), 1217-1219, 2457-2459; **247** (1958), 1083-1085, 1163-1165; **248** (1959), 513-515; MR **20** #2828, 2829; **21** #1645, 1646, 1648]. The investigation is continued in the present paper.

Here are some typical results. Theorem 11: If $f(t)=\exp(2\pi i\varphi([t]))$ and $\gamma(p)=0$ for every positive integer p, then f is pseudo-random. Theorem 12: If φ is a polynomial of degree larger than 1, with irrational leading coefficient, then $f(t)=\exp(2\pi i\varphi([t]))$ is pseudo-random. Theorem 15: If $g(t)$ is periodic, of period 1, and has an absolutely convergent Fourier series, and if φ is a polynomial of degree $\geqq 3$ and the coefficient of at least one power of t higher than the first is irrational, then the function $f(t)=\exp(2\pi i\varphi([t]))\cdot g(t)$ is pseudo-random. Theorem 21: Let φ and ψ be polynomials with irrational leading coefficients, of degrees larger than 2 and 1, respectively. Suppose that the points $(\varphi(n),\psi(n))$ are uniformly distributed (mod 1) in the unit square. Let $r(x)$ be a non-negative bounded integrable function on (0, 1), and define $\{t_n\}$ by $t_{n+1}-t_n=r(\psi(n)-[\psi(n)])$. Then the function f, such that $f(t)=\exp(2\pi i\varphi(n))$ when $t_n<t<t_{n+1}$, is pseudo-random, and its correlation function is

$$\gamma(h)=R^{-1}\int_{r(x)>h}(r(x)-h)\,dx\quad\text{for } h<\sup r(x),$$
$$=0\qquad\qquad\text{for } h>\sup r(x),$$

where $R=\int_0{}^2 r(x)dx$.

{There is some connection between Section V of this

paper and work of the reviewer [Pacific J. Math. **3** (1953), 757–771; MR **15**, 511] on uniform distribution modulo an arbitrary partition of the real axis.}

W. J. LeVeque (Ann Arbor, Mich.)

Citations: MR 15, 511d = K10-14; MR 20# 2828 = K30-18; MR 20# 2829 = K30-19.

Referred to in K05-36, K20-31.

K40-23 (25# 5033)

Kuipers, L.

Continuous distribution modulo 1.

Nieuw Arch. Wisk. (3) **10** (1962), 78–82.

The notion of well-distributed sequences (mod 1) is carried over to well-distributed functions $f(t)$ (of the continuous variable t) (mod 1) and several theorems known in the discrete case [Petersen, Quart. J. Math. Oxford Ser. (2) **7** (1956), 188–191; MR **20** #2313a; Lawton, Proc. Amer. Math. Soc. **10** (1959), 891–893; MR **22** #703; Keough, Lawton and Petersen, Canad. J. Math. **10** (1958), 572–576; MR **20** #2313b] are proved to hold also in the continuous case. Also some earlier results of the author and Meulenbeld [Nederl. Akad. Wetensch. Proc. **53** (1950), 822–827; MR **12**, 15; ibid. **53** (1950), 1038–1048; MR **12**, 489] are reworded: If $f(t)$ is a measurable function for $t \geqq 0$, and if $f'(t) \geqq c$ (constant) as $t \to \infty$, then $f(t)$ is well-distributed (mod 1).

J. F. Koksma (Amsterdam)

Citations: MR 12, 15e = K40-8; MR 12, 489b = K40-10; MR 20# 2313a = K35-9; MR 20# 2313b = K35-9; MR 22# 703 = K35-11.

K40-24 (26# 3692)

Chauvineau, Jean

Suites continûment et périodiquement réparties modulo 1.

C. R. Acad. Sci. Paris **256** (1963), 839–841.

La suite u_n est dite χ_0-c.r. (mod 1) si et seulement si elle est continûment répartie (mod 1) [c.r. (mod 1)] avec χ_0 comme fonction de répartition (mod 1). La suite u_n est dite α, χ_0-c.p.r. (mod 1) si et seulement si elle est c.r. (mod 1) avec une fonction de répartition (mod 1) unique, soit χ, telle que $\chi(a+\alpha)-\chi_0(a+\alpha)=\chi(a)-\chi_0(a)$ pour tout $a \in [0, 1-\alpha]$.

L'auteur a prouvé Théorème 1: Pour que la suite u_n soit α, χ_0-c.p.r. (mod 1), où α est un irrationnel de $I(=[0, 1])$, il faut et il suffit qu'elle soit χ_0-c.r. (mod 1). En le deuxième théorème, il considère le cas où α est rationnel.

G. M. Petersen (Swansea)

K40-25 (26# 4983)

Cugiani, Marco

Successioni uniformemente distribuite nei domini *P*-adici.

Ist. Lombardo Accad. Sci. Lett. Rend. A **96** (1962), 351–372.

The author studies the notion of a uniformly distributed sequence in the compact group D_p of p-adic integers. Among other things, he proves that a sequence of p-adic integers $a_n=\sum_{v=0}^{\infty} a_{n,v}p^v$ $(0 \leqq a_{n,v} < p;\ n=1, 2, \cdots)$ is uniformly distributed in D_p if and only if the corresponding sequence of real numbers $a_n^*=\sum_{v=0}^{\infty} a_{n,v}p^{-v-1}$ is uniformly distributed mod 1.

C. G. Lekkerkerker (Amsterdam)

Referred to in K40-35, K40-48, K40-52, K40-54.

K40-26 (26# 6339)

Kuipers, L.

Continuous distribution mod 1 **and independence of functions.**

Nieuw Arch. Wisk. (3) **11** (1963), 1–3.

In previous papers [Thesis, Free Univ. of Amsterdam, 1947; MR **10**, 235; Nederl. Akad. Wetensch., Proc. **53** (1950), 34–40; MR **11**, 424] the author investigated the behaviour of some real-valued functions $f(t)$ $(t \geqq 0)$ of the form $g(t)+h(t)$ with respect to the (asymptotic) distribution of their values mod 1. As examples he proved that $\alpha t+\sin t$ if $\pi\alpha$ is irrational, $\sqrt{t}+\sin 2\pi(t+t^{-1})$ are uniformly distributed mod 1. In the present paper using his functional version of H. Weyl's criterion for uniform distributivity mod 1 [P1] and H. Steinhaus' independence criterion [H. Steinhaus and M. Kac, Studia Math. **6** (1936), 45–58; ibid. **6** (1936), 59–66; ibid. **6** (1936), 89–97; ibid. **7** (1938), 1–15; ibid. **7** (1938), 96–100; ibid. **9** (1940), 121–132; MR **3**, 2], he gives a simple proof of the independence of the functions $\cos l_1x$, $\cos l_2x$, $\cdots$, $\cos l_nx$ if the real numbers $l_1, l_2, \cdots, l_n$ are linearly (rationally) independent and the uniform distributivity of $\alpha t+\sin t$ if $\pi\alpha$ is irrational.

Chr. Y. Pauc (Nantes)

Citations: MR 3, 2f = K05-3; MR 10, 235f = K40-3; MR 11, 424a = K40-5.

K40-27 (26# 7012)

Cigler, Johann

Folgen normierter Masse auf kompakten Gruppen.

Z. Wahrscheinlichkeitstheorie und Verw. Gebiete **1** (1962/63), 3–13.

The paper deals with the convolution of probability measures on compact topological groups from the point-of-view of the theory of equidistribution. Let G be a compact topological group, $C(G)$ the Banach space of all complex-valued continuous functions on G with the norm $\|f\|=\max_{x\in G}|f(x)|$, $M(G)$ the set of all normed measures on G. If μ is a measure on G, then the mean value of f with respect to μ, $\mu(f)=\int_G f\,d\mu$ is a continuous non-negative linear functional on $C(G)$. A sequence μ_n $(n=1, 2, \cdots)$ of measures on G is said to be distributed according to the measure $\nu \in M(G)$ if $\lim_{N\to+\infty}(1/N)\sum_{n=1}^{N}\mu_n(f)=\nu(f)$ for every $f\in C(G)$.

If ν is the Haar measure, then the sequence μ_n is said to be equidistributed. If $\mu \in M(G)$ let μ^* be the measure defined by $\mu^*(f)=\int_G f(x^{-1})\,d\mu$. The convolutions of two measures $\mu_1 \in M(G)$, $\mu_2 \in M(G)$ is denoted by $\mu_1\cdot\mu_2$. The main theorem of the paper, which is a generalization of a well-known theorem of J. G. van der Corput [Acta Math. **56** (1931), 373–456; see also E. Hlawka, Österreich. Akad. Wiss. Math.-Nat. Kl. Anzeiger **94** (1957), 313–317], is as follows. If $\mu_n \in M(G)$ is a sequence of measures such that the sequence $\mu_{n+h}\mu_n^*$ $(n=1, 2, \cdots)$ is distributed according to a measure ν_h $(h=0, 1, \cdots)$ and the sequence ν_h is equidistributed, then the sequence μ_n itself is also equidistributed.

The paper deals in detail with the study of the sequence $\mu_n=\mu^n$. It is shown that for any $\mu \in M(G)$ the sequence μ^n is distributed according to some measure ν (which is the Haar measure of a certain subgroup of G) and a necessary and sufficient condition in terms of the matrix representation of the group (which exists according to the Peter-Weyl theorem) is given that the sequence μ^n should be equidistributed. Necessary and sufficient conditions are given also for the (weak) convergence of the sequence μ^n itself to the Haar measure. These results were obtained previously by other methods [Y. Kawada and K. Itô, Proc. Phys. Math. Soc. Japan (3) **22** (1940), 977–998; I. Glicksberg, Pacific J. Math. **9** (1959), 51–67; MR **21** #7405; B. M. Kloss, Teor. Verojatnost. Primenen. **4** (1959), 255–290; MR **22** #3761; K. Urbanik, Fund. Math. **44** (1957), 253–261; MR **19**, 1180]. Here they are all derived from the above mentioned main theorem. A further result deals with the equidistribution of the sequence $u^{P(n)}$, where $P(n)$ is a polynomial with integer coefficients. Equidistribution on the additive group R of real numbers is also discussed; R being only locally compact, the definition has to be modified. A sequence μ_n of probability measures on R is called equidistributed if for any bounded almost periodic function f, $\lim_{N\to+\infty} (1/N) \sum_{n=1}^{N} \mu_n(f) = M(f)$, where $M(f) = \lim_{T\to+\infty} (1/T) \int_0 f(t)\,dt$. It is shown that if μ_n is a sequence of measures on R such that the sequence $\mu_{n+h}\mu_n^*$ $(n=1, 2, \cdots)$ is equidistributed for $h=1, 2, \cdots$, then the sequence μ_n itself is also equidistributed. Finally, some examples are given. *A. Rényi* (Budapest)

K40-28 **(27# 1530)**

Helmberg, Gilbert

Eine Familie von Gleichverteilungskriterien in kompakten Gruppen.

Monatsh. Math. **66** (1962), 417–423.

Let X be a compact group with a countable base, containing at least two elements. If $\{x_n\}$ and $\{y_n\}$ are sequences in X, the sequence of all products $x_n \cdot y_m$, in a certain well-defined order, is denoted by $\{x_n\}\times\{y_n\}$. The author shows that $\{x_n\}\times\{y_n\}$ is $N_1 * N_2$-summable if $\{x_n\}$ is N_1-summable and $\{y_n\}$ is N_2-summable ($N_1 * N_2$ designates the convolution of the measures N_1 and N_2). He then proceeds to prove four criteria for the sequence $\{x_n\}$ to be uniformly distributed; this is, e.g., the case if and only if $\{x_n\}\times\{y_n\}$ is summable for every sequence $\{y_n\}$, or also if and only if the sequence $\{x_n\}\times\{x_n^{-1}\}$ is uniformly distributed. *P. C. Baayen* (Amsterdam)

K40-29 **(29# 88)**

Philipp, Walter

Über einen Satz von Davenport-Erdős-LeVeque.

Monatsh. Math. **68** (1964), 52–58.

In the paper of the title [Michigan Math. J. **10** (1963), 311–314; MR **27** #3618], it was shown that if $\{f_k(x)\}$ is a sequence of bounded integrable functions on $[a, b]$, and if $s_n(x) = |n^{-1}\sum_1^n \exp(2\pi i m f_k(x))|^2$ and $I(n) = \int_a^b s_n(x)\,dx$, then $\{f_k(x)\}$ is uniformly distributed (mod 1) for almost all $x\in[a, b]$ if $\sum n^{-1}I(n)$ converges for all integral $m\neq 0$. In the present paper, the author extends this as follows. Let X be a compact space with countable basis, and let $\mathfrak{B}(X)$ be the set of all positive normalized measures on X. Let $\mathfrak{C}(X)$ be the normed (by $\|f\| = \sup_{x\in X}|f(x)|$) linear space of all complex continuous functionals on X, and let $H\subset\mathfrak{C}(X)$ be a countable set spanning a space dense in $\mathfrak{C}(X)$. Let $A = (a_{nk})$ be a complex matrix with $\lim_n \sum_{k=1}^{\infty} a_{nk} = 1$ and $\sup_n \sum_{k=1}^{\infty} |a_{nk}| < \infty$. If $\omega = \{x_k\}$ is an infinite sequence of elements of X, put $\lambda_n(\omega, A, f) = \sum_{k=1}^{\infty} a_{nk} f(x_k)$. The author calls ω strongly A-equidistributed to the measure $\mu\in\mathfrak{B}(X)$ if for each $f\in H$, $\lim s_n = 0$ and $\sum |(s_n - s_{n+1})s_n| < \infty$, where $s_n = |\lambda_n - \mu(f)|^2$. Satz 2: The sequence ω is strongly A-equidistributed to the measure μ if and only if, for each $f\in H$, there is a divergent series $\sum c_n$ with $c_n \geqq 0$, $\sum c_n s_n < \infty$, and $|s_n - s_{n+1}| \leqq c_n$ for $n = 1, 2, \cdots$. Satz 3: Let $\{T_k\}$ be a sequence of measurable operators in X. If to each $f\in H$ there corresponds a divergent series $\sum c_n$ with $c_n \geqq 0$, $\sum c_n \mu(s_n) < \infty$ and $|s_n - s_{n+1}| \leqq c_n$, then the sequence $\{T_k x\}$ is strongly A-equidistributed to the measure μ for μ-almost all $x\in X$. Applications are made to distribution on a compact group with countable basis. *W. J. LeVeque* (Ann Arbor, Mich.)

Citations: MR 27# 3618 = K05-40.

K40-30 **(29# 89)**

Kuipers, L.

Some aspects of the theory of asymptotic distribution modulo 1.

Bull. Soc. Math. Belg. **15** (1963), 379–388.

The first part of this paper has an expository character. It gives the definitions of uniform distribution in the discrete and continuous cases and states the general results obtained by H. Weyl and J. G. van der Corput. Furthermore, some Fejér type theorems are mentioned. Then the author uses the notion of independence in the sense of Kac and Steinhaus to derive results on the uniform distribution of certain sums such as $\alpha t + \sin t$. *J. E. Cigler* (Vienna)

K40-31 **(29# 185)**

Kuipers, L.; Scheelbeek, P. A. J.

Gleichverteilung in kompakten, topologischen Gruppen.

Nederl. Akad. Wetensch. Proc. Ser. A **67** = *Indag. Math.* **26** (1964), 262–265.

Let G be a compact Hausdorff group, and let $\{a_1, a_2, \cdots, a_k\}$ be a subset of G such that $(a_1^n)_{n=1}^{\infty}$ is uniformly distributed in G. {This obviously implies that G is Abelian.} For $i=1, 2, 3, \cdots$, let $\Psi(i)$ be the finite sequence $(a_1^i, a_1^{i-1}a_2, \cdots, a_1^{i-1}a_k, a_1^{i-2}a_2^2, a_1^{i-2}a_2a_3, \cdots, a_2^i, \cdots, a_{k-1}^i, \cdots, a_{k-1}^{i-1}a_k, a_{k-1}a_k^{i-1}, a_k^i)$. Theorem: The sequence $\Psi(1), \Psi(2), \Psi(3), \cdots$ is uniformly distributed in G. *Edwin Hewitt* (Seattle, Wash.)

K40-32 **(29# 422)**

Cigler, Johann

Einige Bemerkungen zur Theorie der fastperiodischen Funktionen.

Arch. Math. **15** (1964), 155–160.

Author's summary: "In der vorliegenden Note werden die Beziehungen zwischen einer topologischen Gruppe $\mathfrak{G}$ und ihrer fastperiodischen Kompaktifizierung $\tilde{\mathfrak{G}}$ vom Standpunkt der Theorie der Gleichverteilung aus untersucht. Es zeigt sich, dass im Falle der Gruppe $\mathfrak{G}$ der ganzen (reellen) Zahlen $\mathfrak{G}$ gleichverteilt in $\tilde{\mathfrak{G}}$ ist. Daraus wird auf einfache Weise die Existenz der Verteilungsfunktion einer reelwertigen fastperiodischen Funktion auf $\mathfrak{G}$ gefolgert. Im Falle einer beliebigen topologischen Gruppe wird die Existenz des Mittelwertes einer fastperiodischen Funktion als Gleichverteilungsaussage gedeutet. Daraus wird wieder die Existenz der Verteilungsfunktion fur eine reelwertige fastperiodische Funktion hergeleitet." *E. Følner* (Copenhagen)

K40-33 **(29# 1223)**

Cigler, Johann

Ein gruppentheoretisches Analogon zum Begriff der normalen Zahl.

J. Reine Angew. Math. **206** (1961), 5–8.

Es sei X eine kompakte abelsche Gruppe mit abzählbarer Basis und T ein steiger Endomorphismus von X auf sich, der bezüglich des Haarschen Maßes μ auf X ergodisch ist. Ein Element $x\in X$ heißt normal bezüglich T, wenn die Folge $\{T^n x\}$ in X gleichverteilt ist, d.h. wenn

$$\lim_{N\to\infty} \frac{1}{N}\sum_{n=1}^{N} f(T^n x) = \int_X f(x)\,d\mu(x)$$

für alle stetigen komplexwertigen Funktionen f auf X gilt. Der Verfasser beweist folgende Verallgemeinerungen bekannter Sätze über (bezüglich einer ganzzahligen Basis $a>1$) normale Zahlen: Fast alle $x\in X$ sind normal bezüglich T. Für jedes $k\geqq 2$ ist das Element $x\in X$ genau dann normal bezüglich T, wenn es normal bezüglich T^k ist. Ist x normal bezüglich T und S ein mit T vertauschbarer stetiger Endomorphismus von X auf sich, dann ist auch Sx normal bezüglich T; ist außerdem der Kern dieses Endomorphismus S endlich, dann ist auch jedes S-Urbild von x normal bezüglich T. Die beim Beweis dieser letzten Behauptung verwendete Existenz der Menge B, deren Rand das Maß 0 hat, scheint dem Referenten nicht voll geklärt zu sein, jedoch läßt sich nach Mitteilung des Verfassers der Beweis auch ohne Verwendung dieser Voraussetzung führen. Durch die Begriffsbildungen des Verfassers werden auch die normalen k-tupel von Maxfield [Pacific J. Math. **3** (1953), 189–196; MR **14**, 851] und Resultate von Rohlin [Izv. Akad. Nauk SSSR Ser. Mat. **13** (1949), 329–340; MR **11**, 40] erfaßt. *G. Helmberg* (Zbl **103**, 19)

Citations: MR 14, 851b = K15-29.
Referred to in K15-75.

K40-34 (30# 1114)

Helmberg, Gilbert
Gleichverteilte Folgen in lokal kompakten Räumen.
Math. Z. **86** (1964), 157–189.

Im ersten Teil dieser Arbeit definiert Verfasser den Begriff der Gleichverteilung einer Folge (x_n) bezüglich eines Wahrscheinlichkeitsmasses auf $[0, \infty)$. Er leitet sodann notwendige und hinreichende Bedingungen dafür ab, stellt den Zusammenhang mit gleichverteilten Folgen auf [0, 1) dar und macht auch quantitative Aussagen über die Genauigkeit der Approximation. Im zweiten Teil werden allgemeine Sätze über Gleichverteilung bezüglich eines regulären Wahrscheinlichkeitsmasses in einem lokalkompakten Hausdorffraum mit abzählbarer Basis hergeleitet. *J. E. Cigler* (Groningen)

K40-35 (30# 3064)

Zaretti, Anna
Equivalenza di due definizioni di successioni equidistribuite nei domini P-adici.
Atti Sem. Mat. Fis. Univ. Modena **13** (1964), 111–118.

Recently M. Cugiani [Ist. Lombardo Accad. Sci. Lett. Rend. A **96** (1962), 351–372; MR **26** #4983] investigated uniformly distributed sequences of p-adic integers. The present author compares Cugiani's results with the general theory of uniform distribution in compact topological groups, such as given by J. Cigler and G. Helmberg [Jber. Deutsch. Math.-Verein. **64** (1961), Abt. 1, 1–50; MR **23** #A2409]. *C. G. Lekkerkerker* (Amsterdam)

Citations: MR 23# A2409 = K02-1; MR 26# 4983 = K40-25.
Referred to in K40-52, K40-59.

K40-36 (31# 2234)

Baayen, P. C.; Hedrlín, Z.
The existence of well distributed sequences in compact spaces.
Nederl. Akad. Wetensch. Proc. Ser. A **68** = *Indag. Math.* **27** (1965), 221–228.

The concept of a well distributed sequence was introduced by E. Hlawka [Rend. Circ. Mat. Palermo (2) **4** (1955), 33–47; MR **17**, 594], and by the reviewer [Quart. J. Math. Oxford Ser. **7** (1956), 188–191; MR **20** #2313a]. In this paper, it is shown that if X is an arbitrary non-void, compact, Hausdorff space satisfying the second axiom of countability, and if μ is an arbitrary Borel measure on X, then there exists a μ well distributed sequence in X. This involves proving that if Φ is an arbitrary real-valued continuous function on X, then there exists an $M_0 = M_0(\Phi, \varepsilon)$ such that

$$|M^{-1} \sum_{m=1}^{M} \Phi(x_{m+k}) - \mu(\Phi)| \leqq \varepsilon$$

for all $M \geqq M_0(\Phi, \varepsilon)$, uniformly in $k \in N$.

The sequence constructed gives an affirmative answer to a question posed in the Colloquium on uniform distribution at the Mathematical Center in Amsterdam, 1963/1964. *G. M. Petersen* (Christchurch)

Citations: MR 17, 594c = K40-17; MR 20# 2313a = K35-9.

K40-37 (31# 3407)

Gotusso, Laura
Successioni uniformemente distribuite in corpi finiti.
Atti Sem. Mat. Fis. Univ. Modena **12** (1962/63), 215–232.

The author extends the concept of uniform distribution to finite fields. Let $\Gamma = \{\gamma_1, \cdots, \gamma_q\}$ be a finite field, of characteristic p, with $q = p^\alpha$ elements. For each sequence $A = \{a_m\}$ of elements of Γ, define $S(\gamma_{s,n})$ as the number of terms among $a_1, a_2, \cdots, a_n$ that satisfy $a_m = \gamma_s$. The sequence A is uniformly distributed with respect to γ_s in case (1) $S(\gamma_{s,n})/n \to 1/q$ as $n \to \infty$. The sequence A is said to be uniformly distributed in Γ if (1) is true for every $\gamma_s \in \Gamma$.

Some elementary theorems about uniform distribution are proved. In particular, the author considers the sequences $\{f(a_m)\}$, where $f(x)$ is a polynomial with coefficients belonging to Γ and $\{a_m\}$ is a uniformly distributed sequence.

The author discusses methods of generating, in Γ, uniformly distributed sequences which may be considered as pseudo-random sequences. A numerical example is given. *J. Kubilius* (Vilnius)

K40-38 (32# 2394)

Holewijn, Petrus Johannes
Contributions to the theory of asymptotic distribution modulo 1.
Proefschrift ter verkrijging van de graad van Doctor in de Technische Wetenschappen aan de Technische Hogeschool te Delft.
Uitgeverij Waltman, Delft, 1965. 49 *pp.* (1 *insert*) 5.90 *gld.*

The author studies some problems about the asymptotic distribution modulo 1 of functions of a real variable with respect to weighted means by reducing this case to the case of arithmetic means and to the discrete case. He also gives two proofs of Weyl's criterion. The first proof is the usual one. In the other, necessity of Weyl's conditions is deduced from the deeper "continuity theorem of probability theory" and sufficiency is proved by using Fourier series. *J. E. Cigler* (Groningen)

K40-39 (32# 5627)

Baayen, P. C.; Helmberg, G.
On families of equi-uniformly distributed sequences in compact spaces.
Math. Ann. **161** (1965), 255–278.

Let X be a compact Hausdorff space, let μ be a regular normed Borel measure on X and let $C(X)$ be the Banach space of all complex-valued continuous functions f on X. Let $\mu(f) = \int_X f(x)\, d\mu(x)$. A family $\mathfrak{S} = \{(x_{\sigma,n})_{n=1}^\infty : \sigma \in S\}$ of sequences in X is called a family of equi-μ-uniformly distributed sequences if for every $f \in C(X)$ and for every real number $\varepsilon > 0$, there exists an integer $N(f, \varepsilon)$ independent of σ, such that $|(1/N) \sum_{n=1}^N f(x_{\sigma,n}) - \mu(f)| \leqq \varepsilon$ for all $N \geqq N(f, \varepsilon)$ and for all $\sigma \in S$.

For the additive group of reals and Lebesgue measure, one has Theorem 1: Let A be a set of irrational numbers such that the sequences $(na)_{n=1}^\infty$ $(a \in A)$ are equi-uniformly distributed mod 1; then the set of residues mod 1, $\hat{A} = \{\hat{a} : a \in A\}$, is nowhere dense in [0, 1[. Theorem 2: Let δ $(0 < \delta < 1)$ be given. Then there exists a closed nowhere dense set A of irrational numbers in $[0, 1[$ such that $\mu(A) \geqq 1-\delta$ and such that the sequences $(na)_{n=1}^{\infty}$ $(a \in A)$ are equi-uniformly distributed mod 1.

If X is also a compact group (μ will now be the normed Haar measure on X), and if there is at least one $a \in X$ (called generator) such that $(a^n)_{n=1}^{\infty}$ is everywhere dense in X, then X is called monothetic. The set of all generators of X is denoted by G. One has Theorem 4: Let X be not totally disconnected and let A be a set of generators of X such that the sequences $(a^n)_{n=1}^{\infty}$ $(a \in A)$ are equi-uniformly distributed; then A is nowhere dense. Theorem 5: The sequences $(a^n)_{n=1}^{\infty}$ $(a \in G)$ are equi-uniformly distributed if and only if X is totally disconnected. Theorem 6: Suppose that X satisfies the second axiom of countability and let ε $(0<\varepsilon<1)$ be given. Then there is a closed nowhere dense subset Q of G such that $\mu(Q) > \mu(G)-\varepsilon$ and such that the sequences $(a^n)_{n=1}^{\infty}$ $(a \in Q)$ are equi-uniformly distributed.

Let X_∞ denote the topological product of countably many copies of X. There are theorems analogous to the above for the general case, for example, Theorem 8: Suppose that X contains at least two points and that $\mathfrak{S}=\{\xi_\sigma \in X_\infty : \sigma \in S\}$ is a family of equi-μ-uniformly distributed sequences; then $\mathfrak{S}$ is nowhere dense in X.

A sequence $\xi \in X_\infty$ is called μ-well distributed if the sequences $P^m\xi$ $(m=1, 2, \cdots)$, where P is the shift transformation, are equi-μ-uniformly distributed. Theorem 10: Suppose that the sequence $\xi \in X_\infty$ is μ-well distributed. Then so is every sequence in the closure of the set $\mathfrak{S}=\{P^m\xi : m=1, 2, \cdots\}$.

The terms almost well, weakly well, and completely uniformly distributed sequences are defined. For example, the sequence is weakly μ-well distributed if for every $f \in C(X)$ one has

$$\lim_{N\to\infty} \limsup_{R\to\infty} \frac{1}{R} \sum_{r=0}^{R-1} \left| \frac{1}{N} \sum_{n=rN+1}^{(r+1)N} f(x_n) - \mu(f) \right| = 0.$$

The following theorem is proved by a neat application of ergodic theory. Theorem 16: Suppose that X satisfies the second axiom of countability. Then μ_∞-almost all sequences $\xi \in X_\infty$ are weakly μ-well distributed.

The paper seems very rich in ideas.

G. M. Petersen (Christchurch)

K40-40 **(32# 5628)**

Helmberg, Gilbert

Abstract theory of uniform distribution.

Compositio Math. **16**, 72–82 (1964).

This paper presents a "survey of typical and basic ideas and results concerning abstract theory of uniform distribution". The survey proceeds at a sophisticated level in the sense that more is offered than a mere list of results with the related bibliography. The author formulates connections between and implications of various results, and thereby gives a valuable overview of the topic.

I. Niven (Eugene, Ore.)

K40-41 **(32# 5629)**

Mück, Rudolf

Über ein Problem in der Theorie der C-Gleichverteilung.

J. Reine Angew. Math. **222** (1966), 201–206.

The real functions f considered are C-equidistributed in the sense that, for each interval $[\alpha, \beta) \subset [0, 1)$, $\lim T^{-1} \int_0^T \chi_{\alpha,\beta}\{f(t)\}\, dt = \beta-\alpha$, where $\chi_{\alpha,\beta}$ is the indicator function of $[\alpha, \beta)$ and $\{f\}=f-[f]$. The author uses some methods of Cassels [Proc. Cambridge Philos. Soc. **46** (1950), 219–225; MR **12**, 162] in the theory of Diophantine approximations to obtain estimates of the rate of convergence in certain cases.

S. J. Taylor (Ann Arbor, Mich.)

Citations: MR 12, 162d = J24-12.

K40-42 **(32# 5639)**

Grandet-Hugot, Marthe

Étude de certaines suites $[\lambda\alpha^n]$ dans les I-adèles.

C. R. Acad. Sci. Paris **261** (1965), 4943–4945.

For any finite set I of valuations of the rational field Q, let $I^+=I \cup \{0\}$ and $I^-=I-\{0\}$, where 0 is the archimedean valuation. Let V_I be the set of adèles $\{x_p\}$ such that $x_p=0$ for all $p \in I$; put $\mathbf{Z}[I]=\{x \in \mathbf{Q} \mid |x|_p \leqq 1$ for all $p \notin I^+\}$, and $F_a(I)=\{x \in V_I \mid |x|_p<1$ for all $p \notin I^-, a \leqq x_0 < a+1\}$ (where a is a fixed real number). It is known that every $x \in V_I$ can be uniquely written in the form $x=E(x)+\mathscr{E}(x)$, where $E(x) \in \mathbf{Z}[I]$ and $\mathscr{E}(x) \in F_a(I)$. If p' is any valuation, and $I'=I \cup \{p'\}$, the author studies the set $s_I^{p'}(a)$ of all $x \in V_I$ such that (i) $|x|_p>1$ for all $p \in I$, and (ii) there is an invertible element $\lambda \in V_I$, for which $\lim_{n\to\infty} \mathscr{E}_{p'}(\lambda x^n)=0$. It is shown that $S_I^{p'}(a)$ is countable. It is then indicated how (ii) can be weakened to the assumption that the sequence $\mathscr{E}_{p'}(\lambda x^n)$ has only a finite number of accumulation points.

B. Gordon (Los Angeles, Calif.)

K40-43 **(33# 114)**

Bertrandias, Françoise

Ensembles remarquables d'adèles algébriques.

Bull. Soc. Math. France Mém. 4 (1965), vi+98 *pp.*

The author generalizes the theory of S-numbers to the ring of adèles of the field Q of rational numbers. The main results are (slightly more general than) the following. For some finite subset I of the set P of all nonequivalent valuations of Q (every $p \in P$ is either the ordinary absolute value or the p-adic valuation for a prime number p), let V_I denote the ring of I-adèles, which can be identified with the completion of Q with respect to the pseudo-valuation $|r|_I=\sup_{p\in I}|r|_p$. Every $x \in V_I$ can uniquely be written in the form $x=E_I(x)+\varepsilon_I(x)$, where $\varepsilon_I(x)$ belongs to the fundamental domain F_I of V_I and $E_I(x)$ belongs to the ring $Z(I)$ of all rational numbers r/s, s containing only primes belonging to I. For every $p' \in P$ the author defines a set $S_I^{p'}$ of algebraic elements $\vartheta \in V_I$ analogous to the definition of S-numbers. She proves two theorems characterizing $S_I^{p'}$, one of which is as follows. An algebraic element $\vartheta \in V_I$, satisfying $|\vartheta|_p>1$ for all $p \in I$, belongs to $S_I^{p'}$ if and only if there exists an invertible element $\lambda \in V_I$ such that $\lim_{n\to\infty} \varepsilon_{p'}(\lambda\vartheta^n)=0$. She then shows that for every ring of algebraic elements in V_I, there exist elements of $S_I^{p'}$ having the degree of the ring. In the next section, sets $\tilde{E}_\xi$ are defined on the fundamental domain of V_I which are generalizations of Cantor sets with constant ratio of dissection $\xi=1/\vartheta$, and the concepts of U- and M-sets (which correspond to sets of uniqueness and multiplicity in the classical case) are introduced. The author gives necessary and sufficient conditions for $\tilde{E}_\xi$ to be a U-set. Finally, a metrical theorem of Koksma on uniform distribution is generalized to the case of functions on V_I.

J. E. Cigler (Groningen)

Referred to in K40-54, R06-32, R06-37, R06-39.

K40-44 **(33# 5597)**

Chauvineau, Jean

Sur la C-équirépartition modulo 1 en valuation p-adique.

C. R. Acad. Sci. Paris Sér. A-B **262** (1966), A557–A559.

A definition for when a function from a measurable set of

p-adic numbers to the p-adic numbers is uniformly distributed modulo 1 is stated. The author shows that the obvious analogues for Weyl's criterion for uniformly distributed modulo 1 real functions hold. Using this criterion he gives several examples of uniformly distributed modulo 1 p-adic functions.

D. J. Lewis (Ann Arbor, Mich.)

K40-45 (33# 7321)

Chauvineau, Jean

Sur la répartition en valuation p-adique.

C. R. Acad. Sci. Paris **259** (1964), 3907–3909.

Several p-adic sequences are shown to be equidistributed or very well distributed (in the sense of Amice [same C. R. **256** (1963), 2742–2744; MR **28** #1194]). Thus, if $p \neq 2$ and $\{x_n\}$ is equidistributed on $\mathbf{Z}_p - \{0\}$, then so also is $\{(x_n |x_n|_p^{1-m})^{1/m}\}$. *W. J. LeVeque* (Ann Arbor, Mich.)

Citations: MR 28# 1194 = Q25-20.

K40-46 (33# 7322)

Decomps-Guilloux, Annette

Ensembles d'éléments algébriques dans les adèles.

C. R. Acad. Sci. Paris **261** (1965), 1929–1931.

The author states necessary conditions that a p-adic number θ be algebraic. Also that an element of a subring of an adele be algebraic. The results are reminiscent of, and the methods related to, those of C. Chabauty [same C. R. **231** (1950), 465–466; MR **12**, 395], C. Pisot [Comment. Math. Helv. **19** (1946), 153–160; MR **8**, 194], and Vijayaraghavan [Proc. Indian Acad. Sci. Sect. A **12** (1940); MR **2**, 33; Proc. Cambridge Philos. Soc. **37** (1941), 349–357; MR **3**, 274; J. London Math. Soc. **17** (1942), 137–138; MR **5**, 35]. They concern the size of the fractional part of $\theta\lambda^n$, for $n \geqq 1$ and some $\lambda \in \mathbf{Q}_p$ or the adele.

D. J. Lewis (Ann Arbor, Mich.)

Citations: MR 2, 33d = K15-3; MR 3, 274c = K25-5; MR 5, 35f = K25-7; MR 8, 194c = K25-9; MR 12, 395d = K40-9.

K40-47 (34# 1297)

Kuipers, L.

Uniform distribution modulo m.

Nieuw Arch. Wisk. (3) **14** (1966), 119–123.

Let $\|E\|$ denote the Lebesgue measure of a measurable set E of real numbers. With $E(u) = E \cap [0, u]$, say that E is relatively measurable if $\lim_{u\to\infty} \|E(u)\|/u$ exists, and this limit is the relative measure, denoted by $\|E\|_R$. Let $f(t)$ be a real function with domain $[0, \infty)$, measurable on every interval $[0, u)$. For a positive integer $m > 1$, let $[f(t)]_m$ denote the greatest multiple of $m \leqq f(t)$ and write $\{f(t)\}_m = f(t) - [f(t)]_m$. For each $j = 0, 1, \cdots, m-1$, let E_j be the set of all nonnegative t such that $j \leqq \{f(t)\}_m < j+1$. Then $f(t)$ is said to be u.d. (uniformly distributed) mod m if $\|E_j\|_R = 1/m$ for $j = 0, 1, \cdots, m-1$. Also, $f(t)$ is said to be u.d. if $f(t)$ is u.d. mod m for all $m = 2, 3, \cdots$. It is proved that if $f(t)/m$ is u.d. mod 1 (in the classical sense) then $f(t)$ is u.d. mod m. If for every $m = 2, 3, \cdots$ the function $mf(t)$ is u.d. mod m, then $f(t)$ is u.d. mod 1. A one-sided analogue of the classical Weyl criterion is given, namely a necessary condition in terms of integrals for $f(t)$ to be u.d. Among other results it is established that $\log t$ is not u.d. mod m. The author points out that his definition of u.d. mod m and some proofs are analogous to those of the reviewer for sequences of integers [Trans. Amer. Math. Soc. **98** (1961), 52–61; MR **22** #10971].

I. Niven (Eugene, Ore.)

Citations: MR 22# 10971 = B48-6.

K40-48 (34# 2555)

Grassini, Elena

Successioni regolarmente equidistribuite negli spazi topologici.

Ist. Lombardo Accad. Sci. Lett. Rend. A **99** (1965), 950–962.

Denote by F a family of sets of points in a compact Hausdorff topological group X. A uniformly distributed sequence $\{a_n\}$ is called "justly distributed" on F if the condition $D_N = \sup_{\sigma\in F} |I_N(\sigma)/N - \mu(\sigma)| = O(1/N)$ is satisfied for $N \to \infty$. Here $I_N(\sigma)$ denotes the number of a_i such that $a_i \in \sigma$, $1 \leqq i \leqq N$, where σ is any set that belongs to F, and $\mu(\sigma)$ the Haar measure of σ. T. van Aardenne-Ehrenfest [Nederl. Akad. Wetensch. Proc. **48** (1945), 266–271; MR **7**, 376; ibid. **52** (1949), 734–739; MR **11**, 336] has proved that a justly distributed sequence cannot exist on F if F is the family of intervals on the real axis.

The reviewer [Ist. Lombardo Accad. Sci. Lett. Rend. A **96** (1962), 351–372; MR **26** #4983] has observed that there exist some examples of justly distributed sequences in the p-adic domain D_p on the family of open spheres of D_p.

The paper under review contains several results concerning this topic. In particular, it is proved that there do not exist justly distributed sequences on the family of closed sets of D_p. The main result proved in the paper is the following: In every separable topological group X, the family F of Jordan measurable closed sets and every family F' of Jordan measurable open sets that constitute a countable basis of X are equivalent; i.e., every sequence justly distributed on F is also justly distributed on F', and vice versa. *M. Cugiani* (Milan)

Citations: MR 7, 376l = K05-6; MR 11, 336d = K05-12; MR 26# 4983 = K40-25.

K40-49 (34# 5799)

Kuipers, L.; van der Steen, P.

Metric theorems in the theory of asymptotic distribution modulo 1.

Nederl. Akad. Wetensch. Proc. Ser. A **69** = *Indag. Math.* **28** (1966), 300–310.

For every interval $[\alpha, \beta) \subset [0, 1)$, let $\theta(u; \alpha, \beta) = 1$ or 0 if u belongs or does not belong to $[\alpha, \beta)$, respectively. Let (x) denote the fractional part of x. A measurable real function $f(t)$ defined for $t \geqq 0$ is called continuously uniformly distributed (c.u.d.) mod 1 if for every interval $[\alpha, \beta) \subset [0, 1)$, $T^{-1} \int_0^T \theta((f(t)); \alpha, \beta)\, dt \to \beta - \alpha$ as $T \to \infty$.

The main result of this paper is the following theorem. Let $f(x, u)$ be a measurable real function of the variable $x \in [a, b]$ $(a < b)$ and $u \geqq 0$. For $h = 1, 2, \cdots$ and $t > 0$, set $p_h(x, t) = t^{-1} \int_0^t e^{2\pi i h f(x,u)}\, du$. If, for every h, $\int_1^T t^{-1} (\int_a^b |p_h(x, t)|^2\, dx)\, dt$ has a finite limit as $T \to \infty$, then for almost all $x \in [a, b]$ the function $f(x, u)$, as a function of u, is c.u.d. mod 1.

The authors deduce from this a theorem of H. Davenport, P. Erdős and W. J. LeVeque [Michigan Math. J. **10** (1963), 311–314; MR **27** #3618] which is the discrete analogue of their theorem. They also give some metrical theorems, as applications of the main theorem, which allow one to determine some classes of functions whose values are u.d. mod 1. These results are too complicated to be stated here. E.g., we mention the function $F(u) = u(2 + \sin \log u - \cos \log u)$, $u > 0$. The function $xF(u)$ is c.u.d. mod 1 for almost all x. The function $x^{F(u)}$ is c.u.d. mod 1 for almost all $x > 1$. *J. Kubilius* (Vilnius)

Citations: MR 27# 3618 = K05-40.

K40-50 **(34# 5963)**

Cigler, J.

Stark gleichverteilte Folgen in kompakten Gruppen.

Monatsh. Math. **70** (1966), 430–436.

Let X be a compact second-countable group, and suppose that $\omega_0=\{x_n\}$ is a uniformly distributed (u.d.) sequence of elements of X. Let $\Omega=\prod X_n$, $X_n=X$, with the usual topology. T is the shift transformation on Ω, $X_{\omega_0}\Omega$ is the closure of the orbit of ω_0 under repeated shifts and V_{ω_0} is the set of condensation points (in the weak*-topology) of the linear functionals μ_n defined by $\mu_n f=n^{-1}\sum_{k=1}^n f(T^k\omega_0)$. A set in X_{ω_0} is called null if its μ-measure is zero for every $\mu\in V_{\omega_0}$. The sequence ω_0 is strongly u.d. if apart from a null set every $\omega\in X_{\omega_0}$ is u.d. Theorem 1: Let X be abelian. Then ω_0 is strongly u.d. if and only if for all $\mu\in V_{\omega_0}$ and all continuous characters $\chi\not\equiv 1$ on X, the average $n^{-1}\sum_1^n\int_{X_{\omega_0}}\chi(\pi(T^k\omega))[\chi(\pi(\omega))]^-\mu(d\omega)$ tends to zero; here π is the projection of Ω on its first coordinate. Van der Corput's theorem is obtained as a corollary. Other corollaries are given and the extension to the non-abelian case is indicated in Theorem 2.

J. G. Wendel (Ann Arbor, Mich.)

K40-51 **(35# 1753)**

Cigler, J.

A characterization of well-distributed sequences.

Compositio Math. **17**, 263–267 (1966).

The author obtains several necessary and sufficient conditions for a sequence to be m-well-distributed in a compact Hausdorff space X satisfying the second axiom of countability, where m is a given probability measure on X. These relate the existence of m-well-distributed sequences $\omega=\{x_1, x_2,\cdots\}$ in X to the properties of invariant probability measures on the space X_ω, the closure in the product space $\prod_1^\infty X$ of the set $\{(x_j, x_{j+1},\cdots)\}_{j=1}^\infty$. They generalise a known theorem of J. C. Oxtoby [Bull. Amer. Math. Soc. **58** (1952), 116–136; MR **13**, 850] on strictly ergodic sets.

B. H. Murdoch (Dublin)

Referred to in K35-33.

K40-52 **(35# 2835)**

Zaretti, Anna Golzi

Sopra un teorema di B. Eckmann riguardante successioni uniformemente distribuite.

Ist. Lombardo Accad. Sci. Lett. Rend. A **100** (1966), 99–108.

The author proves a necessary condition for the uniform distribution of a sequence in a compact topological group. This condition improves an analogous theorem of B. Eckmann [Comment. Math. Helv. **16** (1944), 249–263; MR **6**, 146], giving to it a wider meaning. Further, the author completes a note in a previous paper [Atti Sem. Mat. Fis. Univ. Modena **13** (1964), 111–118; MR **30** #3064] concerning a theorem of the reviewer [Ist. Lombardo Accad. Sci. Lett. Rend. A **96** (1962), 351–372; MR **26** #4983] on a criterion for uniform distribution in the p-adic field.

M. Cugiani (Milan)

Citations: MR 6, 146a = K40-2; MR 26# 4983 = K40-25; MR 30# 3064 = K40-35.

K40-53 **(36# 113)**

de Mathan, Bernard

Sur un théorème métrique d'équirépartition mod 1 **dans un corps de séries formelles sur un corps fini.**

C. R. Acad. Sci. Paris Sér. A-B **265** (1967), A289–A291.

Let N denote the set of positive integers. Let F_q denote the finite field of order q and let $\mathfrak{F}$ denote the field of rational functions $F_q(T)$. A valuation of $\mathfrak{F}$ is defined by means of $|f/g|_0=q^{\deg f-\deg g}$ $(f, q\in F_q[T])$; the completion $\mathfrak{F}_0$ of $\mathfrak{F}$ relative to this valuation consists of the set of formal power series $\sum_{s=-m}^\infty a_sT^{-s}$ $(a_s\in F_q)$. The following theorem is stated.

Let D be a disk of $\mathfrak{F}_0$ of radius q^k and ϕ_n $(n=1, 2, 3,\cdots)$ a sequence of continuous mappings of D in $\mathfrak{F}_0$. Put $\psi_{m,n}=\phi_n-\phi_m$. Suppose that there exists a subset K of $N\times N$, containing the diagonal, and such that $\sum_{N=1}^\infty K_N/N^3=+\infty$, $K_N=\operatorname{card}\{(m, n)\in K:\sup(m, n)\leqq N\}$. Moreover, suppose that for every $(m, n)\notin K$, $\psi_{m,n}$ satisfies

$$|\psi_{m,n}(X)-\psi_{m,n}(Y)|_0 = q^{\lambda_{m,n}}|X-Y|_0$$

for all $X, Y\in D$, with $\lambda_{m,n}\geqq -1-k$. Then the sequence $\phi_n(X)$, $n=1, 2, 3,\cdots$, is uniformly distributed (mod 1) for almost all X belonging to D.

The author also states four corollaries of this theorem.

L. Carlitz (Durham, N.C.)

Referred to in K40-70.

K40-54 **(36# 3734)**

Meijer, H. G.

Uniform distribution of g-adic integers.

Nederl. Akad. Wetensch. Proc. Ser. A **70** = *Indag. Math.* **29** (1967), 535–546.

For any integer $g\geqq 2$, let Q_g denote the g-adic numbers, with the g-adic valuation of a rational number a denoted by $|a|_g$. Let Z_g be the ring of g-adic integers. For any rational integer k and any $d\in Q_g$ define the k-neighborhood $U_k(d)$ of d by $U_k(d)=\{x|x\in Q_g, |x-d|_g\leqq g^{-k}\}$. If $\{x_n\}$ is any infinite sequence in Z_g, let $A(U_k(d), N)$ denote the number of elements x_n satisfying $1\leqq n\leqq N$ and $x_n\in U_k(d)$. The sequence $\{x_n\}$ is said to be k-uniformly distributed in $\boldsymbol{Z_g}$ if as $N\to\infty$, $\lim N^{-1}A(U_k(d), N)=g^{-k}$. The sequence $\{x_n\}$ is said to be uniformly distributed (u.d.) in Z_g if it is k-uniformly distributed in Z_g for all positive integers k. For example, let $x_n=n^r$, where r is a fixed positive integer; then the sequence $\{x_n\}$ is u.d. in Z_g if and only if $r=1$. If $\{x_n\}$ is any u.d. sequence in Z_g and if $a, b\in Z_g$, then the sequence $\{ax_n+b\}$ is u.d. in Z_g if and only if a is a unit in Z_g. If $a, b, c\in Z_g$, where $|a|_g<1$ and b is a unit, then the sequence $\{an^2+bn+c\}$ is u.d. in Z_g; this result is generalized to polynomials of higher degree. The following result connects the theory with that of uniform distribution of real numbers, and so enables the author to derive many corollaries from the real number case. Let $\{x_n\}$ be a sequence of real numbers and let k be a fixed positive integer. If the real sequence $\{g^{-k}x_n\}$ is u.d. modulo 1, then the sequence $\{[x_n]\}$ considered as a sequence in Z_g is k-uniformly distributed in Z_g. Some results of the paper are generalizations of work on p-adic uniform distribution by M. Cugiani [Ist. Lombardo Accad. Sci. Lett. Rend. A **96** (1962), 351–372; MR **26** #4983], F. Bertrandias [Bull. Soc. Math. France Mém. **4** (1965); MR **33** #114] and J. Chauvineau [C. R. Acad. Sci. Paris **260** (1965), 6252–6255; MR **31** #3400]. Also generalized are some results on the uniform distribution of rational integers by the reviewer [Trans. Amer. Math. Soc. **98** (1961), 52–61; MR **22** #10971] and S. Uchiyama [Proc. Japan Acad. **37** (1961), 605–609; MR **25** #1145].

I. Niven (Eugene, Ore.)

Citations: MR 22# 10971 = B48-6; MR 25# 1145 = B48-8; MR 26# 4983 = K40-25; MR 31# 3400 = B48-13; MR 33# 114 = K40-43.

Referred to in K40-57.

K40-55 **(36# 5088)**

Grandet-Hugot, Marthe

Étude de certaines suites $\{\lambda\alpha^n\}$ dans les adèles.

Ann. Sci. École Norm. Sup. (3) **83** (1966), 171–185.

Artin's decomposition of adèles enables the author to define special adèles among the adèles of the field of rationals which have many properties analogous to those of the Pisot-Vijayaraghavan numbers. This analogy is pursued in several theorems about these special adèles. The definitions and theorems are too cumbersome to formulate here. *O. H. Körner* (Marburg)

K40-56 (37# 2726)

Grandet-Hugot, Marthe

Éléments algébriques remarquables dans un corps de séries formelles.

Acta Arith. **14** (1967/68), 177–184.

In 1962 P. T. Bateman and A. L. Duquette [Illinois J. Math. **6** (1962), 594–606; MR **26** #2424] constructed an analogue of the Pisot-Vijayaraghavan numbers in fields of formal power series and proved, among other things, that when k is perfect, an element $\theta \in k\{x^{-1}\}$, $|\theta|>1$, is PV if and only if there exists a non-zero $\lambda \in k\{x^{-1}\}$ such that the "distance" of $\lambda\theta^n$ from the "nearest integer" tends to 0 as $n\to\infty$, and, in that case, $\lambda \in k(x)(\theta)$. (See the review of the above mentioned paper for definitions.) In this paper it is shown that k need not be perfect and a similar result is proved for a slightly wider class of elements of $k\{x^{-1}\}$ than the PV elements. Further, (a) necessary and sufficient conditions for an element of $k\{x^{-1}\}$ to be algebraic and separable over $k(x)$ are given, and (b) it is shown that where k is a finite field the set of all PV elements is not closed. This latter result is in distinction to the theorem of R. Salem [Duke Math. J. **11** (1944), 103–108; MR **5**, 254] in the ordinary case. *J. B. Roberts* (Portland, Ore.)

Citations: MR 5, 254a = R06-1; MR 26# 2424 = R06-21.

K40-57 (37# 4030)

Meijer, H. G.

The discrepancy of a g-adic sequence.

Nederl. Akad. Wetensch. Proc. Ser. A **71** = *Indag. Math.* **30** (1968), 54–66.

For any integer $g\geq 2$, let Z_g denote the g-adic integers. For any infinite sequence $\{x_n\}$ of g-adic numbers, for any positive integer N, denote the discrepancy $\delta(N)$ of the sequence by $\sup|A(U_k(d), N)-g^{-k}N|$, where the supremum is taken over all neighborhoods $U_k(d)$ with $d\in Z_g$ and $k=1, 2, \cdots$. For definitions of neighborhood and the A function, see the earlier paper of the author or the review thereof [same Proc. **70** (1967), 535–546; MR **36** #3734]. Then $\delta(N)\geq 1$ for every sequence in Z_g and any positive integer N. If $x_n=n$ the discrepancy is shown to be 1. Let the g-adic expansion of any g-adic integer a be $\sum a_i g^i$. Define the function ϕ_g, mapping Z_g onto the real interval $[0, 1)$, thus: $\phi_g(a)$ is the fractional part of the real number $\sum a_i g^{-i-1}$. Now consider any sequence $\{x_n\}$ in Z_g with discrepany $\delta(N)$; let $d(N)$ denote the discrepancy in the classical sense of the real sequence $\{\phi_g(x_n)\}$. The main result of the paper is that

$$\delta(N) \leqq d(N) < \delta(N)\{2+2(g-1)\log(N/\delta(N))/\log g\}.$$

It follows that $\{x_n\}$ is uniformly distributed in Z_g if and only if the real sequence $\phi_g(x_n)$ is uniformly distributed in $[0, 1)$. The theory is extended to g-adic sequences in several dimensions. Also an extension is given of a result on Van der Corput sequences. A full bibliography is provided. *I. Niven* (Eugene, Ore.)

Citations: MR 36# 3734 = K40-54.

K40-58 (37# 5159)

Meijer, Hendrik Gerrit

Uniform distribution of g-adic numbers. (Dutch summary)

Doctoral dissertation, University of Amsterdam.

Universiteit van Amsterdam, Amsterdam, 1967. 75 *pp.* (*loose insert*) *n.p.*

Let g be an integer $\geqq 2$ with prime power decomposition $g=\prod_{i=1}^r p_i^{h_i}$. The g-adic pseudo-valuation $|\ \ |_g$ is defined for $a\in\mathbf{Q}$, the field of rational numbers by $|a|_g=\max_{1\leq i\leq r}(|a|_{p_i}{}^{\lambda_i})$, where $|a|_{p_i}$ is the ordinary p_i-adic valuation on $\mathbf{Q}$ and λ_i is that real number for which $|g|_{p_i}{}^{\lambda_i}=1$. The completion of Q under this pseudo-valuation is denoted $\mathbf{Q}_g$ and called the ring of g-adic numbers; Q_g is, of course, the direct sum of the p-adic fields Q_{p_i}. The ring of g-adic integers $\mathbf{Z}_g$ is defined to be the ring of those $a\in Q_g$ satisfying $|a|_g\leqq 1$. The study of diophantine approximation in this ring was initiated by K. Mahler [*Lectures on diophantine approximations. Part 1: g-adic numbers and Roth's theorem*, Univ. Notre Dame Press, Notre Dame, Ind., 1961; MR **26** #78]. The author continues this study by defining uniform distribution in Q_g and investigating some of the consequences of this definition. His work comprehends the results of Niven, Uchiyama, and Van den Enden on the uniform distribution of sequences of integers.

Suppose (x_n) is a sequence in $\mathbf{Z}_g$. Let $A(k, d, N)$ be the number of $i\leqq N$ for which $|x_i-d|_g\leqq g^{-k}$. The sequence (x) is said to be g-adically uniformly distributed if $\lim_{N\to\infty} A(k, d, N)/N=g^{-k}$ for all positive integers k and d. The discrepancy of the sequence (x_n) is defined to be $\delta(N)=\sup_{k,d}|A(k, d, N)-g^{-k}N|_g$. If $a\in Q_g$, then a can be represented uniquely in the form $a=\sum_{j=j_0}^{\infty} a_j g^j$, where each $a_j\in\mathbf{Z}$ and $0\leqq a_j<g$. Define $\psi(a)=\sum_{j=j_0}^{-1} a_j g^j\in\mathbf{R}$, the field of real numbers, $\varphi(a)=\sum_{j=j_0}^{\infty} a_j/g^{j+1}\in\mathbf{R}$, and $\{a\}_g=\sum_{j=0}^{\infty} a_j g^j\in\mathbf{Z}_g$. Finally, define $\chi(a)=e^{2\pi i\psi(a)}$.

The author first proves that if (x_n) is a sequence in $\mathbf{R}$ such that $(g^{-k}x_n)$ is uniformly distributed (mod 1) for every integer $k\geqq 0$, then the sequence $([x_n])$, considered as a sequence in $\mathbf{Z}_g$, is g-adically uniformly distributed. As a consequence such sequences as $[\alpha n]$, where α is real irrational and $[\alpha n^\sigma]$, where α, σ are real, $\alpha\neq 0$ and σ non-integral, are g-adically uniformly distributed.

Next the author proves the analogue of Weyl's classical criterion for uniform distribution (mod 1): The sequence (x_n) in $\mathbf{Z}_g$ is g-adically uniformly distributed if and only if $\lim_{N\to\infty} N^{-1}\sum_{n=1}^{N}\chi(hx_n/g^k)=0$ for all positive integers h and k such that h does not divide g^k. Using this result he proves metrical theorems of which the following is a sample: Let (t_n) be a sequence in Q_g such that $|t_1|_g<|t_2|_g<|t_3|_g<\cdots$. Then for almost all $x\in Q_g$, the sequence $\psi(t_nx)$ is uniformly distributed in $[0, 1]$ and simultaneously the sequence $\{t_nx\}_g$ is g-adically uniformly distributed. The author investigates discrepancy and proves that if (x_n) is a sequence of g-adic integers and $\alpha(N)$ is the discrepancy of the sequence $\varphi(x_n)$ then $\alpha(N)\leqq d(N)$, $d(N)=O(\delta(N)\log(N/\delta(N)))$ and obtains a good estimate for the constant implied by the "O".

Finally these results are generalized to multi-dimensional g-adic sequences.

D. G. Cantor (Los Angeles, Calif.)

Citations: MR 26# 78 = J02-14.

K40-59 (38# 5723)

Beer, Susanne

Zur Theorie der Gleichverteilung im p-adischen.

Österreich. Akad. Wiss. Math.-Natur. Kl. S.-B. II **176** (1967/68), 499–519.

Let D_p be the additive group of p-adic integers. The p-adic valuation $|a|_p$ ($a \in D_p$) induces a metric, under which D_p is a compact topological group; denote by μ the corresponding normed Haar measure. For $E \subset D_p$ with $\mu(\partial E)=0$ and $\omega=\{x_n\}\subset D_p$, set $I_N(E)=\sum_{x_n \in E; n \leqq N} 1$; then the sequence ω is said to be equidistributed if for every closed E, $\lim_{N\to\infty} N^{-1} I_N(E)=\mu(E)$. For m, λ rational integers, $\lambda \geqq 0$, consider the sets $E(m,\lambda)=\{x \in D_p: |m-p|_p \leqq p^{-\lambda}\}$ and let $q_N=q_N(m,\lambda)=N^{-1}I_N(E(m,\lambda))-p^{-\lambda}$; also let $D_N(\omega)=\sup_{E(m,\lambda)}|q_N|$, the discrepancy of ω. It is known that $N^{-1}\leqq D_N \leqq 1$; that $\lim_{N\to\infty} q_N=0$ for every m and λ implies the equidistribution of ω [see A. Zaretti, Atti Sem. Mat. Fis. Univ. Modena **13** (1964), 111–118; MR **30** #3064]; and that the equidistribution of ω is equivalent to $\lim_{N\to\infty} N^{-1}\sum_{n=1}^N \chi(x_n)=0$ for all non-trivial characters χ (Weyl's criterion). The author proves some further properties related to the equidistribution of sequences ω in D_p, among which are the following. If ω is equidistributed, then $\lim_{N\to\infty} q_N=0$ holds uniformly in m and λ. $\omega \subset D_p$ is equidistributed if and only if $D_N(\omega)$ is a null sequence. Contrary to what happens in the real case, where $\limsup ND_N(\omega)=\infty$, there are sequences $\omega \subset D_p$ for which $ND_N(\omega)$ stays bounded. If, in particular, $x_n=na+b$, $|a|_p=1$, one actually has $D_N=N^{-1}$. If $N=kp^\lambda+r$, $0\leqq r<p^\lambda$, and $c_N=c_N(E)$ is defined by $(p^\lambda+r/k)q_N+r(kp^\lambda)^{-1}$, then ω is equidistributed in D_p if and only if $c_N\to 0$ for $N\to\infty$ and every m and λ. If ω is equidistributed and ND_N is bounded, then ω is uniformly equidistributed (i.e., there exists an M such that $k-M<I_N<k+M$ for all N, m and λ); the converse is false. If (x_n), (y_n) are sequences in D_p with $|x_n-y_n|_p<\varepsilon$ for all n and discrepancies D_N and D_N', respectively, then $|D_N-D_N'|<\varepsilon$. If (x_n) is an arbitrary sequence in D_p, of discrepancy D_N, and (y_n) is any sequence dense in D_p, then (y_n) can be reordered so that the discrepancy D_N' of the reordered sequence satisfies $|D_N-D_N'|<N^{-1}\varphi(N)$, where $\varphi(t)$ is any monotonic, strictly increasing function; in particular, any dense sequence can be reordered into an equidistributed one.

E. Grosswald (Philadelphia, Pa.)

Citations: MR 30# 3064 = K40-35.

K40-60 (39# 170)

Lesca, Jacques

Équirépartition dans un anneau d'adèles.

Séminaire Delange-Pisot-Poitou: 1966/67, *Théorie des Nombres, Fasc.* 2, *Exp.* 15, 7 *pp. Secrétariat mathématique, Paris,* 1968.

Let A be the ring of adèles $a=(a_p)$ defined with respect to a given algebraic number field k and a given set S of valuations of k including the archimedean ones. Further, let $\mathcal{O}$ be the subring of integers of k. Put $M=\{a: |a_p|_p\leqq 1$ for all $p \in S\}$. The author expresses in several equivalent ways the condition that a sequence (na), $a\in A$, be uniformly distributed mod $\mathcal{O}$. He also sketches a proof of the following theorem. For each $\lambda \in A$ such that $\lambda_p \neq 0$ for all $p \in S$, and for almost all $x \in A-M$, the sequence (λx^n) is uniformly distributed mod $\mathcal{O}$. This theorem generalizes a classical result by J. F. Koksma [Compositio Math. **2** (1935), 250–258].

C. G. Lekkerkerker (Amsterdam)

K40-61 (39# 5484)

de Mathan, Bernard

Théorème de Koksma dans un corps de séries formelles sur un corps fini.

Séminaire Delange-Pisot-Poitou: 1967/68, *Théorie des Nombres, Fasc.* 1, *Exp.* 4, 12 *pp. Secrétariat mathématique, Paris,* 1969.

Let $\mathfrak{F}_0$ be the set of generalized formal series $\Theta=\sum_{-\infty}^{+\infty} a_nX^n$ ($a_{-n}=0$ for all sufficiently large n) over the finite field $\mathbf{F}_q$ with q elements. Define the valuation $W_0(\Theta)$ of Θ as being the least rational integer n such that $a_n\neq 0$. The space $\mathfrak{F}_0$ together with this valuation is to be thought of as a locally compact additive group on which is defined a Haar measure. Let $\mathscr{H}_0$ be the homomorphism $\mathfrak{F}_0\to\mathfrak{F}_0$ defined by $\sum_{-\infty}^{+\infty} a_nX^n \xrightarrow{\mathscr{H}_0} \sum_1^\infty a_nX^n$ and put $\mathscr{M}_0=\mathscr{H}_0(\mathfrak{F}_0)$. $\mathscr{M}_0$ is thus a compact subgroup of $\mathfrak{F}_0$.

A sequence (Θ_n) of elements of $\mathfrak{F}_0$ is said to be uniformly distributed (mod 1) if the sequence $(\mathscr{H}_0(\Theta_n))$ is uniformly distributed on $\mathscr{M}_0$. The author extends a theorem of Koksma to $\mathfrak{F}_0$ and states two corollaries, namely: (i) Let $\Theta \in \mathfrak{F}_0$ be such that $\mathscr{H}_0(\Theta)\neq 0$ and $W_0(\Theta)\leqq 0$ (the second condition has been omitted by the author). The sequence $(\Lambda\Theta^n)_{n\in N}$ is uniformly distributed (mod 1) for almost all $\Lambda \in \mathfrak{F}_0$. (ii) Let $\Lambda \in \mathfrak{F}_0-\{0\}$. Let J be a sequence of integers which are not divisible by $p=\operatorname{char} F_q$. The sequence $(\Lambda\Theta^n)_{n\in J}$ is uniformly distributed (mod 1) for almost all Θ's such that $\mathscr{H}_0(\Theta)\neq 0$ and $W_0(\Theta)\leqq 0$. The author proves several other related results.

M. Mendès France (Talence)

Referred to in K40-62.

K40-62 (40# 97)

Rhin, Georges

Quelques résultats métriques dans un corps de séries formelles sur un corps fini.

Séminaire Delange-Pisot-Poitou: 1967/68, *Théorie des Nombres, Fasc.* 2, *Exp.* 21, 22 *pp. Secrétariat mathématique, Paris,* 1969.

This article refines some of B. de Mathan's results [*Séminaire Delange-Pisot-Poitou: 1967/68, Théorie des nombres,* Fasc. 1, Exp. 4, Secrétariat mathématique, Paris, 1969; MR **39** #5484].

The author generalizes known results concerning the discrepancy of sequences (mod 1). Here, the sequences are elements of a field of formal series with coefficients in a finite field. Among other results, the author states and proves an analogue of the law of the iterated logarithm.

M. Mendès France (Talence)

Citations: MR 39# 5484 = K40-61.

K40-63 (40# 2614)

Hlawka, Edmund; Niederreiter, Harald

Diskrepanz in kompakten abelschen Gruppen. I. (English summary)

Manuscripta Math. **1** (1969), 259–288.

Authors' summary: "It is the aim of this paper to generalize the notion of discrepancy of a sequence to compact abelian groups satisfying the second axiom of countability. This concept of discrepancy which depends heavily on the study of certain generating subsets of the character group includes the hitherto known case of the n-dimensional unit cube. The definition is subsequently justified by transferring well-known theorems of the classical theory to the general case. This first part mainly deals with the algebraic point of view which doesn't possess an analogy in the theory of uniform distribution mod 1. The proofs involve a lot of group theoretic argument. Theorems concerning distribution and approximation will be presented in the second part [#2615 below]."

J. E. Cigler (Groningen)

K40-64 (40# 2615)

Niederreiter, Harald

Diskrepanz in kompakten abelschen Gruppen. II. (English summary)

Manuscripta Math. **1** (1969), 293–306.

Author's summary: "This is a continuation of the paper reviewed above [#2614]. The concept of discrepancy of a sequence in a compact abelian group with second countability axiom is now investigated in its relation to problems of distribution and approximation. Some fundamental inequalities of the theory of uniform distribution mod 1 can be transferred to the general case, e.g., Koksma's inequality. The crucial point in the generalization of Koksma's inequality is the suitable definition of total variation. Another version of this inequality with an upper bound in terms of the Fourier coefficients is presented. Aardenne-Ehrenfest's theorem is valid for non-totally-disconnected groups." *J. E. Cigler* (Groningen)

K40-65 (41# 5306)

Hlawka, Edmund

Ein metrischer Satz in der Theorie der C-Gleichverteilung.

Monatsh. Math. **74** (1970), 108–118.

Let $f(t)$ be a real-valued, integrable function defined on $t\geqq 0$, and let $a(T,t)\geqq 0$ be defined on $T>0, t\geqq 0$ with $\int_0^\infty a(T,t)\,dt=1$ for each $T>0$. The function f is A-uniformly distributed mod 1 if for each interval $J=[\alpha,\beta)$, $0\leqq\alpha<\beta\leqq 1$, (*) $\lim_{T\to\infty}\int_0^\infty a(T,t)\chi(\{f(t)\},J)\,dt=\beta-\alpha$, where $\chi(x,J)$ is the characteristic function of J and $\{f(t)\}$ is the fractional part of $f(t)$. If (*) holds for a $(T,t)=1/T$ if $0\leqq t\leqq T$, and $a(T,t)=0$ if $t>T$, then f is said to be C-uniformly distributed mod 1. The analogous discrete definition is the following: Let $a(N,n)\geqq 0$ be defined for all positive integers N and n, with $\sum_{n=1}^\infty a(N,n)=1$. A sequence $f(n)$ is A-uniformly distributed mod 1 if $\lim_{N\to\infty}\sum_{n=1}^\infty a(N,n)\chi(\{f(n)\},J)=\beta-\alpha$ for each interval $J=[\alpha,\beta)\subset[0,1]$. In both the discrete and the continuous case, Weyl's criterion holds: f is A-uniformly distributed mod 1 if and only if for each integer $h\neq 0$,

$$\lim_{T\to\infty}\int_0^\infty a(T,t)e^{\pi ihf(t)}\,dt = 0.$$

(In the discrete case the integral is replaced by the analogous sum.) See the survey article by J. C. Cigler and G. Helmberg [Jber. Deutsch. Math. Verein. **64**, Abt. 1, 1–50 (1961); MR **23** #A2409].

Let P be the space of sequences $f(n)$ with $0\leqq f\leqq 1$, and μ the Lebesgue product measure. It is known that almost all sequences are A-uniformly distributed mod 1 if for all $\delta>0$, $\sum_{N=1}^\infty \exp(-\delta^2/a_N)<\infty$, where $a_N=\sum_{n=1}^\infty a^2(N,n)$. The author states and proves the following continuous analogue: Let C_0 be the space of all continuous functions on $t\geqq 0$ with $f(0)=0$, and let μ be Wiener measure. Under certain conditions on the function $a(T,t)$, similar to the ones in the discrete case, almost all $f\in C_0$ are A-uniformly distributed mod 1. In particular, almost all $f\in C_0$ are C-uniformly distributed mod 1.

O. P. Stackelberg (Durham, N.C.)

Citations: MR 23# A2409 = K02-1.

K40-66 (41# 7379)

Baker, R. C.

A diophantine problem on groups. I.

Trans. Amer. Math. Soc. **150** (1970), 499–506.

The "Diophantine problem" of the title is a theorem of Hermann Weyl which says that if $\lambda_1<\lambda_2<\lambda_3<\cdots$ is a sequence of real numbers that does not grow too slowly, then, for almost all real numbers u, $\{\exp 2\pi i\lambda_j u\}_{j=1}^\infty$ is uniformly distributed on the circle. The author proves a similar theorem for compactly generated locally compact Abelian groups G: If $\chi_1,\chi_2,\cdots$ is a sequence of characters of G that satisfy a condition analogous to that of Weyl, then, for almost all $u\in G$, $\{\chi_j(u)\}_{j=1}^\infty$ is uniformly distributed on the circle. *L. Corwin* (New York)

K40-67 (42# 7618)

Descovich, Josef

Zur Theorie der Gleichverteilung auf kompakten Räumen.

Österreich. Akad. Wiss. Math.-Natur. Kl. S.-B. II **178** (1970), 263–283.

In der vorliegenden Arbeit werden Bedingungen dafür abgeleitet, wann eine Folge sich so umordnen läßt, daß die umgeordnete Folge eine vorgegebene Menge von Wahrscheinlichkeitsmaßen als Menge ihrer Verteilungsmaße besitzt. Außerdem wird gezeigt, daß für jede stark reguläre Dreiecksmatrix A zu jedem Wahrscheinlichkeitsmaß μ eine A-μ-gleichverteilte Folge existiert.

J. E. Cigler (Vienna).

K40-68 (42# 7621)

Schmidt, Klaus

Über einen Zusammenhang zwischen gleichverteilten Punkt- und Maßfolgen.

J. Reine Angew. Math. **244** (1970), 94–96.

X sei ein separabler, kompakter Hausdorffraum, und $M(X)$ sei die Menge der Wahrscheinlichkeitsmaße über X in der üblichen schwachen Topologie. Weiter seien $P=\prod_{i=1}^\infty X_i$, $X_i=X$, $i=1,2,\cdots$, und $Q=\prod_{i=1}^\infty M_i$, $M_i=M(X)$, $i=1,2,\cdots$, die Räume der Folgen in X bzw. $M(X)$. Nun wird jedem $\tau=(\mu_i)\in Q$ durch $\tau^+=\prod_{i=1}^\infty\mu_i$ ein Maß auf P zugeordnet.

A sei eine Matrix (a_{nk}), $n,k\doteq 1,2,\cdots$, mit $a_{nk}\geqq 0$, $\sum_{k=1}^\infty a_{nk}=1$, und ferner erfülle A die Hillsche Bedingung. $C(X)$ sei die Menge der stetigen reellwertigen Funktionen über X. Dann sei $\lambda_n(\omega,A;f)=\sum_k a_{nk}f(x_k)$, $\omega=(x_k)\in P$, $\lambda_n(\tau,A;f)=\sum_k a_{nk}\mu_k(f)$, $\tau=(\mu_k)\in Q$, $f\in C(X)$. Sei nun $\tau_0=(\mu_k{}^0)\in Q$ fest gewählt. $\omega\in P$ heißt A-τ_0-gleichverteilt, wenn für alle $f\in C(X)$, $\lim_n(\lambda_n(\omega,A;f)-\mu_n{}^0(f))=0$ ist. $\tau\in Q$ heißt A-τ_0-gleichverteilt, wenn für alle $f\in C(X)$, $\lim_n(\lambda_n(\tau,A;f)-\mu_n{}^0(f))=0$ ist. Die Menge aller A-τ_0-gleichverteilten Punktfolgen werde mit $R(\tau_0)$, die Menge der gleichverteilten Maßfolgen mit $S(\tau_0)$ bezeichnet. Dann gilt Satz 1: Sei $\tau_0\in Q$ beliebig. Dann gilt $\tau\in S(\tau_0)\Leftrightarrow\tau^+(R(\tau_0))=1$. *J. E. Cigler* (Vienna)

K40-69 (44# 2711)

Dijksma, Aalt

Uniform distribution in GF$\{q,x\}$ and GF$[q,x]$. (Dutch summary)

Doctoral dissertation, Technical University of Delft.

Technische Hogeschool Delft, Delft, 1971. iii + 22 pp.

Dans cet exposé l'auteur, après avoir rappelé définitions et résultats généraux sur l'équirépartition dans les groupes topologiques compacts et localement compacts, regroupe les résultats qu'il a obtenus et publiés auparavant (seul ou en collaboration avec H. G. Meijer) relativement à $\Phi=\mathrm{GF}[q,x]$ et $\Phi'=\mathrm{GF}\{q,x\}$, respectivement anneau des polynômes et corps des séries formelles de Laurent sur le corps fini GF(q).

On y trouvera particulièrement l'étude des relations entre l'équirépartition dans Φ et l'équirépartition dans Φ', principalement pour les suites engendrées par des polynômes [l'auteur, Nederl. Akad. Wetensch. Proc. Ser. A **72** (1969), 376–383; MR **40** #2650; ibid. **73** (1970), 187–195; MR **41** #8346; H. G. Meijer et l'auteur, Duke Math. J. **37** (1970), 507–514; MR **42** #1771], des théorèmes sur la répartition presque uniforme [l'auteur, #2713 below] et l'utilisation de la mesure de Banach-Buck l'auteur, Mathematica (Cluj) **11** (**34**) (1969), 221–240; MR **42** #229]. *F. Dress* (Talence)

Citations: MR 40# 2650 = T55-31; MR 41# 8346 = T55-32; MR 42# 229 = T55-43; MR 42# 1771 = T55-44.

K40-70 (44# 2713)

Dijksma, A.

Metrical theorems concerning uniform distribution in GF[q, x] and GF{q, x}.

Nieuw Arch. Wisk. (3) **18** (1970), 279–293.

Soient $\Phi = \mathrm{GF}[q, x]$ l'anneau des polynômes en x sur le corps fini GF(q) et $\Phi' = \mathrm{GF}\{q, x\}$ le corps des séries formelles de Laurent de la forme $\sum_{i=-\infty}^{m} a_i x^i$.

Après avoir rappelé les notions de répartition uniforme sur Φ, de répartition uniforme et presque uniforme sur Φ', l'auteur montre en utilisant un résultat de B. de Mathan [C. R. Acad. Sci. Paris Sér. A-B **265** (1967), A289–A291; MR **36** #113] que sous certaines conditions la suite $\{[\varphi_i(\alpha)]\}$ est équirépartie mod 1 dans Φ pour presque tout α appartenant à un certain disque $\Psi \subset \Phi'$.

Il montre ensuite l'analogue d'un théorème de L. Kuipers et S. Uchiyama sur les entiers [Proc. Japan Acad. **44** (1968), 608–619; MR **39** #148]: si la suite $\{A_i\}$ est équirépartie mod M dans Φ pour une infinité de $M \in \Phi$, la suite $\{A_i\alpha\}$ est presque équirépartie mod 1 dans Φ' pour presque tout $\alpha \in \Phi'$. L'auteur donne enfin un contre-exemple analogue à celui de H. G. Meijer [#2712 ci-dessus]: Il existe une suite $\{A_i\}$ équirépartie dans Φ et un ensemble $\Theta \subset \Phi'$ de mesure positive tels que la suite $\{A_i\alpha\}$ ne soit équirépartie mod 1 dans Φ' pour aucun $\alpha \in \Theta$. *F. Dress* (Talence)

Citations: MR 36# 113 = K40-53; MR 39# 148 = B48-16.

K40-71 (44# 2714)

Rhin, Georges

Généralisation d'un théorème de I. M. Vinogradov à un corps de séries formelles sur un corps fini.

C. R. Acad. Sci. Paris Sér. A-B **272** (1971), A567–A569.

Let F_q be the finite field having q elements, F_0 the field of formal power series of type $\sum_{i=-\infty}^{m} c_i x^i$ ($c_i \in F_q$). L. Carlitz [Trans. Amer. Math. Soc. **72** (1952), 187–208; MR **14**, 23] has defined uniform distribution (mod 1) of a sequence of elements of F_0. Let $\{P_n\}_{n} = 1, 2, \cdots$ be the sequence of polynomials of $F_q[x]$, irreducible and having highest coefficient 1, enumerated according to increasing degree. Theorem 1 states that $\{\alpha P_n\}_{n=1,2,\cdots}$ ($\alpha \in F_0$) is uniformly distributed (mod 1) if and only if α is not a rational function (i.e., $\alpha \notin F_q(x)$). Theorem 2 gives an estimate which is an analogue in $F_q[x]$ of estimates for the number of primes in arithmetic progressions (A. Page et al.), which is needed for the proof of Theorem 1. Theorem 3 gives a condition for the uniform distribution (mod 1) of $\{f(P_n)\}_{n=1,2,\cdots}$ where $f \in F_0[y]$. Only hints for the proofs are given. *I. Danicic* (Aberystwyth)

Citations: MR 14, 23b = K40-14.

K40-72 (44# 3967)

Rhin, Georges

Généralisation d'un théorème de I. M. Vinogradov à un corps de séries formelles sur un corps fini.

Séminaire Delange-Pisot-Poitou: 1969/70, *Théorie des Nombres, Fasc.* 1, *Exp.* 14, 8 *pp. Secrétariat mathématique, Paris*, 1970.

Soit F_q le corps fini à $q = p^r$ éléments de caractéristique le nombre premier $p \neq 2$, et soit $\mathscr{F}$ le corps des fractions $F_q(x)$, qu'on munit de la valeur absolue non archimédienne, dite 0-adique, définie par $|A|_0 = q^{\deg A}$ pour tout polynôme non nul A de $F_q[x]$; le complété $\mathscr{F}_0$ de $\mathscr{F}$ pour cette valeur absolue est le corps des séries formelles $\sum_{-\infty}^{m} c_k x^k$, où $m \in \mathbf{Z}$ et $c_k \in F_q$ pour tout $k \leq m$. L'auteur utilise la définition des suites d'éléments de $\mathscr{F}_0$ équiréparties (e.r.) mod 1 due à L. Carlitz [Trans. Amer. Math. Soc. **72** (1952), 187–208; MR **14**, 23], et le critère de Weyl qui s'y applique, pour généraliser le résultat suivant de I. M. Vinogradov [Izv. Akad. Nauk SSSR Ser. Mat. **12** (1948), 225–248; MR **10**, 599]. où p_n désigne le n-ième nombre premier: La suite $(\alpha p_n)_{n \geq 1}$, où α est un réel, est e.r. mod 1 si et seulement si α est irrationnel. À cet effet, l'auteur définit, à partir d'un ordre choisi sur F_q, un ordre ω sur $F_q[x]^*$, et désigne par P_n le n-ième polynôme unitaire irréductible de $F_q[x]$ suivant l'ordre ω; il démontre alors le théorème suivant: La suite $(\alpha P_n)_{n \geq 1}$, où $\alpha \in \mathscr{F}_0$, est e.r. mod 1 si et seulement si $\alpha \in \mathscr{F}_0 \setminus \mathscr{F}$. La nécessité de cette condition résulte du critère de Weyl; sa suffisance s'établit en adaptant à $\mathscr{F}_0$ la méthode des sommes trigonométriques de Vinogradov [English translation, *The method of trigonometrical sums in the theory of numbers*, Interscience, London, 1954; MR **15**, 941] et en étendant à $\mathscr{F}_0$ un théorème de C. L. Siegel. *J. Chauvineau* (Paris)

Citations: MR 10, 599b = L20-11; MR 14, 23b = K40-14; MR 15, 941b = L02-3.

K40-73 (44# 4187)

Veech, William A.

Some questions of uniform distribution.

Ann. of Math. (2) **94** (1971), 125–138.

A sequence $w_1, w_2, \cdots$ in a compact topological group K, with normalized Haar measure ν, is said to be uniformly distributed in K if for any open set U of K whose boundary has measure 0 the equation $\lim_{N\to\infty} N^{-1} \sum_{n=1}^{N} u(w_n) = \nu(U)$ holds, where u is the characteristic function of U. Definition: A sequence $F = (r_1, r_2, \cdots)$ of positive integers is said to be a uniformly distributed sequence generator (u.d.s.g.) if for any compact group K and any sequence $z_1, z_2, \cdots$ in K that is contained in no proper closed subgroup, the "generated" sequence $w_n = z_{r_1} \cdot z_{r_2} \cdot \cdots \cdot z_{r_n}$ is uniformly distributed in K.

The author is concerned with the existence and construction of u.d.s.g.'s. Theorem 2: Let $R_1, R_2, \cdots$ be an ergodic Markov chain with state space the positive integers and suppose there is a state n such that the one step transition probability from n to any state is positive; then $(R_1, R_2, \cdots)$ is almost surely a u.d.s.g.

Let X be a non-empty open subset of $[0, 1)$. Let x in $0 < x < 1$ be an m-normal number; denote its expansion to the base m by $x = \cdot a_1 a_2 \cdots$, $0 \leq a_j < m$. There exist infinitely many integers $q \geq 2$ such that $x_q \in X$ where $x_q = \cdot a_q a_{q+1} \cdots$. We arrange this sequence $(q_1, q_2, \cdots)$ in increasing order. Let $F(x, X) = (r_1, r_2, \cdots)$ be the sequence $r_1 = q_1 - 1$, $r_2 = q_2 - q_1, \cdots$. Let Y be the boundary of X and for each $\varepsilon > 0$ denote by Y_ε the ε-neighborhood of Y. Write $\delta(Y) = \liminf_{\varepsilon\to 0}(\log \mu_0(Y_\varepsilon))/\log \varepsilon$, where μ_0 is the Lebesgue measure on $[0, 1)$. If $s \geq 1$ is an integer, a sequence of positive integers is said to be an s-u.d.s.g. if the sequence ω_n in the definition above is uniformly distributed whenever $z_1, z_2, \cdots$ is contained in no proper closed subgroup of K. With these notations the author proves Theorem 3: Suppose that either (a) $X = (a, b)$ is an interval or (b) $\delta(Y) > 0$ and X is contained in one of the intervals $(1/m, 1)$ or $(0, 1 - 1/m)$; then there exists an integer $s \geq 1$ such that, for every m-normal number x, $F(x, X)$ is an s-u.d.s.g. *T.-C. Sun* (Detroit, Mich.)

K40-74 **(44# 4796)**

Schmidt, Klaus

Eine Diskrepanz für Maßfolgen auf lokalkompakten Gruppen.

Z. Wahrscheinlichkeitstheorie und Verw. Gebiete **17** (1971), 48–52.

Der Verfasser definiert eine Metrik d_r auf $M(X)$ und benützt diese, um den Begriff der Diskrepanz einer Maßfolge $\{\mu_k\}$ bezüglich eines festen Maßes μ einzuführen. Diese Diskrepanz hat die Eigenschaft, daß sie genau dann gegen 0 konvergiert, wenn die Folge $\{\mu_k\}$ das Verteilungsmaß μ besitzt. Es werden einige wichtige Eigenschaften dieser Diskrepanz bewiesen. *J. E. Cigler* (Vienna)

K40-75 **(44# 4797)**

Schmidt, Klaus

Über die C-Gleichverteilung von Maßen.

Z. Wahrscheinlichkeitstheorie und Verw. Gebiete **17** (1971), 327–332.

Diese Arbeit ist eine Fortsetzung der oben referierten [#4796] dem Fall der C-Gleichverteilung gewidmet. Insbesondere wird auf den Hauptsatz der Gleichverteilung für kompakte und lokalkompakte Gruppen eingegangen. *J. E. Cigler* (Vienna)

K40-76 **(44# 5280)**

Christol, Gilles

Équirépartition dans les séries formelles.

Séminaire Delange-Pisot-Poitou: 1969/70, Théorie des Nombres, Fasc. 1, Exp. 4, 13 pp. Secrétariat mathématique, Paris, 1970.

Let K be a compact set on which is defined a positive normalized measure μ ($\mu(K)=1$). A sequence $u_n \in K$ is μ-distributed if $\lim_{n\to\infty} n^{-1}\sum_{k=0}^{n-1} f(u_k)=\int_K f\,d\mu$ holds for all continuous mappings $f: K\to\mathbf{C}$. The author notices that if $K((X))\sim K^{\mathbf{N}}$ is the set of formal power series on K, then a sequence $\mathbf{u}_n \in K((X))$ is $\mu^{\mathbf{N}}$-distributed if and only if each component is μ-distributed.

Let $K=G$ be a compact group. Define the mapping $U_m: G((X))\to G((X))$ by $U_m(\sum_n a(n)X^n)=\sum_n a(nm)X^n$. An element $\mathbf{a}\in G((X))$ is said to be U_m-normal if the sequence $\mathbf{u}_n=U_m{}^n(\mathbf{a})$ is $\mu^{\mathbf{N}}$-distributed. The author proves that if the group G is profinite, then whatever the positive integer m may be, almost all elements of $G((X))$ are U_m-normal (Théorème 1). {Why should this not be true for any compact group?} The rest of the article is mostly concerned with nonnormal elements. *M. Mendès France* (Talence)

K40-77 **(44# 5288)**

Couot, Jacques

Ergodicité et critères topologiques d'équirépartition d'une famille d'éléments d'un groupe.

C. R. Acad. Sci. Paris Sér. A-B **272** (1971), A1045–A1048.

A sequence $(s_n)_{n\geq 1}$ in the torus $\mathbf{T}$ is called equidistributed if for every $f\in\mathscr{C}(\mathbf{T})$ and $x\in\mathbf{T}$ one has

$$\lim_{N\to\infty} N^{-1}\sum_{n=1}^{N} f(x+s_n) = \int f(x)\lambda_{\mathbf{T}}(dx),$$

where $\lambda_{\mathbf{T}}$ denotes Haar measure on $\mathbf{T}$. Theorem 1 of the paper under review contains a topological criterion for the equidistribution of a sequence $(s_n)_{n\geq 1}$ in $\mathbf{T}$. A family $(s_t)_{t\in\mathbf{R}_+}$ in $\mathbf{T}$ possesses an asymptotic measure μ on $\mathbf{T}$, if for every $f\in\mathscr{C}(\mathbf{T})$ and $x\in\mathbf{T}$, $\lim_{u\to\infty} u^{-1}\int_0^u f(x+s_t)\,dt=\int f(x)\mu(dx)$. The method of the proof of Theorem 1 can be used to establish a topological criterion for the existence of an asymptotic measure for a family $(s_t)_{t\in\mathbf{R}_+}$ in $\mathbf{T}$ (Theorem 2). *H. Heyer* (Tübingen)

K45 PSEUDO-RANDOM NUMBERS, MONTE CARLO METHODS

See also reviews A12-60, A56-39, B40-66, B44-23, K02-2, K05-9, K05-15, K05-22, K05-35, K05-39, K05-44, K05-59, K10-10, K10-11, K15-88, K30-18, K40-22, Q15-11, Z30-67.

K45-1 **(11, 239b)**

Koksma, J. F., and Salem, R. Uniform distribution and Lebesgue integration. Acta Sci. Math. Szeged. **12**, Leopoldo Fejér et Frederico Riesz LXX annos natis dedicatus, Pars B, 87–96 (1950) = Math. Centrum Amsterdam. Rapport **ZW 1949-004**, 9 pp. (1949).

The following is the main result of the paper. Let $f(x)$ be real-valued, of period 1, of mean value zero and of the class L^2. Suppose that the Fourier coefficients c_n of f satisfy the condition $\sum_n^\infty |c_k|^2=O(\log n)^{-\alpha}$, where $\alpha>1$. Let $u_1, u_2, \cdots$ be a sequence uniformly distributed mod 1 and satisfying for k, M, N integral and positive the condition $|\sum_{n=M+1}^{M+N} \exp(2\pi i k u_n)|\leqq A k^\rho N^\sigma (M+N)^\tau$, where A, ρ, σ, τ are constants such that $\sigma+\tau<1$, $\tau<\frac{1}{2}$. Then

$$\lim N^{-1}\{f(x+u_1)+f(x+u_2)+\cdots+f(x+u_N)\}=0$$

for almost every x. An example, due to Erdös and reproduced in the paper, shows that, from a certain point of view, some conditions upon the order of the coefficients c_n seem unavoidable. *A. Zygmund* (Chicago, Ill.).

K45-2 **(13, 495f)**

Lehmer, D. H. Mathematical methods in large-scale computing units. Proceedings of a Second Symposium on Large-Scale Digital Calculating Machinery, 1949, pp. 141–146. Harvard University Press, Cambridge, Mass., 1951. $8.00.

The aim of this paper is to discuss in a general way certain features of the mathematics that are characteristic of the large scale digital computing machines. Special attention is given to the problem of generating random numbers. The author's proposal for generating a sequence of "random" 8 decimal numbers u_n is to start with some number $u_0\neq 0$, multiply it by 23, and subtract the two digit overflow on the left from the two right hand digits to obtain a new number u_1. By repeating this process he generates a sequence u_n which he shows has a repetition period of 5,882,352. An IBM 602A was used to generate 5000 numbers of such a sequence. Four standard tests (unspecified) were applied to check the "randomness" and the sequence passed all four. The author observes, however, that all of the members of the sequence he chose were divisible by 17.

R. Hamming (Murray Hill, N. J.).

K45-3 **(20# 25)**

Franklin, J. N. On the equidistribution of pseudo-random numbers. Quart. Appl. Math. **16** (1958), 183–188.

For real x let $\{x\}$ denote the fractional part $x-[x]$. Many computer programs that use pseudo-random sequences generate them by means of the recurrence $x_n=\{Nx_{n-1}\}$ ($x_0\neq 0$), where N is an integer so chosen, in terms of the word size and number system of the computer, that the period of the sequence should be maximum. The author considers the slightly more general recurrence $x_n=\{Nx_{n-1}+\theta\}$, $0\leqq\theta<1$, in the somewhat theoretical case in which x_0 is irrational, a term not understood by computers. He applies the ergodic theorem to show that for almost all choices of x_0 the values of the x's are equidistributed in the unit interval. The case of $N=1$ is that of the classical paper of H. Weyl [Math. Ann. **77** (1916), 313–352]. If $\nu(n)$ denotes the number of x's among the first n that lie in a given interval, then for almost all x_0 the ratio $\nu(n)/n$ tends to the length of that interval as n increases. The argument uses a very general theorem of F. Riesz [Comment. Math. Helv. **17** (1945), 221–239; MR **7**, 255].

D. H. Lehmer (Cambridge, England)

K45-4 (20# 5169)

Korobov, N. M. Approximate calculation of repeated integrals by number-theoretical methods. Dokl. Akad. Nauk SSSR (N.S.) **115** (1957), 1062–1065. (Russian)

Let $f(x_1, x_2, \cdots, x_s)$ be a function of period 1 in each variable which is expressible in an absolutely convergent s-dimensional Fourier series and let σ be the sum of the moduli of its Fourier coefficients. Let p be a prime greater than s and put $\xi_\nu(k)=\{k^\nu/p^2\}$ (fractional part) for $1\leqq\nu\leqq s$. If the partial derivative

$$\frac{\partial^{2s}f(x_1, x_2, \cdots, x_s)}{\partial x_1^2\partial x_2^2\cdots\partial x_s^2}$$

is continuous and if, for any integers $j_1, j_2, \cdots, j_r$, $1\leqq j_i\leqq s$, $1\leqq r\leqq s$, the partial derivatives $\partial^{2r}f/\partial x_{j_1}^2\partial x_{j_2}^2\cdots\partial x_{j_r}^2$ are bounded in modulus by a constant C, then, for $N=p^2$, the author shows that

$$\left|\int_0^1\cdots\int_0^1 f(x_1, \cdots, x_s)dx_1\cdots dx_s - \frac{1}{N}\sum_{k=1}^N f(\xi_1(k), \cdots, \xi_s(k))\right| \leqq \frac{(s-1)\sigma}{\sqrt{N}}+\frac{sC}{10N}.$$

Here the points $(\xi_1(k), \cdots, \xi_s(k))$ depend upon N. The author indicates how this result can be modified so that the dependence of these points on N is lessened.

R. A. Rankin (Glasgow)

Referred to in K45-7, K45-10, K45-18.

K45-5 (21# 2848)

Korobov, N. M. Approximate evaluation of repeated integrals. Dokl. Akad. Nauk SSSR **124** (1959), 1207–1210. (Russian)

Let $p>3$ be prime. For integral $n\geqq 1$ define $H(z_1, \cdots, z_n)$ as follows.

$$H(z_1, \cdots, z_n) = \sum_{k=1}^{p-1}[1-2\ln(2\sin\pi\{kz_1/p\})]\cdots\times[1-2\ln(2\sin\pi\{kz_n/p\})],$$

where $z_1, \cdots, z_n$ are arbitrary integers of the interval $1\leqq z\leqq p-1$ and $\{kz_\nu/p\}$ is the fractional part of the number kz_ν/p.

The integers $a_1, a_2, \cdots$ will be called optimal coefficients if $1\leqq a_1\leqq p-1$ and for given $a_1, \cdots, a_\nu$ $(\nu\geqq 1)$ the magnitude of $a_{\nu+1}$ equals one of the values of z $(1\leqq z\leqq p-1)$ for which $H(a_1, \cdots, a_\nu, z)$ achieves its minimum.

We consider the function $f(x_1, \cdots, x_s)$ defined on the unit s-dimensional cube $(s\geqq 1)$ which has absolutely convergent Fourier series:

$$f(x_1, \cdots, x_s) = \sum_{m_1,\cdots,m_s=-\infty}^{\infty} C(m_1, \cdots, m_s)e^{2\pi i(m_1x_1+\cdots+m_sx_s)}.$$

Theorem 1: If there exist constants $\alpha>1$ and $C=C(s)$ such that the Fourier coefficients of the function $f(x_1, \cdots, x_s)$ fulfill the condition

$$|C(m_1, \cdots, m_s)| \leqq \frac{C}{[(|m_1|+1)\cdots(|m_s|+1)]^\alpha},$$

then, for each collection of optimal coefficients $a_1, a_2, \cdots$, for $\xi_\nu(k)=\{ka_\nu/p\}$ $(\nu=1, 2, \cdots, s)$, and for any $\varepsilon>0$, the following estimate is valid:

$$\int_0^1\cdots\int_0^1 f(x_1, \cdots, x_s)\,dx_1\cdots dx_s - \frac{1}{p}\sum_{k=1}^p f[\xi_1(k), \cdots, \xi_s(k)] = O\left(\frac{1}{p^{\alpha-\varepsilon}}\right).$$

Theorem 2: For each $\alpha>1$ there exists a function $f(x_1, \cdots, x_s)$ the Fourier coefficients of which satisfy the conditions

$$|C(m_1, \cdots, m_s)| \leqq \frac{C}{[(|m_1|+1)\cdots(|m_s|+1)]^\alpha}$$

such that for any numbers $a_1, a_2, \cdots$ with $\xi_\nu(k)=\{ka_\nu/p\}$ $(\nu=1, 2, \cdots, s)$ we have

$$\left|\int_0^1\cdots\int_0^1 f(x_1, \cdots, x_s)dx_1\cdots dx_s - p^{-1}\sum f[\xi_1(k), \cdots, \xi_s(k)]\right| > p^{-\alpha}.$$

A. H. Stroud (Madison, Wis.)

Referred to in K45-7.

K45-6 (22# 3114)

Korobov, N. M. Approximate solution of integral equations. Dokl. Akad. Nauk SSSR **128** (1959), 235–238. (Russian)

The author considers the multiple integral equation

$$\varphi(P) = \lambda\int_{G_s} K(P, Q)\varphi(Q)dQ+f(P)$$

where P and Q are points in the unit cube G_s in s-dimensional space; the Fourier coefficients of the function f and kernel K are assumed to satisfy certain order conditions, and it is supposed that the Fredholm denominator $D(\lambda)$ is not zero. Let M_i denote the point with coordinates $\{i^r/p\}$ (fractional part), where $1\leqq r\leqq s$ and $1\leqq i\leqq p$, p being a prime greater than s. Then it is shown that a solution of the integral equation can be expressed as

$$\varphi(M_i) = \tilde\varphi(M_i)+O(1/\sqrt{p}) \quad (1\leqq i\leqq p),$$

where $\tilde\varphi(M_i)$ satisfies the system of equations

$$\tilde\varphi(M_j) = \frac{\lambda}{p}\sum_{i=1}^p K(M_j, M_i)\tilde\varphi(M_i)+f(M_j) \quad (1\leqq j\leqq p).$$

Two further approximate results of a more complicated type are proved.

R. A. Rankin (Glasgow)

K45-7 (22# 3115)

Šahov, Yu. N. Approximate solution of second kind Volterra equation by means of iterations. Dokl. Akad. Nauk SSSR **128** (1959), 1136–1139. (Russian)

Number-theoretic methods due to Korobov for the approximate evaluation of repeated integrals [N. M. Korobov, same Dokl. **115** (1957), 1062–1065; **124** (1959), 1207–1210; MR **20** #5169; **21** #2848] are applied to obtain an approximate solution of the Volterra integral equation of the second kind. These methods were previously applied by Korobov [see preceding review] in obtaining an approximate solution of Fredholm's equation of the second kind.

J. F. Heyda (Cincinnati, Ohio)

Citations: MR 20# 5169 = K45-4; MR 21# 2848 = K45-5.

K45-8 **(23# B557)**

Certaine, J.

On sequences of pseudo-random numbers of maximal length.

J. Assoc. Comput. Mach. **5** (1958), 353–356.

The method of obtaining the pseudo-random numbers considered here is that expressed by the recurrence relation $x_{r+1} \equiv ax_r \pmod{m}$. Such a sequence will be cyclic, with a period λ. Classical theorems which determine the maximum attainable value of λ are quoted. A theorem which characterises, and can be used to calculate, one value of a which achieves this maximum period, is proved. Values of a for $m = 10^8$, 10^{10}, 10^{11}, and any power of 2, are given. The aim in this paper is only to maximise the period of the sequence; no consideration is given to the effect that the choice of a has upon the random nature of the sequence.

B. A. Chartres (Providence, R.I.)

K45-9 **(24# B504)**

Bass, J.

Nombres aléatoires, suites arithmétiques, méthode de Monte-Carlo.

Publ. Inst. Statist. Univ. Paris **9** (1960), 289–325.

A sequence of numbers $x_1, x_2, \cdots$ in $[0, 1]$ is said to be uniformly distributed if the number of elements from among $x_1, x_2, \cdots, x_n$ which are less than α is asymptotically equal to $n\alpha$ as $n \to \infty$, for each $\alpha \in [0, 1]$. By a classical theorem of Weyl [Math. Ann. **77** (1916), 313–352], this is the case if and only if (a) for every function f Riemann integrable on $[0, 1]$,

$$(*) \qquad \int_0^1 f(x)\,dx = \lim_{n\to\infty} n^{-1} \sum_{k=1}^{n} f(x_k),$$

or (b) for every integer $h \neq 0$,

$$\lim_{n\to\infty} n^{-1} \sum_{k=1}^{n} \exp(2\pi i h x_k) = 0.$$

The first of these equations suggests a method, analogous to the Monte-Carlo method, for approximate evaluation of integrals; the second provides a powerful tool for proving that specific sequences are uniformly distributed. The present paper, which is mainly expository, is concerned with some of the salient theorems on uniform distribution and their applications to numerical integration.

Let $\{x\} = x - [x]$, where $[x]$ is the largest integer not exceeding x. Weyl showed that the numbers $\{A\}$, $\{2A\}$, $\{3A\}, \cdots$ are uniformly distributed if A is irrational, and more generally, that if $P(x)$ is any polynomial with at least one irrational coefficient which is not the constant term, then $\{P(1)\}, \{P(2)\}, \cdots$ is uniformly distributed. It is shown that if $x_k = \{kA\}$, the difference between the left- and right-hand sides of (*) is $O(n^{-1})$ if $f(x)$ has a Fourier series on $[0, 1]$ which is majorized by the series $\sum k^{-(2+\varepsilon)}$, $\varepsilon > 0$. The case in which A is an algebraic irrationality is given special attention, the investigation being based on theorems of Liouville and Roth concerning the approximability of an algebraic number by rational numbers.

{It seems doubtful to the reviewer that Roth's theorem [Mathematika **2** (1955), 1–20; corrigendum, 168; MR **17**, 242] can have any genuine significance in numerical analysis, since it involves an unspecified constant which is as elusive as any in mathematics. Similarly, most error estimates in the present paper involve unspecified constants, and so are of questionable value in applications.}

W. J. LeVeque (Ann Arbor, Mich.)

Citations: MR 17, 242d = J68-14.

K45-10 **(26# 888)**

Hlawka, Edmund

Zur angenäherten Berechnung mehrfacher Integrale.

Monatsh. Math. **66** (1962), 140–151.

The problem of approximating the integral $\mu(f) = \int_Q f d\bar{x}$ of a (complex-valued) R-integrable function $f(\bar{x})$ of $\bar{x} = (x_1, \cdots, x_s)$ over a parallelepiped Q in R_s is reduced to the case that Q is the unit cube and f is periodic in all x_i $(i = 1, \cdots, s)$ with period 1. Now $\mu(f)$ is replaced by $\Sigma_N = N^{-1} \sum_1^N f(\bar{x}_k)$, where $\bar{x}_1, \cdots, \bar{x}_N$ is a set of N points of Q; this set may be denoted by ω. The error $\Delta_N = |\mu(f) - \Sigma_N|$ depends on (a) the properties of f (bounded variation, continuity, derivability, etc.) and (b) the distribution of ω in Q. If, e.g., the function f is of bounded variation with total variation $V(f)$, one always has the estimate $\Delta_n \leqq V(f) D_N$, where D_N denotes the discrepancy (mod 1) of ω [cf. the author, Ann. Mat. Pura Appl. (4) **54** (1961), 325–333; MR **25** #3029]. Thus the problem, independently from the considered functions f, can be reduced to the selection of convenient sets ω with small D_N $(N \geqq 1)$. Here the author uses a theorem of Erdős, Turán, and the reviewer [see Koksma, Math. Centrum Amsterdam Scriptum, No. 5 (1950); MR **12**, 394] (which is an improvement of an older theorem of van der Corput and Koksma). The merits of this method are discussed and compared with the usual methods: Riemann sums, Simpson's rule, Monte Carlo method and with recent theorems of Korobov [Dokl. Akad. Nauk SSSR **115** (1957), 1062–1065; MR **20** #5169] and Hsu and Lin [Acta Math. Acad. Sci. Hungar. **9** (1958), 279–290; MR **20** #7175]. Finally, a sequence ω is constructed which for $s > 1$ gives a sharper estimate than the known methods: for special values of $N > 1$, viz., for each N which is prime, lattice points $\bar{g} = (g_1, \cdots, g_s)$ ('good' lattice points) exist such that if one puts $\bar{x}_k = (k-1)\bar{g}$ $(k = 1, 2, \cdots, N)$, one will have $\Delta_N \leqq V(f)(80 \log N)^s/N$.

J. F. Koksma (Amsterdam)

Citations: MR 12, 394c = K05-16; MR 20# 5169 = K45-4; MR 25# 3029 = K05-35.

Referred to in K05-39.

K45-11 **(26# 5710)**

Hull, T. E.; Dobell, A. R.

Random number generators.

SIAM Rev. **4** (1962), 230–254.

This is a quite complete and authoritative survey of random number generation. The paper begins with indications of types of problems requiring a large supply of random numbers. This is followed by a careful treatment of the method, called "mixed congruential", in which the nth random number is x_n/m where $x_n \equiv ax_{n-1} + c \pmod{M}$, $0 \leqq x_n < M$.

The case $c = 0$ the authors call the "multiplicative congruential method", introduced as long ago as 1949. Conditions for obtaining maximum periods are discussed in both cases. The general mixed congruential has the advantage that no primitive root of M is required to obtain maximum periods. However, the authors cite cases where the mixed method fails to pass certain statistical tests. The production of non-uniformly distributed random numbers is treated in some detail. Several questions are raised concerning bad local behavior or other statistical blemishes and their dependence on the parameters of the generators. The possible usefulness of a small program to monitor the pertinent characteristics of the produced sequence and to report at the end of the calculation is suggested. The survey concludes with an extensive history of the subject including accounts of all published methods and a masterly bibliography of 148 titles.

D. H. Lehmer (Berkeley, Calif.)

Referred to in K15-80.

K45-12 **(26# 7125)**

Franklin, Joel N.

Deterministic simulation of random processes.

Math. Comp. **17** (1963), 28–59.

If $\{x\}$ denotes the non-negative fractional part of x, the deterministic processes $x_n=\{n\alpha\}$ for integral n and irrational α (Weyl sequences), $x_{n+1}=\{Nx_n+\theta\}$ for integral $N>1$ (multiply sequences), and "polynomial" sequences $x_n=\{n^p\alpha+c_1n^{p-1}+c_2n^{p-2}+\cdots+c_p\}$ for irrational α could (in an idealized non-finite word length digital computer) be used to produce sequences of numbers possessing some properties of sequences of truly random uniformly distributed numbers. The main properties investigated, equipartition by k's and equidistribution by k's, are related to the use of blocks of k consecutive numbers in these sequences to produce k-dimensional vectors. Furthermore, the autocorrelation function's vanishing, $R(\tau)=N^{-1}\sum_1^N(x_n-\frac{1}{2})(x_{n+\tau}-\frac{1}{2})=0$ for every integral $\tau>0$, is taken as the definition of whiteness. The author presents 22 theorems, many, but not all, of which are tantamount to cautions to the unwary who may wish to build uniformily distributed random vectors in a cube or white noise from some of these processes. For example, for every multiply sequence with $\theta=0$, $\text{Prob}\{x_n>x_{n+1}>x_{n+2}\}=\frac{1}{6}(1+N^{-1})$, i.e., the sequence is not equipartitioned by k's for any $k\geqq 3$. Another, there exist multiply sequences which may fail to be equidistributed even if x_0 is irrational.

M. L. Juncosa (Santa Monica, Calif.)

Referred to in K45-19.

K45-13 **(27# 548)**

Hlawka, Edmund

Lösung von Integralgleichungen mittels zahlentheoretischer Methoden. I.

Österreich. Akad. Wiss. Math.-Nat. Kl. S.-B. II **171** (1962), 103–123.

The integral equation

$$(1)\qquad \phi(x)-\lambda\int_0^1 K(x,y)\phi(y)\,dy = f(x)$$

has, for λ sufficiently small, as is well known, a solution of the form

$$\phi(x) = f(x)+\sum_{n=1}^{\infty}\lambda^n\int_0^1 K^{(n)}(x,y)f(y)\,dy,$$

where

$$K^{(n)}(x,y) = \int_0^1\cdots\int_0^1 K(x,x_1)K(x_1,x_2)\cdots K(x_{n-1},y)\,dx_1\cdots dx_{n-1}.$$

The author shows how, by picking a suitable sequence of points,

$$\omega(s(T),T)=\omega(s,T)=\{(y_{11},\cdots,y_{1s}),\cdots,(y_{T1},\cdots,y_{Ts})\},$$

in the s-dimensional cube E_s, (1) has a solution for λ sufficiently small of the form

$$\phi(x) = f(x)+\lim_{T\to\infty}\frac{1}{T}\sum_{k=1}^{T}\sum_{n=1}^{s(T)}\lambda^nK(x,y_{k1})\cdots K(y_{k,n-1},y_{kn})f(y_{kn}),$$

assuming f and $K(x,y)$ are of bounded variation on E_1 and E_2, respectively. The construction of such sequences is effected by number-theoretic methods and depends on the author's previous work on functions of bounded variation and the theory of uniform distribution [Ann. Mat. Pura Appl. (4) **54** (1961), 325–333; MR **25** #3029]. Several specific choices for the sequence $\omega(s(T),T)$ are discussed. The author also applies his method to determining the least positive eigenvalue of the homogeneous integral equation $\phi(x)=\lambda\int_0^1K(x,y)\phi(y)dy$, where the kernel is assumed to be symmetric. The paper closes with a discussion of the reduction of integral equations in several variables to a system of linear equations.

S. L. Segal (Boulder, Colo.)

Citations: MR 25# 3029 = K05-35.

K45-14 **(28# 55)**

De Matteis, A.; Faleschini, B.

Some arithmetical properties in connection with pseudorandom numbers. (Italian summary)

Boll. Un. Mat. Ital. (3) **18** (1963), 171–184.

The authors are concerned with the random number generator in which $\chi_{i+1}=a\chi_i+k \pmod m$ introduced for $k=0$ by the reviewer in 1949. If $k\not\equiv 0 \pmod m$, it is possible with proper choice of a, χ_0 and m to obtain a sequence of χ's of proper period m, while for $k=0$ the period is precisely the exponent of $a \bmod m$ and thus $\leqq\phi(m)$, provided $a\chi_0$ is prime to m.

The authors study those numbers a whose exponents modulo m are the same, especially in the case $m=2^\alpha p^\beta$, p a prime and especially $p=5$. Numerous examples are given, including a table of all numbers a whose exponent is $g \pmod{10^{10}}$ for every divisor g of $\phi(10^{10})/8$. {In the reviewer's opinion it is better not to use a modulus like 2^n or 10^n, but rather a prime modulus quite near by, so that division by m is eliminated.}

D. H. Lehmer (Berkeley, Calif.)

K45-15 **(28# 716)**

Korobov, N. M. [Коробов, Н. М.]

Number-theoretic methods in approximate analysis [Теоретико-числовые методы в приближенном анализе].

Gosudarstv. Izdat. Fiz.-Mat. Lit., Moscow, 1963. 224 pp. 0.59 *r*.

The subject of this book is mostly approximate calculation of integrals over an s-dimensional cube by methods of the form

$$\int_0^1\cdots\int_0^1 f(x_1,\cdots,x_s)\,dx_1\cdots dx_s = \frac{1}{N}\sum_{k=1}^{N} f\left(\left\{\frac{a_1k}{p}\right\},\cdots,\left\{\frac{a_sk}{p}\right\}\right)+R,$$

where estimates are obtained for R, for example in the form $R=O(\ln^\beta N/N^\alpha)$. Here p is a prime, the a_i, $i=1,\cdots,s$, are constants and $\{x\}$ means the fractional part of x. The problem is chiefly to investigate properties of best possible sets of the a_i, and some tables of these are given. Also discussed are applications of this method to interpolation of functions of many variables and to the solution of integral equations.

Most of the results of this book have been obtained in the last few years by Russian authors. Of the 39 references in the bibliography, 10 are papers by Korobov, and among 15 others are papers by N. S. Bahvalov, V. S. Rjaben'kiĭ, S. A. Smoljak and Ju. N. Šakov.

This is an important book since it provides a systematic discussion of significant results not widely known.

A. H. Stroud (Lawrence, Kans.)

Referred to in K45-17, K45-21, K45-28.

K45-16 **(28# 3523)**

Jagerman, David L.

The autocorrelation function of a sequence uniformly distributed modulo 1.

Ann. Math. Statist. **34** (1963), 1243–1252.

Let $\rho(x)=(1/2)-\{x\}$ and let x_j $(j=1(1)\infty)$ be a sequence of real numbers. For the integer τ the autocorrelation function of the sequence x_j is taken to be

$$\psi(\tau) = \lim_{N\to\infty} \frac{1}{N}\sum_{j=1}^{N} \rho(x_j)\rho(x_{j+\tau}).$$

The author proves that for the sequence $x_j=\alpha j^2$, where α is irrational, $\psi(\tau)$ vanishes for all positive integers τ and suggests that $\{\alpha j^2\}$ would make a suitable random number generator. The simpler sequence $x_j=\alpha j$, on the other hand, is less suitable since its autocorrelation function does not vanish identically. In fact, $\psi(\tau)=(1/12)-\int_0^{\alpha\tau}\rho(u)\,du$ so that $-1/24<\psi(\tau)<1/12$. The sequence $x_j=j^\sigma$ $(0<\sigma<1)$, like the two previous examples, is uniformly distributed mod 1. Nevertheless, for this sequence $\psi(\tau)=1/12$ for $\tau\geqq 0$.

{For practical applications to digital computers, the reviewer wonders whether it might be well to point out that irrational numbers do not exist. For a machine with 35 binary digits in its word the irrational $\alpha=(1/2)+2^{-50}\sqrt{2}$ would be taken as $\alpha=1/2$ and then would give highly correlated results for $x_j=\alpha j$ or even αj^2. In practice α would be of the form $N\cdot 2^{-35}$, where N is an integer. The sequence $\{\alpha j\}$ would then be formed by repeated additions of N without attention to overflow. The sequence $\{\alpha j^2\}$ would be given by $x_j\equiv 2x_{j-1}-x_{j-2}+2N \pmod{2^{35}}$. Similar schemes of Fibonacci type, though inexpensive, have not proved satisfactory.}

D. H. Lehmer (Berkeley, Calif.)

K45-17 **(30# 3062)**

Hua, Loo Keng; Wang, Yuan

On Diophantine approximations and numerical integrations. I, II.

Sci. Sinica **13** (1964), 1007–1009; *ibid.* **13** (1964), 1009–1010.

Korobov [*Number-theoretic methods in approximate analysis* (Russian), Fizmatgiz, Moscow, 1963; MR **28** #716] considered quadrature formulas over the s-dimensional unit cube of the form

$$\int_0^1\cdots\int_0^1 f(x_1,\cdots,x_s)\,dx\cdots dx_s = q^{-1}\sum_{t=1}^{q} f\left(\frac{a_1t}{q},\cdots,\frac{a_st}{q}\right)+R.$$

where q and the a's are positive integers and f is supposed periodic of period 1 in each variable. He chose q to be a prime and sought to determine the a's so as to minimize R for a class of functions f whose multiple Fourier coefficients are small. Extending the results of an earlier paper [Sci. Record (N.S.) **4** (1960), 8–11; MR **23** #B1118], the authors indicate two direct methods of determining such formulas, based on units of algebraic fields. In the first method the field is $R(\sqrt{p_1},\cdots,\sqrt{p_t})$, where the p's are distinct primes and the units are the fundamental solutions of the Pell equations $x^2-Dy^2=\pm 4$, where D runs over the 2^t-1 divisors >1 of $p_1p_2\cdots p_t$. Thus for $t=2$, $p_1=2$, $p_2=5$, the authors propose $q=5787$, $a_1=1$, $a_2=2397$, $a_3=1366$, and $a_4=939$. The second method is based on the algebraic field $R(\cos(2\pi/p))$, where p is a prime. Examples are given for $3<p\leqq 13$.

D. H. Lehmer (Berkeley, Calif.)

Citations: MR 28# 716 = K45-15.

Referred to in K45-21.

K45-18 **(30# 5463)**

Hlawka, Edmund

Uniform distribution modulo 1 and numerical analysis.

Compositio Math. **16**, 92–105 (1964).

The author develops a method for numerical calculation of multiple integrals based on uniformly distributed sequences. Whereas the usual Monte Carlo method may fail in certain cases, due to its probabilistic character, this method is always applicable and gives smaller errors.

The main idea is, as in the Monte Carlo method, to approximate integrals by sums of the form $N^{-1}\sum_{n\leqq N} f(x_n)$. The author constructs special sequences (x_n), which are easy to calculate and which show a much better asymptotic behaviour than "almost all" other sequences, on which the Monte Carlo method depends. The author gives explicit error-estimates for a general class of functions and applies his method to the numerical solution of integral equations and the inversion of matrices. The methods developed can also be applied for the calculation of eigenvalues of matrices and differential equations.

A similar method has been constructed by N. M. Korobov [Dokl. Akad. Nauk **115** (1957), 1062–1065; MR **20** #5169]. *J. E. Cigler* (Groningen)

Citations: MR 20# 5169 = K45-4.

K45-19 **(31# 1762)**

Jagerman, David L.

The autocorrelation and joint distribution functions of the sequences

$$\left\{\frac{a}{m}j^2\right\},\ \left\{\frac{a}{m}(j+\tau)^2\right\}.$$

Math. Comp. **18** (1964), 211–232.

The author begins with J. N. Franklin's result [Math. Comp. **17** (1963), 28–59; MR **26** #7125] that the sequence $[\{\alpha j^2\}]_{j=0}^{\infty}$, where $\{\cdots\}$ denotes the fractional part, is equidistributed in $(0, 1)$, and that $[\{\alpha j^2\}]_{j=0}^{\infty}$ and $[\{\alpha(j+\tau)^2\}]_{j=0}^{\infty}$ are uncorrelated when α is irrational and ν is a nonzero integer. He establishes the important result that if α is approximated by a rational number, as it usually is in practical computations, taking the form a/m in lowest terms, then the two sequences above are approximately independently equidistributed in $(0, 1)$. This requires a lengthy number-theoretic argument of some delicacy.

If $(a, m)=1$, $m\geqq 36$, $1\leqq\tau<\sqrt{m}$, and $\lambda=2+\sqrt{2}$, then the following are the main theorems: (I) If $0\leqq\alpha<1$, $0\leqq\beta<1$, and $\nu(m)$ is the number of distinct prime divisors of m, then

$$(1)\quad \left|m^{-1}\sum_{j=0}^{m-1}\left(\frac{1}{2}-\left\{\frac{a}{m}j^2-\alpha\right\}\right)\left(\frac{1}{2}-\left\{\frac{a}{m}(j+\tau)^2-\beta\right\}\right)\right| < m^{-1/2}[(.81)\lambda^{\nu(m)}\ln^2 m+33(2\lambda)^{\nu(m)}\ln m];$$

whence (II) setting $\alpha=\beta=0$ in (1), the autocorrelation function $\psi(\tau)$ of $[\{(a/m)j^2\}]_{j=1}^{\infty}$ satisfies

$$(2)\quad |\psi(\tau)| < m^{-1/2}[(.81)\lambda^{\nu(m)}\ln^2 m+33(2\lambda)^{\nu(m)}\ln m];$$

and (III) if $mG(\alpha,\beta)$ is the number of solutions of

$$(3)\quad 0\leqq\{(a/m)j^2\}<\alpha,\ 0\leqq\{(a/m)(j+\tau)^2\}<\beta,$$

for $0\leqq j<m$, then

$$(4)\quad |G(\alpha,\beta)-\alpha\beta| < m^{-1/2}[(3.24)\lambda^{\nu(m)}\ln^2 m+392(2\lambda)^{\nu(m)}\ln m].$$

The author writes: "The above-enumerated properties show the possible applicability of the sequences as random numbers in Monte Carlo procedures. However, an important question is the behavior, from the viewpoint of random number characteristics, of consecutive portions of the complete sequences." His study of this matter is to be published elsewhere. *J. H. Halton* (Upton, N.Y.)

Citations: MR 26# 7125 = K45-12.

K45-20 **(32# 1878)**

Tausworthe, Robert C.

Random numbers generated by linear recurrence modulo two.

Math. Comp. **19** (1965), 201–209.

The author considers the linear recurrence relation $a_k = c_1a_{k-1} + c_2a_{k-2} + \cdots + c_na_{k-n}$ (mod 2), with $c_n = 1$, where each a_i and c_i is 0 or 1, in the case when the sequence $\{a_k\}$ is of maximum period $p = 2^n - 1$. The recurrence relation generates a sequence of L-bit binary numbers $y_k = \sum_{t=1}^{L} 2^{-t}a_{qk+r-t}$ in (0, 1), with q, r, and L arbitrary, so long as $(p, q) = 1$, $q \geqq L$, and $L \leqq n$. By direct analysis, he shows that, to within errors of the order of 2^{-n}, the mean value, variance, and autocorrelation of the sequence $\{y_k\}$ equal those of a sequence of independent random variables, uniform in (0, 1); and the proportion $\hat{T}$ of N consecutive members of the sequence falling in an interval of length 2^{-d} has the expected value 2^{-d}, to within the same error, indicating uniformity of distribution, while the variance of $\hat{T}$ is given an upper bound close to $1/4N$ (the corresponding variance for a random sequence is $(2^{-d} - 2^{-2d})/N$). Finally, the author shows that the M-dimensional sequence of points $(y_k, y_{k-l_2}, \cdots, y_{k-l_M})$, with $0 < l_2 < \cdots < l_M < n/q$ (mistakenly written $n/q - 1$ in the paper), is also uniform in the unit M-dimensional cube, in the same sense. This would appear to be a well-behaved pseudo-random sequence generator for most purposes, though more searching and extensive testing is indicated. *J. H. Halton* (Upton, N.Y.)

Referred to in A12-60.

K45-21 **(32# 4823)**

Hua Loo-Keng [Hua Lo-keng]; Wang Yuan

On numerical integration of periodic functions of several variables.

Sci. Sinica **14** (1965), 964–978.

The method of approximating an integral

$$(1)\qquad \int_0^1 dx_1 \cdots \int_0^1 dx_s f(x_1, \cdots, x_s),$$

of a function with period 1 in each coordinate, by an average (2) $q^{-1}\sum_{t=1}^{q} f(t\xi_1, \cdots, t\xi_s)$, arose out of the Kronecker-Weyl theory of Diophantine approximation. Korobov and the present authors [e.g., the authors, Sci. Sinica **13** (1964), 1007–1009; ibid. **13** (1964), 1009–1010; MR **30** #3062; N. M. Korobov, *Number-theoretic methods in approximate analysis* (Russian), Fizmatgiz, Moscow, 1963; MR **28** #716], among others, have investigated ways of generating suitable rational $\xi_i = a_i/q$ for this method. The authors claim to have found a class of such numbers which is as good as Korobov's, but requires only $O(q)$ elementary operations for the calculation of the average (2), instead of $O(q^2)$, or at best $O(q^{4/3})$, as in Korobov's method. The difference Δ_f between (1) and (2) is shown to be dominated by $c(\pi^2/6)^s D(q, a_1, \cdots, a_s)$, where $(D+1)$ is the average (2), with $\xi_i = a_i/q$, of the function $3^s \prod_{i=1}^{s} (1 - 2\{x_i\})^2$, $\{x\}$ is the distance from x to the nearest integer, c is an upper bound for $(\overline{m}_1 \cdots \overline{m}_s)^2 c(m_1, \cdots, m_s)$, $\overline{n}$ denotes $\max(|n|, 1)$, and $c(m_1, \cdots, m_s)$ is the coefficient of $\exp\{2\pi i(m_1x_1 + \cdots + m_sx_s)\}$ in the s-dimensional Fourier series for $f(x_1, \cdots, x_s)$.

The authors describe a method for obtaining the integers $q, a_1, \cdots, a_s$ as functions of an independent set of $(s-1)$ units of a suitable real cyclotomic field (other real algebraic number fields are also considered). Since $R(2\cos 2\pi/p)$ is of degree $(p-1)/2$, one must take p greater than $2s$.

The authors claim that the error Δ_f decreases as $q^{-1/2}$; but it must be pointed out that in their principal example, with $s = 11$, the upper bound given is about $55c$, where the integral is less than c (not too impressive an accuracy, especially since $q = 698{,}047$). *J. H. Halton* (Upton, N.Y.)

Citations: MR 28# 716 = K45-15; MR 30# 3062 = K45-17.

K45-22 **(35# 3339)**

Sprindžuk, V. G.

Asymptotic behavior of integrals of quasiperiodic functions. (Russian)

Differencial'nye Uravnenija **3** (1967), 862–868.

Consider the integral $J(t) = \int_0^t F(x, \omega_1 x, \cdots, \omega_n x)\, dx$, where $F(x_0, x_1, \cdots, x_n)$ is a real function 1-periodic in x_j. Let the set of numbers $1, \omega_1, \cdots, \omega_n$ be linearly independent over the field of rational numbers and let $w(\omega)$ be the least upper bound of those $w > 0$ for which there exist infinite number of solutions for the inequality $|a_0 + a_1\omega_1 + \cdots + a_n\omega_n| < h_a^{-w}$, where $h_a = \max_{0 \leqq i \leqq n} |a_i|$, a_i being integers. Suppose that F is continuous together with the partial derivatives $\partial^s F/\partial x_j^s$ $(j = 0, 1, \cdots, n)$ and the integrals

$I_j(F) =$

$$\int_0^1 \cdots \int_0^1 |(\partial^{s+1}/\partial x_j^{s+1}) F(x_0, \cdots, x_n)|\, dx_0 \cdots dx_n < \infty$$

exist, where $s \geqq n+1$ is a fixed number. If $w(\omega_1, \cdots, \omega_n) \leqq s$, then it is shown that $J(t) = C_F t + O(1)$, where $C_F = \int_0^1 \cdots \int_0^1 F(x_0, \cdots, x_n)\, dx_0 \cdots dx_n$. Another set of conditions which implies the same conclusion is also given.

V. Lakshmikantham (Kingston, R.I.)

K45-23 **(36# 4779)**

Coveyou, R. R.; Macpherson, R. D.

Fourier analysis of uniform random number generators.

J. Assoc. Comput. Mach. **14** (1967), 100–119.

The authors' introduction should be quoted, both as describing their intention and as a very pertinent comment on the state of the art of constructing pseudo-random generators: "An indispensable requirement of the Monte Carlo method is a copious and reliable source of 'uniformly distributed random numbers'. More precisely, a facile method is needed for the generation of a sequence which strongly resembles a sample sequence drawn by repeated independent trials from a probability distribution uniform on the unit interval. It is not surprising that methods for the generation of such pseudo-uniform sequences have received much attention. This work has had three major themes: the determination of the periods of the iterative processes used and the selection of those with periods of adequate length; the application of statistical tests to the output sequences of such generators; and the mathematical derivation of the statistical properties of such output sequences.

"Work of the first kind has been well done; work of the second kind has been over-done. The third is rare: known a priori tests so far have been for such weak statistical properties that generators designed to pass such tests have, more often than not, failed for other reasons. This state of affairs is the result of the lack of a unified theory of the statistical behavior of the output sequences of such generators. Such a theory is proposed here."

The authors' novel line of attack considers a sequence $\Xi = [\boldsymbol{\xi}_n]_{n=1}^{\infty}$ of points $\boldsymbol{\xi}_n = [\xi_{ni}]_{i=1}^{k}$, with $\xi_{ni} = P^{-1}q_{ni}$, where

P and q_{ni} are integers and $0 \leq q_{ni} \leq P-1$; yielding a Monte Carlo sum $U = N^{-1} \sum_{n=1}^{N} f(\xi_n)$ as an estimate of the integral $V = \int_0^1 dz_1 \cdots \int_0^1 dz_k f(\mathbf{z})$, or rather, of its Riemann-approximation (with P large) $W = P^{-k} \sum_{\mathbf{q} \in J^k} f(P^{-1}\mathbf{q})$, where $\mathbf{q} = [q_i]_{i=1}^{k}$ is a lattice-point of J^k, with $J = \{0, 1, 2, \cdots, P-1\}$. Working in terms of finite Fourier sums, they write $c(\mathbf{p}) = P^{-k} \sum_{\mathbf{q} \in J^k} f(P^{-1}\mathbf{q}) e^{-2\pi i (\mathbf{p} \cdot \mathbf{q})/P}$, where $\mathbf{p} \in J^k$ (so that $W = c(\mathbf{0})$), whence $f(P^{-1}\mathbf{q}) = \sum_{\mathbf{p} \in J^k} c(\mathbf{p}) e^{2\pi i (\mathbf{p} \cdot \mathbf{q})/P}$. If the "asymptotic frequency" of $P^{-1}\mathbf{q}$ in the sequence is $d(\mathbf{q}) = \lim_{N \to \infty} \nu_N(\Xi, \mathbf{q})/N$, where $\nu_N(\Xi, \mathbf{q})$ is the number of vectors $\boldsymbol{\xi}_n = P^{-1}\mathbf{q}$ for $n = 1, 2, \cdots, N$, and if $b(\mathbf{p}) = \sum_{\mathbf{q} \in J^k} d(\mathbf{q}) e^{2\pi i (\mathbf{p} \cdot \mathbf{q})/P}$, whence $d(\mathbf{q}) = P^{-k} \sum_{\mathbf{p} \in J^k} b(\mathbf{p}) e^{-2\pi i (\mathbf{p} \cdot \mathbf{q})/P}$; then the "asymptotic expectation" of the sum U is $\bar{U} = \sum_{\mathbf{q} \in J^k} f(P^{-1}\mathbf{q})\, d(\mathbf{q}) = \sum_{\mathbf{p} \in J^k} c(\mathbf{p}) b(\mathbf{p})$.

The authors argue that, if the $c(\mathbf{p})$ diminish as $|\mathbf{p}|$ increases, it will suffice that the $b(\mathbf{p})$ should be small for all sufficiently small $|\mathbf{p}| \neq 0$ ($b(\mathbf{0}) = 1$ necessarily, in every case), in order that $|U - W|$ be asymptotically small. They discuss various familiar pseudo-random generators in this light, and make recommendations for the choice of parameters in such sequences.

While this approach is basically sound and valuable, and leads to sensible criteria for rejecting many sequences as obviously unsuitable, the reviewer would wish to see a more quantitative treatment of the criterion that $\sum_{\mathbf{p} \in (J^k - \{0\})} c(\mathbf{p}) b(\mathbf{p})$ be "sufficiently small"—a condition that may turn out to be rather severe—and also a study of how fast $|U - W|$ becomes small, in favorable cases, as N increases. *J. H. Halton* (Madison, Wis.)

Referred to in K45-27.

K45-24 (36# 5098)

Postnikova, L. P.

Sequences behaving like random ones. (Russian)

Proc. Fourth All-Union Math. Congr. (*Leningrad*, 1961) (*Russian*), *Vol. II*, *pp.* 131–134. *Izdat. "Nauka"*, *Leningrad*, 1964.

Author's summary: "This is a survey article, by its theme adhering to A. G. Postnikov's book [Trudy Mat. Inst. Steklov. **57** (1960); MR **26** #6146]. There is some material here which cannot be found in the monograph; one may partly call attention to the exposition of J. Ville's results [*Étude critique de la notion de collectif*, Gauthier-Villars, Paris, 1939]." (RŽMat **1964** #9 A79)

Citations: MR 26# 6146 = K02-2.

K45-25 (36# 7297)

Jansson, Birger

Random number generators.

Almqvist & Wiksell, Stockholm, 1966. 205 *pp.* (*loose errata*) *Kr.* 42.00.

The appearance of this book will please those interested in either the theory or the applications of techniques for generating pseudo-random numbers on digital computers. Although several survey articles on the generation of random numbers have appeared in the past three years, none is as complete as this book. The bibliography consists of 271 entries, and it is the most complete (through 1965) that this reviewer has seen. It is especially rich in references to papers published outside the United States. Much of the value of this book rests on the original contributions which the author himself has made to the subject.

Of all the methods reviewed, the following two well known congruential methods receive the most attention; namely, (a) the multiplicative method $x_{n+1} = \lambda x_n \pmod{P}$, and (b) the mixed method $x_{n+1} = \lambda x_n + \mu \pmod{P}$. The author's contributions both to finding the exact serial correlation coefficient of (x_n, x_{n+k}), for arbitrary lags k, and to understanding the dangers of using λ as a sum of powers of 2 (assuming a binary computer) are presented in considerable detail. {Reviewer's remark: This type of value of λ became popular as a result of the desire to save computer time on those machines for which successive shifting and summing could be done faster than multiplying λ by x_n. Hopefully, the appearance of the author's results will discourage some of the unjustified enthusiasm for this clever trick.}

The book consists of nine chapters. Chapter 1, "Introduction", contains an interesting and useful classification of methods into six classes: (1) Dice-like methods, (2) Tables, (3) Physical devices, (4) Arithmetic procedures applied to random numbers, (5) Arithmetic procedures where $x_{n+1} = f(x_n, x_{n-1}, \cdots, x_{n-m})$, $m \geq 0$, and (6) The digits in transcendental numbers. The introduction also contains a brief discussion of the four criteria which are used throughout the book to determine whether or not a generator is acceptable; namely, (1) simple and short generating procedures, (2) long period, (3) reproducibility, and (4) statistical acceptability. Chapter 2, "Some random number generators", contains a brief and interesting historical summary and description of the known methods which have been considered in the six classes identified above. Chapter 3, entitled "The periods and the generated numbers of some arithmetic generators", begins with a consideration of the middle-square method. Although this method is now well known for its deficiencies, the reader will find some interesting number-theoretic analysis of the method which helps to explain its limitations. The periods for methods based on linear recurrence relations $x_{n+1} = a_0 x_n + a_1 x_{N-1} + \cdots + a_j x_{n-j} + b \pmod{P}$ are reviewed in considerable depth, with particular focus on the Fibonacci generator and the multiplicative and mixed methods. Chapter 4, "Statistical tests", contains a summary of the various statistical tests considered for testing local and global randomness. {Here it is a pity that the author, considering the general depth of the book, did not warn the reader of the dangers or contradictions that can occur from the individual or joint use of some of the tests which have been proposed.} Results of known statistical tests for the Fibonacci, multiplicative, and mixed methods are summarized.

In Chapter 5, entitled "Some number-theoretical sums", the reader will find some algorithms for determining the serial correlation coefficient for arbitrary lags within the sequences of numbers generated by the mixed or multiplicative congruential generators. The algorithms are related to determining sums of the type $S_j(P, a, b) = \sum_{n=0}^{P-1} n^j [(an+b)/P]$. The chapter contains information to relate the present needs for determining such sums with the results known about Dedekind sums ($j = 1$) treated by Rademacher. This chapter is the longest in the book. It should appeal to anyone who has an interest in number theory or who wishes to gain insight into how to determine the low order moments of numbers generated by congruential methods. Chapter 6, "Means, variances and serial correlation coefficients", uses the results of Chapter 5 to provide, as noted earlier, the exact correlation coefficient for arbitrary lags for the mixed and multiplicative methods, assuming that the parameters are chosen to provide a maximum period. The author gives results to show under what choices of parameters the approximations for the correlation coefficients obtained by earlier workers in this field will be in serious error. The author also shows the pitfalls that are possible in the selection of λ as a sum of powers of 2. This reviewer cannot emphasize too strongly the value of the results of this chapter in helping one understand the performance of the mixed and multiplicative generators. Furthermore, one cannot help but be impressed with the complexity associated with achieving and interpreting the results. Much of the material of Chapters 5 and 6, because of its complexity and presentation, may be difficult to interpret and master without great effort; consequently, some of its value may be diminished.

The title of Chapter 7, "Other analytical studies", conveys the intent of this chapter. Attention is devoted to four areas: (1) The review of theoretical results related to distributions of sequences of deterministic numbers reduced modulo 1, when irrational or transcendental numbers are considered; (2) The joint distribution of x_n and x_{n+1}; (3) A limited study of runs up and down for sequences generated by linear congruential generators; and (4) Pseudo-random sequences considered as permutations. J. N. Franklin, in his review of this book in [Math. Comp. **21** (1967), 263, Review 22], noted that "There is little mention of the algebraic theories of random numbers". In Chapter 8, "Pseudo-random numbers from various statistical distributions", there is a summary of known results. This is a very brief discussion, with little explanation. The author does provide in this chapter some interesting results and insight concerning the correlations that can exist between values obtained using sums based on a fixed number of pseudo-random numbers. The last chapter contains a few brief concluding remarks.

This book will undoubtedly stand as a landmark in the field. *M. Muller* (Madison, Wis.)

Referred to in Z30-67.

K45-26 (37# 2701)

Korobov, N. M.

Certain questions of the theory of Diophantine approximations. (Russian)

Uspehi Mat. Nauk **22** (1967), *no.* 3 (135), 83–118.

Let $s \geqq 1$, $f(x_1, \cdots, x_s)$ be a real function defined in the unit s-dimensional cube, $m_1, \cdots, m_s$ be integers. Let

$$C_p(m_1, \cdots, m_s) = p^{-s} \sum_{k_1, \cdots, k_s = 0}^{p-1} f\left(\frac{k_1}{p}, \cdots, \frac{k_s}{p}\right) \exp\left(-2\pi i \frac{m_1 k_1 + \cdots + m_s k_s}{p}\right).$$

A function $f(x_1, \cdots, x_s)$ is said to belong to the class $E_{s,p}(C)$ if $|C_p(m_1, \cdots, m_s)| \leq C(\bar{m}_1 \cdots \bar{m}_s)^{-1}$, where $C > 0$ does not depend on p, $m_1, \cdots, m_s$ and $\bar{m}_j = \max(1, |m_j|)$.

The subject of this paper is an approximate calculation of multiple sums by means of simple sums. Let

$$p^{-s} \sum_{k_1, \cdots, k_s = 0}^{p-1} f(k_1/p, \cdots, k_s/p) = p^{-1} \sum_{k=1}^{p} f(\{a_1 k/p\}, \cdots, \{a_s k/p\}) + R,$$

where $a_1, \cdots, a_s$ are integers, $\{x\}$ is the fractional part of x and R is the error term.

The integers $a_1, \cdots, a_s$ are called optimal coefficients if the estimate $R = O(p^{-1} \ln^\beta p)$ holds for all functions $f \in E_{s,p}(C)$, where β and the constant in the symbol O do not depend on p.

There exists a function $f \in E_{s,p}(C)$ such that for any choice of integers $a_1, \cdots, a_s$, relatively prime to p, the inequality $|R| > C_0 C_p^{-1} \ln^s p$ holds, where $C_0 > 0$ is an absolute constant.

The author discusses properties of optimal coefficients related to the uniform distribution of the linear form $a_1 y_1 + \cdots + a_s y_s$. An algorithm is given for calculating optimal coefficients.

{This article has appeared in English translation [Russian Math. Surveys **22** (1967), no. 3, 80–118].}

J. Kubilius (Vilnius)

K45-27 (38# 3998)

Marsaglia, George

Random numbers fall mainly in the planes.

Proc. Nat. Acad. Sci. U.S.A. **61** (1968), 25–28.

A popularized account of this work has appeared as "Are random numbers really random?" [Sci. Res. **3** (1968), 21].

Let $P > 0$, $0 < \lambda < P$, and $0 < t_0 < P$ be positive integers, λ and t_0 prime to P. Define t_k, $k > 0$, recursively by $t_{k+1} \equiv \lambda t_k \pmod{P}$, $0 < t_{k+1} < P$. Let $U_k = \{t_k/P\} = \{\lambda^k t_0/P\}$, where $\{\cdots\}$ denotes "fractional part of...". Then $(U_0, U_1, \cdots, U_k, \cdots)$ is the output sequence of the multiplicative congruential random number generator (MCRNG) defined by P, λ, and t_0. Also, for fixed $n \geq 0$, and each k, $\mathbf{U}_k = (U_k, U_{k+1}, \cdots, U_{k+n})$ is the kth output $(n+1)$-vector of consecutive RN.

The author has noted and proved that all such $(n+1)$-vectors lie on a finite set of hyperplanes within the $(n+1)$-dimensional unit cube, a fact implicit in the reviewer's paper [the reviewer and R. D. Macpherson, J. Assoc. Comput. Mach. **14** (1967), 100–119; MR **36** #4779]. The "wavelengths" (of the "waves" in the probability distribution within the $(n+1)$ unit cube) discussed there are simply the distances between neighbouring hyperplanes on which the output vectors are to be found. (Much more is true. There is, for each MCRNG and each $n \geq 0$, an (output) lattice in $(n+1)$-space such that the output $(n+1)$-vectors of the random number generator (RNG) fill up that portion of the lattice which lies within the $(n+1)$ unit cube.)

The author then computes upper bounds for the size of the smallest set of such parallel hyperplanes containing all output $(n+1)$-vectors; these numbers are closely related to the reviewer's "smallest wave-numbers". These numbers seem (for practical values of P $(2^{16}-2^{48})$ and rather small (3–10) values of $n+1$) to be quite small. (73, for $P = 2^{16}$ and $n+1 = 3$, for example.)

The author regards these facts as a serious defect of these RNG's, and suggests the possibility that much of the Monte Carlo work done with such RNG's over the last two decades could be in serious error.

{This reviewer does not agree. The statistical performance of RNG's may be assessed either by the evaluation of definite integrals taken over the $(n+1)$-cube or by the calculation of the average (or maximum) distance from a general point to an output vector in the unit $(n+1)$-cube. The results are similar and, in the latter approach, the distances turn out to be of the order of $P^{-1/(n+1)}$ for "most" MCRNG's. It is true that for some choices of the parameters of the RNG (e.g., for computers of short word length), or for computations in which the independence of many successive RN's is required and critical, the distances mentioned above may be much greater and other RNG's may be necessary. But these RNG's must then be designed so that periods much longer than P are assured; the author's suggestions have not been shown to meet this requirement.} *R. R. Coveyou* (Vienna)

Citations: MR 36# 4779 = K45-23.

K45-28 (40# 2232)

Zinterhof, Peter

Einige zahlentheoretische Methoden zur numerischen Quadratur und Interpolation.

Österreich. Akad. Wiss. Math.-Natur. Kl. S.-B. II **177** (1969), 51–77.

Die Rechentechnik stellt häufig die Aufgabe, hochdimensionale Integrale durch gut konvergente Folgen anzunähern, oder gegebene Funktionen in vielen Veränderlichen durch trigonometrische Polynome gut zu approximieren. Es hat sich gezeigt, daß diese Probleme einen deutlich erkennbaren zahlentheoretischen Hintergrund haben: Es kommen vor allem Ergebnisse aus der Zahlengeometrie, der Theorie der diophantischen Näherungen und der Theorie der Gleichverteilung modulo 1 entscheidend in Frage.

Der Verfasser betrachtet Funktionen $f(x)$, die im s-dimensionalen Einheitswürfel G_s definiert sind. Dann ergeben sich Quadraturformeln von der Art: Summe + Restglied R_N = Integral über G_s. Falls $f(x)$ Riemann-integrierbar ist, sichert das Weylsche Kriterium das Verschwinden von R_N. Schärfere Aussagen über $f(x)$ betreffen die Konvergenzgeschwindigkeit der $f(x)$ zugeordneten Fourierreihe, wobei die Klasse aller stetigen Funktionen, die mit dem Einheitsgitter periodisch sind, in der üblichen Weise definiert wird, $E_s^{\alpha}(C)$; die Fourierkoeffizienten genügen einer bekannten Ungleichung.

Bei beliebiger Wahl geeigneter Gewichte ist dann nach N. S. Bahvalov [Vestnik Moskov. Univ. Ser. Mat. Meh. Astronom. Fiz. Him. **1959**, no. 4, 3–18; MR **22** #6077] mindestens $R_N = O(1/N^{\alpha})$ und nach N. M. Korobov [*Number-theoretic methods in approximate analysis* (Russian), Fizmatgiz, Moscow, 1963; MR **28** #716] höchstens $R_N = O(1/N)$. In der daraus folgenden Monte-Carlo-Methode waren jedoch keine explizit angebbaren Punkte Θ bekannt. Benutzt man aber eine diophantische Ungleichung von A. Baker [Canad. J. Math. **17** (1965), 616–626; MR **31** #2204], so folgen die Zahlen Θ einer bestimmten Abschätzung.

Im ersten Teil der Abhandlung werden verschiedene zahlentheoretische Hilfssätze bewiesen, und im zweiten werden Fehlerabschätzungen von Quadraturformeln gegeben. Ähnliche Ergebnisse erhielt auch C. T. H. Baker [SIAM J. Numer. Anal. **5** (1968), 783–804; MR **39** #6512].

E. Karst (Tucson, Ariz.)

Citations: MR 28# 716 = K45-15; MR 31# 2204 = J80-19.

K45-29 **(41# 2924)**

Barrucand, Pierre

Une application de la théorie des nombres à l'intégration numérique.

C. R. Acad. Sci. Paris Sér. A-B **270** (1970), A633–A636.

The author observes that, if $f(x)$ is an analytic function such that, for all $n>0$, $\int_0^{\infty} |f(x)|x^n\,dx < \infty$, and $Sf(x) = \sum_{m=1}^{\infty} f(m) + \frac{1}{2}f(0)$,

$$S_d f(x) = A^{-1}\{\sum_{m=1}^{\infty} a(m, d)f(m) + A_0 f(0)\},$$

then $a^{-1}S_d f(x/a)$ tends towards $\int_0^{\infty} f(x)\,dx$ much faster than $a^{-1}Sf(x/a)$, except when $f(x)$ is even. Here $-d$ ($d=3$, 4 in practice) is the discriminant of the imaginary quadratic field $Q(\sqrt{-d})$, $a(m, d)$ is connected with its zeta function $Z(s)$, namely, $Z(s) = \zeta(s)L_d(s) = \sum_{m=1}^{\infty} a(m, d)m^{-s}$, and $A = L_d(1)$, $A_0 = -L_d(0)\zeta(0)$.

The author's work can be generalized, the consistent principle being to construct a Dirichlet series $D(s)$ having as a unique singularity a simple pole at $s=1$, and such that $F(s)D(s)$ has no other singularity except possibly one, where $F(s) = \int_0^{\infty} f(x)x^{s-1}\,dx$. It would be of interest to compare the numerical efficiency of the summation formulae, for various test functions $f(x)$, with other numerical procedures.

A. M. Cohen (Cardiff)

K45-30 **(43# 3217)**

Budkowski, Stanisław

A certain method of generating the coordinates of uniformly distributed points in an n-dimensional cube. (Polish. Russian and English summaries)

Algorytmy **7**, *no.* 12, 33–61 (1970).

A distribution of points is said to be uniformly distributed in an n-cube if the number of points contained in a subset S of the cube is proportional to the volume of S. The author describes a method of generating point coordinates for such a distribution. He uses both continuous and discrete mappings of a unit segment into the unit n-cube.

E. Jucovič (Košice)

K45-31 **(44# 7700)**

Gustavson, Fred G.; Liniger, Werner

A fast random number generator with good statistical properties. (German summary)

Computing (*Arch. Elektron. Rechnen*) **6** (1970), 221–226.

Authors' summary: "Recently, P. A. Lewis, A. S. Goodman and J. M. Miller carried out extensive statistical tests on a random-number generator of congruential type [IBM Systems J. **8** (1969), 136–146]. They showed that this generator has good statistical properties. Based on a careful analysis of the number-theoretical properties of a class of generators of which the above is a member, we have significantly increased its speed. For example, on the IBM 360/67 the time per calculation of one random number has been reduced from 14.25 μs to 9.25 μs and on the IBM 360/91 from 3.25 μs to 1.3 μs. The modified generator owes its increased speed to a replacement of a division operation by an addition and two shift instructions and the removal of a subsequent overflow test."

K50 METRIC THEORY OF CONTINUED FRACTIONS

See also Sections J24, J60.

See also reviews J24-60, J56-20, K02-2, K02-5, K02-6, K15-47, K55-1, K55-2, K55-29, K55-42, K55-43, Z02-18, Z10-9, Z10-20, Z30-34, Z30-49.

K50-1 **(2, 107e)**

Doeblin, W. Remarques sur la théorie métrique des fractions continues. Compositio Math. 7, 353–371 (1940).

The author proves by using the methods of the theory of probability some of the more important theorems on the metric properties of continued fractions of Borel, Kuzmin, P. Levy, Khintchine and Denjoy. *P. Erdös.*

Referred to in K50-10, K50-33, K50-37.

K50-2 **(3, 75b)**

Good, I. J. The fractional dimensional theory of continued fractions. Proc. Cambridge Philos. Soc. **37**, 199–228 (1941).

The author considers the fractional dimension of sets defined by continued fractions. Among others he proves the following results: (1) The set of continued fractions whose partial quotients tend to infinity has dimensional number $\frac{1}{2}$. (2) The set of continued fractions for which the nth root of the nth partial fraction is unbounded has dimensional number $\frac{1}{2}$. The only previous literature on the subject seems to be a paper by Jarnik [Prace Mat.-Fiz. **36**, 91–106 (1928)]. Jarnik is concerned with the set E of continued fractions whose partial quotients are bounded, and with the sets E_α whose partial quotients are not greater than α. He proves that $\dim E=1$. He also shows that $\dim E_2>\frac{1}{4}$. The author improves this estimate very considerably; he shows that $.5306<\dim E_2<.5320$. *P. Erdös* (Philadelphia, Pa.).

Referred to in K15-58, K50-17, K50-44.

K50-3 (5, 92b)

Dyson, F. J. On the order of magnitude of the partial quotients of a continued fraction. J. London Math. Soc. **18**, 40–43 (1943).

Let ξ be any irrational in the range $0<\xi<1$; consider the regular continued fraction of ξ: $[1/a_n]_1^\infty=\xi$. Given any nondecreasing positive function $\varphi(n)$, write

$$b_n=\max_{\nu\leqq n}\frac{\varphi(n)a_\nu}{\varphi(\nu)}.$$

The following theorem is proved: if $\sum 1/\varphi(n)=\infty$, then $\sum b_n^{-1}=\infty$ for almost all ξ. The author remarks that, in the special case $\varphi(n)=1$, Khintchine obtained a sharper result. *O. Szász* (Cincinnati, Ohio).

K50-4 (13, 757a)

Ryll-Nardzewski, C. On the ergodic theorems. I. Generalized ergodic theorems. Studia Math. **12**, 65–73 (1951).

Let $\mathfrak{E}$ be a σ-field of subsets E of a space X, and let μ be a countably additive measure defined on $\mathfrak{E}$ with $\mu(X)<\infty$. Let φ be a mapping of X into itself such that (i) $E\varepsilon\mathfrak{E}$ implies $\varphi^{-1}(E)\varepsilon\mathfrak{E}$, (ii) $\mu(E)=0$ implies $\mu(\varphi^{-1}(E))=0$, but not necessarily measure preserving ($\mu(E)=\mu(\varphi^{-1}(E))$). Let $L(\mu)$ be the L^1-space on X with respect to $\mathfrak{E}$ and μ. Consider the following conditions: (B) for each $f\varepsilon L(\mu)$ there exists a $g\varepsilon L(\mu)$ such that

$$\lim_{n\to\infty}\frac{1}{n}\sum_{k=0}^{n-1}f(\varphi^k(x))=g(x)$$

μ-almost everywhere on X; (N) for each $f\varepsilon L(\mu)$ there exists a $g\varepsilon L(\mu)$ such that

$$\lim_{n\to\infty}\int_X\left|\frac{1}{n}\sum_{k=0}^{n-1}f(\varphi^k(X))-g(x)\right|\mu(dx)=0;$$

(DM) there exists a constant K such that

$$\frac{1}{n}\sum_{k=0}^{n-1}\mu(\varphi^{-k}(E))\leqq K\mu(E)$$

for all $E\varepsilon\mathfrak{E}$ and for $n=1, 2, \cdots$; (H) there exists a constant K such that

$$\limsup_{n\to\infty}\frac{1}{n}\sum_{k=0}^{n-1}\mu(\varphi^{-k}(E))\leqq K\mu(E)$$

for all $E\varepsilon\mathfrak{E}$.

It is clear that (i) (DM) implies (H). The author proves that (ii) (B) and (H) are equivalent, (iii) (N) and (DM) are equivalent, and (iv) (H) does not necessarily imply (DM). (ii) is the main result of this paper. (In case μ is σ-finite, (H) is replaced by other conditions, for example, by (H_2) there exists a constant K such that

$$\limsup_{n\to\infty}\frac{1}{n}\sum_{k=0}^{n-1}\mu(Y\cap\varphi^{-k}(E))\leqq K\mu(E),$$

for all $E\varepsilon\mathfrak{E}$ and for all $Y\varepsilon\mathfrak{E}$ with $\mu(Y)<\infty$.) The proof of (ii) is based on the idea of Y. N. Dowker [Duke Math. J. **14**, 1051–1061 (1947); these Rev. **9**, 359] of using the Banach limit. (iii) was originally obtained by Dunford and Miller [Trans. Amer. Math. Soc. **60**, 538–549 (1946); these Rev. **8**, 280] who also proved that: (v) (N) implies (B). It is to be observed that (v) is now an immediate consequence of (i), (ii) and (iii). Finally, (iv) is proved by constructing a counter-example which is a modification of that of Y. N. Dowker [Bull. Amer. Math. Soc. **55**, 379–383 (1949); these Rev. **10**, 718]. *S. Kakutani* (New Haven, Conn.).

Referred to in K50-15.

K50-5 (13, 757b)

Ryll-Nardzewski, C. On the ergodic theorems. II. Ergodic theory of continued fractions. Studia Math. **12**, 74–79 (1951).

Let

$$x=\frac{1\ |}{|c_1(x)}+\frac{1\ |}{|c_2(x)}+\frac{1\ |}{|c_3(x)}+\cdots$$

be a continued fraction expansion of an irrational number x $(0<x<1)$, and put

$$\delta(x)=\frac{1\ |}{|c_2(x)}+\frac{1\ |}{|c_3(x)}+\cdots=\frac{1}{x}-\left[\frac{1}{x}\right],$$

where $c_n(x)$ are positive integers and $[\alpha]$ denotes the integral part of α. As a mapping of the set X of all irrational numbers x with $0<x<1$ onto itself, δ is not one-to-one (in fact, δ is an infinity-to-one mapping), but is measure preserving with respect to the measure $\nu(E)$ defined for all Lebesgue measurable subsets E of X by $\nu(E)=(\log 2)^{-1}\int_E(1+x)^{-1}dx$, i.e. $\nu(\delta^{-1}(E))=\nu(E)$ for any Lebesgue measurable subset E of X. The measure $\nu(E)$ satisfies

$$m(E)/2\log 2\leqq\nu(E)\leqq m(E)/\log 2$$

for all E, where $m(E)$ is the Lebesgue measure of E with the normalization $m(X)=1$. (From this follows that ν-integrability is equivalent to Lebesgue integrability.)

The author first shows that δ is indecomposable, i.e. that any Lebesgue measurable subset E of X with $\delta^{-1}(E)=E$ satisfies either $\nu(E)=0$ or $\nu(E)=1$. Essentially the same result was previously obtained by K. Knopp [Math. Ann. **95**, 409–426 (1926)]. From Birkhoff's individual ergodic theorem then follows that for any Lebesgue integrable function $f(x)$ defined on X,

$$\lim_{n\to\infty}\frac{1}{n}\sum_{k=0}^{n-1}f(\delta^k(x))=\frac{1}{\log 2}\int_0^1\frac{f(x)}{1+x}dx$$

for almost all x. The author applies this result to the function $f(x)=\log c_1(x)$ and obtains the result of A. Khintchine [Compositio Math. **1**, 359–382 (1935)]:

$$\lim_{n\to\infty}\sqrt[n]{\prod_{k=1}^{n}c_k(x)}=\prod_{n=1}^{\infty}\left(1+\frac{1}{n(n+2)}\right)^{\frac{\log n}{\log 2}}$$

for almost all x. If $f(x)$ is the characteristic function of the set of all $x\varepsilon X$ such that $c_1(x)=p$, then we have the result of P. Lévy [ibid. **3**, 286–303 (1936)]: for any positive integer p and for almost all $x\varepsilon X$, the frequency of p in the sequence $\{c_n(x)\,|\,n=1, 2, \cdots\}$ exists and is equal to $(\log 2)^{-1}\log((p+1)^2/p(p+2))$.

S. Kakutani (New Haven, Conn.).

Referred to in J04-36, K50-15, K50-30, K55-34.

K50-6 (13, 758a)

Hartman, S. Quelques propriétés ergodiques des fractions continues. Studia Math. **12**, 271–278 (1951).

The author follows the idea of C. Ryll-Nardzewski [see the preceding review; we use the same notation as in the preceding review], and applies Birkhoff's individual ergodic theorem to the case when $f(x)$ is non-negative and non-integrable. In this case, from the indecomposability of δ follows that

$$\lim_{n\to\infty}\frac{1}{n}\sum_{k=0}^{n-1} f(\delta^k(x)) = +\infty$$

for almost all x. If $f(x)=c_1(x)=[1/x]$, then we obtain the result of A. Khintchine [Compositio Math. **1**, 359–382 (1935)]:

$$\lim_{n\to\infty}\frac{1}{n}\sum_{k=1}^{n} c_k(x) = +\infty$$

for almost all x. If $f(x)=c_2(x)/c_1(x)$ or $f(x)=c_1(x)/c_2(x)$, then $f(x)$ is non-integrable and hence

$$\lim_{n\to\infty}\frac{1}{n}\sum_{k=1}^{n}\frac{c_{k+1}(x)}{c_k(x)} = +\infty, \quad \lim_{n\to\infty}\frac{1}{n}\sum_{k=1}^{n}\frac{c_k(x)}{c_{k+1}(x)} = +\infty$$

for almost all x. Further, if $\{\gamma_n\}$ is a sequence of positive integers with $\gamma_n\to\infty$ then, for almost all x, the frequency of integers n for which $c_n(x)\geqq\gamma_n$ exists and is equal to 0.

S. Kakutani (New Haven, Conn.).

K50-7 (13, 758b)

Hartman, S., Marczewski, E., et Ryll-Nardzewski, C. Théorèmes ergodiques et leurs applications. Colloquium Math. **2**, 109–123 (1951).

Exposition of recent results in ergodic theory centering around Birkhoff's individual ergodic theorem and its applications. The mapping φ in question is measure preserving, but not necessarily one-to-one. This makes it possible to apply the individual ergodic theorem to the mapping $\varphi(x)=2x$ (mod 1) or to the shift transformation of a "one-sided" infinite direct product measure space, and to obtain the classical result of Borel concerning normal numbers or the strong law of large numbers of Kolmogoroff. Applications of the individual ergodic theorem to the problems of continued fractions are also discussed. Most of the results discussed in this paper can be found in a paper by F. Riesz [Comment. Math. Helv. **17**, 221–239 (1945); these Rev. **7**, 255] and three papers by C. Ryll-Nardzewski and S. Hartman [see the reviews of these three papers above].

S. Kakutani (New Haven, Conn.).

K50-8 (17, 243a)

Jarník, Vojtěch. Contribution à la théorie métrique des fractions continues. Czechoslovak Math. J. **4(79)** (1954), 318–329. (Russian. French summary)

Let p_n/q_n $(n=0, 1, 2, \cdots)$ be the successive convergents of the continued fraction $\vartheta=(a_1, a_2, a_3, \cdots)$ whose partial quotients a_i are positive integers. For $\delta>0$, let N_δ, or M_δ, denote the set of all numbers such that

$$\limsup_{n\to\infty} a_{n+1}/q_n^{\delta}>0 \text{ or } \liminf_{n\to\infty} a_{n+1}/q_n^{\delta}>0$$

respectively. Let P_∞ be the set of all numbers ϑ for which $\lim_{n\to\infty} a_n=+\infty$, and P_b, for integral $b>1$, the set of all ϑ with $a_n\geqq b$ for all sufficiently large n. It is known that the Hausdorff dimension of N_δ is $2/(2+\delta)$. The author proves that $\dim M_\delta=\frac{1}{2}\dim N_\delta$, $\dim P_\infty=\frac{1}{2}$ and that $\frac{1}{2}<\dim P_b<1$, $\lim_{b\to\infty}\dim P_b=\frac{1}{2}$. *R. A. Rankin.*

K50-9 (17, 466f)

Evans, Arwel. A note on continued fractions and dimension theory. J. Reine Angew. Math. **195** (1956), 102–107 (1955).

If r, s are integers,

$$\varrho=r+\frac{1}{r+}\cdots, \quad \sigma=s+\frac{1}{s+}\cdots,$$

$K(a_1, a_2, \cdots, a_n)$ is the denominator of the continued fraction

$$\frac{1}{a_1+}\cdots\frac{1}{a_n}$$

and if u of the a_i are equal to r, the remaining $v=n-u$ of the a_i are equal to s, then it is shown that $A_1 r^u s^v\leqq K\leqq A_2\varrho^u\sigma^v$. It is deduced that the set $E(r, s)$ of real numbers whose continued-fraction representations have only r and s as partial quotients has Hausdorff dimension d lying between the roots of the equations

$$r^{-2x}+s^{-2x}=1, \quad \varrho^{-2x}+\sigma^{-2x}=1.$$

H. D. Ursell (Leeds).

K50-10 (20# 3594)

Kac, M.; and Kesten, Harry. On rapidly mixing transformations and an application to continued fractions. Bull. Amer. Math. Soc. **64** (1958), 283–287; correction, **65** (1959), 67.

Let Ω be a probability measure space, T a measure-preserving transformation of Ω, B a fixed set of positive measure in Ω. Let $V(\omega)=1$ for ω in B; $=0$ for ω not in B. The authors prove that, under a suitable restriction on T, the distribution of $\sum_{k=1}^{n} V(T^k\omega)$, conditional on ω in B, is asymptotically normal. The restriction on T is that there exist an ε and a sequence of H_m such that for every measurable A

$$\left|\frac{\mu\{T^{-k}A\cap B\cap T^{-m}B\}}{\mu\{T^{-k}A\}\mu\{B\cap T^{-m}B\}}-1\right|\leqq H_m e^{-\varepsilon k}.$$

An application is adduced to the distribution of the integers appearing in continued fractions. (The correction notes that the latter result is contained in Doeblin, Compositio Math. **7** (1940), 353–371 [MR 2, 107].)

S. Katz (New York, N.Y.)

Citations: MR 2, 107e = K50-1.

K50-11 (22# 5842)

Šalát, Tibor. On an application of continued fractions in the theory of infinite series. Časopis Pěst. Mat. **84** (1959), 317–326. (Czech. Russian and English summaries)

Let $\sum_{k=1}^{\infty} u_k$ be a series of real numbers, and $M_k=(c_1^k, c_2^k, \cdots, c_n^k, \cdots)$, $k=1, 2, 3, \cdots$, a sequence of real numbers. Let $A_k=\sup|c_n^k|$, $n=1, 2, 3, \cdots$, and

$$\sum_{k=1}^{\infty} A_k|u_k| < \infty.$$

Let

$$\text{(A)} \qquad x=\frac{1}{n_1+}\frac{1}{n_2+}\frac{1}{n_3+}\cdots$$

be the infinite expansion of the irrational number $x\in(0, 1)$ into a continued fraction. Let the function ϕ be defined as follows: $\phi(0)=0$, $\phi(x)=\sum_{i=1}^{\infty} c_{n_i}{}^i u_i$, if x is irrational with the expansion (A). If x is rational, $0<x\leqq 1$, then its expansion into a continued fraction is finite, of the type (A). In the latter case, $\phi(x)=\sum_{i=1}^{r} c_{m_i}{}^i u_i$. By this definition the set of all values of the function ϕ on the set of all irrational numbers of the interval (0, 1) is identical with the set of all sums of the series $\sum_1^\infty \varepsilon_k u_k$, where ε_k is a member of the sequence M_k. Then the author gives the following theorems: (1) The function ϕ is integrable on (0, 1) in the Riemann sense. (2) Let $\sum_{k=1}^{\infty} u_k$ be a series

of real numbers, let the sequence $\{M_k\}$ be normal with respect to this series. For every $x \in (0, 1)$ except the points of a certain set, $\lim \sup_{k\to\infty} \phi_k(x) = +\infty$, $\lim \inf_{k\to\infty} \phi_k(x) = -\infty$. *E. Frank* (Chicago, Ill.)

K50-12 (24# A3445)

Ibragimov, I. A.

A theorem from the metric theory of continued fractions. (Russian. English summary)

Vestnik Leningrad. Univ. **16** (1961), *no.* 1, 13–24.

For every real number $t \in (0, 1)$ let $q_n(t)$ be the denominator of the nth approximant of the regular continued fraction expansion of t. It is shown that there exist constants a and $\sigma > 0$ such that

$$\lim_{n\to\infty} \operatorname{mes}_{t\in(0,1)} \mathscr{E} \left[\frac{\log q_n(t) - na}{\sigma\sqrt{n}} < z\right] = \frac{1}{\sqrt{(2\pi)}} \int_{-\infty}^{z} e^{-u^2/2}\, du.$$

(Paraphrase of author's summary.)

W. J. Thron (Boulder, Colo.)

K50-13 (26# 1423)

Rohlin, V. A.

Exact endomorphisms of a Lebesgue space. (Russian)

Izv. Akad. Nauk SSSR Ser. Mat. **25** (1961), 499–530.

The paper contains a detailed exposition of the theory of so-called exact endomorphisms, introduced by the author. Let $\mathscr{M}$ be a Lebesgue space with measure μ and T an endomorphism of this space, i.e., a single-valued (but not necessarily one-to-one) transformation $\mathscr{M}$ onto itself such that $\mu(\mathscr{A}) = \mu(T^{-1}\mathscr{A})$ for every measurable $\mathscr{A}$. If $\mathfrak{M}$ is a σ-algebra of all measurable sets, then $T^{-k}\mathfrak{M}$ is a σ-algebra of sets, denoted by $T^{-k}\mathscr{A}$, $\mathscr{A} \in \mathfrak{M}$, or differing from such sets on a set of measure zero. An endomorphism T is called exact if the intersection $\bigcap_{k=0}^{\infty} T^{-k}\mathfrak{M}$ of all σ-algebras $T^{-k}\mathfrak{M}$ is equal to $\mathfrak{N}$, where $\mathfrak{N}$ is a trivial σ-algebra containing only the sets of measure zero or one. Exact endomorphisms are irreducible. The isometry operator adjoint with any exact endomorphism has a countably Lebesgue spectrum: exact endomorphisms are mixing of all degrees. In the same way as for automorphisms, there is an entropy $h(T)$ of endomorphisms T introduced and it is proved that $h(T) \geqq H(\mathfrak{M}|T^{-1}\mathfrak{M})$, where $H(\mathfrak{M}|T^{-1}\mathfrak{M})$ is the mean conditional entropy of the σ-algebra $\mathfrak{M}$ under the condition of $T^{-1}\mathfrak{M}$.

Exact endomorphisms are closely related to the so-called K-automorphisms introduced earlier by Kolmogorov in his fundamental paper on entropy. The author calls such automorphisms "Kolmogorov automorphisms". The investigation of the connections between K-automorphisms and exact endomorphisms is based upon the concept of natural extension of an arbitrary endomorphism. Let $\mathscr{M}'$ be a new measure space, points of which are sequences $(x_0, x_1, \cdots)$, where $x_i \in \mathscr{M}$ and $Tx_{n+1} = x_n$ with measure induced by measure in the original space $\mathscr{M}$. A transformation T' of space $\mathscr{M}'$: $T'(x_0, x_1, \cdots) = (Tx_0, x_0, x_1, \cdots)$ is an automorphism of the space $\mathscr{M}'$. Properties of T define properties of T': if T is irreducible, then T' is also irreducible; if T is mixing then T' is also mixing; entropies of T and T' are equal; there are some relations between the spectra of T and T'; the natural extension of an exact endomorphism is a K-automorphism. So from properties of exact endomorphisms one can deduce corresponding properties of K-automorphisms: the spectrum of K-automorphisms, positiveness of entropy and mixing of all degrees. The first of these properties was pointed out earlier by Kolmogorov.

As an application of the whole theory, the author considers number-theoretic endomorphisms introduced by Rényi. Let a function $\varphi(x)$ be defined on the interval $(0, 1)$. For the transformation $Tx = (\varphi(x))$, where (y) denotes the fractional part of y, Rényi found some conditions under which there exists, for the transformation T, an invariant measure equivalent to Lebesgue measure. In the paper it is proved that under the Rényi conditions the endomorphism T is an exact endomorphism. The proof is based upon some general conditions of exactness. In the end, the value of the entropy for these endomorphisms is found.

There are some misprints in the text.

Ja. G. Sinaĭ (Moscow)

Referred to in J24-45, K50-37, K50-41, K55-44, K55-46.

K50-14 (27# 124)

Szüsz, P.

Über einen Kusminschen Satz. (Russian summary)

Acta Math. Acad. Sci. Hungar. **12** (1961), 447–453.

If t is a number between 0 and 1, let $a_1, a_2, \cdots$ denote the sequence of partial denominators in the regular continued fraction for t, so that $t_0 = t$, $t_n = a_n + 1/t_{n+1}$, $n = 0, 1, \cdots$. If $z_n = 1/t_{n+1}$, then $0 < z_n \leqq 1$. Let $m_n(x)$ denote the Lebesgue measure of the set of all the numbers t for which $z_n \leqq x$ and denote by $k_n(x)$ the difference $m_n(x) - [\log(1+x)/\log 2]$. Kuzmin [Atti Congr. Internat. (Bologne, 1928), 6, pp. 83–89] showed that $k_n(x)$ has the asymptotic form $O(q^{\sqrt{n}})$ as $n \to \infty$, where q is a number between 0 and 1, thus establishing the Gaussian limit $k_n(x) \to 0$ as $n \to \infty$. Lévy [Bull. Soc. Math. France **57** (1929), 178–194], using a quite different method and without knowledge of Kuzmin's work, showed that $k_n(x) = O(q^n)$, where $0 < q < .7$. The present author applies the methods of Kuzmin to show easily that $k_n(x) = O(q^n)$, where $q = 2\zeta(3) - \zeta(2)$ (which is greater than .7), but then obtains a better estimate for q by means of the Riemann zeta-function, so that $0 < q < .4$.

H. S. Wall (Austin, Tex.)

Referred to in K50-29.

K50-15 (27# 2483)

de Vroedt, C.

Measure-theoretical investigations concerning continued fractions.

Nederl. Akad. Wetensch. Proc. Ser. A **65** = *Indag. Math.* **24** (1962), 583–591.

The author continues investigations on metrical properties of regular continued fraction expansions begun in his thesis [Amsterdam, 1960]. His results are related to but different from those of Ryll-Nardzewski [Studia Math. **12** (1951), 65–73; MR **13**, 757; ibid. **12** (1951), 74–79; MR **13**, 757].

W. J. Thron (Boulder, Colo.)

Citations: MR 13, 757a = K50-4; MR 13, 757b = K50-5.

K50-16 (27# 3584)

Szüsz, P.

Verallgemeinerung und Anwendungen eines Kusminschen Satzes.

Acta Arith. **7** (1961/62), 149–160.

Der bekannte Gauss-Kusminsche Satz [R. O. Kusmin, Atti Congr. Internaz. Mat. (Bologna, 1928), T. VI, pp. 83–89, Zanichelli, Bologna, 1932] über Kettenbrüche wird folgendermaßen verallgemeinert: Seien n, r, k_1, k_2 natürliche Zahlen mit (*) $(k_1, k_2, r) = 1$, und sei M die Menge der α, $0 < \alpha < 1$, deren reguläre Kettenbruchentwicklung $\alpha = [0; a_1, a_2, \cdots]$ bei gegebenem $x \in (0, 1)$ der Bedingung $[0; a_{n+1}, a_{n+2}, \cdots] \leqq x$ genügt und bei denen außerdem der $(n-1)$-te bezw. n-te Näherungsnenner die Kongruenzen $B_{n-1}(\alpha) \equiv k_1$, $B_n(\alpha) \equiv k_2 \pmod r$ erfüllen. Dann ist das Lebesguesche Maß von M gleich $\log(1+x)/(c(r)\log 2) + O(q^n)$, wobei $0 < q < 1$ und $c(r)$ die Anzahl aller Restklassenpaare

k_1, k_2 mit (*) ist. Ein entsprechender Satz wird auch für den Fall bewiesen, daß zusätzlich noch die ersten l Kettenbruchelemente fest gegeben sind. Schließlich wird ein Satz angegeben, der für gewisse zahlentheoretische Funktionen $f(k_1, k_2, l)$ garantiert, daß bei fast allen α der Mittelwert $\lim_{n\to\infty}(1/n)\sum_{l=1}^{\infty} f(B_{l-1}, B_l, a_l)$ existiert und den gleichen Wert hat. Dieser Satz gestattet u.a. die Bestimmung der (bei fast allen α gleichen) natürlichen Dichte der Menge der Indizes l, bei denen $B_l(\alpha)$ einer gegebenen Kongruenz genügt. *B. Volkmann* (Mainz)

K50-17 (28# 1420)

Rogers, C. A.
Some sets of continued fractions.
Proc. London Math. Soc. (3) **14** (1964), 29–44.

The fractional dimension of the set E_q of real numbers x, $0<x<1$, having the expansion

$$x = \frac{1}{a_1+}\,\frac{1}{a_2+}\cdots,$$

with all $a_i \leqq q$, was investigated by Jarník [Prace Mat.-Fiz. **36** (1928/29), 91–106]. He proved, for example, that $\frac{1}{4}<\dim E_2<1$. The reviewer [Proc. Cambridge Philos. Soc. **37** (1941), 199–228; MR **3**, 75] showed that $\dim E_q = \lim \sigma_{q,n}$ as $n\to\infty$, where $\sigma_{q,n}$ is the real root of

$$\sum [a_1, a_2, \cdots, a_n]^{-2\sigma} = 1 \qquad (a_1, a_2, \cdots, a_n = 1, 2, \cdots, q),$$

where $[a_1, a_2, \cdots, a_n]$ is the Euler polynomial used in the theory of continued fractions, and that $0.5306<\dim E_2<0.5320$. The author excludes from E_q a set of zero probability, in a natural manner, and shows that this can decrease the fractional dimension considerably. He expresses the dimensional number of the reduced set in terms of the limiting values of the geometric mean and median value of the q^n numbers $[a_1, a_2, \cdots, a_n]^{1/n}$. He also obtains some interesting results concerning the distribution of the Euler polynomial and the joint distribution of $[a_1, a_2, \cdots, a_n]$ and $[a_1, a_2, \cdots, a_{n-1}]$.
I. J. Good (Princeton, N.J.)

Citations: MR 3, 75b = K50-2.

K50-18 (29# 57)

Lochs, Gustav
Vergleich der Genauigkeit von Dezimalbruch und Kettenbruch.
Abh. Math. Sem. Univ. Hamburg **27** (1964), 142–144.

Using a result of A. Khintchine and P. Lévy the author proves the following theorem. Let m be the exact number of partial quotients of the regular continued fraction of a number $x\in[0, 1)$ which can be obtained from the first n decimals of x. Then for almost all x

$$\lim_{n\to\infty}\frac{m}{n} = \frac{6\log 2\log 10}{\pi^2} = 0.97027014\cdots.$$

J. E. Cigler (Vienna)

K50-19 (30# 3058)

de Vroedt, C.
Metrical problems concerning continued fractions.
Compositio Math. **16**, 191–195 (1964).

This is an expository account of some results, including one by the author ["Metrical problems concerning continued fractions", Thesis, Amsterdam, 1960], on the subject of the title. *W. J. Thron* (Boulder, Colo.)

K50-20 (33# 1300)

Chatterji, S. D.
Masse, die von regelmässigen Kettenbrüchen induziert sind.
Math. Ann. **164** (1966), 113–117.

Let $a_i(x)$ be the ith entry in the regular continued fraction representing x^1, and let $A_n(x)/B_n(x)$ be the nth partial fraction. Let h be Lebesgue measure on $[0, 1]$ and P another measure. It is shown that (A) $P\perp h$ if and only if $B_n^2(x)P\{y\,|\,a_i(y)=a_i(x), i=1, \cdots, n\}\to 0$ a.e. (h); (B) if P makes the $a_i(x)$ independent, then $P\perp h$; (C) P is atomic if and only if $\prod_{i=1}^{\infty}\sup_{1<k<\infty} P\{x\,|\,a_i(x)=k\}>0$.

A martingale theorem is used to establish (A). The third contention is established by a zero-one theorem which shows that the condition of (C) forces P to be concentrated on a countable set. The second theorem is shown by establishing that the opposite implies a contradiction to a known theorem of Lévy.
J. R. Kinney (Lexington, Mass.)

Referred to in K50-28.

K50-21 (34# 142)

Schweiger, Fritz
Geometrische und elementare metrische Sätze über den Jacobischen Algorithmus.
Österreich. Akad. Wiss. Math.-Natur. Kl. S.-B. II **173** (1964), 59–92.

Bei der üblichen Darstellung des Jacobialgorithmus [siehe etwa O. Perron, Math. Ann. **64** (1907), 1–76] wird einem reellen Zahlen-n-tupel $\alpha=(\alpha_1, \cdots, \alpha_n)$ eine Folge von n-tupeln $a^{(\mu)}=(a_1^{(\mu)}, a_2^{(\mu)}, \cdots, a_n^{(\mu)})$ ganzer Zahlen und eine Folge $\alpha=\alpha^{(0)}, \alpha^{(1)}, \cdots$ von "transformierten" reellen n-tupeln zugeordnet, die miteinander durch die Beziehungen $[\alpha_\nu]=a_\nu^{(0)}$, $\alpha_1=a_1^{(0)}+(1/\alpha_n^{(1)})$, $\alpha_\nu=a_\nu^{(0)}+(\alpha_{i-1}^{(1)}/\alpha_n^{(1)})$ $(i=2, \cdots, n)$, $\alpha_1^{(\nu)}=a_1^{(\nu)}+(1/\alpha_n^{(\nu+1)})$, $\alpha_i^{(\nu)}=a_i^{(\nu)}+(\alpha_{i-1}^{(\nu)}/\alpha_n^{(\nu+1)})$ $(i=2, \cdots, n;\ \nu=1, 2, \cdots)$ verknüpft sind.

Der Verfasser gibt eine einfachere Darstellung des Algorithmus mit Hilfe projektiver Koordinaten $\beta=(\beta_0, \cdots, \beta_n)$ statt der Zahlen $\alpha_i=\beta_i/\beta_0$, so dass die Folge $\{\alpha^{(\nu)}\}$ in eine Folge $\{\beta^{(\nu)}\}$ übergeht. Es werden Matrizen $\Lambda^{(\nu)}$ angegeben, für die $\beta^{(\nu)}=\Lambda^{(\nu)}\beta^{(\nu+1)}$ $(\nu=0, 1, 2, \cdots)$ gilt. Für die Elemente $\omega_{ij}^{(\nu)}$ der Produktmatrizen $\Omega^{(\nu)}=\Lambda^{(0)}\Lambda^{(1)}\cdots\Lambda^{(\nu-1)}$ werden Rekursionsformeln hergeleitet. Unter Verwendung dieser Elemente werden obere und untere Schranken für das Mass der Mengen $B^{(\nu)}$ aller β im Einheitswürfel gewonnen, bei denen die zugehörigen $a_i^{(\mu)}$ $(i=1, \cdots, n;\ \mu=1, \cdots, \nu)$ fest vorgegebene Werte annehmen.

Mit Hilfe solcher Abschätzungen werden unter anderem folgende Sätze bewiesen. (i) Ist $\varphi(\nu)$ eine positive Funktion, so gibt es für fast alle α unendlich [höchstens endlich] viele ν mit $a_n^{(\nu)}\geqq\varphi(\nu)$, wenn $\sum_{\nu=1}^{\infty}\varphi(\nu)^{-1}$ divergiert [konvergiert]. (ii) Für fast alle α ist die Folge $\{a_n^{(\nu)}\}$ $(\nu=0, 1, \cdots)$ unbeschränkt. (iii) Es gibt eine Konstante C, so dass für fast alle bei hinreichend großem ν die Ungleichung $\omega_{0n}^{(\nu)}<e^{c(\nu-1)}$ gilt. (iv) Für alle α und alle $\nu\geqq n+1$ gilt $\omega_{0n}^{(\nu)}\geqq 2^{(\nu-1)/(n+1)-1}$.

{Bei Satz 1 ist das Wort "1–1-deutig" durch "eindeutig" zu ersetzen; denn trivialerweise gibt es stets unendlich viele mit gleichem $a_i^{(\nu)}$.}

See also #143 below. *B. Volkmann* (Stuttgart)

Referred to in K50-23.

K50-22 (34# 143)

Schweiger, Fritz
Metrische Sätze über den Jacobischen Algorithmus.
Monatsh. Math. **69** (1965), 243–255.

Bezeichnungen wie im vorangehenden Referat [#142]. Der Verfasser setzt die Untersuchungen der vorangehenden Arbeit fort und erhält unter anderem folgende weitere Resultate. (v) Satz (i) wird für beliebiges $i \leqq n$ (statt $i=n$) bewiesen. (vi) Für jedes $\alpha \in B^{(0)}$ sei bei gegebenem n-tupel $s=(s_1, \cdots, s_n)$ von ganzen Zahlen (mit gewissen Zusatzbedingungen) $f(N)$ die Anzahl der $\tau \leqq N$ mit $(k_1^{(\tau)}(\alpha), \cdots, k_n^{(\tau)}(\alpha))=s$. Dann gibt es eine Konstante $\kappa>0$ derart, dass für fast alle α die Ungleichung $\lim\inf_{n\to\infty} f(n)/N \geqq \kappa$ gilt.

B. Volkmann (Stuttgart)

K50-23 **(35# 150)**

Schweiger, F.

Ergodische Theorie des Jacobischen Algorithmus.

Acta Arith. **11** (1966), 451–460.

Bei metrischen Untersuchungen an Kettenbrüchen $x=[a_0, a_1, \cdots]$ ($x>0$, reell) spielt die Verschiebungstransformation $\delta(x)=\{1/x\}$ (mit $\{t\}=t-[t]$) eine Rolle, durch die x in den Kettenbruch $x'=[a_1, a_2, \cdots]$ übergeht. Der Verfasser untersucht beim Jacobischen Algorithmus im R_n (Bezeichnungen wie in einer früheren Arbeit [der Verfasser, Österreich. Akad. Wiss. Math.-Natur. Kl. S.-B. II **173** (1964), 59–92; MR **34** #142]) die analoge Transformation $\delta_n=\delta(x_1, \cdots, x_n)=(\{x_2/x_1\}, \cdots, \{x_n/x_1\}, \{1/x_1\})$ des Einheitswürfels auf sich. Es wird bewiesen, dass die Transformation δ_n ergodisch bezüglich des Lebesgueschen Masses ist.

{See also #151 below.} *B. Volkmann* (Stuttgart)

Citations: MR 34# 142 = K50-21.

Referred to in K50-31.

K50-24 **(35# 151)**

Schweiger, F.

Existenz eines invarianten Masses beim Jacobischen Algorithmus.

Acta Arith. **12** (1966/67), 263–268.

In Fortsetzung einer vorangehenden Arbeit [#150 oben] wird bewiesen, dass es auf der Klasse der Lebesguemessbaren Mengen E im Einheitswürfel des R_n ein Mass μ gibt, das bezüglich der Transformation δ invariant ist. Ferner wird gezeigt, dass zwei positive Konstanten existieren, zwischen denen der Quotient $\mu(E)/\lambda(E)$ ($\lambda=$ Lebesguesches Mass) für alle solchen Mengen E liegt.

B. Volkmann (Stuttgart)

K50-25 **(35# 6201)**

Philipp, Walter

Ein zentraler Grenzwertsatz mit Anwendungen auf die Zahlentheorie.

Z. Wahrscheinlichkeitstheorie und Verw. Gebiete **8** (1967), 185–203.

Let $\{X_n, n=0, \pm 1, \pm 2, \cdots\}$ be a stationary process, and let $\mathfrak{M}_{(a,b)}$ be the σ-algebra generated by the random variables $\{X_n\}$ with $a<n<b$. Various authors have shown that mixing conditions of the kind

$$\sup\{|P(AB)-P(A)P(B)| \leqq \alpha(n): A \in \mathfrak{M}(-\infty, 0), B \in \mathfrak{M}(n, \infty)\}$$

imply the asymptotic normality of $S_n=\sum_{i=1}^n X_i$, under conditions on $\alpha(n)$.

In this paper the author assumes only weak stationarity, but instead assumes the existence of mixed moments of arbitrarily high order. He proves a central limit theorem which is then applied to prove asymptotic normality of certain sums occurring in number theory.

J. R. Blum (Albuquerque, N.M.)

Referred to in K50-36.

K50-26 **(35# 6202)**

Philipp, Walter

Das Gesetz vom iterierten Logarithmus für stark mischende stationäre Prozesse.

Z. Wahrscheinlichkeitstheorie und Verw. Gebiete **8** (1967), 204–209.

Let $\{X_n, n=0, \pm 1, \pm 2, \cdots\}$ be a weakly stationary process with $EX_n=0$, $EX_n^2=\sigma^2>0$. Under certain strong mixing conditions on the process, it is shown that

$$P\{\lim_n \sup|\textstyle\sum_{i=1}^n X_i|/\sqrt{(2\sigma^2 n \log\log n)} = 1\} = 1,$$

i.e., that the law of the iterated logarithm holds.

J. R. Blum (Albuquerque, N.M.)

Referred to in K50-27, K50-33.

K50-27 **(36# 6369)**

Stackelberg, Olaf P.

On the law of the iterated logarithm for continued fractions.

Duke Math. J. **33** (1966), 801–819.

The transformation T given by $x \to 1|x-[1|x]$ is an ergodic transformation of the unit interval into itself. There exists an invariant measure λ equivalent to Lebesgue measure. Let us denote the continued fraction expansion as $x=[a_1(x), a_2(x), \cdots]$; then $a_i(x)=a_1(T^{i-1}x)$. In the paper it is shown that for the sequence of functions $\log a_i(x)$ the law of the iterated logarithm holds. There is a key Lemma 4 which approximates higher moments in order to apply known techniques (e.g., P. Erdős and I. S. Gál [Nederl. Akad. Wetensch. Proc. Ser. A **58** (1955), 65–76; ibid. **58** (1955), 77–84; MR **16**, 1016], the author [ibid. **67** (1964), 48–55; ibid. **67** (1964), 56–67; MR **28** #4558]). The methods presented in the paper have been generalised recently by W. Philipp in Z. Wahrscheinlichkeitstheorie und Verw. Gebiete **8** (1967), 204–209 [MR **35** #6202], where they are applied to stationary processes with strong mixing conditions and applications to "Resttransformation" are given.

F. Schweiger (Vienna)

Citations: MR 16, 1016g = K20-5; MR 28# 4558 = K15-71; MR 35# 6202 = K50-26.

Referred to in K50-33.

K50-28 **(37# 148)**

Schweiger, Fritz

Induzierte Masse und Jacobischer Algorithmus.

Acta Arith. **13** (1967/68), 419–422.

For a thorough introduction to the Jacobi algorithm, see the work of O. Perron [Math. Ann. **64** (1907), 1–76]. To each element $x=(x_1, \cdots, x_n)$ of Euclidean n-space there correspond a sequence of n-tuples of integers $a^{(v)}=a_1^{(v)}, \cdots, a_n^{(v)}$ and a sequence of transformed points $x=x^{(0)}, x^{(1)}, \cdots$, satisfying $x_1^{(v)}=a_1^{(v)}+1/x_n^{(v+1)}$, $x_2^{(v)}=a_2^{(v)}+x_1^{(v+1)}/x_n^{(v+1)}, \cdots, x_n^{(v)}=a_n^{(v)}+x_{n-1}^{(v+1)}/x_n^{(v+1)}$; $a_i^{(v)}=[x_i^{(v)}]$. Convergents $\omega_{in}^{(v)}/\omega_{0n}^{(v)}$ are defined by the recursion $\omega_{in}^{(v)}=\omega_{i0}^{(v)}+a_1^{(v)}\omega_{i1}^{(v)}+\cdots+a_n^{(v)}\omega_{in}^{(v)}$, $i=0, 1, \cdots, n$. The transformation $T: B^{(0)}\to B^{(0)}$ (see below) connected with this algorithm is given by $Tx=(x_2/x_1-[x_2/x_1], \cdots, 1/x_1-[1/x_1])$ if $x_i\neq 0$, and $Tx=0$ if $x_i=0$. $Tx^{(v)}=x^{(v+1)}$. If $n=1$, we have the standard continued fraction situation.

In this work the author generalizes two theorems concerning measures induced by regular continued fractions, proven by S. D. Chatterji [Math. Ann. **164** (1966), 113–117; MR **33** #1300] to apply to the Jacobi algorithm with $n=2$. Let $B^{(0)}$ be the unit square, and $B^{(s)}=\{x\in B^{(0)}|a^{(i)}(x)=k^{(i)}, 1\leqq i\leqq s\}$. For a probability measure P, let $p(k^{(1)}, \cdots, k^{(s)})=P\{x\in B^{(0)}|a^{(i)}(x)=k^{(i)}, 1\leqq i\leqq s\}$. Theorem 1: If P is a measure defined by $P(B^{(s)})=p(k^{(1)}, \cdots, k^{(s)})$, then P is singular with respect to Lebesgue measure L if and only if

$\lim_{s\to\infty}(\omega_{02}^{(s+1)}(x))^3 p(k^{(1)}(x), \cdots, k^{(s)}(x)) = 0$ a.e.-L. Theorem 2: If P is a measure defined by

$$P(B^{(s)}) = p(0, k^{(1)})p(k^{(1)}, k^{(2)})\cdots p(k^{(s-1)}, k^{(s)}),$$

and if P is invariant and ergodic with respect to T, then P is singular with respect to L. Furthermore, P is atomic if and only if $\sup p(k^{(1)}, \cdots, k^{(s)}) \geqq C > 0$ for all s.

O. P. Stackelberg (Durham, N.C.)

Citations: MR 33# 1300 = K50-20.

K50-29 (37# 1336)

Szüsz, P.

On Kuzmin's theorem. II.

Duke Math. J. **35** (1968), 535–540.

In Part I [Acta Math. Acad. Sci. Hungar. **12** (1961), 447–453; MR **27** #124] the author gave a refinement of an assertion of Gauss in the metrical theory of continued fractions (see Part I or the present paper for references and a history of the problem). In this paper the author considers a function theory problem suggested by the proof of the above result. Let $f_0(x)$ be a differentiable function such that $f_0(0)=0$ and $f_0(1)=1$. Define $f_1(x)$, $f_2(x), \cdots$ inductively by $f_{n+1}(x) = \sum_{k=1}^{\infty} \{f_n(1/k) - f_n(1/(k+x))\}$. In Part I the author showed that if $f_0(x)$ is twice continuously differentiable, then $f_n'(x) = (1/\log 2)1/(x+1) + O(q^n)$ $(q<1)$. Here he assumes instead that $f_0'(x) \in \mathrm{Lip}\,\vartheta$ $(\vartheta > 0)$ to draw the same conclusion. Also if $f_0(x)$ is just continuously differentiable then $\lim f_n'(x) = 1/(\log 2)1/(x+1)$ $(n\to\infty)$.

W. W. Adams (Berkeley, Calif.)

Citations: MR 27# 124 = K50-14.

K50-30 (37# 1834)

Šalát, Tibor

Remarks on the ergodic theory of the continued fractions. (Russian summary)

Mat. Časopis Sloven. Akad. Vied **17** (1967), 121–130.

Let x be irrational, $x \in (0, 1)$, and $x = \mathrm{K}_{n=1}^{\infty}(1/c_n(x))$ be the regular continued fraction expansion of x. Let A denote a set of natural numbers and $A(n)$ be the number of $a \in A$ such that $a \leqq n$. $h(A) = \lim n^{-1}A(n)$ is the asymptotic density of A. The following is proved: If $t_n \to +\infty$, then for almost all $x \in (0, 1)$,

$$h(\{n; c_n(x) \geqq t_n\}) = h(\{n; c_n/c_{n+1} \geqq t_n\}) =$$

$$h(\{n; c_{n+1}/c_n \geqq t_n\}) = h(\{n; |c_n(x) - c_{n+1}(x)| \geqq t_n\}) = 0.$$

And if $-1 < \alpha < +1$, then for almost all x the sequence $(c_k(x)/c_{k+1}(x))^\alpha$, $k = 1, 2, 3, \cdots$, is $(C, 1)$ summable to a number independent of x. If $0 \leqq \alpha < 1$, then for almost all x the sequence $|c_k(x) - c_{k+1}(x)|^\alpha$, $k = 1, 2, 3, \cdots$, is $(C, 1)$ summable to a number independent of x.

Finally, if $|\alpha| \geqq 1$, then for almost all $x \in (0, 1)$,

$$\lim n^{-1} \sum_{k=1}^{n} \frac{c_k^{\alpha}(x)}{c_{k+1}^{\alpha}}(x) = \lim n^{-1} \sum_{k=1}^{n} |c_k(x) - c_{k+1}(x)|^\alpha = +\infty.$$

The proofs are based on a theorem by C. Ryll-Nardzewski [Studia Math. **12** (1951), 74–79; MR **13**, 757].

A. Magnus (Ft. Collins, Colo.)

Citations: MR 13, 757b = K50-5.

Referred to in K50-45.

K50-31 (37# 2681)

Schweiger, Fritz

Mischungseigenschaften und Entropie beim Jacobischen Algorithmus.

J. Reine Angew. Math. **229** (1968), 50–56.

Im Zusammenhang mit dem Jacobischen Algorithmus hat der Verfasser in einer früheren Arbeit [Acta Arith. **11** (1966), 451–460; MR **35** #150] eine Verschiebungstransformation δ_n untersucht, durch die der n-dimensionale Einheitswürfel auf sich abgebildet wird. In der vorliegenden Arbeit wird bewiesen, dass δ_n ein exakter Homomorphismus ist. Für die Entropie dieser Abbildung werden eine explizite Integraldarstellung und eine fast überall gültige Grenzwertdarstellung mit Hilfe von Massen der Fundamentalbereiche angegeben.

B. Volkmann (Stuttgart)

Citations: MR 35# 150 = K50-23.

K50-32 (38# 1710)

Marques Henriques, J.

On probability measures generated by regular continued fractions. (Portuguese)

Gaz. Mat. (Lisboa) **27** (1966), *no.* 103–104, 16–22.

The author discusses some familiar concepts and results and proves a necessary and sufficient condition for the measure generated by a continued fraction to be singular with respect to Lebesgue measure.

K50-33 (39# 1423)

Philipp, Walter

Das Gesetz vom iterierten Logarithmus mit Anwendungen auf die Zahlentheorie.

Math. Ann. **180** (1969), 75–94.

The author continues his study of the iterated logarithm law for mixing processes started in Z. Wahrscheinlichkeitstheorie und Verw. Gebiete **8** (1967), 204–209 [MR **35** #6202]. He gives sufficient conditions for the law of the iterated logarithm to hold uniformly for certain families of processes. Among the applications is a generalization of a theorem by S. and L. Gál [Nederl. Akad. Wetensch. Proc. Ser. A **67** (1964), 129–143; MR **29** #392], which concerns the discrepancy of sequences of the form $\{T^{l_n}x\}$, where $\{l_n\}$ is a monotone sequence of integers, and T belongs to a certain class of transformations of the unit interval onto itself which includes the dyadic transformation $Tx = 2x$ (mod 1), and the continued fraction transformation $Tx = x^{-1}$ (mod 1). In the second part of the paper the iterated logarithm law is proven for sums of the form $\sum f(T^n x)$ for certain f (including functions of bounded variation, and functions satisfying a Lip α $(\alpha > 0)$ condition), and for T as above. Taking $Tx = x^{-1}$ (mod 1) generalizes theorems by W. Doeblin [Compositio Math. **7** (1940), 353–371; MR **2**, 107] and the reviewer [Duke Math. J. **33** (1966), 801–819; MR **36** #6369].

O. P. Stackelberg (Durham, N.C.)

Citations: MR 2, 107e = K50-1; MR 29# 392 = K15-74; MR 35# 6202 = K50-26; MR 36# 6369 = K50-27.

Referred to in A54-42, K50-34, K50-36.

K50-34 (43# 4789)

Philipp, Walter

Corrigendum: "Das Gesetz vom iterierten Logarithmus mit Anwendungen auf die Zahlentheorie".

Math. Ann. **190** (1971), 338.

The author makes several corrections in his work in same Ann. **180** (1969), 75–94 [MR **39** #1423].

O. P. Stackelberg (Durham, N.C.)

Citations: MR 39# 1423 = K50-33.

K50-35 (39# 4912)

Philipp, Walter

The remainder in the central limit theorem for mixing stochastic processes.

Ann. Math. Statist. **40** (1969), 601–609.

The author studies the convergence of the df's of normed

sums $s_N^{-1} \sum_{n=1}^N x_n$, $s_N^2 = \sum_{n=1}^N Ex_n^2$, of random variables x_n with $Ex_n = 0$, to the Gaussian d.f. Φ under mixing conditions. Let $\mathscr{M}_{ab}$ be the σ-algebra generated by the x_n for $1 \leqq a \leqq n \leqq b$. The mixing conditions that are used are the following: (I*) $|P(AB) - P(A)P(B)| \leqq \Psi(n)P(A)P(B)$ for $A \in \mathscr{M}_{1t}$, $B \in \mathscr{M}_{t+n,\infty}$, $\Psi(n) \downarrow 0$ and

(II*) $\sup_t \sup_{B \in \mathscr{M}_{t+n}} |P(B|\mathscr{M}_{1t}) - P(B)| \leqq \varphi(n) \downarrow 0$

and slight generalizations of these conditions. Assuming that $\varphi(n) = e^{-\lambda n}$ and that the x_n are a.s. bounded by 1 and $\sum_{n=M+1}^{M+H} E^{1/4} x_n^4 \leqq cE(\sum_{n=M+1}^{M+H} x_n)^2$ uniformly with respect to M, he shows that

$$P[s_n^{-1} \sum_{n \leqq N} x_n < x] = \Phi(x) + O(s_N^{-1/2} \log^3 s_n).$$

Similar results are obtained under other mixing conditions.

H. Bergström (Göteborg)

K50-36 (39# 5503)

Philipp, Walter; Stackelberg, Olaf P.

Zwei Grenzwertsätze für Kettenbrüche.

Math. Ann. **181** (1969), 152–156.

It is shown that for the sequence $\langle q_n(x) \rangle$, where $q_n(x)$ denotes the denominator of the nth approximant in the continued fraction of x, the law of the iterated logarithm and the central limit theorem hold. The proof is an application of results given in former papers by the first author [Z. Wahrscheinlichkeitstheorie und Verw. Gebiete **8** (1967), 185–203; MR **35** #6201; Math. Ann. **180** (1969), 75–94; MR **39** #1423].

F. Schweiger (Vienna)

Citations: MR 35# 6201 = K50-25; MR 39# 1423 = K50-33.

Referred to in K50-50.

K50-37 (39# 6850)

Gordin, M. I.

Random processes produced by number-theoretic endomorphisms. (Russian)

Dokl. Akad. Nauk SSSR **182** (1968), 1004–1006.

The paper deals with the study of the transformations T of the unit interval onto itself defined by $Tx = ((f^{-1}(x)))$ $(0 < x < 1)$, where $((y))$ denotes the fractional part of the number y, while $f^{-1}(x)$ is the inverse function of a monotonic positive function $f(x)$ satisfying the conditions introduced by the reviewer [Acta Math. Acad. Sci. Hungar. **8** (1957), 477–493; MR **20** #3843], who has also shown that under the conditions mentioned there exists a measure P on the Borel sets of the unit interval which is invariant with respect to T and absolutely continuous with respect to the Lebesgue measure; further, that with respect to P, the measure preserving transformation T is ergodic. V. A. Rohlin [Izv. Akad. Nauk SSSR Ser. Mat. **25** (1961), 499–530; MR **26** #1423] has continued the study of such transformations, showing that T is an exact endomorphism. For every $g \in L_2(P)$ the sequence $X_k = g(T_x^k)$ $(k = 1, 2, \cdots)$ is a stationary stochastic process. By introducing further restrictions on the function f, the author shows that this process possesses certain mixing properties which make it possible to apply certain general theorems and to prove, e.g., the central limit theorem for the process $\{X_k\}$. The additional restrictions on f are satisfied, e.g., for $f(x) = 1/x$ (which corresponds to continued fractions); thus some classical results of W. Doeblin [Compositio Math. **3** (1940), 353–371; MR **2**, 107] are obtained as special cases.

{This article has appeared in English translation [Soviet Math. Dokl. **9** (1968), 1234–1237].} *A. Rényi* (Budapest)

Citations: MR 2, 107e = K50-1; MR 20# 3843 = K55-6; MR 26# 1423 = K50-13.

K50-38 (40# 98)

Šalát, Tibor

Bemerkung zu einem Satz von P. Lévy in der metrischen Theorie der Kettenbrüche.

Math. Nachr. **41** (1969), 91–94.

Let X be the set of all irrational numbers of the interval $(0, 1)$. Let $x = [0; c_1(x), c_2(x), \cdots]$ be the continued fraction for x, $x \in X$. P. Lévy [*Théorie de l'addition des variables aléatoires*, Gauthier-Villars, Paris, 1937; second edition, 1954] (cf. also A. Ja. Hinčin [*Continued fractions* (Russian), ONTI NKTP, Moscow, 1935; second edition, GITTL, 1949; MR **13**, 444; German translation, Teubner, Leipzig, 1956; MR **18**, 274; third edition, Fizmatgiz, Moscow, 1961; English translations, Noordhoff, Groningen, 1963; MR **28** #5038; Univ. of Chicago Press, Chicago, Ill., 1964; MR **28** #5037]), in the metric theory of continued fractions, gave a theorem concerning the distribution of natural numbers in the sequences $\{c_j(x)\}_{j=1}^\infty$ of the elements of the numbers $x \in X$. Let $N_n(p, x)$ denote the number of integers p in the finite sequence $c_1(x)$, $c_2(x), \cdots, c_n(x)$. Then for almost all $x \in X$,

(A) $\lim_{n \to \infty} N_n(p, x)/n = (1/\log 2) \log((p+1)^2/(p(p+2)))$,

$p = 1, 2, 3, \cdots$. Let S denote the set of all $x \in X$ for which (A) holds. The set S appears as the analog (for continued fractions) of the set N_g of all g-adic normal numbers of the interval $\langle 0, 1)$, g integral, $g > 1$. The author [Rev. Roumaine Math. Pures Appl. **11** (1966), 53–56; MR **34** #1270] showed that the set N_g from a topological standpoint is only of the first category in $\langle 0, 1)$. The aim of the present study is a similar result for the set S.

E. Frank (Chicago, Ill.)

Citations: MR 13, 444f = Z02-18; MR 18, 274f = Z02-20; MR 28# 5037 = Z02-21; MR 28# 5038 = Z02-22; MR 34# 1270 = K15-85.

K50-39 (40# 4200)

Schweiger, Fritz

Ein Kuzminscher Satz über den Jacobischen Algorithmus.

J. Reine Angew. Math. **232** (1968), 35–40.

In früheren Arbeiten hat der Verfasser die zum Jacobischen Algorithmus gehörende Transformation δ:

$$\delta(x_1, \cdots, x_n) = (\{x_2 x_1^{-1}\}, \cdots, \{x_n x_1^{-1}\}, \{x_1^{-1}\})$$

(mit $\delta(0, \cdots, 0) = (0, \cdots, 0)$) untersucht und die Existenz eines δ-invarianten Maßes μ nachgewiesen, das bezüglich des Lebesgueschen Maßes absolut stetig ist. Obwohl noch keine explizite Form von μ bekannt ist, gelingt es dem Verfasser, für den Jacobischen Algorithmus ein Analogon zum Gauß-Kuzminschen Satz zu beweisen. Dabei wird angenommen, daß die zu μ gehörende Dichtefunktion ρ gleichmäßig beschränkte partielle Ableitungen $\partial\rho/\partial x_i$ hat. Der Beweis lehnt sich methodisch an den von Khintchine behandelten Fall des Kettenbruchalgorithmus an [vgl. A. Ja. Hinčin, *Continued fractions* (Russian), ONTI NKTP, Moscow, 1935; second edition, GITTL, 1949; MR **13**, 444; German translation, Teubner, Leipzig, 1956; MR **18**, 274; third Russian edition, Fizmatgiz, Moscow, 1961; English translations, Univ. of Chicago Press, Chicago, Ill., 1964; MR **28** #5037; Noordhoff, Groningen, 1963; MR **28** #5038].

B. Volkmann (Stuttgart)

Citations: MR 13, 444f = Z02-18; MR 18, 274f = Z02-20; MR 28# 5037 = Z02-21; MR 28# 5038 = Z02-22.

Referred to in K55-43.

K50-40 **(40# 5546)**

Jordan, J. H.

The magnitude of partial quotients.

Amer. Math. Monthly **76** (1969), 800–801.

Let ξ be an irrational number and let $\{p_n/q_n\}$ be the sequence of convergents of its continued fraction expansion. Then the following theorem is proved concerning the expected behavior of its convergents: Let $\{G(n)\}$ be a positive sequence such that $\sum_{n=1}^{\infty}(G(n))^{-1}$ converges, and let C be the set of ξ for which $\lim\sup_{n\to\infty} q_{n+1}/G(q_n)>0$. Then C has measure zero. (The condition that $G(n)$ be monotone increasing is also imposed, but this seems unnecessary.) The proof depends on the fact that a function of bounded variation has a finite derivative almost everywhere. *S. A. Burr* (Whippany, N.J.)

K50-41 **(40# 7219)**

Schweiger, Fritz

Metrische Theorie einer Klasse zahlentheoretischer Transformationen.

Acta Arith. **15** (1968), 1–18.

Sei B der n-dimensionale Einheitswürfel, $\mathfrak{F}$ der σ-Körper der Borelmengen in B, λ das Lebesguesche Maß. Gewisse maßtreue Abbildungen T des Wahrscheinlichkeitsraums $(B, \mathfrak{F}, \lambda)$ in sich werden als "zahlentheoretische" Transformationen bezeichnet, da die in der Zahlentheorie bekanntesten, durch Indexverschiebungen bei Algorithmen definierten Transformationen dieser Klasse angehören. Zu den Eigenschaften, die von T gefordert werden, gehören unter anderem auch Differenzierbarkeits- und Beschränktheitsbedingungen für die Umkehrabbildung.

Es wird gezeigt, daß jede zahlentheoretische Transformation T ergodisch und ein exakter Endomorphismus im Sinne von V. A. Rohlin [Izv. Akad. Nauk SSSR Ser. Mat. **25** (1961), 499–530; MR **26** #1423; translated in Amer. Math. Soc. Transl. (2) **39** (1964), 1–36; MR **28** #3909], also mischend von beliebigem Grad ist, und daß stets ein T-invariantes, zu λ äquivalentes Maß existiert. Als Folgerungen ergeben sich die bekannten Sätze über g-adische Normalität und "Kettenbruch-Normalität" fast aller reellen Zahlen.

Für zahlentheoretische Transformationen T, die zusätzlich einer Art Lipschitzbedingung genügen, wird ein "Kuzminscher Satz" (Satz 5) bewiesen. (Ein Fehler im Beweis ist inzwischen durch ein Corrigendum behoben worden [#7220 unten].)

Weitere Sätze machen Aussagen über die Mischungsgeschwindigkeit einer solchen Transformation, über die zugehörige Entropie $h(T)$ sowie über Verschärfungen des individuellen Ergodensatzes (mit Restglied) in Anlehnung an W. Philipp [Pacific J. Math. **20** (1967), 109–127; MR **34** #5755]. *B. Volkmann* (Stuttgart)

Citations: MR 26# 1423 = K50-13; MR 34# 5755 = J24-45.

Referred to in K55-42.

K50-42 **(40# 7220)**

Schweiger, Fritz

"Metrische Theorie einer Klasse zahlentheoretischer Transformationen": Corrigendum.

Acta Arith. **16** (1969/70), 217–219.

Auf das Corrigendum ist bereits im Referat zu der im Titel erwähnten Arbeit hingewiesen worden [#7219 oben]. *B. Volkmann* (Stuttgart)

Referred to in K55-42.

K50-43 **(40# 7221)**

Schweiger, Fritz

Metrische Theorie einer Klasse zahlentheoretischer Transformationen. II. Hausdorffdimensionen spezieller Punktmengen.

Österreich. Akad. Wiss. Math.-Natur. Kl. S.-B. II **177** (1969), 31–50.

Mit Hilfe der beim Jacobi-Algorithmus im R_n auftretenden Grundintervalle wird eine Hausdorff-Billingsley-Dimension d^* eingeführt. Es wird die Dimension d^* der Menge $E(R)$ aller $x=(x_1, \cdots, x_n)$ bestimmt, bei deren Jacobi-Darstellung alle auftretenden Vektoren $a_v(x)=(k_1^{(v)}, \cdots, k_n^{(v)})$ einer vorgegebenen, endlichen Menge angehören.

Ein zweiter Satz behandelt die Menge $F(\alpha)$ der x, bei denen alle $k_n^{(v)}$ einer Ungleichung $k_n^{(v)}\geqq\alpha$ genügen, und gibt eine untere sowie eine obere Schranke für $d^*(F(\alpha))$ an. Aus einem ähnlichen Satz für den Fall $k_n^{(v)}\leqq\beta$ ergibt sich die Folgerung, daß die Menge E der nicht Jacobi-normalen x die Gleichung $d^*(E)=1$ erfüllt.

Schließlich wird angedeutet, wie sich diese Ergebnisse auf den Fall "zahlentheoretischer Transformationen" [vgl. #7219 oben] übertragen lassen. *B. Volkmann* (Stuttgart)

K50-44 **(41# 3424)**

Hirst, K. E.

A problem in the fractional dimension theory of continued fractions.

Quart. J. Math. Oxford Ser. (2) **21** (1970), 29–35.

Let $S(a, b)$ be the set of real numbers ξ in whose continued fraction development the nth partial quotient a_n satisfies $a_n\geqq a^{b^n}$ for sufficiently large n $(a>1, b>1)$. The author proves that the Hausdorff dimension of $S(a, b)$ is $\geqq 1/(2b)$. It follows from the results of Good that this dimension is $\leqq\frac{1}{2}$ and the author conjectures that it is $<\frac{1}{2}$ [see I. J. Good, Proc. Cambridge Philos. Soc. **37** (1941), 199–228; MR **3**, 75]. *P. Erdős* (Waterloo, Ont.)

Citations: MR 3, 75b = K50-2.

K50-45 **(41# 5322)**

Kostryko, P.

An application of ergodic theory to the theory of continued fractions. **(Russian)**

Acta Fac. Rerum Natur. Univ. Comenian. Math. Publ. 19 (1968), 147–156.

Let $x=[c_1(x), c_2(x), c_3(x), \cdots]$ denote the continued fraction expansion of $x\in(0, 1)$ and put

$$R_k(x) = [c_k(x), c_{k+1}(x), \cdots].$$

Continuing earlier work of T. Šalát [Mat. Časopis Sloven. Akad. Vied **17** (1967), 121–130; MR **37** #1834] the author considers the almost everywhere behaviour of Cesàro means of certain sequences, e.g., $(R_k^{\alpha}(x))_{k=1}^{\infty}$, $(\log R_k^{\alpha}(x))_{k=1}^{\infty}$, $(c_k^{\alpha}(x)/c_{k+1}^{\beta}(x))_{k=1}^{\infty}$, $(|c_k^{\alpha}(x)-c_{k+1}^{\beta}(x)|)_{k=1}^{\infty}$ for real α and β. In most cases this relies on the discussion of summability of certain simple functions. Therefore, the investigation can be traced back to convergence problems of related series. *F. Schweiger* (Salzburg)

Citations: MR 37# 1834 = K50-30.

K50-46 **(41# 5324)**

Ruban, A. A.

Certain metric properties of the p-adic numbers. **(Russian)**

Sibirsk. Mat. Ž. **11** (1970), 222–227.

If $x=a_{-n}\cdots a_{-1}a_0.a_1a_2\cdots=\sum_{i=-n}^{\infty}a_ip^i$ is a p-adic number, the author writes $\{x\}=a_{-n}\cdots a_{-1}a_0.00\cdots$, $[x]=0.a_1a_2\cdots$ and puts X, Y for the totalities of $\{x\}$, $[x]$, respectively. Further, each p-adic x has a continued fraction expansion $x=1/(x_1(x)+1/(x_2(x)+\cdots))$, where each $x_i(x)\in X$. In analogy to the metric theory of numbers for the real domain, the author proves a number of analogous theorems for the p-adic domains. We state only two of his twelve theorems here.

Theorem 7: If $y\in Y$ and $\|y\|=p^k$, then for almost all $x\in X$ the frequency of repetition of y in the continued fraction expansion of x is p^{-2k}. Theorem 8: For almost all $x\in X$, $\lim_{n\to\infty}(\prod_{i=1}^n\|x_i(x)\|)^{1/n}=p^{p/(p-1)}$.

The proofs involve the same sort of techniques used in the reals.

{This article has appeared in English translation [Siberian Math. J. **11** (1970), 176–180].}

J. B. Roberts (Portland, Ore.)

K50-47 (41# 6784)

Monteferrante, Sandra; Szüsz, P.

On a generalization of a theorem of Borel.

Acta Arith. **16** (1969/70), 393–397.

Let $\alpha=[0;a_1,a_2,\cdots]$ be an irrational number expressed by its continued fraction. Let $A_n/B_n=[0;a_1,\cdots,a_n]$ and $D_n=B_n\alpha-A_n$. Then it is known that each τ with $D_1<\tau<1-D_1$ can be represented in the form

$$\tau=\sum_{k=0}^{\infty}C_{k+1}(\tau)D_k,$$

where $C_1(\tau)<a_1$, $0\leqq C_{k+1}(\tau)\leqq a_{k+1}$ and $C_{k+1}(\tau)=a_{k+1}$ implies $C_k(\tau)=0$ [V. T. Sós, Acta Math. Acad. Sci. Hungar. **8** (1957), 461–472; MR **20** #34]. The authors show that for almost all τ, $\sum_{k=0,C_{k+1}=0}^{n}1=\sum_{k=0}^{n}B_k|D_k-D_{k+1}|+o(n)$ and

$$\sum_{k=0,C_{k+1}=r}^{n}1=\sum_{k=0,a_{k+1}>r}^{n}B_k|D_k|+\sum_{k=0,a_{k+1}=r}^{n}B_{k-1}|D_k|+o(n)\quad(r>0).$$

A more precise result is obtained when α is a quadratic irrational, since then the limit of $1/n$ times the right hand side can easily be calculated. They plan to discuss further probabilistic statements on this subject in a future paper.

W. W. Adams (College Park, Md.)

Citations: MR 20# 34 = K10-19.

K50-48 (42# 7620)

Philipp, Walter

Some metrical theorems in number theory. II.

Duke Math. J. **37** (1970), 447–458.

In one of the sections of this paper the author applies his recent work on the law of the iterated logarithm, the central limit theorem and the remainder of the central limit theorem for mixing sequences of random variables to obtain improvements of various known metric theorems for continued fractions. Let $[a_1(x),a_2(x),\cdots]$ be the continued fraction expansion of $x\in(0,1]$. The partial quotients $\{a_n(x)\}$ form a strongly mixing sequence of random variables on the probability space $(I,\mathscr{L},P)$ where $I=(0,1]$, $\mathscr{L}$ are Lebesgue measurable subsets of I, and P is Gauss' measure given by

$$PE=(\log 2)^{-1}\int_E dx/(1+x)$$

for all $E\in\mathscr{L}$. A classical theorem of Bernstein states that the event (*) $[a_n\geqq\varphi(n)]$ occurs infinitely often for almost no [almost all] $x\in(0,1]$ if $\sum 1/\varphi(n)$ converges [diverges]. Let $A(N,x)$ be the number of occurrences of the event (*) above for $1\leqq n\leqq N$. Set $\Phi(N)=(\log 2)^{-1}\sum\log(1+1/\varphi(n))$. Suppose $\varphi(n)\to\infty$ and $\sum 1/\varphi(n)=\infty$. It is proved that

$$\limsup|A(N,x)-\Phi(N)|/\surd(2\Phi(N)\log\log\Phi(N))=1$$

almost surely, and

$$\mu\{x:(A(N,x)-(N))/\surd\Phi(N)<\alpha\}\to(1/\surd(2\pi))\int_{-\infty}^{\alpha}\exp(-t^2/2)\,dt,$$

where μ is a probability measure absolutely continuous with respect to Lebesgue measure. This represents a slight improvement over some results of W. Doeblin [Compositio Math. **7** (1940), 353–371; MR **2**, 107]. Other limit theorems for continued fractions are proved in this section.

In another section of the paper the author establishes an asymptotic estimate of the number of solutions of the system of inequalities $n_kx\in I_k$ (mod 1), where I_k is a sequence of subintervals of (0, 1], and n_k is a gap-sequence of integers with $n_{k+1}/n_k\geqq q>1$ for $k=1,2,\cdots$. This improves a result by W. J. LeVeque [J. Reine Angew. Math. **202** (1959), 215–220; MR **22** #12090].

Finally, a metric theorem concerning uniform distribution of all orders k of sequences on compact groups which are Hausdorff and have a countable base is stated, and a special case proved.

{Part I has appeared in Pacific J. Math. **20** (1967), 109–127 [MR **34** #5755].}

O. P. Stackelberg (Durham, N.C.)

Citations: MR 22# 12090 = J24-25; MR 34# 5755 = J24-45.

Referred to in K50-49.

K50-49 (43# 177)

Philipp, Walter

Errata: "Some metrical theorems in number theory. II".

Duke Math. J. **37** (1970), 788.

The author corrects errors in his paper [same J. **37** (1970), 447–458; MR **42** #7620].

O. P. Stackelberg (Durham, N.C.)

Citations: MR 42# 7620 = K50-48.

K50-50 (43# 1939)

Gordin, M. I.; Reznik, M. H.

The law of the iterated logarithm for the denominators of continued fractions. (Russian. English summary)

Vestnik Leningrad. Univ. **25** (1970), *no.* 13, 28–33.

Let $p_n(t)/q_n(t)$ be the nth convergent of the continued fraction expansion of t. The authors show that for almost all t (with respect to Lebesgue and/or Gauss measure), $\limsup\{|\log q_n(t)-na|/[b\surd(2n\log\log n)]\}=1$ for $n\to+\infty$, where $a=\pi^2/(12\log 2)$ and $b>0$ is a constant. The same result was obtained independently by W. Philipp and O. P. Stackelberg [Math. Ann. **181** (1969), 152–156; MR **39** #5503] (both papers were submitted for publication in 1968). The authors' Lemma 3 is actually a general iterated logarithm theorem for a class of dependent random variables; its conditions are satisfied by the coefficients of continued fractions. Among other things, Lemma 3 involves the assumption that a mixing property is satisfied.

J. Galambos (Philadelphia, Pa.)

Citations: MR 39# 5503 = K50-36.

K55 METRIC THEORY OF OTHER ALGORITHMS AND EXPANSIONS

See also Sections K15, K50.

See also reviews A68-1, A68-2, A68-4, A68-24, B12-16, K02-5, K15-57, K15-58, K40-33, K50-13, K50-25, K50-33, K50-35, K50-37, K50-41, Q15-8.

K55-1 (7, 13c)

Herzog, F., and Bissinger, B. H. A generalization of Borel's and F. Bernstein's theorems on continued fractions. Duke Math. J. **12**, 325–334 (1945).

The classical results of Borel [Rend. Circ. Mat. Palermo **27**, 247–271 (1909)] and F. Bernstein [Math. Ann. **71**, 417–439 (1911)] are generalized to the case of the generalized continued fractions introduced by Bissinger [Bull. Amer. Math. Soc. **50**, 868–876 (1944); these Rev. **6**, 150]. The proofs are elementary but rather intricate.

M. Kac (Ithaca, N. Y.).

Citations: MR 6, 150h = A68-1.

Referred to in K55-2.

K55-2 (7, 434e)

Bissinger, B. H., and Herzog, F. An extension of some previous results on generalized continued fractions. Duke Math. J. **12**, 655–662 (1945).

Generalizations and considerable improvements of some of the recent results by the authors [Duke Math. J. **12**, 325–334 (1945); these Rev. **7**, 13]. *M. Kac.*

Citations: MR 7, 13c = K55-1.

K55-3 (12, 679c)

Sanders, Johannes Marinus. Verdelingsproblemen bij gegeneraliseerde duale breuken. [Distribution Problems for Generalized Binary Fractions]. Thesis, Free University of Amsterdam, 1950. 88 pp.

If the interval [0, 1) is divided into two subintervals $\Delta_{1,0}$ and $\Delta_{1,1}$ and each of these is subdivided into two more, giving $\Delta_{2,0}$, $\Delta_{2,1}$, $\Delta_{2,2}$, $\Delta_{2,3}$, etc., the sequence π of subdivisions V_0, V_1, $\cdots$ so formed is called a distribution process. Let the points of V_n be $0=c_{n,0}<c_{n,1}<\cdots<c_{n,2^n}=1$, so that $\Delta_{n,r}=[c_{n,r}, c_{n,r+1})$, and let the length of $\Delta_{n,r}$ be $d_{n,r}$. Any number $\theta\varepsilon[0, 1)$ lies in just one interval Δ_{n,r_n} for each n, and if we put $c_n=0$ or 1 according as r_n is even or odd, we have a mapping $\theta\to\pi$, $c_1c_2\cdots$ which reduces to the ordinary binary expansion $\theta=0\cdot c_1c_2\cdots$ when $c_{n,r}=r\cdot 2^{-n}$ for $0\leqq r\leqq 2^n$, $n\geqq 0$. This thesis extends many of the known results concerning binary expansions to general processes. A process π is said to be normal when the above correspondence is one-to-one; it is shown that normality is equivalent to the relation $\lim_{n\to\infty}(\text{norm } V_n)=0$. Let $N(n, x)$, for $x\varepsilon[0, 1]$, be the number of points $c_{n,r}$ of V_n for which $0\leqq c_{n,r}\leqq x$. Then $\lim_{n\to\infty} N(n, x)\cdot 2^{-n-1}=\Phi(x)$ is called the distribution function of π, and it is shown that Φ is continuous and non-decreasing in [0, 1], with $\Phi(0)=0$, $\Phi(1)=1$. In going from V_n to V_{n+1}, the interval $\Delta_{n,r}$ is divided into the two intervals $\Delta_{n+1,2r}$ and $\Delta_{n+1,2r+1}$. Write the subratios $d_{n+1,2r}/d_{n+1,2r+1}$ as $\lambda_{n,r}/\mu_{n,r}$, where $\lambda_{n,r}+\mu_{n,r}=1$, and let $V(x, \delta)$ be the set of numbers $\lambda_{n,r}$ corresponding to those subintervals $\Delta_{n,r}$ which lie entirely in the interval $[x-\delta, x+\delta]$. Then $\underline{\lambda}(x)=\lim_{\delta\to 0}(\liminf V(x,\delta))$ and $\bar{\lambda}(x)=\lim_{\delta\to 0}(\limsup V(x,\delta))$ are the lower and upper first ratio functions, respectively. A normal process π is said to be regular if $\underline{\lambda}(x)=\bar{\lambda}(x)$ for $x\varepsilon[0, 1]$. This common first ratio function $\lambda(x)$ is continuous in [0, 1], and has the value $\frac{1}{2}$ at any point of differentiability of $\Phi(x)$. If $\lambda(x)$ is constant, π is called quasiuniform, and π is termed uniform if the subratios themselves are all equal.

Some of the principal results are these: (1) If P_n is the number of occurrences of a given combination of m digits in the first $N+m-1$ digits of the sequence π, $c_1c_2\cdots$ corresponding to θ, and p and q are the numbers of zeros and ones in this combination, then $\lim_{N\to\infty} P_N/N=(\lambda(\theta))^p(1-\lambda(\theta))^q$ for almost all θ if π is regular. This generalizes Borel's theorem on normal numbers. (2) Let $\theta_n\to\pi$, $c_{n+1}c_{n+2}\cdots$ for $n=1, 2, \cdots$. Then (a) the regularity of π is sufficient for the sequence $\{\theta_n\}$ to possess an asymptotic distribution function for almost all θ; (b) if π is regular, then this distribution function is independent of θ for almost all θ if and only if π is quasiuniform; (c) if π is regular, $\{\theta_n\}$ is uniformly distributed (mod 1) for almost all θ if and only if π is uniform. (3) In the case of a uniform process, estimates are given for the quantities $R_{\alpha,\beta}(N)$ and $R^*(N)$ associated with the sequence $\{\theta_n\}$ (for notation, see Koksma [Diophantische Approximationen, Springer, Berlin, 1936]). In particular, it is shown that for fixed α, β, $R_{\alpha,\beta}(N)=O(N^{\frac{1}{2}}\log^{\frac{1}{2}(3+\epsilon)}N)$ for every $\epsilon>0$ and almost all θ, and the same bound obtains for $R^*(N)$ for almost all θ. (4) Two questions asked by C. J. Everett [Bull. Amer. Math. Soc. **52**, 861–869 (1946); these Rev. **8**, 259], concerning real number expansions somewhat similar to those of the present paper, are answered.

W. J. LeVeque (Manchester).

Citations: MR 8, 259c = A68-2.

Referred to in A68-7.

K55-4 (18, 645a)

Berg, Lothar. Allgemeine Kriterien zur Massbestimmung linearer Punktmengen. Math. Nachr. **14** (1955), 263–285 (1956).

As is well-known there exists an extensive metric theory of diophantic approximations [see J. F. Koksma, Diophantische Approximationen, Springer, Berlin, 1936, pp. 43–49, 116–118]. This theory gives information about the measure of sets of real numbers represented by certain classes of continued fractions or by classes of "decimal" representations.

L. Holzer [Akad. Wiss. Wien. Math.-Nat. Kl. S.-B. IIa. **137** (1928), 421–453] extended this theory to the representation of real numbers by Lüroth series. One of his results is the analogue of Bernstein's theorem, that any positive integer k appears an infinity of times in the sequence of the quotients in the continued fraction algorithm of any real number, if we exclude a set of real numbers of measure zero.

In this thesis Berg generalizes Holzer's theory to a class of expansions which include the series of Lüroth, Engel and Sylvester [for these series cf. O. Perron, Irrationalzahlen, 2. Aufl., de Gruyter, Berlin, 1939, pp. 116–127], but not the continued fraction algorithm. His results are in general too complicated to state here, but he shows for example that the analogue of Bernstein's theorem does not hold for the Engel and Sylvester series.

The Engel series are treated in more detail.

J. Popken (Amsterdam).

Referred to in K55-5, K55-17.

K55-5 (20# 94)

Berg, Lothar. Massbestimmung linearer Punktmengen. Math. Nachr. **16** (1957), 195–205.

In his thesis [Math. Nachr. **14** (1955), 263–285; MR **18** 645] the author studied an extensive class of expansions for real numbers, which includes the well-known expansions of Lüroth, Sylvester and Engel. In particular he asked himself whether in certain special cases an analogue of F. Bernstein's theorem for the continued fraction algorithm holds (any positive integer k appears an infinity of times in the sequence of quotients in the continued fraction algorithm of almost any real number).

In this paper the author investigates three special cases of the general theory and for these he solves the above problem. One of his results reads: "Let $x=1/a_1+1/a_2+1/a_3+\cdots+1/a_n+\cdots$ denote the Sylvester expansion of x $(0\leqq x\leqq 1)$, so that $a_n\geqq a_n{}^*$, where $a_n{}^*=a_{n-1}(a_{n-1}-1)+1$. Let M_k for any integer $k\geqq 0$ denote the set of numbers x such that $a_n=a_n{}^*+k$ holds at most for a finite number of values for n. Then M_k has measure 1 for every k." Hence in this case the analogue of Bernstein's theorem does not hold. *J. Popken* (Amsterdam)

Citations: MR 18, 645a = K55-4.

Referred to in K55-17.

K55-6 **(20# 3843)**

Rényi, A. Representations for real numbers and their ergodic properties. Acta Math. Acad. Sci. Hungar. **8** (1957), 477–493.

An "f-expansion" of a real number x is the sequence of integers $\{\varepsilon_n\}$ defined as follows: $\varepsilon_0(x)=[x]$, $r_0(x)=(x)$; $\varepsilon_{n+1}(x)=[f^{-1}(r_n(x))]$, $r_{n+1}(x)=(f^{-1}(r_n(x))$. Here $[z]$ and (z) denote the integral part and fractional part of z, respectively. A real number x is said to have an f-expansion for a particular function f if either $r_n(x)=0$ for some n, so that the f-expansion terminates with ε_n, or the sequence $\{F_n(x, 0)\}$ converges to x. $F_n(x, y)$ is defined as follows: $F_0(x, y)=y$, $F_n(x, y)=F_{n-1}(x, \varepsilon_{n-1}(x)+f(\varepsilon_n(x)+y))$.

The problem of determining for what functions f all real numbers x have an f-expansion had previously been considered for monotone decreasing f by Bissinger [Bull. Amer. Math. Soc. **50** (1944), 868–876; MR **6**, 150] and for monotone increasing f by C. J. Everett [ibid. **52** (1946), 861–869; MR **8**, 259]. The results of the two authors are proved here with somewhat less restrictive assumptions on the function f. The main result of the article is concerned with measure-theoretic properties of f-expansions. It can be summarized as follows: Under certain conditions on f (somewhat more restrictive than those required for the existence of an f-expansion), and for a function g which is L-integrable on (0, 1) and otherwise arbitrary,

$$\lim_{n\to\infty} \frac{1}{n} \sum_{k=0}^{n-1} g(r_k(x))$$

exists for almost all x and is a finite expression depending only on f and g, but not on x. The limit can be written as $\int_0^1 g(x)h(x)dx$, where h depends only on f and is such that $v(E)=\int_E h(x)dx$ is invariant under the transformation $(f^{-1}(x))$. By suitable choice of g and f, known theorems about regular continued fraction expansions, decimal expansions, as well as similar theorems for more general f-expansions, can be obtained.

W. J. Thron (Boulder, Colo.)

Citations: MR 6, 150h = A68-1; MR 8, 259c = A68-2.

Referred to in K50-37, K55-18, K55-21, K55-25, K55-42, K55-44, K55-46.

K55-7 **(20# 6403)**

Turán, Pál. On the distribution of "digits" in Cantor-systems. Mat. Lapok **7** (1956), 71–76. (Hungarian. Russian and English summaries)

The author proves, by using the Cantor-Lebesgue theorem, the following theorem. Let $0\leq x\leq 1$,

$$x=\sum_{r=1}^{\infty} \frac{c_r}{q_1q_2\cdots q_r},$$

$q_r\geq 2$, $0\leq c_r\leq q_r-1$, with c_r, q_1, q_2, $\cdots$ integers. Assume $\sum_{r=1}^{\infty} 1/q_r<\infty$. Then for almost all x

$$\sum_{r=1}^{\infty} \frac{\min(c_r, (q_r-c_r), |c_r-q_r/2|)}{q_r}$$

diverges. The author remarks that one can obtain stronger results by using probability methods.

P. Erdös (Birmingham)

Referred to in K55-9, K55-24.

K55-8 **(20# 6404)**

Rényi, Alfréd. On the distribution of the digits in Cantor's series. Mat. Lapok **7** (1956), 77–100. (Hungarian. Russian and English summaries)

In this paper the author proves that the well-known theorem of Borel on the frequency of the digits of the q-ary expansions can be generalised for Cantor series if $\sum 1/q_n=\infty$. In a previous paper [Acta Math. Acad. Sci. Hungar. **6** (1955), 285–335; MR **18**, 339] the author proved a slightly weaker theorem, but the present proof is simpler. The case $\sum_{n=1}^{\infty} 1/q_n<\infty$ is also considered. [Cf. #6406, #6407 below].

P. Erdös (Birmingham)

Citations: MR 20# 6406 = K55-10; MR 20# 6407 = K55-11.

Referred to in K55-11, K55-24, K55-31, K55-41.

K55-9 **(20# 6405)**

Marczewski, E. Remarks on the Cantor-expansions of real numbers. Mat. Lapok **7** (1956), 212–213. (Hungarian)

The author observes that the result of Turán [#6403] follows from the law of large numbers.

P. Erdös (Birmingham)

Citations: MR 20# 6403 = K55-7.

K55-10 **(20# 6406)**

Szüsz, Péter. Bemerkungen zur Verteilung der Ziffern in der Cantorschen Reihe reeller Zahlen. Mat. Lapok **8** (1957), 68–78. (Hungarian. German and Russian summaries)

Hungarian version of #6407 below.

Referred to in K55-8.

K55-11 **(20# 6407)**

Szüsz, P. Bemerkung über die Verteilung der Ziffern in der Cantorschen Reihe reeller Zahlen. Acta Math. Acad. Sci. Hungar. **8** (1957), 163–168.

Let $q_1, q_2, \cdots$ ($q_n\geq 2$) be an infinite sequence of integers. It is well known that for every x $(0<x<1)$ we have

$$x=\sum_{k=1}^{\infty} \frac{\varepsilon_k(x)}{q_1q_2\cdots q_k},\quad 0\leq\varepsilon_k(x)\leq q_k-1\quad (\varepsilon_k(x) \text{ an integer}).$$

Denote by $N_n(x, r)$ the number of solutions of $\varepsilon_k(x)=r$, $1\leq k\leq n$. Rényi proved [#6404 above] that if $\sum_{k=1}^{\infty} 1/q_k=\infty$ then, if $q_k>\max(r, s)$ for all $k>k_0$, we have for almost all x

$$(*)\qquad \lim_{n\to\infty} N_n(x, r)/N_n(x, s)=1.$$

But he also proved that to every x $(0<x<1)$ there exists a sequence q_k satisfying $\sum_{k=1}^{\infty} 1/q_k=\infty$, $q_n\to\infty$, which does not satisfy (*).

The author sharpens and deepens this last result of Rényi. He proves that for every x $(0<x<1)$ and every sequence $c_1, c_2, \cdots$ ($c_k\geq 2$) of integers there exists a sequence of integers $q_1, q_2, \cdots$ satisfying $0\leq q_k-c_k<16$ and $1\leq\varepsilon_k(x)\leq q_k-1$ (i.e., $\varepsilon_k(x)$ is never 0).

P. Erdös (Birmingham)

Citations: MR 20# 6404 = K55-8.

Referred to in K55-8.

K55-12 **(21# 1288)**

Erdös, P.; Rényi, A.; and Szüsz, P. On Engel's and Sylvester's series. Ann. Univ. Sci. Budapest. Eötvös. Sect. Math. **1** (1958), 7–32.

Every number x with $0<x<1$ can be expanded into an Engel's series

$$x=\frac{1}{q_1}+\frac{1}{q_1q_2}+\cdots+\frac{1}{q_1q_2\cdots q_n}+\cdots,$$

where the $q_n=q_n(x)$ are integers such that $2\leq q_n\leq q_{n+1}$. The authors study the metrical properties of the sequence $q_n(x)$. They consider the interval $0<x<1$ as the space of elementary events and interpret the Lebesgue measure of a measurable subset of (0, 1) as its probability. It is shown that the random variables $x_1=\log q_1$, $x_n=\log(q_n/q_{n-1})$ are "almost independent" and "almost identically distributed". From this the authors derive for $\log q_n=x_1+x_2+\cdots+x_n$ first the central limit theorem (i.e., that the distribution of $n^{-1/2}(\log q_n-n)$ tends for $n\to\infty$ to the normal distribution). Then they prove the strong law of large numbers (i.e., that for almost all x we have $\lim_{n\to\infty} q_n^{1/n} = e$) and, finally, even the law of the iterated logarithm.

The second result has been announced earlier without proof by É. Borel [C. R. Acad. Sci. Paris **225** (1947), 773; MR **9**, 292]; the first and third results are due to P. Lévy [ibid. **225** (1947), 918–919; MR **9**, 292]. However in this paper the authors use a different approach and they give detailed proofs.

They obtain new results by applying their methods to Sylvester's series

$$x = \frac{1}{Q_1} + \frac{1}{Q_2} + \cdots + \frac{1}{Q_n} + \cdots \quad (0 < x < 1),$$

where the Q_n are integers such that $Q_{n+1} \geqq Q_n(Q_n - 1) + 1$. They find that the central limit theorem holds for $\log(Q_n/Q_1Q_2\cdots Q_{n-1})$ and further that $\lim_{n\to\infty} 2^{-n}\log Q_n$ exists and is finite and positive for almost all x in (0, 1).

The paper ends with a discussion of some number-theoretic questions concerning these series; several unsolved problems are mentioned.

J. Popken (Berkeley, Calif.)

Citations: MR 9, 292c = A68-4; MR 9, 292d = A68-5.
Referred to in K55-15, K55-23, K55-38.

K55-13 (21# 1289)

Békéssy, A. Bemerkungen zur Engelschen Darstellung reeller Zahlen. Ann. Univ. Sci. Budapest. Eötvös. Sect. Math. **1** (1958), 143–151.

The author extends some results obtained in the paper reviewed above. Let x be any number in the interval $0 < x < 1$ and let

$$x = \frac{1}{q_1} + \frac{1}{q_1q_2} + \cdots + \frac{1}{q_1q_2\cdots q_n} + \cdots$$

denote its expansion into an Engel's series. Let $p_n(k)$ denote the probability $\mathrm{P}\{q_n(x) = k\}$ and let $p_n(k|j)$ stand for $\mathrm{P}\{q_{n+\nu} = k | q_\nu = j\}$ where $\mathrm{P}\{A|B\}$ denotes the conditional probability of A with respect to the condition B. It follows easily $p_n(k) = p_n(k|2)$. The author derives several asymptotic fomulae for $p_n(k|j)$ and for $W_n(k|j) = \sum_{l=k}^{\infty} p_n(l|j)$ too complicated to reproduce here. From one of his results it follows, e.g., that

$$p_n(k) \sim \Gamma\left(2 - \frac{n-1}{\log k}\right) \frac{(\log k)^{n-1}}{k^2(n-1)!}$$

for $(n-1)/2 \log k \leqq \theta < 1$. Another result (formula (11C)) gives an upper bound for the absolute value of the remainder in the central limit theorem for $\log q_n$ (see the preceding review).

J. Popken (Berkeley, Calif.)

K55-14 (21# 2637)

Rényi, A. On Cantor's products. Colloq. Math. **6** (1958), 135–139.

As a Cantor product expansion of a real number $x > 1$ the author defines the infinite product $\prod (1 + 1/q_v)$, where the q_v are determined by the rule: Let q_1 be the least positive integer for which $1 + 1/q_1 \leqq x$; if $q_1, \cdots, q_{n-1}$ have been chosen, let q_n be the least positive integer such that $\prod_{v=1}^{n} (1 + 1/q_v) \leqq x$. Clearly there is a flaw in this definition, since according to it certain rational numbers (and as a matter of fact all rational numbers) have finite expansions. This leads to a certain amount of confusion in the early part of the paper. A more standard definition is one in which strict inequality is assumed in the relations defining the q_v. Then every number $x > 1$ has an infinite expansion, and a rational number is characterized by the fact that from some n_0 on $q_{n+1} = q_n^2$. The Cantor product expansion of a number $x > 1$ is unique, and thus, for every n, $q_n = q_n(x)$. It is also known that $q_{n+1}(x) \geqq q_n^2(x)$.

For other expansions of real numbers, such as decimal expansions, regular continued fraction expansions, etc., ergodic properties of the coefficients of the expansion have been extensively investigated. This is the first time that ergodic results are obtained for Cantor products. Typical of the results stated (proofs are only sketched) in this article are the following: For almost all $1 < x \leqq 2$

$$\lim_{n\to\infty} (q_{n+1}(x))^{1/2^n} = l(x),$$

where $2 < l(x) < \infty$, and

$$\lim_{n\to\infty} [q_{n+1}(x)/q_1(x)q_2(x)\cdots q_n(x)]^{1/n} = e.$$

W. J. Thron (Boulder, Colo.)

K55-15 (21# 6356)

Erdős, P.; and Rényi, A. Some further statistical properties of the digits in Cantor's series. Acta Math. Acad. Sci. Hungar. **10** (1959), 21–29. (Russian summary, unbound insert)

Let $q_1, q_2, \cdots$ be a fixed sequence of positive integers, with $q_k > 2$ for all k. In an earlier paper [Ann. Univ. Sci. Budapest Sect. Math. **1** (1958), 7–32; MR **21** #1288] the authors, together with P. Szüsz, considered the probabilistic behavior of the "digits" $\varepsilon_n(x)$ in the Cantor expansion

$$x = \sum_{n=1}^{\infty} \frac{\varepsilon_n(x)}{q_1q_2\cdots q_n}$$

where, for each n, $\varepsilon_n(x)$ is an integer with $0 \leqq \varepsilon_n(x) < q_n$. In particular, they generalized a classical theorem of Borel by showing that if

$$f_n(k, x) = \sum_{\substack{1\leqq j\leqq n\\ \varepsilon_j(x)=k}} 1, \quad Q_n(k) = \sum_{\substack{1\leqq j\leqq n\\ q_j>k}} \frac{1}{q_j},$$

then, for all integers $k \geqq 0$ for which $\lim_{n\to\infty} Q_n(k) = \infty$,

$$\lim_{n\to\infty} \frac{f_n(k, x)}{Q_n(k)} = 1 \quad \text{for almost all } x.$$

In the present paper they consider the behavior of $M_n(x) = \max_k f_n(k, x)$, the most frequent number among the first n digits. Put $Q_n = Q_n(1)$. (I) If $\lim_n Q_n/\log n = \infty$, then for a.a. x, $\lim_{n\to\infty} M_n(x)/Q_n = 1$. (II) If $0 < c_2 \leqq q_n/n \leqq c_3$ for $n = 1, 2, \cdots$, and $\lim Q_n/\log n = \alpha > 0$, then, for a.a. x, $\lim_{n\to\infty} M_n(x)/Q_n = y(\alpha)$, where $y(\alpha)$ is the real solution of the equation $y \log y = \alpha^{-1}$. (III) If $q_n/n \to \infty$ but also $Q_n \to \infty$ as $n \to \infty$, then, for a.a. x, $\lim_{n\to\infty} M_n(x)/Q_n = \infty$.

W. J. LeVeque (Ann Arbor, Mich.)

Citations: MR 21# 1288 = K55-12.

K55-16 (22# 702)

Gel'fond, A. O. A common property of number systems. Izv. Akad. Nauk SSSR. Ser. Mat. **23** (1959), 809–814. (Russian)

If $\theta > 1$ is a real number, then any α, $0 < \alpha \leqq 1$, can be represented in a unique way by the series $\alpha = \sum_1^\infty \lambda_n/\theta^n = \sum_1^n \lambda_k/\theta^k + x_{n+1}/\theta^n$, $0 \leqq x_n < 1$, where $x_1 = \alpha, \cdots, x_n = x_n(\alpha) = \{\theta x_{n-1}\}$ and $\lambda_1 = [\alpha\theta], \cdots, \lambda_n = [\theta x_n]$ ($\{x\}$ denotes the proper fractional part of x and $[x]$ the greatest integer $\leqq x$). Further, define: (1) $\psi(x) = 1$ if $0 \leqq x \leqq 1$, $\psi(x) = 0$ if $x < 0$ or $x > 1$; (2) $1 = \sum_1^n \lambda_k/\theta^k + t_{n+1}/\theta^n$, $t_1 = 1$; (3) $\tau = \sum_1^\infty t_k/\theta^{k-1}$; (4) $\sigma_0(t) = \tau^{-1}\sum_1^\infty \psi(1 - t_k + t)/\theta^{k-1}$, $0 < t \leqq 1$. The author establishes the law of distribution of the remainders $x_n(\alpha)$ proving that, if θ is not an integer, then for almost all α

there holds

$$\lim_{N\to\infty} N^{-1} \sum_{k=1}^{N} \psi(1-t+x_k) = \int_0^t \sigma_0(x)dx$$
$$= \tau^{-1} \sum_{k=1}^{\infty} \min(t, t_k)/\theta^{k-1}.$$

[The case where θ is an integer was considered by G. H. Hardy and J. E. Littlewood, Acta Math. **37** (1914), 155–191.] *J. W. Andrushkiw* (Newark, N.J.)

Referred to in K55-18, K55-21, K55-46.

K55-17 (22# 1559)

Berg, Lothar. Massbestimmung linearer Punktmengen. II. Math. Nachr. **17**, 211–218 (1959).

This paper is a continuation of two earlier publications of the author concerning the metrical properties of a certain class of expansions of real numbers [Math. Nachr. **14** (1955), 263–285; **16** (1957), 195–205; MR **18**, 645; **20** #94]. The author first generalizes his "Homogenitätssatz" (Satz 1, p. 270, in the first paper). Then he studies more closely the measure of the set M already introduced in the two former papers (cf. Satz 3, p. 276, in the first paper, and Satz 1, p. 196, in the second). If this measure is zero, then an analogue of the Bernstein-Knopp theorem from the theory of continued fractions holds for the expansions considered. In all the former special cases treated by the author, this set turned out to be homogeneous, hence has measure zero or one. Now it is shown, among other results, that this need not be the case in general. *J. Popken* (Amsterdam)

Citations: MR 18, 645a = K55-4; MR 20# 94 = K55-5.

K55-18 (22# 12092)

Cigler, Johann. Ziffernverteilung in ϑ-adischen Brüchen. Math. Z. **75** (1960/61), 8–13.

Let the transformation $T(x)$ of the interval $0 \leqq x \leqq 1$ onto itself be defined by $T(x) = ((\vartheta x))$, where $\vartheta > 1$ is an arbitrary fixed number and $((y))$ denotes the fractional part of the non-negative number y. It has been proved by the reviewer [Acta Math. Acad. Sci. Hungar. **8** (1957), 477–493; MR **20** #3843] that there exists a measure μ on the Borel sets of the interval $(0, 1)$ which is invariant with respect to T, equivalent with Lebesgue measure, and ergodic. The same result has been independently obtained by A. O. Gel'fond [Izv. Akad. Nauk SSSR. Ser. Mat. **23** (1959), 809–814; MR **22** #702], who gave also an explicit formula for μ: $\mu(B) = \int_B g(t)dt$, where $g(t) = c\cdot\sum_{t<t_k} \vartheta^{-k}$; here $t_k = T^k(1)$ for $k \geqq 1$, $t_0 = 1$, and the constant c is determined by the condition $\int_0^1 g(t)dt = 1$. Starting from this explicit formula for the measure μ, the author gives a new and simple proof of the invariance of the measure μ with respect to T. A proof of the ergodicity of T is also given. Every real number α with $0 \leqq \alpha \leqq 1$ has a representation in the form (*) $\alpha = \sum_{k=1}^{\infty} \lambda_k(\alpha)\cdot\vartheta^{-k}$, where $\lambda_k(\alpha)$ is a non-negative integer defined by $\lambda_k(\alpha) = [\vartheta T^k(\alpha)]$; here $[y]$ denotes the integral part of the number y. The author proves that if ϑ is an algebraic number such that the representation of the number 1 in the form (*) is finite, that is, $1 = \sum_{k=1}^{n+1} a_k\cdot\vartheta^{-k}$, and if α is supposed to be a random variable distributed in $(0, 1)$ according to the measure μ (in which case the $\lambda_k(\alpha)$ are random variables too), then the n-tuples of consecutive "digits" $(\lambda_{\nu+1}(\alpha), \lambda_{\nu+2}(\alpha), \cdots, \lambda_{\nu+n}(\alpha))$, $\nu = 0, 1, 2, \cdots$, form a homogeneous Markov chain. This may be considered as a generalization of the well-known fact that if ϑ is an integer (in which case $n = 0$ and μ is Lebesgue measure) then the random variables $\lambda_k(\alpha)$ are independent and have the same distribution; or of the fact that for $\vartheta = \frac{1}{2}(1+\sqrt{5})$ (in which case $n = 1$) the digits $\lambda_k(\alpha)$ form a Markov chain. *A. Rényi* (Budapest)

Citations: MR 20# 3843 = K55-6; MR 22# 702 = K55-16.

Referred to in K55-21.

K55-19 (22# 12579)

Rényi, A.; Révész, P. On mixing sequences of random variables. Acta Math. Acad. Sci. Hungar. **9** (1958), 389–393.

Using results obtained by Rényi [same Acta **9** (1958), 215–228; MR **20** #4623] the authors extend and apply these results to prove the following theorem: A sequence of random variables $\{\zeta_n\}$ defined on (Ω, A, P) has the mixing property [cf. above reference] if and only if for every $k = 0, 1, 2, \cdots$, there exists a set E_k such that $P(\zeta_k \in E_k) = 1$ and such that if $y \in E_k$ then the distributions $P\{\rho_n < x | \rho_k = y\}$ converge weakly as $n \to \infty$ to a distribution function $F(x)$ independent of y.

Several examples of Markov processes are considered and are shown to have the mixing property. Among these examples is the following: Let the sequence of random variables be $\{q_n(t)\}$, where $q_n(t)$ is an integer not less than 2, and let t be expanded in the Engel series

$$t = \frac{1}{q_1} + \frac{1}{q_1q_2} + \cdots + \frac{1}{q_1q_2\cdots q_n} + \cdots;$$

then it follows from the above result and the paper quoted above that

$$\lim_{n\to\infty} Q\left\{t \left| \frac{\log q_n - n}{\sqrt{n}} < x\right.\right\} = \frac{1}{2\pi}\int_{-\infty}^{x} e^{-u^2/2}\,du,$$

where Q is any probability measure in $(0, 1)$ which is absolutely continuous with respect to the Lebesgue measure. *Y. N. Dowker* (London)

K55-20 (23# A3710)

Erdös, P.; Rényi, A.

On Cantor's series with convergent $\sum 1/q_n$.

Ann. Univ. Sci. Budapest. Eötvös. Sect. Math. **2** (1959), 93–109.

For $0 \leqq x < 1$, let $x = \sum_1^{\infty} \varepsilon_n(x)(q_1 \cdots q_n)^{-1}$, $\varepsilon_n = 0, \cdots, q_n - 1$. Here, $q_n \geqq 2$ are given integers satisfying $\sum q_n^{-1} < \infty$. This paper is concerned with the asymptotic behavior of the sequence $\{\varepsilon_n\}$ of independent random variables. The underlying probability space is $[0, 1]$ with Lebesgue measure. (I) Let ν_k denote the number of ε_n equal to k $(k = 0, 1, \cdots)$. Then $\nu_k < \infty$ with probability 1. The authors determine the exact distribution of lim sup ν_k and further the "best" function $f(k)$ such that, for almost all x, $\nu_k \leqq f(k)$ for $k \geqq k_0(x)$. (II) For almost all x, the set $S(x)$ of integers k equal to some $\varepsilon_n(x)$ has density 0. Let $\{k_j\}$ be a fixed sequence of positive integers having K_n elements $k_j < q_n$. Then for almost all x only finitely many k_j are in $S(x)$ if and only if $\sum K_n/q_n < \infty$. One of the tools used is an extension of the divergence part of the Borel-Cantelli lemma: let $A_1, A_2, \cdots$ be events such that $\lambda_n = \sum_1^n P(A_k) \to \infty$ and $\liminf \lambda_n^{-2} \sum_1^n \sum_1^n P(A_iA_k) \leqq 1$. Then, with probability 1, A_n happens for infinitely many n. *J. H. B. Kemperman* (Rochester, N.Y.)

K55-21 (26# 288)

Parry, W.

On the β-expansions of real numbers. (Russian summary)

Acta Math. Acad. Sci. Hungar. **11** (1960), 401–416.

Let β be an arbitrary positive number greater than 1

which is not an integer. Then every x $(0 \leq x < 1)$ has a "β-expansion" $x = \sum_{k=1}^{\infty} \varepsilon_k/\beta_k$, where the "digits" $\varepsilon_k = \varepsilon_k(x)$ may take on the values 0, 1, $\cdots$, $[\beta]$ ($[\beta]$ stands for the integral part of β), and are obtained by the following algorithm: $\varepsilon_1(x) = [\beta x]$, $\varepsilon_2(x) = [\beta(\beta x)]$, etc. (Here (βx) denotes the fractional part of βx.) It has been shown also [see the reviewer, same Acta **8** (1957), 477–493; MR **20** #3843] that $Tx = (\beta x)$ is an ergodic transformation and for any $g = g(x) \in L(0, 1)$ and for almost all x that $\lim_{n\to+\infty} (1/n) \sum_{k=0}^{n-1} g(T^k x) = M(g)$, where $M(g)$ is a constant depending only on $g(x)$, further, that there exists a unique normalised measure ν_β equivalent to the Lebesgue measure and invariant under T, and that $M(g) = \int_0^1 g\,d\nu_\beta$.

In the present paper the measure ν_β is effectively determined. First it is shown that if $\nu_\beta(E) = \int_E h_\beta(x)dx$, then $h_\beta(x)$ (the Radon-Nikodym derivative of the measure ν_β with respect to the Lebesgue measure) satisfies the functional equation $\beta \cdot h_\beta(x) = \sum_{m=0}^{[\beta-x]} h_\beta((x+m)/\beta)$. It follows that $h_\beta(x) = \sum_{x<T^n1} 1/\beta^n$. Further conditions are given which have to be satisfied by a sequence $\{\varepsilon_k\}$ of integers $(0 \leq \varepsilon_k \leq [\beta];\ k = 1, 2, \cdots)$ in order that the sequence $\{\varepsilon_k\}$ should be the sequence of digits of the β-expansion of a real number x. Finally, those numbers β are investigated in the β-expansion of which the digits form a periodic sequence. Such numbers β are called β-numbers. It is shown that β-numbers are algebraic, and all their conjugates lie in the circle $|z| < 2$. Those β-numbers whose β-expansion is finite are called simple β-numbers. It is shown that the set of simple β-numbers is everywhere dense in $(1, +\infty)$.

It should be added that the measure ν_β has been explicitly determined about the same time by A. O. Gel'fond [Izv. Akad. Nauk SSSR Ser. Mat. **23** (1959), 809–814; MR **22** #702]. The work of the two authors was clearly independent. Recently J. Cigler [Math. Z. **75** (1960/61), 8–13; MR **22** #12092] has given a new and simple proof of the same theorem, and has proved some other related results. *A. Rényi* (Budapest)

Citations: MR 20# 3843 = K55-6; MR 22# 702 = K55-16; MR 22# 12092 = K55-18.
Referred to in K55-46.

K55-22 (26# 4981)

Šalát, Tibor

Cantorsche Entwicklungen der reellen Zahlen und das Hausdorffsche Mass. (Russian summary)
Magyar Tud. Akad. Mat. Kutató Int. Közl. **6** (1961), 15–41.

Let $0 < x < 1$, $q_1, \cdots$ be an infinite sequence of integers. Let further

$$x = \sum_{k=1}^{\infty} \frac{\varepsilon_k(x)}{q_1 \cdot q_2 \cdot \cdots \cdot q_k},\quad 0 \leq \varepsilon_k(x) < q_k,$$

be the development of x into Cantor-series. The author proves (among others) that if $q_k = k+1$, then the Hausdorff dimension of the set in x for which $\varepsilon_k(x) \neq r_i$, $1 \leq k < \infty$, $1 \leq i \leq n$, is always 1.

Let again $q_k = k+1$. Denote by $N_n(r, x)$ the number of k's for which $\varepsilon_k(x) = r$, $1 \leq k \leq n$. The Hausdorff dimension of the set in x with $\limsup_{n=\infty} N_n(r, x)/n \geq d$ equals $1-d$.

Several other related questions are investigated.

P. Erdős (London)

Referred to in K55-33, K55-35.

K55-23 (27# 126)

Rényi, A.

A new approach to the theory of Engel's series.
Ann. Univ. Sci. Budapest. Eötvös Sect. Math. **5** (1962), 25–32.

Every number x in the interval $(0, x]$ can be represented by an Engel's series

$$x = 1/q_1 + 1/q_1q_2 + \cdots + 1/q_1q_2\cdots q_n + \cdots,$$

where the $q_n = q_n(x)$ form a nondecreasing sequence of integers ≥ 2. In a previous paper, jointly written by the author, P. Erdős and P. Szüsz [same Ann. **1** (1958), 7–32; MR **21** #1288], detailed proofs were given for certain probabilistic theorems concerning the sequence $\{q_n\}$, due originally to É. Borel and P. Lévy. These proofs, however, were obtained by a rather difficult technique of handling sums of "almost independent" random variables.

The new approach presented in this paper is much more satisfactory. As basic random variables the author introduces the $\varepsilon_k = \varepsilon_k(x)$ $(k = 2, 3, \cdots)$, defined as the number of times that the integer k occurs in the nondecreasing sequence $\{q_n\}$. It is shown that the ε_k are independent variables and that the distribution of ε_k is given by $P(\varepsilon_k = r) = (k-1)/k^{r+1}$ $(r = 0, 1, \cdots)$. Clearly $\mu_N = \sum_{k=2}^{N} \varepsilon_k$ denotes the number of terms in the sequence $\{q_n\}$ which are $\leq N$. The results of the previous paper, mentioned before, are now obtained as direct consequences of well-known limit theorems of probability theory. For example, Liapounoff's form of the central limit law leads here to

$$\lim_{N\to\infty} P((\log N)^{-1/2}(\mu_N - \log N) < y) = \Phi(y),$$

where $\Phi(y) = (2\pi)^{-1/2} \int_{-\infty}^{y} \exp(-u^2/2)du$, which is equivalent to $\lim_{n\to\infty} P(n^{-1/2}(\log q_n - n) < y) = \Phi(y)$.

The author further proves several new results for Engel's series. For example, he shows that for almost all x the sequence $\{q_n\}$ is strictly increasing for $n \geq n_0(x)$. Finally it is shown that these methods carry over to a certain modification of Engel's representation of real numbers.

A similar approach to the corresponding theory of Sylvester's series is still lacking.

J. Popken (Berkeley, Calif.)

Citations: MR 21# 1288 = K55-12.

K55-24 (28# 1419)

Šalát, Tibor

On the theory of Cantor expansions of real numbers. (Russian. English summary)
Mat.-Fyz. Časopis Sloven. Akad. Vied **12** (1962), 85–96.

Author's summary: "The paper is closely related to the papers of J. D. Hill [Bull. Amer. Math. Soc. **48** (1942), 103–108; MR **3**, 147], P. Turán [Mat. Lapok **7** (1956), 71–76; MR **20** #6403], and A. Rényi [ibid. **7** (1956), 77–100; MR **20** #6404], and is a completion and generalization of them. The main results are contained in the following theorems (Theorem 3 and 4). Theorem: Let $\sum_{k=1}^{\infty} u_k$ be a series with real members and let $\{q_k\}_{k=1}^{\infty}$ be a sequence of natural numbers which are greater than or equal to 2. Let

$$\liminf_{n\to\infty} \sum_{k=1}^{n} (q_k-1)u_k = -\infty,\quad \limsup_{n\to\infty} \sum_{n=1}^{k} (q_k-1)u_k = +\infty.$$

Then for almost all $x \in \langle 0, 1\rangle$,

$$x = \sum_{k=1}^{\infty} \varepsilon_k(x)/(q_1q_2\cdots q_k)$$

(the Cantor expansion of x), the following is true:

$$\liminf_{n\to\infty} \sum_{k=1}^{n} \varepsilon_k(x)u_k = -\infty,$$

$$\limsup_{n\to\infty} \sum_{k=1}^{n} \varepsilon_k(x)u_k = +\infty.$$

Theorem: Let $\sum_{k=1}^{\infty} u_k$ be a series with real numbers and let $\{q_k\}_{k=1}^{\infty}$ have the same meaning as in the preceding theorem. Let the series $\sum_{k=1}^{\infty}(q_k-1)u_k$ be divergent. Then for almost all $x \in \langle 0, 1)$, $x=\sum_{k=1}^{\infty}\varepsilon_k(x)/(q_1q_2\cdots q_k)$, the infinite series $\sum_{k=1}^{\infty}\varepsilon_k(x)u_k$ is divergent."

Citations: MR 20# 6403 = K55-7; MR 20# 6404 = K55-8.

K55-25 (29# 3609)

Parry, W.

Representations for real numbers.

Acta Math. Acad. Sci. Hungar. **15** (1964), 95–105.

While in previous papers on f-expansions [see C. I. Everett, Jr., Bull. Amer. Math. Soc. **52** (1946), 861–869; MR **8**, 259; B. H. Bissinger, ibid. **50** (1944), 868–876; MR **6**, 150; the reviewer, Acta Math. Acad. Sci. Hungar. **8** (1957), 477–493; MR **20** #3843] only sufficient conditions of a metric character on the function f have been given for the validity of the f-expansion, in § 1 of this paper a necessary and sufficient condition (of a topological nature) is given. § 2 deals with the f-expansions corresponding to mod 1 linear functions. Let us consider the linear mod 1 transformation $T(x)=(\beta x+\alpha)$, where $\beta>1$, $0\leqq\alpha<1$, and (y) denotes the fractional part of y. It is shown that, by putting

$$h(x) = \sum_{x<T^n(1)}\frac{1}{\beta^n} - \sum_{x<T^n(0)}\frac{1}{\beta^n},$$

the (signed) measure $\nu(E)=\int_E h(x)\,dx$ is invariant under T. It is shown further that if T is strongly ergodic (i.e., $T^{-1}E\subset E$ implies $l(E)=0$ or $l(E)=1$, where $l(E)$ denotes Lebesgue measure), then $h(x)\geqq 0$. It remains an open question whether $h(x)$ ever assumes negative values. The author mentions also the following further unsolved problems. Are linear mod 1 transformations always ergodic? If there exists a finite measure invariant under T and equivalent to Lebesgue measure, is T strongly ergodic? *A. Rényi* (Budapest)

Citations: MR 6, 150h = A68-1; MR 8, 259c = A68-2; MR 20# 3843 = K55-6.

K55-26 (29# 5789)

Šalát, Tibor

Eine metrische Eigenschaft der Cantorschen Entwicklungen der reellen Zahlen und Irrationalitätskriterien. (Russian summary)

Czechoslovak Math. J. **14 (89)** (1964), 254–266.

Let $\{q_n\}$ be an infinite sequence of integers larger than 1, and for each $x\in[0, 1)$ let

$$x = \sum_{n=1}^{\infty}\frac{\varepsilon_n(x)}{q_1\cdots q_n}$$

be the Cantor expansion of x relative to $\{q_n\}$, so that $\varepsilon_n(x)$ is an integer such that $0\leqq\varepsilon_n(x)\leqq q_n-1$ always, and $\varepsilon_n(x)<q_n-1$ for infinitely many n. It is shown that, for almost all x, $\liminf\varepsilon_n(x)/q_n=0$, and that if $\limsup q_n=\infty$, then for almost all x, the set of limit points of $\{\varepsilon_n(x)/q_n\}$ is the whole interval $[0, 1]$. The proofs use the notion of a homogeneous set and the following condition for homogeneity. Let $Q_n=q_1\cdots q_n$, and let $i_n{}^k$ be the interval $(k/Q_n, (k+1)/Q_n)$ for $0\leqq k<Q_n$. Let B be a measurable set in $[0, 1)$. Then B is homogeneous if $\mathrm{mes}(B\cap i_n{}^k)=\mathrm{mes}(B\cap i_n{}^{k'})$ for all $k, k'<Q_n$.

As an application, the author considers four sufficient conditions for the irrationality of x, in terms of the Cantor expansion of x, given by A. Oppenheim [Amer. Math. Monthly **61** (1954), 235–241; MR **15**, 781] and P. H. Diananda and A. Oppenheim [ibid. **62** (1955), 222–225; MR **16**, 908]. (For example, x is irrational if 1 is a limit point of $\{\varepsilon_n/q_n\}$.) He shows, for each of three of the four cases, that the condition is satisfied by almost all x if $\{q_n\}$ has one kind of behavior, and by almost no x otherwise. The remaining condition is shown to hold for almost all x under certain circumstances.

W. J. LeVeque (Ann Arbor, Mich.)

Citations: MR 15, 781a = J72-26; MR 16, 908g = J72-29.

K55-27 (32# 5626)

Šalát, Tibor

Über die Hausdorffsche Dimension der Menge der Zahlen mit beschränkten Folgen von Ziffern in Cantorschen Entwicklungen. (Russian summary)

Czechoslovak Math. J. **15 (90)** (1965), 540–553.

Let $0<x<1$ and let $q_1, q_2, \cdots$ be an unbounded sequence of integers $\geqq 2$. Let further

$$x = \sum_{k=1}^{\infty}\varepsilon_k(x)(q_1q_2\cdots q_k)^{-1}, \qquad 0\leqq\varepsilon_k(x)<q_{k-1},$$

be the development of x into Cantor series, and let M_∞, M_n be the sets of x for which $\sup\varepsilon_k(x)<\infty$ and $\sup\varepsilon_k(x)\leqq n$, respectively. Write

$$l = \limsup_{k\to\infty}(q_1q_2\cdots q_k)^{1/k}.$$

A simple property is that M_∞ has Lebesgue measure 0. Under a certain additional condition, the present author proves that M_∞ has Hausdorff dimension $\dim M_\infty=0$ if and only if $l=\infty$, and derives estimates for $\dim M_\infty$ in the case $l<\infty$. The proofs are based upon a formula for $\dim M_n$ deduced in an earlier paper by the author [same J. **11 (86)** (1961), 24–56; MR **27** #3763].

C. G. Lekkerkerker (Amsterdam)

K55-28 (32# 7522)

Roos, Paul

Iterierte Resttransformationen von Zahldarstellungen.

Z. Wahrscheinlichkeitstheorie und Verw. Gebiete **4** (1965), 45–63.

A finite real number set $c=\{c_0, c_1, \cdots, c_N\}$ is said to be a set of type A or of type B if $0=c_0<c_1<\cdots<1$ or $0=c_0<c_1<\cdots<c_{N-1}<1<c_N<1+c_{N-1}$, respectively. An infinite set $c=\{c_0, c_1, c_2, \cdots\}$ is a set of type A if $0=c_0<c_1<\cdots<1$, $\lim_{j\to\infty}c_j=1$. For a set c of this kind, denote by $T_1(x)$ a piecewise linear function $(x-x_j)/(c_{j+1}-c_j)$, $x\in[c_j, c_{j+1})$, for $0, 1, \cdots, N-1$ if c is finite, and for $j=0, 1, 2, \cdots$ if c is infinite. Let $T(x)=T_1(x)$, $x\in[0, 1)$, $T(1)=0$ if c is of type A, and $T=T_1(x)$, $x\in[0, 1]$ if c is of type B. The transformation $T(x)$ of the unit interval $[0, 1]$ onto itself is called a remainder-transformation (Resttransformation).

For every $x\in[0, 1)$ in case A and for every $x\in[0, 1]$ in case B, define a sequence of remainders $\{r_\nu\}$ ($\nu=0, 1, \cdots$) by the iteration $r_0=x$, $r_\nu=T(r_{\nu-1})$ ($\nu=1, 2, \cdots$). Let the index $k_\nu=k_\nu(x)$ be such that $r_{\nu-1}\in[c_{k_\nu}, c_{k_\nu+1})$. Then the number x can be represented in the form

$$x = \sum_{\nu=0}^{n-1}c_{k_\nu+1}\prod_{j=1}^{\nu}(c_{k_j+1}-c_{k_j})+r_n\prod_{j=1}^{n}(c_{k_j+1}-c_{k_j}) \qquad (n = 1, 2, \cdots)$$

or

$$x = \sum_{\nu=0}^{\infty}c_{k_\nu+1}\prod_{j=1}^{\nu}(c_{k_j+1}-c_{k_j}).$$

The author proves that the Borel measure is invariant with respect to T of type A. He shows that there exists a measure φ on the Borel sets of the interval $[0, 1]$ which is equivalent to the Borel measure and invariant to T of type B. He gives an explicit formula for φ. The transformation T is mixed (and therefore ergodic) with respect to the

Borel measure in case A and to the measure φ in case B.

If the sequence $\{x_n\}$ of numbers from $[0, 1]$ has a continuous asymptotic distribution function $F(x)$, then the sequence $\{T^k(x_n)\}$ for every $k=1, 2, \cdots$ also has a continuous asymptotic distribution function $F_k(x)$. An explicit formula for $F_k(x)$ is given.

J. Kubilius (Vilnius)

K55-29 **(33# 6670)**

Kinney, John R.; Pitcher, Tom S.

The dimension of some sets defined in terms of f-expansions.

Z. Wahrscheinlichkeitstheorie und Verw. Gebiete **4** (1965/66), 293–315.

Sample theorem: Let $\{k_n\}$ be a sequence of distinct positive integers, and $\{p_n\}$ a probability distribution, with all $p_n>0$. Consider the continued fraction expansion

$$t=\frac{1}{a_1+}\frac{1}{a_2+}\frac{1}{a_3+}\cdots, \qquad 0<t<1,$$

and let E be the set of all t such that, for each j, the event $\{a_i=k_j\}$ has limiting relative frequency p_j. Let μ be the measure induced on $(0, 1)$ by letting the a_i be independent random variables with $P\{a_i=k_j\}=p_j$. Then, provided that the series $\sum p_i \log k_i$, $\sum p_i \log p_i$ converge, the Hausdorff-Besicovitch dimension of E is at least

$$\sum p_i \log p_i/2\int_0^1 \log t\, \mu(dt).$$

{See also #5563 above.}

J. G. Wendel (Ann Arbor, Mich.)

Citations: MR 33# 5563 = J04-53.
Referred to in J04-53.

K55-30 **(35# 2308)**

Rényi, Alfréd

On certain representations of real numbers and on sequences of equivalent events.

Acta Sci. Math. (Szeged) **26** (1965), 63–74.

Let $\{a_n\}$ $(n=0, 1, 2, \cdots)$ be an absolutely monotone sequence of real numbers, that is, a sequence satisfying $\Delta^k a_n>0$ $(k=0, 1, \cdots; n\geqq k)$, where $\Delta^0 a_n=a_n$, $\Delta^1 a_n=\Delta a_n=a_{n-1}-a_n$ $(n\geqq 1)$ and $\Delta^k a_n=\Delta(\Delta^{k-1}a_n)$ $(k\geqq 1, n\geqq k)$. Let $\{a_n\}$ also be normal, i.e. $a_0=1$, and regular, i.e. $a_n\to 0$ and $\Delta^n a_n\to 0$ as $n\to\infty$. The author proves then that any real number $x\in(0, 1]$ can be represented in the form $x=\sum_{k=0}^{\infty}\Delta^k a_{n_k(x)}$, where the increasing sequence of natural numbers $n_k(x)$ is uniquely determined by x.

If x is considered as a random variable uniformly distributed in $(0, 1)$, then the sequence of random variables $n_k(x)$ is a Markov chain with the transition probabilities $P(n_k=n\,|\,n_{k-1}=m)=\Delta^{k+1}a_n/\Delta^k a_m$. If A_n denotes the event that the natural number n is contained in the sequence $n_k(x)$, then the events A_n $(n=1, 2, \cdots)$ are equivalent (symmetrically dependent), and for $1\leqq m_1<m_2<\cdots<m_r$ $(r=1, 2, \cdots)$, one has $P(A_{m_1}A_{m_2}\cdots A_{m_r})=\Delta^r a_r$.

One also has that $k/n_k(x)$ converges almost everywhere to a limit. Consequences for sequences of equivalent events, as well as a construction of a measure-preserving transformation corresponding to the sequence $\{a_n\}$ are given. An example is also discussed.

Y. N. Dowker (London)

K55-31 **(36# 6353)**

Šalát, Tibor

Über die Cantorschen Reihen. (Loose Russian summary)

Czechoslovak Math. J. **18** (**93**) (1968), 25–56.

The author proves many theorems about the distribution of digits in the Cantor series development of real numbers. In this short review we can only give a small sample.

Let $0<x<1$, $x=\sum_{k=2}^{\infty}a_k/k!$, $0\leqq a_k<k$. Put $N_n(r, x)=\sum_{k\leqq n, a_k=r}1$. A. Rényi proved [Mat. Lapok **7** (1956), 77–100; MR **20** #6404] that for almost all x and every r, $N_n(r, x)=\log n+o(\log n)$. The author proves among others that for $\alpha<\frac{1}{2}$, (*) $\limsup_{n=\infty}|N_n(r, x)-\log n|/(\log n)^\alpha=\infty$ $(r=0, 1, \cdots)$. (*) probably holds for $\alpha=\frac{1}{2}$, but undoubtedly fails for $\alpha>\frac{1}{2}$.

He further proves various theorems about the Hausdorff dimension of sets which are defined in terms of the distribution of the digits of their Cantor series development, and he also determines the Baire category of some of these sets.

P. Erdős (Budapest)

Citations: MR 20# 6404 = K55-8.

K55-32 **(37# 1335)**

Pitcher, T. S.

The asymptotic behavior of a decomposition due to Harzheim.

Proc. Amer. Math. Soc. **19** (1968), 938–942.

E. Harzheim [same Proc. **17** (1966), 534–535; MR **32** #5577] has shown the existence for each x in $[0, 1)$ of a sequence of nonnegative integers $\{\alpha_n(x)\}$ with $((x\prod_{k=2}^{n}k^{\alpha_k(x)}))<1/n$, where $((x))=x-[x]$. Conversely, given any sequence $\{\alpha_n\}$ of nonnegative integers, there is an x in $[0, 1)$ with $\alpha_i(x)=\alpha_i$. The author discusses some probabilistic properties of the random variables α_n and x_n, where $x_1(x)=x$ and $x_n(x)=n((x\prod_{k=2}^{n}k^{\alpha_k(x)}))$. For example, he determines the distribution functions of x_n and of α_n, and shows that the α_n are stochastically independent and that x_n is independent of $\alpha_1, \cdots, \alpha_n$.

T. M. Apostol (Pasadena, Calif.)

Citations: MR 32# 5577 = A68-20.

K55-33 **(37# 5179)**

Šalát, Tibor

Zur metrischen Theorie der Lürothschen Entwicklungen der reellen Zahlen. (Russian summary)

Czechoslovak Math. J. **18** (**93**) (1968), 489–522.

Throughout, let x be real and $0\leqq x\leqq 1$. There exist uniquely determined natural integers $d_k=d_k(x)$ such that $x=\sum_{n=1}^{\infty}[(s_1s_2\cdots s_{n-1})(d_n+1)]^{-1}$, where $s_k=d_k(d_k+1)$. This is called the Lüroth representation and the $d_k(x)$ are the Lüroth digits of x. It is classical that the rationality of x is equivalent to the periodicity of the $d_k(x)$. The present paper is a contribution to the metric theory of Lüroth representations of reals, initiated by H. Holzer [S.-B. Akad. Wiss. Wien Math. Natur. Kl. IIa **137** (1928), 421–453]. In spirit, this theory is close to that of the metric theory of g-adic expansions and to that of continued fractions. The paper contains 23 theorems and a large number of lemmas, corollaries, remarks, etc., grouped naturally under four headings: General results, probabilistic methods, Hausdorff measure, and Baire categories. The method of proof is an extension of that of Holzer and of the author [Magyar Tud. Akad. Mat. Kutató Int. Közl. **6** (1961), 15–41; MR **26** #4981; Czechoslovak Math. J. **11** (**86**) (1961), 24–56; MR **27** #3763], but uses also probabilistic ideas and results (see, especially, A. Rényi [*Wahrscheinlichkeitsrechnung. Mit einem Anhang über Informationstheorie*, VEB Deutsch. Verlag Wissenschaft., Berlin, 1962; MR **26** #5597; French translation, Dunod, Paris, 1966; MR **34** #2034]). A systematic presentation is not possible in a short review; instead, a few characteristic results will be quoted. Let φ be a non-negative function on the natural integers and set $T=\lim_{n\to\infty}n^{-1}\sum_{j=1}^{n}\varphi(d_j(x))$. If $\sum_{m=1}^{\infty}\varphi(m)/m(m+1)=+\infty$, then, for almost all x, $T=+\infty$; if $\sum_{m=1}^{\infty}|\varphi(m)|/m(m+1)$ converges, then $T=$

$\sum_{m=1}^{\infty} \varphi(m)/m(m+1)$ for almost all x. Let

$$M_\infty = \{x | \limsup d_n(x) = \infty\};$$

then M_∞ is of (Lebesgue) measure zero. Given a sequence $\{\tau_n\}$, $\tau_n > 0$, the set $\{n | d_n(x) > \tau_n\}$ is finite [infinite] for almost all x, accordingly as $\sum_{n=1}^{\infty} \tau_n^{-1}$ is convergent [divergent]. Let $M_k = \{x | d_n(x) \leqq k$ for every natural integer $n\}$; then the (Hausdorff) dim $M_k = \beta_k$, where β_k is the root of the equation $\sum_{m=1}^{k} (m^2+1)^{-x} = 1$ (one has $\beta_1 = 0 < \frac{1}{2} < \beta_2 < \cdots$, $\beta_k \to 1$). Also, $M_\infty = \bigcup_{k=1}^{\infty} M_k$, and dim $M_\infty = 1$, although M_∞ is of first (Baire) category. Let $L(x)$ be the set of limit points of the sequence $d_{n+1}(x)/d_n(x)$; then $L(x) = [0, +\infty]$, except, possibly, for a set of x of first category. *E. Grosswald* (Philadelphia, Pa.)

Citations: MR 26# 4981 = K55-22.
Referred to in K55-40.

K55-34 (39# 157)

Jager, H.; de Vroedt, C.

Lüroth series and their ergodic properties.

Nederl. Akad. Wetensch. Proc. Ser. A **72** = *Indag. Math.* **31** (1969), 31–42.

If $x \in (1/(n+1), 1/n]$ define $Tx = n(n+1)x - n$ and, for $x \in (0, 1]$, put $\nu_k(x) = T^k x$, $a_{k+1}(x) = [1/\nu_k(x)] + 1$. The numbers $a_k(x)$ define the Lüroth expansion of x. The author obtains several theorems on the properties of this expansion, by exploiting the fact that (1) the a_k are independent random variables, and (2) the mapping $T: (0, 1] \to (0, 1]$ is ergodic. A sample result based on (1) is

$$\lim_{n\to\infty} \text{meas}\left\{x : \left|\frac{a_1(x) + \cdots + a_k(x)}{k \log k} - 1\right| > \varepsilon\right\} = 0.$$

The study of (2) is made along the lines of C. Ryll-Nardzewski's investigation of continued fractions [Studia Math. **12** (1951), 74–79; MR **13**, 757]. *B. Weiss* (Jerusalem)

Citations: MR 13, 757b = K50-5.
Referred to in K55-40.

K55-35 (39# 390)

Wegmann, Helmut

Die Hausdorffsche Dimension von Mengen reeller Zahlen, die durch Zifferneigenschaften einer Cantorentwicklung charakterisiert sind.

Czechoslovak Math. J. **18** (**93**) (1968), 622–632.

T. Šalát [Magyar Tud. Akad. Mat. Kutató Int. Közl. **6** (1961), 15–41; MR **26** #4981] calculated the Hausdorff-Besicovitch dimensionalities of various sets of points x defined in terms of their Cantor expressions $x = \sum_i e_i(x)(q_1 q_2 \cdots q_i)^{-1}$. The author shows that their work can be sharpened and generalized, and the proofs simplified by making use of work of P. Billingsley [Illinois J. Math. **4** (1960), 187–209; MR **24** #A1750; ibid. **5** (1961), 291–298; MR **22** #11094]. *I. J. Good* (Blacksburg, Va.)

Citations: MR 24# A1750 = K15-58; MR 26# 4981 = K55-22.

K55-36 (40# 2616)

Schweiger, F.

Normalität bezüglich zahlentheoretischer Transformationen. (English summary)

J. Number Theory **1** (1969), 390–397.

The author defines a class of measurable transformations on the unit n-cube B, which generalize the g-adic expansion, continued fraction and Jacobi type transformations. Briefly, such a transformation T is defined by describing it on each of at most countably many subsets $B(k) \subset B$ with $\bigcup B(k) = B$ and $\lambda(B(k) \cap B(j)) = 0$; $k \neq j$; λ the n-dimensional Lebesgue measure. T-normality of $x \in B$ is defined, and the question under what conditions on S and T does T-normality imply S-normality for all $x \in B$ is considered.

Write $T \sim S$ if there are positive integers m and n such that $T^m = S^n$. The author proves that $T \sim S$ implies all T-normal x are S-normal. Furthermore, he conjectures [see W. M. Schmidt, Pacific J. Math **10** (1960), 661–672; MR **22** #7994] that $T \nsim S$ implies not all T-normal x are S-normal. This conjecture is known to be true for $Tx = ax \pmod 1$ and $Sx = bx \pmod 1$, a and b integers > 1, and a new example is constructed in this paper.

Finally, the author proves that the set of all non-T-normal $x \in B$ has Billingsley dimension zero. *O. P. Stackelberg* (Durham, N.C.)

Citations: MR 22# 7994 = K15-56.

K55-37 (41# 3423)

Galambos, J.

The ergodic properties of the denominators in the Oppenheim expansion of real numbers into infinite series of rationals.

Quart. J. Math. Oxford Ser. (2) **21** (1970), 177–191.

An expansion (x real, $0 \leqq x \leqq 1$) $x = 1/d_1 + a_1/(b_1 d_2) + \cdots + a_1 a_2 \cdots a_n/(b_1 b_2 \cdots b_n d_{n+1}) + \cdots$ is called a restricted Oppenheim expansion if the following conditions hold: (1) The digits d_n are determined by the algorithm $x = x_1$, $d_n = [1/x_n] + 1$, $x_n = 1/d_n + (a_n/b_n)x_{n+1}$. (2) The positive integers a_n and b_n depend on the last denominator d_n only and the function $h_n(j) = j(j-1)a_n(j)/b_n(j)$ is integer valued. This type of expansion covers many well known special cases, e.g., Engel series, Sylvester series, Cantor product and Lüroth series. The sequence d_n forms a Markov chain the transition probability of which can be calculated easily. In the theorems described below the author assumes additionally (3) $h_{n(j)} \geqq j - 1$ for all n. Theorem 2: The inequality $d_n > h_{n-1}(d_{n-1}) + 1$ holds almost everywhere for all but a finite number of values of n. Theorem 3: $\lim_{n\to\infty} P(h_n(d_n)/(d_{n+1} - 1) < c) = c$ for $0 < c < 1$. Theorem 4: The random variables

$$-\log(h_n(d_n)/(d_{n+1} - 1))$$

are asymptotically normal distributed. Theorem 5: $P(\lim_{n\to\infty} n^{-1} \sum_{n=1}^{\infty} -\log(h_n(d_n)/(d_{n+1} - 1)) = 1) = 1$. The methods used are from probability theory, especially characteristic functions and estimations of the probability of certain elementary events. It would be an interesting question to see to what extent ergodic theory could be used to obtain further information on this kind of expansion. *F. Schweiger* (Salzburg)

K55-38 (41# 3712)

Schweiger, Fritz

Ergodische Theorie der Engelschen und Sylvesterschen Reihen.

Czechoslovak Math. J. **20** (**95**) (1970), 243–245.

P. Erdős, A. Rényi and P. Szüsz have studied the metric theory of the series of Engel and Sylvester [Ann. Univ. Sci. Budapest. Eötvös Sect. Math. **1** (1958), 7–32; MR **21** #1288]. The author comments upon the fact that no ergodic theory is used, although ergodic methods have been used very successfully in closely related problems (e.g., the theory of g-adic expansions and that of continued fractions). The reason for this situation is found in the fact that, although the transformations connected with the two types of series are ergodic (with respect to Lebesgue measure), there exists no absolutely continuous invariant measure. The purpose of the present paper is the proof of these two assertions. The proof is given separately for the series of Engel and for those of Sylvester and in each case two theorems (corresponding to the two assertions) are proven. *E. Grosswald* (Philadelphia, Pa.)

Citations: MR 21# 1288 = K55-12.
Referred to in K55-39.

K55-39 (43# 2190)

Schweiger, Fritz
Addendum zu "Ergodische Theorie der Engelschen und Sylvesterschen Reihen".
Czechoslovak Math. J. **21 (96)** (1971), 165.

The author makes a correction in the proof of Theorem 1, and states that Theorem 2 is still open. The original paper appeared in same J. **20 (95)** (1970), 243–245 [MR **41** #3712].

Citations: MR 41# 3712 = K55-38.

K55-40 (42# 208)

Dávid, A.
Berechnung einer bestimmten Konstante aus der Theorie der Lürothschen Reihen.
Acta Fac. Rerum Natur. Univ. Comenian. Math. Publ. 21 (1968), 15–20 (1969).

Each real number $x \in (0, 1]$ can be uniquely expressed by its Lüroth development

$$x = \frac{1}{d_1+1}+\frac{1}{s_1}\frac{1}{d_2+1}+\cdots+\frac{1}{s_1s_2\cdots s_{n-1}}\frac{1}{d_n+1}+\cdots,$$

where $d_k = d_k(x)$ $(k=1, 2, \cdots)$ are positive integers and $s_k = d_k(d_k+1)$ [cf. O. Perron, *Irrationalzahlen*, pp. 116–122, De Gruyter, Berlin, 1921; fourth edition, 1960; MR **22** #6782]. The reviewer [Czechoslovak Math. J. **18 (93)** (1968), 489–522; MR **37** #5179] and H. Jager and C. de Vroedt [Nederl. Akad. Wetensch. Proc. Ser. A **72** (1969), 31–42; MR **39** #157] have shown that there exists a constant $c_0 > 0$ such that for almost all $x \in (0, 1]$ the relation $\lim_{n\to\infty} \sqrt[n]{(d_1(x)\cdots d_n(x))} = c_0$ holds. The constant c_0 is analogous to Hinčin's constant k_0, well known from the theory of continued fractions. The author uses the automatic computer Odra 1013 and obtains for c_0 the following approximate value: 2.19991175 ± 0.0002571 $(<2.685\ldots = k_0)$. *T. Šalát* (Bratislava)

Citations: MR 22# 6782 = Z02-32; MR 37# 5179 = K55-33; MR 39# 157 = K55-34.

K55-41 (42# 3049)

Schweiger, Fritz
Über den Satz von Borel-Rényi in der Theorie der Cantorschen Reihen.
Monatsh. Math. **74** (1970), 150–153.

Let $\{q_k\}$ be a sequence of integers, $q_k \geqq 2$ $(k=1, 2, \cdots)$. Each real $x \in [0, 1)$ can be uniquely expressed by means of its Cantor development $x = \sum_{k=1}^{\infty} \varepsilon_k(x)/(q_1q_2\cdots q_k)$, $0 \leqq \varepsilon_k(x) < q_k$, $\varepsilon_k(x)$ are integers and for an infinite number of k's we have $\varepsilon_k(x) < q_k - 1$ [O. Perron, *Irrationalzahlen*, pp. 111–116, Verein. Wiss. Verl., Berlin, 1921; second edition, de Gruyter, Berlin, 1939; third edition, 1947; fourth edition, 1960; MR **22** #6782]. The fundamental metric results on the distribution of digits in developments of numbers $x \in [0, 1)$ were given by A. Rényi [Mat. Lapok **7** (1956), 77–100; MR **20** #6404]. These results generalize the well-known classical result of E. Borel on normal numbers. The author, using some results of W. Philipp [Pacific J. Math. **20** (1967), 109–127; MR **34** #5755], proves the following strengthening of the mentioned results of A. Rényi: Put $A(N, a, x) = \sum_{k\leqq N, \varepsilon_k(x)=a} 1$ for $a \geqq 0$ and $\varphi(N) = \sum_{k\leqq N, a<q_k} 1/q_k$. Let $\varepsilon > 0$. Then for almost all $x \in [0, 1)$, the relation $A(N, a, x) = \varphi(N) + O(\varphi^{1/2}(N) \log^{3/2+\varepsilon} \varphi(N))$ holds. *T. Šalát* (Bratislava)

Citations: MR 20# 6404 = K55-8; MR 22# 6782 = Z02-32; MR 34# 5755 = J24-45.

K55-42 (43# 1918)

Guthery, Scott Bates
An inversion algorithm for one-dimensional F-expansions.
Ann. Math. Statist. **41** (1970), 1472–1490.

Let f be a monotone function, φ its inverse function and $TX = \langle\varphi(x_1)\rangle$ the fractional part of $\varphi(x)$. Under certain restrictions T maps $(0, 1)$ into $[0, 1]$, and if $a_1 = a_1(x) = [\varphi(x)]$ and $a_n = a_n(x) = a_1(T^{n-1}x)$, then

$$x = \lim f(a_1+f(a_2+f(a_2+f(a_3+\cdots+f(a_n))\cdots),$$

$x \in (0, 1)$. We then call $\{a_n\}$ the f-expansion of x. Examples: $f = x/10$ on $[0, 10]$, the decimal expansion; $f(x) = 1/x$ on $[1, \infty)$, the continued fraction expansion. General f-expansions have been considered by many authors (see the papers by A. Rényi [Acta Math. Acad. Sci. Hungar. **8** (1957), 477–493; MR **20** #3843] and F. Schweiger [Acta Arith. **15** (1968), 1–18; MR **40** #7219; corrigendum, ibid. **16** (1969/70), 217–219; MR **40** #7220] and the references cited there). If $\mu(dx) = h(x)\,dx$ is a probability measure on $[0, 1]$ that is invariant under T, then it induces in an obvious way a measure P for the sequences $\{a_n(x_1)\} = \{a_1(T^{n-1}x)\}$ that is invariant under the shift ($\mu(dx) = dx$ and $((1+x)\log 2)^{-1}\,dx$ for the decimal and continued fraction expansions, respectively). A pleasant phenomenon occurs when T is ergodic with respect to μ, for then the ergodic theorem can be used to derive a.e. $[P]$ limits for various functionals of sequences $\{a_n\}$. Rényi [loc. cit.] and Schweiger [loc. cit.] gave conditions for the existence of μ and the ergodicity of T. The explicit determination of μ and P for given f is rarely possible. The author treats the following converse problems: (i) Given μ find an f for which f-expansions make sense and is such that μ is invariant with respect to the corresponding f. The solution is not unique in general and the author cleverly gives a class of f's by decomposing $\mu(dx)$ in the form $\sum_{k=0}^{N-1} g(x+k)\,dx$, where N is integral, $g \geqq 0$ and $\int_0^N g(x)\,dx = 1$. He also gives conditions for the resulting T to be ergodic. Several examples are given, among which is an interesting generalization of the continued fraction expansion. (ii) Given a stationary process $(\{a_n\}, P)$ find f and μ that give rise to it. The author gives an explicit construction when $(\{a_n\}, P)$ is a Markov chain of finite order and shows that for a wide class of processes it is possible to choose f so that μ becomes Lebesgue measure. *H. Kesten* (Ithaca, N.Y.)

Citations: MR 20# 3843 = K55-6; MR 40# 7219 = K50-41; MR 40# 7220 = K50-42.

K55-43 (43# 2752)

Gordin, M. I.
Exponentially rapid mixing. (Russian)
Dokl. Akad. Nauk SSSR **196** (1971), 1255–1258.

Let T be a measurable transformation of a probability space; the title refers to quasi-independence among the derived σ-algebras $T^{-n}(M)$, where M is the set of events. Both the hypotheses on T and the consequences for ergodic theory follow the pattern of Kuz'min's theorem in continued fractions; the exact formulations are quite technical. The author mentions applications to other expansions, notably the Jacobi-Perron algorithm. {Compare Chapter III of A. Ja. Hinčin's book [*Continued fractions* (Russian), ONTI NKTP, Moscow, 1935; English translations of the third Russian edition, Univ. Chicago

Press, Chicago, Ill., 1964; MR **28** #5037; Noordhoff, Groningen, 1963; MR **28** #5038] and the paper by F. Schweiger [J. Reine Angew. Math. **232** (1968), 35–40; MR **40** #4200].}

{This article has appeared in English translation [Soviet Math. Dokl. **12** (1970), 331–335].}

R. Kaufman (Urbana, Ill.)

Citations: MR 28# 5037 = Z02-21; MR 28# 5038 = Z02-22; MR 40# 4200 = K50-39.

K55-44 **(44# 173)**

Waterman, Michael S.

Some ergodic properties of multi-dimensional *f*-expansions.

Z. Wahrscheinlichkeitstheorie und Verw. Gebiete **16** (1970), 77–103.

A. Rényi [Acta Math. Acad. Sci. Hungar. **8** (1957), 477–493; MR **20** #3843] studied the *f*-expansions of Everett and of Bissinger and generalised the ergodic theory of Ryll-Nardzewski concerning continued fractions. (For references, cf. the author's paper or Rényi's paper [loc. cit.].) Rényi's main concerns were the validity of *f*-expansions, the ergodicity of their associated transformations of the unit interval, and the existence of finite invariant measures equivalent to Lebesgue measure. To ensure all these properties Rényi proposed a regularity condition on the growth of *f*-expansions. The author is principally concerned with n-dimensional versions of Rényi's results and to facilitate his study imposes a variety of suitable regularity conditions. This non-trivial extension of Rényi's work gives rise to a number of interesting transformations of the n-dimensional cube. Under certain regularity conditions the transformations turn out to be not merely ergodic but exact and therefore mixing of all orders, and the author computes their entropies in terms of their Jacobians. These results generalise V. A. Rohlin's [Izv. Akad. Nauk SSSR Ser. Mat. **25** (1961), 499–530; MR **26** #1423; translated in Amer. Math. Soc. Transl. (2) **39** (1964), 1–36; see MR **28** #3909]. Finally, the author presents a central limit theorem, an iterated logarithm theorem and a discussion of examples. Acknowledgement of recent research in the same direction as his own very interesting work is given, with a full bibliography.

W. Parry (Coventry)

Citations: MR 20# 3843 = K55-6; MR 26# 1423 = K50-13.
Referred to in K55-45.

K55-45 **(44# 5289)**

Waterman, Michael S.

A Kuzmin theorem for a class of number theoretic endomorphisms.

Acta Arith. **19** (1971), 31–41.

Let A be a convex subset of $\mathbf{R}^n$ and $F: A \to]0, 1[^n$ a one-to-one continuous map. The map $T(x) = F^{-1}(x) - [F^{-1}(x)]$, subject to some further conditions, is an ergodic transformation, which was studied extensively in an earlier paper of the author [Z. Wahrscheinlichkeitstheorie und Verw. Gebiete **16** (1970), 77–103; MR **44** #173]. The main purpose of this difficult paper is to prove an extension of the famous Kuzmin theorem on continued fractions. The proof contains some errors which are stated and corrected in a forthcoming paper by the author and the reviewer (to appear in J. Number Theory). It is worth noting that the main incorrect statement traces back to the proof of Kuzmin and can be found in Hinčin's well known booklet as well as in several subsequent papers: The numbers l_i and l_i' depend not only on i, but equally on g_0. Therefore ν_0 depends on g_0 and the choice of ν_0 may vary with g_r. Hence $r \to \infty$ may imply $\nu \to \infty$. Furthermore, the following condition (necessary to handle improper cylinders, but not continued fractions) is implicitly used: $B_\nu \cap A_i \neq \varnothing$ implies that $B_\nu \subseteq A_i$.

F. Schweiger (Salzburg)

Citations: MR 44# 173 = K55-44.

K55-46 **(44# 6938)**

Shiokawa, Iekata

Ergodic properties of piecewise linear transformations.

Proc. Japan Acad. **46** (1970), 1122–1125.

Let $\bar{\beta} = (\beta_0, \cdots, \beta_N)$, $N \geq 1$, be an $(N+1)$-tuple of positive numbers such that $\beta_N(1-(1/\beta_0 + \cdots + 1/\beta_{N-1})) < 1$. On consecutive sub-intervals of $[0, 1)$ of length $1/\beta_0, \cdots, 1/\beta_{N-1}$ define T to be linear, expanding these intervals (by factors $\beta_0, \cdots, \beta_{N-1}$, respectively) onto $[0, 1)$. Define T on the remaining sub-interval by $Tx = \beta_N x - (1/\beta_0 + \cdots + 1/\beta_{N-1})$. These piecewise linear transformations generalise the case $\beta_0 = \beta_1 = \cdots = \beta_N$ studied by A. Rényi [Acta Math. Acad. Sci. Hungar. **8** (1957), 477–493; MR **20** #3843]. The author relates these transformations to Rényi's *f*-expansions and claims that they satisfy Rényi's conditions, although the case $\beta_N < 1$ does not seem immediately obvious. A T-invariant measure is exhibited with a formula similar to the one discovered by A. O. Gel'fond [Izv. Akad. Nauk SSSR Ser. Mat. **23** (1959), 809–814; MR **22** #702] and the reviewer [Acta Math. Acad. Sci. Hungar. **11** (1960), 401–416; MR **26** #288] for constant $\beta_0 = \cdots = \beta_N$, and the transformations are shown to be exact in analogy with V. A. Rohlin's result [Izv. Akad. Nauk SSSR Ser. Mat. **25** (1961), 499–530; MR **26** #1423]. Special cases are related to multiple Markov chains.

W. Parry (Coventry)

K99 NONE OF THE ABOVE, BUT IN THIS CHAPTER

See also reviews B08-4, B08-43, B08-51, B08-59, B08-68, B08-71, B08-77, B08-80, B12-30, B20-3, B20-4, B20-7, B24-11, B24-17.

K99-1 **(30# 4713)**

Martić, Branislav

On a sum of fractional parts.

Mat. Vesnik (N.S.) **1 (16)** (1964), 45–46.

The author provides a new proof and a variation on an elementary number-theoretic problem* posed by the reviewer [Amer. Math. Monthly **64** (1957), 117].

J. L. Ullman (Ann Arbor, Mich.)

[*Namely, if $\Sigma_1^\infty \lambda_i = 1$, then $N^{-1} \Sigma_1^\infty \{N\lambda_i\} \to 0$. Ed.]

K99-2 **(42# 3046)**

Haight, J. A.

A linear set of infinite measure with no two points having integral ratio.

Mathematika **17** (1970), 133–138.

For $x > 0$ and a set E on the positive real line, let $N(x, E)$ denote the number of positive integers n such that $nx \in E$. The set E is said to be of type S if x_1/x_2 is never equal to an integer for any $x_1, x_2 \in E$, $x_1 \neq x_2$. In a paper by the reviewer [Nederl. Akad. Wetensch. Proc. Ser. A **61** (1958), 197–205; MR **24** #A1256a], it is shown that there are Lebesgue measurable sets E which are of infinite measure $|E|$, although $N(x, E) < \infty$ for all $x > 0$. The present author constructs a set E which, in addition, is of type S. The proof uses an inductive process which starts with the construction of a subset E_1 of $[\frac{1}{2}, 1)$ of positive measure possessing the following property T: the complement of $\bigcup_{k=1}^{\infty} kE_1$ contains a sequence of bounded measurable sets B_i, of type S, with $\lim_{i\to\infty} |B_i| = \infty$.

It is also shown that if E is measurable and it is not possible to find a real number $x > 1$ and distinct positive integers l, m with $lx \in E$ and $mx \in E$, the measure of E cannot be greater than nine.

C. G. Lekkerkerker (Amsterdam)

Citations: MR 24# A1256a = H15-71.